GAS CHROMATOGRAPHY

GAS CHROMATOGRAPHY

COLIN F. POOLE
Department of Chemistry, Wayne State University, Detroit, Michigan, USA

ELSEVIER

AMSTERDAM • BOSTON • HEIDELBERG • LONDON
NEW YORK • OXFORD • PARIS • SAN DIEGO
SAN FRANCISCO • SYDNEY • TOKYO

Elsevier
The Boulevard, Langford Lane, Kidlington, Oxford OX5 1GB, UK
Radarweg 29, PO Box 211, 1000 AE Amsterdam, The Netherlands
225 Wyman Street, Waltham, MA 02451, USA

First edition 2012

Library of Congress Cataloging-in-Publication Data
Poole, C. F.
Gas chromatography / Colin F. Poole. -- 1st ed.
p. cm.
ISBN 978-0-12-385540-4
1. Gas chromatography. 2. Chromatographic analysis. I. Title.
QD79.C45P64 2012
543'.85--dc23
2012012782

British Library Cataloguing in Publication Data
A catalogue record for this book is available from the British Library

For information on all Elsevier publications
visit our web site at store.elsevier.com

Printed and bound in USA

12 13 14 15 16 10 9 8 7 6 5 4 3 2 1

ISBN: 978-0-12-385540-4

Contents

For the color version of the figures, the reader is referred to the online version of the book.

Contributors

K. Hüsnü Can Baser Anadolu University, Faculty of Pharmacy, Department of Pharmacognosy, Eskisehir, Turkey

Leonid M. Blumberg Fast GC Consulting, P.O. Box 1243, Wilmington, DE 19801, USA

Joseph A. Caruso Department of Chemistry, University of Cincinnati, Cincinnati, OH 45221-0172, USA

Qilin Chan Department of Chemistry, University of Cincinnati, Cincinnati, OH 45221-0172, USA

Eric Chun Yong Chan Department of Pharmacy, National University of Singapore, Singapore

Cristina Cruz-Hernandez Nestlé Research Center Vers-chez-les-Blanc, P.O. Box 44 1000 LAUSANNE 26, Switzerland

Peter Dawes Chairman, SGE Analytical Science, 7 Argent Place, Ringwood, VIC, 3134, Australia

Bieke Dejaegher Department of Analytical Chemistry and Pharmaceutical Technology, Vrije Universiteit Brussel (VUB), Laarbeeklaan 103, B-1090 Brussels, Belgium

Frédéric Destaillats Nestlé Research Center Vers-chez-les-Blanc, P.O. Box 44 1000 LAUSANNE 26, Switzerland

Frank L. Dorman Associate Professor of Biochemistry and Molecular Biology, Associate Director for Research and Graduate Education, Forensic Science, The Pennsylvania State University, 107 Whitmore Lab, University Park, PA 16802, USA

Susan E. Ebeler Department of Viticulture and Enology, University of California, Davis, CA 95616, USA

Amadeo R. Fernandez-alba University of Almería, Pesticide Residue Research Group, 04120, La Cañada de San Urbano, Almería, Spain

Károly Héberger Research Centre for Natural Sciences, Hungarian Academy of Sciences, H-1025 Budapest, Pusztaszeri út 59/67, Hungary

Yvan Vander Heyden Department of Analytical Chemistry and Pharmaceutical Technology, Vrije Universiteit Brussel (VUB), Laarbeeklaan 103, B-1090 Brussels, Belgium

Walter G. Jennings Food Science, University of California, Davis, California, USA

Leesun Kim Centre for Green Chemistry, School of Chemistry, Monash University, Wellington Road, Victoria 3800, Australia

Matthew S. Klee JAS Inc., Newark, Delaware, USA

Mainak Mal Department of Pharmacy, National University of Singapore, Singapore

Philip J Marriott Centre for Green Chemistry, School of Chemistry, Monash University, Wellington Road, Victoria 3800, Australia

Angeles Martinez-Uroz University of Almería, Pesticide Residue Research Group, 04120, La Cañada de San Urbano, Almería, Spain

M Milagros Mezcua University of Almería, Pesticide Residue Research Group, 04120, La Cañada de San Urbano, Almería, Spain

Jocelyn G. Millar Department of Entomology, University of California, Riverside, CA, 92521, USA

Jeremy S. Nadeau Department of Chemistry, Box 351700, University of Washington, Seattle, WA, 98195, USA

Temel Özek Anadolu University, Faculty of Pharmacy, Department of Pharmacognosy, Eskisehir, Turkey

Kishore Kumar Pasikanti Department of Pharmacy, National University of Singapore, Singapore

Karisa M. Pierce Department of Chemistry, 3307 Third Avenue West, Suite 205, Seattle Pacific University, Seattle, WA 98119, USA

Maria Chiara Pietrogrande Department of Chemistry, University of Ferrara, Via L. Borsari, 46, 44100 Ferrara, Italy

Colin F. Poole Department of Chemistry, Wayne State University, Detroit, Michigan, USA

Eric J. Reiner Senior Mass Spectrometry Research Scientist, Laboratory Services Branch, Ontario Ministry of the Environment, Toronto, ON, M9P 3V6, Canada

Volker Schurig Institute for Organic Chemistry, University Tübingen, Auf der Morgenstelle 18, D 72076 Tübingen, Germany

John V. Seeley Oakland University, Department of Chemistry Rochester, MI 48309, USA

Michael J. Sithersingh Department of Chemistry and Biochemistry, Center for Academic Industry Partnership, Seton Hall University, 400 South Orange Avenue, South Orange, NJ 07079, USA

Johanna Smeyers-Verbeke Department of Analytical Chemistry and Pharmaceutical Technology, Vrije Universiteit Brussel (VUB), Laarbeeklaan 103, B-1090 Brussels, Belgium

Philip A. Smith U.S. Department of Labor Occupational Safety and Health Administration, Health Response Team, Salt Lake Technical Center, 8660 S. Sandy Parkway, Sandy, Utah 84070, USA

Nicholas H. Snow Department of Chemistry and Biochemistry, Center for Academic Industry Partnership, Seton Hall University, 400 South Orange Avenue, South Orange, NJ 07079, USA

Stanley D. Stearns President, Valco Instruments Co. Inc, Houston, USA

Robert E. Synovec Department of Chemistry, Box 351700, University of Washington, Seattle, WA, 98195, USA

Andrew Tipler PerkinElmer Inc., Shelton, CT 06484, USA

Adam Voelkel University of Technology, Institute of Chemical Technology and Engineering, pl. M. Skłodowskiej-Curie 2, 60-965 Poznań, Poland

Thomas P. Wampler CDS Analytical, 465 Limestone Road, Oxford, Pa 19363-0277, USA

Charles L. Wilkins Department of Chemistry and Biochemistry, University of Arkansas, Fayetteville, USA

Elizabeth Woolfenden Markes International Ltd, Gwaun Elai Campus, Llantrisant, RCT, CF72 8XL, UK

CHAPTER

1

Milestones in the Development of Gas Chromatography

Walter G. Jennings, Colin F. Poole

OUTLINE

Gas Chromatography DOI: 10.1016/B978-0-12-385540-4.00001-8

1.1. INTRODUCTION

This article was started by the senior author Walter Jennings who was unable to complete it due to poor health. Sections 1.2–1.6 are a personal account of the early days of gas chromatography seen through the eyes of one of the major pioneers and innovators in this field. With only minor editorial changes made by the junior author, these are presented as Walter intended. As the junior author I am responsible for Sections 1.7 and 1.8. These sections extend Walter's comments on the early days of gas chromatography to the present day.

1.2. THE INVENTION OF GAS CHROMATOGRAPHY

In 1952 A.J.P. Martin and R.L.M. Synge were both awarded Nobel Prizes for their work in the field of liquid/solid chromatography. Martin, in his award address, suggested it might be possible to use a vapor as the mobile phase. Some years later, James and Martin used ethyl acetate vapor to desorb a mixture of fatty acids that had been affixed to an adsorbent, and placed in a tube. The vapor stream eluting from that tube was directed to an automated titration apparatus, resulting in a graph showing a series of "steps" that reflected the sequential additions of base as each eluted acid was neutralized by automated titration [1]. Many practitioners have for far too long considered this as the starting point of gas chromatography.

In 2008 Leslie Ettre published an article in which he stated, "… the activities of Professor Erika Cremer and her students at the University of Innsbruck, Austria, in the years following the Second World War, represented the true start of their continuous involvement in gas chromatography" [2]. After exploring Ettre's arguments and conducting some research myself, I fully agree with his conclusions. I now believe that the theoretical basis for gas chromatography was first conceived by Erika Cremer, an Austrian scientist at the University of Innsbruck, Austria, in the late 1940s during the period of the Second World War. As Ettre points out, this was a period when women, especially in Germanic countries, were expected to confine their activities to "children, church, and kitchen". In spite of her "superb Ph.D. thesis work" she had great difficulty in finding a position. Her opportunity came in 1940 when the war started, and university teachers were drafted. She obtained an academic position at the University of Innsbruck, Austria (then a part of Germany) in the Institute of Physical Chemistry. It was here that she and her students (with major credit to Fritz Prior) constructed the first prototype of a gas chromatograph (Figure 1.1) and, after a long delay that was probably attributable to the war, published the results of their research in 1951 [3,4]. At this time she was promoted to Professor and, some twenty years later, to director of the University's Institute of Physical Chemistry. Professor Dr. Cremer, by all accounts a brilliant woman scientist, died in 1996.

1.3. EARLY INSTRUMENTATION

By 1953, several petroleum companies, primarily in Great Britain and the Netherlands, were exploring this new analytical technique, and in 1954, a few flavor chemists (including this author) were building crude chromatographs, many of them based on an article by N. H. Ray [5]. Ray inserted thermal conductivity cells into a Wheatstone bridge, whose outlet connected to a strip chart recorder, thus generating a Gaussian peak for each eluting solute; this was (to my knowledge) the first gas chromatogram as we know them today. The schematics and chromatograms published by Ray encouraged a number of readers (including the author) to build similar chromatographs. Almost every part of these crude

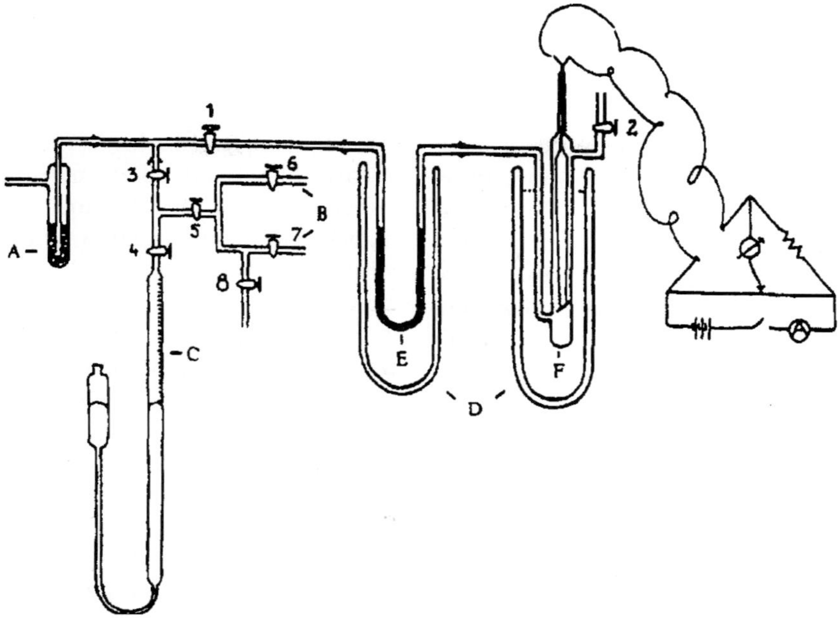

FIGURE 1.1 The gas chromatographic system used in Prior' work, in 1945–1947. A = adsorbent for purification of the carrier gas (hydrogen); B = sample inlet system; C = buret containing mercury with niveau glass for sample introduction; D = Dewar flask; E = separation column (containing silica gel on activated carbon); and F = thermal conductivity detector. *Source: From ref. [4] copyright Friedr. Vieweg & Sohn.*

instruments had to be self-designed and self-made, but this introduced the days of the packed column; supports, such as granules of diatomaceous earth, were coated with a variety of high boiling fluids (e.g. diethylene glycol succinate, DEGS) and, in our earliest efforts, packed into copper columns, typically 3–6 m long with internal diameters of about 6 mm, and (usually) coiled. It was soon recognized that copper columns were quite active and most workers switched, first to stainless steel, and then to coiled glass tubing of similar dimensions. The reminiscences of many of the early pioneers in gas chromatographic instrumentation are summarized in [6].

1.3.1. Early Commercial Instruments

The first companies to manufacture gas chromatographs (GCs) in Europe were Griffin and George (London) and Metropolitan Vickers Electric (Manchester), but the U.S. instrument companies had a greater impact on the development of this new area [7]. Perkin Elmer was one of the first companies to market a gas chromatograph; in May of 1954, they introduced their Model 154 Vapor Fractometer. The temperature of the column oven was adjustable from room temperature to 150 °C, and it offered a "flash vaporizer" with a rubber septum permitting syringe injection into the carrier gas stream. The detector was a thermal conductivity cell. The instrument was a great success and sold widely [8]. In early 1956, Perkin Elmer followed up with their Model 154-B, with a temperature range from room temperature to 225 °C and could be fitted with an optional rotary valve offering a variety of sample loops for the injection of gas samples.

1.4. EARLY COLUMN DEVELOPMENTS

All of the early instruments utilized packed columns, some coiled, some U-shaped. Packed columns all have one thing in common: they possess a high resistance to gas flow, and this limits the practical length of the column, usually to a few meters. In much later days, some packed columns approached an efficiency equivalent to our current 0.32 mm internal diameter capillary columns (ca. 3000 plates per meter), but because of their length limitations, they could achieve perhaps 35,000 theoretical plates while a 30 m open-tubular 0.25 mm internal diameter column should be capable of three times that efficiency, merely because it is three times longer.

It was at the 1958 Second International Symposium on Gas Chromatography in Amsterdam that Marcel Golay, who was then a consultant to Perkin Elmer, introduced the theory of open-tubular capillary columns and demonstrated their superiority to packed columns [9]. (*There is no connection between the old Perkin Elmer referred to here and the Perkin Elmer of today, they are entirely different companies.*) These columns demanded much smaller sample injections, necessitating more sensitive detectors than the thermal conductivity detectors that were in common use at the time. Fortunately, James E. Lovelock had invented an electron-capture detector in 1957 [10], and in 1958 the flame ionization detector appeared; some credit this to Harley, Nel, and Pretorious [11], and others to McWilliams and Dewer [12,13]; these increased detector sensitivities by ca. 10^3–10^6. The invention and development of the flame ionization detector is discussed in more detail in [14]. Early capillary columns were of plastic and copper tubing; the former had serious temperature limitations and the latter was active; this led to stainless steel tubing. Perkin Elmer had filed for and been granted patents on the open-tubular concept, essentially worldwide. I purchased two Perkin Elmer wall-coated open-tubular (WCOT) columns at different times, only to find that the columns available from them produced abominable results. Perkin Elmer apparently recognized this fact, because they essentially abandoned their research efforts on WCOT columns, outsourced their production, and redirected their efforts to columns whose interiors were first coated with a support material (e.g. diatomaceous earth) and then with the stationary phase. These were dubbed "support coated open-tubular (SCOT) columns". Perkin Elmer continued to rigorously enforce the patent on WCOT (and SCOT) columns, but under considerable pressure (especially from the applicant) they were forced to issue one license, to Hansjurg Jaeggi, a former assistant of Kurt Grob in Switzerland. Under Swiss law, if a patent bars a Swiss from conducting his or her business, a license must be issued. Jaeggi had been and still was making excellent glass WCOT columns. The high quality of his columns led Perkin Elmer to later propose that they collaborate, but, probably because of the acrimonious battle he had gone through to obtain his license, he wanted nothing more to do with Perkin Elmer, and refused their offer. His obituary, written by Konrad Grob, makes interesting reading [15].

1.4.1. Do-it-Yourself Glass Capillary Columns

In 1965, Desty et al. invented an elegantly simple machine for drawing long lengths of coiled glass capillaries [16]. Besides the fact that his was much less expensive tubing, it also had a smoother interior *and it was transparent*. For the first time, it was possible to scrutinize the layer of stationary phase as it existed on the column wall; soon it was obvious that when a new unused column exhibiting a thin uniform film of stationary phase was exposed to higher temperatures, the stationary phase collected into beads that were randomly scattered over the inner surface. Almost immediately scores of scientists realized that many of the low viscosity fluids that worked well in packed columns were unsatisfactory for WCOT columns and should be replaced with high viscosity materials that would retain their high viscosity even at higher temperatures. Low-viscosity silicone fluids (e.g. OV 101) were replaced with high viscosity silicones (e.g. OV 30, a viscous paste-like silicone), and experimenters soon began producing much more stable columns.

1.4.2. The Positive Results of Patent Enforcement

The low quality of the Perkin Elmer WCOT columns and enforcement of their patent led many scientists to begin making their own columns for their own use. This caused investigators in many other fields, who would have preferred to purchase usable columns from an outside source and confine their research to some other field, to now have to make columns; scores of scientists were now studying and publishing on methods of pretreating, deactivating, and coating columns. Their combined results were responsible for many of the advances in column improvements, and they soon surpassed the results that had been generated by Perkin Elmer [17].

1.4.3. Mileposts in Coating WCOT Capillary Columns

Golay had coated his original glass open-tubular columns by completely filling them with a dilute solution of stationary phase in a low-boiling solvent, sealing one end, and then drawing the column, open end first, through an oven [18]. As used by Golay, the column could be coiled only after coating, a distinct drawback. The method was improved by Ilkova and Mistrykov [19], who coiled, and then filled the column and sealed one end. The open end of the filled column was fed into a heated oven by supporting the column on a rotating rod fitted with a drive roller at the entrance to the column oven, thus literally screwing the rest of the column into the oven.

Up until just a few years ago, all chemistry departments at the multicampus Universities of California system frowned on applied research, and chemistry department faculties drifting away from pure organic chemistry or pure physical chemistry rarely survived to tenure. Analytical chemistry was essentially forbidden. There were many faculty members scattered through various other departments who were analytical chemists, and eventually they formed a "Group in Analytical and Environmental Chemistry", open to anyone in any department who had interests in the analytical side. At one time I chaired that group for several years and it still exists. At this time, my title was "Professor of Food Science and Technology and Chemist in the Experiment Station". This had several advantages: for one thing, those with just the academic title worked nine months per year, while the Experiment Station operated eleven months a year. My department chairman tolerated what he called my "dabbling" in gas chromatography, but insisted that I should also be working on subjects "more aligned with the food industries"; he suggested "circulation cleaning" which was just emerging. Swallowing what I wanted to say, I assigned a new Ph.D. student (Malcolm Bourne) to the project, built a miniature circulation pipeline, and helped him bake radio-labeled tristearin onto glass microscope slides and strips of stainless steel of similar width and length. These were inserted into the rapid circulating system, and the levels of radioactivity measured every two minutes.

Inevitably, semi-log plots of our cleaning curves showed a rapidly dropping curve that terminated in a straight line with a slight downward slope. All at once Bourne realized that he was looking at two first-order reactions that were occurring simultaneously. The baked film of tristearin existed only at one of two energy levels, a loosely bound system, and a much more tightly bound system – nothing in between. Spatial estimates indicated that the tightly bound form was not a monolayer, but could be up to at least twenty molecular thicknesses [20–22]. He also discovered that by manipulating time and temperature, he could change the ratio of the two species. Bourne went on to study the kinetics of cleaning in much greater depth, and made quite a name for himself. I went back to my "dabbling", then slowly realized that my newly gained knowledge on removal of thin films was just the reverse of achieving more stable coatings of stationary phases to the column surface.

This may have guided me as I attempted to modify our column coating apparatus. After replacing the opaque lid of the oven with a glass lid, it was obvious that the solution was superheated and evaporated in a series of minor explosions, leaving blotches of stationary phase randomly scattered over the surface of the column. I decided to introduce the column into the oven through a preheater made from a short curved length of 1/8 inch stainless steel tubing lagged with an electrically heated wire and heated to 150 °C: the column's passage through the curved preheater was not smooth; indeed, it scraped the walls and vibrated. To avoid breakage, I introduced graphite powder into the loop. The column continued to vibrate, but more gently. In retrospect, that vibration at this point on its passage into the oven would also discourage super-heating. It may have been dumb luck, but from then on, the evaporation of the solvent occurred smoothly. We were routinely producing very stable columns with uniformly thin coatings.

1.4.4. Commercial Column Manufacturers

At this point, one of my Ph.D. students, Robert Wohleb, suggested that we had a marketable product and should establish a column production company. When I broached this suggestion with my department chairman, he called in the dean. After much discussion, they agreed that I could engage in this activity only if I agreed to several restrictions: (1) as long as I was a full time professor, I could receive no remuneration from this venture; (2) there could be no connection between my university research and activities of the company; (3) I could not get involved in the day-to-day activities of the company. I could, *on my own time*, answer trouble-shooting questions and engage in educational activities. J&W Scientific was founded in 1974; I constructed two drawing machines in my machine shop at home, and left them with Wohleb as I departed on my second sabbatical leave, this time in Karlsruhe, Germany. Sandy Lipsky had launched his company, Quadrex, slightly ahead of J&W.

1.4.5. Bonded, Crosslinked, and/or Immobilized Stationary Phases

In 1975, J&W noticed a sudden drop in column sales from several regular customers; fortunately, we had been using (and saving) bar codes to trace every step each column went through. The bar codes on all of the columns those customers had been buying showed they had all gone through the same coating machine in the same general time period. I called several of the buyers and asked if they were having problems with our columns, and was shocked when they replied, "no, the columns are great – they never wear out". This was serious, we were putting ourselves out of business. In checking that coating machine, it was discovered that the thermocouple on the preheater through which the column entered the oven had failed some time ago and was giving much lower readings. Over the past few weeks, the technician on the machine had simply compensated for the low reading by boosting the voltage. Instead of 150 °C, that preheater was close to 300 °C. The stationary phase we were using (SE-54) contained 1% vinyl. Our R&D head, Rand Jenkins, realized that serendipity had smiled on us! He immediately ordered materials and began adding vinyl-containing compounds, to our coating solutions, hoping that the phase would not only crosslink but also connect with some surface hydroxyls. With some pretty harsh testing, we found that the columns exposed to these higher preheater temperatures were much more rugged, and the deposited phase could not be removed with solvent rinsing. At the next "Advances in Chromatography" meeting in Houston, we announced the "Bonded Phase Column". If the technician on that machine had reported the failing thermocouple, we would simply have replaced it and none of this would have happened. It turned out that Kurt Grob in Switzerland had (independently) been working along these same lines. Checking on the dates that both parties had ordered vinyl chemicals, Rand was just about 5 days ahead.

1.4.6. Further Improvements in Stationary Phases

In the late 50s and extending into the 60s, gas chromatographers dealt with a plethora of stationary phases. Petroleum chemists, dealing with nonpolar products, favored phases such as squalane, a fully saturated hydrocarbon ($C_{30}H_{62}$) on the basis of like-attracts-like. The upper temperature limit for Squalane is 125°, and a higher temperature paraffin, Apolane 87 ($C_{87}H_{176}$), was sometimes used instead. Chemists dealing with more polar products needed more polar stationary phases, and they turned to high boiling esters such as diethylene glycol succinate (DEGS). As more and more scientists entered the field, the range of stationary phases exploded to over 200 different phases, most of which were gradually discarded. The polysiloxane stationary phases are now widely used and have been reviewed by Blomberg [23] and Haken [24].

1.5. INTERFACING GLASS CAPILLARY COLUMNS TO INJECTORS AND DETECTORS

Attaching these relatively fragile glass capillaries to injectors and detectors *via* leak-proof connections soon became a problem; some workers punched holes in small disks of silicone rubber which were then substituted for the ferrules supplied with 1/16 inch Swagelock™ fittings, but all too frequently, as the nut on the Swagelock fitting was tightened to the point that there were no leaks, the compressive strain on the glass caused breakage at that point. Others tried lead disks, with similar effects. I was on sabbatical leave at the nuclear research institute in Karlsruhe, Germany, when a clerk in the supply room called my attention to a sheet of plastic heavily infused with graphite. After sliding Swagelock caps onto both column ends, I wound small strips of this around both ends of a glass capillary, removed the stock ferules from the Swagelock connectors to the inlet and detector, and connected the column. As I tightened the connections to a point where there was no leakage and found that the column was still intact, I was delighted. The company that made these graphitized sheets was just a few miles outside of Karlsruhe, and I paid them a visit. They were very cordial and gave me a number of samples. On my return to the U.S., J&W ordered several steel dies dimensioned on the inner measurements of a 1/16 inch Swagelock fitting and began pressing ferrules. These went on the market in 1975; they were a great success, and I have always regretted my failure to patent that product.

1.6. THE HINDELANG CONFERENCES AND THE FUSED-SILICA COLUMN

The first symposium restricted to open-tubular (capillary) columns was organized by Rudolf Kaiser and chaired by Dennis Desty in 1975; this was the first of the four Hindelang conferences and was held in an isolated village in the Bavarian Alps. Ettre visited Kaiser, first to voice his opposition to holding the meeting, and then proposed a delay, arguing that it was being held "far too early", but many other workers in the field were enthused by the idea. I still regard the first Hindelang Conference as the most exciting meeting I have ever attended. It attracted nearly everybody that was working with glass capillary columns. We listened to presentations from our associates in the morning, and spent afternoons and evenings gathered around tables seating perhaps eight participants, drinking strong German brews and exchanging views on chromatography, with the emphasis on WCOT columns. It was then that Golay approached me; he was feeble at the time and steadied by Ettre. He wanted to talk about J&W's method of coating columns. I explained that the first step was his idea of feeding a filled column, open end first, into a heated oven. Ilkova and Mistrykov's idea of screwing the coiled column into the oven showed real ingenuity: my preheater contribution would have been proposed by anyone that could actually see the operation taking place. Now that I have reached (and probably surpassed) Golay's age at the time, I believe I can now understand his curiosity, and I hope he realized that another of his brilliant ideas had borne fruit. He departed courteously, and I was left to continue my attempts to visit every table and meet every participant. I did meet virtually all the experts, absorbed a great deal of knowledge, and left filled with new ideas and a desire to get home quickly where I could get back to work in my own lab.

There are many types of glass, but we will restrict our interests to those studied for their chromatographic interest by Dandeneau and Zerenner: soda lime, borosilicate, uranium, potash soda lead, and fused silica (Table 1.1) [25]. At the third Hindelang conference, I presented a paper on multiple short pass capillary chromatography, in which one of my Ph.D. students (James A. Settlage) had recycled a butane injection sixteen times through two 7.5 m columns to achieve 1170 theoretical plates per second, and generated 850,000 theoretical plates in those sixteen passes in less than twelve

TABLE 1.1 Approximate Compositions for Glass Materials Used for Capillary Column Fabrication

	Metal oxide percent composition (w/w)									
Glass Type	**SiO_2**	**Al_2O_3**	**Na_2O**	**K_2O**	**CaO**	**MgO**	**B_2O_3**	**PbO**	**U_2O_3**	**BaO**
Soda lime	68	3	15		6	4	2			2
Borosilicate	81	2	4				13			
Uranium glass	76	3	4	2			14		1	
Potash soda lead	56	2	4	9				29		
Fused silica*	100									

* *Typically contains less than 1 ppm total metals.*

minutes [26]; as I finished, several members of the audience rushed to the podium with congratulations; at a final synopsis of the meeting, Desty cited also this paper, but I knew they were all wrong. The most significant paper was that of Dandeneau and Zerenner [25], releasing the fused-silica column. When Dandeneau presented his paper, very few people realized its importance. Sandy Lipsky, who had started Quadrex about the same time that J&W was founded, was one, and I was another. Both Lipsky and I had been trying to convert industrial analysts to WCOT glass capillaries, with but scant success. Industrial analysts realized the superiority of results generated by these columns, but they also recognized their fragility. Downtime is rarely critical in academia, but in industry it can be very serious; glass capillaries break easily, packed columns lasted longer. However, this fused-silica column changed everything! Here was a capillary column that was both strong and flexible, but because it was both labor intensive and materials intensive, it was much more expensive.

Fused-silica columns have immense tensile strength, permitting an extremely thin column wall, which in turn makes a flexible column. This permitted drawing long lengths of straight tubing, which can then be coiled. Interfacing these to detectors and injectors was now a simple task; no longer must we flame straighten the ends of the column, it was already inherently straight! But the outer surface of drawn silica requires protection. Like all silica-based glasses, fused silica is subject to flaws at its surface, and flaws grow at a rate determined by stress – and coiling this long straight piece of fused silica places it under stress – as the flaw grows larger, the column snaps at that point. To protect the outer surface, Hewlett Packard first coated the columns with a coating of polysiloxane, but this ended up as a very sticky column. DuPont then came forward with a polyamide coating that was applied during the drawing process. Later they changed to a polyimide; this seals surface flaws in the tubing. It is usually applied in several thin coats during the drawing process.

Not everyone recognized the advantages of the fused-silica column; indeed, in spite of my vigorous protests, my partners at J&W dismissed it as "a gimmick". Two weeks later, it was impossible to give, let alone sell, a glass capillary column, and J&W made the conversion to fused silica. At this point, things became very "touchy". The vast majority of scientists working in gas chromatography had careers based on glass capillary columns – drawing, deactivation, coating, etc. – now those careers were passé. I had a seminar scheduled for Milan, Italy, where I was warmly received – until my first slide was shown. It was a fused-silica column. Several attendees left the room, and the atmosphere took on a chill. Very few in this audience wanted to hear how their futures might be affected. Shortly thereafter, Georges Guiochon organized a chromatography Symposium in Cannes, France. The GC column portion of the meeting was chaired by Gerhardt Schomburg. As he called the meeting to order, he announced that there would be "no discussion of fused-silica columns". As the meeting proceeded, several attendees expressed their objections to the opening restriction and announced that they had come to the meeting just to learn more about the fused-silica column. With obvious distaste, Schomburg turned to me and said, "tell them something about fused silica". At this time, I had been exploring on-column injections with fused silica, sometimes trapping samples within the fused-silica tubing by folding the flexible tube into a "U" shape in a Dewar flask filled with liquid nitrogen, then withdrawing the "U" and re-inserting it into a flask of heated oil for injection. This approach demanded a flexible tube and would have been impossible with glass capillary tubing. An obviously uncomfortable Schomburg interrupted me with the observation that "flexibility appeals only to Americans, and only because they are too clumsy to handle glass". A few months later J&W was selling Schomburg fused-silica columns.

1.7. INCREASING SOPHISTICATION OF INSTRUMENTATION

Table 1.2 provides a timeline for important developments in instrumentation. The essential elements of instruments were developed

TABLE 1.2 Important Advances in Technology that Impacted Gas Chromatography

1955	First commercial instrument (thermal conductivity detection)
1955–1960	Rapid growth in technology
	Invention of ionization detectors
	Interfaces for coupling to mass spectrometry
	Microsyringes produced commercially
	Temperature programming introduced
1960–1970	Period of technical advances
	First open-tubular columns introduced
	Transistors replace vacuum tubes
	Improvements in detector technology
	Stable rubidium sources introduced for TID
	Improved FPD (several designs)
	Pulsed ECD introduced
1970–1980	Period of consolidation and refinement
	Microprocessor-based instruments introduced
	Self-made glass open-tubular columns increasingly used
	Computing integrators introduced for data handling
	Fused-silica open-tubular columns introduced in 1979 and catalyzes further changes in column and instrument technology
1980–1990	Period of technical advances
	Gum and immobilized phases developed
	Columns with thick-film stationary phases developed
	Wide-bore open-tubular columns developed
	Fundamental basis for injection mechanisms developed
	On-column and PTV injection developed
	Large-volume injection described
	Autosamplers for unattended injection developed
1990–2000	Period of consolidation and refinement
	Keyboard instrumentation (PC control of operation and data handling)
	Electronic pneumatic control
	Selectable elemental detection (SED)
	Pulsed flame photometric detector developed (FPD)
	Electron-capture detector with pulsed discharge source developed (ECD)
	More versatile and affordable spectroscopic detectors (MS, FTIR)
	Solid-phase microextraction affords a new approach to sampling and sample introduction
	First microfabricated gas chromatograph on a chip introduced but fails commercially
	Field portable instruments move gas chromatography out of the laboratory
2000–present	Slowing down in the growth of technology
	Micro-GC instruments for fast gas chromatography developed
	Capillary columns with ionic liquid stationary phases introduced
	Gas generators start to replace pressurized gas cylinders
	Programmable robotic systems used to automate sample preparation

by the early 1960s [8], with further developments occurring in short bursts of innovation and advances in technology followed by longer periods of evolutionary changes and consolidation. Many advances were catalyzed by advances in column technology or electronics. For example, the advent of the microprocessor brought about a radical change in instrument design and use. From this point onward, instrument functions were monitored and controlled by networks of circuits communicating with each other and with a central controller. This paved the way for the emergence of the keyboard instrument controlled by software running on a personal computer, which dominated instruments for laboratory use by the early 1990s. A significant milestone in achieving full automation of instrument operation was the introduction of electronic pressure control in the early 1980s. This allowed carrier gas and support gas pressure and flow rates to be set and monitored by electromechanical devices communicating with a central processor. With the introduction of robotic autosamplers at about the same time, the gas chromatograph could now operate without human intervention, 24 h operation became standard practice for routine analysis in high sample throughput environments, and gas chromatographs were deployed to remote locations and monitored electronically with only occasional visits for service and routine maintenance. The complex functions of gas chromatography had been reduced to those of a black box analyzer 50 years after its invention, but still evolutionary changes continue in technology with the view of minimizing the importance of the skill and knowledge of the operator in the production of data [27].

1.7.1. Column Heating

The increasing interest in capillary columns demonstrated the inadequacies of packed column instruments, and the paradigm shift accompanying the introduction of fused-silica open-tubular columns meant significant changes were needed as capillary gas chromatography quickly became normal practice. Apart from the use of liquid thermostats in the early days of gas chromatography, the column heater in a gas chromatograph is typically a forced-circulation air oven, the temperature of which can be changed in a controlled manner with time for temperature programmed separations. Good temperature control is essential to obtain reproducible retention times and to avoid peak distortions associated with temporal and spatial oven temperature gradients. A low thermal mass for the oven is also important since it allows rapid cooling after temperature programming. Typically, forced air-circulation ovens can maintain a higher rate of heating at lower temperatures than higher temperatures due to the greater amount of heat lost to the surrounding environment as the temperature ramp progresses. For fast gas chromatography, some manufacturers have addressed this problem with the oven inside an oven concept or more directly using resistive heating achieved through the use of a metallic heating element to transfer heat to the column by conduction. Fast gas chromatography makes use of short columns of small internal diameter, thin film columns, higher carrier gas flow rates, and fast temperature program rates (Table 1.3) [28]. For the fastest separations the limiting instrument conditions become the available column inlet pressure, maximum oven temperature program rate, maximum detector sampling rate, detector sensitivity and noise level, and sample introduction (initial band width). Typical laboratory instruments can usually meet the requirements for fast gas chromatography (first entry in Table 1.3), but special-purpose instruments are required for very fast gas chromatography (second entry in Table 1.3). The entries in Table 1.3 reflect what is within the capability of commercial instruments with little modification (first entry) and the

TABLE 1.3 Instrumental Parameters for Fast Gas Chromatography

(i)	Fast gas chromatography (separation times <10 min)
	Peak widths 0.5–2 s
	Columns 5–15 m with I.D. 0.25–0.1 mm
	Temperature program rates 20–60 °C/min
	Column inlet pressure <15 bar
	Data acquisition rate 50 Hz
	Mobile phase velocity 1.4 × u_{opt}
(ii)	Very fast gas chromatography (separation time in seconds)
	Peak widths 50–200 ms
	Columns 2–10 m and I.D. 0.1–0.05 mm
	Temperature program rates >60 °C/min
	Data acquisition rates 50–200 Hz

limits of current technology requiring special purposing of instruments (second entry). From the perspective of separation speed, the capability of gas chromatographic instruments limits performance more so than column technology in contemporary practice.

1.7.2. Sample Introduction

The limited sample capacity and low carrier gas flow rates associated with open-tubular columns make sample introduction more difficult than for packed columns. A thermostated flash vaporization chamber in which the evaporated sample is mixed with carrier gas and divided between a stream entering the column (carrier gas flow) and a stream vented to waste (split flow) was the first practical solution to this problem. Split injection can be used to generate extremely small bandwidths for high peak capacity separations but discriminates against less volatile compounds and the quantitative analysis of wide boiling point mixtures is difficult, and for samples present in dilute solution, detectability is limited by the small amount of sample transferred to the column. The splitless injection technique was devised (actually discovered accidentally [29]) to overcome some of the deficiencies of split injection for the analysis of mixtures in dilute solution through the transfer of relatively large sample volumes to the column. The gas flow through a splitless injector is relatively low, and the sample is introduced into the column over a comparatively long time (30–60 s), relying on cold trapping and/or solvent effects to refocus the compounds at the head of the column. The importance of these refocusing mechanisms was not fully understood at first and much of our current knowledge of the injection mechanism owes a great deal to the exemplary studies of Konrad Grob and the publication in 1986 of his classic book on split and splitless injection [29].

The programmed-temperature vaporizer (PTV) injector is perhaps the most versatile injector developed for gas chromatography. It facilitates split and splitless injection as well as new approaches for large-volume injection with solvent elimination [30]. The PTV injector has a low thermal mass to allow rapid heating and cooling. The sample is introduced at a relatively low temperature and then raised ballistically to a temperature sufficient for rapid volatilization of the highest boiling sample components. Slow sample introduction at a low temperature with solvent elimination facilitates the injection of large sample volumes (usually <1 ml) for trace analysis. The PTV injector was originally introduced in Europe in the 1980s but it was over a decade later before it gained traction in the USA. For much of this time it was an option but not heavily promoted or supported by instrument companies based in the USA. This was probably due to a combination of commercial preferences and the not-invented-here syndrome.

The production of wide-bore fused-silica capillary columns coated with immobilized stationary phases in the early 1980s allowed the syringe needle to be introduced directly into the column and eliminated the problem of removal of the stationary phase by the relatively large volume of solvent released inside the column by the syringe [31]. These advances in column technology facilitated the development of cold on-column injection where the sample is introduced as a liquid into the column inlet and subsequently vaporized. Discrimination because of volatility differences was virtually eliminated and the risk of sample decomposition minimized. With secondary cooling of the injector, the oven temperature could be kept well above the boiling point of the solvent while maintaining the column inlet at a much lower temperature. This is important for using on-column injection in high-temperature gas chromatography. Dirty samples present a problem owing to contamination of the sample introduction zone, which leads to poor chromatography and unreliable quantification. Both on-column and programmed-temperature vaporizer injection afford high accuracy and precision. These injectors also facilitated the direct coupling of other chromatographic systems to gas chromatography, such as liquid chromatography and supercritical fluid chromatography [32]. In the 1980s, Carlo Erba manufactured an instrument for on-line liquid chromatography–gas chromatography, but the uptake was poor and the system was discontinued.

The array of sample introduction methods for gas chromatography and an understanding of the mechanistic details on which they rely were complete by the end of the 1980s and modifications since then have been evolutionary. The most visible change is the shift from manual to automated sample introduction systems that accept samples in vials and can be programmed for sequential analysis of each or selected vials in a batch. The introduction of solid-phase microextraction in the early 1990s simplified sample preparation and handling steps using a syringe containing a retractable sorbent fiber compatible with solvent-free sample introduction [33]. By the turn of the century, this had become one of the most popular sampling/sample introduction techniques in gas chromatography. It can be seen as the forerunner of the liquid-phase microextraction techniques developed over the last decade. Together these microextraction formats are responsible for achieving a better scaling of sample size requirements to sample utilization capabilities of capillary columns and are having a considerable impact on laboratory working practices.

1.7.3. Detectors

Gas chromatography is blessed by a number of robust and sensitive detectors based on gas-phase ionization processes (e.g. flame ionization, thermionic ionization, electron capture, and photoionization), bulk property (thermal conductivity), optical (flame photometric, chemiluminescence, and atomic emission), electrochemical (electrolytic conductivity), and advanced spectroscopic (mass spectrometry and infrared spectroscopy) detection principles. Many of these detectors were introduced during the initial phase of the development of gas chromatography and are still used today in a modified form reflecting changes in enabling technology. The exception is the (Hall) electrolytic conductivity detector, which, because of its large detector volume and basis in wet chemical processes, is little used today. Interfacing of modern detectors to capillary columns is not normally difficult except for fast gas chromatography where the detector volume and data acquisition rates limit the use of some detectors. The flame ionization detector facilitated many of the early developments in capillary columns (Section 1.4). This detector has a low dead volume, a high sensitivity for nearly all carbon-containing compounds, and an extremely wide linear response range. Other detectors were developed in response to the need for element-selective or structure-selective detection. These detectors allowed the analysis of target compounds at low concentrations in complex samples, and made an important contribution to the rapid acceptance of gas chromatography in laboratories performing routine analysis as well as research laboratories.

A special mention should be made of the coupling of gas chromatography with mass spectrometry (GC–MS). This coupling dates almost from the beginning of gas chromatography, but in the early days the practical problems and high cost meant that the combination was confined to a few research laboratories. When packed columns dominated the practice of gas chromatography, separators were required to reduce typical column flow rates to the vacuum-handling capacity of the ion source of the mass spectrometer. Separators based on a jet, glass fritted tube, or diffusion membrane design allowed sample enrichment while simultaneously reducing the gas flow to the ion source. Separators disappeared nearly completely with the introduction of fused-silica open-tubular columns, which could be routed from the column oven through a heated transfer line directly into the ion source of the mass spectrometer. During the time that instrumentation for gas chromatography was advancing, so were all aspects of mass spectrometry, which became more powerful, affordable, reliable, and available to laboratories with low-skilled personnel. The great advantage of the mass spectrometer as a detector is its ability to provide structural information for identification using different ionization approaches and through techniques like tandem mass spectrometry. Another advantage is its high sensitivity and selectivity as a detector in the selected ion or high-resolution mode.

1.7.4. Data Handling

In the early days of gas chromatography peak quantification was done by purely manual measurements of peak heights, although it was recognized that peak area measurement was fundamentally better but more difficult. The product of the peak height and the peak width at half height could be used as an approximate area measurement but this was neither fast nor (usually) accurate, especially for narrow peaks. Approaches based on planimetry and cut–and-weigh procedures could afford reasonable accuracy but demanded considerable skill to achieve acceptable results and were slow and tedious. Electromechanical devices such as the ball and disk integrators and integrating amplifiers were early attempts at automating this time-consuming process. These devices were rather limited in range and speed of response. The advent of microprocessors in the early 1970s ushered in the computing integrator which took over the labor-intensive process of data management, eventually resulting in devices that could record a chromatogram simultaneously with data acquisition and, when the separation was complete, instantly generate reports in tabular form of all sorts of descriptive information from the chromatogram, including the calculation of peak areas. It was not long before mainframe computers were in use to handle data produced from a number of instruments simultaneously. The development of the personal computer and the rapid growth in processor speed and memory, however, had a greater impact and resulted in data management being handled by software running on a dedicated personal computer with information being archived locally and/or centrally on a laboratory information network server. Today, the large amount of data produced by capillary columns (especially when coupled to a mass spectrometer) can be handled by a personal computer which simultaneously functions as the system manager controlling and monitoring all aspects of the chromatograph. What has not changed is that the methods used to manipulate the chromatographic data for display and calculation are considered proprietary information, requiring an act of faith that what is provided as a printout or on screen is exactly what happened in the column [34]. What is clear, however, is that it would be nearly impossible to persuade scientists to return to the status quo before the electronic age of data handling as laboratory productivity would be severely limited.

1.7.5. Comprehensive Gas Chromatography

Sometimes the separating power of gas chromatography, even when augmented by the identification power of spectroscopic detectors such as mass spectrometry, is still insufficient to accurately determine the composition of a sample. This problem is tackled in contemporary studies using comprehensive gas chromatography. Comprehensive separations have their origins in multidimensional gas chromatography developed in the early 1960s after the introduction of the Dean's switch for controlling the flow direction of a gas stream at a T-piece using pressure changes. Multidimensional (typically two-dimensional) gas chromatography employs two independent columns connected to each other by an interface. In early versions a fraction of the chromatogram, a heartcut, from the first column was isolated by cold trapping and then released by thermal desorption for separation on the second column. The disadvantage of this approach is that the fraction collected at the interface contains no information about the separation on the first column. In the 1980s, Siemens marketed a remarkable instrument for its time, the SiChromat-2, which had two independently controlled air-circulation ovens, housing the two columns, connected by a T-piece to isolate, focus, and re-inject fractions from the first column to the second by live switching based on the Dean's switching mechanism. This instrument likely contained the most complex pneumatic system of any instrument of its generation and required considerable skill in its operation. It had to compete for preference with a new generation of affordable and easy-to-use gas chromatography—mass spectrometry systems, however, and was not a commercial success [35]. In the early 1990s, Phillips introduced comprehensive gas chromatography in which the whole separation on the first column was transferred to a second column of different selectivity, the two columns being connected through a modulation interface. In this case, the chromatogram is recorded as a two-dimensional contour plot with the planar axes consisting of the independent retention times for each compound on the two columns and the vertical axis is related to the amount of each compound. This technique was not an instant success, but after further refinements in modulator design, a decade later it had entered the main stream of commerce. These instruments are at the cutting edge of instrument development where capabilities are pushed to match column performance. The function of the modulator is to arrest, concentrate, and launch a fraction of the material contained within each peak of the chromatogram from the first column and transfer them as a series of pulses of constant frequency to the second column. To avoid overlap of individual separations on the second column, each chromatogram must be complete in less time than the cycle time of the modulator. Thus, the second-column separations must be very fast with a total separation time measured in seconds. This requires incredibly small injection bandwidths, short second columns of small internal diameter, fast detector response times, and repetitious and fast changes in instrument operating conditions. Comprehensive gas chromatography generates a large amount of data, especially when mass spectrometry is used as a detector, and challenges remain in how best to make these data available to the user in a form convenient for analysis. In these instruments, one can perhaps see a vision of the future of single column gas chromatography (based on technology developed for the second-dimension separation) while at the same time highlighting the limitations of current technology for gas chromatography.

1.8. DECLINE IN THE EXPERTISE OF THE AVERAGE GAS CHROMATOGRAPHER

At the present day the success in the commercial sector in gas chromatography has resulted in a situation not all that different from a narcotic dependency. To be classed as an expert in gas chromatography, all that seems to be necessary is the wherewithal to buy a gas chromatograph and a column. Today, those tasked to operate a gas chromatograph do not need to even make an injection, columns have become classified as laboratory supply items, and data as something that appears on a printout at the end of an experiment. When I was first introduced to gas chromatography in the 1970s, it was necessary to make, modify, and continuously improve columns and instruments to meet the needs of each application. Expert chromatographers of this time were of necessity well versed in troubleshooting and thoroughly knowledgeable of the principles and practice at what often appeared to outsiders as the dark arts of separations. On account of the immense powers of separations virtually any unskilled individual can produce data today; even the village idiot can inject a sample and obtain peaks that are interpreted as "usable data". This has led to a continuing decline in the expertise of the average practicing chromatographer from the mid-1980s to the present time. This can be perilous, because everything from column selection to trouble-shooting skills is based on a fundamental knowledge of chromatographic principles, the absence of which degrades the quality and usefulness of the information acquired by these instruments. To address these problems requires a massive educational effort before the knowledge is lost and the usefulness of gas chromatography to decision makers is called into question. When I first started to make my way in gas chromatography, I never thought that such a reversion would occur so quickly, and as retirement approaches, that intellectually I might leave the field in a poorer state than when I entered it. There is no doubt that advances in technology have delivered better instruments and columns but now in the hands of less qualified operators; this does not always lead to better-quality data.

References

[1] A.T. James, A.J.P. Martin, Gas–liquid partition chromatography: the separation and microestimation of volatile fatty acids from formic to dodecanoic acid, Biochem J 50 (1952) 679–690.

[2] L.S. Ettre, The beginnings of gas adsorption chromatography 60 years ago, LC-GC North America 26 (2008) 48–51.

[3] F. Prior, Determination of adsorption heats of gases and vapors by application of the chromatographic method in the gas phase. Doctoral theses. University of Innsbruck: Innsbruck, Austria; 1947 (in German).

[4] O. Bobleter, Exhibition of the first gas chromatographic work of Erika Cremer and Fritz Prior, Chromatographia 43 (1996) 444–446.

[5] A.N.H. Ray, Gas chromatography, I–II, J Appl Chem 4 (1954). 21–25 and 82–85.

[6] L.S. Ettre, A. Zlatkis (Eds.), 75 years of chromatography – a historical dialogue, Elsevier, Amsterdam, 1979.

[7] H.A. Laitinen, G.W. Ewing, A history of analytical chemistry, division of analytical chemistry, American Chemical Society, Washington, 1977, pp. 303–304.

[8] L.S. Ettre, The early development and rapid growth of gas chromatographic instrumentation in the United States, J Chromatogr Sci 40 (2002) 458–472.

[9] M.J.E. Golay, in: D.H. Desty (Ed.), Gas chromatography 1958 (Amsterdam symposium), Butterworths, London, 1958, pp. 139–143.

[10] J.E. Lovelock, S.R. Lipsky, Electron affinity spectroscopy – a new method for the identification of functional groups in chemical compounds separated by gas chromatography, J Am Chem Soc 82 (1960) 431–433.

[11] J. Harley, W. Nel, V. Pretorious, Flame ionization detector for gas chromatography, Nature (London) 181 (1958) 177–178.

[12] I.G. McWilliam, R.A. Dewer, in: D.H. Desty (Ed.), Gas chromatography 1958 (Amsterdam symposium), Butterworths, London, 1958, p. 142.

[13] I.G. McWilliam, The origin of the flame ionization detector, Chromatographia 17 (1983) 241–243.

[14] L.S. Ettre, The invention, development and triumph of the flame ionization detector, LC-GC (Europe) 15 (2002) 364–373.

[15] K. Grob, Hansjurg Jaeggi – Obituary, J High Resolut Chromatogr 12 (1989) 459.

[16] D.H. Desty, J.N. Haresnape, B.H.F. Whyman, Construction of long lengths of coiled glass capillary, Anal Chem 32 (1960) 302–304.

[17] K. Grob, The role of column technology in capillary gas chromatography, J High Resolut Chromatogr 7 (1984) 252–257.

[18] M.J.E. Golay, in: D.H. Desty (Ed.), Gas chromatography 1958 (Amsterdam symposium), Butterworths, London, 1958, pp. 36–55.

[19] E.L. Ilkova, E.A. Mistrukov, A simple versatile method for coating glass capillary columns, J Chromatogr Sci 9 (1971) 569–570.

[20] M.C. Bourne, W.G. Jennings, Existence of two soil species in detergency investigations, Nature (London) 197 (1963) 1003.

[21] M.C. Bourne, W.G. Jennings, Kinetic studies of detergency. I Analysis of cleaning curves, J Am Oil Chem Soc 40 (1963) 517–523.

[22] M.C. Bourne, W.G. Jennings, Kinetic studies of detergency. II Effect of age, temperature, and cleaning time on rates of soil removal, J Am Oil Chem Soc 40 (1963) 523–530.

[23] L. Blomberg, Current aspects of the stationary phase in gas chromatography, J High Resolut Chromatogr 5 (1982) 520–533.

[24] J.K. Haken, Developments in polysiloxane stationary phases in gas chromatography, J Chromatogr 300 (1983) 1–77.

[25] R.D. Dandeneau, E.H. Zerenner, An investigation of glasses for capillary chromatography, J High Resolut Chromatogr 2 (1979) 351–356.

[26] W. Jennings, J.A. Settlage, R.J. Miller, Multiple short pass glass capillary gas chromatography, J High Resolut Chromatogr 2 (1979) 441–443.

[27] G.M. Gross, V.R. Reid, R.E. Synovec, Recent advances in instrumentation for gas chromatography, Current Anal Chem 1 (2005) 135–147.

[28] C. Cruz-Hernandez, F. Destaillates, Recent advances in fast gas chromatography: application to the separation of fatty acid methyl esters, J Liq Chromatogr Rel Technol 82 (2009) 1672–1688.

[29] K. Grob, Classical split and splitless injection in capillary gas chromatography, 2nd ed., Huethig, Heidelberg, 1986. 1992.

[30] P. Sandra (Ed.), Sample introduction in capillary gas chromatography, Huethig, Heidelberg, 1985.

[31] K. Grob, On-column injection in capillary gas chromatography, 2nd ed., Huethig, Heidelberg, 1987. 1991.

[32] K. Grob, On-line coupled LC–GC, Huethig, Heidelberg, 1991.

[33] J. Pawliszyn, Solid-phase microextraction: theory and practice, Wiley-VCH, New York, 1997.

[34] N. Dyson, Chromatographic integration methods, The Royal Society of Chemistry, Cambridge, UK, 1998.

[35] W. Bertsch, Two-dimensional gas chromatography concepts, instrumentation, and applications – part 2: comprehensive two-dimensional gas chromatography, J High Resolut Chromatogr 23 (2000) 167–181.

CHAPTER

2

Theory of Gas Chromatography

Leonid M. Blumberg

OUTLINE

Gas Chromatography DOI: 10.1016/B978-0-12-385540-4.00002-X

2.1. INTRODUCTION

Analytical gas chromatography (GC) (Figure 2.1) is a technique of separation of components of mixtures (*samples*) with the purpose of obtaining information about their molecular compositions and amounts. The information obtained from a chromatographic analysis can include a chromatogram (a graphical image of a detector output), information regarding the heights and the areas of the *resolved* (adequately separated) peaks in a chromatogram, their molecular identity, etc.

It is assumed in this chapter that the reader is familiar with the basic GC concepts and structures such as the *capillary* i.e. the *open-tubular* column (OTC), *carrier gas* as the *mobile phase* in GC, liquid and solid *stationary phases*, the key mechanisms of the interaction of the solutes migrating (being carried by the carrier gas) through a column with the stationary phase, temperature and/or pressure programming, etc. Several chapters of this volume and other sources [1–15] could help refresh this information.

A theory of GC can be tailored to emphasize different aspects of GC operations. One might be interested, for example, in accurate prediction of retention times and degree of separation of all or some predetermined peaks. As the computerized calculations are usually acceptable for such predictions, complexity of the models becomes a relatively minor issue compared to the accuracy of the model. It might become necessary for the accurate predictions to account for generally minor factors such as nonideal carrier gas, effects of the liquid stationary phase surface, nonuniform column stationary phase thickness, etc. These factors, however, are outside of the main concern of this review. Its main focus is on the effect of the operational parameters of GC analysis on its *general performance*. The theory presented here is designed to address issues such as general effect of the column dimensions, carrier gas type and flow rate, temperature programming, and other factors on the peak retention times, separation, detection limits, and the tradeoffs between these performance factors. To get simple and insightful mathematical descriptions of these relations, simplicity of the basic models becomes more important than their accuracy. As a general trend, this review favors simplicity over unnecessary accuracy. The aforementioned and other similar minor secondary factors complicating the models are typically ignored.

Temperature programming is an indispensible technique in analyses of complex mixtures. The first broad theoretical study of general

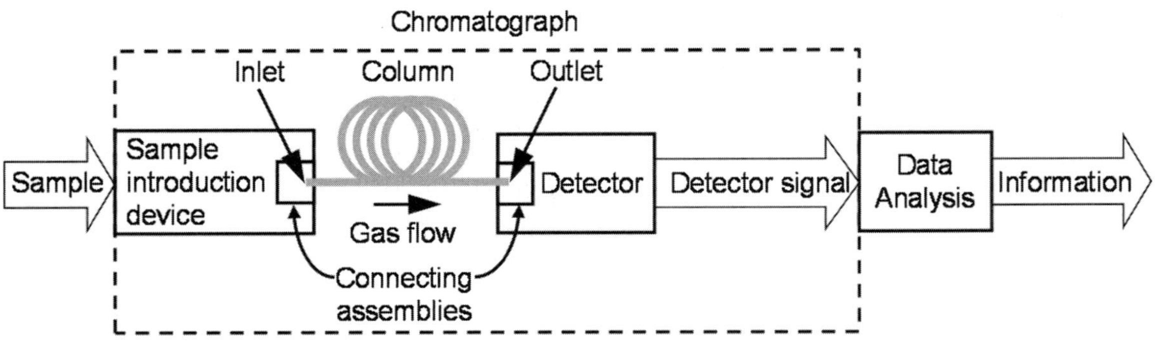

FIGURE 2.1 Block diagram of a chromatographic system.

performance of temperature-programmed GC is known from Giddings who outlined the theory and came close to finding optimal heating rate [16]. Although Giddings did not explicitly express this point of view, he essentially treated the hold-up time as the basic clock in chromatography and found that the optimal heating rate should be inversely proportional to the hold-up time. The view of the hold-up time as the basic clock in chromatography is the basis for the more recently developed techniques of method translation and retention locking in chromatography, as well as for expressing optimal heating rate in terms of temperature per hold-up time. An important contribution to theory of temperature-programmed GC came from Habgood and Harris [12,17,18], who extended the concepts of plate number and plate height previously known only for isothermal GC, to temperature-programmed GC.

This chapter is an overview of the current state of GC theory with the emphasis on temperature-programmed GC. The isothermal GC is viewed as a temperature-programmed GC with zero heating rate. The chapter starts with the review of relevant theoretical information regarding interaction of organic compounds with liquid polymers, properties of ideal gases, and flow of ideal gases. This information is then used as a basis for addressing such core issues of GC theory as formation of retention times and other parameters of eluting solutes; formation of peak spacing and factors affecting reversal of peak order; and formation of peak widths. The overview of all these topics is mostly based on material in a recently published book on temperature-programmed GC [1]. In addition to that, this chapter addresses the issues of column optimization. Optimal flow rate, optimal heating rate, and optimal tradeoff between a column separation performance, analysis time, and detection limits were found from a single perspective. This material is currently available only in periodic literature where it was addressed from different perspectives.

The opening sentence of this introduction implies that only the **analytical** GC is considered in this chapter. The scope of the chapter is primarily limited to the most widely used **capillary** (open-tubular) columns. Other broad constraints highlighted by the **boldface** type are introduced later.

2.2. NOMENCLATURE AND OTHER CONVENTIONS

Abbreviations

DL – detection limit
EOF – efficiency-optimized flow rate
GLC – gas–liquid chromatography
MDA – minimal detectable amount
MDC – minimal detectable concentration
OTC – open-tubular column
PLOT – porous-layer open-tubular (column)
SEOF – specific efficiency-optimized flow rate
SOF – speed-optimized flow rate
SSOF – specific speed-optimized flow rate
WCOT – wall-coated open-tubular (column)

Subscripts

a – asymptotic parameter (parameter of a solute that was highly retained at the beginning of a heating ramp)
init – parameter at initial temperature of a heating ramp
i – parameter at a column inlet
Max – speed-optimized maximum
max – maximum, except for speed-optimized conditions
middle – parameter in the middle of the heating ramp
Min – speed-optimized minimum
min – minimum, except for speed-optimized conditions
Opt – speed-optimized parameter
o – parameter at a column outlet

opt – optimal parameter, except for speed-optimized conditions
R – parameter at retention time
ref – reference parameter
st – parameter at standard temperature and pressure

Symbols (Local Symbols Used only Shortly Before or After Their Definition Are not Listed)

D	solute diffusivity in a gas
$\mathscr{D}$	solute local and/or instant dispersion rate (Eqn (2.94))
D_g	gas self-diffusivity
D_{pst}	solute diffusivity in a gas at standard pressure and arbitrary temperature
D_S	solute diffusivity in stationary phase
E	column efficiency, Eqn (2.127)
d	column internal diameter
d_f	stationary phase film thickness
F	gas flow rate (Eqn (2.40))
f	gas-specific flow rate (Eqn (2.42))
G	Gibbs free energy (Eqn (2.6))
g	dimensionless Gibbs free energy (Eqn (2.6))
G_H	enthalpy (Eqn (2.6))
G_S	entropy (Eqn (2.6))
H	(apparent) plate height (Eqn (2.101))
$\mathscr{H}$	local and/or instant plate height (solute spatial dispersion rate) (Eqn (2.94))
h	dimensionless plate height (Eqn (2.118))
j	James–Martin compressibility factor (Eqn (2.36))

(*Continued*)

Symbols (Local Symbols Used only Shortly Before or After Their Definition Are not Listed) (*cont'd*)

j_G	Giddings compressibility factor (Eqn (2.105))
j_H	Halász–Hartmann–Heine compressibility factor (Eqn (2.47))
k	retention factor (Eqn (2.2))
K_c	distribution constant (Eqn (2.5))
L	column length
ℓ	dimensionless column length (Eqn (2.46))
M	molar mass
N	(apparent) plate number (Eqns (2.100) and (2.121))
$\mathscr{N}$	directly counted plate number (Eqn (2.92))
P	relative pressure (Eqn (2.25))
ΔP	relative pressure drop (Eqn (2.25))
p	pressure
Δp	pressure drop (Eqn (2.25))
p_a	ambient pressure
p_g	gauge pressure (Eqn (2.25))
p_{st}	standard pressure, 1 atm
q_{anal}	throughput (Eqn (2.187))
$\mathscr{R}$	molar gas constant, 8.31447 J/(K mol)
R_T	heating rate (Eqn (2.54))
R_{TM}	normalized heating rate (Eqn (2.63))
r_T	dimensionless heating rate (Eqn (2.59))
s_c	separation capacity of chromatographic analysis (Eqn (2.130))

(*Continued*)

Symbols (Local Symbols Used only Shortly Before or After Their Definition Are not Listed) (*cont'd*)

T	temperature
T_{char}	characteristic temperature (Eqn (2.11))
ΔT_{char}	difference in characteristic temperatures of two solutes
T_{end}	temperature at the end of heating ramp
ΔT_{end}	heating range of a heating ramp (Eqn (2.132))
T_{init}	initial temperature of a heating ramp
T_{norm}	normal temperature, 298.15 K (25 °C)
T_R	solute elution temperature
ΔT_R	difference in elution temperatures of two solutes
T_{st}	standard temperature, 273.15 K (0 °C)
t	time
t_g	gas propagation time (Eqn (2.50))
t_m	dynamic hold-up time (Eqn (2.52))
t_{mol}	mean time between molecular collisions
t_M	hold-up time
$t_{M,R}$	t_M in static analysis under conditions existed at time t_R in dynamic analysis
$t_{M,char}$	hold-up time at characteristic temperature
t_R	retention time
Δt_R	retention time difference of two solutes
$t_{R,stat}$	retention time in static analysis
u	gas velocity (Eqn (2.27))

(*Continued*)

Symbols (Local Symbols Used only Shortly Before or After Their Definition Are not Listed) (*cont'd*)

$\bar{u}$	gas average velocity (Eqn (2.35))
V	volume
v	solute velocity
X_D	gas parameter (Table 2.1)
X_r	sensitivity of ΔT_R to relative change in r_T (Eqn (2.79))
X_T	parameter described in Eqn (2.126)
X_θ	sensitivity of ΔT_R to $\Delta\theta_{char}$ (Eqn (2.77))
X_σ	gas parameter (Table 2.4)
z	distance from a column inlet
β	phase ratio (Eqn (2.7))
ζ	dimensionless distance from column inlet (Eqn (2.31))
η	gas viscosity
θ_{char}	characteristic thermal constant (Eqn (2.11))
$\Delta\theta_{char}$	difference in characteristic thermal constants of two solutes
θ_T	parameter defined in Eqn (2.61)
ϑ_{G1}	parameter defined in Eqn (2.88)
$\mathit{\Lambda}$	linear dynamic range (Eqn (2.187))
λ	mean free path
μ	solute mobility (Eqn (2.2))
μ_{eff}	solute effective mobility (Eqn (2.53))
ξ	gas empirical parameter (Table 2.1)
ρ_n	noise spectral density
σ	peak width (standard deviation) in time domain

(*Continued*)

Symbols (Local Symbols Used only Shortly Before or After Their Definition Are not Listed) (*cont'd*)

σ_T	peak width (standard deviation) in temperature domain
$\tilde{\sigma}$	width (standard deviation) of a solute zone
Φ	function defined in Eqn (2.39)
φ	dimensionless film thickness (Eqn (2.8))
υ	average molecular speed
Ω	gas pneumatic resistance (Eqn (2.30))
ω	solute immobility (Eqn (2.2))
$\overline{\omega}$	distance-averaged immobility (Eqn (2.76))

2.3. GENERAL DEFINITIONS

Conditions of GC analysis are *static* (as in *isothermal isobaric* GC) if they do not change with time. Otherwise, the conditions are *dynamic* (as in temperature- and/or pressure-programmed GC). The conditions could be *uniform* (the same at any place along the column length) or *nonuniform*. Gas decompression along the column is a typical cause of the nonuniform conditions [1].

Throughout this chapter, T is assumed to be the **absolute temperature** (in kelvin, K) unless the contrary is explicitly stated. However, temperature intervals and the differences between two absolute temperatures can be expressed in units °C (degrees in Celsius scale).

Two types of *open-tubular columns* (OTCs) also known as the *capillary columns* are used in GC. The internal walls of a *wall-coated* open-tubular (WCOT) column are coated with absorbing liquid polymer film. The internal walls of a *porous-layer* open-tubular (PLOT) column are covered with a solid porous-layer having large adsorbing surface. Typically, the thickness of a liquid film and a porous solid layer is much smaller than the internal diameter of a column tubing and has a minor effect on a column flow. It is assumed throughout this chapter that the **diameter of the internal open space of an OTC column is the same as the internal diameter of the column tubing**. Both are referred throughout the chapter as a column *internal diameter* (briefly, *diameter*).

2.4. SOLUTE–COLUMN INTERACTION

2.4.1. Solute Distribution Between Mobile and Stationary Phases

A solute migrating through a column might interact with its stationary phase (might be partially absorbed by a liquid stationary phase or adsorbed on a solid stationary phase). As a result, the net *velocity* (v) of migration of a solute zone at some location (z) along the column can be smaller than the carrier gas velocity (u) at the same location. The relationship between v and u can be expressed as

$$v = \mu u \tag{2.1}$$

where μ is the mobile phase fraction of the solute (see below).

Different *solutes* – components of a sample – might be differently distributed between the stationary phase and the inert carrier gas in a given column. In that case, the solutes migrate through the column with different velocities (v) and elute from the outlet at different *elution times* also known as *retention times* – thus being *separated* from each other. This means that

Different distribution of different solutes between the stationary phase and the inert carrier gas resulted from the different interaction of the solutes with the stationary phase in a column is the root cause of the solute separation.

The distribution of a solute between the column phases can be expressed in several ways [1]:

$$k = \frac{a_{\text{stat}}}{a_{\text{mob}}}, \quad \mu = \frac{a_{\text{mob}}}{a}, \quad \omega = \frac{a_{\text{stat}}}{a} \tag{2.2}$$

where $a = a_{\text{mob}} + a_{\text{stat}}$, a_{mob} and a_{stat} are, respectively, the total amount of a solute and its amounts in mobile and stationary phases. Parameters k, μ, and ω are, respectively, a solute *retention factor* [1,15] (*capacity factor* [8,9], *capacity ratio* [15], *partition ratio* [6,10,12]), *mobility* [1] (a parameter originally introduced by Consden et al. [19] and also known as the *retardation factor* [7,15], the *retention ratio* [8], and *the frontal ratio* [9]), and *immobility* [1] (*interaction level* [1]).

It follows directly from their definitions in Eqn (2.2) that parameters k, μ, and ω relate to each other as

$$k = \frac{1-\mu}{\mu} = \frac{\omega}{1-\omega}, \quad \mu = 1-\omega = \frac{1}{1+k}$$
$$\omega = 1-\mu = \frac{k}{1+k} \tag{2.3}$$

and are bound by the conditions

$$0 \le k \le \infty, \quad 0 \le \mu \le 1, \quad 0 \le \omega \le 1 \tag{2.4}$$

Parameters μ and ω are complementary to each other. If $\mu = 0$, then $\omega = 1$. Conversely, if $= 1$, then $\omega = 0$. This justifies the complimentary terms for these parameters such as the immobility (μ) and the immobility (ω). The latter represents a measure of a solute interaction with a column stationary phase (its affinity to the phase). As such measure, the solute immobility (ω) is similar to the retention factor (k). However, ω is a normalized measure (changing from 0 to 1), while k is not.

As parameters k, μ, and ω are interdependent with each other, either one is sufficient for all evaluations that depend on a solute distribution between the phases. However, a need for the simplicity and transparency of interpretation of theoretical results justifies preferential choice of one parameter over the others depending on the type of evaluations. Thus, the mobility (μ) most directly affects (Eqn (2.1)) the solute velocity (v); the immobility (ω) most directly affects the separation [20,21]; and the retention factor most directly relates to the physics of a solute–column interaction – our next topic.

2.4.2. Two Modes of a Solute–Column Interaction

GC utilizing the columns with liquid stationary phase is known as *gas–liquid chromatography* (GLC) and as *partition* GC [22]. This is the most frequently used type of GC by far, and the subject of the key attention in this chapter.

The equilibrium of a solute distribution between two phases (like a liquid polymer phase and a gas phase in partition GC) can be described by the *distribution constant* [1,15] (*partition coefficient* [5,6,8–10,22,23])

$$K_c = \frac{C_{\text{stat}}}{C_{\text{mob}}} \tag{2.5}$$

where C_{mob} and C_{stat} are the concentrations of the solute in the mobile and the stationary phase, respectively. Dependence of the distribution on temperature (T) can be described by the integrated *van't Hoff equation* representing an *ideal thermodynamic model* [8,24] (Figure 2.2.a):

$$K_c = e^g, \quad g = \frac{G}{\mathcal{R}T} = -\frac{G_S}{\mathcal{R}} + \frac{G_H}{\mathcal{R}T} \tag{2.6}$$
(ideal thermodynamic model)

where G, G_S, G_H, and g are the *Gibbs free energy*, the *entropy*, the *enthalpy*, and the *dimensionless Gibbs free energy* [1,21], respectively, of a solute transfer from stationary to mobile phase.

It follows from Eqns (2.2), (2.5) and (2.6) that

$$k = \frac{K_c}{\beta} = \frac{e^g}{\beta}, \quad \beta = \frac{\text{stationary phase volume}}{\text{mobile phase volume}} \tag{2.7}$$

where β is the *phase ratio*.

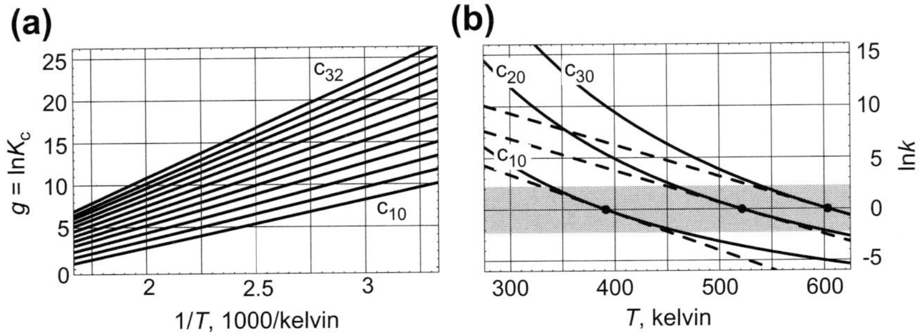

FIGURE 2.2 Thermodynamic properties of *n*-alkanes in 5% DDP (poly(diphenyldimethylsiloxane) containing 5% diphenyl monomer (a) dimensionless Gibbs free energies (g, Eqn (2.6)) of transfer from stationary to mobile phase vs. inverse temperature ($1/T$) and (b) ln$k$ vs. T (Eqn (2.9) at $\varphi = 0.001$). Within shaded area in (b) where $0.1 \le k \le 10$, a tangent line to a ln$k(T)$ curve at $k = 1$ closely approximates the curve itself. *Source: Reconstructed from ref. [1] with permission.*

For a capillary column having its internal walls coated with a liquid stationary phase *film*, β can be expressed as

$$\beta \approx \frac{1}{4\varphi}, \quad \varphi = \frac{d_f}{d} \tag{2.8}$$

where φ is *dimensionless film thickness* [1], d is a column *internal diameter*, and d_f is the stationary phase *film thickness*. Eqn (2.7) becomes (Figure 2.2.b)

$$k = 4\varphi e^g \tag{2.9}$$

which, in the case of the ideal thermodynamic model of Eqn (2.6), becomes

$$\ln k = \frac{G_H}{\mathscr{R}T} - \frac{G_S}{\mathscr{R}} + \ln(4\varphi) \tag{2.10}$$
(ideal thermodynamic model)

Ideal thermodynamic model of Eqn (2.6) is not the only known model of a solute interaction with a column stationary phase in GC [1,8,24–29]. The simpler ones, however, are typically less accurate than others. The model in Eqn (2.6), although relatively simple, has its own shortcomings. Its two parameters, G_H and G_S, change with temperature and solute concentration. The dependence of G_H and G_S on a solute concentration can be negligible if the stationary phase *overloading* is avoided when relatively small sample amounts are analyzed. It is assumed throughout this review that, unless otherwise is explicitly stated, the columns are **not overloaded** and the solute concentration does not affect parameters G_H and G_S. As for the dependence of these parameters on temperature, it is negligible if a solute migration is evaluated within a relatively narrow temperature range (less than ±50 °C [1]) around the temperature at which the parameters were found (measured, calculated, etc.).

Even when its parameters (G_H and G_S) are fixed quantities, the use of the ideal thermodynamic model in theoretical studies of temperature-programmed GC is highly problematic because ln k is not a linear function of T which, in Giddings' words [30], "always leads to rather formidable integrals whose solutions are not easily obtained" [12,16]. Fortunately, within a moderately wide range, $0.1k_0 \le k \le 10k_0$, of retention factors (k) around some predetermined retention factor k_0, the difference between actual ln $k(T)$ curve for a solute and the tangent line to that curve at $k = k_0$ is small and the tangent line can be used as a *linearized model* of a solute–column interaction. This is important because, during a heating ramp covering a reasonably wide temperature range, all solutes elute with approximately equal retention factors (k_R) [1,16] (see also Section 2.7.5). The solutes also migrate through the major portion of a column length with retention factors that are not much larger than $10k_R$ [1]. This suggests that a linearized model can be used for evaluation of the key elution parameters of the solutes in temperature-programmed GC analyses. A choice of $k_0 = 1$ leads to the simplest transformation of parameter G_S and G_H into parameters of a linearized model, and vice versa [1]. It also helps that, at optimal or near optimal heating rate, retention factors of eluting solutes are not far from unity [1].

Let $\ln k(T)$ be some (not necessarily ideal thermodynamic) model of a solute–column interaction where functions $\ln k(T)$ for all solutes are roughly similar to the ones in Figure 2.2.b. The tangent line at $k = 1$ for each solute can be described as

$$\ln k = -\frac{T - T_{char}}{\theta_{char}}, \quad T_{char} = T|_{k=1}, \qquad \theta_{char} = -\left(\left.\frac{d\ln k(T)}{dT}\right|_{k=1}\right)^{-1} \tag{2.11}$$

In other words, the straight line described here can be used as a *linearized model* of a solute–column interaction, i.e.

$$\ln k = -\frac{T - T_{char}}{\theta_{char}} \ (\textit{linearized model}) \tag{2.12}$$

Parameters T_{char} and θ_{char} of this model are defined in Eqn (2.11). As parameters of a straight line, quantities T_{char} and $-1/\theta_{char}$ are its intercept and the slope, respectively.

From another perspective, quantities T_{char} and θ_{char} are the key parameters of each particular model $\ln k(T)$. They can be collectively called as *characteristic* parameters of a solute–column interaction, with T_{char} being *characteristic temperature* (T_{char}) and θ_{char} being *characteristic thermal constant* of the interaction [1]. Both parameters, T_{char} and θ_{char}, are measured in units of temperature. Parameter T_{char} is the temperature at which $k = 1$. It can be any temperature (T) used in GC analyses and beyond. Quantity $(-1/\theta_{char})$ is the slope of $\ln k(T)$ at $k = 1$. An increase in T reduces k. As a result, $\ln k(T)$ has a negative slope, and parameter θ_{char} is a positive quantity typically ranging between 25 °C and 40 °C [1] and describing a measure of change in k due to a change in T. Increasing T by θ_{char} (from T to $T + \theta_{char}$) reduces k by a factor of e (e ≈ 2.72); increasing T by 1 °C reduces k by $1/\theta_{char}$; in order to reduce k by a factor of 2, T should be increased by θ_{char} ln2.

Example 2.1. If $\theta_{char} = 30$ °C then raising T by 30 °C reduces k by a factor of 2.72; raising T by 1 °C reduces k by 3.3%; in order to reduce k by a factor of 2, T should be raised 21 °C [1,30–32].

The linearized model was successfully used for computerized calculations of retention times and peak resolutions in GC applications [28]. However, the model is not sufficiently accurate as a universal basis for such calculations for all GC analyses. Moreover, it is not clear if any known model satisfies this requirement. However, as stated in the Introduction, accurate prediction of chromatograms is not a goal for the theory in this chapter. Rather, the goal is theoretical evaluation of the factors affecting general performance of GC analyses and their optimization. Among these factors are the effects of a column temperature and heating rate not on each particular solute, but general trends in such effects on all solutes. The linearized model of a solute–column interaction is sufficiently accurate for such evaluations and is the only known model that leads to broad theoretical solutions [1]. The theory reviewed in this chapter is based on the **linearized model** of a solute–column interaction.

The usefulness of characteristic parameters (T_{char} and θ_{char}) goes beyond the linearized model in Eqn (2.12). The fact that characteristic parameters T_{char} and θ_{char} have a clear chromatographic interpretation provides a certain insight into chromatographic properties of the solutes (their expected elution temperatures in temperature-programmed analyses, dependence of the temperatures on the heating rate, etc.)

and facilitates further theoretical exploration of those properties. This cannot be said about thermodynamic parameters G_H and G_S in Eqn (2.10). While having a clear thermodynamic interpretation, they tell very little about chromatographic properties of the solute. Thus, the fact that, for C_{20} in Figure 2.2, $G_H \approx 70$ kJ/mol and $G_S \approx 90$ J/mol/K [1], tells very little about chromatographic properties of these parameters. This suggests that characteristic parameters T_{char} and θ_{char} might be useful for expressing not only a linearized model but an ideal thermodynamic model as well.

If parameters, G_H and G_S, of the ideal thermodynamic model in Eqn (2.6) are known, then parameters T_{char} and θ_{char} defined in Eqn (2.11) could be found as [1]

$$T_{char} = \frac{G_H}{G_S - \mathscr{R}\ln(4\varphi)}, \qquad \theta_{char} = \frac{\mathscr{R}\, G_H}{(G_S - \mathscr{R}\ln(4\varphi))^2} \tag{2.13}$$

If necessary, G_H and G_S can be reconstructed from T_{char} and θ_{char} as [1]

$$G_H = \frac{\mathscr{R}T^2_{char}}{\theta_{char}}, \quad G_S = \mathscr{R}\left(\frac{T_{char}}{\theta_{char}} + \ln(4\varphi)\right) \tag{2.14}$$

Substitution of Eqn (2.14) in Eqn (2.10) yields [1]

$$\ln k = \frac{T_{char}}{\theta_{char}}\left(\frac{T_{char}}{T} - 1\right) \tag{2.15}$$

(ideal thermodynamic model)

which is the same ideal thermodynamic model as the one in Eqn (2.10) only differently expressed. It follows directly from Eqn (2.15) that $(-1/\theta_{char})$ is the slope of $\ln k$ at $T = T_{char}$ and that $k = 1$ at $T = T_{char}$.

If two columns have the same stationary phase type of different dimensionless film thicknesses (φ), then Eqns (2.13) and (2.14) can be used for transforming T_{char} and θ_{char} of each solute in one column into these parameters in another column. However, such transformation is rather complex. A simple approximate transformation for T_{char} can be obtained from the approximation [1]

$$\frac{T_{char2}}{T_{char1}} = \left(\frac{\varphi_2}{\varphi_1}\right)^{0.07} \tag{2.16}$$

It shows that T_{char} is a weak function of φ. Thus, doubling φ causes about 5% increase in T_{char}. A transformation for θ_{char} is described later.

2.4.3. Relations Between Characteristic Parameters

As mentioned earlier, a solute characteristic temperature (T_{char}) can be any temperature (T) used in GC analyses and beyond. The values of the characteristic thermal constant (θ_{char}) are typically confined to a 25 °C to 40 °C range [1]. For a given solute, parameters T_{char} and θ_{char} are independent from each other. However, there is a general trend. The solutes having larger T_{char} (and eluting at higher temperatures) have generally larger θ_{char}. In addition to that, dimensionless film thickness (φ) slightly affects θ_{char}. The trend found from examining 2412 solute–liquid pairs involving eight different liquid stationary phase types (Figure 2.3) can be described as [1]

$$\theta_{char} = \theta_{char,st}\left(\frac{T_{char}}{T_{st}}\right)^{0.7}, \qquad \theta_{char,st} = 22°\mathrm{C}\cdot(10^3\varphi)^{0.09} \tag{2.17}$$

Example 2.2. In a column with $\varphi = 0.001$, a solute having $T_{char} \approx 423$ K (150 °C) is expected to have $\theta_{char} \approx 30$ °C. A (possibly different) solute that has the same T_{char} in the same column with the same type but twice as thick stationary phase is expected to have $\theta_{char} \approx 32$ °C.

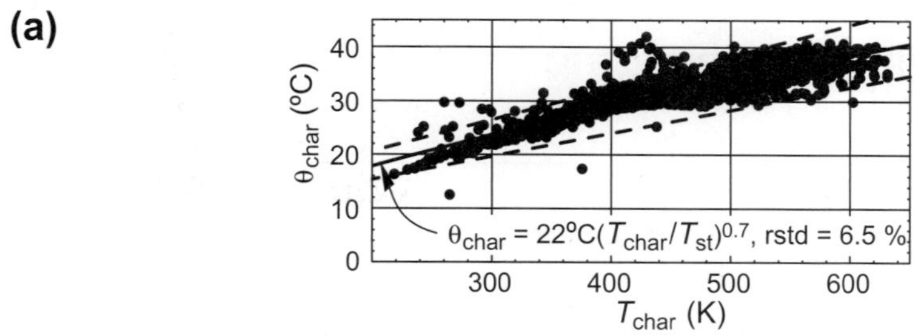

(b) Combinations of liquid polymers and solutes in (a)

Code	Composition of polymer	Number of solutes
1	dimethyl polysiloxane	925 petroleum + 199 pesticides + 213 drugs
5	5 % diphenyl-dimethyl polysiloxane	347 petroleum
1701	14 % cyanopropylphenyl-methyl polysiloxane	150 pesticides
200	35 % trifluoropropylmethyl polysiloxane	100 drugs
50	50 % diphenyl-dimethyl polysiloxane	208 PCB
Wax	polyethylene glycol (PEG)	270 industrial solvents

FIGURE 2.3 (T_{char}, θ_{char}) pairs — the dots in (a) — for 2412 solute–liquid pairs listed in (b) at $\varphi = 0.001$. The solid line is the least-squares fit of the curve in Eqn (2.17), and rstd is the relative standard deviation of vertical departures of all dots from the solid line. The dashed lines mark about $0.3\theta_{char}$-wide stripe around the solid line. Overwhelming majority of the dots is contained within the stripe.

2.5. PROPERTIES OF AN IDEAL GAS

2.5.1. Theoretical Relations

It is assumed throughout this chapter that a pure carrier gas behaves as an **ideal gas** [1,24,33]. This means that the relationship between its mass (m_g), pressure (p), temperature (T), and volume (V) is governed by the *ideal gas law*:

$$pV = \frac{\mathscr{R}\, m_g T}{M} \tag{2.18}$$

where $\mathscr{R} = 8.31447$ J/(K mol) is the *molar gas constant* and M is the gas *molar mass*. Several molecular properties of a gas are important for GC. Among them are *average molecular speed* (υ), *mean free path* (λ), *mean time between collisions* (t_{mol}), *self-diffusivity* (D_g), and *viscosity* (η). They relate to each other and to other gas properties as [1,24,33,34]

$$\upsilon = \sqrt{\frac{8\mathscr{R}T}{\pi M}}, \quad \lambda = \frac{4\upsilon\eta}{5p}, \quad t_{mol} = \frac{\pi\eta}{4p}, \quad D_g = \frac{3\pi}{20}\cdot\frac{\eta\upsilon^2}{p} \tag{2.19}$$

The values of some of these parameters and others, introduced later in this section, are listed in Table 2.1 where temperature- and/or pressure-dependent parameters are quantified at *standard pressure* ($p_{st} = 1$ atm) and *standard temperature* ($T_{st} = 273.15$ K, i.e. 0 °C).

2.5.2. Viscosity

Among the gas parameters listed in Eqn (2.19), only the average molecular speed (υ) can be calculated from *a priori* known molar mass (M) of a gas. Other parameters are expressed via the gas viscosity (η) — a measure of its resistance to flow. Ideally, as is assumed

Table 2.1 Molecular Properties of Several Gases

Gas	He	H_2	N_2	Ar
MOLECULAR PROPERTIES				
M (molar mass), g/mol	4.003	2.016	28.01	39.95
λ_{st} (mean free path at $p = p_{st}$, $T = T_{st}$), μm	0.177	0.112	0.0604	0.0641
υ_{st} (average molecular speed at $T = T_{st}$), km/s	1.20	1.69	0.454	0.38
η_{st} (viscosity at $T = T_{st}$), μPa·s	18.69	8.362	16.84	21.35
$D_{g,st}$ (self-diffusivity at $p = p_{st}$, $T = T_{st}$), cm^2/s	1.26	1.12	0.162	0.144
EMPIRICAL PARAMETERS				
ξ (parameter in Eqn (2.20))	0.685	0.698	0.710	0.750
X_D (parameter in Eqns (2.23) and (2.24))	0.456	0.544	0.693	0.686
COMBINATIONS				
$X_D\ \lambda_{st}$, μm	0.081	0.061	0.042	0.044
$(X_D\ \lambda_{st})/(X_D\ \lambda_{st})_{\text{hydrogen}}$	1.33	1	0.689	0.721
$X_D \upsilon_{st}$, km/s	0.548	0.92	0.315	0.261
$(X_D \upsilon_{st})/(X_D \upsilon_{st})_{\text{hydrogen}}$	0.596	1	0.342	0.284
$X_D^2 D_{g,st}$, cm^2/s	0.262	0.331	0.078	0.068
$D/D_{\text{hydrogen}} = X_D^2 D_{g,st}/(X_D^2 D_{g,st})_{\text{hydrogen}}$ (Eqns (2.23) and (2.24))	0.79	1	0.24	0.21

Notes:
1. *Quantities $D_{hydrogen}$ and D are the diffusivities of the same solute in hydrogen and in another gas, respectively.*
2. *In the source [1], quantity X_D was denoted as γ_D.*
Source: *Reconstructed from ref. [1] with permission.*

throughout this chapter, η is **independent of pressure**. Although theoretical formulas for η are known, they are either complex or not sufficiently accurate for practical applications [24,33,35]. Instead, empirical formulas are used in GC [1,35–37]. The following two are the simplest at their accuracies [1,37]:

$$\eta = \left(\frac{T}{T_{st}}\right)^{\xi} \eta_{st} \tag{2.20}$$

$$\eta = \eta_{st0}\left(\frac{T}{T_{st}}\right)^{\xi(T)}, \quad \xi(T) = \xi_0 + \xi_1 \frac{T - T_{st}}{T_{st}} \tag{2.21}$$

Parameters η_{st} (the gas viscosity at 0 °C) and ξ in Eqn (2.20) are listed for several gases in Table 2.1. Parameters in Eqn (2.21) are listed in Table 2.2. All parameters were obtained by the least-square fittings of Eqns (2.20) and (2.21) to numerical data [35] within 300 K $\le T \le$ 600 K and 250 K $\le T \le$ 750 K, respectively [1]. Eqn (2.20) is simpler than Eqn (2.21), but Eqn (2.21) is more accurate (Table 2.3). The data in Table 2.3 show that, only when Eqn (2.20) is used within the *extended temperature range* (250 K $\le T \le$ 750 K), its errors for viscosity of nitrogen and argon can be larger than the errors in the source of

Table 2.2 Gas Viscosity Parameters in Eqn (2.21)

Gas	He	H_2	N_2	Ar
η_{st0} (μPa·s)	18.63	8.382	16.62	21.04
ξ_0	0.6958	0.6892	0.7665	0.8131
ξ_1	−0.0071	0.005	−0.0378	−0.0426

Source: *Reconstructed from ref. [1] with permission.*

numerical viscosity data [35]. In all other cases compiled in Table 2.3, the errors added by Eqns (2.20) and (2.21) can be ignored because they are smaller [1] than the errors specified in the source [35] for its numerical data.

In the rest of this chapter, **only Eqn (2.20)** is used for the evaluation of gas viscosities. Eqn (2.21) might be useful in designing pneumatic control systems in GC instruments where higher accuracy of viscosity calculation might be required.

It should be mentioned that the numerical values of parameters η_{st} and ξ in Table 2.1 for helium, hydrogen, and nitrogen are different

Table 2.3 The Largest %-Errors in Numerical Viscosity Data in the Source [35] (Primary errors), and Additional Errors Due to Eqns (2.20) and (2.21)

Gas	He	H_2	N_2	Ar
Primary errors	1	2	2	no data
Additional errors from Eqn (2.20), $300\ \text{K} \le T \le 600\ \text{K}$	0.2	0.2	0.8	0.8
Additional errors from Eqn (2.20), $250\ \text{K} \le T \le 750\ \text{K}$	0.5	0.5	2	2.5
Additional errors from Eqn (2.21), $250\ \text{K} \le T \le 750\ \text{K}$	0.1	0.2	0.4	0.4

Note: The additional errors are the largest %-difference between the numerical data in the source [35] and the data obtained from Eqns (2.20) and (2.21).
Source: *Reconstructed from ref. [1] with permission.*

from the previously reported values for these gases [37,38]. As shown elsewhere [1], the viscosity based on the data in Table 2.1 is expected to be more accurate than viscosity based on the previously reported data [37].

2.5.3. Solute Diffusivity in a Gas

Solute *diffusion* in a gas plays a dual – positive and negative – role in GC separation process. On the one hand, the diffusion is the mechanism that transports the solutes from the gas to the stationary phase and vice versa. In this regard, the diffusion is essential for the separation in GC. On the other hand, the diffusion broadens the peaks, thus reducing their separation.

Diffusivity (D) [1,33,39] (*diffusion coefficient* [8,24,33,39–41]) of each solute – a measure of the rate of its diffusion in a gas – can be described by the *Fuller–Giddings empirical formula* [40]:

$$D = \frac{10^{-3}\sqrt{1/M_g + 1/M_{sol}}}{(V_g^{1/3} + (\Sigma V_{sol})^{1/3})^2} \cdot \frac{\text{atm}}{p} \cdot \left(\frac{T}{\text{K}}\right)^{1.75} \cdot \frac{\text{cm}^2}{\text{s}} \tag{2.22}$$

where M_g and M_{sol} are, respectively, molecular weights of a gas and a solute; and quantities V_g and V_{sol} are dimensionless empirical quantities [19] known as the *molecular diffusion volume* of a gas and the *atomic diffusion volume increments* of a solute, respectively. The values of these quantities for some gases and solutes can be found in several sources [40,41].

While Eqn (2.22) is suitable for evaluation of diffusivity of some solutes (those few with known parameters V_g and V_{sol}), it is structurally too complex and not suitable for broad theoretical evaluations. One of the difficulties in using Eqn (2.22) is that it does not provide a simple way of accounting for the fact that the solutes generally tend to elute in the order of increase in the size of their molecules. The following two formulas more directly account for this effect [1]:

$$D = X_D^2 D_{g,st} \cdot (10^3 \varphi)^{0.09} \cdot \frac{p_{st}}{p} \cdot \left(\frac{T_{char}}{T_{st}}\right)^{-1.25} \left(\frac{T}{T_{st}}\right)^{1+\xi} \quad (2.23)$$

$$D = \frac{X_D^2 D_{g,st} \cdot (10^3 \varphi)^{0.09}}{k^{0.1}} \cdot \frac{p_{st}}{p} \cdot \left(\frac{T}{T_{st}}\right)^{\xi-0.25}, \quad (0.1 \le k \le 30) \quad (2.24)$$

where gas parameters $D_{g,st}$ and X_D as well as their products $X_D^2 D_{g,st}$ for several gases are listed in Table 2.1.

Several comments can be made regarding accuracy of Eqns (2.22), (2.23), and (2.24).

1. An error (ΔD) in evaluation of a solute diffusivity (D) eventually leads to the errors in theoretical prediction of peak widths [1] which, in the worst case, can be proportional to $\sqrt{\Delta D/D}$. As long as $\Delta D/D<1$, the relative peak width error is smaller (maybe much smaller) than half of the relative error ($\Delta D/D$) in D.
2. When the effects of operational parameters (column dimensions, carrier gas type, column temperature, heating rate, etc.) on a column performance are considered, the diffusivity errors are typically canceled out, and can be ignored.
3. Eqns (2.23) and (2.24) are the approximations of Eqn (2.22). Therefore, the errors in Eqn (2.22), reported [40] to have 6.7% standard deviation and inaccuracy reaching 40% and beyond in some cases, should be the benchmark for judging the significance of the errors in Eqns (2.23) and (2.24).
4. The additional errors in Eqns (2.23) and (2.24) are expected to be within ±8% at the edges of 25 °C to 325 °C temperature range [1]. This means that the contribution of the additional errors is not significant.
5. Eqn (2.23) describes D as a function of a solute characteristic temperature (T_{char}) and a column temperature (T). In this formula, D is proportional to $T^{1+\xi}$ where, according to Table 2.1, quantity $(1 + \xi)$ ranges between 1.685 and 1.75, i.e. for all gases, is in a close proximity to 1.75th power of T in Eqn (2.22).
6. In the studies of a column performance, it is desirable [1] to express D as a function of measurable parameters such as a column temperature and a solute retention factor (k). Eqn (2.24) is suitable for that. As mentioned earlier, all solutes that were highly retained at the beginning of a heating ramp elute with roughly the same elution retention factor. In this case, k in Eqn (2.24) is a fixed quantity and D is proportional to $T^{\xi-0.25}$ where quantity $(\xi-0.25)$ is about 0.5 for argon and about 0.45 for other gases in Table 2.1. This might appear as a contradiction with Eqn (2.22). However, it is not. According to Eqn (2.22), the diffusivity of a *particular solute* is proportional to about $T^{1.75}$. However, also according to Eqn (2.22), the diffusivity of a solute being a larger molecule is generally lower than the diffusivity of a smaller molecule. During a heating ramp, the solutes having larger molecules generally elute at higher temperatures. Eqn (2.24) reflects the combined effect of these two phenomena which has been experimentally confirmed elsewhere [42].
7. Transition from Eqns (2.23) to (2.24) is based on approximation $T_{char}/T \approx k^{0.08}$ verified for $0.1 \le k \le 30$ and probably valid for retention factors larger than 30 [1].
8. Strictly speaking, Eqns (2.23) and (2.24) are valid only for n-alkanes [1]. However, observation of a large number of GC chromatograms suggests that, whenever there is an n-alkane peak in a chromatogram, its width is not noticeably different

from the widths of its neighbors. As a result, it is reasonable to use Eqns (2.23) and (2.24) for evaluation of diffusivities of all solutes.

2.6. FLOW OF IDEAL GAS IN OPEN CIRCULAR TUBES

From the point of view of a gas flow, a column is a *tube*. It is assumed throughout this chapter that a column (and, therefore, a tube) is **long** ($L \gg d$), has a **uniform** (the same at any location along its length), **circular** *cross section*, and the flow in it is **mass conserving**[1] (no material flows through the tube walls) and **laminar** [1].

Some *pneumatic parameters* of a gas flow in a tube are: gas *flow rate* (F), *pressure* (p), *hold-up time* (t_M) – the time of migration of a narrow packet of the gas molecules from the column inlet to its outlet – etc. [15]. Several pressure parameters are known in chromatography. Among them are *inlet pressure* (p_i) *outlet pressure* (p_o), *ambient pressure* (p_a), *gauge pressure* (p_g), *pressure drop* (Δp), *relative pressure* (P), and *relative pressure drop* (ΔP). Some of these parameters relate to others as

$$\Delta p = p_i - p_o, \quad p_g = p_i - p_a, \quad P = \frac{p_i}{p_o}, \quad \Delta P = \frac{\Delta p}{p_o} \tag{2.25}$$

Due to gas viscosity, the *longitudinal velocity* (u_r) of the gas flow in a tube has a *parabolic profile* (Figure 2.4) [1,8]:

$$u_r = \left(1 - \left(\frac{2r}{d}\right)^2\right) u_{r,\max} \tag{2.26}$$

[1] There is a strong evidence [43] that this condition can be meaningfully violated for helium at 200 °C and higher temperatures.

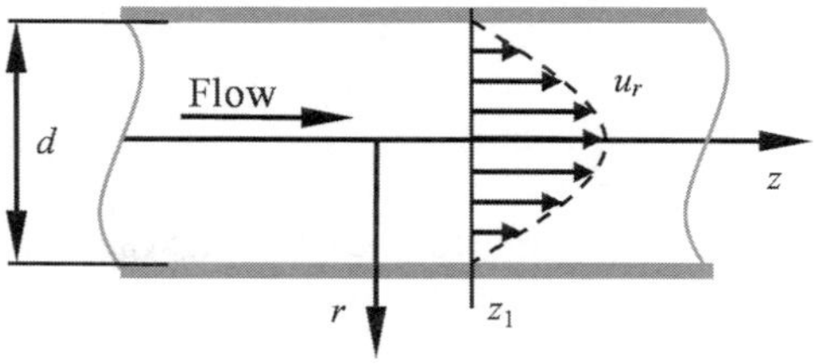

FIGURE 2.4 Parabolic profile (Eqn (2.26)) of longitudinal velocities (u_r) in an open tube with circular cross section.

Let A_{gas} be the total cross-sectional area open to the gas flow, and dA be an area of a small element in that cross section. A *cross-sectional average*,

$$u = u(z) = \frac{1}{A_{gas}} \int_{A_{gas}} u_r \, dA, \quad A_{gas} = \frac{\pi d^2}{4} \tag{2.27}$$

of all longitudinal velocities (u_r) in a cross section located at *longitudinal coordinate* z (the distance from the tube inlet) of the tube is a gas *linear velocity* (briefly, *velocity*) at z. According to *Darcy's law*, gas velocity (u) at location z along a tube can be found as [1,8,9,44,45]

$$u = -\frac{d^2}{32\eta} \cdot \frac{\partial p}{\partial z} \tag{2.28}$$

Ideal gas is a *compressible* fluid. The volume of a fixed mass of a gas is proportional to its temperature (T) and inversely proportional to pressure (Eqn (2.18)). This implies that,

in the mass-conserving gas flow in a uniform, uniformly heated tube, the product pu is uniform (the same at any location along a tube) [1].

Accounting for that, one can describe u in Eqns (2.28) as [1]

$$u = -\frac{1}{\Omega} \cdot \frac{\partial p}{\partial \zeta}, \quad pu = (\text{fixed quantity}) \tag{2.29}$$

where

$$\Omega = \frac{32 L \eta}{d^2} \tag{2.30}$$

$$\zeta = \frac{z}{L} \tag{2.31}$$

Parameters Ω and ζ are, respectively, a tube *pneumatic resistance* and *dimensionless distance* from the inlet. Solving the system in Eqn (2.29), one can express the gas velocity (u) at any location along a tube as a function of pressure (p) at that location, and both parameters as a function of the distance from the tube inlet (Figure 2.5):

$$u = \frac{p_i^2 - p_o^2}{2\Omega p} \qquad (2.32)$$

$$p = p_i\sqrt{1 - \frac{p_i^2 - p_o^2}{p_i^2}\zeta}, \quad u = \frac{u_i}{\sqrt{1 - \frac{p_i^2 - p_o^2}{p_i^2}\zeta}} \qquad (2.33)$$

Gas compressibility complicates relations between pneumatic parameters. To simplify mathematical expressions describing these relations, two extreme levels of the decompression can be considered:

$$\text{The decompression is} \begin{cases} \text{weak,} & \text{when } |\Delta p| \ll p_o \\ \text{strong,} & \text{when } \Delta p \gg p_o \end{cases} \qquad (2.34)$$

When the decompression is *weak* (Figure 2.5.a), the gas pressure (p) and velocity (u) are almost uniform. On the other hand, when the decompression is *strong* (Figure 2.5.c), p and u are substantially nonuniform — the difference in their values at the tube inlet and outlet is large. Under typical GC conditions, gas decompression is weak and can be ignored when relatively short wide-bore columns ($d \geq 0.32$ mm) operate with detectors working at atmospheric pressure. Conversely, the decompression is strong in all cases of GC-MS ($p_o = 0$) and in the analyses of complex mixtures utilizing relatively long narrow-bore columns.

Because the gas velocity (u) can change with the distance from the column inlet, the *time-averaged velocity* (briefly, *average velocity*)

$$\bar{u} = \frac{L}{t_M} \qquad (2.35)$$

of a gas is frequently used in GC as a single gas velocity metric. It relates to the gas outlet velocity (u_o) as [1,7,9,22]

$$\bar{u} = ju_o, \quad j = \frac{3(p_i + p_o)p_o}{2(p_i^2 + p_i p_o + p_o^2)} \approx \begin{cases} 1, & |\Delta p| \ll p_o \\ \frac{3}{2\Delta P}, & \Delta p \gg p_o \end{cases} \qquad (2.36)$$

where j is the *James–Martin compressibility factor.* While presenting a conceptually simple relation of parameters $\bar{u}$ and u_o, the last formula is

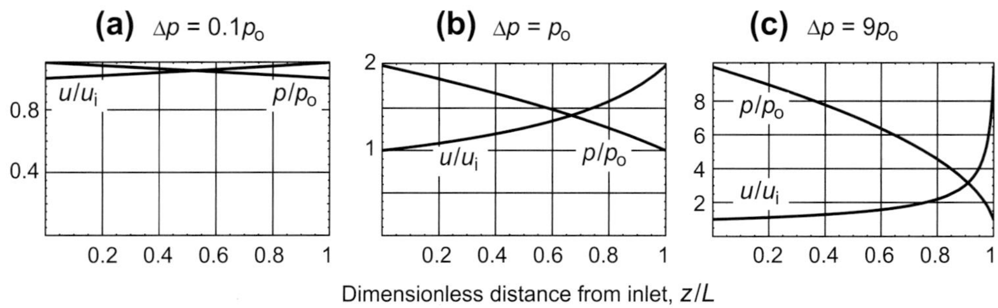

FIGURE 2.5 Pressure (p) and velocity (u) as functions (Eqn (2.33)) of a distance (z) from a tube inlet, and relative pressure drop ($\Delta p/p_o$). The change of quantities p and u along the tube is small when $\Delta p/p_o = 0.1$, gets larger when $\Delta p/p_o = 1$, and is very large when $\Delta p/p_o = 9$. *Source: Reconstructed from [1] with permission.*

incomplete. It describes $\overline{u}$ as a function of three pneumatic parameters, p_i, p_o, and u_o of which only two can be mutually independent [1] while the third is a function of the other two. For example, if quantities p_o and u_o in a given tube are known, then p_i is predetermined by the tube dimensions. This means that Eqn (2.36) is insufficient for finding $\overline{u}$ for a given u_o without additional information about the relationship between p_i, p_o, and u_o, which is described in Eqn (2.32). Solving together Eqns (2.32) and (2.36), one can find that $\overline{u}$ is the function of p_o and u_o [1]:

$$\overline{u} = \frac{3u_o}{2} \cdot \frac{x}{(1+x)^{3/2} - 1}, \quad x = \frac{2\Omega u_o}{p_o} \tag{2.37}$$

Sometimes, it is necessary to express u_o as a function of $\overline{u}$. A solution of Eqn (2.37) for u_o is [1]

$$u_o = \frac{p_o}{2\Omega} \cdot \left(\Phi^2 \left(\frac{\Omega \overline{u}}{p_o} \right) - 1 \right) \tag{2.38}$$

For an arbitrary variable x, function $\Phi(x)$ is defined as

$$\Phi(x) = \frac{1}{9}(-3 + 4x + X_+^{1/3} + X_-^{1/3}) \tag{2.39}$$

where

$$X_+ = 2(X_1 + 27\sqrt{X_2}), \quad X_- = 2(X_1 - 27\sqrt{X_2}),$$
$$X_1 = (3 + 8x)(36 + 3x + 4x^2),$$
$$X_2 = x(3 + 4x)(24 + 3x + 4x^2)$$

In spite of its rather cumbersome definition, $\Phi(x)$ is a monotonic slightly convex function (Figure 2.6).

Two more pneumatic parameters are important in gas chromatography. They are: the gas flow rate and specific flow rate.

A *gas volumetric flow rate* (briefly, *flow rate*) [1]

$$F = \frac{\pi d^2 p u T_{norm}}{4 p_{st} T} = X_{FA} p u, \quad X_{FA} = \frac{\pi d^2 T_{norm}}{4 p_{st} T} \tag{2.40}$$

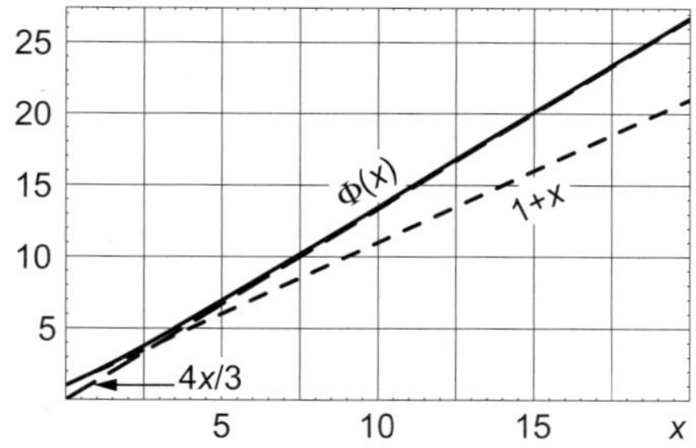

FIGURE 2.6 Function Φ (x) (solid line ———) and its tangent lines: $1 + x$ (short dashes – – –) are the tangent to Φ (x) at $x = 0$, and $4x/3$ (long dashes — — —) is the tangent to Φ (x) at $x = \infty$.

is measured at standard pressure (p_{st}) and at some predetermined temperature regardless of actual pressure (p) and temperature (T) in a tube. It is assumed in Eqn (2.40) and throughout this chapter that F is always measured at *normal temperature*, $T_{norm} = 298.15$ K (25 °C). Substitution of Eqns (2.32) and (2.30) in Eqn (2.40) yields

$$F = \frac{\pi d^4 (p_i^2 - p_o^2) T_{norm}}{256 L \eta p_{st} T} \tag{2.41}$$

In its essence, a gas-*specific flow rate* [1,46]

$$f = \frac{T}{T_{norm}} \cdot \frac{F}{d} \tag{2.42}$$

is the flow rate per unit of a tube diameter (d). Factor (T/T_{norm}) included in the definition of f removes the dependence of f on the specifics of the flow rate measurement in GC. Indeed, due to Eqns (2.40) and (2.32), Eqn (2.42) can be rearranged as [1]

$$f = \frac{\pi d p u}{4 p_{st}} = \frac{\pi d \cdot (p_i^2 - p_o^2)}{8 p_{st} \Omega} = \frac{\pi d^3 \cdot (p_i^2 - p_o^2)}{256 L p_{st} \eta} \tag{2.43}$$

The absence of parameter T_{norm} in Eqn (2.43) is an indication of the independence of quantity f of the temperature conventions in the measurement of F.

One of the key parameters of a mobile phase flow in chromatography (not only in GC) is hold-up time (t_M) which plays a role of a basic time clock in chromatography [1,47] and, from that perspective, is not only a pneumatic parameter but also a chromatographic parameter. Let $t_g = t_g(z)$ be a gas *propagation time* – the time of migration of a narrow gas packet from a column inlet to an arbitrary location z along the column. The *hold-up* time (t_M) is the time of migration of a narrow gas packet through the entire L-long column, i.e.

$$t_M = t_g(L) \tag{2.44}$$

The hold-up time can be found as [1]

$$t_M = \frac{32\,\ell^2 \eta}{j_H \Delta p} \tag{2.45}$$

where

$$\ell = \frac{L}{d} \tag{2.46}$$

$$j_H = \frac{3(p_i + p_o)^2}{4(p_i^2 + p_i p_o + p_o^2)} \approx \begin{cases} 1, & |\Delta p| \ll p_o \\ \frac{3}{4}, & \Delta p \gg p_o \end{cases} \tag{2.47}$$

Parameters ℓ and j_H are the tube *dimensionless length* and the *Halász–Hartmann–Heine compressibility factor* [48]. Eqn (2.45) shows that t_M is proportional to quantity $(L/d)^2$. It should be noticed, however, that this would be the case if Δp was the same for all column dimensions. This is not how the columns actually operate in GC. Lower pressure is typically used for shorter and wider bore columns, and larger pressure is used for longer and narrower bore ones. What typically is the same or more or less the same for all columns is specific flow rate (f) of a given carrier gas [1]. On the other hand, an actually used pressure is a strong function of a column dimension. As a result, t_M in a realistic GC is not proportional to (L/d).^{2} Expressing t_M as a function [1]

$$\begin{aligned} t_M &= t_M(d, L, f) \\ &= \frac{\pi^2 d^4 p_o^3}{1536\ p_{st}^2 \eta f^2} \cdot \left(\left(1 + \frac{256 L p_{st} \eta f}{\pi d^3 p_o^2} \right)^{3/2} - 1 \right) \\ &= \begin{cases} \dfrac{\pi d L p_o}{4 p_{st} f}, & |\Delta p| \ll p_o \\ \sqrt{\dfrac{64 \pi L^3 \eta}{9 d p_{st} f}}, & \Delta p \gg p_o \end{cases} \end{aligned} \tag{2.48}$$

of f exposes a more realistic dependence of t_M on the column dimensions, showing that, under the realistic GC condition of the same f for all column dimensions, t_M is not proportional to the product $(L/d)^2\eta$ as it is in Eqn (2.45). Indeed, at fixed f, quantity t_M is a rather complex function of a tube dimensions (d and L) and gas viscosity (η). The function becomes simpler at the extremes of the gas decompression. Thus, when the decompression is weak, t_M is proportional to the product dL and is independent of η. When the decompression is strong, t_M is proportional to $\sqrt{\eta L^3/d}$. Eqn (2.48) plays a key role in a column optimization [46] discussed in section 2.10.

A large collection of mathematical expressions describing relations between various pneumatic parameters can be found in ref. [1].

2.7. MIGRATION AND ELUTION PARAMETERS OF THE SOLUTES

The main focus of this section is migration and elution of a solute *zone* – a packet of a solute material in a column. Its width is measured in units of distance along the column. Elution of a solute zone out of a column inlet results in a chromatographic *peak* – the rate of flow (the flux) of a solute material at the outlet. The widths of the peaks are measured in units of time. Typically, the peaks are observed as components of visual representation of a detector output. Additional details can be found elsewhere [1,49]. Throughout this chapter, the **first mathematical**

moment of a peak is assumed to be its retention time (t_R), and the **first mathematical moment of a solute zone** (its *center of mass*) is assumed to be the distance (z) of the zone from the column inlet [1]. (The first moments of symmetrical peaks and zones coincide with their apexes.)

In a study of temperature-programmed analyses, it might be convenient to deal with the solute elution temperatures rather than with the retention times (t_R). A solute *elution temperature* (T_R) is a column temperature at the time of the solute *elution* from the column outlet, i.e. at $t = t_R$. The retention time depends on secondary factors of temperature programming such as the column dimensions, carrier gas type, flow rate, etc. Considering the elution temperatures rather than the retention times makes it possible to bypass the effects of the secondary factors.

2.7.1. General Equations of a Solute Migration and Elution

In a simple case of a column operation under uniform, static conditions, a solute retention time can be found as $t_R = L/v$ which, due to Eqns (2.1) and (2.35) can be expressed as

$$t_R = \frac{L}{v} = \frac{L}{\mu u} = \frac{t_M}{\mu}, \text{ (uniform and static conditions)} \quad (2.49)$$

In a more general case of nonuniform and dynamic conditions, the retention time of a given solute can be found from integration of Eqn (2.1) where parameters u and μ, and, therefore, parameter v can be functions $u = u(z,t)$, $\mu = \mu(z,t)$, and $v = v(z,t)$ of time (t) and distance (z) traveled by the solute since its injection in a column inlet.

Due to the gas decompression along a column, gas velocity (u) can be a function (Eqn (2.33)) of z (be nonuniform). Due to the temperature-dependent gas viscosity, u in a temperature-programmed analysis can also change with t (be dynamic). Pressure programming in isothermal GC analysis also causes a dependence of u on t.

A solute mobility (μ) can also be a time-dependent quantity. In fact, the change in μ with t is the prime target of the temperature programming in GC. In addition to being a dynamic quantity, μ can be nonuniform (as it is in gradient-elution LC, for example). The use of a nonuniform μ, where a negative gradient ($\partial\mu/\partial z < 0$) in μ was utilized for the peak sharpening due to what can be called as an *in-column focusing*, has been discussed in the literature [50,51]. It was found [51] that, because the peak sharpening due to the in-column focusing is more than compensated by the peak getting closer to each other due to the same focusing, the focusing could not improve the separation-time tradeoff compared to that obtainable in conventional temperature-programmed GC. It is assumed throughout this chapter that μ is a **uniform** quantity ($\partial\mu/\partial z = 0$) as it is in conventional isothermal and temperature-programmed GC.

The integration of Eqn (2.1) for a given solute can be expressed as the system [1]:

$$t_g = t_g(z) = \int_0^z \frac{dz}{u(z, t(z))}, \text{ (gas } \textit{propagation equation}\text{)} \quad (2.50)$$

$$\int_0^{t(z)} \mu(T(t))dt = t_g, \quad (2.51)$$

(solute *migration equation*)

where $T(t)$ is a temperature program and t_g is a gas *dynamic propagation time*. At $z = L$, gas propagation time becomes *dynamic hold-up time* (t_m) [1], i.e.

$$t_m = t_g(L) \quad (2.52)$$

and Eqn (2.51) becomes a solute *elution equation*.

Quantities t_g and t_M are conceptually similar (but not equal) to their static counterparts t_g and t_M described in section 2.6. Thus, the definition of t_m in Eqn (2.52) is similar to the definition of t_M in Eqn (2.44). However, there are substantial differences between the static t_g and t_M and their dynamic counterparts. The difference comes from the difference between static and dynamic t_g. Under static conditions, t_g is the time of migration of a *given* packet of a gas molecules to an arbitrary location z. The value of parameter t_g at a given z does not depend on the time of its measurement i.e. once a static flow in a column is established, the measurement of t_g at a given z can be made at any time thereafter. On the other hand, under dynamic conditions, parameter $t_g(z)$ is associated with a particular solute (rather than with any arbitrarily chosen gas packet as under the static conditions) and can be different for different solutes. Indeed, according to Eqn (2.50), quantity $t_g(z)$ for a given solute is a sum of all consecutive infinitesimally short time segments (dt_g), representing infinitesimally short static hold-up times in infinitesimally short column segments (dz) centered at all coordinates between 0 and z of a migrating solute. In other words, $t_g(z)$ is the sum of infinitesimally short static hold-up times of all (possibly *different*) gas packets that were overpassed by a given solute during its migration to location z. In a dynamic analysis, function $t_g(z)$ for one solute can be different from that for another solute.

General solution for the system in Eqns (2.50) and (2.51) is unknown. However, the system can be numerically solved for a known model, $\mu(T)$, of a solute–column interaction under any temperature and pressure program [1]. The system can also serve as a starting point for general solutions under some constraints.

2.7.2. Isobaric Analysis

So far, the key constraint to the scope of the study of GC analyses was the requirement that the solute–column interaction was uniform along the column. Here, an additional requirement that a GC analysis should be *isobaric* (conducted at *constant pressure*) is added. Unless otherwise is explicitly stated, the **isobaric** conditions are assumed below.

In an isobaric analysis, the pressure profile $p(z)$ along a column does not change with time (t) even if a column temperature is a function of t. In a temperature-programmed analysis, the temperature dependence of a gas viscosity (η) becomes the only factor causing the change in gas velocity (u) with t. As follows from Eqns (2.32), (2.33), and (2.30), the dependence, $u = u(z,t)$, of u on z and t becomes proportional to the product $p(z)\eta(t)$, where $p(z)$ and $\eta(t)$ are independent of t and z, respectively. The separation of variables z and t in Eqn (2.50) allows one to avoid the need for using the dynamic gas propagation time (t_g) in Eqns (2.50) and (2.51), and to reduce the system to a single migration equation which, at $z = L$, becomes an elution equation [1,47]

$$\int_0^{t_R} \mu_{eff}(T(t))dt = t_{M,ref},$$

$$\mu_{eff}(T(t)) = \frac{\eta_{ref}}{\eta(T(t))}\mu(T(t)) \quad \text{(isobaric analysis)} \tag{2.53}$$

where η_{ref} is a gas *reference viscosity* measured at a predetermined reference temperature (T_{ref}), $t_{M,ref}$ is the reference hold-up time statically measured at T_{ref} (and a fixed pressure of a given isobaric analysis), and μ_{eff} can be interpreted as a solute *effective mobility*.

Equation (2.53) has interesting implications. It follows from Eqn (2.53) that [1], in an isobaric analysis with any temperature program $T(t)$,

The net effect of a temperature-depended solute mobility (μ) and gas viscosity (η) on a solute retention time (t_R) and elution temperature (T_R) can be lumped in a single quantity – the effective solute mobility (μ_{eff}).

The net effect of column dimensions, carrier gas type, and the gas initial conditions on a solute retention time (t_R) can be expressed via a single parameter — the static hold-up time ($t_{M,ref}$) measured at a predetermined fixed temperature (T_{ref}).

Other than through its effect on the hold-up time ($t_{M,ref}$) at some predetermined fixed temperature (T_{ref}), gas decompression along a column has no other effect on a solute retention time (t_R) and elution temperature (T_R).

Equation (2.53) also shows that the hold-up time (t_M) plays a special role in the formation of retention times. It acts as a basic *time clock* of chromatography and a fundamental *scaling factor* of a chromatogram. Thus, it follows from Eqn (2.53) that the effect of a temperature program, $T(t)$, on peak retention times comes not from the absolute time scale of $T(t)$ but only through its relation to t_M. The proportionality of dynamic peak retention times to hold-up time measured at one set of static conditions (the scalability of the peak retention times) is the basis for *method translation* and *retention time locking* (*RTL*) concepts [1,47,52] and techniques [53–58]. It should also be mentioned that, although Eqn (2.53) validates the techniques only in isobaric analyses, the techniques are also valid for *isorheic* (*constant flow*) analyses [59] at a weak or a strong gas decompression [1]. The techniques can be further extended to analyses with arbitrary pressure programs [57].

2.7.3. Linear Heating Ramp and Linearized Model of a Solute–Column Interaction

The results considered so far were valid for any type of solute–column interactions (μ could be any function of T) and for any temperature program, $T(t)$. Here, the scope is narrowed to the linear heating ramp that can be described as

$$T = T(t) = T_{init} + R_T t \tag{2.54}$$

where T_{init} is the *initial temperature* of the ramp and R_T is a fixed *heating rate* (in *isothermal* analysis, $R_T = 0$). The scope is further limited to linearized and ideal thermodynamic models of a solute–column interaction. Due to Eqns (2.3), (2.12), and (2.15), a solute mobility for these models can be expressed as

$$\mu = \left(1 + \exp\frac{T_{char} - T}{\theta_{char}}\right)^{-1}, \quad \text{(linearized model)} \tag{2.55}$$

$$\mu = \left(1 + \exp\left(\frac{T_{char}}{\theta_{char}} \cdot \left(\frac{T_{char}}{T} - 1\right)\right)\right)^{-1} \quad \text{(ideal thermodynamic model)} \tag{2.56}$$

The former is the basis for the closed-form solutions obtained below while the latter can be used for numerical verification of those solutions.

Due to Eqns (2.55) and (2.20), the integrand (μ_{eff}) in Eqn (2.53) for the linearized model can be expressed as

$$\mu_{eff}(T(t)) = \left(\frac{T_{ref}}{T(t)}\right)^{\xi}\left(1 + \exp\frac{T_{char} - T(t)}{\theta_{char}}\right)^{-1} \quad \text{(isobaric analysis, linearized model)} \tag{2.57}$$

Unfortunately, even this integrand with linear temperature program, $T(t)$, does not lead to a closed-form solution for Eqn (2.53). A suitable approximation for Eqn (2.57) exists when a relatively long linear heating ramp (covering more than a 100 °C range) is considered. Under typical conditions, a solute that is not significantly retained at the beginning of a heating ramp (its T_{char} is not significantly larger than T_{init}) does not stay in a column for long time so that the change in T during the solute migration is relatively small. On the other hand, if a solute is significantly retained at the beginning of a heating ramp, then it continues to stay near the column inlet until T gets closer to the solute characteristic temperature (T_{char}). Again, during the solute migration through the major portion of the column length, the change in T is relatively small. In both cases, a meaningful migration of a solute takes place when T is reasonably close to T_{char} [1]. On top of that, factor $(T_{ref}/T)^{\xi}$ in Eqn (2.57) is a relatively weak function of T so that $(T_{ref}/T)^{\xi}$ remains nearly constant when T changes in vicinity of T_{char}. Taking this into account, Eqn (2.53) can be approximated as

$$\left(\frac{T_{ref}}{T_{char}}\right)^{\xi} \int_{0}^{t_R} \left(1 + \exp\frac{T_{char} - T(t)}{\theta_{char}}\right)^{-1} dt = t_{M,ref} \quad (2.58)$$

The integral in this formula has closed-form solutions described shortly.

2.7.4. Dimensionless and Normalized Heating Rates

Before addressing the solutions of Eqn (2.58), it should be noticed that expressing a heating rate in dimensionless form simplifies the solutions and broadens their scope.

It follows from Eqn (2.58) that the significance of a change in a column temperature (T) depends on relation of that change to the solute characteristic thermal constant (θ_{char}), and the rate (R_T) of the change in T is meaningful only in relation to the hold-up time rather than to the absolute time scale. These observations suggest that *dimensionless heating rate* (r_T) can be expressed as [1]

$$r_T = \frac{R_T t_{M,char}}{\theta_{char}} \quad (2.59)$$

where $t_{M,char}$ is the hold-up time statically measured at a solute characteristic temperature (T_{char}). To distinguish it from the dimensionless heating rate (r_T), quantity R_T can be called as the *absoluter heating rate.*

Generally, r_T can be different for different solutes and, therefore, can be a function of T. However, in isobaric analysis with linear heating ramp, this is essentially not the case. Indeed, during the isobaric linear heating ramp, both components of the ratio $t_{M,char}/\theta_{char}$ in Eqn (2.59) change in approximate proportion with $T^{0.7}$. For θ_{char} this follows from Eqn (2.74) where solute elution temperatures are close to their characteristic temperatures and from Eqn (2.17). The proportionality $t_{M,char} \sim T^{0.7}$ follows from Eqn (2.45) where t_M at constant pressure is proportional to gas viscosity (η) which, according to Eqn (2.20), is proportional to T^{ξ} where $\xi \approx 0.7$ (Table 2.1). As a result, $t_{M,char}/\theta_{char}$ at a constant pressure is essentially independent of T, and, therefore, r_T at a fixed R_T is essentially independent of T. One can conclude that

in isobaric analyses with linear heating r amp, all solutes are subjected to approximately the same dimensionless heating rate (r_T).

Due to Eqns (2.45), (2.20), and (2.17), r_T for all solutes can be estimated as [1]

$$r_T \approx \frac{R_T t_{M,ref}}{\theta_{T,ref}} \quad (2.60)$$

where θ_T is defined as

$$\theta_T = \theta_{char,st}\left(\frac{T}{T_{st}}\right)^{0.7} \quad (2.61)$$

and quantities $\theta_{T,ref}$ and $t_{M,ref}$ are calculated at any reference temperature (T_{ref}) as long as it is the same for both parameters.

Example 2.3. If $\theta_{char,st} = 22$ °C, $R_T = 10$ °C/min, $T_{ref} = 423$ K (150 °C), and $t_{M,ref} = 1$ min, then $\theta_{T,ref} \approx 22\ °C \times (273/423)^{0.7} \approx 30$ °C, and $r_T \approx 0.33$ for any solute.

Any temperature can be chosen as T_{ref}. However, for a better consistency of all forthcoming numerical examples and recommendations, it is desirable to base them on the same T_{ref} representing the middle of a typical GC temperature range. $\boldsymbol{T_{ref}}$ **= 423 K** (150 °C) of the previous example will continue to be used for all numerical evaluations in this chapter.

Combining Eqns (2.54) and (2.60), one can express retention time of a peak eluting at temperature T_R as

$$t_R = \frac{\Delta T_R t_{M,ref}}{\theta_{T,ref} r_T}, \quad \Delta T_R = T_R - T_{init} \tag{2.62}$$

In addition to the dimensionless heating rate, also useful is the *normalized heating rate* (R_{TM}) defined as

$$R_{TM} = R_T t_M \tag{2.63}$$

where parameters R_{TM} and t_M are measured at the same temperature (T). Parameter R_{TM} appears to be more practically oriented than parameter r_T. It is known, for example, that the optimal heating rate is about 10 °C per hold-up time ($R_{TM} = 10$ °C) [60,61]. Measured in units of temperature, parameter R_{TM} represents a temperature change during the time interval equal to hold-up time (t_M), and is called sometimes as the *dead temperature* [16], *void temperature* [47], *hold-up temperature* [1], etc. This terminology is avoided here.

To emphasize the fact that R_{TM} and t_M in Eqn (2.63) are measured at the same reference temperature (T_{ref}), the formula can be expressed as

$$R_{TM,ref} = R_T t_{M,ref} \tag{2.64}$$

assuming that T_{ref} can be any temperature as long as it is the same for $R_{TM,ref}$ and $t_{M,ref}$. If $R_{TM,ref}$ for some $t_{M,ref}$ is known, then R_T and r_T could be found from Eqns (2.64) and (2.59) as

$$R_T = \frac{R_{TM,ref}}{t_{M,ref}} \tag{2.65}$$

$$r_T = \frac{t_{M,char}}{\theta_{char}} \cdot \frac{R_{TM,ref}}{t_{M,ref}} \tag{2.66}$$

Equations (2.60) and (2.64) yield

$$r_T \approx \frac{R_{TM,ref}}{\theta_{T,ref}} \tag{2.67}$$

If r_T is known, then $R_{TM,ref}$ can be found from Eqns (2.67) and (2.61) as

$$R_{TM,ref} = r_T \theta_{char,st} \left(\frac{T_{ref}}{T_{st}}\right)^{0.7} \tag{2.68}$$

2.7.5. Migration and Elution Parameters

Let k_{init}, μ_{init}, and ω_{init} be *initial retention factor*, *initial mobility*, and *initial immobility* of a solute, i.e. the solute parameters in Eqn (2.2) at the beginning of a heating ramp. A solute is *highly retained* at the beginning of a heating ramp, or, briefly, *initially highly retained*, if $k_{init} \gg 1$, or equivalently, $|\mu_{init}| \ll 1$, $\omega_{init} \approx 1$. Following are some solutions of Eqn (2.58) [1].

A solute *migration immobility*, $\omega(\zeta)$, can be found as

$$\omega = \omega(\zeta) = \omega_{init}\omega_a(\zeta) \tag{2.69}$$

where ζ is the solute dimensionless distance (Eqn (2.31)) from the column inlet, and

$$\omega_a = \omega_a(\zeta) = e^{-r_T\zeta} \tag{2.70}$$

is the solute *asymptotic immobility* – the immobility it would have if it was highly retained ($\omega_{init} \approx 1$) at the beginning of the heating ramp. At $\zeta = 1$, a solute migration immobility becomes its *elution immobility* (ω_R). The latter can be found from Eqn (2.69) as

$$\omega_R = \omega_{init}\omega_{R,a}, \quad \omega_{R,a} = e^{-r_T} \tag{2.71}$$

where $\omega_{R,a}$ is the solute *asymptotic elution immobility* – the elution immobility it would have if it was highly retained ($\omega_{init} \approx 1$) at the beginning of the heating ramp.

Combining Eqn (2.69) with Eqns (2.3) and (2.12), one can find other *migration parameters* of a solute such as $k(\zeta)$, $\mu(\zeta)$, and $T(\zeta)$ corresponding to the time when the solute is located at $z = \zeta L$ [1]. At $\zeta = 1$, a solute migration parameters become its *elution parameters*. Elution immobility ($\omega_{R,a}$) of an initially highly retained solute is described in Eqn (2.71). Other

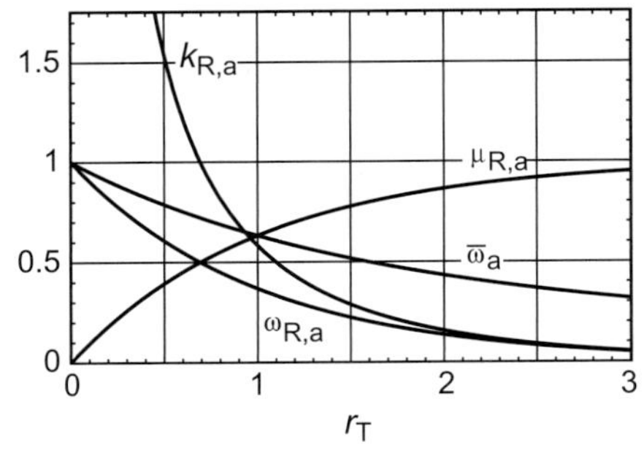

FIGURE 2.7 Asymptotic elution parameters in Eqns (2.71)–(2.73) and (2.76).

parameters of such a solute can be found from Eqns (2.3), (2.12), and (2.71). They are [1] (Figure 2.7):

$$\mu_{R,a} = 1 - e^{-r_T} \approx \begin{cases} r_T, & r_T < 0.3 \\ 1, & r_T > 3 \end{cases} \tag{2.72}$$

$$k_{R,a} = 1/(e^{r_T} - 1) \approx \begin{cases} 1/r_T, & r_T < 0.3 \\ 0, & r_T > 3 \end{cases} \tag{2.73}$$

$$\begin{aligned} T_{R,a} - T_{char} &= \theta_{char} \ln\left(e^{r_T} - 1\right) \\ &\approx \theta_{char} \cdot \begin{cases} \ln r_T, & r_T < 0.3 \\ r_T, & r_T > 3 \end{cases} \end{aligned} \tag{2.74}$$

The last two expressions suggest that $r_T = \ln 2 \approx 0.7$ can be viewed as a *benchmark heating rate*. When $r_T = \ln 2$, each initially highly retained solute elutes at its characteristic temperature ($T_{R,a} = T_{char}$) and the solute elution retention factor is equal to unity ($k_{R,a} = 1$). According to Eqn (2.67), at $r_T = \ln 2$, the normalized heating rate ($R_{TM,ref}$) *measured at $T_{ref} = 150$ °C is about 21 °C per hold-up time which is about two times higher than the optimal heating rate of about 10 °C per hold-up time [60] (see also section 2.10.3). This means that optimal r_T is close to 0.35. This also means that, because the lower end ($r_T < 0.3$) of approximations in Eqns (2.72)–(2.74) is closer to the optimal r_T than is the higher end ($r_T > 3$), the former is practically more important than the latter. Consider, for example, $T_{R,a}$ in Eqn (2.74) illustrated in Figure 2.8. When $r_T < 0.3$ (lower than 9 °C per t_M measured at 150 °C), $T_{R,a}$ is lower than T_{char} and departure of $T_{R,a}$ from T_{char} is proportional to $\ln r_T$. Making r_T 2.72 times smaller subtracts additional θ_{char} from $T_{R,a}$, and making r_T two times smaller subtracts additional $0.7\theta_{char}$ from $T_{R,a}$.

Example 2.4. Let $\theta_{char} = 30$ °C and $r_T = 0.1e \approx 0.272$. It follows from the approximation in Eqn (2.74) that lowering r_T from 0.272 to 0.1 reduces $T_{R,a} - T_{char}$ from −39 °C to −69 °C (a 30 °C change). More accurately, the same change in

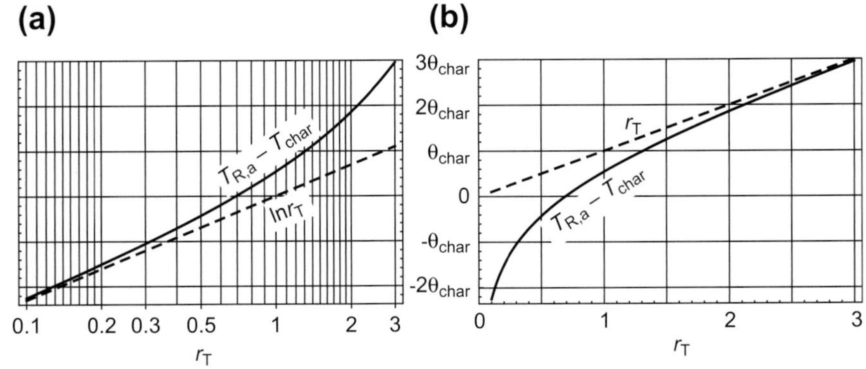

FIGURE 2.8 Graphs (solid lines) of departure ($T_{R,a} - T_{char}$) of a solute elution temperature ($T_{R,a}$) from its characteristic temperature (T_{char}) vs. dimensionless heating rate (r_T). The dashed lines are the tangents to the solid lines (a) at $r_T \to 0$ in logarithmic, and (b) at $r_T \to \infty$ in linear scales for r_T.

r_T reduces $T_{R,a} - T_{char}$ from −34.9 °C to −67.6 °C (a 32.7 °C change).

An important parameter of a column operations is the *distance-averaged immobility*

$$\overline{\omega} = \int_0^1 \omega(\zeta)d\zeta \tag{2.75}$$

which, due to Eqns (2.69), (2.70) and (2.72), can be found as (Figure 2.7)

$$\overline{\omega} = \omega_{init}\int_0^1 e^{-r_T\zeta}d\zeta = \omega_{init}\overline{\omega}_a, \quad \overline{\omega}_a = \frac{1-e^{-r_T}}{r_T} = \frac{\mu_{R,a}}{r_T} \tag{2.76}$$

How accurate are the formulas for the solute migration and elution parameters described here? To answer this question, recall that the formulas were obtained from solving migration integrals under the two approximating assumptions. One was the use of the linearized model of the solute–column interaction. Another was the assumption in Eqn (2.58) that gas viscosity "experienced" by each particular solute remained fixed during its entire migration and equal to $\eta(T_{char})$ where T_{char} is the solute characteristic temperature. The two assumptions lead to two layers of possible errors. The errors in t_R and μ_R relative to the ideal thermodynamic model are illustrated in Figure 2.9 and

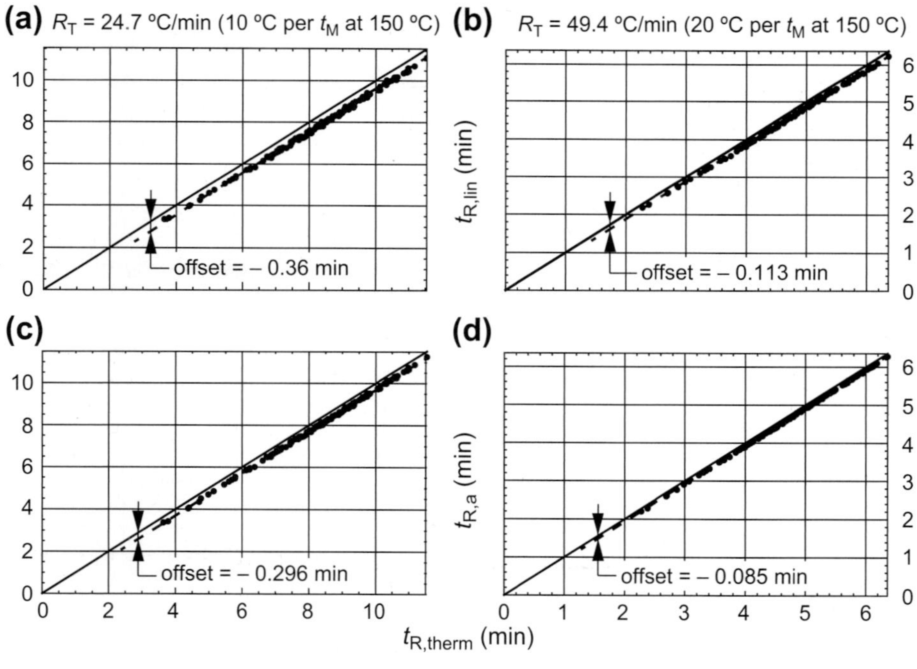

FIGURE 2.9 Retention times (t_R) of 147 pesticides in a column with 1701-type (14% cyanopropylphenyl–methyl polysiloxane) stationary phase polymer. In each graph, a dot represents retention time of the same solute calculated by two of the following three methods: parameters $t_{R,therm}$ and $t_{R,lin}$ were found from numerical integration [1] of Eqns (2.50) and (2.51) with μ from Eqn (2.56) for $t_{R,therm}$ and from Eqn (2.55) for $t_{R,lin}$; parameter $t_{R,a}$ was found from Eqns (2.74) and (2.54). Solid lines are the graphs of $t_{R,lin} = t_{R,therm}$ in (a) and (b), and $t_{R,a} = t_{R,therm}$ in (c) and (d). Dashed lines are the graphs of linear least-squares fits in numerical data. Column: 10 m × 0.1 mm × 0.1 μm. Gas: helium isobaric at $p_o = 1$ atm and initial flow of 0.8 mL/min ($t_{M,init} = 0.32$ min, $t_{M,ref} = 0.405$ min at $T_{ref} = 150$ °C). Temperature program: $T = 300\text{ K} + R_T t$ with R_T shown in the graphs. Characteristic parameters of all 147 pesticides were obtained as described in ref. [1].

Figure 2.10 for two heating rates more or less bracketing practically used optimal or near optimal rates. Parameters t_R and μ_R were chosen for the illustration because the former directly affects the analysis time while the latter directly affects the peak width [1] (discussed later in this chapter). Although the errors are clearly visible, they do not significantly affect evaluation of column performance. The following factors deserve a notice.

1. When compared to the retention time of the last *elute*, the errors in theoretically predicted $t_{R,a}$ do not exceed several percentage points – well below the needed accuracy of theoretical prediction of analysis time in evaluations of column performance.
2. The errors in the average values of μ_R do not significantly exceed the differences in the values of μ_R for particular solutes in the same analysis. These errors are also generally

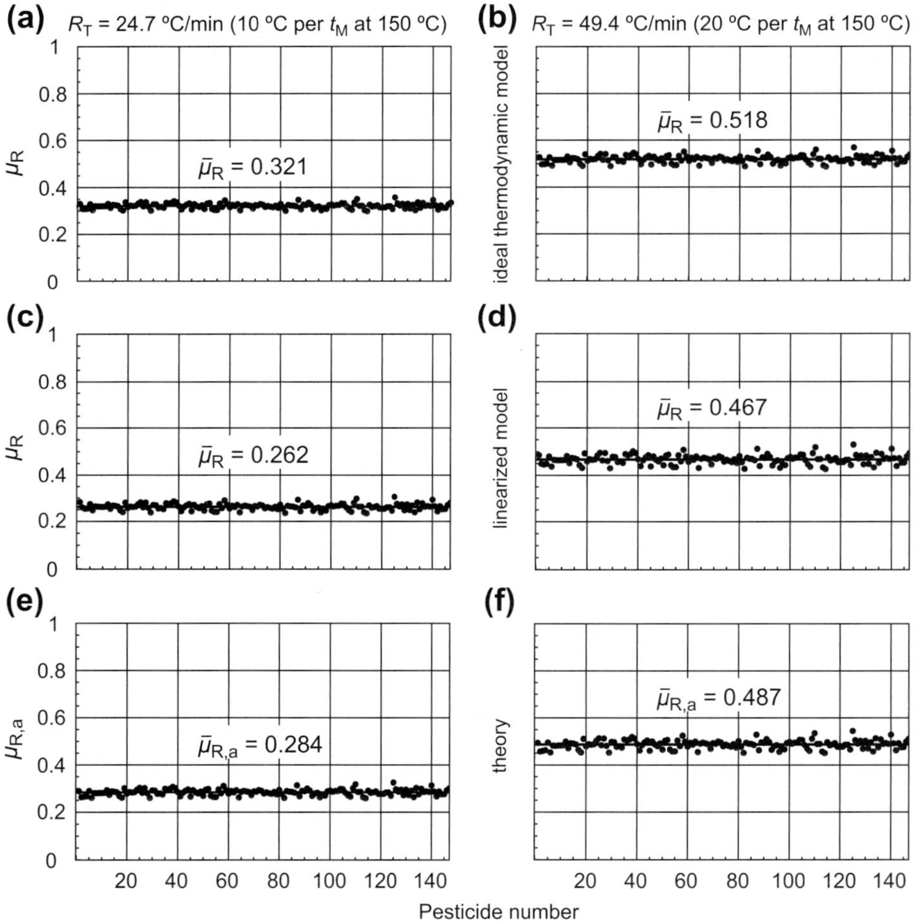

FIGURE 2.10 Elution mobilities (μ_R) of 147 pesticides. In each graph, a dot represents elution mobility of the same solute calculated by one of the following three methods: parameters μ_R were found from numerical integration [1] of Eqns (2.50) and (2.51) with μ from Eqn (2.56) for ideal thermodynamic model and from Eqn (2.55) for linearized model; parameter $\mu_{R,a}$ was found from Eqn (2.72). Quantities $\bar{\mu}_R$ and $\bar{\mu}_{R,a}$ are the averages of the respective numerical values (the dots) within each respective graph. Other conditions are the same as in Figure 2.9.

smaller than the errors in the solute diffusivities (section 2.5.3) and other factors affecting the peak widths.

3. The errors appear as the heating-rate-dependent offsets and can be corrected if necessary.

2.8. PEAK SPACING AND REVERSAL OF PEAK ORDER

Peak *spacing* is the distance between two peaks [28,62]. It can be expressed as *temporal spacing*, i.e. the difference (Δt_R) in the peak retention times (t_R), or as a *thermal spacing*, i.e. the difference (ΔT_R) in the peak elution temperatures (T_R). The thermal spacing, $\Delta T_R = T_{R,B} - T_{R,A}$, of peaks B and A can be estimated as [1]

$$\begin{aligned} \Delta T_R &\approx \Delta T_{char} + X_\theta \Delta\theta_{char}, \\ \Delta T_{char} &= T_{char,B} - T_{char,A}, \\ \Delta\theta_{char} &= \theta_{char,B} - \theta_{char,A} \end{aligned} \quad (2.77)$$

where quantity

$$X_\theta = \frac{r_T}{1 - e^{r_T}} + \ln\left(1 - e^{-r_T}\right) \quad (2.78)$$

is the *sensitivity* of the thermal spacing of two peaks to the difference ($\Delta\theta_{char}$) in their characteristic thermal constants. The sensitivity is a function of dimensional heating rate (r_T); it has a negative value and the magnitude getting monotonically smaller with an increase in r_T (Figure 2.11). These properties of X_θ lead to the following conclusions:

1. The slower the heating, the larger the effect of the difference $\Delta\theta_{char}$ on the peak thermal spacing.
2. When quantities ΔT_{char} and $\Delta\theta_{char}$ for a given solute pair have different signs, the heating rate does not affect the sign of their thermal spacing (ΔT_R) and the heating rate does not affect the solute elution order.
3. When quantities ΔT_{char} and $\Delta\theta_{char}$ for a given solute pair have (a) the same signs and (b) comparable values, then a change in a heating rate can change the sign of ΔT_R and, therefore, the solute elution order.

Example 2.5. Typically, the largest value of the difference $\Delta\theta_{char}$ for two solutes having close T_{char} values does not exceed 30% of the average θ_{char} for that T_{char} (Figure 2.3). For $T_{char} = 423$ K (150 °C), the average θ_{char} is about 30 °C and $|\Delta\theta_{char}| < 9$ °C. For $r_T = 0.35$ (near optimal heating rate), Eqn (2.78) yields: $X_\theta = -2.1$. Therefore, $|X_\theta \Delta\theta_{char}| < 20$ °C. If ΔT_{char} and $\Delta\theta_{char}$ for two solutes have the same sign, then change in a heating rate can change the solute elution order. However, even in that case, the elution order change is unlikely if $|\Delta T_{char}| > 20$ °C.

4. As a general trend, solutes eluting during a heating ramp covering a wide temperature range elute in the order of increase in their characteristic temperatures. Thus, according to the previous example, among two solutes having characteristic temperatures (T_{char}) differing by 20 °C or more, the one that has higher T_{char} will elute after the one that has lower T_{char}.

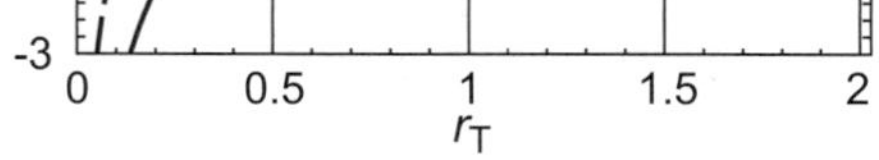

FIGURE 2.11 Sensitivity (X_θ) (Eqn (2.78)) of thermal spacing of two peaks to the difference ($\Delta\theta_{char}$) in their characteristic thermal constants.

Since a change in a heating rate can change the order of two closely spaced peaks, it might also be useful to know the *sensitivity* of the peak spacing to a change in the heating rate. Let us define that sensitivity (X_r) as the change in ΔT_R per unit of relative change (dr_T/r_T) in r_T, i.e. [1],

$$X_r = \frac{d(\Delta T_R)}{dr_T/r_T} \quad (2.79)$$

Equations (2.77) and (2.78) yield (Figure 2.12) [1]

$$X_r = \frac{d(\Delta T_R)}{dr_T/r_T} = \frac{r_T^2 e^{r_T}}{(e^{r_T}-1)^2}\Delta\theta_{char} \quad (2.80)$$

Equation (2.80) can be also expressed as

$$d(\Delta T_R) = X_r\frac{dr_T}{r_T} = \frac{r_T^2 e^{r_T}\Delta\theta_{char}}{(e^{r_T}-1)^2}\frac{dr_T}{r_T} \quad (2.81)$$

Figure 2.12 shows that, in the most practically important region of moderate heating rates, $X_r \approx \Delta\theta_{char}$. As a result,

$$d(\Delta T_R) \approx \Delta\theta_{char}\frac{dr_T}{r_T}, \quad r_T \leq 0.75 \quad (2.82)$$

Example 2.6. Let $\Delta\theta_{char} = 10\ °C$, $r_T = 0.35$, and $dr_T/r_T = 0.1$. The former represents more or less worst-case scenario of nearly the largest possible value of $\Delta\theta_{char}$ (Example 2.5). Eqn (2.82) yields $d(\Delta T_R) = 1\ °C$. This means that, a 10% change in dimensionless heating rate can change thermal spacing of two peaks by up to 1 °C. How significant is this change? Let σ and σ_T be the standard deviations of a peak in, respectively, the time and the temperature domain. They are related as $\sigma_T = R_T\sigma$. Under optimal conditions, σ_T can be estimated to be [1] $\sigma_T = 40\ °C/\sqrt{N}$, where N is a column *plate number* which can be estimated as $N = L/d$. For a conventional 25 m × 0.25 mm column, $\sqrt{N} \approx 320$ and $\sigma_T \approx 0.125\ °C$. This means that a 10% change in the heating rate can cause an $8\sigma_T$ change in ΔT_R. This can definitely cause a reversal of the peak order. Thus, the order of two peaks separated by $4\sigma_T$ could change without changing the absolute value of the peak spacing.

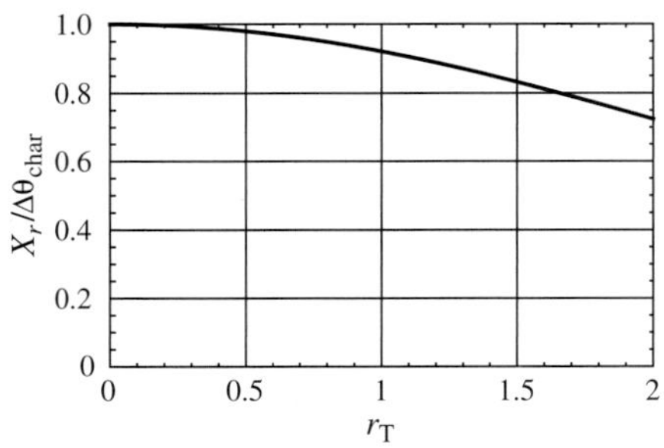

FIGURE 2.12 Normalized sensitivity ($X_\theta/\Delta\theta_{char}$) (Eqn (2.80)) of thermal spacing (ΔT_R) of two solutes to relative change (dr_T/r_T) in dimensionless heating rate.

Equation (2.82) is useful not only for the evaluation of a possibility of reversal of the peak order, it can be also used for evaluation of requirements for the accuracy of method parameters of GC analysis. According to Eqn (2.59), r_T in a given column is proportional to R_T. Therefore, a relative change (dr_T/r_T) in r_T in a given column with a given gas flow is equal to relative change (dR_T/R_T) in R_T. However, a change in R_T is not the only cause of a change in r_T. Eqn (2.59) also shows that r_T is proportional to hold-up time (t_M). Due to Eqn (2.46), t_M is inversely proportional to pressure drop along a column and, when column pressure and temperature are fixed, t_M is proportional to $(L/d)^2$. This means that a column replacement, trimming a column end, etc. can also cause an error in r_T. The relative error in r_T caused by the error in L/d is two times larger than relative error in L/d. This suggests that, to reduce the possibility of the reversal of the peak order, a combined effect of the relative error in heating rate, the relative error in pressure drop, and twice the relative error in the ratio L/d should be tightly controlled.

Example 2.7. Suppose that, to prevent a measurable reversal of a peak order, a possible change in ΔT_R should not exceed $1\sigma_T$. According to Example 2.6, this can be achieved when

$$\frac{dr_T}{r_T} < \frac{40\ ^\circ C}{\Delta\theta_{char}\sqrt{N}} \approx \frac{4}{\sqrt{N}}, \quad (\text{for } d(\Delta T_R) < \sigma_T)$$

For a conventional 25 m × 0.25 mm column, $\sqrt{N} \approx 320$. Therefore, the change in r_T should not exceed 1.25%. For more efficient columns, the acceptable change in r_T should be even smaller, inversely proportional to $\sqrt{N}$.

The last example suggests that the combined effect of the relative errors in all factors affecting r_T should not exceed 1% or be even smaller especially in the analyses using high-efficiency columns. The errors in the column dimensions and sometimes the errors caused by a column trimming are typically much larger than that [63]. As a result, tight control of the heating rate and the column pressure drop combined with retention time locking (RTL) [47] after each column trimming and replacement could be necessary to prevent the reversal of the peak order in analyses of some samples.

2.9. PEAK WIDTH

There are many *peak width* metrics in chromatography. *Standard deviation* (σ) of a peak [1,2,6–10,12,15], *half-height width* (w_h) and *base width* (w_b) [7,10,15], *area-over-height width* (w_A) [1,64], 2σ-width [7], 4σ-width [8], 6σ-width [65], and 8σ-width [66] are some of them. Quantities such as w_h and w_b are easy to measure by "a ruler and a pencil". Metric w_A is typically reported by commercial data-analysis devices (integrators). Most important for theoretical studies is the fact that only σ can be theoretically predicted from parameters of chromatographic analysis if the peak shape is not known *a priori*. As a result, the choice becomes between the metric of σ and its multiples. Each multiple has its own justification. Typical justifications are based on the assumption of *Gaussian peaks* [1,7,8]. Thus, starting from van Deemter et al. [67], the choice of the 4σ-width metric is typically justified by the fact that $w_b = 4\sigma$. However, for non-Gaussian peaks (generated in multidimensional separations and in other practically important applications), the relationship between w_b and σ might be different. Thus, for *exponential peaks*, $w_b = \sigma$ [1]. Other multiples of σ are also questionable for non-Gaussian peaks. In addition to that, the very existence of different peak width metrics based on different multiples of σ is a source of confusing interpretations of some published results. In this chapter, **$\boldsymbol{\sigma}$ is the only peak width metric**.

Peak width (σ) is measured in units of time. In a study of properties of temperature-programmed analysis, it might be convenient to deal with the (earlier considered) *time-domain peak width* (σ_T) defined as

$$\sigma_T = R_T\sigma \tag{2.83}$$

and measured in units of temperature. All known studies of the peak width are based on the studies of evolution of the width ($\tilde{\sigma}$) of a *solute zone* during its migration in a column. This width (the standard deviation of a solute zone) is measured in units of distance along a column.

The width of a solute zone and a peak can be affected by nonideal sample introduction when the *injection pulse* is insufficiently sharp so that its width (σ_i) can broaden the solute zones and, therefore, the peaks. Nonuniform dynamic conditions can further complicate the evaluations. As a starting point, it is convenient to evaluate the widths of solute zones and peaks under the simplest conditions.

2.9.1. Plate Height and Plate Number. Uniform Static Conditions

For the most part, material of this section is suitable not only for GC but also for other separation techniques operating under uniform static conditions.

Let $\tilde{\sigma}_o^2$ be the *variance* of a solute zone (the zone's *second central moment* [1]) eluting from a column outlet. In a column operating under uniform static conditions and *ideal* or nearly ideal sample introduction, $\tilde{\sigma}_o^2$ caused by the solute longitudinal dispersion (*spreading* and *broadening*) due to a solute (a) molecular diffusion in gas, (b) molecular diffusion in stationary phase, (c) retention, and (d) parabolic flow profile can be found from one of the following two formulas (Golay [68]):

$$\tilde{\sigma}_o^2 = \mathscr{H} L \text{ (uniform static conditions)} \tag{2.84}$$

$$\tilde{\sigma}_o^2 = \mathscr{D} t_R \text{ (uniform static conditions)} \tag{2.85}$$

where

$$\mathscr{H} = \frac{2D}{u} + \frac{\vartheta_{G1}^2 d^2 u}{96D} + \frac{2\vartheta_{G2}^2 d_f^2 u}{3D_S} \tag{2.86}$$

$$\mathscr{D} = \mathscr{H} v = \mathscr{H} \mu u \tag{2.87}$$

$$\vartheta_{G1} = \frac{\sqrt{1+6k+11k^2}}{1+k} = \sqrt{11-16\mu+6\mu^2} = \sqrt{1+4\omega+6\omega^2} \tag{2.88}$$

$$\vartheta_{G2} = \frac{\sqrt{k}}{1+k} = \sqrt{(1-\mu)\mu} = \sqrt{(1-\omega)\omega} = \sqrt{\mu\omega} \tag{2.89}$$

Parameter D_S in these formulas is a solute diffusivity in a stationary phase. Other parameters were described earlier.

Quantities $\mathscr{H}$ and $\mathscr{D}$ in Eqns (2.86) and (2.87) are a solute *spatial dispersion rate* and *temporal dispersion rate*. The latter is the rate of increase in $\tilde{\sigma}^2$ per unit of time of a solute migration along a column. The half-value ($D_{eff} = \mathscr{D}/2$) of $\mathscr{D}$ is known in chromatography as *effective diffusion coefficient* [8], *dynamic diffusion constant* [69], *dispersion coefficient* [70], *effective diffusivity* [71,72], and under other names. Both, D_{eff} and $\mathscr{D}$, are measured in units of length^2/time (such as cm^2/s). Quantity $\mathscr{H}$ is the rate of increase in $\tilde{\sigma}^2$ per unit of a solute displacement along a column. It is known as a column *plate height*, and can be measured in units of length^2/length (such as mm^2/m – an increase in $\tilde{\sigma}^2$ in millimeter-squared per one meter of a solute displacement along a column) [1]. However, as a matter of practical convenience, $\mathscr{H}$ is measured in units of length (such as mm) coming from reduction of unit of length^2/length. The basic relation ($\mathscr{D} = \mathscr{H} v$, Eqn (2.87)) between $\mathscr{D}$ and $\mathscr{H}$ is more symmetric than relation $D_{eff} = \mathscr{H} v/2$ between D_{eff} and $\mathscr{H}$. There are other reasons for using parameter $\mathscr{D}$ rather than D_{eff} [1].

Although Golay clearly recognized the greater utility of parameter $\mathscr{H}$ over parameter $\mathscr{D}$ in chromatography, he was notorious in the equal parallel theoretical treatment of both parameters. The goal seems to be to stress conceptual similarity of a mysterious (in his time and even now) parameter $\mathscr{H}$ with the more familiar concept of diffusion as the rate of broadening of a solute zone.

Although Eqns (2.84) and (2.85) have similar structure, and parameters $\mathscr{H}$ and $\mathscr{D}$ represent similar concepts, some differences in components of Eqns (2.84) and (2.85) as well as in parameters $\mathscr{H}$ and $\mathscr{D}$ speak decisively in favor of greater utility of parameter $\mathscr{H}$ over parameter $\mathscr{D}$. First, parameter L in Eqn (2.84) is the same for all solutes while parameter t_R in Eqn (2.85) can be different for different solutes. As for parameters $\mathscr{H}$ and $\mathscr{D}$ in these formulas, both are solute-dependent quantities. However, the values of parameter $\mathscr{H}$ for all solutes are contained within a relatively narrow range. As a result, Eqn (2.84) can be treated as a single formula describing a *column* as a whole (rather than specific solutes), and $\mathscr{H}$ can be (and typically is) treated as a *column parameter* (a *column plate height*). On the other hand, the values of parameter $\mathscr{D}$ have no lower limit and, according to Eqn (2.87), can be zero for highly retained solutes having $\mu = 0$. As both parameters in Eqn (2.85) are solute dependent and each changes within a wide range, Eqn (2.85) is an expression of properties of individual solutes, but not of a column as a whole. These are the main reasons why the plate height is known as one of the key parameters of a column operation while parameter $\mathscr{D}$ is barely known in chromatography.

The name *plate height* for the spatial dispersion rate ($\mathscr{H}$) is rooted in the earlier known concept of H.E.T.P. ("height of equivalent theoretical plate") introduced by Peters in 1922 [73] for packed distillation towers (columns) as a way of comparing them with the towers composed of discrete plates (trays). In 1941, Martin and Singe adopted the concept of H.E.T.P. ("height equivalent to one theoretical plate") for chromatographic columns [23]. It was initially thought that H.E.T.P. represented a *separation stage* and that the stage-to-stage transitioning was the sole reason for the broadening of the solute zones. The inconsistency of this perception became obvious when van Deemter et al. published their celebrated 1956 paper [67] where it was found that H.E.T.P. in a packed column was rooted in a solute diffusion. Shortly after that, a 1957 scientific committee of 10 leading experts in the field-included Martin and Golay agreed that "a theoretical plate is an abstract term with no physical significance other than as a measure of the relative variance of a peak" [74]. A few years later, Giddings suggested that the "current trend is to abandon the postulates that gave birth to the concept of HETP and speak of it as a general parameter for measuring zone spreading" (Giddings and Kelly [75], page 96). "Despite the absence of the theoretical plate model ..., [parameter $\mathscr{H}$] retains the name *plate height* for historical continuity" (Giddings [8], page 97). Additional details regarding the history of the *plate height* concept can be found elsewhere [1].

The width (σ) of a peak produced by elution of a solute zone can be found as

$$\sigma = \frac{\tilde{\sigma}_o}{v_o} \tag{2.90}$$

where v_o is the solute velocity at the column outlet. When condition of a column operation is uniform and static, v_o is the same as v at any location along a column and at any time. As a result, Eqns (2.84) and (2.49) allow one to express Eqn (2.90) as

$$\sigma = \frac{\sqrt{\mathscr{H}\ L}}{v} = \frac{t_R}{\sqrt{\mathscr{N}}} \text{ (uniform static conditions)} \tag{2.91}$$

where [1]

$$\mathscr{N} = \frac{L}{\mathscr{H}} \tag{2.92}$$

is a column *plate number* [1,8,11,76]. Eqns (2.91) and (2.92) show that, all other factors being equal, the larger the plate height the wider the peaks. The plate height has an adverse effect on the peak separation and should be reduced whenever this is possible without compromising other operational parameters.

Equation (2.92) seems to be an intuitively natural definition of the plate number. However, that definition is different from the widely accepted definition recommended by IUPAC [15] and introduced shortly. The difference results from the need to account for the effect of the gas compressibility on the peak width dependence on plate height, plate number, gas velocity, and other parameters.

2.9.2. Nonuniform Dynamic Medium – General Considerations

Gas decompression along a column causes a coordinate-dependent change in a solute density and velocity, i.e. nonuniform conditions of a solute migration. Temperature and/or pressure programming causes a time-dependent change in a solute velocity, i.e. dynamic conditions of a solute migration. Our immediate goal is to find (a) the width of a solute zone migrating in a nonuniform dynamic medium and (b) the width of a corresponding peak.

The variance ($\tilde{\sigma}^2$) of a solute zone migrating in *one-dimensional mass-conserving* nonuniform dynamic medium can be deduced from the following partial differential equation of *mass conservation* in such medium [1,8,71,72]:

$$\frac{\partial a}{\partial t} = \frac{1}{2}\frac{\partial}{\partial z}\left(\mathscr{D}\frac{\partial a}{\partial z}\right) - \frac{\partial}{\partial z}(va), \quad \mathscr{D} = \mathscr{H}v \quad (2.93)$$

where a is *specific amount* of the solute (the solute amount per unit of length), and (as before) parameters $\mathscr{D}$, $\mathscr{H}$, and v are, respectively, the solute temporal dispersion rate, plate height (spatial dispersion rate), and velocity, respectively. All four parameters could be functions of distance (z) and time (t) and, therefore, can be qualified as *local* and/or *instant* quantities. The local values of parameters $\mathscr{D}$ and $\mathscr{H}$ are defined as [1,71,72]

$$\mathscr{H} = \lim_{\tilde{\sigma}\to 0}\frac{\mathrm{d}\tilde{\sigma}^2}{\mathrm{d}z}, \quad \mathscr{D} = \lim_{\tilde{\sigma}\to 0}\frac{\mathrm{d}\tilde{\sigma}^2}{\mathrm{d}t} \quad (2.94)$$

meaning that the *local plate height* ($\mathscr{H}$) is the rate of increase in the variance ($\tilde{\sigma}^2$) of infinitesimally narrow solute zone during its migration along infinitesimally short segment of length dz. Similarly, the *local temporal dispersion rate* ($\mathscr{D}$) is the rate of increase in the variance ($\tilde{\sigma}^2$) of infinitesimally narrow solute zone during infinitesimally short migration time dt. Quantities $\mathscr{H}$ and $\mathscr{D}$ defined in Eqn (2.94) can be found from Eqns (2.86) and (2.87) where all parameters could be functions of distance and time.

Equation (2.93) is not limited to GC, but can be used for addressing the problem of peak broadening in *gradient-elution liquid chromatography* (LC) and beyond.

The distance-dependent changes in parameters of Eqn (2.93) could be abrupt (as in a narrow region around connection of two columns having different diameters, stationary phases, etc.) or gradual (as in the case of a solute velocity change due to the gas decompression). A medium of a solute migration is *smooth* if a local plate height ($\mathscr{H}$) and velocity gradient ($\partial v/\partial z$) do not significantly change within a solute zone. For example, nonuniformity caused by gas decompression along a GC column is smooth [1]. In the case of a smooth medium, Eqn (2.93) can be reduced to an ordinary differential equation [1,71,72]:

$$\frac{\mathrm{d}\tilde{\sigma}^2}{\mathrm{d}z} = \mathscr{H} + \frac{2\tilde{\sigma}^2}{v}\cdot\left.\frac{\partial v(z,t)}{\partial z}\right|_{t=t(z)} \quad (2.95)$$

where a solute velocity gradient $\partial v/\partial z$ is a known function of z and t.

Example 2.8. According to Eqn (2.1), velocity (v) of a solute migrating in a GC column can be expressed as $v = \mu u$, where u is gas velocity and μ is a solute mobility. In a uniform, uniformly heated GC column, μ can change with time, but it is always the same in all locations along a column ($\partial\mu/\partial z = 0$). Due to Eqn (2.33), the gradient in v can be found as

$$\frac{\partial v}{\partial z} = \mu\frac{\partial u}{\partial z} = \frac{\mu u_{\mathrm{i}}}{2}\cdot\frac{X}{(1 - Xz)^{3/2}}, \quad X = \frac{p_{\mathrm{i}}^2 - p_{\mathrm{o}}^2}{Lp_{\mathrm{i}}^2}$$

where parameters p_{i}, p_{o}, u_{i}, and μ could be programmable functions of t.

Other solutions of Eqn (2.93) including the ones that are more general than the solution in Eqn (2.95) can be found elsewhere [1,72,77,78]. However, Eqn (2.95) appears to be a suitable starting point for solving the known problems of the peak width formation not only in nonuniform dynamic GC but also in other separation techniques such as *gradient-elution* liquid chromatography [LC] and beyond.[2] From now on, Eqn (2.95) is treated as the starting point in the study of a solute zone broadening in chromatography. Thus, a known Giddings formula for the apparent plate height in GC and Giddings compressibility factor [79] are derived below from Eqn (2.95). In the case of a uniform medium ($\partial v/\partial z = 0$), Eqn (2.95) converges to the form used by Golay [68]:

$$\frac{d\tilde{\sigma}^2}{dz} = \mathscr{H} \text{ (uniform medium)} \tag{2.96}$$

which, under static conditions, can be expressed in the form of Eqn (2.84).

Back to conditions specific for GC, as mentioned earlier, only the uniform heating of the columns is considered here. In this case, there are no gradients in solute mobility, i.e. $\partial\mu/\partial z = 0$. Eqn (2.95) becomes

$$\frac{d\tilde{\sigma}^2}{dz} = \mathscr{H} + \frac{2\tilde{\sigma}^2}{u} \cdot \left.\frac{\partial u(z,t)}{\partial z}\right|_{t=t(z)} \quad (\text{when } \partial\mu/\partial z = 0) \tag{2.97}$$

In addition to always assumed uniform heating of a column, it will be assumed from now on that, as in Example 2.8, gas **decompression along a column is the only source of nonuniform conditions in GC**. Eqn (2.97), i.e. the starting point of study of a solute zone broadening and a peak width formation in GC becomes

$$\frac{d\tilde{\sigma}^2}{dz} = \mathscr{H} + \frac{\tilde{\sigma}^2}{Y - z}, \quad Y = \frac{Lp_i^2}{p_i^2 - p_o^2} \tag{2.98}$$

where parameters p_i and p_o can be functions of time, and parameter $\mathscr{H}$ can be a function of distance and time.

2.9.3. Plate Height and Plate Number: Gas Decompression in Static GC

Under static (isothermal and isobaric) conditions, p_i and p_o are fixed quantities, and $\mathscr{H}$ at any given location (z) remains fixed through the entire analysis. The following can be noticed regarding the distance-dependent variations in $\mathscr{H}$. As follows from Eqn (2.86), $\mathscr{H}$ is a function of D and u. Both are inversely proportional to local pressure (p) and, therefore, due to the gas decompression along a column, both can change with z. On the other hand, the ratio D/u is independent of z. As a result, only the third term in Eqn (2.86) can change with z. These considerations lead to the following closed-form solution of Eqn (2.98) for the variance (σ_o^2) of the solute zone at the column outlet [1]:

$$\tilde{\sigma}_o^2 = \left(\frac{p_i^2 + p_o^2}{2p_o^2}\left(\frac{2D}{u} + \frac{\vartheta_{G1}^2 d^2 u}{96D}\right) + \frac{2\vartheta_{G2}^2 d_f^2 \bar{u}}{3D_S j^2}\right)L \tag{2.99}$$

The next step should be to find a suitable definition of a column plate height (H) under nonuniform conditions. At the first glance, it might appear that, to be consistent with Eqn (2.84) the entire expression in parentheses of Eqn (2.99) (let us denote it as expression X) should be equated to H. However, this approach leads to substantial inconsistencies. Expression X reflects both, *convective* zone broadening

[2] Because Eqn (2.93) is valid beyond GC, it could be mentioned that a theoretical study of the factors affecting a zone variance ($\tilde{\sigma}^2$) in gradient-elution LC is known from ref. [70] where the results were based on a subset of Eqn (2.93). That subset was inconsistent with the conservation of mass under nonuniform conditions (such as the conditions in gradient-elution LC). For that reason, the boundaries of the solution are unknown even within LC not to speak of other chromatographic techniques, and the solution itself might need further verification. On the other hand, the main result in ref. [70] can be obtained directly from Eqn (2.95) where, due to uniform mobile phase velocity ($\partial u/\partial z = 0$) assumed in ref. [70], $(\partial v/\partial z)/v = (\partial\mu/\partial z)/\mu$.

caused by the gas decompression along the column, and the *dispersive broadening* caused by the solute diffusion. In the case of a deep gas decompression, *convective* zone broadening can be much larger than the dispersive broadening. Thus, in GC-MS where $p_o = 0$, the convective zone broadening is so dominant that expression X becomes infinity. On the other hand, a large convective zone broadening is always accompanied with proportional increase in the gas velocity (u_o) and, therefore, in the solute velocity (v_o) at the column outlet. As a result based on Eqn (2.90), a large increase in the value of expression X has only a minor effect on the peak broadening, and, therefore, the expression itself does not represent a useful concept.

A meaningful extension of the plate height concept to columns with gas decompression is known from Giddings et al. [79]. It can be described as follows. Although Eqn (2.92) provides an intuitive definition of the plate number ($\mathcal{N}$), the latter can also be found from Eqn (2.91). Let us denote so found *plate number* as N, i.e.

$$N = \frac{t_R^2}{\sigma^2} \tag{2.100}$$

With this definition, it is unnecessary to know the plate height in order to find the plate number (N). It allows one to find N from measured (observed) parameters, t_R and σ. Once quantity N is found, the plate height can be found from Eqn (2.92). Let us denote so found *plate height* as H, i.e.

$$H = \frac{L}{N} \tag{2.101}$$

Giddings called this quantity as the *apparent plate height* or *measured plate height* [79,80]. Interestingly, in the case of a nonuniform gas velocity, apparent plate height (H) can be different from local plate height ($\mathcal{H}$) even if the latter is the same at all locations along the column [1,80]. This implies that the apparent plate height is not an average of local plate heights.

Following the logic behind the concept of the apparent plate height, quantity N can be called as the *apparent plate number*. To distinguish quantities H and N from their counterpart $\mathcal{H}$ and $\mathcal{N}$, quantity $\mathcal{H}$ is called from now on as the *local plate height* [1,79] or as the *instant plate height* [1]. Quantity $\mathcal{N}$ is called from now on as the *directly counted plate number* [1]. On the other hand, following accepted conventions, quantities H and N would be called as the plate height and the plate number. Their apparent nature would be only emphasized when necessary. Under uniform conditions, there is no difference between quantities H and $\mathcal{H}$. Similarly, there is no difference between N and $\mathcal{N}$. However, under nonuniform conditions, the differences do exist as will be seen shortly.

The formulas in Eqns (2.92) and (2.100) are known from Glueckauf [76] who treated Eqn (2.92) as the definition of the "number of theoretical plates", and Eqn (2.100) as a way of finding that number from experimental conditions. Shortly after, however, Eqn (2.100) was recommended [74,81] and broadly adopted as the definition of the plate number [2–7,9,10,12–15]. Only a small number of workers (Littlewood [11] and Giddings [8] among them) continued to use formula $N = L/H$ as the definition of N and Eqn (2.100) as a way of finding its value from experimental conditions.

Equations (2.100) and (2.101) define N and H as observable quantities. However, once it is explicitly defined, H can also be found from theoretical considerations, and with that, Eqns (2.100) and (2.101) can be reversed and used for theoretical prediction of the plate number (N) and peak width (σ). One has

$$N = \frac{L}{H} \tag{2.102}$$

$$\sigma = \frac{t_R}{\sqrt{N}} \tag{2.103}$$

Our next immediate goal is to theoretically find the plate height (H).

Substitution of Eqn (2.100) into Eqn (2.101) and accounting for Eqns (2.90), (2.99), (2.1), (2.36), (2.35), and (2.49) yield a known *Giddings formula* for the apparent plate height [79]:

$$H = \frac{L\sigma^2}{t_R^2} = \frac{L\tilde{\sigma}_o^2}{t_R^2 v_o^2} = j_G \cdot \left(\frac{2D}{u} + \frac{\vartheta_{G1}^2 d^2 u}{96D}\right) + \frac{2\vartheta_{G2}^2 d_f^2 \bar{u}}{3D_S} \quad (2.104)$$

where

$$j_G = \frac{9(p_i^2 + p_o^2)(p_i + p_o)^2}{8(p_i^2 + p_i p_o + p_o^2)^2} \approx \begin{cases} 1, & |\Delta p| \ll p_o \\ 1.125, & \Delta p \gg p_o \end{cases} \quad (2.105)$$

is *Giddings compressibility factor* [79]. The latter can change between 1 and 1.125 and, in many practical cases, it can be ignored by assuming that $j_G = 1$ at any pressure.

A column can be viewed as a *thin-film* one if the last term in Eqn (2.104) (the one that depends on the film thickness, d_f) is significantly smaller than the other terms and can be ignored. Due to the relatively fast diffusion through the pores of the porous-layer in PLOT columns, they typically behave as the thin-film ones. Golay proposed these columns [68] exactly because the diffusivity (D_S) of the solutes in gas-filled pores of adsorptive porous-layer is so high compared to that in typical liquid polymers that the third term in Eqn (2.86) and, therefore, in Eqn (2.104) for the PLOT columns is practically negligible compared to the other terms. A column has an *intermediate film thickness* or is a *thick film* one if the last term in Eqn (2.104) is comparable or larger than the other terms, respectively. If the film thickness effect on H is not negligible, then accounting for it is complex [82]. The tendency toward increasing the film thickness usually comes from the need for increasing column *loadability*. This makes it possible to inject larger sample amount and to lower (improve) the *detection limit* for low concentration components without overloading the column by large concentration components present in the same sample. However, increasing the film thickness beyond the level when its effect on H can be ignored substantially reduces a column separation performance. Another way of increasing column loadability is not only to increase the film thickness but to proportionally increase column length, diameter, and film thickness. This also increases H. A comparative analysis (unpublished) of the two approaches shows that an increase in the film thickness alone is only beneficial when the effect of the film thickness on H remains to be relatively small. After that, the proportional increase in column dimensions and film thickness offers a better tradeoff between the column separation performance, analysis time, and detection limit. From now on, **only the thin-film columns are considered.** Eqn (2.104) becomes

$$H = j_G \cdot \left(\frac{2D}{u} + \frac{\vartheta_{G1}^2 d^2 u}{96D}\right) \quad (2.106)$$

Equation (2.106) is simpler than Eqn (2.104). However, it is still not well suited for practical use and practice-oriented theoretical studies. In Eqn (2.106), quantities D and u should correspond to the same pressure. It can be inlet pressure (p_i), outlet pressure (p_o), or pressure (p) at any location along the column. However, if there is a significant difference between p_o and p_i (due to strong gas decompression along a column), then the measurement of outlet (u_o) or inlet (u_i) velocities becomes impractical. This is especially true for GC-MS where u_o approaches infinity. It is easier to measure and to set up inlet pressure (p_i), average gas velocity ($\bar{u}$), flow rate (F), or specific flow rate (f). When expressed via these parameters and outlet pressure (p_o), Eqn (2.106) becomes [1] (Figure 2.13)

$$H = H(f) = j_G \cdot \left(\frac{\pi d D_{pst}}{2f} + \frac{d\vartheta_{G1}^2 f}{24\pi D_{pst}}\right) \quad (2.107)$$

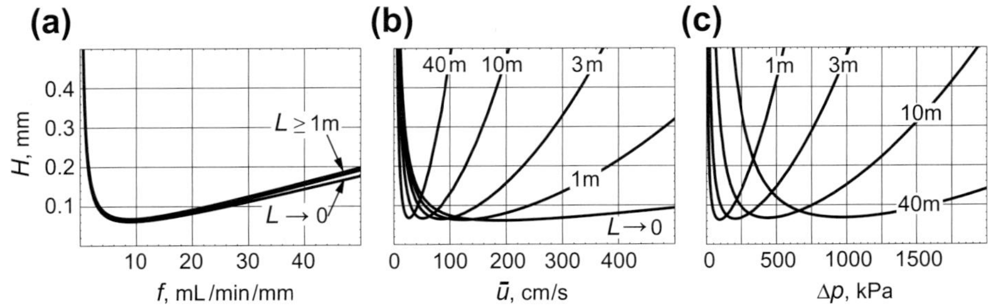

FIGURE 2.13 Plate height (H) as a function of (a) specific flow rate (f) of carrier gas, (b) its average velocity ($\overline{u}$), and (c) pressure drop (Δp) for several column lengths. Conditions: normal decane (C_{10}) in helium, $d = 0.1$ mm, $p_o = 1$ atm, $T = 100$ °C, $k = 1$. Curves at $L \to 0$ correspond to weak (negligible) gas decompression. Column length (L) has practically no effect on $H(f)$ (the slight increase in $H(f)$ with an increase in L comes from the dependence of j_G in Eqn (2.107) on L). On the other hand, L strongly affects the shapes of the curves $H(\overline{u})$ and $H(\Delta p)$, and horizontal positions of their minima. *Source: Reproduced from ref. [1] with permission.*

$$H = H(F) = j_G \cdot \left(\frac{\pi d^2 D_{pst} T_{norm}}{2TF} + \frac{T \vartheta_{G1}^2 F}{24\pi D_{pst} T_{norm}} \right) \quad (2.108)$$

$$H = H(p_o, \overline{u}) = j_G \cdot \left(\frac{128 D_{pst} L p_{st} \eta}{d^2 p_o^2} \cdot \left(\Phi\left(\frac{32 L \eta \overline{u}}{d^2 p_o}\right)^2 - 1 \right)^{-1} + \frac{d^4 p_o^2 \vartheta_{G1}^2}{6144 D_{pst} L p_{st} \eta} \cdot \left(\Phi\left(\frac{32 L \eta \overline{u}}{d^2 p_o}\right)^2 - 1 \right) \right) \quad (2.109)$$

$$H = H(p_o, p_i) = j_G \cdot \left(\frac{128 D_{pst} L p_{st} \eta}{d^2 (p_i^2 - p_o^2)} + \frac{d^4 \vartheta_{G1}^2 (p_i^2 - p_o^2)}{6144 D_{pst} L p_{st} \eta} \right) \quad (2.110)$$

where D_{pst} is a solute diffusivity in the gas at standard pressure ($p_{st} = 1$ atm). Due to Eqn (2.24), quantity D_{pst} can be found as

$$D_{pst} = \frac{X_D^2 D_{g,st} \cdot (10^3 \varphi)^{0.09}}{k^{0.1}} \cdot \left(\frac{T}{T_{st}}\right)^{\xi - 0.25}, \quad (0.1 \le k \le 30) \quad (2.111)$$

where gas-specific quantities $D_{g,st}$, X_D, and ξ are listed for several gases in Table 2.1.

Which formula for H is more suitable for the studies of column performance? To answer this question, several factors should be considered. Among them are simplicity of the formula and simplicity of expressions for optimal pneumatic variables.

Equations (2.107–2.110) and the graphs in Figure 2.13 show that the dependence of H on $\overline{u}$ is the most complex[3] while the dependence of H on f is the simplest ($H(F)$ is not shown (Figure 2.13) because its graphs differ

[3] Eqn (2.109) is different from the widely accepted formula $H = B/\overline{u} + C\overline{u}$ typically attributed to van Deemter et al. [67] or to Golay [68]. However, only the uniform conditions were considered in these works, and neither the time-averaged gas velocity ($\overline{u}$) nor the formula $H = B/\overline{u} + C\overline{u}$ itself have even been mentioned in these works. Furthermore, Golay explicitly stated (page 43) that his results were only "applicable to columns … in which the input to exit pressure ratio is near unity". In the case of a strong gas decompression, formula $H = B/\overline{u} + C\overline{u}$ is substantially incorrect. Thus, it follows from Eqn (2.109) that, at strong gas decompression (all GC-MS applications, and majority of analyses of complex samples with any detector), $H = B/\overline{u}^2 + C\overline{u}^2$ [1]. A history of the formula $H = B/\overline{u} + C\overline{u}$ along with its published criticism and retractions can be found in ref. [1].

from the graphs of $H(f)$ only by horizontal scale).

Equations (2.107–2.110) yield for *optimal flow rate* (F_{opt}), *optimal specific flow rate* (f_{opt}), *optimal average velocity* ($\overline{u}_{opt}$), and *optimal pressure* ($p_{i,opt}$) corresponding to the minimum in H [1]:

$$f_{opt} = \frac{2\sqrt{3}\pi\, D_{pst}}{\vartheta_{G1}} \tag{2.112}$$

$$F_{opt} = \frac{2\sqrt{3}\pi\, dD_{pst}}{\vartheta_{G1}} \cdot \frac{T_{norm}}{T} \tag{2.113}$$

$$\overline{u}_{opt} = \frac{18432 L p_{st}^2 D_{pst}^2 \eta}{d^4 p_o^3 \vartheta_{G1}^2} \cdot \left(\left(\frac{512\sqrt{3} L p_{st} D_{pst} \eta}{d^3 p_o^2 \vartheta_{G1}} + 1 \right)^{3/2} - 1 \right)^{-1} = \begin{cases} \dfrac{8\sqrt{3} D_{pst} p_{st}}{d p_o \vartheta_{G1}}, & |\Delta p| \ll p_o \\ \dfrac{3}{4}\sqrt{\dfrac{\sqrt{3} d D_{pst} p_{st}}{2L\eta\vartheta_{G1}}}, & \Delta p \gg p_o \end{cases} \tag{2.114}$$

$$p_{i,opt} = \sqrt{p_o^2 + \frac{512\sqrt{3} D_{pst} L p_{st} \eta}{d^3 \vartheta_{G1}}} \tag{2.115}$$

Again, the formula for $\overline{u}_{opt}$ is the most complex and the formula for f_{opt} is the simplest. The former depends on column dimensions (L and d), gas viscosity (η), and outlet pressure (p_o) in a rather complex way. What's more, these dependencies at strong gas decompression are different from the ones at weak decompression. The complexity of Eqn (2.114) might be the reason why the values of $\overline{u}_{opt}$ for many useful applications are not known from the literature.[4] On the other hand, f_{opt} is independent of column dimensions, and of parameters η and p_o. Simple recommendations for f_{opt} published in 1999 [83] for several gases are consistently recommended for several generations of commercial GC instruments manufactured by Agilent Technologies (Santa Clara, CA) since 1994.

From now on, Eqn (2.107) is treated as the primary formula for plate height (H) in GC.

In the independence of its optimum (f_{opt}) on a column dimension, quantity f is similar to Giddings' *reduced (dimensionless) gas velocity* [1,8,84]. However, f is more useful in practice because it more directly relates (Eqn (2.42)) to the operator-controllable flow rate (F). The fact that the optimum of quantity f is independent of column dimensions makes it possible to isolate (as it is done later in the chapter) a study of the effects of the column dimensions on the column performance from the effects of the flow rate on that performance.

Equation (2.107) can be described in symmetric form as [1]

$$H = \frac{H_{min}}{2} \cdot \left(\frac{f_{opt}}{f} + \frac{f}{f_{opt}} \right) \tag{2.116}$$

where

$$H_{min} = \frac{j_G d \vartheta_{G1}}{2\sqrt{3}} \tag{2.117}$$

[4] Frequently recommended experimental values of $\overline{u}_{opt}$ (35 cm/sec for helium and 50 cm/sec for hydrogen) are suitable only for conventional (25 m × 0.25 mm) columns. For a 1 m × 0.1 mm column in GC-MS (a popular choice of secondary column in GC × GC), $\overline{u}_{opt} \approx 150$ cm/sec for helium and $\overline{u}_{opt} \approx 250$ cm/sec for hydrogen. A large difference between actual and recommended values of $\overline{u}_{opt}$ is especially harmful in GC-MS where $H = B/\overline{u}^2 + C\overline{u}^2$ [1] and, therefore, the change in H due to departure of $\overline{u}$ from $\overline{u}_{opt}$ is much larger than it is typically recognized.

One can verify by direct substitutions that Eqn (2.116) is equivalent to Eqn (2.107). Introducing *dimensionless plate height* [1,8,84],

$$h = \frac{H}{d} \tag{2.118}$$

one can rearrange Eqn (2.116) as

$$h = \frac{h_{\min}}{2} \cdot \left(\frac{f_{\text{opt}}}{f} + \frac{f}{f_{\text{opt}}} \right) \tag{2.119}$$

where

$$h_{\min} = \frac{H_{\min}}{d} = \frac{j_G \vartheta_{G1}}{2\sqrt{3}} \approx 1 \tag{2.120}$$

2.9.4. Plate Height and Plate Number. Temperature-Programmed GC

The definition in Eqn (2.100) of the plate number in static analysis is not suitable for temperature-programmed analysis. Indeed, all peaks that were highly retained at the beginning of a linear heating ramp have roughly the same widths [1] (see also forthcoming analysis) while their retention times (t_R) gradually increase. As a result, quantity N in Eqn (2.100) tends to increase with time by several orders of magnitude. As a result, N becomes a peak-specific quantity rather than a column parameter and a key factor in evaluation of column performance. The following definition [1] rooted in the one proposed by Habgood and Harris in 1960 [12,18,85] avoids this problem, and defines N as a column-specific quantity similar to the one that comes out of Eqn (2.100) in static analysis [12,18].

Plate number (N) corresponding to a peak having retention time t_R in dynamic analysis is the ratio

$$N = \frac{t_{R,\text{stat}}^2}{\sigma^2} \tag{2.121}$$

where $t_{R,\text{stat}}$ is retention time in static analysis conducted under conditions existing at the time t_R in dynamic analysis, and σ is actual width of a peak having retention time t_R in actual dynamic analysis. Eqn (2.121) is the most general definition of the plate number (N). It incorporates the definition in Eqn (2.100). The following discussion leads to conclusion that the plate number (N) found from Eqn (2.121) for a typical temperature-programmed analysis is numerically close to its counterpart in static analysis [1,12,18].

Due to Eqn (2.49), quantity $t_{R,\text{stat}}$ in Eqn (2.121) can be found as

$$t_{R,\text{stat}} = \frac{t_{M,R}}{\mu_R} \tag{2.122}$$

where $t_{M,R}$ and μ_R are, respectively, the hold-up time statically measured at the conditions existing in dynamic analysis at $t = t_R$, and μ measured at the column temperature (T_R) existing in dynamic analysis at $t = t_R$. Eqn (2.121) becomes

$$N = \frac{t_{M,R}^2}{\mu_R^2 \sigma^2} \tag{2.123}$$

Substitution of this into the definition (Eqn (2.101)) for H yields

$$H = \frac{L \mu_R^2 \sigma^2}{t_{M,R}^2} \tag{2.123a}$$

After finding H from theoretical considerations, one can use it to find N from Eqn (2.102), and for finding the peak width from Eqn (2.123) as

$$\sigma = \frac{t_{M,R}}{\sqrt{N} \mu_R} \tag{2.124}$$

Typically, N in a given analysis is roughly the same for all solutes and can be treated as a fixed quantity. In addition to that, it follows from Eqn (2.72) that μ_R is roughly the same for all solutes that were highly retained at the beginning of a heating ramp in isobaric analysis.

Eqns (2.124), (2.45), and (2.20) together with the data in Table 2.1 suggest that

The width (σ) of a peak generated by a solute that was highly retained at the beginning of a linear heating ramp in isobaric analysis and that elutes during the ramp at temperature T_R is proportional to $T_R^{0.7}$.

Example 2.9. A peak at 473 K (200 °C) should be about 8% wider than the peak at 423 K (150 °C).

Compared to isothermal analysis where peak width can change by more than an order of magnitude, it is reasonable to suggest that,

all peaks generated during a heating ramp have roughly the same width.

In static analysis, $t_{M,R} = t_M$, $t_{R,stat} = t_R$, and $\mu_R = \mu$. As a result, the formulas for dynamic conditions converge to their static counterparts. This means that the formulas provided here for dynamic conditions are, indeed, the generalizations of their static counterparts.

A general exact theoretical formula for H in a temperature-programmed analysis is unknown. Approximate integration of Eqn (2.98) for isobaric temperature-programmed analysis shows that temperature programming causes a minor, hardly measurable increase in minimal plate height and, therefore, in maximal plate number [1]. This is in agreement with known experimental data [12,18]. In addition to that, there is a difference in expression for f_{opt}. The latter can be expressed as [1]

$$f_{opt} = X_T f_{opt,static} = \frac{2\sqrt{3}\pi D_{pst} X_T}{\vartheta_{G1}} \quad (2.125)$$

where $f_{opt,static}$ (Eqn (2.112)) is f_{opt} in static analysis and

$$X_T = \begin{cases} 1, & \text{static conditions} \\ 0.82\sqrt{\dfrac{1 + 4\omega_R + 6\omega_R^2}{1 + 4\overline{\omega} + 6\overline{\omega}^2}}, & \text{isobaric heating ramp} \end{cases} \quad (2.126)$$

where ω_R and $\overline{\omega}$ are the elution and the distance-averaged ω (Eqns (2.71) and (2.76)), respectively.

2.10. OPTIMIZATION

2.10.1. General Considerations

Separation of components of a sample mixture is the primary function of chromatography. An improvement in the separation performance can be considered as the primary goal of column optimization. However, the separation performance is not the only performance factor in chromatography. Other factors are the *analysis time* and the *detection limit* (the lowest concentration of a sample components that a chromatographic system can detect). With no upper limit to the analysis time and no requirement for detection of low concentration components, any separation performance would be possible. Therefore, a column optimization is a matter of obtaining the best tradeoff between the separation, the time, and the detection limit where the separation is the optimization's focal point.

An important column parameter that affects its separation performance is the column *efficiency* [20,46]

$$E = \sqrt{N} \quad (2.127)$$

Replacing a column plate number (N) with the column efficiency (E) in the expressions for the column separation performance leads to more transparent expressions and to more intuitive interpretations of those expressions [20,46].

The tradition of associating a column plate number with various concepts of column efficiency goes back to the origins of the theory of *column chromatography* [23]. In some sources (including earlier publications of this author), the plate number (N) is associated with a column *separation* efficiency. From that perspective, parameter E in Eqn (2.127) might be interpreted as a metric of a column *separation efficiency*. However, this interpretation could be misleading. Indeed, the plate number concept is valid not only for chromatographic columns but also for the inert tubes that are incapable of any separation. Furthermore, according to Eqns (2.88) and (2.117), an inert tube has the lowest plate height, and, therefore, the highest plate number and the highest efficiency (E). However, it would be misleading to say that the tube has the highest *separation efficiency* because it can make no separation. The efficiency that column parameters N and E represent is the *efficiency of delivering* sharp peaks to the outlet of a column or a tube. The larger the parameters N and E, the narrower the peaks in relation to their retention times. It follows from Eqns (2.127), (2.102), (2.118), and (2.124) that E and σ can be found as

$$E = \sqrt{\frac{L}{H}} = \sqrt{\frac{L}{dh}} \tag{2.128}$$

$$\sigma = \frac{t_{M,R}}{E\mu_R} \tag{2.129}$$

As stated earlier, only the general rather than specific separation performance is considered in this chapter. From that point of view, an improvement in separation performance of a chromatographic system means first of all an increase in the total number of peaks that the system can *resolve* (identifiably and quantifiably separate) rather than an improvement in the separation of specific peak pairs. The former depends on the *peak capacity* of a chromatographic system in a given analysis [8,20,46,86]. The peak capacity depends on a column separation capacity in a given analysis, and on the ability of data-analysis system to resolve poorly separated peaks [20,46]. The *separation capacity* of a chromatographic analysis is the number

$$s_c = \int_{t_M}^{t_{anal}} \frac{dt}{\sigma} \tag{2.130}$$

of σ-slots (σ-wide time intervals) in the time interval between the hold-up time and the end (t_{anal}) of the analysis [20,46].

Let t_{anal} be the time at the end of a heating ramp. The separation capacity of a linear heating ramp in isobaric analysis can be estimated as [46]

$$s_c \approx \frac{t_{anal} - t_{M,init}}{\sigma_{middle}} \tag{2.131}$$

where $t_{M,init}$ is the *initial hold-up time* (the hold-up time at the beginning of the heating ramp) and σ_{middle} is the peak width in the middle of the ramp. Due to Eqn (2.62), t_{anal} can be found as

$$\begin{aligned} t_{anal} &= \frac{\Delta T_{end}}{\theta_{T,middle}} \cdot \frac{t_{M,middle}}{r_T}, \\ \Delta T_{end} &= T_{end} - T_{init} \end{aligned} \tag{2.132}$$

where subscript "end" indicates a parameter at the end of the heating ramp. Quantities ΔT_{end} and $\Delta T_{end}/\theta_{T,middle}$ can be interpreted as the *heating range and dimensionless heating range*, respectively.

If the heating range of a linear ramp is relatively wide ($\Delta T_{end} \gg \theta_{T,middle}$), then the majority of the peaks eluting during the ramp including the peaks eluting in the middle of the ramp are the ones that were highly retained at the beginning of the ramp. Parameter σ_{middle} can be found from Eqn (2.129) as

$$\sigma_{middle} = \frac{t_{M,middle}}{E\mu_{R,a}} \tag{2.133}$$

where $t_{M,middle}$ is the hold-up time in the middle of the heating ramp, and quantity $\mu_{R,a}$ is described in Eqn (2.72). Substitution of Eqns (2.132) and (2.133) into Eqn (2.131) and accounting for Eqns (2.72) and (2.76) yields

$$s_c \approx E\overline{\omega}_a \cdot \left(\frac{\Delta T_{end}}{\theta_{T,middle}} - \frac{r_T t_{M,init}}{t_{M,middle}} \right) \tag{2.134}$$

One of the factors affecting s_c is initial hold-up time ($t_{M,init}$) which is typically much smaller than total analysis time (t_{anal}) and, as follows from Eqn (2.131), has minor effect on s_c. On the other hand, the presence of $t_{M,init}$ in Eqn (2.134) obscures its transparency. Ignoring $t_{M,init}$ in Eqn (2.131) transforms Eqn (2.134) into the form

$$s_c \approx E\overline{\omega}_a \frac{\Delta T_{end}}{\theta_{T,middle}}, \quad (\text{at } t_{M,init} \ll t_{anal}) \quad (2.135)$$

The last formula shows that primarily three factors affecting s_c are the column efficiency (E), average immobility ($\overline{\omega}_a$) of eluting solutes, and dimensionless heating range ($\Delta T_{end}/\theta_{T,middle}$). The first two are the functions of method parameters directly controlled by an operator. Thus, $\overline{\omega}_a$ is a function (Eqn (2.76); Figure 2.7) of the heating rate, and E is a function (Eqn (2.128)) of the column dimensions (L and d) and dimensionless plate height (h) which, in turn, is primarily a function (Eqn (2.119)) of the flow rate. On the other hand, dimensionless heating range mostly depends on the sample. It cannot be directly controlled by the operator, and is not a target of a general optimization considered below. Only the column dimensions, the flow rate, and the heating rate are.

The forthcoming analysis addresses two-stage approach to column optimization. First, the *speed optimization* with the goal of obtaining the best tradeoff (the best compromise) among the separation capacity (s_c) and the analysis time (t_{anal}) is considered. This leads to the optimal flow rate of a gas and the optimal heating rate of a column. It also quantifies the effect of column diameter and carrier gas type on the analysis time. Second, the factors affecting the tradeoff between s_c, t_{anal}, and the detection limit are considered.

The problem of speed optimization is formulated as obtaining the shortest analysis time at a *given separation capacity* (at a given number of σ-slots in a chromatogram). Operational parameters optimizing the speed of analysis could be different from the ones that lead to, say, the highest efficiency ($E_{max,orig}$) in a *given* (original) *column*. Thus, the specific flow rate that optimizes the speed of analysis can be different from quantity f_{opt} in Eqn (2.116) corresponding to $E_{max,orig}$. To distinguish the optimal operational parameters leading to the speed optimization from the ones leading to $E_{max,orig}$, the former are denoted by the subscript "Opt" (capital "O" rather than lower case "o" as in f_{opt}). Thus the flow rate (F), the specific flow rate (f), and the heating rate (R_T) that optimize the speed of analysis are denoted as F_{Opt}, f_{Opt}, and $R_{T,Opt}$, respectively. Similarly, the minima and the maxima resulting from the speed optimization are denoted by the subscripts "Min" and "Max". Thus, $t_{M,Min}$ is the shortest hold-up time at a given column efficiency, and $t_{anal,Min}$ is the shortest analysis time at a given separation capacity (in a column with a given diameter and carrier gas).

Due to Eqn (2.42), F_{Opt} can be found from f_{Opt} as

$$F_{Opt} = df_{Opt}\frac{T_{norm}}{T} \quad (2.136)$$

Quantity F_{Opt} is known as the *speed-optimized flow rate* (SOF) [83], while the flow rate $F_{opt} = df_{opt}T_{norm}/T$ maximizing a column efficiency is the *efficiency-optimized flow rate* (EOF) [83]. This terminology can be extended to the specific flow rate by calling quantities f_{Opt} and f_{opt} as *specific speed-optimized flow rate* (SSOF) and *specific efficiency-optimized flow rate* (SEOF), respectively. Only **speed optimization** is considered in the rest of this chapter and, unless otherwise explicitly stated, the terms such as *optimal flow rate*, *optimal specific flow rate*, and others used without additional qualification always mean the speed-optimized parameters.

2.10.2. Optimal Flow Rate

By allowing the gas flow rate, the column heating rate, and the column length, to vary simultaneously one can obtain the shortest t_{anal} at a given s_c in a column with a given diameter and carrier gas. As t_{anal} (of isothermal or temperature-programmed analysis) is proportional to hold-up time (t_M) measured at some static conditions [47] and s_c is proportional to column efficiency (E), solving the problem of obtaining the shortest t_{anal} at a given s_c in dynamic analysis can start with solving the problem of obtaining the shortest t_M at a given E in static analysis [87].

To evaluate t_M at a fixed E and d, one can find L from Eqn (2.128) and substitute it into Eqn (2.48). The latter becomes [46]

$$t_M = \frac{\pi^2 d^4 p_o^3}{1536 p_{st}^2 \eta f^2} \cdot \left(\left(1 + \frac{256 E^2 p_{st} \eta}{\pi d^2 p_o^2} hf \right)^{3/2} - 1 \right) = \begin{cases} \dfrac{\pi d^2 E^2 p_o}{4 p_{st}} \dfrac{h}{f}, & |\Delta p| \ll p_o \\ \dfrac{8dE^3}{3} \sqrt{\dfrac{\pi \eta}{p_{st}} \dfrac{h^3}{f}}, & \Delta p \gg p_o \end{cases} \quad (2.137)$$

2.10.2.1. *Strong Gas Decompression*

It is convenient to start the analysis of gas flow optimization from the case of a strong gas decompression along the column where $\Delta p \gg p_o$. This is probably the most typical case because the decompression is strong in all GC-MS applications and in all analyses of complex mixtures requiring relatively long, narrow-bore columns. It follows from Eqn (2.137) that, at any fixed f (optimal or not)

$$t_M \sim E^3 \quad (\text{at } \Delta p \gg p_o, \text{ fixed } d \text{ and } f) \quad (2.138)$$

$$t_M \sim d \quad (\text{at } \Delta p \gg p_o, \text{ fixed } E \text{ and } f) \quad (2.139)$$

It also follows from Eqn (2.137) at $\Delta p \gg p_o$ that t_M has a minimum ($t_{M,Min}$) at a given E and d when quantity h^3/f has a minimum. Let us consider a normalized version, $h^3 f_{opt}/f$, which has the minimum at the same f as h^3/f has. Eqn (2.119) yields (Figure 2.14.a)

$$\frac{h^3 f_{opt}}{f} = \frac{h_{min}^3}{8} \cdot \frac{f_{opt}}{f} \left(\frac{f_{opt}}{f} + \frac{f}{f_{opt}} \right)^3 \quad (2.140)$$

This quantity and, therefore, t_M have a minimum at $f = f_{Opt}$ where

$$f_{Opt} = \sqrt{2} f_{opt} \quad (\Delta p \gg p_o) \quad (2.141)$$

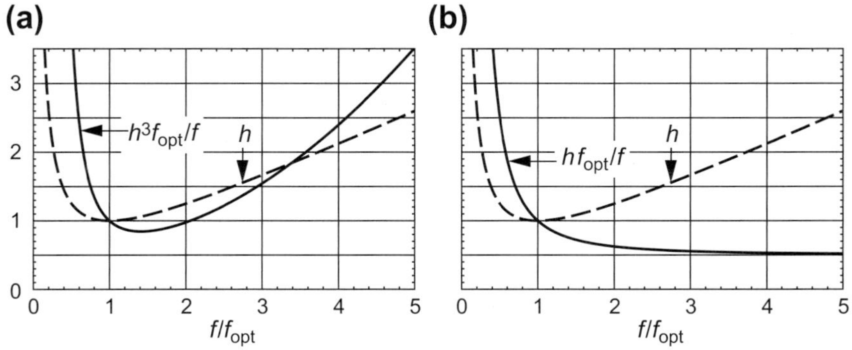

FIGURE 2.14 Graphs of the functions of variable f in Eqns (2.140) and (2.156) — solid lines. The dashed lines representing the graphs of h in Eqn (2.119) are included for the references. Quantity $h^3 f_{opt}/f$ in (a) has a minimum at $f = f_{Opt} = \sqrt{2} f_{Opt}$. Quantity hf_{opt}/f in (b) approaches its lowest level at infinitely large f.

is the *specific speed-optimized flow rate*. The outcome of the speed optimization at strong gas decompression can be interpreted as follows. If a predetermined efficiency or hold-up time can be obtained by choosing a suitable column length, then

At $f = f_{Opt}$, the hold-up time is the shortest for a given efficiency, or, conversely, the efficiency is the largest for a given hold-up time.

For practical applications, one needs to transform f_{Opt} into F_{Opt} (into the speed-optimized flow rate, SOF) which, due to Eqns (2.136) and (2.141), can be expressed as

$$F_{Opt} = \sqrt{2}F_{opt} \quad (\Delta p \gg p_o) \qquad (2.142)$$

where F_{opt} is the efficiency-optimized flow rate. Eqn (2.136) also provides a basis for the transformation of f_{Opt} into F_{Opt}. Assuming that a column temperature in isothermal analysis and initial temperature in a typical heating ramp are not very high, one can approximate Eqn (2.136) by a simple form

$$F_{Opt} = f_{Ref}d \qquad (2.143)$$

It follows from Eqns (2.136), (2.141), (2.125), and (2.111) that parameter f_{Ref} for each carrier gas is proportional to the product $X_D^2 D_{g,st}$ for the gas (Table 2.1). Numerical values of f_{Ref} for several gases are listed in Table 2.4 [46,83].

Several column parameters at $f = f_{Opt}$ are evaluated below.

At $f = f_{Opt}$, Eqn (2.119) yields, for the *optimal dimensionless plate height* (h_{Opt}) in a speed-optimized column,

$$h_{Opt} = \frac{h_{min}}{2}\left(\frac{1}{\sqrt{2}} + \sqrt{2}\right) = \frac{3h_{min}}{2\sqrt{2}} \approx 1.061h_{min} \quad (\Delta p \gg p_o) \qquad (2.144)$$

This means that the speed optimization increases the plate height (h_{Opt}) by about 6% compared to its efficiency-optimized value (h_{min}). This, according to Eqn (2.128), reduces the column efficiency by practically negligible 3% compared to its original maximum ($E_{max,orig}$) at $f = f_{opt}$. However, to sustain the theoretical assumption of a fixed efficiency in Eqn (2.137), the lost efficiency can be regained by using a longer column, i.e. replacing the original L_{orig}-long column with the one having the *optimal length* L_{Opt}, where

$$L_{Opt} = \frac{3L_{orig}}{2\sqrt{2}} \approx 1.061L_{orig} \quad (\Delta p \gg p_o) \qquad (2.145)$$

Let $t_{M,opt}$ and $t_{M,Opt}$ be t_M at $f = f_{opt}$ and $f = f_{Opt}$, respectively. Two versions ($t_{M,Opt,L}$ and $t_{M,Opt,E}$) of $t_{M,Opt}$ are considered below. The former is in a column of a fixed length while the latter is in the column of a fixed efficiency. When quantities $t_{M,Opt,L}$ and $t_{M,Opt,E}$ are compared with each other, it is assumed that all parameters (column diameters, carrier gas types,

Table 2.4 Flow-Related Parameters of Several Gases

Gas	Conditions	Formula	He	H_2	N_2	Ar
f_{Ref} (mL/min/mm)		Eqn (2.143)	8	10	2.4	2.1
X_σ (ms/m)	150 °C	Eqn (2.153)	7.3	4.4	12.4	15.0
$t_{M,Opt,hydrogen}/t_{M,Opt}$	$\lvert\Delta p\rvert << p_o$, same column	Eqn (2.157)	0.8	1	0.24	0.21
$t_{M,Opt,hydrogen}/t_{M,Opt}$	$\Delta p >> p_o$, same column	Eqn (2.148)	0.6	1	0.34	0.28
$t_{M,Opt,hydrogen}/t_{M,Opt}$	same p_o, p_i and E	Eqn (2.158)	0.45	1	0.5	0.4

etc.) other than column length and efficiency are the same in both cases. It can be noticed that $t_{M,Opt,E}$ is also the shortest t_M ($t_{M,Min}$) that can be obtained at a given E in a column of a given diameter with a given carrier gas. However, for better flexibility of forthcoming evaluations, it is more convenient to treat $t_{M,Min}$ as a version ($t_{M,Opt,E}$) of quantity $t_{M,Opt}$.

It follows from Eqn (2.48) at $\Delta p \gg p_o$ that the relative reduction in t_M at a fixed L is proportional to the relative increase in $\sqrt{f}$. As a result,

$$\frac{t_{M,Opt,L}}{t_{M,opt}} - 1 = \sqrt{\frac{1}{\sqrt{2}}} - 1 \approx -0.16 \quad (\Delta p \gg p_o) \tag{2.146}$$

On the other hand, it follows from Eqn (2.137) at $\Delta p \gg p_o$ that the relative reduction in t_M at a fixed E is proportional to the relative reduction in $\sqrt{h^3/f}$. As a result (Figure 2.14.a),

$$\frac{t_{M,Opt,E}}{t_{M,opt}} - 1 = \sqrt{\frac{1}{8}\cdot\frac{1}{\sqrt{2}}\left(\frac{1}{\sqrt{2}} + \sqrt{2}\right)^3} - 1 = \sqrt{\frac{27}{32}} - 1 = -0.0814 \quad (\Delta p \gg p_o) \tag{2.147}$$

The 16% reduction in $t_{M,Opt,L}$ compared to $t_{M,opt}$ comes at a cost of the earlier shown 3% reduction in the column efficiency (E) while the 8% reduction in $t_{M,Opt,E}$ compared to $t_{M,opt}$ comes at no reduction in E.

How does $t_{M,Opt}$ depend on the carrier gas type? Eqns (2.48) and (2.137) describe the dependence of t_M on a gas viscosity (η). Substituting $f = f_{Opt}$ in Eqn (2.48) or Eqn (2.137) and accounting for Eqns (2.141) and (2.112) show how $t_{M,Opt}$ also depends on a solute diffusivity (D_{pst}) at standard pressure. For a given gas, parameters η and D_{pst} are mutually dependent quantities. As a result, dependence of $t_{M,Opt}$ on η and D_{pst} does not provide a complete picture of dependence of $t_{M,Opt}$ on the gas type. Accounting for Eqns (2.111), (2.20), and (2.19) yields

$$t_{M,Opt} = \frac{8\sqrt{5\sqrt{2}L^3\vartheta_{G1}}}{3\sqrt{3\pi\sqrt{3}d}}\frac{k^{0.05}}{(10^3\varphi)^{0.045}}\left(\frac{T}{T_{st}}\right)^{0.125} \times \frac{1}{X_D v_{st}}, \quad (\Delta p \gg p_o) \tag{2.148}$$

This shows that $t_{M,Opt}$ is inversely proportional to the product $X_D v_{st}$ listed for several gases in Table 2.1. The key component in this product is the average molecular speed (v_{st}) of a gas at standard pressure and temperature. Relative values of $t_{M,Opt}$ of hydrogen in relation to $t_{M,Opt}$ of other gases are listed in Table 2.4. One can find from the table that

At strong gas decompression, a column with the speed-optimized flow of hydrogen has 40% shorter hold-up time (and analysis time) than the same column with the speed-optimized flow of helium.

Peak width at strong gas decompression can be expressed as [46]

$$\sigma = \frac{8L}{3\mu_R}\sqrt{\frac{\pi\eta h}{p_{st} f}} \quad (\Delta p \gg p_o) \tag{2.149}$$

which, due to Eqns (2.119), (2.141), and (2.125), yields at $f = f_{Opt}$

$$\sigma_{Opt} = \frac{2L}{\mu_R}\sqrt{\frac{2\pi}{3\sqrt{3}\pi}\frac{h_{min}\vartheta_{G1}\eta}{D_{pst}p_{st}X_T}} \quad (f = f_{Opt}) \tag{2.150}$$

Remarkably [1,46,88],

In a column with strong gas decompression at optimal flow rate, peak width is independent of column diameter and proportional to the column length.

From Eqn (2.149), one can also find σ at $f \to \infty$. This would be the narrowest peak width (σ_{min}) that can be obtained in a column of a given length in isothermal analysis at a given temperature or in temperature-programmed analysis with a given dimensionless heating rate. It is interesting that

$$\frac{\sigma_{\min}}{\sigma_{\mathrm{Opt}}} = \sqrt{\frac{2}{3}} \approx 0.82 \qquad (2.151)$$

In a column with strong gas decompression at optimal flow rate, further flow rate increase up to infinity can only reduce the peak width by less than 20%.

Substitution of Eqns (2.111), (2.20), and (2.19) into Eqn (2.150) expresses σ_{Opt} via a gas-dependent product $X_D v_{\mathrm{st}}$ (Table 2.1) as

$$\sigma_{\mathrm{Opt}} = \frac{4}{3}\sqrt{\frac{10}{\sqrt{3}\pi}}\sqrt{\frac{h_{\min}\vartheta_{G1}}{X_T}}\left(\frac{T}{T_{\mathrm{st}}}\right)^{0.125} \times \frac{Lk^{0.05}}{\mu_R(10^3\varphi)^{0.045}}\frac{1}{X_D v_{\mathrm{st}}} \qquad (2.152)$$

The essence of this formula can be expressed as

$$\sigma = \frac{X_\sigma}{\mu_R}L \qquad (2.153)$$

where the values of X_σ for several gases are listed in Table 2.4 [46].

2.10.2.2. Weak Gas Decompression

It follows from Eqn (2.137) that

$$t_M \sim E^2, \quad (\text{at } |\Delta p| \ll p_o \text{ and fixed } d) \qquad (2.154)$$

$$t_M \sim d^2, \quad (\text{at } |\Delta p| \ll p_o \text{ and fixed } E) \qquad (2.155)$$

According to Eqn (2.137), t_M is the shortest at $|\Delta p| \ll p_o$ and at a given E and d when quantity hf_{opt}/f is the lowest. It follows from Eqn (2.119) that (Figure 2.14.b)

$$\frac{hf_{\mathrm{opt}}}{f} = \frac{h_{\min}}{2}\cdot\frac{f_{\mathrm{opt}}}{f}\left(\frac{f_{\mathrm{opt}}}{f} + \frac{f}{f_{\mathrm{opt}}}\right) = \frac{h_{\min}}{2}\cdot\left(1 + \frac{f_{\mathrm{opt}}^2}{f^2}\right) \qquad (2.156)$$

This quantity and, therefore, t_M do not have a minimum, but approach their lowest levels when f approaches infinity. This means increasing the flow rate and the column length so that the column efficiency remains fixed always reduces t_M (and analysis time). This, however, is not the end of the story of speed optimization at weak gas decompression.

Under realistic conditions, an unlimited increase in the flow rate (F) and in the column length (L) in order to reduce t_M without reducing efficiency (E) inevitably leads to high pressure drop and to strong gas decompression. This means that the weak gas decompression does not have a property of *closure*. Optimization of speed of analysis at weak decompression leads to conditions that are outside of the weak decompression. To find a realistic speed-optimized conditions in a column that, at $f = f_{\mathrm{opt}}$, has weak gas decompression, let us reexamine the dependence of t_M on F and L at a fixed E in the column.

The idea of reducing t_M at a fixed efficiency (E) by simultaneous variation of F and L or, equivalently, f and L, goes back to Scott and Hazeldean's work reported in 1960[5] [89]. The technique is based on the fact that a small increase in f above its value f_{opt} corresponding to the maximum ($E_{\mathrm{max,orig}}$) in the efficiency of a given column reduces the hold-up time (t_M) more rapidly than it reduces E. Even if the lost efficiency is recovered by increasing L compared to its original value (L_{orig}), the final t_M is still shorter than the one in the original

[5] In the original approach [89], simultaneous variation of average gas velocity ($\bar{u}$) and column length (L) was considered and the concept of *optimum practical gas velocity* (OPGV) was introduced. Unfortunately, the original study was based on an incorrect plate height formula $H = B/\bar{u} + C\bar{u}$ (see footnote 3 on page 54). As a result, although the OPGV was recommended without reservations and is widely assumed to be suitable for all GC applications, the concept is valid only at weak gas decompression and cannot be utilized in the applications with strong decompression such as GC-MS [1].

column at $f = f_{opt}$. At strong gas decompression where t_M is inversely proportional to $\sqrt{f}$ (Eqn (2.137) at $\Delta p \gg p_o$), the balance between reducing t_M due to an increase in f and increasing t_M due to an increase in L is reached at $f = f_{Opt} = \sqrt{2}f_{opt}$ (at higher f, the original maximum ($E_{max,orig}$) in E existed in original L_{orig}-long column at $f = f_{opt}$ can be preserved only at the expense of a longer t_M). At weak gas decompression where t_M is inversely proportional to f (Eqn (2.137) at $|\Delta p| \ll p_o$), an increase in f leading to reduction in t_M can go far beyond $f = \sqrt{2}f_{opt}$. As a result, far larger L than L_{Opt} in Eqn (2.145) might be necessary for preserving the original maximal efficiency. However, this works only until the required pressure becomes sufficiently high to cause a meaningful gas decompression. This slows the gas average velocity and the rate of reduction in t_M due to increasing f. Eventually, this leads to a minimum ($t_{M,Min}$) in t_M at a finite f (rather than at $f \to \infty$ as in Eqn (2.156)).

Expressions for the optimal conditions leading to $t_{M,Min}$ at a fixed E in a column having weak gas decompression at $f = f_{opt}$ are relatively complex for practical use. In view of that, it is worth comparing this *complete speed optimization* with a simpler *partial speed optimization* where $f = \sqrt{2}f_{opt}$ regardless of the degree of the gas decompression and where, at a weak decompression, t_M remains higher than $t_{M,Min}$.

Let f^*_{Opt} and L^*_{Opt} be optimal specific flow rate and optimal column length in complete speed optimization. These parameters correspond to the minimum ($t_{M,Min}$) in t_M at a fixed efficiency in a column with an arbitrary degree of gas decompression at $f = f_{opt}$. Parameters f^*_{Opt} and L^*_{Opt}, are shown in Figure 2.15 as functions of relative pressure drop ($\Delta P_{opt,orig}$) at $f = f_{opt}$ in the original L_{orig}-long column. When $\Delta P_{opt,orig}$ is large ($\Delta P_{opt,orig} \gg 1$), parameters f^*_{Opt} and L^*_{Opt} converge to parameters f_{Opt} and L_{Opt} (Eqns (2.141) and (2.145)), minimizing t_M at strong decompression. At small $\Delta P_{opt,orig}$ ($|\Delta P_{opt,orig}| \ll 1$), parameters f^*_{Opt} and L^*_{Opt} could be significantly larger than their strong decompression counterparts f_{Opt} and L_{Opt}.

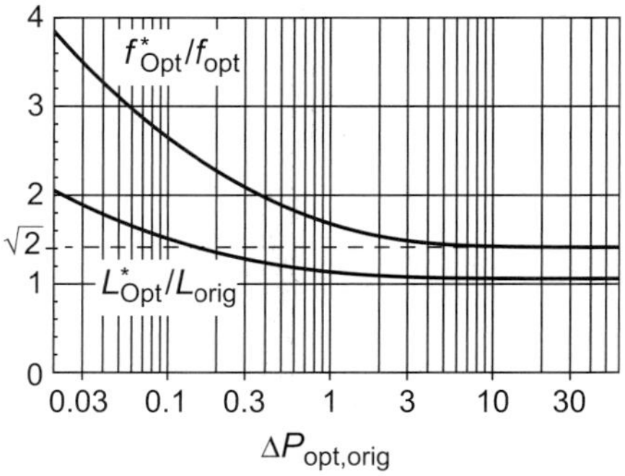

FIGURE 2.15 Optimal specific flow rate (f^*_{Opt}) and optimal column length (L^*_{Opt}) vs. relative pressure drop ($\Delta P_{opt,orig}$) at $f = f_{opt}$ in original L_{orig}-long column. At strong gas decompression (large $\Delta P_{opt,orig}$), f^*_{Opt}, and L^*_{Opt} approach their strong decompression counterparts described in Eqns (2.141) and (2.145).

Example 2.10. Consider a 5 m × 0.53 mm column with hydrogen at 150 °C and at $p_o = 1$ atm. For this column F_{opt} is about 3.75 mL/min [83] (see also Table 2.4 together with Eqns (2.143) and (2.142)). At this flow, $\Delta P_{opt,orig} \approx 0.025$ (one of the lowest pressure drops in practice). It follows from the graph in Figure 2.15 at $\Delta P_{opt,orig} = 0.025$ that $f^*_{Opt} \approx 3.6\ f_{opt}$. $F^*_{Opt} \approx 3.6\,F_{opt} \approx 13.5$ mL/min, respectively. According to Eqn (2.116), this flow rate causes about 1.94-fold increase in a column plate height compared to its minimum (H_{min}) at $F = F_{opt}$. To preserve the column efficiency, the column should be made 1.94 times longer, i.e. $L^*_{Opt} \approx 9.7$ m. Nevertheless, there is a net reduction in t_M compared to the one in the original 5 m-long column at $F_{opt} = 3.75$ mL/min (Figure 2.16).

Hold-up times under complete and partial speed optimization are illustrated in Figure 2.16. Several practical factors can be taken into account when analyzing Figure 2.16.

1. The solid line in Figure 2.16 represents the hold-up time reduction at a partial speed optimization with $f = f_{Opt} = \sqrt{2}f_{opt}$ in

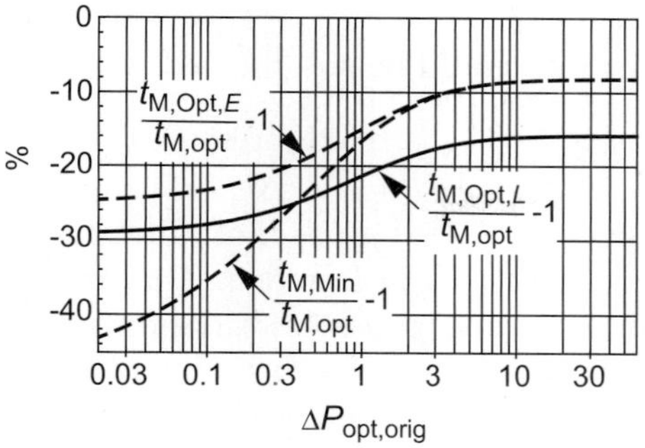

FIGURE 2.16 Relative reductions in hold-up times compared to the hold-up time ($t_{M,opt}$) at $f=f_{opt}$ in original L_{orig}-long column having efficiency $E_{max,orig}$ and relative pressure drop $\Delta P_{opt,orig}$. Parameter $t_{M,Min}$ is the hold-up time in completely optimized column (i.e. at $f = f^*_{opt}$ in L^*_{opt}-long column where both f^*_{Opt} and L^*_{Opt} are found from Figure 2.15). Parameter $t_{M,Opt,E}$ is the hold-up time at $f = f_{Opt} = \sqrt{2}f_{Opt}$ in $1.061L_{orig}$-long column (Eqn (2.145)). Using $L = 1.061L_{orig}$ keeps the column efficiency at the level of $E_{max,orig}$ when f is increased from f_{opt} to $\sqrt{2}f_{opt}$. Parameter $t_{M,Opt,L}$ is the hold-up time at $f = \sqrt{2}f_{opt}$ in original (L_{orig}-long) column. Not increasing the column length when f is raised from f_{opt} to $\sqrt{2}f_{opt}$ reduces the column efficiency by practically negligible 3% compared to its original maximum at $E_{max,orig}$. This case (a solid line) is the most practical implementation of the speed optimization. In majority of practical cases (but not always), it leads to practically the shortest t_M at a given E.

a column of the original length. The reduction (16% at strong decompression and up to 29% at weak decompression) comes at the expense of a practically negligible 3% lower efficiency compared to the highest efficiency ($E_{max,orig}$) that could be obtained in the original column at $f = f_{opt}$.

2. If the relative pressure drop ($\Delta P_{opt,orig}$) at $f = f_{opt}$ in the original column is larger than 0.35 or so, then the time saving described in the previous item is larger (better) than the time saving at a complete speed-optimization using $f = f^*_{Opt}$ and $L = L^*_{Opt}$ where f^*_{Opt} and L^*_{Opt} are taken from Figure 2.15. The region of $\Delta P_{opt,orig} \geq 0.35$ covers practically all columns with $d \leq 0.25$ mm. Thus, in 10 m × 0.25 mm with hydrogen at 150 °C, $p_o = 1$ atm and $F = 1.8$ mL/min (close to F_{opt}), the relative pressure drop is $\Delta P \approx 0.42$.
3. Only in relatively seldom cases of the short wide-bore columns for which $\Delta P_{opt,orig}$ is significantly lower than 0.3, complete speed optimization can result in additional time saving compared to conditions of $f = \sqrt{2}f_{opt}$ and $L = L_{orig}$. If necessary, that additional time saving can be found from Figure 2.16, and the corresponding values of parameters f^*_{Opt} and L^*_{Opt} can be found from Figure 2.15.

Example 2.11. It was found in Example 2.10 that one of a practically lowest values of $\Delta P_{opt,orig}$ was 0.025 (5 m × 0.53 mm column with hydrogen at $F = F_{opt}$). One can find from Figure 2.16 that, at $\Delta P_{opt,orig} = 0.025$, a complete speed optimization leads to 42% reduction in t_M compared to t_M in efficiency-optimized column (t_M at $f = f_{opt}$). This is only 13% better than 29% reduction in t_M at partial speed optimization at a fixed column length.

Due to a relatively modest level of additional time saving in complete speed optimization, and relative complexity of finding its parameters f^*_{Opt} and L^*_{Opt}, only the partial speed optimization is considered below. It is assumed from now on that the optimal specific flow rate, $f_{Opt} = \sqrt{2}f_{opt}$ specified in Eqn (2.141) only for strong gas decompression, is optimal for **any decompression**.

Using the transformations similar to those that led to Eqn (2.148), one can find that $t_{M,Opt}$ corresponding to $f = f_{Opt}$ at a weak gas decompression, can be expressed as

$$t_{M,Opt} = \frac{dLk^{0.1}\vartheta_{G1}}{8\sqrt{6}(10^3\varphi)^{0.09}}\frac{p_o}{p_{st}}\left(\frac{T}{T_{st}}\right)^{0.25-\xi} \times \frac{1}{X_D^2 D_{g,st}}, \quad (|\Delta p| \ll p_o) \tag{2.157}$$

This shows that $t_{M,Opt}$ is inversely proportional to the product $X_D^2 D_{g,st}$ listed for several gases in Table 2.1. Relative values of $t_{M,Opt}$ of hydrogen in relation to $t_{M,Opt}$ of other gases are listed in Table 2.4. One can find from the table that

At weak gas decompression, a column with the speed-optimized flow of hydrogen has 20% shorter hold-up time (and analysis time) than the same column with the speed-optimized flow of helium.

2.10.2.3. *Fixed Pressure*

It has been assumed so far that a GC instrument can provide any pressure necessary for a column optimization. This reflects overwhelming majority of practical applications. It can be also added that, although commercial GC instruments cannot provide unlimited pressure, the practically available pressure is not a fundamental limiting factor [46] and the pressure that a GC instrument can provide can be increased in the next generation of GC instruments if and when higher pressure becomes practically necessary. Nevertheless, speed optimization under a limited pressure deserves consideration. The problem can be formulated as that of the speed optimization under a *fixed pressure*. Conversely, the condition when a pressure required for a column optimization is available can be called as the case of *adjustable pressure*.

Due to Eqns (2.46), (2.128), and Eqn (2.45), t_M at a fixed pressure can be expressed as

$$t_M = \frac{32E^4h^2\eta}{j_H\Delta p} \qquad (2.158)$$

indicating that

$$t_M \sim E^4 \quad (\text{at fixed } \Delta p) \qquad (2.159)$$

$$t_M \sim 1/\Delta p \quad (\text{at fixed } E) \qquad (2.160)$$

It also follows from Eqn (2.158) that if a column pressure is fixed, then t_M is the shortest when h is the smallest i.e. according to Eqn (2.119), at $f = f_{opt}$. This and previous conclusions could be summarized as

$$f_{Opt} = f_{opt} \quad (\text{fixed pressure}) \qquad (2.161)$$

Equation (2.158) does not directly contain column dimensions L and d. This reflects the fact that, at a fixed pressure, variation of a column efficiency (E) by varying both L and d does not affect t_M. However, there is a constraint on these variations coming from other factors. The variations should not affect the specific flow rate which should stay at its optimal level. This according to Eqn (2.43) for a fixed p_o, p_i, and f implies that the ratio d^3/L should be fixed. This, in turn, implies that, in order for two columns having efficiencies E_1 and E_2 to operate at the same pressure each at its optimal flow rate, their diameters (d_1 and d_2) and lengths (L_1 and L_2) should relate to their efficiencies as

$$\frac{d_2}{d_1} = \frac{E_2}{E_1}, \quad \frac{L_2}{L_1} = \left(\frac{E_2}{E_1}\right)^3 \qquad (2.162)$$

This together with Eqn (2.159) implies that

$$t_M \sim d^4 \quad (\text{at fixed } \Delta p) \qquad (2.163)$$

One can also consider a change in a carrier gas while keeping pressure and column efficiency fixed and using optimal flow in both cases. It follows from Eqns (2.43) and (2.19) that, to satisfy these conditions, the column dimensions should be related as

$$\frac{L_2}{L_1} = \frac{d_2}{d_1} = \frac{X_{D,1}\lambda_1}{X_{D,2}\lambda_2} \qquad (2.164)$$

The products of the mean free paths (λ) and empirical parameters X_D for several gases are listed in Table 2.1.

Equation (2.158) shows that, at a fixed pressure and column efficiency, t_M is proportional to gas viscosity (η) listed for several gases in Table 2.1. Relative values of $t_{M,Opt}$ of hydrogen in relation to $t_{M,Opt}$ of other gases are listed in Table 2.4. One can find from the table that

At the same pressure and column efficiency, a column with optimal flow rate of hydrogen has 55% shorter hold-up time (and analysis time) than the same column with optimal flow of helium.

It is interesting that, because η of nitrogen is about 10% lower than η of helium, an analysis with nitrogen should be about 10% faster than analysis with helium if the same efficiency at the same pressure is obtained in both cases.

Example 2.12. The pressure required for operating a 100 m × 0.1 mm column with nitrogen at $f = f_{opt}$ is about 600 kPa – close to the pressure limit in some GC instruments. The same 100 m × 0.1 mm column with helium at $f = f_{opt}$ requires more than twice as high pressure. Instead, one can choose a column with different dimensions that require 600 kPa pressure at $f = f_{opt}$ of helium. According to Eqn (2.164) and Table 2.1, this should be a 193 m × 0.193 mm column (a 180 m × 0.18 mm column can be assembled from commercially available shorter 0.18 mm columns that can be used for experimental verification of this example). The hold-up time in this column with helium at $f = f_{opt}$ is about 10% larger than the hold-up time in 100 m × 0.1 mm column with nitrogen at $f = f_{opt}$. This is in a contrast with the adjustable pressure where, as follows from Table 2.4, an analysis with helium at weak and strong gas decompression is, respectively, almost twice and more than three times as fast as an analysis with nitrogen.

2.10.2.4. *Closing Remarks*

Specific flow rates optimizing speed of analysis at several pressure conditions can be summarized in a single expression

$$f_{Opt} = f_{opt} \cdot \begin{cases} \sqrt{2}, & \text{adjustable pressure} \\ 1, & \text{fixed pressure} \end{cases} \tag{2.165}$$

In a column with a given diameter and a given carrier gas,

Speed optimization of a column flow leads to the shortest analysis times at a predetermined column efficiency and to the largest efficiency at a predetermined analysis time.

When column pressure is fixed, optimal flow for maximal efficiency is the same as that for minimal hold-up time at a given efficiency. Only when a GC instrument can supply pressure required for speed optimization, the latter can reduce the analysis time compared to that at maximal efficiency of a given column. The time savings (16% to 29% at least) are practically useful, but not dramatic. From theoretical perspective, it is not the amount of the time saving that counts the most, but the very fact that the questions of the flow optimization for the best separation-time tradeoff are answered. Even more important is the fact that criteria developed for the speed optimization of the flow rate can be extended to optimization of other parameters such as a column heating rate.

Finally, as mentioned earlier, practical cases when pressure available in a GC instrument is insufficient for the flow optimization are relatively rare. It is assumed from now on that, unless otherwise explicitly stated, **optimal pressure is always available**.

Optimal flow rate depends on temperature. Assuming that parameter ξ (Table 2.1) can be approximated as $\xi = 0.7$, it follows from Eqns (2.136), (2.141), (2.125), and (2.109) that (Figure 2.17.a)

$$F_{Opt} \sim T^{-0.55} \tag{2.166}$$

Experimental verification and additional discussion of this relation can be found elsewhere [42]. It is interesting to compare F_{Opt} with actual flow (F) in a column during a heating ramp. It follows from Eqns (2.43)

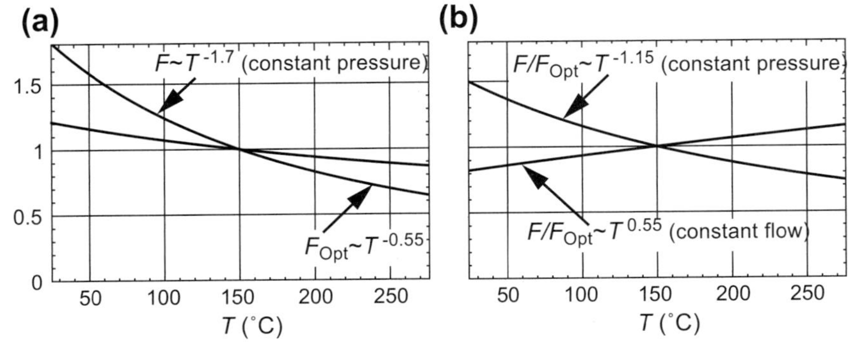

FIGURE 2.17 Normalized (to unity at 150 °C) flow rates in temperature-programmed analyses vs. column temperature (T). (a) Actual flow rate (F) (Eqn (2.167)) in isobaric (constant pressure) analysis and optimal flow rate (F_{Opt}) (Eqn (2.166)). (b) Relative values of F/F_{Opt} in isobaric and isorheic (constant flow) analyses.

and (2.20) that actual flow rate in isobaric analysis is (Figure 2.17.a)

$$F \sim T^{-1.7} \tag{2.167}$$

The last two formulas show that, the ratio F/F_{Opt} changes during a temperature-programmed analysis with both constant pressure and constant flow (Figure 2.17.b). In other words, there is a mismatch of actual and optimal flow in both operational modes. In isorheic analysis, the mismatch is significantly smaller than it is in isobaric analysis. This is not the only and probably not the most important advantage of having a constant column flow. It is practically important that constant gas flow leads to more consistent detector operations.

2.10.3. Optimal Heating Rate

Among factors affecting separation capacity (s_c) of GC analysis is the heating rate (R_T) (Eqns (2.135) and (2.76)). The slower the rate, the larger the s_c. At $R_T = 0$ (isothermal analysis), s_c is the highest. Unfortunately, the highest s_c comes at the expense of a longer analysis time (t_{anal}) which, in the case of analysis of a complex mixture, can become prohibitively long. Temperature programming can substantially reduce the analysis time. However, it can also reduce the separation capacity. These considerations suggest that there can be a point of the best tradeoff for s_c and t_{anal}. Optimization of speed of analysis by using optimal heating rate ($R_{T,Opt}$) that leads to the shortest t_{anal} at a given s_c is another layer (in addition to the flow optimization) of column speed optimization. While the flow optimization leads to the shortest hold-up time (t_M) at a given column efficiency (E), the heating rate optimization leads to the shortest t_{anal} at a given s_c, E, and t_M. The outcome of the heating rate optimization can be interpreted as follows. If a predetermined separation capacity or analysis time in a column with given diameter, carrier gas type, and flow rate can be obtained by choosing a suitable column length, then

The time of analysis of a given sample at optimal heating rate is the shortest for a given separation capacity, or, conversely, the separation capacity is the largest for a given analysis time.

To find the optimal heating rate, let us notice first that, as follows from Eqns (2.137) and (2.158), the relationship between E and t_M can be summarized as

$$t_M = X_i E^i,$$
$$i = \begin{cases} 2, & |\Delta p| \ll p_o, \ p_o \text{ and } \Delta p \text{ can vary} \\ 3, & \Delta p \gg p_o, \ p_o \text{ and } \Delta p \text{ can vary} \\ 4, & \text{fixed } p_o \text{ and } \Delta p \end{cases},$$
$$X_i = \begin{cases} \dfrac{\pi d^2 p_o h}{4 p_{st} f}, & |\Delta p| \ll p_o \\ \dfrac{8d}{3}\sqrt{\dfrac{\pi\eta h^3}{p_{st} f}}, & \Delta p \gg p_o \\ \dfrac{32 h^2 \eta}{j_H \Delta p}, & \text{fixed } p_o \text{ and } \Delta p \end{cases} \tag{2.168}$$

This allows one to transform Eqn (2.132) for the analysis time (t_{anal}) as

$$t_{anal} = \frac{X_i E^i}{r_T} \cdot \frac{\Delta T_{end}}{\theta_{T,middle}}, \quad i = 2,3,4;$$
$$\Delta T_{end} = T_{end} - T_{init} \tag{2.169}$$

showing that

$$t_{anal} \sim E^i, \ i = 2,3,4 \tag{2.170}$$

Equation (2.169) also shows that t_{anal} is proportional to a dimensionless heating range ($\Delta T_{end}/\theta_{T,middle}$) and is inversely proportional to a dimensionless heating rate (r_T). Substituting E from Eqn (2.134) into Eqn (2.169) and accounting for Eqn (2.76) yield (Figure 2.18.b)

$$t_{anal} = \left(\frac{\Delta T_{end}}{\theta_{T,middle}} - \frac{t_{M,init} r_T}{t_{M,middle}}\right)^{-i} \times \frac{\Delta T_{end}}{\theta_{T,middle}} \frac{r_T^{i-1}}{(1-e^{-r_T})^i} X_i s_c^i, \quad i = 2,3,4 \tag{2.171}$$

Optimal dimensionless heating rate ($r_{T,Opt}$) minimizing t_{anal} at a fixed s_c can be found from this formula (Figure 2.18.b). However, the dependence of t_{anal} on r_T in the formula is relatively complex and not transparent. Quantity $r_{T,Opt}$ found from the formula is a function of several parameters (Figure 2.18.b) having a minor effect on $r_{T,Opt}$, but obscuring the key factors affecting $r_{T,Opt}$. Eqns (2.134) and (2.171) can be substantially simplified by ignoring $t_{M,init}$ in Eqn (2.131). This makes sense because typically $t_{M,init}$ is a small fraction of t_{anal}. At $t_{M,init} = 0$, Eqns (2.134) becomes Eqn (2.135), and Eqn (2.171) becomes

$$t_{anal} \approx X_i s_c^i \cdot \left(\frac{\theta_{T,middle}}{\Delta T_{end}}\right)^{i-1} \cdot \frac{r_T^{i-1}}{(1-e^{-r_T})^i}, \tag{2.172}$$
$$i = 2,3,4; \ t_{M,init} \ll t_{anal}$$

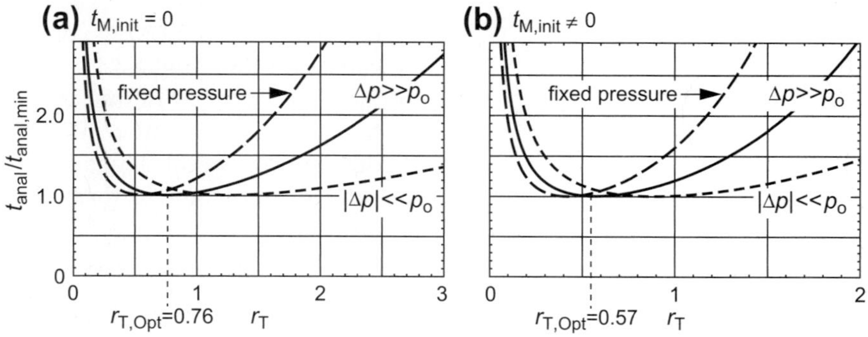

FIGURE 2.18 Analysis time (t_{anal}) as functions of dimensionless heating rate (r_T). Graphs (a) and (b) represent Eqns (2.172) and (2.171), respectively. Conditions for (b): $T_{init} = 50$ °C, $T_{end} = 250$ °C, $t_{M,init} = (T_{init}/T_{middle})^{0.7} t_{M,middle} = 0.828 t_{M,middle}$, $\theta_{T,middle} = (T_{middle}/273)^{0.7} 22$ °C $= 30$ °C. The marked optimal dimensionless heating rates ($r_{T,Opt} = 0.76$ and $r_{T,Opt} = 0.57$) correspond to strong gas decompression (solid-line curves) and adjustable pressure.

The last term in this formula shows how t_{anal} at a given s_c depends on the dimensionless heating rate (r_T). This dependence is illustrated in Figure 2.18.a. Unlike it is in Eqn (2.171), the relationship between t_{anal} and r_T in Eqn (2.172) is independent of all other parameters of GC analysis. Although Eqn (2.172) probably does not yield sufficiently accurate numerical values of optimal dimensionless heating rate ($r_{T,Opt}$) under all typical conditions, it does correctly identify the key factors affecting the dependence of t_{anal} on r_T and their relative effects on that dependence.

Figure 2.18 shows that optimal heating rate depends on pressure conditions in a column. To identify the most practically important conditions, two factors can be considered.

1. Weak gas decompression is typical for analyses of relatively simple samples where temperature programming might not be even necessary. An increase in sample complexity inevitably leads to strong gas decompression.
2. Fixed pressure limited by the maximum pressure available in a GC instrument is also rare in practice [46] because a large inlet pressure is typically associated either with unacceptably long analysis time (2 hours and longer [88]) or with the use of too small column diameters ($d < 0.1$ mm) causing technical difficulties at the current state of the art. The maximum pressure (hundreds of kPa) available in existing commercial GC instrument is more a result of current practical needs, and not a fundamental limitation. It is reasonable to expect that if and when higher pressure becomes practically useful, the GC instruments providing that pressure would be available.

These observations suggest that the strong gas decompression (solid-line curves in Figure 2.18) is a sort of the case of convergence for other pressure conditions in temperature-programmed analyses. In forthcoming discussions of optimal heating rate, **strong gas decompression** is always assumed unless otherwise explicitly stated.

As mentioned earlier, ignoring the fact that no peaks appear prior to the hold-up time ($t_{M,init}$) leads to relatively simple and transparent dependence (Eqn (2.172)) of analysis time (t_{anal}) on the dimensionless heating rate (r_T). As shown in Figure 2.18.a, the optimal dimensionless heating rate ($r_{T,Opt}$) corresponding to the minimum ($t_{anal,min}$) in analysis time in Eqn (2.172) can be estimated as $r_{T,Opt} \approx 0.76$. How does it relate to experimental data?

A limited amount of published experimental data suggest that *optimal heating rate* ($R_{T,Opt}$) is about 10 °C/t_M (10 °C per hold-up time) [60,90]. In the original source [60], t_M was measured at 50 °C (323 K) in a column with dimensionless film thickness (φ) of 0.001. According to Eqns (2.67), (2.61), and (2.17), the experimental optimal dimensionless heating rate ($r_{T,Opt}$) was $r_{T,Opt} \approx 0.4$ which is significantly lower than $r_{T,Opt} \approx 0.76$ in Figure 2.18.a. The following considerations help to reconcile these results.

1. Equation (2.172) yielding $r_{T,Opt} \approx 0.76$ is attractive because it yields $r_{T,Opt}$ that is independent of other parameters of GC analysis. On the other hand, Eqn (2.172) is approximate. It does not account for many experimental factors typically affecting $r_{T,Opt}$ in a relatively minor but complex way. These considerations suggest that Eqn (2.172) is more an identifier of the key factors affecting t_{anal} at a given s_c rather than an accurate description of that dependence.
2. A factor unaccounted for in Eqn (2.172) is the absence of peaks during the hold-up time. Eqn (2.171) that does account for this factor is more complex than Eqn (2.172). Under conditions specified in Figure 2.18, Eqn (2.171) yields $r_{T,Opt} = 0.57$ (Figure 2.18.b) which is closer to experimentally based $r_{T,Opt} = 0.4$ than the result $r_{T,Opt} \approx 0.76$ based on Eqn (2.172).

3. Figure 2.18 as well as the graphs of experimental data in the source [60] suggest that t_{anal} is relatively tolerant to departures of heating rates from their optimal values. In most cases, an increase in a heating rate compared to its optimum by 50% increases t_{anal} by less than 10% – the change that is within the accuracy of the measurement of $r_{T,Opt}$.

These observations suggest that the difference between $r_{T,Opt} = 0.57$ in Figure 2.18.b and the experimentally based

$$r_{T,Opt} = 0.4 \tag{2.173}$$

is within practically acceptable limits, and the latter can be adopted simply because it yields the already known [46,47,60,61,88,90] and easy-to-remember result

$$R_{T,Opt} = 10\,^{\circ}\mathrm{C}/t_M, \quad (\text{at } 50\,^{\circ}\mathrm{C}) \tag{2.174}$$

Numerical values of several elution parameters at this heating rate are listed in Table 2.5.

As mentioned earlier, $r_{T,Opt}$ depends on the pressure conditions of GC analysis. According to Figure 2.18, $r_{T,Opt}$ at weak decompression is about 50% higher than $r_{T,Opt}$ at strong decompression. On the other hand, $r_{T,Opt}$ at a fixed pressure is about 25% lower than $r_{T,Opt}$ at strong decompression with adjustable pressure. These results are summarized in Table 2.6.

Finally, $R_{T,Opt}$ at a given $r_{T,Opt}$ depends on dimensionless film thickness (φ). According to Eqns (2.60), (2.61), and (2.17),

Table 2.5 Asymptotic Elution Parameters (Eqns. (2.71)–(2.73), (2.76), and (2.126)) of the Solutes Eluting During a Linear Heating Ramp Having Dimensionless Heating Rate of $r_T = 0.4$

Parameter	$k_{R,a}$	$\mu_{R,a}$	$\omega_{R,a}$	$\overline{\omega}$	X_T
Value	2	0.33	0.67	0.82	0.71

Note: Asymptotic elution parameters are the ones of the solutes that were highly retained at the beginning of a heating ramp.

Table 2.6 Optimal Dimensionless Heating Rate ($r_{T,Opt}$) and Optimal Heating Rate ($R_{T,Opt}$) of a Linear Heating Ramp in Isobaric Analysis

		Adjustable Pressure		
Parameter	**Units**	$\Delta p \gg p_o$	$\lvert\Delta p\rvert \ll p_o$	**Fixed Pressure**
$r_{T,Opt}$	1	0.4	0.6	0.3
$R_{T,Opt}$	°C/time	$10/t_M$	$15/t_M$	$7.5/t_M$

$$R_{T,Opt} \sim (10^3\varphi)^{0.09} \tag{2.175}$$

When φ is significantly different from its typical values around $\varphi = 0.001$, $R_{T,Opt}$ at a given $r_{T,Opt}$ can be adjusted accordingly.

All conclusions regarding $r_{T,Opt}$ were based on assumption of a linear heating ramp at constant pressure (isobaric analysis). The simplicity of this approach results from the fact that, as observed in the comments to the definition of r_T in Eqn (2.59), the ratio $t_{M,char}/\theta_{char}$, and, therefore, r_T in that analysis do not change during the heating ramp. What about the constant flow mode or other conditions requiring a change in the pressure during a ramp?

When pressure changes during a heating ramp, a change in column temperature (T) might cause a change in parameter $t_{M,char}$ that might no longer be proportional to a change in parameter θ_{char}. It follows from Eqn (2.59) that, because heating rate (R_T) of a linear heating ramp is fixed, r_T can change during the ramp and can significantly depart from $r_{T,Opt}$. To maintain r_T at a fixed optimal level, a nonlinear heating ramp with gradually changing rate (R_T) should compensate for the time-dependent change in the ratio $t_{M,char}/\theta_{char}$ during the ramp.

The temperature dependence of an R_T in practically important isorheic (constant flow) analysis can be found from solving Eqns (2.60), (2.48), (2.42), (2.20), and (2.17) for R_T and assuming that F and r_T are fixed quantities. Assuming also that ξ in Table 2.1 can be approximated as $\xi \approx 0.7$, R_T as a function of T can be expressed as

$$R_T \approx \frac{R_{T,init}(P_{init}^3 - 1)(T/T_{init})^{3.4}}{(1 + (P_{init}^2 - 1)(T/T_{init})^{1.7})^{1.5} - 1} = R_{T,init}\begin{cases}(T/T_{init})^{1.7}, & |\Delta p| \ll p_o \\ (T/T_{init})^{0.85}, & \Delta p \gg p_o\end{cases} \quad (2.176)$$

This heating rate gradually increases with temperature. In practice, it can be implemented as a piecewise linear temperature program consisting of a series of linear heating ramps with increasing rates. Additional details regarding implementation of this and other temperature-dependent heating ramps can be found elsewhere [57].

2.10.4. Detection Limit and Tradeoff Triangle

Three independent factors affect the relationship (Eqn (2.172)) of the analysis time (t_{anal}) and a column separation capacity (s_c): dimensionless heating range ($\Delta T_{end}/\theta_{T,middle}$), dimensionless heating rate (r_T), and parameter X_i described in Eqn (2.168). Typically, the heating range depends on a combination of the properties of a sample and stationary phase and cannot be significantly changed by a method developer. Once the heating rate is set to its optimal value ($r_{T,Opt}$), reducing parameter X_i becomes the only option for further reduction in t_{anal} at a given s_c.

It follows from Eqn (2.168) that, at optimal flow rate and fixed pressure (the latter is an unlikely event), choosing a gas with the lowest viscosity (η) is the only way for further reduction in t_{anal} at a given s_c. When the pressure is not fixed, additional reduction in t_{anal} at a given s_c can be obtained through reducing a column diameter (d). That, however, has negative consequences. Smaller d reduces the sample amount that can be injected in a column without overloading the column. That, in turn, can make it more difficult to detect and to measure low concentration components of a sample. This brings into the issue of the *detection limit* (DL) – the third key factor of column performance tradeoffs.

Following is a brief review of the issues related to DL. Additional information can be found elsewhere [46].

The DL is a metric of the ability of a chromatographic system to detect and to measure small amounts of solutes in a sample. The DL is typically limited by the *baseline noise in* a chromatogram. The larger the noise, the larger (the worse) the DL. The noise can come from carrier *gas impurities, column bleeding* typically increasing with temperature, detector electronics, etc. All noise sources except for the detector noise can be viewed as the external ones. If necessary, the noise coming from these source can be reduced. Thus, the noise due to the gas impurities can be reduced by using purer carrier gas and the column bleeding can be reduced by using *low-bleed columns* and lower temperatures. On the other hand, some factors affecting a detector noise are part of a GC instrument design and cannot be changed by a system operator. In that regard, the detector noise ultimately controls the detection limit of a GC instrument.

Each GC system has one way or another a (digital or analog) *noise filter.* The baseline noise level depends on the width (σ_{filt}) of the filter *impulse response* [91] – the width of the filter's output resulted from an extremely narrow input pulse. Typically, a method developer can control σ_{filt} (by choosing a number of data points per peak, or by other means available in a GC system). However, there are limits to the utility of that control. Increasing σ_{filt} reduces the noise. On the other hand, too wide σ_{filt} can broaden the peaks and make some of them unresolvable. Therefore, it is necessary to maintain in practical applications a certain balance between σ_{filt} and the widths (σ) of the peaks in a chromatogram. Quantitative characteristics of that balance depend on the structure of the baseline noise. For example, the level of the *white noise* [91] – a typical type of detector noise – is inversely proportional to $\sqrt{\sigma_{\text{filt}}}$. Only the **white noise** is considered below. It is also assumed from now on that the ratio $\sigma/\sigma_{\text{filt}}$ is maintained in such a way that any **change in column dimensions does not affect the value of $\sigma/\sigma_{\text{filt}}$**. As a result, the noise level is inversely proportional to $\sqrt{\sigma}$, i.e. [46]

$$(\text{noise leverl}) \sim \sqrt{\frac{\rho_n}{\sigma}} \tag{2.177}$$

where ρ_n is the noise *spectral density* [91]. It is assumed below that the **carrier gas type and flow rate do not affect ρ_n and a detector response**.

Two metrics of detection limit can be considered: the *minimum detectable amount* (MDA) and the *minimum detectable concentration* (MDC) [46,92]. Both metrics are solute specific. The MDA of a given solute is its amount at a predetermined *signal-to-noise ratio*. The MDC of a given solute is its concentration at a predetermined *signal-to-noise ratio*.

Metric MDA is suitable for the situation where the *same amount of sample* can be injected in a column of any dimensions without overloading the column. This might happen where only a small amount of sample is available. In this case, *there is no conflict between the MDA, the separation capacity, and the analysis time* [46]. Thus, when column diameter and length get smaller in proportion with each other, the column maximum efficiency remains unchanged while the analysis time gets shorter. Along with that, the peaks become sharper so that the *signal-to-noise ratio* for the same sample amount increases (improves). This means that MDA gets lower (better) while the analysis time gets shorter and the column separation capacity remains unchanged. If necessary, shorter analysis time can be traded for larger separation capacity. The relationships of MDA and column dimensions can be quantified by the following relations of proportionality [46]:

$$\text{MDA} \sim \sqrt{\frac{p_o\rho_n\sqrt{d^3Lh}}{f\mu_R}} \sim d\sqrt{\frac{Ehp_o\rho_n}{f\mu_R}}, \tag{2.178}$$

$(|\Delta p| \ll p_o)$

$$\text{MDA} \sim \sqrt{\frac{L\rho_n}{\mu_R}\sqrt{\frac{h\eta}{f}}} \sim E\sqrt{\frac{dh^3\eta\rho_n}{f\mu_R}}, \tag{2.179}$$

$(\Delta p \gg po)$

In analyses of complex samples, more typical is a different situation where, in order to prevent column overloading and substantial broadening of larger peaks, a reduction in column dimensions must be accompanied by a reduction in the amount of injected sample. To maintain the same margin of column overloading for all column dimensions, a sample injected in any column should have the *same concentration* (rather than the same amount). MDC becomes a suitable metric of detection limit. The conflict between the need to prevent a column overloading by the high concentration components and to detect low concentration components creates a conflict between separation capacity, analysis time, and MDC. From now on, only the **MDC** is considered as a metric of detection limit.

The *largest sample amount* ($A_{\max}$) that does not overload a column (or causes the same relative level of overloading) is proportional to the product $\varphi d_c^{5/2}\sqrt{L}$ [46], i.e.

$$A_{\max} \sim \varphi d_c^{5/2}\sqrt{L} \tag{2.180}$$

Dependence of MDC on method parameters and properties of a chromatographic system can be described by the following relations of proportionality [46]:

$$\text{MDC} \sim \frac{1}{\varphi}\sqrt{\frac{p_o\rho_n}{f\mu_R\sqrt{d^7hL}}} \sim \frac{1}{d^2\varphi}\sqrt{\frac{p_o\rho_n}{E\mu_R fh}},$$

$(|\Delta p| \ll p_o)$ (2.181)

$$\mathrm{MDC} \sim \frac{1}{\varphi}\sqrt{\frac{\rho_{\mathrm{n}}\sqrt{\eta}}{d^5\mu_{\mathrm{R}}\sqrt{fh}}}, \quad (\Delta p \gg p_{\mathrm{o}}) \qquad (2.182)$$

This formulas describe, among other things, the dependence of MDC on column dimensions (diameter and length), or on column diameter and efficiency. It is interesting that, at strong gas decompression, MDC does not depend on column length or efficiency. This means that the previously considered variations of a column length in order to obtain the shortest analysis time at a predetermined separation performance or to obtain the best separation performance at a predetermined analysis time do not affect MDC.

By excluding column diameter (d) from Eqns (2.182) and (2.172), one has

$$\mathrm{MDC} \sim \sqrt{X_{\mathrm{all}}\frac{s_{\mathrm{c}}^{15}}{t_{\mathrm{anal}}^5}}, \quad (\Delta p \gg p_{\mathrm{o}}) \qquad (2.183)$$

where

$$X_{\mathrm{all}} = \frac{\rho_{\mathrm{n}}}{\varphi^2}\cdot\frac{\eta^3 h^7}{f^3}\cdot\frac{r_{\mathrm{T}}^{10}}{(1-e^{-r_{\mathrm{T}}})^{16}} \qquad (2.184)$$

Equation (2.183) shows how the key performance metrics − the separation capacity (s_{c}), the analysis time (t_{anal}), and the MDC − relate to each other at strong gas decompression. Eqn (2.184) highlights other parameters that affect the tradeoff between s_{c}, t_{anal}, and MDC. Under optimal conditions ($f = f_{\mathrm{Opt}}$, $r_{\mathrm{T}} = r_{\mathrm{T,Opt}}$), the number of parameters affecting the tradeoff gets smaller. First, parameters f, r_{T}, and h become fixed quantities. Second, as shown earlier, the ratio η/f_{Opt} in Eqn (2.184) can be reduced to the gas-dependent product $X_D^2 v_{\mathrm{st}}^2$ (Table 2.1). Eqn (2.183) can be expressed as

$$\mathrm{MDC} \sim \sqrt{X_{\mathrm{all,Opt}}\frac{s_{\mathrm{c}}^{15}}{t_{\mathrm{anal}}^5}}, \quad \Delta p \gg p_{\mathrm{o}} \qquad (2.185)$$

where

$$X_{\mathrm{all,Opt}} = \frac{\rho_{\mathrm{n}}}{(X_D v_{\mathrm{st}})^6\varphi^2} \qquad (2.186)$$

Example 2.13. According to Table 2.1, the product $X_D v_{\mathrm{st}}$ for helium is about 60% of that for hydrogen. Due to the factor $(X_D v_{\mathrm{st}})^6$ in Eqn (2.186), MDC in a system with hydrogen as a carrier gas is about five times lower (better) than MDC in a system with helium that requires the same time for the same separation capacity. To be more specific, consider two GC-MS analyses: one is based on a 10 m × 0.1 mm column with helium and another on a 17 m × 0.17 mm column with hydrogen. Under optimal conditions, both analyses have the same efficiency and the same hold-up time. As a result, the width of a peak corresponding to the same solute and the noise are the same in both cases. However, according to Eqn (2.180), the sample amount that can be injected in a column with hydrogen is about 5 times larger ($1.7^3 \approx 4.9$) than that amount for a column with helium. This means that, to yield the same signal-to-noise ratio in both cases, the analysis with hydrogen could have five times lower concentration of the trace level components than the analysis with helium. In other words, the analysis with hydrogen has five times lower (better) MDC than the analysis with helium. This example can also be viewed as an illustration of trading the advantage of hydrogen over the helium in speed of analysis at the same separation capacity for the advantage in MDC at the same analysis time and separation capacity.

The tradeoff between column separation power, analysis time, and detection limit is sometimes illustrated by a *tradeoff triangle* (*triangle of compromise* [46,61]). Eqn (2.185) is a quantitative description of those triangles under different operational conditions (at strong gas decompression). However, to construct the rectangles, some metrics of column performance should be modified.

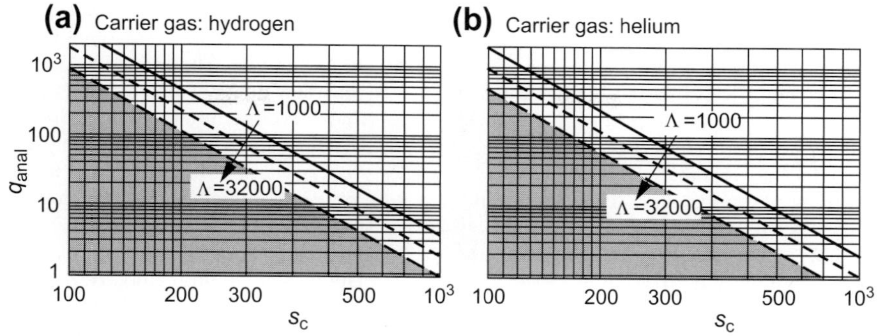

FIGURE 2.19 Tradeoffs between separation capacity (s_c), throughput (q_{anal}) and dynamic range (Λ) (Eqn (2.189)) for hydrogen (a), and helium (b). Three levels of Λ are shown for each carrier gas. Each next level is $4\sqrt{2}$ times larger than the previous one and, as follows from Eqn (2.182), corresponds to twice as large column diameter. Each line in the graphs can be viewed as a side of a triangle of separation-time tradeoff. A triangle corresponding to the same Λ is shaded for each gas.

Quantity s_c can be viewed as a *direct metric* of separation performance in the sense that larger value of the metric corresponds to better separation performance. On the other hand, MDC and t_{anal} can be viewed as *inverted metrics* in the sense that larger values of the metrics correspond to poorer performance. For better visualization of Eqn (2.185), it is desirable that all performance metric in the formula were direct. Such direct counterparts of MDC and t_{anal} are the *linear dynamic range* (Λ) of a chromatogram (the ratio of the largest undistorted *peak height* to the noise), and the *throughput* (q_{anal}) – the number of analyses per time unit. These metrics can be expressed as

$$\Lambda \sim \frac{1}{\mathrm{MDC}}, \quad q_{anal} = \frac{1}{t_{anal}} \tag{2.187}$$

transforming Eqns (2.183) and (2.185) as (Figure 2.19)

$$s_c^{15}\Lambda^2 q_{anal}^5 \sim \frac{\varphi^2}{\rho_n}\frac{f^3}{\eta^3 h^7}\frac{(1-e^{-r_T})^{16}}{r_T^{10}}, \quad \Delta p \gg p_o \tag{2.188}$$

$$s_c^{15} q_{anal}^5 \Lambda^2 \sim \frac{\varphi^2}{\rho_n}(X_D v_{st})^6, \quad (\Delta p \gg p_o, \; f = f_{Opt}, \; r_T = r_{T,Opt}) \tag{2.189}$$

Eqns (2.188) and (2.189) could also be constructed for the weak gas decompression. However, the most demanding tradeoffs between s_c, q_{anal}, and Λ exist in analyses of complex mixtures, typically requiring strong gas decompression. From that perspective, Eqns (2.188) and (2.189) can be viewed as descriptions of the *ultimate tradeoff triangle*.

References

[1] L.M. Blumberg, Temperature-programmed gas chromatography, Wiley-VCH, Weinheim, 2010.
[2] H.M. McNair, J.M. Miller, Basic gas chromatography, third ed., John Wiley & Sons, Inc, Hoboken, New Jersey, 2009.
[3] E.F. Barry, R.L. Grob, Columns for gas chromatography, John Wiley & Sons, Hoboken, New Jersey, 2007.
[4] R.L. Grob, E.F. Barry (Eds.), Modern practice of gas chromatography, fourth ed., John Wiley & Sons, Hoboken, New Jersey, 2004.
[5] C.F. Poole, The essence of chromatography, Elsevier, Amsterdam, 2003.
[6] W. Jennings, E. Mittlefehldt, P. Stremple, Analytical gas chromatography, second ed., Academic Press, San Diego, 1997.
[7] L.S. Ettre, J.V. Hinshaw, Basic relations of gas chromatography, Advanstar, Cleveland, Ohio, 1993.
[8] J.C. Giddings, Unified separation science, Wiley, New York, 1991.

[9] G. Guiochon, C.L. Guillemin, Quantitative gas chromatography for laboratory analysis and on-line control, Elsevier, Amsterdam, 1988.

[10] M.L. Lee, F.J. Yang, K.D. Bartle, Open tubular gas chromatography, John Wiley & Sons, New York, 1984.

[11] A.B. Littlewood, Gas chromatography: principles, techniques, and applications, second ed., Academic Press, New York, 1970.

[12] W.E. Harris, H.W. Habgood, Programmed temperature gas chromatography, John Wiley & Sons, Inc., New York, 1966.

[13] J.H. Purnell, Gas chromatography, John Wiley & Sons, Inc., New York, 1962.

[14] A.I.M. Keulemans, Gas chromatography, second ed., Reinhold Publishing Corp., New York, 1959.

[15] IUPAC, Nomenclature for chromatography, Pure Appl Chem 65 (1993) 819–872.

[16] J.C. Giddings, Theory of programmed temperature gas chromatography: the prediction of optimum parameters, in: N. Brenner, J.E. Callen, M.D. Weiss (Eds.), Gas chromatography, Academic Press, New York, 1962, pp. 57–77.

[17] H.W. Habgood, W.E. Harris, Retention temparature and column efficiency in programmed temperature gas chromatography, Anal. Chem. 32 (1960) 450–453.

[18] H.W. Habgood, W.E. Harris, Plate height in programmed temperature gas chromatography, Anal Chem 32 (1960) 1206.

[19] R. Consden, A.H. Gordon, A.J.P. Martin, Qualitative analysis of proteins: a partition chromatographic method using paper, Biochem. J. 38 (1944) 224–232.

[20] L.M. Blumberg, Metrics of separation performance in chromatography. Part 1. Definitions & application to static analyses, J Chromatogr A 1218 (2011) 5375–5385.

[21] L.M. Blumberg, M.S. Klee, Quantitative comparison of performance of isothermal and temperature-programmed gas chromatography, J. Chromatogr. A 933 (2001) 13–26.

[22] A.T. James, A.J.P. Martin, Gas-liquid partition chromatography: the separation and micro-estimation of volatile fatty acids from formic acid to dodecanoic acid, Biochem. J. 50 (1952) 679–690.

[23] A.J.P. Martin, R.L.M. Synge, A new form of chromatogram employing two liquid phases. 1. A theory of chromatography. 2. Application to the microdetermination of the higher monoamino-acids in proteins, Biochem. J. 35 (1941) 1358–1368.

[24] W.J. Moore, Physical chemistry, fourth ed., Prentice-Hall, Inc., Englewood Cliffs, 1972.

[25] E.C.W. Clarke, D.N. Glew, Evaluation of thermodynamic functions from equilibrium constants, Transactions of the Faraday SocietyTransactions of the Faraday Society 62 (1966) 539–547.

[26] F. Aldaeus, Y. Thewalim, A. Colmsjö, Prediction of retention times of polycyclic aromatic hydrocarbons and n-alkanes in temperature-programmed gas chromatography, Anal Bioanal Chem 389 (2007) 941–950.

[27] G. Castello, P. Moretti, S. Vezzani, Retention models for programmed gas chromatography (Review), J. Chromatogr. A 1216 (2009) 1607–1623.

[28] D.E. Bautz, J.W. Dolan, L.R. Snyder, Computer simulation as an aid in method development for gas chromatography. I. The accurate prediction of separation as a function of experimental conditions, J. Chromatogr. 541 (1991) 1–19.

[29] S.N. Atapattu, K. Eggers, C.F. Poole, W. Kiridena, W. Koziol, Extension of the system constants database for open-tubular columns: system maps at low and intermediate temperatures for four new columns, J. Chromatogr. A 1216 (2009) 1640–1649.

[30] J.C. Giddings, Elementary theory of programmed temperature gas chromatography, J. Chem. Educ. 39 (1962) 569–573.

[31] A.B. Fialkov, A. Gordin, A. Amirav, Extending the range of compounds amenable for gas chromatography – mass spectrometric analysis, J. Chromatogr. A 991 (2003) 217–240.

[32] E. Grushka, Chromatographic peak capacity and factors influencing it, Anal Chem 42 (1970) 1142–1147.

[33] W. Kauzmann, Kinetic theory of gases, W. A. Benjamin, Inc., New York, 1966.

[34] Y.S. Touloukian, S.C. Saxena, P. Hestermans, Thermal conductivity. Nonmetallic liquids and gases, IFI/Plenum, New York, 1970.

[35] Y.S. Touloukian, S.C. Saxena, P. Hestermans, Viscosity, IFI/Plenum, New York, 1975.

[36] J.T.R. Watson, Viscosity of gases in metric units, Her Majesty's Stationery Office, Edinburgh, 1972.

[37] J.V. Hinshaw, L.S. Ettre, The variation of carrier gas viscosities with temperature, J. High Resolut. Chromatogr. 20 (1997) 471–481.

[38] L.S. Ettre, Viscosity of gases used as the mobile phase in gas chromatography, Chromatographia 18 (1984) 243–248.

[39] E. Zauderer, Partial differential equations of applied mathematics, second ed., John Wiley & Sons, New York, 1989.

[40] E.N. Fuller, P.D. Schettler, J.C. Giddings, A new method for prediction of binary gas-phase diffusion coefficients, Ind Eng Chem 58 (1966) 19–27.

[41] V.R. Maynard, E. Grushka, Measurement of diffusion coefficients by gas-chromatography broadening techniques: review, in: J.C. Giddings, E. Grushka, R.A. Keller, J. Cazes (Eds.), Advances in gas

chromatography, Marcel Dekker, Inc., New York, 1975, pp. 99–140.

[42] L.M. Blumberg, W.H. Wilson, M.S. Klee, Evaluation of column performance in constant pressure and constant flow capillary gas chromatography, J. Chromatogr. A 842 (1–2) (1999) 15–28.

[43] J.E. Cahill, D.H. Tracy, Effects of permeation of helium through the walls of fused silica capillary GC Columns, J. High Resolut. Chromatogr. 21 (1998) 531–539.

[44] H. Darcy, Les fontaines publiques de la ville de dijon, Dalmont, Paris, 1856.

[45] H. Darcy, The public fountains of the city of Dijon, Kendall Hunt Publishing Co., Dubuque, IA, 2004.

[46] L.M. Blumberg, Multidimensional gas chromatography: theoretical considerations, in: L. Mondello (Ed.), Comprehensive chromatography in combination with mass spectrometry, Wiley, Hoboken, NJ, 2011, pp. 13–63.

[47] L.M. Blumberg, M.S. Klee, Method translation and retention time locking in partition GC, Anal. Chem. 70 (1998) 3828–3839.

[48] I. Halász, K. Hartmann, E. Heine, Column types in gas chromatography, in: A. Goldup (Ed.), Gas chromatography 1964, The Institute of Petroleum, London, 1965, pp. 38–61.

[49] J.A. Jönsson, Dispersion and peak shapes in chromatography, in: J.A. Jönsson (Ed.), Chromatographic theory and basic principles, Marcel Dekker, New York, 1987, pp. 27–102.

[50] R.W. Ohline, D.D. DeFord, Chromatography, the application of moving thermal gradient to gas liquid partition chromatography, Anal. Chem. 35 (1963) 227–234.

[51] L.M. Blumberg, Limits of resolution and speed of analysis in linear chromatography with and without focusing, Chromatographia 39 (1994) 719–728.

[52] Snyder WD, Blumberg LM, Constant peak elution temperature with GC columns of different diameter. How to increase analysis speed with little or no loss in resolution. In Sandra P, Lee ML editors. Fourteenth international symposium on capillary chromatography, ISCC92, Baltimore, USA, 1992. pp. 28–38.

[53] B.D. Quimby, V. Giarrocco, M.S. Klee, Speed improvement in detailed hydrocarbon analysis of gasoline using 100 μm capillary column. Application Note 228-294, Hewlett-Packard Co, Wilmington, DE, 1995. document (43) pp. 5963–5190E.

[54] V. Giarrocco, B.D. Quimby, M.S. Klee, Retention time locking: concepts and applications. Application Note 228–392, Hewlett-Packard Co, Wilmington, DE, 1997. document (23) p. 5966–2469E.

[55] F. David, D.R. Gere, F. Scanlan, P. Sandra, Instrumentation and applications of fast high-resolution capillary gas chromatography, J. Chromatogr. A 842 (1999) 309–319.

[56] P. Sandra, F. David, High-throughput capillary gas chromatography for the determination of polychlorinated biphenyls and fatty acid methyl esters in food samples, J. Chromatogr. Sci 40 (2002) 248–253.

[57] Blumberg LM, Method translation in gas chromatography, USA patent 6,634,211, 2003. Provisional patent application No. 60/291,406/May 16 2001.

[58] N. Etxebarria, O. Zuloaga, M. Olivares, L.J. Bartolomé, P. Navarro, Retention-time locked methods in gas chromatography, J. Chromatogr. A 1216 (2009) 1624–1629.

[59] C. Costa Neto, J.T. Köffer, J.W. De Alencar, Programmed flow gas chromatography. Part I, J. Chromatogr. 15 (1964) 301–313.

[60] L.M. Blumberg, M.S. Klee, Optimal heating rate in gas chromatography, J Microcolumn Sep 12 (2000) 508–514.

[61] M.S. Klee, L.M. Blumberg, Theoretical and practical aspects of fast gas chromatography and method translation, J. Chromatogr. Sci. 40 (2002) 234–247.

[62] J.C. Giddings, Use of multiple dimensions in analytical separations, in: H.J. Cortes (Ed.), Multidimensional chromatography techniques and applications, Marcel Dekker, Inc., New York and Basel, 1990, pp. 1–27.

[63] W. Jennings, The "replacement" column, a recurring problem in gas chromatography, The Restek Advantage 1 (2006) 2.

[64] B. Yan, J. Zhao, J.S. Brown, J. Blackwell, P.W. Carr, High-temperature ultrafast liquid chromatography, Anal. Chem. 72 (2000) 1253–1262.

[65] J.M. Davis, Statistical theory of spot overlap in two-dimensional separations, Anal. Chem. 63 (1991) 2141–2152.

[66] R.E. Murphy, M.R. Schure, J.P. Foley, Effect of sampling rate on resolution in comprehensive two-dimensional liquid chromatography, Anal. Chem. 70 (1998) 1585–1594.

[67] J.J.J. van Deemter, F.J. Zuiderweg, A. Klinkenberg, Longitudinal diffusion and resistance to mass transfer as causes of nonideality in chromatography, Chem. Eng. Sci. 5 (1956) 271–289.

[68] M.J.E. Golay, Theory of chromatography in open and coated tubular columns with round and rectangular cross-sections, in: D.H. Desty (Ed.), Gas chromatography 1958, Academic Press, New York, 1958, pp. 36–55.

[69] M.J.E. Golay, The dynamic diffusion constant within fluid flow in an open straight tube with an elliptical cross-section, J. Chromatogr. 196 (1980) 349–354.

[70] H. Poppe, J. Paanakker, M. Bronckhorst, Peak width in solvent-programmed chromatography. i. general description of peak broadening in solvent-programmed elution, J. Chromatogr. 204 (1981) 77–84.

[71] L.M. Blumberg, T.A. Berger, Variance of a zone migrating in a non-uniform time invariant linear medium, J. Chromatogr. 596 (1992) 1–13.

[72] L.M. Blumberg, Variance of a zone migrating in a linear medium. II. time-varying non-uniform medium, J. Chromatogr. 637 (1993) 119–128.

[73] W.A. Peters, The efficiency and capacity of fractionating columns, J. Ind. Eng. Chem. 14 (1922) 476–479.

[74] W.L. Jones, S. Dal Nogare, D.H. Desty, M.J.E. Golay, A.I.M. Keulemans, A.J.P. Martin, S.S. Ober, C.S.G. Phillips, J. Thoburn, E. Williams, Standard nomenclature considerations and recommendations, in: V.J. Coates, H.J. Noebels, I.S. Fagerson (Eds.), Gas chromatography, Academic Press, New York, 1958, pp. 315–317.

[75] J.C. Giddings, R.A. Keller, Theoretical basis of partition chromatography, in: E. Heftmann (Ed.), Chromatography, Reinhold Publishing Corp., New York, 1961, pp. 92–111.

[76] E. Glueckauf, Theory of chromatography. The "theoretical plate" concept in column separations, Transactions of Faraday Society 51 (1955) 34–44.

[77] K. Lan, J.W. Jorgenson, Theoretical investigation of the spatial progression of temporal statistical moments in linear chromatography, Anal. Chem. 72 (2000) 1555–1563.

[78] K. Lan, J.W. Jorgenson, Spatial and temporal progressions of spatial statistical moments in linear chromatography, J. Chromatogr. A 905 (2001) 47–57.

[79] J.C. Giddings, S.L. Seager, L.R. Stucki, G.H. Stewart, Plate height in gas chromatography, Anal. Chem. 32 (1960) 867–870.

[80] J.C. Giddings, Plate height of nonuniform chromatographic columns. gas compression effects, coupled columns, and analogous systems, Anal. Chem. 35 (1963) 353–356.

[81] D.H. Desty, E. Glueckauf, A.T. James, A.I.M. Keulemans, A.J.P. Martin, C.S.G. Phillips, Nomenclature recommendations, in: D.H. Desty, C.L.A. Harbourn (Eds.), Vapor phase chromatography, Academic Press, London, 1957, pp. xi–xiii.

[82] L.M. Blumberg, Theory of fast capillary gas chromatography. Part 4: column performance vs. liquid film thickness, J. High Resolut. Chromatogr. 22 (1999) 501–508.

[83] L.M. Blumberg, Theory of fast capillary gas chromatography. Part 3: column performance vs. gas flow rate, J. High Resolut. Chromatogr. 22 (1999) 403–413.

[84] J.C. Giddings, Evidence on thet nature of eddy diffusion in gas chromatography from inert (nonsorbing) column data, Anal. Chem. 35 (1963) 1338–1341.

[85] J.C. Giddings, Plate height theory of programmed temperature gas chromatography, Anal. Chem. 34 (1962) 722–725.

[86] J.C. Giddings, Maximum number of components resolved by gel filtration and other elution chromatography methods, Anal Chem 39 (1967) 1027–1028.

[87] L.M. Blumberg, Theory of fast capillary gas chromatography. Part 2: speed of analysis, J. High Resolut. Chromatogr. 20 (1997) 679–687.

[88] L.M. Blumberg, F. David, M.S. Klee, P. Sandra, Comparison of one-dimensional and comprehensive two-dimensional separations by gas chromatography, J. Chromatogr. A 1188 (2008) 2–16.

[89] R.P.W. Scott, G.S.F. Hazeldean, Some factors affecting column efficiency and resolution of nylon capillary columns, in: R.P.W. Scott (Ed.), Gas chromatography 1960, Butterworth, Washington, 1960, pp. 144–161.

[90] M. van Deursen, H.-G. Janssen, J. Beens, G. Rutten, C.A. Cramers, Design consideration for rapid-heating columns applied in fast capillary gas chromatography, J. Microcolumn Sep. 13 (2001) 337–345.

[91] A. Papoulis, Probability, Random variables, and stochastic processes, McGraw-Hill Book Co., New York, 1965.

[92] T.H.M. Noij, Trace analysis by capillary gas chromatography. Theory and Methods, Ph. D. Thesis, Technical University of Eindhoven, Netherlands, 1988.

CHAPTER

3

Column Technology: Open Tubular Columns

Frank L. Dorman, Peter Dawes

OUTLINE

3.1. INTRODUCTION

In the early days of gas chromatography (GC), practitioners commonly used packed columns that were self-manufactured in their respective laboratories. Early users of the technique had to be capable of not only the instrumental aspects but also the column manufacturing. Commercial manufacturing of GC columns is a relative newcomer to the field, having been in existence

Gas Chromatography DOI: 10.1016/B978-0-12-385540-4.00003-1

for about the last 50 years. When commercial manufacturing first began, packed columns were the only columns used in the practice of GC. This may seem surprising to more recent users of the technique, and packed column GC has almost completely been replaced by wall-coated open tubular (WCOT) capillary columns made from flexible fused-silica tubing. While there have been numerous advances made to the science of GC as a result of the use of these materials, Marcel Golay [1] is largely credited with first considering the benefits of the open tubular column format. In the 1960s and 1970s, these columns were carefully manufactured using glass capillaries, based on the work of Desty [2]. Due to the inflexibility and fragility of these columns, however, they remained beyond the ability of many chromatographers. It was the innovation of drawn fused-silica capillary tubing that really moved the WCOT glass capillary format to commercial success. The work of Dandeneau and Zerenner [3] of Hewlett Packard Corporation as well as of Lipsky [4] made the WCOT capillary format available to all practitioners, and its importance cannot be overlooked. Due to the advantages of the WCOT capillary format, and the ease of use, capillary GC is possibly the most powerful common separation technique readily available to the analytical community. While only a subset of organic and organometallic compounds are amenable to a GC analysis (estimates are generally between 10% and 20%), it is generally the technique of choice due to its high efficiency and resolving power.

The introduction of thin-wall flexible fused-silica capillary tubing to gas chromatography by Dandenau and Zerener in 1979 was the important step that was needed to make capillary gas chromatography the widely utilized analytical technique that it is today.

Conventional glasses that were used up to that point are generally thought of as being inert and unlikely to react with the compounds, but in reality they are highly reactive when passing nanogram or smaller quantities of analytes down a long narrow capillary tube with high residence time at elevated temperatures. In conventional glasses such as soda glass and borosilicate glass, there is a very large percentage of metal oxides and other materials that may interact with the analyte molecules. These metal oxides may also cause catalytic breakdown of the polymer stationary phase at elevated temperatures.

Synthetic fused-silica material is very pure silicon dioxide but it still has to be treated correctly to achieve the levels of inertness required in modern capillary GC columns. Correctly drawn and treated fused-silica tubing is the first essential ingredient of a high-quality fused-silica capillary column.

The inertness that can be achieved with fused-silica capillary columns was the driving force for their introduction but there are other reasons for its successful adoption. Previously, the glass capillary columns that had an inside diameter of 0.25 mm had an outside diameter of around 1.0 mm. This meant that the tubing was not flexible, was fragile, and the curvature meant that it could not be inserted for any distance into a fitting. In addition, the large annular area at the end of the column added a significant dead volume and so real skill was required to be able to make connections without getting some degree of peak tailing and loss of resolving power. Fused-silica capillary material has a thin wall and is flexible, making it very easy to work with compared to the earlier glass columns, and being thin walled it became very easy to make low or even zero dead volume connections.

An additional advantage of fused-silica that the manufacturers understand well was that the glass coils that were previously used for GC columns were extremely slow to draw down to size and then bend it into a coil. The speed of the process was in the order of 1 m/min and it was normal in capillary column production to have banks of glass-drawing machines making the coils very slowly. With thin-wall fused-silica capillary tubing and the techniques used to draw flexible fiber optic material, the draw speed of the tubing can be in the order of hundreds of meters per minute but more usually is at a rate of around 20 m/min. So instead of producing a long 60-meter column in over an hour, enough fused-silica tubing is commonly produced in a few minutes. With the old glass technology, capillary columns would be far more expensive than what analysts are accustomed to with modern fused-silica columns and it is unimaginable how the worldwide demand would have been met (at a reasonable price).

3.2. OVERVIEW OF THE FUSED SILICA DRAWING PROCESS

Companies producing fused-silica capillary tubing consider the techniques as proprietary and so there is very little disclosure of ideas and technology between manufacturers of the tubing. The information given here is somewhat a general overview from the lessons of drawing fused-silica tubing for over 30 years. There will be different views on the appropriate procedures for producing fused-silica tubing.

3.3. THE PREFORM – RAW MATERIAL

The starting material for the fused-silica capillary tubing is known as a preform. The preform is a large high-purity synthetically produced fused-silica tube. There are three common ways of producing the preform material but the most usual is continuous flame hydrolysis. Silane is oxidized to form silicon dioxide and the resulting silicon dioxide dust is fused thermally onto a boule, which is like a big lump of pure fused-silica. The boule can then be melted and formed into the dimensionally very precise preform tube that is required.

In the past, fused-silica was formed by feeding silicon tetrachloride into a hydrogen/oxygen flame in an additive process, forming a boule. The $SiCl_4$ process has fallen out of favor not only because of the large amounts of chlorine and hydrochloric acid that were formed in the process but also because GC columns formed in this manner always tended to have an excessively acidic character.

The hydroxyl concentration in the preform from the preferred hydrolysis process is very low, at below the 1-ppm level, and the metal impurities (metal oxides) are also less than 1 ppm. This is fortuitous because this is exactly what is needed for the production of capillary GC columns; however, it only happens because this is the property required for the much larger fiber optic manufacturing industry where this level of purity is required to achieve the stringent optical properties.

The dimensions of the preforms are typically in the range of 20-mm outside diameter with an inside diameter of around 14 mm, depending on the size required for the finished tubing. The dimension of the preform needs to relate to the ratio of the outside diameter to inside diameter of the drawn capillary tube. Under ideal drawing conditions, the ratio of outside diameter to inside diameter will be similar to the same ratio in the preform.

To achieve the required dimensional tolerances of the finished capillary, the preforms require quite remarkable tolerances, for example, often better than a 10-μm variation on the 20-mm outside diameter.

3.4. SURFACE CHEMISTRY

One of the long-term issues regarding fused-silica capillary tubing is the surface chemistry. There have been many different ways proposed on how to treat the capillary tubing to obtain the ideal surface characteristics for inertness of the finished column, "coatability" of the surface with the stationary phase, and the forming of the covalent bonds between the stationary phase and the glass surface. What is often not discussed is what can be done as part of the drawing process that can have a dramatic impact on the surface chemistry and the quality of the capillary column that is ultimately produced. Without this control and knowledge, it is very difficult to reliably produce many batches of capillary columns even with surface chemistry normalization treatments. A common problem for many capillary column manufacturers is that a column type can be made consistently for a long period of time and then batches of the column fail to coat correctly and meet specifications. Producing capillary columns for many years was perceived as a black art until factors in the drawing process became better understood through careful experimentation with capillary drawing conditions and measurement of parameters such as the wettability of the fused-silica inner wall through contact angle measurements to determine silanol concentrations on the inner surface. In this way, it was eventually possible to understand the seemingly insignificant changes in the drawing process that had such a great impact on the GC capillary column manufacturing process.

3.5. DRAWING OF THE CAPILLARY FROM THE PREFORM

The process of pulling a capillary tube from a large preform tube has been likened to "pulling toffee;" there are no dies to form the fiber. The end of the fused-silica is brought to a very precisely controlled temperature and a molten thread is pulled from the preform. The dimensions of the capillary thread are managed by accurate pulling speed of the capillary combined with precise feed speed of the preform.

The ratio of draw speed to feed speed (draw ratio) predominantly determines the dimension of the capillary tube derived from the larger preform. For typical dimensions of capillary GC columns, a very high draw ratio is used where, from every meter of preform, over 4,000 meters of capillary is produced. As in fiber optic production, it is possible to draw tubing for capillary GC at enormous speeds of hundreds of meters per minute but there are other considerations in chromatography tubing, which will be touched upon later. Typically, draw speeds are in the range of 10–40 m/min.

Because very fine molten glass tubing is being pulled in a steady state, it is easy to imagine that vibrations will be transmitted to the capillary causing imperfections and stress points in the finished tubing. For this reason, it is critical that no vibrations are transmitted into the draw system and so shock-absorbing mounts are used throughout the machine. In addition, motors, gearboxes, and anything that moves on the tower must not introduce a vibration or any unevenness in the force that is applied to the capillary as it is pulled. Long hours have been spent when designing and building the towers to find the source of the most imperceptible unevenness of movement or force.

For conventional glasses such as soda glass or borosilicate glass, the melt temperature is in the range of 600–800 °C. The high-purity fused-silica has a melt temperature in the order of 2,000 °C, which must be very precisely controlled within the range of a few degrees.

Many years ago, a gas/oxygen flame was used for heating or even a hydrogen/oxygen flame. Precise control of the gas flows while monitoring the temperature of the neck down point of the preform enabled a surprisingly accurate temperature control but the fuel/oxygen ratios also had an impact on inner surface chemistry of the capillary tubing (Figure 3.1).

The standard furnace heating system for fused-silica drawing is now an inductively heated graphite furnace. A graphite core in the furnace (termed a susceptor) is heated inductively, which in turn radiates heat to the

FIGURE 3.1 Neck down area on a fused-silica preform.

fused-silica preform. Of course, the heated graphite would ignite in the presence of air and to prevent this as well as to protect the capillary from contamination, the furnace is purged with argon.

To control the level of acid and moisture content inside the preform during the draw process, it is also usual to purge the preform with inert gas. Some also use the purge gas pressure to finely control the dimensions of the resulting capillary.

The dimensional control of fused-silica capillary columns has steadily improved, making it possible to establish more stringent dimensional specifications on the finished capillary. These tight dimensions are maintained through automated feedback loops on the drawing tower. Positioned just below the furnace is a two-axis laser micrometer, which is used to measure the dimensions of the drawn fiber. It is not possible to measure the inside diameter of the tubing but this can be inferred from the outside diameter measurement, provided a narrow range of drawing conditions have been used and the ratio of outside diameter to inside diameter is maintained the same as the preform. The laser micrometer measurement of the outside diameter of the capillary is then used in a feedback loop to control the feed and draw speeds to maintain the required outside diameter. This permits an extremely tight control of the diameter of the capillary tube and consequently the inside diameter.

In gas chromatography, reproducible analyte retention from column to column is required which begins with the fused-silica capillary tubing. The inside diameter is critical for consistent gas flow characteristics of the column, and the smallest difference in the inside diameter of a capillary will have a substantial impact on gas flow or more correctly gas velocity through the column and therefore retention times. The gas flow through the column for a given pressure is proportional to the inside diameter to the power of four (flow $\approx$ ID^4), so on a 0.25-mm ID fused-silica column a 10-μm difference in diameter will give a 17% increase in gas velocity, which is very significant.

Despite statements to the contrary, the inside diameter does not have a significant impact on the phase ratio of the column which is an important parameter in determining retention characteristics. Because most WCOT GC columns are coated statically (described later), the phase ratio is directly related to the concentration of the coating solution. For a given coating solution the phase ratio will stay the same for successfully coated columns irrespective of the inside diameter of the column. The phase thickness may be slighter different if the tubing ID changes but the retention characteristics remain the same.

The two-axis laser micrometer on the tower also measures the ovality of the tubing as it is drawn and a further feedback loop on the preform feed system precisely adjusts the positioning of the preform in the furnace to correct misalignment and ovality that develop during the draw. Ovality must be controlled for the same reasons that consistent inside diameter of the capillary tubing is required. Ovality of the tubing creates additional problems in making connections in analytical chemistry applications, particularly with PressFit connectors and metal ferrules, which are being used more frequently with fused-silica material.

3.6. PROTECTIVE COATING

It is well known that glass fractures easily and yet fine fused-silica tubing is extremely flexible and can even be tied in knots. Glass is inherently strong and flexible provided the surface does not have surface defects. Cracks will nucleate and grow from very small defects on the glass surface. Moisture and even dust particles in the air settling on the unprotected fused-silica are enough to damage the surface and make the tubing extremely fragile. As soon as the drawn capillary exits the furnace and has cooled sufficiently, a protective layer must be applied.

The first fused-silica capillary GC columns had a protective silicone coating, which worked until a higher-temperature capability was required. Polydimethylsiloxane, when used as the stationary phase inside the column where oxygen is excluded, can easily be operated in excess of 300 °C, but the same material on the outside of the column in air fails catastrophically at as low as 220 °C. If the exterior cladding was not thermally stable at the temperatures used in GC separations it would severely restrict the applicability of gas chromatography. Polyimide was quickly settled on as an appropriate material. Polyimide is widely used in the electronics industry and forms a strong coating that hermetically seals and protects the fused-silica. The problem with polyimide is that it can be extremely difficult to work with.

The polyimide resin is applied to the fused-silica through a pressurized die through which the capillary tubing is pulled soon after it exits the furnace. A very uniform and well-controlled coating can be applied but the polyimide must be cured quickly by passing it through a thermal curing system. The manufacturer's material on the polyimide resins suggests curing the material by heating over very long periods of time, for optimum performance. Curing the material too fast at too high-temperature will cause bubbles in the polyimide coating, which of course leads to a weak very-poor-quality fused-silica capillary. In the continuous capillary drawing process, minimal time is available to cure the polyimide before it comes in contact with the take-off wheel pulling the fused-silica capillary. If a tower needs to be run at 20 m/min and it is a 9-meter-high tower, for example, there is less than half a minute to get the polyimide cured hard enough before it must pass over the take-off roller.

With the pressurized coating tips that are now used, thinner better-controlled coatings can be applied which can be more rapidly cured. Multiple coatings can be applied as the capillary is taken through a series of rollers, coating applicators, and curing ovens.

Ultimately, the thickness of the cured polyimide coating needs to be around 15 μm to ensure good protection of the fused-silica. In the SGE process, two coatings are applied as shown in the schematic diagram (see Figure 3.2) but some manufacturers apply more than this.

The final online check in the drawing process is a second 2-axis laser micrometer monitoring the capillary tube before it is wound onto the spool. This checks the thickness and consistency

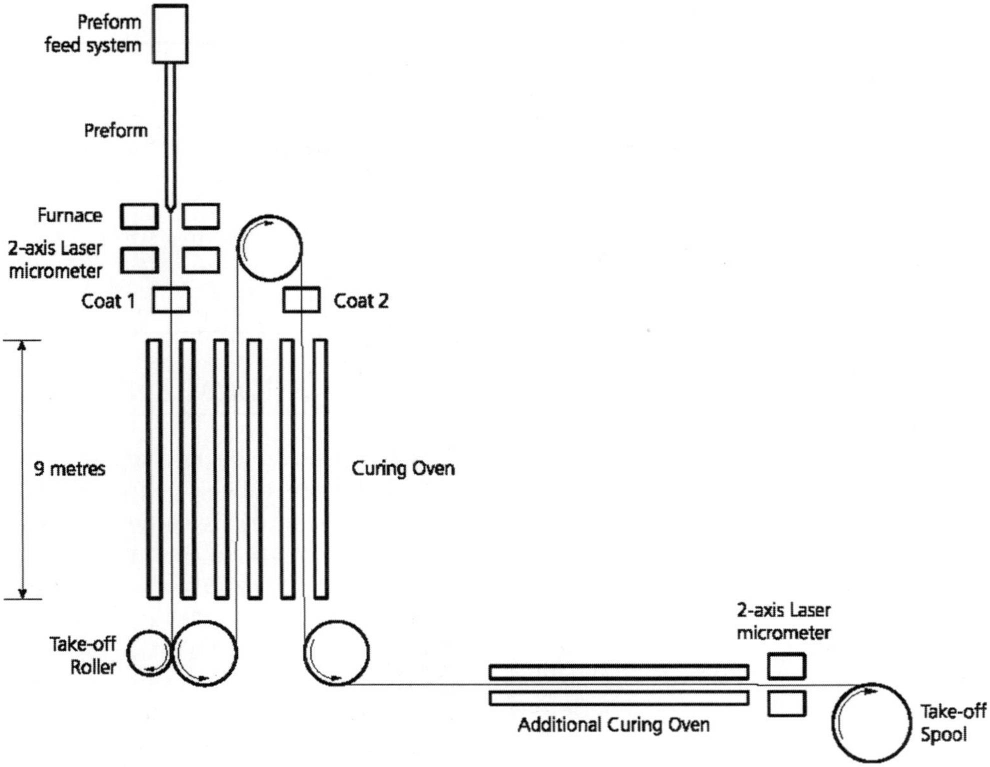

FIGURE 3.2 Fused silica draw tower configuration.

of the polyimide coating. Again, the inline measurement allows feedback into the coating systems during the run (Table 3.1).

3.7. ALTERNATIVE PROTECTIVE COATINGS

One of the limitations in gas chromatography is imposed by the polyimide coating. Polyimide is one of the best polymer materials in terms of its upper temperature limit but conventional polyimide-clad fused-silica will become weak with any prolonged use above 370 °C. Stationary phases for gas chromatography have long been available that will operate at much higher temperatures than this, such as the carborane–siloxane phases that can be operated to 480 °C and silphenylene phases to 400 °C. Several alternatives to deal with these higher required temperatures have been developed.

Aluminum-clad fused-silica will operate to 500 °C for very long periods of time and is produced by quenching molten aluminum onto the fused-silica capillary as it emerges from the furnace. The limitation of this material is that when the capillary is thermally cycled (as happens in gas chromatography), cracks in the aluminum coating develop. The problem comes about because of the mismatch in coefficients of thermal expansion between the fused-silica and the thin aluminum coating. If an aluminum-clad fused-silica column is cooled slowly from the end of a temperature program run (<10 °C/min), it remains strong, but with

TABLE 3.1 Standard SGE Dimensional Specifications for Some of the Different Sizes of Fused Silica

Nominal inside diameter	ID tolerance	Outside diameter	OD tolerance	Ovality	Polyimide coating
0.100 mm	±3 μm	363 μm	±12 μm	≤5 μm	≥15 μm
0.150 mm	±5 μm	363 μm	±15 μm	≤5 μm	≥15 μm
0.250 mm	±5 μm	363 μm	±15 μm	≤5 μm	≥15 μm
0.320 mm	±5 μm	430 μm	±20 μm	≥5 μm	≥15 μm
0.530 mm	±10 μm	680 μm	±25 μm	≤10 μm	≥15 μm

rapid cooling, as used in capillary GC, the thin layer of aluminum cracks causing immediate weakening to the point that it becomes fragile; consequently, the aluminum-clad fused-silica is of limited use in capillary GC.

The most successful solution for extending the temperature limit of fused-silica tubing is a high-temperature polyimide material. It is not disclosed why this material can achieve higher temperatures but it is now used as standard by some manufacturers and does not deteriorate even at temperatures up to 420 °C. The material is 10 times more expensive than traditionally used polyimide resins but the benefits outweigh the costs even for low-temperature applications.

3.8. CLEANROOM ENVIRONMENT

As has already been discussed, the unprotected capillary tube is extremely susceptible to damage from particles and moisture. Some drawing towers are incased in a controlled environment directly around the tower and others are operated in a cleanroom environment (Figure 3.3).

Special attention is given to maintaining low humidity, as moisture, when the capillary is stressed by bending, will cause stress corrosion of the fused-silica and weakening of the tubing. In addition, the conditioned air in the room is filtered to reduce particles that lead to

FIGURE 3.3 Upper level of drawing towers in a controlled environment room.

FIGURE 3.4 Lower level of SGEs 3 drawing tower facility.

point defects in the material and ultimately unexpected breaks in material that otherwise appears strong (Figure 3.4).

3.9. QUALITY MONITORING

Following the drawing process with the tubing on large spools, it must be checked for what are described as "point defects" and general strength. All material is run through a proof tester, which loads the material as it is wound over its entire length. The load should be below the expected failure point. Breakages from this test will lead to the entire batch being rejected. A useful tool to quantitatively monitor the quality of the material being produced is a tensile testing system, which loads samples of the tubing under stress to failure. Again, standards for required failure stress are maintained (Figures 3.5 and 3.6).

Following proof testing, and confirmation of dimensional tolerances from the several-kilometer-long continuous length of fused-silica, the material needs to be wound off into the lengths required for subsequent processing coating, whether it is for producing GC capillary columns, deactivated tubing, or producing into PEEK jacketed material (PEEKsil) for liquid chromatography transfer lines or columns.

It is also normal at this point to verify that the internal surface chemistry of the fused-silica is as specified for the application the material is intended.

3.10. OBSERVATIONS ON HANDLING OF FUSED-SILICA CAPILLARY TUBING

Naturally people are conscious of not damaging the 15-μm-thick polyimide protective coating but consideration must also be given to the inner bore of the tubing. The tubing is inherently straight, but if it is bent, stress is induced in fused-silica which then leads to crack propagation and failure. The greatest point of stress in the tubing is on the outer radius of tubing but there is also stress induced in the bore of the tube, and damage to this inner wall will cause the tubing to be just as weak.

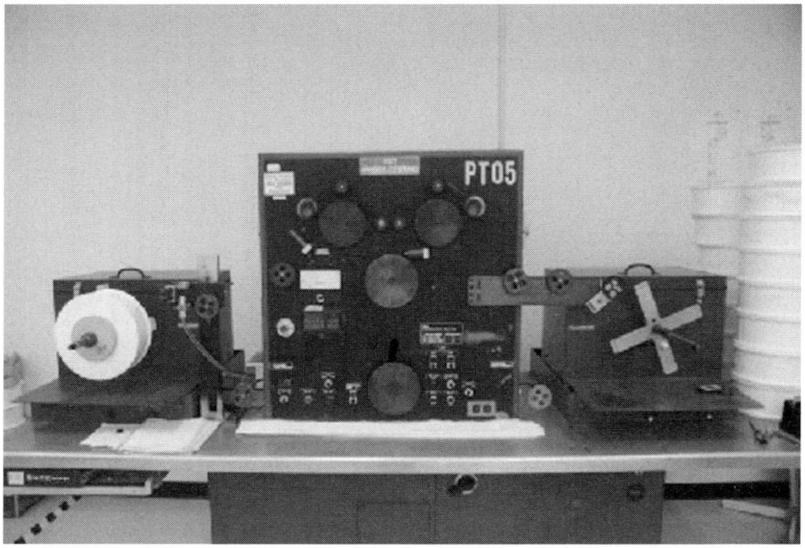

FIGURE 3.5 Continuous proof testing machine.

The types of things that will damage the fused-silica from the bore are as follows:

- Chemically leaching of the bore of fused-silica which should not be necessary if the tubing has been drawn correctly.
- Packing fused-silica with particles will scratch the bore of the material, making the tubing extremely brittle. In LC applications, this is sometimes done but it is essential that the fused-silica is reinforced and cannot be bent. By doing this, it can be possible to work with such columns but the ultimate pressure before the tubing fails is reduced enormously.
- An observation that we do not have an adequate explanation for is when moisture-containing gas is pushed through a fused-silica capillary at a high velocity, even for a few seconds, the material becomes instantly extremely fragile. If gas must be pushed through fused-silica capillary tubing that has not had its inner surface modified in any way, the gas must be moisture free.
- The smaller the radius that fused-silica is coiled, the greater the stress on the material, which is not hard to imagine, but this stress is also highly dependent on the outside diameter of the tubing. For a given radius bend in the tubing, the stress at the surface goes up proportionally to the diameter.

Finally, when it is necessary to cut the fused-silica, never snap it. The stress induced with such a violent action may be transmitted some distance into the tubing, and can cause the material to be very weak for up to a meter from the break. In addition, particles of polyimide and glass will contaminate the bore of the capillary, leading to activity and poor peak-shape issues. The fused-silica should only be cut with a light scratch through the polyimide to the fused-silica surface underneath and gently pulled apart to give a square nonjagged break.

3.11. COLUMN TECHNOLOGY – COATING THE STATIONARY PHASE

3.11.1. Surface Preparations

As discussed in the previous section, surface pretreatment may not always be necessary if control of the chemistry in the bore of the tubing

FIGURE 3.6 Tensile testing of capillaries.

is possible during tubing manufacturing. In general, however, tubing pretreatment is done for several reasons: preparation of the surface for application of the stationary phase, increasing the column surface inertness, dehydration, changing the surface silanol content, or etching to increase roughness or surface area for some specific applications. Depending on what steps are to follow, the fused-silica surface may benefit from further processing or modification.

The only step to make a WCOT GC column that is truly required is the actual coating of the stationary phase. All of the other possible steps are optional, but they can severely impact the overall quality of the column that is produced. It is possible to coat many stationary-phase polymers directly onto untreated fused-silica columns, even if they have a variable surface chemistry. In order to make a high-efficiency, low-bleed column with excellent inertness, it is usually necessary to employ a number of additional steps prior to the actual coating of the stationary phase.

3.11.2. Leaching

Leaching is used to deionize the tubing surface through rinsing with acid solutions. Specifically, leaching solubilizes metal ions and allows them to be removed from the surface of the column bore. Secondary to this removal of ions, leaching also increases the surface silanol concentration, which may be beneficial under certain circumstances as it allows for additional sites which may serve as connection points for deactivation and/or stationary-phase molecules. While leaching could be carried out with caustic solutions, these materials are considerably more aggressive to the glass structure; therefore, leaching is usually carried out using acid solutions and most commonly hydrochloric acid. The concentration of this solution is also variable, but typically 1–5% HCl solutions are used.

Contact time also plays a critical role in the leaching process, and leaching can be done using either static or dynamic techniques, but static is usually preferred. In this process, a plug of the leaching solution is moved through the column using an inert gas. This solution then "wets" the column surface. Once the plug of solution exits the other end of the column, both ends are sealed. Flame sealing is generally considered to provide the highest inertness, but other sealing devices can be used. The column is then heated (typically 200–250 °C) and held at a final temperature for a period of time (typically 1–10 h). It is important to maintain constant temperature over the entire column.

Following leaching, the column will have high water content on the surface, in addition to the HCl. This must be removed prior to deactivation.

3.11.3. Rinsing and Dehydration

Rinsing is necessary to remove the acid solution used in the leaching process. Deionized water may seem an obvious candidate for this procedure, but prolonged exposure to deionized water can cause a column to become very active. For this reason, rinsing should occur with the minimum amount of time and solution necessary to achieve removal of the leached material. Additionally, the same solution used for leaching is typically used due to the activity that arises with the use of deionized water alone. Procedurally, the column ends are reopened, and the column is rinsed with the leaching solution. The rate of rinsing plays an important role, and slower rates often produce a better chromatographic surface. Typically, rinsing rates of 1 cm/sec are reported [5]. Once a plug has been rinsed through the column (typically the plug should be about 25% of the total column volume), the column must then be dehydrated immediately. Dehydration occurs by following the plug with an inert gas, and then heating the column to between 200 and 250 °C and allowing the gas to continue to purge the column bore for approximately 1h.

3.11.4. Surface Deactivations

The deactivation of the column surface primarily serves two main purposes: First, the surface energy can be more closely matched to the stationary-phase molecules that will ultimately be coated. Second, the inertness of the column surface is increased by chemical modification of the acidic silanol sites that are present on either the untreated or acid-leached surface of the tubing. While the stationary phase can be directly coated onto the column surface in many cases, it is generally assumed that in order to produce a column of high inertness and durability, the column surface must be deactivated. This is true even for stationary phases that would normally wet the untreated or undeactivated tubing surface. As discussed in the first part of this chapter, it may be possible to minimize many of these column handling steps if direct control of the tubing chemistry is available. Since most users of GC do not have their own source of tubing, these steps are likely necessary. Additionally, only a few commercial column manufacturers (most notably SGE and Chrompack) actually manufacture their own tubing; so these steps are almost universal.

Deactivation can generally be divided into two classes: pinpoint and polymeric. While the end goal of each of these is the same, the method of reaching that goal is different. The idea behind a pinpoint deactivation is to react each silanol group with a chemical functionality which derivatizes the acid moiety to something more neutral. In order to accomplish this, relatively small reactant molecules should be used to minimize steric hindrance of nearby silanols once a specific point has been reacted. Depending on the surface concentration and position of the silanols, this may be particularly difficult, especially if using highly functionalized reactants that contain larger moieties (e.g., triphenylchlorosilane). In a polymeric deactivation, much larger reactant molecules are used. These molecules are generally designed to react with one silanol and then cover, or mask, adjacent silanols. Mostly based on the research of Lee [6,7] and utilizing hydrosilane monomers or oligomers, polymeric deactivations can basically be described as a very thin layer of stationary-phase-like material that can be linked to the column surface through reaction with a surface silanol. Both techniques have benefits, but the end result can be the same if the chemistry is performed properly.

Pinpoint deactivation is typically carried out using either chlorosilanes or disilazanes. Each reacts with surface silanols, producing either HCl or NH_3 as the leaving group following bonding with the surface. This produces a silyl-ether connection from the surface of the tubing to the silicon of the deactivation reagent. Numerous reagents are commercially available from several sources at purities that typically do not require further processing. In some cases, these materials may be directly synthesized, but due to the wide variety available (Gelest, Silar, etc.), this is rarely necessary. The functionalities used are "tuned" to match the chemistry of the stationary phase that will ultimately be coated. For some manufacturers, a separate deactivation is used for every stationary phase offered, while for others very few different deactivations are employed. It is arguable if a different deactivation is truly necessary for each possible stationary phase, given that most commercially available GC columns are of acceptable quality for most separations. For the most demanding applications, however, there may be an advantage in terms of inertness, bleed level, and lifetime. Finally, the deactivation reagent may also include functionalities that allow for direct chemical bond to the stationary phase. Most commonly this is done using vinyl groups in the deactivation layer and in the stationary phase to allow for bonding following application of the stationary phase.

Pinpoint deactivation is performed by treating the dehydrated tubing with a solution of the deactivation reagent. Since the possible reagents are too numerous to mention here, we will consider the use of a mixture of trimethylchlorosilane and dimethylvinylchlorosilane as a reasonable material. The ratios of these reagents will be adjusted to yield an overall vinyl percentage of typically 1–10%. The vinyl groups are then available for bonding to the stationary-phase polymer, which also likely uses vinyl functionalities for this same purpose. The deactivation solution is made at a concentration ranging typically from 1% to 10% in a suitable solvent (dichloromethane). A plug of this material is then pushed through the column using an inert gas, much like the leaching process described above. When the plug exits the other end of the column, both ends are sealed again, and the column is heated to the temperature necessary to catalyze the reaction with the surface silanols. This temperature is usually in the range of 200 °C to as high as 400 °C, depending on the chemistry of the specific materials. Once the deactivation has occurred (time of heating varies between 2 and 10 h), the column is solvent-rinsed and then is ready for application of the stationary phase. If the column is to be used as a deactivated guard tubing, or for a retention gap, it is now ready for use, following a thermal cycle that runs to maximum operating temperature in an inert gas inner atmosphere.

3.12. STATIONARY PHASES

The coating of the stationary phase is again a similar procedure, regardless of the stationary phase being coated in most cases. In general, a stationary-phase polymer is either purchased from a supplier (Ohio Valley, GE, etc.) or directly synthesized as is typically done by most commercial suppliers of GC columns. A specified amount of the polymer, based on the desired stationary-phase film thickness, is dissolved in a suitable low-boiling solvent (e.g., pentane), forming a "coating solution." The coating solution also likely contains other reagents used for cross-linking catalysis, further column deactivation, and immobilization. The types of compounds used for this are both widely varied and proprietary, but several have been discussed in the open literature such as peroxides and azo compounds [8].

Selection of the best stationary phase is likely the most important parameter in developing a GC separation. The selectivity of the analytes of interest and the stationary phase has a larger impact on the resulting compound resolution than any other parameter. For this reason, many stationary phases are commercially available, yet practitioners usually restrict themselves to very few actual materials. Most separations are performed using either a polydimethylsiloxane (PDMS, or a "-1" column) or a 5%polydiphenyl/95%polydimethylsiloxane (a "-5" column). In terms of selectivity, these stationary-phase moieties are very similar; so it can be concluded that most GC separations are run on essentially the same stationary-phase chemistry − an unfortunate situation. Users of GC often rely on the separation efficiency, or theoretical plates, to obtain resolution of their intended analytes. While this technique does have a very high efficiency, stationary-phase selectivity should still be considered first when choosing a capillary GC column. Modern WCOT GC stationary phases generally fall into one of four categories: Nonpolar polysiloxanes, functionalized polysiloxanes, polar nonsiloxanes and other hybrid materials.

Nonpolar polysiloxanes comprise the largest group of materials, including the 1- and 5-type materials already mentioned. Additional phases employing arylene (or silphenylene) functionalities in the siloxane backbone have also received considerable attention due to their decreased levels of bleed, and increased resistance to damage from oxygen. While substitution of arylene functionalities in place of the oxygen in the polymer backbone does change the stationary-phase selectivity relative to the pendant-substituted materials, the overall polarity is often indistinguishable. Most manufacturers offer both chemistries, and often denote the arylene polymers as "MS" columns to indicate that they are preferred with mass spectrometric detectors which may benefit from reduced bleed levels. Many of these columns use a poly(tetramethyl-1,4-silphenylenepolysiloxane), but many other moieties have been investigated, and many are commercially available under a variety of names [9]. The use of these materials generally improves the performance of the GC column, and, as a result, most new stationary phases that have been commercialized employ this type of chemistry, often in addition to the more typical pendent substitution of the silicon in the polysiloxanes.

Functionalized polysiloxanes, employing both arylene functionalities in the polymer backbone as well as pendent moieties, account for the greatest number of commercially available stationary phases, even if they are not the most used. Substitution of the dimethyl groups in PDMS with high concentrations of diphenyls (20−65%), trifluoropropylmethyl, cyanopropylphenyl, bis-cyanopropyl, and many other moieties make up a large range of possibilities for the analytical chemist to choose from. These materials allow for additional modes of interaction of the stationary phase with the analytes of interest, other than what is offered by PDMS, or -5 phases alone. For functionalized analytes, these materials generally offer improved selectivity, but do so sometimes at the cost of efficiency. Specifically, columns using high cyanopropylphenyl, or bis-cyanopropyl functionalities, often have about 2/3 of the efficiency of a PDMS column with similar dimensions. This is likely due to the polymer structure, and a corresponding reduction in diffusivity as compared to PDMS polymers. Even with this decrease in efficiency, the increase in selectivity makes this a more than beneficial tradeoff, especially in the case of demanding separations such as FAME isomers, dioxins, or PCB congeners.

Polar, nonsiloxanes are a category that includes the waxes. Based on polyethylene glycol (PEG), and similar materials, these materials offer the highest polarity of the common stationary phases. For analytes that can hydrogen-bond to the stationary phase, these materials offer the highest selectivity.

The difficulty with the materials in this category is their thermal limitations. In general, temperatures between 250 and 280 °C represent the maximum isothermal temperatures for these materials. Additionally, these polymers often have increased levels of bleed and decreased efficiency as compared to the siloxanes and are usually not bonded. As a result, they are now viewed as more specialty-purpose materials, whereas a decade or two ago, they were considered general-purpose phases. Often, unless the PEG materials are absolutely necessary, polar-substituted siloxanes may allow for adequate selectivity without the drawbacks of these materials. For example, polytrifluoropropylsiloxane (DB-210, Rtx-200, etc.) are very good choices for the separation of phenolics, and their increased thermal properties and improved efficiency make them a rather good choice as compared to the PEG materials for many applications.

There are also a number of materials that do not easily fit into any of the above categories. These hybrid materials are as diverse as wax/polysiloxanes blends, ionic liquids, liquid crystals, and nonhydrocarbon arylene materials, most notably the polycarboranes. Each of these materials has value to the field of WCOT GC as they allow for selectivity not achieved with the PDMS-like materials. For example, the ionic liquids have received considerable interest for polar compound separations, and as second-dimension GC × GC materials. To date, they have not achieved the efficiency or the thermal stability of many of the polysiloxanes, but they do have very different solubility properties. The polycarboranes have seen application for many high-temperature separations such as high-temperature simulated distillation. These materials can be operated well in excess of 400 °C, but they may exhibit a Lewis acid–base reaction with compounds such as carboxylic acids, so they remain a specialty material. Polycarboranes have also seen use for PCB congener analysis where they exhibit a very different elution order compared to many of the functionalized polysiloxanes, and thus are excellent as primary or confirmatory analysis columns for these separations.

3.13. COATING TECHNIQUES

Modern commercial columns are generally coated using static coating techniques. Static coating implies that the column will be fully filled with a coating solution, and then the solvent will be slowly removed using evacuation under controlled temperature leaving the stationary-phase polymer behind as a thin layer. The benefit of static coating is that it produces the most uniform stationary phase, but the stationary phase must wet the deactivated surface, and stay intact until the solvent has been removed and the column is thermally processed. If this becomes too difficult of a challenge, then the older technique of dynamic coating can be employed.

Dynamic coating requires less time than static coating, and is often useful for scouting new column chemistries and the coating process may only take a few minutes. In dynamic coating, a viscous coating solution (5–25% wt/vol) of the stationary phase is dissolved into a suitable solvent (pentane, dichloromethane, etc.) and an immobilization or a cross-linking reagent is added to this solution. The immobilization reagent, as previously mentioned, is generally in the 1–5% wt/vol concentration range and is used to initiate cross-linking of the polymer molecules and also bonding of the stationary phase to the deactivation layer. The coating solution is then pushed through the tubing using relatively high pressure (due to the viscosity) at a controlled rate. As the plug nears the other end of the tubing, care must be taken to keep the same flow rate of the coating solution, so buffer columns or other devices are used to be able to control this rate. As the concentrated plug moves through the column, a layer of the stationary phase is deposited on the tubing wall. One of the limitations of dynamic coating is that the concentration of the stationary phase in the plug decreases as the plug moves through the column. This has the net effect of producing a gradient of stationary-phase thickness in the column. The stationary phase will be thicker at the end of the column where the plug was introduced and the exit end of the column will have a thinner stationary phase. Ultimately this produces a column that will exhibit different capacity factors for the analytes, depending on the direction of the installation into the GC. Also, if only portions of the column are used, as in GC × GC, each portion will have slightly different capacity factors as well. Finally intended film thickness is not usually determinable without empirical knowledge of the process. While there are formulas for estimation of the stationary-phase thickness that would be produced given the flow rate of the plug and the viscosity of the solution, they are only approximations. If a column is to be coated using dynamic techniques, then the film thickness may only be determined through actual isothermal testing of the final product. Following the coating, the stationary phase is bonded and cross-linked in a similar manner to a column coated by static techniques.

In static coating, a dilute coating solution (ca. 0.1–1.0% wt/vol) is made using a suitable solvent, and immobilization reagent is added. The column is then completely filled with this solution. The concentration of the coating solution is determined by the intended film thickness as based on the following equation [5]:

$$\text{film thickness (um)} = 2.5 \times \text{Column i.d. (mm)} \times \%\ \text{concentration coating solution}$$

As previously mentioned, this coating solution will produce stationary-phase films on dissimilar i.d. columns so that the phase ratio is constant.

Once the coating solution has filled the entire column, one end of the column is sealed and vacuum is applied to the other end following immersion into a constant temperature environment. While the temperature is not necessarily high, it is usually a few degrees above ambient so that an even evaporation rate is maintained as the solvent vaporizes and is pulled away from the bore of the tubing. This process can be quite slow, and represents the major drawback of static coating. Some longer columns with narrow i.d. (ca. 105-M × 0.25 mm i.d.) may take a week or more to be properly evacuated.

There is a real art in sealing the one end of the column before applying the vacuum. It is important to not have a void at this location prior to sealing. Since the solvents used are flammable, flame sealing is not possible; so manufacturers use a variety of materials to make this seal. This is also considered proprietary, and is often the cause for coating failures if the seal is improperly made.

Once the solvent has been completely removed, there will be a thin layer of stationary phase left of the inside surface of the column. The column can now be thermally processed to initiate the cross-linking and bonding.

Depending on the immobilization reagents used, the thermal curing of the column may be done at a variety of temperatures. In general, peroxide catalysts require 250 °C to initiate the cross-linking and bonding reaction, though many different materials are used. The coated column is connected to an inert gas stream, and then heated to this initiation temperature where it is then held for a period of time (1–2 h) to allow the reaction to occur. Depending on the product, the column may be solvent-rinsed a final time to remove unbonded stationary phase. The "rinse-out" during this step should be small (less than 10%) or the column will likely be of questionable quality. Columns with high rinse-out are usually short lived in the analytical laboratory, and repeated thermal cycling will cause the stationary phase to migrate toward the column exit slowly over time. This causes a continual change in capacity factor, and possibly selectivity, depending on the analytes. Rinse-out is often a quality measurement that commercial manufacturers will determine prior to final commercialization of a stationary phase.

3.14. COLUMN TECHNOLOGY – QUALITY EVALUATION

The final step in the manufacturing of capillary GC columns is to ensure the quality of the completed device. These same tests can also be used in the analytical laboratory that ultimately uses the columns to verify initial performance and as continuing verification of column performance as the column ages in normal use. For many years, the most common column performance test mixture was referred to as the "Grob mix" [10]. As developed in 1978, the Grob mix uses the target compounds listed in Table 3.2.

TABLE 3.2 The Grob Test Mixture Compounds

*n*C10-FAME
*n*C11-FAME
*n*C12-FAME
2,3-Butanediol
Dicyclohexylamine
2,6-Dimethylaniline
2,6-Dimethylphenol
2-Ethylhexanoic acid
Nonanal
1-Octanol
Undecane (C11)
Decane (C10)

These compounds are typically injected at higher concentration in split mode, so that injection port inertness is not as critical. For many years, the analysis of this mixture was considered to give a thorough evaluation of column reactivity, probing both acid- and base-reactive sites, as well as allowing for characterization of retention behavior. Due to the range of volatility, temperature-programmed analysis is necessary with this test mixture. When using temperature programming, the solute–stationary phase and solute–column surface interactions are reduced, relative to a lower-temperature isothermal analysis. This also serves to reduce the ability of this test to rigorously test the column inertness beyond a certain point. When it was first developed, column inertness was not nearly as high as it is today, and it was a more appropriate probe, even though it was a temperature-programmed analysis. Despite its limitations, the Grob test is still a viable way to determine column degradation in an analytical laboratory.

More recently, as column technology has continued to advance, a test mixture proposed by Luong, Graz, and Jennings [11] has received attention. Utilizing several more reactive probes, and lower on-column concentrations this approach also uses either a late-eluting solvent or no solvent at all. The theory is that the solvent causes a "masking" of reactive sites, and that by having no solvent in front of the analytes as they move through the column it is possible to get a more accurate test of a column's reactivity. Table 3.3 lists the compounds proposed in this work, but several column-manufacturing companies have adopted variations on this mixture, and the industry seems to be switching to a more rigorous test of column reactivity.

This test mixture has the added advantage of also being able to be analyzed under isothermal conditions due to the more limited range of volatility. This allows, as previously stated, a better interaction of the solute with both the stationary phase and the column surface. It is the combination of late-eluting (or no) solvent,

TABLE 3.3 Test Mix Proposed by Luong, Graz, and Jennings

Propionic acid
Octane
Nitrobutane
4-Picoline
Trimethyl phosphate
1,2-Pentanediol
Propylbenzene
1-Heptanol
3-Octnone
Decane

analyte reactivity, and isothermal analysis that allows this mixture, or similar mixtures, to be better probes of today's higher-inertness columns. These mixtures could also be easily adapted into working analytical laboratories as continuing system control standards so that column degradation through routine use can be measured. This would allow for determination of when system maintenance was necessary, or when the column had to be replaced.

3.15. COLUMN TECHNOLOGY – SUMMARY

Modern WCOT GC columns are highly efficient, robust, and have generally high inertness. Advances in tubing manufacturing, column preparation, deactivation, and stationary-phase chemistry have all led to continued improvements since these formats were first considered. These columns represent what is generally the best separation technique for the compounds which are amenable to this technique, although there are still opportunities for further advancement. Stationary-phase chemistries still need to be developed to allow for increased resolution of many compounds, as efficiency does not solve all of the analytical challenges faced by chromatographers. Even though it is possible to obtain a column with several hundred thousand theoretical plates, selectivity is still the main tool when separating compounds of similar functionality and vapor pressure. Additionally, though not specifically discussed in this chapter, low thermal mass columns and new column formats are being explored which promise to extend the range of analyses that capillary GC is capable of being used for, and increase the ease of use for the practitioner.

References

[1] M.J.E. Golay, in: V.J. Coates, H.J. Noebels, I.S. Fagerson (Eds.), Gas chromatography, Academic Press, New York, NY, 1958.

[2] D.H. Desty, J.N. Haresnape, B.H.F. Whyman, Anal Chem 32 (1960) 302.

[3] R. Dandeneau, E.H. Zerenner, J High Resolut Chromatogr 2 (1979) 2.

[4] S.R. Lipsky, W.J. McMurray, M. Hernandez, J.E. Purcell, K.A. Billeb, J Chromatogr Sci 18 (1980) 1.

[5] K. Grob, Making and manipulating capillary columns for gas chromatography, Huethig, Basel, Heidelberg, New York, 1986.

[6] K.E. Markides, B.J. Tarbet, C.M. Schregenberger, J.S. Bradshaw, M.L. Lee, J. High Resolut Chromatogr 8 (1985) 741.

[7] C.L. Woolley, R.C. Kong, B.E. Richter, M.L. Lee, J. High Resolut Chromatogr 7 (1984) 329.

[8] B.E. Richter, J.C. Kuei, N.J. Park, S.J. Crowley, J.S. Bradshaw, M.L. Lee, J. High Resolut Chromatogr 6 (1983) 371.

[9] P.R. Dvornic, R.W. Lenz, High temperature siloxane elastomers, Huethig & Wepf, Basel, Heidelberg, New York, 1990.

[10] K. Grob Jr., G. Grob, K. Grob, J Chromatogr 156 (1978) 1–20.

[11] J. Luong, R. Graz, W. Jennings, J Sep Sci 30 (2007) 2480–2492.

CHAPTER

4

Packed Columns for Gas–Liquid and Gas–Solid Chromatography

Colin F. Poole

OUTLINE

4.1. INTRODUCTION

Since the early 1980s fused-silica open-tubular columns have dominated the practice of analytical separations by gas chromatography. This was a rather abrupt change in general practice since packed columns were used almost universally up until that time. Fused-silica capillary columns are both flexible and relatively durable unlike other types of glass capillary columns. Open-tubular (or capillary) columns afford a higher peak capacity, higher permeability, and are more inert than packed columns. They are better suited to the requirements for fast separations of both simple and complex mixture and synergistically prompted the development and optimization of the instrumentation for gas chromatography found in most laboratories today. Modern instruments are designed for users of open-tubular columns and require the purchase of specific options for dual use with packed columns.

Packed columns are preferred for applications that are difficult to implement using wall-coated open-tubular (WCOT) columns. For example, only a limited number of liquids can be coated as stable films on the smooth walls of fused-silica capillary columns, while virtually any liquid of

Gas Chromatography DOI: 10.1016/B978-0-12-385540-4.00004-3

low volatility can be handled using packed columns. Consequently, when the purpose is to determine specific physicochemical properties of a liquid employed as a stationary phase, for example, by inverse gas chromatography (Chapter 20), packed columns are generally required. Packed columns can hold a significantly larger volume of stationary phase and are better suited to the needs of process gas chromatography at the pilot plant scale and laboratory preparative gas chromatography when isolation of all but a few milligrams of product is desired. Packed columns are also more forgiving of matrix burden and better suited to handling samples containing a significant amount of involatile or thermally unstable matrix components. These materials and their breakdown products usually accumulate on the packing at the head of the column, which is easily exchanged periodically for fresh packing material. Since a large part of the fundamental understanding of gas chromatography was established using packed columns, it remains necessary for scientists entering this field to be aware of the characteristic properties of packed columns to maintain contact with these early studies.

Some characteristic properties of packed columns are summarized in Table 4.1 [1]. Classical packed columns with an internal diameter >2 mm and packed with particles in the 100–250-μm range are the most widely used and are easily prepared in the laboratory. Micropacked columns have diameters <1 mm and are packed with particles of a similar size to classical packed columns [2,3]. These should be distinguished from packed capillary columns, which typically have an internal diameter <0.6 mm and are packed with particles in the 5–20-μm range [4–6]. Columns with narrow internal diameters facilitate fast temperature programming and particles of small diameters enhance performance. On the other hand, the low permeability of packed capillary columns limits column length and requires instruments modified for high-pressure operation. Micropacked and packed capillary columns are more retentive than WCOT columns and find some applications for the fast separation of gases and similar compounds at above ambient temperatures, but are otherwise not widely used. Support-coated open-tubular (SCOT) columns are capillary columns containing a liquid phase coated on a surface covered with a layer of porous solid material (support) leaving an open passageway through the center of the column. They are a hybrid of packed and open-tubular columns with properties that are a compromise between those of the two column types. As open-tubular column technology evolved, SCOT columns were left with few real advantages and are no longer in general use.

The most significant difference among the various columns in Table 4.1 is their permeability. It is this feature that allows long open-tubular columns to be used for the separation of complex mixtures and restricts the peak

TABLE 4.1 Representative Properties of Different Column Types for Gas Chromatography

Column type	Phase ratio	H_{min} (mm)	u_{opt} (cm/s)	Permeability (10^7.cm^2)
Classical Packed	4–200	0.5–2	5–15	1–50
Micropacked	50–200	0.02–1	5–10	1–100
Packed Capillary	10–300	0.05–2	5–25	5–50
SCOT	20–300	0.5–1	10–100	200–1000
WCOT	15–500	0.03–0.8	10–100	300–20000

H_{min} = minimum plate height at the optimum mobile phase velocity u_{opt}

capacity of packed columns. The intrinsic efficiency (minimum plate height, H_{min}) for packed and open-tubular columns is not that different but the separation performance of WCOT columns is higher because of their greater permeability and a characteristic shift in the optimum mobile phase velocity (u_{opt}) to higher values. Packed columns typically have lower phase ratios (volume of gas phase to liquid phase) than WCOT columns leading to longer separation times at a constant temperature.

4.2. GAS–LIQUID CHROMATOGRAPHY

Classical packed columns for gas-liquid chromatography are prepared by coating a support with the desired liquid phase and transferring the coated support (packing material) to an empty column of appropriate dimensions for the separation. Columns are generally made of stainless steel or passivated nickel tubing [7], glass-lined or silicon-coated stainless steel [8], glass, or Teflon and coiled to fit the column oven and connect with the injector and detector inlets. Metal columns are preferred for applications where mechanical stability and high temperatures are important, but, unless internally coated with a film of glass or silicon, are unsuitable for thermally labile compounds. Hot metal surfaces tend to be catalytically active and degrade labile compounds. Glass columns, with silanized surfaces, are the most inert but are more fragile than metal columns. Teflon columns are limited to low temperatures and are used for the separation of highly reactive chemicals that react with or are degraded by glass and metal surfaces.

4.2.1. Frequently Used Stationary Phases

Since the beginning of gas chromatography thousands of substances have been used as stationary phases [9–15]. Many of these were technical products with a variable or undefined composition, or had similar separation properties to other stationary phases but an inferior temperature operating range, and have long since been abandoned. The development of stationary phase classification schemes, such as the Rorschneider–McReynolds' system of phase constants, played an important role in rationalizing stationary-phase selection during the early development of gas chromatography by identifying redundant stationary phases with (near) identical separation properties. As a result, the number of stationary phases actively used has declined considerably and these can be grouped under a few general headings, namely: (1) hydrocarbons and perfluorocarbons; (2) poly(siloxanes); (3) ethers and poly(esters); (4) ionic liquids; (5) liquid crystals; and (6) chiral stationary phases (see Chapter 22).

It is desirable that the stationary phase has a wide liquid temperature operating range. The lower operating temperature is usually close to the melting point of the stationary phase or glass transition temperature for a polymer. The maximum allowable operating temperature is usually determined by the thermal stability of the stationary phase or its vapor pressure. These considerations tend to favor the use of polymeric materials as stationary phases. Practical considerations dictate that the stationary phase should be uncreative, adequately wet common supports, and have reasonably solubility in some common volatile organic solvent.

High-molecular-weight hydrocarbons such as hexadecane, squalane (2,6,10,15,19,23-hexamethyltetracosane), Apolane-87 (24,24-diethyl-19,29-dioctadecylheptatetracontane), and Apiezon greases have long been used as low-selectivity stationary phases (Table 4.2). All hydrocarbon phases are susceptible to oxidation and should be used with carrier gases having a low oxygen content. Apolane-87 is the most resistant to oxidation of the common hydrocarbon phases because of its low concentration

TABLE 4.2 Structure and Properties of Some Frequently Used Stationary Phases for Packed Column Gas Chromatography

Trade name	Structure and properties
Squalane	2,6,10,15,19,23-hexamethyltetracosane. Obtained by the complete hydrogenation of squalene isolated from shark liver oil. Common impurities squalene and batyl alcohol.
Apiezon	A series of hydrocarbon greases prepared by the high-temperature treatment and molecular distillation of lubricating oils. Ill-defined composition containing unsaturated hydrocarbons and carbonyl- and carboxylic acid-containing impurities. Colored samples should be purified by chromatography over charcoal and alumina before use. Apiezon MH prepared by hydrogenation of purified Apiezon greases.
Apolane-87	24,24-diethyl-19,29-dioctadecylheptatetracontane $(C_{18}H_{37})_2CH(CH_2)_4C(C_2H_5)_2(CH_2)_4CH(C_{18}H_{37})_2$, molecular weight 1222, melting point ≈ 28–34 °C
Fomblin YR	Poly(perfluoroalkyl ether) $-[(OCFCF_3CF_2)_n(OCF_2)_m]$-, average molecular weight 6–7 × 10^3
OV-1	Poly(dimethylsiloxane), average molecular weight >10^6, melting point ≈ 100 °C
OV-101	Poly(dimethylsiloxane), average molecular weight ≈ 3 × 10^4, viscosity 1500 cP
OV-3	Poly(dimethylmethylphenylsiloxane) containing 10 mol % methylphenylsiloxane
OV-7	Poly(dimethylmethylphenylsiloxane) containing nominally 20 mol % methylphenylsiloxane, average molecular weight ≈ 1 × 10^4, viscosity 500 cP
OV-11	Poly(dimethylmethylphenylsiloxane) containing nominally 35% methylphenylsiloxane monomer
OV-17	Poly(methylphenylsiloxane), average molecular weight ≈ 4 × 10^4, viscosity 1300 cP
OV-22	Poly(methylphenyldiphenylsiloxane) containing 65 mol % phenyl
OV-25	Poly(methylphenyldiphenylsiloxane) containing 75 mol % phenyl, average molecular weight 1 × 10^4, viscosity <1 × 10^5
OV-105	Poly(cyanopropylmethyldimethylsiloxane) containing 10 mol % cyanopropylmethylsiloxane
OV-225	Poly(cyanopropylmethylphenylmethylsiloxane) containing 50 mol % cyanopropylmethylsiloxane, average molecular weight 8 × 10^3, viscosity 9 × 10^3 cP
OV-275	Poly(dicyanoalkylsiloxane) containing 70 mol % dicyanopropylsiloxane and 30 mol % Dicyanoalkylsiloxane, average molecular weight 5 × 10^3, viscosity 2 × 10^4 cP
QF-1	Poly(trifluoropropylmethylsiloxane), average molecular weight ≈ 2 × 10^5, viscosity ≈ 1 × 10^4
PPE-5	Poly(phenyl ether) $C_6H_5O(C_6H_5O)_3C_6H_5$, meta linked,
DOP	Dioctyl phthalate $C_6H_4(COOC_8H_{17})_2$, ortho substitution
EGS	Poly(ethylene glycol succinate) $HO(CH_2)_2[OOC(CH_2)_2COO(CH_2)_2]_nOH$
EGAD	Poly(ethylene glycol adipate)
DEGS	Poly(diethylene glycol succinate) $HO(CH_2)_2O(CH_2)_2[OOC(CH_2)_2COO(CH_2)_2\ O(CH_2)_2]_nOH$
CW20M	Carbowax 20 M, a poly(ethylene glycol) with an average molecular weight 14,000, $HO(CH_2CH_2O)_nCH_2CH_2OH$, melting point about 60 °C
FFAP	The product obtained by condensing Carbowax 20 M with 2-nitroterephthalic acid

(Continued)

TABLE 4.2 Structure and Properties of Some Frequently Used Stationary Phases for Packed Column Gas Chromatography (*cont'd*)

Trade name	Structure and properties
TCEP	1,2,3-Tris(2-cyanoethoxy)propane $(CH_2OCH_2CH_2CN)_3$
QBA FOS	Tetra-n-butylammonium perfluorooctanesulfonate
QBA MPS	Tetra-n-butylammonium 4-morpholinepropanesulfonate
QBA OS	Tetra-n-butylammonium octanesulfonate
QBA BES	Tetra-n-butylammonium 2-[bis(2-hydroxyethyl)amino]ethanesulfonate
QBA TS	Tetra-n-butylammonium 4-toluenesulfonate
QBA FMS	Tetra-n-butylammonium trifluoromethansulfonate
QBA PIC	Tetra-n-butylammonium picrate
QBP TS	Tetra-n-butylphosphonium 4-toluenesulfonate
QBP Cl	Tetra-n-butylphosphonium chloride
QEA TS	Tetra-n-ethylammonium 4-toluenesulfonate
DEA TS	Di-n-ethylammoniun 4-toluenesulfonate

of tertiary hydrogen atoms. Both squalane and Apiezon may contain polar impurities that can be removed by column chromatography prior to use. The hydrocarbon stationary phases are primarily used for the separation of hydrocarbons and as low-selectivity reference phases in column classification schemes [9,11,16]. They have poor support deactivating properties and are poor solvents for polar compounds.

Highly fluorinated liquids are used for the separation of reactive compounds and for the separation of other highly fluorinated compounds, such as Freons [9,17]. Volatile metal halides, interhalogen compounds, and the hydrogen compounds of halides, sulfur, and phosphorus tend to destroy conventional phases. The low cohesive energy of highly fluorinated stationary phases results in lower retention compared with the analogous hydrocarbon phases facilitating the separation of thermally labile compounds at lower temperatures. Most highly fluorinated liquids have poor support wetting characteristics leading to low column efficiency and incomplete support coverage at low phase loadings. The most useful stationary phases contain either ether or ester "anchor" functional groups, such as the poly(perfluoroalkyl ether) oil Fomblin YR with a molecular weight between 6,000 and 7,000, to assist in film formation and stabilization. They are generally considered special-purpose stationary phases and little used for general applications.

The poly(siloxanes) used in packed-column gas chromatography are generally linear polymers synthesized from monomers containing methyl, vinyl, phenyl, 3,3,3-trifluoropropyl, or 3-cyanoalkyl substituents. By varying the identity and amount of the substituent groups, polymers with a range of solvent properties can be prepared (Table 4.3) [9–15,18]. The poly(dimethylsiloxanes) are low-selectivity stationary phases, similar to the hydrocarbon phases but with wider liquid temperature ranges. Replacing methyl with phenyl groups increases the dipolarity/polarizability and hydrogen-bond basicity of the poly(siloxanes) with little change

TABLE 4.3 Characteristic Properties of Packed Column Stationary Phases

Stationary phase	Temperature range		System constants (121 °C)				
	Minimum	Maximum	*e*	*s*	*a*	*b*	*l*
Squalane	20	120	0.13	0.01	0	0	0.58
Apolane-87	35	260	0.17	0	0	0	0.55
Fomblin YR	30	250					
OV-1	100	350					
OV-101	<20	350	0.02	0.19	0.13	0	0.50
OV-3	<20	350	0.03	0.33	0.15	0	0.50
OV-7	<20	350	0.06	0.43	0.17	0	0.51
OV-11	<20	350	0.10	0.54	0.17	0	0.52
OV-17	<20	350	0.07	0.65	0.26	0	0.52
OV-22	<20	300	0.20	0.66	0.19	0	0.48
OV-25	<20	300	0.28	0.64	0.18	0	0.47
OV-105	<20	275	0	0.36	0.41	0	0.50
OV-225	<20	250	0	1.23	1.07	0	0.47
OV-275	25	250	0.21	2.08	1.99	0	0.29
QF-1	<20	250	−0.45	1.16	0.19	0	0.42
PPE-5	0	200	0.23	0.83	0.34	0	0.53
DOP	20	130	0	0.80	1.00	0	0.57
EGS	90	220					
EGAD	100	225	0.13	1.39	1.82	0.21	0.43
DEGS	20	200	0.34	1.53	1.75	0.17	0.37
CW20M	60	225	0.32	1.26	1.88	0	0.45
FFAP	50	250					
TCEP	20	170	0.20	1.82	1.79	0.24	0.33
QBA FOS	<20	220	0	1.09	1.62	0	0.40
QBA MPS	<20	180	0	1.75	3.54	0	0.55
QBA OS	<20	180	0.27	1.14	3.53	0	0.50
QBA BES	<20	170	0.25	1.76	3.67	0	0.38
QBA TS	55	200	0.16	1.58	3.30	0	0.46
QBA FMS	112	240	0	1.58	2.14	0	0.42

(Continued)

TABLE 4.3 Characteristic Properties of Packed Column Stationary Phases (*cont'd*)

Stationary phase	Temperature range		System constants (121 °C)				
	Minimum	Maximum	*e*	*s*	*a*	*b*	*l*
QBA PIC	90	200	0.10	1.56	1.42	0	0.45
QBP TS	44	230					
QBP Cl	83	230	0.24	1.85	5.42	0	0.47
QEA TS	85	190	0.33	2.05	3.43	0	0.30
DEA TS	105	210					

in the liquid temperature range (Table 4.3). Poly(siloxanes) with cyano groups on α-carbon atoms and fluorine atoms on α- and β-carbon atoms have low thermal stability and are not used for gas chromatography. Poly(siloxanes) with 2-cyanoethyl, 3-cyanopropyl, and 3,3,3-trifluoropropyl groups have suitable thermal stability and facilitate the synthesis of poly(siloxane) stationary phases with high dipolarity/polarizability and hydrogen-bond basicity. The cyanoalkyl- and 3,3,3-trifluoropropyl-containing poly(siloxanes) have complementary selectivity. The poly(cyanoalkylsiloxane) phases with a high incorporation of cyanoalkyl groups are some of the most cohesive and polar stationary phases in common use. They are also susceptible to oxidation and hydrolysis requiring the use of high-purity carrier gases.

The dialkyl phthalates, poly(ethers), poly (esters), and poly(ethylene glycols) are a further group of stationary phases that cover a wide polarity range, with the poly(ethylene glycols) being the most important because of their complementary properties to the poly(siloxane) stationary phases (Table 4.3). The dialkyl phthalates are moderately polar and weakly hydrogen-bond basic stationary phases [19]. The meta-linked poly(phenyl ethers) with 5- or 6-rings are moderately polar liquids with exceptionally low vapor pressure for their low molecular weight. The poly(ester) stationary phases are represented by a wide range of resinous composite materials derived from the reaction of a polybasic acid with a polyhydric alcohol [9,11,13,20]. The most widely used are the succinate and adipate esters of ethylene glycol, diethylene glycol, and butanediol. The stability of these materials at high temperatures is questionable and they are also slowly degraded by oxygen and water that might be present in the carrier gas or samples. Exchange reactions with samples containing alcohols, acids, amines, and esters are also possible. They have been replaced in many of their applications by the more stable poly(cyanoalkylsiloxanes).

Poly(ethylene glycols) are widely used for the separation of volatile polar compounds and have good support deactivating properties [9–15]. Carbowax 20 M, a waxy solid with a molecular weight of about 16,000, is one of the most popular phases for packed-column gas chromatography. Pluronic phases of lower polydispersity prepared by condensing propylene oxide, ethylene oxide, and propylene glycol have similar properties to the poly(ethylene glycols). Condensing Carbowax 20 M with 2-nitroterephthalic acid produces a new phase, FFAP, recommended for the separation of organic acids. The poly(ethylene glycols) are degraded by oxygen, moisture, strong acids, and Lewis acids at high temperatures.

The ionic liquid stationary phases are organic salts of low melting point with a wide liquid temperature range (Table 4.3) [21,22]. Those

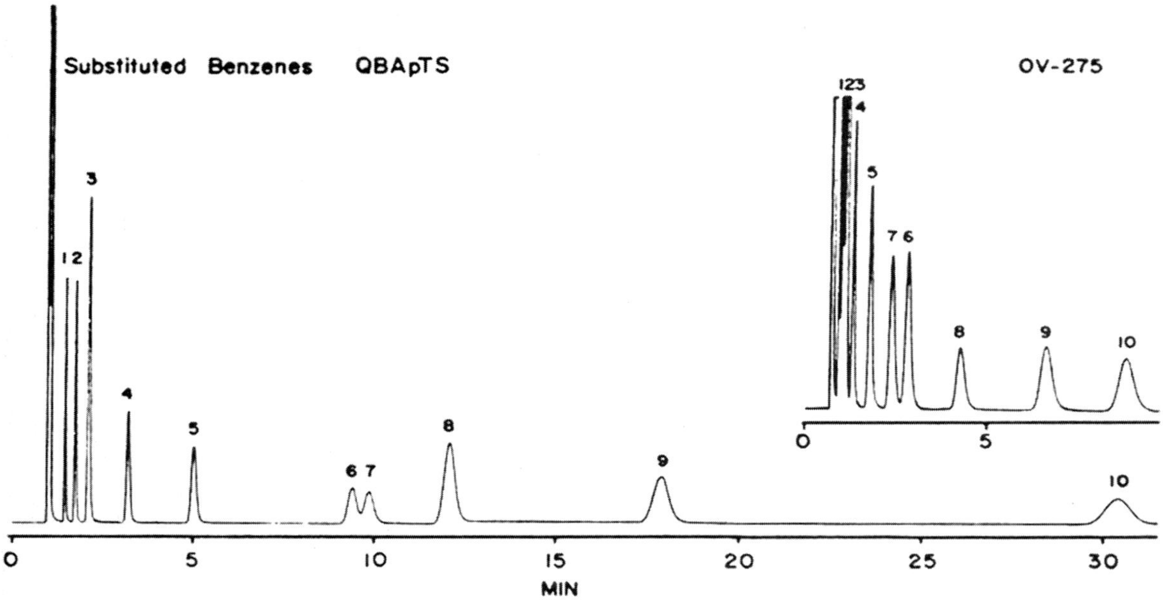

FIGURE 4.1 Separation of a mixture of aromatic compounds on matched packed columns coated with tetra-n-butylammonium 4-toluenesulfonate and OV-275. Each column was 3.5 m × 2 mm I.D. containing 10% (w/w) of stationary phase on Chromosorb W-AW (100–120 mesh) with a nitrogen carrier gas flow rate of 15 mL/min and column temperature 140 °C. Peak assignments: 1 = benzene; 2 = toluene; 3 = ethylbenzene; 4 = chlorobenzene; 5 = bromobenzene; 6 = iodobenzene; 7 = 1,2-dichlorobenzene; 8 = benzaldehyde; 9 = acetophenone; and 10 = nitrobenzene. *From ref [1]; copyright Elsevier.*

used in gas chromatography are mainly alkylammonium, alkylphosphonium, and 1,3-dialkylimidazolium salts with weak nucleophilic anions, such as sulfonate and tetrafluoroborate. The long-range Coulombic forces present in ionic liquids resist the escape of ions into the gas phase resulting in the virtual absence of significant vapor pressure over a wide temperature range. In many cases, the upper temperature limit is established by the thermal stability of the ionic liquid rather than its vapor pressure. Ionic liquids with weak nucleophilic anions possess low chemical reactivity and transformation reactions are relatively rare. Nucleophilic displacement reactions with alkyl halides and proton transfer and other acid/base reactions with amines have been observed in a few cases. The unique selectivity of the ionic liquids is a result of their strong hydrogen-bond basicity and significant capacity for dipole-type interactions (Table 4.3). For example, Figure 4.1 illustrates the separation of aromatic compounds on matched columns of tetra-n-butylammonium 4-toluenesulfonate and the poly(cyanoalkylsiloxane) stationary phase OV-275. The increased retention for the aromatic compounds on the ionic liquid column is indicative of the strong polar interactions between sample and ionic liquid compared to those with OV-275. The hydrogen-bond basicity of the ionic liquids is primarily a property of the anion and is influenced by the size and charge localization on the anion. Ions that can delocalize charge (e.g. picrate and perfluoroalkanesulfonate) are weaker hydrogen-bond bases and less dipolar than other ionic liquids. The halide anions have relatively small atomic radii and no mechanism for charge delocalization. They are the most basic of the ionic liquid stationary phases.

Liquid-crystal stationary phases are used for the separation of positional or geometrical isomers of rigid molecules [23]. Well over two hundred liquid-crystal phases have been used in gas chromatography. They are of a variety of chemical types but all have a markedly elongated, rigid, rod-like structure in common. The most common types are Schiff bases, esters, and azo and azoxy compounds. In most cases, the order of elution is in accord with the solute length-to-breadth ratios with differences in vapor pressure and solute polarity also being important. Long and planar molecules fit better into the ordered structure of the liquid-crystal phase, whereas nonlinear and nonplanar molecules do not permeate so easily between the liquid-crystal molecules of the stationary phase and elute faster from the column.

When single stationary phases fail to provide an adequate separation but the poorly resolved components are different on the individual phases, then the use of a mixture of stationary phases (mixed solvent) or column packings (mixed bed) is a viable approach for the separation. The mixed-bed approach is the most versatile and economical one and, in the absence of specific interactions between stationary phases, results in similar separations to the mixed-solvent approach [1,12,24]. Purnell and co-workers developed the general theory for retention on mixed stationary phases (theory of diachoric solutions) in which the gas–liquid partition coefficient for a solute on a binary mixed bed or solvents column, K_S, is expressed as

$$K_s = K_A + \phi_B(K_B - K_A) \qquad (4.1)$$

where K_A and K_B are the gas–liquid partition coefficients for the solute on the single stationary phases and ϕ_B is the volume fraction of stationary phase B ($\phi_A + \phi_B = 1$). A plot of K_S against ϕ_B affords a series of straight lines with slopes determined by the single-phase partition coefficients plotted as the extreme points on the dual ordinate axis. The elution order of the solutes at any binary phase composition can be read by constructing a vertical line at the composition of interest. The window diagram method affords a convenient approach for identification of the optimum separation conditions [25]. The optimum stationary-phase composition for the separation corresponds to the value of ϕ_B for the highest window provided the most difficult to separate pair is reasonably well retained. In addition to the optimal binary stationary-phase composition and elution order, the window diagram method can also provide the required column length and the total separation time. The above theory assumes a partitioning model for the retention mechanism and predictions will be approximate or poor for solutes retained by a mixed retention mechanism involving interfacial adsorption (Section 4.2.6).

4.2.2. Supports

An ideal support would have sufficient surface energy to cause the stationary phase to wet the surface as a thin, stable film while sufficiently inert to eliminate solute interactions with the surface. It should have a large surface area to weight ratio (facilitates the preparation of columns with a high phase loading), a regular shape with a narrow range of cross-sectional diameters (facilitates the preparation of columns of high efficiency), be mechanically stable (facilitates general handling), and be a good conductor of heat (facilitates rapid thermal equilibrium). No such ideal support exists but the diatomaceous earths represent a reasonable compromise and are the most widely used supports [26,27].

Diatomite (diatomaceous earth) is a natural product composed of the skeletons of single-celled alga found in large beds in various parts of the world. The skeletal material is essentially amorphous silica with small amounts of alumina and metallic oxide impurities. High-temperature processing (>900 °C) is used to agglomerate and strengthen the natural material. The presence of complex iron oxides gives this material its characteristic pink color. Thermal processing of the diatomite in the presence of a small amount of sodium carbonate results in the formation of a white material in which metal impurities are converted to colorless sodium silicate. The bulk chemical composition of the two types of diatomaceous supports is quite similar (SiO_2 88.9–90.6%, Al_2O_3 4.0–4.4%, Fe_2O_3 1.6%, CaO 0.6%, MgO 0.6%, and $Na_2O + K_2O$ 1.0–3.6% m/m). Typical physical properties are summarized in Table 4.4. The pink supports are relatively hard, and have a high packing density, a relatively large surface area, and a high loading capacity. They are used in both analytical and preparative gas chromatographies. The white supports are more friable, less dense, and have a lower surface area and loading capacity. They are used primarily for analytical separations. Supports with a mesh range of 80–100 or 100–120 are a reasonable compromise between column efficiency (proportional to d_p, the average particle diameter) and column pressure drop (proportional to $1/d_p^2$) for analytical separations. Mesh ranges in micrometers are given in Table 4.5. For polar compounds, severe tailing, decomposition, and loss of injected mass by adsorption are observed for some compounds and are associated with the presence of metallic impurities, principally iron, and silanol groups on the support surface.

TABLE 4.4 Characteristic Properties of Chromosorb Supports

Property	P	W	G[a]	A[b]	750[c]
Color	Pink	White	Oyster	Pink	White
Apparent pH	6.5	8.5	8.5	7.1	8.0
Free fall density (g/ml)	0.38	0.18	0.47	0.40	0.37
Packed density (g/ml)	0.47	0.24	0.58	0.48	0.36
Surface area (m^2/g)	4.0	1.0	0.5	2.7	0.5–1.0
Maximum liquid loading (%)	30	15	5	25	12

[a] *A specially prepared white support that is harder, more robust, inert, and has a higher density than typical white supports (a given column volume contains approximately 2.5 times the amount of Chromosorb G, and therefore liquid phase, as Chromosorb W of the same nominal phase loading).*
[b] *Possesses the mechanical strength and high loading capacity of pink supports with a reduced surface activity approaching that of the white supports (used for preparative chromatography).*
[c] *Most inert of the white supports.*

TABLE 4.5 Particle Size Ranges for Diatomaceous Supports

Mesh range	Top screen opening (μm)	Bottom screen opening (μm)	Spread (μm)	Range ratio
10–20	2000	841	1159	2.38
10–30	2000	595	1405	3.36
20–30	841	595	246	1.41
30–40	595	420	175	1.41
35–80	500	177	323	2.82
45–60	354	250	104	1.41
60–70	250	210	40	1.19
60–80	250	177	73	1.41
60–100	250	149	101	1.68
70–80	210	177	33	1.19
80–100	177	149	28	1.19
100–120	149	125	24	1.19
100–140	149	105	44	1.42
120–140	125	105	20	1.19
140–170	105	88	17	1.19
170–200	88	74	14	1.19
200–230	74	63	11	1.19
230–270	63	53	10	1.19
270–325	53	44	9	1.20
325–400	44	37	7	1.19

Acid and/or base washing to remove metal impurities and silanization of silanol groups are the most widely used methods for support deactivation [12]. Silanol groups are converted to silyl ethers by reaction with dimethyldichlorosilane (DMCS), hexamethyldisilazane (HMDS), trimethylchlorosilane (TMCS), or by a combination of these reagents. Highly silanized supports are not wet by polar liquids and are unsuitable for preparing packings with polar stationary phases. For strongly acidic or basic compounds, tailing reducers may be required to improve chromatographic properties. To be effective, the tailing reducer should be a stronger acid or base than the compounds to be separated. For amines, the tailing reducer could be a few percent (w/w) potassium hydroxide or poly(ethyleneimine). For acidic compounds, phosphoric acid or trimer acid is suitable. These active substances also act as subtractive agents (acidic tailing reducers absorb bases, etc.) and must be compatible with the stationary phase. Potassium hydroxide and phosphoric acid, for example, catalyze the depolymerization of poly(ester) and poly(siloxane) stationary phases.

Less commonly used supports include fluorocarbon powders, glass beads, and dendritic salt. The fluorocarbon powders (Table 4.6) are used primarily as supports for highly fluorinated stationary phases required for the separation of corrosive compounds. They have low surface energy and can be difficult to coat, often requiring special techniques [12]. Column efficiencies are often low compared with diatomaceous earth supports. Glass beads with a narrow size distribution can be used to prepare efficient columns but have a low loading capacity (<0.5% w/w). Because of their controlled shape and size, they were used primarily in theoretical studies of band broadening.

TABLE 4.6 Characteristic Properties of Fluorocarbon Powder Supports

Property	Kel-F[a]	Fluoropak-80[b]	Teflon-6[c]
Surface area (m^2/g)	2.2	1.3	10.5
Optimum phase loading (w/w %)	15–20	2–5	15–20
Maximum temperature (°C)	160	275	250

[a] *A hard chlorofluorocarbon powder that can be handled like diatomaceous supports but usually gives columns of low efficiency.*
[b] *A granular fluorocarbon resin with a sponge-like structure with a low loading capacity.*
[c] *Agglomerated poly(tetrafluoroethylene) porous polymer. Not wet by polar stationary phases. Columns packed at low temperatures to minimize buildup of electrostatic charges.*

4.2.3. Coating and Packing Techniques

The preparation of a packed column is a multistep process in which the liquid phase in a suitable solvent is mixed with the support, the solvent removed by evaporation, and the dry packing added to the empty column. No special apparatus, not usually found in chemical laboratories, is required for this process. Column packings are prepared on a weight-for-weight basis and quoted as percent liquid phase (% w/w). The required weight of solid support is added to the required amount of stationary phase dissolved in sufficient solvent to completely cover the support. The solvent should be a good solvent for the stationary phase, free of stabilizers and involatile contaminants, and sufficiently volatile for easy evaporation. The most common coating/evaporation procedures are the rotary evaporator technique, pan-dry method, slurry filtration method, and column adsorption method [12]. Since diatomaceous supports are fragile, they should be handled gently during the coating procedure. After coating the damp packing is air dried, oven dried, or dried in a fluidized bed dryer. The latter is faster and usually gives packings of higher efficiency and permeability, probably due to removal of fine particles by the gas passing through the packing. Mechanical sieving prior to coating and fluidized bed drying after coating are the preferred methods of minimizing the deleterious effects of fines on column performance. If an accurate value for the phase loading is required, it can be determined by either Soxhlet extraction or combustion on a sample of the dried and conditioned packing [12].

Columns of 0.5–3.0-m length and 2–4-mm internal diameter can be packed by the tap-and-fill method, aided by suction. Longer columns are more difficult to pack and require a combination of suction at the free end and pressure applied to the packing reservoir attached to the other end. One end of the column is terminated with a glass wool plug, attached to a water aspirator, and small aliquots of packing added *via* a filter funnel attached to the opposite end. The column bed is consolidated as it forms by gentle tapping of the column sides with a rod or with the aid of an electric vibrator. When the column has been packed to the required length, the funnel is removed and the empty segment of the column (the length of this segment is determined by the design of the column injector) is filled with a glass wool plug. Column preparation is completed by placing the column in the oven of the gas chromatograph and connecting it to the carrier gas flow from the injector end only. The temperature of the oven is then increased to about 20 °C above the highest temperature to be used at for the planned separations (but never above the maximum allowed operating temperature for the stationary phase) and held there for several hours (usually overnight). This conditioning step removes volatile impurities and contaminants from the column packing and ensures that when the column is placed into service a stable detector baseline is obtained.

After column conditioning, a few preliminary tests are performed to ensure that the performance of the column meets expectations. Column efficiency is determined with a simple test mixture containing compounds with weak specific interactions (e.g. n-alkanes and aliphatic ketones), an activity test with a polar test mixture reflecting the functional groups expected to be present in the samples, and a resolution test using a mixture that reflects the intended column use. These tests provide information on column efficiency, absolute retention, peak tailing, separation factors, and resolution of target compounds suitable for demonstrating both column quality and that the column is fit for purpose. Columns with a plate number between 1,500 and 2,500 per meter are suitable for analytical applications (plate number varies with the phase loading and support type). An unusually high column inlet pressure indicates support attrition, possibly as a result of mechanical breakdown during the coating and packing procedure. Excessive tailing or adsorption of polar test compounds indicates residual column activity. Column performance might be restored by on-column silanization of support silanol groups using one of the many available column-conditioning reagents.

4.2.4. Packed Columns for Preparative-Scale Gas Chromatography

Preparative-scale gas chromatography requires the use of larger amounts of stationary phase than analytical separations to achieve high sample throughput [12,28]. For simple separations, short, wide columns are generally used (e.g. 1–3 m × 6–10 cm I.D.) For difficult separations, long, narrow columns (e.g. 10–30 m × 0.5–1.5 cm I.D.) are preferred. Wide-bore columns cannot be coiled and purpose-designed instruments to accommodate long columns are required for their use. A coarser support material with a narrow particle-size distribution, for example, 35–40 mesh, is used to facilitate operation at a reasonable inlet pressure. Reproducibly packing wide-bore columns with reasonable efficiency is more difficult due to the uneven radial packing density resulting from particle-size segregation (the larger particles accumulate closer to the wall). To minimize radial heterogeneity, the shake, turn, and pressure packing method is recommended. The column is shaken in the radial direction and is rotated along its long axis while being packed, and periodically pressurized. Packing with suction and vertical tamping affords a faster approach.

4.2.5. Classification of Stationary Phases

Many methods of classifying stationary phases to facilitate column selection for method development have been proposed and abandoned over the long history of gas chromatography [12,29]. Most early classification scales attempted to rank stationary phases according to their solvent strength (or polarity) and solvent selectivity. Single value scales of solvent strength are unable to provide information that is useful for column selection, since multiple and independent interactions are involved in solute–solvent interactions. Scales of free energy or retention index differences for prototypical compounds were in use for many years but were shown to be unreliable [30]. The system of phase constants introduced by Rohrschneider and further developed by McReynolds are the best known of this type and can be found in some stationary-phase catalogs [11]. The modern approach to stationary-phase classification is based on Abraham's solvation parameter model [1,31,32]. This model assumes that the transfer of a solute from the gas phase to the stationary phase occurs in three steps: the formation of a cavity in the stationary phase of the same size as the solute, transfer of the solute to the cavity with reorganization of the solvent molecules around the cavity, and the setting up of solute–solvent interactions recognized as dispersion, of a dipole type (orientation and induction), and hydrogen bonding. The master equation for gas chromatography is given as

$$\log \mathrm{SP} = c + e\mathrm{E} + s\mathrm{S} + a\mathrm{A} + b\mathrm{B} + l\mathrm{L} \qquad (4.2)$$

where SP is some free energy-related retention property, such as a partition coefficient, retention factor, or relative retention. The uppercase letters are solute descriptors, which define the capability of the solute to participate in the defined intermolecular interactions, but are not important for the present discussion. The capability of the stationary phase to enter into complementary interactions with those of the solute is indicated by the lowercase letters known as system constants. The *e* system constants are determined by electron lone-pair interactions, the *s* system constants by interactions of a dipole type, the *a* system constant by the hydrogen-bond basicity of the stationary phase, the *b* system constant by the hydrogen-bond acidity of the stationary phase, and the *l* system constant by the opposing contributions from cavity formation and dispersion interactions. The equation constant, *c* term, is not a fundamental property of the stationary phase and is determined by a number of factors related to the physical characteristics of the column and statistical contributions from fitting the model to the experimental data.

The system constants for some common stationary phases are summarized in Table 4.3 [1,31]. The system constants are only loosely scaled to each other so that differences in any column can be read directly but small differences along rows must be interpreted cautiously. Two features in this table stand out. None of the common stationary phases are significant hydrogen-bond acids (*b* system constant) and this interaction is unimportant for selectivity optimization. In the case of EGAD, DEGS, and TCEP, the small *b* system constant is more likely a reflection of impurities in the stationary phase produced during synthesis or while in use. The second feature of note is that electron lone-pair interactions (*e* system constant) are generally weak and of limited variation. They afford few opportunities to optimize selectivity. Fluorine-containing stationary phases have negative *e* system constants representing the tighter binding of electron pairs compared with hydrocarbons. The remaining three system constants account for most of the selectivity differences among the stationary phases in Table 4.3. The l system constant is always an important contributing factor to retention and mainly reflects changes in the cohesion of the stationary phases. It has small values for polar stationary phases, which retain compounds of low polarity weakly, and large values for low polarity stationary phases, which retain compounds of low polarity more strongly. The l system constant is also strongly correlated to the partial molar Gibbs free energy of solution for a methylene group and is an indication of the spacing between alternate compounds in a homologous series. An example of this effect is seen in Figure 4.1 for the highly cohesive stationary phase OV-275 (small l value) and the moderately cohesive ionic liquid stationary phase (intermediate l value). The ionic liquids with weakly associating ions have remarkably low cohesion for polar stationary phases due to the effect of the Coulombic fields on interionic distances. Ionic liquids with associating ions have similar cohesion to polar molecular stationary phases. The stationary phases in Table 4.3 differ significantly in their capacity for dipole-type interactions (s system constant). For the poly(siloxane) stationary phases the s system constant changes predictably with the type of substituent (methyl < phenyl < trifluoropropyl < cyanopropyl) attached to the polymer backbone. Poly(siloxane) stationary phases with a high incorporation of cyanoalkyl groups are among the most dipolar stationary phases used in gas chromatography. The poly(ethylene glycol) stationary phases are slightly less dipolar than the poly(ester) stationary phases, and both are intermediate with respect to the dipolarity range for the poly(siloxanes) containing cyanoalkyl substituents. The ionic liquid stationary phases are all dipolar with some examples being the most dipolar of all stationary phases. All of the dipolar stationary phases are also hydrogen-bond bases (a system constant). The ionic liquids possess the widest range of hydrogen-bond basicity and include the highest ranked stationary phases for this interaction (Table 4.7). A notable feature of the polar stationary phases is the characteristic family difference for the s/a system constants ratio. This is the principal reason for retention of the different types of polar stationary phases for selectivity optimization.

A smaller number of stationary phases can be identified from those in Table 4.3 as a starting point for method development (Table 4.8). The system constants (and the intermolecular interactions they represent) are temperature dependent and not highly correlated [33]. Thus, the system constants in Table 4.3 should be considered reliable for selectivity optimization at temperatures in the region of 120 °C but less so at significantly different temperatures. The general strategy for selectivity optimization is

TABLE 4.7 Typical Ranges for the System Constants at 121 °C for Alkylammonium and Alkylphosphonium Ionic Liquids and NonIonic Stationary Phases

	Range	
System constant	**Ionic liquids**	**Nonionic liquids**
e (electron lone-pair interactions)	0.07–0.50	0–0.0.37
s (dipole-type interactions)	1.4–2.1	0–2.1
a (solvent hydrogen-bond basicity)	1.4–5.4	0–2.1
b (solvent hydrogen-bond acidity)	0	0
l (cohesion and dispersion interactions)	0.44–0.55	0.37–0.58
(associated anions)	0.26–0.37	

TABLE 4.8 Classification of Common Stationary Phases into Selectivity Groups for Method Development

Group	Type of stationary phase	Examples of group membership	Basis of selectivity
I	Hydrocarbons and poly(dimethysiloxanes)	Squalane Apolane-87 OV-1	Low cohesion and very weak polar interactions.
II	Poly(methylphenylsiloxanes) <50 % methyphenylsiloxane monomer	OV-3 OV-11 OV-105 PPE-5	Low cohesion and moderate polar interactions ($s/a > 2$)
III	Poly(trifluoropropylmethyl-siloxanes)	QF-1	Moderate cohesion, intermediate dipolarity, low hydrogen-bond basicity, electron lone-pair interactions reduce retention ($s/a > 5$)
IV	Poly(cyanoalkylphenyl-siloxanes)	OV-225 OV-275 TCEP	High cohesion and strong dipolarity and hydrogen-bond basicity ($s/a \approx 1$)
V	Poly(ethylene glycols) Poly(esters)	CW20M DEGS	Intermediate cohesion, dipolarity, and hydrogen-bond basicity ($s/a < 1$)
VI	Ionic liquids	QBA TS QBP Cl	see Table 4.7

to choose a single stationary phase from each group in Table 4.8 for screening of the selectivity groups, and, once a suitable group is identified for the separation, to fine-tune the selection by exploring other phases within the same group. Stationary-phase selection has to take into account the temperature operating range for the stationary phases as well as their selectivity.

4.2.6. Retention Models

Any model devised to explain retention in packed-column gas chromatography in general terms has to take into account the distribution of the stationary phase on the porous support surface. For liquids that wet the support, the stationary phase is first adsorbed as a monomolecular and multimolecular layer over the entire support surface. As the phase loading is increased, it collects in the fine pores initially, and then progressively appears in the large cavities at the same time as the adsorbed layer thickens up. The structured layer formed close to the support surface has more order than in the bulk liquid and different retention properties. Although surface forces are short range, their effect can be transmitted by the successive polarization of adjacent molecules to a considerable depth in the liquid. This structured layer, then, may be of considerable thickness and will dominate the retention mechanism at low phase loadings. At high phase loadings, the presence of bulk liquid will come to dominate the retention mechanism. In addition, variation of the phase loading will result in nonlinear changes in the surface area of the gas–liquid interface as pores of different sizes fill with liquid at different rates. At high phase loadings, it is assumed that the gas–liquid interfacial area approaches a limiting value, approximately equal to the support surface area less the area of its narrow pores and channels. For liquids that do not wet the support surface readily, the stationary phase will be present as droplets primarily at the outer grain surface allowing exposure of the sample to adsorption sites directly on the support surface at low phase loadings. At higher phase loadings, coalescence to a continuous film will occur and interactions with the support surface then depend primarily on the competition between the bulk (or structured) liquid phase and the sample for surface active sites on the support. Support deactivation is then important to minimize these interactions but also to facilitate wetting of the support surface by liquids with poor support wetting characteristics.

Taking the above considerations into account, a general model for retention in gas–liquid chromatography has to consider contributions from interfacial adsorption at the support and/or liquid interfaces, as well as partition at the gas–liquid and bulk liquid-structured liquid interfaces. This results in the general model [34,35]

$$V_N^* = V_L K_L + \delta V_L (K_s - K_L) + (1 - \delta) A_{LS} K_{DSL} + A_{GL} K_{GL} + A_{LS} K_{GLS} \tag{4.3}$$

where V_N^* is the net retention volume per gram of packing, V_L the volume of liquid phase per gram of packing, K_L the gas–liquid partition coefficient, δ a constant constrained to have values of 1 when the film thickness is less than or equal to the thickness of the structured layer (d_S) and zero when the film thickness exceeds the thickness of the structured layer, K_S the gas–liquid partition coefficient for the structured liquid layer, A_{LS} the liquid–solid interfacial area per gram of packing, $K_{DSL} = d_S(K_S - K_L)$, A_{GL} the gas–liquid interfacial area per gram of packing, K_{GL} the adsorption coefficient at the gas–liquid interface, and K_{GLS} the coefficient for adsorption at the liquid–solid interface. Although Eqn (4.3) provides a general description of the retention process, it is rather awkward to use. The equation contains five unknowns (K_L, K_S, K_{DSL}, K_{GL}, and K_{GLS}) requiring data for the packing characteristics (V_L, A_{GL}, and A_{LS}) for at least five columns prepared from the same support at different phase loadings to obtain a numerical solution. In addition, there is no exact method of defining the thickness of the structured layer, which must be established by a trial and error procedure. The general assumption in deriving Eqn (4.3) is that the individual retention mechanisms are independent and additive. This will be true for conditions where the infinite dilution and zero surface coverage approximations apply (i.e. small sample sizes where the linearity of the various adsorption and partition isotherms are unperturbed and solute–solute interactions are negligible).

At intermediate to high phase loadings, the contributions of the structured liquid phase to retention are expected to be small, allowing Eqn (4.3) to be simplified [36,37]

$$V_N^*/V_L = K_L + (A_{GL} K_{GL} + A_{LS} K_{GLS})(1/V_L) \tag{4.4}$$

from which the gas–liquid partition coefficient can be determined by (usually) a linear extrapolation from a plot of V_N^*/V_L against $1/V_L$. This observation is consistent with the view that adsorption at the support–liquid interface is dominant for phases of low polarity and at the gas–liquid interface for polar phases. Some representative examples are shown in Figure 4.2 for several compounds on the polar stationary phases Carbowax 20 M and the poly(dicyanoalkylsiloxane) OV-275. For solutes retained solely by gas–liquid partitioning (e.g. nitromethane, dioxane, and ethanol) on Carbowax 20 M, the plots have a zero slope. The n-alkanes are retained by a mixed retention mechanism indicated by the positive slope and a significant intercept at $1/V_L = 0$. Interfacial adsorption is important for all compounds on OV-275 and is dominant for the n-alkanes, which have a near-zero intercept, indicating that gas–liquid partitioning is of minor importance to their retention.

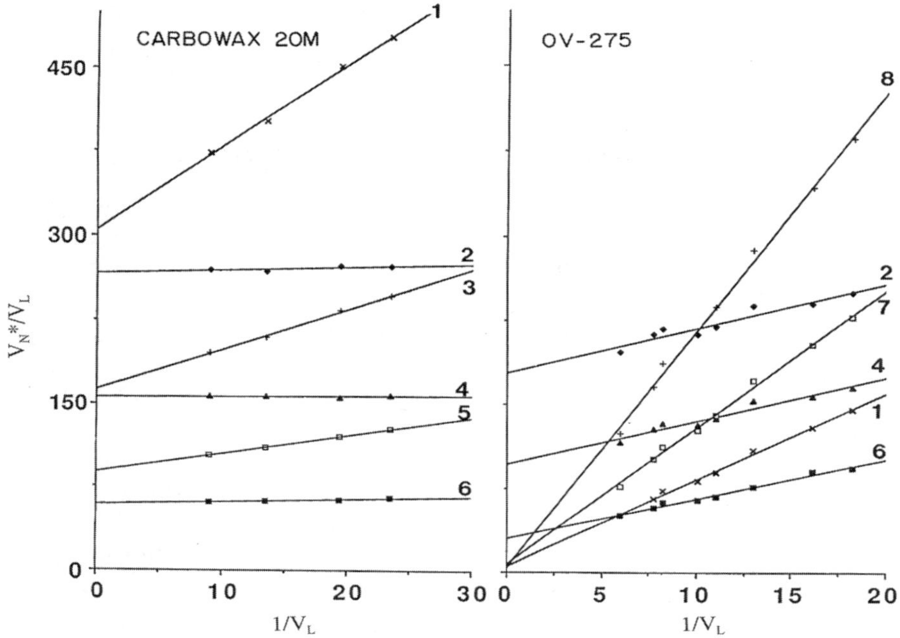

FIGURE 4.2 Plot of V_N^*/V_L against $1/V_L$ for different compounds on poly(ethylene glycol), Carbowax 20 M, and poly-(cyanoalkylsiloxane), OV-275, stationary phases at 80.8 °C. Identification: 1 = dodecane; 2 = nitromethane; 3 = undecane; 4 = dioxane; 5 = decane; 6 = ethanol; 7 = tridecane; and 8 = tetradecane. *From ref. [36]; copyright Elsevier.*

Since there is no good method for measuring A_{GL}, Eqn (4.4) is rarely used to determine K_{GL} but a qualitative assessment of the general contribution of interfacial adsorption to the retention mechanism can be obtained by comparison of the true gas–liquid partition coefficient obtained from Eqn (4.4) by extrapolation and the value measured with the assumption that retention occurs only by gas–liquid partition. These calculations indicate that interfacial adsorption is not a rare retention mechanism, and although gas–liquid partitioning tends to dominate in most cases, interfacial adsorption should not be ignored. Since the contribution from interfacial adsorption depends on the phase loading, temperature, type of stationary phase and support, and properties of the sample, it is impossible to prescribe exact circumstances when it is important; however, the following general comments serve to indicate when it is a likely contributor to retention. Interfacial adsorption tends to be more important at lower temperatures and for a mixed retention mechanism it may fall to zero at a sufficiently high temperature. Interfacial adsorption makes a larger relative contribution to the retention mechanism at lower phase loadings due to a combination of a larger accessible liquid surface area and a smaller bulk liquid volume. For nonpolar phases, interfacial adsorption can generally be related to support properties, and at intermediate temperatures, possibly eliminated by adequate support deactivation. Interfacial adsorption is usually significant for most compounds on cohesive stationary phases and is commonly observed for solutes of low polarity on polar stationary phases. The primary effect of interfacial adsorption on method development is that simple relationships that might be used to estimate retention as a function of phase loading are of limited use because, as well as the volume of liquid phase, different interfacial surface areas need to be considered and their distribution constants, which are generally unknown and do not change linearly with the volume of stationary phase.

4.2.7. Band-Broadening Mechanisms

As a sample migrates along the column its distribution about the zone center increases in proportion to its residence time in the column. The extent of band broadening determines the column efficiency, which is expressed by convention as either the plate number (N) or plate height (H) for the column, and is controlled by a set of kinetic factors affected by column design. The plate number and plate height are calculated from the peak profile as indicated in Table 4.9. These terms have their origin in the plate theory of chromatography, long since abandoned. Modern models of band broadening are based on rate theories that are better able to account for experimental observations as well as facilitating column design [1,4,38–43]. Rate theory considers three general contributions to band broadening, which can be identified as flow anisotropy (eddy diffusion), axial diffusion, and resistance to mass transfer. These terms, except where noted, are considered independent and additive. In general, rate models are able to account for the main features of band broadening but are not exact for packed columns.

TABLE 4.9 Calculation of the Plate Number (N) and Plate Height (H) for a Packed Column Assuming that Peaks can be Approximated by a Gaussian Model

$N = (t_R/\sigma_t)^2 = a(t_R/w)^2$	t_R = solute retention time
	σ_t = peak standard deviation
	a = constant that depends on the peak width
	w = a measure of the peak width
	w_h = peak width at half height and a = 5.54
	w_b = peak width at base and a = 16
$H = L/N$	L = column length

The space between particles in a packed bed is made up of a network of interconnected channels of varied dimensions that depend on the size and shape of the particles and their packing density. For any channel, the mobile phase velocity falls close to zero near the particle surface and increases rapidly toward the channel center. In addition, individual streamlines are not straight since molecules are forced to continually change direction by the obstacles (packing material) in their way. The heterogeneity of the mobile phase velocity and path lengths for different streamlines results in band broadening compared with passage through an open tube. Band broadening resulting from these conditions is relaxed to some extent by radial diffusion, which allows solute molecules to sample multiple streamlines as they migrate along the column. The contribution of flow anisotropy (eddy diffusion) to the plate height, H_E, can be estimated from $H_E = 2\lambda d_p$, where λ is the column packing factor (a dimensionless parameter with typical values between 0.5 and 1.5) and d_p is the average particle size. Band broadening by flow anisotropy can be minimized using a homogeneous bed packed with small particles of a narrow size range at a high packing density. The column pressure drop will ultimately determine the most practical particle size and column length. For typical operating pressures (<10 bar) this corresponds to an average particle size of about 100 μm and columns <5 m long.

The contribution to the plate height from molecular diffusion in the mobile phase arises from the tendency of the solute band to diffuse away from regions of high to lower concentration. Its contribution to the plate height, H_L, is proportional to the diffusion coefficient in the mobile phase, D_M, the tortuosity (obstruction) factor of the column, γ, and the time the solute spends in the mobile phase (inversely proportional to the mobile phase velocity u), $H_L = 2\gamma D_M/u$. The tortuosity factor typically has values of 0.6–0.8 that depend on the packing density over the length of the column. At low mobile phase velocities, its value is averaged over tight and loosely packed domains, while at high velocities, it is weighted in favor of the loosely packed domains where more flow occurs [44]. Diffusion coefficients in the stationary phase are about 10^4 times smaller than in gases and the contribution to band broadening from axial diffusion in the stationary phase can generally be neglected.

During migration along the column, solute molecules are continually and reversibly transferring to and from the stationary phase. This process is not instantaneous because a finite time is required for the molecules to transfer to the interface between phases and enter the stationary phase by diffusion. Those solutes close to the interface enter the stationary phase almost immediately, whereas those some distance away enter sometime later. However, since the mobile phase is moving during this time interval, those molecules that remain in the mobile phase will be swept along the column and dispersed away from those molecules that entered the stationary phase almost immediately. Dispersion occurring in the stationary phase is exactly analogous to that in the mobile phase. Thus, those molecules that were close to the surface will be swept along in the mobile phase and dispersed from those molecules still diffusing to the interface. The combined peak observed at the column outlet is broadened about its band center, which is where it would have been if equilibrium was instantaneous, provided the degree of nonequilibrium is small. The process described above is called resistance to mass transfer, and the mobile phase contribution, H_M, is given by $wudp^2/D_M$ where w is an empirical packing factor with values typically between 0.02 and 5. The stationary-phase contribution to resistance to mass transfer, H_S, is given by $2kd_f^2u/3D_s(1+k)^2$ where k is the retention factor, d_f the film thickness, and D_S the solute diffusion coefficient in the stationary phase. In practice, to account for the influence of convection (band broadening resulting from exchange of solute between streams moving at different velocities), the flow anisotropy term must be coupled with the mobile phase resistance to mass transfer term, as indicated below [38]:

$$H_{MC} = 1/(1/H_E + 1/H_M) \tag{4.5}$$

where H_{MC} is the contribution to the plate height resulting from the coupling of flow anisotropy and resistance to mass transfer in the mobile phase. At typical mobile phase velocities for packed columns, the coupling concept is of minor importance but, at high mobile phase velocities, it is better able to explain experimental observation of the change in the total plate height. In particular, it accounts for the flattening out of the ascending portion of the van Deemter curve observed at high mobile phase velocities compared with the uncoupled approach.

In gas chromatography, the mobile phase is compressible and results in a pressure gradient along the column. As a consequence, diffusion constants in the mobile phase and the mobile phase velocity vary in a nonlinear manner with the fraction of the column length explored. For columns operated at an average mobile phase velocity in the region of the optimum value, a reasonable approximation for the total plate height (obtained by summing the individual contributions discussed above) is given by

$$H = 2\lambda d_p + 2\gamma D_{M,0}/u_0 + [f_g(k)](d_p^2/D_{M,0})u_0 + [f_s(k)d_f^2/D_s u_{AV}] \tag{4.6}$$

$$f_g(k) = (1 + 6k + 11k)/96(1 + k)^2$$
$$f_s(k) = 2k/3(1 + k)^2$$

where the subscript zero is added to indicate that the parameter is specified as the value at the column outlet and the empirical constant w has been replaced by a functional dependence on the retention factor derived exactly for an open-tubular column and used as an approximation for a packed column. All terms containing u_0 should be multiplied by a function dependent on the column inlet/outlet pressure ratio, *P*, but since for the usual range of conditions in gas chromatography it has values between 1 and 1.1, it is often omitted for simplicity [42]. According to Scott the average linear velocity, u_{AV}, can be replaced by $[4u_0/(P+1)]$ in Eqn (4.6) to allow evaluation entirely in terms of the mobile phase velocity at the column outlet [41]. For gas–solid chromatography the stationary-phase mass transfer term is replaced by a function describing the kinetics of adsorption and desorption from a solid surface, which is often the dominant term in the plate height equation for inorganic oxides and chemically bonded adsorbents [4,5,42].

Variation of the column plate height against the mobile phase velocity is a hyperbolic function commonly described by the van Deemter equation [38,39,42,43]

$$H = A + B/u + (C_S + C_M)u \qquad (4.7)$$

The equation coefficients can be superficially assigned to A = flow anisotropy, B = axial diffusion, and the C terms the contributions from resistance to mass transfer in the stationary and mobile phases to the column plate height. The coefficients of the van Deemter equation are empirical, however, and are not simply related to the various terms in Eqn (4.6). An important general contribution of the van Deemter equation was the illustration that an optimum mobile phase velocity existed for a column at which its highest efficiency is realized, H_{min}. Some typical values are summarized in Table 4.1. Since the region around u_{opt} is usually shallow with a fairly flat ascending portion at $u > u_{opt}$, for less demanding separations columns are often operated at $u > u_{opt}$ to reduce the separation time with a modest reduction in column performance.

From Eqn (4.6) it is clear that the efficiency of a packed column is never less than the contribution from flow anisotropy, and columns should have a homogeneous and densely packed bed to minimize this contribution. In addition, columns should be packed with small particles of a narrow size range and coated with a thin, homogeneous film of stationary phase. Possibilities for improving the performance of packed columns by simultaneously increasing the column length and inlet pressure are rather limited. Performance can be substantially improved by decreasing the particle size and increasing the column inlet pressure simultaneously [4,5]. Decreasing the particle size is accompanied by a shift in u_{opt} to higher values requiring a further increase in the column inlet pressure. In practice, the typical range of inlet pressures controls the absolute particle size. Also, typical packing materials have a range of particle sizes and film thicknesses resulting in lower performance than predicted by models assuming these parameters to be constant. For high stationary-phase loadings, 25–35% (w/w), slow diffusion in the stationary phase is the principal cause of band broadening. For lightly loaded columns (<5% w/w), resistance to mass transfer in the mobile phase is generally more important provided that the stationary phase completely wets the support.

4.3. GAS–SOLID CHROMATOGRAPHY

Gas–solid chromatography is used for applications that can broadly be characterized as those difficult to achieve by gas–liquid chromatography above ambient temperatures [9,10,12,42,45–47]. These include the separation of gases, solvents, and volatile hydrocarbons and halocarbons (typically compounds containing <12 carbon atoms and with a boiling point <200 °C). Retention results from adsorption on surfaces with different types and number of active sites providing for complementary selectivity to liquids and an enhanced capability for the separation of isomers and isotopomers. The carrier gas can play a significant role in the separation process by competing with analyte molecules for adsorption at active sites on the stationary phase. Selectivity, therefore, can be modified by the selection of the carrier gas and by the use of gas mixtures [48]. One or more of the following features may be detrimental to applications in gas–solid chromatography: (i) sample-size-dependent retention and asymmetric peak shapes resulting from nonlinear adsorption isotherms; (ii) the presence of chemically active adsorption sites resulting in irreversible binding of some analytes; (iii) the high surface areas and adsorption energies of adsorbents can result in excessive retention for compounds of low volatility and for polar compounds; (iv) column efficiency is often less favorable for adsorbents compared to liquids; (v) some adsorbents are catalytically active; and (vi) the reproducible preparation of adsorbents is generally more difficult than for liquids. Most adsorbents are sufficiently robust to allow dry packing of columns.

4.3.1. Inorganic Oxides

The most important inorganic oxide adsorbents are silica gel and alumina. These are available as spherical porous beads in a wide range of particle sizes with surface areas of 5–500 m^2/g and pore diameters of 8–150 nm. Retention is controlled by the specific surface area, surface deactivation method, the thermal history of the absorbent, and the capability of analytes to participate in specific interactions, such as hydrogen bonding, with surface functional groups. The different surface functional groups (silanol groups in the case of silica and aluminum ions in the case of alumina) result in different selectivities for these adsorbents. Since both adsorbents tenaciously adsorb moisture, reproducible retention requires the rigorous exclusion of moisture from the carrier gas and samples. Mixing these adsorbents with a diatomaceous support or coating a diatomaceous support with a layer of fine-particle adsorbent (dusted columns) reduces retention and simultaneously improves efficiency [49]. A more general approach to control retention and selectivity is to coat the adsorbent with a small amount of liquid stationary phase or an inorganic salt [50]. Alkali metal salts (potassium chloride and carbonate, sodium sulfate, etc.) at loadings of 0.5–30% (w/w) are common salt modifiers. At low loadings of salt or liquid phase, retention is reduced with less effect on selectivity, while at higher loadings, a mixed retention mechanism dominates and selectivity may be changed significantly from that of the bare adsorbent. Inorganic salts have their own adsorption characteristics that are different from silica and alumina, and, in the case of liquid phases, adsorption at the liquid interface and gas–liquid partition are possible contributors to the retention mechanism.

4.3.2. Carbon Adsorbents

Graphitized carbon blacks and carbon molecular sieves are the main types of carbon adsorbents used in gas–solid chromatography [51]. Graphitized carbon blacks have low porosity with surface areas from about 5 to 100 m^2/g. Ideally, they behave as nonspecific adsorbents with retention dominated by dispersion interactions. In reality, the presence of a small number of polar sites associated with surface oxygen complexes introduces strong specific interactions with polar compounds. Low loadings of liquid phase (0.2–5% w/w) are frequently used to mask high-energy adsorption sites and to modify selectivity. Coating graphitized carbon blacks with phosphoric acid or potassium hydroxide facilitates the separation of organic acids and amines. The flat surfaces of graphitized carbon blacks are particularly well suited to the separation of structural and geometrical isomers.

Carbon molecular sieves consist of small graphite crystallites cross-linked to yield a disordered cavity-aperture structure. They are microporous with a large surface area, 200–1,200 m^2/g, and a pronounced retention of organic compounds. Primary applications include the separation of inorganic gases, hydrocarbons containing less than four carbon atoms, and the separation of small polar molecules such as water and formaldehyde. Higher-molecular-weight compounds may be irreversibly bound at available desorption temperatures.

4.3.3. Molecular Sieves

Molecular sieves (zeolites) are artificially prepared alkali metal aluminosilicates. For gas–solid chromatography, the most common types are calcium aluminosilicate (type 5A) with an effective pore diameter of 0.5 nm and sodium aluminosilicate (type 13X) with an effective pore diameter of 1 nm. The molecular sieves

TABLE 4.10 Physical Properties of Porous Polymer Beads for Packed Column Gas Chromatography

Porous polymer	Monomers	Surface area (m^2/g)	Average pore diameter (nm)	Temperature limit (°C)
Chromosorb 101	STY-DVB	30–40	300–400	275
Chromosorb 102	STY-DVB	300–400	8.5–9.5	250
Chromosorb 103	STY	15–25	300–400	275
Chromosorb 104	ACN-DVB	100–200	60–80	250
Chromosorb 105	Polyaromatic	600–700	40–60	250
Chromosorb 106	STY	700–800	5	225
Chromosorb 107	Acrylic ester	400–500	8–9	225
Chromosorb 108	Acrylic ester	100–200	25	225
Porapak N	STY-DVB-VPO	225–500	9	200
Porapak P	STY-DVB-EVB	100–200	7.5–10	250
Porapak Q	EVB-DVB	500–850	7.5–10	250
Porapak R	STY-DVB-VPO	450–600	7.5–10	250
Porapak S	STY-DVB-VP	300–550	7–9	250
Porapak T	EGDMA	250–450	9	200
Porapak PS	Silanized P			
Porapak QS	Silanized Q			250
Tenax-GC	DPO	19–30	25–7500	375

STY = styrene; DVB = divinylbenzene; ACN = acrylonitrile; EVB = ethylvinylbenzene; EGDMA = ethylene glycol dimethacrylate; VPO = vinylpyrolidone; VP=vinylpyridine; DPO = 2,6-diphenyl-p-phenylene oxide
Values for surface area vary widely in the literature

TABLE 4.11 Common Packed Column Applications of Porous Polymers

Polymer	Application
Chromosorb 101 Porapak P and PS	Esters, ethers, ketones, alcohols, hydrocarbons, fatty acids, aldehydes and glycols. Not recommended for amines and anilines.
Chromosorb 102 Porapak Q	Light and permanent gases, volatile carboxylic acids, alcohols, glycols, ketones, hydrocarbons, esters, nitriles and nitroalkanes. Not recommended for amines, anilines. Nitrated by nitrogen oxide gases.
Chromosorb 103 Porapak S	Amines, amides, alcohols, aldehydes, hydrazines and ketones. Not recommended for acids, amines, glycols, and nitriles. Reacts with nitroalkanes.
Chromosorb 104	Nitriles, nitro compounds, sulfur gases, ammonia, carbon dioxide, vinyl chloride, moisture in solvents and xylenols. Not recommended for amines and glycols.
Chromosorb 105 Porapak N	Aqueous mixtures of formaldehyde, acetylene from lower hydrocarbons and most gases. Not recommended for glycols, acids and amines.
Chromosorb 106 Porapak QS	Alcohols, C_2-C_5 carboxylic acids, alcohols and sulfur gases. Not recommended for glycols and amines.
Chromosorb 107 Porapak T	Formaldehyde from water and acetylene from lower hydrocarbons. Sulfur compounds. Not recommended for glycols and amines.
Chromosorb 108	Gases, water, alcohols, aldehydes and glycols.
Porapak R	Esters, ethers, nitriles and nitro compounds. Not recommended for glycols and amines.
Tenax-GC	High boiling polar compounds, diols, phenols, methyl esters of dicarboxylic acids, amines, diamines, ethanolamines, amides, aldehydes and ketones.

are microporous with a tunnel-like pore structure of similar dimensions to small molecules. Retention is primarily controlled by molecular size, whether the analyte can enter the pore structure of the molecular sieve, and by the strength of adsorption interactions with the internal pore surface. They are used primarily for the separation of gases and low-molecular-weight hydrocarbons [52].

4.3.4. Porous Organic Polymers

A number of porous organic polymer beads prepared from different monomers and cross-linking reagents with a range of surface areas and pore sizes are used in gas–solid chromatography (Table 4.10) [12,47,53,54]. Beads with pore diameters <10 nm are used primarily for the separation of gases and those with larger pore diameters for volatile organic compounds (Table 4.11). Compared with other adsorbents their surfaces are relatively inert, facilitating the separation of polar compounds with little difficulty. The natural microporosity of polymer beads results in modest column efficiency. The retention mechanism is not well understood. At low temperatures, adsorption is expected to be the dominant mechanism, while at higher temperatures, it is possible that the porous polymers behave as highly extended liquids with solvation properties.

References

[1] C.F. Poole, The essence of chromatography, Elsevier, Amsterdam, 2003.

[2] T. Herraiz, G. Reglero, M. Herraix, R. Alonso, M.D. Cabuzundo, Micropacked capillary columns: suitable alternative to very thick capillary columns, J. Chromatogr. 388 (1987) 325–333.

[3] M. Inoue, Y. Saito, I. Ueta, T. Miura, H. Ohkita, K. Fujimura, et al., Rapid temperature-programmed separation and retention prediction on a novel packed-capillary column in gas chromatography, Anal. Sci. 26 (2010) 687–691.

[4] R.J. Jonker, H. Poppe, J.F.K. Huber, Improvements of speed of separation in packed column gas chromatography, Anal. Chem. 54 (1982) 2447–2456.

[5] M.M. Robinson, K.D. Bartle, P. Myers, High-pressure microcolumn gas chromatography, J. Microcol. (1998) 115–123. Sep. 10.

[6] M. van Lieshout, M. van Deursen, R. Derks, H-G. Janssen, C. Cramers, A practical comparison of two recent strategies for fast gas chromatography. Packed capillary columns and multicapillary columns, J. Microcol. (1999) 155–162. Sep. 11.

[7] D.C. Fenimore, J.H. Whitford, C.M. Davis, A. Zlatkis, Nickel gas chromatographic columns: an alternative to glass for biological samples, J. Chromatogr. 140 (1977) 9–15.

[8] Y. Takayama, T. Takeichit, Preparation of deactivated metal capillary for gas chromatography, J. Chromatogr. 685 (1994) 61–78.

[9] G.E. Baiulescu, V.A. Ilie, Stationary phases in gas chromatography, Pergamon Press, Oxford, 1975.

[10] K.K. Unger (Ed.), Packings and stationary phases in chromatographic techniques, Dekker, New York, 1990.

[11] H. Rotzsche, Stationary phases in gas chromatography, Elsevier, Amsterdam, 1991.

[12] C.F. Poole, S.K. Poole, Chromatography today, Elsevier, Amsterdam, 1991.

[13] J.A. Yancey, Liquid phases used in packed columns. Part I: polysiloxane liquid phases, J. Chromatogr. Sci. 23 (1985) 161–167.

[14] J.A. Yancey, Liquid phases used in packed columns. Part II: use of liquid phases which are not polysiloxanes, J. Chromatogr. Sci. 23 (1985) 370–377.

[15] E.F. Barry, R.L. Grob, Columns for gas chromatography: performance and selection, Wiley, New York, 2007.

[16] C.F. Poole, R.M. Pomaville, T.A. Dean, Proposed substitution of Apolane-87 for squalane as a nonpolar reference phase in gas chromatography, Anal. Chim. Acta 225 (1989) 193–203.

[17] R.M. Pomaville, C.F. Poole, Thermally stable highly-fluorinated stationary phases for gas chromatography, Anal. Chim. Acta 200 (1987) 151–169.

[18] J.K. Haken, Development in polysiloxane stationary phases in gas chromatography, J. Chromatogr. 300 (1984) 1–77.

[19] G. Park, C.F. Poole, Solvation in weak complexing n-octyl phthalate and n-octyl tetrachlorophthalate solvents by gas chromatography, J. Chromatogr. A 726 (1966) 141–151.

[20] R.F. Koupps, R.S. Henly, Characteristics of EGS and DEGS polyesters in packed GC columns, J. Chromatogr. Sci. 12 (1974) 127–130.

[21] C.F. Poole, K.G. Furton, B.R. Kersten, Liquid organic salt phases for gas chromatography, J. Chromatogr. Sci. 24 (1986) 400–409.

[22] C.F. Poole, S.K. Poole, Ionic liquids as stationary phases for gas chromatography, J. Sep. Sci. 34 (2011) 888–900.

[23] Z. Witkiewicz, J. Oszezudlowski, M. Repelewicz, Liquid-crystalline stationary phases for gas chromatography, J. Chromatogr. A 1062 (2005) 155–174.

[24] G.J. Price, The use and properties of mixed stationary phases in gas chromatography, Adv. Chromatogr. 28 (1989) 113–163.

[25] J.H. Purnell, Window analysis: an approach to total optimization in chromatography, in: F. Bruner (Ed.), The science of chromatography, Elsevier, Amsterdam, 1985, pp. 363–379.

[26] D.M. Ottenstein, Column support material for use in gas chromatography, J Chromatogr Sci 25 (1987) 536–546.

[27] J.F. Paltraman, E.A. Walker, Techniques in gas chromatography part 1. Choice of solid supports, Analyst 92 (1967) 71–82.

[28] A. Zlatkis, V. Pretorius (Eds.), Preparative gas chromatography, Marcel Dekker, New York, 1971.

[29] C.F. Poole, S.K. Poole, Characterization of the solvent properties of gas chromatographic liquid phases, Chem. Revs. 89 (1989) 377–395.

[30] B.R. Kersten, C.F. Poole, K.G. Furton, Ambiguities in the determination of McReynolds stationary phase constants, J. Chromatogr. 411 (1987) 43–59.

[31] C.F. Poole, T.O. Kollie, S.K. Poole, Recent advances in solvation models for stationary phase characterization and the prediction of retention in gas chromatography, Chromatographia 34 (1992) 281–302.

[32] M.H. Abraham, C.F. Poole, S.K. Poole, Classification of stationary phases and other materials by gas chromatography, J. Chromatogr. A 842 (1999) 79–114.

[33] S.K. Poole, T.O. Kollie, C.F. Poole, The influence of temperature on the mechanism by which compounds are retained in gas-liquid chromatography, J. Chromatogr. A 664 (1994) 229–251.

[34] J.R. Condor, C.L. Young, Physicochemical measurements by gas chromatography, Wiley, Chichester, 1979.

[35] R.N. Nikolov, Identification and evaluation of retention mechanisms in gas-liquid chromatographic systems, J. Chromatogr. 241 (1982) 237–256.

[36] C.F. Poole, S.K. Poole, Foundations of retention in partition chromatography, J. Chromatogr. A 1216 (2009) 1530–1550.

[37] B.R. Kersten, C.F. Poole, The influence of concurrent retention mechanisms on the determination of stationary phase selectivity in gas chromatography, J. Chromatogr. 399 (1987) 1–31.
[38] J.C. Giddings, Dynamics of chromatography – part 1, Marcel Decker, New York, 1965.
[39] E. Grushka, L.R. Snyder, J.H. Knox, Advances in band spreading theories, J. Chromatogr. Sci. 13 (1975) 25–37.
[40] J.H. Knox, Band spreading in chromatography: a personal view, Adv. Chromatogr. 38 (1998) 1–49.
[41] R.P.W. Scott, Introduction to analytical gas chromatography, Marcel Decker, New York, 1998.
[42] G. Guiochon, C.L. Guillemin, Quantitative gas chromatography for laboratory analyses and on-line process control, Elsevier, Amsterdam, 1988.
[43] S.J. Hawkes, Modernization of the van Deemter equation for chromatographic zone dispersion, J. Chem. Edu. 60 (1983) 393–398.
[44] P. Thumneum, S. Hawkes, The obstruction factor γ in gas chromatography, J. Chromatogr. Sci. 14 (1981) 576–578.
[45] T. Paryjczak, Gas chromatography in adsorption and catalysis, Wiley, Chichester, 1986.
[46] C.J. Cowper, A.J. DeRose, The analysis of gases by chromatography, Pergamon Press, Oxford, 1983.
[47] L. Henrich, Recent advances in adsorption chromatography for analysis of light hydrocarbons in petrochemical-related materials, J. Chromatogr. Sci. 26 (1988) 198–203.
[48] V.G. Berezkin, I.V. Malyukova, The influence of the carrier gas on the retention parameters and height equivalent to the theoretical plate in gas–solid chromatography, Russ. Chem. Revs. 67 (1998) 761–781.
[49] W.K. Al-Thamir, J.H. Purnell, R.J. Laub, Enhancement of gas–solid chromatographic column performance by inert solid dilution, J. Chromatogr. 188 (1980) 79–88.
[50] K. Naito, M. Endo, S. Moriguchi, S. Takei, Characterization of modified alumina as an adsorbent for gas–solid chromatography, J. Chromatogr. 253 (1982) 205–215.
[51] B. Leboda, A. Lodyga, A. Gierak, Carbon adsorbents as materials for chromatography, I. Gas chromatography, Mat. Chem. Phys. 51 (1997) 216–232.
[52] T.G. Andronikashvili, V.G. Berezkin, N.A. Nadiradze, L.Ya. Laperashvili, Increased effectiveness of chromatographic columns packed with zeolites in the separation of mixtures of some organic compounds, J. Chromatogr. 365 (1986) 269–277.
[53] R. Arshady, Beaded polymer supports and gels. 1 manufacturing techniques, J. Chromatogr. 586 (1991) 181–197.
[54] G. Castello, G. D'Amato, Effect of solute polarity on the performance of Porapak type porous polymers, Chromatographia 23 (1987) 839–843.

CHAPTER

5

Gas–Solid Chromatography (PLOT Columns)

OUTLINE

Since 1960, many laboratories have shown interest in using adsorbents as stationary phases in gas chromatography [1,2]. However, gas–solid chromatography (GSC) was always less attractive than gas–liquid chromatography (GLC). This is probably due to the nonlinear shape of the isotherm of adsorption for many compounds, even in small quantities. Therefore, the peaks obtained by GSC are characterized by larger asymmetry when compared to the same amount of compounds analyzed by GLC. The nonlinear run of this isotherm also causes incomplete recovery of the analyzed compound from the column. Moreover, adsorbents commonly used in GSC have a large surface with much stronger interaction than liquid phases, because of which the columns packed with these adsorbents are, in most cases, useless for analytical purposes when evaluating strongly polar and high-boiling-point compounds. The poor kinetics of transfer of analyzed mass in the columns filled with adsorbents result in lower efficiency compared with the columns of partition chromatography. However, compared with GLC columns, they reveal better selectivity in the case of separation of isotopic and spatial isomers. Most frequently, the GSC is applied to inorganic gases and hydrocarbons of low-molecular-mass analyses, providing by far better separation over the columns with liquid stationary phases.

Gas Chromatography DOI: 10.1016/B978-0-12-385540-4.00005-5

TABLE 5.1 Application Fields of Adsorbents

Adsorbent	Applications
Aluminum oxide with deactivated agent: KCl, Na_2SO_4, mixed salt	Light hydrocarbons C1–C10 isomers, halogen hydrocarbons.
Molecular sieve 5A	Permanent gases, hydrogen isotopes, oxygen, nitrogen, nitrous oxide, carbon monoxide, light hydrocarbons C1–C3.
Molecular sieve 13X	Hydrogen, oxygen, nitrogen, methane, carbon monoxide, paraffins, and naphthenic C1–C12.
DVB porous polymer (type Q)	Hydrocarbon (natural gas, refinery gas, C1–C3 isomers), CO_2, air/CO and water, polar solvents (methanol, acetone, methylene chloride, alcohols, ketones, aldehydes, esters), sulfur compounds (H_2S, mercaptans, COS).
DVB vinylpyridine polymer (type S)	Hydrocarbons and medium polarity volatiles, including hydrocarbons and ketones, esters, halogenated compounds.
DVB ethyleneglycol-dimethylacrylate polymer (type U)	Hydrocarbons (natural gas, refinery gas, C1–C7, all C1–C3 isomers ethane/ethylene, propylene/propane), CO_2, air/CO, water, polar volatiles, nitriles/nitro-compounds, alcohols/aldehydes, sulfur gases.
Carbon	Light hydrocarbons C1–C5, inorganic gases.
Silica	Light hydrocarbons C1–C10, C1–C4 isomers, sulfur gases, CFCs, semi-polar compounds, samples containing moisture.

Yet, in the literature one can find columns coated with polysiloxane phases, frequently used for analysis of volatile compounds. However, due to the development of thick films, the applied efficiency of such columns decreases faster than observed for columns with traditional thickness of the stationary phase. Moreover, wall-coated open-tubular (WCOT) capillary columns used for analysis of this group of compounds have to operate at very low temperatures, requiring the use of coolants such as dry ice or liquid N_2.

The advantage of adsorbents in comparison with liquid stationary phases also stems from their larger resistance for oxygen included in gas samples of compounds to be analyzed. Since columns packed with adsorbents cannot compete with capillary columns, the necessity of the introduction of capillary columns coated with a porous adsorbent became obvious.

As a result, a new type of capillary columns coated with adsorbents, named "porous layer open tubular" (PLOT), became available. As it turned out in practice, due to different properties of adsorbents, methods of coating used thus far required suitable adjustment for usage with a new type of material. In addition, research laboratories intended to develop layers with sufficient durability, which in fact are not easy to produce.

The PLOT columns have a different appearance compared to classical WCOT columns due to the visible presence of adsorbent deposits on the inner walls of the capillaries. Hence, these appear as opaque white or black capillaries depending on the type of adsorbent used.

The most frequently used adsorbents in the GSC method are alumina, molecular sieves, porous polymers, carbon, and silica. Each adsorbent has its own specific application field, as summarized in Table 5.1.

5.1. ALUMINA ADSORBENTS

Alumina is a type of adsorbent showing separation properties that are useful in the analysis of volatile hydrocarbons. For alumina coatings, grains of 2 μm or smaller are used, giving a film of 4–25-μm thickness. The dimensions of columns most frequently applied are 0.25-, 0.32-, and 0.53-mm diameters and 10–50-m length. This adsorbent is prepared by heat-treating hydrated alumina at 300−1000 °C [3−5]. The physical properties of this adsorbent make the preparation of columns coated with a stable layer deposit a difficult operation. Therefore, the earlier capillary columns were often plugged. A scanning electron micrograph of a porous layer containing particles is presented as an example in Figure 5.1.

The other common problem of preparation of PLOT columns coated with alumina was very high activity of the surface of adsorbent, which caused the asymmetry of observed peaks and hence the decrease in the selectivity of the column. Therefore, in this case, the deactivation process was necessary. The preliminary test was carried out with water [1]. To do this, the humid carrier gas was passed through the column. However, this method was limited to the analytical separation processes carried out isothermally at moderate temperatures. As a result the other methods devoid of these inconveniences were appreciated. Schneider, Frone, and Bruderreck [6] proposed a more favorable method of deactivation of the Al_2O_3 surface with potassium chloride. Using this method, the analysis with programmed temperature up to 200 °C was possible. At higher temperatures, this method was useless due to recrystallization of used salt. For the deactivation process, in addition to KCl, other salts such as NaCl, NaI, NaOH, KOH, K_2CO_3, and K_2HPO_4 were applied [4,5,7,8]. In addition, the deactivation of the Al_2O_3 surface was carried out with the liquid stationary phases used in GLC. Most frequently, these were silicone oil, squalene, carbowax, or diphenylphthalate. With such a wide range of deactivation agents, it was possible to make columns coated with different polarities of the Al_2O_3 surface.

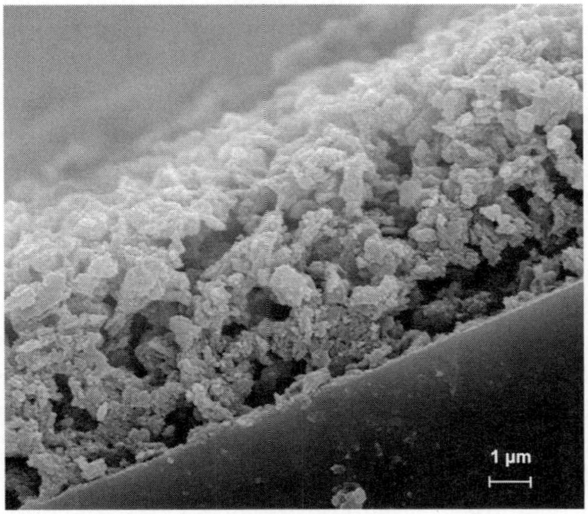

FIGURE 5.1 Scanning electron micrograph of an alumina layer on a 0.53-mm ID fused silica column.

Capillary columns coated with aluminum oxide offer high selectivity for separating ppm levels of C1−C5 hydrocarbons in a main stream of C1−C5 hydrocarbons. These columns analyze more compounds in a single run than packed columns, while still delivering higher resolution and faster analysis times. When compared to liquid stationary phases, the Al_2O_3 PLOT column offers increased selectivity and allows all C1−C5 hydrocarbon isomers to be separated. Al_2O_3 PLOT columns operate without the need for subambient cooling.

An example of this is the separation of impurities in high-purity propylene presented in Figure 5.2. The separation was performed on the column of low polarity, deactivated with KCl. Next, Figure 5.3 presents a separation using the column deactivated with Na_2SO_4. This impacts the separation of the mixture of hydrocarbons C1−C5.

Due to the strong adsorption properties of the column coated with alumina, it is worth noting that, in this case, some admixture, particularly a polar one such as water or carbon dioxide, can cumulate, leading to a change of retention times of analyzed compounds. This phenomenon is easily noticed during operation at low temperatures. Hence, it is necessary to activate the surface of the column at a higher temperature, periodically. It is also worth remembering that the high activity of the adsorbent can decay the separated compounds

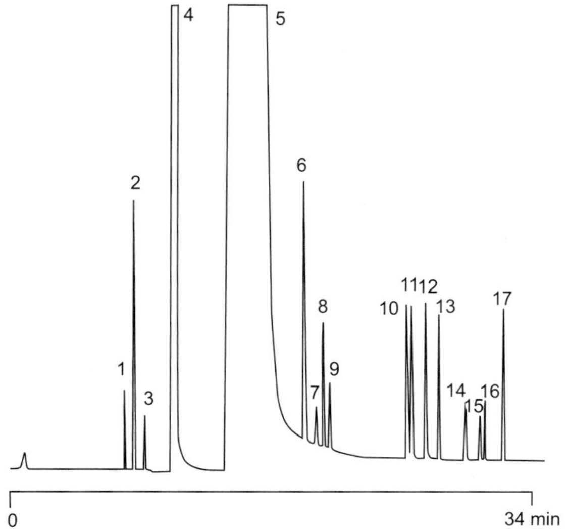

FIGURE 5.2 Separation of impurities in high-purity propylene to proposed ASTM method on the Al_2O_3/KCl PLOT column (application A01312). Column: 50 m × 0.53 mm fused silica, df = 10 μm; temperature: 40 °C (10 min) 5 °C/min to 160 °C; carrier gas: He 80 kPa; injector: valve into splitter, 20:1 (*T* = 200 °C); detector: FID (*T* = 200 °C); sample size: 0.2 μl; peaks: 1 = methane, 2 = ethane, 3 = ethylene, 4 = propane, 5 = propylene, 6 = isobutane, 7 = acetylene, 8 = butane, 9 = propadiene, 10 = *trans*-2-butene, 11 = 1-butene, 12 = isobutene, 13 = *cis*-2-butene, 14 = isopentane, 15 = propyne, 16 = pentane, 17 = 1,3-butadiene. *Source: Reproduced with permission from Agilent, copyright © Agilent Technologies Inc.*

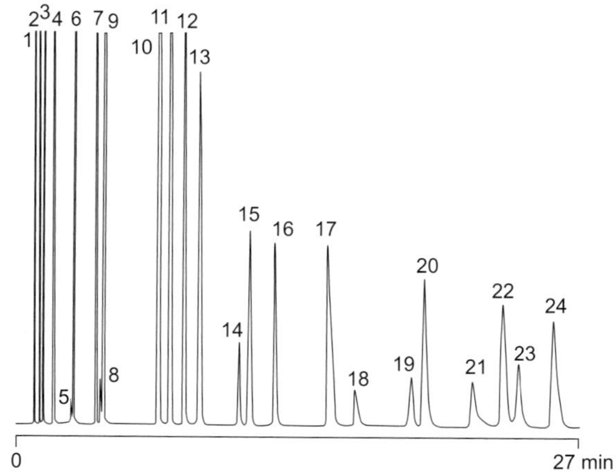

FIGURE 5.3 Separation of C1–C5 hydrocarbons on the Al_2O_3/Na_2SO_4 PLOT column (application A00611). Column: 50 m × 0.53 mm fused silica, df = 10 μm; temperature: 120 °C; carrier gas: He 50 kPa; injector: splitter (*T* = 225 °C); detector: FID (*T* = 250 °C); peak: 1 = methane, 2 = ethane, 3 = ethylene, 4 = propane, 5 = cyclopropane, 6 = propylene, 7 = isobutane, 8 = propadiene, 9 = n-butane, 10 = *trans*-2-butene, 11 = 1-butene, 12 = isobutene, 13 = *cis*-2-butene, 14 = cyclopentane, 15 = isopentane, 16 = n-pentane, 17 = 1,3-butadiene, 18 = propyne, 19 = cyclopentene, 20 = *trans*-2-pentene, 21 = 2-methyl-2-butene, 22 = 1-pentene, 23 = 2-methyl-1-butene, 24 = *cis*-2-petene. *Source: Reproduced with permission from Agilent, copyright © Agilent Technologies Inc.*

during chromatography; in particular, this applies to halogen derivatives. The commercial columns – PLOT, coated with Al_2O_3 – were originally introduced by Chrompack (1983), J&W (1990), Hewlett-Packard (1994), Supelco (1995), and Restek (1995). The deactivation of the surface of these commercial columns was most frequently carried out with KCl or Na_2SO_4.

5.2. MOLECULAR SIEVES

The molecular sieve and zeolite have been used for separation of fixed gases such as CO, CH_4, O_2, and N_2. In GC, type 5A calcium alumino-silicate, with an effective pore diameter of 5A, and type 13X, sodium alumino-silicate, with an effective pore diameter of 10 Å are used. Before use, the molecular sieve requires activation by heat treatment to remove adsorbed water. The activation is realized at 250 °C (for 5A) and 350 °C (for 13X) for 18 h. The layer of the column was made with the grains of 10 nm and a film thickness of 10–50 μm. The molecular sieve PLOT columns are offered in 0.32- and 0.53-mm ID dimension and in length from 5 to 30 m. The scanning electron micrograph of a 5A molecular sieve layer inside a capillary column is presented in Figure 5.4.

Manfred Mohnke [9] presented the first attempts of capillary columns loaded with zeolites, although the first commercially available

FIGURE 5.4 Scanning electron micrograph of a molecular sieve layer on a 0.53-mm ID fused silica column.

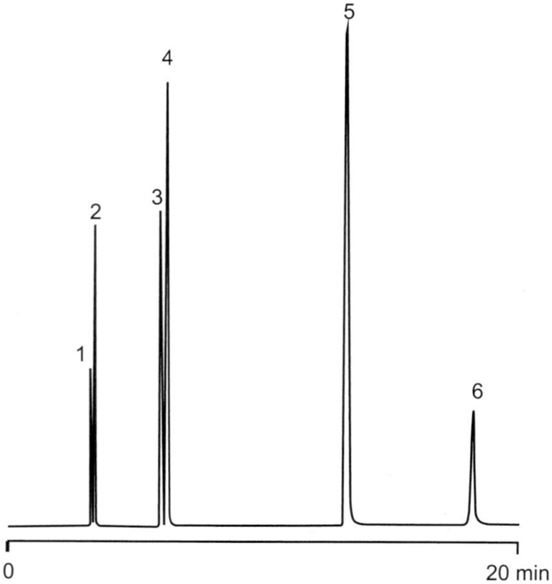

FIGURE 5.5 High resolution of permanent gases on a Molsieve 5A column (5A application A00764). Column: 50 m × 0.53 mm fused silica, df = 50 μm; temperature: 30 °C; carrier gas: H_2 55 kPa; injector: splitter, 200 ml/min (T = 30 °C); detector: TCD (T = 180 °C); peak: 1 = helium, 2 = neon, 3 = argon, 4 = oxygen, 5 = nitrogen, 6 = methane. *Source: Reproduced with permission from Agilent, copyright © Agilent Technologies Inc.*

capillary columns coated with molecular sieves appeared at the end of the 1980s.

The molecular sieves of pore diameters of 5A appear to be ideal adsorbents for the separation of permanent gases. The use of molecular sieves as coatings in capillary columns has the advantage of shortening the time necessary for analysis by as much as 75% when compared with packed columns. However, due to these adsorbing and catalytic properties, they are useless for analytical purposes of hydrocarbons longer than two carbon atoms. A separation of a permanent gas mixture is shown in Figure 5.5. The separation was performed on a column of 0.53-mm diameter and 50-m length coated with molecular sieves of 5A and 50-μm film thickness. This column enables separation of argon and oxygen, as well as helium and neon with baseline, without expensive cryogenic cooling. However, if we do not intend to separate these pairs of gases, then the possible alternative columns are those coated with 13X sieves. These have an efficiency of traditional molecular sieve PLOT columns with a unique selectivity of the 13X molecular sieve. Figure 5.6 shows the rapid and efficient analysis of the permanent gases on the 15-m, 0.32-mm ID Rt-Msieve 13X column. Baseline separation of all compounds is achieved in just over 1.5 min. Moreover, because of their larger pores, the 13X molecular sieves have lower absolute retention, and therefore they are suitable for separation of paraffins and naphthalenes. The introduction of the capillary columns coated with molecular sieves for gas chromatography forced the changes in the design of the TCD detector used for the estimation of fixed gases. Therefore, the TCD detectors were constructed with microcells adapted for carrier gas flow at a level 10 times smaller than that utilized for packed columns.

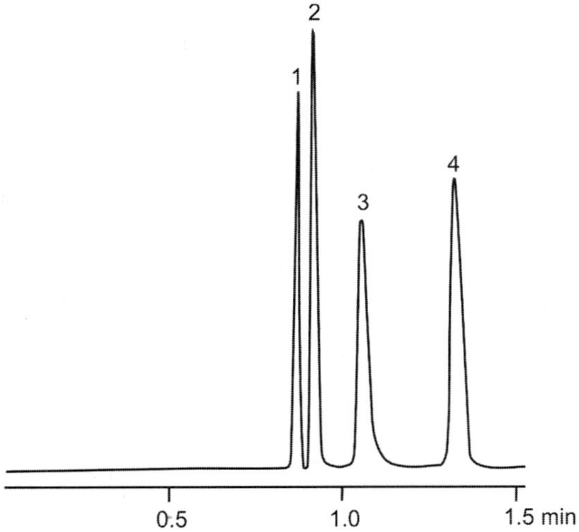

FIGURE 5.6 Analysis of the permanent gases on a Rt-Msieve 13X PLOT column. Column 15 m × 0.32 mm fused silica, temperature: 40 °C; carrier gas: He 1.5 ml/min; injector: splitter 15:1 ($T = 200$ oC); detector: TCD ($T = 200$ °C); sample size: 20 μl; peak: 1 = oxygen, 2 = nitrogen, 3 = methane, 4 = carbon monoxide. *Source: Reproduced with permission from R. Wawrzyniak, W. Wasiak (2003) Journal of Separation Science 26: 1219, copyright © John Wiley & Sons, Inc.*

5.3. POROUS POLYMERS

The porous polymers are useful materials in gas chromatography. These are copolymers of polydivinylbenzene (DVB). DVB polymers, in relation to a highly cross-linked structure, are very porous, ranging from mesoporous to microporous. The trade names include HayeSep, Poropak, and Chromosorb Century. Three kinds of polymers differing in polarity named Q, S, and U are in use. The polymer Q is a DVB–styrene copolymer [10], S is a DVB–pyridine one [11], and finally U is a DVB–ethylene glycol dimethylacrylate copolymer [12]. The polarity of these adsorbents increases in the rank of Q > S > U. Unfortunately, along with the polarity increase, the thermal stability decreases; for the Q polymer it is 250 °C and for U it is only 190 °C. One of the latest developments is the improved stabilization of the styrene–divinylbenzene polymer, which has resulted in the introduction of a high-temperature stable material. This means that CP-PoraBOND Q can be used up to 320 °C without decomposition.

In the above-mentioned set of polymers, the most popular is the Q polymer. This is used most often due to its high retention, inertness, and the selectivity toward estimated compounds. This is also characterized by a high hydrophobic behavior, which produces a weak interaction with highly polar compounds and water. This polymer is a favorite in the analysis of alcohols and water, polar solvents, hydrocarbons, and gases containing sulfur. Again, polymer S is frequently used for analysis of ketones, esters, halogenated compounds, and some hydrocarbons. Next, polymer U is designed for all polar volatiles, nitriles/nitro-derivatives, alcohols/aldehydes, ethane/ethylene, gases containing sulfur, oxygen in the air, and finally the ppm amount of water in samples of different gases. In addition to the above-mentioned polymers, the polymer named "AmiNES" designed for volatile amines, with high retention, is offered by Chrompack. The porous polymer PLOT columns are offered in 0.25-, 0.32-, and 0.53-mm ID dimension with length up to 50 m and film thickness ranging from 8 to 50 μm.

A separation of halogenated hydrocarbons C1–C2 and hydrocarbons C1–C6 on the capillary column coated Q polymer is shown in Figure 5.7. Very sharp peaks of separated compounds make it possible to estimate entities of very low concentration. The results obtained prove well-defined pore-size distribution of the columns used since, in the case of porous polymers, no homogeneous pore-size distribution caused peak broadening occurs and poor detection limits for a number of compounds. Figure 5.8 shows a headspace sampling analysis of volatile compounds at low levels in a complex, "dirty" matrix. The unique selectivity of the polymer U column

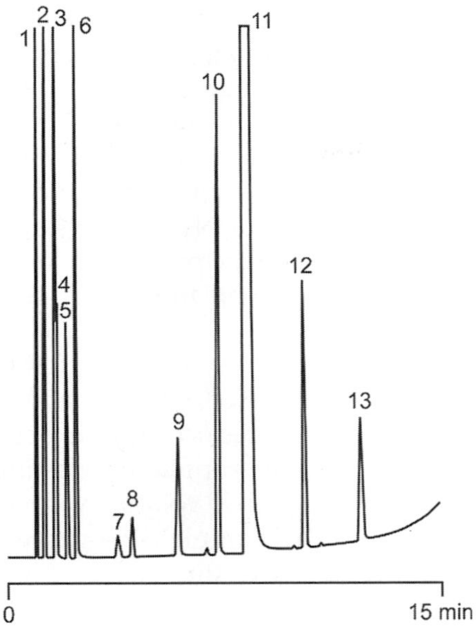

FIGURE 5.7 Separation of CFCs on a CP-PoraBOND Q column (application A01312). Column: 25 m × 0.53 mm fused silica, df = 10 μm; temperature: 100 °C (2 min) 10 °C/min to 250 °C; carrier gas: He 40 kPa; injector: split (T = 250 °C); detector: FID (T = 250 °C); sample size: 50 μl; peak: 1 = methane, 2 = ethane, 3 = CFC 134a, 4 = CFC 22, 5 = propane, 6 = CFC 12, 7 = isobutane, 8 = butane, 9 = CFC 11, 10 = pentane, 11 = CFC 113 + CFC 113a, 12 = hexane, 13 = CFC 112 + CFC 112a. *Source: Reproduced with permission from Agilent, copyright © Agilent Technologies Inc.*

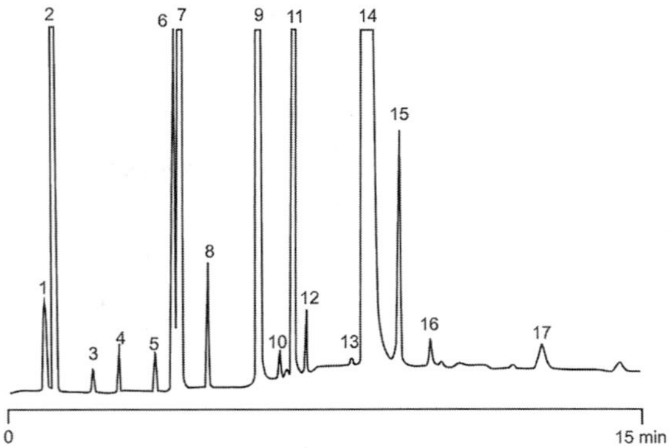

FIGURE 5.8 Impurities in polymer on a CP-PoraBOND U column (application A01338). Column: 10 m × 0.53 mm fused silica, df = 20 μm; temperature: 65 °C (0.1 min) 10 °C/min to 150 °C; carrier gas: He 50 kPa; injector: headspace (T = 40 °C for 30 min); detector: FID (T = 200 °C); peak: 1 = methane, 2 = ethylene, 3 = propylene, 4 = propane, 5 = vinyl chloride, 6 = isobutane, 7 = methanol + acetaldehyde, 8 = n-butane, 9 = ethanol, 10 = 2-methylbutane (isopentane), 11 = acetone, 12 = n-pentane, 13 = 2-propanol, 14 = vinyl acetate, 15 = ethyl acetate, 16 = 1-propanol, 17 = 2-methyl-2-propanol (t-butanol). *Source: Reproduced with permission from Agilent, copyright © Agilent Technologies Inc.*

separates vinyl acetate, vinyl chloride, and other oxygenates and hydrocarbons. Figure 5.10 shows the separation obtained on a capillary column coated S polymer. Thanks to properties of the used polymer, it was possible to have very good separation for polar, as well as nonpolar, compounds.

5.4. CARBON ADSORBENTS

Three types of this kind of adsorbent are commonly used in chromatography: active carbon, graphitized carbon black, and carbon sieve. The columns coated with this adsorbent are called CarbonPLOT, CarboPLOT, or Carbograph.

Bruner and his coworkers at the University of Urbino suggest the name GLOT (graphitized layer open tubular) [13] for these types of columns.

The active carbon is an amorphous form of carbon. It is obtained by destructive distillation of carbonaceous material and their activation at 800–900 °C by carbon dioxide. Active carbon has a porous structure of high adsorptivity properties for many gases and vapors. Graphitized carbon black is a nonporous form of carbon. It has regular crystal, a large surface, and is nonpolar. Carbon sieve is a carbonaceous particle that is inert, is very nonpolar, and has a very high surface area. It is obtained by pyrolysis of a polymer such as divinylbenzene–styrene. The columns coated with this kind of adsorbent are prepared by means of dynamic coating or direct pyrolysis. The dimensions of the columns coated with carbon adsorbent are 10–60 m in length and mostly 0.53 mm in diameter, although 0.32-mm

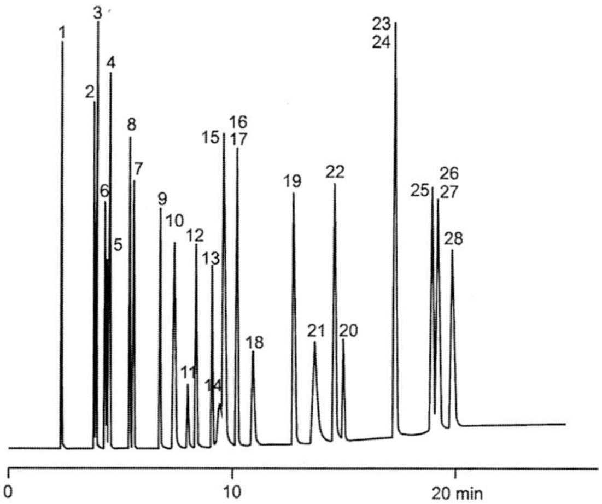

FIGURE 5.9 Separation of polar and nonpolar compounds on an Rt-S-BOND column. Column: 30 m × 0.53 mm fused silica, df = 20 μm: temperature: 120 °C 5 °C/min to 220 °C (5.0 min); carrier gas: H_2 40 cm/s; injector: split (split vent flow 100 ml/min, T = 200 °C); detector: FID (T = 220 °C); sample size: 50 μl; peak: 1 = methanol, 2 = ethanol, 3 = acetonitrile, 4 = acetone, 5 = dichloromethane, 6 = 1,1-dichloroethene, 7 = nitromethane, 8 = *trans*-1,2-dichloroethylene, 9 = *cis*-1,2-dichloroethylene, 10 = tetrahydrofuran, 11 = chloroform, 12 = ethyl acetate, 13 = 1,2-dichloroethane, 14 = 1,1,1-trichloroethane, 15 = benzene, 16 = 1,2-dimethoxyethane, 17 = trichloroethylene, 18 = 1,4-dioxane, 19 = pyridine, 20 = dimethylformamide, 21 = methylcyclohexane, 22 = toluene, 23 = 2-hexanone, 24 = chlorobenzene, 25 = ethylbenzene, 26 = *m*-xylene, 27 = *p*-xylene, 28 = *o*-xylene. *Source: Reproduced with permission from Restek, copyright © Restek Corporation*

diameter columns are also in use. The thickness of the carbon adsorbent layers varies from 1.5 to 25 μm. In the case of thicker films, the problems with their stability are well known, and therefore the maximum temperature that can be applied is 115 °C. Due to poor stability of the layer, the micropacked columns of 0.53-mm diameter are alternatively used with PLOT columns. Columns with thicker films of increased thermal and mechanic stability are available. An example of this group is CP-CarboBOND, which allows for separation processes up to 300 °C.

The columns coated with this type of adsorbent have a unique selectivity for inorganic and organic gases. The active carbon and carbon sieves are used for the estimation of the following gases: hydrogen, nitrogen, oxygen, carbon oxide, carbon dioxide, and hydrocarbons from C1 to C2. Due to the high efficiency of distribution of these columns, the complete separation of CO and N_2 or acetylene and ethylene is possible on the ppm level. Moreover, they are resistant to the moisture present in analyzed samples.

The graphitized carbon black is frequently used in the deactivated form in many liquid stationary phases to reduce the peak tailing. After deactivation, it is applied for estimation of n-alkanes up to C13 volatile vapors and polar solvents. An example of separation on the column coated with this type of adsorbent is given in Figure 5.10. The columns coated with this type of adsorbent are commercially available from Chrompack, J&W, and Supelco. However, the manufacturers do not give the information on which form of "carbon" is used for the preparation of the adsorption layer.

5.4.1. Silica

Silica is a material used from the beginning of a gas chromatography method. For chromatography, two kinds of silica are useful: porous silica and silica gel. These are prepared by the polycondensation method. After drying, silica gel gives a porous material that is available as Porasil and Spherosil. Both kinds of this material are available in varied sizes of pores and also the surface. The grains are porous in their whole volume or in the case of a solid core only on the surface. Silica is a type of material with a large amount of silanols on the surface, which causes a redundant peak tailing effect. The silanols can be removed during a thermal dehydroxylation process [14]. The other deactivation process is the loading of the surface of silica with inorganic salts [14]. The sililation by means of a wide variety of organosilane

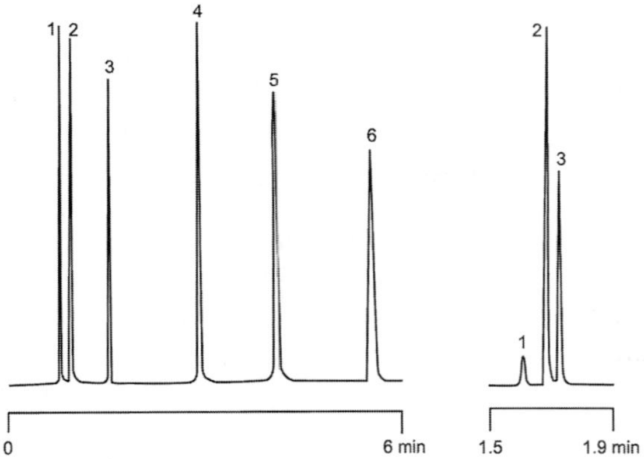

FIGURE 5.10 Analysis of carbon monoxide and carbon dioxide in hydrocarbon streams on a CP-CarboBOND column (application A01431). Column: 50 m × 0.53 mm fused silica, df = 5 μm; temperature: 35 °C (7 min) 30 °C/min 180 °C; carrier gas: H_2 60 kPa; injector: split 1:5 (T = 30 °C); detector chromatogram 1: FID with a Ni-catalyst methanizer, chromatogram 2: TCD (T = 250 °C); sample size: 100 μl; chromatogram 1 peak: 1 = carbon monoxide, 2 = methane, 3 = carbon dioxide, 4 = acetylene, 5 = ethylene, 6 = ethane; chromatogram 2 peak: 1 = helium, 2 = air, 3 = carbon monoxide. *Source: Reproduced with permission from Agilent, copyright © Agilent Technologies Inc.*

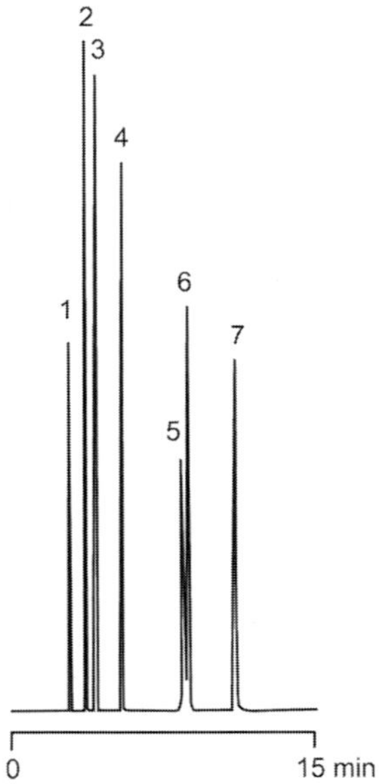

FIGURE 5.11 Separation of halogenated hydrocarbons and C-2 hydrocarbons on a CP-SilicaPLOT column (application A01356). Column: 30 m × 0.32 mm fused silica, df = 4 μm; temperature: 40 °C (2 min) 20 °C/min 200 °C; carrier gas: N_2 50 kPa; injector: split 50 ml/min (T = 200 °C); detector FID (T = 200 °C); sample size: 1 ml in nitrogen; peak: 1 = methane, 2 = ethane, 3 = ethylene, 4 = acetylene, 5 = chloromethane, 6 = vinyl chloride, 7 = chloroethane. *Source: Reproduced with permission from Agilent, copyright © Agilent Technologies Inc.*

compounds [15] is frequently applied with the hope of blocking the silanol groups. Silica in its activated form is seldom used in chromatography.

Silica is a favorable material for analysis of sulfur derivatives such as H_2S, COS, or fluoro and chloro derivatives and ethers. This is also a very useful material in analysis of light hydrocarbons up to four carbon atoms, but, after modification, even up to C10 in a chain. It is very important that the columns coated with silica turn out to be useful for analysis of the samples contaminated with water, since its presence does not influence the retention time of the estimated entities. Deactivated silica is highly inert and so it is possible to estimate halogen compounds with no risk of their decay during the separation process.

The commercial columns PLOT, coated with silica, have been available since 1997. These columns are offered by Chrompack and J&W. An example of a separation of halogenated hydrocarbons and C2 hydrocarbons on a CP-Silica-PLOT column is shown in Figure 5.11. All the C2 isomers are separated with high resolution, which is characterized by sharp and well-separated peaks. This means that the columns are highly selective and inert relative to separated compounds.

5.5. OTHER ADSORBENTS

Frequently, other adsorbents are used to cover the PLOT type of columns, with the most popular being cyclodextrins [16]. The following entities are also used as the coatings of PLOT columns: polyurethanes [17], alkali, and transition metal salts [18–21].

Finally, it is worth mentioning that in addition to the classical capillary system of the columns, some special complex systems including capillary columns exist. For example, analysis in a single-run solution for carbon dioxide and permanent gases. A set of two parallel columns that combine CP-Molsieve 5A for permanent gas analysis and CP-PoraBOND Q for CO_2 analysis was used.

5.5.1. Separation Mechanism

In the case of PLOT columns, the size, strength of dipoles, and polarizability of the adsorbent layer are crucial for the separation of an analyte. The separation made with the columns covered with molecular sieves 5A, 13X, porous polymers, and carbon sieves consists of sieving or sizing. This separation depends on the mechanism in which the analytes smaller than the size of pores can penetrate into the grain of adsorbent, causing relatively larger retention times. However, compounds that are too large to enter the pores will only retain relatively weak adsorption onto active sites on the outside of the particles, thus giving shorter retention times. For example, a pair of isobutane/methane where, on the column coated with molecular sieves, the isobutane is eluted much earlier than methane. The reverse has to be posited for elution of linear n-butane, which follows methane. Separation on the columns coated with aluminum oxide is based on strong dipole interaction. This applies to both active alumina and alumina deactivated with inorganic salts. In the case of nonpolar hydrocarbons, the dipole interactions are evidently weak. However, for hydrocarbons with multiple bonds, the interactions are strong enough that they are not eluted in accordance with their boiling temperatures.

Again, the polar hydrocarbons in the above interactions are so strong that their separation on the column coated with Al_2O_3 is impossible.

Porous polymers provide a separation based on a phenomenon of polarizability, e.g., the inducted dipole at the phenyl group of DVB interacts with the analyte. However, the interactions with the inducted dipoles are much weaker than with the "normal" dipoles mentioned above. Therefore, these types of columns are useful for separation of the polar compounds. On the other hand, the weaker interactions result in the unsuccessful separation of nonpolar compounds compared with the results obtained on the column coated with Al_2O_3.

5.5.2. Column Dimensions

The majority of the PLOT columns commercially available are produced in dimensions of 0.53-mm, 0.32-mm, and 0.25-mm diameter and length up to 60 m. The thickness of the adsorbent layer varies from 5 to 50 μm, depending on the type of adsorbent used. The exact data for given columns are presented in Table 5.2. In addition, the list of commercially available columns supplied by different manufacturers is given in Table 5.3.

5.5.3. Coating Techniques of Adsorbent Layers

The quality of the column strongly depends on the technique of coating. The selected technique influences the capacity factor, coating efficiency, selectivity, and inertness of the prepared column. In the case of PLOT columns, the most popular are static coating, dynamic coating, and *in situ* polarization.

The static coating technique was introduced by J. Bouche and M. Verzele in 1968 [22], and has been used for the preparation of porous polymer, molecular sieve, and silica PLOT columns. In this technique, the column is filled

TABLE 5.2 Parameters of PLOT Columns

Adsorbent	Column parameter			
	Internal diameter (mm)	Length (m)	Film thickness (μm)	Temperature range
Aluminum oxide with deactivated agent: KCl, Na_2SO_4, mixed salt	0.25, 0.32, 0.53	10–50	5–25	To 200
Molecular sieve 5A	0.32, 0.53	10–50	10–50	To 350
Molecular sieve 13X	0.32, 0.53	5–30	–	To 350
DVB porous polymer (type Q)	0.25, 0.32, 0.53	10–50	8–50	To 310
DVB vinylpyridine polymer (type S)	0.25, 0.32, 0.53	10–50	8–20	To 250
DVB ethyleneglycoldimethylacrylate polymer (type U)	0.25, 0.32, 0.53	10–50	8–20	To 190
Carbon	0.32, 0.53	15–60	1.5–25	To 300
Silica	0.32, 0.53	15–30	3–6	To 225

with a stable suspension "solution." In the next step, the solvent is evaporated under vacuum, producing a coating layer on capillary walls. Finally the wet coating layer is dried by a continuous gas flow.

The dynamic coating technique was first described by Dijkstra and DeGoey [23] and has been applied to molecular sieves, zeolite, and alumina columns. In this technique, the suspension solution plug is first inserted into the column using high pressure gas, and then is pushed through the capillary column at a steady speed, leaving a coating layer behind the meniscus of the plug. A buffer capillary is attached to the column end as a restrictor, to avoid acceleration of the solution plug near the exit of the capillary column being coated.

In situ polymerization was described by Hollis in 1973 [24]. This technique is utilized for the preparation of PLOT columns with molecular sieves, zeolite, porous polymer, and silica. For the preparation of the column, a solution of monomer and catalyst is inserted into a column. Next, the column is heated for polymerization, producing a porous layer inside the column. After the reaction, the residue solvents, monomer, and catalyst are removed from the column by gas purging.

Newly prepared columns are required to be conditioned for up to 24 h. This process removes solvent residue and low-molecular-mass compounds of adsorbent, and activates the porous layer by desorption of compounds adsorbed onto the surfaces of adsorbent.

The details concerning the preparation of the solutions used for coating in the above-mentioned techniques are given by Mohnke and Heybey [25].

5.5.4. Application of PLOT Columns

When using PLOT columns, it is worth remembering that they have a smaller sample capacity compared to WCOT columns, i.e., the sample capacity of the PLOT column is only 1% of the sample capacity of the relevant WCOT column. The analytical result is that the tailing peak is observed for the former and the leading peak for the latter columns. Generally, the PLOT columns show smaller efficiency

TABLE 5.3 Manufacturers of PLOT Columns

Porous layer	Agilent/J&W	Alltech	Restek	Supelco	Varian/Chrompack*	Quadrex
Aluminum oxide	GS-Alumina GS-Alumina KCl HP-PLOT Al_2O_3 KCl HP-PLOT Al_2O_3 S HP-PLOT Al_2O_3 M	AT-Alumina	Rt-Alumina BOND/CFC Rt-Alumina BOND/ Na_2SO_4 MXT-Alumina BOND/ Na_2SO_4 Rt-Alumina BOND/KCl	Alumina-PLOT	CP-Al_2O_3 KCl CP-Al_2O_3 Na_2SO_4	–
Molecular sieve 5A	GS-Molsieve HP-PLOT Molsieve	AT-Mole Sieve	Rt-Msieve 5A MXT-Msieve 5A	Molsieve 5A PLOT	CP-Molesieve 5A	PLT-5A
Molecular sieve 13X			Rt-Msieve 13X			
DVB porous polymer	GS-Q HP-PLOT Q	AT-Q	Rt-Q-BOND Rt-QS-BOND MTX-Q-BOND	Supel-Q-PLOT	CP-PoraPLOT Q CP-PoraPLOT Q-HT CP-PoraBOND Q	–
DVB vinylpyridine polymer	–	–	Rt-S-BOND MTX-S-BOND	–	CP-PoraPLOT S	–
DVB ethyleneglycoldi-methylacrylate polymer	HP-PLOT U	–	Rt-U-BOND	–	CP-PoraPLOT U CP-PoraBOND U	–
Carbon	GS-CarbonPLOT	Carbograph VOC	–	–	CP-CarboBOND	–
Silica gel	GS-GasPro	–	–	–	CP-SilicaPLOT	–

* *Actually Agilent.*

than their WCOT equivalents. Yet these columns definitely have better selectivity in relation to fixed gas and light hydrocarbons, which is not attributable to WCOT columns. Moreover, a change of selectivity of the PLOT columns is easy to carry out through the careful selection of the deactivating agent. However, in the case of porous polymers, this is possible by a change in the composition of divinylbenzene copolymer.

Regarding the use of PLOT columns, it is worth noting that they are not inert, which can impinge on the reproducibility of analytical results. So, the adsorption of carbon dioxide or nitrous gases onto the alumina layer is irreversible due to the presence of Lewis acid–base centers. Furthermore, the columns covered with Al_2O_3 can dehydrate the analyzed hydrocarbons. However, in the case of porous polymers, the adsorption of H_2S and hydrocarbons can take place in vestigial amounts, disturbing the analytical results of estimation of small quantities of these compounds. Using molecular sieve 5A we can expect a sorption of hydrogen, but in the case of carbon sieves the adsorption of fixed gases takes place.

When using PLOT columns, we have to keep in mind that mechanical stability of the layer is definitely smaller than that of WCOT columns. A change in the carrier gas flow rate or temperature program rate can result in destruction of the adsorbent layer. The low stability of this layer caused baseline noise and spiking, which generated ghost peaks and a decrease in detection sensitivity; which, in extreme cases, can result in a plugging of the chromatographic system. A solution to this problem is an application of an extension column of 1–2 m length coated with a thick film of polydimethylsiloxane, which acts as the so-called "glue" by capturing up the interrupted particles of a sorbent and protecting the detector from interferences. The retention properties of the used film are definitely weaker than applied adsorbents and so do not influence the retention factors of analyzed compounds. The poor mechanistic stability of the layer of PLOT columns forced manufacturers to develop improved methods for adsorbent immobilization. Currently, two such methods are worked out. The first is based on chemical bonding of adsorbent particles directly to the wall of the column, and the second relies on the physical sticking of adsorbent onto the wall of the column by means of a special "glue" [26] (Figure 5.12).

The PLOT columns with chemically bonded adsorbent are resistant to higher gas flows and head pressures as high as 80 psig. As a result of the application of chemical bonding of an adsorbent, it is possible to regenerate the column layer by rinsing it with a solvent.

Physical sticking of an adsorbent is most commonly used in the case of molecular sieves, zeolites, and carbon sieves. However, this method of immobilization of the layer gives the cover much less stability than the above-described chemical bonding.

5.5.5. Evaluation of Quality and Stability of PLOT Layers

The preparation of PLOT columns coated with equal thickness of a film is not an easy operation.

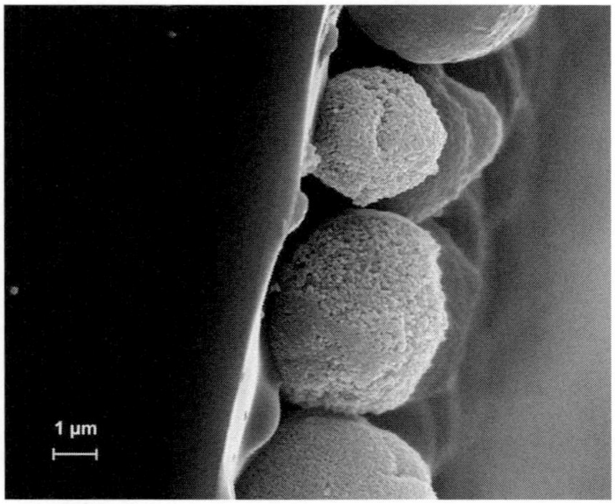

FIGURE 5.12 Scanning electron micrograph of a bonded silica by a special "glue."

The different thicknesses of the obtained stationary layers spoil the laminar flow of the carrier gas. Unfortunately, the usage of PLOT columns allows for the possibility that the thickness of the layer can change during its operation. This is why PLOT columns have lower reproducibility than WCOT columns. In practice, PLOT columns of similar dimensions and column pressure drop can have flow coefficients differing by 4–6 units, which, in the case of any application where the flow rate plays an important role, may cause serious problems. Therefore, Restek Corporation introduced a new factor called "flow restriction factor" (F) [27]:

$F = t_{R1}$ of unretained component (uncoated tubing)/t_{R2} of unretained component (coated column).

This factor is based on the retention time of an unretained marker compound, as measured on both coated and uncoated tubing using the same backpressure setting. In the case of columns coated with adsorbent in this order, methane seems to be most popular. The abovementioned factor of flow restriction – can also be expressed in percentage form. This factor is particularly useful for testing the reproducibility of the coating process of PLOT columns and periodic control of qualities of the layer of the column in use.

References

[1] M.J.E. Golay, Nature 199 (1963) 370–371. doi: 10.1038/199370a0.
[2] J.J. Kirkland, Anal. Chem. 35 (1963) 1295–1297.
[3] J. De Zeeuw, R.C.M. De Nijs, L.T. Henrich, J. Chromatogr. Sci. 25 (1987) 71–83.
[4] K. Naito, R. Kurita, S. Moriguchi, S. Take, J. Chromatogr. 246 (1982) 199–206. doi: 10.1016/S0021-9673(00)95859-X.
[5] K. Naito, R. Kurita, S. Moriguchi, S. Takei, J. Chromatogr. 253 (1982) 205–218. doi: 10.1016/S0021-9673(01)88378-3.
[6] W. Schneider, J. Frone, H. Bruderreck, J. Chromatogr. 155 (1978) 311–327. doi: 10.1016/S0021-9673(00)87992-3.
[7] H. Yang, N. Zou, W.Z. Lu, Sepu 6 (1988) 129–133.
[8] C.G. Scott, C. Phillips, Gas chromatography 1964, Institute of Petroleum, London, 1965.
[9] Mohnke M, Saffert W. Preprints of the 4th international symposium on gas chromatography, Hamburg, Germany, June 13–16 (1962) pp. 214–219.
[10] J. De Zeeuw, R.C.M. De Nijs, J.C. Buyten, J.A. Peene, M. Mohnke, J. High Resolut. Chromatogr. 11 (1988) 162–167. doi: 10.1002/jhrc.1240110204.
[11] Z. Ruan, H. Liu, J. Chromatogr. A 693 (1995) 79–88. doi: 10.1016/0021-9673(94)00951-5.
[12] T.C. Shen, J. Chromatogr. Sci. 30 (1992) 239–240.
[13] E. Bruner, M. Attaran Rezai, L. Lattanzi, Chromatographia 41 (1995) 403–406. doi: 10.1007/BF02318613.
[14] D. Cadogan, D. Sawyer, Anal. Chem. 42 (1970) 190–195.
[15] R. Wawrzyniak, J. Sep. Sci. 32 (2009) 1415–1424. doi: 10.1002/jssc.200800616.
[16] E.J. Smolkova, J. Chromatogr. 251 (1982) 17–34. doi: 10.1016/S0021-9673(00)98506-6.
[17] W.D. Ross, R.T. Jefferson, J. Chromatogr. Sci. 8 (1970) 386–389.
[18] R.L. Grob, E.J. McGonigle, J. Chromatogr. 59 (1971) 13–20. doi: 10.1016/S0021-9673(01)80001-7.
[19] B.H. Gump, J. Chromatogr. Sci. 7 (1969) 755–759.
[20] A.G. Ober, M. Cooke, G. Nickless, J. Chromatogr. 196 (1980) 237–244. doi: 10.1016/S0021-9673(00)80443-4.
[21] W. Wasiak, Chem. Anal. (Warsaw) 33 (1988) 573–580.
[22] J. Bouche, M. Verzele, J. Gas Chromatogr. 6 (1968) 501–505.
[23] D. Dijkstra, J. DeGoey, Gas chromatography 1958, Butterworths, London, 1958, p. 56.
[24] O.L. Hollis, J. Chromatogr. Sci. 11 (1973) 335–342.
[25] M. Mohnke, J. Heybey, J. Chromatogr. 471 (1989) 37–53. doi: 10.1016/S0021-9673(00)94153-0.
[26] R. Wawrzyniak, W. Wasiak, J. Sep. Sci. 26 (2003) 1219–1224. doi: 10.1002/jssc.200301430.
[27] de Zeeuw J, Bromps B, Vezza T, Morehead R, Stidsen G. Advances in porous layer open tubular columns, Technical Article, American Laboratory (2010), http://www.americanlaboratory.com/

CHAPTER

6

Classification and Selection of Open-Tubular Columns for Analytical Separations

Colin F. Poole

OUTLINE

6.1. INTRODUCTION

A simplified blueprint for developing a separation method starts with the principle that the minimum resolution required for the most difficult to separate peak pair in a mixture establishes the difficulty of obtaining a separation and the time required to elute the last peak in the sample under the conditions established by the first criteria fixes the separation time. Peak widths and separation speed are mainly controlled by kinetic parameters while the capability of the stationary phase to differentiate between two compounds (selectivity) is a thermodynamic function. To a large degree, the two contributions to the separation can be treated independently and optimized separately, facilitating different approaches to a partial or complete optimization of a separation. The gas phase is close to ideal for typical separation conditions and selectivity can be considered a property of the stationary phase and column temperature alone.

For any column, resolution is impossible without selectivity or retention. Thus, all separations require a separation factor $\alpha > 1$ and a retention factor $k > 0$. Small changes in the separation factor and retention factor from their minimum values have a significant impact on resolution and the ease of obtaining a useful separation (required plate number, N_{req}) (Table 6.1) [1]. Average retention factors for the most difficult to separate peak pair greater than 5 improve resolution by only a small amount and can lead to longer than necessary separation times. In choosing a separation system, the system with the highest separation factor will facilitate faster separations. The effect of the plate number on a separation is the most predictable separation parameter and can assume a wide range of values in gas chromatography. Its variation by changing the characteristic column dimensions allows many separations to be achieved for moderately complex mixtures under conditions where the separation factor and retention factor are far from optimum. Increasing the column length increases resolution only by the square root of the column length, while the increase in the separation time is proportional to the column length. The maximum column length is ultimately determined by the available column inlet pressure, which increases with the column length. A better strategy to increase the plate number is to reduce the column radius at a fixed column length (Table 6.2) [2]. For the same stationary phase and temperature, observed retention factors are dependent on the column phase ratio (for a partition system the ratio of the volume of gas phase to stationary phase in the column). Thick-film columns, which have a low phase ratio, have a lower intrinsic efficiency (a result of additional band broadening resulting from slow mass transfer in the stationary phase), but thick-film columns then provide greater resolution of volatile compounds through optimization of the retention factor range. Thick-film columns also facilitate the separation of volatile compounds at more convenient temperatures. The opposite argument applies to compounds of low volatility and provides the rationale for the manufacture of columns containing the same stationary phase with different phase ratios.

Nearly all separations in gas–liquid chromatography are achieved with hydrogen, helium, or nitrogen as the carrier gas. Although the choice of carrier gas does not significantly affect the separation factor for typical separation conditions, it can still affect resolution through its effect on efficiency and the optimum mobile-phase velocity for the separation. Differences in gas-phase diffusion coefficients favor the choice of hydrogen and helium for separations on

TABLE 6.1 Plate Count Required for a Peak Resolution of 1 for Different Separation Conditions

Retention factor	Separation factor	N_{req}	Retention factor	Separation factor	N_{req}
3	1.005	1,150,000	0.1	1.05	853,780
3	1.01	290,000	0.2	1.05	254,020
3	1.02	74,000	0.5	1.05	63,500
3	1.05	12,500	1.0	1.05	28,200
3	1.10	3,400	2.0	1.05	15,800
3	1.20	1,020	5.0	1.05	10,160
3	1.50	260	10	1.05	8,540
3	2.00	110	20	1.05	7,780

TABLE 6.2 Characteristic Properties of Some Representative Open-Tubular Columns

Length (m)	Internal diameter (mm)	Film thickness (μm)	Phase ratio	H_{min} (mm)	Column plates number	Plate per meter
30	0.10	0.10	249	0.06	480,000	16,000
30	0.10	0.25	99	0.08	368,550	12,285
30	0.25	0.25	249	0.16	192,000	6,400
30	0.32	0.32	249	0.20	150,000	5,000
30	0.32	0.50	159	0.23	131,330	4,380
30	0.32	1.00	79	0.29	102,080	3,400
30	0.32	5.00	15	0.44	68,970	2,300
30	0.53	1.00	132	0.43	70,420	2,340
30	0.53	5.00	26	0.68	43,940	1,470

H_{min} = minimum plate height.

thin-film columns at mobile-phase velocities at or above the optimum value for the column (Figure 6.1). In addition, the value for the optimum mobile-phase velocity moves to higher values for gases of higher diffusivity favoring faster separations without loss of resolution. For thick-film columns (>0.5 μm), slow diffusion in the stationary phase makes a significant contribution to band broadening compared with mass transfer in the mobile phase. Thick-film columns should be operated close to the optimum velocity, with the choice of carrier gas being less significant and nitrogen often selected because of cost and/or safety considerations.

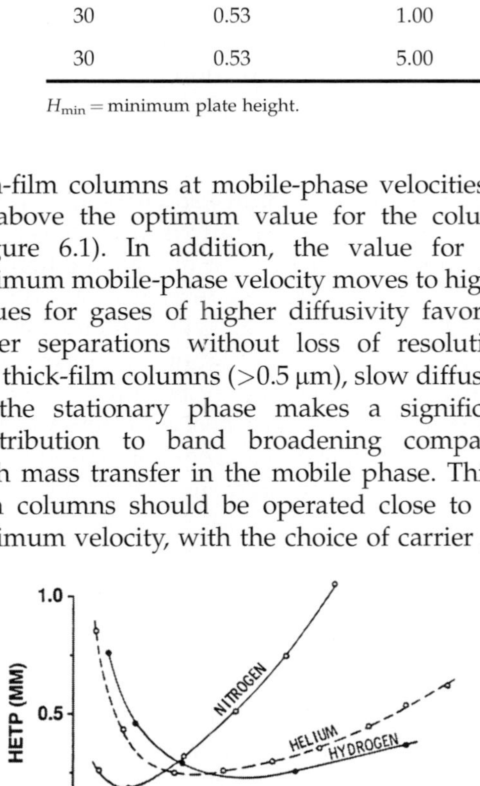

FIGURE 6.1 Plot of the column plate height against the mobile phase velocity (van Deemter plot) for a thin-film column with different carrier gases. *Source: From Ref. [1]; copyright Elsevier.*

6.2. STATIONARY-PHASE CLASSIFICATION

The selectivity of a stationary phase is defined as its relative capacity to enter into specific intermolecular interactions such as dispersion, induction, orientation, and hydrogen bonding. Early attempts to define selectivity scales were based on the system of phase constants introduced by Rohrschneider and subsequently modified by McReynolds, Snyder's solvent selectivity triangle, solubility parameters, and the partial molar Gibbs free energy of solution for functional groups or prototypical compounds among other conceptually similar approaches [3–5]. The Rohrschneider–McReynolds system of phase constants was widely employed for stationary-phase classification and endured until recent times. This method is founded on the assumption that intermolecular interactions are additive, and their individual contribution to retention can be evaluated from the difference in retention index values for a series of prototypical compounds on the stationary phase to be classified and on squalane, adopted as a nonpolar reference phase, with all measurements at a standard temperature of 120 °C. The phase constants (retention index differences) for five prototypical compounds (Rohrschneider) or ten compounds (McReynolds) were used to compare the contribution of different solute interactions with various stationary phases. This approach was abandoned after many years of use because of a number of practical and fundamental problems [6]. Two fundamental issues of particular note are the difficulty of identifying prototypical compounds to represent singular interactions and the demonstration that the retention index differences are composite terms that depend mainly on the choice of standard substances used to determine the retention index values rather than the capability of the prototypical compounds to interact with the stationary phase. To avoid these difficulties, attention was turned to selectivity scales based on the Gibbs free energy of solution for prototypical compounds [7]. These approaches were abandoned after it was demonstrated that the free energy selectivity scales were size dependent and provided anomalous values for solute–solute interactions in cohesive stationary phases as well as after the recognition that the use of prototypical compounds to represent single intermolecular interactions was itself a flawed approach. To overcome these difficulties it was reasoned that the size dependence of the selectivity scales could be removed by separating the free energy into a cavity term and an interaction term containing the contributions from polar intermolecular interactions. A number of models were trialed before settling on Abraham's solvation parameter model, which is currently the approach used nearly universally to classify stationary phases in gas chromatography.

6.2.1. Solvation Parameter Model

The solvation parameter model explains the transfer of a solute from the gas phase to a liquid stationary phase as a three-step process [1,3,8–10]. Initially a cavity is formed in the stationary phase of the same size as the solute. Once the cavity is created, reorganization of the solvent molecules around the cavity occurs to minimize the disruption caused by the solute cavity and to form a more favorable orientation for solute–solvent interactions. Lastly, the solute is inserted into the cavity and sets up solute–solvent interactions indicated as dispersion, orientation, induction, and hydrogen bonding. Cavity formation is an endoergic process that opposes transfer from the gas phase and is expected to be less favorable for polar stationary phases due to stronger solvent–solvent interactions. Reorganization of solvent molecules around the cavity is a comparatively low-energy process in free energy terms and can usually be neglected. The strength of solute–solvent interactions favors transfer of the solute from the gas phase and must exceed the free energy required for cavity formation. Retention in gas–liquid chromatography, therefore, will depend on the cohesive energy of the stationary phase, represented by the free energy required for cavity formation, the formation of additional dispersion interactions of a solute–solvent type, and on solute–solvent interactions that depend on the complementary character of the polar properties of the solute and the stationary phase. To model retention in gas–liquid chromatography requires parametrization of the cavity model to quantitatively identify the individual contributions to the retention process and to develop a series of scales for stationary-phase selectivity. The model proposed by Abraham, using the modern symbols for the descriptors, is set out below:

$$\log K = c + e\mathrm{E} + s\mathrm{S} + a\mathrm{A} + b\mathrm{B} + l\mathrm{L} \qquad (6.1)$$

The gas–liquid partition coefficient, K, is the relevant free energy parameter for modeling retention but is more difficult to determine than the retention factor, k. The two terms are connected through the phase ratio, β, by the relationship $K = k\beta$. When the retention factor is used in Eqn (6.1), the phase ratio is subsumed by the model intercept term, c, which is not assigned any chemical significance.

The model is made up of a series of product terms that define the contribution of specific intermolecular interactions to the solvation process. The uppercase letters are solute descriptors that define the contribution of the solute to participate in defined intermolecular interactions and the lowercase letters in italics are the system constants that define the complementary capability of the stationary phase to participate in the same interactions. The specific interactions are identified as follows: eE is the contribution from electron lone-pair interactions that occur between polarizable molecules; sS is the contribution from interactions of a dipole-type that occur between molecules with a permanent dipole moment and/or with polarizable molecules due to interactions between a permanent dipole and an induced dipole; and the aA and bB terms represent the contributions of hydrogen bonding with the solute behaving as either a hydrogen-bond acid, A descriptor, or a hydrogen-bond base, B descriptor. Since a solute hydrogen-bond acid will interact with a solvent hydrogen-bond base, the a system constant is a measure of the hydrogen-bond basicity of the stationary phase, and in the same way, the b system constant is a measure of the hydrogen-bond acidity of the stationary phase. The lL term is a measure of the opposing contributions of cavity formation and dispersion interactions to the retention process. The sign of the lL term (positive or negative) provides an indication of the relative importance of cavity formation (breaking solvent–solvent interactions) and dispersion interactions (formation of solute–solvent dispersion interactions).

To characterize the selectivity of a stationary phase, that is to calculate the values of the system constants in Eqn (6.1), requires the experimental determination of the retention factors at a constant temperature for a varied group of compounds with known descriptor values. A collection of compounds with optimized descriptor values for characterizing gas chromatographic stationary phases over the temperature range of 60–240 °C is available [8,11]. For each column and temperature, a sufficient number of solutes are required to provide statistically meaningful values for the system constants and to explore a wide-enough descriptor space for the models to have useful predictive properties. The descriptor values for a collection of compounds with acceptable retention characteristics for convenient measurement on the selected column must also have low cross-correlation to facilitate extraction of the true system constants by multiple linear regression analysis. Rather than using the same compounds for the classification of all stationary phases, the above criteria that require different compound collections are used for columns with different polarities and/or evaluated at different temperatures. This allows all column types to be classified at any temperature and is more flexible than methods that employ a group of prototypical compounds to define each interaction. More details of the selection of solutes for stationary-phase classification using Eqn (6.1) are given elsewhere [8–11] and a brief description of the methods used to determine descriptors is given in Table 6.3.

6.2.2. System Constants Database for Open-Tubular Columns

The system constants fully describe all stationary-phase interactions and provide a logical basis for stationary-phase classification. A database of system constants determined at 20 °C intervals over the temperature range of 60–140 °C was established for 56 open-tubular columns [12–14] and the temperature range extended to include 160–240 °C for 14 columns [11,13,14]. The system constants facilitate the

TABLE 6.3 Identity of the Solute Descriptors Used in the Solvation Parameter Model and Methods for their Determination [8–11]

Descriptor	Method of determination
E	Excess molar refraction ($cm^3mol^{-1}/10$). For liquids, it can be calculated from the refractive index and characteristic volume. For solids, computer-estimated values for the refractive index can be used. For solids may be determined together with the other descriptors using chromatographic methods.
S	Dipolarity/polarizability descriptor usually determined by experiment. Initially estimated by gas chromatography on polar stationary phases but more commonly by liquid–liquid partition together with chromatographic methods today.
A	The effective or summation hydrogen-bond acidity descriptor. Originally determined from complexation constants in inert solvents for monofunctional hydrogen-bond acids and bases. The effective values broadened to allow for multiple solute–solvent interactions. They are commonly determined by liquid–liquid partition and chromatographic measurements today.
B	The effective or hydrogen-bond basicity descriptor. Determined as described for the A descriptor.
L	The gas–liquid partition coefficient for the solute on n-hexadecane at 298 K. For volatile compounds, it can be determined directly. For compounds of low volatility, it is determined by back calculation from gas chromatographic retention measurements on nonpolar stationary phases.

comparison of the separation characteristics of different stationary phases, afford a link between chromatographic selectivity and monomer composition for polymeric stationary phases, and enable the identification of selectivity-equivalent stationary phases with different or unknown compositions. In addition, the database can be used to assist in column selection, identification of initial separation conditions, and method development. The columns and their general properties included in the system constants database are summarized in Table 6.4. The system constants for each column at 120 °C are summarized in Table 6.5. The system constants in Table 6.5 are only loosely scaled to each other so that changes in any column can be read directly but changes across rows must be interpreted cautiously. Most stationary phases possess some capacity for electron lone-pair interactions (*e* system constant), but selectivity for these interactions is rather limited except for fluorine-containing stationary phases (PMTS) and poly(ethylene glycols). Even in this case, interactions are weak. None of the stationary phases in Table 6.5 are significant hydrogen-bond acids (*b* system constant is zero) [15]. Some stationary phases have small *b* system constants, but this is a result of the method employed for deactivation and is a column property but not a stationary-phase property [13]. In any case, these interactions are too weak to be important for selectivity optimization. In practice, the important considerations for selectivity optimization are differences in cohesion/dispersion properties, dipole-type interactions, and stationary-phase hydrogen-bond basicity. These intermolecular interactions are temperature dependent and differences in selectivity at one temperature, as illustrated by Table 6.5, are not necessarily the same as those at a different temperature. Plots of the system constants as a continuous function of temperature are called-system maps and allow a straightforward evaluation of the variation of temperature on selectivity. An example of a system map is shown in Figure 6.2 for the poly(cyanopropylphenyldimethylsiloxane) stationary phase DB-1701 [11]. Individual system constants exhibit a nonlinear change with temperature with the cohesion of the stationary phase and its capacity for polar interactions being smaller at higher temperatures. The shape of the plots is an indication that polar interactions persist to quite high temperatures and

TABLE 6.4 Stationary-Phase Chemistries and Column Identities for the System Constants Database

Column	Type	Identity
PDMSO	Poly(dimethylsiloxane)	DB-1, SolGel-1
PMOSO	Poly(methyloctylsiloxane)	SPB-Octyl
PMPS-5	Poly(dimethyldiphenylsiloxane)	HP-5, DB-5, OV-5, SPB-5, PTE-5
	5% diphenylsiloxane monomer	
PMPS-20	20% diphenylsiloxane monomer	Rtx-20
PMPS-35	35% diphenylsiloxane monomer	DB-35
PMPS-50	50% diphenylsiloxane monomer	HP-50$^+$, Rxi-17
PMPS-65	65% diphenylsiloxane monomer	Rtx-65
PMPS	Poly(methylphenylsiloxane)	Rtx-50, Rxi-50
SASO-5	Silarylene–siloxane copolymer	DB-5ms, ZB-5ms, HP-5TA,
		Rxi-5Sil MS
SASO-35		DB-35ms
SASO-50		DB-17ms
SASO-XLB		DB-XLB
CBSO-5	Carborane–siloxane copolymer	Stx-500
PMTS-35	Poly(dimethylmethyltrifluoropropylsiloxane)	DB-200
PMTS-50	Poly(methyltrifluoropropylsiloxane)	DB-210
PCPM-06	Poly(cyanopropylphenyldimethylsiloxane)	DB-1301
	6% cyanopropylphenylsiloxane monomer	
PCPM-14	14% cyanopropylphenylsiloxane monomer	DB-1701
PCPM-50	50% cyanopropylphenylsiloxane monomer	DB-225
Rtx-440		Rtx-440
PCM-50	Poly(cyanopropylmethylsiloxane)	DB-23
PCPS	Poly(biscyanopropylsiloxane)	SP-2340
SACPS-88	Bis(cyanopropylsiloxane)-*co*-methylsilarylene	HP-88
	88% bis(cyanopropylsiloxane) monomer	
SACPS-90	90% bis(cyanopropylsiloxane) monomer	BPX90
PEG	Poly(ethylene glycol)	HP-20M, HP-INNOWax, EC-Wax,
		DB-WAXetr, SolGel-Wax,
		DB-FFAP
PAG	Poly(ethylene glycol)-*co*-propylene oxide	PAG

(Continued)

TABLE 6.4 Stationary-Phase Chemistries and Column Identities for the System Constants Database (*cont'd*)

Column	Type	Identity
Rtx-Volatiles	Application specific	Rtx-Volatiles
Rtx-VGC		Rtx-VGC
DB-608		DB-608
DB-624		DB-624
Rtx-Dioxin		Rtx-Dioxin
Rtx-Dioxin2		Rtx-Dioxin2
Rtx-OPPesticide		Rtx-OPPesticides
Rtx-CLPesticides		Rtx-CLPesticides
Rtx-TNT		Rtx-TNT
Rtx-TNT2		Rtx-TNT2
DX-1	PDMSO + 10% PEG	DX-1
DX-3	PDMSO + 50% PEG	DX-3
DX-4	PDMSO + 85% PEG	DX-4
Cyclodex-B[1]		Cyclodex-B
CycloSil-B[2]		CycloSil-B

[1] *10.5% permethylated β-cyclodextrin dissolved in a 14% poly(cyanopropylphenyldimethylsiloxane) solvent (DB-1701).*
[2] *30% heptakis(2,3-di-O-methyl-6-O-t-butyldimethylsilyl)-β-cyclodextrin dissolved in a 14% poly(cyanopropylphenyldimethylsiloxane) solvent (DB-1701).*

TABLE 6.5 Representative System Constants at 120 °C from the System Constants Database

		System constants				
Column	**Identity**	*e*	*s*	*a*	*b*	*l*
PDMSO	DB-1	0	0.215	0.175	0	0.507
PMOSO	SPB-Octyl	0.182	0.081	0	0	0.585
PMPS-5	HP-5	0.004	0.306	0.199	0	0.518
PMPS-20	Rtx-20	0.030	0.535	0.245	0	0.545
PMPS-35	DB-35	0.074	0.640	0.277	0	0.531
PMPS-50	Rxi-17	0.140	0.759	0.317	0.050	0.553
PMPS-65	Rtx-65	0.120	0.828	0.330	0	0.528
PMPS	Rtx-50	0.085	0.787	0.331	0	0.522
SASO-5	DB-5ms	0.007	0.315	0.205	0	0.548
SASO-35	DB-35ms	0.074	0.683	0.300	0	0.561

(*Continued*)

TABLE 6.5 Representative System Constants at 120 °C from the System Constants Database (*cont'd*)

Column	Identity	System constants				
		e	*s*	*a*	*b*	*l*
SASO-50	DB-17ms	0.134	0.769	0.330	0	0.558
SASO-XLB	DB-XLB	0.011	0.396	0.227	0	0.561
CBSO-5	Stx-500	−0.042	0.520	0.129	0	0.570
PMTS-35	DB-200	−0.335	1.041	0.147	0	0.465
PMTS-50	DB-210	−0.417	1.304	0.174	0	0.433
PCPM-06	DB-1301	−0.049	0.467	0.455	0	0.542
PCPM-14	DB-1701	−0.101	0.710	0.619	0	0.517
PCPM-50	DB-225	−0.035	1.232	1.176	0	0.457
Rtx-440		0.022	0.429	0.280	0	0.562
PCM-50	DB-23	−0.034	1.524	1.464	0	0.453
PCPS	SP-2340	0.058	1.874	1.873	0	0.391
SACPS-88	HP-88	0.015	1.820	1.724	0	0.424
SACPS-90	BPX90	0.041	2.034	1.811	0	0.393
PEG	HP-20M	0.223	1.364	1.969	0	0.477
PAG	PAG	0.188	1.238	1.928	0	0.510
Rtx-Volatiles		0.022	0.455	0.235	0	0.541
DB-608		0.132	0.748	0.328	0	0.540
DB-624		−0.054	0.484	0.399	0	0.522
Rtx-Dioxin		−0.049	0.523	0.102	0	0.592
Rtx-Dioxin2		0.011	0.438	0.257	0	0.582
Rtx-OPP		−0.289	0.924	0.206	0	0.485
Rtx-CLPesticides		−0.220	0.743	0.162	0	0.520
Rtx-TNT		−0.048	0.274	0.225	0	0.505
Rtx-TNT2		−0.256	0.879	0.175	0	0.449
DX-1		0.032	0.360	0.501	0	0.530
DX-3		0.166	1.113	1.671	0	0.536
DX-4		0.194	1.205	1.838	0	0.516
Cyclodex B		−0.071	0.737	1.000	0	0.541
Cyclosil B		−0.062	0.641	1.040	0	0.545

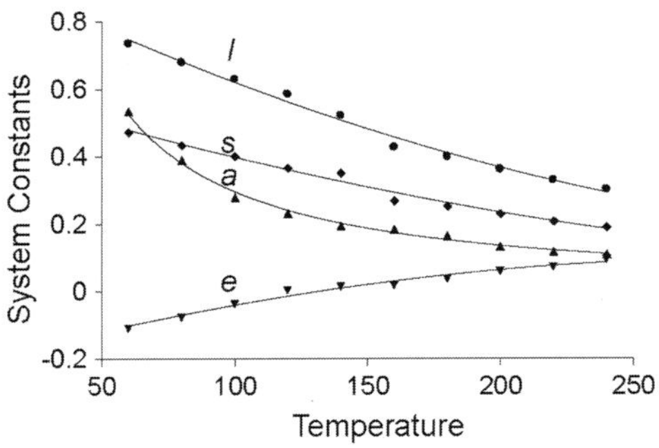

FIGURE 6.2 System map for the poly(cyanopropylphenyldimethylsiloxane) stationary phase DB-1701 containing 14% cyanopropylphenylsiloxane monomer. *Sourec: From Ref. [11]; copyright Elsevier.*

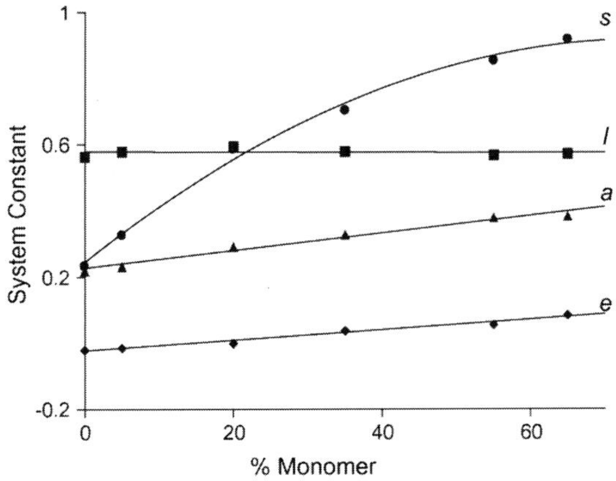

FIGURE 6.3 Plot of the system constants for the poly(dimethyldiphenylsiloxane) stationary phases as a function of the diphenylsiloxane monomer composition at 100 °C. *Source: From Ref. [12]; copyright Elsevier.*

also that changes in selectivity due to polar interactions are larger at lower temperatures. The additional dispersion interaction that polarizable molecules experience on account of their loosely held lone-pair electrons represented by the *e* system constant is the only interaction that increases retention at higher temperatures. In several cases, the sign of the *e* system constant is negative at low temperatures and positive at higher temperatures. Although electron lone-pair interactions are relatively more important at higher temperatures, they remain comparatively weak interactions in most cases and only important for optimizing the selectivity of low polarity and polarizable molecules, such as polycyclic aromatic hydrocarbons.

The poly(dialkylsiloxane) stationary phases (PDMSO and PMOSO) are low-selectivity stationary phases characterized by low cohesion and weak polar interactions. Separations are governed mainly by dispersion interactions with good selectivity for the separation of nonpolar compounds. Replacing dimethylsiloxane monomers by diphenylsiloxane monomers for poly(dimethyldiphenylsiloxane) stationary phases (PMPS) results in an orderly change in the capacity of the stationary phase for dipole-type interactions and in their hydrogen-bond basicity (Figure 6.3). The changes in the system constants are approximately linear up to about 50% diphenylsiloxane monomer. Given that the cavity/dispersion term, *l* system constant, changes only slightly with diphenylsiloxane monomer composition, the principal selectivity difference among these phases is due to the variation in the ratio of dipole-type to hydrogen-bond base interactions. There are small selectivity differences for stationary phases prepared from methylphenylsiloxane monomer and from dimethylsiloxane and diphenylsiloxane monomers of the same nominal composition [14]. Stationary phases prepared from diphenylsiloxane monomers are more thermally stable and are generally preferred for most applications. The poly(dimethyldiphenylsiloxane) stationary phases are general-purpose low-polarity stationary phases with many applications for compounds covering a wide polarity range. Incorporation of silarylene or carborane monomers into the backbone of poly(siloxanes) enhances their thermal stability by inhibiting the formation of cyclic siloxanes. These phases, designated SASO and CBSO in Table 6.4, generally have the extension ms added to the name of the nominally equivalent poly(dimethyldiphenylsiloxane) stationary phase and are designed for demanding applications where low column bleed is important. Although possessing similar separation characteristics to the nominally equivalent poly(dimethyldiphenylsiloxane) stationary phases, they are not selectivity equivalent and the separation of critical compound pairs may be different. The stationary phase DB-XLB has an undisclosed composition but is said to have similar separation properties to PMPS-5 stationary phases. Careful analysis of the system constants indicates that this phase is significantly more dipolar/polarizable (larger *s* system constant) than the SASO-5 stationary phases and separation differences at temperatures <200 °C are significant [16]. The SASO-5 stationary phases are generally based on silphenylene monomers and DB-XLB likely contains a different silarylene monomer. This is also seen in the carborane–siloxane stationary phases which have similar separation properties to the nominally equivalent PMPS-5 stationary phases but are not selectivity equivalent.

The incorporation of 3,3,3-trifluoropropylmethylsiloxane monomers into a poly(dimethylsiloxane) stationary phase results in characteristic changes in their selectivity that makes these phases a useful alternative to the poly(dimethyldiphenylsiloxane) stationary phases. They are significantly more dipolar/polarizable and weaker hydrogen-bond bases than the poly(dimethyldiphenylsiloxane) stationary phases with a similar incorporation of polar monomer groups. In addition, electron lone-pair interactions are important and repulsive in type compared with the poly(dimethyldiphenylsiloxanes).

The incorporation of 3-cyanopropylphenylsiloxane and 3-cyanopropylmethylsiloxane monomers into a poly(dimethylsiloxane) stationary phase, indicated as PCPM, PCM, and SACPS in Table 6.4, results in significant changes in selectivity with those phases containing a high incorporation of polar monomer being among the most polar of the common stationary phases. These stationary phases are cohesive and strongly dipolar/polarizable and hydrogen-bond basic. They have different selectivity to the PMTS stationary phases containing 3,3,3-trifluoropropylmethylsiloxane monomers with a similar incorporation of polar monomer. For the PCPM and PCM stationary phases, the ratio of the s/a system constants is close to 1 while it is much less than 1 for the PMTS stationary phases. In addition, electron lone-pair interactions are weak for the PCPM and PCM stationary phases but more significant for the PMTS stationary phases. The PCPM and SACPS stationary phases with a high incorporation of polar monomer are among the most cohesive stationary phases in common use with small *l* system constants resulting in low methylene group selectivity. The composition of the Rtx-440 stationary phase is undisclosed but stands out in this general class with separation properties similar to PCPM-6 stationary phases for compounds which are weak hydrogen-bond acids. Hydrogen-bond acids are selectively displaced to shorter retention times compared with a PCPM-6 stationary phase due to characteristically smaller *a* system constants.

Poly(ethylene glycol) stationary phases are more hydrogen-bond basic and nearly as dipolar/polarizable as the SACPS stationary phases indicated in Table 6.4. In addition, they are generally less cohesive, and electron lone-pair interactions, while weak, are more important and of opposite sign. The methods used to stabilize PEG stationary phases employ different chemistries, but the variation in separation properties for all column types in this group is quite small. Incorporation of propylene oxide into the polymer backbone, PAG phase, results in small changes in selectivity. The PAG column is slightly less dipolar/polarizable, cohesive, and electron lone-pair attractive compared with the PEG stationary phases. It is expected to provide greater separation between members of a homologous series than PEG but diminished separation of compounds that differ mainly in their dipolarity or polarizability. A more significant variation in the properties of the PEG phases can be obtained by dilution of PEG with PDMSO to prepare columns with a mixed stationary-phase composition indicated as DX-1, DX-2, and DX-3 in Table 6.4.

A number of application-specific columns are included in Table 6.4. The stationary-phase compositions for these columns are generally considered proprietary information, but in several cases can be identified from a comparison of their system constants with the stationary phases of known composition in the database and confirmed by comparison of retention factors for varied compounds on the stationary phases identified as selectivity equivalent. The stationary phases identified in this way are summarized in Table 6.6. Most application-specific columns are prepared from common stationary phases and optimized for the specified separation by selection of the appropriate column dimensions and phase ratio. In a few cases, for example, Rtx-Volatiles and Rtx-CLPesticides, common polymers are used but with a monomer composition that is different from standard phases. These columns might be useful for other applications as they extend the range of monomer compositions available for common types of stationary phases.

TABLE 6.6 Identification of the Stationary Phases Employed for Common Application-Specific Columns

Column identification	Selectivity-equivalent stationary phase
DB-VRX	PMPS-5
Rtx-Volatiles	PMPS containing about 16% diphenylsiloxane monomer
Rtx-VGC	No selectivity equivalent stationary phase in the database
DB-608	PMPS-50
DB-624	PCPM-06
Rtx-Dioxin	CBSO-5
Rtx-Dioxin2	SASO-XLB
Rtx-OPPesticides	PMTS-35 probably based on silarylene–siloxane chemistry
Rtx-CLPesticides	PMTS-28 containing between 25–30% polar monomer and probably based on silarylene–siloxane chemistry
Rtx-TNT	PMPS-5
Rtx-TNT2	Rtx-OPPesticides

6.2.3. Classification of Stationary Phases

The system constant ranges for the ten temperatures used to construct the database are summarized in Table 6.7. These ranges describe the selectivity space available for separations using the open-tubular columns contained in the system constants database. An obvious deficiency is the lack of a stationary phase with significant hydrogen-bond acidity. In addition to new stationary phases with a significant *b* system constant, other stationary phases with system constants either higher or lower than the ranges shown in Table 6.7 would also be useful. Principal component analysis of the system constants provides a useful tool to visualize how individual columns occupy the selectivity space, Figure 6.4. The large central cluster in the score plot contains the poly(dimethylsiloxane), poly(dimethyldiphenylsiloxane), phenyl-containing silarylene–siloxane copolymers, and poly(3-cyanopropylphenyldimethylsiloxane) stationary phases. Figure 6.4 shows that the monomers used to prepare these stationary phases have a rather limited capability to explore the full selectivity space. There are also many stationary phases with similar properties in this group, albeit, a common group of stationary phases used for gas chromatography. The two

TABLE 6.7 Range of System Constants for Wall-Coated Open-Tubular Columns in the System Constants Database

	System constant							
	e		*s*		*a*		*l*	
Temperature (°C)	High	Low	High	Low	High	Low	High	Low
60	0.22	−0.61	2.34	0.07	2.92	0	0.78	0.52
80	0.22	−0.50	2.17	0.07	2.54	0	0.71	0.47
100	0.23	−0.46	2.08	0.07	2.28	0	0.65	0.43
120	0.23	−0.42	1.87	0.08	2.00	0	0.59	0.39
140	0.24	−0.33	1.80	0.09	2.00	0	0.54	0.37
160	0.19	−0.13	1.31	0.05	1.55	0	0.43	0.27
180	0.21	−0.10	1.23	0.05	1.33	0	0.39	0.24
200	0.22	−0.07	1.15	0.04	1.19	0	0.35	0.22
220	0.22	−0.05	1.09	0.04	1.03	0	0.32	0.19
240	0.18	−0.03	1.03	0.04	0.89	0	0.29	0.16

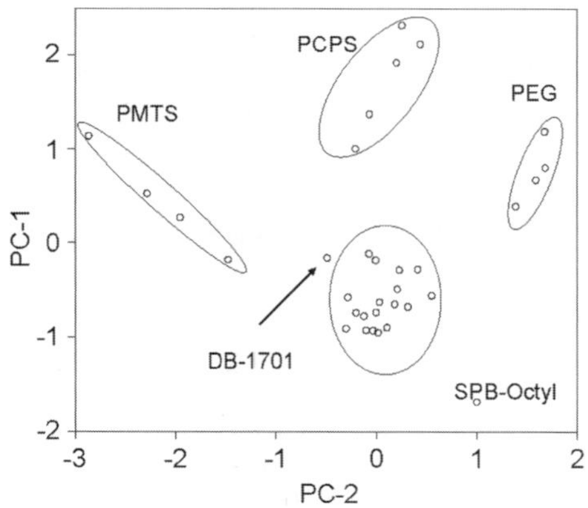

FIGURE 6.4 Score plot for principal component factor analysis (varimax rotation) of the system constants for the temperature range 60–140 °C. To simplify the plot, all duplicate entries were removed by using a single average for selectivity equivalent stationary phases. The first two principal components account for 93.3% of the variance and so represent the whole selectivity space quite well. Principal component 1 (PC-1) is dominated by contributions from the *l* and *s* system constants and principal component 2 (PC-2) the *e* system constant. The *a* system constant is almost equally loaded on both components. *Source: From Ref. [12]; copyright Elsevier.*

stationary phases SPB-Octyl (least selective of all stationary phases in the database) and DB-1701 are located at a remote position from the central cluster and have rather singular properties for method development. The importance of the poly(dimethyltrifluoropropylmethylsiloxane), poly(biscyanopropylsiloxane) and related stationary phases based on silarylene chemistry (PCPS and SACPS phases in Table 6.4), and poly(ethylene glycols) and their mixtures with poly(dimethylsiloxane) (DX-1, DX-2, and DX-3 in Table 6.4) is illustrated by their position on the score plot relative to the central cluster. The PMTS, PCPS, and PEG phases occupy different regions of the selectivity space but as relatively compact groups reflecting distinct character but limited coverage. The space between groups marked in Figure 6.4 is an indication of the selectivity space that is not accessible by the available stationary phases in the database. This is not surprising given the limited number and type of monomers used to prepare contemporary stationary phases.

A general method development strategy is to select the minimum number of stationary phases that span the selectivity space as evenly as possible (Table 6.8). After trial separations on each stationary phase, the stationary phase that separates the greatest number of peaks is selected for further optimization. This is achieved by evaluating its near neighbors in the selectivity space. In other cases, there may be a great deal of empirical information available and the decision becomes how to select the best stationary phase from a group of stationary phases considered suitable for a particular application. For example, the separation of volatile organic compounds of regulatory concern can be achieved to different extents on the columns identified in Figure 6.5 [17]. These columns have been classified using hierarchical cluster analysis. In choosing stationary phases for this application, the dendrogram indicates the close similarity in properties of three groups of stationary phases (DB-624 and DB-1301, HP-5 and DB-VRX, and Rtx-50 and Rtx-65). These three groups offer different selectivities and to investigate the use of one phase from each group would be more sensible as a screening tool than choosing several phases from a single group. On the other hand, DB-1, Rtx-Volatiles, DB-35, DB-1701, and Rtx-VGC are indicated as stationary phases with significant differences in selectivity and provide complementary separation characteristics for this application. The connections in the dendrogram indicate the degree of similarity and facilitate the selection of stationary phases with a small or relatively large selectivity difference. Empirical and historical information indicates that more polar stationary phases contained in the database are rarely successful for this application and the approach here is to blend established practice with a rationale approach for column selection.

6.2.4. Ionic Liquids

After a long period of largely evolutionary progress in stationary-phase chemistry a new class of stationary phases, the ionic liquids, was introduced recently [18,19]. The discovery of the dialkylimidazolium ionic liquids was an important catalyst in furthering interest in ionic liquid stationary phases since these ionic liquids and their analogous alkylpyrolinium, *N*-alkylpyrrolidinium, *N*-alkylpyridinium, and alkylphosphonium salts with bulky and or charge-delocalized anions have provided

TABLE 6.8 System Constants for Columns Selected from Different Selectivity Groups at 100 °C ($b = 0$ for all Column Types)

Column type	% Polar monomer	System constants			
		e	s	a	l
Poly(methyloctylsiloxane)		0.175	0.067	0	0.647
Poly(dimethyldiphenylsiloxane)	5	−0.020	0.332	0.247	0.572
Poly(dimethyldiphenylsiloxane)	50	0.054	0.851	0.377	0.566
Poly(methyltrifluoropropylsiloxane)	50	−0.460	1.377	0.195	0.455
Poly(cyanopropylphenyldimethylsiloxane)	50	−0.062	1.334	1.321	0.510
Poly(biscyanopropylsiloxane)	100	0.027	2.044	1.947	0.427
Poly(ethylene glycol)	100	0.205	1.407	2.117	0.511

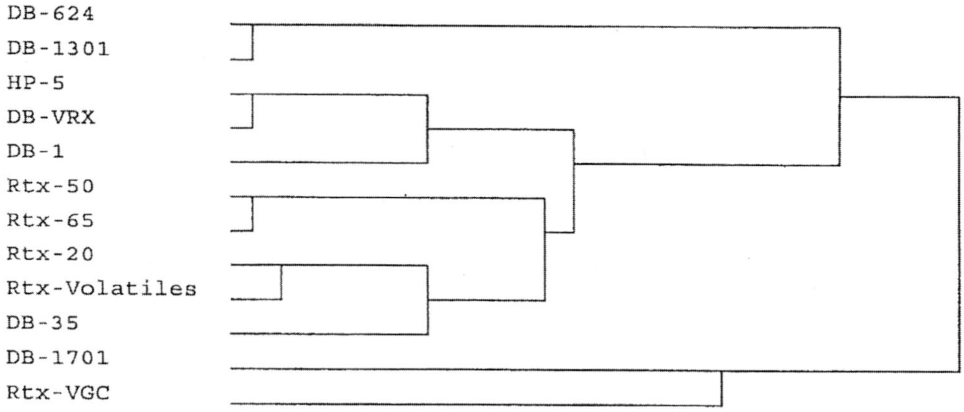

FIGURE 6.5 Dendrogram for the classification of columns for the separation of volatile organic compounds using the nearest neighbor agglomeration method for columns identified in Table 6.4. Input data are the system constants for the temperature range 60–140 °C. *Source: From Ref. [17]; copyright Elsevier.*

the greatest number of low-melting-point materials of high thermal stability. The long-range Coulombic interactions, characteristic of ionic liquids, results in stationary phases with very low vapor pressure. The key to the development of ionic liquid stationary phases for use in open-tubular columns was the identification of organic salts of high thermal stability, high viscosity, and with moderate surface tension. Stationary phases that wet glass surfaces adequately and simultaneously possess high viscosity form stable thin films that resist droplet formation with temperature variations. Ionic liquids with 1-vinyl-3-alkylimidazolium cations can be cross-linked using a free radical catalyst to obtain stationary phases of enhanced thermal stability. Compared with the poly(siloxanes) these stationary phases are not immobilized by reaction with the column wall, but cross-linking increases the apparent viscosity of the ionic liquids resulting in more robust and stable films at higher temperatures. Ionic liquids with a multication core also possess favorable properties for gas chromatography and provide stationary phases with the highest thermal stability, up to 350 °C in favorable cases. Some columns of the above type are commercially available but the identity of the stationary phase composition is often unnecessarily withheld as proprietary information.

The solvation parameter model has been used to characterize a number of the new ionic liquid stationary phases at two or more temperatures in the range of 40–110 °C [19]. Some representative examples are given in Table 6.9. A notable feature is the significant hydrogen-bond acidity of some of these ionic liquids with *b* system constants between 0.50 and 1.62. Since none of the commonly used poly(siloxane) and poly(ethylene glycol) stationary phases are hydrogen-bond acids, this feature provides an additional possibility for selectivity optimization using ionic liquids. For the imidazolium-based ionic liquids, the C-2 hydrogen is one source of hydrogen-bond acidity. The ionic liquids with the largest *b* system constant contain cations with additional amide or hydroxyl substituents and anions with weak hydrogen-bond basicity or structural features that minimize ion–ion association. For the other system constants, there is significant overlap for the ionic liquids and the polar stationary phases in the system constants database (Section 6.2.2). The 1-butyl-3-methylimidazolium

TABLE 6.9 System Constants for Some Representative Ionic Liquid Stationary Phases at 100 °C

Ionic liquid	System constants				
	e	*s*	*a*	*b*	*l*
(i) Monocations					
1-Butyl-3-methylimidazolium					
Bis(trifluoromethylsulfonyl)imide	0	1.60	1.55	0.24	0.49
Trifluoromethanesulfonate	0	1.39	2.35	0	0.49
Hexafluorophosphate	0	1.54	1.37	0	0.44
1-Butyl-2,3-dimethylimidazolium bis(trifluoromethylsulfonyl)imide	0.09	1.58	1.57	0.11	0.48
1-Hexyl-3-methylimidazolium tris(pentafluoroethyl) trifluorophosphate	0	1.46	0.49	0.61	0.42
1-(4-Methoxyphenyl)-3-methylimidazolium Trifluoromethanesulfonate	0.28	2.05	2.03	0.16	0.38
1-Butyl-1-methylpyrrolidinium					
Bis(trifluoromethylsulfonyl)imide	0	1.44	1.55	0	0.48
Tris(pentafluoroethyl)trifluorophosphate	0.19	1.46	0.60	0.60	0.44
1-(6-Aminohexyl)-1-methylpyrrolidinium tris(pentafluoroethyl) trifluorophosphate	0.21	1.73	1.66	0.29	0.38
1-(2-Hydroxyethyl)-1-methylpyrrolidinium					
Tris(pentafluoroethyl)trifluorophosphate	0.22	1.59	0.65	1.62	0.37
Bis(trifluoromethylsulfonyl)imide	0.22	1.81	1.65	0.97	0.37
1-(2-Hydroxyethyl)-3-methylimidazolium					
Tris(pentafluoroethyl)trifluorophosphate	0	1.79	0.71	1.51	0.33
Bis(trifluoromethylsulfonyl)imide	0.13	1.99	1.81	1.03	0.37
Trihexyl(tetradecyl)phosphonium tris(pentafluoroethyl) trifluorophosphate	−0.31	1.30	0.45	0.27	0.62
(ii) Dications					
1,4-Di(3-methylimidazolium)butane bis(triflioromethylsulfonyl)imide	0.20	1.69	1.57	0.33	0.37
1,9-Di(3-methylimidazolium)nonane bis(trifluoromethylsulfonyl) imide	0.11	1.64	1.50	0.15	0.43
1,12-Di(3-benzylimidazolium)dodecane bis(trifluoromethylsulfonyl) imide	0	1.47	1.44	0.52	0.46
1,11-Di(3-methylimidazolium)-2,6,9-trioxaundecane					
Bis(trifluoromethylsulfonyl)imide	0.12	1.68	1.65	0	0.44
Trifluoromethanesulfonate	0.45	1.95	2.72	0.22	0.31

(*Continued*)

TABLE 6.9 System Constants for Some Representative Ionic Liquid Stationary Phases at 100 °C (*cont'd*)

	System constants				
Ionic liquid	*e*	*s*	*a*	*b*	*l*
1,9-Di(3-hydroxyethylimidazolium)nonane Bis(trifluoromethylsulfonyl)imide	0.29	1.44	1.34	0.76	0.40
(iii) Tricationic (all bis[trifluoromethylsulfonyl]imides)					
Core = mesitylene R = 3-benzyl-1-imidazolium	0.10	1.87	1.61	0.58	0.39
Core = benzene R = 3-methylimidazolium	0.18	1.51	1.42	0	0.42
R = 3-butylimidazolium	0.09	1.56	1.57	0.15	0.48
R = 3-benzylimidazolium	0	1.97	1.78	0.39	0.43
Core = tris(2-hexanamido)ethylamine					
R = 3-methylimidazolium	0.16	2.10	2.50	0.17	0.37
R = 3-benzylimidazolium	0	1.69	1.93	0	0.42
R = tri(n-propyl)phosphonium	0.14	1.72	2.17	0	0.44
(iv) Polymeric					
Poly(1-nonyl-3-vinylimidazolium-*co*-di-1,9-[3-vinylimidazolium] nonane) bis(trifluoromethylsulfonyl)imide	0	1.54	1.41	0.31	0.54
Poly(1-hexyl-3-vinyl)imidazolium bis(trifluoromethylsulfonyl)imide	−0.35	1.76	1.38	0.95	0.49

trifluoromethanesulfonate ionic liquid and the tricationic ionic liquid with a tris(2-hexanamido) ethylamine core, tri(n-propyl)phosphonium substituents, and bis(trifluoromethylsulfonyl) imide anions possess similar separation properties to the poly(ethylene glycol) stationary phases. The ionic liquids 1-(4-methoxyphenyl)-3-methylimidazolium trifluoromethanesulfonate, 1-butyl-2,3-dimethylimidazolium bis(trifluoromethylsulfonyl)imide, 1,11-di(3-methylimidazolium)-2,6,9-trioxaundecane bis(trifluoromethylsulfonyl) imide, and the tricationic ionic liquid with a benzene core, 3-butylimidazolium substituents, and bis(trifluoromethylsulfonyl)imide anions have similar separation properties to the poly-(3-cyanopropylphenyldimethylsiloxane) and poly (biscyanoalkylsiloxane) stationary phases. An advantage of the ionic liquids is their higher-temperature operating limits compared with conventional polar stationary phases. The lack of anion diversity for the ionic liquids shown in Table 6.9 [most contain either the bis(trifluoromethylsulfonyl)imide, trifluoromethanesulfonate, or tris (pentafluoroethyl)trifluorophosphate anions] suppresses the selectivity space and leaves plenty of room for future development.

A notable feature of the ionic liquids in Table 6.9 is the similar magnitude of the *l* system constant for ionic liquids with weakly associated anions and the low-polarity conventional stationary phases. The *l* system constant provides an indication of the capability of a solvent to dissolve higher members of a homologous series and in gas chromatography an indication of the peak spacing between homologs. For ionic liquids, the equilibrium distance between ions is controlled primarily by Coulombic forces. These distances are comparatively large for ionic liquids containing bulky ions resulting in relatively low cohesion forces and separations of low-polarity compounds resembling those obtained on nonionic stationary phases of low polarity. When either or both ions in an ionic liquid are capable of strong ion–ion intermolecular interactions, such as association by hydrogen bonding, there is an increase in the cohesive energy of the ionic liquid resulting in retention properties for low-polarity solutes similar to those observed for polar nonionic solvents. In addition, for polar solutes additional intermolecular interactions with individual ions are possible and result in increased retention and selectivity.

Ionic liquids that have been more of an academic curiosity than a bedrock of application-based research now seem on the verge of entering the main stream. The commercial availability of columns containing ionic liquid stationary phases was an important step, and the appearance of an increasing number of applications, an indication of wider interest and activity. There are also possibilities for further improvements in both high-temperature operation and extension of the selectivity range beyond those of molecular liquids to sustain current interest.

6.3. POROUS-LAYER OPEN-TUBULAR COLUMNS

Porous-layer open-tubular columns (PLOT columns) are used for a narrow range of separations based on gas–solid chromatography. On account of the high characteristic retention of adsorbents, typical applications include the separation of gases, volatile hydrocarbons, and organic solvents. Table 6.10 provides a general applications-based guide for column selection [1,20]. PLOT columns require more careful use, the average efficiency may be lower, the sample capacity lower, and chemical activity higher than for columns used for gas–liquid chromatography. The choice of carrier gas can affect retention and selectivity by competing with the sample for active sites on the adsorbent

TABLE 6.10 General Applications of PLOT Columns in Gas Chromatography

Stationary phase	Maximum operating temperature (°C)	Typical applications
Alumina oxide	200	Alkanes, alkenes, alkynes, and aromatic hydrocarbons from C_1 to C_{10}. C_1 and C_2 halocarbons
Silica gel	250	Hydrocarbons (C_1 to C_4), inorganic gases, volatile ethers, esters, and ketones
Carbon	350	Inorganic gases, hydrocarbons (C_1 to C_5) and oxygenated
Carbosieves	150	C_1 to C_6 compounds
Molecular sieves (5A and 13X)	350	Hydrogen, oxygen, nitrogen, methane, and noble gases. Particularly the separation of He/Ne and Ar/O_2. Hydrocarbons (C_1 to C_3) on 5A with higher alkanes on 13X (up to C_{12}) but not isomer separations
Cyclodextrins		Fixed gases, halocarbons, hydrofluorocarbons, C_1 to C_{10} hydrocarbons
Porous polymers		
Q	310	Hydrocarbons (C_1 to C_{14}), halocarbons (C_1 and C_2), volatile
S	250	Oxygenated solvents (C_1 to C_6), thiols, amines, nitro
U	190	Compounds, nitriles, water, and inorganic gases

Q = Poly(divinylbenzene–styrene),
S = poly(divinylbenzene–vinylpyridine) and
U = poly(divinylbenzene–ethylene glycol dimethacrylate).

surface. They should be considered complementary rather than competitive to liquid-phase columns.

6.4. TEMPERATURE-PROGRAMMED SEPARATIONS

In gas chromatography, there is an approximate exponential relationship between retention time and solute vapor pressure for separations at a constant temperature. Consequently, it is impossible to establish a suitable compromise temperature for the separation of mixtures with a boiling point range exceeding about 100 °C. For mixtures that span a wide boiling point range, isothermal separations are characterized by long separation times, poor separations of early-eluting peaks, and poor detectability of late-eluting peaks due to zone broadening. There are many separation problems that fall into this category handled by gas chromatography by employing temperature programming, or less often because of its greater limitations, flow programming [21]. Stationary phases of high thermal stability allow wide temperature ranges to be used. Neither constant nor programmed-temperature modes are superior to each other; in an individual case, the best approach is defined by the properties of the sample. Temperature-programmed techniques are the most useful approach for scouting the properties of an unknown sample.

Column selection for temperature-programmed separations is more challenging than for isothermal separations because the column selectivity is not constant throughout the temperature range of the separation. The influence of temperature on selectivity can be observed from the shape of system maps as shown, for example, in Figure 6.2. In a temperature program, each substance interacts with the stationary phase only over the change in temperature while it is resident in the column. Separated substances are associated with different temperature ranges, and, therefore, different selectivity ranges. For separations with large differences in temperature ranges, a typical situation for mixtures with a wide boiling point or polarity range, temperature-induced changes in selectivity are likely to be significant. The notion of a column having an assignable selectivity, therefore, has no chemical meaning for temperature-programmed gas chromatography. The properties of the sample dictate the selectivity space for the column, and there is no general possibility of a column classification system for temperature-programmed conditions. To obtain columns likely to provide different selectivities throughout, a programmed separation selection is based on the general characteristics observed for isothermal separations at an average temperature for the programmed temperature range.

Leaving apart the selection of the stationary phase, many other column characteristics can be selected by computer simulation using method translation software (or calculations) [21–23]. Column characteristics within a method that can be optimized by method translation are summarized in Table 6.11. For

TABLE 6.11 Translatable Parameters in Temperature-Programmed Gas Chromatography

Translatable (do not affect peak elution patterns)
Column dimensions (length and internal diameter)
Carrier gas (identity of carrier gas)
Pressure conditions (column inlet and outlet pressure, carrier gas flow rate)
Temperature program conditions (proportional changes in the duration of program steps and isothermal plateaus)
Nontranslatable (can effect peak elution patterns)
Stationary-phase type
Column phase ratio
Set temperatures for isothermal segments of a temperature program

speed optimization, method translation facilitates the simultaneous optimization of the carrier gas and its flow rate, column dimensions (length and internal diameter), stationary-phase film thickness, system operating pressure (inlet or outlet pressure), and temperature program rate to rescale an obtained separation to a new set of speed-optimized parameters in which the elution order for the original separation is maintained. The basis of method translation is that the column hold-up time can be viewed as a fundamental parameter to express any time-related parameter for a separation. Two methods are usually translatable if they have identical non-translatable parameters and the same normalized temperature program rate. Studies in method translation indicate that an approximate optimum linear temperature program rate of about 10 °C/(column hold-up time) exists for all separations [24]. For translatable methods, the speed gain is given by the ratio of the column hold-up times for the two methods and the change in resolution by the square root of the ratio of the plate numbers for the two methods. Faster separations usually result in the choice of shorter columns of smaller internal diameters. Practical bounds to speed optimization are usually set by inadequate column sample capacity, required column inlet pressure, or unrealistically high temperature program rates for standard column ovens. Retention time locking used to minimize differences in retention times with variation of operating conditions and to standardize retention times for the same method with different systems is based on the method translation approach [23,25].

6.5. STATIONARY-PHASE SELECTIVITY TUNING

The common approach to method development in gas chromatography is to choose a stationary phase (column) that provides the best separation from a group of available stationary phases. The selected column, however, is simply the best available for the fixed stationary-phase compositions sampled and not necessarily the true optimum composition for the separation. Two general approaches are available for optimizing the stationary-phase composition for a separation: coupling of two or more columns containing different stationary phases in which the residence time is varied in each column to tune the overall selectivity for a separation [26,27] and computer-assisted stationary-phase design using initial experimental data on individual stationary phases of different composition to determine thermodynamic parameters that facilitate the simulation of the separation for different possible stationary-phase compositions to identify an optimum composition for a particular separation [27,28]. The latter approach was used to develop application-specific stationary phases of the type indicated in Table 6.6. Neither approach, however, is widely used in the current practice of laboratory methods. The emergence of comprehensive gas chromatography over the last decade is an exception.

Comprehensive multidimensional techniques employ two columns separated by a modulation interface. To achieve a useful increase in peak capacity with respect to a single column, the coupled columns should be of different selectivity. This selectivity difference is usually described in terms of "orthogonality," but in reality stationary phases differ in the intensity of specific intermolecular interactions, and even the most different stationary phases cannot be stated to be "orthogonal" [29]. Orthogonality is also a useful concept for the selection of separation systems for conventional separations, since the inclusion of systems with the widest possible separation properties ensures adequate sampling of the available separation space facilitating column selection. Using the system constants of the

TABLE 6.12 The Orthogonality of Different Column Types to SPB-Octyl (as reference phase) at 100 °C (Column Stationary Phases are Identified in Tables 6.4 and 6.9)

Column	*D*-parameter	cos θ
DB-1	0.348	0.862
DB-VRX	0.374	0.850
PMPS-5	0.401	0.831
DB-5ms	0.436	0.810
DB-XLB	0.498	0.785
Rtx-Volatiles	0.548	0.748
Rtx-440	0.554	0.754
Stx-500	0.578	0.739
Rtx-20	0.622	0.711
DX-1	0.700	0.655
DB-35	0.730	0.656
DB-1301	0.753	0.624
DB-35ms	0.779	0.652
DB-17ms	0.878	0.617
Rtx-50	0.882	0.589
Rtx-CLP	0.908	0.523
Rtx-65	0.937	0.576
Rtx-VGC	0.989	0.501
DB-1701	1.036	0.503
Rtx-OPP	1.103	0.432
DB-200	1.221	0.385
DB-210	1.481	0.296
1-Butyl-1-methylpyrrolidinium tris(pentafluoroethyl) trifluorophosphate	1.644	0.352
DB-225	1.851	0.312
1,9-Di(3-hydroxyethylimidazolium) nonane bis(trifluoromethylsulfonyl) imide	2.081	0.278
DX-3	2.111	0.319
1,9-Di(3-methylimidazolium)nonane bis(trifluoromethylsulfonyl)imide	2.190	0.266
DB-23	2.256	0.266
DX-4	2.315	0.292
PAG	2.539	0.270
1-(2-Hydroxyethyl)-3-methylimidazolium bis(trifluoromethylsulfonyl)imide	2.604	0.222
PEG	2.629	0.258
HP-88	2.641	0.232
SP-2340	2.759	0.223
BPX90	2.787	0.217
Tricationic ionic liquid	3.279	0.184
Core = tris(2-hexanamido) ethylamine		
Substituent = 3-methylimidazolium		
Anion = bis(trifluoromethylsulfonyl) imide		

solvation parameter model, two scales were used to predict the orthogonality of a wide range of stationary phases at 100 °C: the Euclidean distance between stationary phases in multidimensional hyperspace (*D*-parameter) and the angle θ between the linear vectors calculated from the system constants arranged in five-dimensional space. A *D*-parameter less than about 0.5–0.8 is a good indication that the compared systems are similar in terms of their capability for intermolecular interactions. The closer cos θ is to zero, the greater the "orthogonality" of the compared stationary phases and the closer cos θ is to 1, the greater their similarity. Taking the poly(methyloctylsiloxane) stationary phase as a reference point, since it has the lowest capacity for polar interactions, the orthogonality of a number of common stationary phases is summarized in Table 6.12 [29]. Stationary phases with similar solvation properties are next to each other in the table and those most different to the poly (methyloctylsiloxane) reference phase are found at the bottom of the table. These include the poly(biscyanoalkylsiloxane), poly(3,3,3-trifluoropropylmethylsiloxane), poly(ethylene glycol), and some of the ionic liquid stationary phases. These stationary phases are expected to be the most effective when combined with the poly(methyloctylsiloxane) stationary phases for comprehensive gas chromatography. The orthogonality represented by cos θ corresponds to angles between about 70 and 80° for the above-mentioned stationary phases. Although orthogonality maximizes the peak capacity, it does not necessarily result in better separations unless the polarity of the sample covers a sufficiently wide range to spread across the selectivity space. If not, large areas of the retention space remain empty. Many practical separations of complex mixtures of a narrow polarity range, therefore, will require columns with appropriate selectivity differences for the sample type, which do not necessarily correspond to the most orthogonal stationary phases available.

References

[1] C.F. Poole, The essence of chromatography, Elsevier, Amsterdam, 2003.

[2] C. Cruz-Hernandez, F. Destaillates, Recent advances in fast gas chromatography: application to the separation of fatty acid methyl esters, J. Liq. Chromatogr. Rel. Technol. 82 (2009) 1672–1688.

[3] M.H. Abraham, C.F. Poole, S.K. Poole, Classification of stationary phases and other materials by gas chromatography, J. Chromatogr. A 842 (1999) 79–114.

[4] C.F. Poole, S.K. Poole, Characterization of the solvent properties of gas chromatographic liquid phases, Chem. Rev. 89 (1989) 377–395.

[5] L. Rohrschneider, Characterization of GC stationary phases in multilinear retention model, Chromatographia 48 (1998) 728–738.

[6] B.R. Kersten, C.F. Poole, K.G. Furton, Ambiguities in the determination of McReynolds stationary phase constants, J. Chromatogr. 411 (1987) 43–59.

[7] C.F. Poole, T.O. Kollie, S.K. Poole, Recent advances in solvation models for stationary phase characterization and the prediction of retention in gas chromatography, Chromatographia 34 (1992) 281–302.

[8] C.F. Poole, S.N. Atapattu, S.K. Poole, A.K. Bell, Determination of solute descriptors by chromatographic methods, Anal. Chim. Acta. 652 (2009) 32–53.

[9] M.H. Abraham, A. Ibrahim, A.M. Zissimos, Determination of sets of solute descriptors from chromatographic measurements, J. Chromatogr. A 1037 (2004) 29–47.

[10] M. Vitha, P.W. Carr, The chemical interpretation and practice of linear solvation energy relationships in chromatography, J. Chromatogr. A 1126 (2006) 143–194.

[11] S.N. Atapattu, C.F. Poole, Solute descriptors for characterizing retention properties of open-tubular columns of different selectivity in gas chromatography at intermediate temperatures, J. Chromatogr. A 1195 (2008) 136–145.

[12] C.F. Poole, S.K. Poole, Separation characteristics of wall-coated open-tubular columns for gas chromatography, J. Chromatogr. A 1184 (2008) 254–280.

[13] S.N. Atapattu, C.F. Poole, Selectivity equivalence of two poly(methylphenylsiloxane) open-tubular columns prepared with different deactivation techniques for gas chromatography, J. Chromatogr. A 1185 (2008) 305–309.

[14] S.N. Atapattu, K. Eggers, C.F. Poole, W. Kiridena, W.W. Koziol, Extension of the system constants database for open-tubular columns: system maps at low and intermediate temperatures for four new columns, J. Chromatogr. A 1216 (2009) 1640–1649.

[15] S.D. Martin, C.F. Poole, M.H. Abraham, Synthesis and gas chromatographic evaluation of a high-temperature hydrogen-bond acid stationary phase, J. Chromatogr. A 805 (1998) 217–235.

[16] W. Kiridena, C.C. Patchett, W.W. Koziol, H. Ahmed, C.F. Poole, Separation characteristics of phenyl-containing stationary phases for gas chromatography based on silarylene-siloxane copolymer chemistries, J. Sep. Sci. 29 (2008) 211–217.

[17] C.F. Poole, J. Qian, W. Kiridena, C. DeKay, W.W. Koziol, Evaluation of the separation characteristics of application specific (volatile organic compounds) open-tubular columns for gas chromatography, J. Chromatogr. A 1134 (2006) 284–290.

[18] C. Yao, J.L. Anderson, Retention characteristics of organic compounds on molten salts and ionic liquid-based gas chromatography stationary phases, J. Chromatogr. A 1216 (2009) 1658–1712.

[19] Poole CF, Poole SK. Ionic liquid stationary phases for gas chromatography. J. Sep. Sci. 34 (2011) 888–900.

[20] Z. Ji, R.E. Majors, E.J. Guthrie, Porous layer open-tubular capillary columns; preparations, applications and future directions, J. Chromatogr. A 842 (1999) 115–142.

[21] L.M. Blumberg, Temperature-programmed gas chromatography, Wiley-VCH, Weinheim, Germany, 2010.

[22] M.S. Klee, L.M. Blumberg, Theoretical and practical aspects of fast gas chromatography and method translation, J. Chromatogr. Sci. 40 (2002) 234–247.

[23] L.M. Blumberg, M.S. Klee, Method translation and retention time locking in partition GC, Anal. Chem. 70 (1998) 3828–3838.

[24] L.M. Blumberg, M.S. Klee, Optimal heating rate in gas chromatography, J. Microcolumn. Sep. 12 (2000) 508–514.

[25] N. Etxebarria, O. Zuloaga, M. Olivares, L.J. Bartolome, P. Navarro, Retention-time locked methods in gas chromatography, J. Chromatogr. A. 1216 (2009) 1624–1629.

[26] R. Sacks, C. Coutant, A. Grall, Advancing the science of column selectivity, Anal. Chem. 72 (2000) 524A–533A.

[27] F.L. Dorman, P.D. Schettler, L.A. Vogt, J.W. Cochran, Using computer modeling to predict and optimize separations for comprehensive two-dimensional gas chromatography, J. Chromatogr. A 1186 (2008) 196–201.

[28] F.L. Dorman, P.D. Schettler, C.M. English, D.V. Patwardhan, Predicting gas chromatographic separation and stationary-phase selectivity using computer modeling, Anal. Chem. 74 (2002) 2133–2138.

[29] S.K. Poole, C.F. Poole, The orthogonal character of stationary phases for gas chromatography, J. Sep. Sci. 31 (2008) 1118–1123.

CHAPTER

7

Multidimensional and Comprehensive Gas Chromatography

John V. Seeley

OUTLINE

7.1. INTRODUCTION

A modern gas chromatograph equipped with a split/splitless inlet, a capillary column, and a fairly universal detector (e.g., a flame ionization detector) can be used to generate high-resolution separations of a wide range of samples. Unfortunately, experience and theory [1] have shown that complex samples, such as petroleum-based fuels or environmental

Gas Chromatography DOI: 10.1016/B978-0-12-385540-4.00007-9

extracts, produce chromatograms with significant peak overlap. The limited ability of single-column gas chromatography (GC) to isolate analytes in complex mixtures has led to the adoption of selective detectors such as the electron-capture detector or scanning detectors such as the mass spectrometer. Such detectors supply additional resolving power but often decrease generality or increase cost. It is also possible to improve performance by increasing the resolving power of the chromatographic separation. Changing stationary phases can separate a critical pair of components, but it frequently leads to the creation of new overlapping pairs. Switching to a longer and/or narrower column can generate more theoretical plates, but this also increases analysis time or decreases sample capacity.

Introducing an additional chromatographic stage is an effective approach for increasing the resolving power of a separation. This chapter examines methods that employ two sequential GC separations. Such analyses fall into the category of two-dimensional gas chromatography (2D GC). In addition to increased peak capacity, obtaining retention information on two different stationary phases can help distinguish functional classes of components. This effect can be observed by examining the retention indices of components on two different stationary phases. For example, the Kovats retention indices of alkyl esters and aliphatic alcohols [2] are interspersed on both the DB-5 and DB-Wax stationary phases (see Figure 7.1a and 7.1b). However, when a two-dimensional plot of DB-Wax and DB-5 retention indices is constructed, the alkyl esters are fully resolved from the aliphatic alcohols (see Figure 7.1c). Combining two separations has the potential to increase the resolution and information provided by GC analyses, but the challenge is to develop practical laboratory methods capable of exploiting this potential.

Two-dimensional GC was first reported over 50 years ago [3] and numerous improvements have been made since. For a more

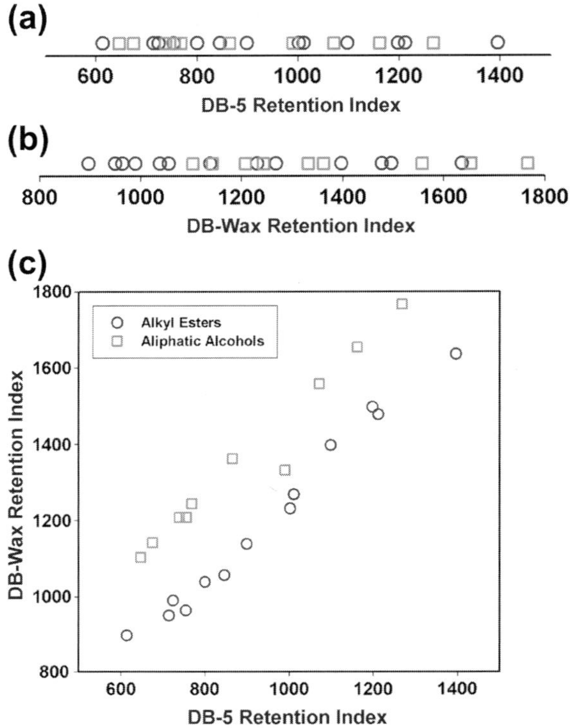

FIGURE 7.1 Kovats retention indices of C_4–C_{12} alkyl esters (blue square) and C_4–C_{10} aliphatic alcohols (red squares) on the DB-5 and DB-Wax stationary phases. (a) DB-5 retention indices. (b) DB-Wax retention indices. (c) A 2D plot of DB-5 and DB-Wax retention indices.

comprehensive account of the development of multidimensional GC, the reader is referred to several excellent review articles and book chapters published over the past two decades [4–9]. This chapter focuses on the chromatographic characteristics required for an effective 2D GC separation and some of the instrument designs that are being actively developed and applied, including heartcutting 2D GC and comprehensive two-dimensional gas chromatography (GC × GC).

Unlike 2D gel electrophoresis or 2D thin layer chromatography, 2D GC is not performed with a planar separation medium. Instead, two capillary columns containing different stationary phases are used. The majority of the instrument

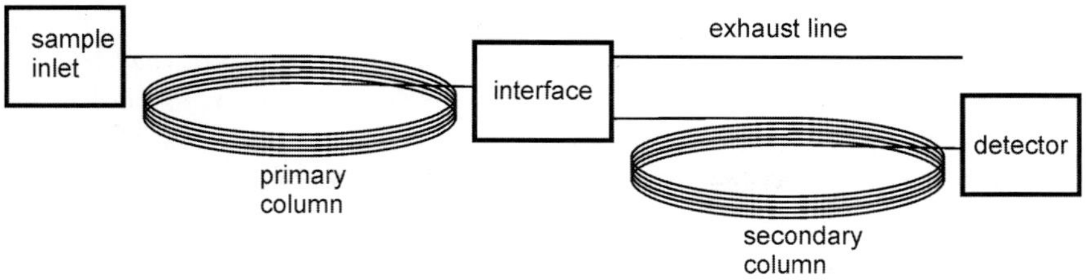

FIGURE 7.2 The basic construction of a 2D GC instrument. Samples are injected through a standard inlet and separated by the primary column. Upon eluting from the primary column, components pass through an interface that controls entry into a secondary column. Components are detected as they elute from the secondary column.

components associated with 2D GC are identical to those used for conventional GC. However, 2D GC analyses require a device for coupling the pair of capillary columns and in some cases special software for analyzing the resulting data. A schematic of a basic gas chromatograph capable of performing a 2D separation is shown in Figure 7.2. The sample is injected through an inlet located at the head of the primary column. The mixture components pass through the primary column until they reach an interface that controls transfer from the primary column to the secondary column. Depending on the mode of operation, the interface may send the component to an exhaust line, temporarily accumulate the component in a storage region, or send the component to the head of the secondary column. Components that enter the secondary column are transported down its length until they reach a detector.

Two-dimensional GC separations combine three time-dependent processes: transport through the primary column, transport through the interface, and transport through the secondary column. As such, there are many more degrees of freedom present in 2D GC than conventional GC. This increased flexibility provides the opportunity to tune separations for a specific set of analytes, and it also introduces more sources of error. Successful separations require precise control of many parameters including the carrier flow rates, the temperatures of heated zones, and the timing of interfacial events. Fortunately, GC instruments are now equipped with high-precision electronic pneumatics, multiple independently heated regions, and programmable electronic outputs. Stationary-phase selection is a critical part of the development of a successful 2D GC analysis. The primary and secondary stationary phases should have complementary selectivities that maximize the contrast between the analytes and the sample matrix. The ease of identifying promising stationary-phase pairs has been improved with the introduction of simple models for quantitatively predicting chromatographic retention [10].

7.2. A GRAPHICAL REPRESENTATION OF 2D GC SEPARATIONS

Component transport in the primary column, the interface, and secondary column all occur simultaneously in a 2D GC separation. In this chapter, a graphical approach will be used to represent these concurrent processes. Figure 7.3 is a plot that represents the retention of a single component in a coupled-column separation. In this case, the component flows directly from the exit of the primary column to the head of the secondary column (i.e., the interface acts as a simple connecting union). The primary retention time t_1 of the component is represented by the horizontal position of the peak in the 2D plot and the cumulative retention time t_c (i.e., the sum of the primary and secondary retention times) is indicated by its vertical position. For simplicity, this analysis assumes that the residence time in the interface is negligible. The secondary retention time t_2 can be determined by the difference of the cumulative retention time and the primary retention time ($t_2 = t_c - t_1$). Graphically, this corresponds to the vertical distance between the 2D peak and the diagonal line representing $t_c = t_1$. The time-dependent concentration of the component at the exit of the primary column is determined by integrating the 2D plot in the vertical direction, whereas the concentration at the exit of the secondary column is determined by integrating the 2D plot in the horizontal direction. The width of a vertical slice of the 2D peak is proportional to the standard deviation of the secondary retention times. For example, if the peaks do not broaden appreciably on the secondary column, the 2D peak will appear as a slender ellipse.

Figure 7.4 shows a simulation of the retention of 20 components when the primary retention time was randomly assigned within a range of

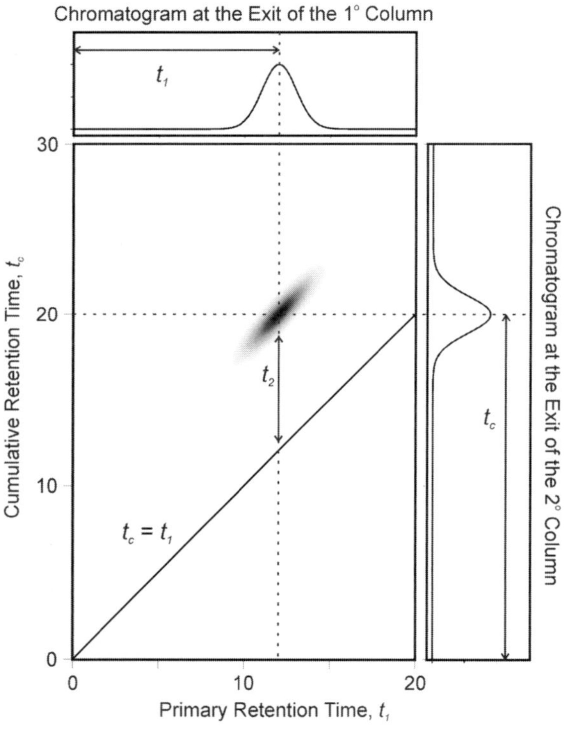

FIGURE 7.3 A graphical representation of a 2D GC separation. The primary retention time t_1 of a component is represented by its horizontal position, and the cumulative retention time t_c (i.e., the sum of t_1 and the secondary retention time t_2) is represented by its vertical position. The chromatogram at the end of the primary column is found by integrating the 2D plot in the vertical direction, and the chromatogram at the end of the secondary column is found by integrating the 2D plot in the horizontal direction.

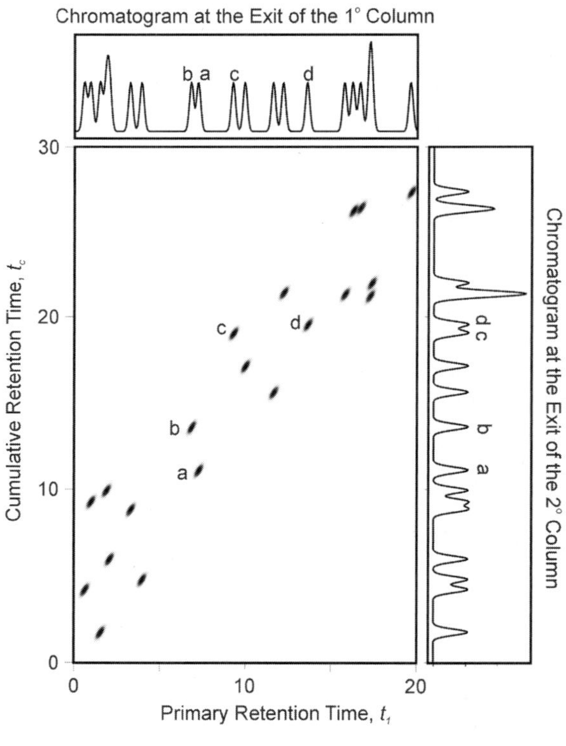

FIGURE 7.4 Simulation of a coupled-column separation of 20 components. All of the components are allowed to pass directly from the primary column to the secondary column. The chromatogram at the top of the figure shows that 8 of the 20 components are separated at the end of the primary column. The chromatogram on the right shows that 7 of the 20 components are separated at the end of the secondary column.

0–20 min and the secondary retention time within a range of 0–10 min. Note that this simulation assumes that components flow directly from the primary column to the secondary column. Within the 2D plot, 16 of the 20 components are fully isolated. This means that 80% of the components have a primary retention time and cumulative retention time combination that is substantially different from those of the other components. The 2D plot shown in Figure 7.4 is useful in understanding the dynamics of coupled-column separations, but it is not experimentally observable with an apparatus like that shown in Figure 7.2. This plot could be constructed in this theoretical exercise because the primary and secondary retention times and peak shapes are known for each component. In an experimental setting, only the chromatogram generated at the end of the secondary column (i.e. the chromatogram at the right of Figure 7.4) would be available. If a second detector was added to monitor a portion of the primary column effluent, then the chromatogram at the top of Figure 7.4 would also be available. Unfortunately, it would be impossible to use these two 1D chromatograms to unambiguously assign the primary and secondary retention times to each of the 20 components when they are analyzed simultaneously. As will be shown in subsequent sections, the ability to experimentally generate an unambiguous 2D representation of primary and secondary retentions for the chromatographic analysis of a complex mixture is only possible under highly controlled experimental conditions.

These simulations demonstrate that merely adding a secondary column to the end of the primary column does not drastically increase resolving power. For example, the chromatogram at the top of Figure 7.4 shows that 8 of the 20 components are resolved from their neighboring peaks (i.e., $R > 1$ on both sides) at the end of the primary column, and the chromatogram at the right of Figure 7.4 shows that 7 of the 20 components are fully isolated at the end of the secondary column. One might predict that the chromatogram obtained at the end of the secondary column would have significantly greater resolution because components that coelute on the primary column can be separated in the secondary column. However, components resolved by the primary column can also be recombined in the secondary column. In the absence of action taken at the interface between the columns, these two processes are nearly balanced. For example, components *a* and *b* in Figure 7.4 coelute on the primary column, but they are separated when they reach the end of the secondary column as component *b* experiences much greater secondary retention than component *a*. In contrast, components *c* and *d* are separated on the primary column with component *d* being more strongly retained. However, component *c* has much greater secondary retention causing the pair to recombine as they exit the secondary column.

To maximize resolving power, 2D GC systems use the action of the interface along with carefully controlled retention in the secondary column to (1) decrease the probability that components resolved in the primary column are recombined in the secondary column and (2) increase the probability that components that coelute from the primary column are separated in the secondary column. The remainder of this chapter will examine the main forms of 2D GC currently being employed or developed, with particular attention paid to the conditions that must be met to make a 2D GC separation successful.

7.3. BACKFLUSHING 2D GC

Backflushing 2D GC is the simplest form of 2D GC. It allows a class of analytes to be fully resolved from sample matrix components when the analytes are fairly volatile and substantially more polar than the matrix components.

7.3.1. Basic Mode of Operation

A common implementation of backflushing 2D GC is shown in Figure 7.5. In this case, the interface is just a tee union that introduces an auxiliary flow of carrier gas to the junction between the primary and secondary columns. The apparatus can be placed in two different states: foreflush and backflush. In the foreflush state (Figure 7.5a), the carrier gas flows from the sample inlet through the primary column to the tee union. The auxiliary flow mixes with the primary column effluent in the tee union and the combined stream moves through the secondary column toward the detector. In the backflush state (Figure 7.5b), the pressure in the sample inlet is reduced significantly while the pressure in the tee union is held constant. This causes the flow in the primary column to reverse direction while the flow in the secondary column is unchanged.

A backflushing 2D GC analysis involves a single switch from the foreflush state to the backflush state. The apparatus is initially in the foreflush state when the sample is injected through the split/splitless inlet. Compounds that are weakly retained on the primary column pass through the tee and reach the secondary column. Just after the last analyte reaches the secondary column, the apparatus is placed in the backflush state. This reverses the flow in the primary column and causes any sample matrix components that are still in the primary column to be backflushed out of the split vent, while the components in the secondary column continue traveling toward the detector.

The retention characteristics and timing required for an effective backflushing 2D GC analysis are demonstrated in the simulation shown in Figure 7.6. The analytes are circled in

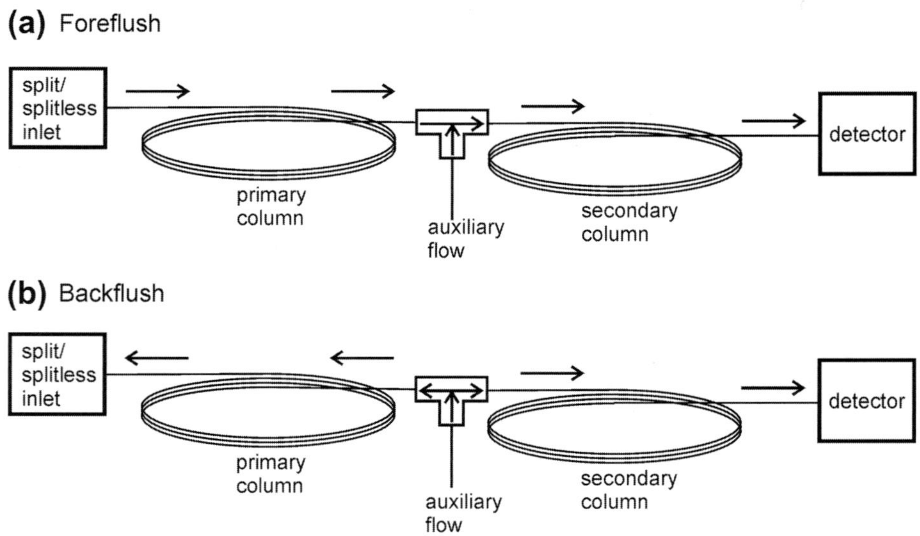

FIGURE 7.5 A backflushing 2D GC apparatus. (a) The instrument is initially in the foreflush state where the sample passes from the inlet toward the interface through the primary column. (b) After the last analyte enters the secondary column, the instrument is placed in the backflush state where the sample matrix components that are still in the primary column are backflushed out of the split vent and thereby prevented from reaching the secondary column.

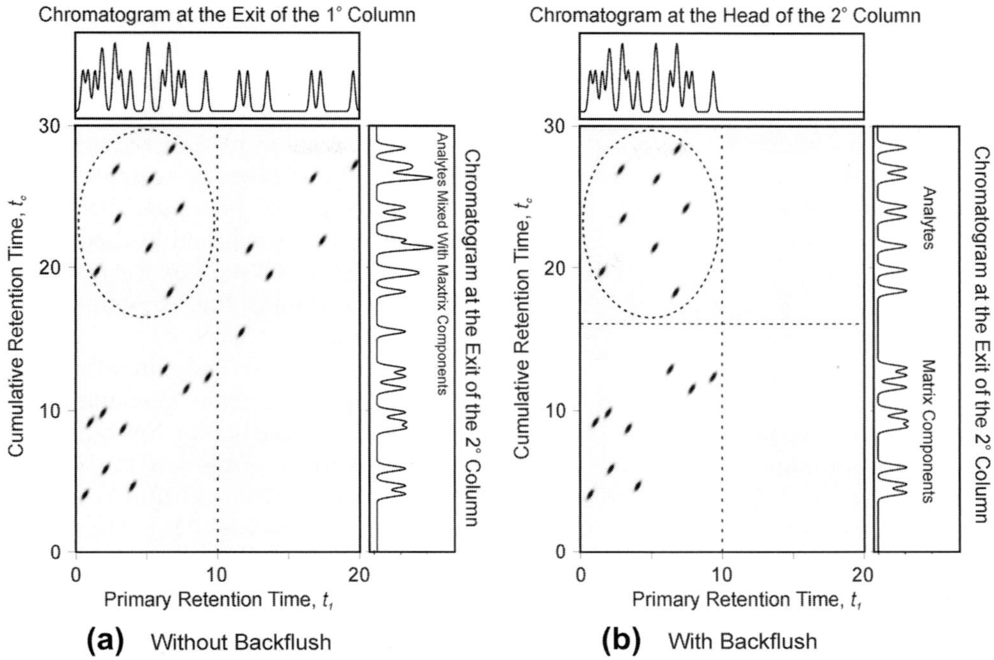

FIGURE 7.6 2D plots of a simulated backflushing 2D analysis. (a) Without backflushing, the analytes (circled in the 2D plot) coelute with sample matrix components at the end of the primary column and at the end of the secondary column. (b) Backflushing prevents the components that elute from the primary column after the dashed vertical line from entering the secondary column. This causes the analytes to be fully separated from the sample matrix components at the end of the secondary column.

the 2D retention plot and the remainder of the peaks represent sample matrix components. The analytes have very large secondary retention times. In the absence of backflushing, the analyte peaks have primary retention times that overlap with the early-eluting matrix peaks (see the top of Figure 7.6a) and cumulative retention times that overlap with the late-eluting matrix peaks (see the right of Figure 7.6a). Thus, neither the chromatogram generated at the end of the primary column nor the chromatogram generated at end of the secondary column would fully isolate the analytes from the matrix. The backflush time is represented by the vertical dashed line in Figure 7.6a. Peaks that are to the right of the backflush line are not allowed to enter the secondary column and thus are removed by the backflush process.

The effective 2D retention plot when backflushing is performed is shown in Figure 7.6b. The chromatogram generated at the end of the secondary column successfully separates the analyte components from the sample matrix components. In this scenario, a single event in the interface, the backflush, allows the analytes to be fully resolved from the matrix components.

Backflushing works when the analytes are weakly retained on the primary column and much more strongly retained on the secondary column as compared to the matrix components. Under such conditions, the 2D retention plots shown in Figure 7.6 can be divided into quadrants with the upper-left quadrant containing only analyte components. The matrix components that overlap with the analytes at the end of the secondary column (i.e., the components in the upper-right quadrant) do not reach the head of the secondary column until after the last analyte peak has been loaded onto the secondary column. Thus, a backflush immediately after the last analyte peak has reached the secondary column removes these interfering matrix components.

7.3.2. An Example of Backflushing 2D GC: The Analysis of Oxygenates in Gasoline

The separation of oxygenates from gasoline hydrocarbons is a common application of backflushing 2D GC. Gasoline contains hundreds of hydrocarbons with carbon numbers ranging from C_5 to C_{12}. Gasoline can also contain oxygenated compounds. The most prevalent oxygenates are C_1–C_5 aliphatic alcohols and C_5–C_6 aliphatic ethers. The concentrations of many oxygenates are at parts per million (ppm) levels or below; however, ethanol and methyl *tert*-butyl ether (MTBE) can be present at levels above 10%. The main analytical goal is to separate the entire oxygenate class from the hydrocarbons. Unfortunately, there is no single capillary column that can generate such a separation. This is because oxygenates overlap with the low-molecular-weight hydrocarbons on nonpolar columns and oxygenates overlap with the high-molecular-weight hydrocarbons on polar columns. However, backflushing 2D GC with a poly(dimethylsiloxane) primary column and a CP-Lowox secondary column can fully separate the oxygenates from the hydrocarbons.

An example of an instrument used for separating oxygenates from gasoline is shown in Figure 7.7. This particular configuration, developed by Andrew Tipler of PerkinElmer, Inc., employs a miniature column coupling device known as an S-Swafer [11]. The S-Swafer has minimal unswept volume, low thermal mass, and provides leak-free connections for the primary column, secondary column, and auxiliary flow. Figure 7.7 also shows a flow restrictor that leads to an additional detector that allows a chromatogram of the components eluting from the primary column to also be obtained. The chromatogram of gasoline eluting from the primary column without backflushing is

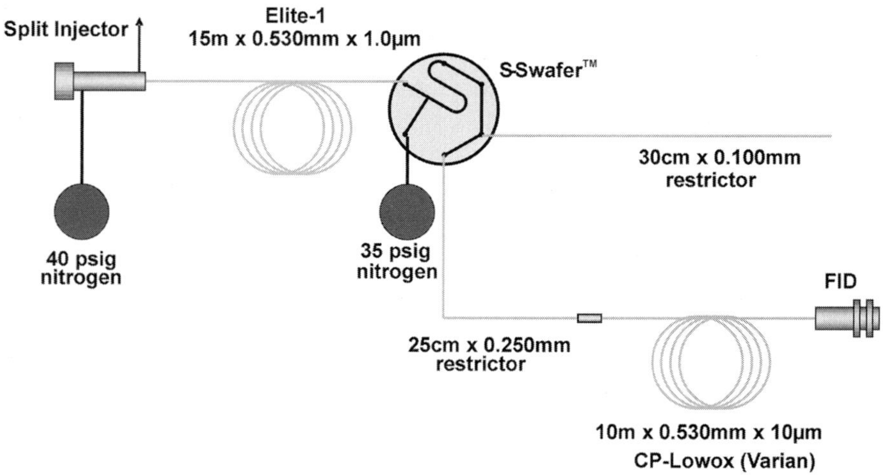

FIGURE 7.7 Backflushing 2D GC apparatus used to analyze oxygenates in gasoline. *Source: Reprinted with permission from Ref. [11]. Copyright 2010, PerkinElmer, Inc.*

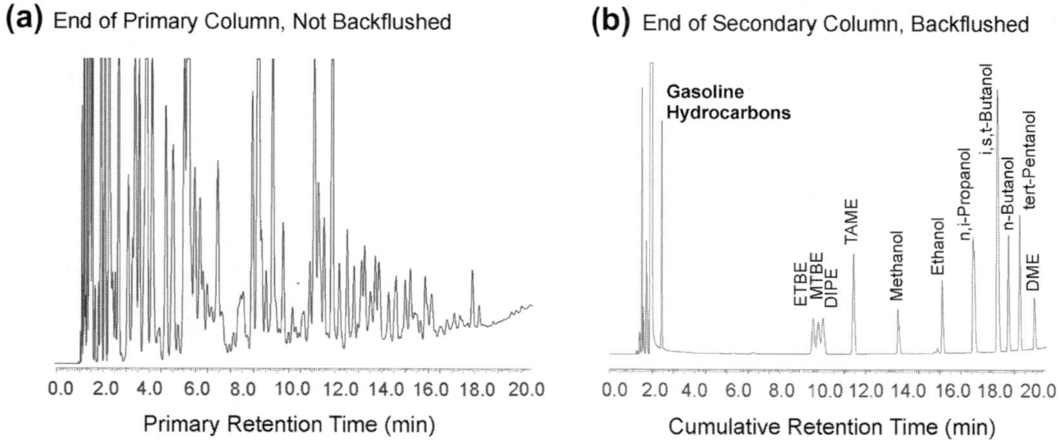

FIGURE 7.8 (a) Gasoline chromatogram obtained on the nonpolar primary column. The oxygenates elute between 1.0 and 1.5 min along with numerous hydrocarbons. (b) Chromatogram obtained at the end of the CP-Lowox secondary column for gasoline containing percent levels of oxygenates. When the primary column is backflushed at 1.52 min, the chromatogram obtained at the end of the secondary column fully separates the oxygenates from the hydrocarbons. The names of the alcohols are listed in the figure. The names of the ethers are abbreviated: ETBE = ethyl *tert*-butyl ether; MTBE = methyl *tert*-butyl ether; DIPE = diisopropyl ether; TAME = *tert*-amyl methyl ether; DME = 1,2-dimethoxyethane. *Source: Reprinted with permission from Ref. [11]. Copyright 2010, PerkinElmer, Inc.*

shown in Figure 7.8a. The oxygenates elute between 1.0 and 1.5 min and, when present in low concentrations, are obscured by the hydrocarbons. Under normal conditions, backflushing is initiated at 1.52 min when the last oxygenate, *tert*-amyl methyl ether (TAME), enters the secondary column. All compounds with retention times on poly(dimethylsiloxane) greater than that of TAME are not transferred to the CP-Lowox secondary column and are backflushed out of the split inlet. The chromatogram observed at the end of the CP-Lowox secondary column when backflushing is employed is shown in Figure 7.8b. The high affinity of the CP-Lowox stationary phase for oxygenates causes the ethers and alcohols to have significantly higher cumulative retention times than the hydrocarbons. A complete separation of the oxygenate class from gasoline hydrocarbons is achieved. The high resolution of this method allows ppm levels of oxygenates to be accurately quantified. Backflushing improves the speed of the analysis as the run can be concluded shortly after the last oxygenate elutes from the secondary column, and there is no need to wait for the low-volatility hydrocarbons to elute from the secondary column because they have been backflushed off the primary column. Another advantage of backflushing is that it protects the high-polarity CP-Lowox secondary column from being exposed to low-volatility hydrocarbons that could otherwise lead to contamination.

Backflushing 2D GC analyses can also be performed with multiport rotary valves in place of the coupling tee union. While such valves are generally considered more appropriate for packed-column analyses, they can generate a more diverse set of flow patterns than fluidic devices. For example, ASTM method D 4815 [12] analyzes oxygenates in petroleum-based fuels using a micropacked primary column containing 1,2,3-tris-(2-cyanoethoxy)propane (TCEP) followed by a poly-(dimethylsiloxane) secondary column. The sample is initially injected into the TCEP column where the oxygenates and heavy hydrocarbons are strongly retained while the light hydrocarbons are only weakly retained. The rotary valve is initially in a position that causes the light hydrocarbons to be directed to an exhaust line as they elute from the primary column. When all of the light hydrocarbons have fully eluted from the TCEP column, the valve is rotated and the trapped oxygenates and heavy hydrocarbons are backflushed onto the poly(dimethylsiloxane) secondary column. The oxygenates elute individually from the secondary column and are detected with an FID, while most of the heavy hydrocarbons are strongly retained at the head of the nonpolar secondary column. The rotary valve is actuated back to its original position and the heavy hydrocarbons are backflushed out of the secondary column to the FID.

While these two methods for analyzing oxygenates employ different hardware and different polar stationary phases, they exploit the same fundamental principle found in all backflushing 2D GC analyses: A precisely timed change in flow direction in the primary column removes sample matrix components that would otherwise coelute with the analytes at the exit of the secondary column.

7.4. HEARTCUTTING 2D GC

Heartcutting 2D GC is a simple approach for isolating a few selected analytes from a large number of sample matrix components. Within the field of analytical chemistry, the generic term "two-dimensional gas chromatography" is often used to denote what will be called heartcutting 2D GC in this chapter.

7.4.1. Basic Mode of Operation

Heartcutting 2D GC divides the primary column effluent into a series of fractions. Fractions that contain the targeted analyte components along with any coeluting sample matrix components are directed to the secondary column, whereas fractions that contain only sample matrix components are directed to an exhaust flow restrictor. The general schematic of a 2D GC instrument shown in Figure 7.2 can also serve as a schematic for a heartcutting 2D GC instrument. The interface is normally in a bypass state where the primary column effluent is directed to an exhaust flow restrictor. The interface is switched from the bypass state to an inject state immediately before an analyte elutes from the primary column. In the inject state, the primary effluent is directed to the head of the secondary column. As soon as the analyte has been loaded onto the secondary column, the interface is switched back to the bypass state.

Figure 7.9 shows a simulation that demonstrates the key retention features associated with a successful heartcutting 2D GC analysis of two analytes in a mixture containing 21 sample matrix components. In the absence of heartcutting (see Figure 7.9a), the analytes coelute with sample matrix components at the exits of the primary and secondary columns (see the top and right of Figure 7.9a). However, if none of the individual interfering compounds coelutes with the analytes on *both* the primary and secondary columns (i.e., share the same position in the 2D plot), then they can be resolved by heartcutting. The beginning and ending of two heartcuts for the simulation are shown in Figure 7.9a with the dashed vertical lines. The effective 2D retention plane after performing two heartcuts is shown in Figure 7.9b. Heartcutting removes the sample matrix components that would potentially recombine with the analytes in the secondary column (i.e., removes the matrix components that may be to the left or right of the analytes in the 2D retention plot). Thus, the analytes are fully isolated in the chromatogram acquired at the exit of the secondary column (see the right side of Figure 7.9b).

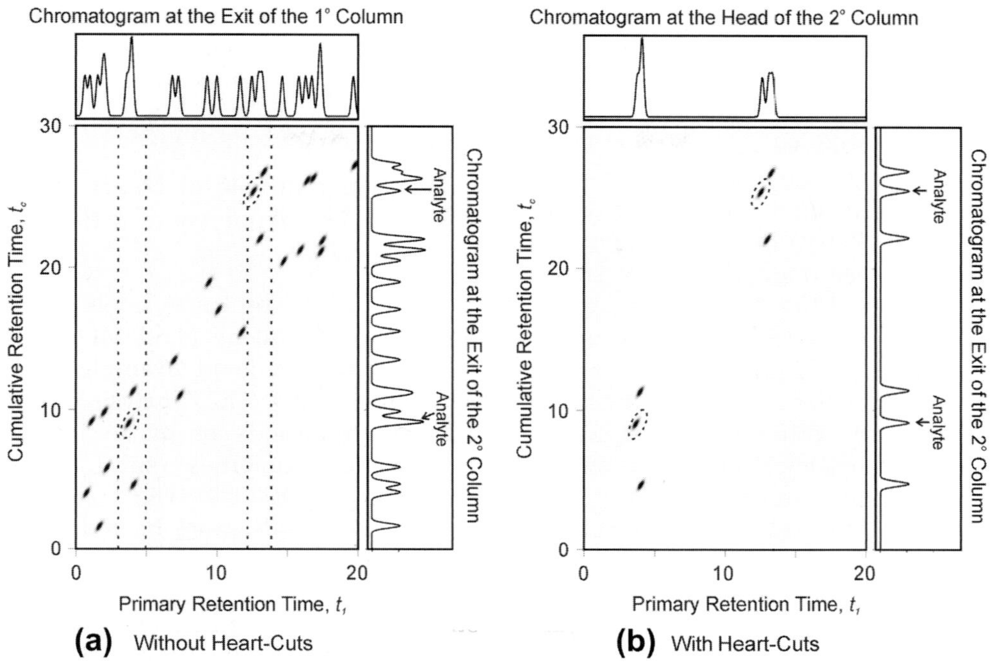

FIGURE 7.9 2D plots of a simulated heartcutting 2D analysis. (a) Without heartcutting, the analytes (circled in the 2D plot) coelute with sample matrix components at the end of the primary column and at the end of the secondary column. (b) Two separate heartcuts are performed to prevent numerous sample matrix components from entering the secondary column. This allows the analytes to be fully separated from the sample matrix components at the end of the secondary column.

There are numerous interfaces that can be used to execute heartcuts. Multiport valves have been used extensively in the past, but many current methods employ a Deans switch. The Deans switch is a fluidic device introduced 34 years ago [13] that uses an auxiliary flow to control the direction of the primary column effluent. A Deans switch is constructed from a two-way, three-port solenoid valve and an assembly of tee junctions. A schematic of a simple Deans switch is shown in Figure 7.10. The exit of the primary

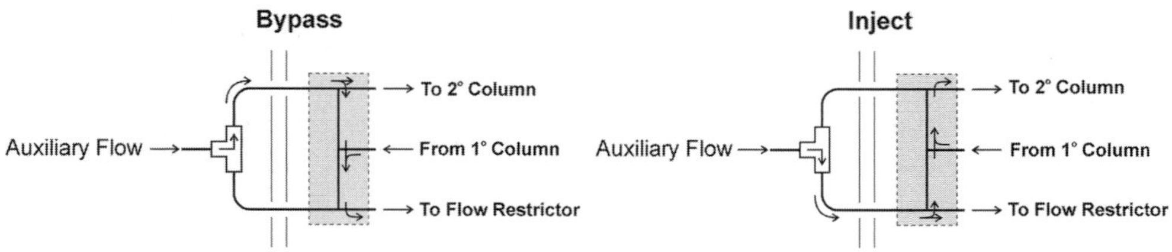

FIGURE 7.10 A simple Deans switch. A Deans switch is constructed from a solenoid valve placed outside the oven and an assembly of tee junctions placed inside the oven. The left side of the figure shows the Deans switch in the "bypass" state where the primary column effluent is directed away from the secondary column and into the flow restrictor. The right side shows the switch in the "inject" state where the primary column effluent is directed into the secondary column.

column is connected to the center tee junction, while the secondary column and an exhaust flow restrictor are connected to the peripheral tee junctions. The auxiliary flow passes through the solenoid valve and then to one of the two peripheral tee junctions. The bypass state is generated by having the solenoid valve introduce the auxiliary flow to the tee junction connected to the secondary column (see the left part of Figure 7.10). The vast majority of the auxiliary flow goes to the secondary column while a smaller portion moves toward the center tee junction where it directs the incoming primary column effluent to the exhaust flow restrictor. The inject state is generated by having the solenoid valve introduce the auxiliary flow to the tee junction connected to the exhaust restrictor (see the right part of Figure 7.10). A small portion of the auxiliary flow goes to the center tee junction where it directs the primary column effluent to the secondary column.

Deans switches have several advantages over multiport valves. The only moving part of the Deans switch, the solenoid valve, is not in the sample path; thus, it can be placed outside of the oven. The portion of the device that contacts the sample, the tee junction assembly, is a static device that can be assembled from inert materials that are capable of operating over a broad range of temperatures. Tee junction assemblies can be constructed by combining three individual tee unions, but now several manufacturers make assemblies that integrate all of the necessary flow paths and connections into a single device [14]. This reduces dead volume within the device and decreases the likelihood of leaks. Transitioning a Deans switch between the bypass and inject state does not significantly disturb the flow in the primary or secondary columns. Thus, heartcuts can be performed without affecting the primary retention times of later-eluting analytes. This allows multiple heartcuts to be made with high precision. In contrast, multiport valves require mechanical activation and cause brief but significant disturbances to the primary and secondary flows.

7.4.2. An Example of Heartcutting 2D GC: The Analysis of 4,6-DMDBT in Diesel Fuel

The analysis of trace levels of 4,6-dimethyldibenzothiophene (4,6-DMDBT) in diesel fuel serves as a good example of a simple heartcutting 2D GC analysis. Diesel fuel contains thousands of hydrocarbon components and numerous sulfur compounds including 4,6-DMDBT. The concentration of sulfur compounds must be reduced to ppm levels to meet regulatory requirements. Refiners determine the efficacy of their sulfur mitigation strategies by monitoring the levels of 4,6-DMDBT because it is one of the most difficult compounds to remove. The analysis is normally done with GC and a sulfur-selective detector; however, McCurry and Quimby [15] have demonstrated that heartcutting 2D GC with FID detection is also a viable strategy. They used a 15 m × 0.25 mm × 0.25 μm HP-5MS (Agilent Technologies) primary column followed by a 30 m × 0.25 mm × 0.25 μm Innowax (Agilent Technologies) secondary column. Preliminary studies with high levels of 4,6-DMDBT spiked into diesel fuel showed that 4,6-DMDBT had a primary retention time near 6.5 min. Thus, low levels of 4,6-DMDBT in diesel fuel were analyzed by using a Deans switch to heartcut the components eluting from the primary column between 6.40 and 6.65 min (see Figure 7.11a). The Innowax secondary column was selected because it retained 4,6-DMDBT more than the hydrocarbons that coeluted on the HP-5MS primary column (see Figure 7.11b). The 4,6-DMDBT is fully isolated from the diesel fuel hydrocarbons and the quantitative result is in excellent

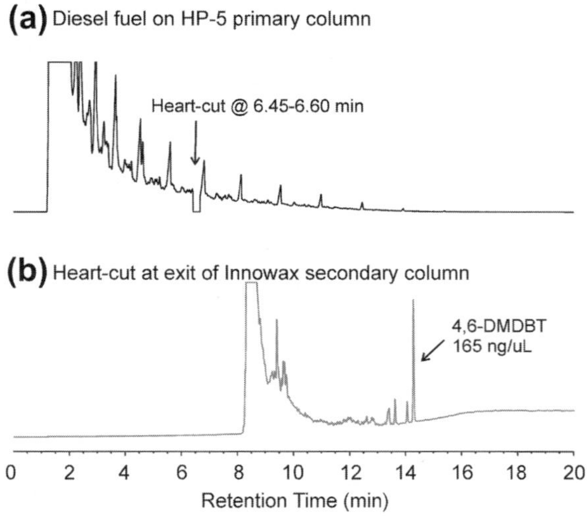

FIGURE 7.11 Heartcutting 2D GC analysis of 4,6-DMDBT in diesel fuel. (a) The chromatogram obtained at the end of the primary column. A heartcut is performed from 6.40 to 6.65 min to direct the 4,6-DMDBT to the secondary column. (b) Chromatogram of the heartcut obtained at the end of the secondary column. The 4,6-DMDBT is fully separated from the hydrocarbons that coeluted on the primary column. *Source: Reprinted with permission from Ref. [15]. Copyright 2003, Preston Publications.*

agreement with GC results obtained with an atomic emission detector. This method was found to be capable of accurately measuring 4,6-DMDBT levels down to 2 ppm.

7.4.3. Advanced Applications: Multiple Heartcuts and Independent Column Heating

The current generation of heartcutting 2D GC instruments is capable of making numerous, precisely timed heartcuts. For example, Gras et al. [16] have shown that a polyethylene glycol primary column coupled to a CP-PoraBond Q column can be used to analyze alkyl mercaptans in natural gas. Four heartcuts were used to individually isolate the C_1 through C_4 mercaptans. Sciarrone et al. [17] have recently described the use of 13 heartcuts to evaluate the quality of Australian Tea Tree oil. A 5% diphenyl/95% dimethyl polysiloxane primary column was coupled to a polyethylene glycol secondary column. A mass spectrometer was used to monitor the secondary column effluent. With this approach, they were able to isolate and identify 18 critical components within the complex mixture.

It is important to note that if multiple heartcuts are performed, care must be taken to ensure that the peaks from the adjacent fractions do not remix on the secondary column. This can be prevented by keeping the range of secondary retention times within each fraction less than the time between heartcuts. This measure ensures that a slow-moving component from a prior heartcut does not recombine with a fast-moving component from a subsequent heartcut. Thus, multiple heartcutting 2D GC requires that additional attention be paid to the chromatographic conditions including the column dimensions, carrier gas flow rates, and the column temperatures. Fortunately, GC instrumentation is constantly improving and new technologies are giving analysts unprecedented control. For example, direct heating of the capillary columns [18] allows the analyst to run independent temperature programs on the primary and secondary columns. This ability provides greater control over the range of secondary retention times and allows complicated heartcutting strategies to be successfully implemented.

7.5. COMPREHENSIVE 2D GC

Heartcutting 2D GC is effective for isolating a few important components, but much less effective for a broad-based characterization of sample composition. Comprehensive two-dimensional gas chromatography (GC $\times$ GC), a technique introduced 20 years ago by John B. Phillips and Zaiyou Liu [19], is the only 2D GC method suited for a complete analysis of the sample composition. In contrast to backflushing and heartcutting,

GC × GC passes each sample component through both the primary and secondary columns. The principles and applications of GC × GC have been the subject of numerous reviews over the past 10 years [6,20–25].

7.5.1. Basic Mode of Operation

GC × GC analysis combines a standard GC separation in the primary column with a series of high-speed separations in the secondary column. In GC × GC, the interface between the primary and secondary columns is called a modulator. The modulator repetitively transfers samples of primary column effluent to the secondary column at a constant interval called the modulation period. The modulation period is set to be less than the width of the peaks emerging from the primary column so that the primary separation is preserved. Each modulation event introduces primary effluent to the head of the secondary column as a narrow pulse. These narrow pulses are separated on the secondary column and passed on to the detector. The secondary separations in a GC × GC analysis are approximately three orders of magnitude faster than the primary separation. The high speeds of the secondary separations are achieved by using a short narrow-bore column (e.g., 1 m × 0.10 mm × 0.10 μm) or a high carrier gas flow. At the completion of the run, the detector signal is divided into segments, with each segment representing a secondary separation. The segments are stacked side by side to generate a two-dimensional chromatogram.

The retention characteristics required for a GC × GC analysis are demonstrated in

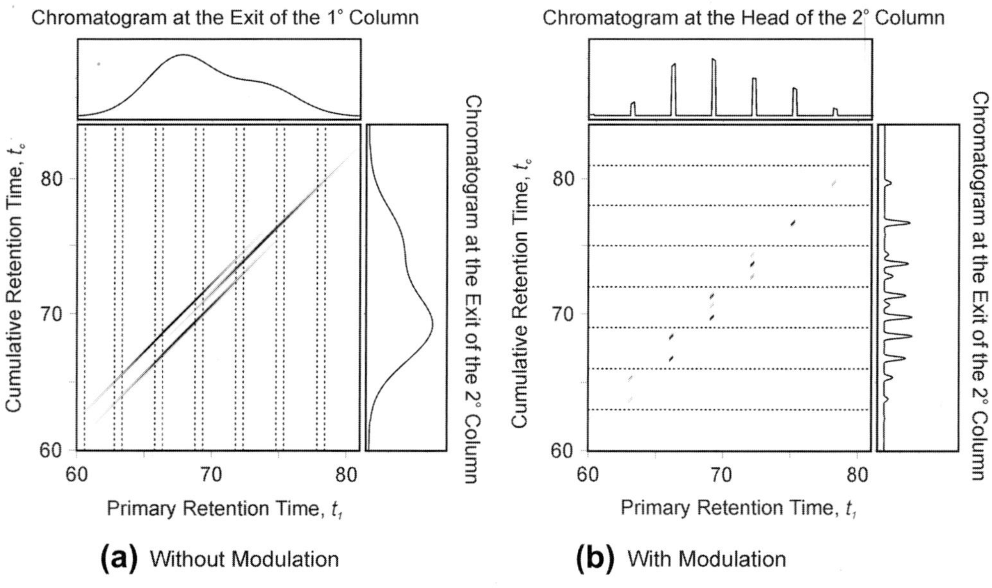

FIGURE 7.12 2D plots of a simulated GC × GC analysis of three components. (a) Without modulation, the components appear as diagonal streaks in the 2D plot. This is because the short, highly efficient secondary column leads to secondary retention times that are less than the widths of the peaks along the primary dimension. The sampled sections of the 2D retention plot when modulation is performed are shown with the vertical dashed lines. (b) With modulation, the broad component peaks are replaced by a series of narrow pulses that enter the secondary column at the beginning of each modulation period (see the top 1D chromatogram) and exit the secondary column as a series of separated peaks (see the right 1D chromatogram).

Figure 7.12 for a three-component mixture. The components are assumed to have primary retention times between 60 and 80 s with primary widths of 10 s and secondary retention times between 0 and 3 s. Figure 7.12a represents the case where the modulator allows primary column effluent to flow uninterrupted to the secondary column (i.e., the modulator is "turned off"). Due to the short secondary retention times, the peaks in the 2D retention plot appear as a series of long, narrow, diagonal streaks situated very close to the $t_c = t_1$ line. The chromatograms generated at the end of the primary column (see the top of Figure 7.12a) and at the end of the secondary column (see the right of Figure 7.12a) look very similar when modulation is not performed due to the small retention time shifts generated by the secondary column.

In order to simulate modulation, it was assumed that the primary column effluent enters the secondary column for the first 10% of each 3.0-s modulation period. The primary column effluent is sent to an exhaust vent for the remaining 90% of the modulation period. The sampled slices of primary effluent are shown between the vertical dashed lines in Figure 7.12a. The modulated 2D plot is shown in Figure 7.12b. The act of modulation causes a series of 300-ms pulses to enter the secondary column every 3.0 s (see the top of Figure 7.12b). The secondary separation is conducted within the baseline spaces between these input pulses. The components are retained to varying degrees in the secondary column and reach the detector as numerous individual peaks (see the right side of Figure 7.12b). The high efficiency of the secondary column allows the peaks to retain much of the sharpness of the original input pulses.

When the range of secondary retention times is kept below the modulation period, the components injected in a particular modulation cycle do not overlap with components from previous or subsequent modulation cycles. This is demonstrated in the 2D graph in Figure 7.12b by the absence of horizontal overlap between the groups of peaks representing individual modulation cycles. Under such conditions, the chromatogram obtained at the end of the secondary column is composed of segments that map directly to discrete regions of the 2D retention plane. This allows a 2D chromatogram to be constructed from the 1D chromatogram. This process is shown in Figure 7.13. The first step is to divide the modulated chromatogram into a series of segments that have the length of the modulation period

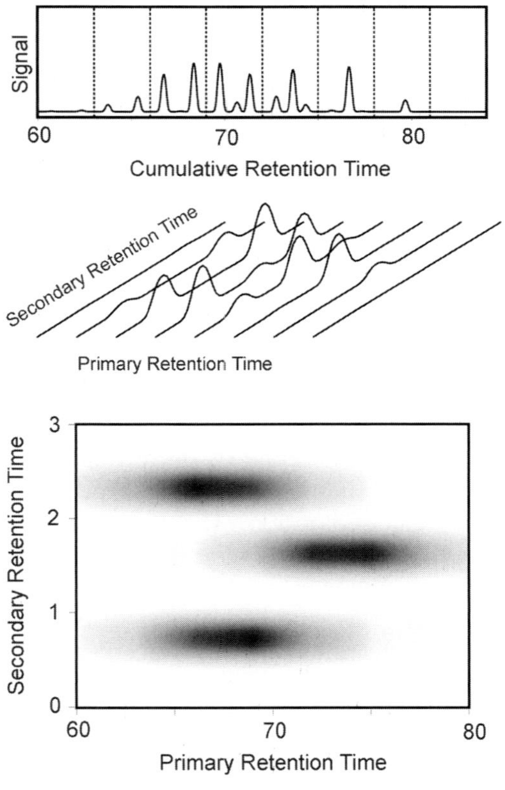

FIGURE 7.13 The process of converting the signal array into a 2D chromatogram. The signal array is first divided into segments with widths given by the modulation period (see top graph). The individual segments are stacked side-by-side to generate a 2D chromatogram (see middle graph). 2D chromatograms are most often displayed as a contour plot (see bottom graph).

(see the top of Figure 7.13). These segments are then rotated 90° and stacked side by side to generate a two-dimensional chromatogram (see middle of Figure 7.13). In practice, the 2D chromatograms generated by GC × GC are usually displayed as a linearly interpolated contour plot (see bottom of Figure 7.13). The primary retention time is displayed on the horizontal axis and the secondary retention time is displayed on the vertical axis. The construction of 2D chromatograms requires software that is normally not included with the software used to operate conventional gas chromatographs.

If the range of secondary retention times exceeds the modulation period, then an unambiguous construction of the 2D chromatogram cannot be produced from the 1D chromatogram. In practice, it is common for a few components to experience high secondary retention and to fall outside the preferred range. Such components appear to be "wrapped around" in the 2D chromatogram. As long as only a few components are wrapped around, the resulting 2D chromatograms are still useful.

7.5.2. GC × GC Modulators

The modulator is a piece of hardware that is unique to GC × GC, and its performance plays a critical role in the resolution produced by a GC × GC separation. The development of GC × GC modulators continues to this day. Numerous reviews have described the operating principles and relative merits of the modulator technology introduced over the past two decades [20–22,24–25]. Two classes of modulators have been developed: valve-based modulators and thermal modulators. Valve-based modulation strategies will be considered first in this chapter as they have much in common with the flow switching technology used in backflushing and heartcutting 2D GC. However, it should be kept in mind that thermal modulation has been used more frequently than valve-based modulation.

Valve-based modulation was introduced by Bruckner et al. [26] seven years after the original GC × GC work of Phillips and Liu. Bruckner et al. used a high-speed diaphragm valve to produce a series of pulses at the head of the secondary column. Their modulator transferred only a small fraction of the primary column effluent to the secondary column and thus is classified as a low-duty cycle modulator. The study of Bruckner et al. showed that a modulator constructed with off-the-shelf components could be combined with a standard GC to create an effective, high-speed GC × GC instrument. Several other low-duty cycle modulators have been introduced [27–29] since the original work of Bruckner et al.

Low quantitative precision is possible when employing a low-duty cycle modulator. If the modulation period is too large, then the total amount of the component transferred to the secondary column depends on the position of the primary peak within the sequence of modulation events [30]. For example, if the modulation period is substantially larger than the width of the peak emerging from the primary column, then the entire peak would be missed if it eluted between injections into the secondary column. The ratio of the primary peak width (4s) to the modulation period is known as the modulation ratio [31]. Theoretical and experimental studies [29,30,32] have shown that when the modulation ratio is kept above 3.0 the fraction of effluent transferred to the secondary column is essentially constant. This means that if the modulation produces three or more significant pulses per primary peak, then the quantitative precision is not diminished by employing a low-duty cycle modulator.

While low-duty-cycle modulators can produce precise quantitative measurements, they are not optimized for high sensitivity as most of the primary effluent does not reach the detector. The majority of GC × GC separations have been performed with modulators that transfer 100% of the primary column effluent to the secondary column (i.e., they have duty cycles of 1). These modulators work by accumulating the primary effluent throughout the modulation period and then introducing the effluent as a pulse at the end the of modulation period. Full transfer modulators are well suited for analyses requiring high sensitivity. In addition, these devices do not cause a decrease in quantitative precision when operated at a large modulation ratio; however, large modulation ratios lead to diminished chromatographic resolution along the primary axis [33]. Two classes of full transfer modulators have been developed: differential flow modulators and thermal modulators.

Differential flow modulation is a valve-based technique first demonstrated by Seeley et al. [30] using a diaphragm valve fitted with a sample loop. The primary effluent is collected in the sample loop. Near the end of the modulation period, the valve is actuated and the loop contents are flushed into the secondary column. The flow rate of carrier gas is much higher in the secondary column than in the primary column (e.g., 15 mL/min secondary flow and 1 mL/min primary flow). Thus, the time required to flush the loop is much less than the time required to fill the loop. Alternating between the fill and flush states of the valve leads to a series of high-intensity pulses from each primary peak. Several designs of differential flow modulators have been introduced including a simple fluidic device that samples all of the primary column effluent [34]. Agilent Technologies now offers a commercial version of a fluidic differential flow modulator based on this design [14].

The main limitations of differential flow modulation are caused by the high secondary flow rate. High carrier gas velocities in the secondary column lead to increased plate heights. Fortunately, longer secondary columns can be used to partially offset the loss in secondary resolution [30]. Narrow-bore columns (i.e., diameter <0.25 mm) are rarely used in differential flow modulation GC × GC because of the extremely high head pressures that would be required to generate high flow rates. Perhaps the biggest drawback of differential flow modulation is that the high flows do not allow the secondary column to be directly coupled to a mass spectrometer. However, Poliak et al. [35] have shown that mass spectrometers with differentially pumped vacuum chambers are compatible with differential flow modulation.

Thermal modulators are the most commonly used GC × GC modulators. They employ precisely timed temperature changes to convert primary peaks into concentrated pulses. The original GC × GC modulator of Phillips and Liu [19] was a thermal modulator that accumulated components in a thick-film capillary column situated between the primary and secondary column. Concentrated pulses were generated by the two-stage heating of the trapping capillary. In 1998, Marriot and Kinghorn [36] introduced the longitudinally modulated cryogenic system that is still in use. This system employs a cryogenic fluid to cool the capillary joining the primary and secondary columns. Components eluting from the primary column condense inside the capillary at the location of the cooled region. Precise movement of the cooled region along the length of the capillary segment produces extremely sharp pulses of primary effluent.

Today most commercially produced thermal modulators are two-stage devices that use jets of gas to cool or heat two small sections of a capillary joining the primary and secondary columns [37–39]. Figure 7.14 shows the cryogenic modulator developed and marketed by Zoex Corporation [40]. The modulator is situated inside the main GC column oven. A jet of cold nitrogen is sprayed onto a capillary loop. The looped configuration causes the cold jet to hit the capillary in two places. Primary column effluent accumulates in the cold spots. A jet of hot gas is pulsed in a precise manner to heat the cold spots to release the accumulated components.

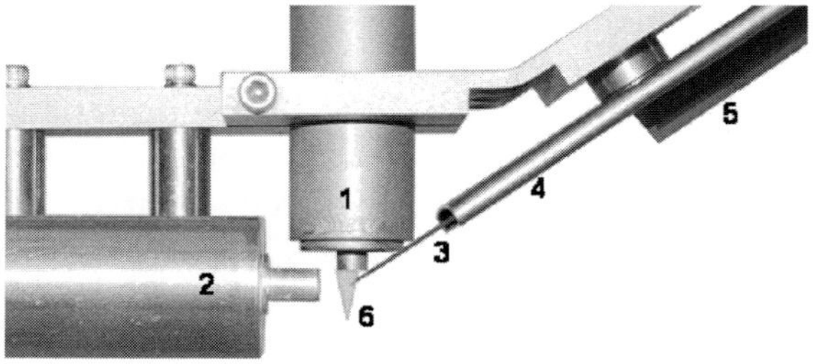

FIGURE 7.14 The cryogenic loop modulator from Zoex: (1) cold jet assembly; (2) hot jet assembly; (3) capillary loop; (4) loop clip; (5) loop retainer; (6) cold jet. *Source: Copyright Zoex Corporation.*

The basic mechanism for a two-stage thermal modulation is illustrated in Figure 7.15. The plots represent the spatial concentration profile of a component as it moves down the length of the capillary joining the primary and secondary columns. The capillary is cooled in an upstream location and a downstream location. Both locations are shown with arrows in Figure 7.15. The sequence of events leading to the formation of a single pulse is described below: As a component elutes from the primary column, it moves through the capillary from left to right. Figure 7.15a shows the central portion of the primary peak as it encounters the upstream cold spot shown with the left arrow. The low temperature condenses the component and forms the concentration pulse highlighted with a star. The component continues to flow into the upstream cold spot causing the pulse amplitude to grow as shown in Figure 7.15b. At a specified time, the cooling is replaced by heating as shown in Figure 7.15c. The rapid increase in temperature causes the accumulated pulse to be released into the carrier gas and then move downstream (i.e., to the right in Figure 7.15c). During this brief heating period, unfocused material passes through the upstream zone producing a tail on the left side of the focused pulse. After a brief period, the cold jet is reapplied as shown in Figure 7.15d, while the pulse and shoulder move downstream. The pulse and shoulder enter the downstream cold spot where they are focused further into a single pulse

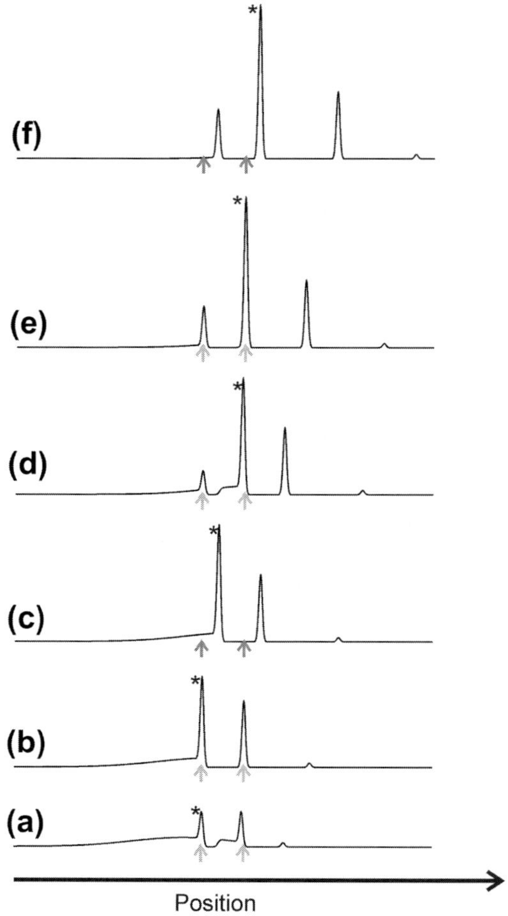

FIGURE 7.15 Two-stage thermal modulation. The carrier gas moves from left to right bringing a single component from the exit of the primary column into the modulator. The position of the two heating/cooling spots are designated with arrows. This analysis follows the formation of a single pulse designated with the star: (a) A portion of a component emerging from the primary column is accumulated in the upstream cold spot. (b) The amount of collected material increases with time. (c) The hot jets heat the capillary column and causes the focused material to be released. (d) The hot jets are turned off and the pulse moves into the downstream cold spot where it is focused further. (e) The pulse is fully focused with no tail. (f) The hot jets heat the capillary column to release a sharp pulse into the carrier stream that heads to the secondary column.

as shown in Figure 7.15e. Finally, a sharp symmetric pulse is released to the secondary column by heating the downstream cold spot as shown in Figure 7.15f. This entire sequence is repeated throughout the chromatographic run and, in the process, converts primary column peaks into a series of highly focused pulses that enter the secondary column for further separation. Two-stage thermal modulation can produce pulses with widths less than 50 ms and concentrations that are increased by nearly two orders of magnitude [38].

In terms of chromatographic performance, thermal modulation is superior to valve-based strategies. However, thermal modulation GC × GC is more expensive to implement. Most thermal modulators use large amounts of liquid cryogen and working fluid. This represents a substantial increase in consumable costs when compared to conventional GC and mostly limits thermal modulation GC × GC to well-funded R&D laboratories. However, a few research groups have recently developed thermal modulators that do not require liquid cryogen [41,42]. It remains to be seen if these low-resource modulators can match the outstanding performance of cryogenic modulators.

7.5.3. Detection and Quantitation in GC × GC Separations

The peaks exiting the secondary column in a GC × GC analysis often have widths on the order of 100 ms. Thus, it is imperative that the detector has a fast response time, low internal volume, and 50 Hz or greater sampling rates. To date, the flame ionization detector (FID) has been the most common GC × GC detector. The stable, fairly uniform response of the FID makes it particularly well suited for the comprehensive analysis of complex organic mixtures. Studies have also been performed with several selective detectors such as the electron-capture detector and the sulfur chemiluminescence detector [24].

Mass spectrometry (MS) is an especially effective GC × GC detector. Mondello et al. have recently reviewed the history of GC × GC–MS [43]. Numerous studies have been performed with quadrupole mass spectrometers, but the scanning rates of these analyzers (<20 Hz) make them incompatible for GC × GC separations that generate especially narrow peaks. Time-of-flight mass spectrometers (TOFMS) are ideally suited for GC × GC analysis as they can easily generate full scans at a rate ≥100 Hz. Several companies now market fully integrated GC × GC–TOFMS systems.

Quantitation in GC × GC is complicated by the fact that a single component produces several 1D peaks. However, software is now available that automatically groups the peaks according to their position in the 2D plane and provides precision that matches that of conventional single-column GC [44]. Furthermore, the detection limits of GC × GC are often better than those of GC because peak overlap is greatly reduced and peak intensities are increased when thermal modulators or differential flow modulators are used. Chemometric data analysis strategies can often greatly increase the resolution and sensitivity of GC × GC. Two reviews of recent developments in the application of chemometrics to GC × GC have been published [45,46].

7.5.4. Example GC × GC Application I: The Aromatic Composition of Gasoline

GC × GC is particularly well suited to analyze the composition of complex petrochemical mixtures. Seeley et al. [29] used a GC × GC instrument equipped with a Deans switch modulator (see Figure 7.16) to determine the aromatic composition of gasoline. A 15 m × 0.25 mm × 0.50 μm poly(dimethylsiloxane) primary column was coupled to a 2.5 m × 0.25 mm × 0.25 μm polyethylene glycol secondary column. A modulation period of 1.0 s was employed with the modulator in the inject state for 0.07 s at the beginning of each

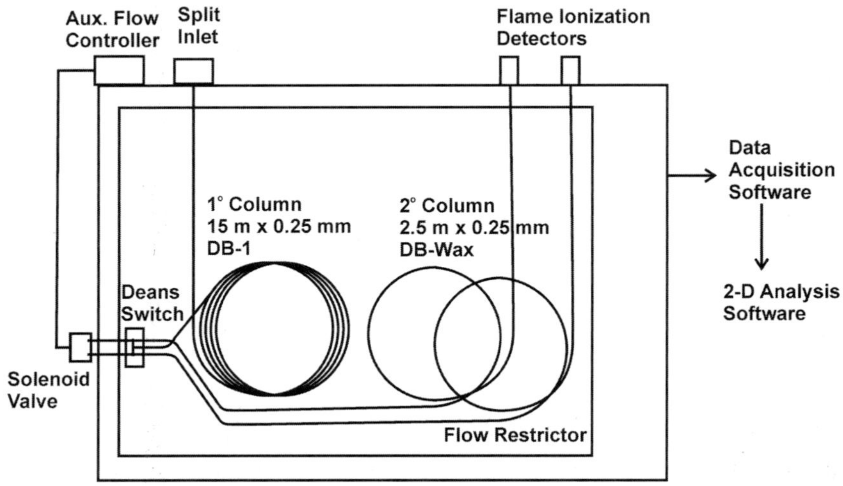

FIGURE 7.16 GC × GC apparatus employing a Deans switch modulator. Gasoline samples were injected into a split inlet and then passed through a nonpolar primary column. A Deans switch was used to transfer the primary column effluent as a series of pulses to the polyethylene glycol secondary column. Gasoline components were detected with an FID. *Source: Reprinted with permission from Ref. [29]. Copyright 2007, American Chemical Society.*

modulation period. Thus, only 7% of each component was passed from the primary column to the secondary column. A typical GC × GC chromatogram of gasoline is shown in Figure 7.17. The primary retention times ranged from 60 to 500 s and the secondary retention times ranged from 1.2 to 2.2 s. Individual peaks were observed to have 100-ms widths along the secondary axis.

Adding modulation and a secondary separation greatly increased the peak capacity of the analysis. This can be observed in Figure 7.17 by noting the number of peaks that share the same primary retention time but are separated

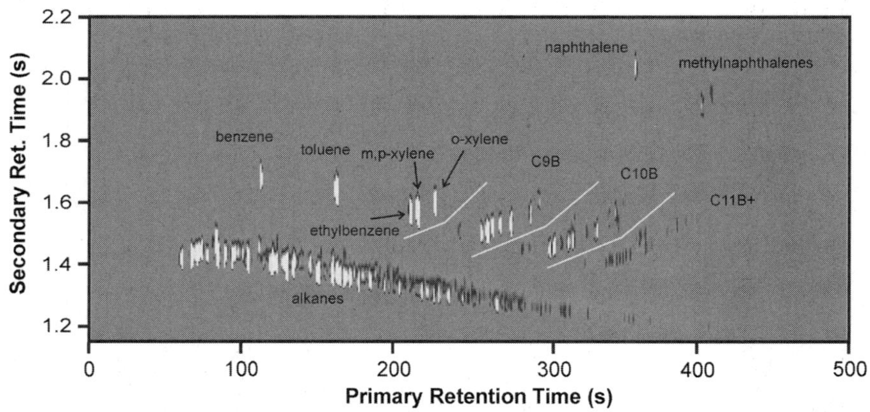

FIGURE 7.17 Deans switching GC × GC analysis of gasoline. Compound classes were observed to form distinct peak clusters in the chromatogram. The alkylbenzene classes are highlighted in the chromatogram. *Source: Reprinted with permission from Ref. [29]. Copyright 2007, American Chemical Society.*

in the secondary dimension. In addition to greater peak capacity, GC × GC analysis often produces "structured" chromatograms. This is demonstrated in the gasoline chromatogram by the formation of peak clusters representing specific compound classes. Saturated hydrocarbons form a narrow band with low secondary retention running horizontally across the base of the 2D chromatogram. The alkylbenzenes form a series of "roof-tile" bands with moderate secondary retention. Each roof-tile corresponds to a group of alkylbenzene isomers having the same number of carbon atoms. The diaromatic compounds (i.e., naphthalene and methylnaphthalene) have a large primary and secondary retention and are located in the upper-right corner of the chromatogram. Seeley et al. determined the concentration of each compound class from the total peak areas within each group. Excellent agreement was observed between the concentrations determined with Deans switching GC × GC and GC-MS.

7.5.5. Example GC × GC Application II: GC × GC-MS Analysis of Yeast Extracts

GC × GC–TOFMS was used by Mohler et al. [47] to identify chemical differences in the metabolite extracts of yeast cells. The extracts were methoximated and trimethylsilylated prior to analysis. Mohler et al. used a GC × GC instrument equipped with a quad-jet cryogenic modulator and a time-of-flight mass

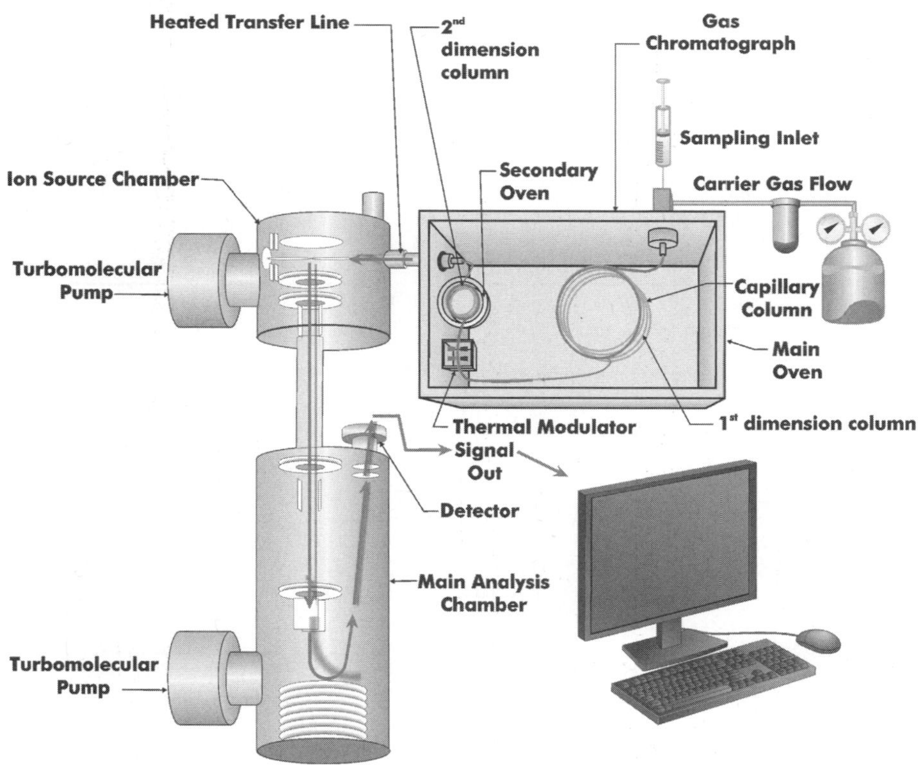

FIGURE 7.18 Leco GC × GC – TOFMS system. This system uses a quad-jet thermal modulator and a high speed time-of-flight mass spectrometer. *Source: Copyright Leco.*

spectrometer that were both manufactured by Leco (see Figure 7.18). A 20 m × 0.25 mm × 0.50 μm 5% phenyl/95% dimethyl polysiloxane primary column was coupled to a 2.0 m × 0.18 mm × 0.20 μm poly(trifluoropropylmethylsiloxane) secondary column. Extracts were analyzed with a 1.5-s modulation period and a total run time of 37.75 min. The mass spectrometer produced 100 spectra/s with a mass range of 40–600 m/z.

A typical 2D chromatogram of a sample extract is shown in Figure 7.19a. This particular chromatogram was obtained for a mass channel, $m/z = 73$, that represents the trimethylsilyl functional group. Over 2500 individual peaks can be observed in the chromatogram. Figure 7.19b is a zoomed-in portion of the 2D chromatogram that demonstrates the high resolution and complexity of the chromatogram. The mass spectra produced by the instrument allowed full 2D chromatograms to be generated for each mass channel. This provides the ability to produce 2D chromatograms that are selective for a particular compound class. For example, the $m/z = 205$ mass fragment is known to be indicative of trimethylsilyl derivatives of carbohydrates. Thus, the 2D chromatogram of this channel (see Figure 7.19c) is representative of carbohydrate composition. Similarly, the 2D chromatogram obtained at $m/z = 387$ (see Figure 7.19d) represents the trimethylsilyl derivatives of sugar phosphates. The key point

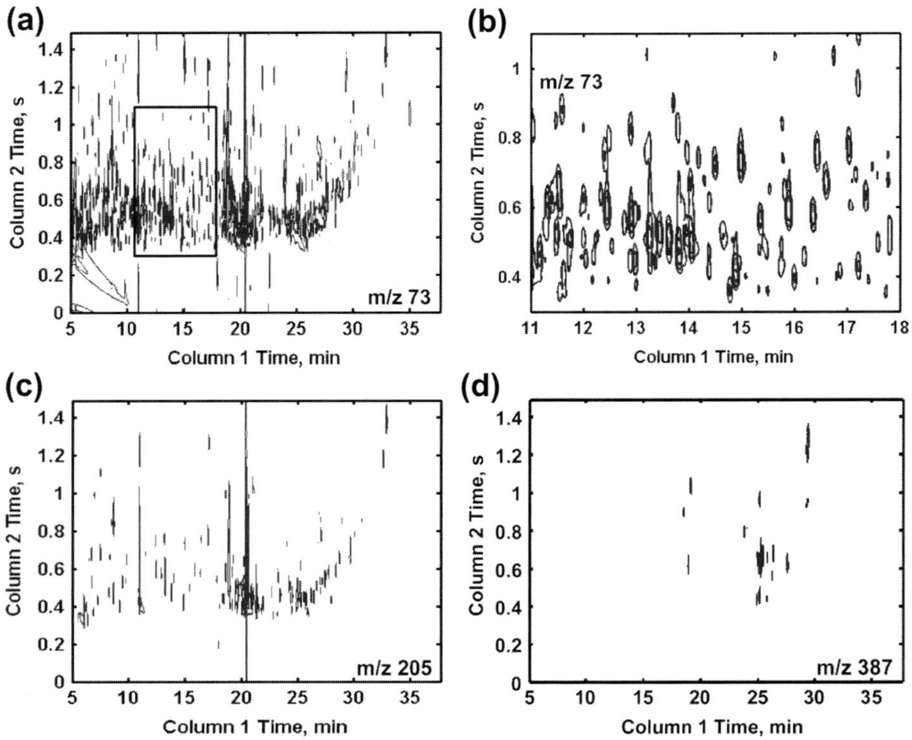

FIGURE 7.19 GC × GC–TOFMS chromatograms of yeast metabolite extract. (a) Chromatogram obtained at $m/z = 73$ indicative of trimethylsilyl derivatives. (b) Zoomed-in version of the chromatogram shown in panel A. (c) Chromatogram obtained at $m/z = 205$ indicative of carbohydrates. (d) Chromatogram obtained at $m/z = 387$ indicative of sugar phosphates. *Source: Reprinted with permission from Ref. [47]. Copyright 2006, American Chemical Society.*

is that "real world" samples such as yeast metabolite extracts produce extremely complex chromatograms even when GC × GC is used. The resolution gained by adding time-of-flight mass spectrometry can both greatly reduce the complexity of the 2D chromatograms and greatly increase the amount and quality of information produced.

7.6. CONCLUSIONS

This chapter has described several basic approaches for combining two GC separations. Backflushing 2D GC separates an entire class of analyte components from the sample matrix by employing a single change in primary-column flow direction. To be successful, the polarity difference between the analyte class and the sample matrix must be quite large. Heartcutting 2D GC separates targeted analytes from the sample matrix. Subtle retention differences can be exploited to fully resolve individual analytes. GC × GC combines a standard GC separation with a series of high-speed secondary separations. GC × GC can be used for the comprehensive analysis of the sample composition, and it is also effective at class separations and monitoring individual analytes. All three of these 2D GC approaches have been successfully applied to the analysis of complex mixtures such as petroleum-based fuels, flavors and fragrances, and environmental samples.

It is important to note that the individual forms of 2D GC separations examined in this chapter can be combined. For example, Maikhunthod et al. [48] have constructed an apparatus that can perform both GC × GC and heartcutting 2D GC analyses. An overview of sample composition is first obtained with GC × GC and then heartcutting 2D GC is used to isolate a few particularly important analytes. Several groups have also shown that additional GC stages can be added to generate multidimensional GC separations. For example, Watson et al. [49] have recently reported the successful implementation of comprehensive three-dimensional gas chromatography (GC^3). A third dimension of separation is added by injecting the secondary effluent into a tertiary column every 200 ms. It is clear that future research efforts will continue to discover effective ways of exploiting the benefits of combining GC separations.

References

[1] J.M. Davis, J.C. Giddings, Statistical theory of component overlap in multicomponent chromatograms, Anal. Chem. 55 (1983) 418–424.

[2] K.L. Goodner, Practical retention index models of OV-101, DB-1, DB-5, and DB-Wax for flavor and fragrance compounds, LWT - Food Sci. Technol. 41 (2008) 951–958.

[3] M.C. Simmons, L.R. Snyder, Two-stage gas-liquid chromatography, Anal. Chem. 30 (1958) 32–35.

[4] J.C. Giddings, Use of multiple dimensions in analytical separations, in: H.J. Cortes (Ed.), Multidimensional chromatography: techniques and applications, M. Dekker, New York, 1990, pp. 1–27.

[5] W. Bertsch, Two-dimensional gas chromatography. Concepts, instrumentation, and applications – Part 1: Fundamentals, conventional two-dimensional gas chromatography, selected applications, J. High Resolut. Chromatogr. 22 (1999) 647–665.

[6] W. Bertsch, Two-dimensional gas chromatography. Concepts, instrumentation, and applications – Part 2: comprehensive two-dimensional gas chromatography, J. High Resolut. Chromatogr. 23 (2000) 167–181.

[7] A.C. Lewis, Multidimensional high resolution gas chromatography, in: L. Mondello, A.C. Lewis, K.D. Bartle (Eds.), Multidimensional chromatography, Wiley, New York, 2002, pp. 47–75.

[8] P.J. Marriot, P.D. Morrison, R.A. Shellie, M.S. Dunn, E. Sari, D. Ryan, Multidimensional and comprehensive two-dimensional gas chromatography, LCGC Europe 16 (2003) 23–31.

[9] L. Ramos, U.A.T. Brinkman, Multidimensionality in gas chromatography: general concepts, in: L. Ramos (Ed.), Comprehensive two dimensional gas chromatography, Elsevier, Amsterdam, 2009, pp. 3–14.

[10] C.F. Poole, S.K. Poole, Separation characteristics of wall-coated open-tubular columns for gas chromatography, J. Chromatogr. A 1184 (2008) 254–280.

[11] A. Tipler, Application note: determination of low-level oxygenated compounds in gasoline using the Clarus 680 GC with S-Swafer micro-channel flow technology, PerkinElmer, Inc., Shelton, CT, 2010.

[12] Standard test method D4815 for determination of MTBE, ETBE, TAME, DIPE, tertiary-amyl alcohol and C1 to C4 alcohols in gasoline by gas chromatography, ASTM International, West Conshohocken, PA, 2009.

[13] D.R. Deans, A new technique for heart cutting in gas chromatography, Chromatographia 1 (1968) 18–22.

[14] B.D. Quimby, J.D. McCurry, W.M. Norman, Capillary flow technology for gas chromatography: reinvigorating a mature analytical discipline, LCGC The Peak (2007) 7–15.

[15] J.D. McCurry, B.D. Quimby, Two-dimensional gas chromatography analysis of components in fuel and fuel additives using a simplified heart-cutting GC system, J. Chromatogr. Sci. 41 (2003) 524–527.

[16] R. Gras, J. Luong, V. Carter, L. Sieben, H. Cortes, Practical method for the measurement of alkyl mercaptans in natural gas by multi-dimensional gas chromatography, capillary flow technology, and flame ionization detection, J. Chromatogr. A 1216 (2009) 2776–2782.

[17] D. Sciarrone, C. Ragonese, C. Carnovale, A. Piperno, P. Dugo, G. Dugo, et al., Evaluation of tea tree oil quality and ascaridole: a deep study by means of chiral and multi heart-cuts multidimensional gas chromatography system coupled to mass spectrometry detection, J. Chromatogr. A 1217 (2010) 6422–6427.

[18] J. Luong, R. Gras, G. Yang, H. Cortes, R. Mustacich, Multidimensional gas chromatography with capillary flow technology and LTM-GC, J. Sep. Sci. 31 (2008) 3385–3394.

[19] J.B. Phillips, Z. Liu, Comprehensive two-dimensional gas chromatography using an on-column thermal modulator interface, J. Chromatogr. Sci. 29 (1991) 227–231.

[20] P. Marriott, R. Shellie, Principles and applications of comprehensive two-dimensional gas chromatography, TrAC Trends Anal. Chem. 21 (2002) 573–583.

[21] J. Dallüge, J. Beens, U.A.T. Brinkman, Comprehensive two-dimensional gas chromatography: a powerful and versatile analytical tool, J. Chromatogr. A 1000 (2003) 69–108.

[22] T. Górecki, J. Harynuk, O. Pani&cacute, The evolution of comprehensive two-dimensional gas chromatography (GC × GC), J. Sep. Sci. 27 (2004) 359–379.

[23] J.M.D. Dimandja, GC × GC, Anal. Chem. 76 (2004) 167A–174A.

[24] M. Adahchour, J. Beens, U.A.T. Brinkman, Recent developments in the application of comprehensive two-dimensional gas chromatography, J. Chromatogr. A 1186 (2008) 67–108.

[25] H.J. Cortes, B. Winniford, J. Luong, M. Pursch, Comprehensive two dimensional gas chromatography review, J. Sep. Sci. 32 (2009) 883–904.

[26] C.A. Bruckner, B.J. Prazen, R.E. Synovec, Comprehensive two dimensional high-speed gas chromatography with chemometric analysis, Anal. Chem. 70 (1998) 2796–2804.

[27] J.F. Hamilton, A.C. Lewis, K.D. Bartle, Peak amplitude and resolution in comprehensive gas chromatography using valve modulation, J. Sep. Sci. 26 (2003) 578–584.

[28] A.E. Sinha, K.J. Johnson, B.J. Prazen, S.V. Lucas, C.G. Fraga, R.E. Synovec, Comprehensive two-dimensional gas chromatography of volatile and semi-volatile components using a diaphragm valve-based instrument, J. Chromatogr. A 983 (2003) 195–204.

[29] J.V. Seeley, N.J. Micyus, S.V. Bandurski, S.K. Seeley, J.D. McCurry, Microfluidic Deans switch for comprehensive two-dimensional gas chromatography, Anal. Chem. 79 (2007) 1840–1847.

[30] J.V. Seeley, Theoretical study of incomplete sampling of the first dimension in comprehensive two-dimensional chromatography, J. Chromatogr. A 962 (2002) 21–27.

[31] W. Khummueng, J. Harynuk, P.J. Marriott, Modulation ratio in comprehensive two-dimensional gas chromatography, Anal. Chem. 78 (2006) 4578–4587.

[32] W.C. Siegler, B.D. Fitz, J.C. Hoggard, R.E. Synovec, Experimental study of the quantitative precision for valve-based comprehensive two-dimensional gas chromatography, Anal. Chem. 83 (2011) 5190–5196.

[33] R.E. Murphy, M.R. Schure, J.P. Foley, Effect of sampling rate on resolution in comprehensive two-dimensional liquid chromatography, Anal. Chem. 70 (1998) 1585–1594.

[34] J.V. Seeley, N.J. Micyus, J.D. McCurry, S.K. Seeley, Comprehensive two-dimensional gas chromatography with a simple fluidic modulator, Am. Lab. 38 (2006) 24–26.

[35] M. Poliak, A.B. Fialkov, A. Amirav, Pulsed flow modulation two-dimensional comprehensive gas chromatography-tandem mass spectrometry with supersonic molecular beams, J. Chromatogr. A 1210 (2008) 108–114.

[36] P.J. Marriott, R.M. Kinghorn, Longitudinally modulated cryogenic system. A generally applicable approach to solute trapping and mobilization in gas chromatography, Anal. Chem. 69 (1997) 2582–2588.

[37] E.B. Ledford Jr., C.A. Billesbach, J.R. Termaat, Inventors; assignee Zoex Corporation. Transverse thermal modulation. US patent 6547852; April 15, 2003.
[38] J. Beens, M. Adahchour, R.J.J. Vreuls, K. van Altena, U.A. Th, Brinkman, simple, non-moving modulation interface for comprehensive two-dimensional gas chromatography, J. Chromatogr. A 919 (2001) 127–132.
[39] J. Harynuk, T. Górecki, New liquid nitrogen cryogenic modulator for comprehensive two-dimensional gas chromatography, J. Chromatogr. A 1019 (2003) 53–63.
[40] Ledford Jr EB, Inventor; assignee Zoex Corporation. Method and apparatus for measuring velocity of chromatographic pulse. US patent 7258726; Aug. 21, 2007.
[41] M. Libardoni, J.H. Waite, R. Sacks, Electrically heated, air-cooled thermal modulator and at-column heating for comprehensive two-dimensional gas chromatography, Anal. Chem. 77 (2005) 2786–2794.
[42] O. Panic, T. Górecki, C. McNeish, A.H. Goldstein, B.J. Williams, D.R. Worton, et al., Development of a new consumable-free thermal modulator for comprehensive two-dimensional gas chromatography, J. Chromatogr. A 1218 (2011) 3070–3079.
[43] L. Mondello, P.Q. Tranchida, P. Dugo, G. Dugo, Comprehensive two-dimensional gas chromatography-mass spectrometry: a review, Mass Spectrom Rev 27 (2008) 101–124.
[44] S.E. Reichenbach, M. Ni, V. Kottapalli, A. Visvanathan, Information technologies for comprehensive two-dimensional gas chromatography, Chemometrics and Intelligent Laboratory Systems 71 (2004) 107–120.
[45] R.E. Synovec, J.C. Hoggard, Chemometric approaches, in: L. Ramos (Ed.), Comprehensive two dimensional gas chromatography, Elsevier, Amsterdam, 2009, pp. 3–14.
[46] Z.D. Zeng, H. Hugel, P. Marriott, Chemometrics in comprehensive multidimensional separations, Anal. Bioanal. Chem. (2011) 1–14.
[47] R.E. Mohler, K.M. Dombek, J.C. Hoggard, E.T. Young, R.E. Synovec, Comprehensive two-dimensional gas chromatography time-of-flight mass spectrometry analysis of metabolites in fermenting and respiring yeast cells, Anal. Chem. 78 (2006) 2700–2709.
[48] B. Maikhunthod, P.D. Morrison, D.M. Small, P.J. Marriott, Development of a switchable multidimensional/comprehensive two-dimensional gas chromatographic analytical system, J. Chromatogr. A 1217 (2010) 1522–1529.
[49] N.E. Watson, W.C. Siegler, J.C. Hoggard, R.E. Synovec, Comprehensive three-dimensional gas chromatography with parallel factor analysis, Anal. Chem. 79 (2007) 8270–8280.

CHAPTER

8

Sample Introduction Methods

Andrew Tipler

OUTLINE

8.1. INTRODUCTION

The means of introducing samples into a gas chromatograph (GC) remains one of the processes most critical to its successful operation. It does not matter how good the rest of the system is (column, oven, detector, etc.); if the sample cannot be introduced into the system reliably, then the results are not going to be reliable either. The old computer software adage 'garbage in, garbage out' is particularly relevant within this context.

So, what do we mean by a 'good sample introduction technique' within the field of gas chromatography?

Gas Chromatography DOI: 10.1016/B978-0-12-385540-4.00008-0

In simple terms, we want the analytical results we get from whatever is introduced into the GC to be representative of the original sample.

In practice, we want the following:

- narrow symmetrical peaks to be able to fully exploit the efficiency of the GC column,
- good-sized peaks to be able to discern them from the background (noise) signal,
- no losses of analytes due to adsorption or chemical breakdown,
- good quantitative repeatability to provide confidence in results,
- easy to understand and use, and
- compatible with autosampler injection for throughput and performance.

Typical GC columns and GC detectors are really only suited to handle a few micrograms of a particular sample component at the most; hence, rarely do we introduce the whole sample into the system.

Many samples exist in the liquid form – either as a solution in a suitable solvent (for example, a plant extract) or because they are naturally liquid (for instance, gasoline). For such samples, it is convenient to use a microliter syringe to take a small amount of the sample and introduce it into a device called a *'liquid injector'*. This liquid injector provides the interface between the sample in the syringe and the GC column. There are several different types of liquid injector, which may cause confusion to the novice gas chromatographer. Each injector has its place; each has its advantages and idiosyncrasies that will make it particularly suited to certain types of sample and column.

This chapter focuses mainly on the design and operation of liquid injectors; however, we will also discuss the use of gas and liquid mechanical valves in making sample introductions of gases and pressurized liquids.

There are also more specialized sample introduction systems such as those for headspace sampling, thermal desorption, and pyrolysis sampling that are covered elsewhere in this book (see Chapters 9, 10, and 11).

8.2. CHOOSING A SAMPLE INTRODUCTION SYSTEM

In this section, we summarize the various injection devices and techniques discussed in this chapter and suggest how they may be used. These are intended to be just guidelines – one of the advantages of gas chromatography is learning how to bend the rules for a particular need or application.

In this table, we classify the injectors into two groups:

- Vaporizing – the sample is injected into a hot zone so that vaporization starts immediately. Typically, this will be set to 25–50 °C above the maximum programmed column temperature.
- Nonvaporizing – the sample is injected into a cold zone where it remains as a liquid until the temperature is increased at some point after the injection. Typically, this will be set to 10–20 °C below the boiling point of the solvent.

Table 8.1 summarizes the injector types, their modes of operation, and guidelines on their application.

8.3. SUPPORTING DEVICES

Before we discuss the injectors and injection techniques, it would be useful to consider some of the other technology needed to support the injectors.

8.3.1. Syringes

A microliter syringe is usually used to make liquid sample injections into GC injectors. For

TABLE 8.1 Summary of Injectors and Injection Techniques Covered in this Chapter

Injector	Technique	Type	Column type	Concentrated samples	Dilute samples	Trace samples	Labile components	Volatile compounds	Heavy compounds	Gases	Pressurized liquids
Cold on column	Cold on-column	Non-vaporizing	Capillary	Poor	Best	Good[1]	Best	Poor	Best	Poor	Poor
Flash vaporization	Flash injection	Vaporizing	Packed and 0.53 mm capillary[2]	Good[3]	Fair[3]	Poor	Poor	Fair[3]	Fair	Fair[3]	Fair[4]
	Hot on-column	Vaporizing	Packed	Poor	Fair	Poor	Poor	Poor	Fair	Poor	Poor
Split/splitless	Classical split	Vaporizing	Capillary	Good	Poor	Very poor	Poor	Good	Fair	Fair	Fair[4]
	Classical splitless	Vaporizing	Capillary	Poor	Good	Fair[1]	Poor	Poor	Fair	Poor	Poor
Programmable split/splitless	Programmed split	Non-vaporizing	Capillary	Best	Poor	Very poor	Good	Good	Good	Fair	Fair[4]
	Programmed splitless	Non-vaporizing	Capillary	Poor	Good	Fair[1]	Good	Poor	Good	Poor	Poor
	Vaporizing split	Vaporizing	Capillary	Fair	Poor	Very poor	Poor	Good	Fair	Fair	Poor
	Vaporizing splitless	Vaporizing	Capillary	Poor	Fair	Fair[1]	Poor	Poor	Fair	Poor	Poor
	Cold On-column	Non-vaporizing	Capillary	Poor	Best	Good	Best	Poor	Best	Poor	Poor
	Hot on-column	Vaporizing	Capillary	Poor	Fair	Poor	Poor	Poor	Fair	Poor	Poor
	Large volume injection	Non-vaporizing	Capillary	Poor	Good	Best	Good	Poor	Good	Poor	Poor
Gas sampling valve	Gas loop injection	Not applicable	Packed and capillary[5]	Good[3]	Good[3]	Fair[3]	Poor	Best	Poor	Best	Poor
Liquid sampling valve	Pressurized liquid injection	Non-vaporizing	Packed & capillary[5]	Good[3]	Good[3]	Poor	Poor	Best	Poor	Poor	Best

[1] *With a retention gap*
[2] *At high flow rates*
[3] *With packed columns*
[4] *With valved syringe*
[5] *With splitter at column inlet*

most applications, the injection volume is in the range 0.1–10.0 μL. The needle is normally fabricated from hypodermic stainless-steel capillary tubing and the body from glass. The plunger will normally be stainless steel and may have a PTFE tip for better sealing. The length of the needle needs to be compatible with the type and design of injector being used. The outer diameter is normally 0.63 mm, although a smaller diameter (of 0.47 mm) will be needed for on-column injection into 0.53-mm tubing. The tip of the needle may have a variety of geometries, conical, tapered, flat, etc., to suit different injector seals and septa.

Two types of syringe are available:

- Plunger in barrel – this is the most familiar type of syringe in which the tip of the plunger resides inside the visible barrel of the syringe. This is the type normally used for samples of 1 μL or above. For large-volume injection applications (see Section 8.7.5), the syringe capacity may be 50 μL or even more.
- Plunger in needle – in this type of syringe, the plunger extends down into the needle itself. This is suitable for small-volume injections of 0.5 μL or less.

The two types of syringe are illustrated in Figure 8.1.

A larger-capacity, gastight syringe may be used to inject gas samples either directly into a liquid injector or into a gas sampling valve. It is normally made from glass and has a gastight plunger to prevent sample leakage when the needle is inserted into a pressurized injector. Some syringe designs have a sealing valve at the point where the needle joins the barrel. This effectively prevents sample vapors from escaping from the syringe barrel or ambient air entering the syringe while it is being handled prior to injection.

Smaller-capacity syringes with a sealing valve may be used to sample and inject pressurized liquids such as aerosol propellants or liquid petroleum gases into a liquid injector.

8.3.2. Liners

Most liquid injectors use some form of liner into which the sample is injected by syringe. These are normally fabricated from glass or quartz tubing. For some injectors, the liner may need to be packed with glass or quartz wool.

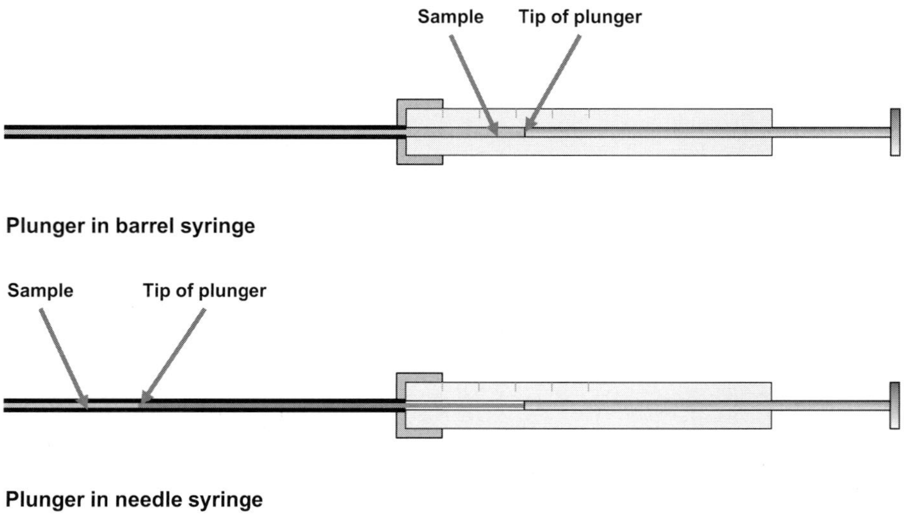

FIGURE 8.1 Types of syringe used for liquid injections.

It is usual to deactivate the liner with a suitable chemical agent to prevent breakdown or adsorption of sensitive analytes particularly at low concentrations.

The geometry of the liner depends on the type of injector and the injection technique, which will be discussed further in the sections on injectors.

For vaporizing injectors, it is important that the internal capacity of the liner is sufficient to hold the volume of vapor generated after a liquid sample injection. This will depend on the following factors:

- volume of sample injected,
- number of moles in the injected sample,
- the liner temperature,
- the carrier gas pressure inside the liner, and
- the injection speed.

The potential vapor volume from a given solvent can be calculated using a variant of the ideal gas law given in Eqn (8.1):

$$V_v = \frac{V_l \cdot \rho \cdot R \cdot T_v}{M \cdot P_v} \quad (8.1)$$

where V_v is the potential volume of solvent vapor generated in the liner (μL), V_l is the volume of liquid solvent injected (μL), ρ is the density of the liquid solvent (g mL^{-1}), R is the ideal gas constant (8314 mL kPa K^{-1} mol^{-1}), T_v is the absolute temperature of the solvent vapor inside the liner (K), M is the molar mass (molecular weight) (g), and P_v is the absolute pressure of the solvent vapor inside the liner (kPa) (1 psi is equivalent to 6.894 kPa).

The potential vapor volumes for typical solvents used in GC are given in Table 8.2. This table also contains boiling-point information, which is discussed later in this chapter.

Table 8.2 clearly shows that liner capacity will be more of a concern for polar solvents such as water and alcohols and less of a concern with higher hydrocarbons. The solvent with the lowest potential vapor volume in this list is isooctane (and so has greater tolerance for larger-volume liquid injections). This is a popular solvent for many applications.

TABLE 8.2 Boiling Points and Vapor Volumes of Common Solvents Used in GC

Solvent	Normal boiling point (°C)	Vapor volume from 1 μL of liquid injected at 250 °C and 15 psig (μL)
Acetone	56	290
Carbon disulfide	46	352
Chloroform	61	265
Cyclohexane	81	197
Dichloromethane	40	332
Ethanol	78	364
Ethyl acetate	77	217
Isooctane	99	124
Isopropanol	82	278
Methanol	65	526
n-Pentane	36	185
n-Hexane	69	162
n-Heptane	98	145
n-Octane	126	131
n-Nonane	151	119
n-Decane	174	109
Toluene	111	200
Water	100	1180

8.3.3. Septa

There have been many developments to seal the syringe inlet of liquid injectors and yet facilitate the introduction of the needle into the liner below. None has yet surpassed the simplicity and ruggedness of a simple rubber septum. Most injectors still make use of this type of seal – it is cheap, reliable, easy to replace, and generally gives good performance. The advent of modern silicone polymers has improved their performance significantly – they last much longer than earlier versions, and issues with septum bleed have been largely addressed.

8.3.4. Autosampler

Modern autosamplers automate the process of syringe injection. They save time for the operator as they will function continuously even at times when the operator is not present. The other great advantage is the precision of the injection process. Every setting in the autosampler method is executed with high repeatability, giving excellent precision in the chromatography. With a good manual injection technique, the expected quantitative precision of an analysis will be in the order of 2–4% relative standard deviation; with an autosampler this should improve to 1% or even lower.

Purchase of an injector autosampler is a highly recommended option for any GC – it will quickly repay the initial investment with much higher sample throughput, better quality data, and give operators more time for other duties.

8.3.5. Pneumatic Systems

The design of carrier gas control systems is a complex subject and difficult to cover as part of a single chapter in a general book on GC such as this. However, pneumatic system design and configuration are critical parts of GC injector operation and so we will cover the basics here.

Several types of pneumatic components may be used to control GC carrier gases:

- Forward pressure regulator – this is a pressure-reducing device that adjusts a variable restrictor to maintain a constant pressure at its outlet.
- Backpressure regulator – this is a pressure-reducing device that adjusts a variable restrictor to maintain a constant pressure at its inlet.
- Mass flow controller – this device adjusts a variable restrictor to maintain a constant mass flow rate of gas through it.
- Needle valve – this device is a variable restrictor.
- Pressure gauge or transducer – a device that displays or reads the pressure at a given point.
- Flow transducer – a device that reads the mass flow rate through a given point.

In the early days of GC, all the controllers were mechanical devices that required the user to turn a rotary knob until the required pressure or flow rate was achieved. These days, most GCs use microprocessor-controlled electro-mechanical modules to perform these functions. These devices offer significant advantages over their mechanical counterparts. They not only make method setup easier and more precise but also enable flow rates and pressures to be dynamically programmed during the course of an analysis. Some modes of injection are not possible without this pneumatic programmability.

Details of specific pneumatic requirements for each injector will be made in the following sections, but one topic that will be covered here concerns the control of the carrier gas flow rate through open-tubular capillary columns.

Many capillary injectors have multiple gas outlets. These may include the column, septum purge, and split vent. Traditional mass flow controllers will regulate the flow rate of carrier gas going *into* the injector. As often only a small fraction of this gas will make it into the column, the flow rate of carrier gas through the column cannot be controlled directly in this way.

Most programmable pneumatic systems control the carrier gas flow rate through the column by applying a carrier gas *pressure* calculated to give the required flow rate through the column. In this way, the column flow rate is completely unaffected by septum purge, splitting, or even gas leaks.

The calculation used for this purpose is usually based on a variant of the Hagen–Poiseuille relationship [1] developed in the mid-nineteenth century and modified here for calculations with compressible gases and is shown in Eqn (8.2):

$$F_o = \frac{\pi \cdot d_c^4 \cdot (p_i^2 - p_o^2)}{256 \cdot L \cdot p_o \cdot \eta} \quad (8.2)$$

where F_o is the flow rate at the column outlet at the temperature and pressure at the outlet (mL.min^{-1}), d_c is the internal diameter of the column (cm), L is the length of the column (cm), p_i is the absolute carrier gas pressure at the column inlet (kPa), p_o is the absolute carrier gas pressure at the column outlet (kPa), and η is the dynamic viscosity of the carrier gas which varies with column temperature (kPa s).

If the pneumatic control system has knowledge of the column geometry and type of carrier gas (which would be entered by the user), it can adjust the column inlet pressure, p_i, to deliver a set flow rate, F_o, at any set column temperature.

It is important when using carrier gas flow control to understand what the pressure will be inside the injector. Some of the advanced injection techniques discussed in this chapter rely on pressure changes.

8.4. THE COLD ON-COLUMN INJECTOR

The cold on-column injection (COCI) technique is, in concept, the simplest injection technique in capillary GC and yet it is probably the least used.

Essentially, it involves the deposition of a liquid sample directly into the inlet of the capillary column at a low temperature as shown in Figure 8.2. The syringe needle is withdrawn and the sample in the column is heated and vaporized during the column oven temperature program [2,3]. In this way, once it has left the syringe, the sample only makes contact with the surface of the capillary column as it makes its way to the detector. Because the injection takes place at a low temperature, there is low risk of thermolysis of labile compounds on a hot metallic syringe needle and almost no risk of the analyte mass discrimination from preinjection or postinjection vaporization effects like those seen with vaporizing injectors.

If applied correctly, cold on-column injection will outperform all other injection techniques. It is a technique not suited to all samples, however (especially for concentrated or very volatile analytes), and care must be taken in choosing the conditions in the analytical method.

The main challenge to the user is the need to insert the syringe needle directly into the column inlet. When such injectors were first introduced, this was achieved by using syringes with a fine delicate length of fused silica tubing that was manually threaded through a guide and complex seal inside the injector and directly into column inlets down to 0.25 mm in internal diameter. This was not for the fainthearted.

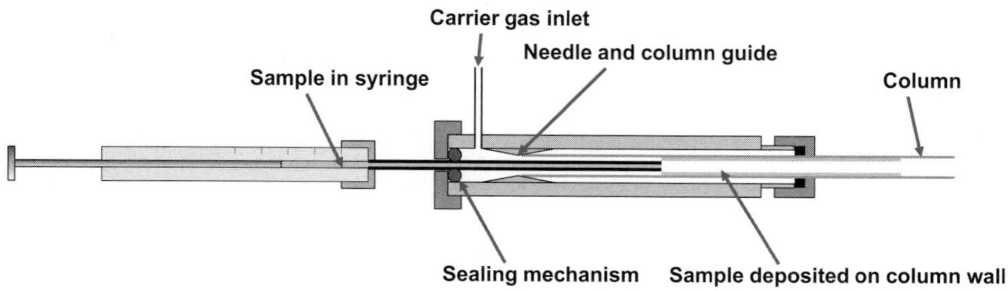

FIGURE 8.2 The basics of cold on-column injection.

The advent of retention gaps (see Section 8.4.3) enabled the use of syringes with more robust metal needles that could also be driven by a standard autosampler and make use of a standard septum as a seal. Although this has made the prospect of COCI more approachable to many users, there is still an element of skill and uncertainty in the installation and alignment of the column within the injector. Many users prefer to opt for the higher robustness of the classical split/splitless or the PTV injection systems.

8.4.1. Sample Considerations

Because the whole sample injection is made directly into the column, care must be taken not to overload the column with too much analyte. Overloaded peaks cause characteristic 'fronting' peaks on columns as shown in Figure 8.3. Overloaded peaks are no taller than regular peaks but they are much broader and may give apparent changes in retention time and so should be avoided if possible.

The amount of analyte that will overload a capillary column will depend on the geometry of that column, the stationary-phase thickness, and the applied conditions, but a value of 50–100 ng would be fairly typical. For COIC, a typical injection volume would be 1–5 μL, and so maximum analyte concentrations would need to be in the order of 10–100 μg mL^{-1}. Thus, COCI is only suited to very dilute liquid

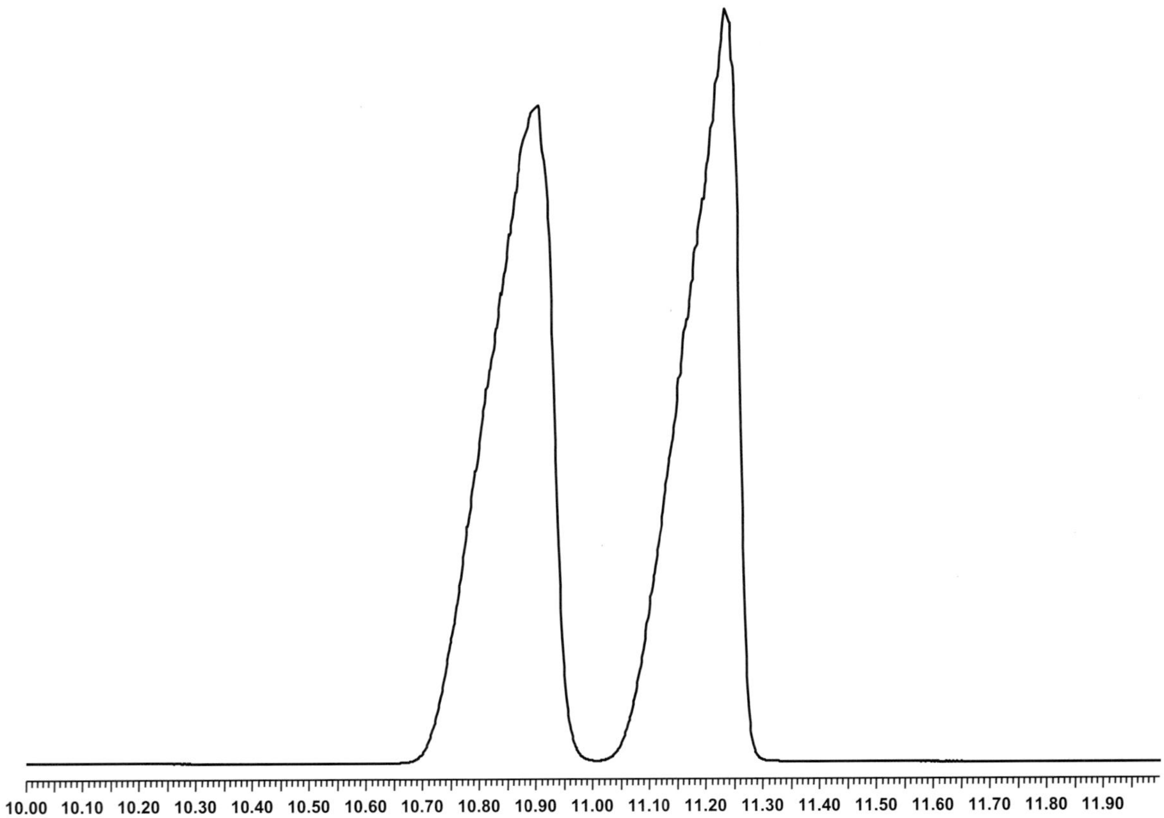

FIGURE 8.3 Example of overloaded peaks.

samples and, in most applications, a suitable solvent must be added to ensure that analyte concentrations are compatible with the column being used.

Another point that should be considered is that because the sample is being deposited directly into the column, any involatile or reactive components will also be introduced into that column. Over a period of time, there may be sufficient buildup of residue or attack of the stationary phase to cause significant degradation of the column. This may need the removal of a short section of the column from the inlet (or exchange of the retention gap) but a better solution would be to ensure that such materials are not present in the sample at all. To be robust, COCI needs samples to be reasonably clean. If this is not possible, then an injection technique that uses a replaceable liner may be preferable.

8.4.2. Role of the Solvent

In order for COCI to work, the sample must essentially remain in the liquid state before, during, and after injection. This requires careful consideration of the sample solvent and the temperature of the GC column during injection. Generally, the column oven temperature is chosen to be approximately 10–20 °C lower than the boiling point of the solvent at the column inlet pressure. If the temperature is too high, the solvent will boil within the confined space within the column causing 'flashback', leading to poor chromatography, poor quantitative performance, and a high risk of system contamination. Table 8.2 lists the normal boiling points for some of the most common solvents used in GC. Note that, because the column inlet is at an elevated pressure because of the applied carrier gas, the solvent boiling points will also be elevated.

The solvent itself may play an active part in obtaining good chromatographic peak shapes especially for the early-eluting peaks. In COIC, it is normally (but not exclusively) assumed that the solvent will be the first component that will elute from the column. At the point of injection, the solvent should distribute itself as a liquid layer on the inside of the column walls near the inlet. This 'pseudo' stationary phase serves to refocus the analyte vapors that follow it out of the injector and so concentrate them as a narrow band at the column inlet. The column oven is then temperature-programmed to first vaporize the solvent and then the refocused analytes. This ensures that the early-eluting peaks are sharp and symmetrical. This effect is termed the 'solvent effect' [4] and requires the careful choice of a solvent that will both adhere to the walls of the column (wettability) and have an affinity (partitioning) for the analytes.

If the solvent does not adhere to the walls of the column, then it will form droplets of the solvent that will absorb analyte vapors. These droplets are easily driven down the column sometimes for a significant distance by the carrier gas, which has a velocity normally in the range of 20–50 cm s^{-1}. As the droplets travel along the column, they will slowly evaporate. This has the effect of depositing patches of analytes randomly along the column. When temperature programming is initiated and these analytes vaporize, the chromatographic peaks may become highly distorted, as the molecules for each analyte will have different distances to travel through the column. This *'solvent flooding effect'* [5] is usually seen with midrange-volatility components. An example of a severe case is given in Figure 8.4.

If the analytes do not sufficiently partition into liquid solvent, then there will be no focusing effect and so the early peaks may be broad or even tailing.

If analytes precede the solvent as it passes from the injector and through the column, they may undergo what is known as the *'reverse solvent effect'* [6]. In such instances, the solvent may actually displace the analyte molecules from the stationary phase and so undermine

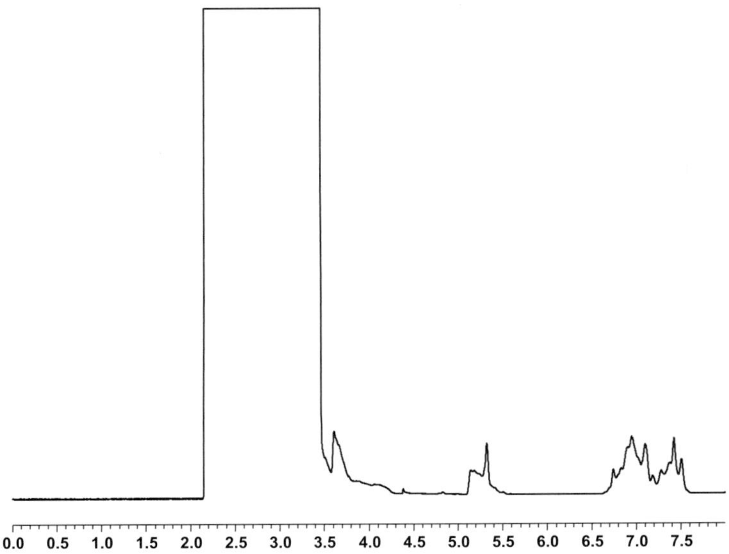

FIGURE 8.4 Example of the solvent flooding effect caused by using methanol as a solvent with a thin-film nonpolar column.

the whole chromatographic process and the resultant peaks may be broad and poorly separated (Figure 8.5).

As should now be apparent, choosing a suitable solvent is one of the most critical decisions when developing a COCI method. Figure 8.6 shows an example of good chromatography obtained with COCI.

Although the solvent effect is an important technique for volatile analytes in a sample, it may not be needed for heavier (semivolatile) components (although these will still be at risk

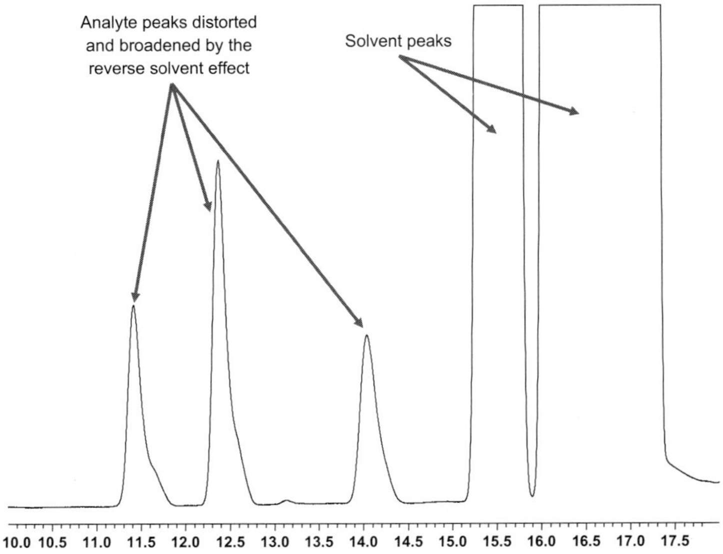

FIGURE 8.5 Example of the reverse solvent effect caused by using a less volatile solvent (hexane isomer mixture) to chromatograph light hydrocarbons.

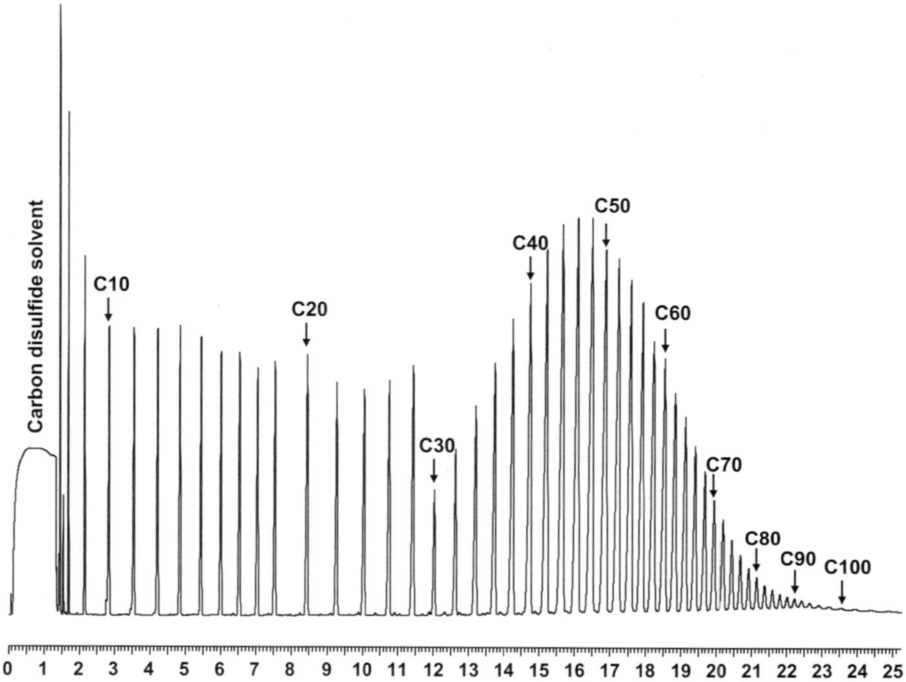

FIGURE 8.6 Example of a chromatography obtained with cold on-column injection. This is an injection of a high-temperature simulated distillation calibration mixture containing hydrocarbons up to 100 carbon atoms. The column was a 10 m × 0.53 mm × 0.1 μm methyl silicone high-temperature metallic column and the solvent was carbon disulfide. The temperature program was −20 °C to 425 °C at 20 °C min^{-1} and held for 4 min.

from the solvent flooding effect). The heavier components will tend to sit as a narrow band on the column at the point that they were initially deposited during injection. The solvent will be long gone before they start to vaporize during a temperature program. They will vaporize as a narrow plug of vapor and generally give narrow and symmetrical peaks.

8.4.3. Using a Retention Gap

In many instances, choosing a suitable solvent is not easy and compromises must be made. The use of a technique called the '*retention gap*' [7] enables COCI to be far more forgiving of the solvent being used.

A retention gap is simply a length of uncoated and inert tubing connected between the injector and the column. It is normally made from chemically deactivated fused silica tubing. It may be of a wider internal diameter than the column itself and the length is chosen to suit the solvent being used and the injection volume. Many modern columns may be purchased with a retention gap prebonded by the vendor to the column inlet.

The main function of the retention gap and the reason why it is so named are to act as an unretentive buffer between the injector and the column. The length of the retention gap should be sufficient to ensure that all the liquid solvent is retained within it and only enters the separation column as vapor. The solvent effect is effectively reduced but, more importantly, so is the solvent flooding effect. This enables selection from a wider range of solvents that would not normally work with COCI. The loss of the solvent effect, which would have helped keep the chromatographic peaks sharp, is compensated by another effect. The stationary-phase gradient that is created at the union between the retention gap and the column acts as an analyte focusing point as illustrated in Figure 8.7.

A retention gap provides several other benefits besides improving the chromatographic peak shape. It is such a powerful technique that many experienced chromatographers will insist on its use with most columns.

Because the retention gap is normally removable and replaceable, it can be seen as a guard column to protect the main (and much more expensive) separation column from dirty or reactive samples. When the chromatography degrades, it is a relatively simple matter to replace the retention gap and return to the original chromatography.

One of the main reasons a retention gap is popular is because, with the wider internal diameter normally used, it is possible to insert a robust metal syringe needle directly into it.

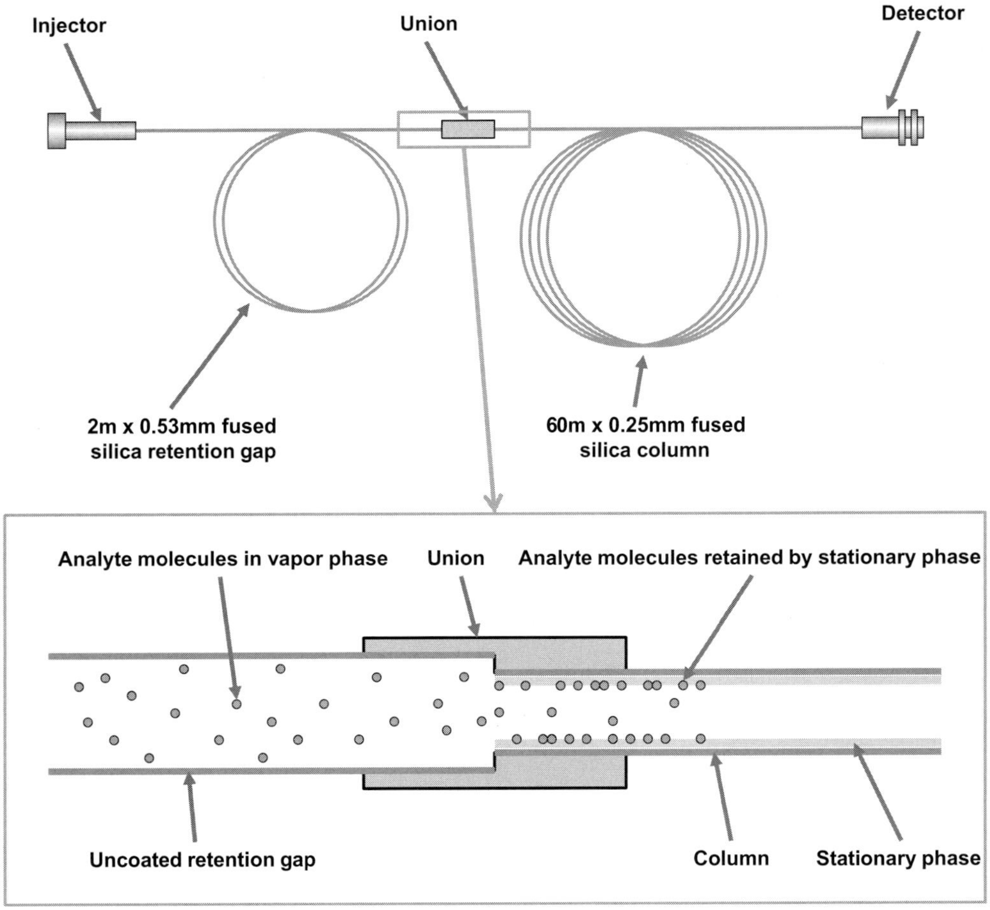

FIGURE 8.7 The retention gap principle.

This enables a regular autosampler to automate cold on-column injections.

Finally, as mentioned earlier, the retention gap is able to overcome the solvent flooding effect that would otherwise cause severe peak distortion. By using a long retention gap, large sample volumes may be injected. As a general rule, approximately 0.2–0.5 m of 0.53-mm internal-diameter retention gap is required for each μL of sample injected, although this will depend on the potential vapor volume of the solvent as indicated in Table 8.2. Some workers have demonstrated liquid sample injections of 1 mL using a 100-m retention gap. These injection volumes generate a lot of solvent vapor to move through a capillary column at a typical carrier gas flow rate and so this is often not a very practical technique.

One variant of the retention gap technique that overcomes the problems of venting large volumes of solvent vapor through a capillary column is the *'vented retention gap'* technique [8] as shown in Figure 8.8. In this case, a split line is opened for a short time after injection to vent the solvent. The split vent is closed and a column temperature program is initiated. This keeps the bulk of the solvent out of the column and so helps to protect it and minimizes the time that would otherwise be required to vent the solvent vapor through the column.

Although the use of a long retention gap and a vented retention gap has facilitated large-volume sample injections, the use of a programmable temperature vaporizing injector (see Section 8.7) for this purpose is usually more robust and easier to set up and operate.

8.5. THE FLASH VAPORIZATION INJECTOR

This is perhaps the simplest and most robust means of injecting liquid samples into a GC column. Sample is injected into a heated (normally) glass liner, and it is vaporized and carrier gas carries it into the GC column. Method development is very straightforward: the liner temperature must be high enough to vaporize the whole sample. Figure 8.9 illustrates the basics of flash vaporization injection.

Unfortunately, despite its simplicity, flash vaporization does suffer from some significant

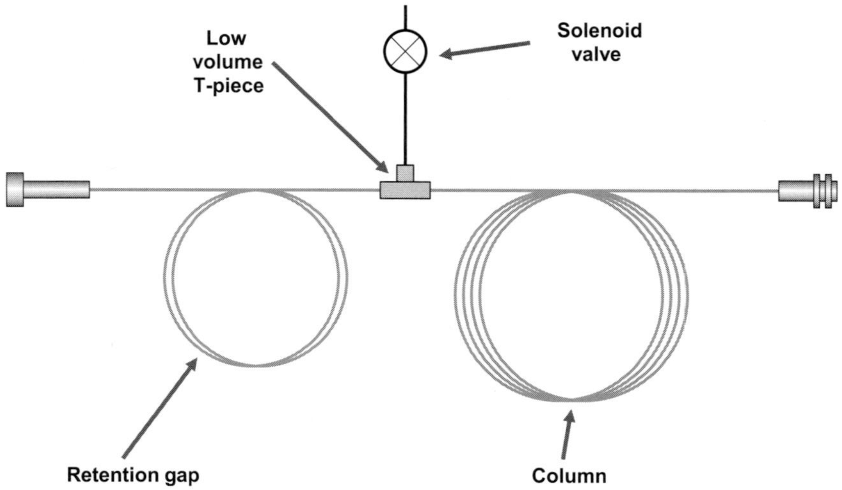

FIGURE 8.8 The vented retention gap technique.

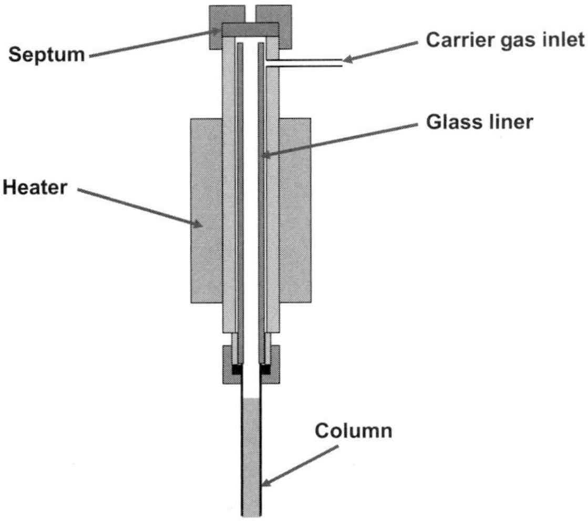

FIGURE 8.9 The flash vaporization injector.

disadvantages. The liner capacity has to be sufficient to retain the volume of sample vapor generated, as the sample is injected and vaporized. An injection volume of 1 μL may generate as much as 0.5 mL of vapor with some solvents as shown in Table 8.2 and so require a liner with a capacity of between 0.5 and 1 mL. For a capillary column with a carrier gas flow rate of, say, 1 mL min^{-1}, it would take over 5 min to sweep the sample vapor into the GC column. Thus, the chromatography would start with a very broad tailing solvent peak. An example of this is shown in Figure 8.10.

For this reason, the flash vaporization technique is normally restricted to packed columns where the much higher carrier gas flow rates are more efficient at sweeping the sample vapor from the liner. For packed columns, the flash vaporization injection technique performs very well.

One exception to the rule that says that this technique is unsuitable for capillary columns is in the case of 0.53-mm internal-diameter (wide-bore) capillary columns at high carrier gas flow rates (e.g. 10–20 mL min^{-1}) and especially those with thick stationary-phase coatings. Although this will not be using the column at its optimum chromatographic efficiency, it will allow the use of a simple flash vaporizing injector as shown in Figure 8.11.

Some flash vaporization injector designs will support the technique of 'hot on-column injection'. This was popular in the days of mainly packed-column chromatography as injecting a sample directly into the packing of a glass packed column appeared to be more inert than injecting into a liner. To use this technique, the liner would be removed and the inlet end of a specially designed glass packed column with an extended inlet would be pushed right up through the injector until it was immediately below the injector septum. These days, if inertness is sought, then a suitable fused silica capillary column is nearly always a better option.

Hot on-column injection may also be used with 0.53-mm i.d. capillary columns. The inlet end of the column is inserted into a special adapter that is installed in the injector and guides the syringe needle directly into the column. The injection site is within the heated injector and so there is rapid vaporization within a confined space. Because of this, the technique is really only suitable for a small injection volume of 0.5 μL or less. This is very inferior to the other capillary column injection techniques but it does retain the robustness and simplicity offered by the flash vaporization injector.

8.6. THE SPLIT/SPLITLESS INJECTOR

The classical split/splitless injector is the most popular GC injector currently in use. It is reasonably rugged as well as being easy to use, and there are hundreds of papers describing its operation and application. It is normally operated in one of two modes: split injection or splitless injection. These two 'classical' injection modes

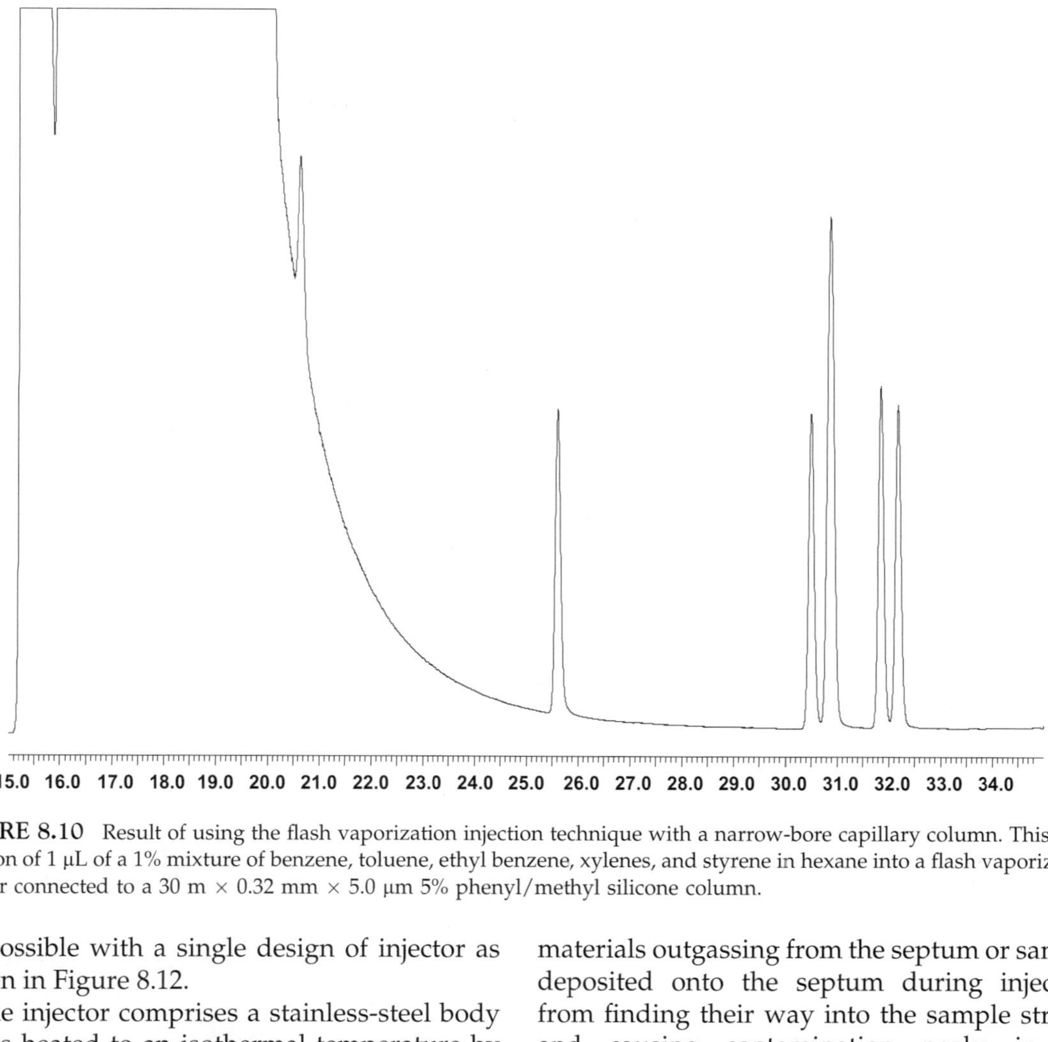

FIGURE 8.10 Result of using the flash vaporization injection technique with a narrow-bore capillary column. This is an injection of 1 μL of a 1% mixture of benzene, toluene, ethyl benzene, xylenes, and styrene in hexane into a flash vaporization injector connected to a 30 m × 0.32 mm × 5.0 μm 5% phenyl/methyl silicone column.

are possible with a single design of injector as shown in Figure 8.12.

The injector comprises a stainless-steel body that is heated to an isothermal temperature by a surrounding heater block. A glass (or quartz) liner resides inside the heated area and is sealed around its outer circumference to ensure that gas only passes through the center of the liner and not between it and the body. Carrier gas enters the top of the liner and exits into the column and, optionally, a split vent at the bottom. The split vent is enabled by activating a solenoid valve. A septum is used to seal the interior at the top of the injector and enables liquid sample injection by syringe into the liner. A septum purge is provided to prevent materials outgassing from the septum or sample deposited onto the septum during injection from finding their way into the sample stream and causing contamination peaks in the chromatography.

We will consider the two injection modes separately here.

8.6.1. Classical Split Injection

The sample is injected into a heated liner where it undergoes rapid vaporization. A high flow rate of carrier gas is applied through the liner during injection (typically 25–500 mL/min). The flow rate through the column is much less and is chosen to suit that

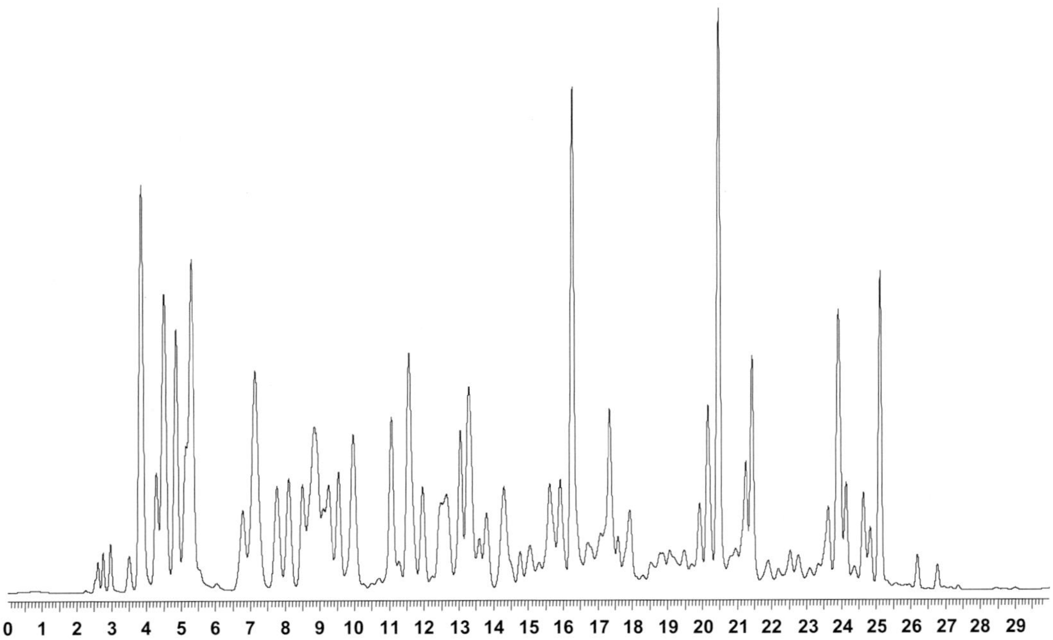

FIGURE 8.11 Flash vaporization injection of a gasoline sample with a 30 m × 0.53 mm × 5.0 μm methyl silicone capillary column. The helium carrier gas flow rate was 12 mL min^{-1} and the temperature program was 35 °C for 5 min., then 5 °C min^{-1} to 160 °C.

column. The excess carrier gas is vented from the injector through the split vent. This technique achieves two things: first, it reduces the sample vapor residence time in the liner and delivers a sharp plug of vapor to the column inlet and, second, it delivers only a small fraction of the injected sample to the column inlet. In summary, split injection gives very sharp peaks but is only suitable for samples with high concentrations of analytes; it is not suitable for trace-level analyses.

Split injection has a wide tolerance for different solvents – they can elute before or after the analytes without significantly affecting chromatographic performance. It is even possible to inject some samples without the use of an added solvent, for example, in the analysis of gasoline.

Almost any column oven temperature (isothermal or programmed) may be used with

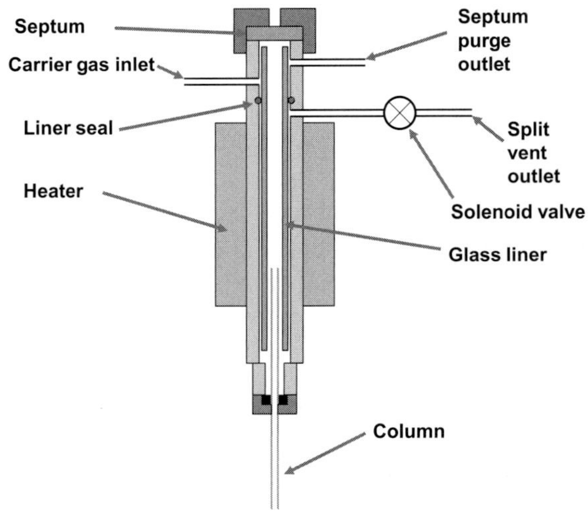

FIGURE 8.12 The split/splitless injector.

split injection. The injector liner temperature, however, must be sufficiently high to vaporize efficiently the whole sample.

Although split injection is conceptually simple, the dynamics occurring inside the heated liner during the injection are very complex and there are a number of considerations that must be taken into account when developing methods.

8.6.1.1. The Role of the Syringe

To make an injection, the syringe needle must first be inserted into the heated liner. Thus, the needle heats up before the syringe plunger is depressed. Any sample that resides inside the needle at this point is going to become vaporized and will enter the GC column prematurely. This 'preinjection effect' may cause apparent peak distortion or even splitting especially with early-eluting components and may degrade the quantitative performance. This effect would be more apparent with plunger in-needle style syringes.

Similarly, after the plunger is depressed to inject the sample, the residual sample left in the needle may become vaporized and enter the GC column. This 'post-injection effect' may give rise to peak distortion again and cause an effect called mass discrimination, where more of the volatile sample components will enter the GC column than they should. This effect would be more apparent with plunger in-barrel style syringes.

Figure 8.13 shows an example of poor syringe technique. In this instance, care was not taken in the way the syringe needle was inserted or withdrawn. The peaks are larger and broader than they should be and the earlier peak has become split.

One of the most successful ways of overcoming preinjection and postinjection effects is to use a fast injection technique. This requires an autosampler designed to support this, as it would be almost impossible to perform this manually because of the speeds involved. The autosampler inserts the syringe needle into the injector liner, depresses the plunger to effect the sample injection, and then withdraws the needle from the injector. This complete process takes 100 ms or even less. The syringe needle is within the injector liner for such a short time that it does not really have the chance to heat up to produce either the preinjection or postinjection effects. A plug of glass wool is needed within the liner to wipe the syringe needle during its withdrawal or else it is likely that some of the liquid sample may be pulled back onto the base of the septum and lost from the analysis. Fast injection will provide more accurate injection volumes, sharp peaks, and low mass discrimination of wide-volatility-range samples. Figure 8.14 shows how fast injection can improve the chromatography. Compare this chromatogram against Figure 8.14 that was from the same sample run under identical chromatographic conditions and plotted with the same scaling.

Another approach to overcome preinjection effects with a plunger in barrel syringe is to pull back the plunger prior to injection so that all sample is withdrawn from the needle. It does not work with plunger in-needle syringes.

The postinjection effects may be addressed on a plunger in-barrel syringe by using either the 'air plug' or the 'solvent plug' technique. Prior to taking up sample into the syringe, a fixed volume of air or solvent is loaded into the syringe. The required volume of sample is then drawn into the syringe. The plunger may then be further withdrawn to empty the needle as described above. A small volume of air or solvent now sits between the sample and the end of the syringe plunger. When the syringe needle is inserted into the heated liner and the plunger is depressed, the air or solvent plug sweeps the whole sample out of the needle so that there is no sample in the needle at the end of the injection. This effectively eliminates the postinjection

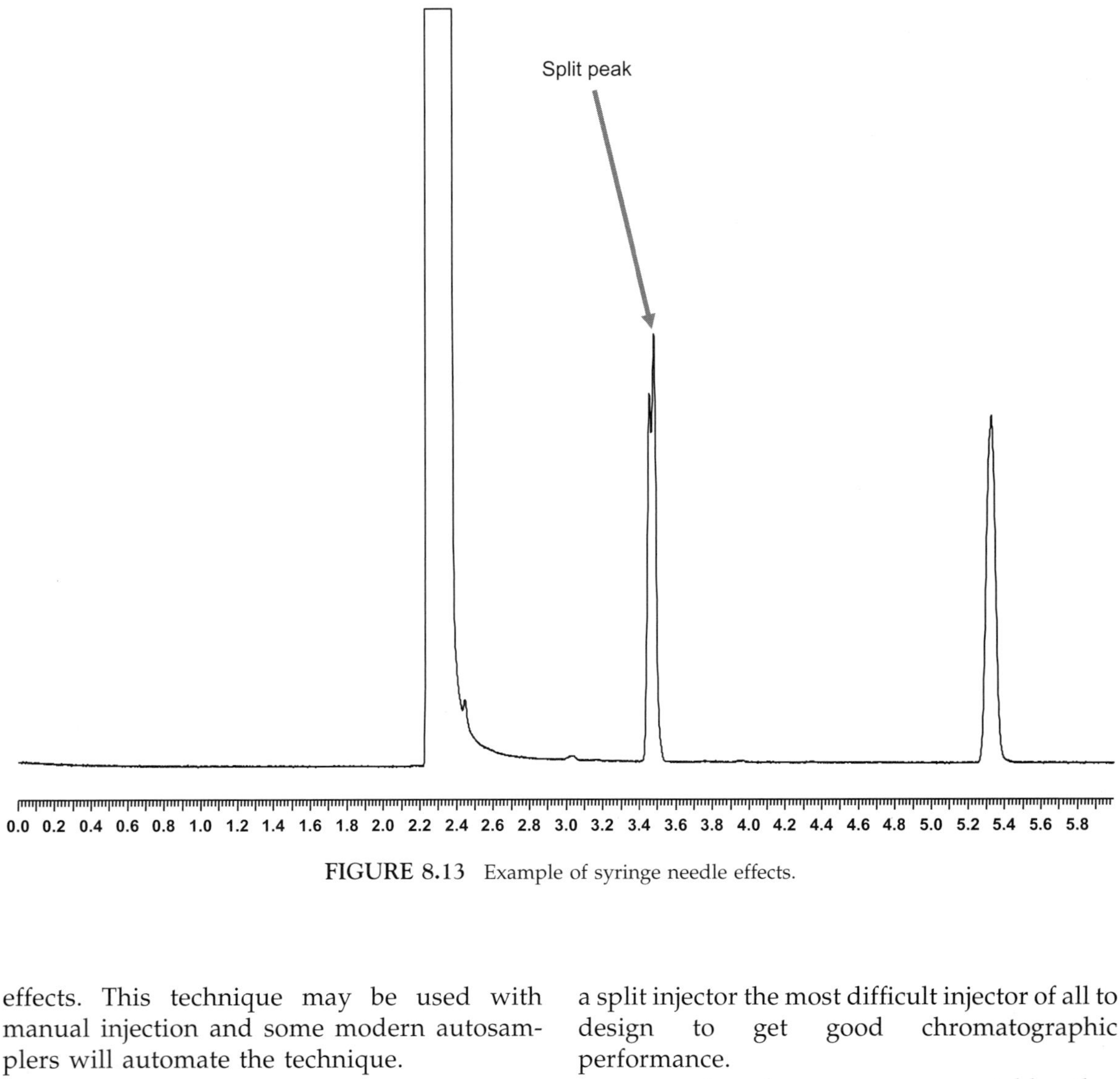

FIGURE 8.13 Example of syringe needle effects.

effects. This technique may be used with manual injection and some modern autosamplers will automate the technique.

8.6.1.2. The Role of the Injector Liner

The geometry of the injector liner is a critical factor with classical split injection. The role of the liner is twofold. First, it is to keep the sample vapor within the confines of the liner and deliver a fraction of the sample into the column that is truly representative of the sample injected. Second, it is to do this in a repeatable manner. This is what makes a split injector the most difficult injector of all to design to get good chromatographic performance.

In practice, there are many variables that conspire to affect performance and so make the liner choice difficult. Different manufacturers will give apparently conflicting advice and different users seem to have very different opinions as to what liner is best; there is possibly an element of black magic, too, in getting the best out of a given liner. In reality, there is probably no liner design that is good for all applications.

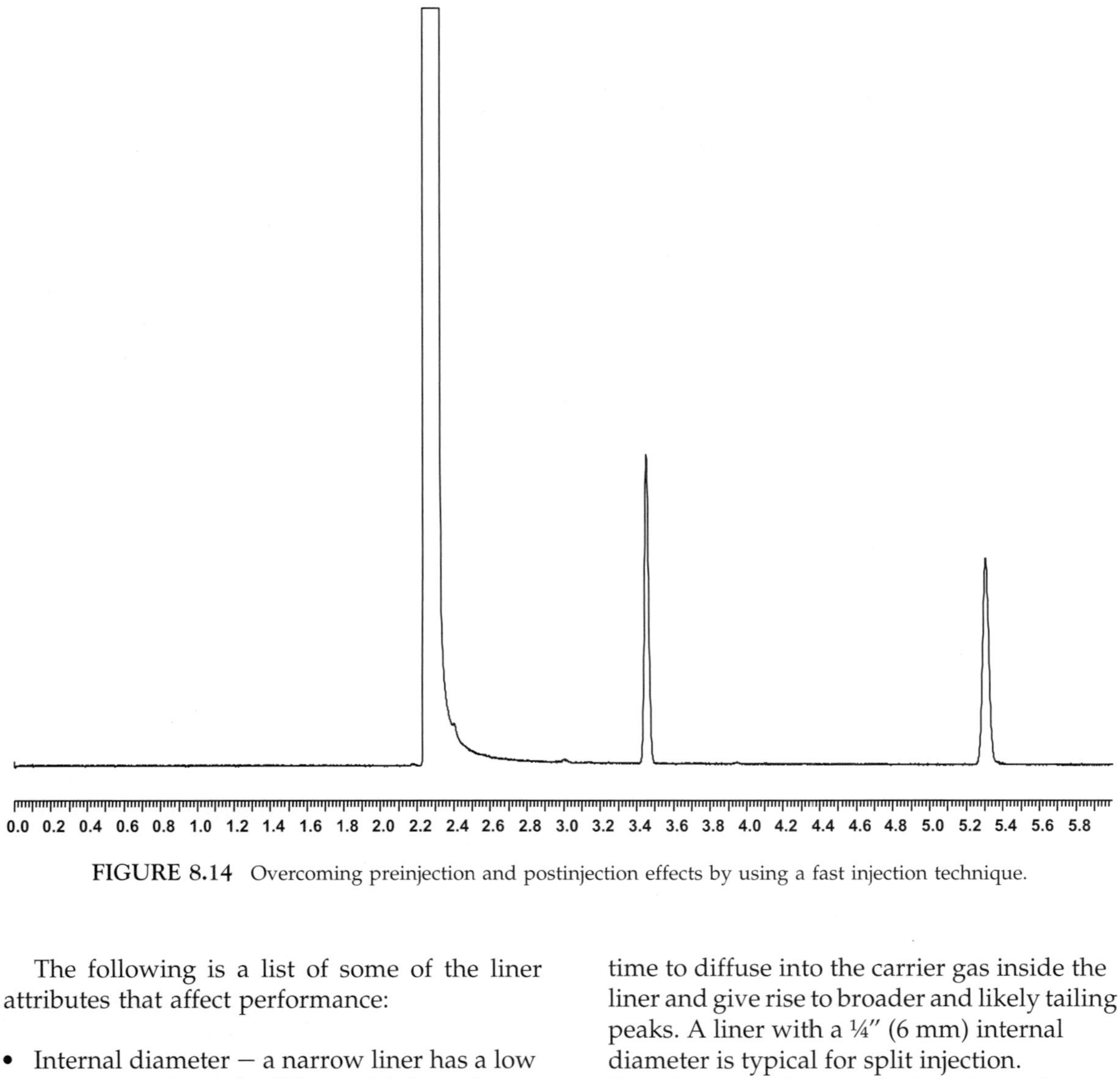

FIGURE 8.14 Overcoming preinjection and postinjection effects by using a fast injection technique.

The following is a list of some of the liner attributes that affect performance:

- Internal diameter – a narrow liner has a low vapor capacity and will have a high carrier gas velocity flowing through it. This will limit the volume of sample injected and reduce the time available for sample vaporization and homogeneous mixing with the carrier gas before it reaches the column inlet. Conversely, a wide-bore liner will be able to accept larger sample injections and have more time for the vapor to homogenize with the carrier gas. However, the sample vapor will have more time to diffuse into the carrier gas inside the liner and give rise to broader and likely tailing peaks. A liner with a ¼″ (6 mm) internal diameter is typical for split injection.
- Sample size – the sample injection volume must be kept sufficiently low to prevent the generated vapor from escaping from the sample liner (see Section 8.3.2). Typically, 2 μL is the maximum sample volume that should be injected with split injection.
- Thermal mass – as the injected sample vaporizes, the energy needed for this vaporization will be drawn from the immediate surroundings, in this case the

walls of the liner. If these walls are thin, then this would cause a significant temperature drop causing inconsistent vaporization and possible mass discrimination in the chromatography of wide-volatility-range samples.

- Packing – most split injector liner designs require some form of packing material to be present. Normally this will comprise a plug of glass or quartz wool. This provides a suitable bed on which the sample may be initially deposited so that it is ideally exposed to the carrier gas flowing through it and aids the vaporization process. It also assists with the mixing of the sample vapor with the carrier gas so that a homogeneous mixture is presented to the column inlet. Finally, with high-speed injection (see Section 8.6.1.2), the packing will wipe the end of the syringe needle and help prevent unevaporated liquid sample from being pulled back (and effectively lost) on the outside of the needle and be deposited on the underside of the septum.
- Internal geometry – although a plug of packing material will greatly assist with sample vapor homogenization, the flow path of sample vapor and carrier gas through the liner will also have a significant impact on performance. Liners have been developed with sophisticated internal geometry such as the Jennings inverted cup and various baffles.
- Inertness – for many applications, the liner should be chemically deactivated. Different types of samples will need different types of deactivation chemistries. The most important part of the deactivation process is to ensure that any glass or quartz wool plug is fully deactivated – this is the material in closest contact with the sample during injection. Liner deactivation is a complex process and many users will purchase liners that have already been fully deactivated from vendors.

8.6.1.3. *The Split Ratio*

An important factor in split injection is the split ratio, which may be calculated for a given set of conditions from Eqn (8.3). The two flow rates apply to movement of the sample vapor in the carrier gas at the split point (i.e. the end of the column in the injector). In practice, the flow rates are measured (or calculated) at the split vent and at the end of the column where the temperature and pressure are at ambient conditions. As long as the two measurements are made under the same conditions of temperature and pressure, the result should remain the same:

$$\text{Split ratio} = \frac{\text{Flow rate into split line} + \text{Flow rate into column}}{\text{Flow rate into column}} \qquad (8.3)$$

The split ratio is used as a measure of how much of the sample injected into the injector liner actually enters the capillary column. A small ratio will place more sample in the column, whereas a large ratio will reduce it.

There are practical limits on the split ratio for any split injector; at low ratios, peaks become broad and tailing and, at high ratios, insufficient time for mixing may occur.

8.6.1.4. *Pneumatics*

The design of the pneumatics system is critical to split injection performance. The flows and pressures must be carefully controlled to deliver a stable split ratio and hence good quantitative performance and to retain the sample vapors within the liner during a fairly explosive injection process as liquid sample hits a hot liner.

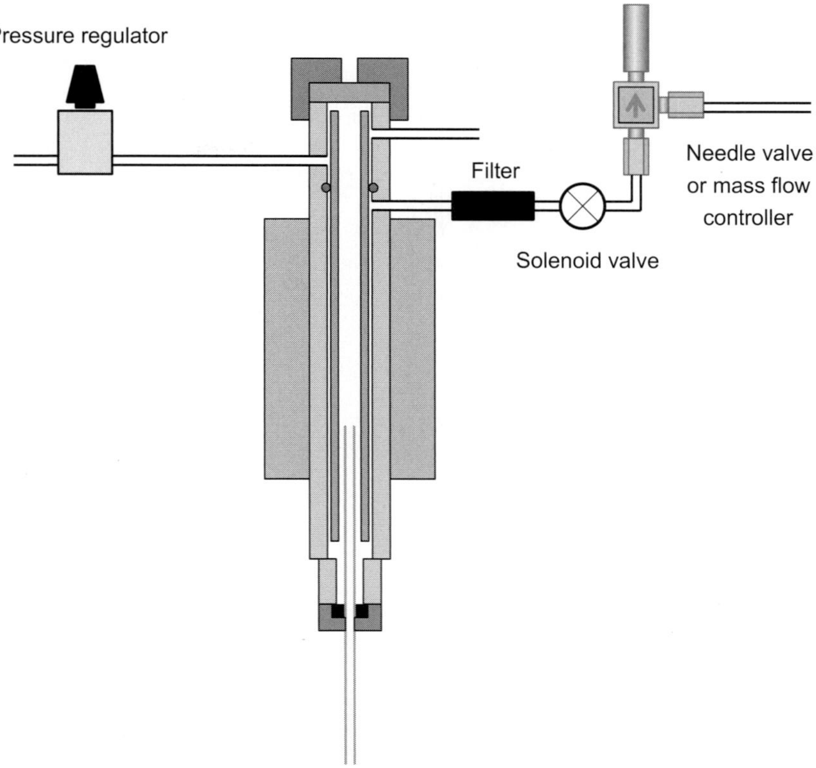

FIGURE 8.15 Forward pressure regulation in split injection.

There are essentially two approaches to the pneumatic system that are discussed in the following:

Forward pressure regulation as shown in Figure 8.15 is the traditional design. This may be implemented using mechanical or programmable electronic controllers. The pressure (or flow rate) of carrier gas into the column is set by the pressure regulator and the split flow rate is set by the needle valve or mass flow controller on the split line. The solenoid valve is normally left on during split injection although with programmable electronic pneumatics, the split flow rate may be later reduced to conserve carrier gas. The main disadvantage with this approach is that the pressure is controlled upstream of the injector and so any pressure drop through the plumbing or through the liner will cause a reduction in flow rate through the GC column affecting retention times and split ratios.

A better approach is to use backpressure regulation as shown in Figure 8.16. At first, this looks similar to forward pressure regulation but it is fundamentally different. Carrier gas is supplied to the injector by means of a mass flow controller. This controls the total flow into the injector. A backpressure regulator maintains the pressure inside the injector by regulating the passage of carrier gas through it until the set pressure is attained. Some of the carrier gas will flow into the column, some will exit from the septum purge vent, and the rest will be released through the split vent. Because the pressure is regulated

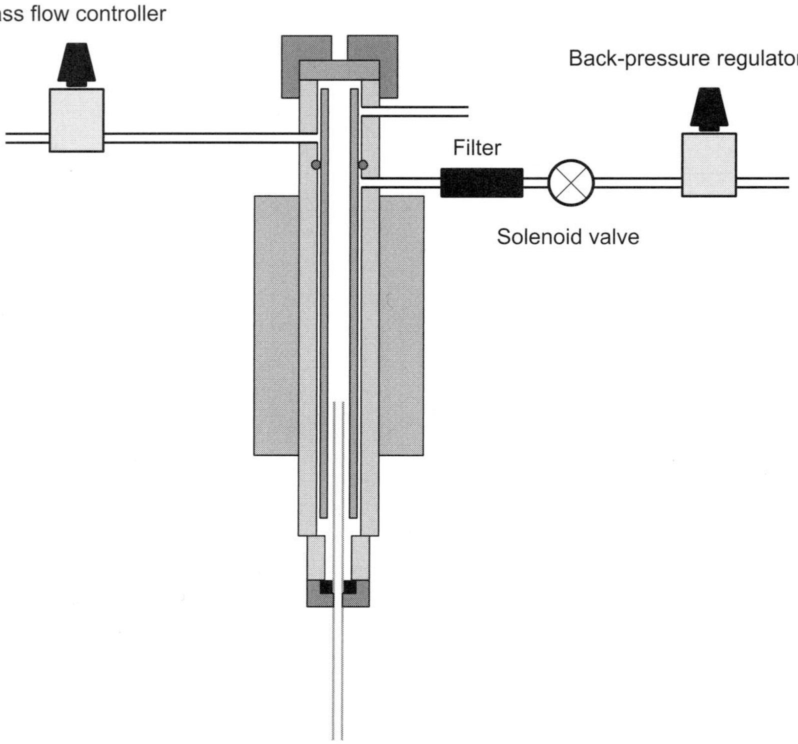

FIGURE 8.16 Backpressure regulation in split injection.

downstream of the liner, the pressure applied to the head of the column is more tightly controlled and better retention time repeatability and split ratio linearity are observed.

Backpressure regulation is the approach adopted in most modern GCs and especially those with programmable electronic controllers.

8.6.2. Splitless Injection

Classical splitless injection [9] must be the most difficult of all the liquid injection techniques to master. There are so many critical variables that affect performance that a good understanding of the principles is essential before embarking on method development using this technique. Yet, it remains one of the most popular injection techniques – mainly because of its robustness and because the same injector can usually be used for split injection too (see Section 8.6.1).

The aim of splitless injection is to allow sample injection into a glass or quartz liner and then transfer the whole sample into the GC column. The use of a liner overcomes several of the key concerns with cold on-column injection (see Section 8.3) but raises the question of how to get several hundred microliters of sample vapor into a narrow-bore capillary column and still get sharp peaks.

Like COCI, the key to successful splitless injection lies with the solvent.

In splitless injection, the sample is injected into a heated liner where it is immediately vaporized. There is no split applied at this stage, so the carrier gas will exit the liner through the column alone. The flow rate through the column will be low (perhaps only 1.0 mL min^{-1}) and so it will take several minutes for the sample to be fully purged into the column (and so produce peaks several minutes wide at the start of the chromatography). However, if the column is held at a temperature that is less than the boiling point of the solvent, then most of the solvent will recondense in the column. This condensation induces a rapid drop in pressure within the column inlet and serves to pull the rest of the sample vapor out of the liner and into the column. In this way, a very rapid transfer of the sample into the column as a narrow band can be achieved.

Once in the column, the recondensed liquid sample is in a very similar state to that of COCI, and so the discussion on solvent effect, solvent flooding, and retention gaps applies here too (see Section 8.3.3). It also means that splitless injection normally requires a column temperature program to be effective.

At some point after injection (normally about 30 s), the split vent is opened to flush out any remaining sample residue from the liner (and any other parts of the injector into which it may have diffused); this creates a small but unavoidable loss of the sample from the column. The benefit of doing this is that it cleans up the solvent peaks and sharpens many of the early-eluting peaks in the chromatography. The split flow is set to typically 25 mL min^{-1}. At the point that the split vent is opened, there will be a momentary drop in pressure inside the liner, which will serve to backflush any sample close to the end of the column – this will help to remove any peak tailing.

8.6.2.1. *The Role of the Syringe*

Classical splitless injection is, like its split equivalent, a vaporizing injection technique. A syringe deposits the liquid sample into a heated liner. It will have the same issues and solutions with the syringe technique that have already been discussed for split injection (see Section 8.6.1.1).

The volume of liquid sample that may be injected is much greater with splitless injection than with split injection. There are two things to consider:

- The injection rate – this should not be greater than the rate of recondensation in the capillary column or else vapor will exit the liner in other ways than into the column. The rate at which the syringe plunger is depressed will control this effect. A typical rate for large-volume injection is 0.5–1.0 μL s^{-1}. This technique has been termed *'total concurrent vaporization'* [10].
- The buildup of liquid in the column – like cold on-column, too much solvent causes problems with solvent flooding. The use of a retention gap is recommended for large-volume injections.

Although some methods do use classical splitless injection for large sample volumes, the use of a PTV is often a better option as there is dynamic control of the liner temperature (see Section 8.7.5).

8.6.2.2. *The Role of the Injector Liner*

Unlike split injection, splitless injection does not need to concern itself about homogeneous mixing of the sample vapor with the carrier gas in the liner; it just needs to get the whole sample into the column as efficiently as possible. Because of the solvent recondensation effect, the splitless liner can be much narrower than that used for split injection; a 2-mm-internal-diameter liner is typical. It is likely that much of the sample is not fully vaporized as it enters the column – this would improve transfer efficiency. The exotic baffles that are used in some split injector liners have no place in splitless injection.

For the same reason, packing the liner with glass or quartz wool is normally not required except to wipe the liquid sample off the end of the syringe needle during high-speed liquid autosampler injection.

Liner deactivation is much more critical with splitless injection because of the much lower analyte concentrations in the samples and because the residency time in the liner will be greater.

8.6.2.3. *Pneumatics*

The split injection mode pneumatics are normally also deployed with the splitless injection technique.

With forward pressure regulation (see Figure 8.15), the solenoid valve on the split line is simply closed during injection and opened later in the run.

With backpressure regulation (see Figure 8.16), matters are more complicated because a high flow rate is still needed for the mass flow controller and backpressure regulator to function. Either the system should revert to forward pressure regulation (which is possible with programmable electronic controllers) or the flow from the flow controller should bypass the injector and be fed directly to the backpressure regulator.

8.6.2.4. *Pressure-Pulsed Injection*

Although a good splitless injector and method will get most of the sample into a capillary column, it will not get it all. Compounds will be lost through adsorption or diffusion into other parts of the injector.

Clearly, the flow rate of the carrier gas that is passing through the liner into the column at the point of injection is going to have a very direct effect on the transfer efficiency. Temporarily increasing this flow rate during the injection can have a significant effect on the peak size in the resultant chromatography. This technique has been termed *'pressure-pulsed injection'* [11] and is normally only possible with a GC that has a programmable electronic pneumatic system. Figure 8.17 shows a typical carrier gas pressure program for pressure-pulsed injection. The value of pressure pulsed injection will

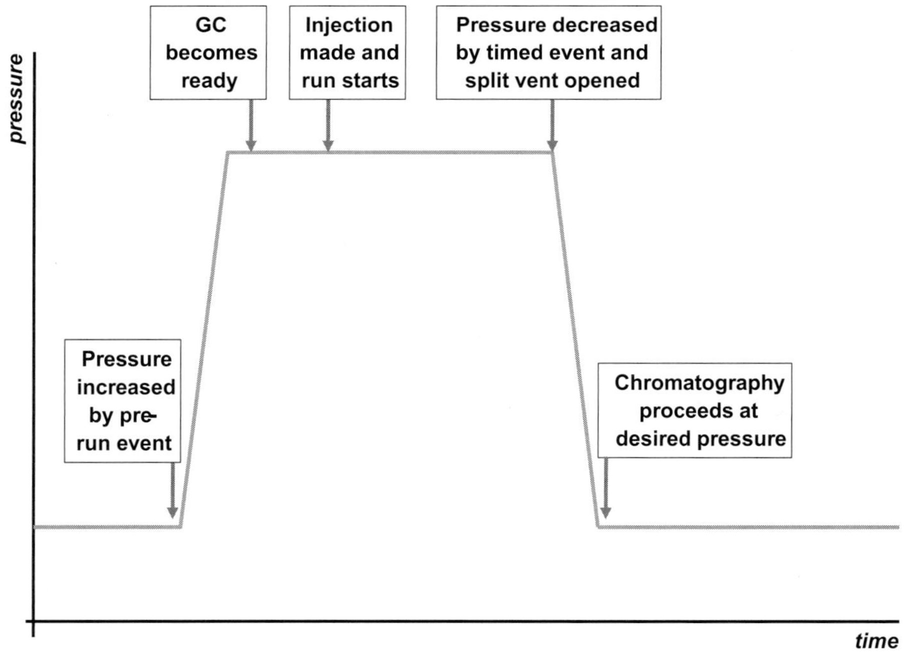

FIGURE 8.17 The principle of pressure-pulsed injection.

depend on the application and the design of the injector being used.

8.7. THE PROGRAMMABLE TEMPERATURE VAPORIZING (PTV) INJECTOR

This injector is the most versatile of all the GC liquid injection systems. It has the convenience of a classical split/splitless injector and yet approaches the performance of a cold on-column injector. It also is able to offer a number of additional techniques that are not possible with any other injector type.

The PTV may be called different names by different instrument vendors: the temperature vaporizing injector (TPI), the programmable split splitless injector (PSS), the septum-equipped temperature programmable capillary injector (SPI), the brightly enhanced sample transfer (BEST) injector, and the OPTIC.

Despite the confusing array of acronyms, all these injectors essentially perform the same basic functions.

A PTV injector differs from other injectors in that the injector liner inside the injector is able to be programmatically heated and cooled [12]. Figure 8.18 shows a typical PTV design in which the heater block has fins to enable

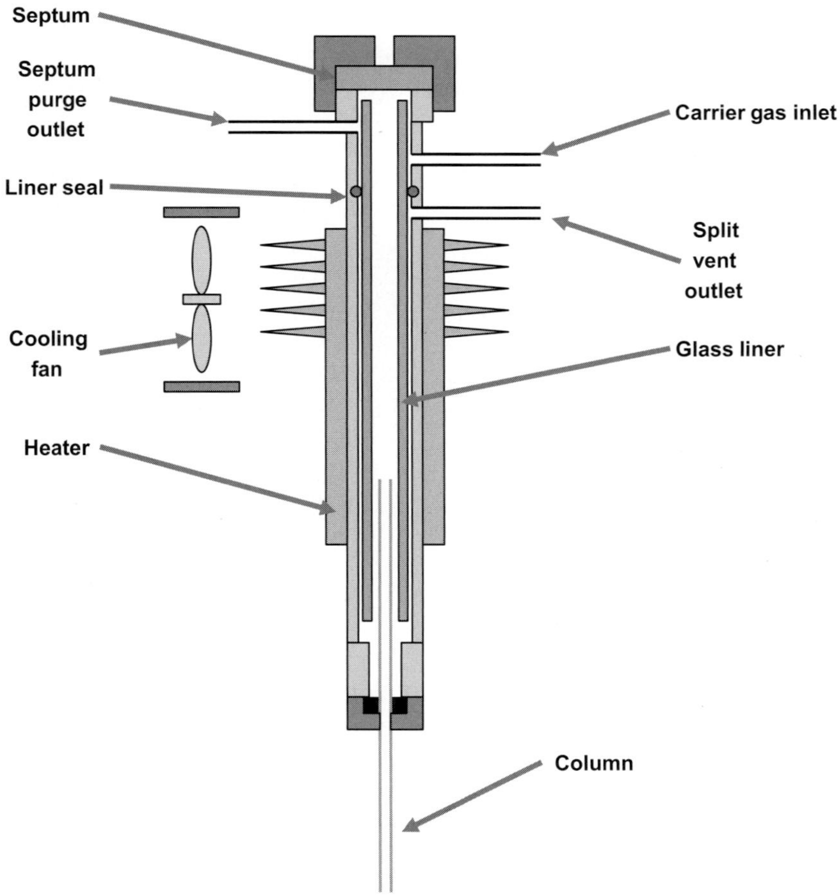

FIGURE 8.18 The programmable temperature vaporizing injector.

TABLE 8.3 Injection Modes that may be Supported by a PTV Injector

Mode
Programmed split
Programmed splitless
Vaporizing split
Vaporizing splitless
Cold on-column
Hot on-column
Large-volume injection

fast cooling from a fan directed at it. The body of the injector will normally be narrower and lighter than its vaporizing equivalent to facilitate rapid heating and cooling. Normally there is a split vent to enable both split and splitless modes of injection. Table 8.3 shows the various injection modes achievable from this injector.

8.7.1. Programmed Split Injection

This mode is very similar to the split injection technique described earlier (see Section 8.6.1). The main difference is that the sample is injected by syringe into a cooled liner. The syringe needle is then withdrawn and only then is the liner heated to vaporize the sample and transfer the vapors to the GC column.

This two-step injection technique provides several advantages over traditional heated liner injection:

- The sample largely remains as a liquid as it is deposited into the liner. This eliminates the explosive vaporization of the sample solvent seen with heated vaporizing injection, which can cause the sample to be expelled from the liner resulting in an apparent loss of sensitivity, discrimination effects, and possible cross-contamination between samples. With PTV injection, the solvent vaporization occurs at a more gradual rate as the liner temperature is increased which helps keep vapors within the confines of the liner.
- The (normally) metal syringe needle is not exposed to high temperatures while in contact with the sample. Such contact with hot metal may promote the chemical breakdown of sensitive components. Injection into a cold liner ensures that there is no hot metal present during sample vaporization, so breakdown of sample components is much less likely.
- The syringe needle is not exposed to heat and so the preinjection and postinjection effects experienced with heated injectors are not seen.
- Each sample component, as soon as it is vaporized as the liner temperature is increased, will be swept out of the liner by carrier gas into the column and out through the split vent. Thus, no component should stay in the liner long enough to be exposed to any temperatures higher than necessary for its vaporization. This is one of the key reasons why PTV injection is so suited for the injection of samples containing thermally labile compounds.
- Once the sample components of interest have vaporized, the liner may be increased to a much higher temperature than would be used in methods with other injectors. This serves to bake-out traces of less volatile sample residue, which would otherwise build up in the liner. The PTV liner is much easier to keep clean than with other injectors.

The injector liner is normally narrower (typically 2-mm internal diameter) than for classical split injection because there is no longer the explosive vaporization that occurs with that technique during sample injection. The lower-capacity liner assists in getting the sample out of the liner and into the column, as the temperature is increased. The walls of the liner tend to be much thinner to aid heat transfer from the heater into the liner interior. The liner is usually a simple glass or quartz cylinder. Because this is a nonvaporizing injection technique and the sample will be deposited as a liquid into the liner, it is important to have a plug of glass or quartz wool to wipe traces of liquid sample off the syringe needle during injection to prevent being pulled back onto the septum.

Despite the apparent complexity of a PTV injector, it is a much more straightforward prospect to design than a classical split/splitless injector because the dynamics of rapid sample vaporization no longer have to be considered.

Because of the narrower liner there will be a significant pressure drop across it at high split flow rates. This pressure drop will increase as the liner is temperature-programmed due to the viscosity of the carrier gas increasing. This will have the effect of changing the split ratio during the liner program, giving rise to significant mass discrimination effects. Fortunately, by using backpressure regulation (see Section 8.6.2.3) the effect is largely eliminated.

8.7.2. Programmed Splitless Injection

The principle of programmed splitless injection is very similar to programmed split injection: the sample is injected into a cooled liner; the syringe needle is removed and only then is the liner heated to affect controlled sample vaporization. All the benefits of programmed split injection apply to programmed splitless injection too.

As with classical splitless injection, the split vent is opened shortly after injection to purge out residual traces of sample from the injector. This should only be done after the liner has reached its final temperature or else loss of heavier components may result. Typically, the vent will be opened with a flow rate of, say, 25 mL min^{-1} 0.5–1.0 min after the liner has reached its top temperature.

For programmed splitless injection, the liner may be very narrow (down to 1.0 mm) to ensure rapid transfer of the desorbed vapors into the GC column during temperature programming of the liner.

8.7.3. Vaporizing Split and Splitless Injection

These are essentially the same techniques as those described for classical split and splitless injection (as described in Sections 8.6.1 and 8.6.2). The main difference is that the liners for the PTV techniques are usually smaller and of lower mass than those used in their vaporizing injector counterparts. This may mean that the maximum volume of sample that can be handled will be much less. For a 2-mm liner, this may limit the sample volume to 0.3 or even 0.2 μL.

It is not possible to design an injector that is optimized for both vaporizing and nonvaporizing injection techniques without making compromises in the performance or application. However, a good PTV design is as close as it can get.

8.7.4. Cold and Hot On-Column Injection

A PTV injector may be easily adapted for on-column injection. A special liner may be installed as shown in Figure 8.19 The end of a 0.53-mm column or retention gap is pushed right up into this liner until it is located within the hourglass constriction in the liner. The other side of the hourglass constrictor acts as a guide for a metal syringe needle, which can now be fully inserted into the end of the 0.53-mm tubing. A small hole may be required in the side of the liner for coupling to pneumatic components in the split line.

The sample is injected into the column or retention gap with the liner set to a low temperature that is just below the boiling point of the solvent. After the needle is withdrawn, the liner may be rapidly heated to a high temperature or programmed to track the temperature of the column during a temperature program. In other respects, the cold on-column and hot on-column techniques are as described in Sections 8.3 and 8.5, respectively.

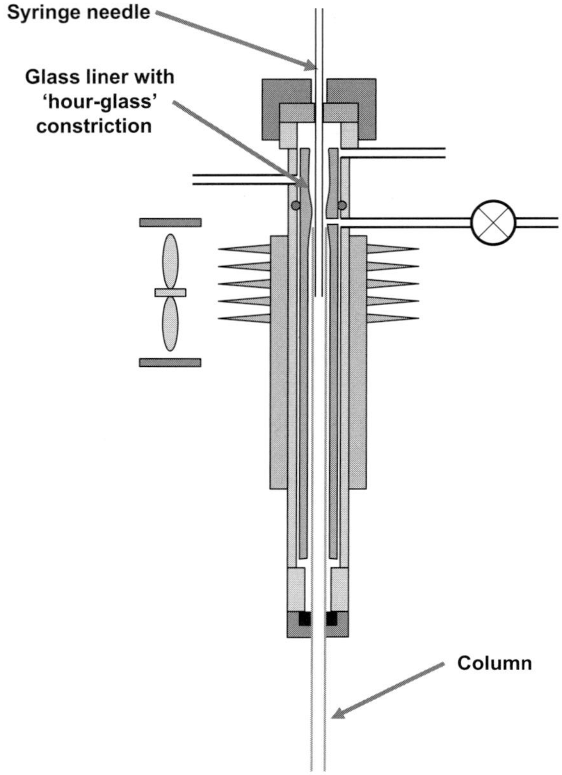

FIGURE 8.19 Cold on-column injection with a PTV injector.

8.7.5. Large-Volume Injection

For some samples, lower detection limits are needed that can only be achieved by injecting more sample into the column. The possibility of using cold on-column or splitless injection techniques to do this has been discussed earlier. None of these is very easy or practical as some effective way of dealing with large volumes of sample liquid and vapor is needed.

The PTV injector allows a mode of operation that simplifies large-volume injection (LVI) and yet addresses the issues with the large solvent volume.

The LVI technique may be viewed as a combination of both split and splitless techniques applied to the same sample injection at different temperatures. Figure 8.20 illustrates the principle. A suitable syringe injects a large volume (typically 50–100 μL) of sample into a cold PTV injector liner. The split vent is open and the carrier gas flow rate will be set to 100 mL min^{-1} or above. The injector is left under these conditions for a few minutes. During this time, the solvent will become vaporized and be carried out of the liner through the split vent. A small amount of the solvent vapor will enter the column. At the end of this '*solvent purge*' process, the less volatile components in the original sample should still remain in the liner. The split vent is then closed and the liner temperature is raised to effect the vaporization of the sample residue and its splitless transfer by carrier gas into the GC column.

For this technique to work, there must be a sufficient difference between the volatility of the solvent and the analytes. For n-alkane solvents, this normally means a 4-carbon number difference or a boiling-point difference of 100 °C. The liner must be able to retain the large volume of liquid without it spilling over into the column or into the far recesses of the injector interior. A tight plug of glass or quartz wool will help with the liquid retention. Some liners have an etched surface to help with the liquid retention. Figure 8.21 shows an example of a chromatogram produced using this technique with a chlorinated solvent on an electron capture detector.

A 2 mm × 50 mm liner bed will have a maximum volumetric capacity of just 160 μL (leaving no space for carrier gas); so such a liner would comfortably handle injection volumes of 50 μL but 100 μL would be too much.

For larger volumes, there are two approaches that could be considered:

- Multiple injections – for example, a 50-μL injection could be made and then solvent purged and then further injection/purge cycles could be applied until the desired

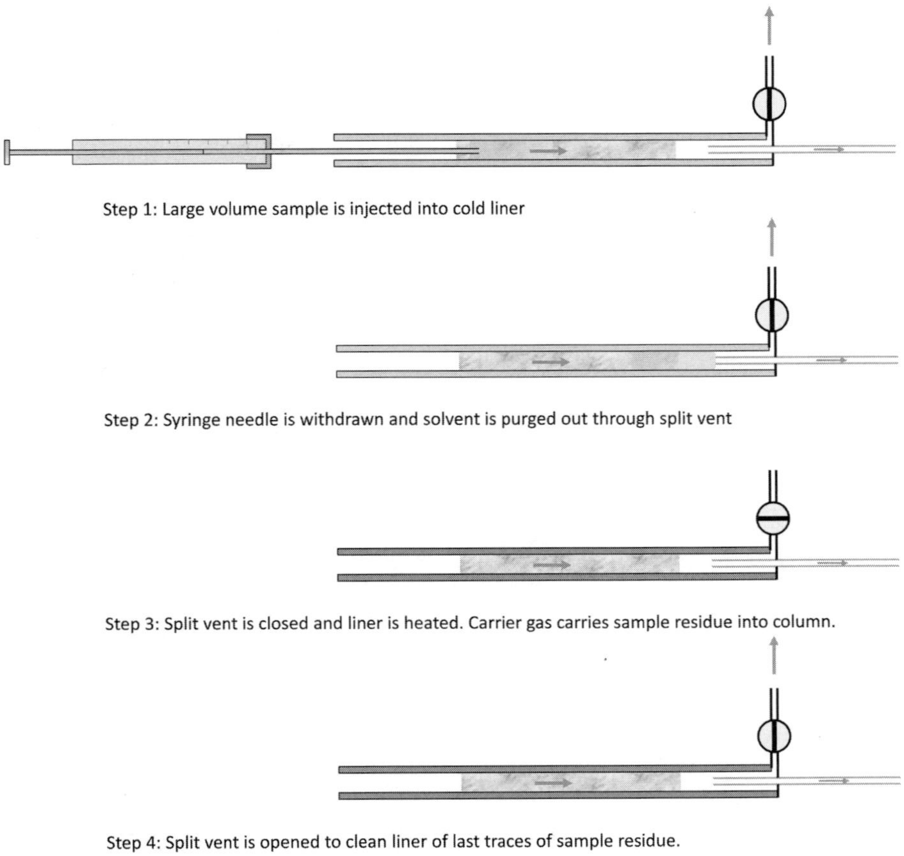

FIGURE 8.20 Large-volume injection with a PTV injector.

volume is reached. After the final cycle, the split vent is closed and the liner temperature is increased to affect the injection.

- Partial concurrent vaporization [13–15] – the liner temperature is set to a value that is found to vaporize the sample solvent at a reasonable rate. A large syringe injects the sample at a slow rate so that the total volume of liquid inside the liner at any time does not exceed its capacity. This needs some trial and error to establish satisfactory conditions.

8.8. THE GAS SAMPLING VALVE

The previous injectors discussed are primarily designed for liquid injection. It is possible to use some of these injectors with a gastight syringe to make injections of gas samples; however, this is a very manual process and subject to errors resulting mainly from temperature and pressure changes occurring inside the syringe during sample loading and injection.

In most instances, much better performance will be obtained through the use of a gas sampling valve. Although there is not real scope

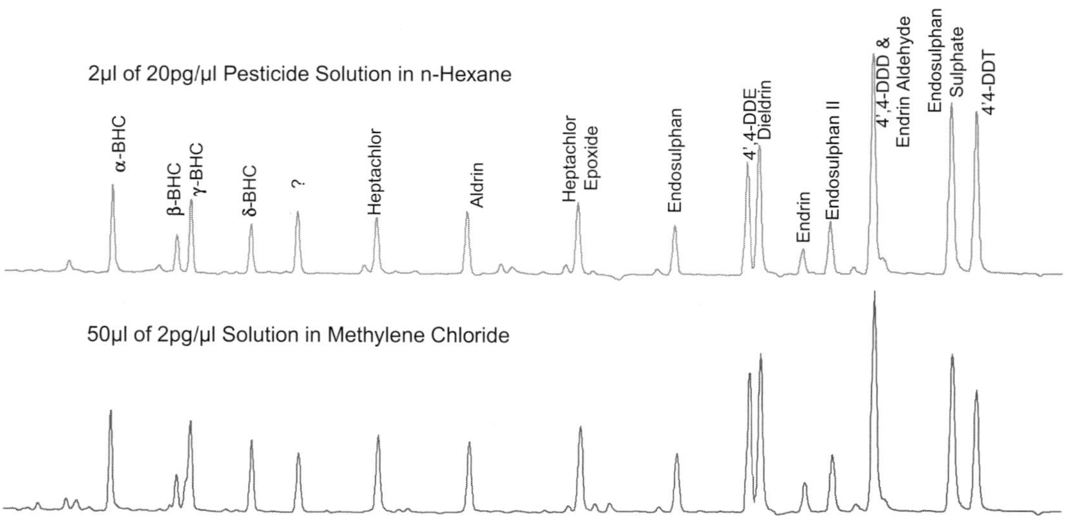

FIGURE 8.21 Example of a large-volume injection with a PTV injector. This compares the chromatography obtained on an electron capture detector from a solution of pesticides in a regular solvent like n-hexane against a dilute solution of the same compounds in methylene chloride using the LVI technique.

to cover this topic in detail in this section, the basics will be discussed.

A gas sampling valve is a mechanical device comprising a rotor with etched channels in it rotating between a series of ports that connect it to other components in a GC. There are a variety of valve designs with different numbers of ports in them that may be configured either independently or with other valves in a large number of ways for many different applications.

For injecting a gas sample into a GC column, we will just consider the classic 6-port valve configuration given in Figure 8.22.

The principle of operation is straightforward. Prior to injection, sample gas flows through a valve loop as shown in Figure 8.22. This stream may be flowing from a pressurized source, it may be introduced by a gas syringe, or it may be drawn through by a vacuum pump. Although the loop will have a fixed and normally calibrated capacity, the amount of sample injected into the column will also be directly dependent on the temperature and pressure of the sample in the loop at the point of injection. For all gas analyses to be accurate, the samples and the calibration standard mixtures must all use the same loop and at the same temperature and pressure. If these requirements are met, a GSV is capable of producing the highest precision of any injection technique in GC.

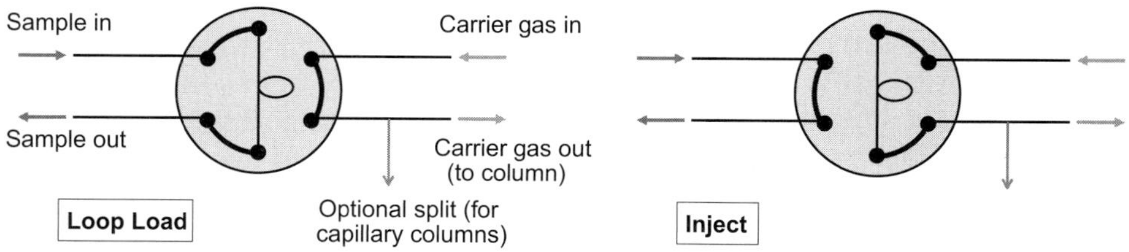

FIGURE 8.22 Example of a gas sampling valve configured for injection of gas samples into a GC column.

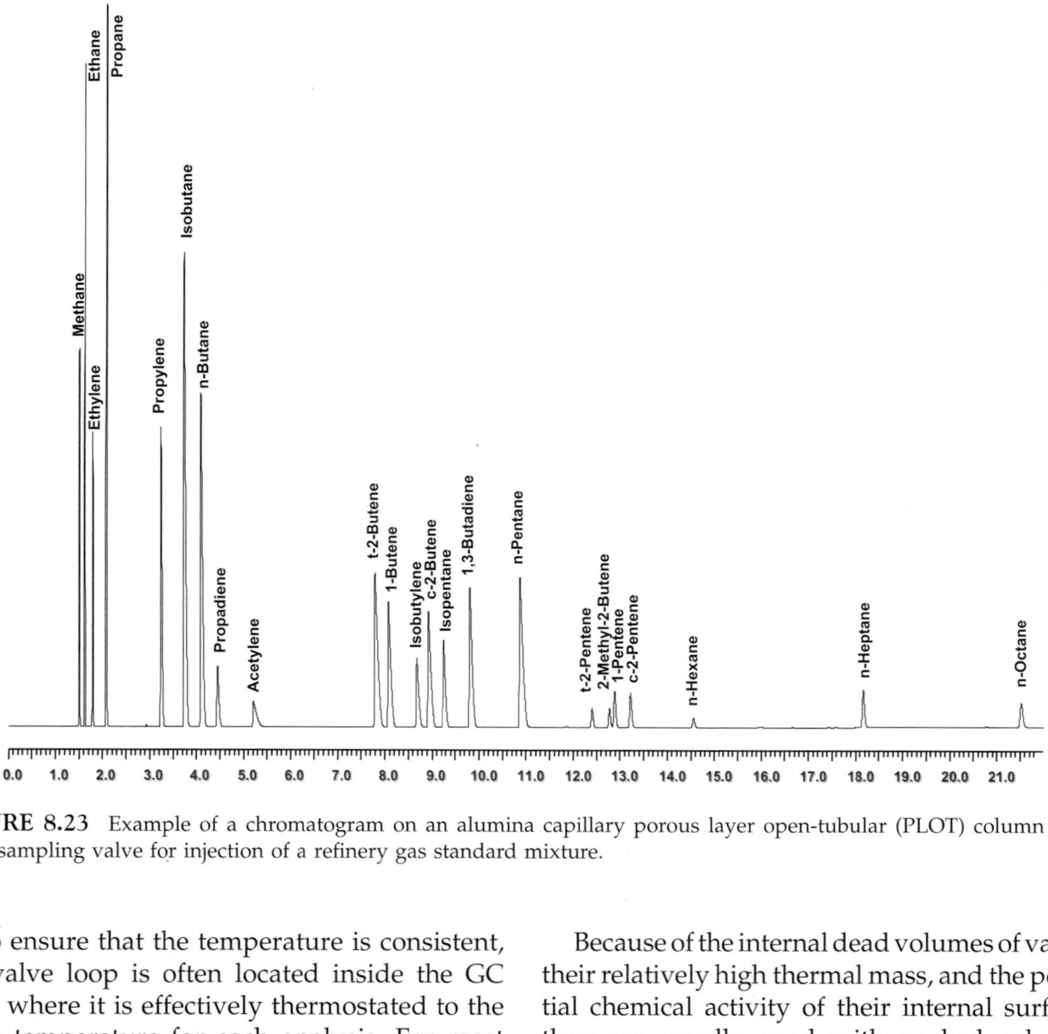

FIGURE 8.23 Example of a chromatogram on an alumina capillary porous layer open-tubular (PLOT) column using a gas sampling valve for injection of a refinery gas standard mixture.

To ensure that the temperature is consistent, the valve loop is often located inside the GC oven where it is effectively thermostated to the same temperature for each analysis. For most applications, the loop is held in a semisealed location so that it is effectively insulated from changes in the temperature of the ambient air.

For consistent sample pressure inside the loop, two approaches may be considered:

- using a backpressure regulator on the valve sample vent or
- turning off the sample supply (or reduce the flow rate) immediately prior to injection to allow the pressure inside the loop to decay to atmospheric pressure.

Because of the internal dead volumes of valves, their relatively high thermal mass, and the potential chemical activity of their internal surfaces, they are usually used with packed columns. Generally, this is not a problem as most gas analyses will require packed columns anyway to get the necessary retention and dynamic range.

There are methods, however, that use capillary columns especially with some of the modern porous layer open-tubular (PLOT) columns. In such instances, a splitter is often deployed between the valve and column to get the peak widths sufficiently sharp for good chromatography.

An example of a gas analysis using GSV injection is given in Figure 8.23.

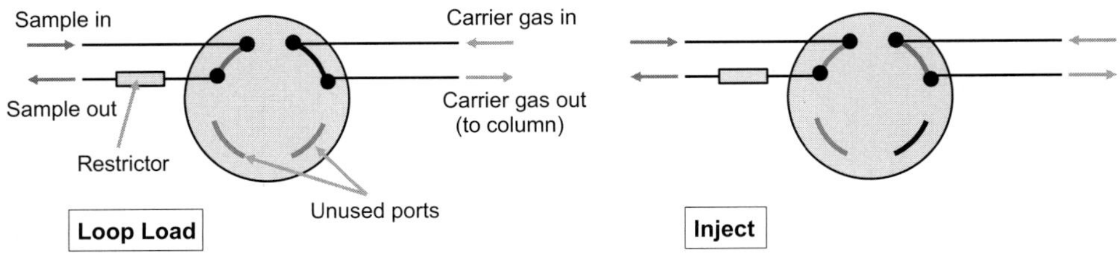

FIGURE 8.24 Example of a liquid sampling valve with internal loop built into rotor for injection of liquid samples into a GC column.

8.9. THE LIQUID SAMPLING VALVE

Like the GSV, the liquid sampling valve (LSV) is a mechanical rotary valve with a fixed sample capacity. It is used to inject a pressurized liquid into a GC column. Instead of an external loop that is connected to ports on the valve, the sample is loaded into a cavity on the rotor that is typically just a few microliters in capacity. The

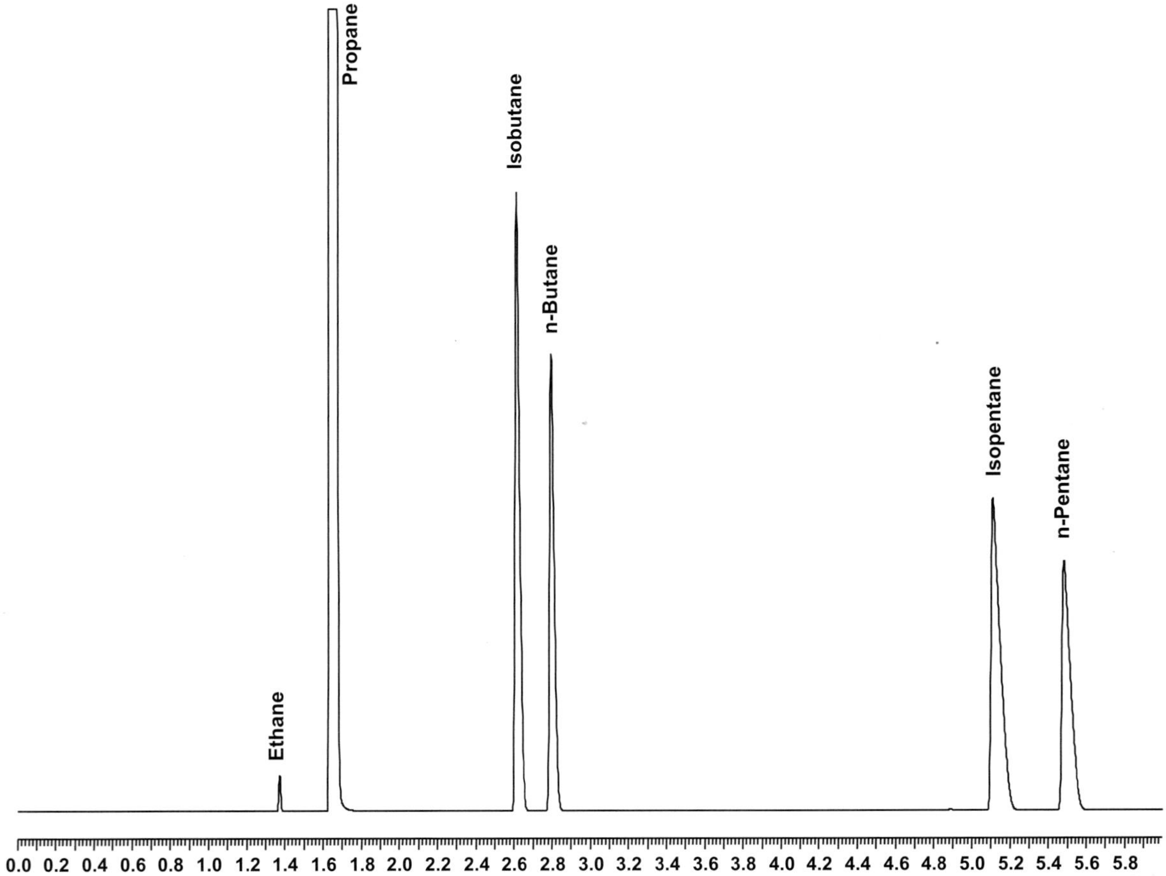

FIGURE 8.25 Example of a chromatogram using a liquid sampling valve for injection of a liquid petroleum gas standard mixture into an adsorbent packed column.

sample is fed, under pressure into the valve – a restrictor on the vent port maintains the sample in the liquid state within the valve.

Once the sample is loaded, the rotor rotates and the pressurized liquid rapidly vaporizes and it is carried to the column by a flow of carrier gas. A schematic diagram of a typical LSV installation is given in Figure 8.24 and an example application is given in Figure 8.25.

Acknowledgment

The author would like to thank PerkinElmer for its support in the writing of this chapter and for colleagues who were so forthcoming with their help and advice.

References

[1] S.P. Sutera, R. Skalak, Annu. Rev. Fluid. Mech. 25 (1993) 1–20. [Hagen–Poisseuille].
[2] K. Grob, K. Grob Jr., J. Chromatographr. 151 (1978) 311.
[3] K. Grob, J. High Resolut. Chromatogr. 1 (1978) 263.
[4] Solvent effect – reference required.
[5] K. Grob, J. Chromatogr. A 213 (1) (21 August 1981) 3–14. [Solvent flooding].
[6] Grob K. Chromatographia, 17(7) 357–360. [Reverse solvent effect].
[7] K. Grob, G. Karrer, M.L. Riekkola, J. Chromatogr. 334 (1985) 129. [Retention gap].
[8] K. Grob, H.-G. Schmarr, A. Mosandl, J. High Resolut. Chromatogr. 12 (1989) 375. [Vented retention gap].
[9] K. Grob, K. Grob Jr., J. Chromatogr. Sci. 7 (1969) 584.
[10] Total concurrent vaporization – reference required.
[11] P.L. Wylie, R.J. Phillips, K.J. Klein, M.Q. Thomas, B.W. Hermann, J. High Resolut. Chromatogr. 14 (1991) 649. [High pressure injection].
[12] F.S. Poy, F. Visani, F. Terrosi, J. Chromatogr. 217 (1981) 81. [PTV].
[13] W. Vogt, K. Jacob, A.B. Ohnesorge, H.W. Obwexer, J. Chromatogr. 186 (1979) 197.
[14] W. Vogt, K. Jacob, H.W. Obwexer, J. Chromatogr. 174 (1982) 437.
[15] F. Munari, A. Trisciani, G. Mapelli, S. Trestianu, K. Grob Jr., J.M. Colin, J. High Resolut. Chromatogr. 8 (1985) 601.

CHAPTER

9

Headspace-Gas Chromatography

Michael J. Sithersingh, Nicholas H. Snow

OUTLINE

9.1. INTRODUCTION AND HISTORY

The term "headspace" in gas chromatography denotes the vapor phase within a sealed container also containing a liquid or solid. Usually, the vapor is above the liquid or solid, at the top of the container, hence the term "headspace." Headspace extraction refers to the collection and analysis of the vapor phase in the container. Generally, the vapors are sampled from a system that is brought to equilibrium prior to the sampling.

In a typical headspace extraction, a liquid or solid sample is placed in a sealed vial and heated to a predetermined temperature until equilibrium is reached, providing a constant composition mixture of gases in the vapor

Gas Chromatography DOI: 10.1016/B978-0-12-385540-4.00030-4

phase. An aliquot of this vapor-phase mixture is withdrawn from the container and transferred to a gas chromatograph for separation and analysis of the components.

Since a defined amount of the equilibrium vapor is taken and carried to the column in the gas chromatograph for the analysis, only volatile substances reach the gas chromatographic column, with nonvolatile matrix components remaining in the sample container. Complex sample matrices, especially those that include volatile analyte(s) in a less volatile matrix, which would otherwise require sample extraction or preparation, or be difficult to analyze directly, are ideal candidates for headspace extraction, since they can be placed directly in a vial with little or no preparation.

9.1.1. History

The precise origin of the idea that analyzing the headspace above a sample in a sealed container would provide useful information about the composition of the sample is unclear; however, the use of headspace extraction with gas chromatography is almost as old as gas chromatography itself, dating to the 1950s. Early work in the late 1950s and early 1960s included analysis of nontraditional samples, such as food in flexible packaging, soda in cans, and ethanol in blood and breath, and led to the introduction of the first automated instrumentation for headspace-gas chromatography in the late 1960s.

The popularity of headspace-gas chromatography has grown over the past 40 years due to automation of the instruments and the seeming fundamental simplicity of the technique [1–5]. Some common fields of application include analysis of polymers [6], volatile components in drinks and foodstuffs [7], blood alcohol [8], water and environmental analysis [9], and fragrances in perfumes and cosmetics [10]. In the pharmaceutical industry, headspace-gas chromatography is widely used to determine residual solvents in active pharmaceutical ingredients (APIs) and drug substances. The United States Pharmacopeia (USP) incorporated the technique into its General Chapter <467> "Residual Solvents" to determine the most common residual solvents in drug substances using headspace chromatography [11,12].

9.1.2. Print and Online Resources

There is a very large literature of headspace-GC applications, although it can be difficult to access a specific headspace symposia, books, journals, and reviews have been relatively few. The theory and practice of static headspace extraction are thoroughly described in three texts, by Kolb and Ettre, Ioffe and Vitenberg, and Hachenberg and Schmidt [13–15]. Headspace extraction in its various forms is often covered as a chapter or as part of the sample preparation chapter in general textbooks about gas chromatography [16,17]. When searching literature databases, such as SciFinder™ or Science Direct™, care should be taken to be as specific as possible about the use of keywords, as general keywords, such as "headspace and water," can produce thousands of references. Instrument vendor websites and general chromatography websites may also provide information and application notes to assist in developing headspace-based methods and in general gas chromatographic theory and practice [18–20]. As a general resource for static headspace extraction, the text by Kolb and Ettre, mentioned above, is the best starting point.

9.2. FUNDAMENTALS OF HEADSPACE EXTRACTION

There are several means by which the vapor component in a sealed container may be collected and transferred to a gas chromatograph. In static headspace extraction (SHE), analyte vapors may be transferred using gas tight syringe, a capillary or a sample loop placed between the vial and the gas chromatographic inlet or vapors may be trapped on a sorbent, followed by desorption into the gas chromatograph. In dynamic headspace extraction (purge and trap), a gas is passed through the liquid phase, evaporating analytes, which are then collected on a sorbent and transferred to the gas chromatograph. While the solution–vapor phase equilibrium upon which headspace extraction is based is generally straightforward, there are subtleties in both theory and instrumentation that make headspace extraction similar to chess: easy to learn, yet challenging to master.

9.2.1. Static Headspace Extraction (SHE)

Figure 9.1 shows a schematic representation of a vial containing the sample phase (S) in equilibrium with the headspace of the vial, gaseous phase (G). The sample phase (S) contains the analytes and matrix. The volatile analytes, originally present in the sample phase (S), are distributed between the two phases at equilibrium. Most extraction and chromatographic theory assumes that the system is at equilibrium. Failure to bring the system to equilibrium is one of the most common causes of problems with reproducibility in extractions.

The chemical equation and equilibrium constant expression leading from this configuration are given below in Eqns (9.1) and (9.2), which describe the phase transfer of an analyte from the sample phase in the bottom of the vial into the vapor phase above it:

$$A(\text{sample}) \leftrightarrow A(\text{headspace}) \tag{9.1}$$

$$K = \frac{[A(\text{headspace})]}{[A(\text{sample})]} \tag{9.2}$$

In their text, Kolb and Ettre use mass balance to derive an expression relating the chromatographic peak area (A) in SHE-GC to the analyte concentration in the vapor phase (C_G), the initial analyte concentration of a liquid sample (C_o), the analyte's solution–vapor partition coefficient (K), and the phase ratio (β) in the sealed vial. This is the fundamental relationship used by most method developers in SHE that provided the basis for related techniques:

$$A \propto C_G = \frac{C_o}{K + \beta} \tag{9.3}$$

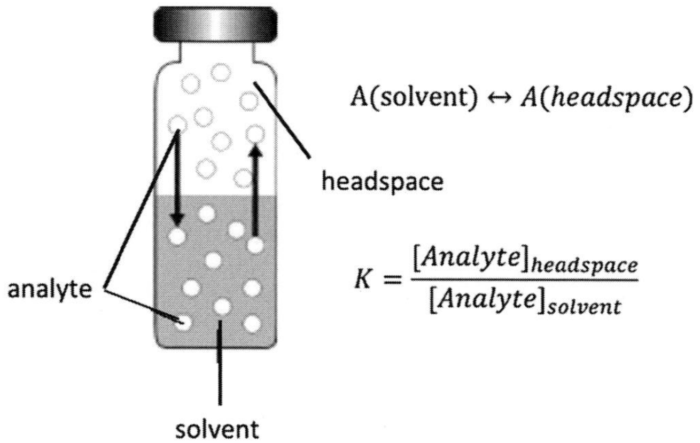

$$K = \frac{[Analyte]_{headspace}}{[Analyte]_{solvent}}$$

FIGURE 9.1 Schematic of the system for static headspace extraction inside a sealed vial with equilibrium constant expressions.

It is interesting to note that the partition coefficient in Eqn (9.3) is for the reverse of the reaction shown in Eqn (9.1), so most partition coefficients determined in SHE systems are expressed with the larger values favoring the condensed phase. The fundamental importance of β and K are described subsequently.

9.2.1.1. *Phase Ratio* (β)

The phase ratio is defined as the ratio of the volume of gaseous phase, V_G, to the volume of the sample liquid phase, V_S, in a vial of total volume V_v. It allows definition of the mass balance of the distribution of a dissolved analyte between the two phases in the vial:

$$\beta = V_G/V_S \text{ and } V_V = V_G + V_S \quad (9.4)$$

A known, reproducible sample volume (careful delivery of the sample into the vial) and reproducible vial volume are both necessary for effective quantitative analysis. Although the vial is usually heated following sample delivery, the change in the volume of the sample phase is generally insignificant. Since the vial is a closed system, the vapor moving from the sample phase to the gaseous phase does not change the vapor phase volume although it can change the pressure. Therefore, the initial volume of the sample added to the headspace vial remains the same after equilibrating the vial to the desired condition.

9.2.1.2. *Partition Coefficient*

The partition coefficient or distribution constant is the equilibrium constant of an analyte partitioning between two phases at equilibrium as described in Eqns (9.1) and (9.2). In this chapter, K is defined as described by Kolb and Ettre: the ratio of the concentration of the analyte in the sample phase (C_S) to that of the analyte in the gaseous phase (C_G) at equilibrium, the inverse of Eqn (9.2):

$$K = C_S/C_G \quad (9.5)$$

The sensitivity of an analysis in headspace gas chromatography depends on the partition coefficient of the analyte. A lower partition coefficient indicates a higher concentration of the analyte in the vapor phase and higher sensitivity. Method development efforts such as increasing the vial temperature, salting out, and adjusting the pH are often geared toward reducing the partition coefficient. K is related to more than the vapor pressure of the analyte; its solubility in the sample phase is also important. Lower solubility can assist in driving the analyte into the vapor phase. These ideas will be discussed later in this chapter.

As seen in the denominator of Eqn (9.3), the relative magnitude of the partition coefficient and phase ratio is a critical aspect of method development, discussed in detail in Section 9.4. The measured chromatographic peak area is further directly proportional to the concentration of the analyte in the gas phase, through instrument response-related factors such as injection technique and detector response, which is clearly proportional to the initial concentration of the analyte in the sample.

SHE is an equilibrium-based technique; generally, the analyte is not exhaustively removed from the sample matrix. Rather, a portion of the total amount of the analyte, based on the partition coefficient and factors in the discussion above, is collected and analyzed. This can be problematic, as much method development time and effort involves defeating the problem of reaching equilibrium in the sample vial, with a reproducible partition coefficient, even as the matrix may change from sample to sample.

9.2.1.3. *Multiple Headspace Extraction*

In multiple headspace extraction (MHE), analytes are extracted multiple times from the sample vial, providing the possibility for exhaustive extraction or derivation of the instrument response that exhaustive extraction would provide. MHE is analogous to extracting the compounds of interest from a sample multiple times using a separatory funnel with fresh aliquots of solvent each time and combining the extracts. After several extraction steps, the analytes of interest may be exhaustively extracted from the original sample.

Using the same instrumentation as SHE, the- concentrations of the analytes are determined at each extracted step. After each extraction, the analytes in the headspace vial are re-equilibrated between the sample phase and the gaseous phase. The decrease in concentration of the compound in subsequent extraction steps follows a first-order exponential decay with the time variable being the number of extractions and the exponential term k being determined experimentally in the same fashion as a first-order rate constant:

$$C = C_o e^{-kt} \quad (9.6)$$

In order to be useful experimentally, it is convenient to express Eqn (9.6) in terms of the chromatographic peak area, replacing the time t with the number of extraction steps performed, n. The initial concentration C_o is replaced with the peak area A_1 from the first extraction step which occurs at $n-1$:

$$A_n = A_1 e^{-k*(n-1)} \quad (9.7)$$

A_n is the peak area of the analyte at the number of extraction steps, n, and k^* is the new constant obtained by including instrument parameters. The parameters may be obtained by expressing Eqn (9.7) in a linear form and plotting the natural logarithm of the obtained peak areas versus the number of extractions, shown in Eqn (9.8):

$$\ln A_n = -k^* (n-1) + \ln A_1 \quad (9.8)$$

By summing the peak areas from each extraction step, the total peak area of any volatile analyte in the sample can be determined as seen in Eqn (9.9):

$$\sum A_n = A_1/(1 - e^{-k*}) \quad (9.9)$$

An extrapolated total peak area which corresponds to the total amount of the analyte present in the sample can be obtained by applying the values determined using regression analysis, with Eqns (9.8) and (9.9). MHE allows exhaustive extraction using the same instrumentation as SHE; although with multiple extraction steps, it can be time consuming.

9.2.2. Sorptive Extraction

Many SHE techniques involve the addition of a sorbent to trap analytes for transfer to the GC rather than the direct transfer of the vapor phase. A sorbent may be placed directly into the sample vial, as in headspace solid-phase microextraction (HS-SPME) [21], stir-bar sorptive extraction (SBSE) [22], or single-drop microextraction (HS-SDME, HS-LPME) [23], in which a liquid drop is placed in the headspace, or the sorbent can be placed directly in the transfer line (HS-trap) [24]. HS-SPME is the most developed of these techniques, so it is discussed in more detail here; the other methods rely on similar principles.

Figure 9.2 gives a schematic presentation of an HS-SPME sampling system. Note the similarity to Figure 9.1 except that the system now includes the SPME fiber, exposed to the headspace. This generates a second equilibrium, besides the sample-vapor-phase equilibrium that must be considered: the vapor-phase–fiber equilibrium. By manipulating both equilibrium constants, HS-SPME has proven highly sensitive and versatile.

In HS-SPME, the analytes are partitioned between three phases, liquid phase, gaseous phase, and the fiber phase, so the mass balance is more complex. The liquid phase and the fiber coating are connected by the gaseous phase.

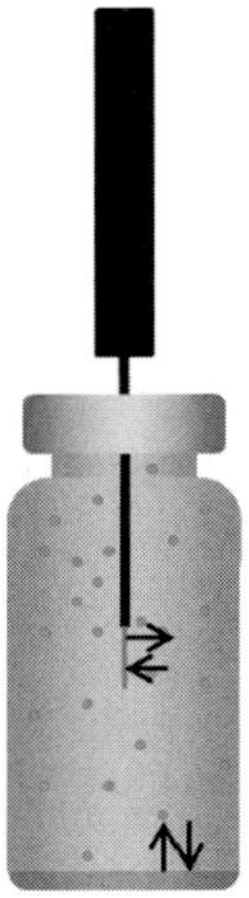

$$K_{FG} = \frac{[Analyte]_{fiber}}{[Analyte]_{headspace}}$$

$$K_{SG} = \frac{[Analyte]_{headspace}}{[Analyte]_{sample}}$$

FIGURE 9.2 Schematic of the system for static headspace extraction inside a sealed vial with a sorbent, an SPME fiber in this example, collecting analyte vapors from the headspace. Note the presence of two equilibria.

The amount of a volatile analyte distributed between these three phases is shown in Eqns (9.10) and (9.11):

$$C_o \times V_s = (C_s \times V_s) + (C_G \times V_G) + (C_F \times V_F) \quad (9.10)$$

$$A(sample) \leftrightarrow A(headspace) \leftrightarrow A(fiber) \quad (9.11)$$

where C_o is the concentration of the analyte in the original sample with volume V_S and C_S, C_G, and C_F are the concentrations of the analyte at equilibrium in liquid phase, gaseous phase, and fiber with volumes V_S, V_G, and V_F, respectively. As seen above, there are two sets of equilibrium constants, $K_{G/S}$ and $K_{F/G}$, which are involved in this system and can be written as follows:

$$K_{G/S} = C_G/C_S \quad \text{or} \quad C_S = C_G/K_{G/S} \quad (9.12)$$

$$K_{F/G} = C_F/C_G \quad \text{or} \quad C_G = C_F/K_{F/G} \quad (9.13)$$

Equations (9.10)–(9.13) can be combined and rearranged to give the mass of the analyte extracted into the fiber, shown in Eqn (9.14):

$$W_F = (C_o \times V_S \times V_F \times K_{F/G} \times K_{G/S})/[(K_{F/G} \times K_{G/S} \times V_F) + (K_{G/S} \times V_G) + V_G] \quad (9.14)$$

The mass of the analyte extracted into the fiber, which is then injected into the GC, is a function of nearly all variables that may occur in the vial. If the small fiber volume along with possible cases of $K_{G/S}$ is considered, Eqn (9.14) simplifies to two cases:

$$K_{G/S} \ll 1 : W_F = C_o(1/\beta)V_F K_{F/G} K_{G/S} \quad (9.15)$$

$$K_{G/S} \gg 1 : W_F = C_o(1/\beta)V_F K_{F/G} \quad (9.16)$$

When $K_{G/S}$ is large (low volatility or semivolatile analytes), the mass extracted is essentially related to the vapor–fiber partition coefficient which is likely to also be large; when $K_{G/S}$ is small (volatile analytes), the mass is related to both partition coefficients. For example, heating the vial may drive more analyte into the vapor phase from the sample, but reduce extraction from the vapor to the fiber. In any event, both equilibria should be considered during method development.

The other sorbent-based techniques work on the same principles. HS-SBSE is essentially the same as HS-SPME except that the stir-bar coating has a much larger volume than the SPME fiber (although still much smaller than the sample and vapor volumes), allowing a larger mass of the analyte to be extracted, although with slower kinetics. In HS-SDME and HS-LPME, a small drop of traditional liquid solvent is suspended in the headspace phase; therefore, extraction theory is almost exactly the same as for SPME, except that the liquid solvent is less viscous and has lower surface tension than the polymeric phases used in SPME, allowing for possibly faster equilibration. In HS-trap systems, analyte vapor is trapped on a sorbent following extraction in a manner similar to purge and trap, described next.

9.2.3. Dynamic Headspace Extraction (Purge and Trap)

Dynamic headspace extraction, also called "purge and trap," provides a means for exhaustive extraction of the analyte from the matrix. An inert gas is bubbled through a liquid sample, continuously evaporating and removing volatile analyte(s) form the less volatile solvent. This system does not reach equilibrium in the sample vial, as the evaporation process is continuous; however, the drive toward equilibrium generated by removing the vapor-phase analyte(s) and continuously refreshing the gas stream with fresh gas enables exhaustive extraction. The ability and time required to extract analytes are dependent on the equilibrium constant and the volume of gas passed through the sample. Following extraction, the flowing gas containing extracted analyte(s) is passed over a sorbent trap to collect them. They are subsequently desorbed into the inlet of a gas chromatograph.

As in SHE with a sorbent, purge and trap involves two steps driven by their equilibrium constants. In the purge step, an inert gas is passed through the sample to evaporate and carry volatile analytes into the vapor phase. In the second step, the gas is passed over a sorbent trap that extracts the analytes from the vapor phase. Most fundamental development in purge and trap has focused on developing sorbents that can extract a wide array of volatile analytes form the vapor phase. Often, a multilayer sorbent that consists of a combination of weak, moderate, and strong materials, such as Tenax, silica gel, and charcoal, respectively, is used.

9.3. INSTRUMENTATION AND PRACTICE

The basic principle underlying all instrumentation for headspace-gas chromatography is that an aliquot of the analyte from the vapor phase above a liquid or solid sample in a sealed vial or container must be reproducibly and effectively transferred to the inlet of a gas chromatograph. There are several means for accomplishing this, including: gas tight syringe, transfer line, sample loop and collection on a sorbent. The most important challenges are to ensure that the sample composition that reaches the gas chromatograph is truly representative of the composition of the headspace vapor in the vial and that the headspace vapor is representative of the composition of the original sample. Fundamentals of each instrumental approach are described in the following sections.

9.3.1. Gas Tight Syringe

Classically and most simply, headspace may be sampled using a gas tight syringe of appropriate volume. This is also the most convenient and inexpensive method for sampling vapors from nontraditional containers such as cans or bags. The syringe is inserted into the headspace and an aliquot of appropriate volume is removed. Although seemingly simple, there are several challenges associated with this method:

- If the syringe volume is similar to the headspace volume, equilibrium in the vial may be disrupted. In effect, insertion of the syringe and withdrawal of the syringe plunger cause an increase in the vapor-phase volume, disrupting equilibrium.
- If the syringe is at a lower temperature than the vial, analyte(s) or matrix components may condense in the syringe.
- The syringe volume and concentration of analytes may necessitate significant optimization of the injection conditions on the gas chromatograph.
- The syringe may require cleaning or purging with inert gas between each analysis.

Heated and/or purged syringes are available to reduce these challenges. Although reproducibility may be difficult to obtain, even using an automated system, a gas tight syringe is the least expensive and simplest way to begin using or to test the feasibility of a proposed static headspace extraction method.

9.3.2. Transfer-Line-Based Systems

The most common method for directly transferring an aliquot of the headspace within a vial to a gas chromatograph is by using a capillary transfer line. Because the gas chromatographic inlet is pressurized, the vial must be pressurized to a pressure higher than the inlet prior to sampling. The pressure drop from the vial to the inlet, the sampling time, and the transfer-line dimensions determine the amount of vapor that is transferred from the vial to the inlet. This also means that there are several steps in a transfer-line-based analysis that must be optimized during method development. A schematic diagram of these steps is shown in Figure 9.3. The same gas flow is used both for pressurizing the vial and as a carrier gas for the GC.

9.3.2.1. Thermostatting and Pressurizing

During the thermostatting phase, the sampling needle is in the upper position. The carrier gas flows through the solenoid valve V1 to the column; at the same time, the needle cylinder is purged by a small cross-flow vented through solenoid/needle valve V2. After thermostatting, the two phases attain equilibrium. At this stage, the sampling needle pierces the septum and the tip of the needle is positioned in the headspace of the vial. The carrier gas flows into the vial headspace, pressurizing the vial for a preset time, usually a few minutes, to achieve homogenization of the volatile analyte and the incoming carrier gas in the headspace of the vial.

9.3.2.2. Injection

After pressurization of the vial, the solenoid valves are closed, stopping the carrier gas flow from the headspace sampler. Now, the needle inside the headspace vial is opened up to the injection port of the GC system via the transfer

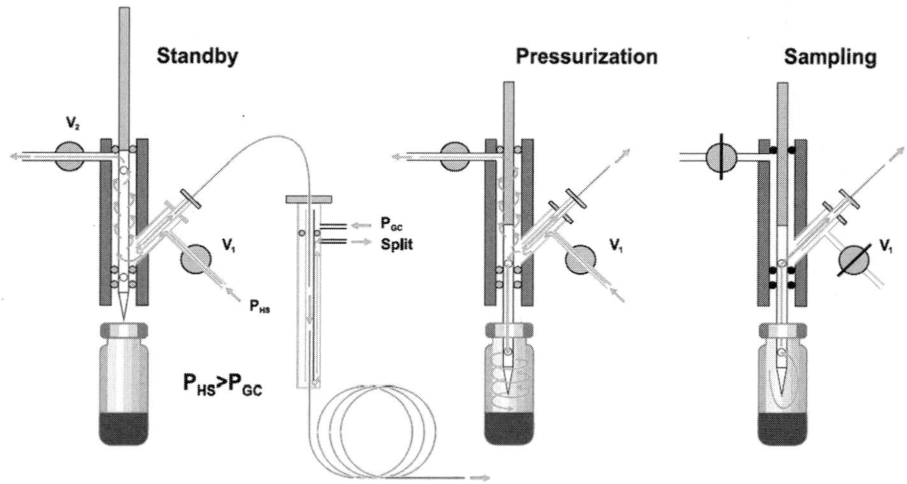

FIGURE 9.3 Steps in static headspace extraction with a transfer-line-based system. First, the vial is heated with no connection to the inlet (standby). Second, the vial is pressurized with carrier gas (pressurization). Third, the pressurized vial is opened to the transfer line and inlet (sampling). *Source: Courtesy Perkin Elmer Instruments.*

line. The pressurized gas expands and flows into the injection port through the transfer line and then to the column for a preselected injection time. At the end of the injection time, the solenoid valves are opened, preventing any of the vial headspace from being additionally transferred through the needle.

This technique is highly reproducible and uses the injection time and pressure to determine the volume of the sample transferred to the GC, so these are generally electronically controlled. It should be noted that when the headspace vial is pressurized during the pressurization mode, the volatile analyte in the gaseous phase is diluted by the addition of the carrier gas and also the equilibration achieved after thermostatting the vial gets disturbed. It is possible that a small amount of the analyte might move back to the liquid phase, but the effects on quantitation are negligible as these factors are constant throughout multiple injections. Since pressurization of the vial following equilibration may disrupt the equilibrium, this type of SHE is not the best for determining physical constants such as the partition coefficients themselves.

9.3.3. Sample-Loop System

The sample-loop system is another injection mode used in SHE and MHE, depicted schematically in three steps in Figure 9.4. This system encompasses a sample loop connected to the headspace vial and transfer line to the GC system through a six-port valve. The headspace sample vial is pressurized to a preset value after thermostatting the vial. The vial is then opened to the loop for a short time. The sample thus collected in the loop is swept by the carrier gas to GC injection port through a transfer line. Figure 9.4(a)–(c) depict the various modes in the loop sampling system.

During pressurization, the heated needle pierces through the septum of the headspace vial and is positioned in the headspace of the vial. The vial is pressurized using the carrier gas for a preset time in order to obtain a uniform mixture of volatile components in the headspace. After pressurization, the headspace of the vial is opened through the needle to the heated sample loop through the six-port valve. The sample from the headspace of the vial moves into the sample loop and purges the loop. The sample loop is opened to the atmosphere during filling and then closed to allow it to reach equilibrium. During injection, the sample collected in the loop is swept into the inlet through the heated transfer line for a preset duration. After the injection time, the system goes back to a standby condition in which the carrier gas bypasses the loop and flows to the inlet through the transfer line.

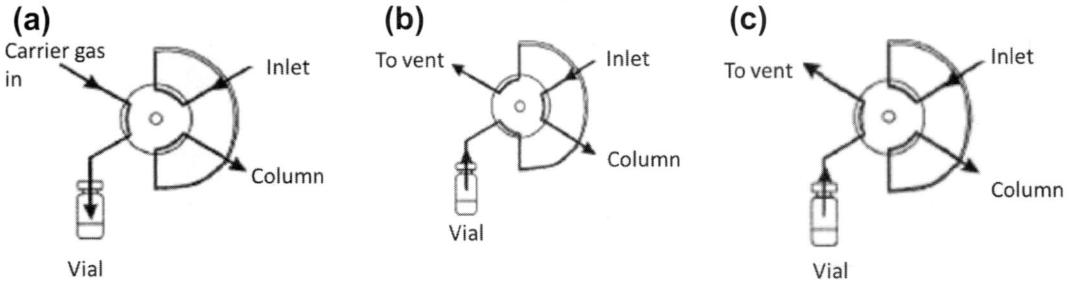

FIGURE 9.4 Steps in static headspace extraction with a sample-loop-based system. (a) The vial is pressurized with carrier gas while not connected to the sample loop. (b) The vial is connected to the sample loop to fill the loop while the loop is vented. (c) The loop is connected to the column.

9.3.4. Purge and Trap

Like SHE, purge and trap relies on the volatility of the analytes to achieve extraction and release from the matrix. However, the volatile analytes and matrix are not allowed to reach a state of equilibrium. This is accomplished by continually sweeping the carrier gas across the headspace of the sample matrix, thus providing a continuous concentration gradient, which aids in the more exhaustive extraction of the analytes by the Le Chatelier's principle. Once in the carrier gas, the analytes are swept from the vial and trapped on a sorbent prior to analytical analysis. Figure 9.5 shows a schematic representation of a typical purge-and-trap system. The carrier gas is passed through the sample in a vial and is vented to the trap using a six-port valve. This allows control over variables such as the time and flow rate of the gas. The valve is then switched to desorb the trap into the GC. Note that this is a multistep process: first analytes are evaporated from the matrix; they are then adsorbed on the trapping material and finally desorbed into the gas chromatograph.

There are several methods of analyte trapping: cryogenic, sorbent, and column focusing. Each of these methods has advantages and disadvantages. In general, the trap should do the following: retain the analytes of interest, not introduce impurities, and allow rapid injection of the analytes to the column. Once the analytes are sorbed on the trap, they must be desorbed into the gas chromatograph. This is most often accomplished by passing the carrier gas over the trap, combined with heating to facilitate desorption.

9.3.5. SPME, SBSE, and SDME

The instrumentation for SPME, SBSE, and SDME is very straightforward. In SPME, a coated fused-silica fiber is contained within a syringe needle as seen in Figure 9.2. The fiber is retracted into the needle for storage and transport and is extended from the needle for extraction in a vial and desorption into a gas chromatograph. SPME can be operated in both manual and automated instruments. In SBSE,

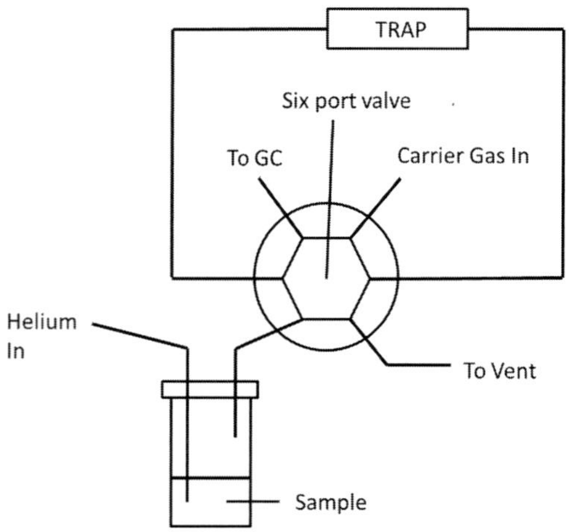

FIGURE 9.5 Schematic of a purge-and-trap system. First, carrier gas is passed through a liquid sample in a sealed vial and passed over the trap. Second, the six-port valve is moved and the trap is connected to the inlet.

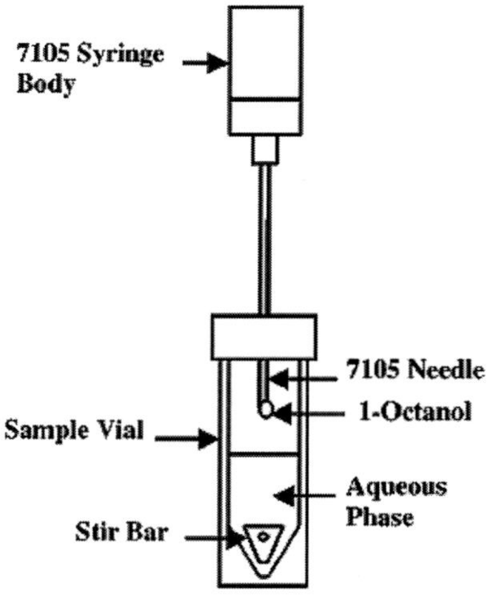

FIGURE 9.6 Schematic of a headspace-SDME system with the liquid drop suspended in the headspace of the vial. The drop would be retracted into the needle for transfer to the gas chromatograph. *Source: Reproduced with permission from Ref. [25]. Copyright 2001, American Chemical Society.*

the coated stir bar is suspended in the headspace, while in SDME, a liquid drop is suspended from the tip of a standard microsyringe needle, as seen in Figure 9.6, which shows a schematic of an HS-SDME setup with a drop of 1-octaonol suspended in the headspace of a glass vial [25].

9.4. METHOD DEVELOPMENT CONSIDERATIONS

Method development for headspace extractions is based on similar considerations for all analytical methods. There are compromises between sensitivity, resolution, and ease of use that must be made. In all cases, the mass of the analyte that can be extracted is governed by initial concentration in the original sample, the partition coefficient, and the phase ratio, as shown in Eqn (9.3). As seen in the instrumentation discussion, the simplest instruments, such as a gas tight syringe, may be the most difficult to obtain figures of merit such as adequate reproducibility and wide linear range. In each case, method development efforts focus on driving analytes into the vapor phase within a sealed vial, as seen in Figures 9.1 and 9.2.

9.4.1. Effect of Vial Temperature

In general, the relationship between solution–vapor equilibrium constant and temperature is given by classical thermodynamics, with K increasing exponentially with temperature. However, as written in Eqn (9.5) for SHE, K decreases with temperature. According to Eqn (9.3), the effect of a change in temperature on the sensitivity of an analyte depends on the sum of the partition coefficient (K) and phase ratio (β). This generates three cases relevant to developing methods:

- $K \ll \beta$: In this case, a change in K has little effect on Eqn (9.3); therefore a change in T has little effect.
- $K = \beta$: In this case, a change in K will be somewhat offset by not changing β, but there will be an effect from changing T. Increased T will lower K and increase A.
- $K \gg \beta$: In this case, a change in T may have a dramatic effect on sensitivity. Increased T will lower K and increase A.

9.4.2. Effect of Sample Volume

Changing the sample volume would result in change in the phase ratio (β). Again, the combination of phase ratio and equilibrium constant determines the effect of changing sample volume and phase ratio. As in the effect of temperature, the effect of sample volume can be illustrated in three cases:

- $K \ll \beta$: In this case, the change in b dominates; more sample means lower β and more sensitivity.
- $K = \beta$: In this case, more sample will generate more sensitivity, but not as dramatically.
- $K \gg \beta$: In this case, additional sample volume has little effect.

9.4.3. Effect of Sample Solvent

Selecting the right solvent or solvent mixture can be a challenging method development step. The ideal solvent would dissolve the analytes and matrix components but not with such high solubility that it hinders vaporization. It also would be nonvolatile and not react with sample components. These requirements often lead to unusual solvent choices, such as the use of dimethyl sulfoxide (DMSO), *N,N*-dimethyl acetamide (DMAC), and water in pharmaceutical headspace methods. For example, ethanol has a partition coefficient (K) of ~7000 at room temperature in water, but benzene which is not soluble in water has a partition coefficient (K) of only 7 in water, although the boiling points are similar. It is important to know as much as possible about the solubility of analytes and matrix components in the solvent. In addition to traditional solvents, there has been recent interest in the use of ionic liquids as diluents for SHE [26]. An ionic liquid is an organic salt that is a liquid at room temperature. Many ionic liquids can dissolve a wide range of materials and have very low vapor pressure, making them ideal solvents for SHE.

Beyond the solvent choice, there are several methods for reducing *K*. Temperature increases, as discussed above, should be used carefully, as increasing the temperature close to or above the boiling point of the solvent increases the amount of solvent vapor in the headspace of the vial which in turn can flood the inlet with solvent. This is one reason for the use of DMSO and DMAC, mentioned above, with boiling points of 189 °C and 165 °C, respectively. By using solvents with higher boiling points, the thermostatting temperature can be increased without generating much vapor pressure from the solvent and the sensitivity of the analytes may be increased. One drawback in using these solvents at elevated temperature is that any residual volatile impurities present in these solvents will also be detected and may interfere with the chromatography.

Salting out is also commonly used to assist in driving analytes into the vapor phase. This is the addition of a large (often 1 M or more) concentration of a common salt such as sodium or potassium chloride to the samples. Addition of the salt is believed to disrupt hydration of the analytes by the solvent, making them more available to vaporize and reducing *K*. There are many examples of salting out being used in the literature, but there is little systematic information on how to use it with specific analyte and solvent combinations. Salting-out conditions must be studied by experimentation.

9.5. CONCLUSIONS

Headspace extraction combined with gas chromatography is a powerful separation and analysis technique. While the theory of headspace extraction is straightforward, there are subtleties that often make method development challenging. Static headspace extraction techniques including SHE, HS-SPME, HS-SBSE, and HS-SDME are all equilibrium-based techniques which do not generally perform exhaustive extraction. For exhaustive extraction, MHE and purge and trap are used. Instrumentation for all of these techniques is generally automated, although simple sampling with a gas tight syringe remains an excellent choice for method feasibility testing. Factors such as vial temperature, solvent choice, phase ratio, and instrumental conditions may have a profound effect on analytical results.

Acknowledgments

The authors gratefully acknowledge the Sanofi Aventis Foundation for providing funding to the Center for Academic Industry Partnership. Perkin Elmer Instruments has provided instrumental and technical support to our headspace work through an ongoing collaboration.

This chapter is dedicated to the memory of Dr. Leslie S. Ettre, who may be considered as the father of modern headspace extraction and analysis. His insight, dedication, and humor reflect a spirit that we may hope to continue.

References

[1] N.H. Snow, G.C. Slack, Headspace analysis in modern gas chromatography, TrAc Trends Anal. Chem. 21 (2002) 608–617.
[2] S. Mitra (Ed.), Sample preparation in analytical chemistry, John Wiley and Sons, New York, 2003.
[3] B. Kolb, Chromatography: gas: headspace gas chromatography, encyclopedia of separation science, Elsevier, 2007, pp. 489–496.
[4] J.D. Green, Headspace analysis: static, encyclopedia of analytical science, Elsevier, 2005. 229–236.
[5] N.H. Snow, G. Bullock, Novel techniques for enhancing sensitivity in static headspace extraction gas chromatography, J. Chromatogr. A 1217 (2010) 2726–2735.

[6] N. Delaunay-Bertoncini, F.W.M. van der Wielen, P. de Voogt, B. Erlandsson, P.J. Schoenmakers, Analysis of low-molar-mass materials in commercial rubber samples by Soxhlet and headspace extractions followed by GC–MS analysis, J. Chromatogr. A 35 (2004) 1059–1073.

[7] M. Mestres, O. Busto, J. Guasch, Chromatographic analysis of volatile sulphur compounds in wines using the static headspace technique with flame photometric detection, J. Chromatogr. A 773 (1997) 261–269.

[8] R.G. Gullberg, Alcohol: breath alcohol analysis, encyclopedia of forensic and legal medicine, Elsevier, 2005. 21–29.

[9] H. van der Jagt, H.F. Reijnders, Water analysis: overview. Encyclopedia of analytical science (2005), pp. 233–252.

[10] F. Augusto, A.L. e Lopes, C. Zini, Sampling and sample preparation for analysis of aromas and fragrances, TrAc Trends Anal. Chem. 22 (2003) 160–169.

[11] General chapter <467> Residual solvents, USP 34-NF 29, United States Pharmacopeia, 2011.

[12] L. Dai, A. Quiroga, K. Zhang, H.B. Runes, D.T. Yazzie, K. Mistry, et al., LC-GC North America 28 (2010) 54–66.

[13] B. Kolb, L.S. Ettre, Static headspace-gas chromatography theory and practice, second ed., John Wiley and Sons, 2006.

[14] B.V. Ioffe, A.G. Vitenberg, Headspace analysis and related methods in gas chromatography, John Wiley and Sons, 1984. Translated by I.A. Mamantov.

[15] Translated by D. Verdin H. Hachenberg, A.P. Schmidt, Gas chromatographic headspace analysis, Hayden and Son, 1977.

[16] R.L. Grob, E.F. Barry (Eds.), Modern practice of gas chromatography, fourth ed., John Wiley and Sons, 2003.

[17] H.M. McNair, J.M. Miller, Basic gas chromatography, second ed., John Wiley and Sons, 2009.

[18] http://www.chromedia.org [accessed July, 2011].

[19] http://www.chromatographyonline.com [accessed July 2011].

[20] http://www.separationsnow.com [accessed July 2011].

[21] D.A. Lambropoulou, I.K. Konstantinou, T.A. Albanis, Recent developments in headspace microextraction techniques for the analysis of environmental contaminants in different matrices, J. Chromatogr. A 1152 (2007) 70–96.

[22] F. David, P. Sandra, Stir bar sorptive extraction for trace analysis, J. Chromatogr. A 1152 (2007) 54–69.

[23] M.A. Jeannot, A. Przyjaznyb, J.M. Kokosac, Single drop microextraction—development, applications and future trends, J. Chromatogr. A 1217 (2010) 2326–2336.

[24] K. Schulz, J. Dreßler, E. Sohnius, D.W. Lachenmeier, Determination of volatile constituents in spirits using headspace trap technology, J. Chromatogr. A 1145 (2007) 204–209.

[25] A.L. Theis, A.J. Waldack, S.M. Hansen, M.A. Jeannot, Headspace solvent microextraction, Anal. Chem. 73 (2001) 5651–5654.

[26] G. Laus, M. Andreb, G. Bentivoglio, H. Schottenberger, Ionic liquids as superior solvents for headspace gas chromatography of residual solvents with very low vapor pressure, relevant for pharmaceutical final dosage forms, J. Chromatogr. A 1216 (2009) 6020.

CHAPTER

10

Thermal Desorption for Gas Chromatography

Elizabeth Woolfenden

OUTLINE

Gas Chromatography DOI: 10.1016/B978-0-12-385540-4.00010-9

10.1. GENERAL INTRODUCTION TO THERMAL DESORPTION

Thermal desorption (TD) is arguably the most powerful and versatile of all sample introduction technologies for gas chromatography (GC). It is readily automated and serves to combine sampling/sample preparation, selective concentration, and efficient GC injection in one fully automated procedure. It is compatible with sampling and analysis of gas- (vapor-) phase organics trapped on sorbent media, allowing concentration factors up to 10^6 to be comfortably achieved. It can also be used for direct gas extraction of volatiles and semivolatiles from solid or liquid matrices.

Thermal desorption is generally, but not exclusively, used with GC or GC/mass spectrometer (GC-MS) technology. Alternative vapor-phase analytical options include process mass spectrometry and sensors ('Enose' technology). TD also provides the basis for many other GC sampling procedures — most notably purge and trap, sorptive extraction, some forms of large-volume injection, and head-space trap.

The development of thermal desorption was fundamentally driven by the limitations and complexity of conventional GC sample preparation methods – liquid extraction, steam distillation, etc. It held out the promise of an alternative high-sensitivity/solvent-free gas extraction process that could be fully automated.

10.1.1. General Principles

In its simplest form, thermal desorption is a straightforward extension of gas chromatography. It involves heating sample materials or sorbents in a flow of inert 'carrier' gas, such that retained organic volatiles and semi-volatiles 'desorb' and are transferred or injected into the GC column in the carrier gas stream. As in GC, key method parameters include temperature, carrier gas flow rate, desorption time, and sorbent (stationary phase) selection.

However, as soon as you move away from the simplest form of TD, the power and potential of the technique expand rapidly. It is possible, for example, to configure TD technology in multiple stages such that analytes are repeatedly extracted/desorbed into smaller and smaller volumes of gas, thus concentrating the compounds of interest and enhancing sensitivity/detection limits – see Figure 10.1. Figure 10.1 illustrates a relatively conventional monitoring procedure whereby 100 L of air or sample gas are pumped through a sorbent sampling tube over a period of say 12 h. Retained vapors are then desorbed in approximately 100 ml of carrier gas and subsequently refocused on a smaller ('cold') sorbent trap. Depending on the trap design, this in turn can be quantitatively desorbed, with the analytes eluting in as little as 100 μl of gas, thus providing a million-fold enhancement in vapor concentration overall.

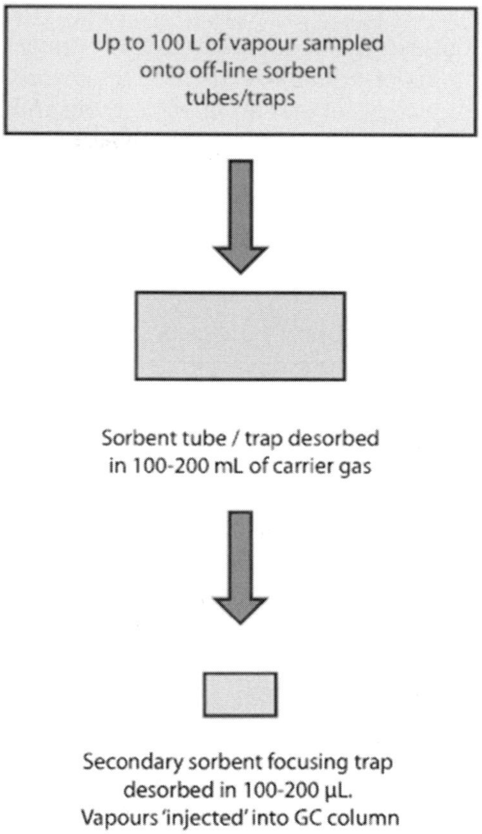

FIGURE 10.1 An illustration of the concentration potential of multistage thermal desorption.

Another benefit of thermal desorption is that it is often possible to quantitatively retain target compounds during one or more of the trapping stages while unwanted interferents, such as water and/or permanent gases, are selectively purged to vent. This allows compounds of interest to be transferred/injected into the GC analytical column with minimal interference.

10.1.2. Comparing Thermal Desorption with Solvent Extraction

The first and most obvious advantage of thermal desorption *versus* solvent extraction is that it is possible to transfer 100% of retained analytes to the analytical system, whereas solvent extraction generally involves microliter injections of milliliter extracts. This means that TD typically offers at least a thousand-fold enhancement in sensitivity *versus* equivalent solvent extraction procedures for volatile organic compounds (VOCs). Other significant advantages are discussed in the following.

10.1.2.1. Enhanced Desorption Efficiency

Assuming appropriate selection of sampling and analytical conditions (sorbent, temperature, and flow), it is usually straightforward for TD methods to exceed 95% desorption efficiency [1]: In other words, for retained compounds to be stripped completely from the sorbent tube or trap and transferred quantitatively to the analytical system. This is because TD is a dynamic process, with gas continually purging compounds away from the sorbent or sample matrix as soon as they are released into the vapor-phase by the rising temperature. In contrast, typical solvent extraction procedures are static, with analytes partitioning between the sorbent, solvent, and vapor (headspace) phases. This limits desorption efficiency, and standard methods for solvent extraction therefore typically attain only 75% recovery [2].

10.1.2.2. Reliable Extraction Efficiency

Static partitioning systems such as most solvent extraction procedures are also subject to increased variability of analyte recovery depending on the nature of the compounds of interest and the presence of interferences. The desorption efficiency of methods specifying charcoal sample tubes with CS_2 extraction, for example, has been shown to fall as low as 20 or 30% for polar compounds in the presence of water [3]. This uncertainty is particularly problematic for air-monitoring methods or measurements of industrial VOC emissions, as the analyst may not be aware of field/sample conditions such as high water content; moreover, poor recovery may lead to significant under-reporting.

10.1.2.3. Enhanced Automation

Thermal desorption is inherently less labor-intensive, requiring little or nothing in the way of manual sample preparation.

10.1.2.4. Reduced Analytical Interference

Solvent interference can be a major consideration for liquid extraction methods. One of the reasons CS_2 was originally selected as the preferred solvent for many charcoal-based air-sampling methods was that it gives little or no signal on a GC flame ionization detector. However, nowadays, with the preference for MS detection, this advantage no longer holds. Common solvent interference issues include masking of peaks of interest, signal quenching (for components coeluting with the solvent), and baseline disturbances. All these make peak integration difficult and more prone to error. TD is inherently free of solvent interference.

10.1.2.5. Selective Elimination of Interferents

Depending on the volatility of the compounds of interest, thermal desorption often facilitates selective purging of sample interferences such as water or ethanol prior to analysis. Applications as diverse as monitoring VOC emissions from paint and characterizing the aroma of whisky benefit from the selection of sorbents which quantitatively retain compounds of interest while allowing water and, in the latter case, ethanol, to purge to vent. Selectivity is usually only possible for solvent extraction procedures when there is a very significant volatility difference between the compounds of interest and the interferences.

10.1.2.6. Reduced Exposure Risk

Many common extraction solvents, such as CS_2, are toxic, odorous, and present a significant potential health and safety hazard. Thermal desorption is inherently safer in this respect. TD–GC(MS) systems can generally be installed without fume hoods or other extraction equipment, provided all outlet points, including sample split lines, are configured with appropriate filters. In TD operation, wet chemistry procedures are confined to the preparation of liquid standards for spiking tubes when gas standards are not available – see the section below on calibration.

10.1.2.7. Reusable Samplers and Lower Cost Per Analysis

Solvent extraction invariably involves destruction of the sorbent sampler used for vapor collection. For example, the charcoal tubes traditionally used for industrial hygiene comprise glass tubes with drawn/sealed ends that are broken during sampling and analysis. Typically, prepacked TD tubes (glass, stainless, or coated steel) are in the order of 10 times more expensive than charcoal tubes, but can then be reused at least a hundred times. They are also automatically cleaned by the TD process. This reduces the sampling costs of TD methods to roughly a tenth of equivalent solvent extraction procedures.

10.2. BRIEF HISTORY OF THERMAL DESORPTION – ESSENTIAL FUNCTIONS AND PERFORMANCE CHARACTERISTICS

The history of thermal desorption can be traced back to the mid-1970s. Scientists struggling with the limitations of conventional GC sample preparation and extraction techniques began to experiment by packing conventional GC injector liners with sorbent material. These sorbent-packed injector liners were used to sample a fixed volume of air or gas and were then quickly inserted back into the GC inlet for desorption and transfer of analytes to the analytical column. The limitations of these primitive adaptations of conventional GC injectors are many and obvious (air ingress, volatile losses, variability, contamination from the outer surfaces of the liner, single stage, etc.), but the fact that it was attempted at all is testament to the fundamental need for this technology.

Another early incarnation of thermal desorption was in purge-and-trap technology. The US Environmental Protection Agency (EPA) rushed to develop purge-and-trap/GC-based test methods to measure VOCs in drinking water in the late 1970s in response to serious environmental incidents such as the Love Canal tragedy. Love Canal refers to an area of housing in Niagara City in New York State, which had been developed on land contaminated by the chemical industry in the 1940s and 1950s. Work to develop test methods began in earnest after an unusually high incidence of serious birth defects was reported in the neighborhood. Ultimately, it was found that underground chemical waste had been seeping into the drinking-water supply and causing a number of serious human health concerns. The 500-series methods, produced by EPA in response to this and similar issues, relied on volatiles being purged/sparged from the water in a stream of pure nitrogen and trapped on a tube packed with sorbents. This was subsequently heated in a reverse stream of carrier gas to thermally desorb the organic chemicals of interest and transfer them to the GC analytical system in a standard TD-type procedure.

The first early commercial configurations of dedicated general purpose TD technology were invariably based around desorption of a single tube or badge. The 'Coker cooker', designed by Environmental Monitoring Systems Ltd. (UK) in the mid-1970s [4], was a popular example and accommodated samples or sorbents contained in ¼-inch O.D. tubes. These early desorbers were very primitive by modern standards, typically offering only single-stage desorption and without any of the functions that would now be regarded as standard such as leak testing or prepurging of air from the tube. However, they usually operated sufficiently well, within specific constraints (e.g. packed column only, stable compounds only, and narrow concentration-and-volatility ranges) to provide a useful tool for straightforward applications such as workplace air monitoring in the petrochemical industry [4].

Perhaps the most important early technical breakthroughs came from 'Working Group 5' (WG5) of the UK Health and Safety Executive's (HSE) 'Committee on Analytical Requirements' (CAR). HSE/CAR WG5 began with a chance meeting at a conference on workplace air monitoring in the late 1970s. Scientists including Dr Kevin Saunders (then of BP), Dr Richard Brown (then of HSE), and Jack Charlton and Brian Miller (then of ICI) found that they had a common interest in both diffusive (passive) sampling and thermal desorption.

This group believed that passive (diffusive) sampling would allow quantitative air monitoring without the complications and expense of personal sampling pumps [5]. At a series of meetings over the next couple of years, various other experts joined the team including Peter Hollingdale-Smith of Porton Down [6], David Coker of Exxon, and Nico van den Hoed of Shell. Between them the working group evaluated the available forms of diffusive monitor, and homed in on axial samplers, based on the ¼-inch O.D. sorbent tubes used in the 'Coker cooker'. These were identified as a practical size, being least susceptible to air-speed limitations [5,7] and having the most flexibility (notably they are suitable for both passive and pumped sampling).

WG5 saw thermal desorption as an enabling technology for passive sampling because it offered approximately 1000 times better sensitivity than solvent extraction which was more than enough to offset the slow sampling (uptake) rate of axial-form diffusive samplers (typically around 1 ml/min). They also realized that TD overcame the toxicity and variability issues inherent in the charcoal/CS_2 extraction methods in use at that time (see above). As soon as an agreement was reached on the sampler, WG5 therefore set about outlining a specification for the world's first automated thermal desorber to accommodate it. The TD functionality requirements that came out of these discussions in the late 1970s are still relevant today. They include a two-stage desorption, the necessity of certain predesorption checks (stringent leak testing and prepurging of air to vent), and automation.

10.2.1. Two-Stage Desorption

With standard sampling tubes containing 100 mg to 1 g sorbent (depending on density), WG5 realized that single-stage thermal desorption (Figure 10.2) was inherently limited – tens of milliliters of gas are required for complete extraction of a standard tube. Volumes like these are not compatible with capillary chromatography and compromise analytical resolution/sensitivity even with packed columns.

Initial attempts to address this issue utilized capillary cryofocusing (either on-column or in cooled GC inlets such as 'programmable temperature vaporizers' – PTVs) positioned

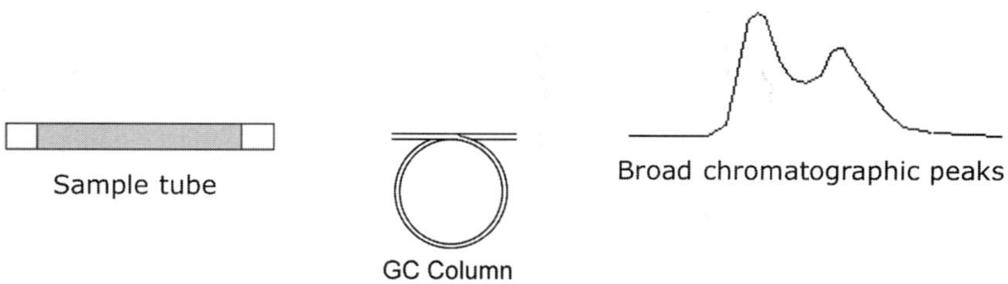

FIGURE 10.2 Single-stage thermal desorption.

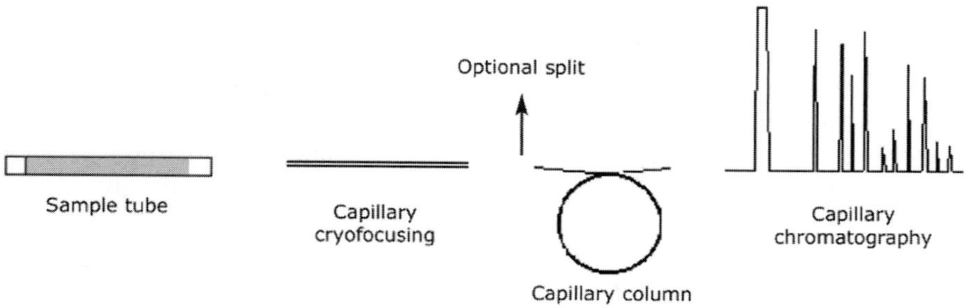

FIGURE 10.3 Two-stage thermal desorption incorporating capillary cryofocusing.

in between the sorbent tube and analytical system – see Figure 10.3. In this case, analytes desorbed from the primary sorbent tube were refocused/concentrated in a short length of capillary or narrow-bore tubing (typically 1 mm internal diameter or less) cooled with liquid cryogen. Heat was then applied to release the compounds into the analytical system in a small volume of carrier gas.

Early TD configurations which harnessed capillary cryofocusing included the Chrompack CTC unit designed in the Netherlands. The peak shape produced by this system was excellent even under splitless conditions and it therefore offered exceptional sensitivity. However, the limitations of capillary cryofocusing quickly became apparent. Key concerns included ice blockage, incomplete retention of very volatile compounds [8], loss of high boilers due to aerosol formation [9], and high running costs (systems consumed up to 6 L of liquid nitrogen per hour in operation) [10]. More importantly, because capillary cryofocusing devices were connected directly to the GC column, it made it difficult to implement the essential predesorption checks specified by WG5.

Ultimately, the PerkinElmer ATD 50 unit, developed around the WG5 specification, addressed all these issues. Designed by Dr Peter Higham and introduced in 1981, the ATD 50 incorporated a small, Peltier- (electrically)cooled, sorbent-packed focusing trap – see Figure 10.4. The combination of sorbent packing and modest focusing temperatures (minimum: −30 °C) was a real breakthrough. It offered quantitative retention of a wide range of compounds including

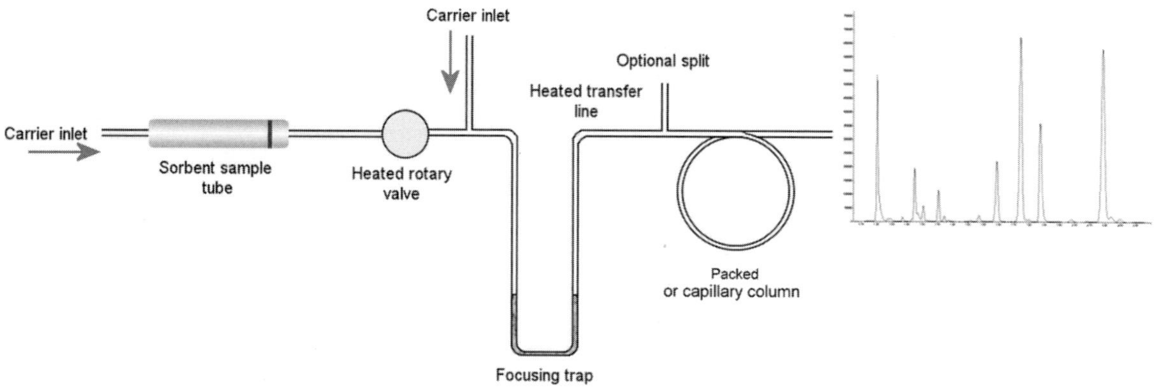

FIGURE 10.4 Two-stage thermal desorption incorporating sorbent focusing trap and heated valve.

very volatile species, such as the lightest gasoline components [11], SF_6 and N_2O [12], without the cost and inconvenience of liquid cryogen. The steel focusing trap also had a wide enough internal diameter (~3 mm) to prevent ice plug formation, yet could be heated at rates approaching 60 °C/s to allow rapid (capillary-compatible) desorption/injection with minimal split and good sensitivity. Another breakthrough was the inclusion of a rotary valve in the flow path of the desorber – see Figure 10.4. This isolated the sorbent tube from the GC, allowing both stringent 'stop-flow' leak testing and prepurging of air to vent prior to desorption of every tube.

10.2.2. Automation

Automating thermal desorption requires sealing sample tubes both before and after desorption to minimize the risk of artifact ingress from laboratory air and to prevent loss of analytes over periods of say 1 or 2 days. Caps are essential because sample losses from poorly sealed tubes can be very significant. In the end, this issue was addressed on the ATD 50 by using stainless steel end caps which incorporated a ball valve. They were not ideal, involving several o-rings in direct contact with the sample flow path, but they could remain on the tubes throughout an entire sequence (no uncapping/recapping required) and provided a robust solution for the time.

Introduction of the ATD 50 led to a rapid expansion of TD–GC applications. Over and above personal exposure assessment [11–13], these included residual solvents in drugs [14], and ambient air monitoring [15–17] – see in more detail below.

10.2.3. Double Splitting

Double splitting was not part of the original WG5 specification, but was introduced as an enhancement to the ATD 50 in around 1985 (Figure 10.5). It allows the transfer of analytes from the tube to the trap to be carried out split or splitless and likewise the subsequent transfer/injection of analytes from the secondary (focusing) trap to the GC analytical column.

Double splitting brought with it some significant benefits in terms of application versatility. Overall split ratios as high as 10,000:1 could accommodate milligrams of individual analytes while ng- or pg-level samples could still be analyzed with negligible split or no split at all. While it might seem strange to want to implement such a high split ratio for thermal desorption applications, it must be remembered that sensitivity is not the only advantage that TD

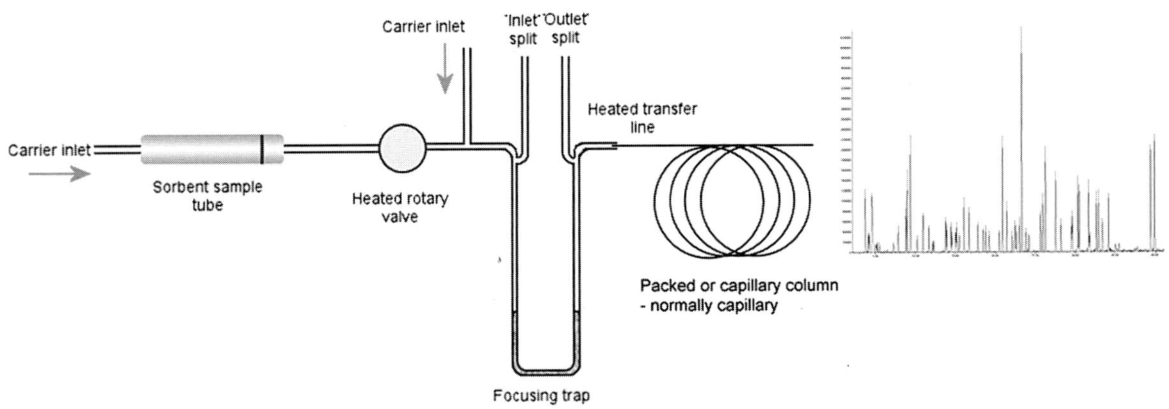

FIGURE 10.5 Two-stage thermal desorption, as shown in Figure 4, but incorporating double splitting capability.

offers – many laboratories simply preferred to use thermal desorption because it eliminated the hazard of CS_2 and was less labor-intensive. A surprising number of workplace air or industrial emission monitoring applications benefit from (or even require) a significant split ratio. For example, when sampling compound X (nominal molecular weight 100) at 50 ml/min over a full 8-hour shift (sample volume 24 L), a vapor concentration of 5 ppm would mean collecting ~500 μg of X. An overall split ratio of at least 1000:1 would be advisable in this case in order to prevent overload of high-resolution capillary GC columns and detectors.

Over and above application flexibility, the introduction of double splitting also enhanced operation of the two-stage thermal desorption process itself. Analytical objectives during primary (tube) desorption are complete removal (extraction) of retained vapors from the sample tube combined with quantitative trapping of the compounds of interest on the secondary (focusing) trap. If the sample is sufficiently large to allow implementation of a split during primary desorption, this benefits both objectives – it allows application of a relatively high carrier gas flow through the hot sample tube during desorption while at the same

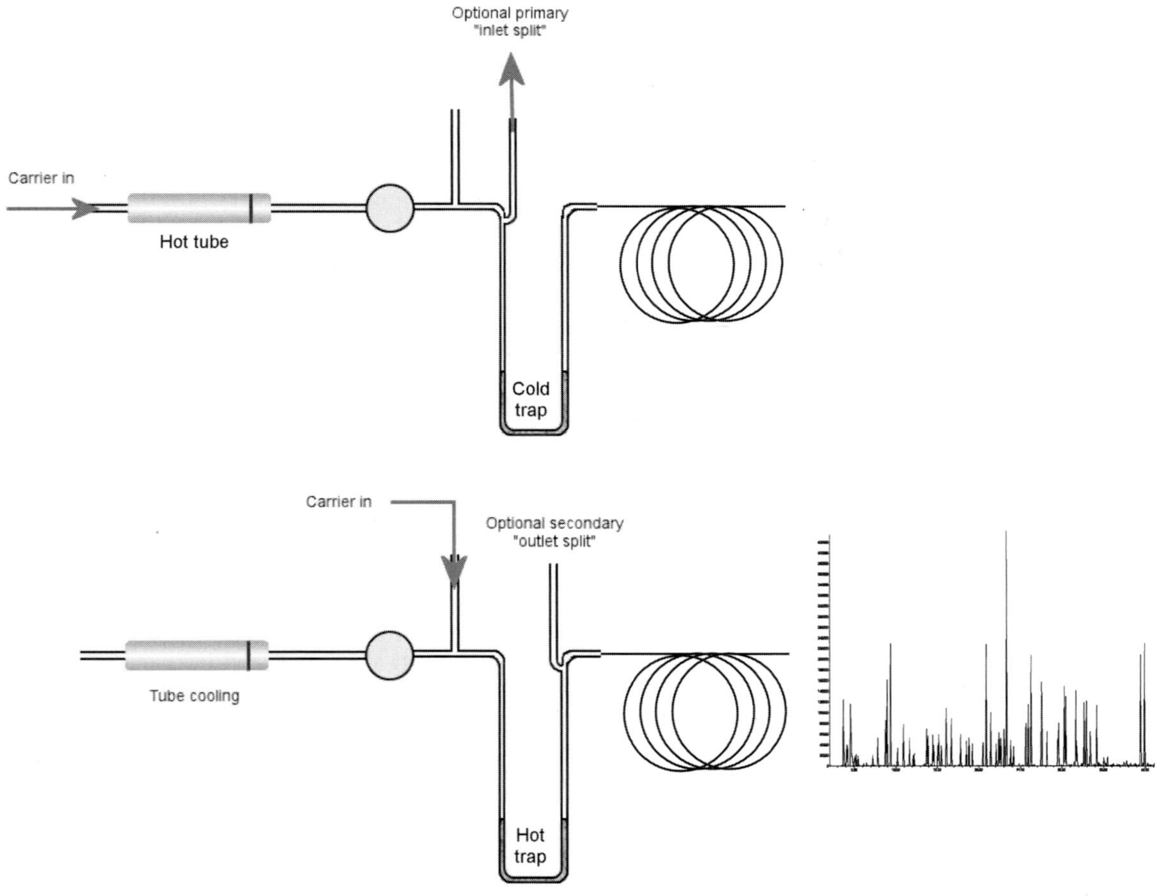

FIGURE 10.6 An illustration of gas flow during two stages of desorption with double splitting.

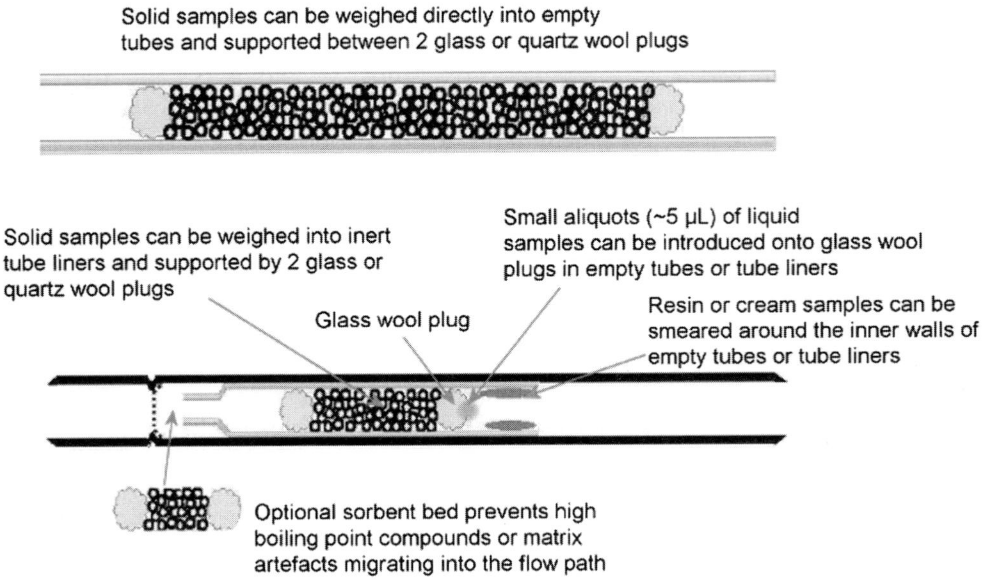

FIGURE 10.7 An illustration of sample preparation for direct desorption of materials.

time allowing a low flow to be maintained through the focusing trap, thus aiding analyte retention. Application of a second split during subsequent trap desorption then increases the gas flow through the trap when it is being heated and desorbed – Figure 10.6.

Implementation of double splitting enabled the first serious expansion of TD into direct desorption of materials – Figure 10.7. Relevant application examples include solvents in paint, residual monomer in polymer, essential oils, and volatiles in dried vegetable products such as tobacco or spices [14,18,19] – Figure 10.8.

10.3. THE EVOLUTION OF TD TECHNOLOGY

10.3.1. The Limitations of Early Systems

While the ATD 50 represented the absolute state of the art for TD instrumentation in its day, new thermal desorption application requirements began to emerge which highlighted the limitations of this first-generation technology. Key concerns included the following:

- Forward-flow trap desorption which meant analytes had to pass through the entire sorbent bed. This limited the volatility range of components that could be analyzed simultaneously.
- Flow path and desorption temperatures: The ATD 50 operated with a flow path maximum of 150 °C and maximum desorption temperatures of 250 °C and 300 °C for the tube and trap, respectively. This was hot enough to allow complete recovery of compounds up to n-C_{26} (b.p. ~400 °C) [20]. However, interest was already growing in measuring the vapor concentration of higher-boiling semivolatiles such as PCBs, phthalates, and multiring PAHs.
- Inertness: the predominantly stainless steel flow path of early systems, such as the ATD 50, caused degradation of the most reactive VOC species.

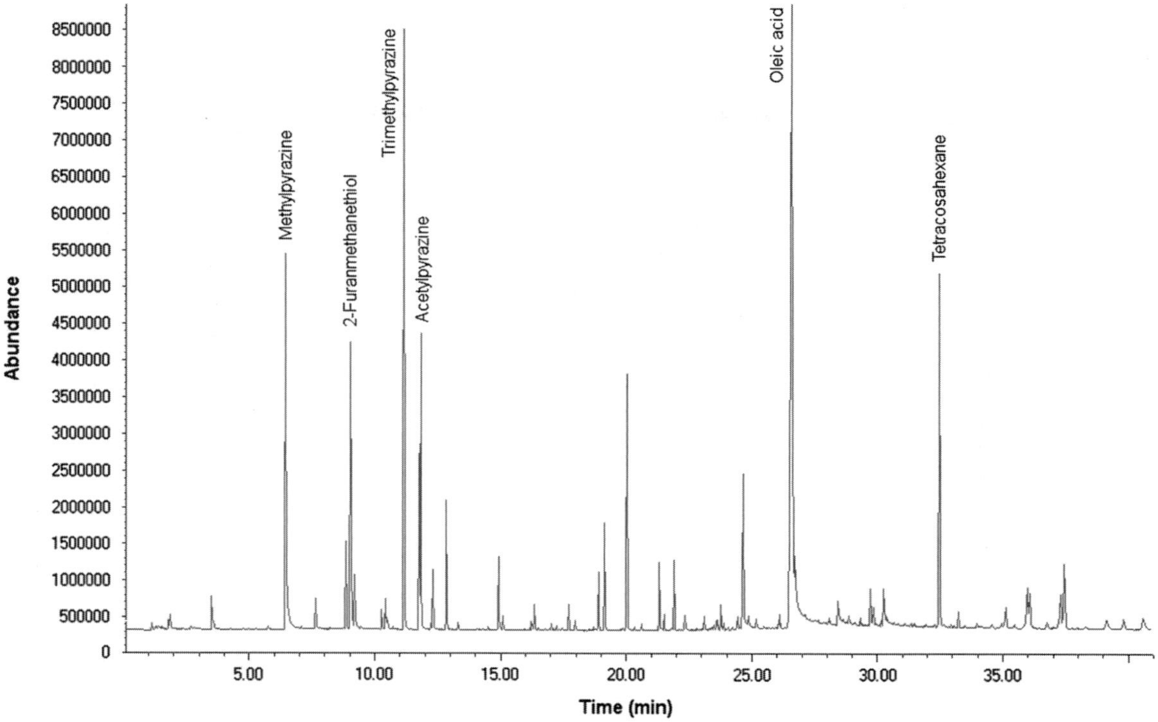

FIGURE 10.8 Direct thermal desorption with GCMS analysis of aroma constituents from powdered dry mushroom. Experimental conditions: TD-100 automated thermal desorber (Markes International Ltd., UK) combined with a 7890-5975 GCMS system (Agilent Technologies). Desorption: 5 min at 50 °C into a multisorbent ('air toxics') trap at 0 °C. Trap desorption: 300 °C with 25 ml/min split flow. Flow path: 140 °C. Column: 30 m × 0.25 mm I.D. × 0.25 μm film HP-Innowax. GC oven: 50 °C to 220 °C at 3 °C/min. Scan range: 35–300 amu.

- Internal standard addition had become an accepted part of automated GC procedures generally, and thermal desorption users were beginning to demand this option.
- Whole-air sampling – there was increasing interest in ultra-volatile compounds such as C_2 hydrocarbons and freons, which cannot be quantitatively retained by conventional sorbent tubes at ambient temperatures. Demand was also growing for semicontinuous, near-real-time monitoring of urban air pollutants with known adverse health effects – specifically C_2–C_9 hydrocarbons originating primarily from vehicle exhaust emissions [21,22] – so-called 'ozone precursors'. This led to thermal desorption technology being adapted to accommodate the controlled introduction of whole-air or gas samples directly into the cooled focusing trap [23] – see below.

To address these limitations and respond to the new demands, thermal desorption technology began to evolve rapidly from the early 1990s. The most significant changes were made in the following areas.

10.3.2. Optimization of the Focusing Trap

While electrically cooled/sorbent-packed focusing traps remain the most robust and versatile platform for two-stage thermal

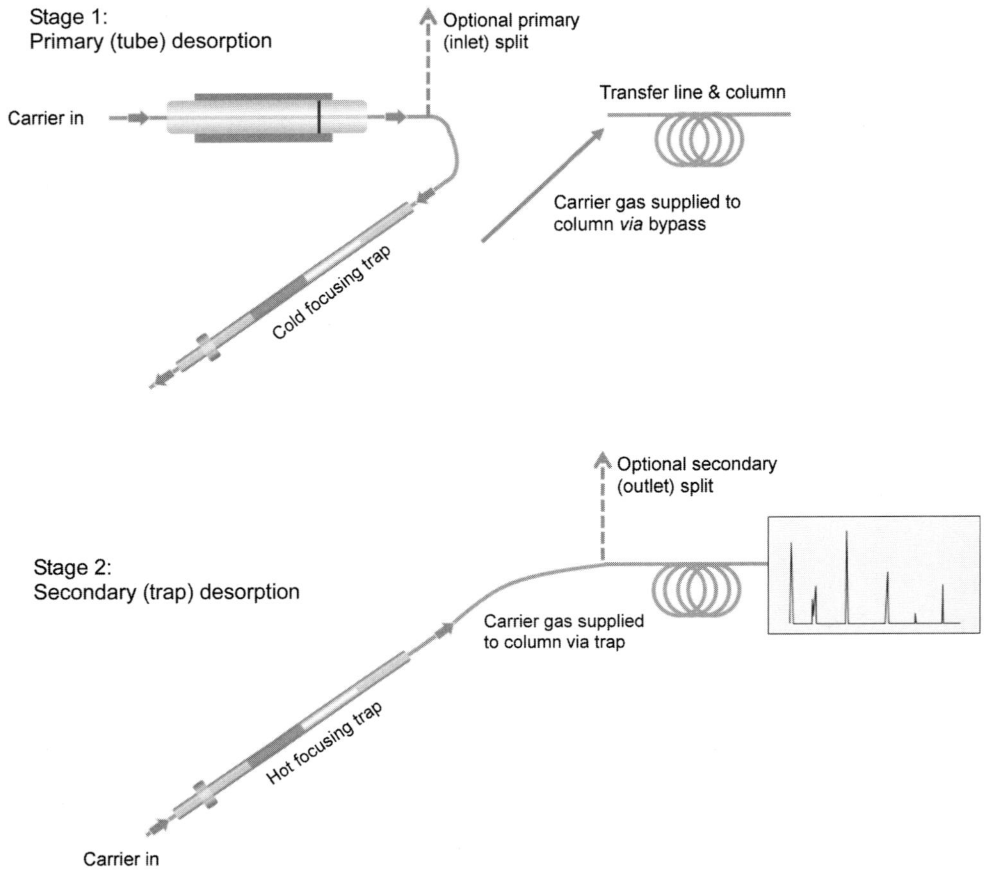

FIGURE 10.9 Illustration of two-stage thermal desorption, incorporating backflush of the focusing trap.

desorption, the technology has been refined considerably since 1981. Modern traps are typically constructed of inert materials such as quartz, can be heated at rates up to 100 °C/s, and can be desorbed efficiently at flows down to 1.5 ml/min to optimize method sensitivity. They are also invariably configured in 'backflush' mode, such that analytes enter and leave the trap from the same end – see Figure 10.9. This allows the trap to be packed with a series of sorbents of increasing strength and extends the analyte volatility range [7]. Examples of results that can be routinely obtained for air monitoring using the latest TD trapping technology are presented in Figures 10.10 and 10.11.

Trapping performance has also been optimized considerably over recent years. It is now possible to quantitatively retain ultra-volatile organics from air/gas samples without resorting to liquid cryogen cooling – see Figure 10.12 and Table 10.1.

10.3.3. The Evolution of Heated Valve Technology for Thermal Desorption

The essential functions of leak testing and prepurging of air to vent, plus the more recent

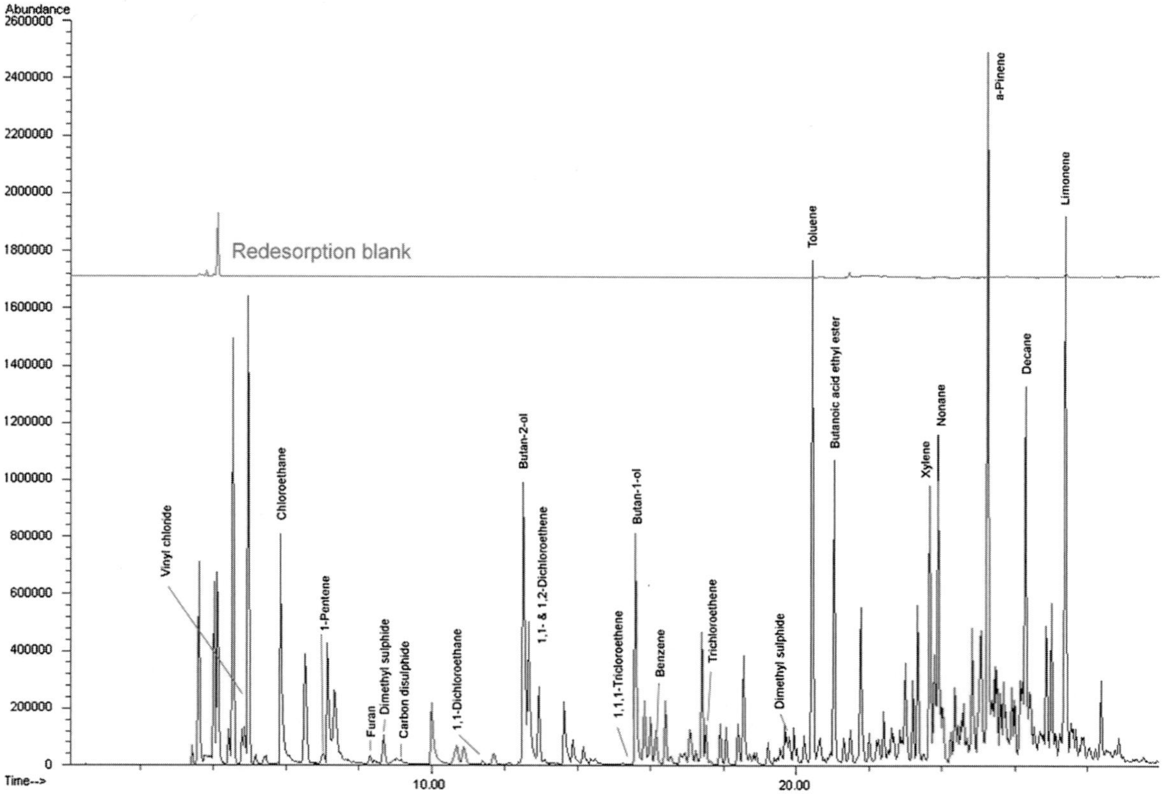

FIGURE 10.10 100 ml landfill gas with trace target analytes and many major components identified. Experimental conditions: ULTRA-UNITY automated thermal desorber (Markes International Ltd., UK) combined with a 6890-5975 GCMS system (Agilent Technologies). Silcosteel tube trap flow packed with Tenax/Unicarb. Tube desorption: 5 min at 200 °C into a multisorbent ('sulfur') trap at −15 °C, split 20 ml/min:20 ml/min. Trap desorption: 220 °C with 80 ml/min split flow. Flow path: 120 °C. Column: 60 m × 0.25 mm I.D. × 1.4 μm film DB-VRX (Agilent Technologies). GC oven: 40 °C to 225 °C at 10 °C/min. Scan range: 35–260 amu.

requirements for backflush desorption, dry purging, and internal standard addition (to the sampling end of the tube), have all re-enforced the need for valving in the TD sample flow path. Rotary valves are still widely used in commercial TD systems, but concerns regarding cold spots and temperature limitations have also led to the development of inert, low-volume TD-specific valving. Systems incorporating the new TD valves are compatible with high-boiling semivolatiles (e.g. up to n-$C_{40/44}$ – see Figure 10.13) yet still allow flow path temperatures to be set low enough for quantitative recovery of the most labile species. Thiols and CS (tear) gas, for example, work best with TD flow path temperatures at or below 125 °C [24].

10.3.4. Tube Sealing for Automation

Early attempts to overcome the sorption and artifact limitations of the original ATD 50 analytical tube seals involved caps that were removed and replaced by the thermal desorber

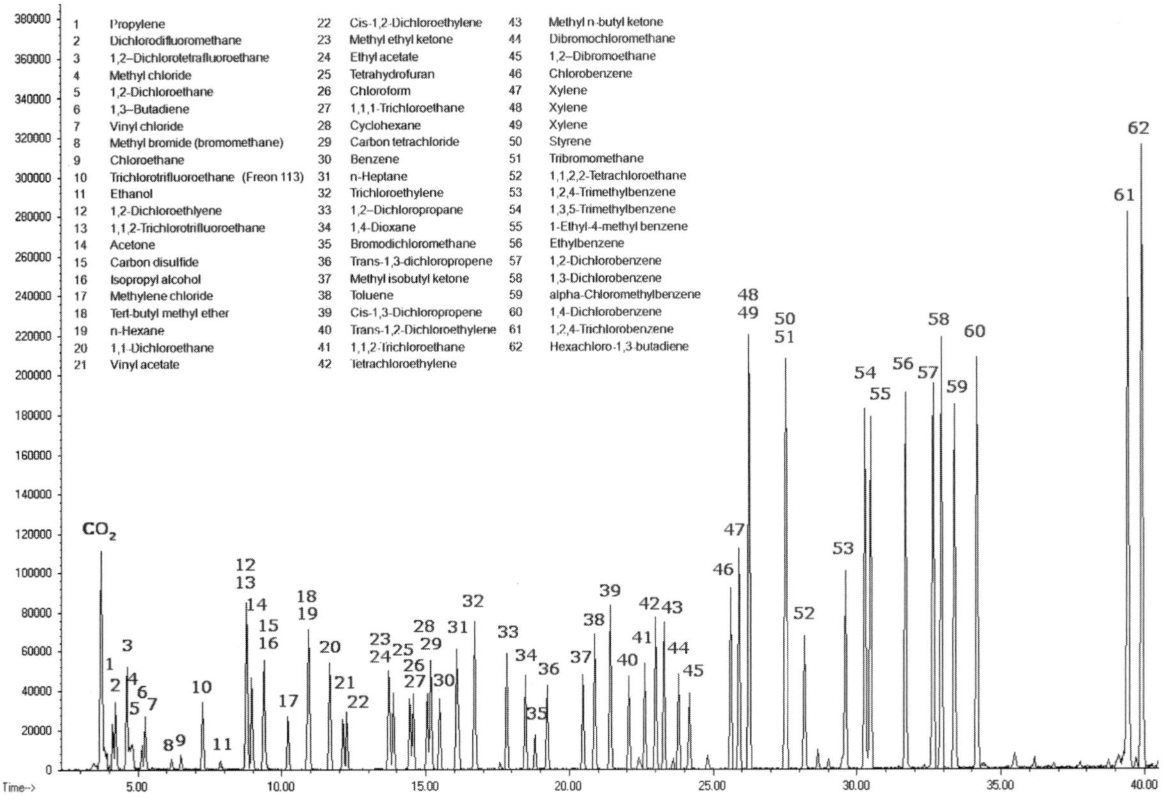

FIGURE 10.11 Splitless analysis of 1 L × 1 ppb 62-component air toxics standard in a canister. Experimental conditions: UNITY-CIA system (Markes International Ltd., UK) combined with a 7890-5975 GCMS system (Agilent Technologies). Multisorbent ('air toxics') trap at 25 °C. Trap desorption: 40 °C/s to 320 °C. Flow path: 140 °C. Column: 60 m × 0.32 mm I.D. × 1.8 μm film DB-624 (Agilent Technologies). GC oven: 40 °C to 230 °C at 5 °C/min. Scan range: 35–300 amu.

TABLE 10.1 Sampling and Detection Limit Data, Determined in SIM Mode, for Low-Concentration Standards of Potent Greenhouse Gases in Nitrogen and Real Air

Compound	Sample volume (mL)	Sampling flow (mL/min)	Lowest measured concentration (std) (ppt)	RMS S:N (std)	Estimated minimum detection limit (std) (ppt)	Lowest measured concentration (air) (ppt)	RMS S:N (air)	Estimated minimum detection limit (std) (ppt)
CF_4	25	10	70	40:1	<10	300	20:1	50
C_2F_6	1000	50	6	80:1	0.2	100	2000:1	0.2
SF_6	1000	50	1.5	80:1	0.05	100	5500:1	0.05

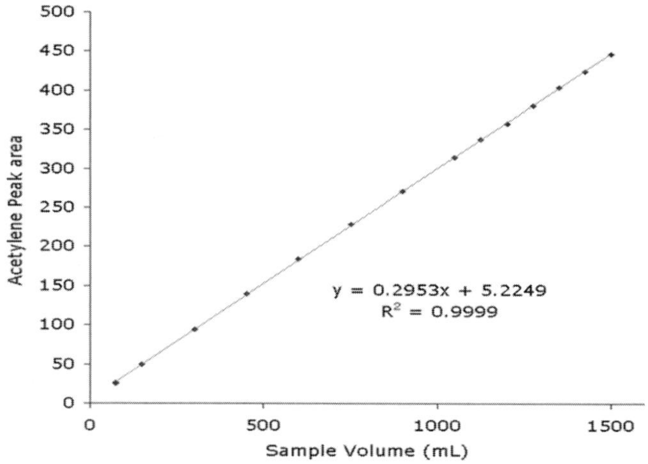

FIGURE 10.12 Cryogen-free retention of ultra-volatile analytes from large volumes of air/gas: Illustration with acetylene.

during automatic operation. These incorporated PTFE-coated o-rings to reduce friction and make the automatic uncapping/recapping processes as reliable as possible. However, while a major step forward, these caps were shown to be prone to significant loss of volatiles over time [25]. A more robust subsequent solution was the development of caps that incorporated a very long, narrow gas flow path through the cap. 'Diffusion-locking' mechanisms like these could reduce analyte loss and artifact ingress to negligible levels, even over extended periods (e.g. a week) but still allow gas to flow unimpeded when pressure was applied. This meant that the integrity of sampled and clean (desorbed) tubes could be rigorously maintained without complicating the TD automation process – i.e. without the need to uncap/recap tubes.

10.3.5. Re-Collection of Split Flow

An inherent drawback of all early thermal desorption systems was that the technique was '1-shot' – if anything went wrong during the analytical procedure, no sample remained to repeat the test. Quantitative re-collection of TD sample split flow was first reported by Dr Jan Kristensson in 1988 [14,26]. His initial implementation involved adapting a standard TD system of the time and was therefore unavoidably cumbersome. Nevertheless, it showed the potential of split flow re-collection for overcoming the inherent one-shot limitation of traditional TD technology and for confirming analyte recovery and test results.

The first commercial implementation of quantitative split flow re-collection for TD was developed in 1998 by a team led by Dr Peter Higham, then lead mechanical engineer at Markes International. Harnessing a new TD-specific heated valve (pioneered a few months earlier by the same team), the TD flow path was configured such that both primary tube desorption ('inlet') split flow and secondary trap desorption ('outlet') split flow were directed to the same 're-collection' tube – Figure 10.14. This overcame the old one-shot limitation of thermal desorption for the vast majority of applications. Furthermore, because re-collection involved analytes passing through an extended version of the TD flow path, carrying out a short sequence of repeat analyses on a single standard allowed selective losses of

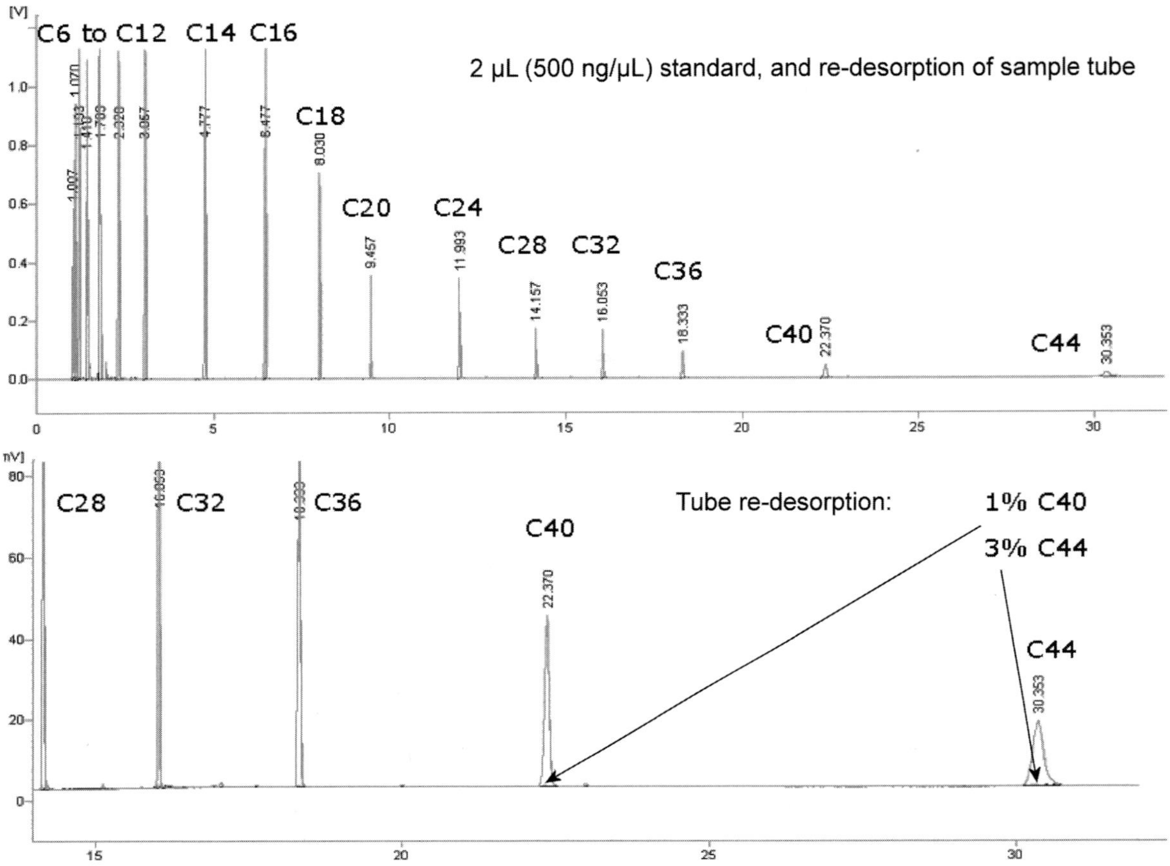

FIGURE 10.13 Analysis of a sorbent tube loaded with 2 μl of an n-C_6 to n-C_{44} hydrocarbon standard (500 ng/μl) using 2-stage TD with a valve and with trap desorption in backflush mode. Experimental conditions: UNITY 2 thermal desorber (Markes International Ltd., UK) combined with a Trace GC/FID (Thermo Electron Inc). Stainless-steel tube packed with quartz wool and graphitized carbon black. Tube desorption: 10 min at 370 °C with 40 ml/min flow into a multisorbent (quartz/carbon black) trap at 0 °C. Trap desorption: 370 °C with 30 ml/min split flow and 3 ml/min constant column flow. Flow path: 210 °C. Column: 30 m × 0.25 mm I.D. x 0.25 μm film DB 5 (Agilent Technologies). GC oven: 100 °C (1 min hold) to 325 °C at 15 °C/min.

one or more analytes (relative to split ratio or to other more stable/volatile compounds in the mix) to be readily identified.

Quantitative sample re-collection has now been automated and implemented on several different commercial systems. It is one of those TD functions that has become accepted as a standard requirement and is now referenced in many international standards [27,28].

10.3.6. Electronic Control of Flows and Pressures

While GC systems with conventional liquid injectors have benefited from electronic pneumatic control of carrier gas and split flow for many years, most two-stage TD procedures present a significantly more difficult technical challenge. This is because the route (flow path)

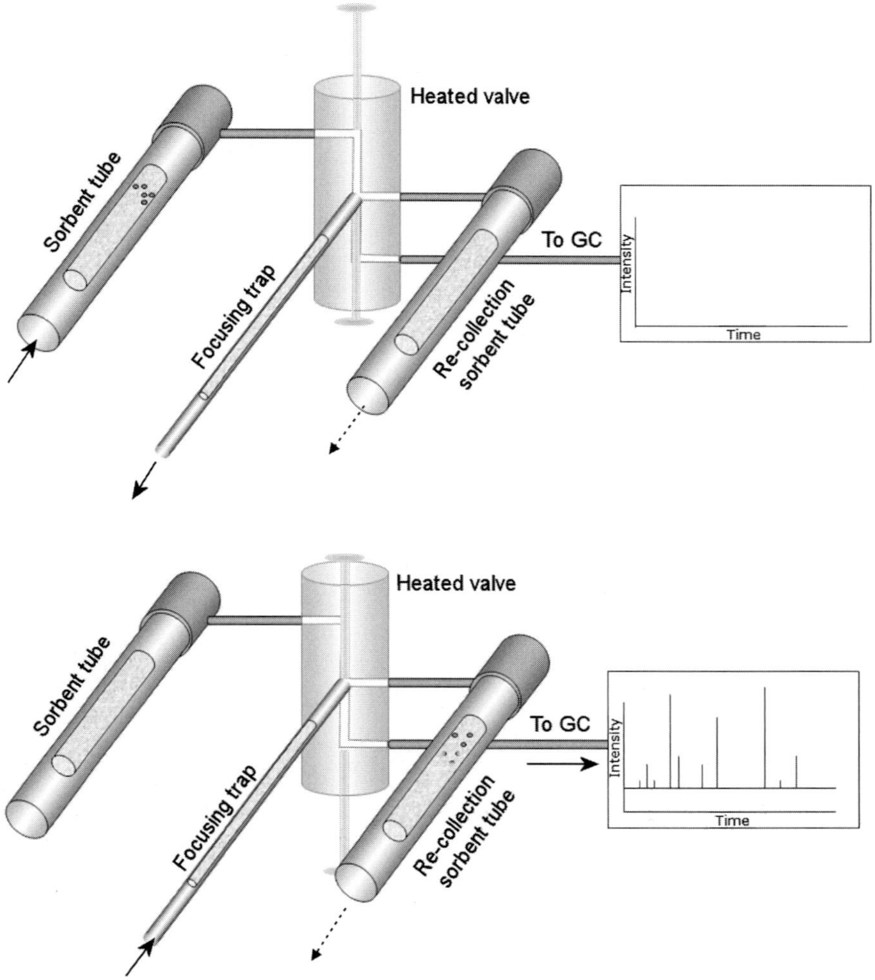

FIGURE 10.14 Operation of two-stage thermal desorber with integrated low volume, TD-specific valve and the capability to re-collect primary and/or secondary split flow on the same tube.

by which carrier gas is supplied to the column alters depending on the phase of operation. For example, at the start of trap desorption, the flow to the column switches from a simple bypass line to pass through the very different impedance of a sorbent focusing trap. The trap then heats, thus changing the impedance once again. Maintaining a stable electronically controlled carrier gas pressure or flow to the head of the GC column under these conditions requires a very robust closed-loop feedback.

Despite this complexity, several modern TD systems have now implemented both precise electronic pneumatic control of carrier gas flow/pressure (either *via* the attached GC or *via* independent TD modules) and electronic mass flow control of desorption and split flows. This enhances complex analyses by stabilizing ('locking') peak retention times independent of split flow, trap impedance, etc. – Figure 10.15. Without electronic pneumatic control of carrier gas, late-eluting components would be subject

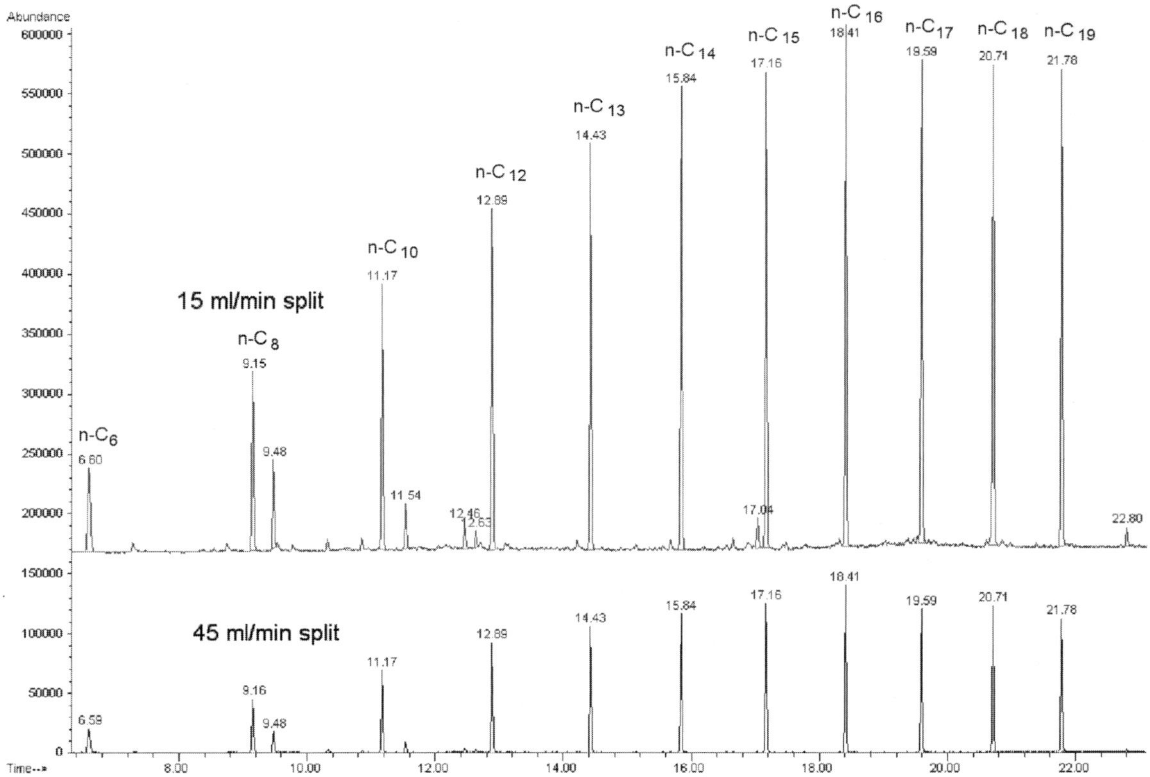

FIGURE 10.15 Using electronic carrier gas control to stabilize TD-GC(MS) retention times under different analytical conditions.

to significant retention time variations if the analytical conditions changed.

10.3.7. TD Innovations for Whole-Air/Gas Sampling (Canisters/Bags and On-line Monitoring)

Recent advances in thermal desorption technology for whole-air/gas monitoring have included optimization of focusing trap technology for retention of cryogen-free ultra-volatile components, the development of reciprocal twin-trap system configurations for continuous on-line operation, extended sequencing capabilities (more channels), and improved general analytical performance (linearity, reproducibility, reduced carryover, etc.).

On-line air/gas monitoring systems tend to be installed in industrial plant or in environmental field monitoring stations and are thus often required to operate unattended for extended periods of time. High liquid cryogen consumption was a major limitation for the earliest on-line TD–GC systems [10], but this was addressed as soon as the first Peltier-cooled systems became available for on-line work [24]. More recently, the optimization of trap performance has played a major part in extending the compatibility of cryogen-free on-line TD technology even further, to include ultra-light perfluorinated compounds (see Table 10.1).

Systems incorporating two reciprocally operated traps offer on-line monitoring without any sampling 'blind spots'. These are often targeted at very hazardous applications, for example, in counter-terrorism or for monitoring particularly dangerous chemical processes – see below.

The first widely used standard method for air monitoring using canisters was published by the EPA for consistency with earlier use in 1991 and involved cryofocusing [29]. Automated analytical technology for whole-air/gas containers such as canisters or bags has been slow to evolve since then. This situation is finally beginning to change, and most modern canister standards [30,31] now specify (or at least include) the option of cryogen-free sorbent trapping. Improved trap performance, better-designed (more uniformly heated) sample flow paths, and extended purging have also served to reduce carryover [32] and make modern canister-based autosamplers more versatile and productive than their predecessors.

10.4. SAMPLING OPTIONS AND THE ROLE OF THERMAL DESORPTION AS A FRONTEND TECHNOLOGY FOR GC

Sampling options associated with thermal desorption have conventionally included vapor monitoring, *via* sorbent tubes/traps, canisters, and bags, [7] and direct desorption of homogeneous liquids or solids (Figure 10.7). However, thermal desorption is also the primary interface for many other GC 'frontend' technologies, namely purge and trap [33,34], large-volume injection, sorptive extraction, and headspace trap or headspace-TD (HS-TD).

10.4.1. Headspace-TD

Traditional equilibrium headspace technologies for GC are fundamentally static systems. They rely on target compounds partitioning reproducibly between the sample matrix and the vapor (headspace) phase under fixed conditions of temperature and pressure, such that analyte concentrations in the headspace are representative of analyte concentrations in the sample matrix. Aliquots of headspace vapor in the order of 1-ml volume are typically transferred to the GC either *via* a simple syringe or using rather more sophisticated mechanisms such as gas loop or pressure balance.

The addition of a focusing trap turns headspace into what is essentially a step-wise dynamic process. It allows larger volumes of headspace vapor to be collected/focused over a longer period of time. Moreover, many HS-TD systems allow the headspace vials to be repressurized and resampled repeatedly (in multiple cycles) before the focusing trap is finally thermally desorbed to 'inject' all the retained vapors in one go. The main advantages of HS-TD *versus* conventional static headspace procedures are a 10–100-fold increase in sensitivity (depending on analyte volatility and the number of cycles), an extended volatility range (conventional static headspace is intended to preferentially increase the concentration of volatiles at the expense of higher-boiling matrix components) – Figure 10.16 – and selective prepurging of interferences (water, ethanol, etc.). In effect, it allows headspace procedures to approach purge-and-trap sensitivity levels but with the practical advantages of disposable vials, no foaming, and minimal risk of aerosol formation [35].

10.4.2. Solid Phase (micro-)Extraction (SP(M)E) or Sorptive Extraction (SE) for GC

SP(M)E/SE is available in various commercial formats* comprising fibers, tubes, or stir bars. All are coated with sorbent or stationary

* Commercial implementations of sorptive extraction technology for TD–GC include coated fibers from Sigma-Aldrich Inc., PDMS-coated stir bars (Gerstel-Twister® – Gerstel GmbH and Co. KG), and SPE-tD™ cartridges (Markes International Ltd, UK).

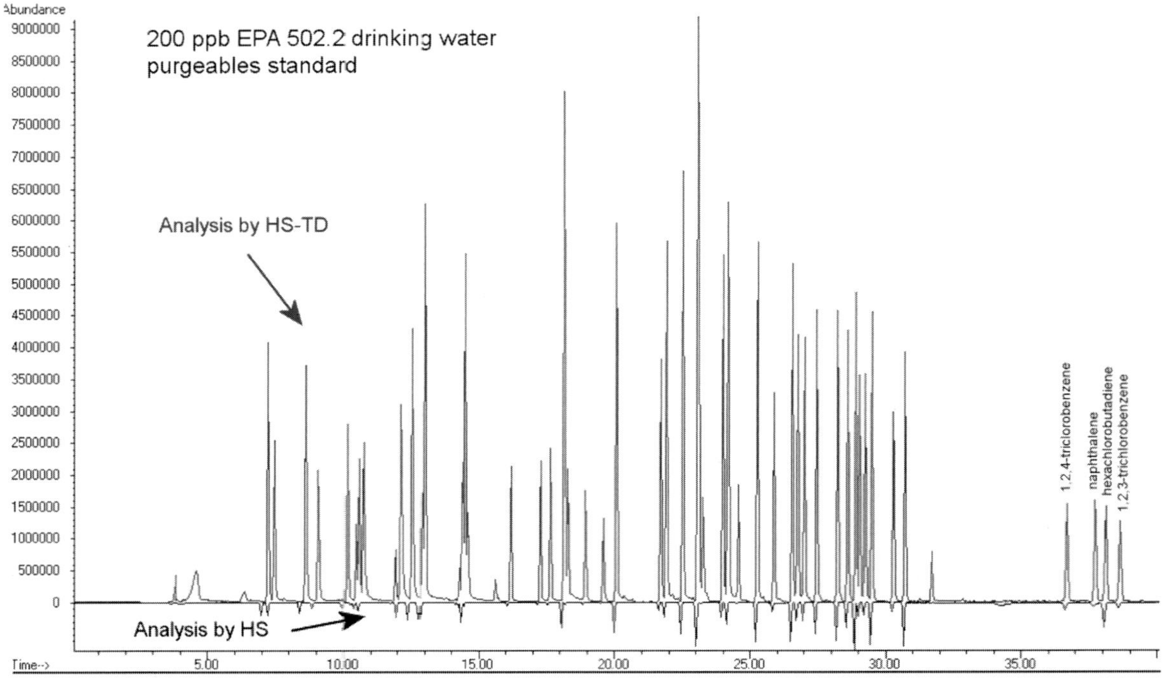

FIGURE 10.16 200 ppb purgable VOC standard in drinking water analyzed by conventional HS-GCMS (lower inverted chromatogram) and HS-TD-GCMS (top chromatogram). Experimental conditions: HS-5TD system (Markes International Ltd., UK) combined with a 6890-5975 GCMS system (Agilent Technologies). Multisorbent ('air toxics') trap at 25 °C. Vial temp: 40 °C. 10 HS vial sampling cycles at 50 ml/min. 3 min trap purge. Trap desorption: 40 °C/s to 320 °C with 5 ml/min split. Flow path: 150 °C. Column: 60 m × 0.25 mm I.D. x 1.4 μm film DB-VRX (Agilent Technologies). GC oven: 40 °C (5 min), 5 °C/min to 180 °C (10 min) then 20 °C/min to 240 °C. Scan range: 29–400 amu.

phase – most commonly polydimethylsiloxane (PDMS). The technique is principally applied to higher-boiling organics in the liquid phase and is thus a good complement to headspace trap – Figure 10.17. Organic compounds partition between the sample matrix and the stationary phase on the SP(M)E/SE device at a given temperature. The sampled cartridge/fiber is then rinsed, dried, and analyzed using thermal desorption (or solvent extraction) with GC(MS) [36].

The approach is most commonly used for screening (rather than absolute quantitation) because the limited range of predominantly nonpolar sorptive coatings is not compatible with all analytes. Furthermore, the partition system can be very sensitive to variations in sample conditions – humidity, analyte concentration, matrix composition, time, temperature, etc. Nevertheless, SP(M)E/SE provides a useful extraction tool for complex samples and is widely used for routine drug screening and for monitoring persistent organic pollutants in foods, beverages, and other natural products [37].

10.4.3. Large-Volume Injection

While rarely thought of as an extension of thermal desorption technology, large-volume injection systems are also frequently based on

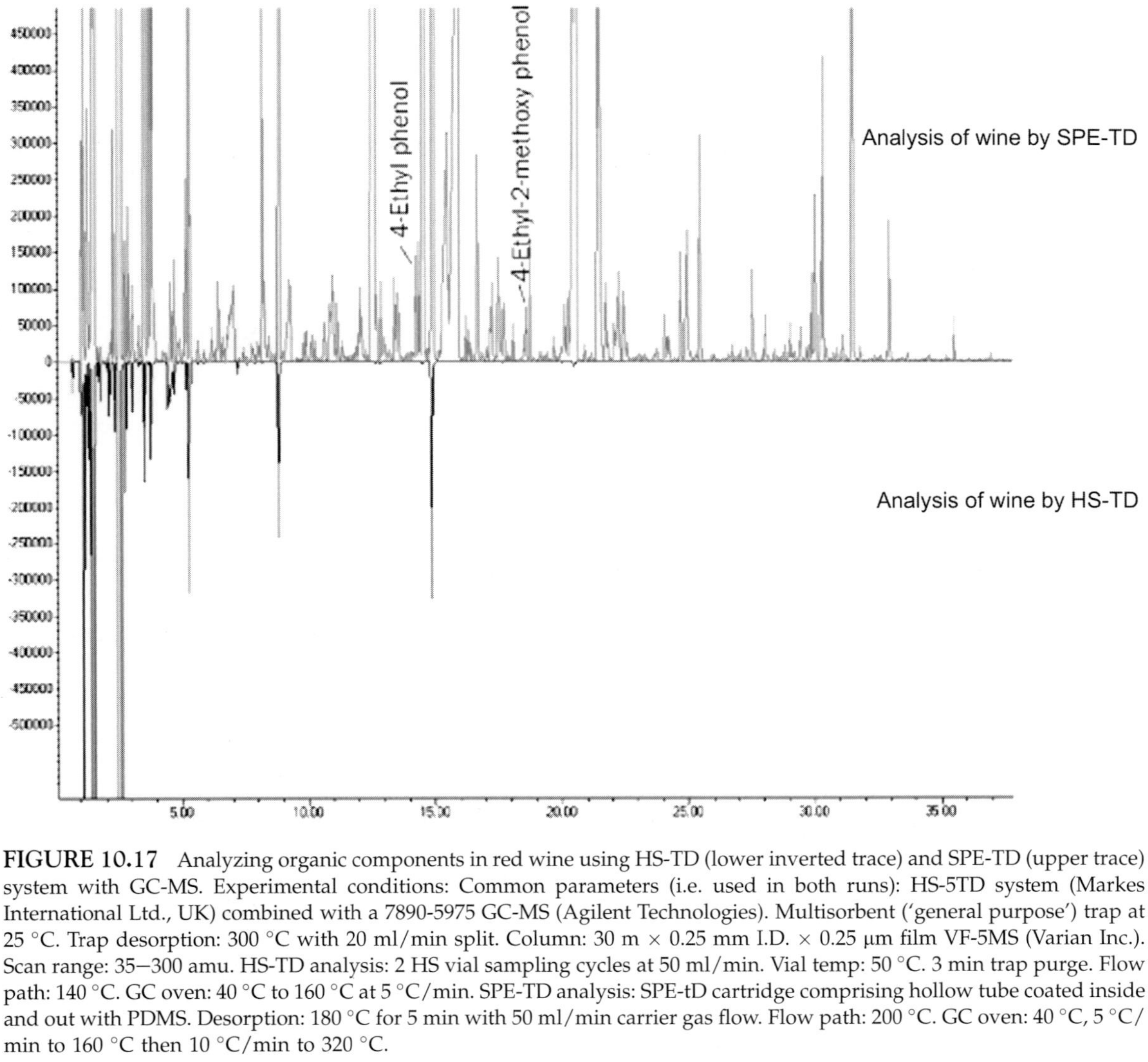

FIGURE 10.17 Analyzing organic components in red wine using HS-TD (lower inverted trace) and SPE-TD (upper trace) system with GC-MS. Experimental conditions: Common parameters (i.e. used in both runs): HS-5TD system (Markes International Ltd., UK) combined with a 7890-5975 GC-MS (Agilent Technologies). Multisorbent ('general purpose') trap at 25 °C. Trap desorption: 300 °C with 20 ml/min split. Column: 30 m × 0.25 mm I.D. × 0.25 μm film VF-5MS (Varian Inc.). Scan range: 35–300 amu. HS-TD analysis: 2 HS vial sampling cycles at 50 ml/min. Vial temp: 50 °C. 3 min trap purge. Flow path: 140 °C. GC oven: 40 °C to 160 °C at 5 °C/min. SPE-TD analysis: SPE-tD cartridge comprising hollow tube coated inside and out with PDMS. Desorption: 180 °C for 5 min with 50 ml/min carrier gas flow. Flow path: 200 °C. GC oven: 40 °C, 5 °C/min to 160 °C then 10 °C/min to 320 °C.

TD. Some of the earliest studies of groundwater and soil contamination, for example, described 'adsorption/thermal desorption' and involved injecting several milliliters of water through Tenax tubes before they were subsequently dried and analyzed by TD–GC(MS) [38]. Modern implementations are typically built on similar principles, albeit in more automated and integrated configurations.

10.4.4. 'Stand-Alone' Sampling Accessories which Extend the TD Application Range

Numerous specialist 'stand-alone' sampling devices have also been introduced for thermal desorption over recent years [39]. Key examples include alveolar breath samplers [40,41], material emission testing equipment [42–44], and soil probes [45,46] – some examples are described in detail in ref [39] and relevant example applications are presented below.

10.4.5. Summary of the Versatility of TD as a Frontend Technology for GC

The multiple roles played by thermal desorption in GC sample introduction are probably best illustrated diagrammatically – see Figure 10.18. In its various manifestations, TD offers compatibility with gas-, liquid-, and solid-phase samples and with GC-compatible organic analytes ranging in volatility from C_2 hydrocarbons and freons to n-C_{40} compounds and 6-ring PAHs. Key TD applications include air/gas monitoring (including fugitive industrial emissions), odor and aroma profiling, materials characterization, civil defense, product quality control, and testing chemical emissions from everyday products to indoor air.

In many respects, high-performance TD systems can be thought of as versatile, readily automated, programmable split/splitless GC injectors. The desorption efficiency of the focusing trap should equate to that of a well-designed liquid inlet for GC in terms of peak shape, compatible boiling range, stability, longevity, etc. In other words, the performance of the thermal desorber should allow it to be interfaced directly to the analytical column, using a simple heated transfer line, without any additional form of injector or focusing device.

10.5. METHOD DEVELOPMENT AND OPTIMIZATION

The fundamentals of TD method development and optimization have remained

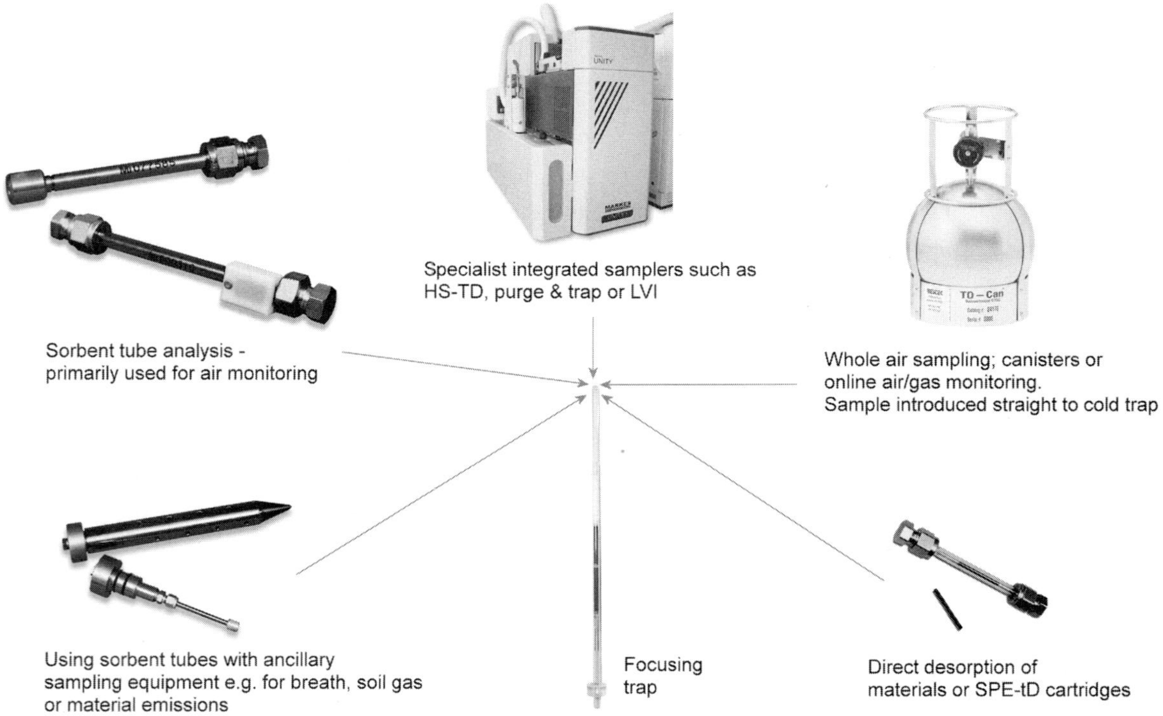

FIGURE 10.18 GC(MS) sample introduction options supported by thermal desorption.

essentially the same since the advent of back-flush desorption of the focusing trap. While this chapter is not an appropriate forum for a detailed tutorial on TD–GC(MS) method development, the following general guidelines may be useful.

10.5.1. Sampling Considerations for Successful Analysis

As with any measurement procedure, it is essential to make sure that the samples are collected correctly. Considerations for optimizing sorbent selection for on- or off-line collection of air/gas samples, with and without specialist sampling accessories (breath, material emissions, etc.), have been well documented [7,39]. While the selected sorbents must be strong enough to prevent breakthrough and loss of any target compounds during sampling, they must also be weak enough to allow quantitative recovery of all the analytes of interest at safe temperatures during the desorption phase – i.e. at temperatures that do not exceed the stability limits for the sorbent or compounds of interest. It is also an advantage if retention of interfering compounds, such as permanent gases and water, can be minimized during the sampling process – e.g. by selecting hydrophobic sorbents when sampling humid air/gas streams. Similarly, the use of canisters, bags, or unheated on-line air/gas manifolds must be restricted to vapors that can readily be recovered from ambient temperature containers/streams – this typically means compounds more volatile than n-$C_{9/10}$.

Headspace-TD and sorptive extraction applications have been extensively reported [35,37], and associated optimized sampling conditions are presented in the literature and in application notes from relevant manufacturers. The direct desorption of materials also requires some specific considerations. General guidelines are given below.

10.5.2. Optimizing the Analytical Procedure

10.5.2.1. General Considerations

As described above, TD is essentially an extension of gas chromatography with key parameters including temperature, gas flow, time, and sorbent (stationary phase) selection. While thermal desorption parameters vary widely from application to application, it is possible to apply some general rules that aid the development of robust methods.

It is usually best to start by considering the analysis as a whole: What are the objectives? What are the target compounds and likely interferences? What is the expected concentration range? What range of analyte masses should be allowed to reach the column and detector to best ensure optimal GC performance and required detection levels?

In the case of sorbent tubes, it is usually (but not always) necessary to desorb everything that has been trapped/collected as the primary sample while at the same time quantitatively retaining the compounds of interest on the focusing trap. Key considerations at this stage include the volatility range of compounds of interest, the maximum temperature of the sorbents, and the likely presence of interferences. If it is possible to adjust the trapping/focusing parameters such that compounds of interest will be quantitatively retained for the duration of primary (tube) desorption while unwanted interferences (CO_2, water, and ethanol, for example) are purged to vent, this is a big bonus.

Once primary desorption is complete, the objective of secondary (trap) desorption is invariably to completely desorb everything retained by the trap and transfer it to the GC analytical column, usually in as small a band (volume) of carrier gas as possible.

10.5.2.2. Temperature

Selection of optimum desorption and flow path temperatures is usually straightforward, taking into account both the volatility and thermal stability of the compounds of interest and the temperature limits of the tube and/or trap sorbents concerned. One commonly misunderstood factor is that the energy required to break the sorbent–sorbate bond and release retained analytes into the gas stream is much higher than that required to keep analytes in the vapor-phase as they pass through the empty narrow-bore tubing that comprises the rest of the TD system flow path. In fact, provided the rest of the TD flow path is uniformly heated, inert, and narrow bore, desorbed compounds will remain comfortably in the vapor-phase in the stream of carrier gas at temperatures well below that required for tube or trap desorption. This is intuitive to most gas chromatographers: just as analytes would be expected to elute from a 30- m coated capillary column at a temperature well below their boiling point, so will compounds elute very readily from the short, uncoated, narrow-bore internal flow path of a TD at surprisingly moderate temperatures. For example, re-collection and repeat analysis (Figure 10.14) was used to validate recovery of the hydrocarbon standard shown in Figure 10.13 through an automated TD system. The experiment was carried out with the TD flow path and transfer line set at only 210 °C. Results demonstrate negligible losses of n-C_{40} (b.p ~550 °C) and are shown in Figure 10.19 [47,48]. It is important to understand this, because it allows relatively low flow path and transfer line temperatures to be set for most applications, thus enhancing recovery of reactive species [25].

10.5.2.3. Flow and Split Ratio Selection

The power of gas flow to enhance the thermal desorption process is commonly underestimated. Increasing flow can be a very useful

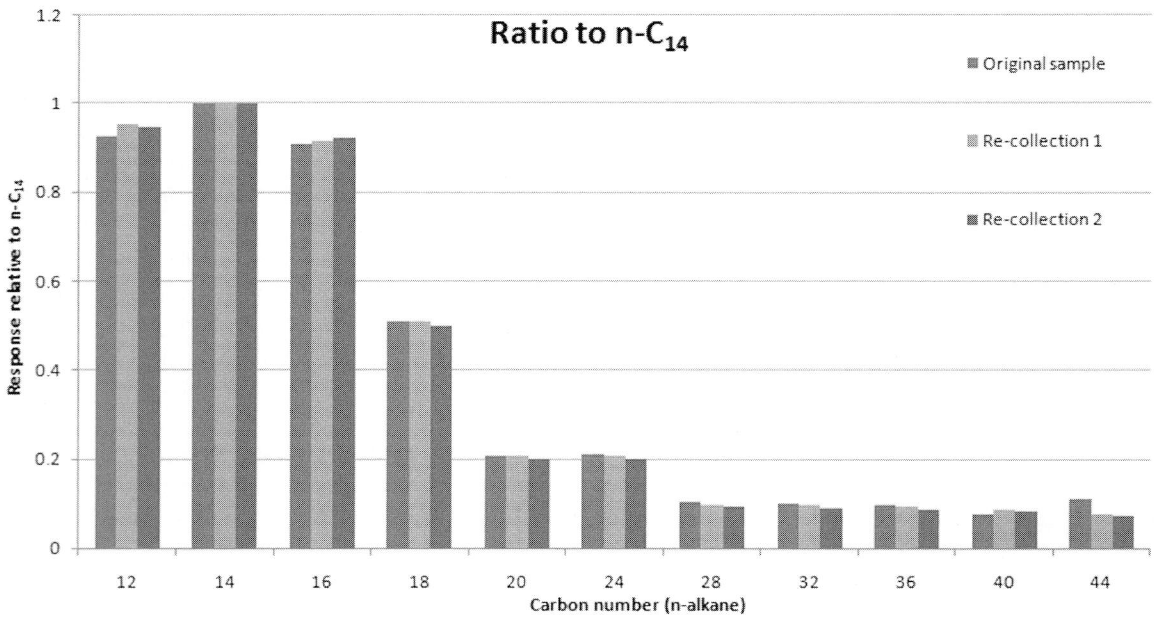

FIGURE 10.19 Quantitative recovery of high-boiling hydrocarbons validated using quantitative re-collection of thermal desorption split flow and repeat analysis. Experimental conditions as for Figure 10.13.

alternative to increasing temperatures – for example, when analyzing reactive compounds such as explosives and/or to minimize artifacts when using less stable sorbents such as Chromosorb 106 or Porapak N. As a general rule, doubling the desorption flow approximately halves the desorption time.

Typical flow rates used in thermal desorption are in the order of 20–200 ml/min for tube desorption, 10–100 ml/min for whole-air sampling, 2–50 ml/min through the cold trap during focusing, and 2–100 ml/min during secondary (trap) desorption.

As described above, implementation of sample splitting often directly benefits the thermal desorption process by allowing tube and/or trap desorption flows to be set higher than (and independently of) focusing or GC column flows, respectively. It also extends the applicable concentration range. Whole-air/gas sampling methods, headspace-TD, and other procedures that transfer unconcentrated analytes directly into the focusing trap are all generally restricted to selection of a single split i.e. during desorption of the focusing trap. This limits split ratios to between zero and approximately 200:1 in these cases. However, all two-stage TD procedures, including tube desorption, direct desorption, large-volume injection, and sorptive extraction-TD, benefit from the option of double splitting. In this case, the overall split ratio is the product of the two individual split ratios and milligram level samples can be reduced to a few 100 ng on-column [19].

10.5.2.4. Minimizing Interferences

As described above, thermal desorption offers several opportunities for selective elimination of common GC interferences such as permanent gases, water [49], and, when applicable, volatiles such as methanol, ethanol, and acetic acid. Common options include selective retention during sampling/focusing, dry purging or pre-purging prior to desorption, and, occasionally, the use of specific devices such as a Nafion dryer; e.g. for on-line monitoring of very volatile nonpolar compounds such as freons or 'ozone precursors' [50].

10.5.2.5. Direct desorption of materials or sorptive extraction cartridges:

Key considerations for direct desorption include the following:

- Ensuring that the flow path through the tube is not blocked by sample material or the SP(M)E/SE sampling device.
- Ensuring that the material/device is positioned securely in the central, heated portion of the sample tube not in the relatively cooler zones at either end of the tube.
- Understanding the objective of the process – i.e. Is it the intention to carry out complete extraction of the volatiles of interest (e.g. for residual solvent studies) or simply to obtain a representative VOC profile (e.g. for characterizing the aroma of natural products)?
- Understanding the nature of the sample matrix – maximum temperature, water content, etc.

For many solid and liquid samples, direct thermal desorption allows both interfering volatiles and unwanted higher-boiling matrix components to be excluded from the analysis. In these cases, TD effectively combines sample preparation/cleanup and selective extraction into one fully automated process, thus extending the lifetime of capillary columns and other GC(MS) system components.

10.6. CALIBRATION AND VALIDATION

Thermal desorption procedures are generally calibrated using external standard methods with the optional addition of a gas-phase internal standard such as deuterated toluene or bromofluorobenzene.

Theoretically, gas-phase standards should be used for calibrating vapor-monitoring applications, but they can be expensive and/or difficult to obtain at appropriate concentrations. Certified cylinders of ppb-level hydrocarbons for calibrating ozone precursor systems, for example, retail at upwards of US$ 3,000.

Reliable gas standards are also notoriously difficult to generate. Static standard atmospheres are prone to analyte losses through surface sorption, condensation, and dissolution into any liquid water present inside the container – even the thinnest surface film. (N.B. Many of these same issues afflict air/gas samples collected in modern canisters.) Dynamic standard-atmosphere generation methods are much more reliable and are described in several papers [51–53]. However, there are very few laboratories in the world that have sufficiently sophisticated apparatus, including continuous monitoring equipment, to produce low-concentration (low ppb) standard atmospheres that are truly traceable to primary standards.

For these reasons, the most important international standard thermal desorption methods describe external standard calibration using liquid solutions [27,28,54,55]. Standard solutions are prepared such that injection of small volumes (typically 0.2–2 μL) introduces the same analyte masses that will be retained by (or loaded into) the TD tube during sampling. The preferred injection method involves introducing the liquid standards through what is essentially a simple unheated GC injector which is connected to the sampling end of the sorbent tube as if it was the injector end of a ¼-inch packed column. Carrier gas flows are typically set to something in the order of 100 ml/min, and the syringe is usually inserted through the septum and into the 'injector' such that it just touches the sorbent-retaining material within the tube – gauze, frit, quartz/glass wool, etc. Analytes vaporise rapidly in the fast flow of carrier gas and reach the sorbent in the vapor phase, as they do during sample collection. After the aliquot of standard has been introduced in this way, tubes are typically left *in situ* with the carrier gas flowing for up to 5 minutes to allow selective purging of solvent (typically methanol) if applicable. Calibration of very light components is similar but involves gas standards and gas syringes. If calibration is required over a wide volatility range using multiple standards, liquid standards containing higher-boiling stable components are introduced first with gas standards of the lighter compounds being loaded last.

Note that with two-stage thermal desorption, it is not usually important to load standards onto tubes packed with the same sorbents that will be used for vapor sampling. This is because the critical 'analytical' injection is the second (trap) desorption. It is more important to make sure that the standard loading and calibration procedure is simple/robust and to use selective purging if possible to eliminate the carrier solvent.

For obvious practical reasons, if the nature of the analytes means it is not possible to use a solvent that can be selectively purged prior to analysis, it is usually best to minimize injection volumes (i.e. to <0.5uL) and to choose a solvent that gives a sharp, well-resolved peak under the analytical conditions selected.

Traditional methods for validating analyte recovery through a thermal desorption system were complex and tedious. Users were generally recommended to carry out a multilevel calibration using the thermal desorber and then set up an equivalent liquid injection system (same column, carrier gas flow, split ratio, etc.) and repeat the process [54]. Aside from the time required, it is difficult to exactly match column and split flows for two such different GC injection systems, especially in the case of double-split TD methods. The latest international standard methods for thermal desorption have therefore begun to recommend the use of quantitative re-collection for repeat analysis as an alternative approach – see above and Figures 10.14 and 10.19.

10.7. AN INTRODUCTION TO THERMAL DESORPTION APPLICATIONS

Having concentrated thus far on describing the necessary functions for thermal desorption and how to optimize methods, we can now begin to address the most important and interesting aspect – what GC applications benefit from thermal desorption?

Before looking at this in detail, it is prudent to start with a reality check and consider what *does not* work well.

Most thermal desorption systems do not provide the best GC interface option for the following:

- Inorganic gases such as O_2, O_3, N_2, CO_2, NO_x, SO_x, and NH_3 (exceptions include N_2O, H_2S, and SF_6, which can all be conveniently monitored using TD methods).
- Methane: This is hard to trap quantitatively, even on the most efficiently cooled sorbent traps. Moreover, it is often present in such abundance (relative to other organic vapors) that it can usually be monitored without preconcentration, e.g. using conventional flame ionization detectors.
- Organic compounds that are not compatible with conventional GC analysis.
- Nonvolatiles – i.e. compounds less volatile than n-$C_{40/44}$, didecyl phthalate or 6-ring polyaromatic hydrocarbons (PAHs).
- Organics retained in a thermally unstable sample matrix.

Apart from this limited number of exceptions, thermal desorption has the versatility to benefit a wide range of GC applications especially those that present a 'challenge' to conventional inlet technology, i.e. liquid autosamplers, static headspace, or as sample valves. Suitable analytes include any GC-compatible organic compounds within the constraints listed above.

10.8. AIR MONITORING

Air monitoring is the application everyone first thinks of in connection with thermal desorption, but it is a broad field in its own right and can be subdivided into several distinct areas [7,39]. The following text summarizes these different areas and presents some topical examples.

10.8.1. Industrial (occupational) Hygiene or Workplace Air Monitoring

Monitoring personal (inhalation) exposure to toxic chemicals for compliance with workplace health and safety regulations was one of the very first applications of thermal desorption. Samples are typically collected using pumped or diffusive sorbent tubes according to various national and international standard methods [27,54,56–58]. Time-weighted average measurements, e.g. over an 8-h shift, are then assessed against threshold limit values (TLVs), sometimes called 'occupational exposure limits' (OELs), to check compliance. While vast improvements in workplace health and safety have been implemented in most industrialized countries in recent years, the developing world still struggles to keep personal exposure levels below safe limits. Additional information on chemical toxicity has also led to the continued reevaluation and reduction of many limit levels, thus driving the ongoing need to monitor personal exposure at lower concentrations.

There are many excellent publications in the literature reporting on the use of analytical thermal desorption for occupational hygiene [59,60]. One point of note is that best practice typically requires average monitoring results to be well below prescribed limit levels (e.g. one-tenth). This is because differences in behavior between individuals generate such a wide range of results (for example, over 1 or 2 orders of magnitude [1]) that it is only when the average falls well below the OEL that it is safe to assume no workers are being exposed to unsafe levels.

Related thermal desorption applications include biological monitoring – i.e. measuring chemical concentrations in the blood, urine, or breath of workers as a means of assessing their chemical exposure *via* all possible routes (ingestion and skin absorption, as well as inhalation [41,61,62]). Generally speaking, subjects prefer to provide a breath sample, rather than blood or urine, and another benefit of this noninvasive approach is that it does not require trained medical personnel [41]. With sufficient data, guideline 'acceptable breath concentration' levels can be set for specific processes or tasks. However, results are more typically interpreted in relative terms – e.g. for checking mean exposure levels are not increasing over time or for identifying anomalies across a group of workers all supposedly doing the same job. Guidance notes are available to help interpret breath-monitoring data for some common skin-absorbed compounds [63].

10.8.2. Other TD Applications Relating to Breath Sampling

While perhaps not a mainstream air-monitoring application, there is also increasing interest in the diagnostic potential of VOCs in breath [40,64]. Many biological processes and medical conditions produce specific indicative VOCs or patterns of VOCs – e.g. diabetes and stress. The diagnostic potential of breath has been extensively studied for lung/respiratory diseases (lung cancer, asthma, TB, etc.) [40,64–67] and has even been investigated for gut disorders and mental health conditions.

Additional applications for monitoring VOCs in breath have included the following:

- investigations of the permeability of human skin to volatile halogenated species at elevated (bath/shower) temperatures [62];
- studies of halitosis or breath odor [68]; and
- identification of compounds that can be used as reliable indicators of smoking [69,70].

10.8.3. Ambient Outdoor and Indoor Air Monitoring

TD–GCMS has been the analytical method of choice for ambient air-monitoring applications for at least two decades. A wide variety of sampling options are applied to this field depending on monitoring objectives. Online/near-real-time monitoring systems are used for round-the-clock monitoring of key pollutants such as ozone precursors [22–24,71,72] in urban air (Figure 10.20) and odorous sulfur compounds (hydrogen sulfide, methane thiol, dimethyl sulfide, and dimethyl disulphide) near landfill sites and sewage treatment works [73,74]. Off-line sampling options include both sorbent tubes [75,76] (Figure 10.10) and whole-air sampling into canisters [77] (Figure 10.11). Applications for whole-air sampling options such as canisters include the more volatile 'air toxic' [78] species and ultra-volatile trace greenhouse gases – CF_4, C_2F_6, SF_6, etc. [79]. Generally speaking, however most indoor and outdoor air-monitoring applications are more conveniently sampled using pumped sorbent tubes [17,27,56,78,80–83]. Some detailed studies of air pollution in metropolitan areas have been carried out using diffusive sampling onto sorbent tubes. The low cost of passive (diffusive) sampling facilitates the collection of large numbers of samples, allowing accurate mapping of pollution isopleths – see Figure 10.21 [84].

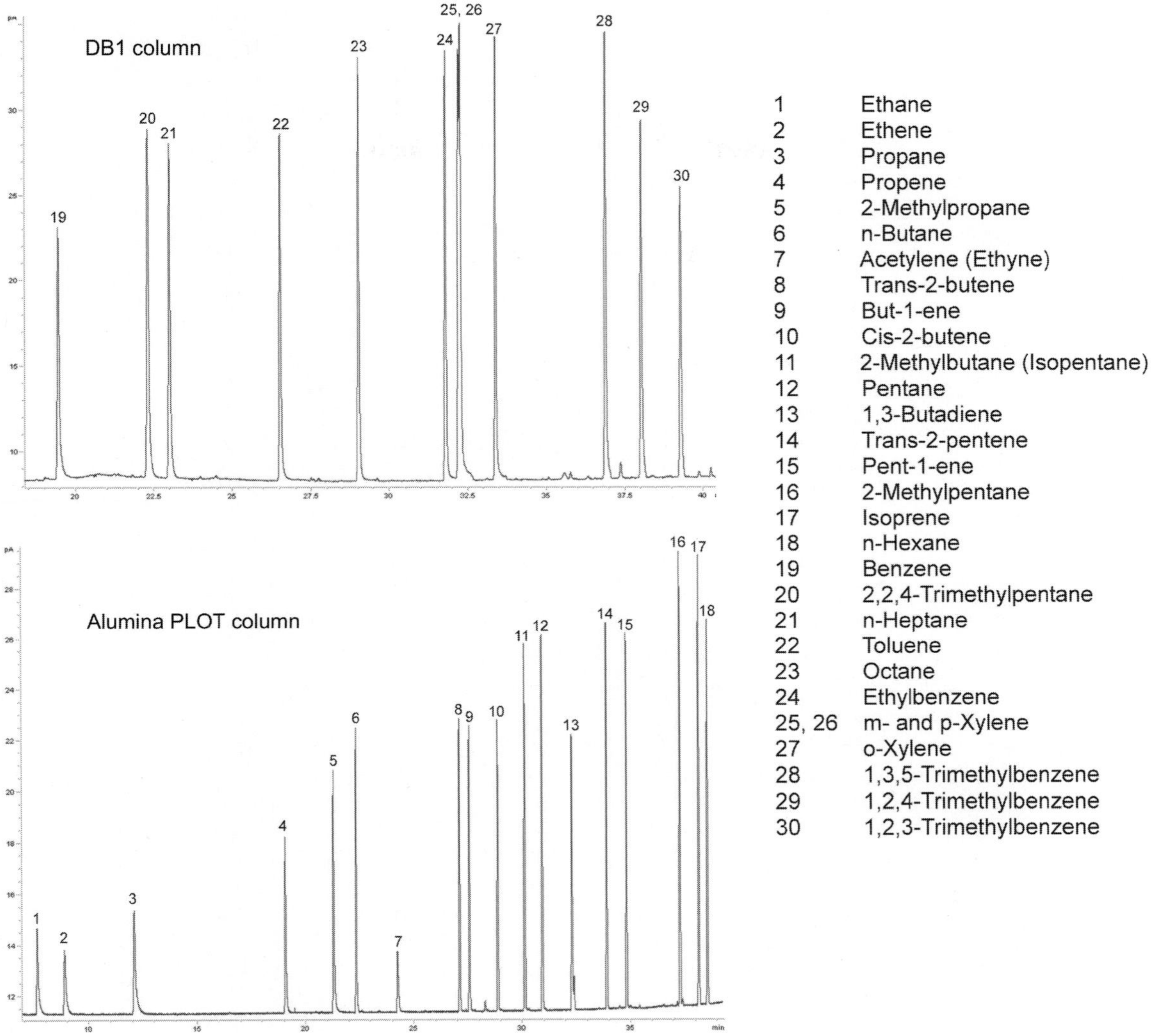

FIGURE 10.20 On-line monitoring of 'ozone precursors' – C_2 to C_9 hydrocarbons – using TD-GC with dual column / Deans switch configuration and dual FID. Experimental conditions: UNITY-Air Server on-line thermal desorption system (Markes International Ltd., UK) combined with a 7890 GC configured with Deans switch, 'Micro-fluidics' (Agilent Technologies). Multisorbent ('ozone precursor') trap at −30 °C. Trap desorption: 40 °C/s to 325 °C. Flow path: 80 °C. Columns: 60 m × 0.25 mm I.D. × 1.0 μm film DB1 (top trace) and 50 m × 0.32 mm × 8μm Alumina PLOT (Na_2SO_4 wash) (Agilent Technologies) (bottom trace). GC oven: 30 °C (12 min) 5 °C/min to 170 °C then 15 °C/min to 200 °C. Deans switch at 17.5 min.

10.8.4. Industrial (fugitive) Emissions – Stack (flue) Gases and Perimeter Monitoring

Thermal desorption has always been popular for monitoring around the perimeter of industrial plants as a check on the impact of industrial emissions. Again the simplicity of diffusive monitoring mean 10 or 20 samplers can be cost-effectively deployed around a site, allowing accurate mapping of pollution/contamination levels under different wind and weather conditions [85].

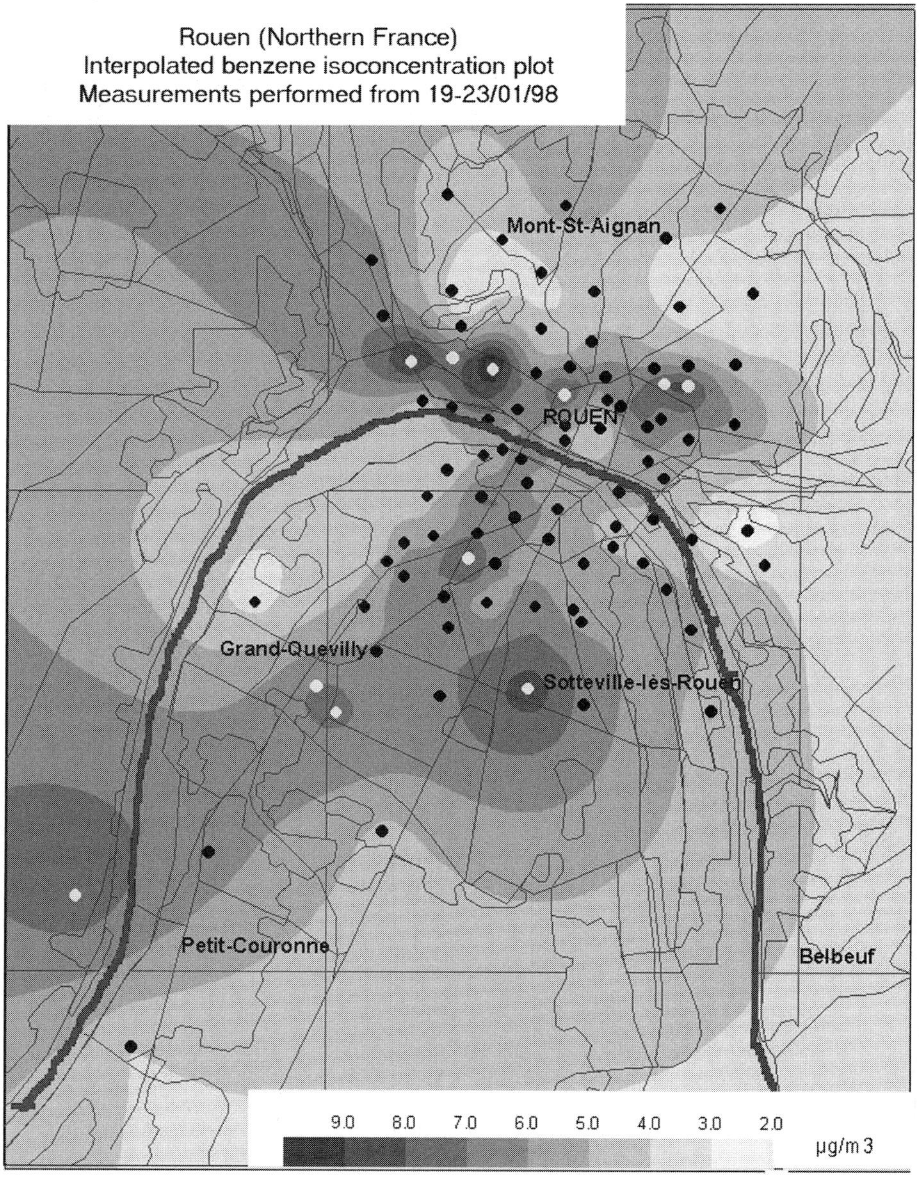

FIGURE 10.21 Mapping criteria pollutants in urban air using diffusive sampling.

While most bulk organic vapor measurements in industrial emissions (stack or flue gases) are made using sensors (continuous emission monitoring (CEM) technology), lower-level toxic organics are measured using sorbent tubes with either solvent extraction or thermal desorption. Key test methods have included EPA Methods 0030, 0031, and 5041A in the US and EN 13649 in Europe [86]. While originally restricted to solvent extraction only,

EN 13649 is currently being revised to include thermal desorption in response to demand from industry.

10.8.5. Atmospheric Research

Recent fears about the impact of air pollution on climate and other finely balanced natural systems such as the ozone layer, have led to numerous national and international research projects into atmospheric pollution. Investigated issues have included global background pollution (monitoring some of the cleanest atmospheres on earth [87,88]) long range pollution transport, biogenic emissions, atmospheric chemistry [89,90], and measurements of air–seawater interactions [91].

These studies typically involve vapor concentration measurements at ppt or even ppq concentrations requiring the best available thermal desorption-GC technology coupled with high-sensitivity MS detection (negative-ion chemical ionization (NCI), time-of-flight, etc.). Preferred sampling options include pumped sorbent tubes or canisters [88], depending on target analyte range.

10.8.6. Soil Gas and Vapor Intrusion into Buildings

As the human population continues to expand, there is increasing pressure to redevelop disused industrial land rather than build on 'green field' sites. However, as in the case of the infamous Love Canal incident mentioned earlier, there is always concern that the residue from chemical processes originally carried out on a site, or leaks from disused chemical/fuel storage tanks, may remain in the soil. Redevelopment of industrial land therefore requires detailed assessment and remediation of any identified contamination. Even after remediation (cleanup), long-term monitoring of soil gas, or air from inside buildings developed on the site, may be required to make sure that the risk from any residual pollution remains low. Occasionally it is necessary to adapt building construction to minimize the risk of vapor intrusion – for example, by installing impermeable membranes in the basement.

The US has led the way in this field, and many related monitoring methods are now under development, for example, within ASTM. Cited soil gas (or 'under slab') sampling approaches include using canisters (limited to compounds more volatile than n-$C_{9/10}$) and active or passive sampling onto sorbent tubes (compatible with a wider volatility range). Analysis is by thermal desorption in either case [7,92–94].

10.8.7. Water Odor

A few specific odorous organic compounds – e.g. geosmin, methyl-i-borneol, and trichloranisole – are responsible for a high proportion of consumer complaints about drinking-water quality. They are typically detectable down to 10 ppt by the human nose and, while they do not present an actual health hazard at these levels, their musty, earthy smell is a real concern to members of the public. Conventional GC sampling methods such as static headspace or purge and trap do not offer sufficient sensitivity for this demanding application and it is one of the areas where sorptive extraction or HS-TD (HS-trap) has recently been found to present a potentially useful automatic alternative. Detection limits in the order of 1–2 ppt have been reported [35] – see Figure 10.22.

10.8.8. Monitoring Tracer Gases, for Example, in Studies of Building Ventilation

Tracer gases typically comprise perfluorinated compounds (sulfur hexafluoride or perfluoromethylcyclohexane) which do not occur naturally and are readily detected at low levels by GC methods (e.g. by using electron-capture detection or GC-MS in NCI mode). Sources of tracer gases are placed in various locations in

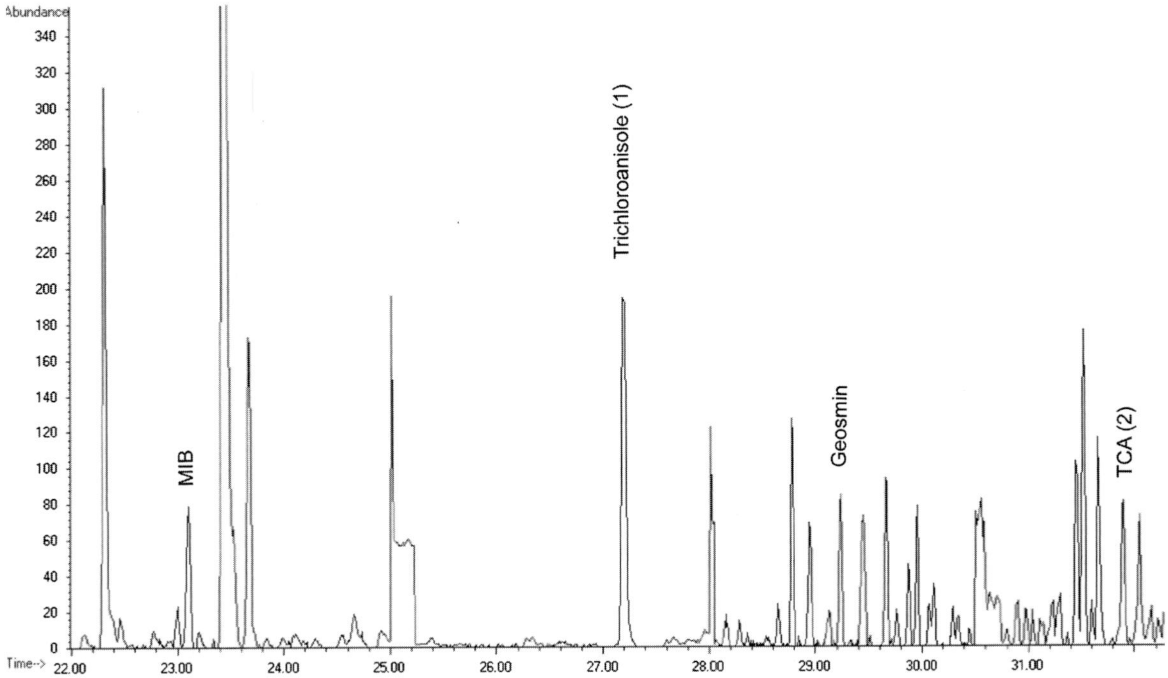

FIGURE 10.22 5 ppt level odorants in drinking water analyzed by TD-GC-MS (SIM). Experimental conditions: HS-5TD system (Markes International Ltd., UK) combined with a 6890-5975 GC-MS system (Agilent Technologies). Multisorbent ('general purpose') trap at 25 °C. Vial temp: 45 °C. 10 HS vial sampling cycles at 50 ml/min. 2 min trap purge. Trap desorption: 50 °C/s to 300 °C splitless. Flow path: 160 °C. Column: 60 m × 0.25 mm I.D. x 0.25 μm film DB-1701 (Agilent Technologies). GC oven: 50 °C (2 min), 5 °C/min to 175 °C then 20 °C/min to 260 °C. SIM ions: Gp 1 – 95, 108, 135, Gp 2 – 195, 197, 210, 212, Gp 3 – 112, 182.

buildings or vehicles. On- or off-line TD–GC methods are then used to monitor the concentrations as they change over time, thus allowing ventilation efficiency to be evaluated. Both passive and active sampling methods have been applied [49,95].

In one interesting aside, halogenated organic compounds are now known to play a key role both in ozone depletion (i.e. in damaging the ozone layer) and to have very high global warming potential (5,000–10,000 times more than that of CO_2). They also have a long half-life in the atmosphere. SF_6, for example, can now be detected around the planet at ~6 ppt and Freon 113 at ~75 ppt – see Figure 10.23. While these levels do not present any immediate or significant environmental risk, scientists are nevertheless concerned to make sure their research does not contribute to global pollution levels. The use of tracer gases has therefore been minimized in recent years.

10.9. CHEMICAL EMISSIONS FROM EVERYDAY PRODUCTS TO INDOOR AIR

Fears relating to global warming have also driven new regulations relating to the energy performance of buildings [96]. Unfortunately, this well-intentioned legislation has had the unwanted side effect of significantly reducing building ventilation levels and impacting indoor air quality. Whereas air turnover in

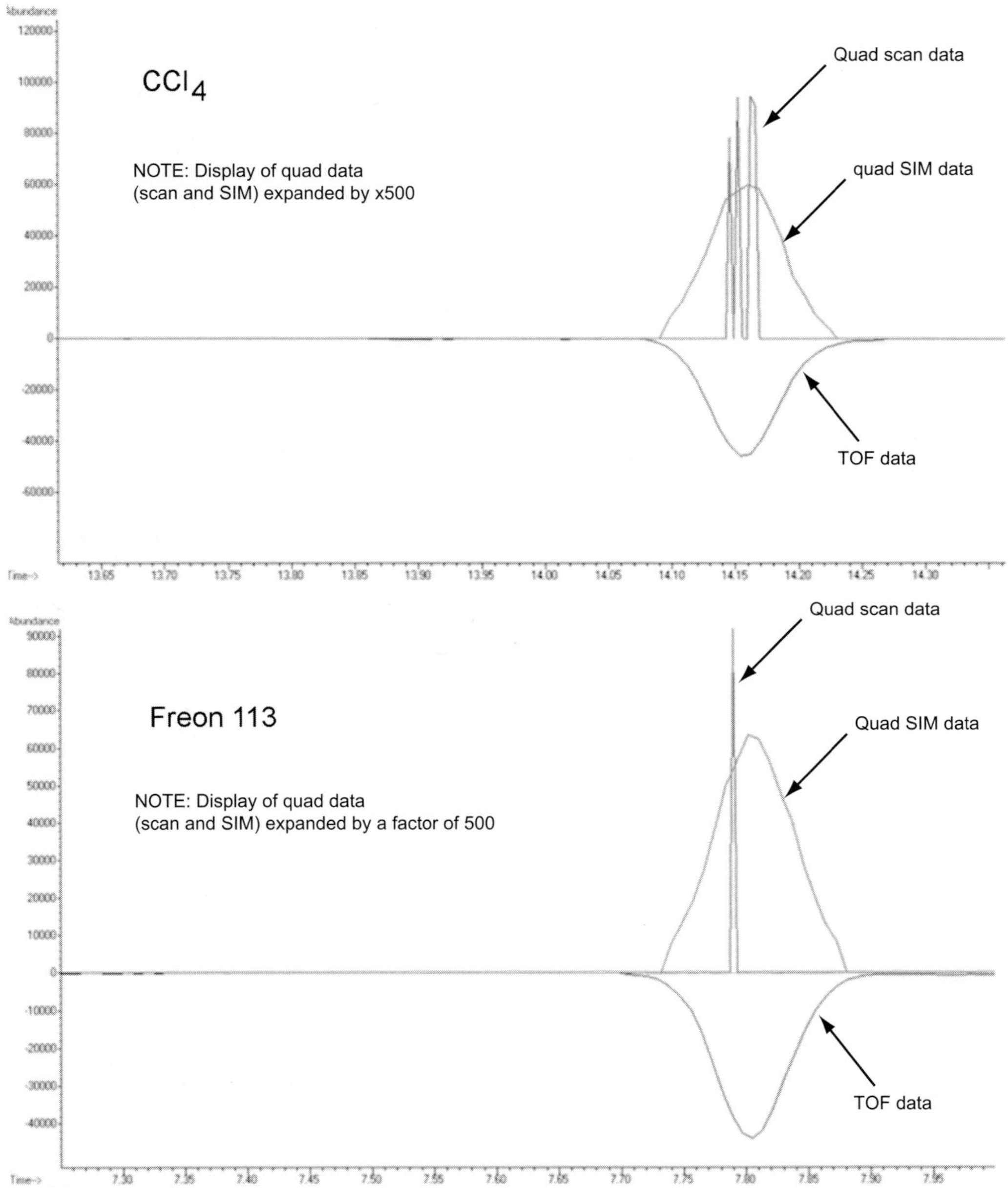

FIGURE 10.23 Comparing TD-GCMS (quad) with TD-GC-MS (TOF) results for the analysis of carbon tetrachloride and Freon 113 in 200 ml forest air. Experimental conditions: TD-100 (Markes International Ltd., UK) linked to a 7890-5975 GCMS system (Agilent Technologies) and a 7890 GC with a BenchTOF-dx system (ALMSCO International Ltd.). Common parameters: Stainless steel tube packed with carbon black and carbon molecular sieve. Tube desorption: 10 minutes at 320 °C into a multisorbent ('air toxics') trap at +25 °C. Trap desorption: 320 °C at 40 °C/s with 5 ml/min split. Flow path: 140 °C. Column: 60 m × 0.32 mm I.D. x 1.8 μm film DB-624 (Agilent Technologies). GC oven: 35 °C to 230 °C at 5 °C/min. Quad MS: 20–300 amu (scan) and SIM groups: 1) 56, 67, 93, 151; 2) 57, 78, 106; 3) 58, 91, 117. Masses acquired using TOF MS: 20–300 amu. Note: Display of SIM and SCAN data from quad MS expanded by factor of 500 relative to TOF data.

North European or US housing stock was traditionally in the order of 1 or 2 air changes per hour, this has been reduced to 0.2 air changes per hour or even lower in some new homes/offices [97,98]. Reports of adverse health effects have therefore led to increased focus on controlling the sources of indoor pollutants, including chemical emissions from products used indoors [99]. Construction products, decorative materials, car interior trim components, furniture, and cleaning products have all come under the spotlight. Even natural materials that have been used for centuries in traditional housing may compromise air quality when installed in modern, airtight dwellings.

New regulations [100–102], test protocols [103–105], and analytical methods [106–108] have been developed to address these concerns. Those relating to vapor-phase organic chemicals predominantly specify the use of sorbent tubes with subsequent TD–GC-MS analysis. [28,109].

Reference methods for material emission testing generally specify small environmental chambers (typically 20–1000 L volume) or test cells, both of which can be used to evaluate chemical emissions from products and materials under simulated real-use conditions. Samples are usually prepared such that only the surface exposed to the indoor environment is exposed in the test chamber or cell. Pure humidified air is then driven into the chamber under specified conditions of temperature, humidity, time, etc. After a given period, the exhaust air is sampled and analyzed as described above to measure the area-, mass-, or component-specific chemical emission rate. Tests are usually carried out over an extended period – typically 3, 7, 14, or 28 days – to simulate airborne concentrations soon after product installation or building occupation.

Faster emission screening methods using microchambers have been developed recently [112,113] to complement long-term reference tests and to provide industry with a practical tool for routine in-house use. Microchambers are also used in combination with sorbent tubes and TD–GC-MS to measure emitted organic vapors – see Figure 10.24 – or with DNPH cartridges and subsequent HPLC analysis for formaldehyde emissions.[†]

There are numerous publications and reports describing emission tests from construction products and car trim [114,115]. Other everyday products, which are commonly subjected to emission testing, include furniture, furnishings, and toys – see Figure 10.25. Note that even though the toy emission profiles shown in Figure 10.25 were obtained at relatively low temperatures (40 and 90 °C, respectively), they still show emission of significant levels of key SVOCs – including phthalates several of which are now designated 'substances of very high concern' under REACH[‡], and bisphenol A which is a known endocrine disruptor.

Emission testing is now being implemented more widely within manufacturing industry in the interests of consumer safety and to take advantage of market demand for low-emission products. Specific applications include quality control of production, development of new low-emission product ranges, monitoring raw materials, and comparison against best-in-class competitors. Similar procedures are used for quality control of volatile and semivolatile chemical emissions from sensitive electronic components such as PC hard-drives.

[†] N.B. Formaldehyde is the primary exception to this. It can be analyzed by TD–GC(MS) but is so reactive and prone to hydrolysis that it is very difficult to store in its free state. Most reference methods therefore specify sampling using cartridges impregnated with dinitrophenylhydrazine (DNPH). This reacts with the formaldehyde to form a more stable derivative which is then analyzed by solvent extraction and HPLC [110,111].

[‡] REACH: European Directive [2006/121/EC] on the Registration, Evaluation, Authorization and restriction of CHemicals.

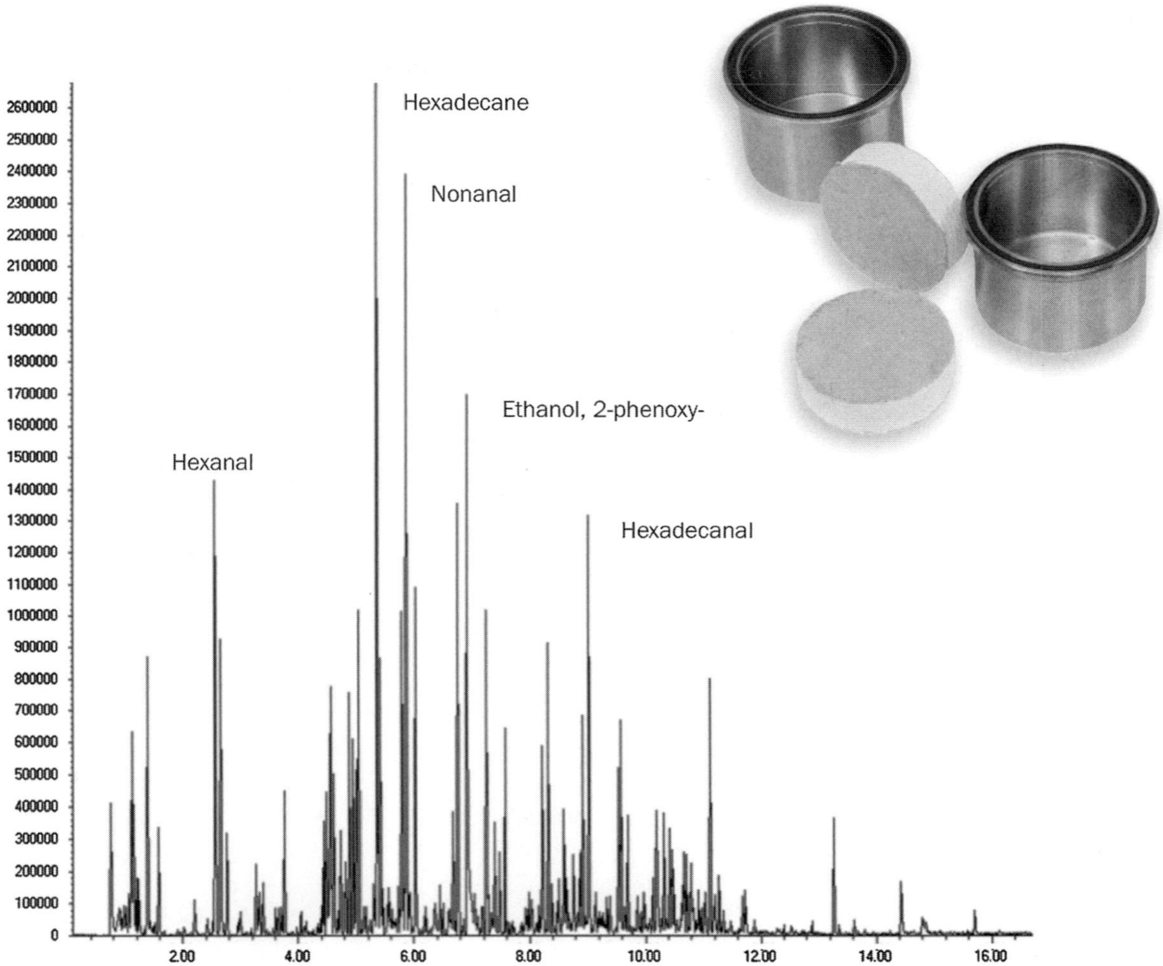

FIGURE 10.24 Screening chemical emissions from plasterboard (dry wall) using test chambers / microchambers or emission cells with sorbent tubes and TD-GCMS. Experimental conditions: Micro-Chamber/Thermal Extractor (Markes International Ltd) set at 50 °C with a flow of 100 ml/min dry air. Equilibration time: 20 min, vapor sampling time: 15 min. Sampling onto stainless steel tube packed with Tenax TA. TD-100 automated thermal desorber (Markes International Ltd., UK) combined with a 7890–5975 GC-MS system (Agilent Technologies). Tube desorption: 5 minutes at 280 °C into a multi-sorbent qtz/Tenax/Carbograph 5 TD trap at +25 °C. Trap desorption: 300 °C with 30 ml/min split flow. Flow path: 150 °C. Column: 60 m × 0.25 mm I.D. × 0.5 μm film DB5 (Agilent Technologies). GC oven: 40 °C to 225 °C at 10 °C/min. Scan range: 35–300 amu.

10.10. TOXIC CHEMICAL AGENTS AND CIVIL DEFENSE

While in many ways a simple extension of air monitoring, the use of thermal desorption for detection of extremely toxic compounds, such as chemical warfare agents, is now considered a field in its own right. Relevant example applications include the following:

- battlefield protection – using sorbent tubes with TD–GC(MS) to evaluate protective equipment – e.g. monitoring the

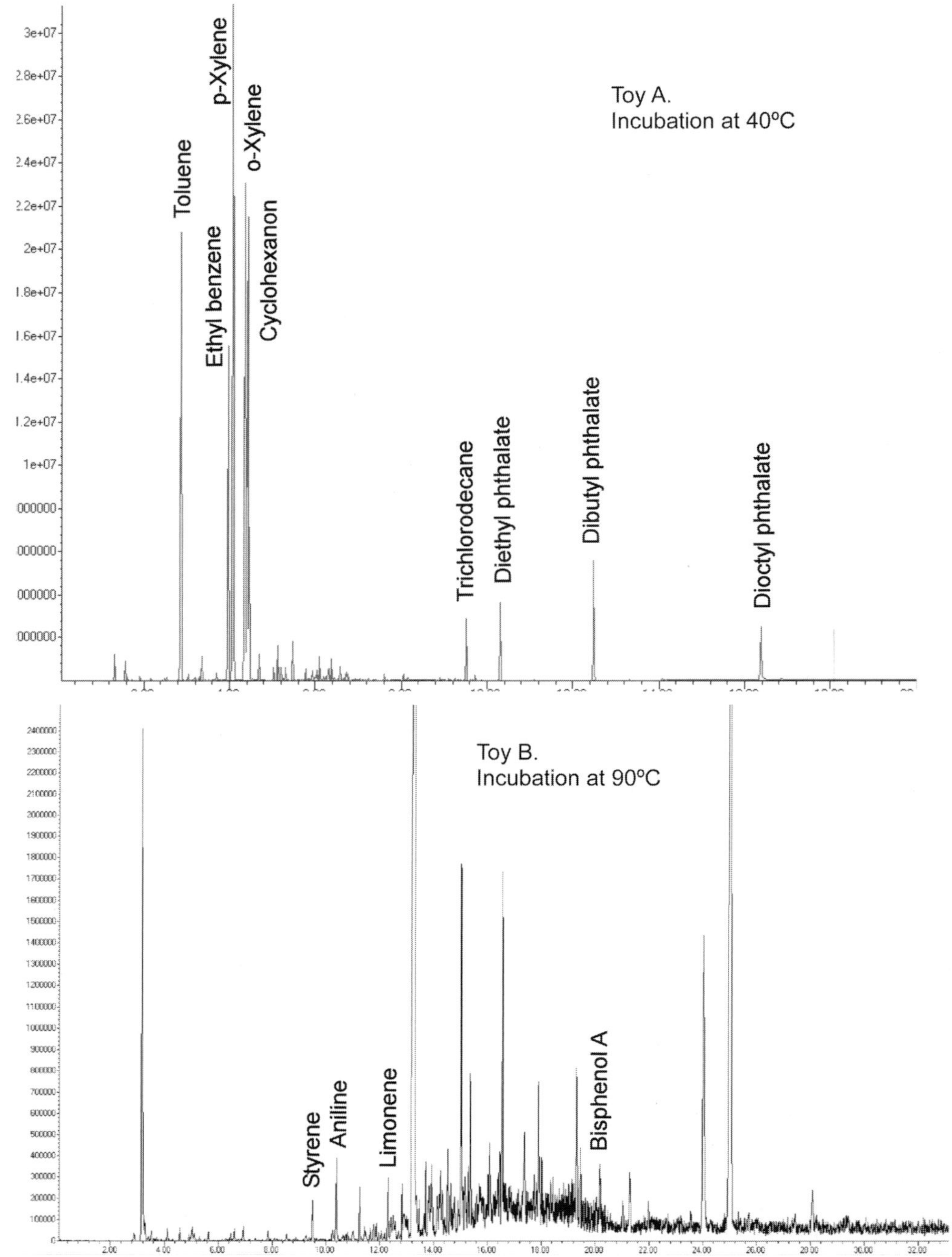

FIGURE 10.25 Screening volatile and semivolatile chemical emissions from children's toys Experimental conditions: Micro-Chamber/Thermal Extractor (Markes International Ltd) set at 40 °C and 100 ml/min flow (top chromatogram) or 90 °C/50 ml/min flow (bottom chromatogram). Equilibration time: 20 min, vapor sampling time: 15 min in each case. Similar analytical conditions to Fig. 10.24.

permeation of agent through masks and clothing.

- first responders – teams equipped with mobile laboratories and trained to be the first on the scene in the event of a major chemical incident.
- monitoring agent storage and destruction facilities – to ensure the safety of personnel and protect the nearby environment.
- studies of decontamination technology and protective coatings – e.g. paints designed to prevent agent from penetrating into the fabric of buildings or machinery to simplify decontamination in the event of a chemical attack.
- continuous monitoring of critical civilian locations – transport hubs, key government buildings, etc.

Many of the most toxic chemical warfare agents present a major analytical challenge – because of (1) the low detection limits required (e.g. 0.0006 μg/m^3 for general population exposure) and (2) the nature of the chemicals themselves. Early thermal desorption technology was not compatible with many of the highest boiling or most reactive CW agents. VX, for example, was traditionally monitored by sampling the air through silver fluoride pads to produce the more volatile 'G-analog' [116]. However, this conversion process was rarely 100% efficient and the performance tended to diminish as the pads aged, leading to risk of under-reporting. The latest on- and off-line thermal desorption technology is compatible with free-VX even at the lowest levels (Figure 10.26), thus removing the need

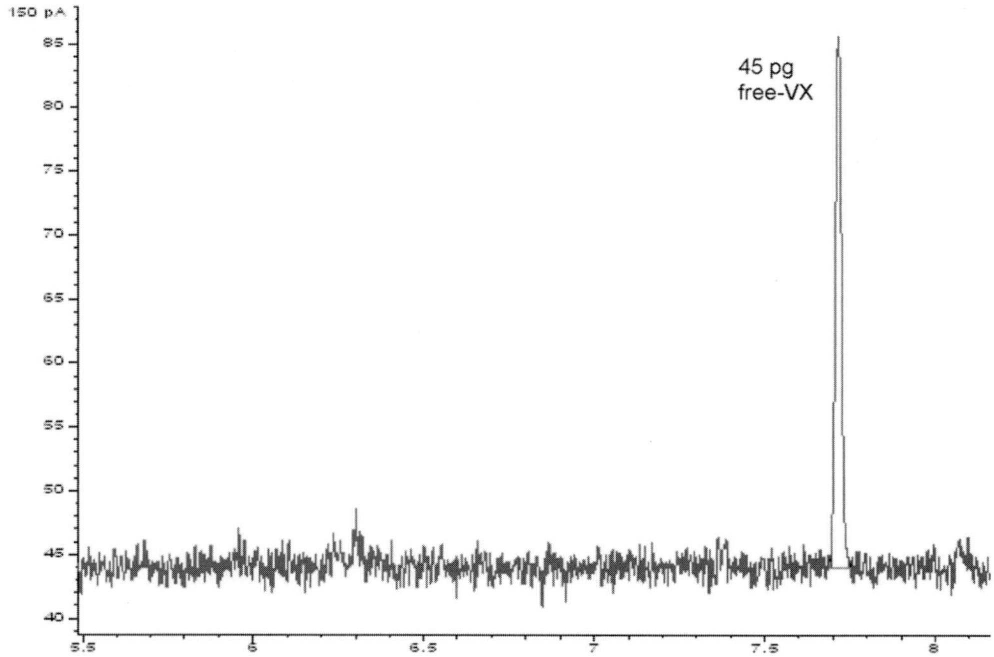

FIGURE 10.26 TD-GC-FPD analysis of free-VX at 45 pg level on tube Experimental conditions: Silcosteel tube packed with Tenax TA. UNITY 2 thermal desorber (Markes International Ltd., UK) combined with a 6890 GC/FPD system (Agilent Technologies). Tube desorption: 8 min at 300 °C into a multisorbent qtz/Tenax trap at +20 °C. Trap desorption: 300 °C splitless. Flow path: 200 °C. Column: 30 m × 0.25 mm I.D. × 0.5 μm film HP-5 (Agilent Technologies). GC oven: 60 °C to 250 °C at 20 °C/min. FPD: 250 °C, H_2 150 ml/min, air 110 ml/min, N_2 55 ml/min.

for derivatization and reducing analytical uncertainty.

Much of the impetus for developing twin-trap TD configurations also came from chemical-agent-monitoring applications. Twin-trap systems allow air or gas streams to be sampled continuously, thus generating near-real-time data without any unsampled time or 'blind spots'. While air is being drawn into trap A, trap B is desorbed and analyzed. Once trap B has cooled, it can be switched to sampling, thus allowing trap A to be desorbed. Data shown in Figure 10.27 illustrate continuous monitoring of the chemical agent Sarin (GB) using one such system. Typical applications for this kind of technology include continuous monitoring of government buildings against terrorist attack and process monitoring at agent destruction facilities to ensure the safety of site personnel.

10.11. TD–GC(MS) ANALYSIS OF RESIDUAL VOLATILES

Direct thermal desorption is now widely applied for measurement of residual volatiles and VOC content. As described above, many relatively dry and homogeneous materials can be conveniently weighed into empty sample tubes or tube liners (Figure 10.7) for what is effectively a gas extraction or dynamic head-space process. When operated in this mode the thermal desorber combines sample preparation/extraction and GC injection into one fully automated procedure.

The technique works best if samples present a high surface area to mass ratio (powders, granules, fibers, and films). The power and simplicity of direct thermal desorption relative to conventional static (equilibrium) headspace

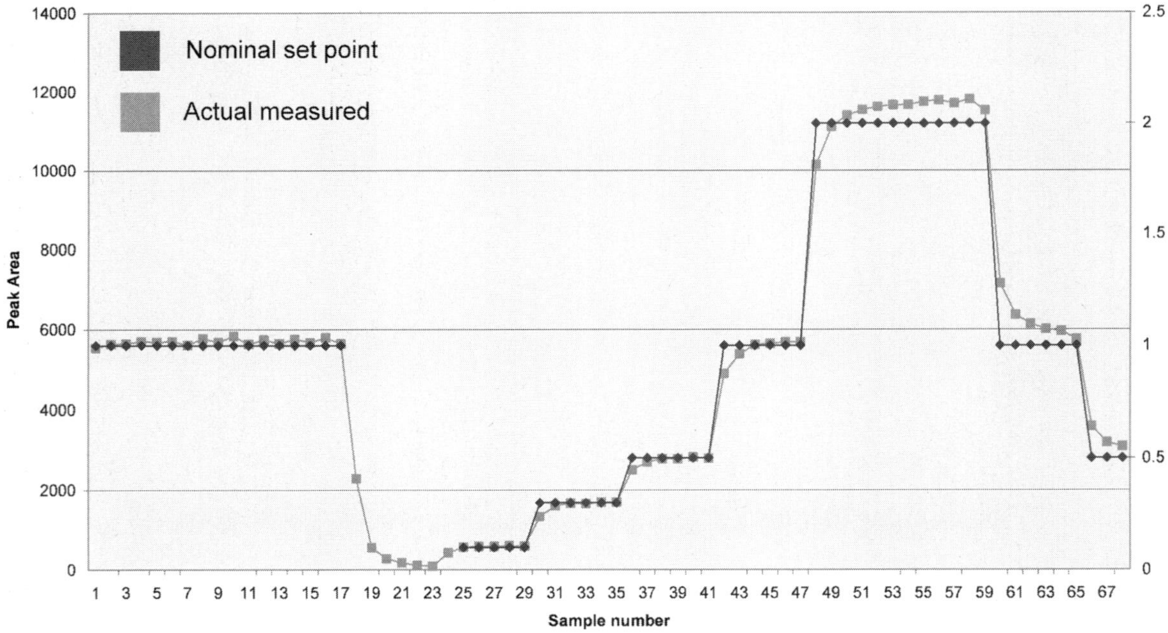

FIGURE 10.27 Using twin-trap thermal desorption equipment TT24-7 (Markes International Ltd., UK) for continuous, near-real-time monitoring of the chemical warfare agent GB (Sarin) in a dynamically generated standard atmosphere *(see reference [54])*.

methods are that no additional dissolution or salting-out steps are required and it does not rely on partition coefficients or equilibria. Moreover, complete (or nearly complete) extraction is often possible in one run, thus simplifying calibration. Another advantage of direct thermal desorption is that it can be applied to much smaller sample sizes (e.g. 2 or 3 mg rather than the 2 or 3 g often required for conventional headspace). This makes it suitable for some forensic applications (see below) and for measuring residual solvents when there is a limited supply of sample material e.g. prototype pharmaceuticals.

Many direct desorption applications fall under the general header of manufacturing QA/QC. Materials suitable for this approach include the following:

- pharmaceutical powders and medicinal ointments – see Figure 10.28 [15];
- soap powders;
- textiles and treated leather – see Figure 10.29;
- paints and adhesives [117];
- polymers in powder, granule, film, or fiber form [15];
- packaging materials [118]; and
- house dust [119].

10.12. FLAVOR, FRAGRANCE, AND ODOR PROFILING

Direct thermal desorption can also be applied to the characterisation of flavor and fragrance profiles. Relevant applications include the profiling of dried foodstuffs – spices, instant coffee, cocoa powder, tobacco, etc. In this case, the intention is not to get complete extraction but to obtain a representative odor/aroma profile. Temperatures are typically kept low to prevent denaturing the sample. Direct TD (dynamic headspace) fulfills a different purpose than conventional equilibrium headspace for this type of work. Whereas standard HS selectively concentrates the most volatile constituents, thermal desorption allows analysis of a wider, more representative profile of the overall aroma (odor), including both volatile and less volatile components. The technique is also very sensitive, accentuating even minor differences in composition – 5% less of a given ingredient, 10% more of another – thus allowing very precise quality control [120,121].

HS-TD and sorptive extraction (see above) complement direct thermal desorption by providing tools that can be used to study organic volatiles and semivolatiles in aqueous samples such as wine, beer, fruit juices [35,122–124], and food extracts. When applied together these two approaches provide a comprehensive profile of organic chemicals in natural products – Figure 10.17.

The conventional on- or off-line sorbent sampling modes of thermal desorption are also extensively applied to monitoring fragrance, aroma, and odor profiles in air.

Overall, this is a growing field with many interesting applications and research opportunities. Highlights include the following:

- accelerated food shelf-life studies – sometimes using similar microchamber technology to that applied to material emission testing [125];
- the composition of air freshener profiles and the kinetics of fragrance decay;
- identification of crop emissions, plant species, and plant health [126–128];
- monitoring flavor/aroma quality in genetically modified foods [129];
- studies of insect/plant interactions – pheromones, etc. [130];
- identifying moulds, fungi, and bacteria. [131];
- tracking sources of taint and off-odor; and
- studies of human body odor and bad breath [132,133].

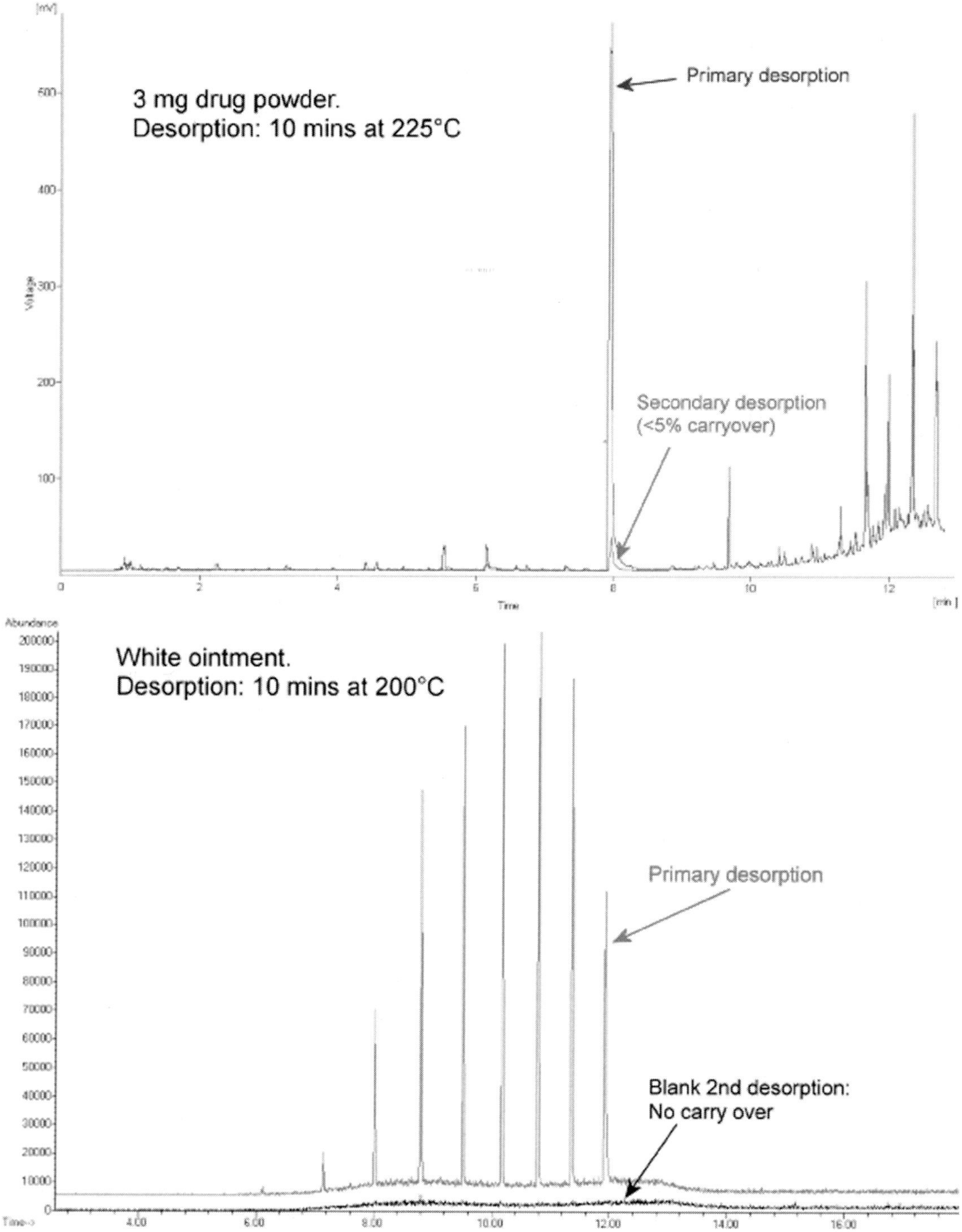

FIGURE 10.28 Direct desorption of residual solvents and/or active ingredients from pharmaceutical preparations.

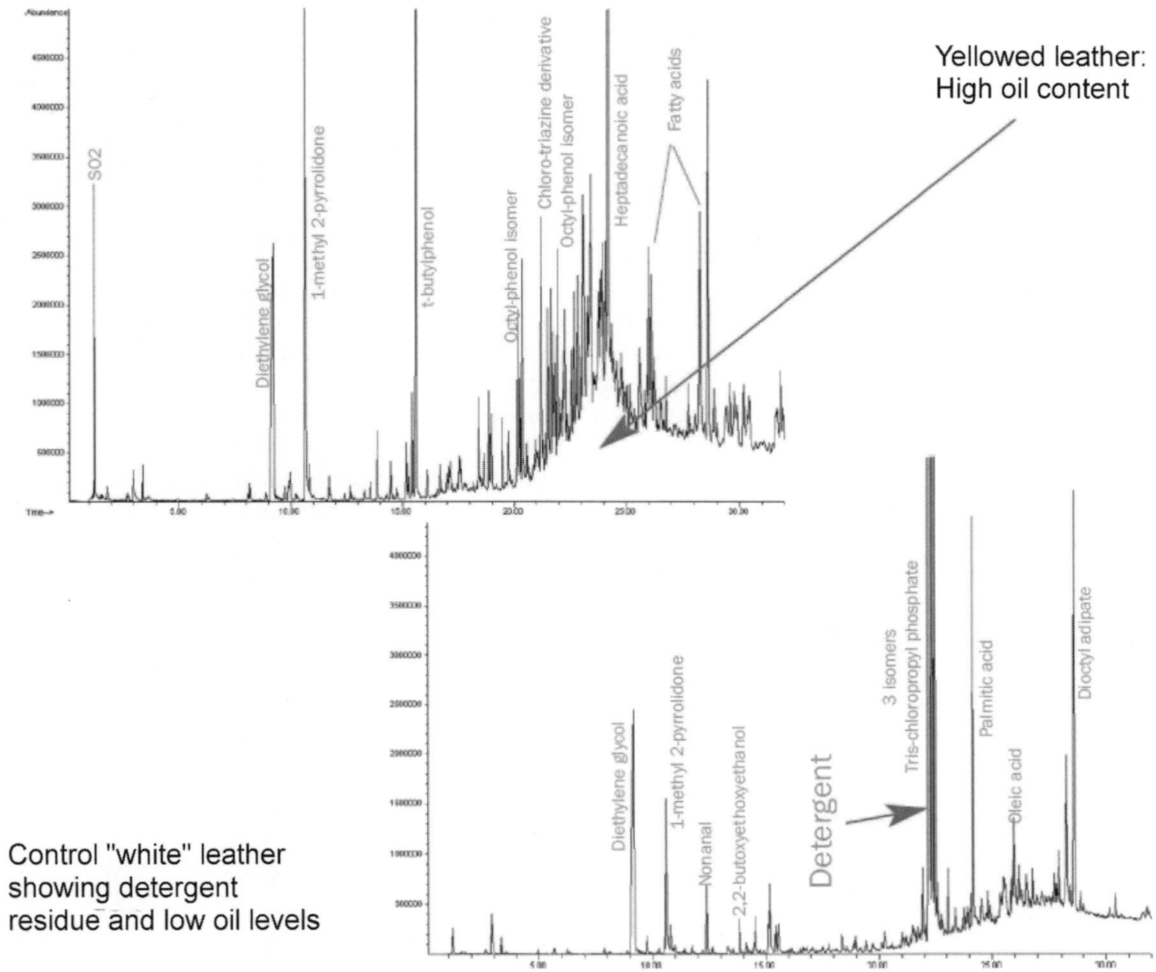

FIGURE 10.29 Direct desorption of discolored and control samples of white leather. The control sample shows detergent residue and much lower levels of natural oils. Experimental conditions: 1.5 mm × 10 mm sections of control and discolored leather were desorbed at 150 °C for 5 min in a flow of 60 ml/min carrier gas (split 50:50) using a UNITY TD system (Markes International Ltd., UK). Trap: packed with Tenax/Carbograph 1 held at −10 °C. Trap desorption: 300 °C with 30 ml/min split flow and 2 ml/min column flow. Flow path: 200 °C. Column: 30 m × 0.32 mm I.D. × 1.0 μm film DB 1 (Agilent Technologies). GC oven: 60 °C (5 min hold) to 280 °C at 10 °C/min. Scan: 45−350 amu.

10.13. FORENSIC APPLICATIONS

The general advantage of thermal desorption as a GC sample introduction technology for forensic applications is that it requires little or no sample preparation – i.e. it involves very few manual steps. This is important for forensic science because it means that there is less risk of compromising data and thus of evidence being challenged in court. Relevant application examples are discussed in the following.

10.13.1. Accelerants in Fire Debris

TD is often used to determine the presence or absence of 'accelerants' (i.e. fuels such as gasoline or kerosene) in burned residue from the scene of suspected cases of arson [134]. Typically, the fire debris is collected in inert containers such as metal cans or nylon bags. Pumps or large gas syringes are then used to draw the headspace vapors through well-conditioned sorbent tubes. The volume of vapor sampled varies considerably because the only objective in this case is to confirm or exclude the presence of accelerants. Small volumes (20–50 ml) are sufficient if the sample smells of fuel.

Thermal desorption offers several advantages over conventional static headspace methods for this application. First, it is not limited to specific container sizes – e.g. headspace vials; thus, it allows larger, more representative samples of the fire debris to be analyzed, giving enhanced sensitivity and better reliability. Moreover, the presence of large quantities of liquid water in many fire residue samples can present a challenge to detection of trace levels of higher-boiling accelerants using conventional HS. In comparison, the dynamic sampling process of TD allows both selective elimination of water and enhanced detection of less volatile compounds.

10.13.2. Drugs of Abuse

Many drug-related applications benefit from the versatility of thermal desorption. Examples include direct desorption of house dust from the scene of a crime (Figure 10.30), direct desorption of bank notes [135], and detection of amphetamine 'factories' by monitoring the air for indicative solvents. The example shown in Figure 10.30 is interesting, not only because the levels of drugs were so high in this case (the dust was found to be nearly 3% heroin/cocaine) but also because phenobarbital – an anti-epileptic treatment for dogs – was identified at the same time. Were the police able to use this finding to narrow their investigations to drug dealers who had recently visited the local vet?!

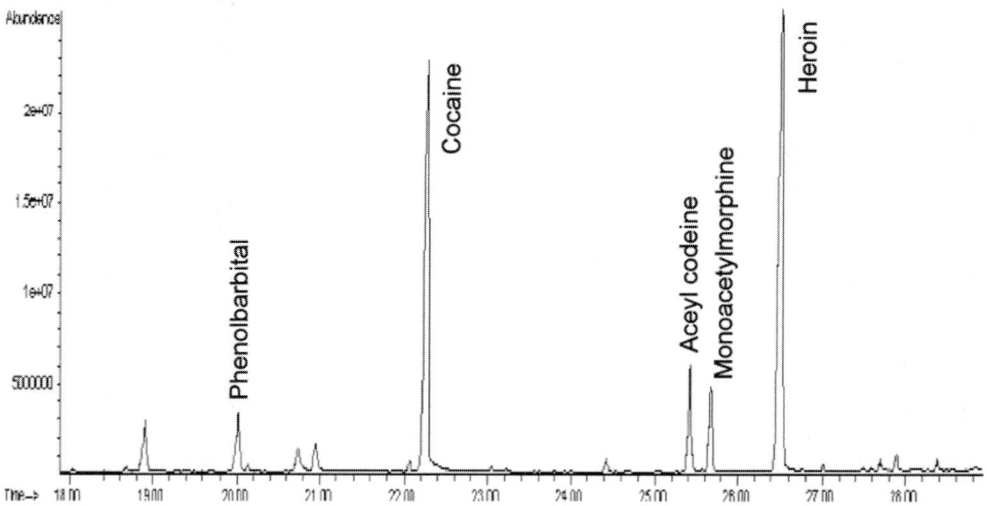

FIGURE 10.30 Direct desorption of proscribed drugs in house dust. Experimental conditions: ~5 mg dust weighed into empty glass tubes supported by quartz wool plugs. Desorption: 150 °C for 10 min in a flow of 50 ml/min carrier gas using an ULTRA-UNITY automated TD system (Markes International Ltd., UK).Trap: packed with quartz/Tenax held at +20 °C. Trap desorption: 40 °C/s to 250 °C with 30 ml/min split flow and 2 ml/min column flow. Flow path: 150 °C.

Further, sorptive extraction with TD provides a convenient analytical approach for detecting proscribed substances in biological fluids [136].

10.13.3. Explosives and Shotgun Propellant Residues

Explosive vapors present a significant challenge even to conventional GC analysis. These highly reactive compounds are very sensitive to any slight deterioration in system performance, for example, injector activity, column age, and detector contamination. It is therefore a testament to the quality of modern TD technology that two-stage thermal desorption of trace-level DNT and TNT is now considered routine and that even more challenging compounds such as RDX and PETN can be detected at trace levels – see Figure 10.31. The use of thermal desorption to detect and characterize ('fingerprint') residual propellant on spent shotgun cartridges has also been reported and can be used to link individual cartridges or weapons to specific crimes.

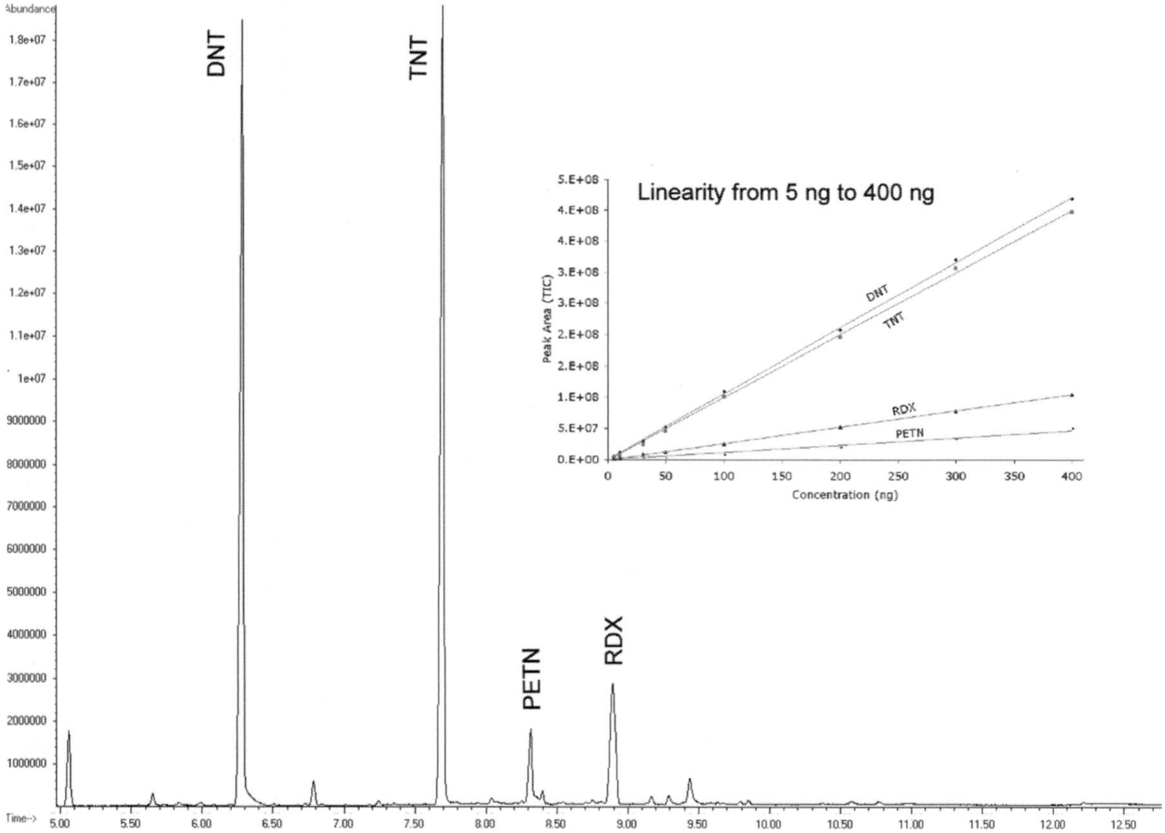

FIGURE 10.31 Using TD-GC-MS for detection of trace explosive vapors. Experimental conditions: Air sampled into Silcosteel tube packed with quartz wool and Tenax TA. UNITY 2 thermal desorber (Markes International Ltd., UK) combined with a 6890 GC/FPD system (Agilent Technologies). Tube desorption: 3 min at 180 °C followed by 2 min at 210 °C into a multisorbent qtz/Tenax trap at +20 °C. Trap desorption: 300 °C at 20 °C/s with 50 ml/min split flow and 3 ml/min column flow. Flow path: 180 °C. Column: 30 m × 0.25 mm I.D. × 0.5 μm film HP-5 (Agilent Technologies). GC oven: 60–250 °C at 20 °C/min. Scan range: 35–300 amu.

10.13.4. Forensic (characterisation) of Inks, Paper, and Other Materials

Many other materials can be reliably characterized from their VOC profile. Key examples include paper, ink, and natural materials such as plants and their fossilized derivatives [126,137].

Figure 10.32 shows direct desorption of organic volatiles/semivolatiles from paper with and without writing. Investigations like this may be used to link the document to a particular paper source and specific ink/pen. The extent of selective losses of the most volatile constituents in the ink can also be useful in estimating the age of a document.

10.14. MONITORING MANUFACTURING AND OTHER INDUSTRIAL CHEMICAL PROCESSES

The concentration power of thermal desorption makes it invaluable for detecting leaks in dangerous industrial chemical processes. Key examples include monitoring chemical agent destruction facilities and chemical syntheses that generate lethal byproducts such as bis(chloromethyl) ether [138]. In each case, continuous or very regular monitoring is necessary to ensure the safety of plant personnel.

TD–GC-MS systems are also increasingly used for routine product quality control and development of low-emission materials. Relevant industries include construction products, electronic components, car interior trim, carpeting, furniture, decorative materials, and consumer products (air fresheners, domestic cleaning agents, etc.) – see above.

Other examples of TD implementation for industrial/process applications include the following:

- monitoring trace impurities in CO_2 supplying the food and beverage industries,
- monitoring tracer gases for leak detection in critical fuel pipes or lines, and
- control of monoethylene glycol (MEG) and other additives to domestic fuel gas supplies [139].

When used to monitor trace organic impurities in process gas streams such as CO_2, thermal desorption may sometimes be coupled directly to real-time detectors such as sensors ('e-nose' technology) or process mass spectrometry. The GC can be eliminated in these cases provided the process gas stream is well characterized, and as long as the range of failure modes and potential contaminants are well known. Elimination of the GC, where applicable, minimizes cycle times and allows rapid detection/notification of any contaminants exceeding control levels.

10.15. NEW GC-RELATED TECHNOLOGY DEVELOPMENTS WHICH BENEFIT THERMAL DESORPTION

The concentration power and application range of thermal desorption are complemented by a number of recent developments in other GC-related fields. Some of the most important examples are discussed in the following.

10.15.1. Mass Spectrometry

Many of the recent developments in mass spectrometry for GC are directly relevant to thermal desorption. Triple quadrupole mass spectrometers, for example, are well suited to TD applications that require detection of specific target compounds at ultra-low levels [79,88,91]. However, given that TD is so often used for uncharacterized atmospheres/materials and/or for screening large numbers of compounds (see general references), the development of GC-compatible time-of-flight (TOF)

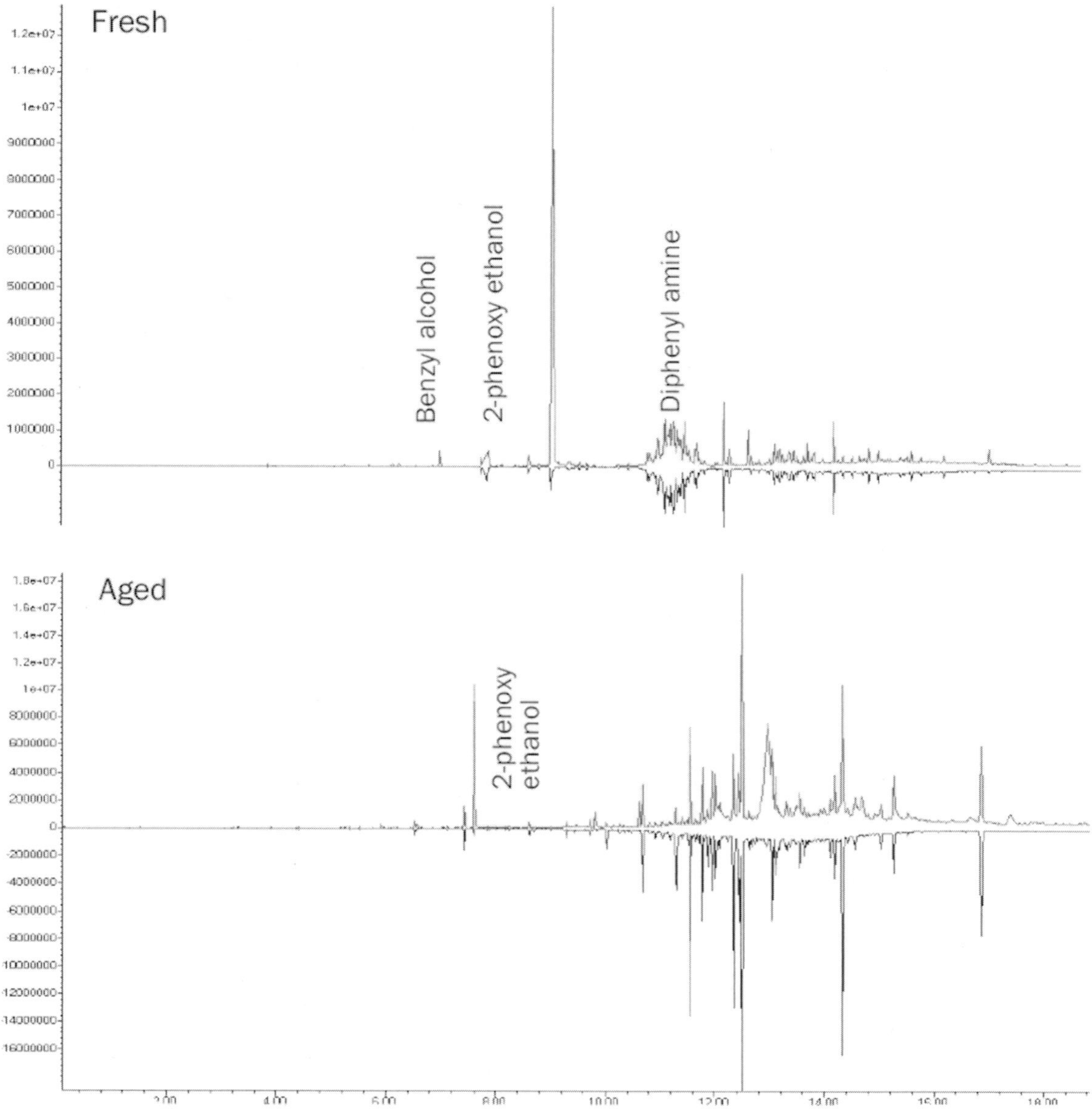

FIGURE 10.32 Direct desorption of documents for forensic characterisation of paper and inks. Experimental conditions: Direct desorption of ~2 mm × 20 mm sections of paper (with and without ink) at 100 °C for 5 min in a flow of 30 ml/min carrier gas using a UNITY 2 thermal desorber (Markes International Ltd., UK). Trap: packed with Tenax/Carbograph 1 held at +25 °C. Trap desorption: 300 °C with 20 ml/min split flow. Flow path: 200 °C. Column: 30 m × 0.32 mm I.D. × 1.0 μm film DB 1 (Agilent Technologies). GC oven: 60 °C (5 min hold) to 280 °C at 10 °C/min. Scan: 45–350 amu.

mass spectrometers, which offer high sensitivity together with full spectral information, is particularly exciting.

10.15.1.1. *Sensitivity*

As TOF technology does not involve scanning, it should theoretically offer 2 or 3 orders of magnitude better sensitivity than conventional quadrupole MS systems operating in SCAN mode (depending on mass range). Actual GC–TOF sensitivity varies significantly from instrument to instrument, but some modern systems do live up to their billing, and are able to provide full spectral information at levels approaching single-ion detection limits on a regular quad-MS system – see Figure 10.23. This aids quantitative and qualitative analysis of toxic or odorous analytes at the lowest possible levels and, from a TD perspective, is an obvious advantage for screening/characterizing unknown samples and for atmospheric research.

Using high-sensitivity MS detectors can also benefit routine air/gas monitoring applications by allowing collection of smaller samples without compromising method detection limits. For example, 100 ml of air analyzed by TD–GC–TOF may provide as much qualitative and quantitative information as 10 L of air analyzed using a conventional quad-MS based system. Quick/easy low-volume 'grab' sampling options [140] and short-term diffusive monitoring can both significantly simplify air monitoring in this case.

Another advantage of high-sensitivity TOF MS is that it facilitates implementation of more robust and TD-relevant configurations of comprehensive GC (GC × GC). While the immense resolving power of GC × GC has already been usefully applied to particularly complex thermal desorption applications such as breath and atmospheric research [90,141], most system configurations have, until recently, been based on thermally modulated GC × GC technology. This uses intermittent cooling and heating to achieve a 2-dimensional separation and works best in expert hands [142]. More robust flow-modulated GC × GC technology is better suited to thermal desorption because it offers enhanced compatibility with volatile compounds [143]. However, as it requires flows in the order of 20 ml/min, implementation with MS has previously been limited by sample splitting. Given that the latest GC–TOF systems offer roughly 100 times better sensitivity than regular benchtop systems operating in SCAN mode, the necessity to split is no longer so much of an issue – flow-modulated TD–GC × GC–TOF systems can still readily exceed standard TD–GC-MS method performance requirements without limiting the collection of full spectral data.

10.15.1.2. *Speed*

TOF is the preferred MS technology for fast GC and GC × GC operation because it generates in the order of 10,000 full spectra per second (typically from 10 to 1500 amu and above) without scanning. This eliminates the spectral skew that can sometimes be observed when using conventional MS systems to scan fast peaks and improves signal to noise. The high data collection rate of TOF MS also allows more efficient application of 'data mining' algorithms, such as spectral deconvolution, which are increasingly important for TD applications – see below.

10.15.1.3. *Mass Resolution and Stability*

Some of the most recently introduced GC–TOF technologies offer high mass resolution (e.g. to 5 ppm) such that individual compounds can be identified from their accurate mass. These systems are typically at the high end of the cost scale and may not offer the best available sensitivity, but they do provide an invaluable tool for some important environmental applications such as distinguishing the most toxic PCBs or dioxins within a complex mixtures of congeners.

However, while accurate mass is rarely a critical issue for the primarily volatile-related applications of thermal desorption, mass resolution better than 1 amu (e.g. 0.1 or 0.01 amu) can be a real advantage, provided it is combined with mass stability. This is because subunit mass resolution allows selective elimination of bulk interferences (Figure 10.33), thus enhancing identification of trace toxic or odorous compounds. Relevant thermal desorption

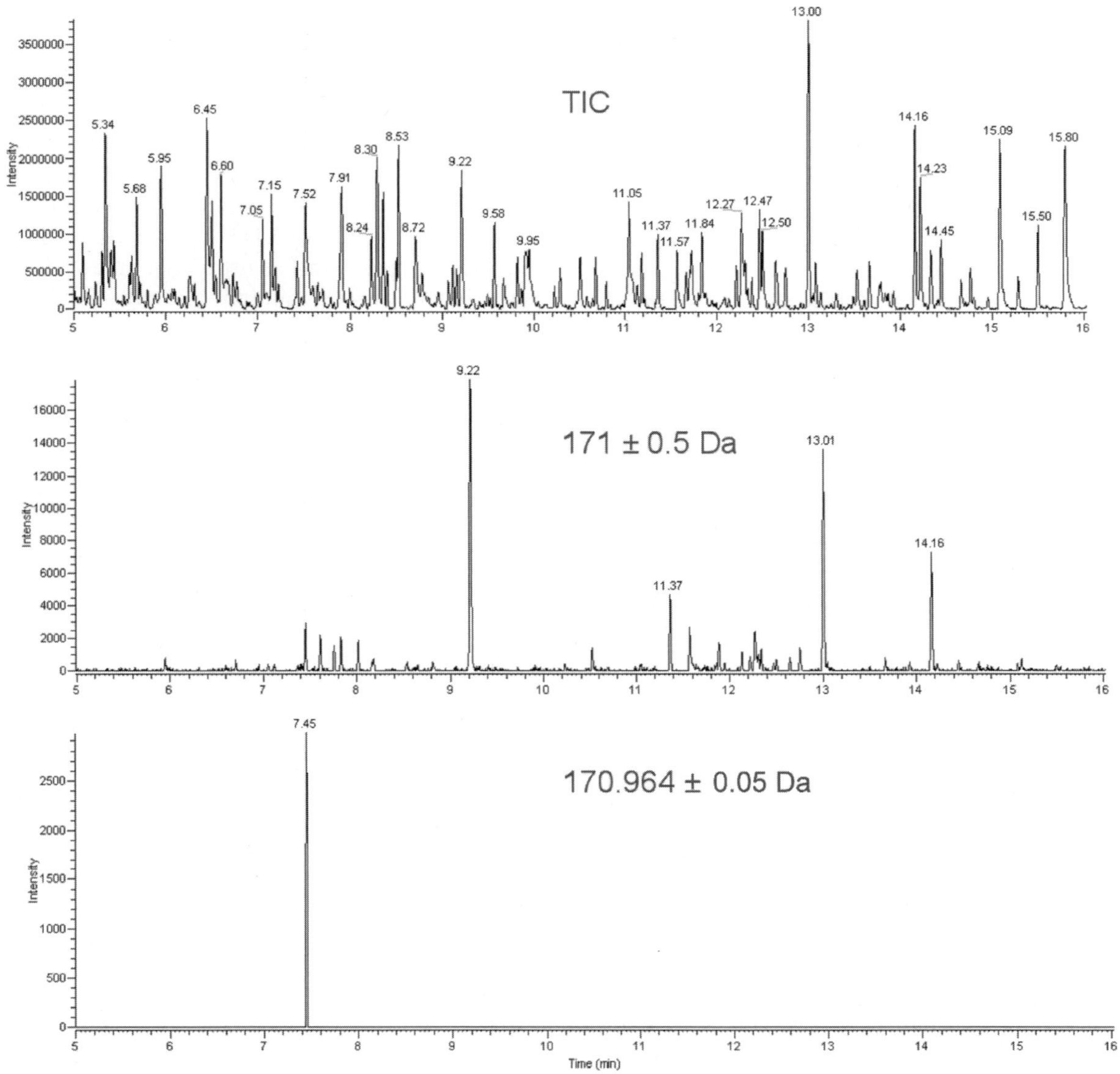

FIGURE 10.33 Applying subunit-mass resolution to minimize chromatographic background and enhance signal to noise in TD-GC-TOF analysis of trace components. Detection of 1 pg dichlobenil in unprocessed apple extract using an 7890 GC (Agilent Technologies) combined with BenchTOF-dx (ALMSCO International, UK).

applications include environmental research, chemical agent detection, and odor profiling.

10.15.2. Real-Time Organic Vapor Monitors such as Sensors or Process Mass Spectrometry

On-line TD configurations are occasionally coupled directly to technology for direct measurement of organic vapors – i.e. without a GC interface. Key examples include process mass spectrometry or organic vapor sensors (sometimes called 'e-noses'). In both cases, thermal desorption improves detection limits (e.g. from ppm to ppb) and allows selective elimination of bulk interferences such as air, water, and ethanol, which could otherwise swamp (mask) detector response to the compounds of interest.

Near-real-time TD configurations involving direct read-out detectors work best for monitoring stable, well-characterized samples such as industrial gas streams or processes. They are well suited for ensuring that vapor composition is within control limits and for detecting significant variations. That said, when a deviation is detected, conventional TD–GC-MS invariably offers the most flexible and powerful analytical option for investigating the cause.

10.15.3. GC-MS Data-Mining Software

Many TD applications result in complex organic profiles – see Figure 10.24 – with the odorous or toxic compounds of interest frequently comprising the smallest components in a chromatogram. New MS technologies such as time of flight can help by virtue of generating accurate (classical) spectra at increased data acquisition rates and with better than unit-mass resolution to eliminate interferences – see above. However, without the benefit of advanced/automated processing functionality such as spectral deconvolution, data analysis remains a skilled and time-consuming task.

Many powerful and automated 'data mining' software packages are now available commercially and can be applied to enhance TD–GC-MS data analysis postrun. They typically

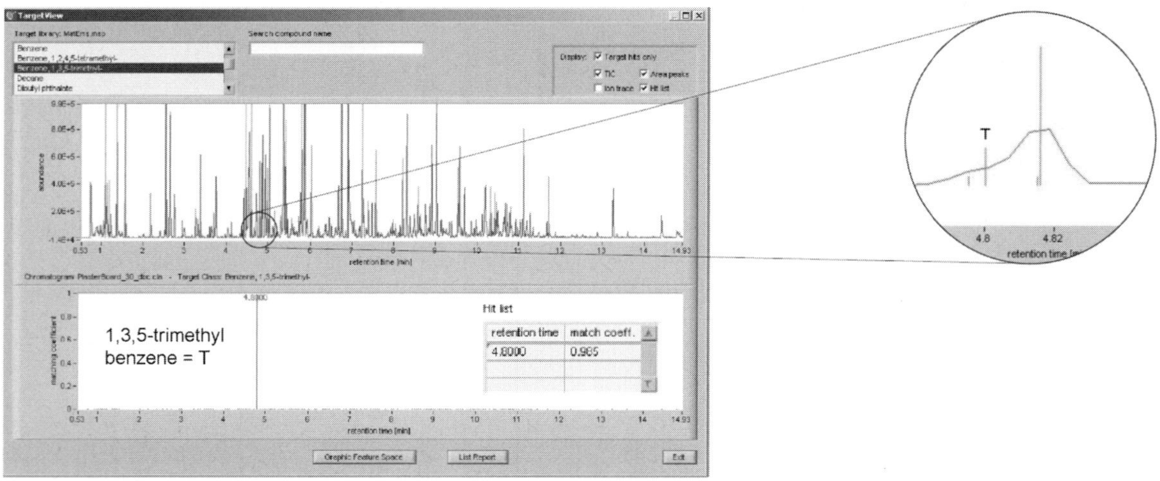

FIGURE 10.34 Application of data-mining software combining spectral deconvolution and principal component analysis to enhance RT-independent detection of trace target compounds in complex chromatographic profiles. EXAMPLE: Application of TargetView (ALMSCO International) to detect trace levels of 1,3,5-trimethylbenzene in a composite peak within the complex TD-GC-MS emission profile of plasterboard sample (see Figure 24).

comprise spectral deconvolution algorithms which 'separate' (deconvolute) coeluting compounds by evaluating the rate of change of individual mass ions. This allows the software to identify the numbers of components present in a single chromatographic peak and assign individual mass ions and peak profiles to respective constituents.

One of the original and most widely used spectral deconvolution packages for GC-MS was developed by NIST in the US and is now available free of charge under the trade name AMDIS (Automated Mass Spectral Deconvolution and Identification System). Other more automated options have since become available that combine AMDIS-type or proprietary spectral deconvolution algorithms with additional screening parameters such as retention time locking or principal component analysis. These can be applied to screen complex TD–GC-MS total ion data sets (such as that shown in Figure 10.24) for multiple trace target analytes – Figure 10.34. Results are produced automatically within minutes, and should be at least as reliable as those that can be generated manually by a GC-MS expert over several hours.

10.16. CONCLUDING REMARKS

The versatility and power of thermal desorption are unequaled among GC frontend technologies, and adoption of TD–GC(MS) methods is accelerating – both generally and in specific application areas such as material emission testing. The question remains, however, as to why it has taken almost 30 years since the introduction of the first reliable commercial units for this expansion to really get going. That is to say, while you might expect at least one in ten GC/GC-MS systems to benefit significantly from interfacing to TD given the application range (considerably more if you consider all the TD-associated procedures such as purge and trap, headspace trap, and sorptive extraction) why is actual take-up still only a fraction of this?

Probably the most significant limiting factor historically has been availability. Modern, high-performance thermal desorption technology was not supplied by many of the mainstream GC(MS) manufacturers until very recently. Now that this situation has changed, this is beginning to lead to a better general understanding of how to get the most out of the technique and a much broader appreciation of its versatility and application potential – gone are the days when thermal desorption was pigeonholed into a narrow range of VOC air-monitoring applications.

Recent technical advances such as addressing the one-shot limitation of earlier systems and implementing robust automation (cryogen-free operation, reliable tube sealing, etc.) have also made it possible for regulators to rely more heavily on TD-GC-based methods – both for the service laboratory sector and, more recently, for quality control of manufacturing. Going forward, it is possible to see that growth in TD–GC-MS applications will accelerate even faster – both in the area of prepackaged 'ruggedized' systems for regulated industrial applications and, in combination with some of the new GC-MS technologies, for advanced research.

References

[1] E.A. Woolfenden, Ch. 5: Practical aspects of monitoring volatile organics in air, in: G. Subramanian (Ed.), Quality Assurance in Environmental Monitoring: Instrumental Methods, VCH, 1995.

[2] ISO 16200-1 Workplace air quality – Sampling and analysis of volatile organic compounds by solvent desorption/Gas chromatography. Part 1: Pumped sampling method.

[3] B. Callan, K. Walsh, P. Dowding, Industrial hygiene VOC measurement interference, Chem. Ind. 5 (April, 1993) 250–252.

[4] D. Coker, Personal monitoring techniques for gases and vapours, Int. Environ. Saf. J. (April, 1979) 43–44.

[5] Annual report of the UK Health and Safety Executive. 1978 ISBN 0-11-883285-9.

[6] P.A. Hollingdale-Smith, A. Bailey, Passive sampling and dosimetry. In 'Trace-organic sample handling. Volume 10 of Methodological Surveys (A): Analysis. Published by Ellis Horwood Ltd, UK, Distributed by Halstead Press a division of John Wiley & Sons.

[7] E.A. Woolfenden, A review of sorbent-based sampling methods for volatile and semi-volatile organic compounds in air. Part 1 – Sorbent-based air monitoring options, J. Chromatogr. A 1217 (2010) 2674–2684.

[8] J. Manura, Selection of GC guard columns for use with GC cryo-trap, Sci. Instrument Services Inc., 1999. Application Note 24a.

[9] B. Kolb, J. Chromatogr. A 842 (1–2) (1999) 163.

[10] MW. Holdren, DL. Smith, Performance of automated gas chromatographs used in the 1990 Atlanta ozone study. Proceedings of the 1991 US EPA/AWMA International Symposium, Measurement of Toxic and Related Air Pollutants; Air and Waste Management Assoc. Pittsburgh, USA; 1991.

[11] D.T. Coker, N. van den Hoed, K.J. Saunders, P.E. Tindle, A monitoring method for gasoline vapour giving detailed composition, Ann. Occup. Hyg. 33 (1) (1989) 15–26.

[12] J. Kristensson, M. Widen, Development and evaluation of a diffusive sampler for measurements of anaesthetic gases, in: A. Berlin, R.H. Brown, K.J. Saunders (Eds.), Diffusive sampling – an alternative approach to workplace air monitoring, RSC Publication, 1987, pp. 423–426.

[13] J. Kristensson, Diffusive sampling and GC analysis of volatile compounds. PhD Thesis, Dept. of Analytical Chemistry, Arrhenius Laboratory, Stockholm University, 1987.

[14] Woolfenden EA, A novel approach to the determination of volatile organics in pharmaceuticals, polymers and food stuffs. Proceedings of the Pitsburgh Conference, New York; 1990.

[15] A.P. Bianchi, M.S. Varney, Sampling and analysis of volatile organic compounds in estuarine air by gas chromatography and mass spectrometry, J. Chromatogr. 643 (1993) pp. 11–23.

[16] V.M. Brown, D.R. Crump, D. Gardiner, Measurement of volatile organic compounds in indoor air by a passive technique, Environ. Tech. 13 (1992) 367–375.

[17] R. Janson, J. Kristensson, Sampling and analysis of atmospheric monoterpenes, Dept. of Meteorology, Stockholm University, 1991. Report CN-79.

[18] PerkinElmer Thermal Desorption Application Note N. 20: The determination of residual Freon 11 in dried vegetable matter. PerkinElmer Corp: USA.

[19] PerkinElmer Thermal Desorption Application Note No. 26: The thermal desorption of volatiles from food packaging film. PerkinElmer Corp: USA.

[20] E.A. Woolfenden, G.M. Broadway, An overview of sampling strategies for organic pollutants in ambient air, LC-GC Int 5 (12) (1986) 28–35.

[21] L. Purdue, Technical assistance document for sampling and analysis of ozone precursors, US Environmental Protection Agency, 1991. EPA/600-8-91/215.

[22] D. Kotzias, J. Hjorth, M. Duane, J.V. Eijk, Sampling and analysis of selected volatile organic compounds (VOC) relevant for the formation of photochemical oxidants. in: Proceedings of the conference: 'Reactivite chimique de l'atmosphere et measure des polluants atmospheriques' Grenoble, France; 1990.

[23] J. Gibich, L. Ogle, P. Radenheimer, Analysis of Ozone Precursor Compounds in Houston, Texas Using Automated, Continuous Gas Chromatographs. In: Proceedings of the AWMA Conference: 'Measuring Toxic and Related Air Pollutants', US, May; 1994. pp. 164–191.

[24] R. Muir, W.A. Carrick, D.B. Cooper, Application of central composite design in the optimisation of thermal desorption parameters for the trace level determination of the chemical warfare agent chloropicrin, Analyst 127 (2002) 1198–1202.

[25] P. Perez-Ballesta, Losses from ATD-400. 'The Diffusive Monitor', Issued by HSE/CAR WG5, Issue 9; 1997.

[26] J. Kristensson, Repeat analysis: the diffusive monitor, 1, UK HSE/CAR WG5, 1988, p. 3.

[27] ASTM D6196 Standard practice for selection of sorbents, sampling and thermal desorption analysis procedures for VOCs in air (and material emissions chambers).

[28] FDIS 16000-6. Indoor air – Part 6: Determination of VOCs in indoor and test chamber air by active sampling on Tenax TA sorbent, thermal desorption and gas chromatography using MS/FID.

[29] US EPA Compendium Method TO-14: The determination of volatile organic compounds in ambient air using Summa passivated canister sampling and gas chromatographic analysis; 1988.

[30] US EPA Compendium Method TO-15: Determination of volatile organic compounds in air collected in specially-prepared canisters and analyzed by gas chromatography/mass spectrometry; 1999.

[31] ASTM D5466: Standard test method for determination of volatile organic chemicals in atmospheres (canister sampling methodology).

[32] S. Davies, M. Bates, D. Wevill, L. Kelly, K. Thaxton, One system for trace and high level air monitoring. The future of ambient air and soil gas analysis. Proceedings of Pittcon; 2011.

[33] US EPA Method 524.2: Measurement of purgable organic compounds in water by capillary column GCMS; 1995.

[34] US EPA Method 8260B: Volatile organic compounds in solid waste by GCMS; 1996.

[35] L. Kelly, E.A. Woolfenden, Enhanced GC–MS aroma profiling using thermal desorption technologies, J. Sep. Sci. 1 (2008) 16–23.

[36] A. Prieto., O. Basauri, R. Rodil, A. Usobiaga, L.A. Fernandez, N. Etxebarria, et al., Stir-bar sorptive extraction: a view on method optimization, novel applications, limitations and potential solutions, J. Chromatogr. A 1217 (2010) 2642–2666.

[37] N. Bukowski, Accurate and reliable analysis of beer using time-of-flight technology for gas chromatography. American Laboratory; 2010.

[38] J.F. Pankow, M.P. Ligocki, M.E. Rosen, L.M. Isabelle, K.M. Hart, Adsorption/thermal desorption with small cartridges for the determination of trace aqueous semivolatile organic compounds, Anal. Chem. 60 (1987) 40–47.

[39] E.A. Woolfenden, A review of sorbent-based sampling methods for volatile and semi-volatile organic compounds in air. Part 2 – sorbent selection and other aspects of optimizing air monitoring methods, J. Chromatogr. A 1217 (2010) 2685–2694.

[40] M. Philips, Breath tests in medicine, Scientific American, July, 1992. 74–79.

[41] D. Dyne, J. Cocker, H.K. Wilson, A novel device for capturing breath samples for solvent analysis, Sci Total Environ 199 (1997) 83–89.

[42] L. Gunnarsen, P.A. Nielsen, P. Wolkoff, Design and characterization of the CLIMPAQ chamber for laboratory investigations of materials, pollution and air quality, Indoor Air 4 (1994) pp. 56–62.

[43] P. Wolkoff, An emission cell for measurement of volatile organic compounds emitted from building materials for indoor use – the field and lab. Emission cell, Gefahrstoffe-Reinhaltung der Luft 56 (1996) 151–157.

[44] T. Schripp, B. Nachtwey, J. Toelke, T. Salthammer, E. Uhde, M. Wensing, et al., A microscale device for measuring emissions from materials for indoor use, Anal. Bioanal. Chem. 387 (5) (2007) 1907–1919.

[45] K.J. Saunders, Air monitoring goes underground, The diffusive monitor, vol. 3, HSE/CAR WG5, 1989.

[46] Kristensson, Soil-probe. The diffusive monitor, vol. 4, HSE/CAR WG5, 1991.

[47] E. Wauters, P. Van Caeter, G. Desmet, F. David, C. Devos, P. Sandra, Improved accuracy in the determination of polycyclic aromatic hydrocarbons in air using 24 h sampling on a mixed bed followed by thermal desorption capillary gas chromatography-mass spectrometry, J. Chromatogr. A 1190 (2008) 286–293.

[48] M. Bates, P. Bruno, M. Caputi, M. Caselli, G. de Gennaro, M. Tutino, Analysis of polycyclic aromatic hydrocarbons (PAHs) in airborne particles by direct sample introduction thermal desorption GC/MS, Atmos. Environ. 42 (2008) 6144–6151.

[49] D. Helmig, L. Vierling, Water adsorption capacity of the solid adsorbents tenax TA, tenax GR, carbotrap, carbotrap C, carbosieve SIII, and carboxen 569 and water management techniques for the atmospheric sampling of volatile organic tracer gases, Anal. Chem. 67 (23) (1995) 4380–4386.

[50] A. Tipler, Water management in capillary gas chromatographic air monitoring systems. Proceedings of the AWMA conference: 'Measuring toxic and related air pollutants', US; May (1994). pp. 624–630.

[51] UK Health and Safety Executive, Methods for the determination of hazardous substances No. 4: Generation of test atmospheres of organic vapours by the permeation tube method. Apparatus for laboratory use; 1981.

[52] T. Hafkenscheid, F. Langellan, Dynamically generated standard atmospheres: a support for air monitoring, Proceedings of Conference 'Measuring air pollution by diffusive sampling', Montpellier; 2001.

[53] J.H. Buchanan, L.C. Buettner, A.B. Butrow, D.E. Tevault, Vapor pressure of VX (Research and Technology Directorate). Edgewood Chemical Biological Center: Report No. ECBC-TR-068; 1999.

[54] EN ISO 16017. Air quality – Sampling and analysis of volatile organic compounds in ambient air, indoor air and workplace air by sorbent tube/thermal desorption/capillary gas chromatography. (Part 1: Pumped sampling. Part 2: Diffusive sampling).

[55] US EPA Compendium Method TO-17: Determination of volatile organic compounds in ambient air using active sampling onto sorbent tubes; 1999.

[56] NIOSH 2549 Volatile organic compounds – (screening) using multibed sorbent tubes, thermal desorption, gas chromatography and mass spectrometry.

[57] MDHS 72 Volatile organic compounds in air. Laboratory method using pumped solid sorbent tubes, thermal desorption and gas chromatography; February 1992.

[58] MDHS 80 Volatile organic compounds in air. Laboratory method using diffusive solid sorbent tubes, thermal desorption and gas chromatography; August 1995.

[59] Clean air at work: New trends in assessment and measurement for the 1990s. in: RH. Brown, M. Curtis, KJ. Saunders, S. Vandendriessche. (Eds.), Proceedings of the Luxembourg Symposiy, organized by the Health and Safety Directorate of the European Commission. RSC Publication No. 108; Sept 1991, ISBN: 0-85186-217-9.

[60] A.A. Grote, E.R. Kennedy, Workplace monitoring for VOCs using thermal desorption-GC-MS, J. Environ. Monit. 4 (2002) 679–684.

[61] K. Jones, J. Cocker, L.J. Dood, I. Fraser, Factors affecting the extent of dermal absorption of solvent vapours: a human volunteer study, Ann. Occup. Hyg. 47 (2) (2003) 145–150.

[62] S.M. Gordon, L.A. Wallace, P.J. Callahan, D.V. Kenny, M.C. Brinkman, Effect of water temperature on dermal exposure to chloroform, J. National Ins. Environ. Health. Sci. 106 (6) (1998) 337–345.

[63] Suite of UK Health and Safety Laboratory Guidance Notes for biological monitoring via breath.

[64] S.M. Gordon, J.P. Szidon, B.K. Krotoszynski, R.D. Gibbons, H.J. O'Neill, Volatile organic compounds in exhaled air from patients with lung cancer, Clinical Chem. 31 (8) (1985) 1278–1282.

[65] J.W. Dallinga, C.M.H.H.T. Robroeks, J.J.B.N. van Berkel, E.J.C. Moonen, R.W.L. Godschalk, Q. Jöbsis, et al., Volatile organic compounds in exhaled breath as a diagnostic tool for asthma in children, Clin. Exp. Allergy 40 (2010) 68–76.

[66] M. Phillips, R.N. Cataneo, R. Condos, G.A.R. Erickson, J. Greenberg, V.L. Bombardi, et al., Volatile biomarkers of pulmonary tuberculosis in the breath, Tuberculosis 87 (2007) 44–52.

[67] J.J.B.N. van Berkel, J.W. Dallinga, G.M. Möller, R.W.L. Godschalk, E. Moonen, E.F.M. Wouters, et al., A profile of volatile organic compounds in breath discriminates COPD patients from controls, Respir. Med. 104 (2010) 557–563.

[68] S.K. Pandey, K.H. Kim, A review of human body odor components and their determination. Trends Anal. Chem. in press.

[69] S.M. Gordon, Identification of exposure markers in smokers' breath, J. Chromatogr. 511 (1990) 291–302.

[70] S.M. Gordon, Application of continuous breath sampling to determine VOC dose and body burden: some VOC Markers of ETS Exposure, EPA, 1998. Contract: 68-D4-0023.

[71] J. Roukos, H. Plaisance, T. Leonardis, M. Bates, N. Locoge, Development and validation of an automated monitoring system for oxygenated volatile organic compounds and nitrile compounds in ambient air, J. Chromatogr. A (2009).

[72] H.T. Nguyen, K.-H. Kim, M.-Y. Kim, Volatile organic compounds at an urban monitoring station in Korea, J. Hazard Mater. 161 (2009) 163–174.

[73] K.-H. Kim, Some insights into the gas chromatographic determination of reduced sulfur compounds (RSCs) in air, Environ. Sci. Technol. 39 (2005) 6755–6769.

[74] K.P. Song, J.S. Han, M.D. Lee, D.W. Ju, M.S. Im, T.H. Kim, et al., A Study of quality assurance/quality control between institutions for reduced sulfur compounds in the ambient air using cryofocusing thermal desorber with GC/PFPD, Kor J. Odor. Res. Eng. 6 (1) (2007) 33–39.

[75] N.T. Plant, M.D. Wright, European diffusive sampling initiative: World survey for BTX by diffusive sampling. Rpt: IACS 97/16, UK Health and safety laboratory, 1998.

[76] M.R. Ras, R.M. Marcé, F. Borrull, Characterization of ozone precursor volatile organic compounds in urban atmospheres and around the petrochemical industry in the Tarragona region, Sci. Total Environ. 487 (2009) 4312–4319.

[77] S.A. Batterman, G.-Z. Zhang, M. Baumann, Analysis and stability of aldehydes and terpenes in electropolished canisters, Atmos. Environ. 32 (10) (1998) 1647–1655.

[78] E.H. Daughtrey, K.D. Oliver, J.R. Adams, K.G. Kronmiller, W.A. Lonneman, W.A. McClenny, A comparison of sampling and analysis methods for low-ppbC levels of volatile organic compounds in ambient air, J. Environ. Monit. 3 (2001) 166–174.

[79] D. Wevill, The use of a thermal desorption system as a cryogen-free method for the monitoring of trace greenhouse gases in air, Labmate, UK, September, 2009.

[80] M.D. Wright, N.T. Plant, R.H. Brown, Storage stability study of TO-14 compounds on single and multi-bed carbon thermal desorption tubes. Rpt: IACS 98/02, UK Health and safety laboratory, 1999.

[81] S. Baek, Y. Kim, R. Perry, Indoor Air quality in homes, offices and restaurants in Korean urban areas-indoor/outdoor relationships, Atmos. Environ. 31 (4) (1997) 529–544.

[82] V.M. Brown, S.K.D. Coward, D.R. Crump, J.W. Llewellyn, H.S. Mann, G.J. Raw, Indoor air quality in English homes – VOCs. in: Proceedings: Indoor Air; 2002. pp. 477–482.

[83] P.A. Clausen, P. Wolkoff, Evaluation of automatic thermal desorption – capillary gc for determination of semivolatile organic compounds in indoor air, J. High Resolut. Chromatogr. 20 (1997) 99–108.

[84] E. De Saeger, P. Perez-Ballesta, BTX monitoring campaign in Brussels. The diffusive monitor, vol. 7, HSE/CAR WG5, 1995. 7–8.

[85] Energy Institute: Protocol for the determination of the speciation of hydrocarbon emissions from oil refineries New ISBN 9780852934050 Old ISBN 085293405X; 2004.

[86] prEN 13649 Stationary source emissions – Determination of the mass concentration of individual gaseous organic compounds; 2011.

[87] P. Ciccioli, E. Brancaleoni, A. Cecinato, R. Sparapani, Identification and determination of biogenic and anthropogenic volatile organic compounds in forest areas of Northern and Southern Europe and a remote site of the Himalaya region by high-resolution gas chromatography-mass spectroscopy, J. Chromatogr. 643 (1993) 55–69.

[88] N. Schmidbauer, M. Oehme, Comparison of solid adsorbent and stainless steel canister sampling for very low ppt-concentrations of aromatic compounds (>C6) in ambient air from remote areas, Fresenius Z Anal. Chem. 331 (1988) 14–19.

[89] D. Helmig, Air analysis by gas chromatography, J. Chromatogr. A 843 (1999) 129–146.

[90] X. Xu, L.L.P. van Stee, J. Williams, J. Beens, M. Adahchour, R.J.J. Vreuls, et al., Comprehensive two-dimensional gas chromatography (GC x GC) measurements of volatile organic compounds in the atmosphere, Atmos. Chem. Phys. 3 (2003) 665–682.

[91] R. Chance, A.R. Baker, F.C. Küpper, C. Hughes, B. Kloareg, G. Malin, Release and transformations of inorganic iodine by marine microalgae, Estuar Coast Shelf Sci 82 (3) (2009) 406–414.

[92] H. Hayes, D.J. Benton, S. Grewal, N. Khan, Evaluation of sorbent methodology for petroleum-impacted site investigations. A & WMA Conference: "Vapour Instrusion: Learning from the Challenges" September 26–28, Providence, RI, USA; 2007.

[93] ASTM WK 23766: Standard practice for active soil gas sampling for direct-push or manual-driven hand sampling equipment.

[94] ASTM WK 20609 Standard practice for passive soil gas sampling in the vadose zone for source identification, spatial variability assessment, monitoring and vapour intrusion evaluations.

[95] H.J.Th. Bloemen, T.T.M. Balvers, A.P. Verhoeff, J.H. Van Wijnen, P. Van der Torn, E. Knol, Ventilation rate and exchange of air in dwellings – development of a test method and pilot study. Report; 1992.

[96] EC directive on Energy Performance of Buildings (EPBD) [2002/91/EC]

[97] California air resources board: ventilation and indoor air quality in new homes. CEC-500-2009-085; 2009.

[98] P. Wargocki, J. Sundell, W. Bischoff, G. Brundrett, P.O. Fanger, F. Gyntelberg, et al., Ventilation and health in non-industrial indoor environments: report from a European, multi-disciplinary scientific consensus meeting (EUROVEN), Indoor Air 12 (2002) 113–128.

[99] M. Larson, J. Sundell, B. Kolarik, L. Hagerhed-Engman, C.G. Bornehag, The use of PVC flooring material and the development of airway symptoms among young children in Sweden. Proceedings of Indoor Air '08, Copenhagen, Denmark. Paper 862; 2008.

[100] European Construction Product Regulation, EC 305/2011.

[101] International Green Construction Code, Public Version 2; 2010.

[102] Japanese Ministry of health Labour and Welfare – Recommendations of the Committee on Sick House Syndrome.

[103] AgBB/DIBt, 2000 (last updated 2008) Health-related evaluation procedure for Volatile organic compound emissions from building products, DIBt-Mitteilungen 1/2001, 3–12. http://www.umweltbundesamt.de/building-products/agbb.htm.

[104] ANSI/ASHRAE/USGBC/IES Standard 189.1, Standard for the design of high performance green buildings; 2009.

[105] BIFMA Standard M7.1, Revised standard test method for determining VOC emissions from office furniture systems, components and seating; 2010.

[106] Japanese Industrial Standard A 1901: Determination of the emission of volatile organic compounds and aldehydes for building products – Small chamber method.

[107] California Department of Public Health: Standard method for the testing and evaluation of VOC emissions from indoor sources using environmental chambers. Version 1.1. (February 2010). Reference CA Spec 01350, CA/DHS/EHLB/R-174.

[108] TC351 WG2 Construction products – Assessment of emissions of regulated dangerous substances from construction products – Determination of emissions into indoor air.

[109] ECA IAQ 18.

[110] ISO 16000-3. Indoor air – Part 3: Determination of formaldehyde and other carbonyl compounds – Active sampling method.

[111] ASTM D5172: Test method for determination of formaldehyde and other carbonyl compounds in air (active sampler methodology).

[112] ASTM D7706 Standard practice for rapid screening of VOC emissions from products using micro-scale chambers; 2011.

[113] ISO DIS 12219–3 Indoor Air of road vehicles – Screening method for the determination of the emissions of VOCs from vehicle interior parts and materials – Micro-scale chamber method; 2011.

[114] M. Lor, K. Vause, K. Dinne, E. Goelen, F. Maes, J. Nicolas, A.C. Romain, C. Degrave, Final Report – Horizontal evaluation method for the implementation of the construction products directive HEMICPD, Belgium. Final Report; 2010.

[115] M. Pharaoh, Final report: work on the correlation between the VDA 276 test and micro-chamber testing. PARD Extension Report: Warwick Manufacturing Group, University of Warwick, 2009.

[116] R.M. Black, B. Muir, Derivatisation reactions in the chromatographic analysis of chemical warfare agents and their degradation products, J. Chromatogr. A 1000 (2003) 253–281.

[117] Markes international thermal desorption technical support Note No. 57: Characterisation of paint samples by direct thermal desorption with GC/MS. Markes International Ltd., UK.

[118] M. Ezrin, G. Lavigne, Analysis of organic compounds in recycled dairy-grade HDPE by thermal desorption with GC/MS, SPE Recycling Division 2nd Annual Recycling Conference; 1995. pp. 104–110.

[119] P. Wolkoff, C.K. Wilkins, Indoor VOCs from household floor dust: comparison of headspace with desorbed VOCs; method for VOC release determination, Indoor Air 4 (1994) 248–254.

[120] E.A. Woolfenden, Controlling quality, Food Process (Jan. 1989) 33–35.

[121] S. Eri, B.K. Khoo, J. Lech, T.G. Hartman, Direct TD-GC and GC/MS profiling of hop (Humulus Lupulus L) essential oils in support of varietal characterisation, J. Agric. Food Chem. 48 (4) (2000) 1140–1149.

[122] Markes international thermal desorption technical support note no. 94: Using thermal desorption technology to automate high/low analysis of a complex beer sample. Markes International Ltd., UK.

[123] N. Watson, E.A. Woolfenden, Complementary techniques used for enhancing GC/MS analysis of flavour and fragrance components in consumer beverages. Proceedings of the HTC-11 Conference, Be.

[124] H.S. Lee, H.J. Lee, H.J. Yu, D.W. Ju, Y. Kim, C.T. Kim, et al. A comparison between high hydrostatic pressure extraction and heat extraction of ginsenosides from ginseng (Panax ginseng CA Meyer). J. Sci. Food Agric., 91, (0) in press.

[125] Markes international thermal desorption technical support note no. 95: Food decomposition analysis using the micro-chamber/thermal extractor with TD-GCMS. Markes International Ltd., UK.

[126] J. Barberio, J. Twibell, Chemotaxonomy of plant species using headspace sampling, thermal desorption and capillary GC, J. High Resolut. Chromatogr. 14 (1991) 18–20.

[127] G.W. Robertson, D.W. Griffiths, W. MacFarlane Smith, R.D. Butcher, The application of thermal desorption-gas chromatography-mass spectrometry to the analyses of flower volatiles from five varieties of oilseed rape (Brassica naous spp. Oleifera), Photochem. Anal. 4 (1993) 152–157.

[128] R.M.C. Jansen, J.W. Hofstee, J. Wildt, F.W.A. Verstappen, H.J. Bouwmeester, M.A. Posthumus, et al., Health monitoring of plants by their emitted volatiles: trichome damage and cell membrane damage are detectable at greenhouse scale, Ann. Appl. Biol. (0003-4746) (2009) 1–12.

[129] D.W. Griffiths, G.W. Robertson, A.N.E. Birch, R.M. Brennan, Evaluation of thermal desorption and solvent elution combined with polymer entrainment for the analysis of volatiles released by leaves from midge (dasineura tetensi) resistant and susceptible Blackcurrant (*Ribes Nigrum L.*) cultivars, Phytochem. Anal. 10 (1999) 328–334.

[130] A. Kessler, I.T. Baldwin, Defensive function of herbivore-induced plant volatile emissions in nature, Science 291 (2001) 2141–2144.

[131] K. Wilkins, K. Larsen, M. Simkus, Volatile metabolites from mould growth on building materials and synthetic media, Chemosphere 41 (2000) 437–446.

[132] S.K. Pandey, K.H. Kim, Human body-odor components and their determination. Trends Anal. Chem. in press.

[133] S. van den Velde, M. Quirynen, P. van Hee, D. van Steenberghe, Halitosis associated volatiles in breath of healthy subjects, J. Chromatogr. B 853 (2007) 54–61.

[134] G.P. Jones, Evaluation of a fully automated thermal desorption device for the headspace screening of fire debris. Science Miscellany & Forensic Fillips; 1986, pp. 141–148.

[135] J.F. Carter, R. Sleeman, J. Parry, The distribution of controlled drugs on banknotes via counting machines, Forensic Sci. Int. 132 (2003) 106–112.

[136] B. Tienpont, F. David, A. Stopforth, P. Sandra, Comprehensive profiling of drugs of abuse in biological fluids by stir-bar sorptive extraction-thermal desorption-capillary GCMS. LC-GC Europe, December; 2003. pp. 2–10.

[137] M. Vîrgolici, C. Ponta, M. Manea, D. Negut, M. Cutrubinis, I.S. Moise, et al., Thermal desorption/

gas chromatography/mass spectrometry approach for characterization of the volatile fraction from amber specimens: a possibility of tracking geological origins, J. Chromatogr. A 1217 (2010) 1977–1987.

[138] R.P. Galvin, M. House, Atmospheric monitoring of bischloromethylether at low ppb levels using an automated system, Environ. Tech. Lett. 9 (1988) 563–570.

[139] Markes international thermal desorption technical support note no. 43: Large-scale monitoring of mono-ethylene glycol vapour in natural gas using pumped sampling onto Tenax tubes followed by TD-GC analysis. Markes International Ltd., UK.

[140] J. Dallüge, L.L.P. van Stee, X. Xub, J. Williams, J. Beens, R.J.J. Vreuls, et al., Unravelling the composition of very complex samples by comprehensive gas chromatography coupled to time-of-flight mass spectrometry cigarette smoke, J. Chromatogr. A 974 (2002) 169–184.

[141] N. Watson, Making SORBENT Tube Sampling Easier; the Development of a New Type of 'Grab' Sampler', Proceedings of the Air & Waste Management Association Conference, Beyond all Borders; June 2011.

[142] R.B. Gaines, G.S. Frysinger, Temperature requirements for thermal modulation in comprehensive two-dimensional gas chromatography, J. Sep. Sci. 27 (2004) 380–388.

[143] G. Semard, C. Gouin, J. Bourdet, N. Bord, V. Livardis, Comparative study of differential flow and cryogenic modulator systems for comprehensive two-dimensional GC systems for the detailed analysis of light cycle oil, J. Chromatogr. A (2010). doi; 10.1016/j.chroma.2010.08.082.

CHAPTER

11

Pyrolysis Gas Chromatography

Thomas P. Wampler

OUTLINE

11.1. THERMAL SAMPLING GC

11.1.1. Desorption

The principle behind all thermal sample preparation methods for GC is that the volatile analytes are delivered to the GC column by controlling the sample temperature rather than by injecting them in a solvent [1]. Since there is no large solvent peak at the beginning of the GC run, these techniques are both sensitive and ideal for determining the presence of trace levels of solvents in products. In many cases, the volatile organics are first adsorbed onto a trapping material, and then thermally desorbed in a GC carrier. However, whether desorbed from a sorbent or directly from a solid or liquid sample, the volatiles that constitute the GC analysis are compounds that are isolated from the matrix and transported to the GC intact. This is more a physical technique in that the analytes have not been chemically

Gas Chromatography DOI: 10.1016/B978-0-12-385540-4.00011-0

altered; hence, the identification of a particular compound as a peak in the chromatogram means that that compound was present in the sample originally. Typical applications of this kind of thermal sampling include the determination of residual solvents and monomers in polymers, air-quality monitoring using sorbent tubes, the identification of aroma compounds in foods, and the identification of plasticizers.

11.1.2. Pyrolysis

Analytical pyrolysis-GC differs from other thermal techniques in one important way: the temperatures used are intentionally high enough to cause chemical changes in the sample material [2]. This is specifically avoided in thermal desorption work; in fact, the temperatures must be modified and carefully controlled to prevent chemically altering the analytes. At low temperatures, the intention is to demonstrate that the peaks identified in the chromatogram indicate that those exact compounds were present in the sample before processing. The purpose of pyrolysis, however, is to create smaller, volatile molecules from a large molecule so that GC may be used to study the macromolecule [3], with the understanding that it is a destructive technique. The result is similar to mass spectrometry in that a large molecule is intentionally fragmented and information about the original molecule is inferred from the collection and identity of the fragments made. It is therefore important to understand the chemistry involved when large molecules are thermally fragmented in interpreting the analytical results of pyrolysis-GC and Py-GC-MS. The reactions include dehydration, decarboxylation, and, to a large extent, bond dissociation to form free radicals, with the typical reactions and rearrangements of such entities.

11.2. CHEMICAL THEORY

Analytical pyrolysis is, strictly speaking, the study of chemical changes that occur to a molecule at high temperatures under vacuum or in an inert atmosphere. Although these reactions are also involved in analyses such as hydrogenation and oxidation, which may also be performed with pyrolysis instruments, the theory here will assume that the sample is being heated in GC carrier gas, typically helium at relatively low pressure (about one atmosphere). The reactions involved have been categorized into several schemes, including depolymerization, random scission, and side-group elimination, but for practical purposes in studying polymers, it is convenient to limit consideration to just two cases: things that make oligomers and things that do not.

11.2.1. Bond Dissociation and Free Radicals

The molecular fragmentation caused by pyrolysis – and the ultimate products formed and identified in the analysis – depends on the relative strengths of the bonds found in the molecule and how the free radicals formed when the bonds dissociate stabilize to make products. As the sample is heated, the weakest bonds break first. In the case of most polymers, there is a molecular chain and then additional atoms or groups attached to the chain that make one chain different from another. For the sake of using a GC, the best case is if the chain bonds are broken first, since this reduces the size of the molecule, eventually to a point at which the pieces are small enough to be volatile and can go through the GC column. On the other hand, if the side groups are attached with bonds that are weaker than the chain bonds, they will be removed from the chain, which generally then becomes unsaturated before degrading.

Once the free radicals have been formed by bond dissociation, they behave in normal fashion. Primary free radicals are more energetic, and tertiary ones are more stable. Primary free radicals will stabilize through reactions that result in the formation of secondary or tertiary free radicals, including rearrangements, especially moving a hydrogen atom from a carbon four atoms removed from the free radical (a 1–5 hydrogen shift). In general, molecules in which the chain bonds are the weakest break apart to form oligomers, including monomers, and polymers with weakly attached side groups do not.

11.2.2. Oligomer Formation

Many synthetic polymers are essentially carbon–carbon chains with side groups attached, and look like scheme 1 in Figure 11.1. Here, two adjacent carbons each have two side groups, shown as R1–R4. If all four R groups are hydrogens, then the polymer is polyethylene. If one is a methyl group, it is polypropylene. If one is benzene, the polymer is polystyrene, and so on. Carbon–hydrogen bonds are stronger than carbon–carbon bonds; hence, these polymers generally fragment by chain-bond dissociation, producing two free radicals. If R1 and R2 are both hydrogen, then the resulting free radical is primary, and very reactive. If one of the R groups is not hydrogen, then it is a secondary free radical, and if neither R1 nor R2 is hydrogen, the result is a tertiary free radical.

Line 2 of Figure 11.1 shows the bond between the two chain carbons dissociating to make the free radicals. One path is available for the free radical to take a hydrogen from another molecule, making a saturated end. This may or may not make a volatile product, depending on the size of the rest of the molecule. Another mechanism is β-scission, shown in line 3 of Figure 11.1. In this case, the unpaired electron and one of the electrons from the bond between the second and

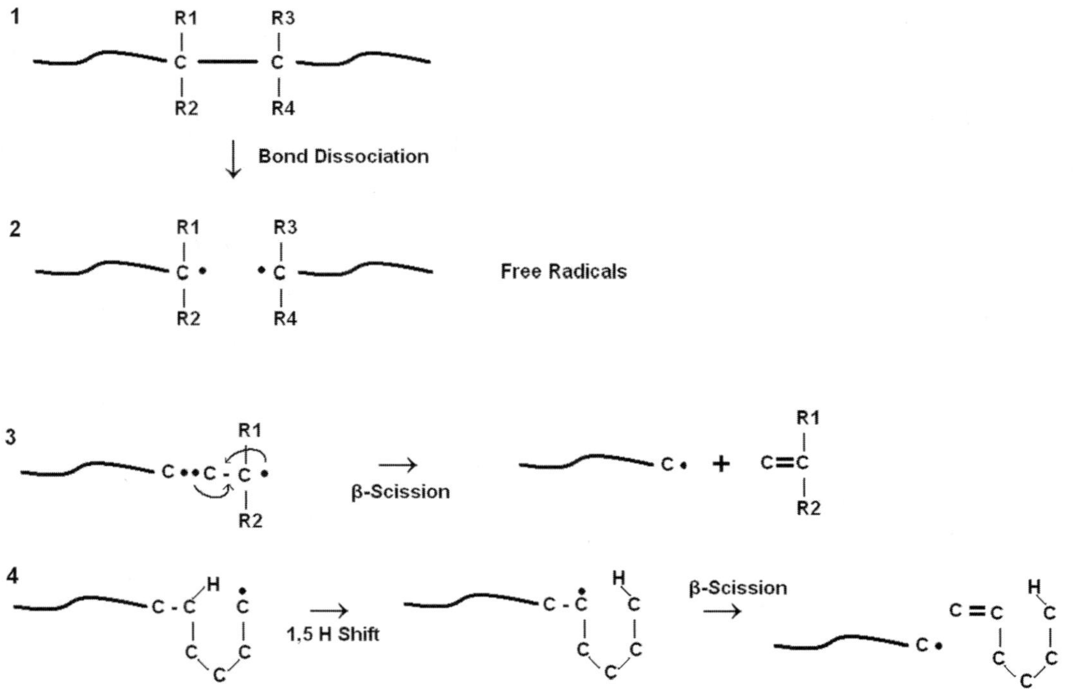

FIGURE 11.1 Free radical generation with production of monomer and trimer.

third carbons in the chain form a bond, making a double bond between the first two carbons and breaking the bond to the rest of the molecule. This produces a new free radical and a stable product of two carbons with R1 and R2 still attached. This product is frequently a molecule of the monomer of the polymer, and may be the primary degradation pathway, that is, a molecule of the monomer is split off, creating a new free radical, which undergoes β-scission again, eventually unzipping the whole polymer chain. Methacrylate polymers and polytetrafluoroethylene are examples of polymers that essentially unzip to monomer through pyrolysis.

If the free radical is primary or secondary, as shown in line 4 of Figure 11.1, it can become more stable by rearranging. By forming a six-membered ring, with five carbons and a hydrogen, the hydrogen can be transferred from the fifth carbon to the first. The unpaired electron is now on the fifth carbon; therefore, the free radical is secondary or tertiary. This free radical can now undergo β-scission, again making a new, terminal free radical, and a stable product. However, this product is a six-carbon fragment of the polymer, with the six carbons still connected to each other as they were originally. This makes it a trimer molecule (the oligomer with three monomers), which contains significant microstructural information. The sequence of bond dissociation, then 1,5 hydrogen shift to make a more stable free radical, and then β-scission to make a trimer is a significant pathway for many polymers. In fact, in some cases, it produces the main or largest peak in the pyrogram, and the information it contains can be very valuable in distinguishing various, similar polymers.

Some polymers, especially polyolefins, produce a very wide range of oligomers when pyrolyzed, including molecules with thirty-five or more carbons. Oligomers beyond the trimer generally result from breaking bonds at two different places in the molecule. If the section between the bond-breaking sites is volatile enough to leave the system, it becomes part of the pyrogram. The free radical at each end may abstract a hydrogen, producing a saturated end to the fragment, or the end may be unsaturated. Polyethylene, for example, generates fragments that may have a double bond at each end, only one double bond or no double bonds; hence, the oligomers consist of alkanes, alkenes, and dienes (see Figure 11.5).

11.2.3. No Oligomers

For some polymers, the bonds attaching side groups are weaker than the bonds holding the chain together; hence, during pyrolysis, the weaker bonds break first and the side groups are removed. In this case, it is not possible to make monomers since the identity of the polymer chain has been altered. Instead, small molecules from the side groups are created, and then the remaining chain is fragmented. This is shown in Figure 11.2 using polyvinyl chloride as an example. The C—Cl bond is weaker than either the C—C or the C—H bonds; hence, at a relatively low temperature the C—Cl bond breaks. The chlorine takes a hydrogen atom from the neighboring carbon and forms HCl, leaving behind a chain with many double bonds. At higher temperatures, this chain fragments, making aromatics, including benzene, toluene, and naphthalene. No vinyl chloride is formed, but the pyrogram, which includes HCl and the aromatics, is still characteristic of the polymer and can be used to identify it. Peaks 1—8 in Figure 11.6 are typical products from the pyrolysis of PVC.

11.2.4. Copolymers

In practice, it is more likely that a material is a mixture, blend, or copolymer than that it is a pure, single polymer. Copolymer systems are frequently divided into two classes: random and block. In block copolymers, and also in

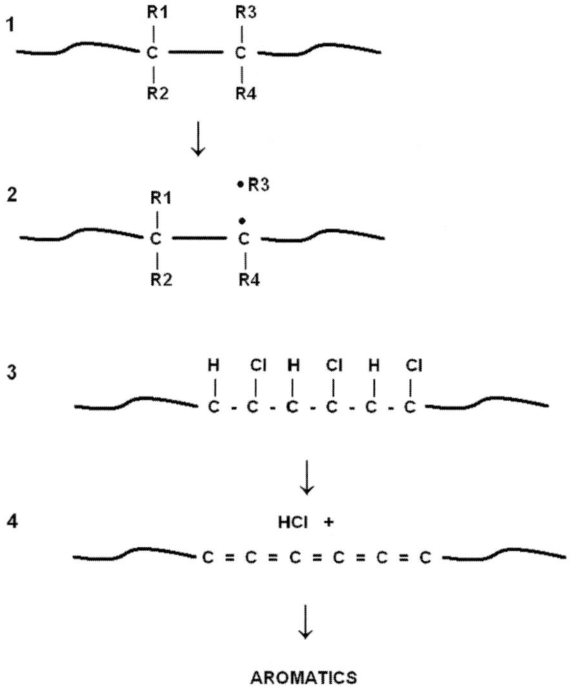

FIGURE 11.2 Side-group elimination with production of aromatics.

blends, mixtures, laminates, and layered materials, there are discrete regions of the individual polymers. In random copolymers, the different monomers are actually part of the polymer molecule and are chemically bound to each other in a random distribution. Therefore, a section of a block copolymer of monomers A and B can be represented like this:

- - - -A-A-A-A-A-A-A-A-A-A-A-A-A-A-A-A-B-B-B-B-B-B-B-B-B-B-B-B-B-B-B-BB- - - -

and a section of a random copolymer like this:

- - - -A-A-B-A-B-B-A-B-A-B-B-A-A-B-A-B-A-B-B-A-B-A-A-B-B-A-B-A-B-A- - - - - -

making two very different polymer molecules. If the polymers form oligomers, then the block sample (or a blend or mixture) will make oligomers of just A and just B, including the trimers A-A-A and B-B-B. The random copolymer will make additional mixed trimers like A-B-A, A-A-B, B-A-B, B-B-A, and so on; hence, the pyrograms will be substantially different. The relative concentrations of A and B will also create differences, as well as the actual degree of randomness in the sample thought to be completely random.

11.3. INSTRUMENTATION

For typical pyrolysis-GC and pyrolysis-GC-MS, the sample is placed in the carrier gas stream of the GC and heated to a high temperature – usually 600–800 °C – rapidly, causing the material to volatilize and go straight onto the GC column. Sample size is dictated by the capacity of the chromatographic system, but is generally about 10–100 μg. This is the equivalent of injecting 1 μl of a standard, that is 1–10% analyte. Since there is no solvent used in pyrolysis, the sample is pure analyte, and even if the pyrogram contains 100 peaks, each one will still represent about 1 μg. Even with such a small sample, split ratios of 50:1 or higher are typical. Dead volume should be as small as possible so that the peak resolution is not affected. Cold spots are of particular concern since many polymers produce oligomers that are fairly large and likely to condense on cool surfaces, producing contamination peaks in later runs.

Although many pyrolysis-GC runs consist of just a single, high-temperature analysis, the technique and the instrumentation used have expanded to include slow, programmed runs, multiple-step analyses taking samples to sequentially higher temperatures, and analyses performed in reactive atmospheres, including hydrogen and air.

11.3.1. Pulse Pyrolysis

The ideal for a pulse pyrolysis is to take a small sample to a high temperature instantaneously. In this way, it will degrade all at once and the resulting peaks will be sharp and well resolved. Commercially available pyrolyzers for pulse pyrolysis come close to achieving this ideal. With heating rates in the tens of degrees per millisecond, they reach a final temperature in less than a second, producing very rapid pyrolysis. There are basically three types of instruments designed for pulse pyrolysis, Curie-point, microfurnace, and heated filament. Each has its own specific characteristics, but all are capable of producing a satisfactory pyrogram in a way that is well controlled and reproducible [4]. The ways in which they differ involve how the sample is introduced and how the heating is applied.

11.3.1.1. *Curie-Point*

In a Curie-point pyrolyzer, a ferromagnetic wire or foil is heated inductively by placing it into a high-frequency coil. The wires are made of various combinations of nickel, cobalt, and iron. The current induced into the wire causes it to heat to a point at which it is no longer ferromagnetic – the Curie-point – at which the temperature stabilizes. Each different alloy has a different Curie-point; hence, the temperature is actually controlled by the choice of alloy rather than by a setpoint on the instrument. The heating is very fast, and the sample, usually applied directly to the wire or foil, pyrolyzes quickly.

Curie-point systems are simple, heat quickly, and the temperature is reproducible within a given batch of alloy. Although the final pyrolysis time is selectable, the heating rate is always ballistic; hence, programmed heating is not an option. The choice of analytical temperature depends on the alloys that are available. Most instruments supply a range of ten to twenty different alloys, and therefore temperatures to choose from.

11.3.1.2. *Microfurnace*

Microfurnaces are generally used isothermally, inserting the sample into a preheated pyrolysis zone. The temperature is usually selectable in 1 °C increments to about 900 °C. Samples may be placed into small cups, which are dropped into the furnace, or inserted using a special syringe that can accommodate powders and fibers. Some furnaces have a separate, lower temperature zone that permits analysis of the sample first at a desorption temperature followed by a second run for pyrolysis.

In addition to being used isothermally, a microfurnace can usually be programmed to heat in °C/minute for time- or temperature-resolved runs. Because of the difficulty in loading very volatile samples onto a filament or Curie-point wire, microfurnaces are preferred for the pyrolysis of such samples. Since a microfurnace generally has a larger heated surface area than a filament or Curie-point system, secondary pyrolysis is a concern, and it is typical to operate them at lower temperatures to compensate for this.

11.3.1.3. *Heated Filament*

In a resistively heated filament pyrolyzer, the sample is placed either onto the surface of a cool filament or into a small quartz tube which is then fitted into a coil of heating wire. The filament is usually platinum, and the temperature is controlled by the voltage applied across the wire or strip. The mass of the filament is very small, so they heat at very fast rates, but can also be heated at slow or programmed rates. Temperature selection is generally in 1 °C increments to about 1400 °C, and the run may include an initial temperature, heating rate, and final temperature and time.

Since the filament temperature is controlled by the instrument setpoint, it is easier to perform multistep sequences, unlike a Curie-point system, which requires a different wire for each temperature. Sample heating is very fast when the sample is placed directly onto the filament, but slower if the sample is contained inside a quartz tube.

11.3.2. Programmed Pyrolysis

Rapid heating for a short time is preferred for a simple pyrolysis-GC analysis since the quick production of volatiles insures narrow peaks and good resolution. There are, however, other analyses that require slower, or programmed heating. These may involve interfacing the pyrolyzer directly to the mass spectrometer, reproducing thermogravimetric programs or simulating industrial processes. If the sample is to be heated slowly, but the analysis is conducted using gas chromatography, the evolved products must be trapped before introduction to the GC. This is done using either a sorbent or a cryogenic trap so that the sample introduction to the GC is rapid even if the analyte production extended over a period of several minutes or longer. Programmed heating is usually not possible with a Curie-point pyrolyzer, but microfurnaces may be programmed to heat in degrees per minute, and filaments, because of their small mass, may be programmed to heat in degrees per minute, second, or millisecond.

11.4. APPLICATIONS

Pyrolysis-GC has been used in laboratories for decades in the analysis of rubber [5], forensic samples [6], coatings [7], artwork [8], and many other kinds of solid materials. The examples shown in this applications section have been chosen both to show the range of materials analyzed using pyrolysis-GC and to illustrate the chemical behaviors described in the theory section. Most of the examples show "real world" materials, and are consequently not pure polymers but blends, copolymers, and composite materials such as paints, adhesives, textiles, and so on.

11.4.1. Polystyrene

As shown in lines 3 and 4 of Figure 11.1, if a polymer structure has a hydrogen atom on the 5th carbon from the bond breaking, it can rearrange and make significant trimer in addition to monomer. This is true of the acrylates (as opposed to the methacrylates) and for other polymers, including polystyrene. Figure 11.3 shows a pyrogram of polystyrene at 750 °C, in which about 70% of the pyrolysate is comprised of the monomer, styrene (peak 2). There is also considerable dimer (peak 4) and trimer (peak 5). The relative amount of the trimer is sensitive to the temperature, so that at cooler temperatures the trimer peak is larger, and at hotter temperatures it decreases. Small peaks are also present for toluene and alpha-methylstyrene, which are generally present in the pyrolysis of polystyrene.

11.4.2. Acrylics

Acrylic polymers include the acrylates such as ethyl acrylate and butyl acrylate, and the methacrylates, such as methyl methacrylate and butyl methacrylate. Looking at line 1 in Figure 11.1, for acrylates, R1, R3, and R4 are all hydrogen, and R2 is the ester group. This means that there is a hydrogen on the 5th carbon from the bond breaking that can be transferred; hence, the trimer can be formed by the mechanism shown in line 4. For the methacrylates, R1 is a methyl group; therefore, the bond breaking makes a tertiary free radical. There is also a methyl group on the 5th carbon; hence, there is no hydrogen to transfer, and the polymer simply unzips to monomer.

The acrylic material shown in Figure 11.4 contains both acrylic and methacrylic monomers, specifically methyl methacrylate, butyl methacrylate, and butyl acrylate. Consequently, there is a large peak for methyl methacrylate monomer (peak 1) and a peak for butyl methacrylate (peak 3). There is also a considerable amount of butyl acrylate, which appears as the monomer, dimer, and trimer (peaks 2, 4, and 5). The trimer peak is the largest, which is typical, and there are actually two dimer peaks,

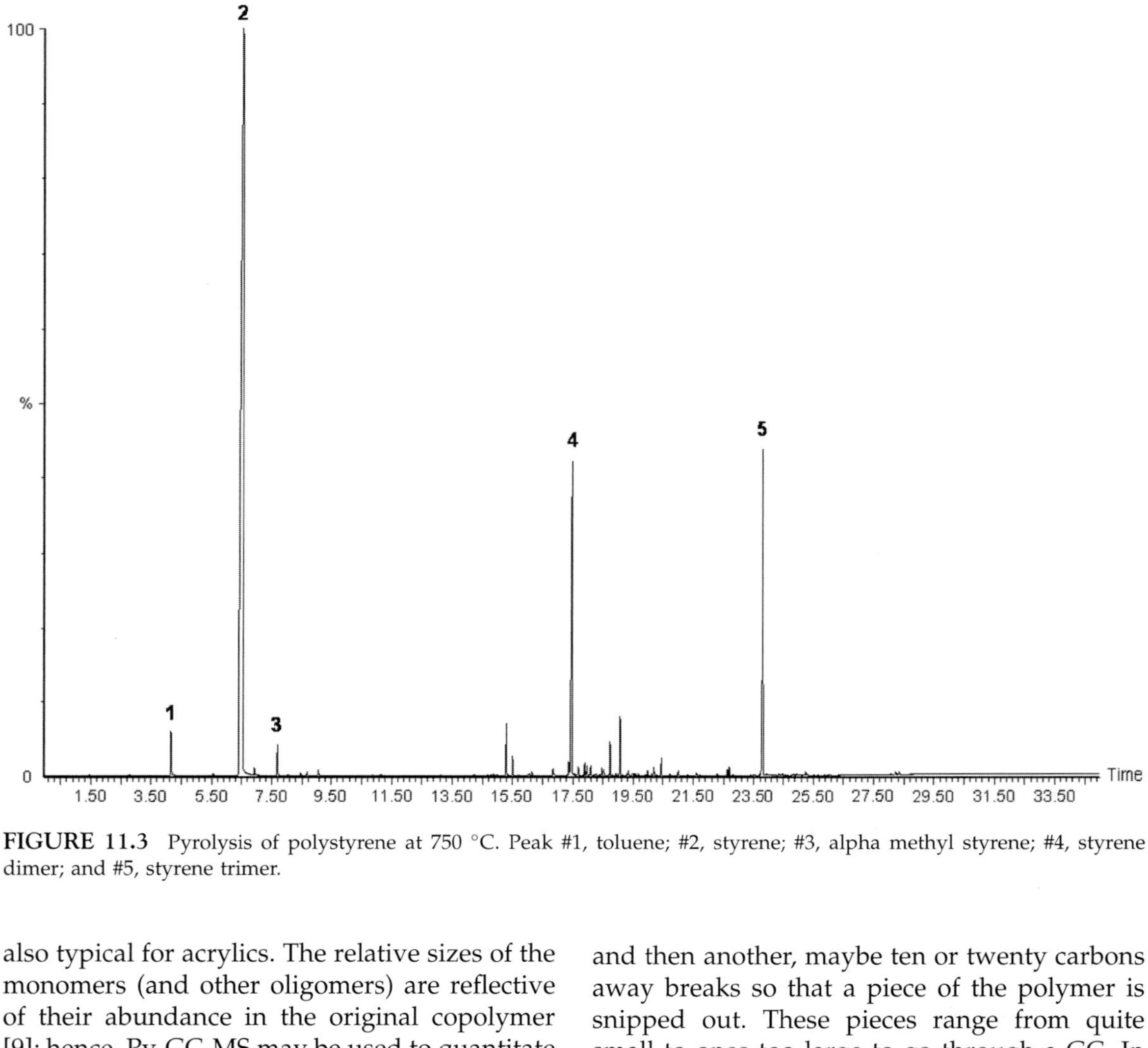

FIGURE 11.3 Pyrolysis of polystyrene at 750 °C. Peak #1, toluene; #2, styrene; #3, alpha methyl styrene; #4, styrene dimer; and #5, styrene trimer.

also typical for acrylics. The relative sizes of the monomers (and other oligomers) are reflective of their abundance in the original copolymer [9]; hence, Py-GC-MS may be used to quantitate the monomer ratios in copolymers like these.

11.4.3. Polyolefins

Polyolefins, including polyethylene, polypropylene, polyisobutylene, and polybutene, form a wide array of fragments when pyrolyzed, making pyrograms with many – frequently hundreds of – peaks. In general, it takes two bond scissions to create a pyrolysis fragment for the polyolefins; hence, one bond breaks, and then another, maybe ten or twenty carbons away breaks so that a piece of the polymer is snipped out. These pieces range from quite small to ones too large to go through a GC. In each case, a series of pyrolysis products makes a pattern specific for the polyolefin pyrolyzed. Polypropylene produces a series of oligomers with each group of peaks having three more carbons than the one that eluted previously. For polyisobutylene, the groups contain four more carbons. For polyethylene, the groups of peaks increase by one carbon; hence, hydrocarbons of each carbon number are present. In all cases, the end of the pyrolysate molecule (where the bond broke) may be a double bond or may

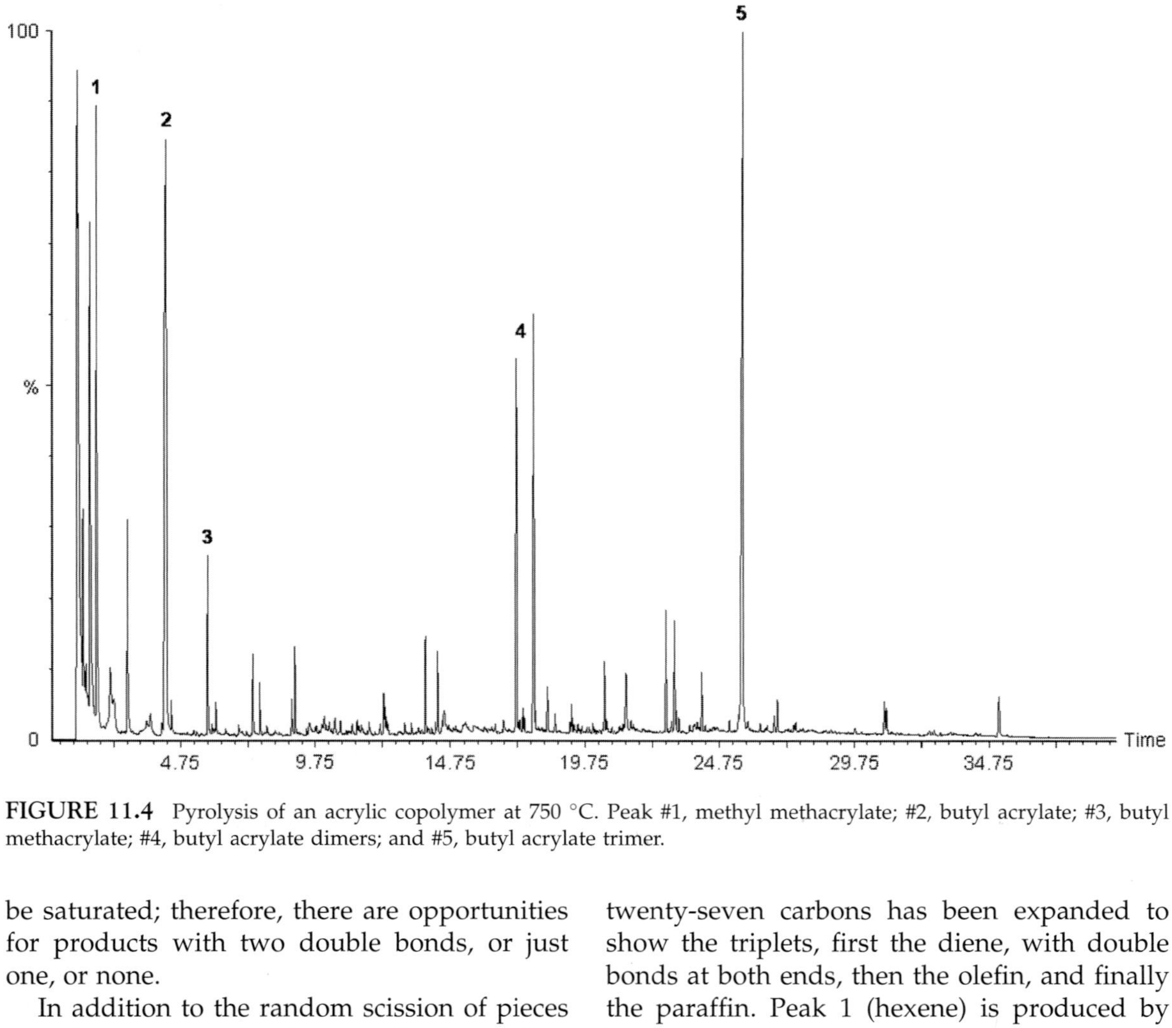

FIGURE 11.4 Pyrolysis of an acrylic copolymer at 750 °C. Peak #1, methyl methacrylate; #2, butyl acrylate; #3, butyl methacrylate; #4, butyl acrylate dimers; and #5, butyl acrylate trimer.

be saturated; therefore, there are opportunities for products with two double bonds, or just one, or none.

In addition to the random scission of pieces out of the macromolecule, once the free radical has been formed, it may still rearrange as in Figure 11.1 line 4; hence, there is a second pathway for some of the products, especially the trimer.

All of this is demonstrated in the pyrolysis of polyethylene, like the piece of a plastic bag pyrolyzed to make what is depicted in Figure 11.5. The overall result is a chromatogram of a series of triplet peaks, all normal hydrocarbons. The group with peak 1 (hexene), has six carbons, and are followed by groups with seven, then eight carbons, and so on. The group with twenty-seven carbons has been expanded to show the triplets, first the diene, with double bonds at both ends, then the olefin, and finally the paraffin. Peak 1 (hexene) is produced by breaking two bonds six carbons apart, and also by the 1–5 hydrogen shift free radical rearrangement that makes trimer. Consequently, there are more of the C6 compounds than of the C5 or C7. This rearrangement can take place more than once. A second shift moves the unpaired electron to carbon 9, and β-scission produces decene. A third shift produces tetradecene. These are marked as peaks 2 and 3 in the pyrogram, each of which is present at greater abundance than the neighboring compounds because of the rearrangement. In general, for the polyolefins, the trimer is the largest

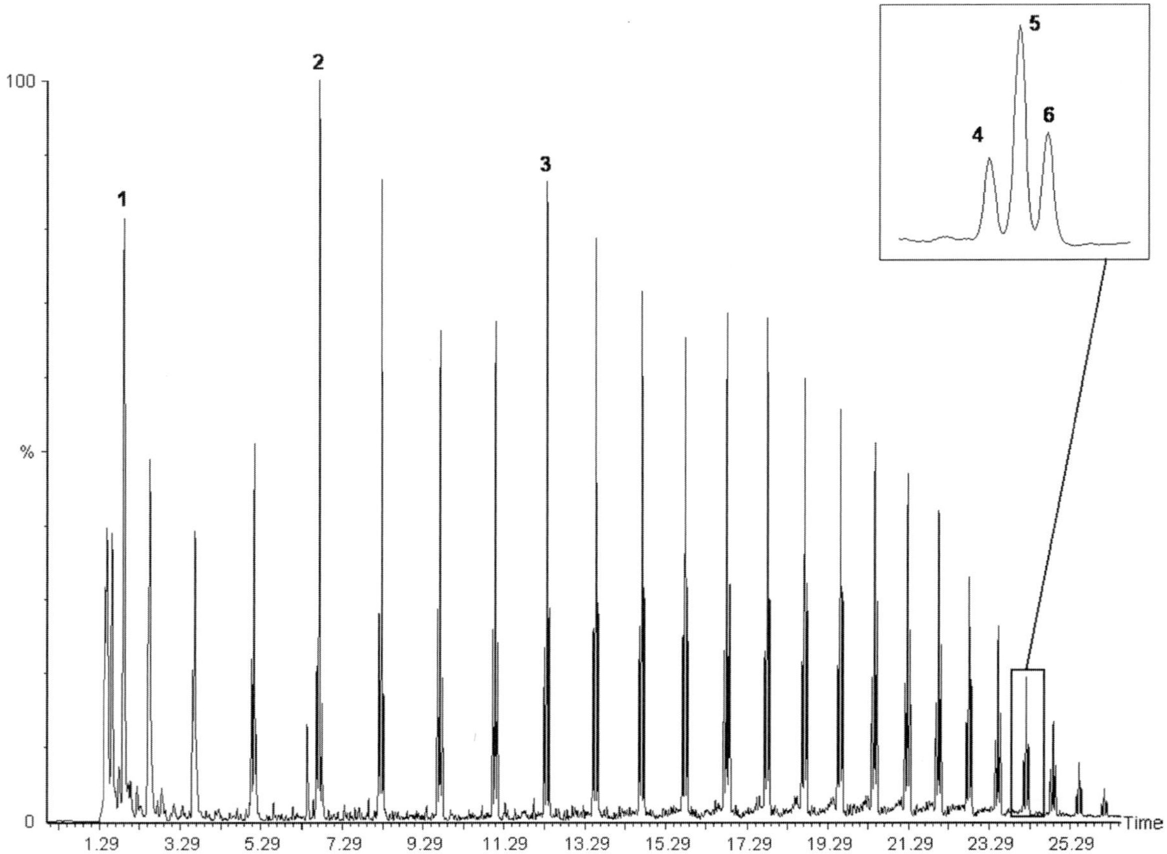

FIGURE 11.5 Pyrolysis of polyethylene at 750 °C. Peak #1, hexene; #2, decene; #3, tetradecene; #4, C27 Diolefin; #5, olefin; and #6, paraffin.

peak – for example, with polypropylene, this is dimethyl heptene.

11.4.4. Vinyl Polymers

Figure 11.2 shows the degradation scheme for polymers that contain a side-group bond weaker than the carbon chain bonds. In these cases, the side group is removed first, and then the chain breaks apart to form smaller molecules, but not monomer. Polyvinylchloride (PVC) and polyvinylacetate (PVA) both behave in this way. In each case, the side group breaks off, producing acetic acid from PVA and HCl from PVC, and leaving a polyunsaturated chain, which then forms many aromatics.

Figure 11.6 shows a polymer packaging material that is largely PVC, with other polymers as well. Peaks 1 through 8 all result from pyrolyzing the PVC portion of the material. Peak 1 is HCl, and the aromatics benzene (2) and naphthalene (7) are significant in the pyrogram. There is also a large peak for methyl methacrylate (9) and for styrene (10). As with most polymers, the presence of several materials being pyrolyzed at the same time does not necessarily confuse things. The peaks 1–8 for PVC would look the same if the sample were pure PVC with no other compound

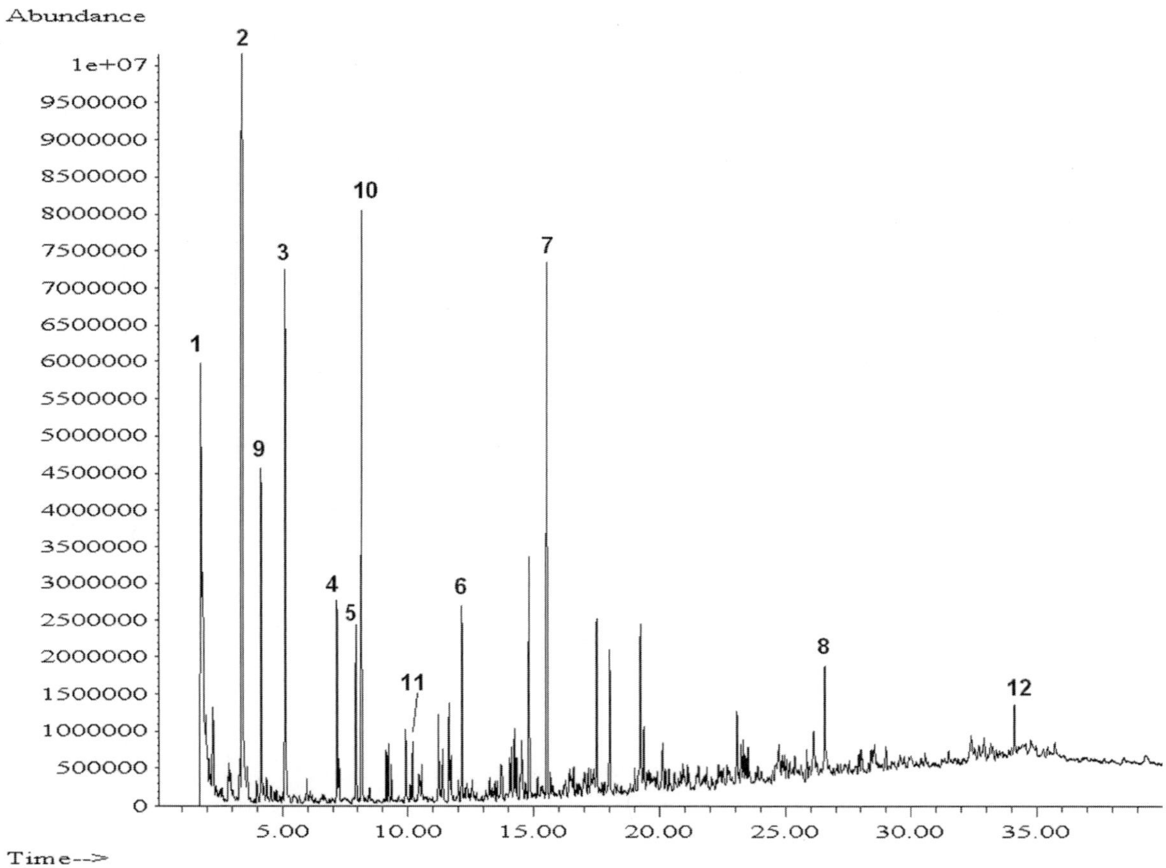

FIGURE 11.6 Pyrolysis of PVC with polystyrene and acrylic at 750 °C. Peak #1, HCl; #2, benzene; #3, toluene; #4, ethyl benzene; #5, xylene; #6, indene; #7, naphthalene; #8, anthracene; #9, methyl methacrylate; #10, styrene; #11, alpha methyl styrene; and #12, styrene trimer.

present. The styrene and methyl methacrylate productions are likewise not affected by the products from PVC. This means that it is unlikely that the pyrolysis products from one material react with those from some other material and make new compounds. This is especially important when considering copolymers. Mixed dimers and trimers in random copolymers are not the result of monomers reacting with each other after pyrolysis, but rather indicate the original structure of the molecule before pyrolysis. In this example, there is a small peak for styrene trimer, which confirms the presence of polystyrene.

11.4.5. Polyurethanes

As a group, polyurethanes are one of the easiest to recognize using Py-GC-MS. They are produced from a diisocyanate and a polyol, and the polyol may be a polyester or a polyether. The diisocyanates are regenerated and generally identified by a mass spectrometer. In addition, fragments from the polyols are frequently simple and indicative of the type used. Many consumer goods called polyurethanes, especially paints, actually contain only a small amount of polyurethane, with other, cheaper ingredients. However, even if the polyurethane content is quite small, the characteristic peaks are present and identifiable.

Figure 11.7 shows an interesting polyurethane made with two different diisocyanates. Peak 3 is actually a double peak for toluene diisocyanate (TDI), which exists as two isomers, both of which appear in commercial products. Isophorone diisocyanate also exists as two isomers, marked in the pyrogram as peaks 4 and 5. The isomeric identity of the diisocyanate is not altered in pyrolysis; hence, the isomers reappear in the pyrogram in the relative abundance they had in the polymer. That is, if only one isomer is used, only one will be made in pyrolysis, and if two are present in a specific ratio in the polymer, that ratio is preserved in the chromatogram.

Peaks 1 and 2 are small ether compounds, indicating that the polyol part of the polyurethane was a polyether. Polyester polyols frequently use adipic acid (also used in some polyamides and other condensation polymers), which produces cyclopentanone when pyrolyzed. Consequently, it is usually possible to determine the diisocyanate and the type of polyol used in a polyurethane from the identity of just a few peaks in the pyrogram.

11.4.6. Polyamides

Condensation polymers, including the nylons [10], have been extensively studied using PY-GC-MS. Nylons are made either from a single monomer, with the acid group at one end and

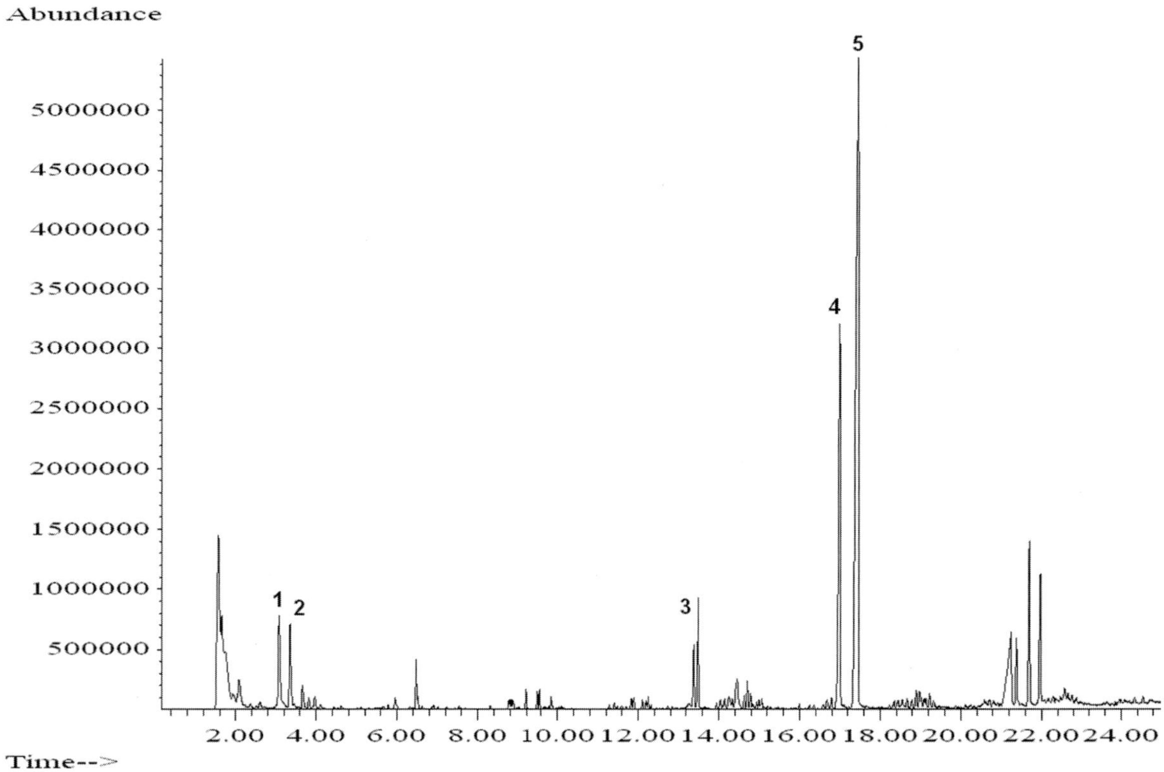

FIGURE 11.7 Pyrolysis of a polyurethane at 750 °C. Peak #1, 2-ethoxypropane; #2, 1-(1-methylethoxy)propane; #3, toluene diisocyanate; #4 and 5, isophorone diisocyanate.

the amine at the other, or from two monomers, one a diacid and the other a diamine. The most commonly used are Nylon 6, which is polycaprolactam, and Nylon 6.6, which combines a six-carbon diacid (adipic acid) and hexanediamine. Although similar in molecular structure, analytically Nylon 6 and Nylon 6.6 are very easy to discriminate. Nylon 6 essentially reverts to monomer; hence, the pyrogram has a large peak for caprolactam, while Nylon 6.6 produces many fragments. One of the most abundant peaks in the pyrogram of Nylon 6.6 (and any polymer made with adipic acid) is cyclopentanone.

Figure 11.8 was produced from carpet fibers, which contained mostly Nylon 6.6. The cyclopentanone peak (1) is prominent, but there are also peaks from the hexanediamine constituent, including hexanenitrile (2) and hexanedinitrile (3). There is also a small peak for caprolactam (4), which is the major product created from Nylon 6. There are larger fragments of the molecule eluting later in the pyrogram as well.

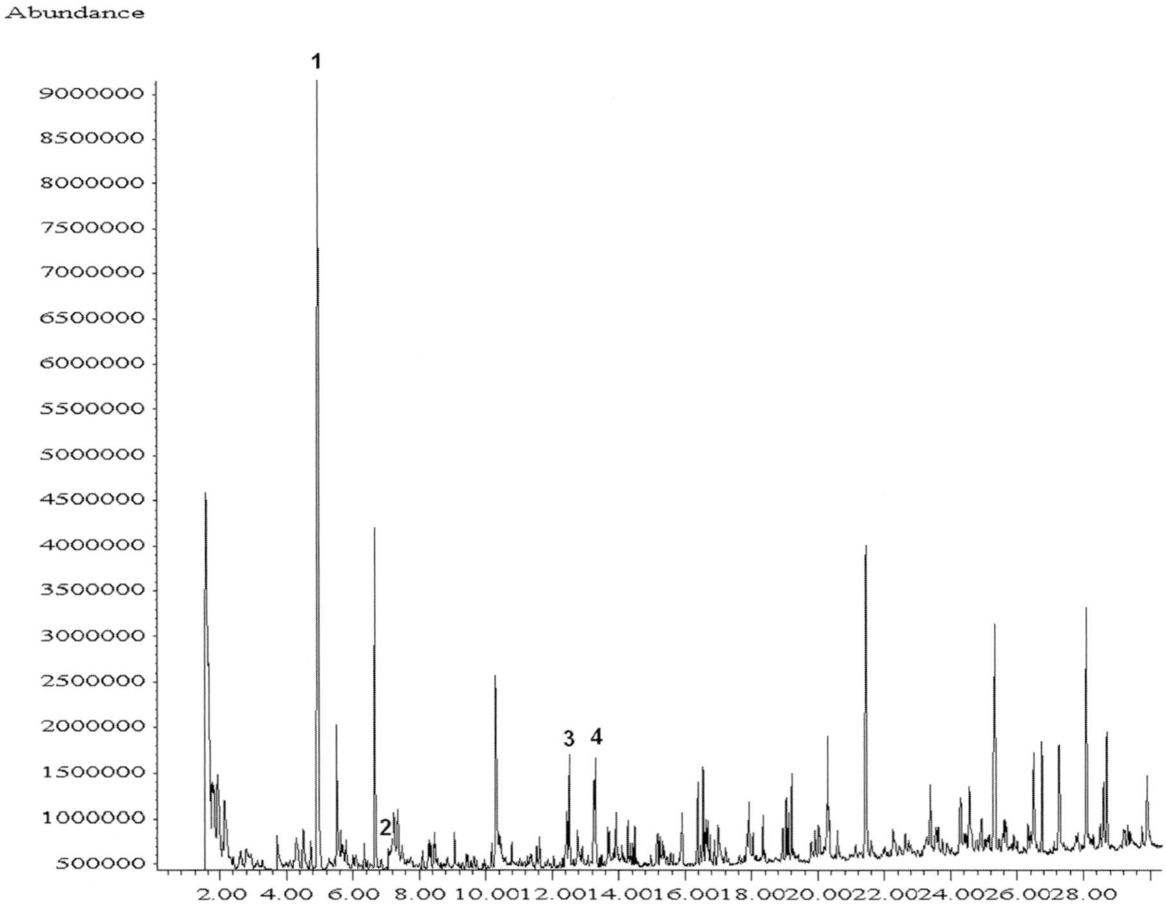

FIGURE 11.8 Pyrolysis of nylon carpet fiber at 800 °C. Peak #1, cyclopentanone; #2, hexanenitrile; #3, hexanedinitrile; and #4, caprolactam.

11.4.7. Epoxies

Epoxies are generally formulated using Bisphenol A and once compounded seem so rugged that chemical analysis would be difficult. They are, however, organic polymers and as such respond to pyrolysis quite readily. Epoxies are increasingly being used in coatings in the form of powdercoats. These can be pure epoxy or hybrids using epoxies and other polymers.

The pyrogram in Figure 11.9 was produced by pyrolyzing the powdercoat finish used on a metal hook. It is a simple epoxy finish, with no other polymers; hence, it is typical of epoxy materials. The large peak (6) at about 25 min is Bisphenol A, which is generally present for epoxies but not actually specific since polycarbonates also use, and produce, Bisphenol A.

Hybrid powdercoats incorporating polyesters or polyurethanes, for example, will still make the same distribution of phenolic compounds seen for a pure epoxy, but will also show the specific pyrolysis products (benzoic acid and diisocyanates) from the additional polymers.

11.4.8. Biomass

Most plant material, including wood, grasses, and leaves, are lignocellulosic [11], meaning that they are comprised of two biopolymers – lignin

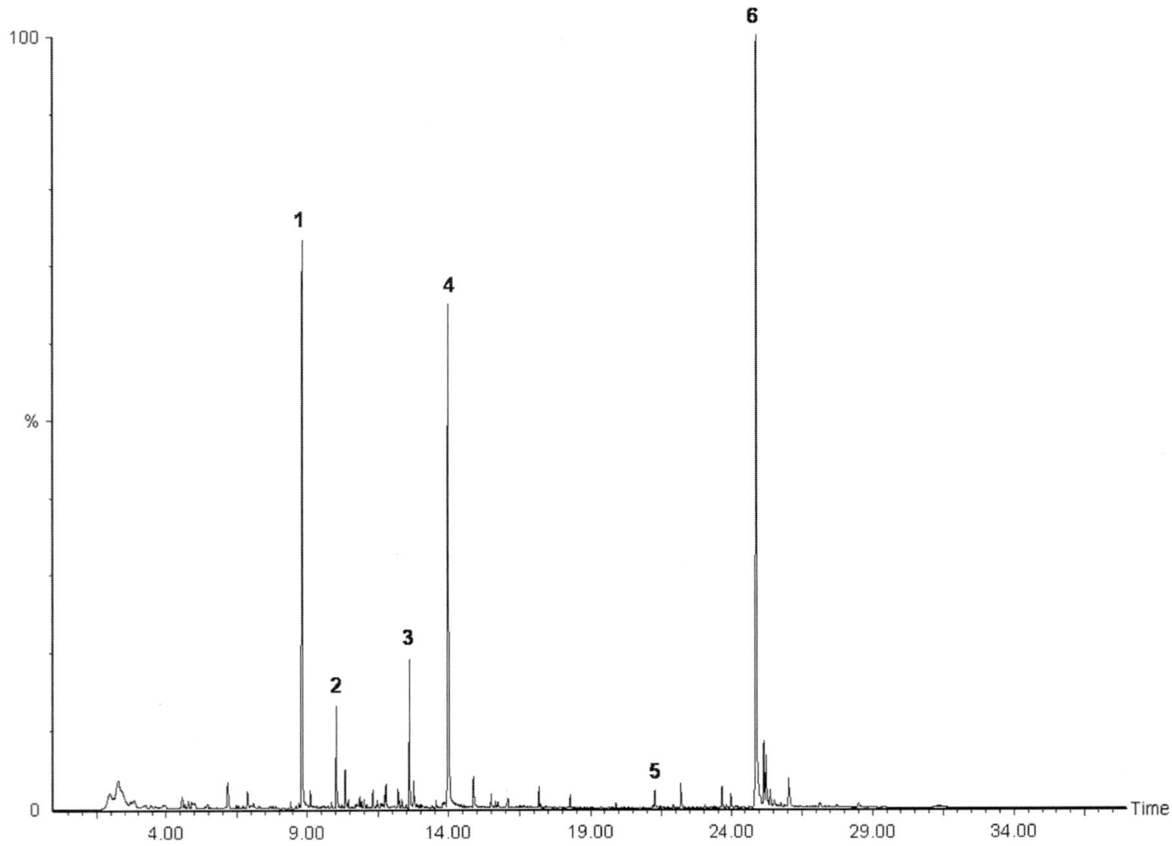

FIGURE 11.9 Pyrolysis of an epoxy powdercoat at 750 °C. Peak #1, phenol; #2, methylphenol; 3, 4-(1-methylethyl)-phenol; #4, p-isopropenylphenol; #5, 4-(1-methyl-1-phenylethyl)phenol; and #6, bisphenol A.

and cellulose. Lignin is a phenolic polymer, and cellulose is a polymer of glucose; hence, their structures, and consequently their pyrolysis products, are quite different. When pyrolyzed, cellulose makes considerable water and carbon dioxide, and also organics that are suited to GC analysis. These include many furans, alcohols, acetic acid, and levoglucosan. Different plants make different lignins, but they are all phenolic, and produce substituted phenols when pyrolyzed.

Like other copolymer systems, if a plant material that contains both lignin and cellulose is pyrolyzed, the resulting pyrogram contains compounds from each of the individual polymers. Figure 11.10 shows the pyrogram made from a sample of wood. Peaks 1, 2, and 10 (acetic acid, furancarboxaldehyde, and levoglucosan) together are characteristic of cellulose pyrolysis. Just after peak 3 at about 9 min is a small peak for phenol, but the more characteristic methoxyphenols, like peaks 4 and 5, are both larger and

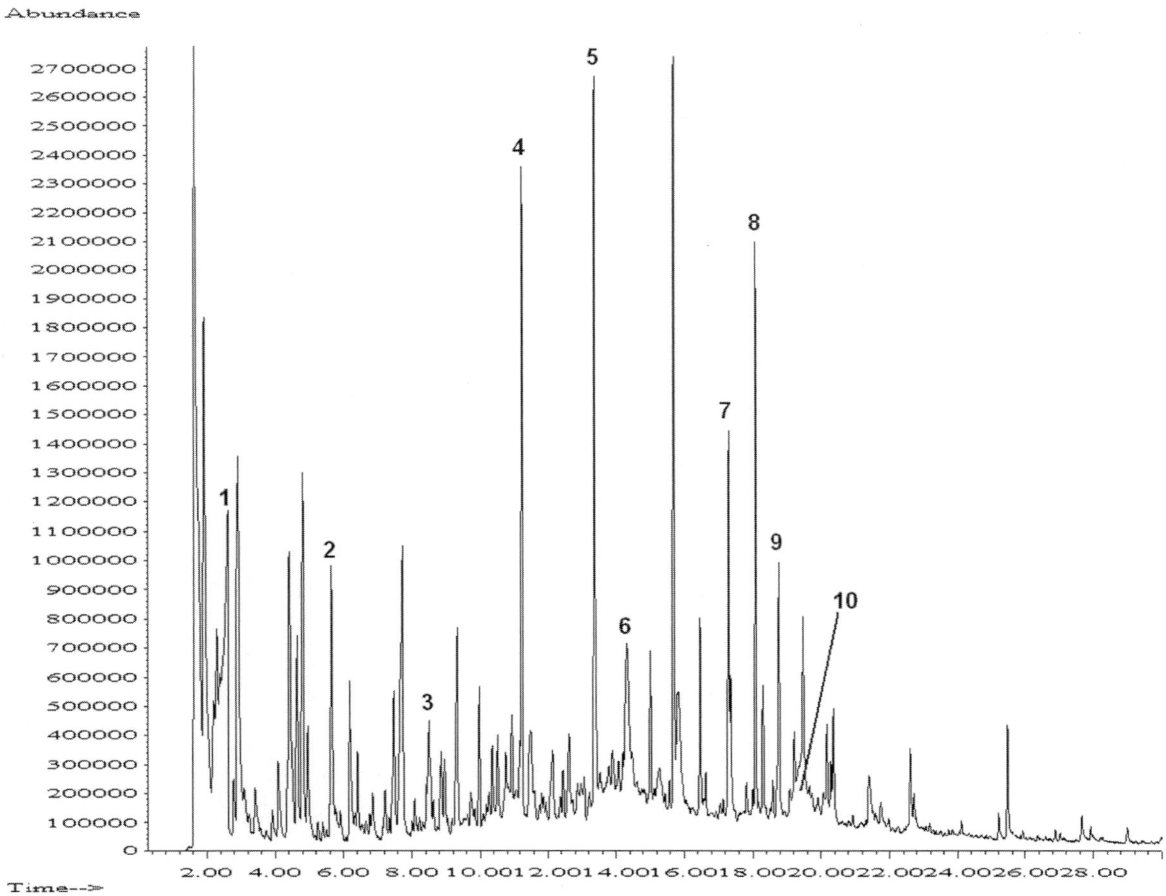

FIGURE 11.10 Pyrolysis of wood at 600 °C. Peak #1, acetic acid; #2, 2-furancarboxaldehyde; #3, 5-methyl-2-furancarboxaldehyde; #4, 2-methoxyphenol; #5, 2-methoxy-4-methylphenol; #6, 5-(hydroxymethyl)-2-furancarboxaldehyde; #7, vanillin; #8, eugenol; #9, 1-(4-hydroxy-3-methoxyphenyl)ethanone; and #10, levoglucosan.

more informative of the type and structure of the lignin involved.

11.4.9. Advanced Applications

In addition to traditional pulse pyrolysis applications for gas chromatography, pyrolysis instruments may be used for a variety of more complex sample treatments. Simultaneous derivatization with a methylating [12] or silylating [13] agent has been applied to many samples, especially those that make polar products, particularly free fatty acids. By adding a methylating agent, such as tetraethylammonium hydroxide, directly to the sample before pyrolyzing, the polar compounds are produced and are methylated, improving the chromatography. Samples may also be processed at low temperatures or using slow heating rates over a protracted time, with the collection of the products onto a trap before transfer to the GC for analysis. The use of a trap also permits heating in a reactive atmosphere such as air or oxygen [14] to study oxygenation without introducing air into the GC system. Systems are also capable of performing multiple step analyses [15] and operating at elevated pressures. Some systems have incorporated catalytic reactors [16] to process the pyrolysate and modify it after pyrolysis, to permit reforming or hydrogenation.

References

[1] T.P. Wampler, Temperature as a sample preparation tool in the analysis of materials by GC-MS, LC GC 17 (2S) (1999) 14–17.

[2] I. Ericsson, Pyrolysis nomenclature, J. Anal. Appl. Pyrolysis 14 (1989) 219–221.

[3] S. Tsuge, H. Ohtani, Structureal characterization of polymeric material by pyrolysis-GC/MS, Polym Degrad. Stab. 58 (1–2) (1997) 109–130.

[4] B.A. Stankiewicz, P.F. van Bergen, M.B. Smith, J.F. Carter, D.E.G. Briggs, R.P. Evershed, Comparison of the analytical performance of filament and Curie-point pyrolysis devices, J. Anal. Appl. Pyrolysis 45 (1998) 133–151.

[5] T.P. Wampler, Pyrolysis techniques in the analysis of polymers and rubbers, in: Encyclopedia of analytical chemistry, John Wiley & Sons, Pubs, 2000.

[6] B.K. Kochanowski, S.L. Morgan, Forensic discrimination of automotive paint samples using pyrolysis-GC/MS with multivariate statistics, J. Chromatogr. Sci. 38 (3) (2000) 100–108.

[7] K.D. Jansson, T.P. Wampler, C.P. Zawodny, Analysis of coatings using pyrolysis-GC/MS, Paint Coat Ind. Mag. (February 2009).

[8] T. Learner, The analysis of synthetic paints by Py-GC/MS, Stud. Conservat. 46 (4) (2001) 225–241.

[9] H. Matsubara, A. Yoshida, H. Ohtani, S. Tsuge, Compositional analysis of UV-cured acrylic ester resins by pyrolysis-gas chromatography in the presence of organic alkali, J. Anal. Appl. Pyrolysis 64 (2) (2002) 159–175.

[10] R.S. Lehrle, I.W. Parsons, M. Rollinson, Thermal degradation mechanisms of nylon 6 deduced from kinetic studies by pyrolysis-GC, Polym. Degrad. Stab. 67 (1) (2000) 21–33.

[11] D. Meier, I. Fortmann, J. Odermatt, O. Faix, Discrimination of genetically modified poplar clones by analytical pyrolysis-gas chromatography and principal component analysis, J. Anal. Appl. Pyrolysis 74 (2005) 129–137.

[12] J.M. Challinor, Review: the development and applications of thermally assisted hydrolysis and methylation reactions, J. Anal. Appl. Pyrolysis 61 (2001) 3–34.

[13] L. Osete-Cortina, M.T. Domenech-Carbo, Characterization of acrylic resins sued for restoration of artworks by pyrolysis-silylation-gas chromatography/mass spectrometry with hexamethyldisilazane, J. Chromatogr. A 1127 (1–2) (2006) 228–236.

[14] T.P. Wampler, E.J. Levy, Effect of heating rate on oxidative degradation of polymeric materials, J. Anal. Appl. Pyrolysis 8 (1985) 153–161.

[15] T.P. Wampler, C.P. Zawodny, K.D. Jansson, Multistep thermal characterization of polymers using GC-MS, Am. Lab. 39 (6) (2007) 16–19.

[16] D.A. Gunawardena, S.D. Fernando, Deoxygenation of methanol over ZSM-5 in a high-pressure catalytic pyroprobe, Chem. Eng. Technol. 34 (2) (2011) 173–178.

CHAPTER

12

Detectors

Matthew S. Klee

OUTLINE

12.1. INTRODUCTION

One of the key strengths of gas chromatography over other forms of chromatography is its detectors. Because the carrier gases used in GC are transparent to most GC detectors, background levels and interferences are very low. This transparency provides choices in detectors and design simplicity that are not possible with other separation techniques such as HPLC and SFC.

Some common GC detectors that generate a univariate (single) signal in response to an eluting solute are listed below:

- thermal conductivity detector (TCD),
- flame ionization detector (FID),
- electron-capture detector (ECD),
- alkali bead or nitrogen–phosphorus detector (NPD),
- flame photometric detector (FPD),
- electrolytic conductivity detector (ELCD),
- photoionization detector (PID), and
- chemiluminescent detector (CLD).

There are also GC detectors that generate multivariate (multiple, multiplex) response, such as

- atomic emission detector (AED),
- infrared detector (IRD), and
- mass spectrometer (MSD).

Gas Chromatography DOI: 10.1016/B978-0-12-385540-4.00012-2

In this chapter, the popular univariate detectors as well as the AED will be discussed.

Detectors fall into two general categories: *universal* and *selective*. Universal detectors are able to detect all compounds (or most compounds) that elute. A selective detector detects only specific classes of compounds based on some molecular, elemental, or physical property. A subset of selectivity is *specificity*, that is, the ability to detect a specific atom or functional group to the exclusion of others. Table 12.1 lists GC detectors in terms of their general attributes, including if they are considered universal, selective, or specific.

Another characteristic of detector response is whether it responds to the concentration of an analyte passing through it or to the mass flow of the analyte passing through it. A *concentration-sensitive* detector will generate a signal proportional to the concentration of an analyte in the carrier gas as it passes through. The thermal conductivity detector (TCD) is the most common concentration-sensitive detector. If, for example, a makeup gas flow rate equal to the column flow rate was added at the end of the column prior to the detector, the concentration of an eluting analyte will be cut in half and so would its signal. In contrast, if all else were equal and the column flow rate was doubled, the detector response would not change because the concentration of analyte passing through the detector would remain the same. LOD of concentration-sensitive detectors is most often stated in concentration terms (e.g., ng/mL).

Mass-sensitive detectors respond to the mass of analyte passing through the detector in a given time. The *flame ionization detector* (FID) is an example of a mass-sensitive detector and its response is stated in mass per time (pg/s). Adding makeup gas at the end of the column would not change the detector response because the same mass of analyte will pass through the detector in the same time (it will just be more dilute). However, if the flow of the column was reduced to ½, the same amount of analyte would generate ½ the peak height because the same mass of analyte would take twice the time to pass through the detector. The peak area would remain essentially the same; however, the LOD would be less since it is dependent on peak height, not area. For both mass-sensitive and concentration-sensitive detectors, narrower peaks yield better detection limits because both mass/time and concentration are higher.

An additional important characteristic of GC detectors is their destructivity. Non-destructive detectors are typically also concentration-sensitive detectors such as the TCD and ECD. Non-destructive detectors can detect solutes without changing them chemically. This is a benefit if you are interested in fraction collection, smelling (olfactory detection), or especially when putting detectors in tandem (e.g., passing effluent through a non-destructive detector and then through a second detector) to get the best detection limits from each.

To understand and contrast GC detectors, we must first define a few metrics related to detection in general. The first is the *detection limit*. This is the minimum quantity of material that can be distinguished from background. The detection limit or *limit of detection* (LOD) is a measure of the ability of the detector to differentiate the signal generated by an eluting compound from the neighboring background noise that is usually characterized as "high-frequency noise" relative to chromatographic peak widths. The numerical measure of this is known as the *signal-to-noise ratio* (S/N). The larger the S/N for a given amount of solute and set of conditions, the better the detector for detection of minute quantities of solute. The LOD is usually specified as the amount (concentration or mass per time) of compound

TABLE 12.1 Comparison of GC Detector Attributes

Detector	Response	Detects	Typical LOD[1]	Linear dynamic range	Selectivity	Concentration or mass sensitive	Destructive?
FID	Universal	Carbon	2 pg C/s	10^7	n/a	Mass	Yes
TCD	Universal	Anything with thermal conductivity different from carrier	400 pg/ml carrier	10^5	n/a	Concentration	No
ECD	Selective	Gas-phase electrophores	50 fg/ml (Varies significantly with structure)	10^4	Up to 10^6	Concentration	No
NPD	Selective	N, P hetroatomas	0.4 pg N/s	10^5	25,000 gN/gC	Mass	Yes
			0.2 pg P/s	10^5	75,000 gP/gC	Mass	Yes
FPD, pFPD	specific	P, S	4,1 pg S/s	10^3	10^6 gS/gC	Mass	Yes
			60, <100 fg P/s	10^4, 10^3	10^6 gP/gC	Mass	Yes
PID	Universal/ selectivity	Compounds ionized by UV light	10 pg C/s	10^6	∞ versus those with IP > lamp	Concentration	No
ELCD	Selective	Halogens, N, S	0.5 pg CI/s	5×10^6	10^6 gCI/gC	Mass	Yes
			2 pg S/s	10^5		Mass	Yes
			4 pg N/s	10^5		Mass	Yes
CLD	Specific	S, N	0.5 pg/s/s	10^4	10^7 gS/gC	Mass	Yes
			3 pg N/s	10^4	10^7 gN/gC	Mass	Yes
AED	Universal and Specific	Atomic Emission (e.g., C, S, N, H, N, CL, O .)	1–150 pg/s (element dependent)	10^3– 10^4	10^4– 10^5 vs C	Mass	Yes
FTIR	Universal/ Specific	Molecular vibrations	100 pg/ml of strong absorber	10^3	10^4 depending on functional group	Concentration	No
MSD	Universal and specific	Tunable for any species	10 pg to ng (depending on SIM)	10^5	∞ depending on functional group	Mass	Yes

[1] *Assuming a 1 μL splitless injection, 1 mL/min flow rate and 5 s peak width.*

that can be detected with an S/N equal to 2 or 3. The lower the amount of analyte that generates this S/N, the better the detector for low-level detection. An illustration of a signal near the LOD is shown in Figure 12.1.

Noise can come from many sources. All detectors have a characteristic noise that comes from random and/or systematic events in the detection process, electronics, thermal, vibration, pressure sensitivities, etc. Some of this high-frequency noise can be reduced by analog and/or digital detector electronics. Detector designers make choices in how much filtering is done in analog and digital sections of the signal processing circuitry. Compromises are made to balance detector response time and noise suppression.

However, the major source of noise in most real analyses comes from chemical noise. This manifests as "low-frequency noise". Impurities in carrier gas, stationary-phase bleed, and carryover from prior sample injections conspire to produce rolling baselines, exaggerated baseline rise with oven temperature programming, ghost peaks, and elevated overall background that

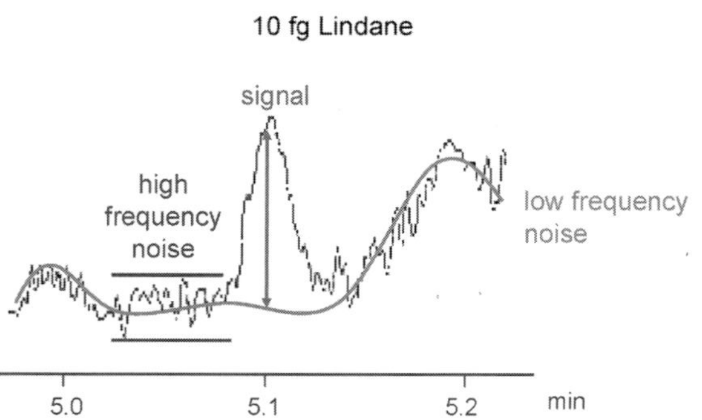

FIGURE 12.1 Example near the detection limit of an ECD. Limits of detection (LOD) are typically stated as some multiple of high-frequency noise, such as S/N = 3 times RMS noise. However, in practice with real samples and an instrument that has been in use for some time, the noise is dominated by chemical noise (low frequency) and method detection limits (MDL) can be 10 times higher than the (LOD). *Source: Redrawn from original courtesy of Agilent Technologies, Inc.*

complicate analyte signal detection. The more sensitive a detector is, the more sensitive it is to chemical noise as well. So, the probability of being able to reach a detector's specified LOD is very low except with new systems that have not been exposed to "real samples". More practical metrics such as *method detection limit* (MDL) or *method limit of quantitation* (LOQ) reflect more realistic minimum quantities of an analyte that can be determined analytically because they take into account additional sources of analytical variability and noise. The MDL is a method that is analyte specific and is often 10–20 times higher than the LOD. Therefore, LOD numbers should be considered as a means of comparing detectors to each other on paper but not as a realistic measure of what one can be routinely achieved in analysis of samples.

Another metric relating to detector performance is its *dynamic range*. This is the usable (operating) range over which the detector will generate a changing signal as the amount of analyte changes. The low end of the dynamic range is limited by the LOD. The upper range is limited by *saturation* of the detector. Saturation is routinely observed for solvent peaks, where the top of the peak appears to roll off, becoming flat and perhaps noisy. Saturation can sometimes go unnoticed for large analyte peaks. Therefore, it is important to calibrate analyses over the full analytical range of interest and look for "roll off" of the calibration curve, as illustrated in Figure 12.2.

Another aspect related to dynamic range is the range over which the detector response is linear (signal increases proportionally to the amount or concentration of analyte). The *linear dynamic range* is a characteristic that was more important in the early days of chromatography when data analysis tools were less sophisticated. Modern data systems can easily deal with nonlinear calibrations. Nevertheless, a large linear dynamic range simplifies calibration procedures and data reduction and therefore it is still a useful attribute for a detector to have. The larger the linear

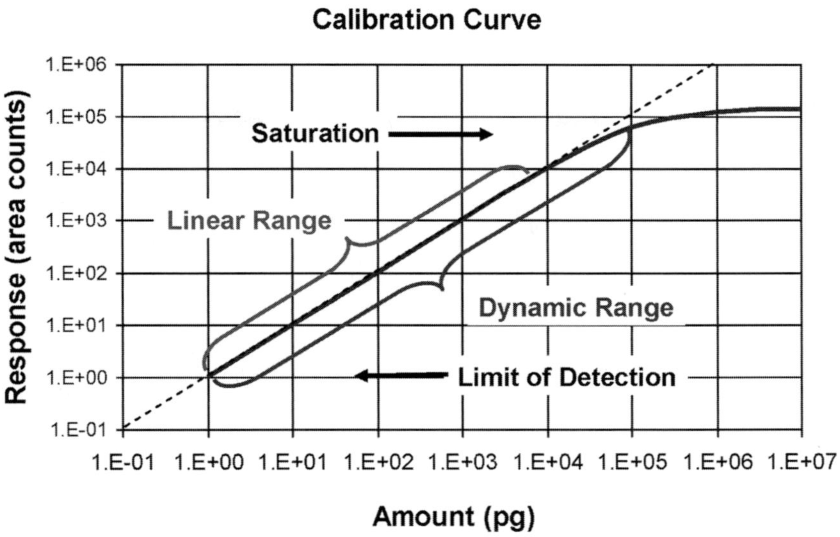

FIGURE 12.2 Generic calibration curve in log(response) vs. log(amount) showing detection limit and comparing linear and dynamic response ranges.

dynamic range, the better. Dynamic and linear dynamic ranges are usually stated as powers of 10 (orders of magnitude). The linear dynamic range of GC detectors is often several orders of magnitude larger than typical LC detectors. Typical dynamic and linear operating ranges are listed in Table 12.1 for common detectors.

Figure 12.3 illustrates another way of displaying linear detector responses, a response factor curve. The response factor (response/sample amount or concentration) is constant over the linear response range but then falls off at the onset of saturation.

When comparing *selective* detectors, one is interested in the ability to detect a characteristic of interest while ignoring everything else. For example, one may be interested in selectively measuring compounds containing sulfur. In comparing a standard *flame photometric detector* (FPD) to a pulsed FPD, sulfur chemiluminescence

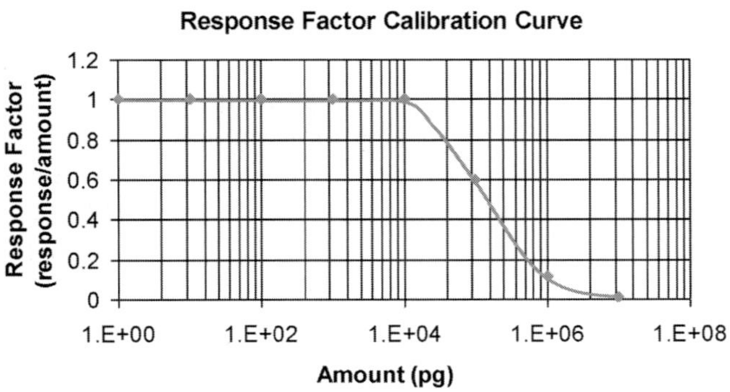

FIGURE 12.3 Calibration curve shown in normalized response factor format.

detector (SCD), or an atomic emission detector (AED), all of which have selective sulfur responses, an important metric is how selective each detector is for sulfur-containing compounds compared to compounds containing no sulfur. *Selectivity* is usually stated as the ratio of (c), the mass of a nontarget interfering species (usually carbon) that generates equivalent signal, to (x), the mass of a target species (e.g., sulfur). So, if 1 mg (10^{-3} g) of carbon in a reference standard generates the same response as 100 pg S (10^{-10} g) in a sulfur-containing reference, then the *S* selectivity ratio over carbon would be

$$c/x = 10^{-3}/10^{-10} = 10^{7}$$

That means that it would take 10^{7} more of the sulfur-free compound to yield a peak of the same size as the sulfur-containing compound. For selectivity, the larger the number, the better.

Some selective detectors are selective for a category or class of compounds, as is the case for an *electron-capture detector* (ECD). The ECD is selective for compounds containing atoms or functional groups that are electronegative (capture electrons). These functional groups include halogens (group 7 in the periodic table of the elements: Cl, Br, and I) and oxygen-containing functional groups. With these types of detectors, values for LOD, dynamic range, and selectivity are usually different for each of the functionalities, and in the case of the ECD, they can vary significantly based on how many active groups are on each molecule. The differences can be very dramatic. Because of this, manufacturers often try to use the same compound and conditions as their competitors when stating performance numbers so that realistic comparisons can be made.

Selective detectors usually provide better MDLs for target compounds because they are able to reject more chemical noise than can universal detectors (similar signal with lower background chemical noise = better MDL).

A detector metric that often gets misconstrued is *sensitivity*. Sensitivity is the change in response per change in solute amount or concentration. The higher the sensitivity, the better one can differentiate between small changes in analyte. Although a highly sensitivity detector can also have a very low LOD, this is not always the case.

Specificity is the ability to detect a specific atom or functionality. For example, an atomic emission detector (AED) can monitor emission lines from specific atoms. A flame photometric detector (FPD) can measure the emission specific to the HPO* species. A mass spectrometer can measure a specific ion m/z.

Table 12.1 compares the common GC detectors in terms of the above metrics. From this comparison, one can see why GC detectors are key differentiators of GC over other separation techniques. Figure 12.4 graphically compares typical operating ranges of GC detectors.

12.2. THERMAL CONDUCTIVITY DETECTOR

The thermal conductivity detector (TCD) is a universal detector that is non-destructive. It is most commonly used for the analysis of light and permanent gases with packed columns or capillary PLOT columns. It is one of the least sensitive GC detectors, but it is rugged, has a fairly wide linear dynamic range, requires no fuel gases, has no flame (it is intrinsically safe), requires little power, and is inexpensive. This combination of attributes often makes the TCD a popular detector and often the only choice for some applications.

TCDs are especially well suited for use in portable and micro gas chromatographs not only due to their low power consumption but also because they are easily miniaturized and can be micromachined and integrated into chip formats [1,2].

The TCD incorporates a relatively simple Wheatstone bridge circuit to measure resistance

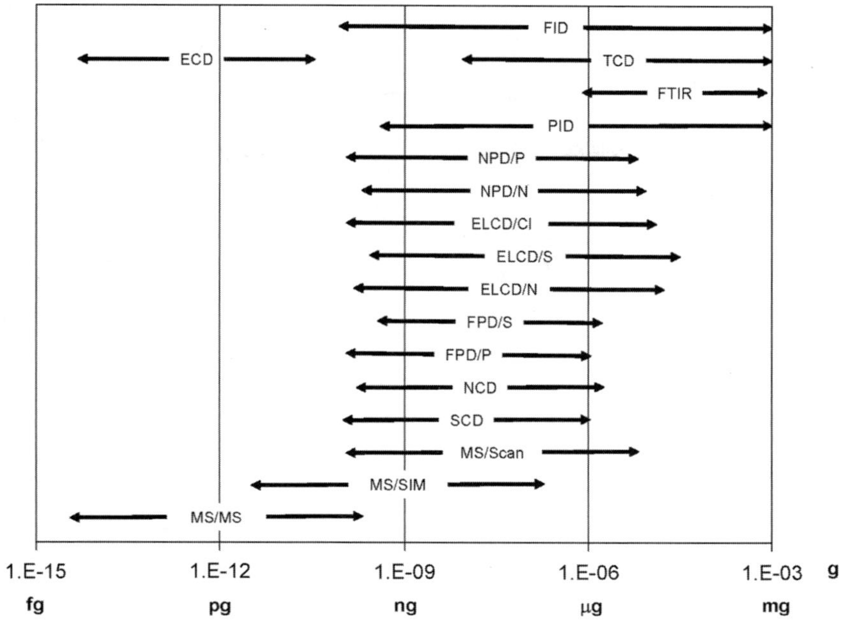

FIGURE 12.4 Typical operating ranges of GC detectors.

changes in a heated filament as sample passes over it. In the simplest design, as illustrated in Figure 12.5, there are two heated filaments. Column effluent passes over one filament and reference gas (clean carrier gas, or effluent from a second unused column) flows over the other. A constant voltage or current is applied to the circuit and the balanced current or voltage between opposite legs of the bridge is monitored.

The reference side in original TCD designs was used primarily to cancel baseline rises due

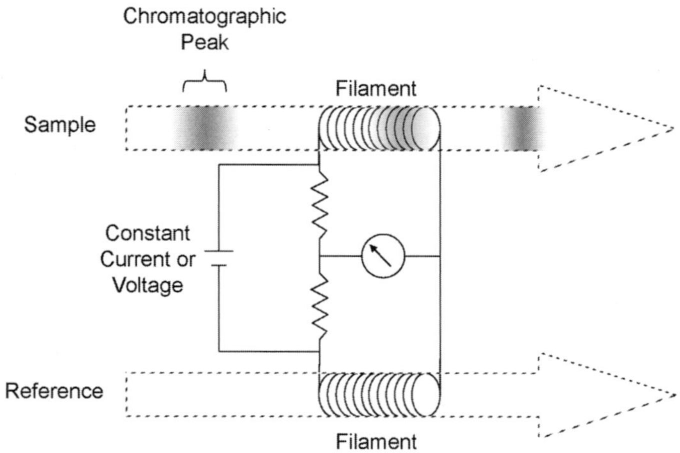

FIGURE 12.5 Schematic of typical TCD Wheatstone bridge.

to stationary phase bleed during the run. A duplicate column was attached to the reference side with the same column flow as the analytical side. The assumption was that the reference column would bleed the same as the analytical column and the signal would cancel the rise occurring on the analytical side. With capillary column and stationary phase advances over the decades, stationary phase bleed is not as much a factor as it used to be. The main function of the reference side of TCDs today is minimizing drift due to temperature and flow variation of the cell and flow rate changes (constant pressure mode) during the run. TCDs are much more sensitive to temperature variations than most other GC detectors.

A clever approach to dual column TCD analyses is to use one side of the TCD for channel 1 and the other for channel 2 detection. Peaks appear positive for channel 1 and negative for channel 2. Of course, one must develop method conditions so that the peaks from each channel elute at different times and one must have data analysis tools that can handle positive and negative peaks in the same chromatogram.

Thermal conductivity is a physical characteristic of molecules that relates to their ability to conduct heat. GC carrier gases have fairly high thermal conductivity, as shown in Table 12.2, which means that they can drain heat away from a heated filament relatively efficiently. As a solute elutes and passes over the detector's sample filament, the filament temperature will rise relative to the reference filament because analytes have lower thermal conductivity and are therefore less efficient at draining the heat. The increase in temperature and therefore the signal are a function of solute concentration and its thermal conductivity relative to that of the carrier gas.

To get the highest sensitivity out of TCD, one should use a carrier gas with as different a thermal conductivity as possible from the lowest concentration analyte of interest in the sample. For example, if one wanted to measure

TABLE 12.2 Comparison of the Thermal Conductivities of Common Carrier Gases and Several Analytes. Units are mW per meter Kelvin (mW/m K)

Formula	Name	Thermal conductivity at 400 K	Relative to He
H_2	Hydrogen	230.4	1.209
He	Helium	190.6	1.000
Ne	Neon	60.3	0.316
CH_4	Methane	49.1	0.258
H_3S	Ammonia	37.4	0.196
C_2H_6	Ethane	35.4	0.186
C_2H_4	Ethylene	34.6	0.182
O_2	Oxygen	33.7	0.177
C_2H_2	Acetylene	33.3	0.175
NO	Nitric oxide	33.1	0.174
N_2	Nitrogen	32.3	0.169
CO	Carbon monoxide	32.3	0.169
C_3H_8	Propane	30.6	0.161
C_4H_{10}	Butane	28.4	0.149
H_2O	Water	27.1	0.142
CH_4O	Methanol	26.2	0.137
N_2O	Nitrous oxide	26	0.136
C_2H_6O	Ethanol	25.8	0.135
CO_2	Carbon dioxide	25.1	0.132
C_5H_{12}	Pentane	24.9	0.131
CF_4	Tetrafluoromethane	24.1	0.126
C_6H_{14}	Hexane	23.4	0.123
Ar	Argon	22.6	0.119
H_2S	Hydrogen sulfide	20.5	0.108
C_3H_6O	Acetone	20.2	0.106
CCl_2F_2	Dichlorodifluoromethane	15	0.079
SO_2	Sulfur dioxide	14.3	0.075
Kr	Krypton	12.3	0.065
$CHCl_3$	Trichloromethane	11.1	0.058
Xe	Xenon	7.3	0.038

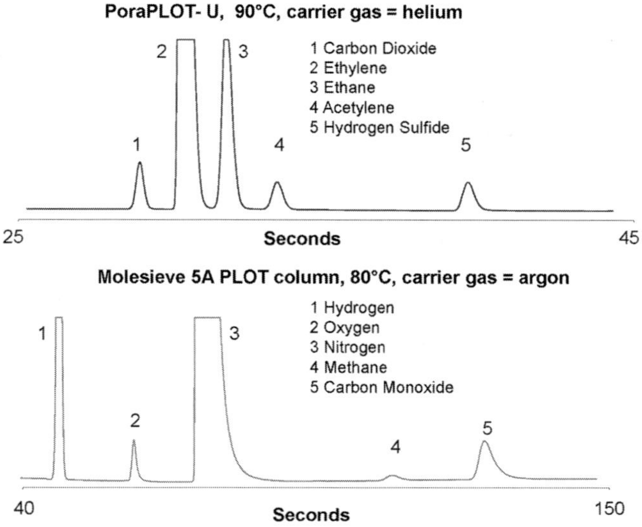

FIGURE 12.6 Dual channel configuration with different carrier gases maximizes TCD sensitivity for target compounds. Due to similarity of thermal conductivities, H_2S would not have been easily detected with Ar carrier gas. Conversely, H_2 would not have been sensitively detected with He carrier gas. *Source: Courtesy of Agilent Technologies, Inc.*

N_2, He or H_2 carrier gases would be good choices. Conversely, one would use N_2 as the carrier gas to purposely hide the N_2 peak (perhaps because it is a major peak and obscures a minor peak of interest).

If one wanted the lowest detection limits for H_2, one would be better served to use N_2 (or Argon) than He as the carrier gas. In addition, there is an anomalous non-linearity in the thermal conductivity of H_2/He gas mixtures. At approximately 7.5% H_2 in He, the mixture has the same thermal conductivity as pure He. At concentrations higher than approximately 7.5%, response is linear but the peak is negative so calibration curves that cover the full concentration range are problematic. At lower concentrations, response is linear but positive. Depending on the amount of H_2 in the sample, its peak can end up looking like a W. To avoid these problems, most people analyzing H_2 in light gas mixtures will use a two-column approach, with one column using N_2 or Ar carrier gas for quantification of H_2 and a second column with He carrier gas for more sensitive quantification of the remaining analytes, as illustrated in Figure 12.6, wherein detection of H_2S would have been very insensitive with Ar due to the similarity of their thermal conductivities.

As with most detectors, it is advisable to operate the TCD at a temperature that matches the isothermal column temperature or the highest column temperature in a temperature-programmed method in order to minimize solute condensation, detector contamination, and peak tailing due to solute interaction with the filament and/or flow path surfaces. Excessive temperature should be avoided with the TCD because it is less sensitive at higher temperatures. Figure 12.7 illustrates the decrease in thermal conductivity of several analytes relative to He as a function of temperature.

A popular alternative TCD design illustrated in Figure 12.8 uses a single filament and fluidic switching to avoid some of the problems associated with two-filament designs. A solenoid

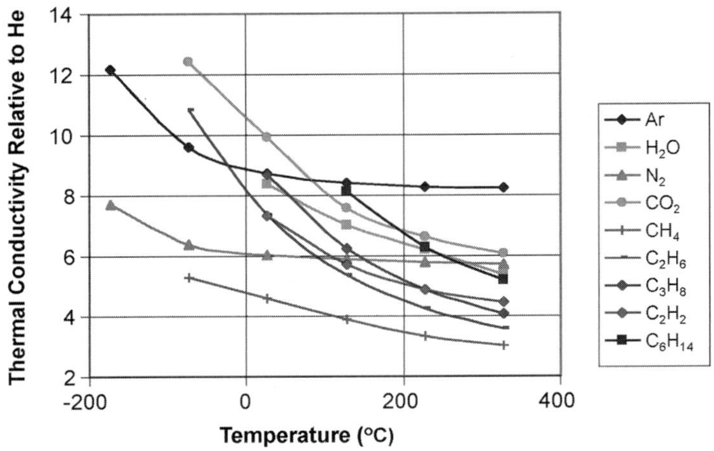

FIGURE 12.7 Example of TCD response dependence on detector temperature. As temperature increases, thermal conductivity of most analytes decreases relative to typical carrier gas (He), decreasing sensitivity. By selecting the lowest reasonable detector temperature one maintains the highest sensitivity.

operating at 10 Hz switches flow between clean reference gas and column effluent. Detector electronics gate detection of the signal and reference flows synchronously with the solenoid switch time (with an appropriate delay time to allow the cell to be cleared and filled with the switched gas stream). Although this provides a very stable response, the downside of the

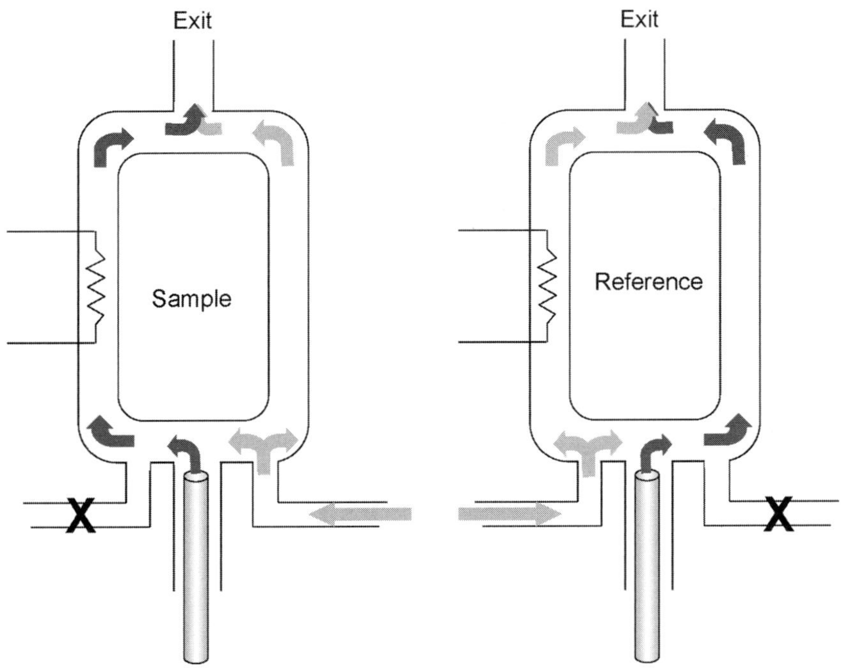

FIGURE 12.8 Microfluidically switched single-filament TCD design. Sample and reference streams are alternately switched across a single filament.

design is that it was designed for packed-column flow rates >20 mL/min. It does not work as well with capillary columns with narrow peaks or low flow rates (effluent has to be diluted by 20 mL/min, or so, makeup flow necessary for the detector to work).

TCD filaments do burn out eventually with use. The lifetime is somewhat related to temperature (higher temperature decreases lifetime), corrosivity or reactivity of gases being analyzed, and exposure to vibration/shock (e.g., in plant installations). Most TCD designs allow filaments to be easily replaced.

With most TCD designs, solutes flow directly over the filament. Some polar or reactive solutes interact with filaments (usually tungsten), causing peak tailing. Acidic compounds are particularly notorious for tailing badly with exposed filaments. To minimize this, filaments are typically coated to reduce interaction with the metal filament. Over time, the protective coating can crack or wear, exposing the filament to sample. When polar peaks start to tail more than they had previously, then replacing the filaments should help. Some TCD designs incorporate indirect heating (filament is isolated from the sample path by a thin membrane) to avoid any interaction with solutes, but the response and detection limits can be negatively impacted.

All in all, the TCD has proven over the decades to be a reliable, rugged, and useful GC detector, especially for the analysis of light and permanent gases.

12.3. FLAME IONIZATION DETECTOR

The flame ionization detector (FID) is the premier detector in gas chromatography. It has unique properties and performance that puts it above and beyond all other general-use detectors in gas chromatography (or any other form of chromatography, for that matter).

Topping the list of unique attributes are *unit carbon response* and wide *linear operating range* (up to 7 orders of magnitude). When combined with its other attributes of low cost, ease of use, speed of response, and ruggedness, it is not surprising why the FID is the premier univariate detector of choice for GC.

The FID is a destructive *mass-sensitive* detector, which means that its response is proportional to the mass of carbon that passes through it in unit time. In this regard, FID response is stated in terms of picograms carbon per second (pg/s). Detection limits for FIDs are in the low pg C/s.

Unit carbon response means that FID responds linearly to the mass of carbon flowing through it, independent of compound structure. The FID gives unit response for most hydrocarbons within a couple percent error. This attribute of unit carbon response allows one to quantify components in mixtures without having calibration standards for every component. Amounts of components in a sample will be proportional to their peak areas. So, a simple area percent report will fairly closely reflect the mass percent of each component in a mixture. This is extremely useful when analyzing complex samples such as those in the petroleum industry wherein samples can contain well over 1000 components. In the same vein, the relative ratio of the area of one peak to another (e.g., the peak of an unknown component relative to a calibrated reference peak) closely reflects its relative amount in the sample. This is useful when estimating concentration levels of components in a sample when identities are unknown or when standards are not available for calibration.

The FID was first described by two independent groups at approximately the same time [3,4]. FIDs were commercially available soon thereafter (the early 1960s). Most of the developments since the original functional designs have been primarily in the areas of usability, adaptation for capillary instead of packed columns,

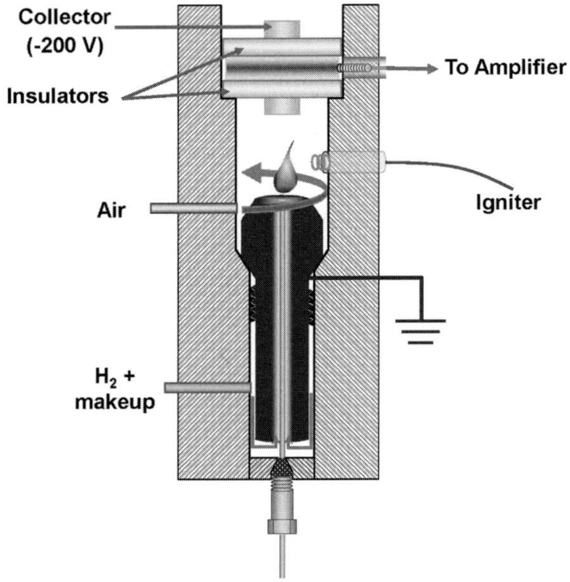

FIGURE 12.9 Schematic of typical FID. H_2 (and sometimes N_2 or He makeup gas) enters the bottom of the FID jet and mixes with column effluent prior to exiting the jet. Air is added above the jet and flame is established. Charged combustion products are accelerated toward and collected by the collector electrode creating the signal.

improvements in associated electronics, and signal processing. A general diagram of an FID is shown in Figure 12.9.

In a hydrogen flame, hydrogen gas (H_2) reacts exothermically with the oxygen (O_2) to form water. A hydrogen/air flame temperature burns at 2210 °C [*note: the hydrogen flame is light blue and often very difficult to see, so open hydrogen flames are quite hazardous; fortunately, in an FID, the flame is contained inside the FID*]:

$$2H_2 + O_2 \rightarrow 4H\cdot + 2O\cdot \rightarrow 2H_2O + \text{heat}$$

In the above reaction, one part of oxygen reacts with two parts of hydrogen. This ratio (1:2) is called the *stoichiometric ratio*. Since air is approximately 20% oxygen (O_2), a stoichiometric mixture would require an air/hydrogen gas flow ratio of 2:5; 2.5 times more air than hydrogen. Although stoichiometric ratios of oxygen to hydrogen can provide a reasonable dynamic range for the FID, experience has shown (in part due to the complex combustion processes described below) that an excess flow of air is required to ensure complete combustion, unit carbon response, and the widest linear dynamic range. In addition, higher than stoichiometric air flow helps to avoid carbon deposits from forming in the jet when high concentrations of analyte or solvent pass through.

A ratio of at least 6:1 of air to hydrogen has empirically been found necessary to achieve the widest dynamic range possible with the FID. Many manufacturers recommend ratios of 10:1 or more air/hydrogen flows. The higher the sample load to the FID, the more flame gases are required to prevent blowout and carbon formation. That is why gas flows used with packed columns (higher sample loads) generally need to be higher than when capillary columns are used. Exact flow rates are somewhat instrument specific and also relate somewhat to carrier gas flow rates, so following manufacturer recommendations are wise.

12.3.1. Sample Combustion and Signal Generation

A very good explanation of FID flame chemistry has been provided by Holm [3]. Therein he supports a mechanism whereby most organic compounds are reduced to saturated counterparts in the initial portion of the flame, where temperatures are lower. As they continue up the flame, these saturated counterparts then continue to react with hydrogen atoms to form methane, as illustrated in the following equations.

$$CH_3{-}(CH_2)_n{-}CH_3 \xrightarrow{H\cdot} CH_4$$
$$+\,CH_3{-}(CH_2)_{n-1}{-}\dot{C}H_3 \xrightarrow{\ldots H\cdot} \ldots(n+2)CH_4$$

Still higher up in the flame, methane is combusted to form formylium ion CHO^+, the primary FID signal-producing ion:

$$CH_4 \rightarrow H_2 + :CH \xrightarrow{O\cdot} CHO^* \rightarrow CHO^+ + e^-$$

Other reactions can of course occur, such as ones that form other positive ions such as hydronium ion:

$$CHO^+ + H_2O \rightarrow H_3O^+ + CO$$

All positive ions are collected by a negatively biased collector causing a current to flow, which is then electronically amplified and digitized (Figure 12.9). The current is proportional to the number of ions collected. The yield of ions from carbon passing through the detector is on the order of one ion per 10^6 carbon atoms [4]; yet this is still enough for the FID to give picogram-level detection. Electrons flow in the opposite direction and are grounded out on the FID jet (Figure 12.10). By biasing the collector electrode high enough relative to the jet (e.g., -200 V), recombination of positively charged ions and electrons is minimized, and signal maximized.

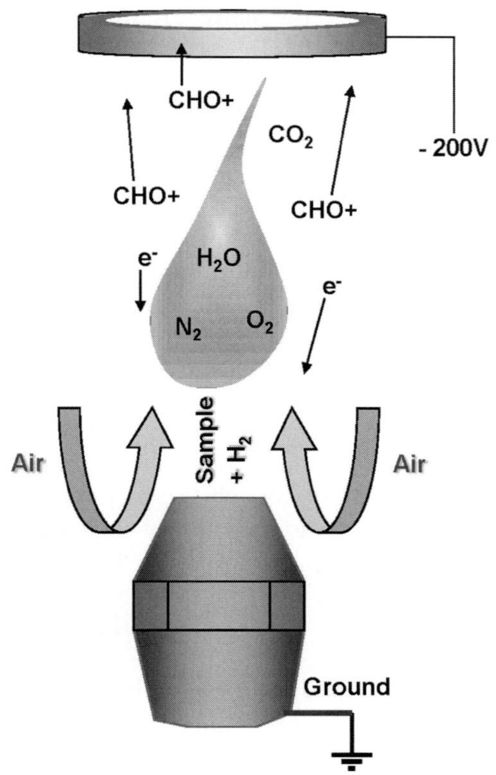

FIGURE 12.10 Sample components are combusted in the flame, creating positive ions and electrons. The positive ions are attracted to the negatively biased collector while the (negatively charged) electrons are repelled toward the jet.

Through the combustion process, hydrocarbons are oxidized in the flame, eventually forming CO_2 and H_2O. The charged intermediate species (primarily CHO^+) give rise to the FID response. If a molecule already incorporates some oxygen, it is farther along the oxidation process and will yield a lower proportion of ions and therefore generate a lower response per molecule or mass of eluting analyte. This means that the same amount of an oxygenated compound will yield a lower peak of less area than would its saturated counterpart. In a similar fashion, the more oxygen per molecule, the less the response.

Response for oxygenated compounds generally decreases in the following order: alcohols, ethers, aldehydes, ketones > esters > acids. The deviation from unit carbon response becomes less significant; the more carbons are in a molecule (the lower the proportion of oxygen per total carbon). Table 12.3 shows the trend of response for a homologous series of oxygenated compounds.

Deviations from unit carbon response are only an issue if one does not calibrate peak responses of individual components. For example, if one were to use an area percent report as an estimate of relative amounts of sample components, then the low mass oxygenated compounds would be under-reported.

When using uncalibrated reports from an FID to estimate weight percent composition, one must also keep in mind that even when

TABLE 12.3 FID Response Relative to Carbon Content of Oxygenated Compounds [5,6]

	Relative response
ALCOHOLS	
Methanol	0.23
Ethanol	0.46
n-Propanol	0.6
n-Hexanol	0.74
n-Decanol	0.84
KETONES	
Acetone	0.49
Methylethylketone	0.61
Dibutylketone	0.72
ESTERS	
Methyl acetate	0.2
Ethyl acetate	0.38
Isopropyl acetate	0.49
ACIDS	
Formic	0.01
Acetic	0.24
Propanoic	0.40
Octanoic	0.65

a compound generates close to theoretical response per mass of carbon in the molecule, the molecular mass of the component may be much higher, especially if there are a significant number of heteroatoms. This would lead to under reporting of the mass percent of these compounds. For example, one might get a similar response from an equal number of molecules of ethane and dichloroethane, but a 50/50 molar mixture of those gases would actually contain 77 wt% dichloroethane (molecular weights of ethane = 30, dichloroethane = 98). Stated another way, the peak size of ethane for equal masses of dichloroethane and ethane would be approximately 3.3 times larger than that for dichloroethane.

Other functional groups such as nitrogen and sulfur can also cause a less-than-theoretical carbon response. All that being said, for the majority of compounds, FID responses are close enough to accurately estimate composition of individual components within 10% of their actual amounts (often within 2% error). This is a unique and powerful attribute of the FID.

12.4. ELECTRON-CAPTURE DETECTOR

The electron-capture detector (ECD) is the detector of choice for many environmental GC methods due to its inherent sensitivity and selectivity for halogenated compounds. It is non-destructive, concentration sensitive, and is the most sensitive non-hyphenated GC detector. The invention of the ECD in the late 1950s is considered by many to be the catalyst for the environmental revolution that soon followed its commercial availability and rapid subsequent improvements [7–9]. Due to its unique attributes, the ECD allowed low-level analyses of chlorinated pesticides and PCBs to be performed on a routine cost-effective basis. Governmental agencies such as the US EPA promulgated test methods and regulations leading to the Contract Lab Program and remediation of Superfund sites in the United States. In the pinnacle of this period, thousands of samples were being analyzed per day by a network of labs that years earlier had not existed.

A key inventor and champion, James E. Lovelock, contributed much to the understanding of the detection principles involved in electron-capture detection [8,9]. Although the details are a bit more complex – highly dependent on specific detector design, experimental setpoints, and solute characteristics – the basic mechanism of detection is described

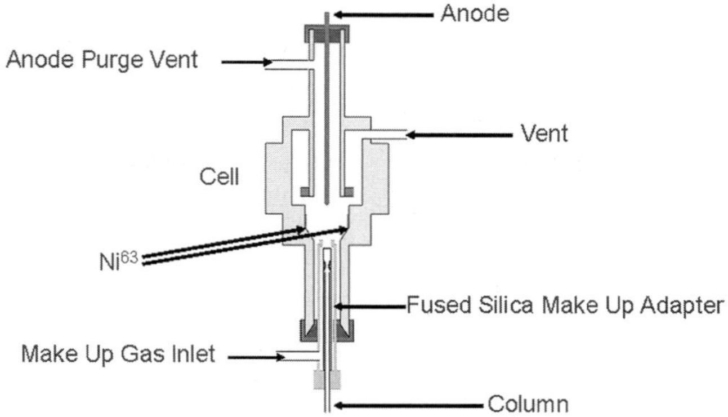

FIGURE 12.11 Example of an ECD. A purge gas protects the anode from excessive interaction with analytes. Turbulent mixing at the entrance to the cell ensures mixing of carrier gas effluent with reagent gas (N_2 or Ar/CH_4). A fused silica sleeve minimizes exposure to metal surfaces. *Source: Copyright Agilent Technologies, Inc. [10].*

below. A schematic of an ECD design is shown in Figure 12.11 [10].

An electron plasma is established in the detector through the ionization of a continuously fed "reagent" gas of relatively low ionization potential such as N_2, or a few percent of CH_4 in He. β-Particles emitted from a radioactive source are used to ionize the reagent gas, releasing electrons and sustaining the plasma. Typically, radioactive sources such as ^{63}Ni (65.9 keV max) or ^{3}H (18.6 keV max) are used.

The resulting background abundance of electrons is measured, usually by using a pulsed circuit with 30–50 V bias and μs pulses. During each pulse, the background of electrons is completely collected and the current measured as illustrated in Figure 12.12. The plasma quickly reestablishes at the conclusion of each pulse.

One approach in ECD detector electronics is to integrate the current measured during the pulses, and output the result as a continuous DC signal. Another, more common approach that provides a wider linear response range and more reliable measurement, is to use a feedback circuit wherein frequency of the pulses is adjusted to maintain a constant current [7]. In such a constant-current design, the frequency then becomes the monitored signal. Most users are not aware of the design specifics of their detector since in all cases the detector output appears as a continuous signal.

As compounds with significant electron-capturing ability pass through the ECD cell,

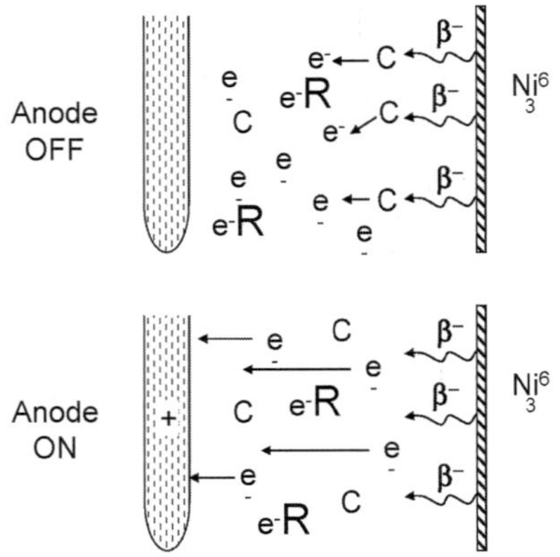

FIGURE 12.12 Simplified representation of the mechanism of ECD response. Background levels of electrons are quickly established by impact of beta particles (β-) with ionizable carrier gas (C) or carrier gas dopant. As an electrophilic analyte (R) enters the cell, it scavenges some of the electrons, lowering the background. Electrons are collected by pulsing the anode.

they scavenge some of the free electrons, lowering the collected current (or increasing frequency necessary to maintain a constant current). The magnitude of the reduction is proportional to the concentration of the solute in the cell, so the ECD is a concentration-sensitive detector [11].

The response of the ECD is highly dependent on the molecular structure of the analytes, as can be seen in Table 12.4. Saturated hydrocarbons have little ability to capture electrons. Pi bonds and electronegative atoms such as the larger halogens are more able to do so and therefore are easier to detect. Multiple electronegative groups within a single molecule yield a highly enhanced response. A comparison of response is shown in Figure 12.13 where the saturated hydrocarbon solvent, octane, yields a peak similar in size to the organochlorine pesticides that are present at over 7 orders of magnitude lower concentration.

TABLE 12.4 Approximate Relative ECD Responses Factors

Compound	**Formula**	**Response relative to benzene**
Benzene	C_6H_6	1
Toluene	C_7H_8	3
Acetone	C_3H_6O	8
n-Butanol	C_4H_9OH	17
1-Chlorobutane	C_4H_9Cl	17
Chlorobenzene	C_6H_5Cl	1,200
1-Bromobutane	C_4H_9Br	5,000
Bromobenzene	C_6H_5Br	7,600
2,3-Butanedione	$C_4H_6O_2$	800,000
Chloroform	$CHCl_3$	1,000,000
1-Iodobutane	C_4H_9I	1,500,000
Carbon tetrachloride	CCl_4	6,600,000

The linear dynamic range of an ECD is relatively narrow (3–4 orders of magnitude); however, due to the widely varying response based on molecular structure, the ECD working range is quite large; some compounds can be analyzed in the mg levels, some at ng levels, and others at pg levels. It is therefore important to establish individual calibration curves for each compound of interest.

Since the ECD is not as selective as many selective GC detectors, it is not able to reject sample matrix as well and so ECD methods are quite reliant on effective sample cleanup in order to benefit from its high sensitivity. In addition, ECDs can be contaminated by dirty samples, causing significant drifts in baseline and seriously degrading the sensitivity due to local charging effects on the surface of the cell. Should that happen, the cell needs to be cleaned, often by the manufacturer (because of the radioactive material inside).

Some countries have laws that regulate the use of ECDs because they contain radioactive material such as ^{63}Ni. The regulations are often tied to the level of radioactivity of the particular ECD design (e.g., cells with >10 mCi might be regulated, whereas those <10 mCi are not). If regulated, owners might need to obtain a license or other compliance document, and may need to periodically (e.g., every 6 months) establish that the ECD is not leaking by doing a wipe test. Instrument manufacturers will usually provide the pertinent compliance information to their customers.

12.5. ALKALI BEAD DETECTOR

The alkali bead detector, also known as a thermionic ionization detector, is also commonly known as a nitrogen/phosphorus detector (NPD). The NPD is a highly selective, destructive detector that finds wide use in foods, flavors, forensics, and pharmaceutical and pesticide analysis applications. However,

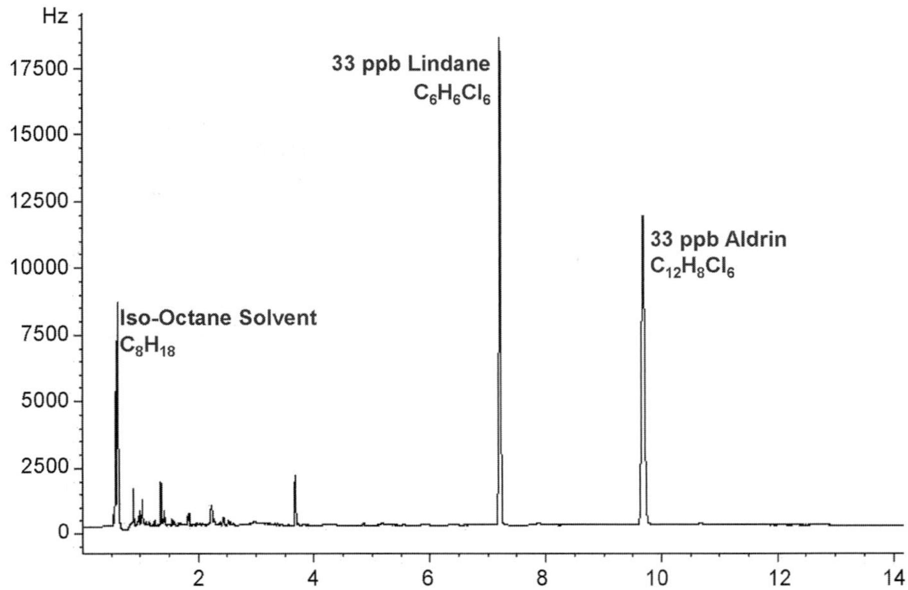

FIGURE 12.13 A checkout standard run by ECD shows the high selectivity for compounds with multiple halogens compared to saturated hydrocarbon. *Source: Copyright Agilent Technologies, Inc. [10].*

it is a finicky detector with drifting response and easily can be damaged if the user is not careful.

Most modern NPDs are based on a heated bead design described by Kolb and Bischoff [12]. Response of the NPD is generally thought to rely on a selective moderated surface reaction between nitrogen- or phosphorus-containing compounds and certain alkali atoms in a cool plasma. Salts of rubidium (Rb) and cesium (Ce) have been used successfully for both N and P selective detection. Salts of these alkali metals can be either coated onto or integrated into ceramic or glass beads that are in turn attached (e.g., melted) onto a wire that is used to heat them. A basic NPD design is illustrated in Figure 12.14.

The nature of NPD response is complicated and dynamic. Column effluent mixes with a few mL/min of H_2 and a few hundred mL/min of air as it leaves the detector jet. With the low flow of H_2, there is actually no flame at the tip of the detector jet. In fact, it is important to keep the bead cool for maximum sensitivity, especially for N-containing compounds. A flame would overheat the bead, causing similar problems as running the bead with too much current. Without a flame, H_2 is combusted just at the surface of the heated bead, forming a localized plasma wherein alkali salts in the bead are reduced (neutralized) and evaporated so that the reduced alkali metals can then catalytically react with N- and P-containing compounds. These ions are then attracted to and collected by a collector electrode that is held at a few hundred volts relative to the jet, as illustrated in Figure 12.15.

The ability of an NPD to work in a repeatable way relies on reaching a steady state of available alkali metal near the surface and replenishment of the depleted alkali by alkali from the bulk of the bead or coating.

As the alkali metals are depleted from the surface, they are replenished from the bulk of

FIGURE 12.14 Cross-sectional view of a typical NPD. *Source: Copyright Agilent Technologies, Inc. [10].*

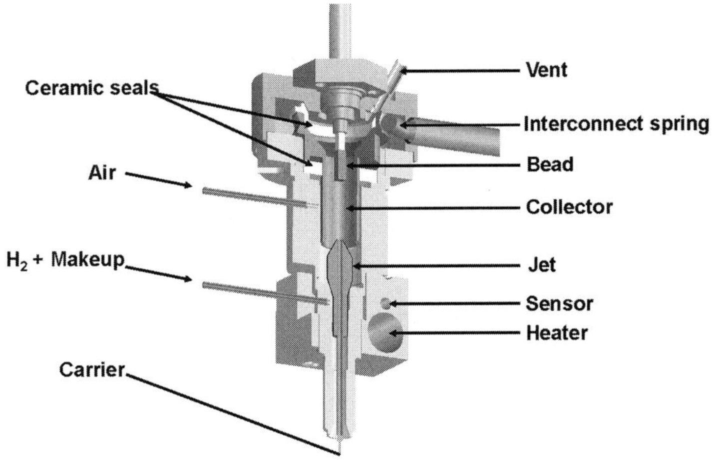

the bead or coating material. Bead temperature certainly has a bearing on the rate of replenishment. The hotter the bead, the faster the replenishment, the more sensitive the detector, and the more able to maintain a linear response for higher analyte concentrations. However, excessive bead temperature leads to premature bead failure, unstable baseline, higher noise, and continuously drifting response factors. Each bead composition and style will have its own "sweet spot" for operation.

There are many variables associated with response of the NPD [13–16] that conspire to yield a different sensitivity, selectivity, and lifetime based on bead type and operating conditions:

- characteristics of the bead: bulk composition, size, porosity, nature, and concentration of the alkali salt;
- operating variables: bead temperature (related to bead current), air/H_2 gas ratio, detector temperature, carrier gas, and makeup gas flow rates;
- detector design: voltage bias on the collector electrode and proximity of the jet to the bead surface;
- Bead history (e.g., age, exposure to poisons, and maximum temperature); and
- Startup procedure (e.g., gradual vs. ballistic).

Startup of the bead to initially establish the plasma can also be a bit tricky. To "turn the bead on," a higher current is often required at first to get things going. The power is then backed off after the plasma has been established and H_2 is burning at a steady state at the surface.

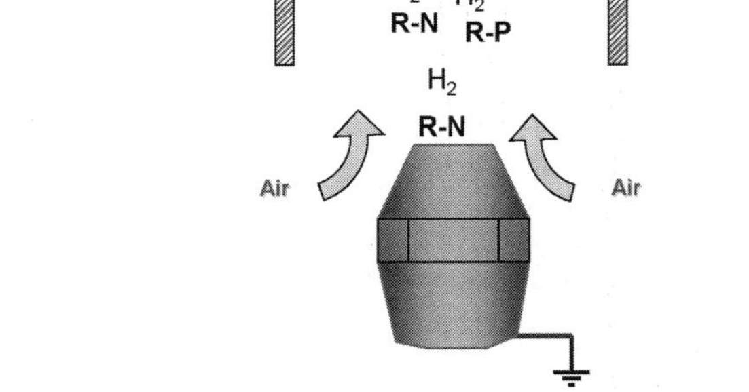

FIGURE 12.15 Simplified schematic diagram for the catalytic process on the surface of a hot bead. Temperature not only drives the reaction but also influences replenishment of Rb to the surface.

If the power is not backed off after this initial period, bead lifetime is shortened due to premature depletion of alkali from the bulk of the bead or coating.

There are several different bead types available for any given instrument design. Proper selection depends on the sample type and goals of the analysis. One design might be of general use for analysis of both N- and P-containing compounds, but would not have the highest selectivity or sensitivity for either. Another bead might be optimized for nitrogen but it likely would give poor selectivity and sensitivity for phosphorus compounds and would most likely cause them to tail as well. Still another bead might be optimized for "ruggedness" and maximum lifetime, but would certainly not give the best detection limits. Figure 12.16 compares FID and NPD chromatograms of an extract of blood from a cigarette smoker.

Certain compounds in high concentrations can kill an NPD bead response by depleting the plasma completely of alkali metal or by irreversibly changing the nature of the surface (poisoning). Chlorinated solvents are notorious for this. Some detector designs allow the H_2 flow to be temporarily suspended during aggressive solvent elution, effectively turning off the bead plasma and suppressing the interaction with aggressive solvents. When the H_2 flow is resumed after the solvent peak passes, the

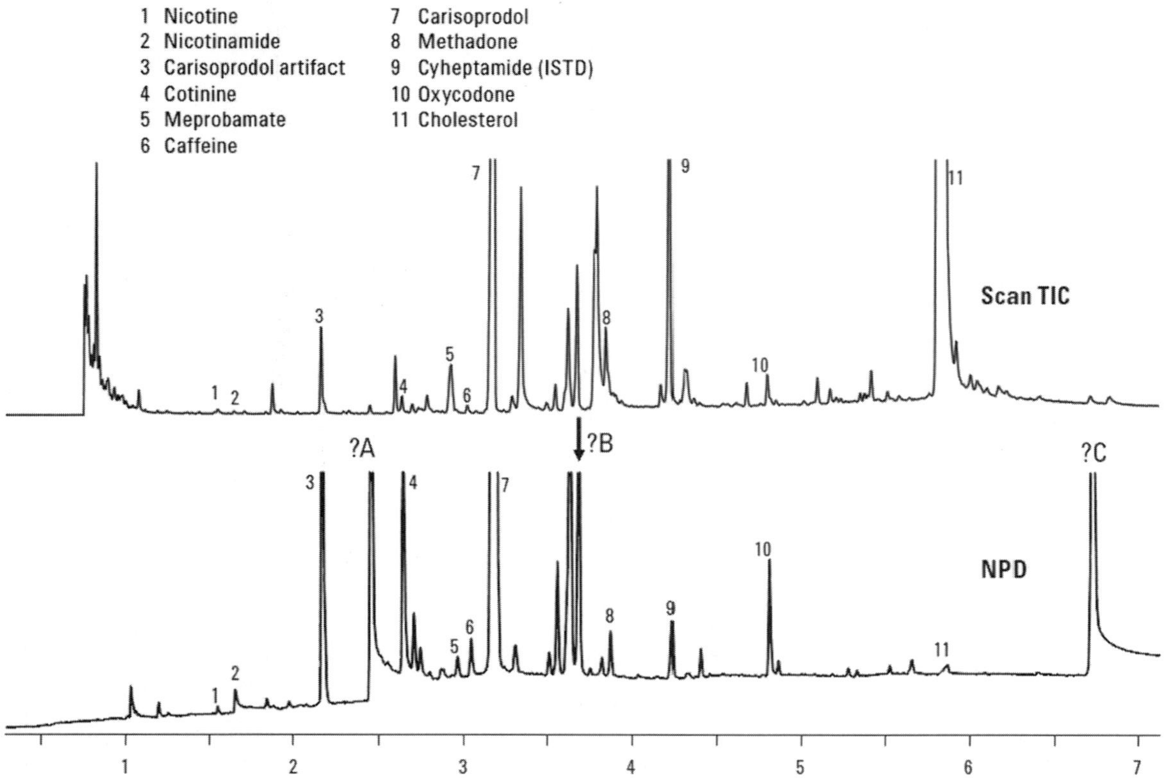

FIGURE 12.16 Comparison of MS total ion and NPD chromatograms for a smoker's blood extract analysis. *Source: Copyright Agilent Technologies, Inc. [10].*

plasma is quickly reestablished (stabilizes much quicker than initial startup). An effective alternative that allows the plasma to be maintained is to direct column effluent away from the detector during solvent elution using a technique such as Deans switching. Although these clever and useful techniques can help when aggressive solvents must be used, it is better to avoid the use of aggressive solvents whenever possible when using an NPD.

The NPD user is advised to be aware of the compromises and read the specific instrument manufacturers' literature in order to select appropriate bead and conditions.

Response of the NPD is mass sensitive for given compounds. However, due to the nature of the ionization process, each compound will have a different mass response. The difference is not as dramatic as detectors like the ECD, but it is enough to require separate calibration curves for each analyte of interest.

12.6. FLAME PHOTOMETRIC DETECTOR

The flame photometric detector (FPD) is a popular selective detector used, among other things, for the specific detection of sulfur in petroleum and petrochemical samples and for the specific detection of phosphorus-containing pesticides in foods and the environment (Figure 12.17). The FPD is a destructive detector that uses a flame, but, instead of measuring the ions formed from one process or another, it measures the light emitted from phosphorus or sulfur combustion products.

During the combustion process of organic molecules in a hydrogen flame, many different forms of excited fragments and recombination products are formed. When these excited ions recombine and relax into stable forms, they emit light. The spectral and temporal characteristics of the emission are species specific and can be

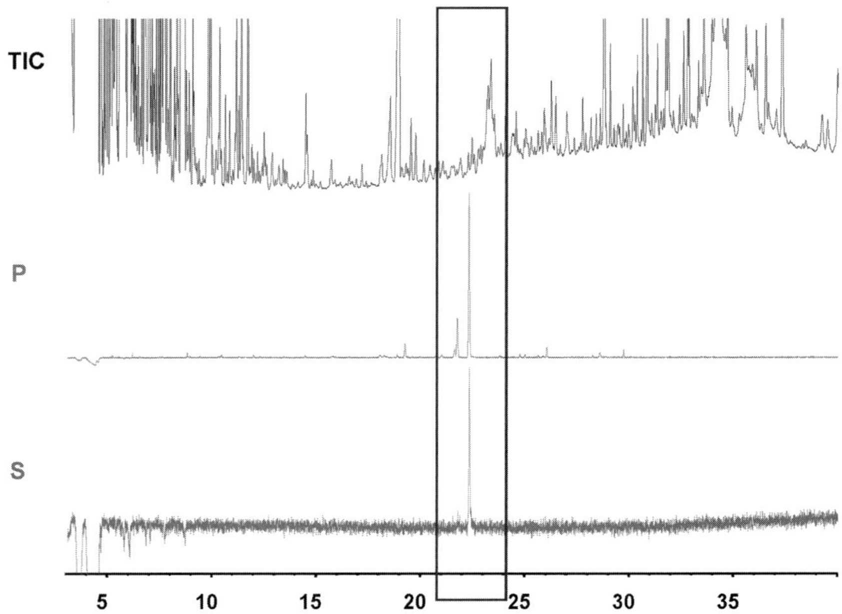

FIGURE 12.17 Demonstration of FPD selectivity for P- and S-containing pesticides in a lemon oil. *Source: Copyright Agilent Technologies, Inc. [10].*

exploited for selective detection. Typical excited species that emit light include those from carbon (CH^*, C_2^*), oxygen (OH^*), phosphorus (HPO^*), and sulfur (S_2^*). Figure 12.18 [17] compares the emission spectra of these species.

The response for phosphorus is predominately from the HPO^* species and is linear with mass for each compound, although, due to different combustion and emission efficiencies, different compounds can yield different responses per mass of P.

Since the emission from sulfur combustion and recombination is from the S_2^* species, response has a quadratic relationship with mass of sulfur in a given compound. This is not a problem with modern data analysis systems wherein it is usually a simple matter to select a quadratic curve fit of the calibration data. However, to simplify matters further, some GC manufacturers incorporate a quadratic correction circuit in their FPD detector electronics for sulfur detection so that the observed response appears linear.

Typical FPD designs, as illustrated in Figure 12.19, monitor emission using a photomultiplier tube (PMT) positioned above the flame (in the cool region where recombination and emission occur). An optical notch filter is positioned between the emission zone and the PMT to select a narrow wavelength region (window) corresponding to where the sulfur or phosphorus emission is highest relative to background (usually carbon) emission. As indicated in Figure 12.18, filters of $\approx$394 nm for S and $\approx$525 for P are typically used. To measure both sulfur and phosphorus simultaneously, dual detector designs are used with two filter/PMT combinations as illustrated in Figure 12.20.

Unfortunately, there is no spectral region where S_2^* or HPO^* emission is completely isolated; there are always interfering emissions

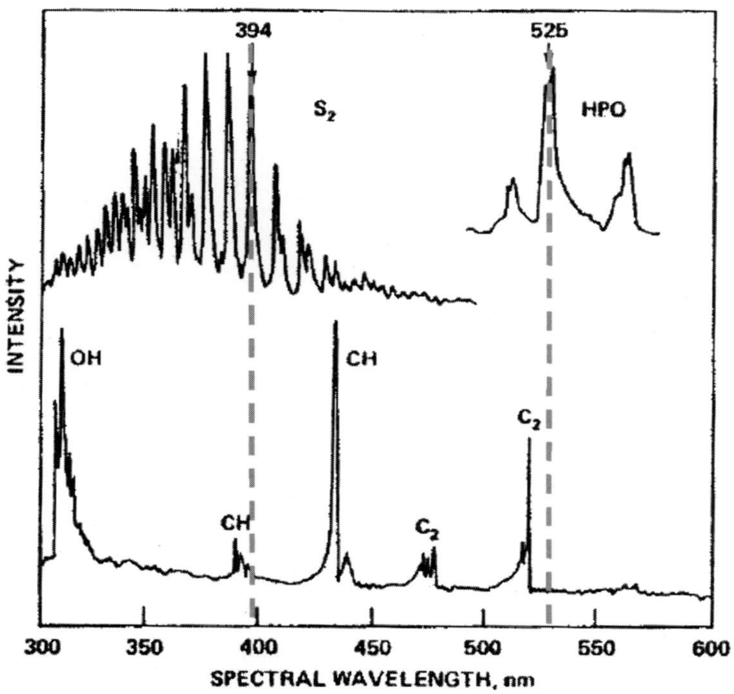

FIGURE 12.18 Emission spectra for carbon, phosphorus, and sulfur combustion products. Dotted lines represent typical spectral filters used for selective detection with FPDs. *Source: Copyright ASTM International [17].*

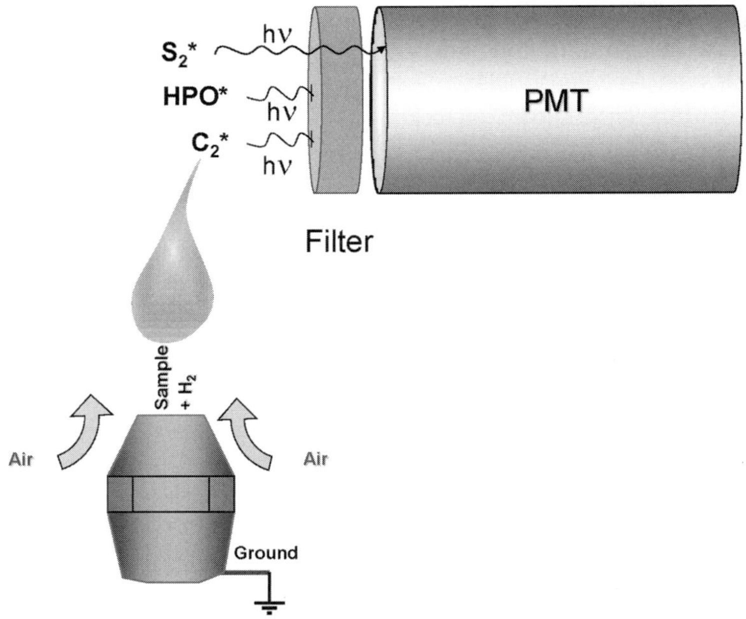

FIGURE 12.19 Illustration of the mechanism of detection of an FPD. Sample components are combusted in the flame, creating positive ions and electrons. The positive ions are attracted to the negatively biased collector, while the (negatively charged) electrons are repelled toward the jet.

from other species. Even if the strength of a target emission is 1000 times that of the background, matrix components in actual samples are often many orders of magnitude higher in concentration than the analytes of interest and their emission can exceed that of the analytes of interest. In these cases, analysis of low-level P- or S-containing compounds can be problematic and a detector with higher selectivity is required.

Probably more problematic than interferences, however, is that the emissions of P and S can be quenched. High concentrations of coeluting matrix (as is typical when measuring sulfur in petroleum samples) can react or interact with the excited S_2* or HPO* species, providing other relaxation pathways short circuiting the emission of light. This reduces the response for a given amount of sulfur or phosphorus in samples compared to standards; so concentrations end up significantly under-reported.

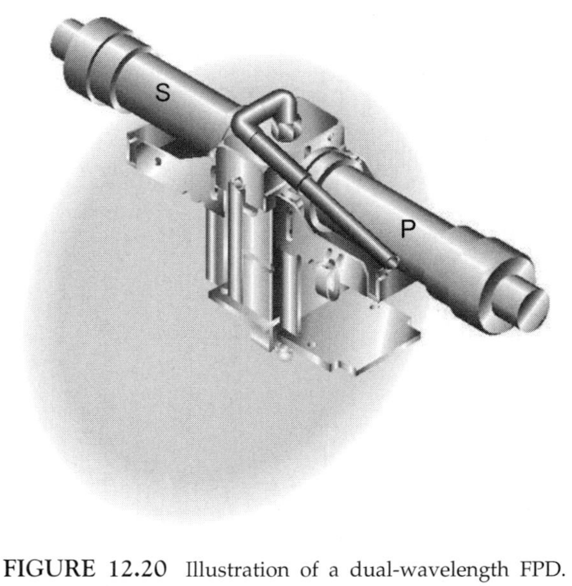

FIGURE 12.20 Illustration of a dual-wavelength FPD. The opposing configuration of filter/PMT combinations allows simultaneous detection of both P and S. *Source: Copyright Agilent Technologies, Inc. [10].*

Quenching is sample dependent and not always obvious because S or P signals can be suppressed without a significant rise in background signal.

Amirav et al. exploited the fact that there is a kinetic aspect to the emission of light [18,19]. Emission from the excited carbon, phosphorus, and sulfur species happens at slightly different rates after the combustion products are formed. As illustrated in Figure 12.21 [20], by starving the flame of hydrogen fuel gas in the presence of a heated wire, the H_2 level will eventually reach a critical concentration, the flame will ignite, quickly depleting all available fuel and therefore extinguishing (happens within a few ms). A few tenths of a second later, the process repeats. This form of FPD is called a "pulsed" FPD (pFPD). By gating detection at the onset of each flame ignition, one can record/select emission as a function of time.

A typical time-gated emission is shown in Figure 12.22. C* emission is fairly fast, occurring just after ignition. Phosphorus emission peaks approximately 5 ms after ignition; S_2* emission happens over a very wide time frame from approximately 5 to 25 ms after ignition.

By gating detection of sulfur and phosphorus emissions after that of carbon, most of the spectral interference from carbon can be eliminated. However, this approach is still susceptible to quenching by high concentrations of coeluting solutes. Figure 12.23 illustrates the excellent selectivity for low levels of sulfur compounds at the higher boiling point region of a diesel fuel chromatogram.

The pFPD has some other limitations in use as well. The detector pulses occur only a few times each second and only for a narrow operating range of column and fuel gas flows. In addition, the pulse rate is flow related; so in methods where column flow rate changes (e.g., constant pressure temperature programmed runs), the pulse rate will change throughout the run. Since the self-ignition of the flame is a function of the proper mix of fuel and oxidant, the flame will not ignite when solvent or high concentrations of analyte pass through the detector (not enough oxidant). This is sometimes not a problem if analytes of interest do not coelute with the solvent or other high-level components, but one is not always so fortunate and important peaks can be missed. Due to these limitations, the range of applications in which a pFPD can be used is narrower than that of the standard FPD.

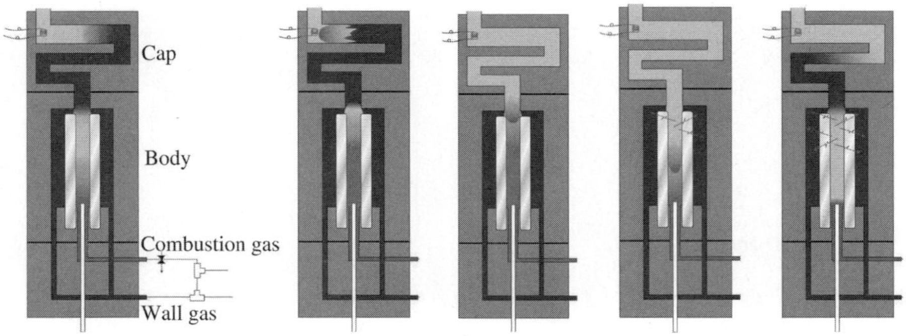

FIGURE 12.21 Combustible mixture of H_2 and air (shown in blue and purple) fills detector body, combustor, and cap from the bottom up, eventually reaching the glow plug and initiating combustion. The flame begins to travel back along the original pathway, burning the combustion gas and eventually the column effluent. Excited combustion products are formed (e.g., HPO* and S_2*) and begin to fluoresce for up to 25 ms after the flame has extinguished. *Source: Copyright OI Corporation [20].*

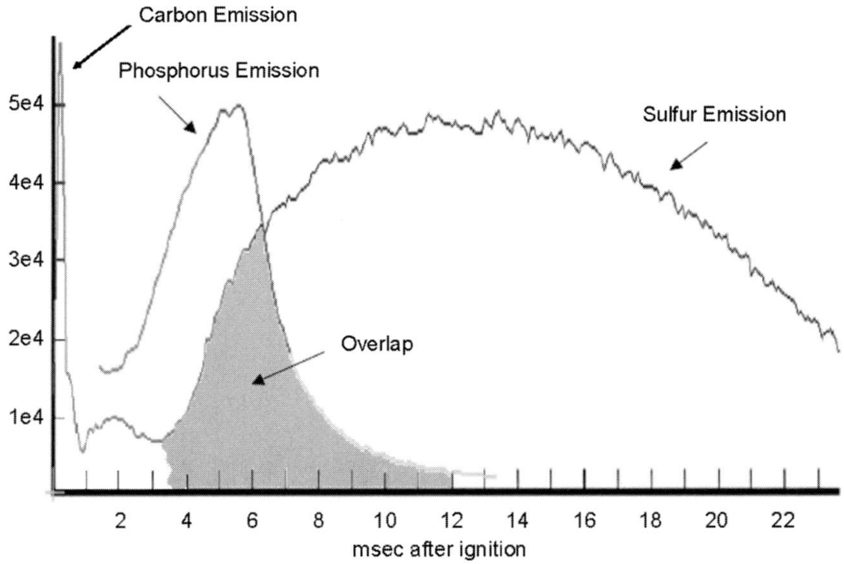

FIGURE 12.22 Emission by combustion products in a pulsed FPD is time related. Carbon species emit quickly, phosphorus more slowly, and sulfur slower still. By using time-gated detection, interferences from coeluting species can be reduced. *Source: Copyright OI Corporation [20].*

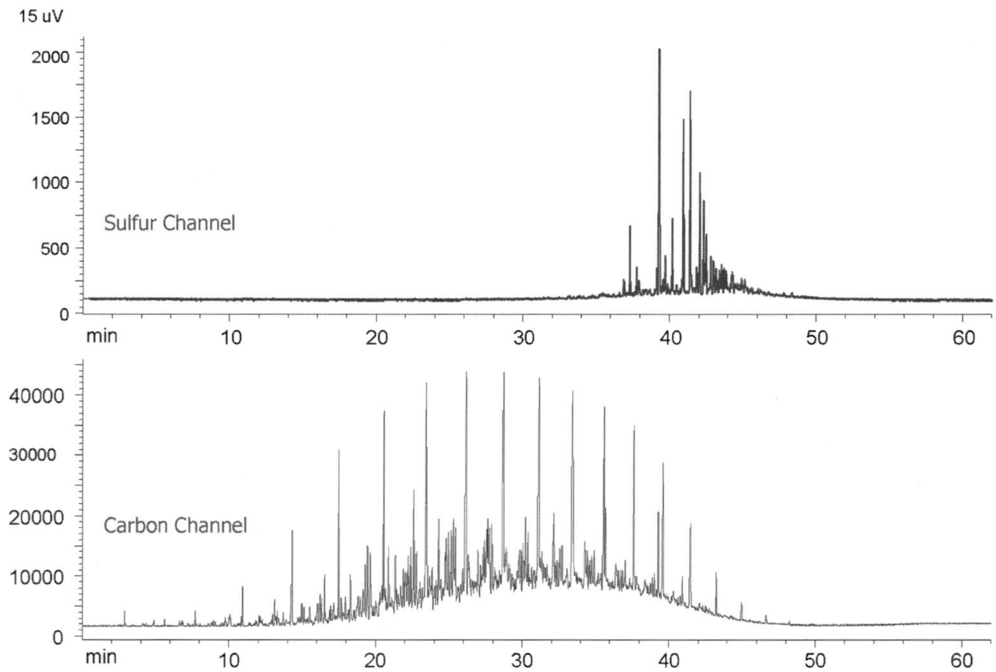

FIGURE 12.23 Comparison of carbon and sulfur signals from a pulsed FPD for a diesel fuel containing 302 ppm total sulfur. *Source: Copyright OI Corporation [20].*

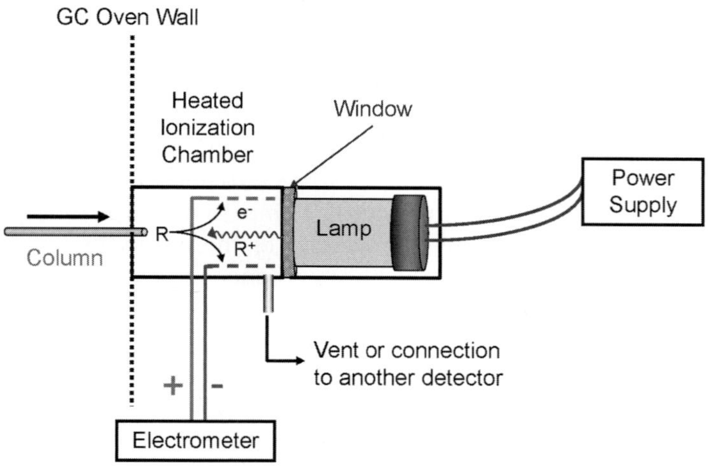

FIGURE 12.24 Illustration of a basic photoionization detector design. UV light ionizes compounds that have lower ionization potential than that of the lamp. Ions are collected and neutralized. Resulting current is proportional to number of ions formed.

12.7. PHOTOIONIZATION DETECTOR

The photoionization detector (PID) is non-destructive, partially selective, and concentration sensitive. It has no flame, is fairly simple, safe, and rugged, and can be effectively integrated into portable gas chromatographs for field use. It is often the detector of choice for trace levels of aromatic compounds of environmental or health concern.

Figure 12.24 illustrates the basic PID design.

The PID creates molecular ions using high-energy photons from a sealed light source. Even though strong enough to ionize molecules, the energy from the lamp usually does not cause molecules to fragment. The molecular ions are attracted to a cathode and are neutralized, yielding the original intact molecule. The current resulting from neutralization is measured and represents the detector signal. The current is proportional to the number of ions neutralized, which is, in turn, proportional to the concentration of the compound in the cell:

$$R + \text{photon} \rightarrow R^* + e^- \xrightarrow{i} R$$

Since the PID is non-destructive, it can be used effectively in series with other detectors.

The selection of light source (usually a noble gas discharge lamp) affords some general

TABLE 12.5 Characteristic Lines Found in Vacuum UV Output of HNu Lamps

Lamp designation (eV)	Major emission line wavelengths (nm)	Energy (eV)	Output (%)	Lamp type
8.3	147	8.44	100	Xenon
9.5	114	10.88	0.03	Xenon
	117.2	10.58	0.01	
	119.3	10.4	0.18	
	125	9.92	0.05	
	139.6	9.57	2.1	
	147	8.44	97.6	
10.2	116.6	10.64	17.1	Krypton
	123.6	10.03	82.9	
11.7	104.9	11.82	26.2	Argon
	106.6	11.62	71.8	
	121.6	10.2	2	
N/A	80.5	15.5	100	Helium

Adapted from [21].

TABLE 12.6 Some Materials used in the Construction of UV-Lamp Windows [23]

Material	Permeability (nm)	Limit (eV)
LiF	104	11.9
MgF_2	112	11.1
CaF_2	122	10.3
NaF	132	9.4
BaF_2	134	9.2
Sapphire	143	8.7

selectivity based on the lamp chosen. Only compounds with lower ionization energies than that of the lamp will be detected. Table 12.5 shows the nominal energy designation of common PID lamps as well as the underlying noble gas emission lines and their relative abundances.

Most PID lamps are sealed to contain approximately 1 atm of noble gas inside and, therefore, must have a window through which the resulting light can pass. There are only a few materials that have the necessary combination of ruggedness, manufacturability, and transparency to UV radiation to be suitable for this use. Table 12.6 lists several of these materials and their effective wavelength cutoffs. Lamp lifetimes are often limited by degradation of these windows due to exposure to the high-energy photons, especially those capable of passing the highest energy emission lines.

Figure 12.25 [22] illustrates discharge spectral and corresponding energy emission curves for three noble gases. Referring back to Table 12.6, note that there is no material that can pass the highest energy He emission lines or some of the higher-energy lines of argon.

In designing a lamp, the combination of discharge gas and window dictates the nominal output energy of the lamp, with the window acting as a cutoff filter. The highest energy emission line that can pass through

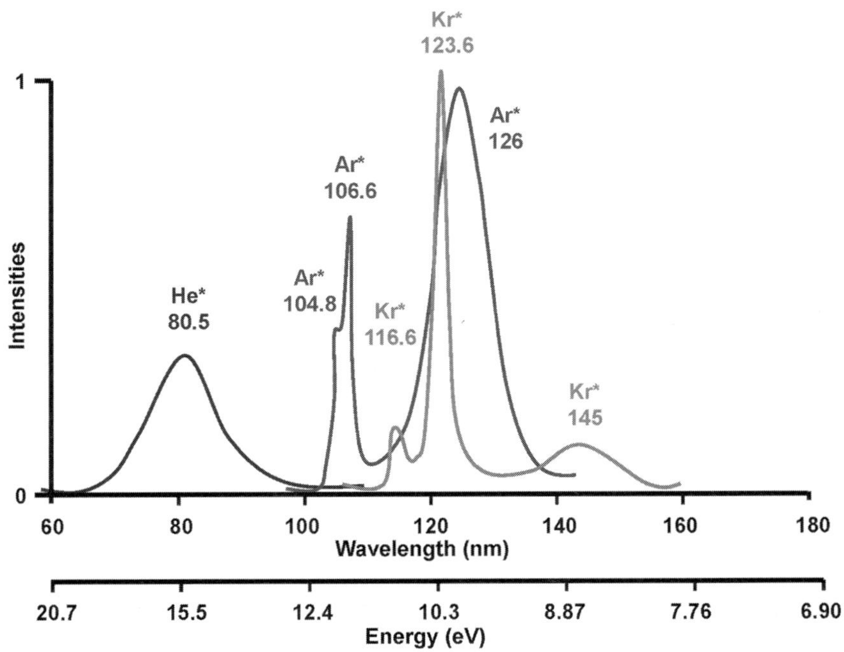

FIGURE 12.25 Emission spectra of noble gas discharges. *Source: Redrawn from [22].*

TABLE 12.7 Ionization Potentials of Compounds

Saturated alkanes		*Nitrogen containing*	
Methane	12.98	Ammonia	10.15
n-Butane	10.63	Methyl amine	8.97
Cyclohexane	9.88	i-propyl amine	8.72
		Diethyl amine	8.01
		Nitromethane	11.08
Oxygenated		Acetonitrile	12.22
Water	12.59	n-butyronitrile	11.67
Methyl alcohol	10.85	Acrylonitrile	10.91
n-butyl alcohol	10.04		
Hydrogen Sulfide	10.46	*Aromatic*	
Methanethiol	9.44	Benzene	9.245
1-butanethiol	9.14	Toluene	8.82
Diethyl sulfide	8.43	Ethyl benzene	3.76
Carbon Dioxide	13.79	o-xylene	8.56
Formaldehyde	10.87	Naphthalene	8.12
n-butyraldehyde	9.86	Chloro-benzene	9.07
Acetone	9.69	o-chlorotoluene	8.83
Methyl ethyl ketone	9.53		
Carbon Dioxide	13.79	*Halogenated*	
Formic acid	11.05	1-chlorobutane	10.67
n-butyric acid	10.16	1-bromobutane	10.13
Ethylene oxide	10.565	2-iodobutane	9.09
p-dioxane	9.13	CF_3CCl_3 (Freon 113)	11.78
		Phosgene	11.77

Excerpted from [25].

the window without significant attenuation (for example, the wavelength corresponding to 50% transmission) is usually stated as the lamp's nominal energy since this is what dictates which molecules can be ionized. For example, a sapphire window would block any xenon emission lines more energetic than 8.7 eV; so based on the closest emission line

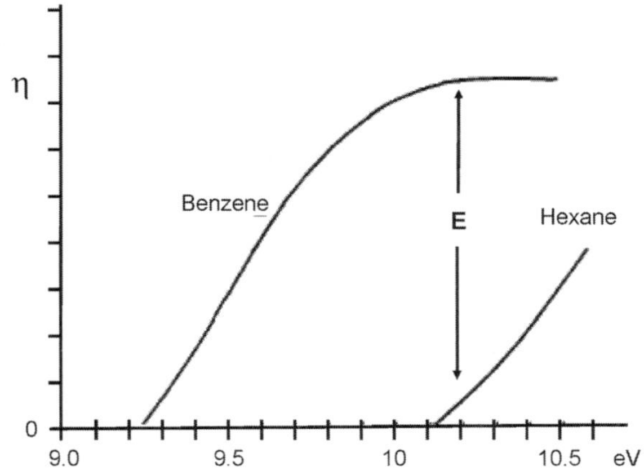

FIGURE 12.26 Relationship between the photoionization efficiency (η) and the photon energy. *E* indicates the energy of the UV lamp. *Source: Redrawn from [24].*

with lower energy (higher wavelength) the lamp would have effective energy near 8.4 eV. In contrast, the same lamp with a calcium fluoride window would pass the 9.57 eV xenon emission line, so its effective lamp energy would be approximately 9.5 eV.

Table 12.7 lists several classes of molecules. The weaker the molecular bonds, the easier an electron can be ejected, ionizing the molecule. The statistical probability that a given molecule will be ionized relates to the difference in the ionization potential of the molecule and the lamp. The probability function reaches a maximum when the difference is several eV higher than the ionization energy of the molecule, as illustrated in Figure 12.26 [24] for benzene.

As long as the lamp energy is above the ionization potential of the molecule it will be reproducibly detected, even though with less sensitivity than with a more energetic lamp. So, compounds will have a difference in response factor that relates to their ionization potential relative to the lamp, as illustrated in Figure 12.27 [26] and Table 12.8. This can be used to advantage if the compounds of analytical interest have lower ionization potentials than sample matrix components. By proper selection of lamp, one can

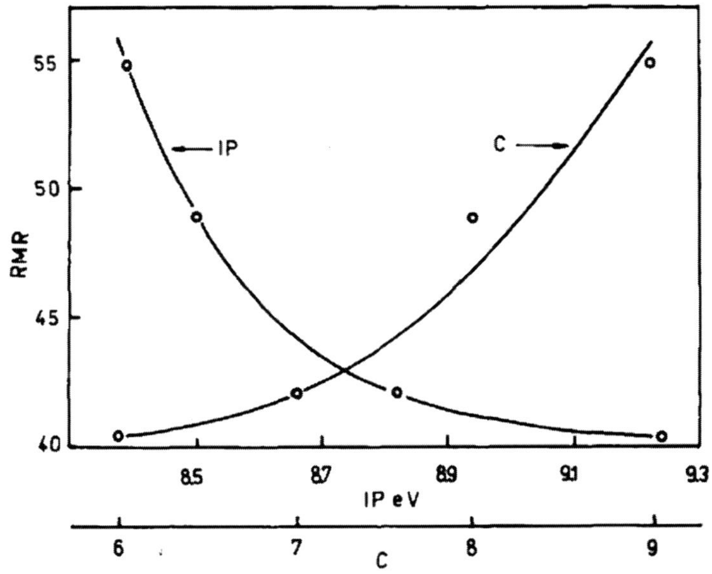

FIGURE 12.27 Relative molar response (*RMR*) of aromatic hydrocarbons as a function of the ionization potential (curve IP) and the number of carbon atoms (curve C) in a molecule RMR of benzene = 1. C = number of carbon atoms. *Source: Redrawn from [26].*

TABLE 12.8 Relative Photoionization Sensitivities for Gases

Chemical	Relative sensitivity	Examples
Aromatic	10	Benzene, Toluene, Styrene
Aliphatic acid	10	Diethylamine
Chlorinated unsaturated	5–9	Vinyl Chloride, Vinylidene, Chloride, Trichloroethylene MEK, MiBK, Acetone,
Carbonyl	7–9	Cyclohexanone Acrolein, Propylene
Unsaturated	3–5	Cyclohexanone, Allyl alcohol
Sulfide	3–5	Hydrogen sulfide, Methyl Mercaptan
Paraffin (C5-C7)	1–3	Pentane, Hexane, Heptane
Ammonia	0.3	

Excerpted from [25].

minimize the response of matrix components, while still detecting target compounds of interest. In the example of Figure 12.28 [24], by choosing a lamp with an energy below 10 eV, saturated hydrocarbons will not be ionized

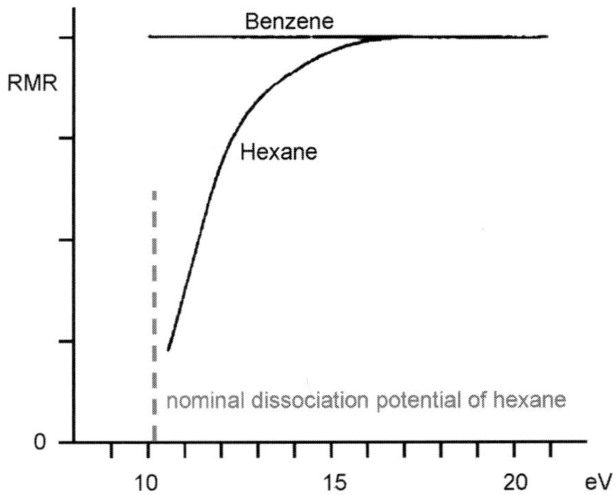

FIGURE 12.28 Relative molar responses per mole of carbon (*RMR*) of an alkane (hexane) and an aromatic hydrocarbon (benzene, dissociation energy 9.25 eV) as a function of the photon energy. *Source: Redrawn from [24].*

while aromatics will be. Figure 12.29 [21] illustrates the selectivity of a PID with 9.5 eV lamp compared to that of an FID in gasoline chromatograms.

In order to make use of the more energetic emission lines in the vacuum UV, there are windowless lamps and detector designs. A simplified schematic of a helium ionization discharge detector that operates as a windowless discharge source is shown in Figure 12.30. In windowless designs, the discharge gas needs to be continuously fed to the integrated lamp/detector. The most energetic discharge is with He alone (approximately 15.5 eV mean emission), which has enough energy to ionize most molecules. In this mode, the He ionization detector has almost a universal response. By doping the He with other noble gases, the energy from He ionization is transferred to the dopants and the output emission is then more characteristic of the dopants, as illustrated in Figure 12.31 [27]. This approach provides some flexibility in tuning selectivity for specific methods without having to change lamps.

The same basic design of windowless PID can be operated in an electron capture mode [28]. Since the He plasma generates free electrons, appropriate positioning and bias of electrodes and addition of appropriate dopant can establish a suitable electron-rich zone near the

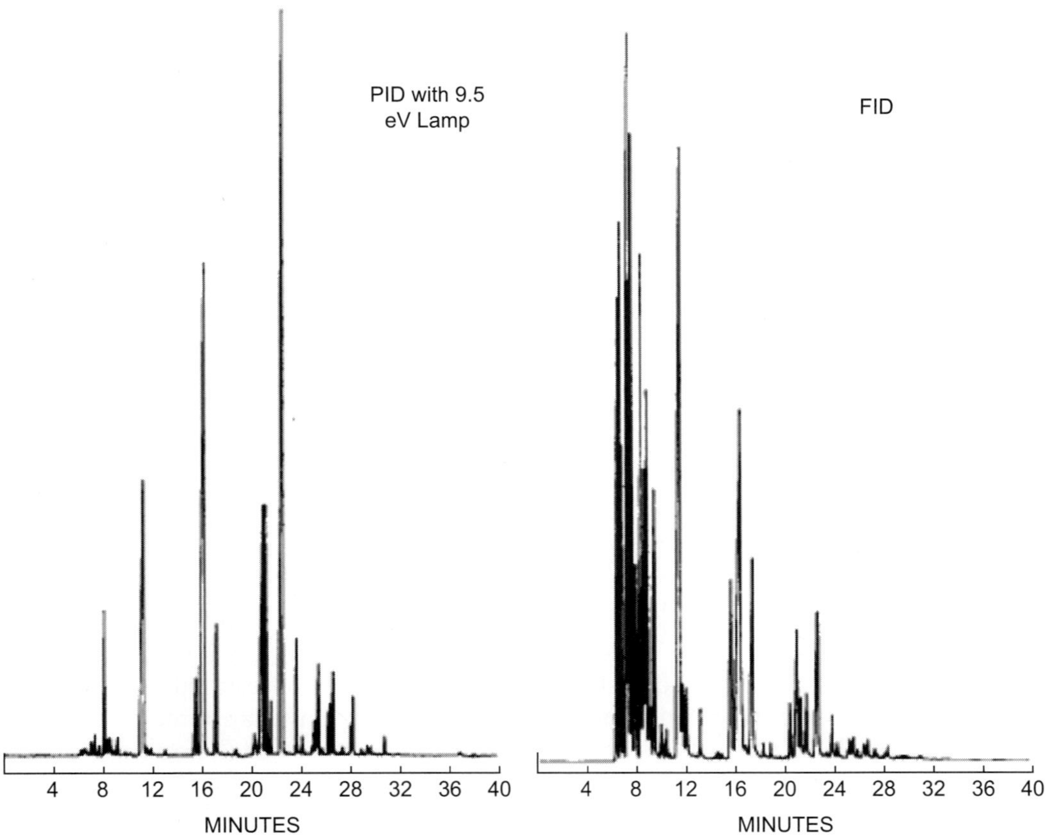

FIGURE 12.29 Comparison of response of PID with 9.5 eV lamp and FID for gasoline sample. *Source: Redrawn from [21].*

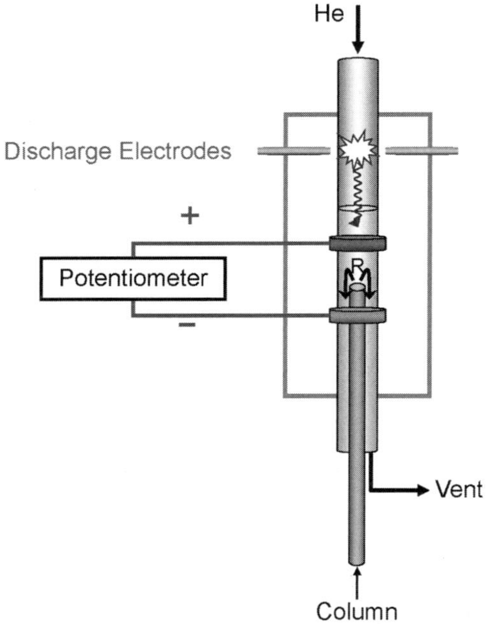

FIGURE 12.30 Simplified illustration of a discharge ionization detector. Highest energy photons can reach the sample since there is no window between the discharge light source and the sample. The higher energy photons create a higher overall sensitivity and less selective response.

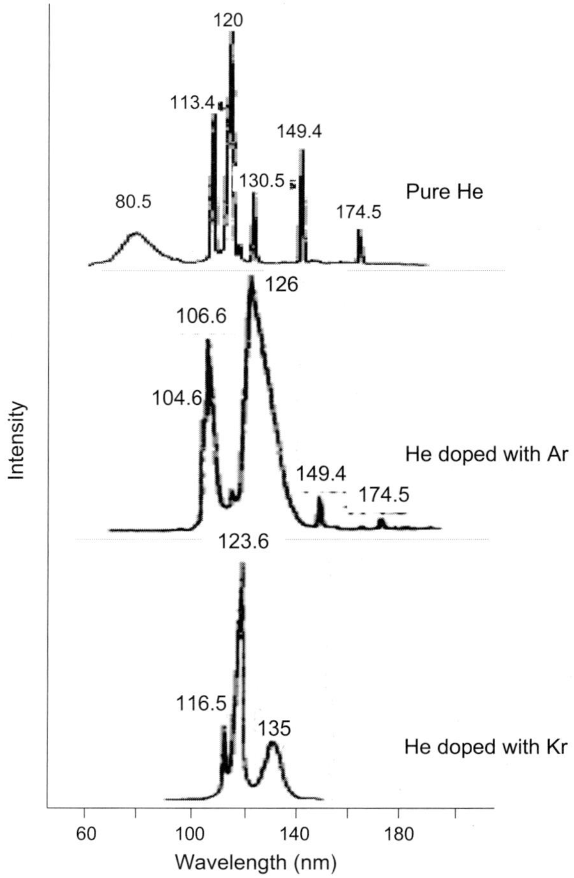

FIGURE 12.31 Spectra of pure helium, argon-doped helium, and krypton doped helium. *Source: Redrawn from [27].*

exit of the column. By measuring the change in electron flux as electron-capturing analytes elute, one can achieve a response similar to that of an ECD (Figure 12.32).

12.8. ELECTROLYTIC CONDUCTIVITY DETECTOR

The electrolytic conductivity detector (ELCD) is a destructive, mass-sensitive selective detector. Its main use is for regulated methods designed for selective detection of halogen-containing compounds. The ELCD consists of three principal components: the reactor assembly, the cell-solvent assembly, and the detector controller. Although the principal mode of operation of the ELCD is the halogen mode (X), sulfur (S) and nitrogen (N) modes are also possible. Each detection mode requires a specific reactor, resin cartridge, and solvent. The ELCD can be used as a stand-alone detector or in tandem following a photoionization detector (PID) or other non-destructive detectors.

The ELCD converts eluting compounds with the target heteroatom (halogens, S or N) to an ionizable gas (HX) using reductive conditions at temperatures from 800 to 1,100 °C in a catalytic micro reactor, as illustrated in Figures 12.33 and 12.34. The gaseous reaction products

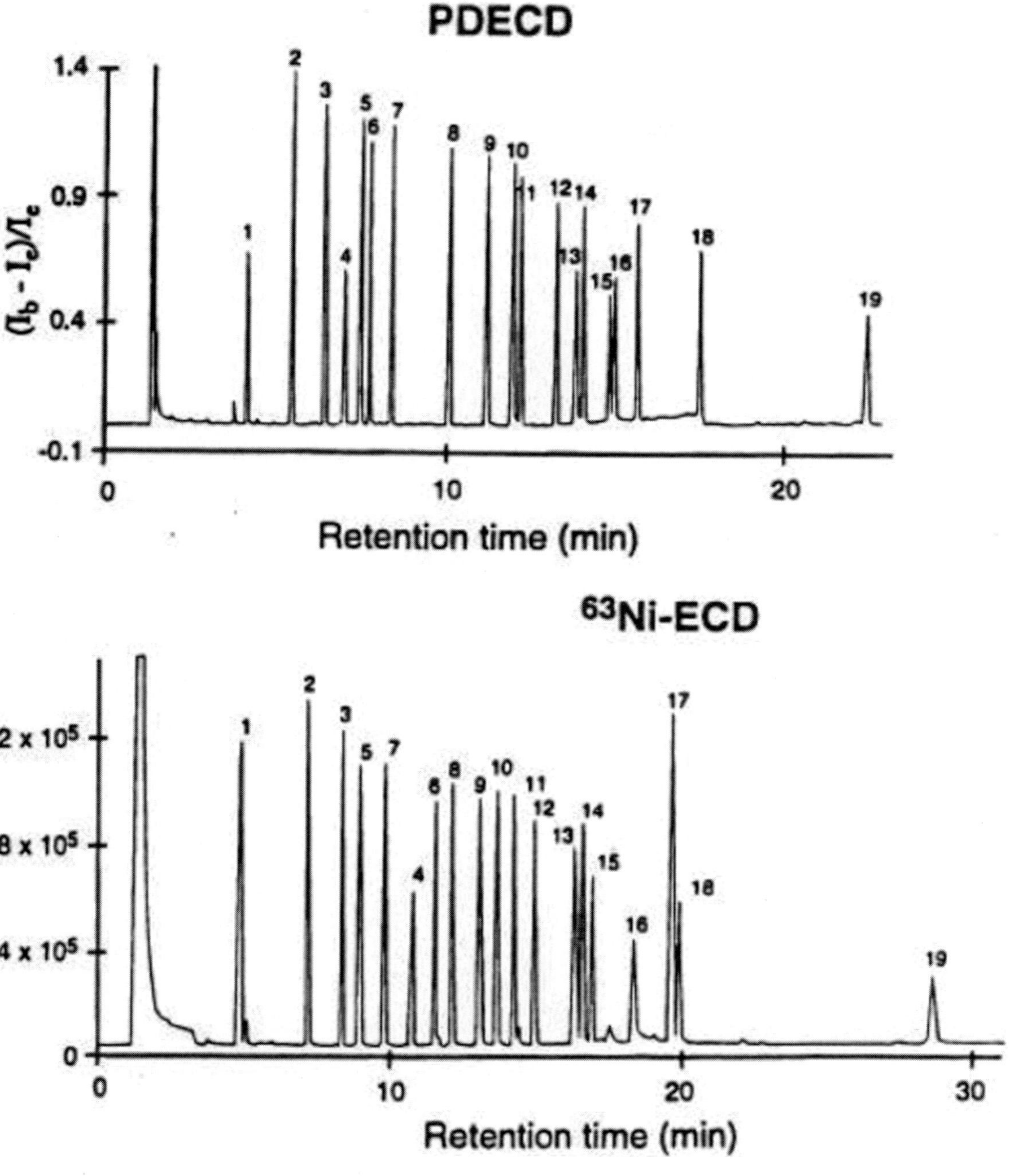

FIGURE 12.32 Comparison of pesticide analysis using the PDECD and 63Ni−ECD. (1) TCMX (surrogate), (2) a-BHC, (3) g-BHC, (4) b-BHC, (5) heptachlor, (6) d-BHC, (7) aldrin, (8) heptachlor epoxide, (9) endosulfan I, (10) 4,49-DDE, (11) dieldrin, (12) endrin, (13) 4,49-DDD, (14) endosulfan II, (15) 4,49-DDT, (16) endrin aldehyde, (17) endosulfan sulfate, (18) methoxychlor, and (19) DCB (surrogate). [28].

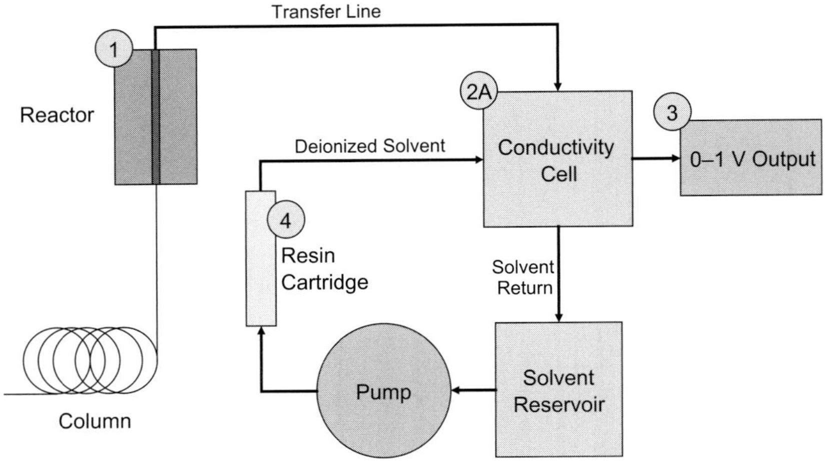

FIGURE 12.33 Eluting analytes are catalytically pyrolyzed and reduced within the reactor (1). Reaction products transfer to the conductivity cell and dissolve in an appropriate deionized solvent (2), increasing solvent conductivity. Solvent is continuously deionized, filtered, and recycled. *Source: Copyright OI Corporation [20].*

continue to the detector cell where they quickly dissolve in a flowing deionized solvent stream, increasing its electrolytic conductivity. A simple conductivity detector amplifies the change in conductivity, producing a signal proportional to the mass of target species. The solvent stream then flows through a deionizing resin bed and filter as it is continuously recycled.

High levels of hydrocarbon solvents cause elemental carbon buildup in the reaction tube

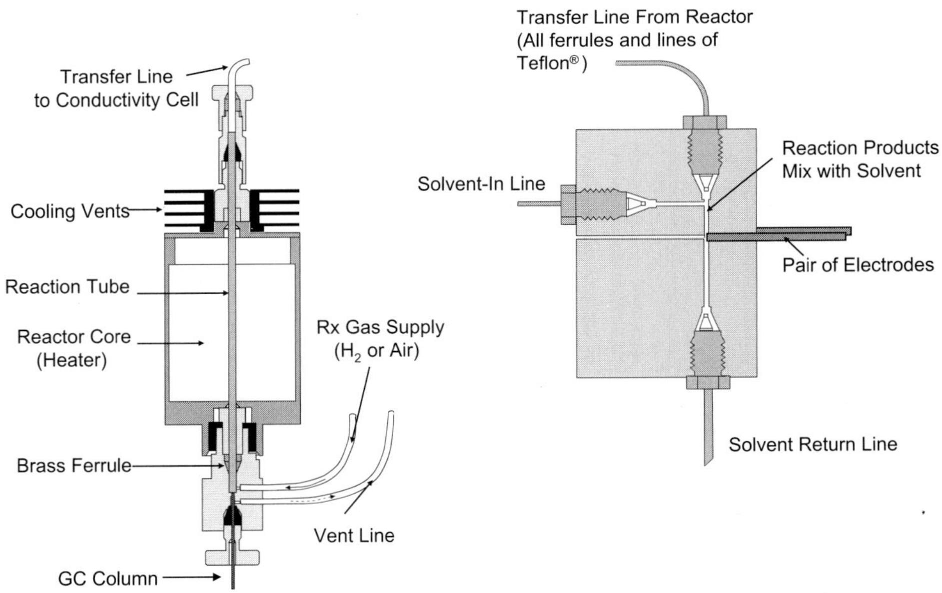

FIGURE 12.34 More detailed view of the reactor and conductivity cell of a typical ELCD. *Source: Copyright OI Corporation [20].*

under certain conditions, decreasing the performance of the detector and requiring more frequent maintenance. Oxygenated solvents can immediately and irreversibly poison the catalyst. High levels of halogenated solvents can tax the deionizer. To prevent detector overload by high amounts of solvent, solvent venting is done by time-programmed control of using a vent valve. This approach can also be used to vent column effluent at other operator-specified times during the run, such as when high-level analytes of no analytical interest elute.

Figure 12.35 compares the selectivity of an ELCD to that of a PID for a spiked soil sample. The PID, although sensitive for the PCBs, is not selective enough to avoid responding to other matrix components. On the other hand, the ELCD is quite selective for the chlorine-containing PCBs, improving method detection limits, improving integration precision and quantitative accuracy, and significantly simplifying data interpretation.

12.9. ATOMIC EMISSION DETECTOR

The atomic emission detector (AED) is both a universal and a specific detector. It is a destructive detector that responds to the light emitted from atoms as they relax to the ground state after having been dissociated and exited in a high-energy microwave plasma:

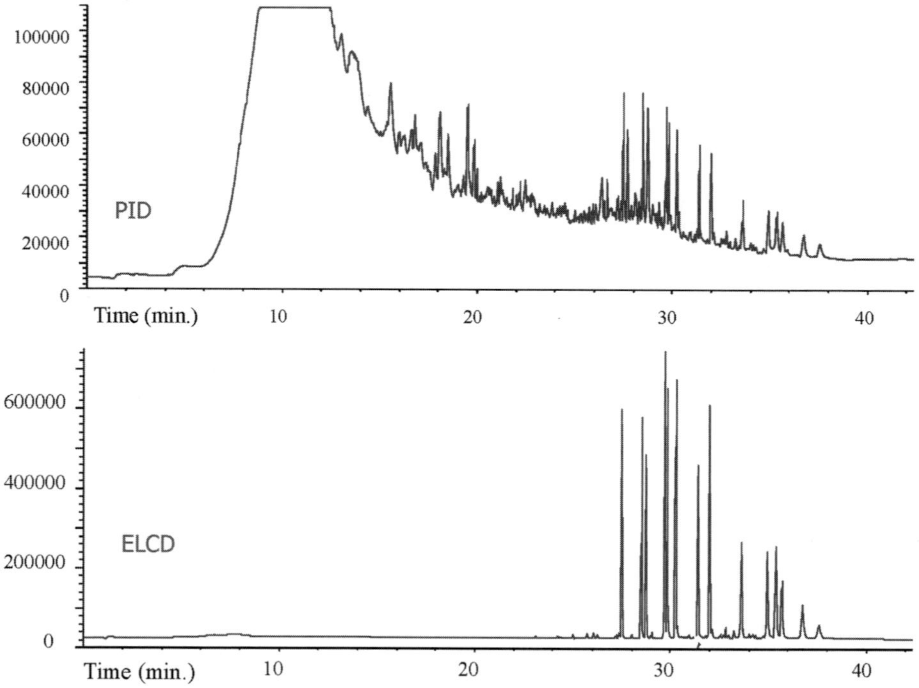

FIGURE 12.35 Comparison of PID and ELCD signals for an extract of soil spiked with 5 ppm mixture of PCBs. The PID is selective for aromatics, but not selective enough to ignore the high concentrations of interferences in the matrix. The ELCD is more selective and more specific for halogen-containing compounds. This simplifies identification and quantification of the target species. *Source: Copyright OI Corporation [20].*

$$C_XH_YS_Z \rightarrow xC + yH + zS \rightarrow xC^* + yH^* + zS^* \xrightarrow{\text{Reagent\_Gases}} h\nu + CO_2 + H_2O + SO_2$$

Helium plasmas are energetic enough to totally dissociated molecules into their composite atoms, totally eliminating any molecular history of origin. This atomization process leaves the atoms in an excited state. Upon relaxing to more stable states, the atoms emit light at discrete wavelengths. Each atom has characteristic atomic emission lines (as discussed in the PID lamp section). A few common elements are listed in Table 12.9 with emission lines commonly used in AED "recipes." In addition, as the exited atoms interact with other atoms in proximity to form stable molecules, the resulting exited molecules can also emit light at characteristic wavelengths. By spectroscopically selecting and monitoring these atomic or molecular product emission lines, one can selectively detect atoms of interest in eluting compounds. Table 12.9 lists common atomic and molecular emission lines used to monitor common atoms by AED and the associated analytical figures of merit.

One of the most significant advantages of an AED over other GC detectors is that the calibration curves for each atom are essentially compound independent. The benefit is that one can develop elemental calibration curves based on easily available standards of known elemental composition and concentration. This allows fairly accurate quantification of target compounds without having a standard for that exact compound. In addition, if calibration curves are developed for each element in the unknown, one can deduce an elemental formula from the relative calibrated peak areas [30]. This is a powerful capability, especially when coupled with mass spectral information.

A basic block diagram of an AED is shown in Figure 12.36. Figures 12.37 and 12.38 [31] further illustrate the atomic emission detection processes. Emitted light is separate and detected by a purged spectrometer capable of monitoring emission from the vacuum UV (near 150 nm) to approximately 800 nm. A photodiode array detector is limited to a narrow spectral range at any given time; so in order to cover a wider range of atomic emissions, multiple analyses of the same sample are required. "Recipes" are the combination of emission lines and reagent gases used for given sets of elements that can be monitored in a single run (a few examples

TABLE 12.9 Representative Figures of Merit for Elemental Detection by a Commercial AED [29]

Group	Element(s)	Wavelength (nm)	Minimum detectable level (pg/s)	Selectivity over carbon	Dynamic range	Measurement compound
1	Carbon	193	1	–	1×10^4	t-Butyl disulfide
	Sulfur	181	2	10000	1×10^4	t-Butyl disulfide
	Nitrogen	174	30	2500	2×10^4	Nitrobenzene
2	Hydrogen	486	4	–	5×10^3	t-Butyl disulfide
	Chlorine	479	30	3000	1×10^4	1,2,4-Trichlorobenzene
3	Phosphorus	178	2	5000	1×10^3	Triethyl phosphate
4	Oxygen	171[1]	150	5000	5×10^3	Nitrobenzene

[1] *Uses molecular band instead of atomic emission line.*

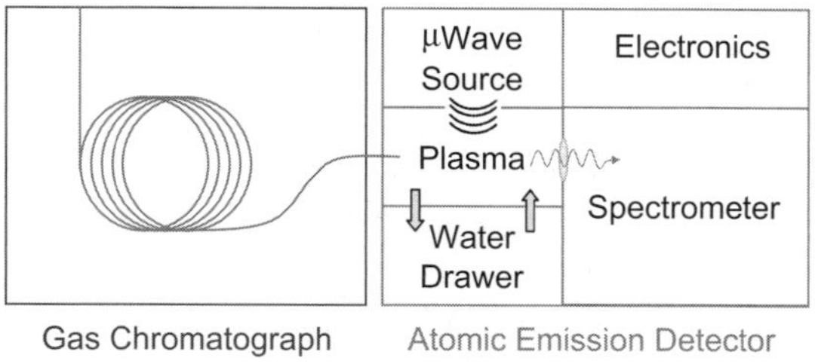

FIGURE 12.36 Block diagram of basic AED design. Microwave energy creates a high-energy He plasma inside a water cooled fused silica discharge tube. Emitted light passes to a spectrometer for separation and detection.

are listed in Table 12.10). Method development involves selecting the desired groups of elements to be monitored, and the minimum number of runs required to monitor these target elements.

In order to prevent formation of deposits in the exit of the source and to minimize peak tailing due to interaction with surfaces, "reagent gases" are added to the column effluent. These gases are chosen specifically to help form inert "friendly" recombination gaseous products by reacting with the atoms as they cool and exit the source. For example, hydrogen is added when monitoring metals because it forms volatile metal hydrides. Oxygen is added when monitoring carbon because it encourages the formation of CO_2.

The benefits of element-specific detection can be seen by comparing element-specific chromatograms, as shown in Figure 12.39. By monitoring carbon emission, one gets a chromatogram very similar to that of an FID. Unlike the ECD, one can monitor halogens separately and with high selectivity. Unlike an FPD, response for sulfur is linear with mass and quenching is not an issue. In addition, the

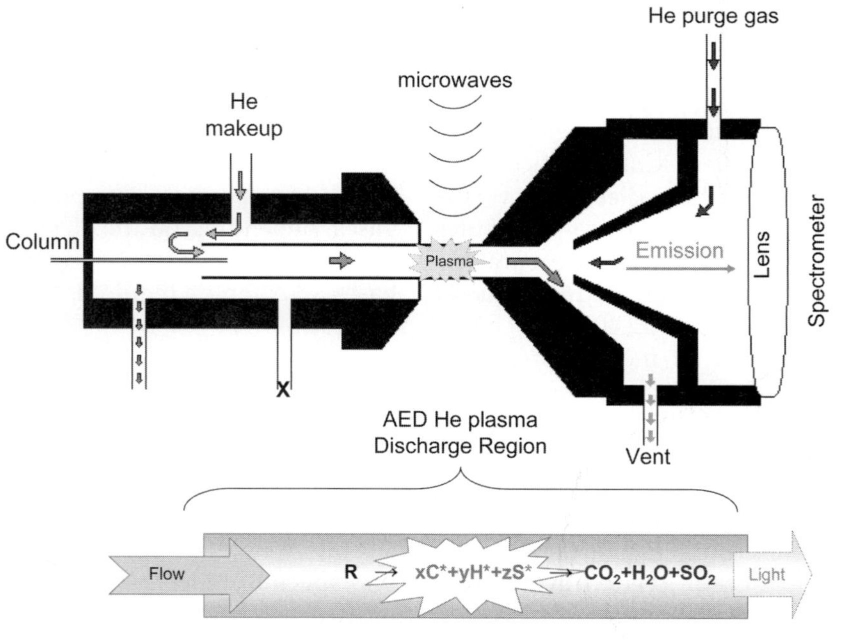

FIGURE 12.37 More detailed look at AED detection processes. Column effluent combines with He makeup gas and is carried into the microwave induced plasma zone where compounds are atomized, atoms are excited then emit light. Light is focused by the entrance lens to the spectrometer. A purge gas of N_2 blocks air from the light path to prevent absorption of low UV emission lines.

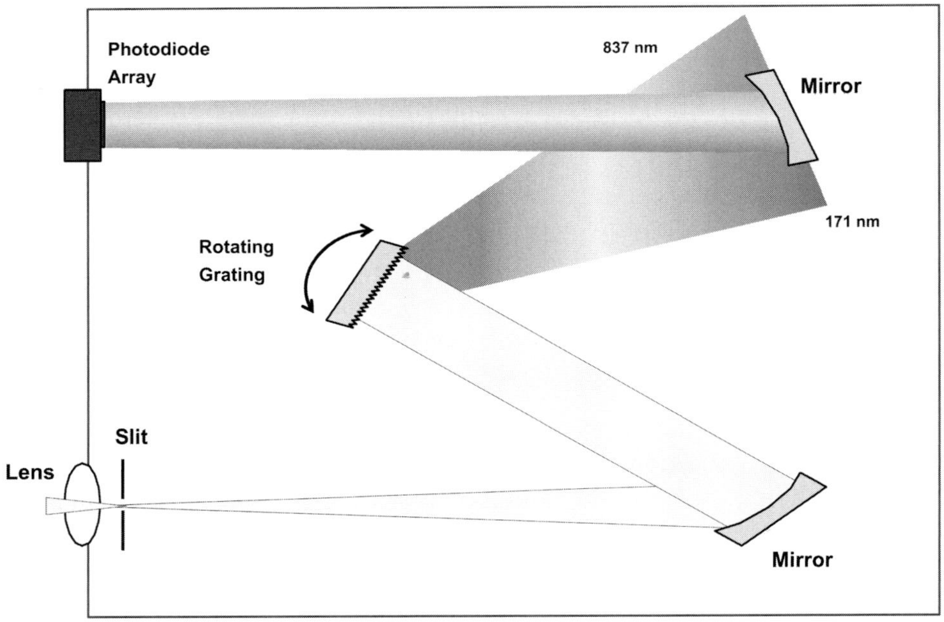

FIGURE 12.38 The purged Czerny–Turner spectrometer focuses diffracted light onto a photodiode array detector. Only a narrow range of wavelengths can be monitored by the PDA at a given time; so to cover multiple elements, several runs of the sample are sometimes required. *Source: Copyright Joint Analytical Systems [31].*

AED can produce unique element-specific signals that other simpler GC detectors cannot. Coupled with reasonable selectivity versus high carbon background, these attributes make the AED a nice tool to measure the boiling point distribution of sulfur in diesel fuel, as illustrated in Figure 12.40.

The AED requires a bit more maintenance than univariate GC detectors, but a similar level to that of an MSD. The He microwave plasma is contained in a fused silica tube (few materials have the necessary characteristics of high-temperature stability, low level of background impurities, low surface interaction with analytes, and reasonable cost). Even with its high-temperature resistance, the tube must be cooled externally by re-circulating water pumped from a reservoir. The tube has a finite lifetime and needs to be replaced after approximately every 6 weeks of use.

The AED also requires more gases than any other GC detector: He for the plasma, N_2 to purge the spectrometer, plus several reagent gases, depending on the elements being monitored (Table 12.10). Of course, the reagent gases are only required used during the analysis.

12.10. CHEMILUMINESCENT DETECTOR

The chemiluminescent detector (CLD) is a mass-sensitive, destructive detector used for specific detection of sulfur (SCD) or nitrogen (NCD). Even though the detection processes are similar for N and S, the hardware and experimental setpoints are different enough that the NCD and SCD should be considered separate detectors from a practical

TABLE 12.10 Selected Elements and Conditions for AED Analysis

Group	Element(s) or isotope	Wavelength (nm)	Reagent gas(es)[3]	Other
1	Carbon	193	O_2 and H_2	
	Iodine	183		
	Sulfur	181		
	Carbon	179[1]		
	Nitrogen	174		
2	Carbon	496	O_2	
	Hydrogen	486		
	Chlorine	479		
	Bromine	478		
4	Phosphorus	178	H_2	He high flow[2]
	Oxygen	171[1]	H_2 and 10% CH_4/90% N_2	
13	Tin	303	O_2 and H_2	He high flow[2]
	Iron	302		
	Nickel	301		
	Tin	301		
	Vanadium	292		

[1] *Uses molecular band instead of atomic emission line.*
[2] *"He High Flow" – For certain element groupings, additional helium carrier makeup gas flow is added to minimize peak tailing due to interaction with the discharge tube surface*
[3] *Performance for elements monitored in multiple groups may vary depending on the reagent gas(es) used, neighboring spectral interferences, and relative S/N for the given emission line.*

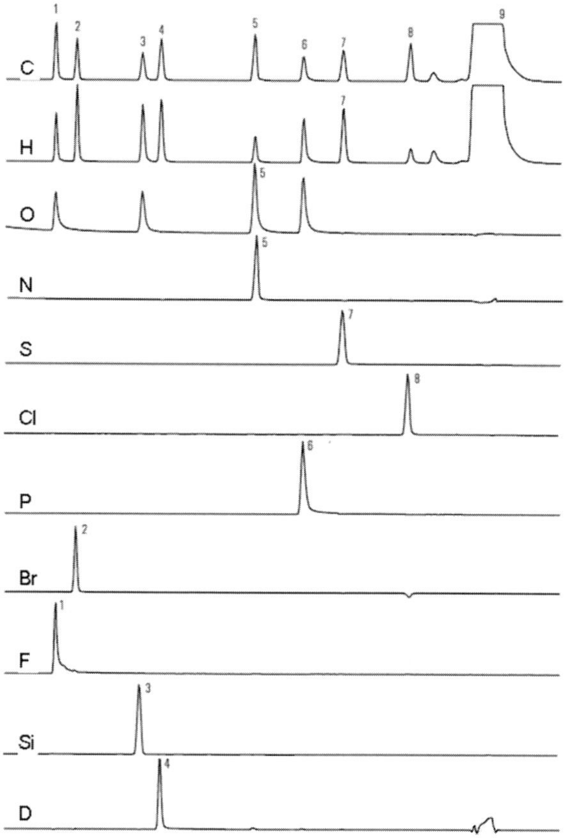

FIGURE 12.39 Element specific AED chromatograms of (1) 4-flouroanisole , (2) 1-bromohexane , (3) tetraethylorthosilicate, (4) n-decane (perdeuterated), (5) nitrobenzene, (6) triethylphosphate, (7) tert-butyl disulfide, (8) 1,2,4-trichlorobenzene, and (9) n-dodecane. [29].

standpoint. The CLD has excellent selectivity and dynamic range, with little quenching issues compared to other N or S selective detectors. However, it is a bit more complicated to run and maintain than most other GC detectors. The sensitivity and performance of the CLD can also degrade with use – dramatically so if improperly set up or if contaminated by high sample loads. The mechanism of detection is based on a two-step process: initial combustion followed by low-pressure reaction with ozone. In the original design [32], combustion was accomplished using a typical hydrogen flame, as is used with an FID. More recent designs use a ceramic combustion chamber [33] and seem to have better overall performance.

The block diagram of a typical flameless CLD is illustrated in Figure 12.41. Column effluent passes through a high temperature (≈800 °C

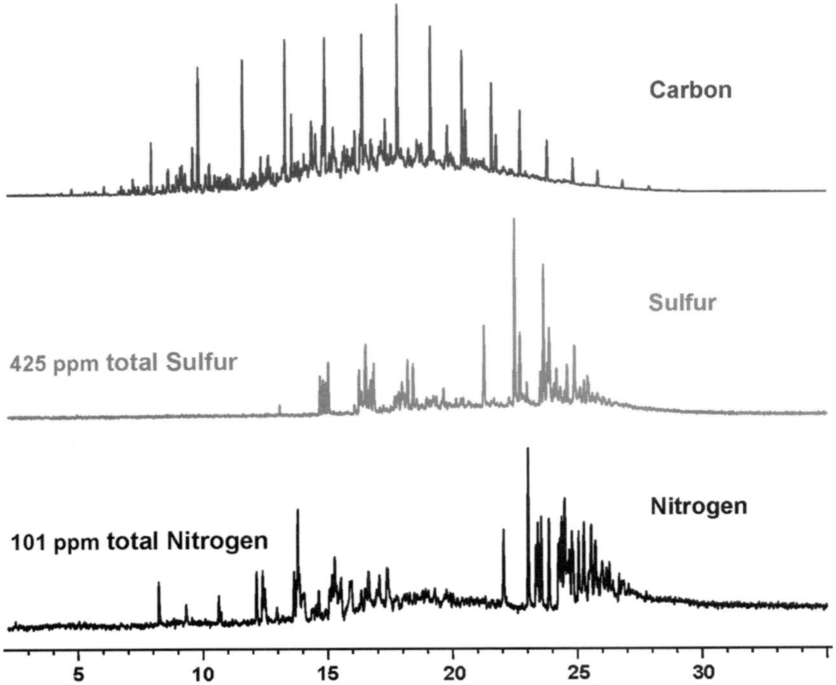

FIGURE 12.40 Comparison of C-, S-, and N-specific chromatograms for the analysis of NIST SRM 2724 low-level sulfur in diesel fuel standard by GC-AED. *Source: Copyright Agilent Technologies, Inc. [10].*

for SCD, ≈925 °C for NCD) burner where components are combusted:

$$R - N \xrightarrow{\Delta + \text{catalyst} + O_2} NO + CO_2 + H_2O + \ldots$$

$$R - S \xrightarrow{\Delta + \text{air}} SO + CO_2 + H_2O + \ldots$$

The combustion chambers used for N and S detection are different. The largest difference is that a catalyst is needed to quantitatively convert N combustion products to NO. Oxygen is typically used for the NCD and air is used

FIGURE 12.41 Block diagram of one design of a CLD. Column effluent is initially combusted in a burner and then is drawn by vacuum into a low-pressure reactor where NO or SO reacts with ozone and emits light. The light is detected by photomultiplier (PMT).

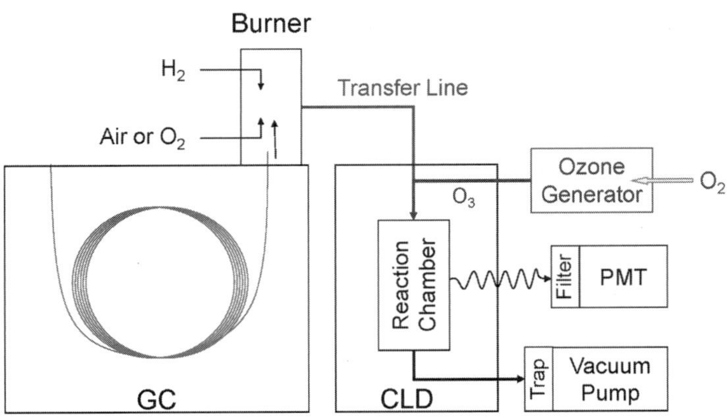

for the SCD. Hydrogen is added near the exit of both burners to ensure that all final combustion products are volatile, reducing fouling of the burner.

There is a possibility of irreversibly poisoning the catalyst in the NCD burner if it is exposed to H_2; so care has to be taken to ensure a positive upward flow of O_2 + column effluent at all times.

A vacuum is used to draw the combustion products from the burner to a separate reactor. Ozone is added, further oxidizing and exciting NO or SO species at a reduced pressure (<12 torr). As the excited oxidation products relax, they emit characteristic light:

$$NO + O_3 \rightarrow NO_2^*(+O_2) \rightarrow NO_2 + h\upsilon \quad (>800\ \text{nm})$$

$$SO + O_3 \rightarrow SO_2^*(+O_2) \rightarrow SO_2 + h\upsilon \quad (<300\text{–}400\ \text{nm})$$

The emitted light ($h\upsilon$) is filtered and detected by a photomultiplier tube and amplified, yielding linear response relative to the mass of S or N.

The CLD has several advantages relative to other N and S selective detectors:

- linear response,
- low (or no) quenching because of the highly efficient combustion processes, and
- highly selective and specific response because of the highly selective reactions with O_3 and wavelength specific emission of light.

Figure 12.42 compares the response of an SCD to other sulfur selective detectors near their detection limits.

Figure 12.43 [34] compares SCD and FID signals for the analysis of a jet fuel. In this example, a special adapter was used to allow the SCD burner to mount in series on the FID. The adapter contains a restrictor between the

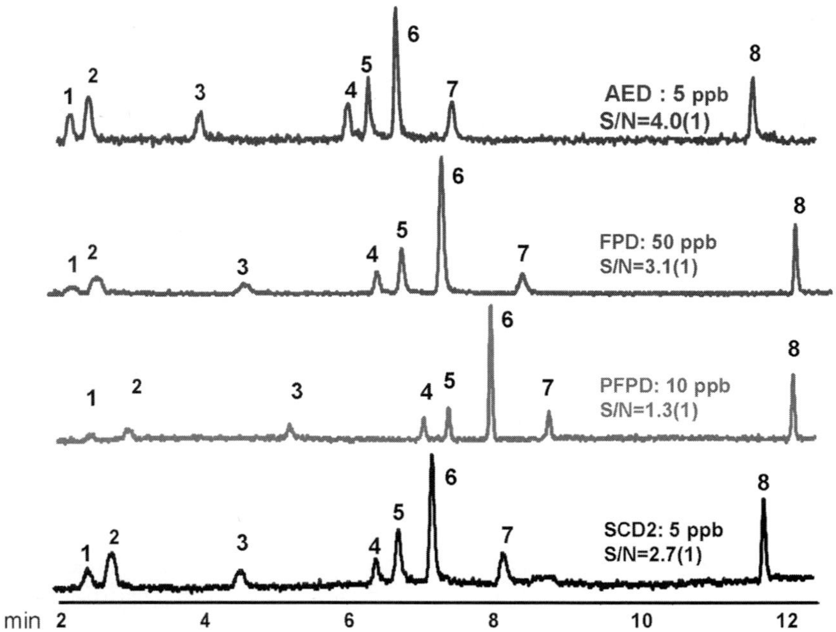

FIGURE 12.42 Sulfur detector comparison 8 component sulfur mix in helium. 1, hydrogen sulfide; 2, carbonyl sulfide; 3, methyl mercaptan; 4, ethyl mercaptan; 5, dimethyl sulfide; 6, carbon disulfide; 7, t-butyl mercaptan; 8, tetrahydrothiophene.

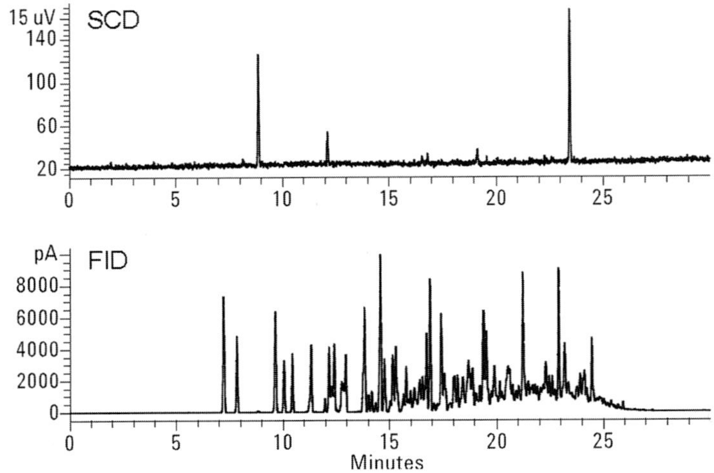

FIGURE 12.43 Comparison of SCD and FID signals for a military JP-4 jet fuel, using an SCD operated in "dual mode". Initial combustion is accomplished by an FID, then combustion products transfer to the reaction cell of an SCD for selective S detection. Source: Copyright Agilent Technologies, Inc [10].

FID and the SCD burner that allows the FID to operate at atmospheric pressure, yielding a traditional FID response. FID combustion products are drawn by vacuum through the restrictor, through the SCD burner, and eventually to the low-pressure SCD reaction chamber. Although this configuration simplifies acquisition of simultaneous signals, the performance of the SCD is degraded compared to an SCD-only configuration.

References

[1] R.R. Reston, E.S. Kolesar, Silicon-micromachined gas chromatography system used to separate and detect ammonia and nitrogen dioxide. I. Design, fabrication, and integration of the gas chromatography system, IEEE/ASME J. Microelectromech. Syst. 3 (1994) 134–146.

[2] E.S. Kolesar, R.R. Reston, Silicon-micromachined gas chromatography system used to separate and detect ammonia and nitrogen dioxide. II. Evaluation, analysis, and theoretical modeling of the gas chromatography system, IEEE/ASME J. Microelectromech. Syst. 3 (1994) 147–154.

[3] T. Holm, Aspects of the mechanism of the flame ionization detector, J. Chromatogr. A 842 (1999) 221–227.

[4] D.K. Bohme, in: P.J. Ausloss (Ed.), Kinetics of ion–molecule reactions, Plenum Press, New York, 1979.

[5] W.A. Dietz, Response factors for gas chromatographic analyses, J. Gas Chromatogr. 5 (1967) 68.

[6] J.T. Scanlon, D.E. Willis, Calculation of flame ionization detector relative response factors using the effective carbon number concept, J. Chrom. Sci. 23 (1985) 333–340. [and references therein].

[7] J.E. Lovelock, S. Wilts, A.J. Davies, F.R. Ferris, Method and apparatus for linearity measuring electron capture with an electron capture detector. US patent 3634 (1972) 754.

[8] W.E. Wentworth, E.C.M. Chen, J.E. Lovelock, The pulse-sampling technique for the study of electron-attachment phenomena, J. Phys. Chem. 70 (1966) 445.

[9] J.E. Lovelock, The electron capture detector: theory and practice, J. Chromatogr. 99 (1974) 3–12.

[10] Agilent Technologies, Inc., 5301 Stevens Creek Blvd., Santa Clara, 95051 USA.

[11] J.E. Lovelock, A.J. Watson, Electron-capture detector: theory and practice II, J. Chromatogr. 158 (1978) 123–138.

[12] B. Kolb, J. Bischoff, J. Chromtogr. Sci 12 (1974) 625.

[13] J.A. Lubkowitz, J.L. Glajch, B.P. Semonian, L.B. Rogers, Study of the nitrogen response mode of the thermionic rubidium silicate detector, J. Chromatogr. 133 (1977) 37–47.

[14] B.P. Semonian, J.A. Lubkowitz, L.B. Rogers, Phosphorus response mode of the thermionic rubidium silicate detector, J. Chromatogr. 151 (1978) 1–10.
[15] P.L. Patterson, Selective responses of a flameless thermionic detector, J. Chromatogr. 167 (1978) 381–397.
[16] P.L. Patterson, Recent advances in thermionic ionization detection for gas chromatography, J. Chromatogr. Sci. 24 (1986) 41–52.
[17] ASTM E840 - 95 standard practice for using flame photometric detectors in gas chromatography, ASTM International, 100 Barr Harbor Drive, West Conshohocken, PA, 19428 USA, 2005.
[18] E. Atar, S. Cheskis, A. Amirav, Pulsed flame - a novel concept for molecular detection, Anal. Chem. 63 (1991) 2061–2064.
[19] S. Cheskis, E. Atar, A. Amirav, Pulsed-flame photometer: a novel gas chromatography detector, Anal. Chem. 65 (1993) 539–555.
[20] OI Analytical Corporation, 151 Graham Road, College Station, TX 77845 USA.
[21] J.N. Davenport, E.R. Adlard, Photoionization detectors for gas chromatography, J. Chromatogr. 290 (1984) 13.
[22] W.E. Wentworth, Nadege Helias, Albert Zlatkis, E.C.M. Chen, Stanley D. Stearns, Multiple detector responses for gas chromatography peak identification, J. Chromatogr. A 795 (1998) 319–347.
[23] P. Verner, Photoionization detection and its application in gas chromatography, J. Chromatogr. 300 (1984) 249–264.
[24] J.N. Driscoll, J. Ford, L.F. Jaramillo, E.T. Gruber, Gas chromatographic detection and identification of aromatic and aliphatic hydrocarbons in complex mixtures by coupling photoionization and flame-ionization detectors, J. Chromatogr. 158 (1978) 171.
[25] Photoionization detector (PID) HNU, US EPA SOP # 2114, 1994.
[26] A.N. Freedman, The photoionization detector: theory, performance and application as a low-level monitor of oil vapour, J. Chromatogr. 190 (1980) 263.
[27] G. Gremaud, W.E. Wentworth, A. Zlatkis, R. Swatloski, E.C.M. Chen, S.D. Stearns, Windowless pulsed-discharge photoionization detector application to qualitative analysis of volatile organic compounds, J Chromatogr. A 724 (1996) 235–250.
[28] W.E. Wentworth, Ju Huang, Kefu Sun, Yu Zhang, Lei Rao, Huamin Cai, et al., Non-radioactive electron-capture detector, J. Chromatogr. A 842 (1999) 229–266.
[29] Agilent G2350A atomic emission detector (AED) specifications, Agilent Technologies, Inc. publication 5966–4141EN.
[30] N.R. Hardas, P.C. Uden, Empirical formulae studies of chlorofluorocarbons using gas chromatography coupled to atomic emission detection, J. Chromatogr. A 844 (1999) 271–281.
[31] Joint analytical systems, GmbH, Carl-Zeiss-Str. 49, D-47445 Moers, Germany.
[32] R.L. Benner, D.H. Stedman, Universal sulfur detection by chemiluminescence, Anal. Chem. 61 (1989) 1268–1271.
[33] R.L. Shearer, Development of flameless sulfur chemiluminescence detection: application to gas chromatography, Anal. Chem. 64 (1992) 2192–2196.
[34] Agilent 355 sulfur chemiluminescence detector (355 SCD): sulfur compounds in destilate fuels, Agilent Technologies, Inc. Publication 5989–6790EN.

CHAPTER

13

Hyphenated Spectroscopic Detectors for Gas Chromatography

Charles L. Wilkins

OUTLINE

13.1. INTRODUCTION

From the earliest days of time-of-flight mass spectrometry, it was recognized that the characteristic rapid scanning capability of such instruments qualified them for a key role in conjunction with gas liquid partition chromatography (GLPC). Specifically, it is possible to use TOF-MS for on-line monitoring of GC effluents. Accordingly, in 1959 Golke at Dow reported demonstration of on-line GC-MS utilizing a packed gas chromatography column [1]. Relatively soon after that, he demonstrated extension of the technique to capillary column GC [2]. Dorsey shortly thereafter contributed an additional feasibility study demonstrating its applicability to petroleum hydrocarbons [3]. These early efforts clearly established the feasibility of coupling a spectroscopic detector with gas chromatography, while falling short of providing convincing evidence of the advantages of the methods. Thus, it was not until about 10 years later that more demanding applications of capillary GC and theoretical analysis of coupling open-tubular columns with a mass spectrometer [4] were published. Hirschfeld provided a more general detailed summary of spectroscopic detector possibilities in an early 1980 article on possible hyphenated separation/spectroscopy systems [5]. In his article, two hyphenated methods, GC-mass spectrometry (MS) and GC-infrared spectroscopy (IR), were identified as successful and 6 other types of GC-spectroscopy detectors were identified as feasible for

 DOI: 10.1016/B978-0-12-385540-4.00013-4

state-of-the-art versions of those instruments. The use of spectroscopic detectors continued to evolve from that time forward as documented in a 1983 article [6] and later articles in 1989 [7] and 2002 [8]. The 2002 article cited was devoted exclusively to the GC/infrared/MS combination. It was optimistically concluded in that chapter "...that continuing advances in the field of chromatography and spectral detectors will lead to improved GC/IR/MS systems in the future" [8]. However, subsequent developments have not borne that out. In fact, a volume on Hyphenated Methods in the Encyclopedia of Mass Spectrometry, published only 4 years later includes no mention of GC-IR-MS in its index, nor indeed in any of the chapters, except for the entry defined as GC-isotope ratio-(IR)-MS, that is discussed in some depth in Chapter 2 of that book [9].

At present, the primary hyphenated spectroscopic GC technique in routine use is GC-MS and its variants include GC-MS-MS and GC-MS_n, where n indicates multiple stages of MS in either space (e.g. the use of triple quadrupole MS or Q-TOF instruments) or time (e.g. the use of ion trapping instruments such as a quadrupole ion trap or a Fourier transform ion cyclotron resonance instrument) [10]. It is beyond the scope of this chapter to discuss the very extensive and more prevalent LC-MS theory and applications that form the majority of the topics in Volume 8 (Hyphenated Methods) of the Encyclopedia of Mass Spectrometry. However, it is significant that Chapter 2 on GC-MS Principles and Instrumentation constitutes 95 pages of this 1068-page volume (approximately 9 percent of the coverage in this book). However, that is not surprising, in that there appears to be significantly more contemporary interest in analysis of nonvolatiles which are the strength of LC-MS and other methods. Before surveying the current state of the art with respect to GC-spectroscopic detectors, it is worthwhile to quickly review the series of relevant manufacturer changes and their corresponding impact on the field. In late 1980s, Hewlett-Packard (that part of HP was later acquired by Agilent) announced an integrated system consisting of a GC, an infrared detector (IRD), and a quadrupole mass spectrometer (MSD). However, sales were disappointing and, as a result, HP divested itself of this system by selling it to Bio-Rad in about 1995. Bio-Rad briefly continued to market that system and to support the previous Hewlett-Packard systems. However, they too discontinued this product. An excellent article reviews the current state of the art for GC/FTIR as it stood at the end of 1998 [11]. Currently, to the present author's knowledge, there is no commercial source of integrated GC-FTIR-MS systems.

13.2. GC INTERFACES

Regardless of what spectroscopic detector is used, any hyphenated system must address the mismatch between a gas chromatograph and whatever detector is employed. It is useful to review how these requirements are dealt with in the more complicated case where several detectors are involved. As mentioned above, the gas chromatography-infrared spectrometry-mass spectrometry combination is attractive for a number of reasons. In this case, the infrared spectrometer is at a serious disadvantage with respect to relative detection sensitivity, when compared with mass spectrometry. This can be compensated in essentially two different ways. One can use a serial flow arrangement, taking advantage of the nondestructive nature of infrared spectrometry. This approach routes the GC effluent through the infrared detector (IRD) and splits a portion to the mass spectrometer (because its sample requirements are much lower), as in the Hewlett-Packard IRD design. In this way, sample requirements for the IR are achieved by roughly matching a gold-plated lightpipe volume to the peak volume of the emerging GC effluent for maximum IR detection efficiency. One early paper dealt with lightpipe design and how it affected sensitivity [12]. The lower sample requirement of the mass spectrometer is thus nicely accommodated. Even though commercial GC-FTIR-MS systems are not currently available, some scientists continue their use [13–15]. A second method is to use a sample trapping approach. This permits enhanced detection limits for infrared detection in an effort to bring these closer to those of MS. For example, as diagramed in Figure 13.1, a commercial system for matrix isolation GC/MI-FTIR was linked to a Hewlett-Packard MSD to be used for heartcut analysis [16]. In this system, the Mattson Cryolect spectrometer mixed xenon with the emerging GC effluent, which was subsequently deposited upon a cryogenically cooled surface. Mattson Instruments was founded in 1983 by David Mattson, and the company acquired by Orion Research in 1989. Subsequently, Orion Research was acquired by Analytical Technology, Inc. that in turn was purchased by Thermo Electron in 1995. Mattson's 100-person Madison operation was then merged into Thermo Electron's Nicolet Instrument Division. The Mattson Cryolect method mentioned above resulted in separated mixture components being matrix isolated in deposited xenon. By careful adjustment of the operating parameters, the scientist could accumulate the contents of an entire GC run upon a gold-plated surface, which was slowly translated and rotated under computer control during a GC separation. Following the separation and storage of the MI separated components, to the surface could be systematically interrogated with an infrared spectral beam to acquire the MI-infrared spectra of separate components. For the MS analysis, the sample was split, prior to the matrix isolation step. One difficulty of this method is the experimental need to adequately exclude water and carbon dioxide, which are significant infrared interferents, even in low quantities.

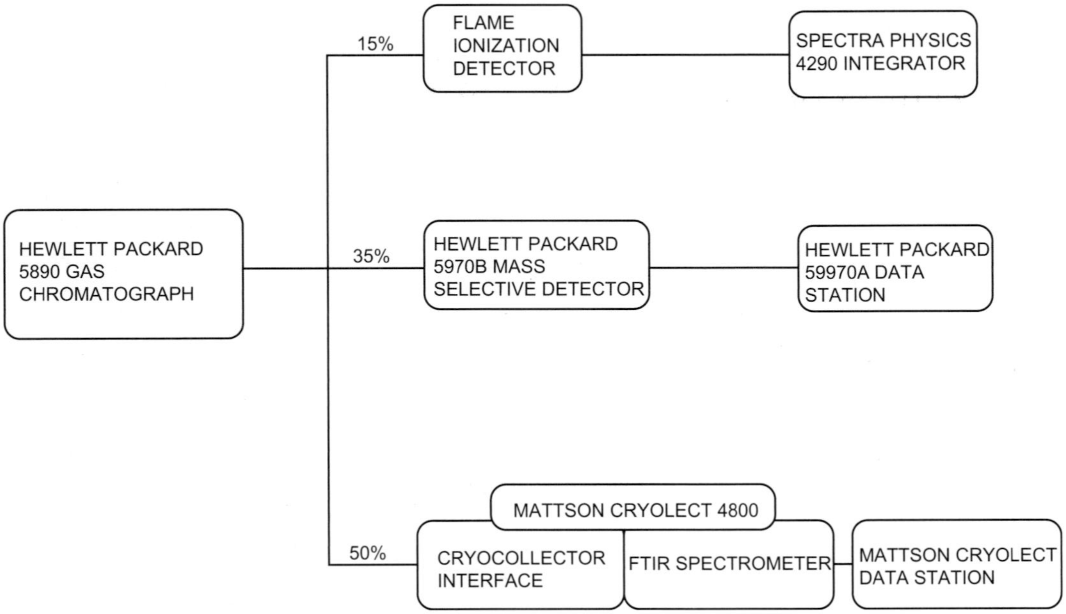

FIGURE 13.1 Schematic diagram of a combined GC-MS- cryogenic FTIR system. *Reproduced from the Journal of Chromatographic Science by permission of Preston Publications, A Division of preston Industries, Inc.*

An alternative to the Cryolect design was the lower cost cryogenic deposition FT-IR offered by Bio-Rad in the late 1980s. However, this instrument also suffered from the same interferences.

In spite of the fact that combined GC-FTIR-MS systems are not currently available, most Fourier transform infrared manufacturers still can supply GC-FTIR systems that are readily interfaced with present-day mass spectrometers, using the lightpipe-based approach mentioned above and an appropriate sample splitter to route a portion of the sample to the mass spectrometer. Thus, the capability is not difficult to implement, if the analytical application warrants it. Here, as in GC-MS, the practical need is for appropriate software and spectral databases. However, the authors of a very ambitious paper describing a reversed-phase high-performance liquid chromatography, ultraviolet diode array, FTIR, ^{1}H NMR-MS system aptly concluded (quoting Hirschfeld [5]) "merely because something is possible does not mean it represents a good solution to a problem" [17]. In that particular example, the cost of such an array of spectroscopic detectors probably outweighs the possible benefits. However, the fact that the research described could be accomplished once again clearly demonstrates the feasibility of joint use of multiple spectroscopic detectors. For GC-MS One interface to the mass spectrometer is even simpler. Using modern vacuum systems, it is easily possible to adopt a splitless injection system that can be used with direct insertion of a capillary GC column into the mass spectrometer source. This technique employs a makeup gas and results in somewhat decreased sensitivity, because although a fixed quantity of GC effluent is delivered to the mass spectrometer, it is lower than the total amount. A typical interface might deliver about 80 percent of the effluent.

Historically, when packed GC columns were interfaced with MS a jet separator was used [18]. One of the limitations of this type of separator is that minimum input separator flow rates of 20–25 ml/minute must be maintained. Therefore, if a capillary column is interfaced to an MS using such a separator, makeup gas must be added to provide the required flow rate. Finally, the membrane separator depends upon the use of a membrane that differentially responds to organic compounds (readily passing them to the MS) and carrier gas, normally helium or hydrogen (discharged and not passed to the MS). Although this approach was once in widespread use, this is no longer the case.

13.3. DATA ANALYSIS

Given the popularity of GC-MS for applications involving analysis of volatile samples, there are numerous commercial sources of GC-MS systems. Most use either splitless injection or open split interfaces. It is most common that manufacturers provide proprietary software for the data analysis task and restrict the use of that software to a single computer. As a typical example, a Varian 350 Triple Quadrupole GC-MS (now supported by Bruker as a result of their purchase of the Varian GC-MS operations) requires separate licenses for the National Institute of Standards and Technology (NIST) database and the GC-MS software to permit the operation with a single computer. Purchases of additional workstation software and NIST database library are charged separately and two licenses are required, one for the NIST library and one for the GC-MS software to enable library searching on each workstation added. Thus, costs can easily become prohibitive. As mentioned above, this is a typical pricing scheme for GC-MS vendor supplied library search and GC-MS software. Of course, there are also a number of third-party software sources, in the event that the features supported by the GC-MS manufacturer are not satisfactory. With such sources, a variety of options are available, and features vary, reflective of the fact that such packages often result from perceived inadequacies of the standard software provided.

13.4. GC-ATOMIC EMISSION-MASS SPECTROMETRY

In this application area, the direct connection of a gas chromatograph and an atomic emission detector (AED, to use the previous Hewlett-Packard designation) is commonly done in parallel with a second GC-mass spectrometer connection rather than directly linking the two spectroscopic detectors to one GC. This approach, as explained in an excellent recent review article [19], is a consequence of the much different pressure requirements of the AED detector (slightly above atmospheric) and the MS which is operated at vacuum. As these authors also noted, results are best if a single column with eluent splitting at the GC column outlet is employed, resulting in GC-AED-MS, obviating the necessity of trying to correct for different retention times occasioned by the dual GC approach. The comprehensive review cited explains that there have been many hundreds of papers on GC-AED since the technique was first demonstrated in 1965. Accordingly, the review of this topic does not attempt a comprehensive coverage of that literature. Even so, 146 references are cited in this more selective treatment of the subject (intended to show "... the versatility and practicability of GC-AED to solve a wide variety of problems ...")!

13.5. SPECTROSCOPIC DETECTORS FOR GC

As mentioned above for the case of GC-FTIR-MS, this combination has generally been relatively neglected in recent years, although a review of the entire topic of 'Gas Chromatography with Spectroscopic Detectors' did appear in 1999 [20]. That review article was a fairly comprehensive discussion of that topic, beginning with consideration of sample introduction techniques, including purge-and-trap methods as well as spray-and-trap methods. GC-GC with various detectors was also included in the discussion, even though that is a technique that has been addressed in many papers over the years and is sufficiently important to justify its own review. The review also discusses a number of different approaches to GC-MS analysis, along with selected examples of their use. GC-FTIR was also covered in some detail with a number of examples of the advantages of this mode of GC detection and discussions of the advantages of linked GC-IR-MS systems, which, in retrospect, have not led to commercially viable solutions. After treating various data analysis methods in some detail, the review concludes that "... recent developments in multispectral detection systems for gas chromatographic effluents have provided unrivaled analytical capabilities to analytical chemists ..." This review, with its 233 references [20], in addition to the more recent review of GC-AED cited earlier [19] provides an excellent snapshot of the current status of spectroscopic detectors for GC and leads to the general conclusion that the field is currently dominated by GC-MS, in spite of the proven advantages of linking additional spectroscopic detectors. This situation can be understood by realizing that use of complementary detectors is often justified on the basis of specific application needs, but that the sensitivity and specificity advantages inherent in mass spectrometry have largely outweighed the advantages that complementary detectors confer. One could also speculate that the numerous changes in companies in the spectroscopic detection field, with corresponding reevaluation of company direction and goals must inevitably have had significant influence on this dynamic field. It is certainly true that a significant number of creative, innovative, analytical chemistry approaches have resulted from scientists' interest in the possibilities so persuasively discussed in Hirschfeld's seminal 1980 article [5].

References

[1] R.S. Gohlke, Time-of-flight mass spectrometry and gas-liquid partition chromatography, Anal. Chem. 31 (4) (1959) 535–541.

[2] R.S. Gohlke, Time-of-flight mass spectrometry. Application to capillary column gas chromatography, Anal. Chem. 34 (10) (1962) 1332–1333.

[3] J.A. Dorsey, R.H. Hunt, M.J. O'Neal, Rapid-scanning mass spectrometry. Continuous analysis of fractions from capillary gas chromatography, Anal. Chem. 35 (4) (1963) 511–515.

[4] M. Novotny, Coupling of open tubular columns with a mass spectrometer through the jet-type molecule separator, Chromatographia 2 (1969) 350–353.

[5] T. Hirschfeld, The Hy-phen-ated methods, Anal. Chem. 52 (1980) 297A–312A.

[6] C.L. Wilkins, Hyphenated techniques for analysis of complex organic mixtures, Science 222 (1983) 291–296.

[7] C.L. Wilkins, Advances in hyphenated analytical chemistry instrumentation, Science 243 (1989) G14–G22. Part II (Guide 1989).

[8] C.L. Wilkins, Directly-linked gas chromatography-infrared-mass spectrometry (GC-IR-MS), in: J.M. Chalmers, P.R. Griffiths (Eds.), Handbook of vibrational spectroscopy, John Wiley and Sons, Ltd, Chichester, West Suffix, UK, 2002, pp. 1627–1633.

[9] W. Kulik, Principles and applications of GC-isotope-ratio-MS, in: W.M.A. Niessen (Ed.), The encylopedia of mass spectrometry, Elsevier, Oxford, UK, 2006, pp. 105–115.

[10] W.M.A. Niessen, Principles and instrumentation of MS-MS, in: W.M.A. Niessen (Ed.), Encyclopedia of mass spectrometry, John Wiley and Sons, Ltd., Oxford, UK, 2006, pp. 13–19.

[11] B. Erickson, Is there new life ahead for hyphenated IR, Anal. Chem. 70 (1998) 801A–805A.

[12] G.N. Giss, C.L. Wilkins, Effects of LIghtpipe dimensions of gas chromatography/fourier transform infrared sensitivity, Appl. Spectrosc. 38 (1984) 17–20.

[13] P. Christen, S. Bieri, O. Munoz, Characterization of positional and configurational tropane alkaloid isomers by combining GC with NPD, MS, and FTIR, Natural Prod. Commun. 4 (2009) 1341–1348.

[14] V. Basiuk, J. Douda, Analysis of less-volatile products of poly-L-valine pyrolysis by gas chromatography/fourier transform infrared spectroscopy/mass spectrometry, J. Anal. Appl. Pyrolysis 60 (2001) 27–40.

[15] A.M. Vinggaard, W. Körner, K.H. Lund, U. Bolz, J.H. Petersen, Identification and quantification of estrogenic compounds in recycled and virgin paper for household use as determined by an in vitro yeast estrogen screen and chemical analysis, Chem. Res. Toxicol. 13 (12) (2000) 1214–1222.

[16] E.R. Baumeister, L. Zhang, C.L. Wilkins, Determination of C-7 isomer distribution in commercial feedstocksby GC-Matrix Isolation-FTIR-MS, J. Chromatographic Sci. 29 (1991) 331–338.

[17] D. Louden, A. Handley, S. Taylor, E. Lenz, S. Miller, I.D. Wilson, et al., Reversed-phase high-performance liquid chromatography combined with on-line UV diode array, FT infrared, and 1H nuclear magnetic resonance spectroscopy and time-of-flight mass spectrometry: application to a mixture of nonsteroidal antiinflammatory drugs, Anal. Chem. 72 (16) (2000) 3922–3926.

[18] R. Ryhage, Use of a mass spectrometer as a detector and analyzer for effluent emerging from high temperature gas liquid chromatography columns, Anal. Chem. 36 (4) (1964) 759–764.

[19] L.L.P. van Stee, U.A.T. Brinkman, Developments in the application of gas chromatography with atomic emission (plus mass spectrometric) detection, J. Chromatography A 1186 (1-2) (2008) 109–122.

[20] N. Ragunathan, K.A. Krock, C. Klawun, T.A. Sasaki, C.L. Wilkins, Gas chromatography with spectroscopic detectors, J. Chromatography A 856 (1999) 349–397.

CHAPTER

14

Plasma-Based Gas Chromatography Detectors

Qilin Chan, Joseph A. Caruso

OUTLINE

14.1. INTRODUCTION TO PLASMA-BASED DETECTORS

Plasma is a state of matter, other than solids, liquids, or gases, containing positive ions and negative electrons. A gas can be turned into plasma when losing or gaining an electron ionizes the gaseous molecules. The presence of high electron density and positive ions makes plasma a good ionization source. Their capability to fully or partially ionize is related to their electron density of ca. $10^{15}/cm^3$. There are three major plasma-based ionization sources, inductively coupled plasma (ICP), microwave-induced plasma (MIP), and glow discharge (GD). More recently, additional intriguing

Gas Chromatography DOI: 10.1016/B978-0-12-385540-4.00014-6

plasma sources have been introduced, but these have not yet received the popularity for GC detection and are not included in this chapter.

ICP is created by igniting plasma gas, typically argon, with a spark, and the plasma is sustained under a high radio frequency electromagnetic field. ICP temperatures range between 6000 and 10,000 K [1]. The samples introduced to ICP are instantly dried, vaporized, atomized, and finally ionized by the plasma. ICP was first used as an excitation source for atomic emission spectroscopy in the 1960s [2], but commercial inductively coupled plasma-atomic emission spectrometry (ICP-AES) instrumentation was not available until 10 years later. Inductively coupled plasma mass spectrometry (ICPMS) emerged commercially in the 1980s and has been growing rapidly. With the unmatched advantages of producing singly charged positive ions for most elements in the periodic table, ICP-AES and ICPMS quickly surpassed other techniques of elemental detection, such as atomic absorption spectrometry, and have become the dominant techniques for trace-level elemental analysis.

Both ICP-AES and ICPMS have been used as atomic detectors for gas chromatography (GC) since the 1980s. However, ICPMS has been the first choice when an online atomic detection is needed for GC, because of the advantages of ICPMS over ICP-AES, such as higher sensitivity, lower detection limits, higher selectivity, and the ability for multi-isotope detection. The development of GC-ICPMS has been ascribed to the continuous growth of interest in elemental speciation. In addition to GC-ICPMS, the emergent needs of elemental speciation also drive the growth of other hyphenated techniques, such as HPLC-ICPMS [3,4] and capillary electrophoresis (CE-ICPMS) [5].

Microwave-induced plasma (MIP), an alternative plasma-based ionization source, uses microwaves produced from a magnetron (a microwave generator) to oscillate the electrons in the plasma gas. The oscillating electrons collide with atoms in the flowing plasma gas to create and maintain a high-temperature plasma. MIP typically uses helium (He) as a plasma gas, which leads to one advantage of MIP over the ICP, namely the capability to ionize all elements in the periodic table with an ionization potential of 24.5 eV, which is significantly higher than Ar at 15.7 eV. Therefore, MIP can easily excite nonmetals, which are poorly ionized in Ar plasma. However, it is worth noting that electron densities and electron energies are typically lower in the analytical MIP – important parameters for efficient ionization. Another benefit of using He is that it reduces isobaric interferences since mono-isotopic ^{4}He forms fewer polyatomic ions than multi-isotopic Ar. Despite the advantages of MIP, ICP is the dominant plasma source on the market. The main drawbacks of MIP include difficulties with plasma ignition and sustainment as well as sample desolvation and atomization due to its limited thermal energy. Therefore, the amount of sample introduced into MIP may be limited. However, this is not an issue for the hyphenated technique GC-MIP-AES/MS because GC is also a great matrix removal technique allowing the plasma more energy to excite or ionize particular analyte ions.

GD, another type of plasma, is developed in a cell with two separate electrodes with a potential difference from a few hundreds to thousands of volts [6]. The plasma gas, argon, or helium typically, fills the cell and is excited by electrons from the cathode. The generated positive ions are accelerated toward the cathode, and secondary electrons are released at the electrode. The first and secondary electrons make GD a self-sustained plasma. For surface analytical applications, the cathode is the subject of analysis, but for GC, it is the plasma (GD) that is the compartment of interest. GD is typically powered by direct current (dc) and radio frequency (rf). The dc-GD has been the most commonly used, but the rf-GD is increasingly popular due to its capability of ionizing nonconductive cathode materials. GD has been used in glow discharge atomic emission spectrometry (GD-AES) and glow discharge-mass spectrometry (GD-MS), which are widely used for bulk solids analysis and depth profiling of solid conducting materials as discussed by Marcus [6]. However, Olson et al. were first to show the power of the GD as a GC detector, illustrating both soft and hard ionization by coupling GC to dc-GD and rf-GD [7,8]. In addition to inorganic analysis, a versatile GD capable of both atomic and molecular ionization has been developed [9].

Increasingly strong interest in elemental speciation has been driving fast growth of hyphenated techniques, which enable online atomic detection for various separation approaches. Figure 14.1 shows the comparison among the three major GC-hyphenated atomic detections, GC-ICP, GC-MIP, and GC-GD. All three were actively studied in the 1990s, but GC-ICP was clearly the most studied. GC-ICP has become the dominant online atomic detection for GC since 2000, and over 90% of the GC-ICP applications are performed on GC-ICPMS. ICPMS is the most favorable atomic detector for GC in the last ten years.

14.2. GC-ICPMS

14.2.1. Brief Introduction to ICPMS

Inductively coupled plasma mass spectrometry is currently one of the most powerful and popular multielement atomic detectors. The first ICPMS, an ICP source coupled to a quadrupole-based mass analyzer, was introduced by Houk et al. in 1980 in the USA while British scientists Gray and Date were making excellent progress [10]. The introduction of collision/reaction cells in 2001 increased the potential of ICPMS for

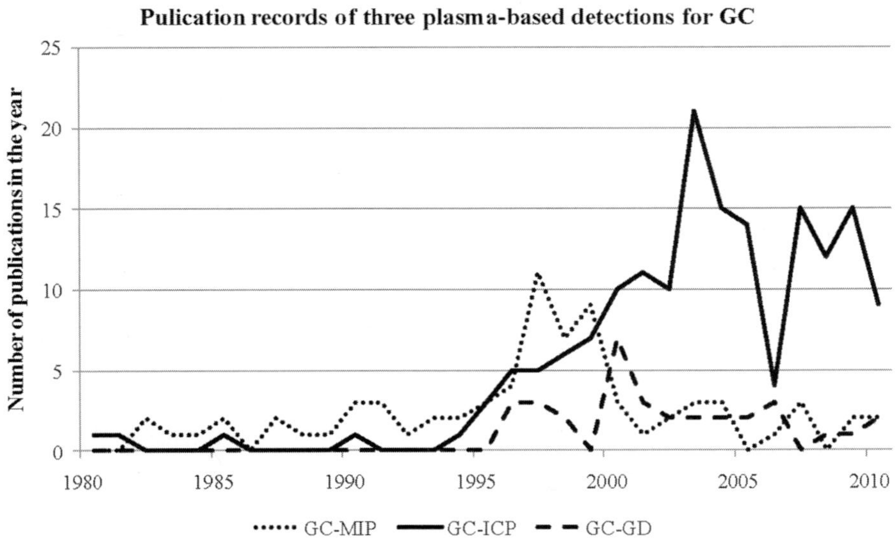

FIGURE 14.1 This graph shows the abundance of publications on each of the three plasma-based detections for GC in an individual year since the first online plasma-based atomic detection for GC in 1980. Above 90% of the GC-ICP publications are from GC-ICPMS. The data were obtained from Scopus by searching the corresponding keywords. The graph is for a trend analysis of these hyphenated techniques, rather than an accurate representation of their actual number of publications.

high matrix samples [11,12]. In the ensuing years, ICPMS has become one of the most important techniques for the detection of more than 70% of the elements in the periodic table. Since the development of laser ablation (LA) and LA-ICPMS [13], ICPMS has been a versatile tool for any type of samples, gaseous, liquid, and solid. In addition to total concentration measurement, it is also a useful online detection tool for common separation techniques, GC, LC, and CE. More than 5000 ICPMS instruments have been installed in various institutions, academic labs, companies, and government agencies all over the world [1].

ICPMS is typically made of six compartments: sample introduction, ICP, ion extraction, ion optics, mass analyzer, and detector (Figure 14.2). Liquid samples are usually introduced into the sample introduction system in which the nebulizer/spray chamber and high-pressure gas (Ar) convert liquid into fine aerosol. The sample is then carried through the ICP torch and enters the center of the plasma, about 6000−8000 K temperature. The liquid sample is instantly dried, vaporized, atomized, and finally ionized in the plasma. The ions are extracted by sampler and skimmer cones, and accelerated and focused by the ion optics. If a collision/reaction cell (usually octopole or hexapole between ion optics and mass analyzer) is installed, the collision/reaction gas can be introduced in the cell to minimize or eliminate the polyatomic interference ions by collision, reaction, or energy discrimination. The mass analyzer and detector collect the ions of different m/z values in rapid sequence (essentially simultaneously). There are four types of commercially available mass analyzers for ICPMS: quadrupole, time of flight, double-focusing sector field, and multicollector, which are nicely reviewed by Becker [1]. For the detector, an electron multiplier is usually used for quadrupole and time-of-flight mass analyzers.

14.2.2. Advantages and Limitations of ICPMS Detection

As a "hard" ion source, ICP breaks down most molecules to atoms or atomic ions, making ICPMS a robust elemental detection technique for complex matrices. The atoms of the same element share the same ionization energy and the difference of their ionization efficiency is negligible; hence, ICPMS is also an exceptional tool for elemental quantification. On the other hand, the ionization efficiency of molecular mass spectrometry heavily depends on the specific structure of the analytes and the matrix. A slight modification of the structure, for example, replacing a hydroxyl group with an ether group, can cause substantial alteration of ionization potential in ESI-MS, even if the elemental composition remains unchanged. The differences in ionization efficiency make

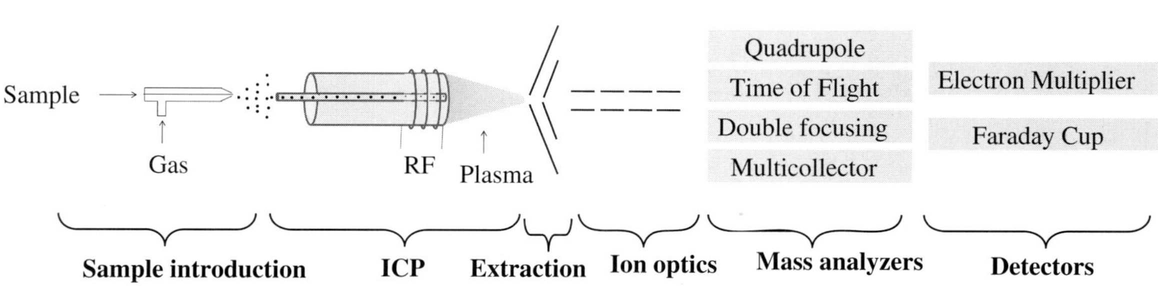

FIGURE 14.2 A simplified schematic of ICPMS.

quantification problematic although it is done. ICPMS has been proposed as a complementary tool for molecular mass spectrometry when both identification and quantification are necessary and is known as a metallomics approach. This has been done by introducing an ICPMS-sensitive element such as Eu through derivatization, or simply using a "natural" elemental label or tag, such as P.

Another advantage of ICPMS is the accurate measurement of the abundance of a single isotope, which cannot be achieved by atomic emission. Due to this capability GC-ICPMS with postcolumn isotope dilution can be used for species-unspecific quantification when the standard is not available. An isotope-enriched elemental standard is introduced to ICPMS as a spike to the eluent coming from a GC column [14,15]. The concentration and isotopic abundance of the element are usually known and the sample's concentration depends on the ratio of the two isotopes used for calculation. Post-column isotope dilution has also been used widely for species-specific quantification, in which the isotope-enriched specific compound was spiked. Isotope dilution is considered as the most accurate method of quantification because the analytes and spikes share the same matrix and interferences.

One of the major limitations of ICPMS is that it is not capable of molecular structure characterization, because molecular structures are destroyed in plasma due to the harsh ionization conditions. In order to obtain structural information, molecular mass spectrometry as a complementary technique for ICPMS is more and more often used in elemental speciation and now has become associated with "metallomics studies." Recently, scientists have started thinking of a versatile mass spectrometry that is capable of both atomic and molecular mass spectrometry. The first mass spectrometry of "dual-source," ICP and ESI, was designed, constructed, and characterized by the Hieftje group [16,17]. Another strategy to achieve both atomic and molecular detection was initially developed by the Sanz-Medel group by utilizing a versatile ionization source, microsecond-pulsed direct current GD [9,18]. Both these two designs were able to provide element-specific detection and molecular structural identification simultaneously or close to simultaneously, suggesting that the study of elemental speciation in the absence of standards has a promising future.

14.2.3. Interfacing GC to ICPMS

An interface from GC to ICPMS needs to efficiently transport the effluent from the end of the GC column to the inlet of the ICP torch. There have been a few interface designs and they have been comprehensively reviewed by Lobinski et al. [19] and Caruso et al. [20]. The main challenge of these designs is to avoid condensation, which causes sensitivity loss or peak broadening. Two strategies have been commonly adopted in an interface design. One is isothermal heating of the transfer line to eliminate hot or cold spots and the other is introducing a flow of make-up carrier gas at the end of the GC column, which effectively dilutes the analyte and better transports it to the plasma. With the exceptional detection limits of ICPMS, this dilution is not a problem, for there is the great advantage with GC of gaseous sample introduction and matrix removal. Further, the plasma prefers a gaseous sample, thereby eliminating the energy it must provide for desolvation, vaporization, and atomization, resulting in far better analyte ionization. As shown in Figure 14.3, the current GC-ICPMS interfaces integrate both the strategies and offer very efficient analyte transfer at temperatures up to 325 °C (some commercial vendors indicate even higher temperatures are possible). Interfacing of MIP and GD to GC is similar to that of ICPMS – all use heated transfer lines to direct GC eluent to the plasma.

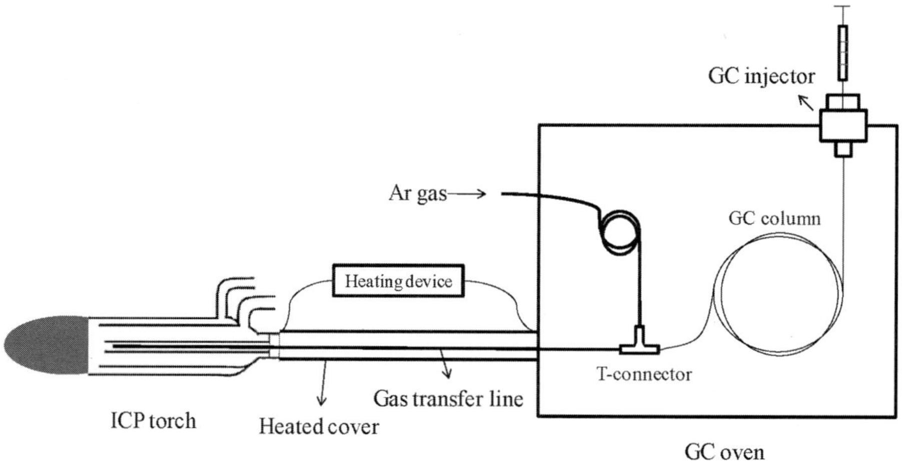

FIGURE 14.3 Schematic representation of the GC-ICPMS interface. The effluent from a GC column is mixed with preheated Ar gas at the T-connector, and the Ar gas carries the effluent through the heated transfer line till the center of plasma.

14.2.4. ICPMS as a Detector for GC

GC-ICPMS is a very effective hyphenated system for speciation analysis. Its extraordinary sensitivity of attogram (10^{-18}) to femtogram (10^{-15} g) levels of detection for Hg, Sn, Pb, Se, etc. is because of the superb sensitivity of both GC and ICPMS enhanced by avoiding condensed mobile phase. GC-ICPMS also has very good separation efficiency due to the high resolving power of GC capillary columns. Therefore, GC-ICPMS is the most sensitive speciation technique. However, GC-ICPMS usage is limited solely for volatile and thermally stable species, or for some compounds that can be converted to such by derivatization. Macromolecules or certain nonvolatile compounds are best suited to other techniques, such as HPLC-ICPMS or CE-ICPMS.

The extraordinary sensitivity of ICPMS is enhanced when it is used as a detector for GC. First, GC introduces relatively "clean" sample to the plasma compared to direct aspiration or LC. Second, the gas-phase compounds consume less energy from ICP with the benefit of higher ionization efficiency. Third, the sample is totally consumed with 100% loaded onto a GC column carried to the ICP, rather than the 2% with a standard nebulizer/spray chamber. GC-ICPMS has been established for about 30 elements, including the ones of most common interests in elemental speciation, such as Hg, Se, and As Table 14.1 lists the elements that have been studied by GC-ICPMS post-2001.

Isotope dilution analysis, an alternative to standard quantification with calibration curves, is used frequently in GC-ICPMS. About 24% of the total GC-ICPMS studies involve isotope dilution techniques, and the percentage is still increasing. Species-specific isotope dilution with GC-ICPMS has been markedly increasing since 2003 and accounts for more than half of the total isotope dilution studies. Species of Hg, Sn, Se, and S are accurately quantified with this technique from a variety of matrices, such as soil, water, petroleum, and biological tissues. Researchers also have started developing methods for multi-isotope GC-ICPMS detection using multiple spikes. Double spike and triple spike have been reported for simultaneous detection of Sn [21].

TABLE 14.1 The List of Elements Analyzed by GC-ICPMS Since 2001

Element	No. of publications	Species determined	Sample matrices	Detection limits (ng L^{-1} or ng kg^{-1})	Sample treatment/ Comment	Representative references
Sn	32	MBT, DBT, TBT, TPhT, and DPhT	Sediment and water	0.18–0.25 [21]	Isotope dilution with single or multiple spikes was well developed for organotin quantification.	[50,69]
Hg	25	Inorganic Hg, methylmercury, ethylmercury, and Hg-S compounds	Seafood, water, blood, plant, and sediment	4 [70]	Isotope dilution has been often used.	[64,71–73]
Se	14	DMSe, DMDSe, DEtSe, SeMet, SeEt, SeCys, and Se-S compounds	Urine, breath, plant, microbe, and cell	7–300 [29]	SPME has been often used for sampling. Selenoamino acids need to be derivatized [74].	[60,75]
S	8	Thiophene derivates, pesticides, and sulfur gases	Petroleum, water, breath, and landfill gas emission	800–3300 [76]		[77–79]
P	8	Pesticides, nerve agents, other organophosphates, and phosphine	Water, blood, flame retardant, and industrial gas	0.09–143 [46]	SPME or derivatization [36] is often needed.	[46,53,80]
Br	8	PBDE and brominated volatile organic compounds	Flame retardant, blood, soil, and water	15.2–40.5 [81]		[82,83]
As	6	MAs, TMAs, TPhAs, TEtAs, DMA, MMA, and TMAO	Soil, natural gas, and gas condensate	2–10 [84]	Hydride generation is needed for DMA, MMA, and TMAO [84].	[32,85]
Pb	4	Trimethyl-lead, $PbEt_4$	Standard solution	60 [86]		[86]
Sb	4	SbH_3, monomethyl-, dimethyl-, and trimethylantimony	Sediment	310 [87]	Hydride generation was used for sampling.	[88,89]
I	3	Iodinated phenols and iodinated volatile organic compounds	Water and air	0.07–0.12 [90]	SPME was used for sampling.	[83,90]

14.3. GC-MIP AND GC-GD

14.3.1. GC-MIP-AES

GC-MIP-AES has been a well-established technique since the 1990s [22]. The wide spread of GC-MIP-AES is due to its low cost as well as relatively high performance. The cost of an ICPMS is much higher than an MIP-AES. Although GC-MIP-AES is not as sensitive as GC-ICPMS, it offers lower detection limits than GC-ICP-AES, for nonmetals. The advantage of high sensitivity for nonmetals is because He is used as a plasma gas instead of Ar. However, GC as sample introduction is favored by MIP-AES because the low load produced from gaseous samples enhances plasma performance as noted years ago by the Cincinnati group [23]. Elemental speciation studies using GC-MIP-AES are summarized in Table 14.2.

14.3.2. GC-GD-MS and GC-GD-AES

The versatility and stability of GD make it a constantly attractive ionization source. The primary application of GD is for direct analysis of solid samples, both conductive and nonconductive [6]. However, GD has been used for gaseous analytes, and it has been combined with GC for elemental speciation since the 1990s [7,8]. The system construction cost is less for GC-GD-AES than for GC-MIP-AES, as well as lower consumption of plasma gas, but higher sensitivity by GC-GD-AES was reported for mercury [24]. However, the applications of GC-GD-AES are limited; only a few elements have been studied by this technique so far. GD-MS has been coupled to GC for elemental speciation studies as well [25].

As an online atomic detection for GC, either GD-AES or GD-MS plays a minor role compared to ICPMS and MIP-AES. In addition GC-GD-MS has been used for elemental speciation with both atomic and molecular detections which occur virtually simultaneously [9]. This unique feature gives GC-GD-MS possible future advantages over the other hyphenated techniques.

14.4. SAMPLE PREPARATION FOR GC-PLASMA SPECTROSCOPY

Gaseous analytes are often at ultratrace levels in the ambient environment. Although the detection limits of GC-ICPMS can go as low as 0.01 ng L^{-1}, a preconcentration step is sometimes required. For volatile or semivolatile species, the preconcentration is typically done with extraction or cryotrapping. There are many different types of extraction methods for GC-ICPMS, such as solid-phase microextraction (SPME) [26], stir bar sorptive extraction (SBME) [27], and single-drop liquid-phase microextraction (SDME) [28]. SPME has been the most popular one because of its advantages such as small sample size, low cost, solventless nature (although it can be used immersed in solvent), and full automation of the entire process. The in-needle fiber can be directly inserted into the GC injection port [29]. The applications of SPME in elemental speciation were comprehensively reviewed by Diez et al. [30]. In cryogenic trapping, another option for preconcentration in GC-ICPMS, the volatile analytes are trapped just by condensation at very low temperatures followed by instantaneous evaporation [31]. Cryotrapping using chromatographic packing followed by thermal desorption into GC-ICPMS was studied by Feldmann et al. [32–34].

Because most organometallic species lack volatility, a derivatization step is often mandatory prior to the extraction. With derivatization, the analytes are converted to volatile or semivolatile compounds in alkylation and hydride-forming reactions. It has been applied to elemental speciation for tin [35], phosphorus [36], mercury [37], lead [30], etc. Hydride generation has been employed to convert nonvolatile species of

TABLE 14.2 The List of Elements Analyzed by GC-MIP-AES

Element	No. of publications	Species determined	Sample matrices	Detection limits	Sample treatment/ comment	Representative references
F	2	Fluoroethers, 2-(Heptadecafluoroctyl) ethanol	Rat blood plasma	0.82 ng [91]		[91]
Br and Cl	4	$CH_xBr_yCl_z$	Seawater, motor car exhaust gas	0.01–0.12 ng [47]		[47,92]
I	2	CH_xI_y and iodopropane	Seawater	11–257 pg for molecules [47]		[47,93]
Hg	13	MeHg, Me_2Hg, EtHg, PhHg, MePhHg, Ph_2Hg, Hg^{2+}	Seafood, water, and sediment	0.8–1.1 ng L^{-1} [94]		[94–99]
Pb	5	Me_xPb, Et_xPb	Water, sediment, and gasoline	43–83 pg [100]	Derivatization	[94,101–103]
Se	2	Me_2Se, Me_2Se_2, Et_2Se, and selenite	Plants	0.19–0.57 μg L^{-1}	SPME, derivatization (selenite)	[97,104]
Sn	7	MBT, DBT, TBT, MPhT, TPhT, and DPhT	Sediment, seawater, seafood, fresh water, and urine	0.4–0.6 ng L^{-1} [94]		[43,96,99,105–107]
Mn	1	Cyclopentadienyl-manganese tricarbonyl and (methylcyclopentadienyl) manganese tricarbonyl	Sediment and seawater	0.62–0.65 pg L^{-1} [108]		[108]
As	1	Arsenic-containing hydrocarbons	Seafood	0.05–0.13 ng μL^{-1} for molecules [109]	Parallel analysis with GC-ICPMS	[109]

Se, As, and Sb to gas-phase hydride followed by a cryotrapping step [38].

14.5. ADVANCES IN APPLICATIONS OF GC-PLASMA SPECTROSCOPY

GC-plasma spectroscopy is one of the most powerful tools for elemental speciation because of its unmatched advantages of high-resolution separation and sensitive detection. This hyphenated technique has been employed for numerous applications in a variety of fields. In this book chapter, the most recent and significant applications of GC-ICPMS are presented in the following two main categories, environmental and biological applications.

14.5.1. Environmental Applications

14.5.1.1. Water

Elemental speciation has been a critical field for environmental studies since many elements are of high environmental concern [39]. Water pollution is a major problem in local and global contexts. It has been suggested that water pollution is the leading worldwide cause of deaths and diseases and that it accounts for thousands of deaths daily [40]. In order to evaluate water safety, many studies of elemental speciation have been carried out for all types of water samples, including wastewater, drinking water, seawater, and groundwater.

Some species of interest in water are at an ultratrace level that often requires excellent sensitivity and detection capabilities for the speciation methods. The sensitivity can be increased by a preconcentration step like SPME or an enhanced detection like hydride generation (HG)-ICPMS. Due to the widespread use of tri-butyltin (TBT)-based paints on all vessel types since the 1970s, organotins are priority substances (Decision 2455/2001/EC) [41] that have to be taken into account, in evaluating water chemical pollution according to the European Water Framework Directive 2000/60/EC. A rapid and sensitive determination was achieved for mono-, di-, and tri-butyltin (MBT, DBT, and TBT) in seawater by GC-ICPMS with ^{119}Sn species-specific isotope dilution [42]. The detection limits are in the range of 0.04–0.27 ng L^{-1} for Sn. The ultralow detection limits for organotins were also achieved by GC-MIP-AES [43]. The method was applied to various organotin species in urine samples, and the detection limits were reported as 0.42–0.67 μg L^{-1} for tin.

Organotins are usually derivatized before they are extracted and analyzed by GC-ICPMS. Xiao et al. determined butyltin compounds in seawater with GC-ICPMS [44]. The organotins were derivatized by adding sodium tetraethylborate ($NaBEt_4$) and sodium tetrahydroborate ($NaBH_4$) and the products, volatile tin species, were extracted by single-drop microextraction (SDME). In SDME, a microdrop of solvent suspended at a syringe needle tip was exposed to the headspace of the sample vial, in which the derivatized butyltins were extracted into the microdrop. SDME is a simple, fast, inexpensive, and effective method of extraction, but its application in elemental speciation is rare. The LODs were 1.4 ng L^{-1} for MBT, 1.8 ng L^{-1} for DBT, and 0.8 ng L^{-1} for TBT, and relative standard deviations (RSDs) were in the range of 1.1–5.3% ($c = 1$ μg/L, $n = 3$).

Mercury contamination in water, sediment, and seafood raises many concerns for human health. GC-ICPMS and GC-MIP-AES are primary techniques for mercury speciation. Methylmercury is the primary species of interest. The sensitivity, accuracy, and precision of the analytical method for measuring methylmercury have been significantly improved by innovative methods such as isotope dilution. Jackson et al. reported a GC-ICPMS method offering excellent figures of merit for methylmercury quantification in Lake Champlain, VT, in USA [45]. The analysis utilized species-specific isotope dilution, purge-and-trap GC-ICPMS. The instrument detection limit was about 0.3 fmolar (0.06 pg L^{-1}), and the method detection limit was 15 fmolar (0.003 ng L^{-1}). The method was accurate even at low concentrations of 0.025 ng L^{-1}. This combination of precision, accuracy, and sensitivity allows for quantification of significant differences in methylmercury concentration between different locations and over time, within the same spot. GC-ICPMS is also a good choice for characterizing organophosphorus in water samples. A few studies were carried out to evaluate the water contamination caused by usage of organophosphorus pesticides. Fidalgo et al. developed an SPME-GC-ICPMS method to measure these in river water samples [46]. In this study, the experimental conditions for SPME of organophosphorus pesticides, such as type of fiber, extraction mode, extraction time, extraction temperature, salt additives, and pH, were optimized.

With the advantage of better excitation efficiency on nonmetals, GC-MIP-AES has been often used for speciation of F, Cl, Br, I, Se, etc. Halogenated compounds in seawater were well characterized by GC-MIP-AES [47]. Br and Cl were monitored for the volatile halogenated compounds such as CH_2BrCl and CH_2Br_2. The detection limits was 0.01–0.04 ng for bromine and 0.03–0.12 ng for chlorine.

14.5.1.2. *Soil*

Soil contamination is another big threat to the environment, food safety, and sustainable agriculture. The leachate of the contaminated area can cause cross-contamination of water within and underlying it, and the emission from the soil can cause cross-contamination of the surrounding atmosphere. Elemental speciation in soil is as important as that in water, but it is a more challenging task. The complexity of the matrix and variation from sample to sample in soil make the sample preparation for speciation a difficult step. It often requires a complex sequential extraction procedure to recover analytes from the soil sample. Microwave extraction was employed to increase the speed and efficiency of extraction [48]. GC-ICPMS and GC-MIP-AES are extensively used for elemental speciation in soil. The elements of interest include arsenic, mercury, tin, antimony, selenium, tellurium, chromium, iodine, etc.

Tin, mercury, arsenic, and antimony in soil are frequently speciated by GC-ICPMS. The use of tributyltin (TBT) as a marine antifouling agent has led to its near-global dispersal, and it is still prominent in coastal seawaters. The sediment is a good indicator for environmental changes in water, and thus TBT and its degradation products, di-butyltin (DBT) and mono-butyltin (MBT), in sediment are often analyzed by GC-ICPMS to evaluate TBT pollution in the water environment. Extracting organotins from sediments with high recovery and low degradation is not a simple task. Kumar et al. evaluated five extraction protocols and two derivatization methods [49]. The recovery rate varied significantly from one tin species to another. Microwave and ultrasonic radiation were introduced to enhance the extraction [50]. Different SPME procedures were also evaluated, and the best results were obtained by using the divinylbenzene/carboxen/polydimethylsiloxane 2 cm 50/30 μm fiber with the optimal extraction conditions as the following: $T = 30\,°C$, $t = 40$ min, and pH = 4 [51]. Isotope dilution has been well established for accurate GC-ICPMS analysis of organotins, including species-specific isotope dilution with a single spike solution of ^{118}Sn-enriched butyltins [52], as well as a triple spike solution containing each butyltin species enriched with a different tin isotope [30]. The results demonstrated that multiple spikes could provide information on possible extraction-derived rearrangement reactions when optimizing and selecting the always-critical solid–liquid extraction procedure of the species from a solid sample. Once it has been demonstrated that the selected extraction procedure provides quantitative extraction without promoting degradation of the species, less sophisticated single isotope spikes can be routinely applied for the determination of the amount of butyltin in a given matrix.

14.5.1.3. Air

Indoor air pollution and urban air quality have been listed as two of the world's worst pollution problems. Some volatile metal(loid) compounds are very toxic; hence, their species and concentration in air can be of high concern. The gas-phase analytes are collected by either absorbing stationary phases such as SPME fibers [31] or cryotrapping with liquid nitrogen [21]. A cryosampling system together with a low temperature GC-ICPMS method was developed for the determination of phosphine emission from a tobacco factory [53]. CO_2 was eliminated by a cartridge filled with NaOH during the analytical desorption step. Phosphine recovery was above 98%, and its concentration in ambient air was about 1 ng m^{-3}.

To enhance the sensitivity and accuracy of gas characterization, a GC-ICPMS method based on species-specific isotope dilution in combination with online derivatization was developed for the simultaneous determination of volatile organometallic compounds and reactive gaseous mercury-related species, such as Hg^0, $(CH_3)_2Hg$, CH_3HgX, and HgX_2 (X = counter ions), in gaseous samples [54]. Elevated concentrations of these mercury compounds have been found in, for example, emissions from contaminated sediments, landfills, and sewage treatment plants. In this work, a mixture of developing Hg species and isotope-enriched standards was derivatized by 1% sodium tetraethylborate solution, and the derivatives were collected with Tenax TA and Au–Pt tubes. After preheating the Tenax TA tube to 250 °C for 45 s, it was purged with 100 ml min^{-1} of He for 5 s, allowing efficient injection of the vapor enriched with Hg species from the Tenax TA tubes to the GC for separation. Determination of Hg^0 in natural samples based on collection on Au–Pt tubes in series with Tenax TA provided a detection limit of 0.8 ng m^{-3}, which is about two times lower than the reported atmospheric background concentrations of total gaseous mercury. The developed method was successfully applied for measuring gaseous mercury in the emission from sediment samples.

14.5.2. Biological Applications

14.5.2.1. Plants and Microbes

Plants and microbes have been very popular materials for analysis in the field of elemental speciation. A number of studies have been devoted to elucidating the biotransformation of various elements whose species are of great interest in toxicology, nutrition, biology, and environmental science [55]. Furthermore, plants and microbes are relatively easy to grow and the growth conditions can be readily manipulated. For instance, nutrient supplementation in the soil of plants is a common strategy for making certain element-enriched samples.

Selenium speciation in plants and yeast has been a popular topic since the 1990s when Se was suggested as a chemopreventive agent because of Se supplementation reducing cancer incidence in a clinical trial [56]. Although Se's anticancer effect is not totally understood, the study of Se biotransformation in plants and yeasts has been well established. Previous studies have shown that plants and microbes are able to make volatile Se species, selenoamino acids, selenopeptides, and some high-molecular-weight selenium from inorganic selenium supplementation [57]. GC-ICPMS has been a common technique for studying selenium transformation in plants and microbes. Usually, the samples are grown in Se-enriched media, allowing them to uptake selenium. The selenium species are extracted and then determined by GC or LC-ICPMS. B'Hymer et al. reviewed these studies [58]. Headspace-SPME GC-ICPMS was used for determining volatile selenium compounds (DMSe and DMDSe), whose vapors are emitted from plants such as *Brassica juncea* [59]. The same selenium volatiles were measured in microbes by a cryotrapping cryofocusing GC-ICPMS system (CT-CF-GC-ICPMS) [60]. For nonvolatile selenium metabolites in plants, ion-pairing RP is the dominant separation method and was shown to provide high-resolution separations as well as high sensitivity (LOQ 2–50 μg L^{-1}) when coupled to ICPMS [61]. Recently, structural characterization with tandem mass spectrometry (MS, MS^2, and MS^3) was achieved for selenium metabolites in plants: selenomethionine (SeMet) in soybean [62] and methylselenocysteine (MeSeCys) in kale [62]. The spectra obtained with ESI-ITMS offered a clear identification, which is essential for analysis of complicated samples like plant extracts.

14.5.2.2. *Body Fluids*

Speciation of trace elements in body fluids not only facilitates elucidation of their metabolic or detoxification pathways, but also helps in identifying biomarkers or certain proteins that are related to some diseases. For example, the species of Hg and As in blood cause much concern because of their toxicity. On the other hand, glutathione peroxidase (a selenium-containing protein) is important to the functioning of the immune system and to the inflammatory process, and its level in blood is related to some diseases.

GC-ICPMS has also been applied to blood samples for speciation of elements such as mercury and phosphorus. The blood samples from a group of mercury-exposed workers were analyzed for methylmercury using a combined GC–EI-MS/ICPMS system, allowing the sensitive detection of mercury (ICPMS) and simultaneous structure characterization by EI-MS [63]. Methylmercury was extracted and then derivatized with sodium tetra-n-ethylborate solution. The limit of quantification was 300 ng L^{-1} for ICPMS detection, lower than 500 ng L^{-1} for EI-MS. Phosphoric acid triesters in human plasma samples were determined by monitoring phosphorous with GC-ICPMS [64]. Alkyl and aryl phosphates were extracted and preconcentrated by direct immersion SPME with 65 μm poly(dimethylsiloxane/divinylbenzene) fiber. The detection limits from blood plasma were 17–240 ng L^{-1} for tripropyl phosphate, tributyl phosphate, tris(2-chloroethyl) phosphate, and triphenyl phosphate, whose molecular structures were elucidated by GC-TOF-MS.

14.5.2.3. *Urine*

Many metabolites of trace elements passed through body fluids, such as blood, are finally excreted by urination; hence, determining the elemental species present in urine can provide valuable metabolic information in terms of their toxicity or health benefits. LC-ICPMS has been commonly used for elemental speciation in urine samples.

As an essential element for human health, selenium has caught much attention. Elucidation of selenium metabolism is of high importance to understanding Se's health effects as well as identifying better Se supplementation compounds. Three stable-isotope labeled Se compounds, ^{76}Se-selenomethionine (SeMet), ^{77}Se-methylselenocysteine (MeSeCys), and ^{82}Se-methylseleninic acid (MSA), were orally administrated to rats, and their metabolites in urine and breath were determined by LC-ICPMS and GC-ICPMS [65]. By comparing the quantity of different isotopes in the metabolites, this study revealed that the three Se species produced similar amount of selenosugars but SeMet generated much less dimethylselenide (DMSe) in breath and trimethyl selenium (TMSe) in urine than the other two. Since DMSe and TMSe are the main metabolites of methylselenol, which is considered to be responsible (at least as a precursor) for the cancer chemopreventive effects of Se, the authors suggest urinary TMSe and exhaled DMSe as biomarkers for production of effective anticancer Se species. The authors were able to distinguish exogenous isotope-enriched Se from endogenous natural Se, and the quantitative speciation of Se was achieved by isotope pattern deconvolution.

14.5.2.4. *Food*

Many elements are known as essential nutrients while others are considered as toxicants. Regardless of whether an element is nutritional or toxic or both, the actual nutritional value or toxicity highly depends on its specific chemical form(s) because they vary a lot from one species to another. For example, As(III) and As(V) are carcinogens while arsenobetaine (AsB) is relatively nontoxic. Therefore, a proper analysis with elemental speciation is necessary to evaluate food safety or nutritional values. The speciation studies have been carried out with GC-ICPMS, GC-MIP-AES, and other hyphenated techniques.

Seafood provides a prime source of high-quality protein: 14–16% of the animal protein consumed worldwide; over one billion people rely on seafood as their primary source of animal protein. The bioaccumulation of arsenic and mercury in seafood has been a big concern in terms of food safety. Because the toxicity of arsenic and mercury is tightly associated to their particular chemical forms, their speciation in seafood has been a hot topic for decades. The common arsenic species found in seafood include arsenite (As(III)), arsenate As(V), monomethylarsonate (MMA), dimethylarsinate (DMA), arsenobetaine (AB), and arsenosugars. The primary mercury species include inorganic mercury and methylmercury. The bioaccessibility of the accumulated arsenic and mercury compounds was also evaluated by simulating the digestion in a gastrointestinal environment [66].

GC-ICPMS and GC-MIP-AES offer excellent sensitivity for mercury speciation in fish tissues. The main analytes of interests are the common toxic mercury compounds, methylmercury, and inorganic mercury. Either species-specific or species-unspecific isotope dilution is usually employed to obtain accurate quantification [67]. Yang et al. determined methylmercury in fish tissues using GC-ICPMS with species-specific isotope dilution, in which an in-house synthesized ^{198}Hg-enriched methylmercury was used as the spike [68]. The samples were digested with methanolic potassium hydroxide, derivatized in aqueous solution with sodium tetrapropylborate and headspace sampled with a polydimethylsiloxane-coated, SPME-fused silica fiber. The isotope dilution method was compared with a standard addition method and it showed nearly 4-fold precision enhancement over the other method.

14.6. CONCLUSIONS AND PERSPECTIVES

GC and plasma spectroscopy are both well-developed techniques for elemental speciation. They offer unmatchable detection limits, as low as ppt level for species of a variety of elements, as well as high-resolution separation. The analytes vary from volatile gases to semivolatile compounds in a wide variety of matrices, such as water, soil, plants, and animals. This chapter has demonstrated the rapid growth of the GC-ICPMS hyphenated technique in the science community. This advancement includes better sensitivity, new sample preparation procedures, as well as innovative detection methods. Although GC-ICPMS has become a powerful technique for elemental speciation, the new challenges rising from novel applications often require complementary techniques such as molecular mass spectrometry. GC with both detections of ICPMS and molecular MS, either offline or parallel, has been increasingly popular in modern elemental speciation and is dubbed by some as "a metallomics approach."

References

[1] J.S. Becker (Ed.), Inorganic mass spectrometry: principles and applications, John Wiley & Sons, Ltd, Chichester, UK, 2007.

[2] K. Iwasaki, H. Uchida, K. Tanaka, Rapid determination of the alloying elements in titanium alloys by inductively-coupled plasma emission spectrometry, Anal. Chim. Acta. 135 (2) (1960) 369–372.

[3] M. Montes-Bayón, K. DeNicola, J.A. Caruso, Liquid chromatography-inductively coupled plasma mass spectrometry, J. Chromatogr. A 1000 (1–2) (2003) 457–476.
[4] K.L. Sutton, J.A. Caruso, Liquid chromatography-inductively coupled plasma mass spectrometry, J. Chromatogr. A 856 (1–2) (1999) 243–258.
[5] S.S. Kannamkumarath, et al., Capillary electrophoresis-inductively coupled plasma-mass spectrometry: an attractive complementary technique for elemental speciation analysis, J. Chromatogr. A 975 (2) (2002) 245–266.
[6] R.K. Marcus, J.A.C. Broekaert (Eds.), Glow discharge plasmas in analytical spectroscopy, John Wiley & Sons Inc., West Sussex, England, 2003.
[7] L.K. Olson, M. Belkin, J.A. Caruso, Radiofrequency glow discharge mass spectrometry for gas chromatographic detection: a new departure for elemental speciation studies, J. Anal. At. Spectrom. 11 (7) (1996) 491–496.
[8] N.G. Orellana-Velado, R. Pereiro, A. Sanz-Medel, Glow discharge atomic emission spectrometry as a detector in gas chromatography for mercury speciation, J. Anal. At. Spectrom. 13 (9) (1998) 905–909.
[9] A. Solà -Vázquez, et al., Tuneable microsecond-pulsed glow discharge design for the simultaneous acquisition of elemental and molecular chemical information using a time-of-flight mass spectrometer, Anal. Chem. 81 (7) (2009) 2591–2599.
[10] A.L. Gray, A.R. Date, Plasma source mass spectrometry of inorganic samples - recent developments of the technique, Int. J. Mass Spectrom. Ion. Phys. 46 (1983) 7–10.
[11] S.F. Boulyga, J.S. Becker, Improvement of abundance sensitivity in a quadrupole-based ICP-MS instrument with a hexapole collision cell, J. Anal. At. Spectrom. 17 (9) (2002) 1202–1206.
[12] S.D. Tanner, V.I. Baranov, D.R. Bandura, Reaction cells and collision cells for ICP-MS: a tutorial review, Spectrochim Acta Part B At. Spectrosc. 57 (9) (2002) 1361–1452.
[13] B. Hattendorf, C. Latkoczy, D. Günther, Laser ablation-ICPMS, Anal. Chem. 75 (15) (2003) 341A–347A.
[14] J.R. Encinar, J.I. García Alonso, A. Sanz-Medel, Synthesis and application of isotopically labelled dibutyltin for isotope dilution analysis using gas chromatography-ICP-MS, J. Anal. At. Spectrom. 15 (9) (2000) 1233–1240.
[15] J.P. Snell, et al., Species specific isotope dilution calibration for determination of mercury species by gas chromatography coupled to inductively coupled plasma- or furnace atomisation plasma ionisation-mass spectrometry, J. Anal. At. Spectrom. 15 (12) (2000) 1540–1545.
[16] D.A. Rogers, S.J. Ray, G.M. Hieftje, An electrospray/inductively coupled plasma dual-source time-of-flight mass spectrometer for rapid metallomic and speciation analysis: instrument design, Metallomics 1 (1) (2009) 67–77.
[17] D.A. Rogers, S.J. Ray, G.M. Hieftje, An electrospray/inductively coupled plasma dual-source time-of-flight mass spectrometer for rapid metallomic and speciation analysis. Part 1. Molecular channel characterization, Metallomics 2 (2010) 271–279.
[18] A. Sol-Vzquez, et al., Gas chromatography coupled to tunable pulsed glow discharge time-of-flight mass spectrometry for environmental analysis, Analyst 135 (5) (2010) 987–993.
[19] B. Bouyssiere, J. Szpunar, R. Lobinski, Gas chromatography with inductively coupled plasma mass spectrometric detection in speciation analysis, Spectrochim Acta Part B At. Spectrosc. 57 (5) (2002) 805–828.
[20] J.C.A. Wuilloud, et al., Gas chromatography/plasma spectrometry - an important analytical tool for elemental speciation studies, Spectrochim Acta Part B At. Spectrosc. 59 (6) (2004) 755–792.
[21] G. Centineo, et al., Isotope dilution GC-MS routine method for the determination of butyltin compounds in water, Anal. Bioanal. Chem. 384 (4) (2006) 908–914.
[22] R. Lobinski, F.C. Adams, Recent advances in speciation analysis by capillary gas chromatography-microwave induced plasma atomic emission spectrometry, Trends Analyt. Chem. 12 (2) (1993) 41–49.
[23] W. Robbins, J. Caruso, F. Fricke, Determination of Ge, As, Se, Sn, and Sb in complex samples by hydride generation microwave induced plasma atomic emission spectrometry, Analyst 104 (1979) 3535.
[24] N.G. Orellana Velado, R. Pereiro, A. Sanz-Medel, Mercury speciation by capillary gas chromatography with radiofrequency hollow cathode glow discharge atomic emission detection, J. Anal. At. Spectrom. 15 (1) (2000) 49–53.
[25] M.A. Belkin, L.K. Olson, J.A. Caruso, Radiofrequency glow discharge as an ion source for gas chromatography with mass spectrometric detection, J. Anal. At. Spectrom. 12 (11) (1997) 1255–1261.
[26] S. Risticevic, et al., Recent developments in solid-phase microextraction, Anal. Bioanal. Chem. 393 (3) (2009) 781–795.
[27] F. Sánchez-Rojas, C. Bosch-Ojeda, J.M. Cano-Pavón, A review of stir bar sorptive extraction, Chromatographia 69 (Suppl. 1) (2009).

[28] Y. Wang, J. McCaffrey, D.L. Norwood, Recent advances in headspace gas chromatography, J. Liq. Chromatogr. Relat. Technol. 31 (11–12) (2008) 1823–1851.
[29] J. Meija, et al., Monitoring volatile selenium and sulfur species from Se accumulating plants (wild type and genetically modified) by GC-MS and GC-ICP-MS using SPME for sample introduction, Anal. Chem. 74 (2002) 5837–5844.
[30] S. Díez, J.M. Bayona, Trace element determination by combining solid-phase microextraction hyphenated to elemental and molecular detection techniques, J. Chromatogr. Sci. 44 (7) (2006) 458–471.
[31] B. Kolb, Headspace sampling with capillary columns, J. Chromatogr. A 842 (1–2) (1999) 163–205.
[32] E.M. Krupp, et al., Investigation into the determination of trimethylarsine in natural gas and its partitioning into gas and condensate phases using (cryotrapping)/gas chromatography coupled to inductively coupled plasma mass spectrometry and liquid/solid sorption techniques, Spectrochim Acta Part B At. Spectrosc. 62 (9) (2007) 970–977.
[33] D. Kremer, G. Ilgen, J. Feldmann, GC-ICP-MS determination of dimethylselenide in human breath after ingestion of ^{77}Se-enriched selenite: monitoring of in-vivo methylation of selenium, Anal. Bioanal. Chem. 383 (3) (2005) 509–515.
[34] S. Wehmeier, A. Raab, J. Feldmann, Investigations into the role of methylcobalamin and glutathione for the methylation of antimony using isotopically enriched antimony(V), Appl. Organomet. Chem. 18 (12) (2004) 631–639.
[35] R. Morabito, P. Massanisso, P. Quevauviller, Derivatization methods for the determination of organotin compounds in environmental samples, TrAC - Trends Anal. Chem. 19 (2–3) (2000) 113–119.
[36] D.D. Richardson, J.A. Caruso, Derivatization of organophosphorus nerve agent degradation products for gas chromatography with ICPMS and TOF-MS detection, Anal. Bioanal. Chem. 388 (4) (2007) 809–823.
[37] M. Dzurko, D. Foucher, H. Hintelmann, Determination of compound-specific Hg isotope ratios from transient signals using gas chromatography coupled to multicollector inductively coupled plasma mass spectrometry (MC-ICP/MS), Anal. Bioanal. Chem. 393 (1) (2009) 345–355.
[38] P. Pohl, Hydride generation - recent advances in atomic emission spectrometry, TrAC - Trends Anal. Chem. 23 (2) (2004) 87–101.
[39] M. Popp, S. Hann, G. Koellensperger, Environmental application of elemental speciation analysis based on liquid or gas chromatography hyphenated to inductively coupled plasma mass spectrometry-a review, Anal. Chim. Acta 668 (2) (2010) 114–129.
[40] UN, The 2nd United Nations world water development report: water, a shared responsibility, 2006.
[41] T.F.M. Etty, et al., Yearbook of European environmental law, vol. 5, Oxford University Press, 2005.
[42] P. Rodríguez-González, et al., Determination of butyltin compounds in coastal sea-water samples using isotope dilution GC-ICP-MS, J. Anal. At. Spectrom. 17 (8) (2002) 824–830.
[43] G.A. Zachariadis, E. Rosenberg, Speciation of organotin compounds in urine by GC-MIP-AED and GC-MS after ethylation and liquid-liquid extraction, J. Chromatogr. B Analyt. Technol. Biomed Life Sci. 877 (11–12) (2009) 1140–1144.
[44] Q. Xiao, B. Hu, M. He, Speciation of butyltin compounds in environmental and biological samples using headspace single drop microextraction coupled with gas chromatography-inductively coupled plasma mass spectrometry, J. Chromatogr. A 1211 (1–2) (2008) 135–141.
[45] B. Jackson, et al., Low-level mercury speciation in freshwaters by isotope dilution GC-ICP-MS, Environ. Sci. Technol. 43 (7) (2009) 2463–2469.
[46] N. Fidalgo-Used, et al., Determination of organophosphorus pesticides in spiked river water samples using solid phase microextraction coupled to gas chromatography with EI-MS and ICP-MS detection, J. Anal. At Spectrom. 20 (9) (2005) 876–882.
[47] S. Slaets, F. Laturnus, F.C. Adams, Microwave induced plasma atomic emission spectrometry: a suitable detection system for the determination of volatile halocarbons, Fresenius J. Anal. Chem. 364 (1–2) (1999) 133–140.
[48] G.M.M. Rahman, H.M. Kingston, Development of a microwave-assisted extraction method and isotopic validation of mercury species in soils and sediments, J. Anal. At. Spectrom. 20 (3) (2005) 183–191.
[49] S.J. Kumar, et al., A simple method for synthesis of organotin species to investigate extraction procedures in sediments by isotope dilution-gas chromatography-inductively coupled plasma mass spectrometry Part 2. Phenyltin species, J. Analyt. At. Spectrom. 19 (3) (2004) 368–372.
[50] K. Inagaki, et al., Certification of butyltins and phenyltins in marine sediment certified reference material by species-specific isotope-dilution mass spectrometric analysis using synthesized ^{118}Sn-enriched organotin compounds, Anal. Bioanal. Chem. 387 (7) (2007) 2325–2334.
[51] F. Bianchi, et al., Optimization of the solid phase microextraction procedure for the ultra-trace determination of organotin compounds by gas chromatography- inductively coupled plasma-mass spectrometry, J. Anal. At. Spectrom. 21 (9) (2006) 970–973.

[52] K. Inagaki, et al., Species-specific isotope dilution analysis of mono-, di, and tri-butyltin compounds in sediment using gas chromatography-inductively coupled plasma mass spectrometry with synthesized ^{118}Sn-enriched butyltins, Analyst 128 (3) (2003) 265–272.

[53] M.P. Pavageau, et al., Phosphine emission measurements from a tobacco factory using cryogenic sampling and GC-ICP-MS analysis, J. Analy. At. Spectrom. 18 (4) (2003) 323–329.

[54] T. Larsson, E. Bjorn, W. Frech, Species specific isotope dilution with on line derivatisation for determination of gaseous mercury species, J. Anal. At. Spectrom. 20 (11) (2005) 1232–1239.

[55] S. Husted, et al., Review: the role of atomic spectrometry in plant science, J. Anal. At. Spectrom. 26 (1) (2011) 52–79.

[56] L.C. Clark, et al., Effects of selenium supplementation for cancer prevention in patients with carcinoma of the skin: a randomized controlled trial, J. Am. Med. Assoc. 276 (24) (1996) 1957–1963.

[57] N. Terry, et al., Selenium in higher plants, Annu. Rev. Plant Physiol. Plant Mol. Biol. 51 (2000) 401–432.

[58] C. B'Hymer, J.A. Caruso, Selenium speciation analysis using inductively coupled plasma-mass spectrometry, J. Chromatogr. A 1114 (1) (2006) 1–20.

[59] K.M. Kubachka, et al., Selenium volatiles as proxy to the metabolic pathways of selenium in genetically modified *Brassica juncea*, Environ. Sci. Technol. 41 (6) (2007) 1863–1869.

[60] M. Peitzsch, D. Kremer, M. Kersten, Microfungal alkylation and volatilization of selenium adsorbed by goethite, Environ. Sci. Technol. 44 (1) (2010) 129–135.

[61] M. Kotrebai, et al., Selenium speciation in enriched and natural samples by HPLC-ICP-MS and HPLC-ESI-MS with perfluorinated carboxylic acid ion-pairing agents, Analyst 125 (1) (2000) 71–78.

[62] Q. Chan, S.E. Afton, J.A. Caruso, Selenium speciation profiles in selenite-enriched soybean (Glycine max) by HPLC-ICPMS and ESI-ITMS, Metallomics 2 (2) (2010) 147–153.

[63] J. Hippler, et al., Comparative determination of methyl mercury in whole blood samples using GC-ICP-MS and GC-MS techniques, J Chromatogr B Analyt. Technol. Biomed. Life Sci. 877 (24) (2009) 2465–2470.

[64] S. Mounicou, et al., Localization and speciation of selenium and mercury in Brassica juncea - implications for Se-Hg antagonism, J. Anal. At. Spectrom. 21 (4) (2006) 404–412.

[65] Y. Ohta, et al., Speciation analysis of selenium metabolites in urine and breath by HPLC- and GC-inductively coupled plasma-MS after administration of selenomethionine and methylselenocysteine to rats, Chem. Res. Toxicol. 22 (11) (2009) 1795–1801.

[66] A.I. Cabañero, Y. Madrid, C. Cámara, Selenium and mercury bioaccessibility in fish samples: an in vitro digestion method, Anal. Chim. Acta. 526 (1) (2004) 51–61.

[67] J.I. García Alonso, et al., Determination of butyltin compounds in environmental samples by isotope-dilution GC-ICP-MS, Anal. Bioanal. Chem. 373 (6) (2002) 432–440.

[68] L. Yang, Z. Mester, R.E. Sturgeon, Determination of methylmercury in fish tissues by isotope dilution SPME-GC-ICP-MS, J. Anal. At. Spectrom. 18 (5) (2003) 431–436.

[69] P. Rodrigucz-Gonzalcz, J.I.G. Alonso, A. Sanz-Medel, Development of a triple spike methodology for validation of butyltin compounds speciation analysis by isotope dilution mass spectrometry. Part 2. Study of different extraction procedures for the determination of butyltin compounds in mussel tissue CRM 477, J. Anal. At. Spectrom. 19 (6) (2004) 767–772.

[70] L. Lambertsson, E. Bjorn, Validation of a simplified field-adapted procedure for routine determinations of methyl mercury at trace levels in natural water samples using species-specific isotope dilution mass spectrometry, Anal. Bioanal. Chem. 380 (7–8) (2004) 871–875.

[71] R.C. Rodriguez Martin-Doimeadios, et al., Using speciated isotope dilution with GC-inductively coupled plasma MS to determine and unravel the artificial formation of monomethylmercury in certified reference sediments, Anal. Chem. 75 (13) (2003) 3202–3211.

[72] D. Point, et al., Simultaneous determination of inorganic mercury, methylmercury, and total mercury concentrations in cryogenic fresh-frozen and freeze-dried biological reference materials, Anal. Bioanal. Chem. 389 (3) (2007) 787–798.

[73] T. Karlsson, U. Skyllberg, Bonding of ppb levels of methyl mercury to reduced sulfur groups in soil organic matter, Environ. Sci. Technol. 37 (21) (2003) 4912–4918.

[74] M.V. Pelaez, et al., Comparison of different derivatization approaches for the determination of selenomethionine by GC-ICP-MS, J. Anal. At. Spectrom. 15 (9) (2000) 1217–1222.

[75] H.G. Infante, et al., Investigation of the selenium species distribution in a human B-cell lymphoma line by HPLC- and GC-ICP-MS in combination with HPLC-ESIMS/MS and GC-TOFMS after incubation with methylseleninic acid, J. Anal. At. Spectrom. 22 (8) (2007) 888–896.

[76] J. Rodriguez-Fernandez, et al., Gas chromatography double focusing sector-field ICP-MS as an innovative tool for bad breath research, J. Anal. At. Spectrom. 16 (9) (2001) 1051–1056.

[77] J. Heilmann, K.G. Heumann, Development of a species-specific isotope dilution GC-ICP-MS method for the determination of thiophene derivates in petroleum products, Anal. Bioanal. Chem. 390 (2) (2008) 643–653.

[78] D. Profrock, et al., Sensitive, simultaneous determination of P, S, Cl, Br and I containing pesticides in environmental samples by GC hyphenated with collision-cell ICP-MS, J. Anal. At. Spectrom. 19 (5) (2004) 623–631.

[79] S. Junyapoon, et al., Analysis of malodorous sulfur gases and volatile organometalloid compounds in landfill gas emissions using capillary gas chromatography with programmed temperature vaporization injection and atomic emission detection, Int. J. Environ. Anal. Chem. 82 (2) (2002) 47–59.

[80] M. Garcia-Lopez, et al., Determination of organophosphate flame retardants and plasticizers in sediment samples using microwave-assisted extraction and gas chromatography with inductively coupled plasma mass spectrometry, Talanta 79 (3) (2009) 824–829.

[81] Q. Xiao, et al., Analysis of PBDEs in soil, dust, spiked lake water, and human serum samples by hollow fiber-liquid phase microextraction combined with GC-ICP-MS, J. Am. Soc. Mass Spectrom. 18 (10) (2007) 1740–1748.

[82] R.F. Swarthout Jr., J.R. Kucklick, W.C. Davis, The determination of polybrominated diphenyl ether congeners by gas chromatography inductively coupled plasma mass spectrometry, J. Anal. At. Spectrom. 23 (12) (2008) 1575–1580.

[83] A. Schwarz, K.G. Heumann, Two-dimensional on-line detection of brominated and iodinated volatile organic compounds by ECD and ICP-MS after GC separation, Anal. Bioanal. Chem. 374 (2) (2002) 212–219.

[84] T. Guerin, et al., Arsenic speciation in some environmental samples: a comparative study of HG-GC-QFAAS and HPLC-ICP-MS methods, Appl. Organomet. Chem. 14 (8) (2000) 401–410.

[85] B. Bouyssiere, et al., Investigation of speciation of arsenic in gas condensates by capillary gas chromatography with ICP-MS detection, J. Anal. At. Spectrom. 16 (11) (2001) 1329–1332.

[86] N. Poperechna, K.G. Heumann, Simultaneous multi-species determination of trimethyllead, monomethylmercury and three butyltin compounds by species-specific isotope dilution GC-ICP-MS in biological samples, Anal. Bioanal. Chem. 383 (2) (2005) 153–159.

[87] Z. Mester, R.E. Sturgeon, J.W. Lam, Sampling and determination of metal hydrides by solid phase microextraction thermal desorption inductively coupled plasma mass spectrometry, J. Anal. At. Spectrom. 15 (11) (2000) 1461–1465.

[88] R. Miravet, et al., Speciation of antimony in environmental matrices by coupled techniques, TrAC - Trends Anal. Chem. 29 (1) (2010) 28–39.

[89] S. Wehmeier, J. Feldmann, Investigation into antimony mobility in sewage sludge fermentation, J. Environ. Monit. 7 (12) (2005) 1194–1199.

[90] R.G. Wuilloud, et al., Determination of iodinated phenol species at parts-per-trillion concentration levels in different water samples by solid-phase microextraction/offline GC-ICP-MS, J. Anal. At. Spectrom. 18 (9) (2003) 1119–1124.

[91] D.F. Hagen, J. Belisle, J.S. Marhevka, Capillary GC/helium microwave emission detector characterization of fluorine containing metabolites in blood plasma, Spectrochim Acta Part B At. Spectrosc. 38 (1–2) (1983) 377–385.

[92] H. Baumann, K.G. Heumann, Analysis of organobromine compounds and HBr in motor car exhaust gases with a GC/microwave plasma system, Fresenius Z. Anal. Chem. 327 (2) (1987) 186–192.

[93] J.P.J. van Dalen, P.A. de Lezenne Coulander, L. de Galan, Optimization of the microwave-induced plasma as an element-selective detector for non-metals, Anal. Chim. Acta 94 (1) (1977) 1–19.

[94] J. Carpinteiro Botana, et al., Fast and simultaneous determination of tin and mercury species using SPME, multicapillary gas chromatography and MIP-AES detection, J. Anal. At. Spectrom. 17 (8) (2002) 904–907.

[95] W. Frech, J.P. Snell, R.E. Sturgeon, Performance comparison between furnace atomisation plasma emission spectrometry and microwave induced plasma-atomic emission spectrometry for the determination of mercury species in gas chromatography effluents, J. Anal. At. Spectrom. 13 (12) (1998) 1347–1353.

[96] S. Tutschku, M.M. Schantz, S.A. Wise, Determination of methylmercury and butyltin compounds in marine biota and sediments using microwave-assisted acid extraction, solid-phase microextraction, and gas chromatography with microwave-induced plasma atomic emission spectrometric detection, Anal. Chem. 74 (18) (2002) 4694–4701.

[97] C. Dietz, et al., Volatile organo-selenium speciation in biological matter by solid phase microextraction-moderate temperature multicapillary gas chromatography with microwave induced plasma atomic emission spectrometry detection, Anal. Chim. Acta 501 (2) (2004) 157–167.

[98] A. Delgado, et al., Production of artifact methylmercury during the analysis of certified reference sediments: use of ionic exchange in the sample treatment step to minimise the problem, Anal. Chim. Acta 582 (1) (2007) 109–115.

[99] M. Zabaljauregui, et al., Fast method for routine simultaneous analysis of methylmercury and butyltins in seafood, J. Chromatogr. A 1148 (1) (2007) 78–85.

[100] M. Heisterkamp, F.C. Adams, Simplified derivatization method for the speciation analysis of organolead compounds in water and peat samples using in-situ butylation with tetrabutylammonium tetrabutylborate and GC-MIP AES, Fresenius J. Anal. Chem. 362 (5) (1998) 489–493.

[101] I.R. Pereiro, R. Łobiński, Fast species-selective screening for organolead compounds in gasoline by multicapillary gas chromatography with microwave-induced plasma atomic emission detection, J. Anal. At. Spectrom. 12 (12) (1997) 1381–1385.

[102] M. Heisterkamp, F.C. Adams, In situ propylation using sodium tetrapropylborate as a fast and simplified sample preparation for the speciation analysis of organolead compounds using GC-MIP-AES, J. Anal. At. Spectrom. 14 (9) (1999) 1307–1311.

[103] J.R. Baena, et al., Comparison of three coupled gas chromatographic detectors (MS, MIP-AES, ICP-TOFMS) for organolead speciation analysis, Anal. Chem. 73 (16) (2001) 3927–3934.

[104] E. Dimitrakakis, et al., Solid-phase microextraction-capillary gas chromatography combined with microwave-induced plasma atomic-emission spectrometry for selenite determination, Anal. Bioanal. Chem. 379 (5–6) (2004) 842–848.

[105] S. Tutschku, S. Mothes, K. Dittrich, Determination and speciation of organotin compounds by gas chromatography - microwave induced plasma atomic emission spectrometry, J. Chromatogr. A 683 (1) (1994) 269–276.

[106] S. Girousi, et al., Speciation analysis of organotin compounds in Thermaikos Gulf by GC-MIP-AED, Fresenius J. Anal. Chem. 358 (7–8) (1997) 828–832.

[107] S. Aguerre, et al., Speciation of organotins in environmental samples by SPME-GC: comparison of four specific detectors: FPD, PFPD, MIP-AES and ICP-MS, J. Anal. At. Spectrom. 16 (3) (2001) 263–269.

[108] N. Campillo, R. Peñalver, M. Hernández-Córdoba, Solid-phase microextraction combined with gas chromatography and atomic emission detection for the determination of cyclopentadienylmanganese tricarbonyl and (methylcyclopentadienyl)manganese tricarbonyl in soils and seawaters, J. Chromatogr. A 1173 (1–2) (2007) 139–145.

[109] U. Arroyo-Abad, et al., Detection of arsenic-containing hydrocarbons in canned cod liver tissue, Talanta 82 (1) (2010) 38–43.

CHAPTER

15

Field and Portable Instruments

Stanley D. Stearns

OUTLINE

Gas Chromatography DOI: 10.1016/B978-0-12-385540-4.00015-8

15.1. HISTORY

Although space instrumentation is discussed in a separate chapter, the history of portable GCs cannot be discussed without referencing America's pioneering space programs. In 1963, W.F. Wilhite (with acknowledgments to V.I. Oyama of the Jet Propulsion Laboratory, Dr. S.R. Lipsky of Yale University, and Dr. J.E. Lovelock of Baylor University) reported the results of a prototype GC designed to be soft-landed on the surface of the moon as part of the Surveyor scientific payload [1].

Wilhite followed this up in 1966 with a subminiaturized GC weighing approximately 100 g, capable of separating gases in a few seconds with a carrier gas flow of 1 ml/min. This GC was designed to "analyze the atmosphere of Mars in a few seconds during the descent of a landing capsule" [2].

In 1969, Josias, Bowman, and Lovelock applied for a patent on a portable GC using an electron-capture detector [3]. At about the same time, Wilhite founded Aptech, incorporating JPL technologies in the first commercial micro-GC. The instrument (Figure 15.1) used micropacked columns with an early Valco 10 port valve and a TCD with a .0001" straight filament. After the company was sold to a larger firm more interested in older technology, the techniques and components employed in the Aptech GC led, in the early 1970s, to the production of a high-speed in vivo blood-gas analyzer that sampled microliter quantities of gas from an interarterial silicone catheter (Figures 15.2 and 15.3).

In 1972, a group at the Jet Propulsion Laboratory demonstrated a completely portable, self-contained GC that generated its own H_2 carrier gas from a water supply [4]. But not all of the ongoing research and development was related to the nation's space efforts. Valco Instruments delivered its first portable GC, the Model 1000 Halocarbon Monitor, in 1972 (Figure 15.4). The Model 1000 measured the concentration of

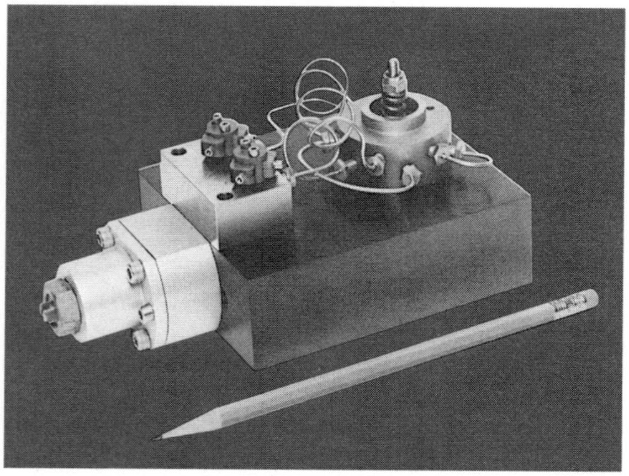

FIGURE 15.1 Aptech micro-gas-chromatograph. *(Courtesy of Frank Wilhite.)*

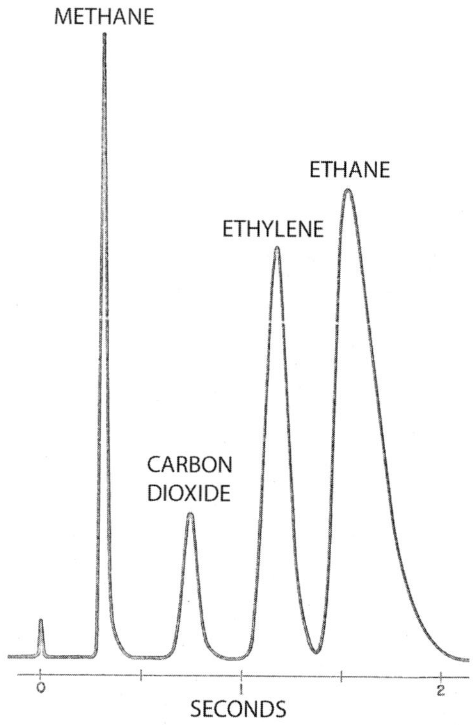

FIGURE 15.2 Aptech micro-gas-chromatograph analysis of light gases. Column: 3" × 0.023" ID, 40–56 μ Porapak S, ambient temperature; carrier: helium at 6.7 mL/min; sample size: 1 μL. *(Courtesy of Frank Wilhite.)*

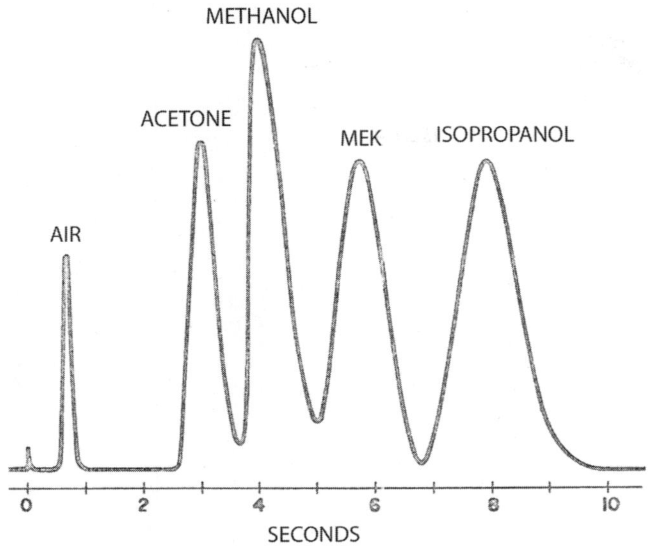

FIGURE 15.3 Aptech micro-gas-chromatograph analysis of oxygenates. Column: 5″ × 0.023″ ID, 36–75 μ Porasil 60, 75 °C; carrier: helium; sample size: 1 μL (gas); inlet pressure: 40 psi. *(Courtesy of Frank Wilhite.)*

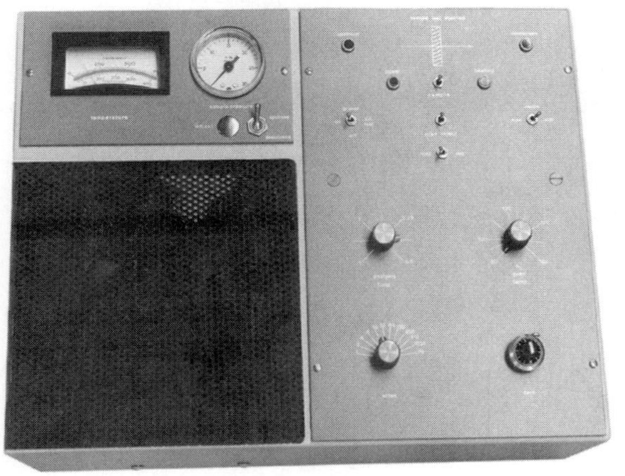

FIGURE 15.4 Valco Instruments Model 1000 Halocarbon Monitor for charcoal filters. *(Courtesy of Valco Instruments.)*

a chlorofluorocarbon (CFC) tracer gas introduced upstream and downstream of a charcoal filter bed in a nuclear reactor facility. A ppm level of the tracer challenged the filter with the downstream flow measured at the ppb level. The linear range of the H_2^3 source was extended by sample μl's upstream and cc's downstream. The same peak height could be found with a 50 ppm upstream concentration and 50 ppb downstream concentration.

A 1973 publication cites the Analytical Instrument Development Model 510, featuring an ECD with 5% methane in argon as the carrier gas. The Model 510 had rechargeable batteries sufficient to operate the electrometer and detector for more than 10 h. Line voltage was required for the Varian 7-port solenoid-actuated sampling valve, Honeywell recorder, and positive-displacement sampling valve [5].

Figure 15.5 shows Valco's 1975 prototype flyable GC for the detection of stratospheric CFCs and N_2O. The instrument was packaged in a purged enclosure with an absolute pressure regulator. The very low N_2O response with N_2 the carrier led to the conclusion that impurities in carrier gas and lab GC system leaks accounted for the N_2O response.

James Morgan patented a portable GC for transformer gas analysis in 1977 [6] (Figure 15.6). Polymer tubes isolated dissolved gases from transformer oil, which were then injected into a packed column using air as a carrier gas. H_2

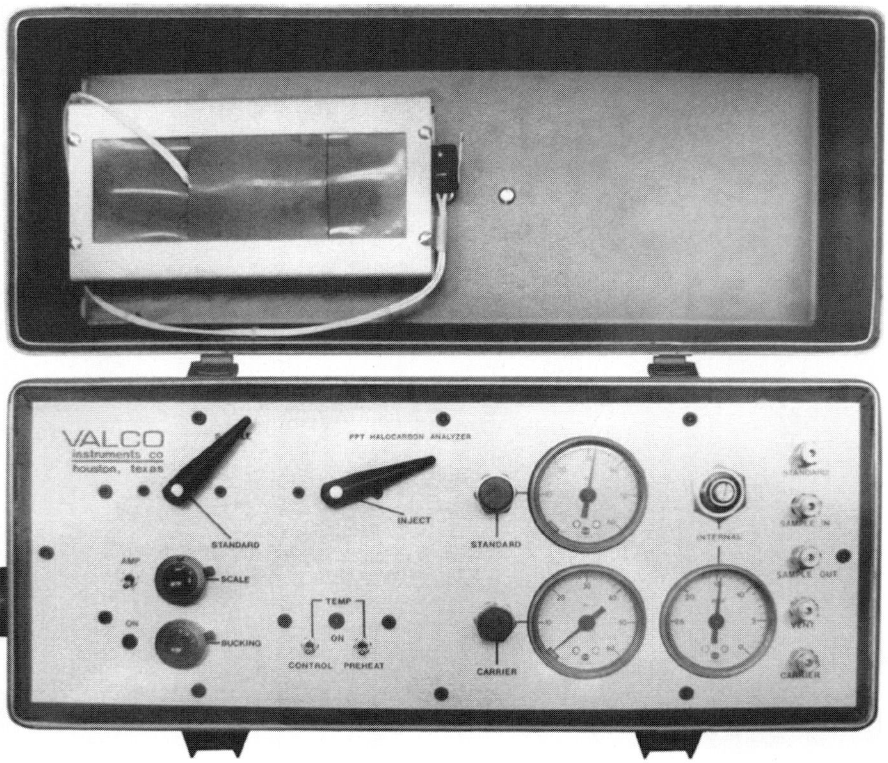

FIGURE 15.5 Valco Instruments prototype flyable CFC and N_2O analyzer, designed for NOAA. *(Courtesy of Valco Instruments.)*

was measured with a thermal conductivity detector (TCD). The first commercial version was manufactured for more than 25 years.

Also in the late 1970s, Photovac introduced the first portable GC with a photoionization detector (PID), followed by HNu, whose founder, Jack Driscoll, was the inventor of the lamp-type PID.

15.2. DESIGN CHALLENGES

One of the major limitations of portable gas chromatography is imposed by electrical requirements. To minimize power consumption (and battery size), overhead from control circuitry must be minimal, and heated columns and detectors must have the lowest possible mass.

The second major limitation is carrier gas. Ambient air can be used; however, maximum column temperature is reduced, and the choice of detectors is limited. There is also a tradeoff in power consumption, since power is required for air compression and purification. A bottled carrier gas such as He does not have these limitations, but again, there is a tradeoff: gas bottles and regulators take up a lot of space. Small-diameter columns and minimum-flow detectors will allow use of the smallest possible bottle or cartridge, while miniature pressure regulators (Figure 15.7) are available for the two-stage regulation required for constant flow as the bottle pressure drops. If a flame ionization

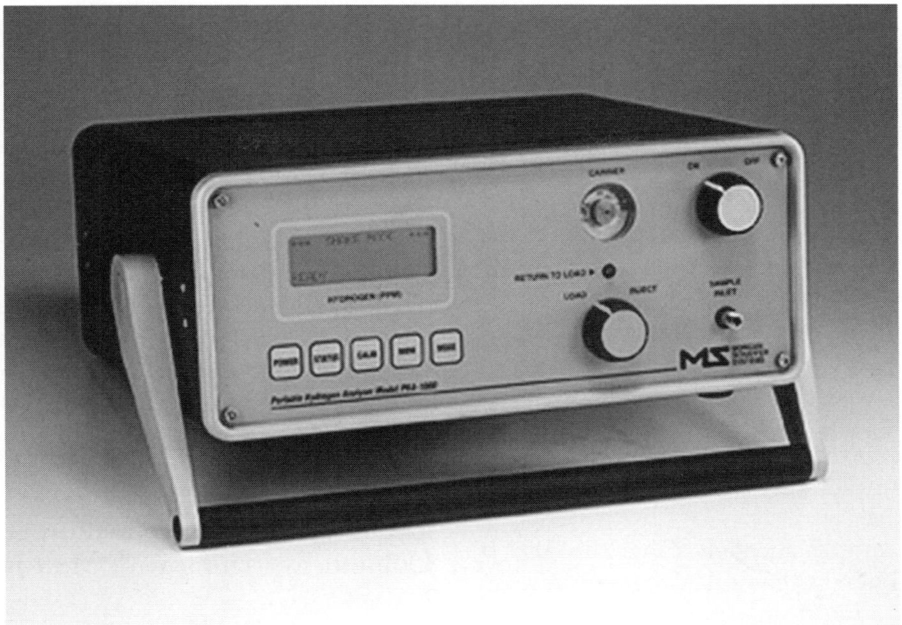

FIGURE 15.6 Morgan Shaffer Model PFGA-P200 Portable Fault Gas Analyzer. *(Courtesy of Morgan Shaffer.)*

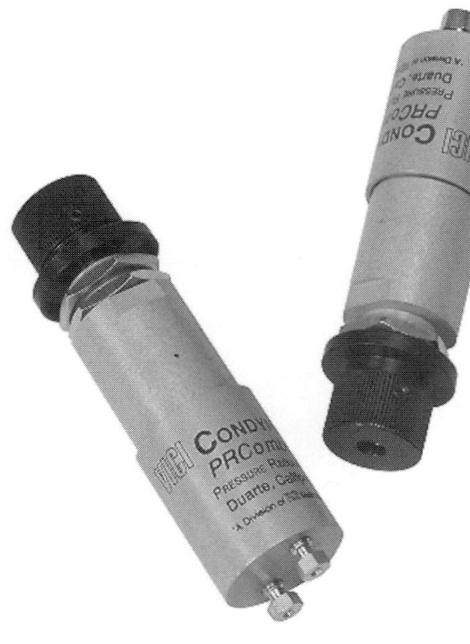

FIGURE 15.7 Miniature regulators. *(Courtesy of Valco Instruments.)*

detector is used, hydrogen is required in addition to air, although an FID has been reported that generates both required gases from the electrolysis of water [7].

The original "lab on a chip" concept [8] has evolved to include an array of on/off valves, separately heated columns, and an independent thermal conductivity detector. While small size and low power consumption have been achieved, separations are limited to low-molecular-weight compounds. In the last several years, C2 V (Thermo) has introduced a microscale GC with a column programmable to 180 °C.

15.3. SAMPLE INTRODUCTION

The simplest means of introducing a sample is with a syringe through a septum inlet. For gaseous samples, the inlet can be unheated, but liquid samples may require heat. Field

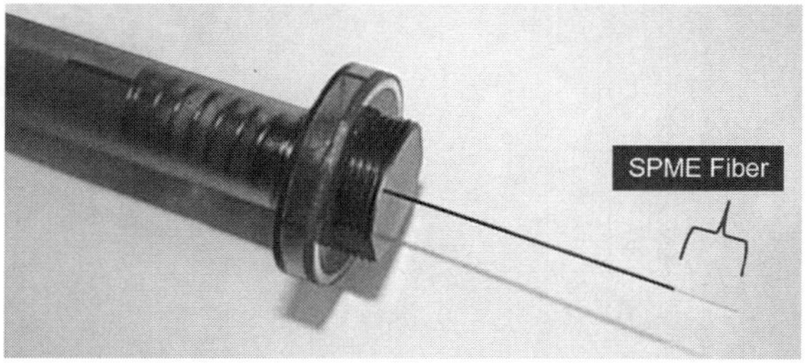

FIGURE 15.8 Custodian SPME syringe from Torion Technologies. *(Courtesy of Torion Technologies.)*

syringe sampling can be of a liquid or of a headspace gas from a sealed container. Purge and trap samplers are also widely used with portable GCs.

A basic sample injection valve can also be used as a liquid or gas inlet. Sample injection valves measure the sample with a slot engraved on the rotor (4-port internal sample injector) or with an external loop (6-port external loop sample injector). An 8- or 10-port design can perform useful column-switching functions as well as sample injection (see Section 15.4.1).

An alternative sample introduction method uses a silica fiber treated with a bonded phase like those used for coating capillary columns. The fiber is exposed to the air or liquid to be analyzed, then inserted into an inlet heated to desorb the retained compounds [9]. A current version of such a sampling device is the CUSTODION® SPME syringe manufactured by Torion Technologies (Figure 15.8). It is used as the preferred sample concentration/injection device for the Torion portable GC mass spectrometer.

15.3.1. Sample Introduction and Column Switching with Multiport Valves

If an isothermal column is used and the sample contains slowly eluting compounds, two columns in series may be attached to

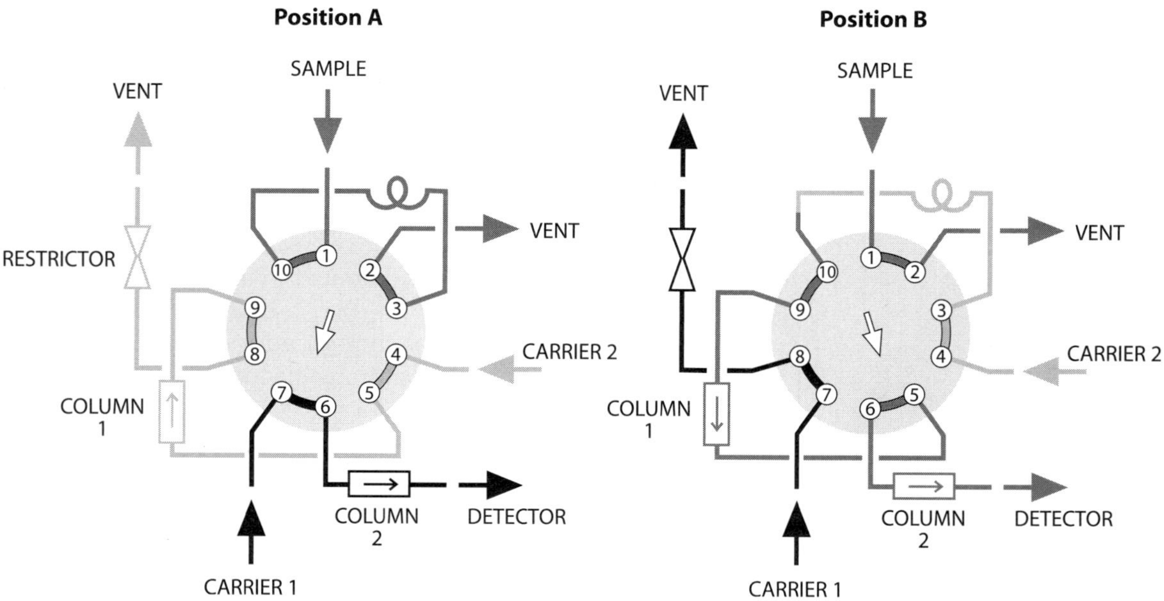

FIGURE 15.9 Loop sampling with back flush of precolumn to vent. *(Courtesy of Valco Instruments.)*

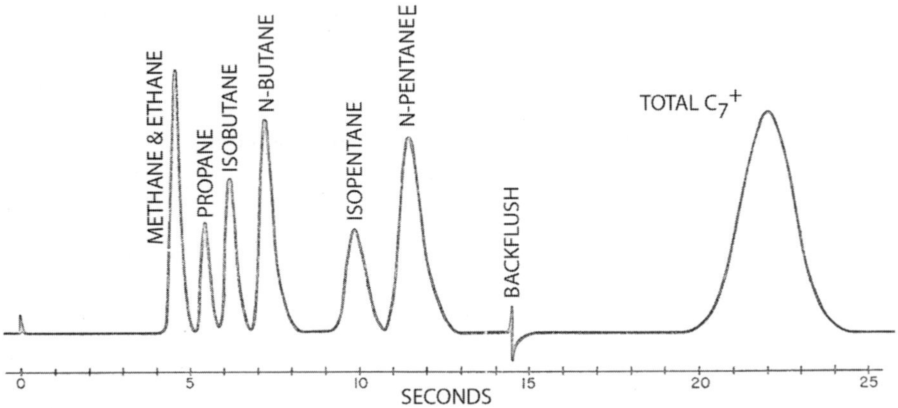

FIGURE 15.10 Natural gas analysis demonstrating back flush. Column: 30″ × 0.043″ ID, 14.4% bis(2-methoxyethyl) adipate on 100–120 mesh chromosorb P, AW, DMCS; column temperature: 44.6 °C; carrier gas: helium; inlet pressure: 40 psi; sample size: 3 μL (gas). *(Courtesy of Frank Wilhite.)*

the valve. After the last component of interest has eluted from the first column, returning the valve to the loading position backflushes the late eluting components into the detector or vent (Figures 15.9 and 15.10). Applications shown with multiport valves can also be done with individual on/off valves and tee fittings, but with greater system complexity.

The use of a trapping or concentrating inlet greatly lowers the minimum detectable level of compounds of interest. If the loop on a standard 6-port sample injector is replaced with a trap

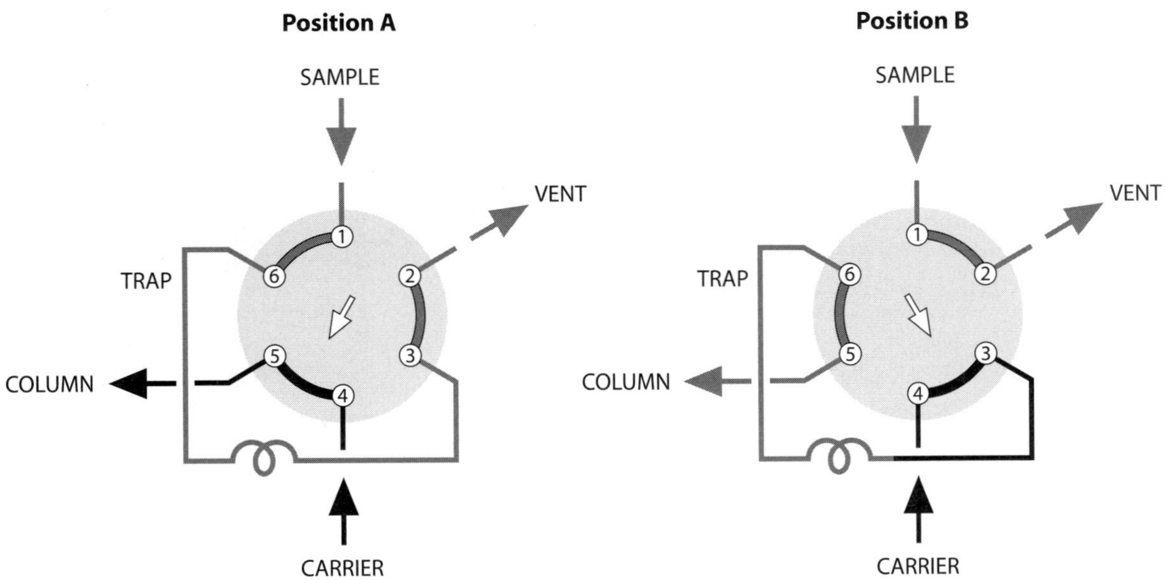

FIGURE 15.11 Six-port valve used to concentrate sample. *(Courtesy of Valco Instruments.)*

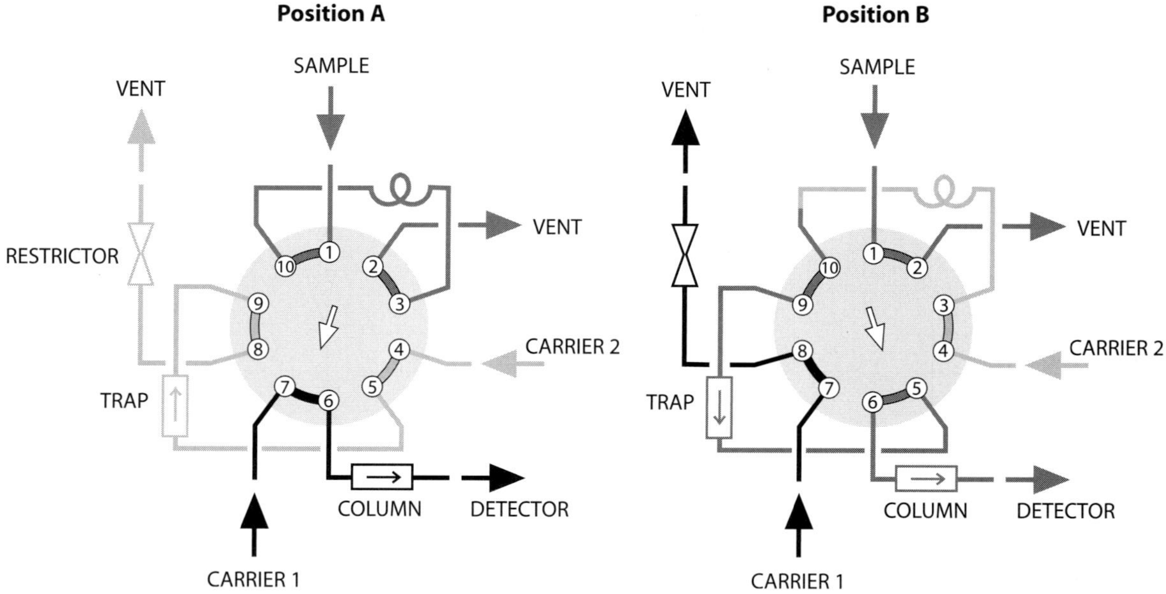

FIGURE 15.12 Volumetric sample injection with concentrating column or trap. *(Courtesy of Valco Instruments.)*

packed with a porous polymer such as Tenax® or Hayesep®, a volume of sample may be pulled through the trap to effect a significant concentration of the sample before the trap is backflushed and heated to transfer the compounds of interest into the separation column (Figure 15.11).

In the ten-port configuration shown in Figure 15.9, if column 1 is replaced with a trap, column switching as described in the previous paragraph may be used to separate a fraction of interest – early or late eluting – from the bulk of concentrated sample (Figure 15.12).

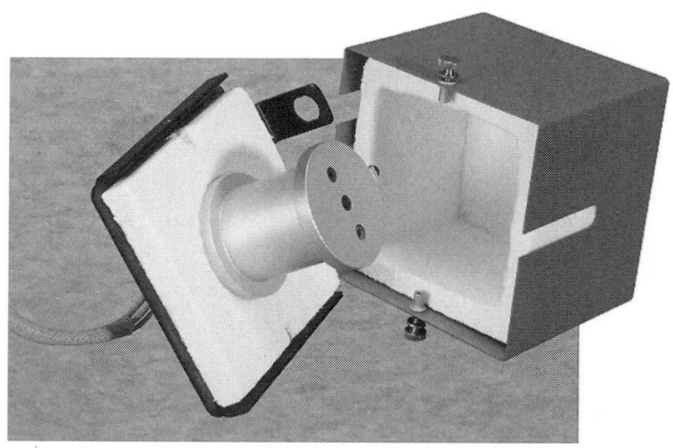

FIGURE 15.13 Column mandrel and insulated enclosure.

15.4. COLUMN CONFIGURATIONS

While air bath ovens are the norm for lab GCs, their size and power consumption mandate a different approach for portable instruments. Packed columns can be heated by winding them on conductive mandrels and surrounding them with low k factor insulation (Figure 15.13).

15.4.1. Isothermal Operation

A micropacked or capillary column is effectively heated by winding the column into a helix and holding it to a polyimide or polyimide film heater by compression or cement.

15.4.2. Temperature Programming

The use of conventional packed columns larger than 1/8" makes little sense if temperature programming is required; capillary and micropacked columns provide greater efficiency and lower power consumption. Columns can be programmed at >1000 °C per minute with low power.

Resistance heating may be accomplished by applying voltage directly to the ends of an electrically insulated column and measuring its temperature with a low mass sensor. Variation in column temperature will occur if the column is not wound or bundled carefully.

15.4.3. Isothermal Packed Columns

The application that has done more than any other to stimulate the production of portable GCs is the separation and quantitation of hydrocarbons at the sites of underground storage tanks (USTs). The most important compounds, from the perspective of the Environmental Protection Agency (EPA), are the aromatics from gasoline: benzene, toluene ethylbenzene, and xylene (BTEX). These can be separated with a capillary column or a packed column at room temperature or slightly above (75 °C for a 6′ × 1/8" packed column) and detected by a PID (Figure 15.14). The presence of higher-molecular-weight compounds may require backflushing or elevating the column temperature periodically. In some instruments, line power is used for the significant increase in column temperature required to elute the high-boiling compounds.

15.4.4. Multiple Columns

Many separations, such as the separation of N_2O and CFCs, require more than one column for a complete analysis. Sample is injected into a porous silica column upstream of a porous polymer column. After the N_2O enters the porous polymer column, the sequence of the columns is reversed as the valve returns to the load position. The CFCs, which would be retained much longer on the polymer column, elute directly to the detector, while the N_2O makes a second pass through the silica column before detection (Figure 15.15).

15.5. DETECTORS

15.5.1. Photoionization Detectors (PIDs)

While generally any detector used in a laboratory GC can be used in a portable GC, some types are used more often than others. The PID, with its insensitivity to air and its ppb sensitivity to important pollutants, is an excellent detector for field instrumentation. The selectivity of the PID can be enhanced by the choice of lamps.

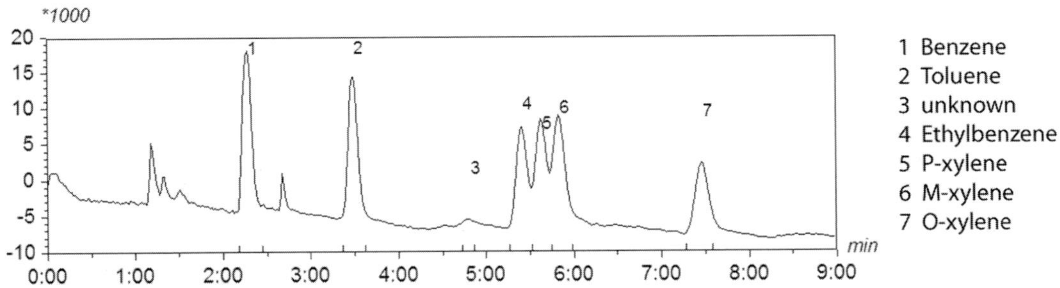

FIGURE 15.14 BTEX chromatogram separated on a packed column. *(Courtesy of PID Analyzers.)*

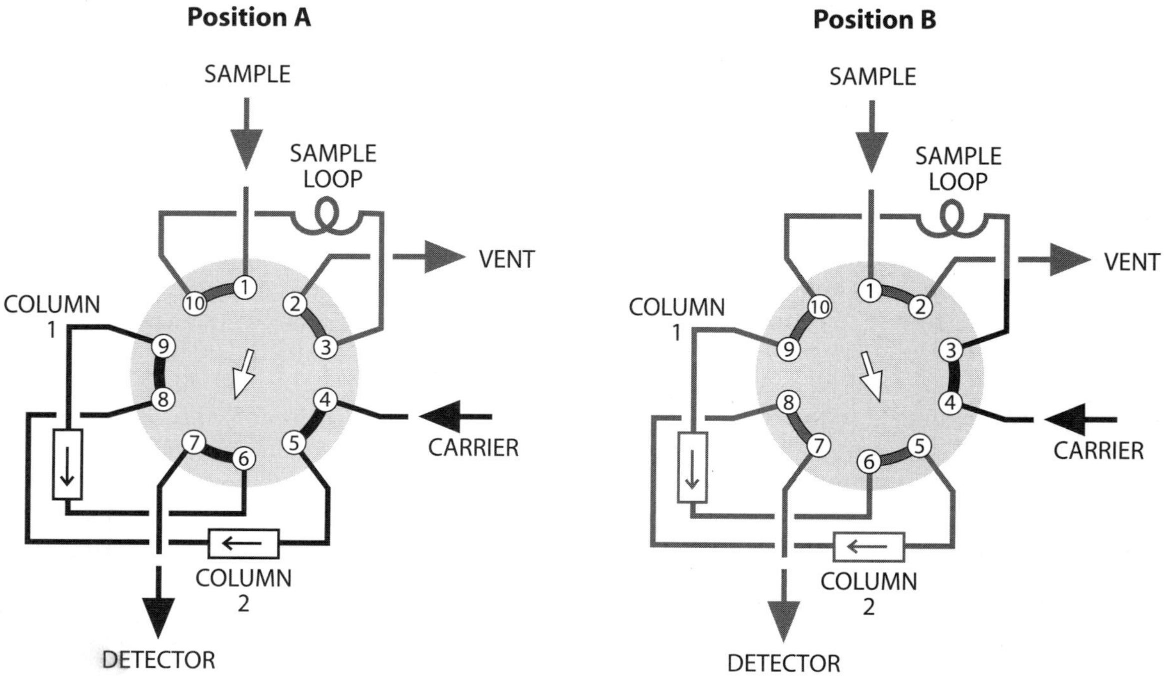

FIGURE 15.15 Loop sampling with two-column sequence reversal. *(Courtesy of Valco Instruments.)*

15.5.2. Pulsed Discharge Photoionization Detectors (PDDs)

The PDD is a universally responsive helium photoionization detector with high ionization efficiency. In Ar-, Kr-, or Xe-doped modes, it responds similarly to a PID, but with less selectivity since there is no window to exclude the high-energy emissions from He. Compared to a PID, it requires higher purity carrier gas, but functions at much lower flow (<5 cc/min for mini- and microversions).

15.5.3. Electron-Capture Detectors (ECD and PDECD)

The electron-capture detector is useful for portable chromatography, as it has unparalleled sensitivity to strongly electronegative compounds. ECDs require high-purity gases, N_2 or Ar/5% CH_4, for radioactive source detectors and He for pulsed discharge photoionization detectors. Both the ECD and PDECD have drawbacks. Radioactive ECDs, which have a Ni^{63} or H^3 source, require licensing, and there may be restrictions on transport. The nonradioactive PDECD requires very high purity He for carrier and discharge gases; however, a small, low-power active metal purifier is available for portable use.

15.5.4. Flame Ionization Detectors (FIDs)

The FID is used in portable GCs when universal sensitivity to hydrocarbons is desired. Unlike the lamp-type PID, the FID is sensitive to CH_4. It is not affected by water or air, so air may be used as a carrier gas. FIDs require H_2 as fuel for the flame; therefore, it must be supplied from a gas bottle or generated within the instrument.

15.5.5. Thermal Conductivity Detectors (TCDs)

The TCD is universally responsive, requires little power, and is simple to incorporate. However, its ppm sensitivity may require sample concentration for effective use in portable GCs.

15.5.6. Nitrogen-Phosphorous Detectors (NPDs)

The NPD is a modified FID that has a selective response for nitrogen and phosphorus compounds, halogens, and some oxygenates.

15.5.7. Surface Acoustic Wave Detectors (SAWs)

SAW detectors are piezoelectric crystals with a coating that is selected for response to specific classes of compounds. They may be used alone or in arrays. Their principal use has been the detection of chemical warfare agents.

15.5.8. Mass Spectrometers (MSs)

The mass spectrometer as a detector for portable gas chromatography adds considerable complexity, cost, weight, and support requirements, but offers unparalleled power to detect and identify compounds in the field. From the first efforts to construct a space mission GC-MS, advances in design and separation have made possible truly portable systems.

15.5.9. Other Detectors

Emission detectors using discharge sources can be successfully used in portable instrumentation [10].

15.6. GAS SUPPLY

Lecture bottles and similar, small, pressurized cylinders are convenient sources for portable GCs. For flow stability consistent with maximum content usage, miniature two-stage pressure regulators, two separate regulators, or a combination of a primary pressure regulator and a flow controller can be used. Electronic pressure control (EPC) is used in many current instruments.

When air is the (carrier gas), it may be from a cylinder or from a nonlubricated diaphragm-type miniature air compressor within the analyzer. Although diaphragm compressors are less likely to add contaminants, most miniature versions are limited to <2 atm. Filtration/purification can eliminate virtually all impurities except methane, which can only be removed with a heated (power-consuming) purifier. Nonheated purifiers can be used with PIDs, since PIDs do not respond to methane.

If air-actuated components are used, actuation pressure comes from the primary-stage regulator, with carrier flow coming from the second-stage regulator.

15.7. POWER MANAGEMENT

Perhaps the greatest challenge in designing a portable GC is providing adequate electricity for standalone, independent operation. If an instrument can be tethered to an auto battery or line supply until the actual time of use, portable use time will be maximized. Establishing a budget for electrical use is often critical.

A wide variety of battery sizes and capacities is available with the most efficient being, not surprisingly, the most expensive (Table 15.1).

15.7.1. Column Power (Isothermal)

Obviously, minimum power is required for column heat if the column(s) can provide the required separation at room temperature. If the sample contains slow-eluting, high-boiling compounds, analyses can be performed until the baseline drifts excessively, at which point extra power is needed to purge the column

TABLE 15.1 Comparison of the Properties of NiMH, Lead Acid, and Li-Ion Batteries at Similar Energy Level

Battery type	Weight	Volume	Energy density by weight/size		Cost ($/wh)
SLA (3 × 12 v 10aH)	10 kg	3270 cm^3	36 wh/kg	0.11 wh/cm^3	$0.19/wh
NiMH (36 V 10aH)	5.5 kg	2430 cm^3	65 wh/kg	0.15 wh/cm^3	$1.00/wh
Poly Li-ion (8aH)	1.75 kg	1340 cm^3	170 wh/kg	0.23 wh/cm^3	$1.25/wh
Poly Li-ion (10aH)	2.15 kg	1613 cm^3	170 wh/kg	0.23 wh/cm^3	$1.20/wh

Reproduced with the permission of BatterySpace.com.

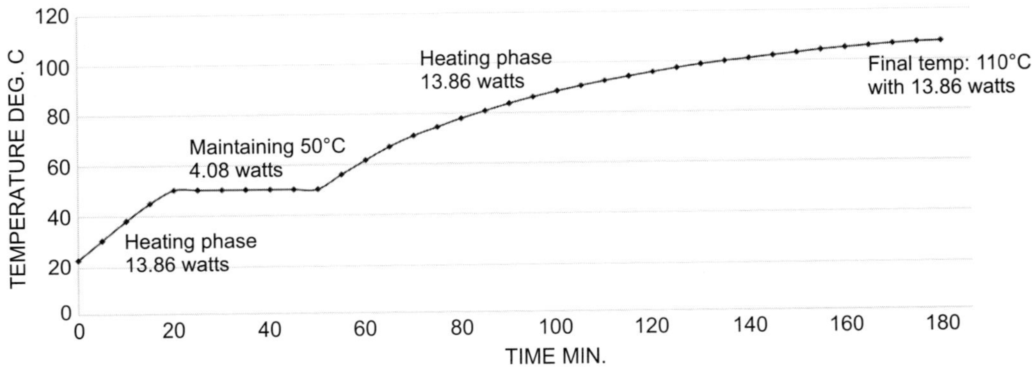

FIGURE 15.16 Column power requirements. Column: 10′ Hayesep DB in .125″ × .085″ SS; flow: 60 cc/min He; ambient temperature: 22.4 °C; power input: 63 VDC @ 220 mA–13.86 VA; power to maintain 50 °C: 34 VAC @ 120 mA–4.08 VA. *(Courtesy of Valco Instruments.)*

(Figure 15.16). As previously described, if the column is connected to a multiport valve or divided into two sections, column and precolumn, regular backflushing or column sequence reversal to vent can discharge the higher boiling compounds without contaminating the detector and without using extra power for column heat.

15.7.2. Column Power (Programmed)

Programming of capillary (Figure 15.17) or micropacked columns may be accomplished with minimum power by closely associating the column with a low-mass heat source, typically with resistance wire or by plating a conductive coating on the column [11,12]. The temperature of the column may be measured for control via the use of a separate sensing wire [13,14,15] or a single wire can be bundled along the length of the column and serve as both heater and sensor [16] (Figure 15.18).

Plating a fused silica column with nickel allows the ends of the column to be connected to a common ground with voltage applied to a center electrical tap, allowing even 30 m columns to be heated rapidly with less than 50 V. Efficiencies and repeatability that approach those of an air bath column oven have been measured (Figures 15.19 and 15.20, Table 15.2). To achieve this level of efficiency and repeatability, cold spots in the flow path must be eliminated. Fused silica columns have

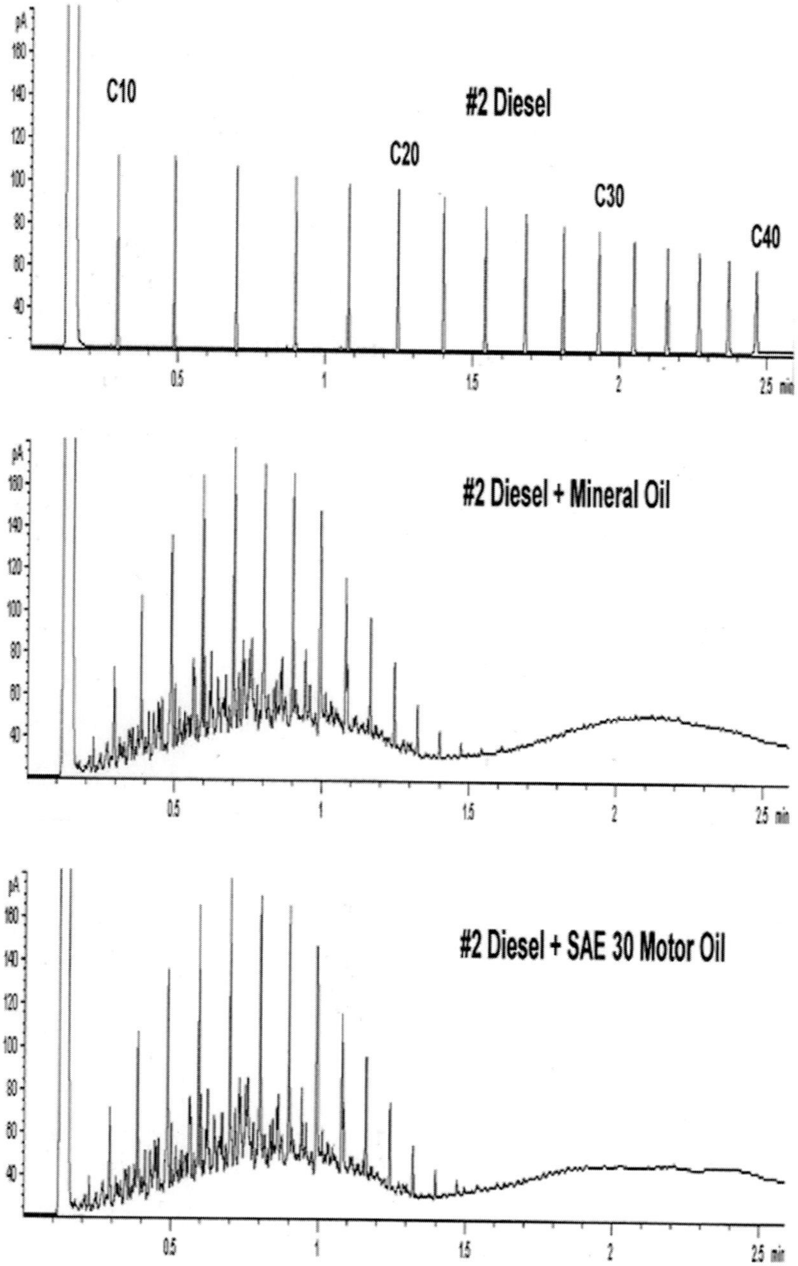

FIGURE 15.17 2.5 min analysis with a temperature-programmed capillary column. Column: VB-5, 5 m × 0.10 mm × 0.10 µm, 3.2 mL/min; detector: FID, 330 °C; column heating: 70 °C to 340 °C at 120 °C/min. *(Courtesy of Valco Instruments.)*

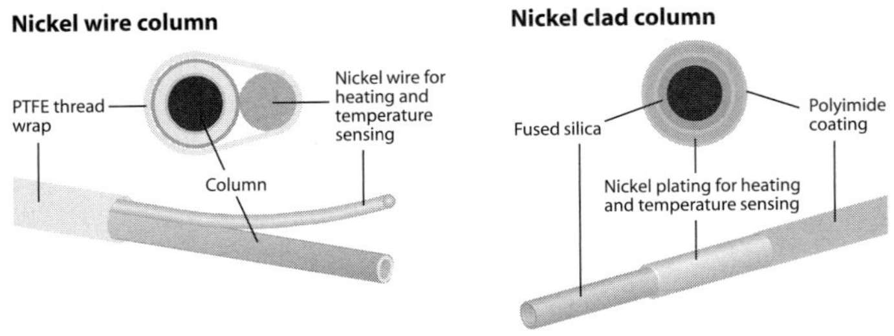

FIGURE 15.18 Resistively heated low mass columns. *(Courtesy of Valco Instruments.)*

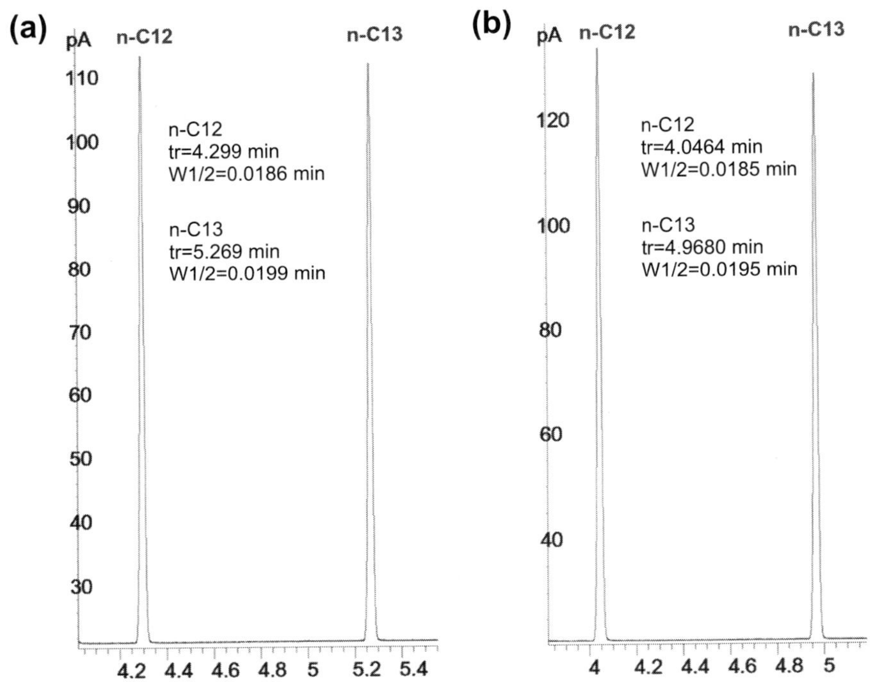

FIGURE 15.19 Heating efficiency of oven-heated column (a) compared to Ni wire type resistively heated column (b), programmed. Column: VB-5, 15 m × 0.25 mm × 0.25 μm, 70 °C to 220 °C at 15 °C/min. *(Courtesy of Valco Instruments.)*

also been heated by jacketing them with stainless steel tubing [17,18].

15.7.3. Pumps: Pressure or Vacuum

DC low-voltage pumps can deliver up to 24 psi or 20" Hg vacuum at a maximum of 6 W.

15.7.4. Additional

Electric precision controllers, which replace mechanical pressure regulators, can operate with less than 3 W. Some applications may also require a heated gas purifier. Miniature models are available that consume less than 10 W.

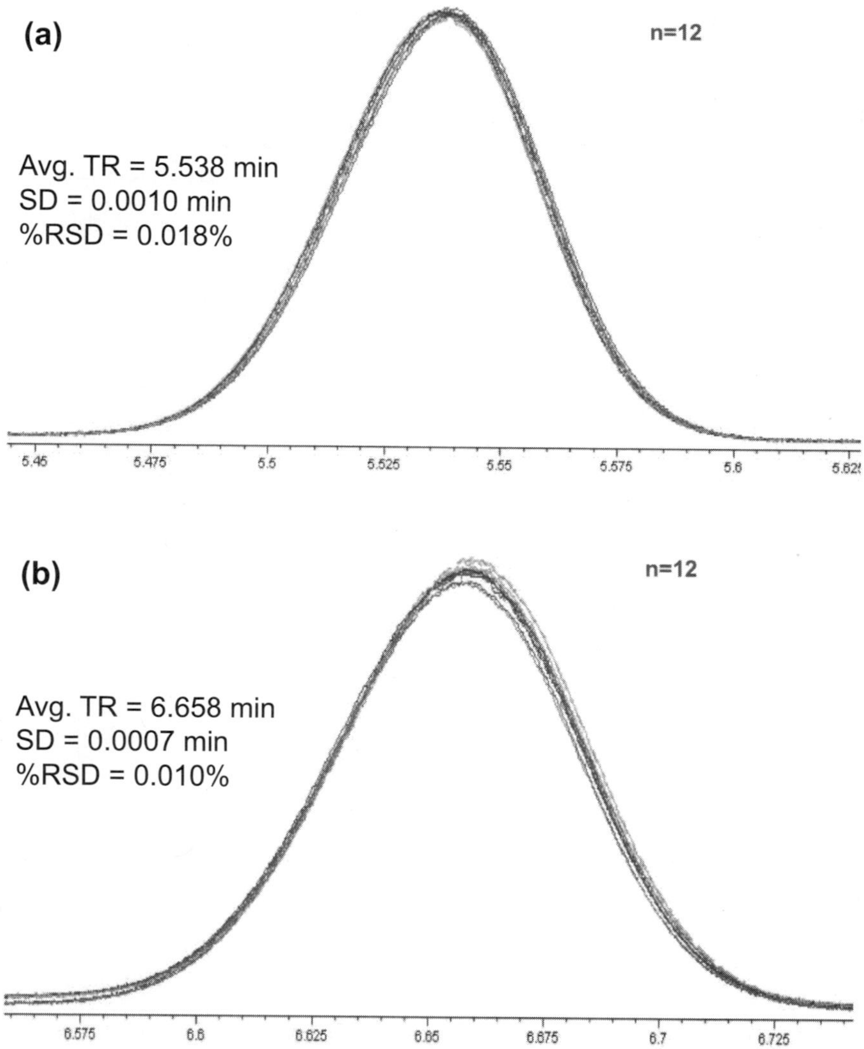

FIGURE 15.20 Repeatability of oven-heated column (a) compared to Ni wire type resistively heated column (b), isothermal. Column: VB-5, 15 m × 0.25 mm × 0.25 μm at 110 °C. *(Courtesy of Valco Instruments.)*

15.8. PROTOTYPING

The assembly of a portable GC is basically a matter of connecting components on a panel as closely as possible without causing thermal zone interference. Transfer lines, injectors, and other components may be heated with high-temperature, insulated resistance wire. Polyimide foam or low-K-factor fiber insulation is helpful in eliminating cold spots and helps reduce power consumption and weight. Figure 15.22 shows an assembly (covered and uncovered) with a diaphragm sample valve, a resistance heated column, and a micro PDD.

TABLE 15.2 Heating Efficiency of Resistive Heating Compared to Air Bath Oven Heating

	Isothermal Plates loss (%)	Programmed SN loss (%)
Ni wire	4.8	4.1
Ni clad	1.7	1.0
Ni tubing	4.5	6.6

Courtesy of Valco Instruments.

In this unit, the valve and detector are heated as well as the column. The chromatogram of chemical weapon surrogates in Figure 15.21 was done on this platform.

The light gases in Figure 15.23 were separated on a somewhat cruder arrangement of similar components (Figure 15.24). At the completion of the analysis, the small fan is turned on to cool the column. As an aid to cooling, the lid of the cardboard box holding the column is blown open by the fan (Figure 15.24).

15.9. FUTURE TRENDS

While the majority of this chapter is dedicated to discussion of practical solutions to field and portable GCs using components that are either readily available or manufacturable, researchers continue to explore the possibility of incorporating semiconductor fabrication techniques and micromachining in the development of compact, low-power, high-speed instruments [19,20].

The original "GC on a chip" became commercial only when the injector, column, and detector were separated. Conventional fused silica columns simply offer much better performance, even considering the difficulty of intracomponent connections.

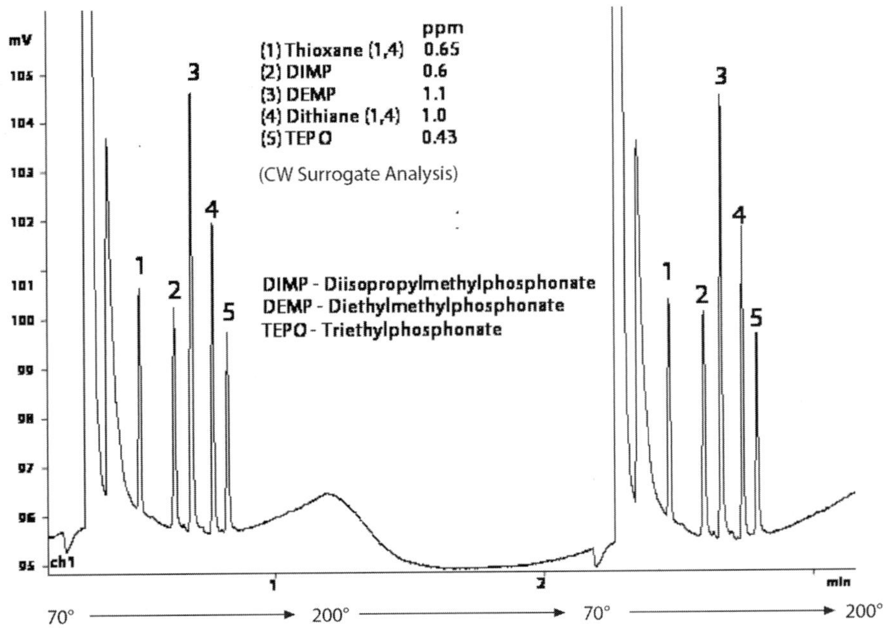

FIGURE 15.21 Less than 1 min. analysis of chemical weapon surrogate. *(Courtesy of Valco Instruments.)*

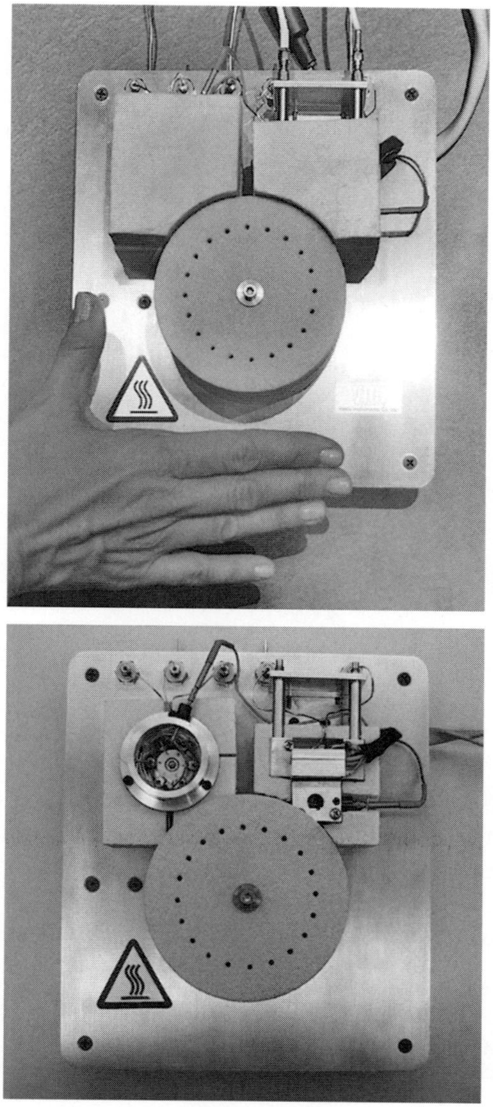

FIGURE 15.22 Prototype portable GC with a diaphragm valve, micro PDD, and resistance heated column. *(Courtesy of Valco Instruments.)*

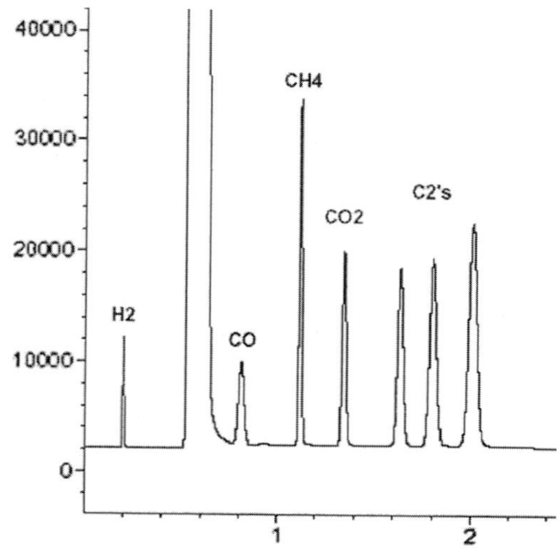

FIGURE 15.23 2 min gas mixture analysis with a temperature-programmed micropacked column. Column: Shin carbon column, 8 mL/min; column heating: 30 °C (0.9 min) to 230 °C at 500 °C/min, 1 min cooling time. *(Courtesy of Valco Instruments.)*

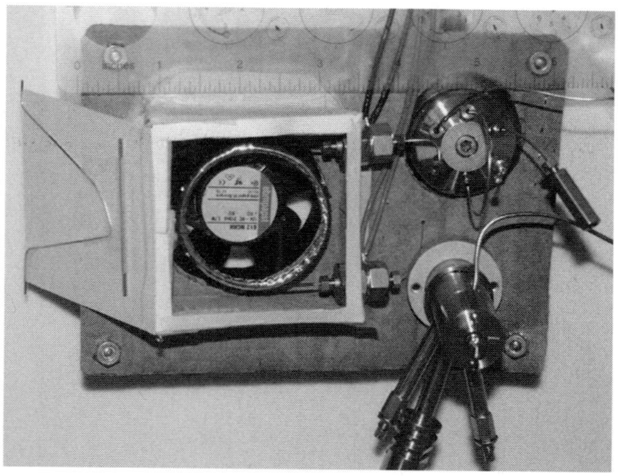

FIGURE 15.24 A second prototype portable GC with a diaphragm valve, micro PDD, and resistance heated column. *(Courtesy of Valco Instruments.)*

There are now manufacturers of micromachined injectors and detectors, making the assembly of micro GCs feasible for specific applications. Manufacturers of microcomponents for injection and detection include Seyonic in Switzerland (www.seyonic.com), Xensor in the Netherlands (www.xensor.nl), and Neroxis, also in Switzerland (www.neroxis.ch).

Currently, the GC-MS instruments from INFICON (Figure 15.25), Griffin/FLIR, and

FIGURE 15.25 INFICON HAPSITE® technology in Afghanistan. *(Courtesy of Inficon.)*

FIGURE 15.26 Tridion® 9 GC-MS from Torion Technologies. *(Courtesy of Torion Technologies.)*

Torion (Figure 15.26) allow a user to bring virtually an entire laboratory to the field, offering a balance of power and performance.

Acknowledgments

Thanks to W.F. Wilhite for providing early portable GC data, to Dr. Huamin Cai of Valco Instruments for demonstrating separation with minimal complexity, and to Dr. John Driscoll of PID Analyzers for providing crucial insights and data. Thanks also to David Salge of VICI Gig Harbor Group, without whose efforts in layout and editing this chapter would not have been completed.

References

[1] W.F. Wilhite, The development of the Surveyor gas chromatograph, technical report no. 32–425, Jet Propulsion Laboratory, California Institute of Technology, May 15, 1963.

[2] W.F. Wilhite, Subminiaturized gas chromatograph gives fast, efficient analysis, NASA Tech Brief (May 1966) 66–10182.

[3] C.S. Josias, L.D. Bowman, J.E. Lovelock. 1973. Gas detector and analyzer. US Patent 3,714,421, filed May 29, 1969, and issued January 30, 1973.

[4] M.R. Stevens, C.E. Giffin, G.R. Shoemake, P.G. Simmonds, A portable self-contained gas chromatograph, Rev. Sci. Instrum. 43 (10) (October 1972) 1530.

[5] R.N. Dietz, E.A. Cote, Tracing atmospheric pollutants by gas chromatographic determination of sulfur hexafluoride, Environ. Sci. Technol. 7 (4) (1973) 338–342.

[6] J.E. Morgan. Transformer Fault Detection. US Patent 4,112,737, filed April 27, 1977, and issued September 12, 1978.

[7] N. Tzanani, A. Amirov, Electrolyzer powered flame ionization detector, Anal. Chem. 69 (1997) 1218–1255.

[8] J.B. Angell, J.H. Jerman, S.C. Terry, S. Saadat, Prototype gas analysis system using a miniature gas chromatograph, interagency energy/environment R&D program report, NIOSH/EPA, December, 1980.

[9] C.L. Arthur, J. Pawliszyn, Solid phase microextraction with thermal desorption using fused silica optical fibers, Anal. Chem. 62 (19) (1990) 2145–2148.

[10] W. Wentworth, K. Sun, D. Zhang, J. Madabushi, S. Stearns, Pulsed discharge emission detector: an element-selective detector for gas chromatography, J. Chromatogr. A 872 (1–2) (2000) 119–140.

[11] R.A. Yost, M.E. Hall, Compact gas chromatograph probe for gas chromatography/mass spectrometry utilizing resistively heated aluminum-clad capillary columns, Anal. Chem. 61 (21) (1989) 2410–2416.

[12] R.A. Yost, M.E. Hall, Direct resistive heating and temperature measurement of metal-clad capillary columns in gas chromatography and related separation techniques. US Patent 5, 114,439, filed September 25, 1990, and issued May 19, 1992.

[13] R.V. Mustacich, 1998. Gas chromatography column assembly temperature control system. US Patent 5,782,964, filed January 27, 1997, and issued July 21, 1998.

[14] R.V. Mustacich, J.F. Everson, 2001, Reduced power consumption gas chromatograph system. US Patent 6,217,829, filed January 27, 1997, and issued April 17, 2001.

[15] R.V. Mustacich J.F. Everson 2004. Reduced power consumption gas chromatograph system. US Patent 6,682,699, filed March 2, 2001, and issued January 27, 2004.

[16] S.D. Stearns, H. Cai, J.A. Koehn, M. Brisbin, C. Cowles, C. Bishop. Direct resistively heated columns for fast and portable gas chromatography, Pittcon oral session 1130-3, March 4, 2008.

[17] D.P. Rounbehler, E. Hainsworth, 1992. Selective detection with high speed gas chromatography. US Patent 5,099,743, filed July 8, 1987, and issued March 31, 1992.

[18] W.A. Rubey, 1991. Gas chromatography methods and apparatus. US Patent 5,028,243, filed March 5, 1990, and issued July 2, 1991.

[19] A. Peters, M. Klemp, L. Puig, C. Rankin, R. Sacks, Instrumentation and strategies for high speed gas chromatography, Analyst 116 (1991) 1313–1320.

[20] V. Reid, M. Stadermann, O. Bakajin, R. Synovec, High-speed temperature programmable gas chromatography utilizing a microfabricated chip with an improved carbon nanotube stationary phase, Talanta 77 (4) (2009) 1420–1425.

CHAPTER

16

Preparative Gas Chromatography

Leesun Kim, Philip J. Marriott

OUTLINE

16.1. INTRODUCTION

Preparative gas chromatography (prep-GC) has a rich and well-established history. This ranges from the use of GC as a bulk isolation method for compounds in mixtures that are well known, but simply cannot be otherwise separated from mixture components in large-scale amounts, to applications where much smaller quantities of compounds that require more substantial structure elucidation than available with classical GC detectors.

Gas Chromatography DOI: 10.1016/B978-0-12-385540-4.00016-X

The characterization of trace components, unknown analytes, and complex mixtures of components has always been a challenge in organic analytical chemistry, such as in the flavor/fragrance [1], pheromone [2], environmental [3], pharmaceutical [4], metabolomics [5], and petrochemical areas [6]. As modern techniques or technical innovations in analysis methods increasingly deal with trace compounds, often this simply pushes the boundary of sample characterization to focus on components at even lower abundance. This is well demonstrated in the field of metabolomics, where there is a continual demand for lower detection limits, and a corresponding need to accurately characterize more compounds at trace levels in order to cover a wider scope of the metabolome [7]. There are many strategies for the reliable measurement of components at a low level. Samples can be concentrated using prior sample preparation for the subsequent analytical measurement. A large volume of sample can be applied to the analytical instrumental step (e.g. large-volume sample introduction; LVSI) in gas chromatography (GC) [8,9] or a chromatographic preparative step can be an integral part of the analysis method. It is also possible to remove, reduce, or eliminate matrix in a sampling step, so that the sensitivity of target analytes can be increased in the detection step and a concomitant reduction of unwanted or potentially interfering material can be excluded for the 'analytical finish'.

Prep-GC includes the elegant approach of incorporating a sample preparative step as an integral part of the analysis step, allowing the component of interest to be isolated from complex matrices and then to be collected after GC analysis. As a result of this operation, the analytes can be characterized further to provide improved identification. This allows an expanded range of spectroscopic detection techniques which are not normally available as hyphenated chromatographic methods. Thus, the components collected can be studied by other analysis techniques, to be more suitably characterized using mass spectrometry (MS), Fourier transform infrared spectroscopy (FTIR), nuclear magnetic resonance (NMR) spectroscopy, X-ray, accelerator MS (AMS), or other spectroscopic tools [10–13]. Traditionally, GC hyphenated with MS has been essential to identify compounds from a variety of matrices [14]. However, significant problems exist when some compounds cannot be identified from their mass spectra and retention data alone. This is due to high similarities between isomeric compounds, and the lack of specificity of the derived mass spectrum for similar compounds or lack of specific fragmentation patterns. The (over-) reliance of many researchers on the mass spectrum to deduce the molecular structure can often cause erroneous identification.

In consequence, prep-GC may play an important role, when compounds either need to be enhanced in abundance (enriched) or need to be isolated due to inadequate elucidation of structure with available on-line detection methods. Prep-GC can conveniently be classified as a large-scale or analytical scale methodology, depending on the sample and component properties and the purpose of using this technique. This chapter will use the following differentiation between these two goals, outlined in Section 16.2.

16.2. APPLICATION SCALE OF PREPARATIVE GAS CHROMATOGRAPHY

Prep-GC can range from large-scale prep-GC methods capable of producing kg/h of substance, to the analytical scale preparatory method, where it is only necessary to isolate sufficient material for subsequent characterization techniques e.g. off-line spectroscopic or microscale reaction methods.

16.2.1. Large-Scale Preparative GC

We may define large-scale prep-GC as a technique which is used as a purification step for a compound, often synthetic, but sometimes natural, that is needed for either large-scale commercial production or subsequent synthesis reaction [15]. Generally, this means that the target component is well characterized and therefore is a known compound, and just needs to be produced in large amounts of pure compound. The scale can therefore be in the kg or greater quantity, and GC may be suitably scaled up to provide the necessary purity of the compounds. In this case, it is simply the act of recovering the compound in a large amount that is the goal of the prep-GC step.

This can be summarized as follows:

Known synthetic or natural sample → large-scale prep-GC → pure compound in large amount

The present review will not consider this role of prep-GC here, as it has been the focus of recent publications, such as the book by Schmidt-Traube [15]. As this citation illustrates, scaling up of the prep-GC method operates according to well-established physical and chemical separation principles, and the reader is directed to this and other references for further information.

16.2.2. Analytical-Scale Prep-GC

Prep-GC on the analytical scale plays a significant role, when it is difficult to identify a molecule by using just the retention time or retention index, or by correspondence of retention with a coinjected authentic compound, or when MS is not sufficient to provide adequate characterization of the structure of the compound. Under these circumstances, alternative methods for compound identity are needed – above and beyond that available from classical GC detectors. This requires a collection method to isolate and transport the compound to other detector(s) or techniques. The standard proof of identity relied on by chemists include techniques such as UV, NMR, and FTIR, and, where possible, X-ray analysis; so it is useful to apply this knowledge and this degree of identification for separated/collected GC peaks where necessary.

In the analytical-scale process, prep-GC has been utilized for a wide range of applications, including compounds such as sex pheromones, essential oil compounds, environmental residues, polycyclic aromatic hydrocarbons, petroleum components, and chiral drugs. These applications can be categorized into the following situations (note that some of these target individual compounds, and others target zones of compounds or indeed the full range of volatile compounds):

1. Analytes in complex samples
   a. Natural products such as pheromones [16–19] and essential oils [20]
   b. Synthetic products such as nonylphenol (NP) [3,21]
2. "Simple" samples but with impurities that require precise structural characterization [10]
3. Samples with trace-level compounds requiring further concentration or enrichment in order to meet the needs of the detection limits for quantitative and qualitative analysis.

The general need and justification for analytical-scale prep-GC therefore can be readily articulated, but there is still scope for improvement in systematic approaches to the general problem of enhancing the identification of components at trace levels and in complex sample matrices. The following section briefly highlights various separation and collection strategies that illustrate approaches to analytical prep-GC taken by researchers, in order to provide elucidation of the structure of components in a variety of sample types. Discussion in Section 16.4 focuses on case studies of selected applications.

16.3. EXPERIMENTAL TECHNIQUES FOR ANALYTICAL-SCALE PREP-GC

16.3.1. 1D GC

The implementation of prep-GC with a single packed column format is the most productive [22–24] in terms of recovering a large mass of compound with few injections. However, it does not provide the best separation for a complex sample with many potentially interfering peaks. Many applications will require a higher-resolution separation method. In this case, higher-resolution capillary columns which may be 'megabore' dimensions –0.53 mm i.d. – or narrower bore 0.25–0.32 mm i.d. may be used. In all cases, a smaller phase ratio will reduce the effect of peak broadening due to overloading effects if larger amounts of analyte are injected. A typical schematic of a prep-GC instrument is shown in Figure 16.1(a). The method incorporates a suitable collection or trapping device T which often may be operated at subambient or cold temperature by using a cooling system. A detector and a switching system are also incorporated here to monitor the progress of the GC analysis. The switching device, which is directed to the detector in normal operation, allows selection of the component(s) to be transferred to the trapping system by switching the flow to the trapping channel. The trap may also be an adsorbent cartridge, a phase-coated capillary trap, or a solvent-filled collector, which will be described further in Section 16.3.3.

16.3.2. Capillary Multidimensional Gas Chromatography (MDGC)

Since MDGC was first proposed by Simmonds and Snyder in 1958 [25], it has been mostly implemented using a wide range of switching devices and/or valves. Multidimensional chromatography (MDC) can be defined as the coupling of two (or more) different separation stages, with independent elution through

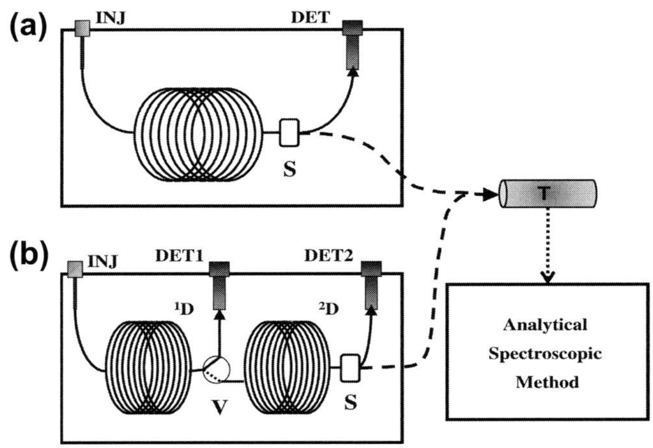

FIGURE 16.1 Schematic arrangements of one-dimensional (a) and multidimensional (b) separation systems, comprising on-line detector(s) with switching to a preparative sample collection system followed by off-line detection. The switching device (S) directs flow either to the detector or to collection. The trapping system (T) can be any mode of collection used for the prep-GC fractions. For MDGC, a valve or interface (V) is used for the heart-cutting process to transfer compounds to the ^{2}D column.

each stage. Therefore, it is imperative that the chromatographic dimensions not only should be connected directly (e.g. two GC columns) but there also should be an interface, such as a valve or switching system between dimensions. The valve or switching system allows a small zone of effluent from the first dimension (^{1}D) to be directed into the second dimension (^{2}D) by changing the flow of gas from the detector channel to the ^{2}D channel as shown in Figure 16.1(b). In this way, improved resolution in the ^{2}D column can be achieved by MDGC, which can benefit a wide range of applications areas already alluded to such as petroleum, environmental pollutants, and essential oil components, all of which demonstrate a degree of complexity that often precludes separation of individual peaks on a single column [26,27].

16.3.3. Trapping Systems

Trapping systems are critical to the success of the prep-GC method. Most studies using prep-GC have utilized preparative fraction collection into dedicated collection devices, or capillary tubes as a trapping method. Both systems can incorporate an additional cooling system, which can increase the recovery efficiency of the semi-volatile or volatile target components [11,17,28]. Sorbent traps are alternatives to the above, and liquid traps are also described.

16.3.3.1. Gerstel Preparative Fraction Collector

The Gerstel preparative fraction collector (PFC) has been used in various studies using prep-GC [19,29,30]. This sample collection system, equipped with six sample traps and one waste trap, automatically collects individual compounds, a series of compounds, or specific classes of compounds after GC separation. Trap tubes are available in 1 μL or 100 μL volumes. To increase recovery efficiency, PFC can be equipped with optional N_2 (liq) or cryostatic trap cooling systems. Microprocessor control allows the trap switching times to be selected to within 0.01 min, which permits reliable collection of individual compounds that are closely resolved. The reliability and reproducibility of the system make it possible to trap compounds over the course of hundreds of injections [10,31]. This allows further analyses of the fractions by techniques requiring larger sample mass such as NMR or IR. In order to overcome the technical limitation to only six collectable fractions, Meinert et al. [28] introduced a new technique combining two PFCs through a special zero-dead volume effluent splitter, shown in Figure 16.2. The same group [31] optimized the performance of the prep-GC system by identifying the best operating parameters for the two PFC system using six environmentally significant target analytes: phenol, naphthalene, acenaphthene, methyl parathion, methoxyclor, and benzo[a]pyrene. Solvent-filled traps were used instead of temperature-controlled trapping. Dichloromethane (DCM) proved to be the most suitable solvent for a large range of compounds. Recoveries were in the range of 50–70% for all compounds apart from benzo[a]pyrene with a recovery of 94% using one PFC (not two PFCs) and DCM-filled traps at a trapping temperature of −10 °C. Transfer line and PFC switch temperatures of 300 °C for phenol, 400 °C for benzo[a]pyrene, and 320 °C for the rest of the analytes gave the best results. Highly heat-tolerant capillaries for the transfer line were used in order to prevent significant capillary porosity and breakdown [31].

16.3.3.2. Capillary Trapping

Eyres et al. [11,12] and Rühle et al. [32,33] used an Agilent G2855B microfluidic Deans switch (DS) to isolate the peak of interest eluting from the outlet of the column, directing the flow to either a detector (flame ionization detector (FID)) or the external trapping assembly (xTA) via deactivated fused silica (DFS) transfer lines. A length of megabore DFS tubing (trapping

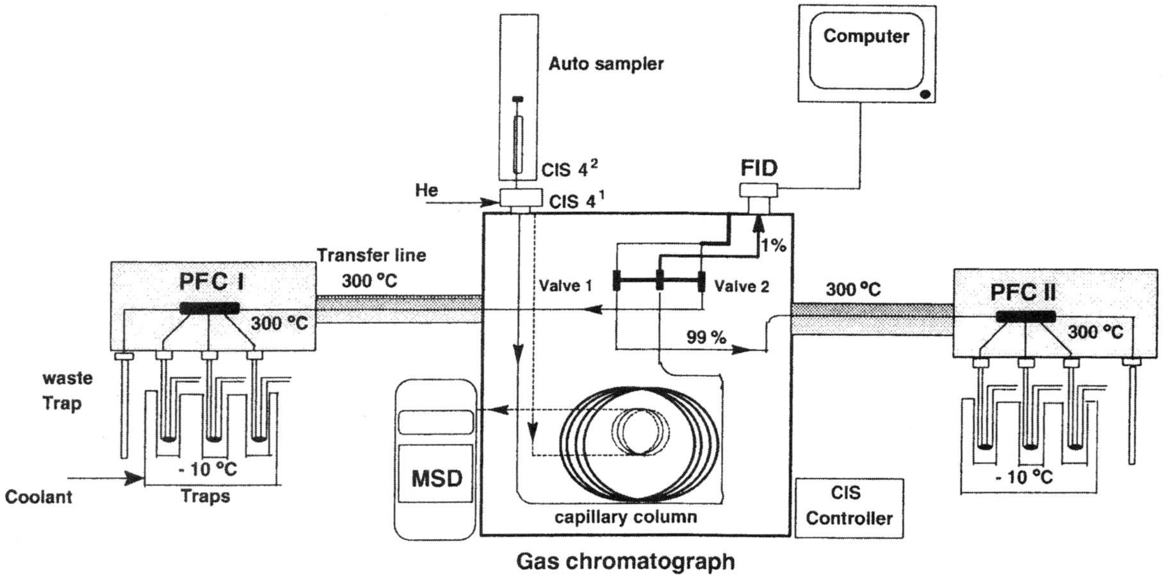

FIGURE 16.2 Schematic illustration of an analytical scale prep-GC system consisting of two prep-fraction collectors (PFCI and PFCII) modules and a waste trap, as described for the Gerstel system in [28]. *Used with Permission of the Copyright Holder.*

capillary) was used within a cryotrap cooled with liquid CO_2 to collect the transferred (heart-cut) peak. The simple heart-cutting procedure is shown in Figure 16.3. After multiple injections, the capillary was removed from the cryotrap and eluted with an appropriate solvent for characterization using spectroscopic techniques including NMR or X-ray.

In a series of prep-GC studies, Nojima et al. [13,17,18,34] described the use of a short collection tube attached to the end of the analytical capillary column. The tube was pulled through a heated exit and passed out of the GC just before the elution time window of the target solute, to collect only the component of interest. A sheath that generated a cooling zone was attached to the tube and assisted sample trapping. After the sample collection, the sheath was removed and the collection tube was withdrawn from the analytical column, with collected compound eluted and taken for NMR analysis. This method is described in Figure 16.4. Nojma et al. [34] investigated the performance of various types of capillary column traps (deactivated, methyl polysiloxane, and polyethylene glycol) using the prep-GC technique and under various collection conditions. The model compounds were C4−C20 normal alkanes, esters, and alcohols. Above a critical Kovat's index, recovery efficiencies of traps with methyl polysiloxane films were 80−100% for a wide range of injected sample mass.

16.3.3.3. Sorbent Trapping Method

In order to enrich many key odor compounds which can occur at a very low level in a complex sample such as wine, Ochiai et al. [35] introduced a single PFC module consisting of a heated transfer line, an additional heater, an adsorbent packed tube (e.g. Tenax tube), and a PFC pneumatic box, shown in Figure 16.5. The adsorbent packed tube was used to enrich the target components from multiple injections. The trapped and enriched compounds were thermally desorbed and subsequently analyzed

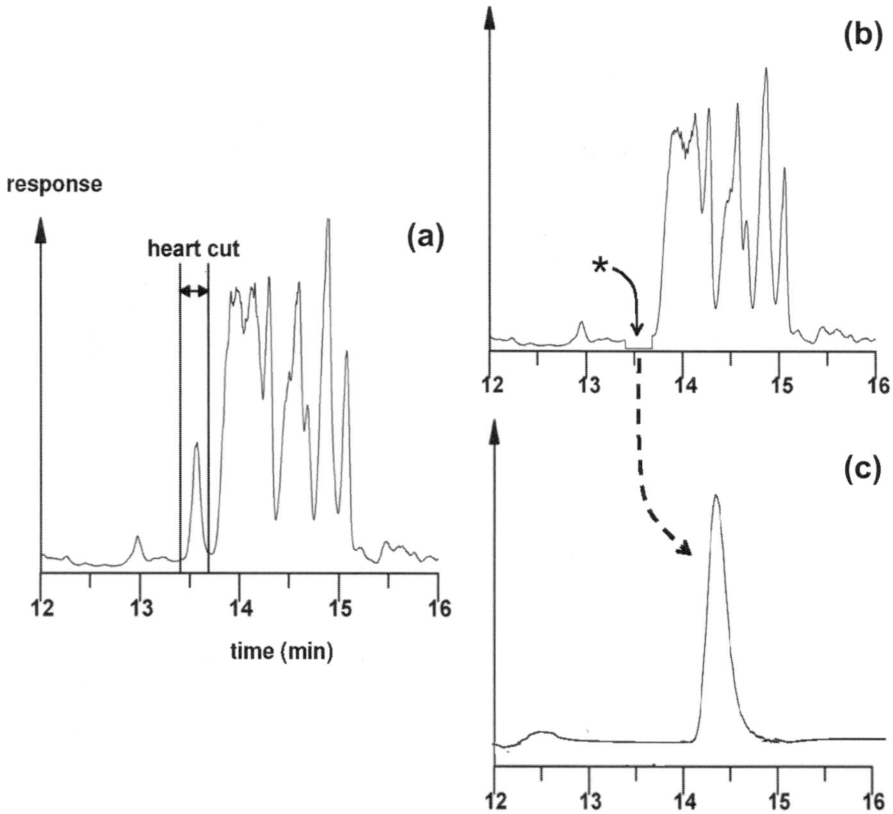

FIGURE 16.3 Demonstration of the heart-cutting procedure using a Deans switch (a) before heart-cutting and (b) after heart-cutting of a target compound – here, a nonylphenol isomer of a polyethoxylate sample. (c) The subsequent GC analysis of the single isolated component peak.

by GC-olfactometry (GC-O)/MS for identification. Stir-bar sorptive extraction (SBSE) was used to extract components from white wine (Sauvignon Blanc and Chardonnay) and placed in a glass thermal desorption liner. Subsequently, the glass liner was placed in the thermal desorption unit. More than 30 extractions could be performed with the same bar, which allowed enriching the target components into the adsorbent packed tube. They evaluated PFC recovery of a single injection of the 15 model flavor compounds at 10 ng each, including alcohols, aldehydes, esters, lactones, and phenol. Recovery in the range of 85–98% was obtained [35]. This application is typical of the general sorbent-trapping procedure.

16.3.4. Spectroscopic Methods for Use with Prep-GC

Once the target compound(s) is (are) trapped in adequate amount, they can be taken – usually manually – to any spectroscopic instrument for subsequent study. This is limited only by either the needs or availability of the appropriate technique, or the ingenuity of the researcher to devise suitable analytical approaches for identification. Generally, this constitutes off-line hyphenation

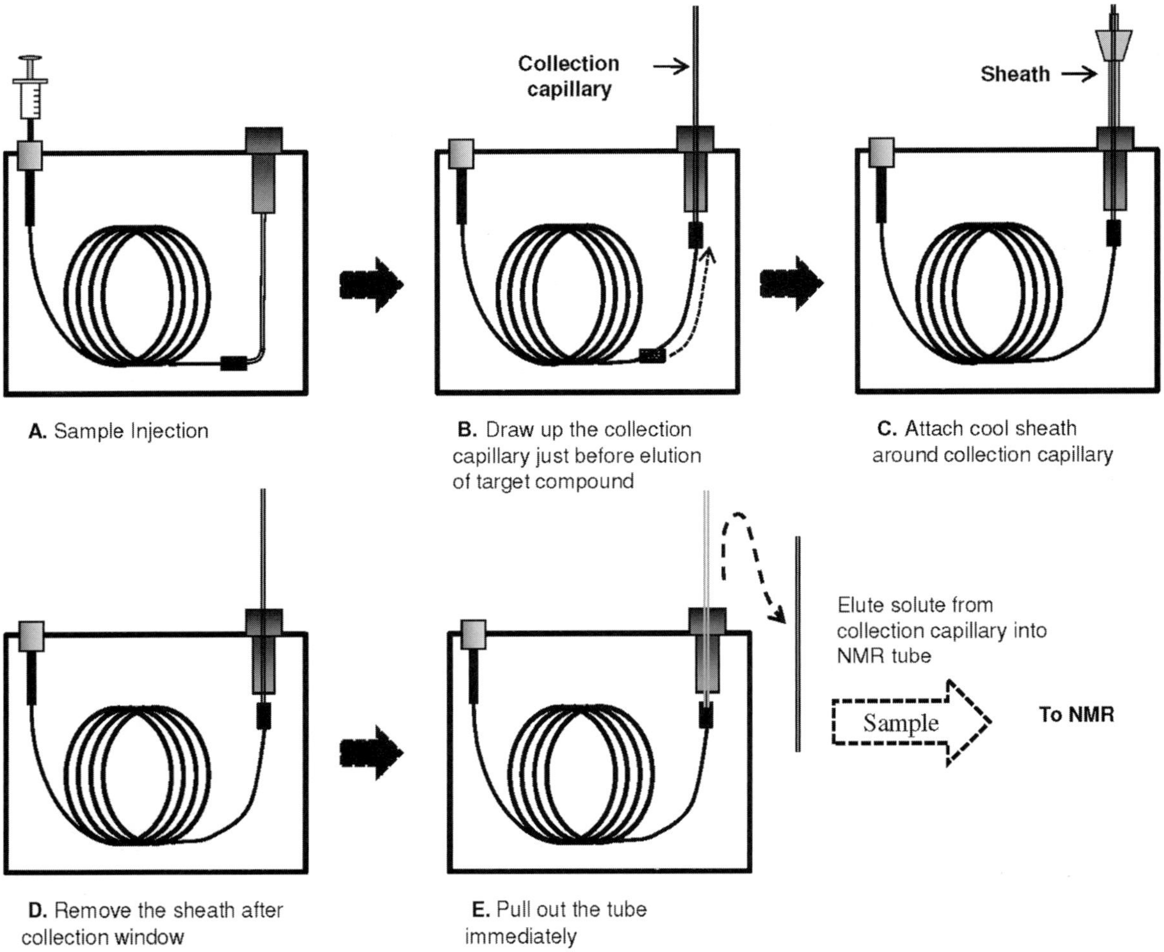

FIGURE 16.4 Sample collection procedure using prep-GC with a trapping capillary method adopted by Nojima. *Source: Adapted from Nojima et al. [34].*

(denoted by the double-forward slash mark chromatography//spectroscopy), where the collected fraction is remote from the subsequent molecular characterization tool. With fraction in hand, therefore, any tool can be used to provide identification, subject to sensitivity requirements and physical or chemical considerations. This section is not intended to provide an exhaustive survey of the various techniques which have been employed as the '//spectroscopy' component of the system, but, rather, in conjunction with the applications outlined Table 16.1 and those covered in Section 16.4, will illustrate a selection of typical (mainly recent) studies that have been conducted with prep-GC.

16.4. CASE STUDIES: APPLICATIONS

Many studies, which have been recently carried out using prep-GC, are summarized in Table 16.1. Some of these will be discussed further below.

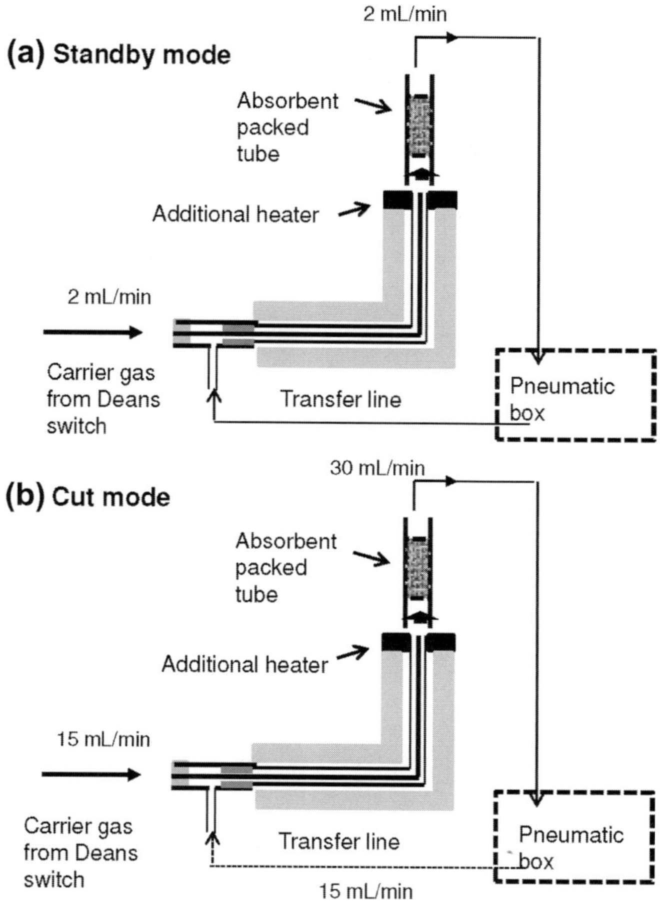

FIGURE 16.5 Sorbent-trapping method introduced by Ochiai and Sasamoto [35] for a switchable 1D–2D GC method. (a) Standby mode; (b) Cut mode. The absorbent packed tube (Tenax TA) after trapping and enriching the target components can be thermally desorbed using thermal desorption (TD)-GC-O/MS for identification. Diagram redrawn based on [35].

16.4.1. Natural Products: Pheromones

Many pheromones, which often are multi-component mixtures and present in nanogram or picogram amounts, require structure elucidation for a variety of purposes such as pest control. Nojima et al. [17] developed a simple and inexpensive approach to sub-micro-scale prep-GC of volatile compounds for use with off-line NMR. The recovery efficiency of volatile compounds using cryogenic trapping with a short section of a deactivated megabore capillary tube was >80% with sample sizes of 0.05–0.5 μg. The purity of the collected samples was enough for structure elucidation by high-sensitivity NMR analyses including 2D NMR experiments. Subsequently using the same technique, Nojima et al. [18] described a method to isolate a thermally and chemically unstable sex pheromone from the female German cockroach (*Blattella germanica* (L.)). The gland was collected and extracted, followed by preliminary high-performance liquid chromatography (HPLC) fractionation in a biological-activity-directed

TABLE 16.1 Selected Reports of Analytical Prep-Scale GC

	Study objective	GC conditions	Trapping method/collected quantity	Analytical methods	Ref
1	Chiral inhalation anesthetics isoflurane and desflurane	**Prep-GC**: FID, traps **Column**: Packed column octakis(2,6-di-0-n-pentyl-3-O-butanoyl)-cyclodextrin in SE-54 (10%, w/w), coated on Chromosorb PAW DMCS (20.3%, w/w)	**Collector**: traps cooled with liquid nitrogen	NMR X-ray crystallography	Schurig [24]
2	Degradation product of inhalation anesthetic sevoflurane	**Prep-GC**: FID, traps **Column**: Packed column (1 m × 18 mm i.d.)	**Collector**: same as case 1	NMR X-ray crystallography	Schmidt [22]
3	Volatile composition of Cognac and Calvados	**Prep-GC**: Thermal conductivity detector **Column**: Packed column (4 m × 5.2 mm i.d.) with a 5% SE-30 (100% dimethylpolysiloxane on Chromosorb PWA 100 mesh)	**Collector**: manual collector-connected glass tubes in liquid nitrogen	GC/MS (CI, EI)	Ledauphin [36]
4	Aromatic hydrocarbons from the UCM recovered from a biodegraded crude oil.	**Prep-GC**: FID, Gerstel PFC **Column1**: BP-5 (15 m × 0.53 mm i.d.× 1.0 μm d_f) **Column2**: SolGel-Wax (30 m × 0.25 mm i.d.× 0.25 μm d_f)	**Collector**: Gerstel PFC	GC/MS	Sutton [37]
5	14 fractions from a commercial nonylphenol(NP)	**Prep-GC**: FID with PTV-LVI with PFC **Column**: DB-5 (60 m × 0.25 mm i.d.× 0.25 μm d_f)	**Collector**: same as case 4	GC/MS, NMR recombinant yeast screen	Kim [3]
6	Gentisyl quinone isovalerate, sex pheromone of German cockroach, *Blattella germanica*	**Prep-GC**: GC-EAD system converted to a prep-GC **Column:** Non-polar Equity-1 (5 m × 0.53 mm i.d.× 1.5 μm d_f)	**Collector**: Injection port modified as a collection port with cryogenic trapping, a detachable sheath with a reservoir for a refrigerant to cool the collection traps	GC/MS NMR	Nojima [18]
7	Fractionation of nonylphenol isomers into 11 fractions	**Prep-GC**: mps2-Auto-injector, PFC with CIS 4 **Column**: HP-5MS, 30 m × 0.32 mm i.d., 0.5 μm d_f	**Collector**: two Gerstel PFC with CIS 4. **Recovery**: 77–552 μg of isomers after 600 single injections	GC/MSD	Meinert [28]

8	Congeners isolated from several widely produced PCB commercial/technical mixtures	**Prep-GC**: FID and a 7683 Series autoinjector in conjunction with a septumless CIS and Gerstel PFC **Column**: a VF-5MS "megabore" fused silica capillary column ($60\ m \times 0.53\ mm$ i.d. $\times 0.5\ \mu m\ d_f$)	**Collector**: Gerstel PFC with glass traps with CIS was operated in "solvent vent" mode **Recovery**: Individual PCBs (or coeluting congeners groups) ranged from 90% to 100%	Chlorine isotope analysis, sealed-tube combustion with TIMS	Mandalakis [29]
9	Sex pheromone component of the Japanese mealybug	**Prep-GC**: PFC with two columns **Column1**: TC-FFAP ($15\ m \times 0.53\ mm$ i.d. $1\ \mu m\ d_f$) **Column2**: TC-1 ($15\ m \times 0.53\ mm$ i.d.)	**Collector**: Gerstel PFC	GC/MS NMR	Sugie [19]
10	Investigate capillary column traps under various collection conditions for different compounds	**Prep-GC**: FID, outlet port assembly **Column**: nonpolar EC-5 megabore capillary column ($30\ m \times 0.53\ mm$ i.d. $\times 1.0\ \mu m\ d_f$) **Deactivated column (with a press fit connector)**: $2\ m \times 0.53\ mm$ i.d.	**Collector**: same as case 6 **Collection traps**: a DB-1 $1.5\ \mu m\ d_f$ **Recovery**: 84% of injected compounds mass with injected sample mass (from 10 to 1,000 ng of each)	NMR	Nojima [34]
11	Volatile components (geraniol) from 15 partially coeluting compounds from the first column	**Prep-MDGC**: two columns in an oven, FID1, 2, a three-channel EPC module, DS and LMCS. **1D column**: DB-5 ($15\ m \times 0.32\ mm$ i.d. $\times 1\ \mu m\ d_f$) **2D column**: DB-Wax ($15\ m \times 0.32$ i.d. $\times 0.5\ \mu m\ d_f$)	**Collector**: External trapping assembly (xTA), trapping capillary ($100\ mm \times 0.53\ mm$ i.d.) in a cryotrap (xCT) **Recovery**: from 10 (8.6 μg) and 100 injections (77.6 μg; purity >99%) 12.3 μg/mL (10 injections) for proton and gCOSY 110.8 μg/mL (100 injections) gHSQC and gHMBC	2D NMR gCOSY NMR gHSQC and gHMBC NMR	Eyres [11]
12	Isolation of 1,4-dimethoxybenzene from an essential oil, isomers of methylnaphthalene from a crude oil	**Prep-MDGC**: Same as case 10 **1D column**: Same as case 10 **2D column**: Same as case 10	**Collector**: same as case 11 **Recovery**: 1,4-dimethoxybenzene (5.2 μg, 10 injections) 1- and 2-methylnaphthalene (3.1 μg, 38 injections and 5.0 μg, 35 injections)	^{1}H-NMR	Eyres [12]

(*Continued*)

TABLE 16.1 Selected Reports of Analytical Prep-Scale GC (*cont'd*)

	Study objective	GC conditions	Trapping method/collected quantity	Analytical methods	Ref
13	Characterization of menthol, linalyl acetate, carvone and geraniol from essential oil	**Prep-MDGC**: Same as case 10 **1D column**: Same as case 10 **2D column:** same as case 10	**Collector, DFS transfer lines**: Same as case 10 **Recovery**:	MS NMR	Rühle [32]
14	Identification of (*E,E*)-2,4-undecadienal from coriander	**Prep-GC**: FID **Column**: Supelco-wax (60 m × 0.75 mm i.d.)	**Collector**: Short capillary column (25 cm × 0.53 mm i.d. × 2.0 μm d_f)	GC/MS	Ikeura [38]
15	Multidimensional method to assess quality of pure tea tree oil and presence of allergenic agents	**Prep-GC**: FID with a fractionation system View Prep Station (VPS278) **Column**: A wide bore non-polar (30 m × 0.53 mm i.d. × 5.0 μm d_f)	**Collector**: VPS278 through software, VPS Control. Seven microtubes (5 °C) (six × collection, one × waste) An additional gas flow (He) was used, with reversed gas flow to avoid cross contamination inside collection tubes	GC/MS ^{1}H-NMR	Sciarrone [20]
16	Potentially mutagenic components	**Prep-GC**: Same as case 7 **Column**: Same as case 7	**Collector**: Same as case 7 **Auto-sampler mps2 (Gerstel) and a speed of injection**: 100 μL/s	GC/MS	Meinert [30]
17	Best parameters for PFC using six test analytes with different physicochemical properties	**Prep-GC**: zero-dead volume effluent splitter to FID, two Gerstel PFC with CIS 4. **Column**: HP-5MS (30 m × 0.32 mm i.d. × 0.5 μm d_f)	**Collector**: two Gerstel PFC with CIS 4. **Two columns**: 0.87 m × 0.32 mm i.d. **Deactivated FS-Phenyl-Sil column**: 0.5 m × 0.1 mm i.d. (for effluent splitter) **Recovery**: 50–70% for all compounds except benzo[a]pyrene (94%)	GC/MS	Meinert [31]
18	Reaction of phenylacetylene with para-substituted aryl-iodides	**Prep-GC**: FID, xTA with cryotrap **Column**: a DB-5 (15 m × 0.32 mm i.d. × 1 μm d_f)	**Collector**: same as case 10 **DFS transfer lines**: 2 m × 0.18 mm i.d.	GC/MS NMR X-ray	Rühle [33]

19	Volatile impurities at major and minor percentage levels in a pharmaceutical matrix by NMR and MS	**Prep-GC**: FID, a CIS 4 PTV and Gerstel PFC **Column**: a DB-5 (30 m × 0.53 mm i.d. × 5 μm d_f) **PTV**: at 20 °C for the injection	**Collector**: Gerstel PFC with traps containing 200 μL of methanol-d_4 and a small (1.7 mm) cryogenic probe **Conditions**: trapping optimized using liquid sorbents **Recovery**: total amount of one of the isolated impurities <60 nmol	GC-APCI-TOF MS 1D NMR 2D NMR-COSY HSQC, HMBC DQF, CASE	Codina [10]
20	Selectable 1D or 2D GC–O/MS with PFC. Off-flavor compounds	**1D Column**: DB-Wax (30 m × 0.25 mm i.d., × 0.25 μm d_f) **2D Column**: DB-5 (10 m × 0.18 mm i.d. × 0.40 μm d_f)	**Collector**: heated transfer line, additional heater, adsorbent packed tube, PFC pneumatic box **Selectable 1D/2D GC–Olfactometry/MS with PFC**: 1D or 2D GC-MS, DS2, PCM 2, single PFC module. 1D and 2D column outlets (connected to DS2) **Transfer capillary**: 1 m × 0.32 mm i.d. d_f **Recovery**: PFC enrichment with 20 injection of 15 model compounds; 500 pg each (98–116%) **SBSE-PFC enrichment efficiency**: 71–78%	2DGC–O/MS	Ochiai [35]
21	Isolation, purification of semiochemicals (i.e. insect pheromones) and NMR analysis of <1 μg of material	**Prep-GC**: FID, outlet port assembly **Column**: either a non-polar EC-5 or polar EC-WAX megabore capillary column, 30 m × 0.53 mm i.d. × 1.0 μm d_f	**Collector**: Injection port modified as a collection port **Collection traps**: DB-1 (2 cm × 0.53 mm i.d. × 1 or 5 μm d_f) **Recovery**: 50 ng of geranyl acetate for NMR 250 ng of geranyl acetate for H-H COSY NMR	^{1}H-NMR H-HCOSY NMR	Nojima [13]

isolation step to discern which LC fraction contained the active component. The active fraction was then further separated by analytical-scale GC with antennography, followed by prep-GC of the active region using multiple injection steps to finally isolate the individual active compound of interest. However, it seems that this system is not suitable for sequential fractionation of multiple chemicals with a single GC run.

Nojima et al. [13] described simple off-line integration of prep-GC and capillary NMR analysis to facilitate the identification of minute amounts of small volatile compounds. ^{1}H-NMR spectra were obtained with 50 ng of geranyl acetate, a model compound, and reasonable H-H COSY NMR spectra were obtained with 250 ng of geranyl acetate. This is a significant development since it pushes the amount required to be isolated by using GC to considerably smaller quantities than in prior work. Thus, either fewer repeat injections are needed or the method can be extended to much smaller abundance components.

16.4.2. Impurities in Pharmaceutical Samples

Impurity profiling in modern pharmaceutical products analysis is an important but challenging task to improve the quality of drugs as well as the safety and efficacy of drug therapy. In an editorial, Görög [4] described a wide range of techniques for impurity profiling with the dominance of HPLC and GC with MS technologies acknowledged. While many other separation and bulk analysis methods are available, they tend to be less commonly used. Since off-line NMR (both as a bulk analytical method and with fraction collection with LC) provides considerable structural information, its complementary role to MS is valuable. Therefore, the HPLC/NMR/MS coupled instrument has been attracting increased attention. However, Codina et al. [10] reported prep-GC with fraction collection and NMR precisely for the task of impurity identification. Various computer-aided structural elucidation (CASE) tools for molecular formula assignment supported the task. Three impurities (A, B, and C) were required to be identified in a pharmaceutical matrix with known main compound band (MB). However, the least concentrated sample (impurity D) was identified as well, with sufficient amount collected for 1D and 2D homonuclear (^{1}H) and possibly heteronuclear (^{1}H, ^{13}C) NMR experiments. The trapping efficiencies were 63% for MB and 39, 63, and 54% for impurities A, B, and C respectively [10]. Their structures are shown in Figure 16.6. Codina also observed that the prep-GC method saved considerable time and streamlined the analysis procedure for identification of impurities in the drug synthesis they investigated. This was from the perspective of researchers in the pharmaceutical industry.

16.4.3. Environmental Pollutant Studies

Using a combination of LC fractionation and prep-GC, Kim et al. [3] obtained 14 purified fractions from a commercial nonylphenol (NP). The isomeric structure of the nonyl group was confirmed by GC-MS and NMR analysis. Meinert et al. [28] fractionated technical p-NP into 11 fractions by using prep-GC, with a total of 600 injections made. Being a technical sample, large quantities of sample were available. The collected amounts of fractionated NP isomers ranged from 77−552 μg of isomer, which was sufficient to allow subsequent biotesting in the screen assay used. In the absence of pure isolates of each isomer, it is not possible to uniquely allocate the bioactivity associated with each isomer.

Using combined methodologies including genotoxicity testing and reversed-phase liquid chromatography (RP-LC) with prep-GC, Meinert et al. [30] tentatively identified a total

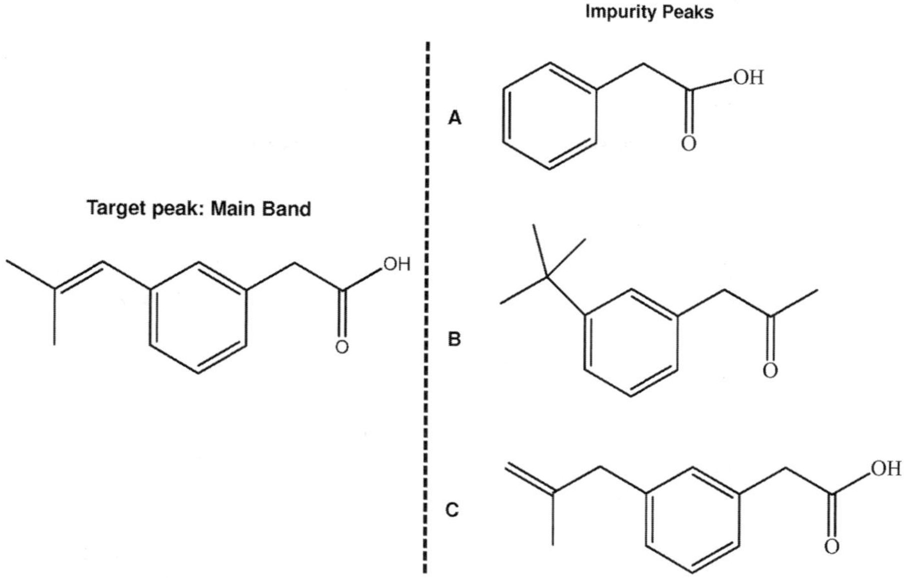

FIGURE 16.6 Main compound and impurity peak structures identified in a pharmaceutical matrix by Codina [10].

of 62 components in genotoxic fractions from a contaminated groundwater. For structure characterization of unknown chemicals, a computer-based structure generation tool called MOLGEN-MS was added to spectral library information from the NIST05 library [30].

In order to study the heterogeneity arising from both ethoxylate and nonylphenol groups, Wu et al. [21] fractionated a mixture of nonylphenol polyethoxylate (NPnEO; n = number of ethoxylate groups), nonionic surfactants, using normal-phase liquid chromatography (NPLC). Collected fractions were then analyzed by GC and GC-MS to allow the nonyl isomeric distribution to be displayed. Since mass spectra obtained did not give enough information about the chemical structures of the well-resolved NPnEO isomers, absolute structural identification was not achieved. Therefore, one of the isomers of an NP2EO sample was heart-cut and collected by prep-capillary-scale GC. GC-MS was used to confirm its purity and the structure of the isomer was elucidated by ^{1}H-NMR spectroscopy.

16.4.4. Chiral Studies and X-Ray Analysis

Schurig et al. [23] reported the preparative enantiomer separation of the inhalation anesthetics, enflurane and isoflurane, in high chemical purity (>99.5%) and enantiomeric purity (>99.9 and 99.4%, respectively). The prep-GC method used a packed column with octakis(2,6-di-O-n-pentyl-3-O-butanoyl)-cyclodextrin dissolved in SE-54 (10% w/w) and coated on Chromosorb P-AW-DMCS (20.3%, w/w). This approach showed the first preparative racemate resolution of both anesthetics by chromatography. Subsequently, Schurig et al. [24] characterized isoflurane and desflurane, which are normally clinically administered as a racemic mixture, by using prep-GC and crystal structure determination.

Schmidt et al. [22] described an experiment in which the enantiomers of an inhalation anesthetic compound were resolved on a chiral packed column in order to isolate the individual isomers of the R and S forms of sevoflurane.

The enantiomers were reported to have an unprecedented high separation factor on the enantioselective phase used, and so this allows considerable overloading without compromising the purity of separation. Not only was sufficient of each of the enantiomers collected to permit NMR analysis, but also crystal structures were possible by careful crystallization of the enantiomers. This provided absolute structural assignment of each isomer.

Marriott and co-workers [33] studied GC separation of a mixture of Sonogoshira catalyst products, which could not be separated by classical liquid chromatographic approaches. The products of interest were well separated with one-dimensional GC with a thick-film phase column, but gave equivalent mass spectral data. Therefore, the MS data were not conclusive for structural details. Using prep-GC, up to about 100 μg of pure product was collected, sufficient to conduct NMR and allow crystallization of the bulky aromatic compound. The small crystals were subsequently analyzed by using synchrotron radiation. The first of the peaks had a distinctive NMR result that was similar for all of the different compounds, suggesting a similar structural environment. However, the NMR spectrum of the second isomer was quite different. The primary product was found to be a symmetric tetra-aryl benzene compound.

16.4.5. Petroleum Studies

The unresolved complex mixture (UCM) or GC 'humps', which is present to various degrees in almost all crude oil samples, has for decades been a difficult task for analysts to identify using one-dimensional GC. Recently, it has been studied by comprehensive two-dimensional gas chromatography (GC×GC) which has proved capable of decoding its compositional complexity [39].

Sutton et al. [37] isolated hydrocarbon fractions from a biodegraded crude oil by using prep-GC. This followed a sequence of prep-open column chromatography, prep-HPLC (using 3 columns in series), prep-GC on an a polar phase into 10 s cuts, and finally prep-GC on a polar phase to recover the final component. They reported a series of $C_{26}-C_{28}$ triaromatic steroids, and tentatively identified a novel C_{26} 17-desmethyl triaromatic steroid. They suggested that the UCM might comprise 250,000 compounds, which is a rationale for none of them being resolvable by using a 1D chromatographic analysis. With a novel-capillary multidimensional gas chromatography (MDGC) approach, Eyres et al. [12] isolated the isomers of 1- and 2-methylnaphthalene (MN) from a complex crude oil. A total of 3.06 μg 1-MN was collected from 38 injections to give a concentration of 5.07 μg/mL for NMR analysis. A total of 5.00 μg 2-MN was collected from 35 injections to give a concentration of 8.22 μg/mL for NMR.

16.4.6. MDGC Methods

In addition to the above study, Eyres et al. [11] had earlier introduced a novel prep-cap MDGC method to resolve geraniol from an essential oil mixture made up of lavender and peppermint oils. Geraniol, a target component, was coeluted with at least six other compounds on the first column. Using a longitudinally modulated cryotrap as a cryogenic storage system, the coeluted zone was completely resolved on the ^{2}D column. A Deans switch connected to the end of the second column was used to heart-cut only the geraniol into an external trapping assembly. This system is shown in Figure 16.7. All other eluted compounds were directed to a monitor FID. From 50–100 injections were made to collect sufficient geraniol for 1D and 2D NMR analysis. The 800 MHz NMR instrument provided improved S/N for collected analyte.

In a follow-up study using the same sample matrix as that used by Eyres above [11], Ruhle et al. [32] proposed that by targeting a range of different compounds to be heart-cut to an

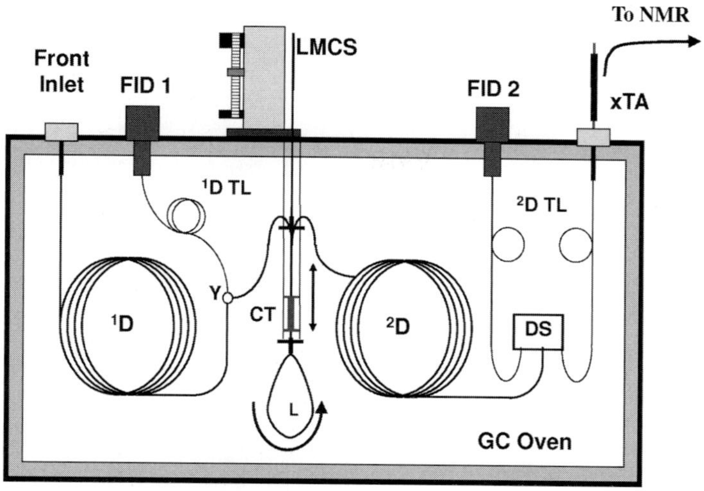

FIGURE 16.7 Preparative capillary MDGC method described by Eyres et al. [11] incorporating a trapping loop segment (L) and cryotrap (CT) that isolates a target region of a ^{1}D elution and effectively separates it from earlier and later peak regions. Separation of the target region on ^{2}D yields single peaks that can be heart-cut using a Deans switch (DS) through a transfer line (TL) into an external trapping assembly (xTA). *Used with Permission of the Copyright Holder.*

external trapping assembly, it would be possible to prepare novel mixtures of compounds from a sample. If this was a natural sample, such as an essential oil, it could theoretically be possible to create a new mixture that is commercially unavailable, or even one that is 'impossible' to synthesize due to the natural complexity of some compounds. In the case referred to, various combinations of linalyl acetate, carvone, geraniol, and menthol were collected into mixtures comprising different components, simply as a proof of concept of the method. However, if a specific compound were desired, it could be isolated, and its NMR and mass spectra recorded (and if possible, an X-ray structure obtained) to obtain a pure microquantity reference standard.

16.4.7. Prep-Scale Enrichment of Bulk Compounds for Further GC Analysis

The prep-GC method can be used to enrich bulk volatile compounds from a sample, in order to reapply the collected mixture to a further analysis – and usually this will be by GC/MS. This was used by Ledauphin et al. [36] for cognac and calvados to increase component abundance and achieve greater detectability for many minor components in the samples. In this case, trace compounds present in both spirits were characterized, aided by prep-GC separations. A multiple analytical approach was used to tackle this problem: Initially, groups of compounds were separated by prep-GC, and the fractions were analyzed on a polar stationary phase by GC/MS. Silica gel fractionation was then used to separate prep-extracts by polarity with further GC/MS analysis. A total of 331 compounds were characterized in both freshly distilled cognac and calvados, with 162 considered to be trace compounds, which would have been difficult to identify in the absence of the prep-GC method. Of the trace compounds, the authors reported 39 to be common to both spirits, 30 specific to cognac, and 93 present in calvados but not cognac. Clearly, the prep method proved to be a useful adjunct approach for the discovery process.

16.4.8. Prep-GC via Multiple Injections with Prep-Collection in the GC Column

By using a cryofocus method and switching devices (e.g. Deans switch) within one GC oven, it is possible to scale up individual or zones of compounds injected, in order to collect larger amounts of compound at the cryotrap [40]. Elution of the cryotrap then allows the multiple-collected compound(s) to be eluted into a second column (in an MDGC method) or directly to a detector or an external trap to give higher mass of material. This was proposed to be of relevance to give a better response for the compound where the injected sample had a low detection limit, or could give a greater degree of spectral confirmation by increasing mass in a spectroscopic detector, or allow the use of a detector that has a poor detection limit by also increasing the mass delivered to the detector. All of these meet the scope of a prep-GC method, but it is unique in being an on-line method.

16.5. CONCLUSIONS

Analytical-scale preparative GC plays a niche role that is largely associated with the requirement to characterize a compound's structure. This is especially applied to the flavor and fragrance areas, where discovery of new compounds is still directed toward natural products, and the perception of desirable characteristics. As this discovery step is pushed to compounds with ever-lower abundance, there remains a need to provide surety of structure. Mass spectrometry does not always provide this level of assurance; so alternative molecular spectroscopic methods are required. This necessitates isolation via prep-GC approaches. NMR has been a mainstay tool for this process, and becomes the 'gold standard' for characterization. However, in some cases, prep-GC has allowed sufficient material to be collected to permit crystals to be grown, with absolute crystal structure providing unambiguous identity.

As long as there remains a need to fully characterize compounds, where they are present in mixtures such that pure materials cannot be generated by any other means other than by using GC, or where standard GC detectors such as MS are limited in their ability to fully interpret the compound structure, there will be a role to play for analytical-scale prep-GC, as described here.

Acknowledgments

LSK thanks Monash University for the provision of a Dean's International Postgraduate Research Scholarship. This research was supported under Australian Research Council's *Discovery Projects* funding scheme (project number DP1095335). This work was conducted as part of our affiliation with the Australian Centre for Research on Separation Science.

References

[1] A. Wanikawa, K. Hosoi, T. Kato, K. Nakagawa, Identification of green note compounds in malt whisky using multidimensional gas chromatography, Flavour Fragrance J. 17 (3) (2002) 207–211.

[2] G.R. Jones, N.J. Oldham, Pheromone analysis using capillary gas chromatographic techniques, J. Chromatogr. A 843 (1–2) (1999) 199–236.

[3] Y.S. Kim, T. Katase, M. Makino, T. Uchiyama, Y. Fujimoto, T. Inoue, et al., Separation, structural elucidation and estrogenic activity studies of the structural isomers of 4-nonylphenol by GC-PFC coupled with MS and NMR, Aust J. Ecotoxicol. 11 (3) (2005) 137–148.

[4] S. Görög, The importance and the challenges of impurity profiling in modern pharmaceutical analysis, TRAC Trends Anal. Chem. 25 (8) (2006) 755–757.

[5] J.R. Idle, F.J. Gonzalez, Metabolomics, Cell Metab. 6 (5) (2007) 348–351.

[6] J. Blomberg, P.J. Schoenmakers, U.A.T. Brinkman, Gas chromatographic methods for oil analysis, J. Chromatogr. A 972 (2) (2002) 137–173.

[7] Y. Li, T. Pang, Y. Li, X. Wang, Q. Li, X. Lu, et al., Gas chromatography-mass spectrometric method for metabolic profiling of tobacco leaves, J. Sep. Sci. 34 (12) (2011) 1447–1454.

[8] J.C. Bosboom, H.-G. Janssen, H.G.J. Mol, C.A. Cramers, Large-volume injection in capillary gas chromatography using a programmed-temperature vaporizing injector in the on-column or solvent-vent injection mode, J. Chromatogr. A 724 (1–2) (1996) 384–391.

[9] K. Grob, M. Biedermann, Vaporising systems for large volume injection or on-line transfer into gas chromatography: classification, critical remarks and suggestions, J. Chromatogr. A 750 (1–2) (1996) 11–23.

[10] A. Codina, R.W. Ryan, R. Joyce, D.S. Richards, Identification of multiple impurities in a pharmaceutical matrix using preparative gas chromatography and computer-assisted structure elucidation, Anal. Chem. 82 (21) (2010) 9127–9133.

[11] G.T. Eyres, S. Urban, P.D. Morrison, J.-P. Dufour, P.J. Marriott, Method for small-molecule discovery based on microscale-preparative multidimensional gas chromatography isolation with nuclear magnetic resonance spectroscopy, Anal. Chem. 80 (16) (2008) 6293–6299.

[12] G.T. Eyres, S. Urban, P.D. Morrison, P.J. Marriott, Application of microscale-preparative multidimensional gas chromatography with nuclear magnetic resonance spectroscopy for identification of pure methylnaphthalenes from crude oils, J. Chromatogr. A 1215 (1–2) (2008) 168–176.

[13] S. Nojima, D.J. Kiemle, F.X. Webster, C.S. Apperson, C. Schal, Nanogram-scale preparation and NMR analysis for mass-limited small volatile compounds, Plos One 6 (3) (2011) 7.

[14] N. Ragunathan, K.A. Krock, C. Klawun, T.A. Sasaki, C.L. Wilkins, Gas chromatography with spectroscopic detectors, J. Chromatogr. A 856 (1–2) (1999) 349–397.

[15] H. Schmidt-Traub (Ed.), Preparative chromatography of fine chemicals and pharmaceutical agents, Wiley-VCH Weinheim, 2005.

[16] D.A. Carlson, F. Mramba, B.D. Sutton, U.R. Bernier, C.J. Geden, K. Mori, Sex pheromone of the tsetse species, Glossina austeni: isolation and identification of natural hydrocarbons, and bioassay of synthesized compounds, Med. Vet. Entomol. 19 (4) (2005) 470–479.

[17] S. Nojima, D.J. Kiemle, F.X. Webster, W.L. Roelofs, Submicro scale NMR sample preparation for volatile chemicals, J. Chem. Ecol. 30 (11) (2004) 2153–2161.

[18] S. Nojima, C. Schal, F.X. Webster, R.G. Santangelo, W.L. Roelofs, Identification of the sex pheromone of the German cockroach, *Blattella germanica*, Science 307 (5712) (2005) 1104–1106.

[19] H. Sugie, M. Teshiba, Y. Narai, T. Tsutsumi, N. Sawamura, J. Tabata, et al., Identification of a sex pheromone component of the Japanese mealybug, Planococcus kraunhiae (Kuwana), Appl. Entomol. Zool. 43 (3) (2008) 369–375.

[20] D. Sciarrone, C. Ragonese, C. Carnovale, A. Piperno, P. Dugo, G. Dugo, et al., Evaluation of tea tree oil quality and ascaridole: a deep study by means of chiral and multi heart-cuts multidimensional gas chromatography system coupled to mass spectrometry detection, J. Chromatogr. A 1217 (41) (2010) 6422–6427.

[21] Z.Y. Wu, C.P.G. Ruhle, P.J. Marriott, Liquid chromatography fractionation with gas chromatography/mass spectrometry and preparative gas chromatography-nuclear magnetic resonance analysis of selected nonylphenol polyethoxylates, J. Chromatogr. A 1218 (26) (2011) 4002–4008.

[22] R. Schmidt, M. Roeder, O. Oeckler, A. Simon, V. Schurig, Separation and absolute configuration of the enantiomers of a degradation product of the new inhalation anesthetic sevoflurane, Chirality 12 (10) (2000) 751–755.

[23] V. Schurig, H. Grosenick, Preparative enantiomer separation of enflurane and isoflurane by inclusion gas chromatography, J. Chromatogr. A 666 (1–2) (1994) 617–625.

[24] V. Schurig, M. Juza, B.S. Green, J. Horakh, A. Simon, Absolute configurations of the inhalation anesthetics isoflurane and desflurane, Angew Chem. Int. Ed. 35 (15) (1996) 1680–1682.

[25] M.C. Simmons, L.R. Snyder, Two-stage gas-liquid chromatography, Anal. Chem. 30 (1) (1958) 32–35.

[26] W. Bertsch, Two-dimensional gas chromatography. Concepts, instrumentation, and applications - part 1: fundamentals, conventional two-dimensional gas chromatography, selected applications, J. High Resolut. Chromatogr. 22 (12) (1999) 647–665.

[27] L. Mondello, A.C. Lewis, K.D. Bartle, Multidimensional chromatography, John Wiley & Sons Ltd, Chichester, 2002.

[28] C. Meinert, M. Moeder, W. Brack, Fractionation of technical *p*-nonylphenol with preparative capillary gas chromatography, Chemosphere 70 (2) (2007) 215–223.

[29] M. Mandalakis, H. Holmstrand, P. Andersson, O. Gustafsson, Compound-specific chlorine isotope analysis of polychlorinated biphenyls isolated from Aroclor and Clophen technical mixtures, Chemosphere 71 (2) (2008) 299–305.

[30] C. Meinert, E. Schymanski, E. Kuster, R. Kuhne, G. Schuurmann, W. Brack, Application of preparative capillary gas chromatography (pcGC), automated structure generation and mutagenicity prediction to improve effect-directed analysis of genotoxicants in a contaminated groundwater, Environ. Sci. Pollut. Res. Int. 17 (4) (2010) 885–897.

[31] C. Meinert, W. Brack, Optimisation of trapping parameters in preparative capillary gas chromatography for the application in effect-directed analysis, Chemosphere 78 (4) (2010) 416–422.

[32] C. Rühle, G.T. Eyres, S. Urban, J.-P. Dufour, P.D. Morrison, P.J. Marriott, Multiple component isolation in preparative multidimensional gas chromatography with characterisation by mass spectrometry and nuclear magnetic resonance spectroscopy, J. Chromatogr. A 1216 (30) (2009) 5740–5747.

[33] C.P.G. Ruhle, J. Niere, P.D. Morrison, R.C. Jones, T. Caradoc-Davies, A.J. Canty, et al., Characterization of tetra-aryl benzene isomers by using preparative gas chromatography with mass spectrometry, nuclear magnetic resonance spectroscopy, and X-ray crystallographic methods, Anal. Chem. 82 (11) (2010) 4501–4509.

[34] S. Nojima, C. Apperson, C. Schal, A simple, convenient, and efficient preparative GC system that uses a short megabore capillary column as a trap, J. Chem. Ecol. 34 (3) (2008) 418–428.

[35] N. Ochiai, K. Sasamoto, Selectable one-dimensional or two-dimensional gas chromatography-olfactometry/mass spectrometry with preparative fraction collection for analysis of ultra-trace amounts of odor compounds, J. Chromatogr. A 1218 (21) (2011) 3180–3185.

[36] J. Ledauphin, J.F. Saint-Clair, O. Lablanquie, H. Guichard, N. Founier, E. Guichard, et al., Identification of trace volatile compounds in freshly distilled calvados and cognac using preparative separations coupled with gas chromatography-mass spectrometry, J. Agric. Food Chem. 52 (16) (2004) 5124–5134.

[37] P.A. Sutton, C.A. Lewis, S.J. Rowland, Isolation of individual hydrocarbons from the unresolved complex hydrocarbon mixture of a biodegraded crude oil using preparative capillary gas chromatography, Org. Geochem. 36 (6) (2005) 963–970.

[38] H. Ikeura, K. Kohara, X.X. Li, F. Kobayashi, Y. Hayata, Identification of (E, E)-2,4-Undecadienal from coriander (Coriandrum sativum L.) as a highly effective deodorant compound against the offensive odor of porcine large intestine, J. Agric. Food Chem. 58 (20) (2010) 11014–11017.

[39] T.C. Tran, G.A. Logan, E. Grosjean, D. Ryan, P.J. Marriott, Use of comprehensive two-dimensional gas chromatography/time-of-flight mass spectrometry for the characterization of biodegradation and unresolved complex mixtures in petroleum, Geochim. Cosmochim. Acta 74 (22) (2010) 6468–6484.

[40] S.T Chin, B. Maikhunthod, P.J. Marriott, Universal method for on-line enrichment of target compounds in capillary gas chromatography using in-oven cryotrapping. Anal. Chem. 83 (17) (2011) 6485-6492.

CHAPTER

17

Data Analysis Methods

Karisa M. Pierce, Jeremy S. Nadeau†, Robert E. Synovec†*

OUTLINE

17.1. INTRODUCTION

In chromatography, the detected signal of a resolved analyte is usually directly proportional to the concentration of the analyte. Analyte quantification is traditionally achieved either by peak height or by integrating chromatographic peaks, while using the internal standard method [1,2]. Signal integration is generally achieved by baseline-correcting and automatically summing the signals of a peak, often while using Gaussian statistics to define the retention time limits of each peak. Prior to automated peak integration, quantification was occasionally achieved by manually cutting out peaks from a chart recording, and observing the masses of the cutouts. Since those days, analysis methods have

Gas Chromatography DOI: 10.1016/B978-0-12-385540-4.00017-1

evolved from simple integration to powerful chemometric analysis of massive volumes of multidimensional chromatographic data, whereby the term chemometrics refers to applying advanced mathematical methods and algorithms to chemical data. Excellent chemometric textbooks have been written by Massart [3], Brereton [4], Beebe et al. [5], and Sharaf et al. [6]. Analysts can apply chemometrics to both objectively and automatically convert chromatographic data into useful knowledge while reducing subjectivity and reducing manual intervention in the analysis. Chemometric methods should maintain chromatographic data integrity by preserving peak resolution and precision throughout data exportation, data compression, and data reduction steps, with all of these steps tempered to the extent necessary to provide the desired chemical information. Since the more traditional data analysis methods for chromatography are well known, this chapter focuses on chemometric advances for chromatography, and loosely categorizes chemometric methods into the following four procedural categories: preprocessing, pattern recognition, calibration, and experimental method optimization.

The analyst's chemometric options are defined and limited by the dimensions of the data; so within each of the four chemometric categories, this chapter will categorize the chemometrics methods based on the data dimensionality. Data of a variety of dimensions are depicted in Figure 17.1. A one-dimensional (1D) gas chromatograph coupled to a univariate detector, such as a flame ionization detector (GC-FID), continually records detector signal as a function of time, yielding a 1D data vector as shown in Figure 17.1a. Likewise, a univariate detector coupled to a comprehensive two-dimensional (2D) gas chromatograph (GC × GC-FID) continually records detector signal as a function of time, again yielding a 1D data vector as shown in Figure 17.1b. However, this vector is composed of consecutive second-dimension separations defined by the modulation period that can be reshaped or *folded* into a 2D matrix as shown in Figure 17.1c to match the physical process of the modulator in a GC × GC separation. In this 2D matrix, the peak modulations ("slices") of a single chemical component will exhibit a similar first-dimension retention time and identical second-dimension retention time. Thus, it is necessary to combine multiple 1D peak modulations into a single 2D peak that ideally has an integrated signal volume that is proportional to analyte concentration. These 2D chromatograms are frequently depicted either as a surface plot (Figure 17.1d) or as a contour plot (Figure 17.1e), where the contours represent signal magnitude. Thus, for GC × GC data, chemometric options are related to whether or not the data have been folded into a 2D matrix. When multiple 2D chromatograms are combined into a single 3-way array, then a new sample dimension is added to the data (Figure 17.1f). Again, this added dimensionality affects chemometric options.

While an instrument with two separation dimensions and a univariate detector will generate 2D data, a comprehensive 2D gas chromatograph coupled to a multivariate detector, such as a time-of-flight mass spectrometer (GC × GC-TOFMS), will generate three-dimensional (3D) data, which again affects the analyst's chemometric options. When multiple 3D chromatograms are combined into a single 4-way array, then a fourth dimension (the sample dimension) is added and the data are 4D. Figure 17.2 depicts the data structures that are most common in gas chromatography. Figure 17.2 shows how analysts often combine multiple chromatograms into a higher-order array, thus increasing dimensionality in order to use chemometric methods that are only available for high-order data structures. Figure 17.2 also depicts how analysts often *unfold* data and purposely decrease dimensionality in order to

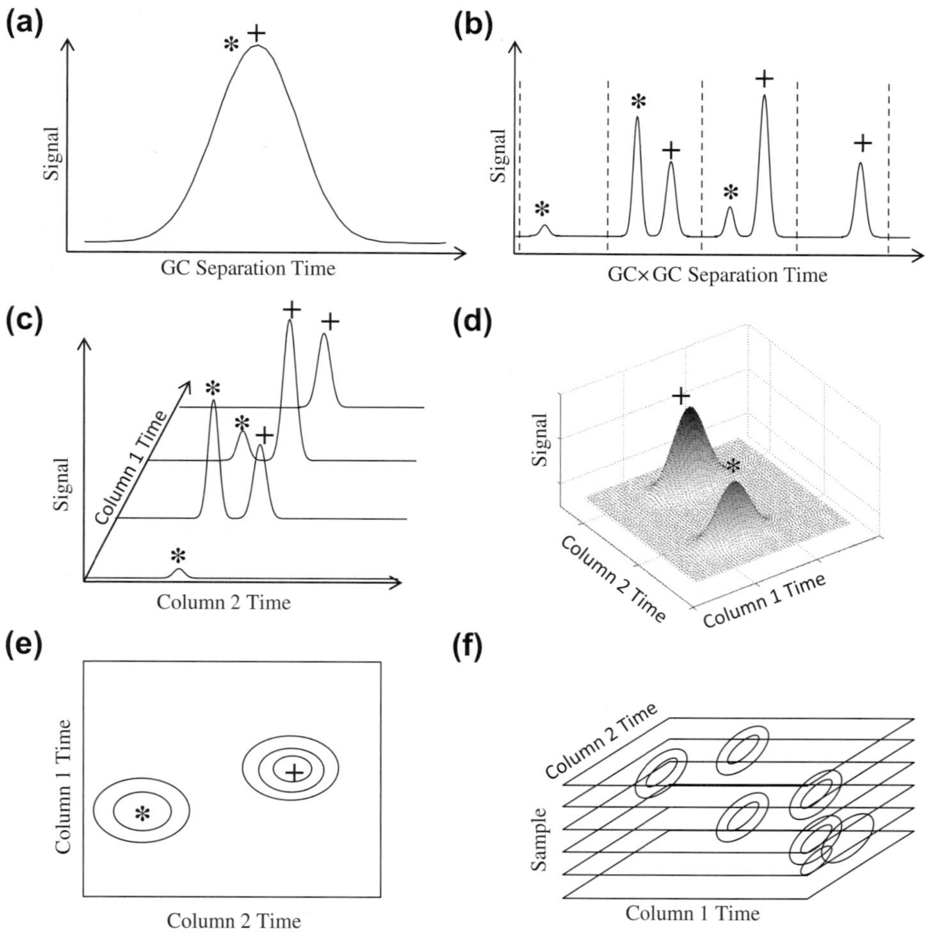

FIGURE 17.1 (a) A univariate detector on a 1D chromatographic instrument collects a 1D data vector as a function of time. (b) A univariate detector on a 2D chromatographic instrument collects a 1D data vector composed of consecutive second-dimension separations, that (c) can then be reshaped into a 2D matrix where the peak modulations ("slices") that elute at a similar column 1 retention time and identical column 2 retention time are considered to be a single compound. Thus, multiple 1D peak modulations are actually a single analyte peak with an integrated signal volume that is proportional to analyte concentration. (d) These 2D chromatograms are frequently depicted as surface plots, or (e) as a contour plot where the contours represent signal magnitude. Two chemical components represented by an asterisk or cross are shown in these figures. (f) When multiple 2D chromatograms are combined into a single 3-way array, then another dimension (the sample dimension) is added to the data.

use chemometric methods that are only available for lower-order data structures. Some chromatographers choose to analyze "peak-level" data (tables of peak data often provided by native instrument software) rather than "pixel-level" data (raw chromatographic data points exported out of native instrument software). The dimensions of peak-level data are generally lower than the dimensions of pixel-level data and so this dimensionality also affects chemometric options. If appropriate preprocessing is applied and if data integrity is maintained

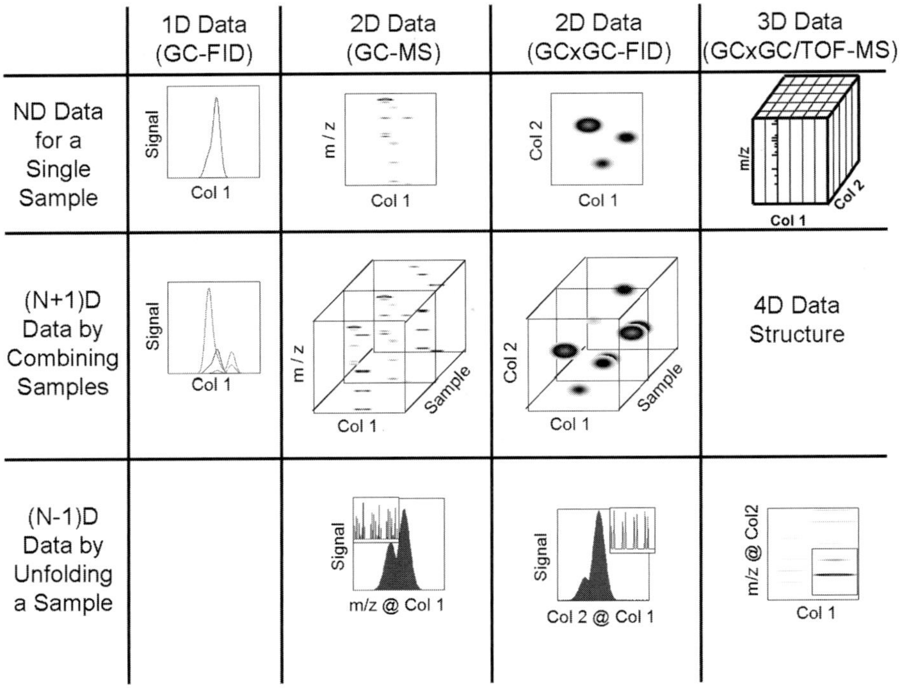

FIGURE 17.2 The analyst's chemometric options are defined by the dimensionality of the data set. If it is appropriate, multiple pixel-level 1D chromatograms can be combined into a 2D matrix that is suitable for certain chemometric methods. Likewise, multiple pixel-level 2D chromatograms can be combined into a 3D array, and 3D chromatograms can be combined into a 4D array, thus increasing dimensionality in order to use desired chemometric methods that are only available for high-order data structures. Analysts can also *unfold* a pixel-level chromatogram and purposely decrease dimensionality in order to use desired chemometric methods that are only available for lower-order data structures.

when raw data in its native file format are converted into a compatible format, then analyzing entire chromatograms at the pixel level should return results that are equally as accurate as analyzing comprehensive peak-level data. However, when the analyst has access to pixel-level data, then this opens up new opportunities to advance the field of novel data analysis software development. Indeed, often when data are output at the peak level, many important data analysis software decisions have been made by the native instrument software that can impact analytical results. This can be an advantage if the analyst is satisfied with the peak-level outputs. However, this can be a disadvantage if the analyst feels the peak-level outputs exhibit shortcomings for their application.

17.2. PREPROCESSING

Interesting chemical variations that reveal important information in chromatographic data are often obscured by chemically irrelevant variations. Preprocessing chromatographic data reduces chemically irrelevant variations and improves results of qualitative and quantitative analyses. The major preprocessing steps that may be required are baseline correction, noise reduction, normalization, and retention time alignment.

17.2.1. Baseline Correction

In chromatographic data analysis at the pixel level, baseline correction is typically the first preprocessing step. Baseline-correction procedures are designed to correct drifting baselines and reduce low-frequency baseline signal variations that arise due to uncontrollable column bleeding, background ionization, and low-frequency detector variations. Figure 17.3

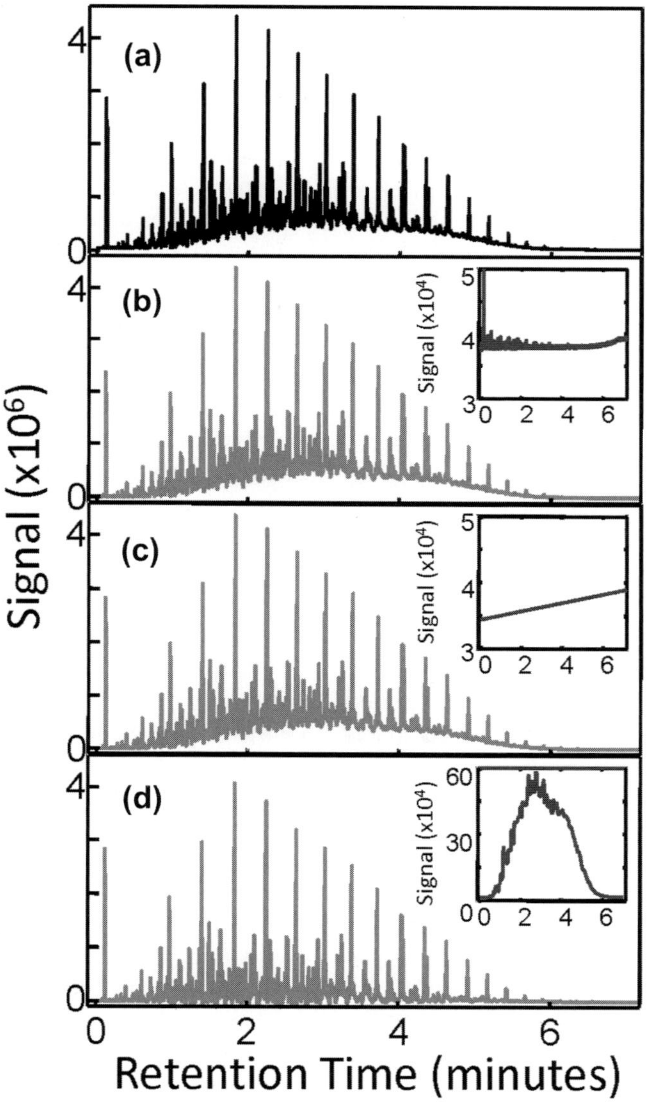

FIGURE 17.3 A single diesel fuel chromatogram is baseline corrected using various baseline-correction procedures. The raw chromatogram in (a) is baseline corrected in (b) using blank run subtraction and in (c) using linear fit subtraction. Both techniques produce almost identical results in this example, while in (d) the higher-order baseline subtraction algorithm removes much of the chemical information. The calculated baselines are shown in the insets.

shows GC chromatograms of a diesel fuel before and after submission to three different baseline-correction procedures. The raw chromatogram is shown in Figure 17.3a. The simplest baseline-correction procedure is to subtract a "blank" chromatogram from the sample chromatogram, as shown in Figure 17.3b. The next simplest baseline-correction procedure is to identify the regions of noise in a sample chromatogram, fit a line through that noise, and subtract the best-fit line from the sample chromatogram as shown in Figure 17.3c. In Figure 17.3c, the beginning and end of the chromatogram were the fitted noise regions. More complex baseline-correction procedures exist to correct wider (more severe) variations [7]. One approach is to subtract the least-squares fit of a high-order polynomial from the sample chromatogram. Another approach is to directly fit a collection of points calculated from the actual chromatogram using a piecewise cubic Hermite interpolation, as shown in Figure 17.3d [7]. In this case, the parameters for piecewise cubic Hermite interpolation were not optimized for the diesel chromatogram, and the high-order baseline was overfit. This illustrates that choosing an inappropriate baseline-correction procedure that overfits the data could remove important chemical variations. The baselines calculated by the different algorithms are shown as insets in Figure 17.3b–d. It is important to understand the chemical nature of the samples and the achieved degree of chromatographic resolution when choosing a baseline-correction procedure because inadvertently fitting a baseline to chemical signals rather than the targeted baseline signal will remove important chemical variations along with the baseline variations. Similar baseline correction procedures can be used with GC × GC and GC–MS chromatograms. These first three baseline-correction procedures perform well for correcting drifting baselines and low-frequency noise, but other noise-reducing procedures are necessary to reduce high-frequency variations.

17.2.2. Noise Reduction

For chromatographic data analysis at the pixel level, several approaches exist to improve the signal-to-noise ratio (S/N). Noise-reduction procedures are designed to reduce uncontrollable instrumental noise and high-frequency noise that may result from flow variations, particulates, or stationary phase escaping the column, external frequencies entering into the electronic signals, and thermal noise. Figure 17.4a shows a raw chromatogram from an isothermal GC separation of a 10-component sample. The end of the raw chromatogram has a very low S/N, which should be improved upon submission to noise-reduction algorithms. A popular noise-reduction procedure is boxcar filtering, which replaces each data point with the average of a certain number of data points surrounding and including that data point [1,2]. The analyst chooses the number of data points that are averaged (boxcar window size) and it is important to choose an appropriate boxcar size that is sufficiently smaller than the peak widths so that noise is minimized while preserving the chemical information present in the analyte peaks. The result of boxcar filtering the raw chromatogram is shown in Figure 17.4b, and it is apparent that the S/N at the end of the chromatogram improves. Median filtering is complementary to boxcar filtering if any outlier signals exist in a chromatogram because the mean is influenced significantly by the outlier itself, while the median is not [8]. The next degree of complexity in noise reduction is using a low-pass filter. This takes advantage of analyte peaks that have a lower-frequency-signal content relative to much of the detected noise. For example, after submitting the chromatogram to the fast Fourier transform (FFT) function, the analyte peaks with 50 ms width at base are located well below 60 Hz in the frequency domain. The signal content defined at frequencies greater than about 60 Hz can be eliminated by replacing the values with zeroes in the frequency domain. The zero-filled frequency domain chromatographic data

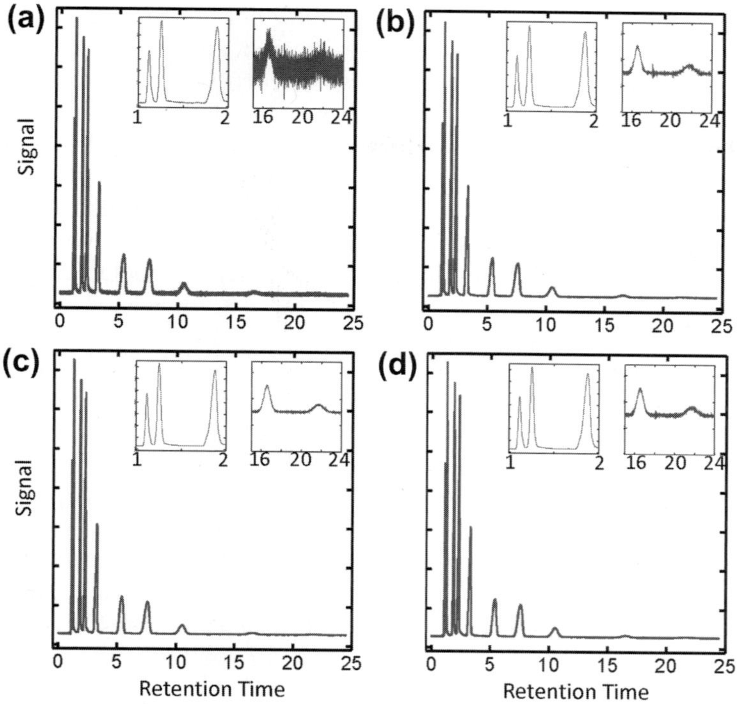

FIGURE 17.4 Different noise-reduction procedures are applied to an isothermal separation. (a) The raw data are (b) boxcar filtered, (c) low-pass filtered with a fast Fourier transform (FFT), and (d) filtered with a Savitsky–Golay smoothing algorithm. The window size for the boxcar filter was 9, the low-pass filter was set to 59 Hz, and the window for the Savitsky–Golay smoothing algorithm was 25.

are then submitted to the inverse fast Fourier transform (IFFT) function to convert the chromatogram back into the time domain. This significantly improves the S/N of the raw chromatogram shown in Figure 17.4c.

Savitzky–Golay smoothing is another popular noise-reduction algorithm that replaces each data point with a polynomial fit to a certain number of data points surrounding the central point within a given window [9]. Again, as with all noise-reduction procedures, the analyst must choose an appropriate window size. Figure 17.4d shows the chromatogram after being filtered by the Savitzky–Golay function using a second-order polynomial fit using an appropriate window size of 25. Care must be taken to not overly smooth the data and hence adversely impact peak shapes and/or chromatographic resolution. In the example presented in Figure 17.4, the resolution has not been compromised at the front end of the isothermal chromatogram, while the noise-reduction procedures improved S/N for the wider peaks at the end of the chromatogram. Other noise-reduction procedures include applying power functions or wavelets to model and eliminate noise. In all of the noise-reduction procedures, the analyst must pick a boxcar size or window that is small enough to preserve the original peak shape and peak area, but large enough to reduce chemically irrelevant noise. Boxcar averaging is the simplest and provides sufficiently improved S/N; hence it is the most commonly applied noise-reduction procedure.

17.2.3. Normalization

During data analysis, chromatograms should be normalized to correct for variation introduced during the injection process, and to correct variations inadvertently introduced during manual sample preparation procedures. The two major forms of normalization are normalizing to an internal standard or, in appropriate cases, using the sum-normalization method. For the simplest internal standard normalization procedure, as per the experimental design the analyst adds a constant amount of one or more non-native standard(s) to every sample. Every sample is then chromatographically separated, ensuring that the internal standard is completely resolved. Each data point in each chromatogram is then divided by the signal of the internal standard in that particular chromatogram. Ultimately, all the normalized chromatograms in the data set have equal internal standard signals, corresponding with the experimental design. This minimizes variation introduced due to the injection process. However, new problems may arise such as undetected coelution with a native component, difficulty in choosing a non-native standard that is completely resolved from all other components of complex unknown samples, and ensuring that the non-native chemical is inert and will not react with components of unknown samples. Another potential shortcoming of the internal standard method is that variation introduced by the additional sample handling could be worse than the variation in the injection process. For instance, it is difficult to precisely manipulate small volumes of viscous or volatile standards. When it is not practical to use the internal standard method, some analysts use the sum-normalization method. The sum-normalization method uses the total sum of all baseline-corrected signals as the normalization factor, so each data point in a chromatogram is divided by the total sum of all the data points in that chromatogram, and thus all chromatograms in a sum-normalized data set would sum to unity. The major assumption here is that within the constraints of the instrumental error, the samples are sufficiently similar such that equal volumes of the samples should have equal total signals at the detector. Indeed, different chemicals have different response factors, and so the major assumption is not strictly true; however, if there is less variation in total response factors than the variation introduced by injector volume discrepancies, then the sum-normalization method can be supported and applied with confidence. The sum-normalization method is described in chromatography textbooks [10,11].

17.2.4. Retention Time Alignment

Retention time variation obscures important chemical variations when comparing one chromatogram to the next; therefore, retention time alignment is a critical preprocessing step for nearly all chemometric methods. Retention time alignment algorithms shift peak positions, so each peak has an accurate retention time. These algorithms should also preserve the accuracy of peak areas or peak volumes. Alignment algorithms can be classified into four major categories: simple scalar shift, alignment to select target peaks, local alignment algorithms, and globally optimized alignment algorithms. The simplest alignment algorithms apply a scalar shift to the entire sample chromatogram. Scalar shift algorithms quickly use a cross-correlation coefficient (or other similarity metric) to calculate the shift that minimizes the difference between the sample and target chromatograms. The problem is that the similarity metrics are usually heavily influenced by the largest peak in the chromatogram. For more complex shifting, more sophisticated alignment algorithms can align selected standard peaks in a sample chromatogram to the same standard peaks in a target chromatogram. However, if the samples are complex and shifting is present, it is difficult to algorithmically identify the standard peaks without manual intervention. For cases when selected standard peaks are not available, and reducing manual intervention is necessary, local alignment algorithms are useful. Local alignment algorithms iteratively shift subregions of the sample chromatogram across the target chromatogram until a matching metric, such as a correlation coefficient, is maximized, thus yielding the shift correction that is applied to each subregion [12–15]. The most sophisticated, robust, and powerful alignment algorithms are globally optimized alignment algorithms that can handle severe and dynamic shifting [16–18]. The globally optimized algorithms can make use of dynamic programming to find the locally and globally optimized shift for every window in the chromatogram. The correlation optimized warping (COW) algorithm is the most popular alignment method produced for globally optimized alignment. The algorithm was introduced in 1998 by Nielsen et al. [16]. It has been improved by various chemometricians to be applied to GC-FID [16,17], GC–MS [19], GC × GC-TOFMS [20], and GC with diode array detection (GC-DAD) [16], but all of the algorithms are very similar. The chromatograms are separated into windows, sometimes called segments, and each window is shifted and warped until the maximum correlation between the sample window and the target is determined, yielding the locally optimized shift. Dynamic programming is used to find the globally optimized shift. Starting at one end of the chromatogram, the locally optimized correlation paths are added together until the opposite end of the chromatogram is reached. At this point, the globally optimized shifts are the maximum values for each of the locally optimized shifts creating one globally optimized path for all the shifts for each window. This is clearly outlined in the pioneering report by Nielsen et al. [16] and also in Massart's insightful comparison of alignment algorithms [21]. The piecewise alignment (PWA) algorithm also uses correlations between windows of a sample chromatogram and a target chromatogram to calculate optimal shifts for a window [18]. By using the locally optimized shifts and comparing the correlations of surrounding windows, a globally optimized path can be calculated in a similar way to the COW algorithms. The alignment of the data using PWA is much faster than the COW algorithms, because there is no warping of the data. Because the window sizes for the globally optimized PWA algorithm and the segment sizes for COW are approximately the size of a peak width, both algorithms accurately align chromatograms as well as accurately preserve peak areas [18]. By combining a fast cross-correlation coefficient algorithm with the COW algorithm, the ChromAlign software reduces the number of iterative calculations required by the COW algorithm, and yields a fast alignment algorithm for GC–MS chromatograms [22]. For the local alignment algorithms and globally optimized alignment algorithms, one must optimize certain parameters such as window length, selecting a target chromatogram, and sometimes even selecting a maximum shift. Analysts typically want parameter optimization to be objective, automated, and reduce manual intervention [23–25].

17.2.5. Software Platforms

In order to apply preprocessing procedures and advanced chemometric methods, the analyst must be able to import pixel-level or peak-level data into user-friendly data analysis platforms. Commercially available software is critical in this regard. Common data analysis platforms are MATLAB (Mathworks), Excel (Microsoft), and SAS (SAS Cary, NC). These platforms can process both pixel-level and peak-level data and they allow the analyst to develop algorithms and visualize their data. The most common software packages that provide pixel-level visualization are ChromaTOF (LECO), ChemStation (Agilent, Santa Clara, CA, USA), GC Solution or GCMS Solution from Shimadzu (Columbia, MD, USA), Xcalibur from Thermo Fisher Scientific (Waltham, MA, USA), and the Reichenbach et al. GC Image GC × GC Software for 3D chromatographic data (http://www.gcimage.com) [26].

ChromaTOF and ChemStation software programs provide easily exportable peak-level data with preprocessing, calibration, and/or pattern recognition capabilities. GC Image software provides pixel-level analysis options such as baseline correction, alignment, blob detection, and differential analysis, and it converts the data into easily exportable formats. The most common chemometrics software packages for processing pixel-level or peak-level data are the PLS Toolbox (Eigenvector Research, Inc), Pirouette (Infometrix), Statistica (Statsoft Inc), and SIMCA (Umetrics, Umean, Sweden).

17.3. PATTERN RECOGNITION

Pattern recognition is a very important chromatographic data analysis tool, especially to study complex samples, and/or when it may be impractical or insufficient to identify and quantify all of the peaks present. Pattern recognition generally takes advantage of the reproducible "chemical fingerprint" provided by a sample in a chromatographic context. Chemometric pattern recognition methods can be grouped into two categories: unsupervised or supervised. Unsupervised pattern recognition methods are helpful when the analyst desires to discover the class membership of a sample. Supervised pattern recognition methods are helpful when the analyst desires to discover the chemical components that distinguish sample classes, so the algorithm requires class information to be input by the analyst. Mathematical resolution of unresolved chromatographic peaks is also a form of a pattern recognition method that can be either supervised or unsupervised. Resolution algorithms are expected to recognize typical peak shapes and sometimes specific spectral profiles; so in this chapter resolution algorithms are categorized as pattern recognition algorithms.

17.3.1. Hierarchical Cluster Analysis

Hierarchical cluster analysis (HCA) is a common unsupervised pattern recognition method. HCA is one of many clustering algorithms that generally function by calculating the distance between samples in the original independent variable space, where distance can be defined as the Euclidean distance or Mahalanobis distance among all samples from the centroid or origin. Classification is achieved by assuming close samples are chemically similar to each other while distant samples are chemically different from each other. Figure 17.5 illustrates how a chromatogram can be thought of as a single point in independent variable space. A pixel-level chromatogram that may contain thousands of data points is described by a single vector endpoint that is positioned by the chromatographic signal (dependent variable) at each retention time data point (independent variable) in independent variable space where each independent variable has its own axis. Dendrograms are often the graphical output containing the distance in independent variable space among the samples. HCA is applicable to multiple pixel-level or peak-level 1D chromatograms that are combined into a 2D matrix. HCA can be applied to multiple pixel-level 2D or 3D chromatograms, if the 2D or 3D chromatograms are each unfolded into 1D data vectors and then combined into a single 2D matrix, as shown in Figure 17.2. An interesting law enforcement application of HCA was reported to model and accurately classify GC × GC-FID and GC × GC-TOFMS separations of illicit drug samples that had been seized by the police [27].

17.3.2. Principal Component Analysis

Principal component analysis (PCA) is a common unsupervised pattern recognition method. Just as in HCA, the first step in PCA is to plot each mean-centered 1D data vector in independent variable space, so each

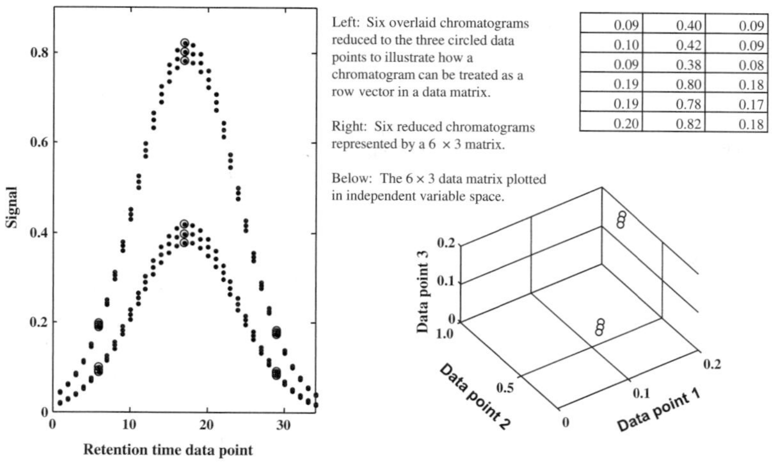

0.09	0.40	0.09
0.10	0.42	0.09
0.09	0.38	0.08
0.19	0.80	0.18
0.19	0.78	0.17
0.20	0.82	0.18

FIGURE 17.5 Illustration of how a chromatogram can be thought of as a single point in independent variable space. A 1D pixel-level chromatogram is described by a single vector endpoint that is positioned by the chromatographic signal (dependent variable) at each retention time data point (independent variable) in independent variable space wherein each data point has its own axis. Chemically similar chromatograms will cluster together in independent variable space. This example shows two clusters, so two classes of samples must be present in the original data set of six chromatograms. This simple data set can be manually classified by visual observation of the overlaid chromatograms, but the same idea applies to large data sets of complex chromatograms where visually observing the chromatograms is not effective.

chromatogram is represented by a single vector endpoint. Then, a line of finite length called a principal component (PC) is fit so that it captures the greatest variance in those vector endpoints. A second principal component (PC 2) that is orthogonal to the first PC (PC 1) is fit such that it captures the next greatest variance in those vector endpoints. More orthogonal PCs are fit and ordered based on percent variance captured until 100% of the variance is captured. This allows the analyst to see the vector endpoints in PC space, rather than in the original independent variable space. Since the PCs are ordered and nested, the analyst can truncate the later PCs and focus on the vector endpoints projected onto the primary PCs that captured the most variation (i.e. the most useful information). The distance between each vector endpoint and each PC is called a score. Samples that are similar to each other will have scores that cluster together in PC space. Samples that are different from each other will have scores that are further apart in PC space. Independent variables (retention time data points) that have the most influence on the clustering observed in the scores will be the most positively loaded and most negatively loaded variables because the loadings for each variable are the cosine of the angle between each PC and each original independent-variable axis. If a retention time data point has a substantial amount of signal variation over the data set, then it will be highly loaded (very negatively or very positively for mean-centered data) in the primary PCs. Chemical similarities (or differences) among samples can be deduced from the clustering in the scores plot and by identifying the retention time data points that are highly loaded in the loadings plots for each PC. Noise, which may have obscured important chemical information in the raw data, is removed when the scores are visualized in the truncated PC space. Chemometric texts provide detailed explanations of the linear algebra behind all the chemometrics described in this chapter [3–6,28]. Data are usually mean-centered and sometimes

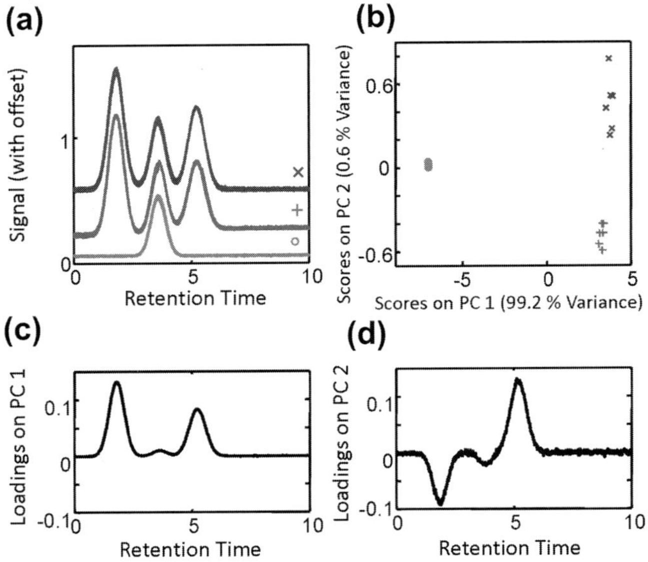

FIGURE 17.6 Illustration of how PLSDA models show covariation between given quantitative information and chromatographic peaks for three simulated chromatograms. A PLSDA loading plot would have positive loadings for the middle peak, negative loadings for the last peak, and little to zero loadings for the regions that have little to no correlation with the given quantitative information (first peak and noise regions).

autoscaled prior to PCA. Figure 17.6 illustrates PCA applied to 18 simulated chromatograms containing thousands of data points. The 18 overlaid chromatograms in Figure 17.6a are actually 6 replicates of 3 classes of samples. For this simple example, manual intervention such as offsetting overlaid chromatograms allows the analyst to determine how many classes there are in the data set and which chromatograms belong to each class. However, for truly complex sample analysis, manual observation and manual intervention generally do not yield class information. To provide class information, PCA fits PC 1 and PC 2 to capture the greatest variations of the 18 vector endpoints in variable space, and the projections of these endpoints onto the PCs are the scores on PC 1 and PC 2, which are shown in Figure 17.6b, where the clustering indicates that there are indeed 6 replicates of 3 classes of samples. The highly loaded variables on PC 1 are shown in Figure 17.6c, where it is apparent that the earliest eluting component peak and the latest eluting component peak were responsible for distinguishing two of the classes (symbolized by × and +) from the third class (symbolized by o). The variables with little to zero loadings on PC 1 are also shown in Figure 17.6c, indicating that the component with the middle elution time did not differentiate the samples significantly, and the noise regions in the chromatograms did not play a role in differentiating the components at all. The variables with largest loadings on PC 2 are shown in Figure 17.6d, indicating that the latest eluting component was responsible for differentiating the two classes that PC 1 did not previously differentiate (symbolized by × and +). PCA is applicable to 1D chromatograms that are combined into a 2D matrix. PCA can be applied to multiple pixel-level 2D or 3D chromatograms if the multiple 2D or 3D chromatograms are each unfolded into 1D vectors and then combined into a single 2D matrix, as shown in Figure 17.2. PCA is applicable to peak-level data as long as the 1D data vectors for each sample are combined into a 2D matrix. An interesting application of PCA was reported to model and classify GC−MS separations of oil spill samples and oil source samples, demonstrating the ability to accurately determine the source of spill samples gathered from the coastal environment [29].

17.3.3. Discriminant Analysis

Partial least squares discriminant analysis (PLSDA) is another common pattern recognition method, but, unlike PCA, PLSDA is supervised. In a PCA model, the PCs (latent variables) are defined by the signals that mostly vary, but the latent variables in a PLSDA model are defined by the signals that both vary and correlate (i.e. covary) with given quantitative information about each sample class. The analyst mines the data and inspects PLSDA loading plots of primary latent variables to discover chemicals that covary with the given quantitative information [30,31]. Figure 17.7 illustrates how PLSDA models are based on covariations between given quantitative information and chromatographic data. Three simulated chromatograms with given quantitative levels (90 units, 60 units, and 30 units) are shown. PLS loadings for latent variable 1 would positively load the middle chromatographic peak because its signal variations positively correlate with the given quantitative levels. PLS would negatively load the latest eluting peak because its signal variations negatively correlate with the given quantitative levels. The PLS loadings for the earliest peak would be zero because there is no correlation between the signals for that peak and the given quantitative levels. Likewise, noise regions in the chromatograms would have little to no correlation, so zero loadings are observed. Like PCA, PLSDA can be applied to 2D or 3D chromatograms as long as the chromatograms are unfolded and combined into a single 2D matrix and as long as retention time precision is adequate. An interesting application of PLSDA was reported to

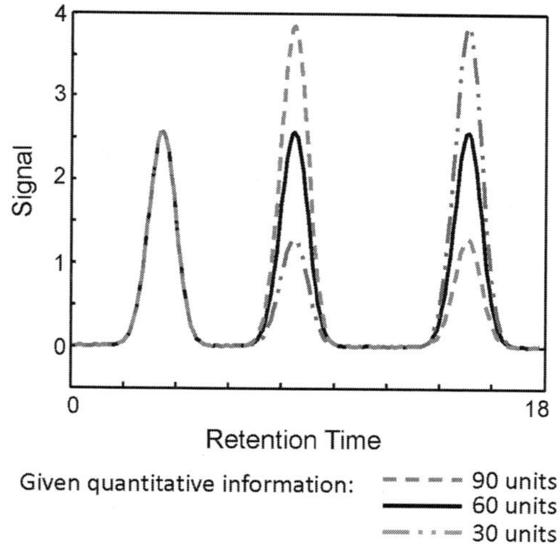

FIGURE 17.7 Illustration of how PLSDA models show covariation between given quantitative information and chromatographic peaks for three simulated chromatograms. A PLSDA loading plot would have positive loadings for the middle peak, negative loadings for the last peak, and little to zero loadings for the regions that have little to no correlation with the given quantitative information (first peak and noise regions).

mine GC−MS separations of urinary metabolites from cancer patients, yielding confirmation of two known biomarkers and three new potential biomarkers for certain cancers [32].

Supervised pattern recognition methods require an experimental design where sample class membership is known, sufficient replicates are obtained, and important sources of variation are modeled. In this regard, the Fisher criterion is a statistic that calculates the ratio of class-to-class signal variance (information) relative to the within-class signal variance (noise), as a function of an independent variable such as the data point location along the retention time dimension [3,33]. A variety of supervised algorithms exists that use the Fisher ratio criterion (or another discriminatory criterion) to do a point-by-point calculation at the pixel level (or at the peak level) in a data set of multiple chromatograms [34−37]. Chromatographic features with large Fisher ratios (or other discrimination metric) reveal the chemicals that significantly differentiate the sample classes. This is sometimes called supervised feature selection. Algorithms based on this sort of point-by-point comparison are applicable to 2D, 3D, or 4D data as long as retention time precision is adequate. The point-by-point comparison basically means the method is univariate. Linear discriminant analysis (LDA) is a multivariate algorithm that uses the Fisher criterion to seek linear combinations of variables that distinguish classes as long as the classes are normally distributed [37]. Whether univariate or multivariate, feature selection algorithms can greatly reduce data density by filtering out retention time data points that do not contain relevant chemical information as long as retention time precision is adequate. Orthogonal signal correction (OSC) is another supervised feature selection method that identifies variations in the chromatographic data that are orthogonal to given quantitative information and then removes irrelevant signals, so the important chemical variations are more likely to be captured and modeled by early latent variables. For the same purpose, some analysts use unsupervised pattern recognition algorithms such as PCA to filter interesting chromatographic features from noise prior to submission to other chemometric methods [38].

17.3.4. Resolution Methods

Mathematical resolution methods are able to resolve overlapping peaks in a chromatogram. In this chapter, mathematical resolution methods are loosely classified as pattern recognition methods because they are often expected to recognize particular peak shapes or spectral profiles. Multivariate curve resolution (MCR) and the generalized rank annihilation method (GRAM) are mathematical resolution methods applicable to 2D chromatograms that are bilinear, meaning the chromatographic profile of the pure compound can theoretically be defined by the outer product of two vectors (the chromatographic peak profile on each dimension). This bilinearity is instrumentally achieved if retention time precision is adequate and if signals of coeluting components are linearly additive. MCR uses chemically selective portions of a chromatogram where there is only one component to obtain the correct shapes of each fully resolved peak. Then, that shape is used as a constraint, along with a nonnegativity constraint, to linearly combine estimated pure profiles until the calculated chromatogram approximates the given unresolved chromatogram via alternating least squares fitting. This reveals the pure peak profiles and peak area information [39]. GRAM requires that both a sample chromatogram and a standard chromatogram be augmented together to form a single matrix in order to resolve overlapping peaks in the sample chromatogram [40]. As such, in addition to being a preprocessing resolution algorithm, GRAM could also be considered a single point external standard calibration method based on the four categories defined earlier in this chapter. Many analysts use rank minimization prior to GRAM in order to improve data structure bilinearity and retention time precision. The rank minimization algorithm seeks to maximize the information content in the lower singular values using singular value decomposition (SVD) to calculate loadings for the original independent variables that are consistent with the previous description of PCA loadings. An interesting application of GRAM was reported to mathematically resolve subregions of GC × GC-FID separations containing overlapping alkyl benzenes, demonstrating the ability to accurately quantify low concentrations of overlapping analytes even in regions with a higher concentration of interfering compounds [41].

Parallel factor analysis (PARAFAC) and trilinear decomposition (TLD) are mathematical resolution methods that are applicable to either a single 3D chromatogram (such as GC × GC-TOFMS) or multiple 2D chromatograms that have been combined into a 3D matrix as shown in Figure 17.2. The 3D data must exhibit a sufficiently trilinear structure, meaning the signal for a pure compound can theoretically be defined by the outer product of three vectors, which is instrumentally achieved if retention time precision is adequate and if signals of coeluting components are linearly additive. Impressively, PARAFAC back-calculates the outer product of the three vectors and predicts pure component concentrations, pure spectral profiles, and pure chromatographic peak profiles for analyte signals that were originally overlapping using alternating least squares with unimodality and nonnegativity constraints to converge upon a best fit. PARAFAC greatly improves S/N by removing noise as an independent component separate from the chemical signals present [42]. TLD estimates the profiles of each component by eigenvector decomposition and then estimates the component concentrations by least squares fitting [43,44]. As long as the data are sufficiently trilinear, as is generally the case for GC × GC-TOFMS, for example, and there is adequate selectivity for each of the coeluting components on two of the three dimensions, then PARAFAC and TLD can resolve overlapping peaks [45,46]. PARAFAC and TLD algorithms are usually only applicable to manually selected chromatographic subregions where a maximum of about 6–7 chemical compounds are present in the subregion, with a smaller rank preferred. However, to avoid the step of selecting a subregion, a version of PARAFAC that automatically resolves all peaks in an *entire* GC × GC-TOFMS chromatogram has been reported [47]. The key to this was an algorithm that automatically selects an appropriate number of factors for PARAFAC modeling (i.e. automated rank determination), thus reducing the number of required analyst inputs and reducing the amount of manual intervention [48,49]. A derivative of PARAFAC called PARAFAC2 was developed for data that severely deviate from trilinearity due to insufficient retention time precision or a drifting retention time in the second dimension of GC × GC. PARAFAC2 can improve trilinearity of appropriate data structures, thus improving resolution results [50,51].

Nontarget algorithms that are designed to qualitatively identify all sample components using mass spectral data can also be considered pattern recognition algorithms because they are expected to recognize similarities between observed spectra and library mass spectra. The publicly available US National Institute of Standards and Technology (NIST) MS Search for electron impact spectra allows analysts to input observed mass spectra and the algorithm outputs a best match to identify analytes. Many instrument manufacturers have incorporated the NIST MS Search algorithm as a data processing option in their native instrument software. The automated mass spectral deconvolution and identification system (AMDIS) also matches observed mass spectra with library mass spectra for identification, and it also has a mathematical resolution (deconvolution) capability [52]. Both the MS Search and AMDIS algorithms can be applied to mass spectra from GC–MS or GC × GC–MS chromatograms.

17.4. CALIBRATION

Calibration methods are used to predict a quantitative property of interest for an unknown sample by regression. A training set of chromatograms with known quantitative properties is used to build the regression model. Models are generally evaluated using an independent test set or by a leave-one-out cross-validation procedure. Classical least squares (CLS) is a calibration method that is used to mathematically resolve and quantify chromatographic peaks since, in chromatography, pure component signals are generally additive and linearly related to concentration [53,54]. To construct a CLS model, the chromatograms of pure standards must be available and the algorithm calculates a standard mixture chromatogram that is the linear combination of the pure component peaks with known concentration coefficients. An unknown sample chromatogram is regressed onto this model by least squares fitting, yielding the pure resolved profiles of each component. A shortcoming of CLS is the presence of completely unresolved and thus undetected interfering components. CLS is applicable to pixel-level 1D chromatograms or spectra. An interesting application of CLS was reported to resolve mass spectra of overlapping GC × GC-TOFMS peaks that were initially a combination of a spiked ^{13}C-labeled analyte peak and the unlabeled analyte peak. CLS provided the pure mass spectra, the $^{12}C/^{13}C$ ratio, and, thus, absolute amounts of ^{12}C and ^{13}C in the unknown samples [53].

17.4.1. Partial Least Squares Regression

Partial least squares (PLS) is a calibration method that highly loads chromatographic features that covary with the given quantitative information [54,55]. The loadings are linearly combined into latent variables, with the most highly loaded independent variables (retention time data points) captured in the primary latent variables. The analyst chooses a subset of the latent variables to build the calibration model. The PLS algorithm then estimates the quantitative property for an unknown chromatogram by regression. Prior to applying PLS, the matrix of chromatograms and the quantitative properties are both usually mean-centered. PLS is designed for multiple 1D chromatograms combined into a 2D matrix. It is also applicable to data sets of 2D or 3D chromatograms as long as they are unfolded into 1D chromatograms and combined into a single 2D matrix. However, keeping data in their original structure without unfolding may improve accuracy; so N-PLS was developed and is applicable to even higher dimensions of data [56]. N-PLS can accept an ND matrix composed of multiple (N-1)D chromatograms. An interesting application of PLS was reported to model and accurately predict quantities of naphthalenes in various jet fuel samples using GC × GC-FID data [57].

17.4.2. Principal Component Regression

Principal component regression (PCR) is a calibration method that functions by building a model that regresses given quantitative values onto the PCA scores for a training set, and then the scores for an unknown chromatogram are submitted to that model to quantify the unknown. Mean centering and sometimes autoscaling the data are common procedural steps prior to PCR. An interesting application of PCR was reported to model and classify food packaging materials according to emissions of volatile chemicals from the wrappers [58].

Overall, calibration models perform best when future samples are similar to the samples that were in the training set. All calibration models decline when forced to extrapolate beyond the range of the variations in the modeled training set. Therefore, all calibration models should be evaluated by checking the accuracy of predicted values for an independent test set, or if the size of the data set is limited, then by using the leave-one-out cross-validation technique. Leave-one-out cross-validation is a traditional method whereby, one at a time, each chromatogram is pulled out of the training set, the model is built with the remaining chromatograms, and then that left-out chromatogram is submitted to the model to predict that sample's quantitative property. This is repeated for every chromatogram in the training set and yields a leave-one-out cross-validation plot of predicted quantity versus known quantity (typically obtained from a benchmark method). The calibration model is evaluated based on the slope, R^2, and relative root-mean-squared-error of cross-validation (RMSECV) values of this plot. When the model is robust and useful, the covariations between quantitative property and chromatographic signal are strong, the slope = 1.00, ideal R^2 values are close to 1.0, and the ideal RMSECV value approaches 0. Such a robust model will then yield reliable quantitative predictions for unknown samples. The underlying assumption is that the benchmark method to provide the known quantity information is sufficiently accurate and precise for modeling purposes. A typical application of this approach is to provide a fast method via the predictive modeling to replace a slow (often more traditional) method used for benchmarking.

17.5. EXPERIMENTAL METHOD OPTIMIZATION

Chromatographic method development often requires that a large number of parameters be considered and optimized: choice of stationary phase(s), column dimensions, flow rate, temperatures, run time, detector settings, types of replicates, numbers of replicates, and a host of sample preparation options. Optimization of these parameters can be determined by assessing a variety of properties: resolution of key analytes, S/N, response factors, magnitude of variance, sources of variance, use of the separation space, and 2D separation orthogonality. Statistical experimental design methods provide mathematical optimization of combinations of these continuous experimental variables. The classic statistical designs are factorial designs, response surface designs, mixture designs, and simplex optimization [4,59]. Classic chemometric texts provide additional information about each of these statistical designs [4]. It is likely that industrial researchers who are doing large-scale and/or long-term research applications are using the classic statistical designs to make intelligent decisions about efficient sampling options, instrumental options, and data analysis methods, all to improve analytical performance, to save time, and to lower expense. Less commonly, individual analysts doing individual research will apply classic statistical design strategies, but individual analysts do monitor experimental response as a function of experimental conditions, and they do carefully consider obtaining appropriate replicates, using appropriate calibration methods, and modeling appropriate sources of variation, all in order to properly support their results. There are generally two approaches that individual researchers use to seek optimal parameters: methods that assess experimentally obtained chromatographic data and output the optimal parameters [60,61] or methods that use given parameter ranges to simulate chromatograms and then assess the simulated chromatograms to determine the optimal parameters [62,63]. An interesting method that uses experimental chromatographic data is a closed-loop machine-learning program that repeatedly runs samples on a GC $\times$ GC-TOFMS, iteratively analyzes the collected data, and adjusts instrument parameters until the separations are at an optimum based on optimizing resolution, S/N, and run time [60]. Experimentally obtained chromatographic data were also used to thoroughly and definitively optimize the experimental design for extraction and GC−MS analysis of human blood plasma metabolites based on PLS results [64].

17.6. CONCLUSION

This chapter introduced basic chemometric methods for chromatographic data analysis while loosely categorizing them as preprocessing, pattern recognition, calibration, or experimental optimization methods. There are many derivatives of these chemometrics that were not mentioned. Some of these derivatives actually cause the chemometric method to shift categories. For instance, when PCA is used as a noise-reduction method prior to other chemometrics, it could fit in the preprocessing category instead of the pattern recognition category. Another example is TLD applied prior to PARAFAC to improve trilinearity, so this method could fit in the preprocessing category as well as the pattern recognition category [46].

Some of the derivatives allow the algorithms to relax their requirements and still yield accurate results, such as fuzzy methods that are robust against shifting or indirect classical least squares (ICLS) that does not require given standards (as CLS requires) to resolve overlapping peaks, and all the nonlinear chemometric methods. For information about these and more, the classic chemometric textbooks are excellent resources [3–6]. In this chapter, we also refrained from explaining the linear algebra driving the chemometrics. For descriptions of the math behind the chemometrics, the classic chemometric textbooks and the PLS Toolbox user's manual are excellent resources [28]. For information about recent developments in chemometrics, review articles covering chemometrics applicable to gas or liquid chromatography are good sources [65–69]. Overall, it is indeed exciting to see how the field of gas chromatography is being advanced by the emergence, acceptance, and application of chemometric data analysis methods.

References

[1] D.A. Skoog, F.J. Holler, T.A. Nieman, Principles of instrumental analysis, vol. 5, Sanders College Publishing, San Francisco, 1998.

[2] D.C. Harris, Quantitative chemical analysis, vol. 8, W. H. Freeman and Company, New York, 2010.

[3] D.L. Massart, Chemometrics: a textbook, Elsevier Sciences Ltd., New York, 1988.

[4] R.G. Brereton, Chemometrics: data analysis for the laboratory and chemical plant, Wiley, New York, 2003.

[5] K.R. Beebe, R.J. Pell, M.B. Seasholtz, Chemometrics: a practical guide, Wiley-Interscience, New York, 1998.

[6] M.A. Sharaf, D.L. Illman, B.R. Kowalski, Chemometrics, John Wiley & Sons, New York, 1986.

[7] L.F. Zhu, R.G. Brereton, D.R. Thompson, P.L. Hopkins, R.E.A. Escott, On-line HPLC combined with multivariate statistical process control for the monitoring of reactions, Anal. Chim. Acta. 584 (2) (2007) 370–378.

[8] S.C. Wang, S.M. Chiang, C.M. Huang, Parametric studies of matched filters to enhance the signal-to-noise ratios of LC-MS-MS peaks, Anal. Chim. Acta. 556 (1) (2006) 201–207.

[9] A. Savitzky, M.J.E. Golay, Smoothing and differentiation of data by simplified least squares procedures, Anal. Chem. 36 (8) (1964) 1627–1639.

[10] K. Robards, P.R. Hadad, P.E. Jackson, Principles and practice of modern chromatographic methods, Academic Press, New York, 1994.

[11] M.L. Lee, F.J. Yang, K.D. Bartle, Open tubular column gas chromatography, John Wiley & Sons, New York, 1984.

[12] C.G. Fraga, B.J. Prazen, R.E. Synovec, Comprehensive two-dimensional gas chromatography and chemometrics for the high-speed quantitative analysis of aromatic isomers in a jet fuel using the standard addition method and an objective retention time alignment algorithm, Anal. Chem. 72 (17) (2000) 4154–4162.

[13] C.G. Fraga, B.J. Prazen, R.E. Synovec, Objective data alignment and chemometric analysis of comprehensive two-dimensional separations with run-to-run peak shifting on both dimensions, Anal. Chem. 73 (24) (2001) 5833–5840.

[14] B.J. Prazen, R.E. Synovec, B.R. Kowalski, Standardization of second-order chromatographic/spectroscopic data for optimum chemical analysis, Anal. Chem. 70 (2) (1998) 218–225.

[15] K.J. Johnson, B.W. Wright, K.H. Jarman, R.E. Synovec, High-speed peak matching algorithm for retention time alignment of gas chromatographic data for chemometric analysis, J. Chromatogr. A 996 (1–2) (2003) 141–155.

[16] N.P.V. Nielsen, J.M. Carstensen, J. Smedsgaard, Aligning of single and multiple wavelength chromatographic profiles for chemometric data analysis using correlation optimised warping, J. Chromatogr. A 805 (1–2) (1998) 17–35.

[17] G. Tomasi, F.v.d. Berg, C. Andersson, Correlation optimized warping and dynamic time warping as preprocessing methods for chromatographic data, J. Chemometr. 18 (5) (2004) 231–241.

[18] J.S. Nadeau, B.W. Wright, R.E. Synovec, Chemometric analysis of gas chromatography-mass spectrometry data using fast retention time alignment via a total ion current shift function, Talanta 81 (1–2) (2010) 120–128.

[19] M. Chae, R.J. Shmookler Reis, J.J. Thaden, An iterative block-shifting approach to retention time alignment that preserves the shape and area of gas chromatography-mass spectrometry peaks, BMC Bioinformatics 9 (Suppl. 9) (2008) S15.

[20] D. Zhang, X. Huang, F.E. Regnier, M. Zhang, Two-dimensional correlation optimized warping algorithm for aligning GC × GC-MS data, Anal. Chem. 80 (2008) 2664–2671.

[21] V. Pravdova, B. Walczak, D.L. Massart, A comparison of two algorithms for warping of analytical signals, Anal. Chim. Acta. 456 (1) (2002) 77–92.

[22] R.G. Sadygov, F.M. Maroto, A.F.R. Huhmer, ChromAlign: a two-step algorithmic procedure for time alignment of three-dimensional LC-MS chromatographic surfaces, Anal. Chem. 78 (24) (2006) 8207–8217.

[23] K.M. Pierce, B.W. Wright, R.E. Synovec, Unsupervised parameter optimization for automated retention time alignment of severely shifted gas chromatographic data using the piecewise alignment algorithm, J. Chromatogr. A 1141 (1) (2007) 106–116.

[24] T.I. Dearing, J.S. Nadeau, B.G. Rohrback, L.S. Ramos, R.E. Synovec, Real-time target selection optimization to enhance alignment of gas chromatograms, Talanta 83 (3) (2011) 738–743.

[25] T. Skov, F.V.D. Berg, G. Tomasi, R. Bro, Automated alignment of chromatographic data, J. Chemometr. 20 (11–12) (2006) 484–497.

[26] S.E. Reichenbach, X. Tian, Q. Tao, E.B. Ledford, Z. Wu, O. Fiehn, Informatics for cross-sample analysis with comprehensive two-dimensional gas chromatography and high-resolution mass spectrometry (GC × GC-HRMS), Talanta 83 (4) (2010) 1279–1288.

[27] T. Groger, M. Schaffer, M. Putz, B. Ahrens, K. Drew, M. Eschner, et al., Application of two-dimensional gas chromatography combined with pixel-based chemometric processing for the chemical profiling of illicit drug samples, J. Chromatogr. A 1200 (1) (2008) 8–16.

[28] B.M. Wise, N.B. Gallagher, R. Bro, J.M. Shaver, W. Windig, S.R. Koch, PLS Toolbox 3.5 for Use with Matlab™ ISBN:0-97611840-8. 2004.

[29] J.H. Christensen, G. Tomasi, A.B. Hansen, Chemical fingerprinting of petroleum biomarkers using time warping and PCA, Environ. Sci. Technol. 39 (1) (2005) 255–260.

[30] D.J. Crockford, J.C. Lindon, O. Cloarec, R.S. Plumb, S.J. Bruce, S. Zirah, et al., Statistical search space reduction and two-dimensional data display approaches for UPLC-MS in biomarker discovery and pathway analysis, Anal. Chem. 78 (13) (2006) 4398–4408.

[31] S.J. Dixon, Y. Xu, R.G. Brereton, H.A. Soini, M.V. Novotny, E. Oberzaucher, et al., Pattern recognition of gas chromatography mass spectrometry of human volatiles in sweat to distinguish the sex of subjects and determine potential discriminatory marker peaks, Chemometr. Intell. Lab. Syst. 87 (2) (2007) 161–172.

[32] H.M. Woo, K.M. Kim, M.H. Choi, B.H. Jung, J. Lee, G. Kong, et al., Mass spectrometry based metabolic approaches in urinary biomarker study of women's cancers, Clin. Chim. Acta. 400 (2009) 63–69.

[33] R.O. Duda, P.E. Hart, Pattern classifications and scene analysis, vol. 1, Wiley, New York, 1973.

[34] K.J. Johnson, R.E. Synovec, Pattern recognition of jet fuels: comprehensive GC × GC with ANOVA-based feature selection and principal component analysis, Chemometr. Intell. Lab. Syst. 60 (2002) 225–237.

[35] K.M. Pierce, J.C. Hoggard, J.L. Hope, P.M. Rainey, A.N. Hoofnagle, R.M. Jack, et al., Fisher ratio method applied to third-order separation data to identify significant chemical components of metabolite extracts, Anal. Chem. 78 (2006) 5068–5075.

[36] T. Kind, V. Tolstikov, O. Fiehn, R.H. Weiss, A comprehensive urinary metabolomic approach for identifying kidney cancer, Anal. Biochem. 363 (2) (2007) 185–195.

[37] D.R. Burgard, J.T. Kuznicki, Chemometrics: chemical and sensory data, CRC Press, Boca Raton, 1990.

[38] M. Jalali-Heravi, A. Kyani, Use of computer-assisted methods for the modeling of the retention time of a variety of volatile organic compounds: a PCA-MLR-ANN approach, J. Chem. Inf. Comput. Sci. 44 (4) (2004) 1328–1335.

[39] R. Gargallo, R. Tauler, F. Cuesta-Sanchez, D.L. Massart, Validation of alternating least squares multivariate curve resolution for chromatographic resolution and quantitation, Trends Anal. Chem. 15 (7) (1996) 279–286.

[40] L.S. Ramos, E. Sanchez, B.R. Kowalski, Generalized rank annihilation method. II. Analysis of bimodal chromatographic data, J. Chromatogr. A 385 (C) (1987) 165–180.

[41] C.G. Fraga, B.J. Prazen, R.E. Synovec, Enhancing the limit of detection for comprehensive two-dimensional gas chromatography (GC × GC) using bilinear chemometric analysis, J. High Resolut. Chromatogr. 23 (3) (2000) 215–224.

[42] J.C. Hoggard, J.H. Wahl, R.E. Synovec, G.M. Mong, C.G. Fraga, Impurity profiling of a chemical weapon precursor for possible forensic signatures by comprehensive two-dimensional gas chromatography/mass spectrometry and chemometrics, Anal. Chem. 82 (2010) 689–698.

[43] E. Sanchez, B.R. Kowalski, Tensorial resolution: a direct trilinear decomposition, J. Chemometr. 4 (1) (1990) 29–45.

[44] K.S. Booksh, Z. Lin, Z. Wang, B.R. Kowalski, Extension of trilinear decomposition method with an application to the flow probe sensor, Anal. Chem. 66 (15) (1994) 2561–2569.

[45] P. Hindmarch, K. Kavianpour, R.G. Brereton, Evaluation of parallel factor analysis for the resolution of kinetic data by diode-array high-performance liquid chromatography, Analyst 122 (9) (1997) 871–877.

[46] A.E. Sinha, C.G. Fraga, B.J. Prazen, R.E. Synovec, Trilinear chemometric analysis of two-dimensional comprehensive gas chromatography-time-of-flight mass spectrometry data, J. Chromatogr. A 1027 (1–2) (2004) 269–277.

[47] J.C. Hoggard, W.C. Siegler, R.E. Synovec, Toward automated peak resolution in complete GC × GC-TOFMS chromatograms by PARAFAC, J. Chemometr. 23 (7–8) (2009) 421–431.

[48] J.C. Hoggard, R.E. Synovec, Parallel factor analysis (PARAFAC) of target analytes in GC × GC-TOFMS data: automated selection of a model with an appropriate number of factors, Anal. Chem. 79 (2007) 1611–1619.

[49] J.C. Hoggard, R.E. Synovec, Automated resolution of nontarget analyte signals in GC × GC-TOFMS data using parallel factor analysis, Anal. Chem. 80 (2008) 6677–6688.

[50] T. Skov, J.C. Hoggard, R. Bro, R.E. Synovec, Handling within run retention time shifts in two-dimensional chromatography data using shift correction and modeling, J. Chromatogr. A 1216 (18) (2009) 4020–4029.

[51] R. Bro, C.A. Andersson, H.A.L. Kiers, PARAFAC2 - Part II. Modeling chromatographic data with retention time shifts, J. Chemometr. 13 (3–4) (1999) 295–309.

[52] J.M. Halket, A. Przyborowska, S.E. Stein, W.G. Mallard, S. Down, R.A. Chalmers, Deconvolution gas chromatography/mass spectrometry of urinary organic acids – potential for pattern recognition and automated identification of metabolic disorders, Rapid Commun. Mass Spectrom. 13 (4) (1999) 279–284.

[53] E.M. Humston, J.C. Hoggard, R.E. Synovec, Utilizing the third order advantage with isotope dilution mass spectrometry, Anal. Chem. 82 (2010) 41–43.

[54] D.M. Haaland, E.V. Thomas, Partial least-squares methods for spectral analyses. 1. Relation to other quantitative calibration methods and the extraction of quantitative information, Anal. Chem. 60 (11) (1988) 1193–1202.

[55] P. Geladi, B.R. Kowalski, Partial least-squares regression: a tutorial, Anal. Chim. Acta. 185 (C) (1986) 1–17.

[56] R. Bro, Multiway calibration. Multiway PLS, J. Chemometr. 10 (1) (1996) 47–61.

[57] K.J. Johnson, B.J. Prazen, D.C. Young, R.E. Synovec, Quantification of naphthalenes in jet fuel with GC × GC/Tri-PLS and windowed rank minimization retention time alignment, J. Sep. Sci. 27 (5–6) (2004) 410–416.

[58] M. Frank, H. Ulmer, J. Ruiz, P. Visani, U. Weimar, Complementary analytical measurements based upon gas chromatography-mass spectrometry, sensor system and human sensory panel; a case study dealing with packaging materials, Anal. Chim. Acta. 431 (1) (2001) 11–29.

[59] S.L.C. Ferreira, R.E. Bruns, E. Galvão, P.d. Silva, W.N.L.d. Santos, C.M. Quintella, et al., Statistical designs and response surface techniques for the optimization of chromatographic systems, J. Chromatogr. A 1158 (2–14) (2007) 2–14.

[60] S. O'Hagan, W.B. Dunn, J.D. Knowles, D. Broadhurst, R. Williams, J.J. Ashworth, et al., Closed-loop, multiobjective optimization of two-dimensional gas chromatography/mass spectrometry for serum metabolomics, Anal. Chem. 79 (2) (2007) 464–476.

[61] V.M. Morris, J.G. Hughes, P.J. Marriott, Examination of a new chromatographic function, based on an exponential resolution term, for use in optimization strategies: application to capillary gas chromatography separation of phenols, J. Chromatogr. A 755 (1996) 235–243.

[62] J. Beens, H.-G. Janssen, M. Adahchour, U.A.Th. Brinkman, Flow regime at ambient outlet pressure and its influence in comprehensive two-dimensional gas chromatography, J. Chromatogr. A 1086 (1–2) (2005) 141–150.

[63] D. Ryan, P. Morrison, P. Marriott, Orthogonality considerations in comprehensive two-dimensional gas chromatography, J. Chromatogr. A 1071 (1–2) (2005) 47–53.

[64] A. Jiye, J. Trygg, J. Gullberg, A.I. Johansson, P. Jonsson, H. Antti, et al., Extraction and GC-MS analysis of the human blood plasma metabolome, Anal. Chem. 77 (24) (2005) 8086–8094.

[65] R.E. Synovec, B.J. Prazen, K.J. Johnson, C.G. Fraga, C.A. Bruckner, Advances in chromatography, Marcel Dekker, Inc., New York, 2003.

[66] J.M. Amigo, T. Skov, R. Bro, Chromatography: solving chromatographic issues with mathematical models and intuitive graphics, Chem. Rev. 110 (8) (2010) 4582–4605.

[67] K. Pierce, J. Hoggard, R. Mohler, R. Synovec, Recent advancements in comprehensive two-dimensional separations with chemometrics, J. Chromatogr. A 1184 (1–2) (2008) 341–352.

[68] H.J. Cortes, B. Winniford, J. Luong, M. Pursch, Comprehensive two-dimensional gas chromatography review, J. Sep. Sci. 32 (5–6) (2009) 883–904.

[69] O. Amador-Muñoz, P.J. Marriott, Quantification in comprehensive two-dimensional gas chromatography and a model of quantification based on selected summed modulated peaks, J. Chromatogr. A 1184 (2008) 323–340.

CHAPTER

18

Validation of Gas Chromatographic Methods

Bieke Dejaegher, Johanna Smeyers-Verbeke, Yvan Vander Heyden

OUTLINE

18.1. INTRODUCTION

Method validation is the last step in the development of an analytical method (Figure 18.1), for example a gas chromatographic method [1]. During method development, a method is selected, e.g. based on the literature or on an already known method. This method is then optimized for the intended application. This can, for example, be necessary because the matrix is different from that analyzed in the original method (e.g. blood versus urine), because another but similar component is determined (e.g. propranolol versus alprenolol in blood), because the chromatographic column is not present in the laboratory and another but similar column will be used, etc. After optimization and selection of the calibration scheme, a candidate method is obtained, which is expected to comply with the requirements set for the method.

Gas Chromatography DOI: 10.1016/B978-0-12-385540-4.00018-3

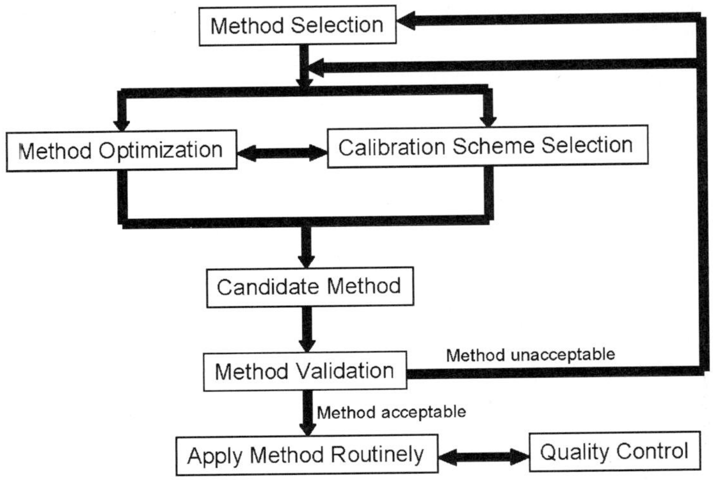

FIGURE 18.1 Different steps in the lifetime of an analytical method.

This candidate method should then be validated to ensure its suitability for the intended purpose(s). The method validation should indicate that the candidate method complies with the requirements, i.e. that the method is able to measure a given component with a suitable precision, trueness, detection limit, etc.

After validation, there are two possible outcomes (Figure 18.1). A first is that the method performs acceptably and can be used routinely. Such a method is then regularly submitted to a quality control, to verify whether the method still performs acceptably in routine analysis. A second outcome is that the validation indicates that the method does not perform acceptably well. Then either the method is re-optimized or another method is selected. The above steps (Figure 18.1) are then repeated until validation indicates that the new candidate method performs acceptably well.

During validation, some performance criteria are evaluated and documented. The requirements for these criteria are specified and then experimentally verified [1]. The characteristics possibly tested are the linearity, precision, trueness, detection limit, quantification limit, range, specificity, and robustness.

There are three golden rules for method validation, i.e. validate the whole method, validate it over the range of concentrations, and validate it over the range of matrices [1].

In this chapter, first, some regulatory aspects concerning method validation are overviewed. Then, the different method validation items are discussed and illustrated with some examples. Finally, the application of an accuracy profile is discussed shortly.

18.2. REGULATORY ASPECTS

During validation, it is examined whether the method provides acceptable analytical results when measuring samples in the expected concentration range and matrices. For instance, drug substance, drug product, blood, plasma, urine, food, water, soil, organs, air samples, etc., are possible matrices. Method validation is most elaborated and the requirements are most strict in pharmaceutical analysis, as can be seen below.

Different guidelines on validation, describing the requirements for analytical methods, for example, guidelines from the International Conference on Harmonisation of Technical

Requirements for the Registration of Pharmaceuticals for Human Use (ICH) [2], the United States Food and Drug Administration (FDA) [3,4], Eurachem [5], the International Union of Pure and Applied Chemistry (IUPAC) [6,7], the European Medicines Agency (EMA) [8], the International Organization for Standardization (ISO) [9], the Association of Official Analytical Chemists (AOAC) [10,11], the Societé Française des Sciences et Techniques Pharmaceutiques (SFSTP) [12], and the European Commission [13] are available. These guidelines indicate which validation characteristics should be evaluated, i.e. linearity, precision (repeatability and time-different intermediate precision), trueness, specificity, limits of detection (LOD) and quantification (LOQ), range, and/or robustness, for a given type of method (see further). The requirements might slightly vary depending on the guideline considered. The FDA, for instance, also recommends evaluating the sample stability.

According to the ICH guidelines [2], which are adopted by the EMA [8], drug analysis methods can be divided into identification tests (qualitative methods), assays (quantitative methods), or testing for impurities (quantitative methods or limit tests) (Table 18.1a). The FDA [3,4] added specific test methods to test the drug substance, excipient, or drug product, e.g. methods as particle size analysis, droplet distribution, spray pattern, and dissolution. Depending on the type of method, a number of characteristics need to be validated.

The AOAC guideline [10,11] divides the methods into identification tests, qualitative methods, and limit tests and quantitative methods for a component with either a low or a high concentration (Table 18.1b).

18.3. METHOD VALIDATION ITEMS

As mentioned in the Introduction, one of the golden rules of method validation is to perform the validation over the whole concentration range. This means that the validation items such as linearity, precision, accuracy and/or trueness should be evaluated in that range. In this section, each item is briefly explained and illustrated with some examples from the literature.

The examples discuss the validation of assays to determine finasteride in tablets [14] and antidepressants in their pharmaceutical preparations [15] and in human urine [16]. Furthermore, validation is discussed for GC assays to determine 3,4-methylenedioxymethamphetamine and its metabolites in plasma and urine [17], estrogens and testosterone in human urine [18], free and total bisphenol A and B in human urine [19], sterol oxidation products in serum [20], cocaine and its metabolites in human primary cultured renal cells [21] and in hair [22], amphetamine-type stimulants and their metabolites in human urine [23], benzene in urine [24], the insecticide lufenuron in wheat flour [25], flumethrin in honey [26], pesticide residues in food samples [27], acrylamide in heat-processed starchy foods [28], phthalates in wine [29], and pesticides in vegetables [30,31], fruits [31], and baby food [31]. However, this list is not exhaustive.

Some papers [16,20,21] follow the ICH guidelines, one [23] follows the FDA guidelines, one [22] the SFSTP guidelines, one [27] the European Commission guidelines , and one [30] the Eurachem guidelines. Most papers, however, do not refer to any guideline.

18.3.1. Linearity

In different guidelines, **linearity** defines *"a method's ability to obtain test results proportional to the concentration of analyte in a sample"* [2,5], and most often refers to the linearity of the calibration line. Most analytical techniques use some type of calibration scheme to estimate the (unknown) concentration of a given compound in a sample. Most often, a straight line relationship between response and concentration is aimed at. Sometimes, a mathematical transformation is needed

TABLE 18.1 Recommended (+) Validation Characteristics for Different Types of Analytical Methods, Specified by: (a) Different Guidelines [2–4,8], and (b) the AOAC Guidelines [10,11]

(a)

Validation characteristic	Type of analytical method				
	Identification test	Assay, dissolution, content/potency	Testing for impurities		Specific test
			Limit test	Quantitative method	
Linearity		+		+	
Precision					
Repeatability		+		+	(+)
Time-different intermediate precision		+		+	(+)
Accuracy		+		+	(+)
Specificity	+	+	+	+	(+)
Detection limit			+	(+)	
Quantification limit				+	
Range		+		+	
Robustness		+	(+)	+	(+)

(b)

Validation characteristic	Type of analytical method				
	Identification test or qualitative test	Component with a low concentration		Component with a high concentration	
		Limit test	Quantitative method	Limit test	Quantitative method
Linearity			+		+
Precision					
Repeatability			+	+	+
Time-different intermediate precision			+	+	+
Accuracy/Trueness			+	+	+
Specificity	+	+	+	+	+
Detection limit		+	+		(+)
Quantification limit			+		(+)
Range			+		+
Robustness			+		+

to obtain a linear relation [1]. After determining a straight line model, its adequacy should be evaluated.

Commonly, calibration approaches are the classic calibration with or without internal standard (IS), and the standard addition also with or without IS. These approaches are discussed below. In GC, very often the classic calibration with IS is used [14–23,29]. In [24], a classic approach with IS to quantify benzene in urine was compared to a standard addition one with IS, and the latter was preferred because it allows compensation for differences in urinary matrices. Some publications [27,30] also describe the application of the standard addition with IS as the calibration scheme to correct for matrix effects. Rarely, classic [25,26] or standard addition calibration [28,31] without IS is applied in GC. The different approaches to assess linearity are described below.

18.3.1.1. Classic Calibration

The simplest calibration procedure is the classic approach, in which a calibration line is constructed with standards of different concentrations. For each standard, a GC chromatogram is recorded and least-squares regression is used to establish a linear model between the dependent variable, the response, y, and the independent variable, the concentration, x, of a given compound [1]. Usually, the considered response in GC methods is the peak area of the component, $\text{Area}_{\text{component}}$. When sharp symmetrical Gaussian peaks are obtained, also the peak height, $H_{\text{component}}$, might be considered a response.

However, most often in GC an IS is added to each standard and sample in order to correct for random errors, such as the variability in injection volume. The IS is added at the start of the procedure in a constant concentration. Consequently, the IS is subject to the same treatments as the component to be determined. The IS preferably should fulfill some requirements: have a similar structure as the component to be determined, not interfere with the component, be well separated from the other peaks, and have a signal of the same order of magnitude as the calibration standards. For each solution, a chromatogram is recorded and least-squares regression is used to estimate the linear model between the response, here the ratio of the peak area of the component, $\text{Area}_{\text{component}}$, to that of the IS, Area_{IS}, i.e. $\text{Area}_{\text{component}}/\text{Area}_{\text{IS}}$, and the concentration of a given compound.

The true linear relation between the x and y variables is given by

$$\eta = \beta_0 + \beta_1 x \tag{18.1}$$

where η is the real dependent variable of which y is an estimation, and β_0 and β_1 are the unknown true intercept and slope of the calibration line. They are estimated by the intercept b_0 and the slope b_1 of the calibration line. The least-squares estimation of the calibration line is given as follows:

$$\hat{y} = b_0 + b_1 x \tag{18.2}$$

where $\hat{y}$ is the response predicted by the model. The least-squares line minimizes the sum of the squared residuals ($\min \sum e_i^2$) [1]. For each point i of the calibration line, the residual e_i is

$$e_i = y_i - \hat{y}_i \tag{18.3}$$

where y_i is the measured response and $\hat{y}_i$ the response predicted by the model.

For the sample(s), from the response y_{sample}, i.e. either $\text{Area}_{\text{component}}/\text{Area}_{\text{IS}}$ or $\text{Area}_{\text{component}}$ depending on whether or not an IS is used, and the estimated calibration line, the concentration of the component in the sample x_{sample} is determined:

$$x_{\text{sample}} = \frac{y_{\text{sample}} - b_0}{b_1} \tag{18.4}$$

18.3.1.2. Standard Addition

In standard addition, the concentration of a component in a sample is determined by adding to aliquots of the sample several known

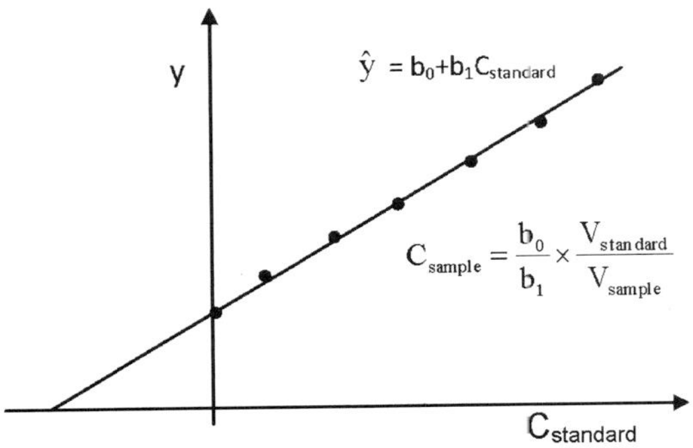

FIGURE 18.2 Standard addition approach.

quantities of the component ($C_{standard}$). $C_{standard}$ represent the added concentrations without taking into account the dilution. For example, the different standard addition solutions are prepared by adding to each "s" mL of the sample, each time "t" mL of a standard with a different concentration, $C_{standard}$.

The response for the different solutions is then measured and the standard addition calibration line calculated using least squares. For GC, the response can again be either $Area_{component}/Area_{IS}$ or $Area_{component}$.

The response plotted as a function of $C_{standard}$ is given in Figure 18.2.

The concentration of the sample is then estimated as

$$C_{sample} = \frac{b_0}{b_1} \times \frac{V_{standard}}{V_{sample}} \quad (18.5)$$

Standard addition approaches are often used in spectroscopic methods to correct for matrix interferences. In GC methods, they are applied to correct for matrix-induced response enhancement [32]. This enhancement occurs when the chromatographic response of a compound in a matrix is larger than that of the same compound with the same concentration in a matrix-free solution. Matrix-induced response enhancement often occurs, e.g. in the analysis of pesticide residues. Ref. [32] provides a table with compounds typically susceptible to matrix enhancement. Different correction methods, including standard addition, are suggested [32]. Standard addition will usually be used if only a few samples are to be analyzed because of its elaborate and time-consuming procedure for each individual sample.

18.3.1.3. *Assessment of Linearity*

To assess the linearity of the calibration line, i.e. to evaluate whether a straight line model fits the experimental results, a number of standards, usually at least 5 [2] or 6 [5,6] and covering the expected concentration range, are prepared. After determining the signal for each standard, the calibration line (intercept and slope) is estimated with least-squares regression. In order to apply least-squares modeling, two requirements should be fulfilled: (i) the x values are exactly known, i.e. errors in x are considered negligible, and errors occur only in the responses, and (ii) the measurements are homoscedastic, i.e. the responses have a constant variance, independent of x. The latter can be verified statistically using a Cohran's test or graphically by evaluating a residual plot [1]. A Cohran's test compares n variances s_i^2, each based on n_k replicates, by calculating $C = s_{max}^2 / \sum_{i=1}^{n} s_i^2$, with s_{max}^2 being the maximum variance, and comparing C to a tabulated value $C_{tab(\alpha,n_k,n)}$ at a given significance level α, usually $\alpha = 0.05$. When $C > C_{tab}$, the variances are considered heteroscedastic, otherwise homoscedastic. A residual plot shows the residuals of the calibration standards, e_i, as a function of their concentrations, x_i. For homoscedastic data, the residuals are randomly distributed, while for heteroscedastic data, they show a trend and increase in absolute value with increasing concentration. For heteroscedastic data, either a data transformation (e.g. a square root transformation or log transformation) can be applied to obtain constant variances or weighted least-squares regression can be used to estimate the parameters of the calibration line [1]. In the latter regression, a higher weight is assigned to the standards with a small variance and a lower weight to those with a high variance.

To assess linearity, the data are plotted and their linearity is first inspected visually to check for outliers or a deviation from the straight line model. Outliers can also be observed graphically by evaluating a residual plot. They further can be detected with statistical diagnostics, such as the evaluation of the standardized residuals, Cook's squared distance, the Mahalonobis distance, or the leverage measure, or with robust methods, such as the single median method, the repeated median method, or the least median of squares method [1].

The adequacy of the straight line model can be verified statistically using either an analysis of variance (ANOVA) lack-of-fit (LOF) test or a test for the significance of the quadratic coefficient b_2 in a second-order model fitted to the data [1,6]. The LOF test requires replicate measurements. The total residual sum of squares (SS), SS_R, is divided into the SS due to LOF of the applied model, SS_{LOF}, and the SS due to the pure experimental error, SS_{PE} (Table 18.2). By dividing these SSs by their corresponding number of degrees of freedom (df), the corresponding mean squares (MS) are obtained, MS_{LOF} and MS_{PE}. The lack-of-fit test is a one-sided F-test at significance level α, often $\alpha = 0.05$, to compare MS_{LOF} and MS_{PE}, i.e., $F = MS_{LOF}/MS_{PE} \leftrightarrow F_{\text{tab}(\alpha;\text{df}_{LOF},\text{df}_{PE})}$.

If the variance due to LOF is found to be significantly larger than that due to experimental error, the straight line model is considered to be inadequate to describe the data. If the variance due to LOF is not significantly larger, the straight line model is considered adequate.

An alternative approach to verify linearity of the data, for example when no replicate measurements are available, is to fit a second-order polynomial model, $y = b_0 + b_1x + b_2x^2$, to the data and to test the significance of the quadratic coefficient b_2 by means of the 95% two-sided confidence interval (CI) around β_2 (Eqn (18.6)) or by means of a two-sided t-test (Eqn (18.7)) [6].

$$95\%\text{CI}_{\beta_2} = b_2 \pm t_{\text{tab}(\alpha,\text{df}=n-3)} s_{b_2} \tag{18.6}$$

$$|t| = \left|\frac{b_2}{s_{b_2}}\right| \leftrightarrow t_{\text{tab}(\alpha,\text{df}=n-3)} \tag{18.7}$$

where s_{b_2} is the standard deviation of b_2 (and obtained from the variance–covariance matrix of the regression coefficients) and t_{tab} the tabulated t-value, based on the significance level α, usually $\alpha = 0.05$, and the number of degrees of freedom, $n - 3$, with n being the number of calibration standards, including the blank. When zero is not included in the 95%CI or when $|t| \geq t_{\text{tab}}$, b_2 is significantly different from zero and the linear model is considered inadequate. Otherwise, the linear model is considered appropriate. However, this latter conclusion is not always correct. Indeed, a quadratic coefficient that is not significantly different from zero means that the quadratic model does not fit the data better than

TABLE 18.2 ANOVA Lack-of Fit Test

Source of variation	Sum of squares (SS)	Degrees of freedom (df)	Mean squares (MS)	F-value
Residual	$SS_R = SS_{LOF} + SS_{PE}$	$n_{Tot} - 2$	$MS_R = \dfrac{SS_R}{df_R}$	
Lack of fit	$SS_{LOF} = \sum_i^k n_i(\bar{y}_i - \hat{y}_i)^2$	$k - 2$	$MS_{LOF} = \dfrac{SS_{LOF}}{df_{LOF}}$	$F = \dfrac{MS_{LOF}}{MS_{PE}}$
Pure error	$SS_{PE} = \sum_i^k \sum_j^{n_i} (y_{ij} - \bar{y}_i)^2$	$n_{Tot} - k$	$MS_{PE} = \dfrac{SS_{PE}}{df_{PE}}$	

y_{ij} = one of the n_i replicate measurements at x_i, $\sum_{i=1}^{k} n_i = n_{Tot}$ = total number of observations, k = number of different x values (including the blank), $\bar{y}_i$ = average of the replicates y_{ij} at x_i, $\hat{y}_i$ = predicted value of y at x_i, and $\bar{y}$ = average of all observations [1].

the straight line, but this does not imply that there is still no LOF from the straight line model.

In the literature, linearity is often verified visually [14–31]. The statistical ANOVA lack-of-fit test was used in [26,27]. Although a correlation or determination coefficient is not appropriate to evaluate linearity [1,6], many authors still report it as a proof of linearity.

The **range**, i.e. the concentration interval in which the method possesses acceptable linearity, precision, and trueness [2–4], was determined in [14–31].

18.3.2. Limits of Detection and Quantification

The **limit of detection (LOD)** of an analytical procedure is defined by the ICH as "*the lowest amount of analyte in a sample that can be detected but not necessarily quantitated as an exact value*" [2]. ICH also states that the **limit of quantification (LOQ)** of an analytical procedure is "*the lowest amount of analyte in a sample which can be quantitatively determined with suitable precision and accuracy*" [2]. In fact, this definition corresponds to the lower limit of quantification (LLOQ). Besides LLOQ, also an upper limit of quantification (ULOQ) exists, which can be defined as the highest concentration of a compound that can be quantified with an acceptable precision and accuracy applying the calibration line. According to the ICH and IUPAC guidelines [2,6,7], the LOD and LOQ can be determined from the standard deviation of replicated measurements of the blank, and the slope of the calibration line:

$$L_d = \mu_{bl} + k'_d\sigma_{bl} \text{ with } k'_d = k_c + k_d \quad (18.8)$$

$$X_d = (L_d - \mu_{bl})/b_1 = k'_d\sigma_{bl}/b_1 \quad (18.9)$$

$$L_q = \mu_{bl} + k_q\sigma_{bl} \quad (18.10)$$

$$X_q = (L_q - \mu_{bl})/b_1 = k_q\sigma_{bl}/b_1 \quad (18.11)$$

where L_d and X_d are the limits of detection, expressed as response/signal and as concentration, respectively. L_q and X_q are the corresponding limits of quantification. μ_{bl} and σ_{bl} are the true response and the standard deviation of the blank, and b_1 is the slope of the calibration line. μ_{bl} and σ_{bl} are estimated experimentally as $\overline{y}_{bl}$ and s_{bl}, often determined from 6 [6] or 10 [5] blank measurements (see further). The multiplication factors k_c, k_d, k'_d, and k_q are constants and their values determine the risk of making a wrong decision one is willing to take. There are two types of wrong decisions, i.e. false-positive decisions related to the type I or α-error, and false-negative decisions related to the type II or β-error. For $\alpha = \beta = 5\%$, $k_c = k_d = 1.645$ and $k'_d = 3.29$. With $k_q = 10$, one expects at the quantification limit a precision, expressed as relative error, of 10% [7].

To ensure realistic LOD and LOQ estimates, it is of the utmost importance to select an appropriate blank [1]. For example, an analytical blank, containing all reagents and analyzed in the same way as the samples, can be considered appropriate. Ideally, a matrix blank, having exactly the same composition as the sample except for the compound to be analyzed, is used. In the absence of an appropriate blank, a standard with a low concentration near the expected LOD can also be used. Alternatively, the residual standard deviation of the calibration line, $s_e = \sqrt{\sum(y_i - \hat{y}_i)^2/n - 2}$, or the standard deviation of the intercept of the calibration line, $s_{b_0} = s_e\sqrt{\sum x_i^2/n\sum(x_i - \overline{x})^2}$, can also be used as an estimate of σ_{bl} in the estimation of LOD and LOQ. The mean blank response, $\overline{y}_{bl}$, is then replaced by the intercept of the calibration line. The above approach (Eqns (8–11)) was used in GC to estimate LOD in [15,16,29] and LOQ in [15,16].

The SFSTP guidelines [12] suggest an alternative approach to estimate LOD and LOQ for chromatographic methods. A matrix blank is injected and the maximum amplitude, h_{max}, in the chromatogram is determined over a distance equal to 20 times the width at half height of the peak of the compound around its retention time. LOD is then defined equal to 3 $h_{max}R$ and LOQ to 10 $h_{max}R$, with R the response factor, i.e. the component concentration/peak height (C/H) ratio. The latter implies that this approach is only used for responses expressed as peak height.

LOD and LOQ can also be determined based on the signal-to-noise ratio, $S/N = 2H/h_{max}$, where H is the height of a low concentration standard [2,33,34]. A concentration corresponding to an S/N ratio of 3 is then considered the LOD, and one with an S/N ratio of 10 the LOQ. The thus-obtained LOD and LOQ values are a factor two lower than those estimated according to the SFSTP guidelines, and are therefore more optimistic estimations. However, these latter approaches can only be applied to analytical procedures that exhibit baseline noise, such as chromatographic methods. Moreover, this approach is only appropriate when the baseline noise is obtained from the injection of an appropriate blank sample, for example, a matrix blank. If the baseline noise is determined in a region where no peaks elute from the injection of a standard or a sample (which often is done), it is assumed and should be verified that the baseline noise is constant in the entire chromatogram.

No GC example that applied the SFSTP procedure was found. The signal-to-noise ratio procedure was used in [14,17,21,26,28,30] to estimate both LOD and LOQ, and in [18,19,22,25,29] for LOD. However, none of the papers specified how the baseline noise was determined. In [22,23,25], LOQ was determined as the lowest point of the calibration line with adequate precision (e.g. $\%RSD_{repeatability} < 15\%$ [22] or <20% [23], and $\%RSD_{intermediate\ precision} < 25\%$ [22]) and trueness (e.g. %recovery = 100% ± 20% [22,23]).

18.3.3. Precision

The **precision** of an analytical method *"expresses the closeness of agreement (degree of scatter) between a series of measurements obtained from multiple sampling of the same homogeneous sample under the prescribed conditions"* [2]. Precision is related to random errors [1]. According to the ICH, FDA, Eurachem, ISO, and AOAC guidelines [2–5,9,11], precision comprises repeatability, intermediate precision, and reproducibility, depending on the conditions for the repeated experiments. Repeatability is the precision estimated under the most optimal operating conditions (same testing material, same laboratory, same analyst, same instrument, and short time interval, e.g. same day). Reproducibility is the precision estimate obtained under the most diverse operating conditions (same testing material, different laboratories, different analysts, different instruments, and different days) and is only to be determined when different laboratories are involved in the routine analysis. Between both extreme operating conditions, intermediate precision conditions can be considered. These result in the precision estimated in the same laboratory, but where larger time intervals (different days), different analysts, and/or different instruments are considered. The M-factor intermediate precision is defined by ISO [9], where M refers to the number of factors (time, analyst, or equipment) that differs ($M = 1, 2$, or 3). It is recommended to estimate always the time-different intermediate precision.

The precision can be expressed as a standard deviation s, variance s^2, or percentage relative standard deviation %RSD. It was reported as s in [14,17,18,21,25] and as %RSD in [14–27]. To estimate the repeatability, time-different intermediate precision, or reproducibility, a number of samples covering the whole range of concentrations and matrices are replicatedly measured under the specified circumstances. Repeatability was assessed in [14–30] and time-different intermediate precision in [14–24,26, 28,29]. When possible, the samples should be real homogeneous samples. If this is not possible, artificial samples can be used [2].

The precision is estimated for each concentration and then it is evaluated whether a mathematical relation exists between the standard deviation (as precision estimate) and the concentration. Such a relation allows estimating the precision for concentrations intermediate to those tested [9].

The ICH guidelines advise using an experimental design approach (full factorial design) to simultaneously determine the repeatability and time-different intermediate precision [2]. In this approach, n replicates (e.g. $n = 2$) are analyzed during p days (e.g. $p = 6$, often $5 \leq p \leq 8$) at each concentration level [1]. Then, ANOVA is used to estimate the variance components, i.e. the repeatability, s_r, and the between-days variance, s^2_{between}, at each concentration level. The time-different intermediate precision is then estimated as $s^2_{I(t)} = s^2_r + s^2_{\text{between}}$. However, no GC examples were found applying this approach.

18.3.4. Trueness

The **trueness** is *"the closeness of agreement between an average value from a number of test results and the accepted reference value"* and is only related to systematic errors [5–9,11]. Two types of systematic errors can be distinguished. The first type is a constant or absolute systematic error, which does not depend on the reference value. The second type is a proportional or relative systematic error, which is proportional to the reference value. To evaluate trueness, parameters, such as bias $= \overline{x} - \mu_0$, %bias $= \dfrac{\overline{x} - \mu_0}{\mu_0} \times 100\%$, and %recovery $= \dfrac{\overline{x}}{\mu_0} \times 100\%$, can be used, with $\overline{x}$ the average value and μ_0 the accepted reference value.

To estimate the trueness, different setups can be applied, depending on the concentration range to be evaluated, on the availability of reference material, on the availability of blank material that can be spiked with the compound to be determined, on the possibility to reconstitute the sample, or on the possibility to add the compound in a representative way to the sample [1].

When blank matrix that can be spiked with known amounts of the compound is available or when the sample can be reconstituted, the following approaches may be used. In the examined concentration range, a number of replicates are measured at each concentration tested. At each level, a t-test is performed to compare the average measured concentration, $\overline{x}$, with the true concentration, μ_0. This procedure can be applied if at the most 3 or 4 levels are evaluated [1]. When more levels are measured, two options are possible: either the above t-test is performed at each level using the Bonferroni correction (i.e. a correction where α and thus t_{tab} are adapted depending on the number of t-tests performed) or a regression analysis to relate the measured to the known concentrations [1].

In the literature on GC analysis, to evaluate the trueness, spiking of placebos of pharmaceutical formulations [15], urine [16–18,23,24], plasma [17], renal cells [21], hair [22], wheat flour [25], blank honey [26], and synthetic wine [29], with different concentrations of the compound(s) to be analyzed, was performed. In none of the studies were t-tests or a regression approach applied to statistically evaluate the trueness. The trueness was simply evaluated calculating and reporting the parameters %bias [18,23,24] and %recovery [14–17,21–23, 25,26,29].

When a blank matrix is not available, but when representative addition of the compound to the sample is possible, one can proceed as follows. In the evaluated concentration range, a one-point standard addition is performed at each concentration level. At each level, a t-test is performed to compare the absolute difference between the average concentrations measured after and before the addition, $|\overline{x}_2 - \overline{x}_1|$, with the known added concentration. This procedure can be applied, provided that at the most 3 or 4 levels are considered [1]. With more levels, the two above options again are possible: the latter described *t*-tests with the Bonferroni correction or a regression analysis [1].

In [28], the trueness was evaluated using a one-point standard addition at only one concentration level for each of four food matrices, and the parameter %recovery was reported.

18.3.5. Specificity

The **specificity** is *"the ability of a method to assess unequivocally the analyte in the presence of components which may be expected to be present, such as e.g. impurities, degradants, matrix, etc"* [2]. If one analytical method is insufficient to indicate method specificity, the lack of specificity may be illustrated using other supporting analytical methods [1–5].

Specificity in GC was evaluated in [14,15,17,18,20,21,23,25,26] by analyzing possible interfering substances, potentially occurring in blank plasma or urine, or in the sample matrix in general. In fact, the compound(s) to be determined should not coelute or overlap with other substances occurring in the samples.

18.3.6. Robustness

Robustness is sometimes also called **ruggedness**. Several definitions are available. Some only use the term ruggedness [10], some distinguish between robustness and ruggedness [34], while others consider them synonyms [2–4]. Youden and Steiner [10] used the term **ruggedness** test for a setup in which the influences of minor but deliberate and controlled changes in the method parameters or factors are evaluated applying an experimental design, in order to detect non-robust factors. In the United States Pharmacopeia (USP) [34], ruggedness is defined as *"The **ruggedness** of an analytical method is the degree of reproducibility of test results obtained by the analysis of the same sample under a variety of normal test conditions, such as different laboratories, different analysts, different instruments, different lots of reagents, different elapsed assay times, different assay temperatures, different days, etc."* In this approach, no deliberate changes in method parameters are introduced and the method is executed under different test conditions. This definition is equivalent to that of intermediate precision or reproducibility, depending on whether or not the test is performed in one laboratory. Detailed ISO (International Organization for Standardization) guidelines exist for these precision estimates [9].

The USP definition of robustness is the same as that of the ICH [2], i.e.: *"The **robustness** of an analytical procedure is a measure of its capacity to remain unaffected by small, but deliberate variations in method parameters and provides an indication of its reliability during normal usage"*. This definition is most widely applied for robustness and similar to the "older" ruggedness definition of [10].

A robustness test is an experimental setup applied to evaluate the robustness of the method. Although robustness tests are not mandatory in the ICH guidelines [2], they are demanded by the US Food and Drug Administration (FDA) for the registration of drugs in the United States of America [3,4]. Robustness can be evaluated using a one-variable-at-a-time (OVAT) approach. In [14], an OVAT procedure in GC analysis was used to examine the effects of injector-, detector-, initial oven-, and final oven temperatures, and injection volume on the finasteride concentration in tablets. No significant effect was found on the results, and the method was considered robust. However, an OVAT approach is not recommended [35,36].

A more appropriate approach is to apply a multivariate approach, i.e. using an experimental design. Such robustness tests were performed on GC assays to determine antidepressants in pharmaceutical preparations [15] or in human urine [16]. Examined factors were column head pressure, injector temperature, time and temperature of the splitless step, detector temperature, oven temperature programs, voltage of the MS detector, steps of the SPE procedure, SPE cartridge lots, time (days), and analysts. Examined responses were peak resolutions, efficacy (expressed as plate count *N*), and relative peak areas of compounds. However, the most important responses to examine should be related to the content or concentration of the compound(s). In [15], a reflected two-level Plackett–Burman (PB) design was executed to examine the effects of 7 factors at three levels in 15 experiments. The results were analyzed by comparing, for each response, the main factor effects with an error estimate, according to the procedure described in [10]. Graphically, bar plots were drawn. In [16], a PB design was executed to examine the effects of 11 factors at two levels in 12 experiments. Graphically, plots of the ranked factor effects were drawn.

18.3.7. Sample Stability

Some publications on GC assays [15,17,18,23, 24,26,27] also examine the sample **stability**, which is required by the FDA [3,4].

In [15], the stability of standard solutions and of spiked placebo solutions stored in darkness at 4 °C was verified over time by comparing the response factors or relative peak areas, respectively, with those of the corresponding freshly prepared solutions. In [17], the stability of spiked matrix solutions at 4 °C was evaluated as a function of time. In another paper [18], the stability of the composition of a mixture of standard compounds was tested by measuring the mixture before and after the urinary work-up and comparing the results. A freeze–thaw (freezing during 24 h at −20 °C and thawing at room temperature), a short-term (8 h at room temperature), and a long-term (3–6 months at −20 °C) stability study of spiked samples was performed in [23]. The influence of freezing and thawing on the stability of spiked samples was also evaluated in [24]. Zhou et al. [26] evaluated the stability of stock solutions after 12 h at room temperature, after 4 weeks at 4 °C, and after 2 months at −18 °C. A long-term stability study was also performed in [27] to evaluate the stability of the calibration solutions.

18.4. ACCURACY PROFILES

Method validation is most often performed as described in section 18.3. However, the parameters and approaches described there are used for a diagnostic purpose, e.g. the trueness and precision parameters are individually estimated and evaluated.

An alternative approach consists of determining accuracy profiles. An accuracy profile is a visualization that allows determining the **method capability**. Such profiles are based on two-sided β-expectation tolerance intervals of validation standards for the measurement of the total error, i.e. including both bias and precision. They are developed in Ref. [37–39]. This statistical tool is used for decision purposes. The use of this approach reduces the risks of accepting an unsuitable assay or to reject a suitable one. In [37], method validation is critically reviewed in order to develop a more harmonized approach. In [38], several experimental protocols are proposed, presenting the types of calibration and validation standards, the concentration levels to use, the number of replicates, and the number of series to be performed. A decision tree (Figure 18.3) is presented to allow selecting a suitable protocol (Figure 18.4) needed to obtain all required information to demonstrate the reliability of the method. In [39], the statistical methodology used to produce the accuracy profiles is described (Figure 18.5).

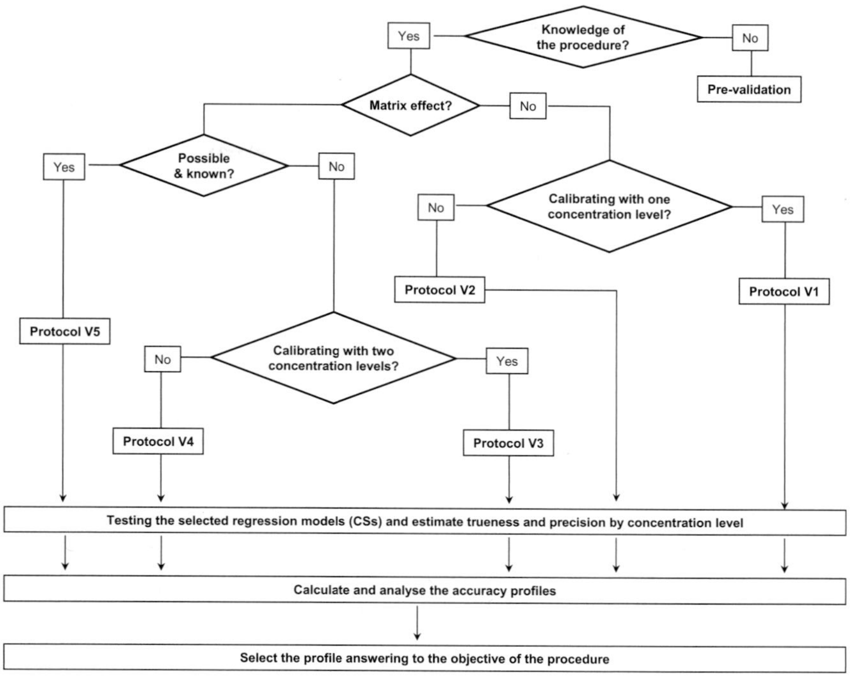

FIGURE 18.3 Determining accuracy profiles: Decision tree to select a validation protocol.

Standards	*Concentration levels*	*Protocol* *V1*	*V2*	*V3*	*V4*	*V5*
CSs without matrix	Low		2		2	
	Mid	2	(2)[a]	2	(2)[a]	
	High	(2)[b]	2	(2)[b]	2	
CS within matrix	Low				2	2
	Mid			2	(2)[a]	(2)[a]
	High			(2)[b]	2	2
	Additional					(2)[c]
VSs within matrix	Low	3	3	3	3	3
	Mid	3	3	3	3	3
	High	3	3	3	3	3
Minimum number of series		3	3	3	3	3
Minimum total number of experiments		33	45	39	63	45

[a] Considering the regression model selected (ex.: simple regression line), the possible suppression of the mid range concentration level depending on the regression
model considered to express the response function (for example: model as the simple regression line). In this case, there are 39 experiments for the protocols V2 (without matrix) and V5 (within matrix). There are 51 experiments for the protocol V4.
[b] Selection of a concentration level higher than the target concentration in order to calibrate (for example: 120% of target concentration).
[c] Addition of a concentration level for a more complex response function (for example: 4-parameter logistic regression).

FIGURE 18.4 Determining accuracy profiles: Validation protocols

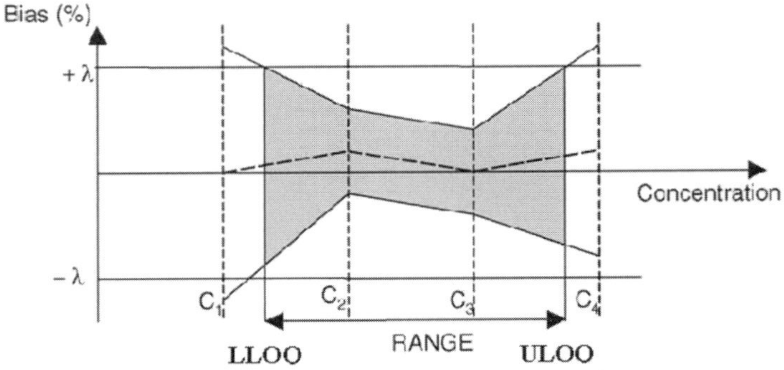

FIGURE 18.5 Determining accuracy profiles: Accuracy profile (gray area describing the range in which the procedure is able to quantify with a known accuracy and a risk *a priori* fixed by the analyst). λ = acceptance limit, LLOQ = lower limit of quantification, and ULOQ = upper limit of quantification. *Source: Reproduced with Permission from [37,38].*

18.5. CONCLUSIONS

This chapter describes method validation aspects, with a focus on gas chromatographic methods. After an overview of the regulatory aspects, the different method validation items were discussed and illustrated with some examples from the literature. The considered parameters were the linearity and the range of the calibration line, the detection limit, the quantification limit, the precision, the trueness, the specificity, and the robustness. The evaluation of the sample stability was also described. Some methods described in the literature were critically reviewed. Finally, the accuracy profile is briefly discussed.

Acknowledgments

Bieke Dejaegher is a postdoctoral fellow of the Fund for Scientific Research (FWO) – Vlaanderen, Belgium.

References

[1] D.L. Massart, B.G.M. Vandeginste, L.M.C. Buydens, S. De Jong, P.J. Lewi, J. Smeyers-Verbeke, Handbook of chemometrics and qualimetrics: part A, Elsevier, Amsterdam, 1997.

[2] Guidelines prepared within the International Conference on Harmonisation of Technical Requirements for the Registration of Pharmaceuticals for Human Use (ICH), Validation of analytical procedures: text and methodology, Q2(R1) 2005; p. 1–13, http://www.ich.org/, [accessed on 28.07.11].

[3] Food and Drug Administration (FDA), Department of health and human services, Guidance for industry: analytical procedures and methods validation: chemistry, manufacturing, and controls documentation, 2000; 1–37, http://www.fda.gov/, [accessed on 28.07.11].

[4] Food and Drug Administration (FDA), Department of health and human services, Guidance for industry: bioanalytical method validation. 2001; 1–25, http://www.fda.gov/, [accessed on 28.07.11].

[5] Eurachem. The fitness for purpose of analytical methods: a laboratory guide to method validation and related topics. 1998; pp. 1–75.

[6] M. Thompson, S.L.R. Ellison, R. Wood, Harmonized guidelines for single laboratory validation of methods of analysis, International Union of Pure and Applied Chemistry (IUPAC), Pure Appl. Chem. 74 (2002) 835–855.

[7] L.A. Currie, Nomenclature in evaluation of analytical methods including detection and quantification capabilities, Pure Appl. Chem. 67 (1995) 1699–1723.

[8] The European Medicines Agency, Note for guidance on validation of analytical procedures: text and methodology, CPMP/ICH/381/95, http://www.ema.europe.eu/, [accessed on 28.07.11].

[9] International Organization for Standardization (ISO), Statistical methods for quality control. vol. 2, fourth ed., Accuracy (trueness and precision) of measurement methods and results – part 1–6, ISO 1994(E), 5725 – 1 till 6, http://www.iso.org/, [accessed on 28.07.11].

[10] W.J. Youden, E.H. Steiner, in: Statistical manual of the Association of Official Analytical Chemists,

The Association of Official Analytical Chemists, Arlington, 1975.
[11] Association of Official Analytical Chemists, How to meet ISO 17025 requirements for method verification. 2007; pp. 1–18, http://www.aoac.org/, [accessed on 28.07.11].
[12] Commission Societé Française des Sciences et Techniques Pharmaceutiques (SFSTP), Guide de validation analytique, Rapport d'une commission SFSTP I Méthodologie, STP Pharma Pratiques 2 (1992) 205–226.
[13] European Commission, Method validation and quality control procedures for pesticide residues analysis in food and feed, Document No. SANCO/2007/3131, October 31st 2007, Brussels, http://www.crl-pesticides.eu/library/docs/srm/AqcGuidance.pdf, [accessed on 28.07.11].
[14] S. Sağlık, S. Tatar Ulu, Anal. Biochem. 352 (2006) 260–264.
[15] J.J. Berzas, C. Guiberteau, M.J. Villaseñor, V. Rodríguez, Anal. Chim. Acta. 519 (2004) 219–230.
[16] J.J. Berzas Nevado, M.J. Villaseñor Llerena, C. Guiberteau Cabanillas, V. Rodríguez Robledo, J. Chromatogr. A 1123 (2006) 130–133.
[17] D. Gomes da Silva, P. Guedes de Pinho, H. Pontes, L. Ferreira, P. Branco, F. Remião, et al., J. Chromatogr. B 878 (2010) 815–822.
[18] P. Hoffmann, M.F. Hartmann, T. Remer, K.-P. Zimmer, S.A. Wudy, Steroids 75 (2010) 1067–1074.
[19] S.C. Cunha, J.O. Fernandes, Talanta 83 (2010) 117–125.
[20] M. Menéndez-Carreño, C. García-Herreros, I. Astiasarán, D. Ansorena, J. Chromatogr. B 864 (2008) 61–68.
[21] M. João Valente, F. Carvalho, M. Lourdes Bastos, M. Carvalho, P. Guedes de Pinho, J. Chromatogr. B 878 (2010) 3083–3088.
[22] E. Cognard, S. Rudaz, S. Bouchonnet, C. Staub, J. Chromatogr. B 826 (2005) 17–25.
[23] W.A. Wan Raihana, S.H. Gan, S.C. Tan, J. Chromatogr. B 879 (2011) 8–16.
[24] P. Basilicata, N. Miraglia, M. Pieri, A. Acampora, L. Soleo, N. Sannolo, J. Chromatogr. B 818 (2005) 293–299.
[25] K.C. Ahire, M.S. Arora, S.N. Mukherjee, J. Chromatogr. B 861 (2008) 16–21.
[26] J. Zhou, X. Xue, Y. Li, J. Zhang, L. Wu, L. Chen, J. Zhao, J. Sep. Sci. 30 (2007) 1912–1919.
[27] A. Garrido Frenich, J.L. Martínez Vidal, J.L. Fernández Moreno, R. Romero-González, J. Chromatogr. A 1216 (2009) 4798–4808.
[28] Y. Zhu, G. Li, Y. Duan, S. Chen, C. Zhang, Y. Li, Food Chemistry 109 (2008) 899–908.
[29] J.D. Carrillo, M.P. Martínez, M.T. Tena, J. Chromatogr. A 1181 (2008) 125–130.
[30] A. Garrido Frenich, M.J. González-Rodríguez, F.J. Arrebola, J.L. Martínez Vidal, Anal. Chem. 77 (2005) 4640–4648.
[31] P. Sandra, B. Tienpont, F. David, J. Chromatogr. A 1000 (2003) 299–309.
[32] C.F. Poole, J. Chromatogr. A 1158 (2007) 241–250.
[33] European Pharmacopoeia 6.2, Council of Europe, Strasbourg, France, 2008.
[34] United States Pharmacopeia 30, National Formulary 25, General chapter 1225, Validation of compendial methods, The United States Pharmacopeial Convention, Rockville, Maryland, USA, 2007.
[35] B. Dejaegher, Y. Vander Heyden, J. Chromatogr. A 1158 (2007) 138–157.
[36] B. Dejaegher, A. Durand, Y. Vander Heyden, Experimental design in method optimization and robustness testing, in: G. Hanrahan, F.A. Gomez (Eds.), Chemometric methods in capillary electrophoresis, chapter 2, John Wiley & Sons, New Jersey, 2010, pp. 11–74.
[37] Ph. Hubert, J.-J. Nguyen-Huu, B. Boulanger, E. Chapuzet, P. Chiap, N. Cohen, et al., J. Pharm. Biomed. Anal. 36 (2004) 579–586.
[38] Ph. Hubert, J.J. Nguyen-Huu, B. Boulanger, E. Chapuzet, P. Chiap, N. Cohen, et al., J. Pharm. Biomed. Anal. 45 (2007) 70–81.
[39] Ph. Hubert, J.-J. Nguyen-Huu, B. Boulanger, E. Chapuzet, N. Cohen, P.-A. Compagnon, et al., J. Pharm. Biomed. Anal. 45 (2007) 82–96.

CHAPTER

19

Quantitative Structure—Retention Relationships

Károly Héberger

OUTLINE

19.1. INTRODUCTION

The search for quantitative structure—(chromatographic) retention relationships (QSRRs) is a very useful approach for discussing the retention mechanism and for prediction of retention data in gas chromatography. Although QSRR models provide insight into the mechanisms of retention on a molecular level, the present chapter focuses on prediction of retention data, and the mechanistic aspects [1] are covered in chapter 6 (pp. 137—159).

QSRR models can be applied [2] to: (i) identify the most informative structural descriptors

Gas Chromatography DOI: 10.1016/B978-0-12-385540-4.00019-5

(features), (ii) quantitatively compare separation properties of individual types of chromatographic columns, and (iii) determine physicochemical properties (this aspect is partly covered by chapter 20, Physicochemical Measurements (Inverse Gas Chromatography) pp. 477–494 novel aspects are outlined in Ref. [3]: QSRR models can be used for successful classification of drugs of various compound classes and for testing various chemometric methods. As the number of descriptors encoding the molecular structure is enormous, descriptor selection and building of predictive models while avoiding chance correlation is not a trivial task.

This review summarizes some of the cornerstones in developing today's QSRR approach (historical part) and gives practical advice regarding how to avoid common problems and how to validate the models (parts: Most frequent errors, and Correct validation, respectively). Since the period of 1996–2006 is covered comprehensively in Ref. [3], recent papers (during last five years) are gathered in the final part (Recent developments).

Core journals for QSRR in gas chromatography disseminating around 50% of scientific information [3] are as follows: J. Chromatogr. A, Chromatographia, Anal. Chim. Acta, Talanta, J. Chemometr., and QSAR and Combinatorial Sciences. However, the other half is dispersed in hundreds of different journals, making a survey particularly difficult.

The basic QSSR formula was conceived by Dimov and Osman decades ago [4]:

$$I^{\mathrm{Stat.Phase}}_{\mathrm{temperature}} = b_0 + \sum_{i=1}^{n} b_i X_i + \sum_{j=n+1}^{n+k} b_j Z_j \quad (19.1)$$

where I is the isotherm Kováts retention index measured on a given stationary phase, b's are the parameters to be fitted, X_i's are the basic molecular descriptors (predictors) encoding the molecular structure (measured or calculated); and Z_i's are the so-called tuning variables, which help to achieve calculated I's within the interlaboratory reproducibility of the experimental retention indices, whereas all terms in Eqn (19.1) are still statistically significant. The task is to select statistically significant, and at the same time physically relevant, best combinations of basic and tuning descriptors. Naturally the distinction between basic and tuning descriptors is not sharp; however, generally, the tuning variables themselves are not significantly correlated with the retention data.

There are other ways of giving retention data (e.g., relative retention times). Then, the logarithm of adjusted retention time is used in Eqn (19.1). However, the accuracy and precision of Kováts indices are much better than any other alternatives; therefore, this index is the most frequently predicted one. Some other options unite the advantages of Kováts index with the peculiarities of the topic, i.e., nonisotherm retention indices and other reference scales [as n-alkanes, e.g., Lee scale for polycyclic aromatic hydrocarbons (PAHs)] are also used.

This multilinear approach remained dominant until recently. Multiple linear regression (MLR) is the most frequently applied chemometric method today. Even if better descriptions are produced by nonlinear methods, such as artificial neural networks, support vector machines, and other techniques, MLR remains a kind of standard for comparison.

19.2. HISTORICAL PERSPECTIVE

The roots of QSRR can be traced back to a historical paper by James and Martin [5], who first observed the log-linear relation for adjusted retention time and carbon numbers. Similarly, Ervin sz. Kováts utilized this relation and attributed the central role of boiling point (Trouton's rule) to retention correlations [6]. However, as QSAR is generally due to the early trials of Hansch and Fujita's multilinear approach, QSRR can only be attributed to searching multivariate functionality (c.f. Eqn (19.1)). It is difficult to establish the author and the time of the first multivariate relationship concerning chromatographic retention data. The pioneering work can be rendered to Kaliszan, who gathered and systematized a large number of sources [7,8]. There is no possibility to survey all efforts, but some important references should be mentioned here (expressing the author's personal preferences). Perhaps the first bilinear equation was suggested by Gassiot et al. [9]; later they summarized the results in a comprehensive way [10]: bilinear equation predicting retention indices could explain 99.9% of the total variance.

Some authors have suggested multilinear relationships as early as 1979 and 1981 [11,12], whereas others prefer the bilinear one [13,14].

The first explicit mentioning of the linear Kováts index–boiling point relation was done by Calixto et al. [15]. The perfect linear equation has been found for *n*-alkylbenzenes as a function of logarithm of vapor pressure [16]. Bermejo and Guillén cleared the basic role of boiling points in descriptive equations and they observed that a better description can be achieved if the tuning variables (molar refraction, molar volume, van der Waals volume, etc.) are used in reciprocal form [17,18]. Rohrbaugh and Jurs developed a QSRR model with four descriptors for 86 alkenes [19].

Parallel to these multivariate approaches, nonlinear functions were built in the models for prediction of retention data. Golovnya and Grigoryeva defined a universal equation for calculation of retention parameters such as specific retention volume and retention index, and tested them for O-, N-, and S-containing substances of different classes (alcohols, esters, amines, and sulfur-containing, not cyclic, compounds). Their equation could be used successfully in homologous series [20,21]. Although the equation provides a perfect description of the measured data, the inclusion of the last two terms is questionable; they are not statistically significant, if $n > 7$ [7].

Systematic model building was carried out for relative retention times and Kováts indices, keeping boiling point as the basic variable, introducing an exponential function for it, and reciprocal function for the tuning variables [22]. Later it was possible to prove the general character of the combined equation (combined from an exponential and a linear part) on various stationary phases [23]. Independent from these efforts, the same function was discovered and described in logarithmic form [24]. The nonlinear function of adjusted retention times versus carbon number has been confirmed again [25].

To illustrate the predictive power of multivariate approach and the proper nonlinear one, Figure 19.1 of Ref. [26] is redrawn here. The residual error (middle curve, middle *Y*-axis) can be diminished to around 3 (index unit) after incorporating 28–29 variables (!) into the multilinear model. For comparison, an exponential–linear relation provides less residual error for the same alkylbenzenes using just two descriptors [23], one of which is the boiling point (a cross indicates the residual error for the combined equation on the figure).

Kaliszan's two books [7,8] summarize the early efforts in the QSAR field very well. A thorough review collects the literature sources from 1996 until 2006 August [3]. Some of the not-yet-surveyed or interesting papers are discussed below.

Kováts indices and relative retention times were predicted from the molecular structure for various compound classes: alkanes [27],

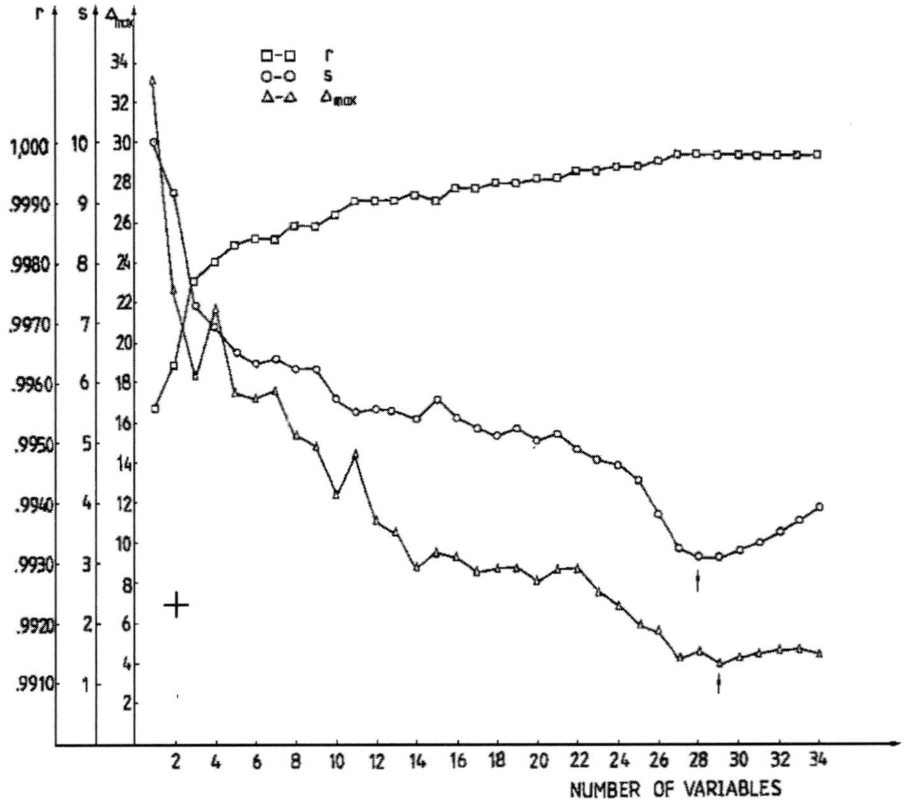

FIGURE 19.1 Effect of including more and more variables into the model (correlation coefficient, residual error, and maximum deviations (measured-predicted)). *Source: Copyright: Elsevier with Permission License Number: 2677131330496.*

anabolic steroids [28], acyclic C5–C8 alkenes [29], alkylbenzenes and naphthalenes [30], dialkylhydrazones [31], substituted aromatics [32], amine–metal complexes [33], PAHs [34], substituted phenols [35], etc.

The time period from 2006 August till 2011 July is covered in the Recent developments section.

19.3. MOST FREQUENT ERRORS

Anonymous referee reports (87) are available courtesy of Elsevier, Springer, and Wiley scientific publishing houses. Some general tendencies can be revealed and given below.

It is a peculiarity that a QSRR paper is seldom written by scientists, who are experts from both a chromatographic and a modeling point of view. Many manuscripts have all the attributes of a paper: introduction, computational part, complicated formulas, results and discussion, conclusion, references, tables, and figures. However, even perfect partitioning is not sufficient for acceptance. The results should be useful for the scientific community. Usefulness is meant in a broad sense, for chromatographers, chemometricians, analysts, statisticians, scientists working on the machine learning field, etc.

19.3.1. Novelty

The reviewers frequently mention the lack of novelty. Rationalization of existing data is justified only if a significantly better fit is achieved or a novel chemometric method has been introduced. The novelty is questionable if reliable measured retention data are available for the given compounds in table form and these data are derived from the literature; no new data were measured and evaluated. In most cases, the modeling methods are similarly well known, and no new techniques were developed. In particular cases, though novelty can be acknowledged, the significance of findings is questionable and the results are subordinate. Many compound classes were studied and straightforward QSRR models were developed 20–30 years ago. Recent investigations ignore the old results though the predictive performance could not be surpassed. The usefulness of the models is doubtful; it is not clear as to retention data of which compounds are to be predicted.

19.3.2. Applicability Domain

Generally, the applicability domain is not given. It is questionable whether the models can be used for prediction purposes and, if so, for which compounds. Unfortunately, the models do not provide a deeper insight into the retention phenomena in many cases. Sometimes the applicability domain has been studied, but no new compounds were to be predicted. The authors frequently claim that QSRR models can be used to estimate the retention indices for new compounds whose experimental values are unknown, but they forget to mention what kinds of compounds are suitable for prediction of their retention data. No author can seriously believe that their models are able to predict retention indices of the 40–50 millions of compounds known nowadays.

It is a serious deficiency when the authors do not provide the applicability domain; increasing evidences [36,37] suggest the necessity of that in the QSAR field. There is no reason to avoid a similar standard in the QSRR modeling.

19.3.3. Problems with Predictive Performance

Novelty, applicability domain, and predictive performance are all closely related issues. Indicators for the model performance (n, R, q, F, s, etc.) are n, number of solutes involved; R, multiple correlation coefficient; q, cross-validated correlation coefficient (adjusted R and q are not used, but recommended); F, overall Fisher statistics; and, s, the residual error or RMSE (root mean squared error). R and F are indeed linear indicators, but they can be calculated for the I(measured) versus I(calculated) linear relationship even if the calculated I was derived from a nonlinear model (artificial neural networks, ANN; support vector machines, SVM; I is the notation for Kováts index but any form of retention data, response factor, can be used instead). To predict Kováts indices with RMSE between 50 and 300 i.u. is not much gain. The predicted values cannot be used for identification. A low level of cross-validated correlation coefficients is a strong indicator about the nonapplicability of the models (q^2 < 0.3, $n < 30$). Prediction error is strongly preferred to be within the interlaboratory reproducibility. This cannot be realized in many cases. Therefore the models should give deeper insight into the retention phenomena, or novel chemometric techniques should be developed.

19.3.4. Deficient Validation

The validation is indeed a complex problem, and the scientific community has only started to accept standards for that. Moreover, the views are diverging about the usefulness of internal and external validations [38–40]. Leave-one-out (LOO), leave-many-out (LMO) cross-validations (CVs) show the consistency and robustness of the model building. The predictive power of LOO CV is questionable [41]. While CV almost unbiasedly estimates the prediction error when no model selection (i.e., variable selection) has been performed, it is heavily biased when considerable model selection is applied [42]. More and more scientists require external test set validation [37,40–42]. There are other resampling methods for validation: randomization test (Y-scrambling), bootstrap, and Monte Carlo simulations – all are suitable to seek distribution to the existing data.

The internal validation methods are thought to be overoptimistic [42]. However, internal indicators of model performance can signal wrong modeling and fits; it is just that there is no guarantee that their high value involves good predictive performance. Therefore, internal validation is a necessary, but not a sufficient form, of validation.

In many cases, the splits into training, validation, and test sets are far from being satisfactory, with the consequence of obtaining nonoptimal and overtrained models. The number of solutes included is often too low, especially in the validation and test sets.

19.3.5. Descriptors

There are some descriptor calculation programs (Dragon, Codessa, Hyperchem, etc.), but it is advisable to search for physically relevant descriptors (especially the usage of boiling point, which proved to be necessary). Calculation of the descriptors is simple but their meaning is often obscure [2], i.e., their application should be avoided. Three-dimensional descriptors are not better than two-dimensional ones [42]. Rendering physical relevance to descriptors enhances the predictive ability.

Linearly selected descriptors are always used in nonlinear methods, e.g., to feed ANN inputs, but this is not advisable. Vanyur et al. proved that linear descriptors should be used for linear modeling techniques and nonlinear ones for nonlinear techniques [43].

Heat of formation (generally denoted by ΔH_f) frequently appears in QSRR models, but erroneously. It is true that ΔH_f can easily be calculated by quantum chemical programs, but they have nothing to do with the solution phenomenon. Heat of formation does not couple to solute–solvent interactions. Moreover, a difference between heat of formation for surface and for analyte does not correlate with chromatographic retention. The correlation of retention data with ΔH_f is simply a consequence of increasing ΔH_f values with increasing molecular mass (and related magnitudes, molar volume, van der Waals volume, molar refraction, etc.).

Several studies are limited to models with one or two descriptors without determining the optimal number of descriptors: a stepwise forward selection, or fairly a best subset selection, would give the optimal numbers of descriptors in the models, if the performance were checked on a separate (test) set. In such cases, better models can be developed if more descriptors are used. Similarly, the optimal number of latent variables is not always determined in case of partial least squares regression (PLS) or determined incorrectly.

Calculations of some parameters (e.g., log$P$ and quantum-chemical) are not reproducible; they depend on the software used, the parametrization, etc. Semiempirical methods can give fairly good and rather bad results and nobody knows when.

19.3.6. Techniques Used

MLR is commonly chosen for the model building algorithm nowadays and is considered a standard approach. PLS and ANN have also been applied; chance correlation and overfitting have not been considered properly in many situations. A descriptor selection is not absolutely necessary before model building, especially in the case of PLS. ANN and SVM are prone to overfitting like kernel methods, as well. The results of these techniques are not always reproducible, which is a crucial problem. If nobody can reproduce the findings, the contribution cannot be considered a scientific one. A simple 3-2-1 ANN architecture (input(3)-hidden(2)-output(1) layer) is sufficient to describe retention data and predicts them better than complicated architectures. Empirical rules suggest using less hidden neurons than inputs (square root of the number of input neurons is a good guess).

Reduction of the number of descriptors might be advisable to eliminate constant and highly correlated variables, but a simple univariate elimination based on correlation coefficients might be a dangerous tactic: two (or more) descriptors together can be highly predictive, whereas none of them is predictive alone. It is safer to follow an algorithm, e.g., forward stepwise, all subset regression, etc.

19.3.7. Comparison with Literature Data is Missing

A comparison with earlier findings is essential. Some compound classes, such as alkylbenzenes, polycyclic aromatic hydrocarbons (PAHs), and polychlorinated biphenyls (PCB), are modeled frequently. It should be noted that if ignoring the predecessors, the novelty of new models cannot be evaluated. At least five models built using various chemometric techniques should be compared, according to the relevant literature. The small number of compared models seriously limits the possibility to draw relevant conclusions.

Abundant prediction papers are available for alkylbenzenes [44], oxygen-containing compounds: aldehydes and ketones [45], alcohols [46], PAHs [47], and oxygen, nitrogen, and sulfur-containing heterocycles [48].

19.3.8. Number of Compounds

Many authors are not aware that finding QSRR between three-four compounds is an illusion. Generally, the number of compounds is too small in the test set, but most reviewers think prediction of 14 retention indices on two stationary phases is not sufficient for a publication.

19.3.9. Congener Series Violation

Some authors are oblivious to the basic assumption of QSRR; it is valid for congener series only. Diverse compounds do not have the same retention mechanism. The distribution phenomena should be the same (or at least similar) to all of the compounds. Therefore, most reported correlations between the molecular structure and GC retention indices either relate to narrow classes of compounds or utilize other experimentally determined physicochemical properties [49].

Various volatile compounds (alkanes, alkenes, ethers, amines, alcohols, alkylbenzenes, and alkylhalides) do not realize one congener series [50], i.e., a QSRR study cannot be based on a common mechanism. The same can be said for drug-like compounds.

Even the assignment of essential oil components to one congener series is dubious, though practically useful information can be obtained [51,52].

19.3.10. Insufficient Conclusions

Several authors, while conducting a correct modeling study, form trivial or misleading conclusions. There is no use stating "The ANN model ... gave a significantly better performance than the other models." Who has ever doubted statements such as "ANN can be used as an alternative modeling tool for quantitative structure–property/retention relationship (QSPR/QSRR) studies"? This has been known for a long time and has been confirmed by thousands of papers. Nonlinear models fit better. Such statements are commonplace and have nothing to do with science.

19.3.11. Inadequate References

On the one hand, the references are too numerous and, on the other hand, at the same time deficient. Many references are superfluous in the Introduction part; references are frequently missing for correct validation and for comparison with similar models. Naturally it is impossible to cite all relevant QSRR investigations, but the efforts concerning the studied compound class should be enumerated. Many authors forget to mention earlier findings, especially from several decades ago. As a rule of thumb, never cite a paper if you have not read it completely.

Similarly, it is not appropriate to cite a source not yet published or not retrievable.

19.4. RECOMMENDATIONS TO AVOID THE MOST COMMON ERRORS

It should be totally clear from the abstract, and/or from the introduction, what the novel contributions are in the paper, and which ideas already exist in the literature, and the authors build on them.

Unfortunately, determination of thermodynamic quantities is out of the forefront of the research nowadays. However, I think, theoretical studies are essential to understand the solution phenomenon better.

One should ensure that the new compounds predicted will fall between the validity ranges of models, within the chemical domain, and within ranges of physicochemical parameters.

The optimal split would be 1/3 for training, 1/3 for validation, and 1/3 for test sets. The splits should be done randomly, and performance parameters (R^2, q^2 for LOO, for LMO, etc., and *RMSE*) should be calculated and given for all three sets.

In the case of MLR, the descriptors should be selected on the training set, the regression coefficients should be calculated on the validation set (calibration), and the overall performance is checked on an independent test set (independent from any part of the modeling). In the case of PLS or PCR, the number of latent components should be determined on the training set and using the validation set. The usage of validation set for ANN calculations is trivial.

The notation "RI" is not a proper abbreviation; IUPAC recommends Capital-(italic)-I-subscript-temperature-and-superscript-stationary phase for Kováts retention indices.

The newly developed models should be superior to the earlier ones and the predictive performance should be compared with similar models published in the literature.

19.5. CORRECT VALIDATION

It is a principal issue how to complete a correct validation of QSRR models. Although the standards for validation are not yet fixed, an increasing number of experts suggest external validation as the necessary validation step. The chromatographic measurements are relatively cheap, and the purity of substances is not critical. Hence, completion of new measurements is strongly encouraged.

If this is not feasible, rigorous numerical validation should be carried out. CV variants might show the insufficient model performance, though not necessarily. One aspect of CV has been critically evaluated by Bro et al. [53]. As a rule of thumb, the more variants applied, the better. Randomization test, LOO, LMO, bootstrap, and Monte Carlo simulations all "map" the data structure in a different way. However, there exists another possibility to check the consistency of models and examine the correctness of splits for training, calibration, and test sets.

There is an easy test; a *de facto* standard for comparison of the modeling methods applied. A novel procedure based on sum of ranking differences (SRD) [54] will show: (i) which modeling method is the best; and (ii) whether the splits are correct (nonconsistent rankings of methods will show problematic splits). The theoretical distribution for SRD values is given in Ref. [55]; validation of SRD ordering by comparison of ranks by random numbers can also be found in Ref. [55]. An MS Excel macro (computer code) for SRD ranking can be downloaded together with sample input and output files from the Internet: http://knight.kit.bme.hu/CRRN or http://goliat.eik.bme.hu/~kollarne/CRRN.

As an illustrative example, the Sutter et al. models [56] are compared.

Figure 19.2a shows the model comparison for the training set. The distribution for random ranking is omitted for clarity (the number of objects and the fitted parameters of the approximate Gaussian distribution for random numbers are given in the header of the figure). As a point

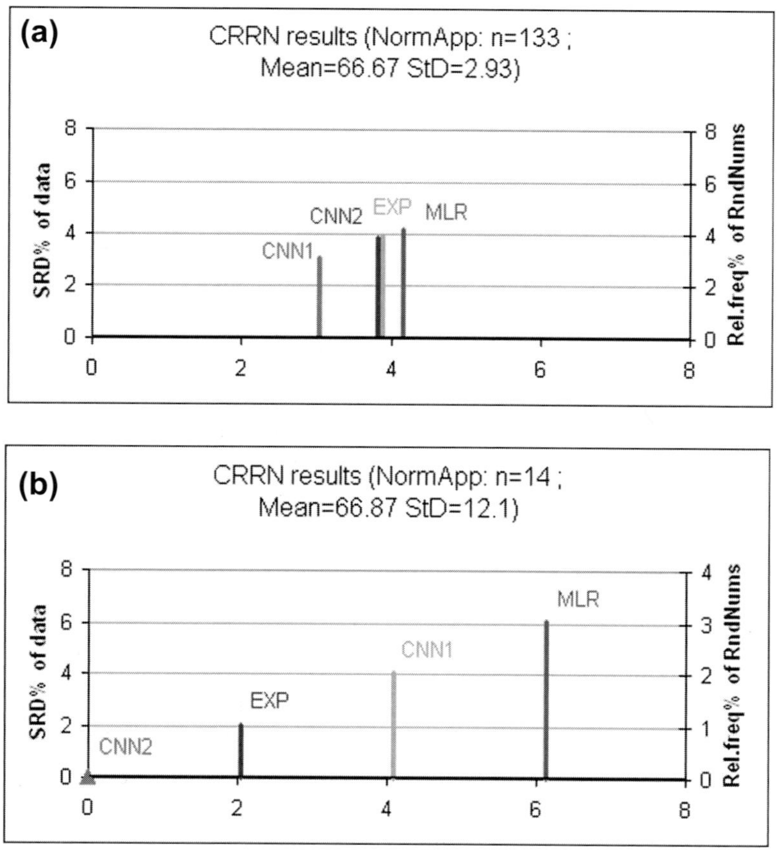

FIGURE 19.2 (a) Ranking of models and experimental values of retention indices for *training set* (alkylbenzenes). Values for sum of ranking differences (SRD) were scaled in between 0 and 100. CNN1: computational neural network model 1, CNN2: computational neural network model 2, MLR: model built by multiple linear regression, Exp: measured retention indices. (b) Ranking of models and experimental values of retention indices for *test set* (alkylbenzenes). Values for SRD were scaled in between 0 and 100. CNN1: computational neural network model 1, CNN2: computational neural network model 2, MLR: model built by multiple linear regression, Exp: measured retention indices.

of reference, the average values of retention indices have been chosen as biases of various calculation and measurement methods cancel each other. Even if this is not the case, we may define a hypothetical average model and look for the similarity to this by SRD ranking.

The first conclusion is that CNN model rationalizes the data better than the experimental values; however, MLR provides worse position in ranking. The second conclusion is that the models (and experimental data) involve much smaller SRD values than the random ranking (c.f., the mean and standard deviation for the Gaussian fit of random numbers in the header of Figure 19.2a).

Figure 19.2b shows the model comparison for the test set. The distribution for random ranking is again omitted for clarity. As a point of reference, the average values of retention indices have been chosen as earlier. A computational neural network (CNN) model 2 rationalizes the data exactly as the average of all models for the test set. Experimental values are in the second position in ranking. Other modeling methods cannot rationalize the data better than the experimental ones. Still the SRD ranking is much better than the random ranking (c.f., the mean and standard deviation for the Gaussian fit of random numbers in the header of Figure 19.2b).

Comparing Figure 19.2a and b one can immediately see that the ranking is different for the two sets. The SRD values for the training set are more compact and the Gaussian standard deviation is much smaller ($StDtrain = 2.93 \ll StDtest = 12.1$), but these are the outcomes of the larger number of solutes included in the training set. SRD rankings show unambiguously that the split for training and test sets was not consistent.

SRD rankings do not utilize information about the predictive performance. Another possibility for verification is whether all terms in the model are significant. In some cases, the theoretical purity outweighs the statistical considerations. Keeping some insignificant terms in the model will not significantly deteriorate the predicted values [34].

Another unsolved problem is the validation of recurrent relationships. It is simply not allowed to sacrifice known values from among the small number of predicted values [57].

The correlation coefficient is not a particularly good indicator for the model performance. Its value says nothing without the degrees of freedom. Phrases such as 'satisfactory' or even 'excellent' correlation should be avoided if the degree of freedom is not given. The readers should evaluate the performance and not the authors themselves.

Residual analysis is often missing; though the adequateness of models can be checked easily: curvature and trend should not be seen in the residua plot(s) (Y(measured) minus Y(calculated)) against Y(calculated).

A good example of how to conduct a proper QSRR study is given in Ref. [58].

Ojha et al. have recently developed a new metric for validation of quantitative structure–property relationships [59]. The average of correlation coefficients with and without intercept might be a useful tool empirically, but statisticians have known for a long time that the two measures cannot be compared directly.

An alternative way of model comparison has been suggested by Todeschini et al. [60]: a new distance measure has been defined between two models. The distance allows us to find clusters of similar models and the most diverse models can also be determined while preserving maximum information and diversity.

19.6. RECENT DEVELOPMENTS

Table 19.1 summarizes the QSRR efforts in gas chromatography between 2006 August and 2011 July. The author is indebted to those who point out missing sources. The time period of 1996–2006 has been covered in Ref. [3].

TABLE 19.1 QSRR in Gas Chromatography 2006 Aug – 2011 July

Solutes	Descriptors	Model building	Stationary phase (SP)	Validation	Source
Substituted benzenes, benzaldehydes and acethophenones	Calculated descriptors, a polarity term for stationary phase (M)	*I*, MLR, BP-ANN	Six OV stationary phases with different phenyl percentages	Cross-validation, test set	[61]
PCN	MEDV	RRI, MLR, additivity	DB-5	LOO	[62]
Alkyl pyridines (18)	Valence molecular connectivity indices (7), steric parameter s,	*I*, ΔH, ΔS, MLR	Apolar (C78) and polar (POH) Kováts' phases	No	[63]
PCDF (115), polychlorinated dibenzo-p-dioxins (41), polychlorinated naphthalenes (62) PCBs (210)	Generalized correlative index (GCI)	Various retention data	Different	Cross-validation	[64]
Molecules in fuel	Topological descriptors	Programmed-temperature *I*, PCR		LOO	[65]
Sulfide series	New topological index NPm (m = 1,2,3) based on distance matrix and branch vertex of atoms	*I*, MLR			[66]
Substituted benzenes and naphtalenes, etc. (109)	LSER descriptors	*k* – adjusted retention factor. MLR, PCA, radar plots	Alkylsiloxane-bonded stationary phases, polar phases	No	[67–69]
Aliphatic alcohols and benzene homologues	Inclusion complex properties	*I*, Simple linear	Calixarenes	No	[70]
PCBs	Constitutional, topological, geometrical, electrostatic, and quantum chemical (236)	Retention time, PCA, MLR, PCR	HT-8/BPX-50	LOO, training-test sets	[71]

(*Continued*)

TABLE 19.1 QSRR in Gas Chromatography 2006 Aug – 2011 July (*Cont'd*)

Solutes	Descriptors	Model building	Stationary phase (SP)	Validation	Source
Terpenes	Constitutional descriptors and topological indices	*I*, MLR, ANN		Random split, external test set	[72]
PBDE-s (126)	Molecular descriptors (17) HOMO, LUMO, ΔH_f, LogP etc.	RRT, MLR	DB-1, DB-5, HT-5, DB-17, DB-XLB, HT-8, CP-Sil 19	LOO, training-test sets	[73]
Alcohols, ketones, esters (86+20)	Modified molecular polarizability effect index, modified inner molecular polarizability index, topological indices	*I*, MLR, ANN	OV-1 and SE-54	LOO, training-test sets	[74]
Chlorinated pesticides, herbicides, and organohalides (38)	ChemSAR (physicochemical, thermodynamic, electronic and spatial) descriptors	Retention times MLR, PLS	Relatively non-polar	LOO, external set	[75]
Pesticides and toxicants (110)	Constitutional, topological, geometrical, electrostatic, quantum chemical descriptors	Retention times , MLR, SVM	DB-35MS	LOO of the whole training set	[76]
Saturated esters (90)	Topological descriptors	*I*, MLR	SE-30, OV-7, DC-710, OV-25, XE-60, OV-225 and Silar-5CP	LOO	[77]
Polychlorinated dibenzothiophenes (135)	Mean molecular polarizability and entropy	*I*, simple linear			[78]
Complex Petroleum Fractions	Dragon descriptors	RT, MLR	PONA	No	[79]
Essential-Oil Components	Topological	*I*, stepwise and GA-MLR	DB-5MS	LMO: training (33) prediction (9), randomly	[80]

Psychiatric drugs (124)	Topological, geometric, and electronic descriptors (699)	RT, CART, ANFIS	Crosslinked methylsilicone	Calibration and prediction set	[81]
Sulphonamides (32)	LSER	*k*, MLR,	2-ethylpyridine (2-EP)	No	[82]
Alkyl substituted cyclic hydrocarbons (174)	Charged Partial Surface Area (CPSA) 56 VolSurf descriptors	*I*, SVM, RBF-NN, MLR	Squalane	LOO	[83]
Fatty acid methyl esters (130+37)	Zero-, one- and two-dimensional descriptors	*I*, MLR, PLS, Various descriptor selection schemes	Standard non-polar polydimethyl siloxane	LMO 3-fold cross-validation	[84]
Hydrocarbons, fluorocarbons, chlorocarbons, bromocarbons, iodocarbons, alcohols, acids, ketones, aldehydes, ethers, esters, amines, nitriles. (208)	Quantumchemical and CODESSA	*I*, MLR, RBF-NN	Absorbent material	LOO training (165) and test (43) sets	[85]
Halogenated aliphatic compounds (141)	Semi-empirical topological index, I-ET,	T_b, melting point, log*P*, *I*, simple linear		LOO, test set (8)	[86]
Alkanes, alkenes	Semi-empirical electrotopological index	*I*, additivity			[87]
Phenol - pentafluorobenzyl (PFB) derivatives (37)	Topological, geometric, and hybrid (geometric and topological) descriptors	RT, stepwise MLR, BP-ANN	DB-1701 (14% cyanopropylphenyl 86% dimethyl-polysiloxane)	LOO, LMO and Y-randomization	[88]
Polycyclic aromatic sulfur heterocycles (114)	Quantumchemical, V_m, HOMO, surface area	*I* (Lee scale), stepwise MLR,		LOO	[89]

(*Continued*)

TABLE 19.1 *QSRR in Gas Chromatography 2006 Aug – 2011 July* (Cont'd)

Solutes	Descriptors	Model building	Stationary phase (SP)	Validation	Source
Alkylbenzenes (122)	Semi-empirical topological index, I-ET,	*I*, additivity, T_b, R_m, melting point, log*P*,	Squalane, SE-30, Carbowax 20 M		[44]
Saffron aroma constituents (43)	Constitutional, topological, geometrical, electrostatic, quantum-chemical descriptors	RT, MLR, Projection Pursuit Regression	ZB-5 MS	Training (34) test (9) sets	[90]
Aliphatic alcohols and benzene homologues	Inclusion complex properties	*I*, Simple linear	Calixarenes	No	[91]
Monoterpenes (17), sesquiterpenes (30)	Mass spectra and *I*	MLR	GFA models	Random selection	[92]
Complex petroleum condensate fractions	Components molecular descriptors, e.g. total path counts and T_b	Adjusted RT, non-linear model, MLR			[93]
PCBs (208)	Quantum chemical, HOMO, V_m, polarizability, ΔH_f	RRT, MLR	DB1, SPB-Octyl, SPBOctyl, CP-Sil5-C18, DB5-MS, RTX-5, CP-Sil-13, SPB-20,HP-35, RTX-35, DB-17, HP-1301, DP-XLB, DB-35-MS, HT-8, Apiezon L, CNBP#2,	LOO, Y-randomization, external validation, training (160)	[94]
Pesticides (168)		Stepwise MLR, Kernel-PLS		Monte Carlo cross validation, Y-randomization	[95]
Chlorinated cyclohexane isomers	No of Carbon atoms, R_m, M_w	*I*, additivity, bilinear T_b	OV-101	No	[96]
Acetone, ethyl alcohol, toluene chloroform, dioxane, *n*-hexane, etc. (21)	Steric substituent constants, R_m	ΔH, linear quadratic	Mesoporous carbon adsorbent	No	[97]

McReynolds test solutes	Quantum chemical, topological descriptors (276)	*I*, MLR	37 stationary phases	LOO, validation and external set	[98]
Components of essential oils (100)	Dragon descriptors (325)	*I*, GA-MLR, poly-PLS, SVM	HP-5	LOO, training and test set	[52]
Various compounds: benzene, *n*-butanol, ethyl formate, 2-butanone, aniline (52)	M_w, T_b, R_m, Branching index etc.	*I*, MLR	7 poly(3-cyanopropyl-methylsiloxane) phases	No	[99]
Trimethylsilylated anabolic androgenic steroids	Topological (molecular connectivity, shape indices, path), 3D, physicochemical descriptors (81)	RRT, MLR, PLS	Methylsilicone gum	Internal cross validation, Y-scrambling.	[100]
Aldehydes (15), ketones (42)	Semi-empirical electrotopological index, I-SET	MLR			[101]
Diverse compounds (94)	LSER descriptrs	MLR	SPB-Octyl, HP-5, Rtx-440, Rxi-50, Rtx-OPP, DB-1701, DB-225, HP-Innowax, HP-88	No	[102]
Volatile compounds (71)	M_W, Density Hbond-donor AlogP98	*I*, MLR	DB-5	LOO	[103]
Various groups	Chemical domain	*I*, linear		No	[104]
Alkylphenols	Number of H atoms, relative number of O atoms, Balaban index, partial charges hydrogen bond donor atoms HDCA (2) index	*I*, SVM, MLR		Training and test sets	[105]
Essential oil components	Nonzero E-state indexes (13), TSAR descriptors (56)	RT, MLR, PCR, PLS	HP5-MS	LOO, external set	[51]

(*Continued*)

TABLE 19.1 *QSRR in Gas Chromatography 2006 Aug – 2011 July* (*Cont'd*)

Solutes	Descriptors	Model building	Stationary phase (SP)	Validation	Source
Saturated esters (100)	Semi-empirical electrotopological index, I-SET, McReynolds constants	*I*, MLR	SE-30, OV-7, DC-710, OV-225 and XE-60	LOO, external set	[106]
Diverse forensic compounds (846)	Dragon descriptors (529)	*I*, GA-MLR, ANN	Methyl polysiloxanes	LOO, LMO, external set	[107]
Polybrominated diphenyl ethers (180)	Numbers of total, ortho, meta and para substituents	RRT, MLR	Rtx-1614	No	[108]
Volatile substances (127)	Various *I*	*I*, PCA	C78-paraffin, its polar derivatives	No	[109]
Glycerol Esters	Number of carbon atoms	*I*, simple linear	OV-101	No	[110]
Hydrocarbons (134)	Dragon descriptors (>400)	*I*, MLR, ANN	PONA methyl silicone	Training set (126) test (35) set	[111]
Petroleum hydrocarbons (73)	Total path count (TPC), sum of atomic polarizabilities (SP)	*I*, MLR, PCA, projection pursuit	Dimethyl-polysiloxane	No	[112]
Diverse compounds in essential oils		*I*, GA, MLR, PLS kernel PLS, ANN		Training, test set	[113]
Benzene, toluene, xylenes, ethylbenzene, substituted benzenes	Quantumchemical descriptors, *V*m, LUMO, radius, HOMO, Mullikan charges, dipole moment	RT, MLR	DB-1, DB-624, DB-wax		[114]
Substituted aromatics (92)	Mean structural increment (MSI)	Additivity, MLR	DB-1		[115]
Cyclic compounds in rosemary and sage essential oil (40)	Quantumchemical, LUMO, HOMO, polarizability, dipole moment, Dragon descriptors	GA-MLR, GA-PLS, GA-ANN	HP 5MS	Training (24) calibration (8) test (8) sets	[116]

PCBs (209)	22 molecular descriptors	RRT, MLR	See ref. [96]	LOO, Y-randomization, external validation, training (160)	[117]
Illicit drugs (two sets)	Codessa descriptors	RT, MLR		LOO	[118]
Aroma components of essential oils	one- and two-dimensional descriptors	*I*, GA, MLR, PLS kernel PLS, ANN		LMO, external set	[119]
Volatile organic compounds (132)	Topological descriptors (30) from DRAGON program	*I*, MLR, PLS, PCR	Apolar (C78) and polar (POH, PCN) Kováts' phases	Y-randomization LOO, external set	[50]
Phenol derivatives, saturated esters	Topological descriptors,	RF, MLR, ANN			[120]
Ketones, aldehydes and esters	Semi-empirical electrotopological index I-SET	*I*, simple linear	SE-30, OV-3, OV-7, OV-11, OV-17, OV-25	External set	[121]
Adamantane, its alkyl derivatives, mono- and bicyclic hydrocarbons $C_{10}H_{(n)}$	Excess thermodynamic functions of mixing		Squalane	No	[122]
Linear and branched aliphatic hydrocarbons (22), cyclic and aromatic hydrocarbons (20), halogenated, ethers, etc.	Bond types	Ordering of molecules by interaction strength	Cu(II)-, Co(II) Cyclam complexes	No	[123]
Ketones, alcohols	Group contribution, No of carbon atoms	RT, thermodynamic functions	SLB5ms (5% phenyl), SPB50 (50% phenyl), Supelco Wax (polyethylene glycol)	No	[124]
PAHs	Polarizability, the second order connectivity Kier and Hall index	*I*, MLR	SE-52	LOO, LMO	[47]

(*Continued*)

TABLE 19.1 *QSRR in Gas Chromatography 2006 Aug – 2011 July (Cont'd)*

Solutes	Descriptors	Model building	Stationary phase (SP)	Validation	Source
Esters, ketones, aldehydes, alcohols (106)	Hydrogen-association classified molecular electronegativity-distance vector (H-MEDV)	*I*, MLR	OV-1 dimethylpolysiloxane	LOO, training and test sets	[125]
Alkylbenzenes (22)	M_w, T_b, HOMO, LUMO, dipole moment, etc.	*I*, MLR, PCR	Cu(II)-, Co(II) Cyclam complexes	Cross-validation for determination of PCs	[126]
Fatty acid methyl esters (49+15)	Dragon descriptors	RT, MLR, PLS, BP-ANN	HP-88, (biscyanopropyl polysiloxane)	Traning, test sets	[127]
Alcohols, alkanes	He gas effect	Thermodynamic functions	See ref. [126]	No	[128]
Nitrobenzene derivatives (35)	Multivariate image-, + Dragon descriptors	RT, MLR	DB-5	LOO, LMO	[129]
Diverse compounds (230), achiral solutes	LSER descriptors	*k*, MLR, PCA	Chiral phases, tris-(3,5-dimethylphenyl-carbamate) amylose and cellulose	No	[130]
Alkanones and chlorinated counterparts	ΔI,	Additivity,	OV-101	No	[131]
Phenol derivatives	Electronic and quantum chemical descriptors	MLR, PLS, ANN	Rtx-200	LMO	[132]
Polycyclic aromatic sulfur heterocycles (PASHs), (114)	MEDV	*I*, MLR		Calibration and test sets	[133]
Diverse compounds (116) in essential oils		*I*, GA, MLR, PLS, kernel-PLS, ANN		LMO	[134]

PCBs (209)	Dragon dscriptors	RT, GA, MLR, ANN, PLS	DB-1/HT-8, DB-XLB/HT-8, DB-XLB/BPX-50, HT-8/BPX-50	Calibration (70) test (139)	[135]
Unsubstituted and substituted PAHs	Volume, dispersion and hydrophobic interactions	*I*, GA, MLR		Internal and external	[139]

Notations
ANN - artificial neural network
ANFIS - Adaptive Neuro-Fuzzy Inference System
BP - back-propagation
CART - Classification and Regression Tree
DB-1 - 100% dimethylpolysiloxane
DB-5 - 5% diphenyl and 95% dimethylpolysiloxane
DB-210 - trifluoropropylmethyl polysiloxane
DB-wax - polyethyleneglycol
GA - genetic algorithm
HP-1 - 100% dimethylpolysiloxane,
HP-5 - 5% diphenyl and 95% dimethylpolysiloxane
HP-50 - 50% diphenyl and 50% dimethylpolysiloxane
HP-Innowax - polyethyleneglycol
I - Kováts retention index
k - *retention factor or coefficient, (capacity factor)*
LOO - leave-one-out (internal) cross-validation
LMO - leave-multiple-out (internal) cross-validation also known as leave-many-out or leave-group-out
MEDV - Molecular Electronegativity Distance Vector(s)
PAH - polycyclic aromatic hydrocarbon
PBDEs - polybrominated diphenyl ethers
PCA - principal component analysis
PCN - polychlorinated naphthalenes
PCR - principal component regression
PCBs - polychlorinated biphenyls
PCDF - polychlorinated dibenzofuran
PLS - Partial least squares regression
PP - projection pursuit
RBF-NN - radial basis function neural network
RF - response factors
R_m - *molar refraction*
RRT - relative retention time
RRI - relative retention index
RT - retention time
SD - standard deviation
SVM - support vector machine
T_b - *boiling point*
V_m - *molar volume*

There are not many reviews about QSRR in gas chromatography. However, Dorman et al. mention [137]: "The final area of empirical modeling, quantitative structure–retention relationships, continues to be developed."

References

[1] S.N. Atapattu, C.F. Poole, Solute descriptors for characterizing retention properties of open-tubular columns of different selectivity in gas chromatography at intermediate temperatures, J. Chromatogr. A 1195 (2008) 136–145.

[2] R. Kaliszan, QSRR, quantitative structure-(chromatographic) retention relationships, Chem. Rev. 107 (2007) 3212–3246.

[3] K. Héberger, Quantitative structure-(chromatographic) retention relationships, J. Chromatogr. A 1158 (2007) 273–305.

[4] N. Dimov, A. Osman, Selection of molecular descriptors used in quantitative structure-gas chromatographic retention relationships II. Isoalkanes and alkenes, Anal. Chim. Acta. 323 (1996) 15–25.

[5] A.T. James, A.J.P. Martin, Gas–liquid partition chromatography: the separation and micro-estimation of volatile fatty acids from formic acid to dodecanoic acid, Biochem. J. 50 (1952) 679–690.

[6] E. sz. Kováts, Gas-chromatographische charakterisierung organischer verbindungen: Teil 1. Retentionsindices aliphatischer halogenide, alkohole, aldehyde und ketone, Helv. Chim. Acta. 41 (1958) 1915–1932.

[7] R. Kaliszan, Quantitative structure-chromatographic retention relationships, Wiley, New York, 1987.

[8] R. Kaliszan, Structure and retention in chromatography - a chemometric approach, Harwood, Amsterdam, 1997.

[9] M. Gassiot, E. Fernández, G. Firpo, R. Carbo, M. Martin, Empirical quantum chemical approach to structure gas chromatographic retention index relationships 1. Sterol acetates, J. Chromatogr. 108 (1975) 337–344.

[10] M. Gassiot-Matas, G. Firpo-Pamies, Relationships between gas chromatographic retention index and molecular structure, J. Chromatogr. 187 (1980) 1–19.

[11] G. Dahlmann, H.J.K. Köser, H.H. Oelert, Multiple correlation of retention indices, Chromatographia 12 (1979) 665–671.

[12] L. Buydens, D.L. Massart, Prediction of gas chromatographic retention indexes from linear free energy and topological parameters, Anal. Chem. 53 (1981) 1990–1993.

[13] E.E. Kugucheva, V.I. Mashinskii, Retention indices of aromatic hydrocarbons on capillary columns with squalane and poly(phenyl ether), J. Anal. Chem. USSR 38 (1983) 1558–1561. Zh. Anal. Khim. 38 (1983) 2023–2026 (in Russian).

[14] J. Bermejo, M.D. Guillén, Biparameter equations for calculating Kováts retention indices of hydrocarbons, Int. J. Environ. Anal. Chem. 23 (1985) 77–86.

[15] F.S. Calixto, A.G. Raso, P.M. Déya, Capillary gas chromatography: a correlation between retention indices and boiling temperatures for alcohols, aldehydes, ketones and esters, J. Chromatogr. Sci. 20 (1982) 7–10.

[16] V.A. Gerasimenko, V.M. Nabivach, Relationships between molecular structure and gas chromatographic retention of C_6–C_{12} alylbenzenes on polydimethylsiloxane, Zh. Anal. Khim. 37 (1982) 110–116 (in Russian).

[17] J. Bermejo, M.D. Guillén, A study of Kováts retention indices of aliphatic saturated esters and their relation to the polarity of the stationary phase, J. Chromatogr. 318 (1985) 187–194.

[18] J. Bermejo, M.D. Guillén, Prediction of Kováts retention index of saturated alcohols on stationary phases of different polarity, Anal. Chem. 59 (1987) 94–97.

[19] R.H. Rohrbaugh, P.C. Jurs, Prediction of gas chromatographic retention indexes of selected olefins, Anal. Chem. 57 (1985) 2770–2773.

[20] R.V. Golovnya, D.N. Grigoryeva, The reason for nonlinear variation of specific retention volumes and retention indices for members of homologous series of organic compounds, Chromatographia 17 (1983) 613–622.

[21] R.V. Golovnya, D.R. Grigoryeva, Violation of the linearity principle of additivity of sorption energy in chromatography, J. High Resolut. Chromatogr. 9 (1986) 584–589.

[22] K. Héberger, Empirical correlations between gas-chromatographic retention data and physical or topological properties of solute molecules, Anal. Chim. Acta. 223 (1989) 161–167.

[23] K. Héberger, Discrimination between linear and nonlinear models describing retention data of alkylbenzenes in gas-chromatography, Chromatographia 29 (1990) 375–384.

[24] I.G. Zenkevich, L.M. Kuznetsova, A new approach to the prediction of gas chromatographic retention indices from physicochemical constants, Collect Czechoslov. Chem. Commun. 56 (1991) 2042–2056.

[25] J.E. Quintanilla-López, R. Lebrón-Aguilar, J.A. García-Domínguez, Hold-up time in gas chromatography V. Dependence of the retention of *n*-alkanes on the

chromatographic variables in isothermal gas chromatography, J. Chromatogr. A 878 (2000) 125–135.

[26] O. Mekenyan, N. Dimov, V. Enchev, Methodology for deriving quantitative structure–retention relationships in gas chromatography, Anal. Chim. Acta. 260 (1992) 69–74.

[27] T. Körtvélyesi, M. Görgényi, L. Seres, Estimation and prediction of GC retention applied quantum-chemical calculations: alkanes and azo compounds, Chromatographia 41 (1995) 282–286.

[28] C.G. Georgakopoulos, O.G. Tsika, J.C. Kiburis, P.C. Jurs, Prediction of gas chromatographic relative retention times of anabolic steroids, Anal. Chem. 63 (1991) 2025–2028.

[29] Y. Ren, H. Liu, X. Yao, M. Liu, Three-dimensional topographic index applied to the prediction of acyclic C5–C8 alkenes Kováts retention indices on polydimethylsiloxane and squalane columns, J. Chromatogr. A 1155 (2007) 105–111.

[30] N. Dimov, A. Osman, O. Mekenyan, D. Papazova, Selection of molecular descriptors used in quantitative structure – gas chromatographic retention relationships: I. Application to alkylbenzenes and naphtalenes, Anal. Chim. Acta. 298 (1994) 303–317.

[31] Z. Király, T. Körtvélyesi, L. Seres, M. Görgényi, Unusual retention relations in the gas chromatography of N, N′-dialkylhydrazones, Chromatographia 42 (1996) 653–659.

[32] B. Lučić, N. Trinajstić, S. Sild, M. Karelson, A.R. Katritzky, A new efficient approach for variable selection based on multiregression: prediction of gas chromatographic retention times and response factors, J. Chem. Inf. Comput. Sci. 39 (1999) 610–621.

[33] I. Rykowska, W. Wasiak, Comparative studies of gas chromatographic properties of new packings with chemically bonded complexes, J. Chromatogr. Sci. 39 (2001) 313–320.

[34] S.S. Liu, C.S. Yin, S.X. Cai, Z.L. Li, Molecular structural vector description and retention index of polycyclic aromatic hydrocarbons, Chemometr. Intell. Lab. Syst. 61 (2002) 3–15.

[35] R. Kaliszan, H.D. Höltje, Gas-chromatographic determination of molecular polarity and quantum chemical calculation of dipole moments in a group of substituted phenols, J. Chromatogr. 234 (1982) 303–311.

[36] P. Gramatica, P. Pilutti, E. Papa, Validated QSAR prediction of OH tropospheric degradation of VOCs: splitting into training-test sets and consensus modeling, J. Chem. Inf. Comput. Sci. 44 (2004) 1794–1802.

[37] P. Gramatica, Principles of QSAR models validation: internal and external, QSAR Comb. Sci. 26 (2007) 694–701.

[38] A. Tropsha, P. Gramatica, V.K. Gombar, The importance of being earnest: validation is the absolute essential for successful application and interpretation of QSPR models, QSAR Comb. Sci. 22 (2003) 69–77.

[39] D.M. Hawkins, Assessing model fit by cross-validation, J. Chem. Inf. Comput. Sci. 43 (2003) 579–586.

[40] K.H. Esbensen, P. Geladi, Principles of proper validation: use and abuse of re-sampling for validation, J. Chemometr. 24 (2010) 168–187.

[41] A.M. Doweyko, 3D-QSAR illusions, J. Comp. Aid. Mol. Des. 18 (2004) 587–596.

[42] K. Baumann, Chance correlation in variable subset regression: influence of the objective function, the selection mechanism, and ensemble averaging, QSAR Comb. Sci. 24 (2005) 1033–1046.

[43] R. Vanyúr, K. Héberger, I. Kövesdi, J. Jakus, Prediction of tumoricidal activity and accumulation of photosensitizers in photodynamic therapy using multiple linear regression and artificial neural networks, Photochem. Photobiol. 75 (2002) 471–478.

[44] L.C. Porto, E.S. Souza, B.D.S. Junkes, R.A. Yunes, V.E.F. Heinzen, Semi-empirical topological index: development of QSPR/QSRR and optimization for alkylbenzenes, Talanta 76 (2008) 407–412.

[45] K. Héberger, M. Görgenyi, M. Sjöström, Partial least squares modeling of retention data of oxo compounds in gas chromatography, Chromatographia 51 (2000) 595–600.

[46] O. Farkas, K. Héberger, Comparison of ridge regression, partial least squares, pair-wise correlation, forward- and best subset selection methods for prediction of retention indices for aliphatic alcohols, J. Chem. Inf. Model. 45 (2005) 339–346.

[47] J.C. Drosos, M. Viola-Rhenals, R. Vivas-Reyes, Quantitative structure–retention relationships of polycyclic aromatic hydrocarbons gas-chromatographic retention indices, J. Chromatogr. A 1217 (2010) 4411–4421.

[48] O. Farkas, K. Héberger, I.G. Zenkevich, Quantitative structure–retention relationships XIV. Prediction of gas chromatographic retention indices for saturated O-, N-, and S-heterocyclic compounds, Chemometr. Intell. Lab. Syst. 72 (2004) 173–184.

[49] A.R. Katritzky, U. Maran, V.S. Lobanov, M. Karelson, Structurally diverse quantitative structure-property relationship correlations of technologically relevant physical properties, J. Chem. Inf. Comput. Sci. 40 (2000) 1–18.

[50] R. Ghavami, S. Faham, QSRR models for Kováts' retention indices of a variety of volatile organic compounds on polar and apolar gc stationary phases using molecular connectivity indexes, Chromatographia 72 (2010) 893–903.

[51] L.-T. Qin, S.-S. Liu, H.-L. Liu, J. Tong, Comparative multiple quantitative structure–retention relationships modeling of gas chromatographic retention time of essential oils using multiple linear regression, principal component regression, and partial least squares techniques, J. Chromatogr. A 1216 (2009) 5302–5312.

[52] S. Riahia, E. Pourbasheer, M.R. Ganjali, P. Norouzi, Investigation of different linear and nonlinear chemometric methods for modeling of retention index of essential oil components: concerns to support vector machine, J. Hazard Materials 166 (2009) 853–859.

[53] R. Bro, K. Kjeldahl, A.K. Smilde, H.A.L. Kiers, Cross-validation of component models: a critical look at current methods, Anal. Bioanal. Chem. 390 (2008) 1241–1251.

[54] K. Héberger, Sum of ranking differences compares methods or models fairly, Trends Anal. Chem. 29 (2010) 101–109.

[55] K. Héberger, K. Kollár-Hunek, Sum of ranking differences for method discrimination and its validation: comparison of ranks with random numbers, J. Chemometr. 25 (2011) 151–158.

[56] J.M. Sutter, T.A. Peterson, P.C. Jurs, Prediction of gas chromatographic retention indices of alkylbenzenes, Anal. Chim. Acta. 342 (1997) 113–122.

[57] I.G. Zenkevich, Application of recurrent relations in chemistry, J. Chemom. 23 (2009) 179–187. DOI: 10.1002/cem.1214.

[58] B. Ren, Atom-type-based AI topological descriptors for quantitative structure–retention index correlations of aldehydes and ketones, Chemometr. Intell. Lab. Syst. 66 (2003) 29–39.

[59] P.K. Ojha, I. Mitra, R.N. Das, K. Roy, Further exploring r_m^2 metrics for validation of QSPR models, Chemometr. Intell. Lab. Syst. 107 (2011) 194–205.

[60] R. Todeschini, V. Consonni, M. Pavan, A distance measure between models: a tool for similarity/diversity analysis of model populations, Chemometr. Intell. Lab. Syst. 70 (2004) 55–61.

[61] P. Ebrahimi, M.R. Hadjmohammadi, Simultaneous modeling of the Kovats retention indices on phenyl OV stationary phases with different polarity using MLR and ANN, QSAR Comb. Sci. 25 (2006) 836–845.

[62] Y. Zhou, L. Sun, H. Mei, S.Z. Li, Estimation and prediction of relative retention indices of polychlorinated naphthalenes in gc with molecular electronegativity distance vector, Chromatographia 64 (2006) 565–570. DOI: 10.1365/s10337-006-0054-0.

[63] P. Tulasamma, K.S. Reddy, Quantitative structure and retention relationships for gas chromatographic data: application to alkyl pyridines on apolar and polar phases, J. Molec. Graph. Model. 25 (2006) 507–513.

[64] Z. Peng, M. Hu, F.F. Tian, Z.L. Li, Novel generalized correlative index used to carry out research in quantitative structure–retention relationship for some persistent organic pollutants, Chinese J. Anal. Chem. 34 (2006) 1096–1100.

[65] T. Zhang, Y.Z. Liang, C.X. Zhao, D.L. Yuan, Prediction of temperature-programmed retention indices from molecule structures, Chinese J. Anal. Chem. 34 (2006) 1607–1610.

[66] C.M. Nie, G.W. Peng, F.Z. Xiao, S. Li, X.M. He, Z.H. Li, et al., Study on topological chemistry of gas chromatography retention index for sulfide, Chinese, J. Anal. Chem. 34 (2006) 1560–1564.

[67] C. West, E. Lesellier, Characterisation of stationary phases in subcritical fluid chromatography by the solvation parameter model I. Alkylsiloxane-bonded stationary phases, J. Chromatogr. A 1110 (2006) 181–190.

[68] C. West, E. Lesellier, Characterisation of stationary phases in subcritical fluid chromatography by the solvation parameter model, II. Comparison tools, J. Chromatogr. A 1110 (2006) 191–199.

[69] C. West, E. Lesellier, Characterisation of stationary phases in subcritical fluid chromatography with the solvation parameter model III. Polar stationary phases, J. Chromatogr. A 1110 (2006) 200–213.

[70] M. Bialon, J. Hetper, Investigation of inclusion properties of p-tert-butylcalixarenes by inversion gas chromatography method, Polimery 52 (2007) 274–279 (in Polish).

[71] Y. Ren, H. Liu, X. Yao, M. Liu, An accurate QSRR model for the prediction of the GC×GC-TOFMS retention time of polychlorinated biphenyl (PCB) congeners, Anal. Bioanal. Chem. 388 (2007) 165–172.

[72] B. Hemmateenejad, K. Javadnia, M. Elyasi, Quantitative structure–retention relationship for the Kováts retention indices of a large set of terpenes: a combined data splitting-feature selection strategy, Anal. Chim. Acta. 592 (2007) 72–81.

[73] H.Y. Liu, S.S. Liu, L.T. Qin, Semi-empirical topological method for prediction of the gas chromatographic relative retention times of polybrominated diphenyl ethers (PBDEs), J. Mol. Model. 13 (2007) 611–627.

[74] F. Liu, Y. Liang, C. Cao, N. Zhou, Theoretical prediction of the Kováts retention index for oxygen-containing organic compounds using novel topological indices, Anal. Chim. Acta. 594 (2007) 279–289.

[75] J. Ghasemi, S. Asadpour, A. Abdolmaleki, Prediction of gas chromatography/electron capture detector retention times of chlorinated pesticides, herbicides, and organohalides by multivariate chemometrics methods, Anal. Chim. Acta. 588 (2007) 200–206.

[76] X. Li, F. Luan, H. Si, Z. Hua, M. Liu, Prediction of retention times for a large set of pesticides or toxicants based on support vector machine and the heuristic method, Toxicol. Lett. 175 (2007) 136–144.

[77] F. Liu, Y. Liang, C. Cao, N. Zhou, QSPR study of GC retention indices for saturated esters on seven stationary phases based on novel topological indices, Talanta 72 (2007) 1307–1315.

[78] S.D. Chen, H.X. Liu, Z.Y. Wang, Study of quantitative structure–retention relationship (QSRR) of gas chromatography for polychlorinated dibenzothiophenes on non-polar columns, QSAR Comb. Sci. 26 (2007) 889–896.

[79] N.E. Moustafa, Prediction of GC retention times of complex petroleum fractions based on quantitative structure–retention relationships, Chromatographia 67 (2008) 85–91. DOI: 10.1365/s10337-007-0467-4.

[80] S. Riahi, M.R. Ganjali, E. Pourbasheer, P. Norouzi, QSRR study of GC retention indices of essential-oil compounds by multiple linear regression with a genetic algorithm, Chromatographia 67 (2008) 917–922. DOI: 10.1365/s10337-008-0608-4.

[81] M. Jalali-Heravi, P. Shahbazikhah, A. Ghadiri-Bidhendi, QSRR study of psychiatric drugs using classification and regression trees combined with adaptive neuro-fuzzy inference system, QSAR Comb. Sci. 27 (2008) 729–739.

[82] A. Cazenave-Gassiot, R. Boughtflower, J. Caldwell, R. Coxhead, L. Hitzel, S. Lane, et al., Prediction of retention for sulfonamides in supercritical fluid chromatography, J. Chromatogr. A 1189 (2008) 254–265.

[83] H.-F. Chen, Quantitative predictions of gas chromatography retention indexes with support vector machines, radial basis neural networks and multiple linear regression, Anal. Chim. Acta. 609 (2008) 24–36.

[84] O. Farkas, I.G. Zenkevich, F. Stout, J.H. Kalivas, K. Héberger, Prediction of retention indices for identification of fatty acid methyl esters, J. Chromatogr. A 1198–1199 (2008) 188–195.

[85] F. Luan, H.T. Liu, Y. Wen, X. Zhang, Quantitative structure–property relationship study for estimation of quantitative calibration factors of some organic compounds in gas chromatography, Anal. Chim. Acta. 612 (2008) 126–135.

[86] A.C.S. Arruda, B. da S. Junkes, E.S. Souza, R.A. Yunes, V.E.F. Heinzen, Semi-empirical topological index to predict properties of halogenated aliphatic compounds, J. Chemometr. 22 (2008) 186–194.

[87] E.S. Souza, C.A. Kuhnen, R.A. Yunes, V.E.F. Heinzen, On a new semi-empirical electrotopological index for QSRR models, J. Chemometr. 22 (2008) 378–384.

[88] K. Asadpour-Zeynall, N. Jalili-Jahani, Modeling GC-ECD retention times of pentafluorobenzyl derivatives of phenol by using artificial neural networks, J. Sep. Sci. 31 (2008) 3788–3795.

[89] H.Y. Xu, J.W. Zou, Y.J. Jiang, G.X. Hu, Q.S. Yu, Quantitative structure-chromatographic retention relationship for polycyclic aromatic sulfur heterocycles, J. Chromatogr. A (2008) 1198–1199. 202–207.

[90] H. Du, J. Wang, Z. Hu, X. Yao, Quantitative structure–retention relationship study of the constituents of saffron aroma in SPME-GC-MS based on the projection pursuit regression method, Talanta 77 (2008) 360–365.

[91] M. Bialon, J. Hetper, Study of the effect of substituent size on the inclusion properties of calixarenes by inverse gas chromatography, Polimery 54 (2009) 216–220 (in Polish).

[92] Z.G. Li, H. Cao, L.L. Wang, W.M. Mo, D.L. Shen, Quantitative structure–retention relationship for terpenes and their application in structure characterization, Acta. Chimica Sinica 67 (2009) 289–294 (in Chinese).

[93] N.E. Moustafa, Gas chromatographic retention times prediction for components of petroleum condensate fraction, Chemical Papers 63 (2009) 608–612.

[94] R. Ghavami, F. Sadeghi, QSRR-based evaluating and predicting of the relative retention time of polychlorinated biphenyl congeners on 18 different high resolution GC columns, Chromatographia 70 (2009) 851–868. DOI: 10.1365/s10337-009-1233-6.

[95] M. Jalali-Heravi, H. Ebrahimi-Najafabadi, A. Khodabandehloo, Use of Kernel orthogonal projection to latent structure in modeling of retention indices of pesticides, QSAR Comb. Sci. 28 (2009) 1432–1441.

[96] I.G. Zenkevich, E.V. Eliseenkov, A.N. Kasatochkin, Chromatographic identification of cyclohexane chlorination products by an additive scheme for the prediction of retention indices, Chromatographia 70 (2009) 839–849.

[97] M. Snauko, D. Berek, T. Cserháti, Determination of the adsorption energy of some volatile solvents on the surface of a mesoporous carbon adsorbent by gas-chromatography, Croat. Chem. Acta. 81 (2008) 409–412.

[98] E.A. Hoffmann, Z.A. Fekete, R. Rajkó, I. Pálinkó, T. Körtvélyesi, Theoretical characterization of gas-liquid chromatographic stationary phases with quantum chemical descriptors, J. Chromatogr. A 1216 (2009) 2540–2547.

[99] A.M. Tello, R. Lebrón-Aguilar, J.E. Quintanilla-López, J.M. Santiuste, Isothermal retention indices on poly(3-cyanopropylmethylsiloxane) stationary phases, J. Chromatogr. A 1216 (2009) 1630–1639.

[100] A.G. Fragkaki, A. Tsantili-Kakoulidou, Y.S. Angelis, M. Koupparis, C. Georgakopoulos, Gas chromatographic quantitative structure–retention relationships of trimethylsilylated anabolic androgenic steroids by multiple linear regression and partial least squares, J. Chromatogr. A 1216 (2009) 8404–8420.

[101] E.S. Souza, C.A. Kuhnen, B.D.S. Junkes, R.A. Yunes, V.E.F. Heinzen, Modeling the semi-empirical electrotopological index in QSPR studies for aldehydes and ketones, J. Chemometr. 23 (2009) 229–235.

[102] C.F. Poole, S.N. Atapattu, S.K. Poole, A.K. Bell, Determination of solute descriptors by chromatographic methods, Anal. Chim. Acta. 652 (2009) 32–53.

[103] Z.-G. Li, H. Cao, M.-R. Lee, D.-L. Shen, Analysis of volatile compounds emitted from *Chimonanthus praecox* (L.) link in different florescence and QSRR study of GC retention indices, Chromatographia 70 (2009) 1153–1162.

[104] I.G. Zenkevich, "Chemical" domains of definition of mathematic relations in organic chemistry, Russ. J. Gen. Chem. 79 (2009) 2164–2174.

[105] M.H. Fatemi, E. Baher, M. Ghorbanzade'h, Predictions of chromatographic retention indices of alkylphenols with support vector machines and multiple linear regression, J. Sep. Sci. 32 (2009) 4133–4142.

[106] E.S. Souza, C.A. Kuhnen, B.D.S. Junkes, R.A. Yunes, V.E.F. Heinzen, Quantitative structure–retention relationship modelling of esters on stationary phases of different polarity, J. Mol. Graph. Model. 28 (2009) 20–27.

[107] Z. Garkani-Nejad, Use of self-training artificial neural networks in a QSRR study of a diverse set of organic compounds, Chromatographia 70 (2009) 869–874.

[108] H. Wei, R. Yang, A. Lia, E.R. Christensen, K.J. Rockne, Gas chromatographic retention of 180 polybrominated diphenyl ethers and prediction of relative retention under various operational conditions, J. Chromatogr. A 1217 (2010) 2964–2972.

[109] E. sz. Kováts, S. Morgenthaler, The choice of polar stationary phases for gas-liquid chromatography by statistical analysis of retention data, Chromatographia 70 (2009) 831–837.

[110] A.S. Leol'ko, E.L. Krasnykh, S.V. Levanova, Retention indices of glycerol esters, J. Anal. Chem. 64 (2009) 1126–1130. Zh. Anal. Khim. 64 2009; 1154–1158.

[111] X. Zhang, L. Ding, Z. Sun, L. Song, T. Sun, Study on quantitative structure–retention relationships for hydrocarbons in FCC gasoline, Chromatographia 70 (2009) 511–518. DOI: 10.1365/s10337-009-1174-0.

[112] N.E. Moustafa, K. E. k. F. Mahmoud, Classification and prediction of retention indices in one-dimensional capillary gas chromatographic separation of petroleum hydrocarbons, Chromatographia 72 (2010) 905–912. DOI: 10.1365/s10337-010-1734.

[113] H. Noorizadeh, A. Farmany, A. Khosravi, Investigation of retention behaviors of essential oils by using QSRR, J. Chinese Chem. Soc. 57 (2010) 982–991.

[114] A. Morsali, S.A. Beyramabadi, M.R. Bozorgmehr, M. Raanaee, B. Keyvani, G.R. Jafari, Prediction of gas chromatography retention of BTEX and other substituted benzenes based on quantum chemical parameters, Scientific Research and Assays 5 (2010) 349–351.

[115] J. Acevedo-Martínez, I.G. Zenkevich, R. Carrasco-Velar, Use of a simple additive scheme to predict the GC retention indices of aromatic compounds with different structures, Chromatographia 71 (2010) 881–889. DOI: 10.1365/s10337-010-1587-9.

[116] H. Noorizadeh, A. Farmany, QSRR models to predict retention indices of cyclic compounds of essential oils, Chromatographia 72 (2010) 563–569. DOI: 10.1365/s10337-010-1660-4.

[117] R. Ghavami, S.M. Sajadi, Semi-empirical topological method for prediction of the relative retention time of polychlorinated biphenyl congeners on 18 different HR GC columns, Chromatographia 72 (2010) 523–533. DOI: 10.1365/s10337-010-1696-5.

[118] B.B. Xia, Y.J. Wang, R.Q. Yang, X.Y. Zhang, Quantitative structure–retention relationship study on the GC-MS retention time of illicit drugs, Chinese J. Struct. Chem. 29 (2010) 1879–1885.

[119] H. Noorizadeh, A. Farmany, Exploration of linear and nonlinear modeling techniques to predict of retention index of essential oils, J. Chinese Chem. Soc. 57 (2010) 1268–1277.

[120] Z. Garkani-Nejad, Quantitative structure–retention relationship study of some phenol derivatives in gas chromatography, J. Chromatogr. Sci. 48 (2010) 317–323.

[121] E.S. Souza, C.A. Kuhnen, B.D.S. Junkes, R.A. Yunes, V.E.F. Heinzen, Development of semi-empirical electrotopological index using the atomic charge in QSPR/QSRR models for alcohols, J. Chemometr. 24 (2010) 149–157.

[122] S. Yashkin, Equilibrium parameters of a liquid-vapor system and thermodynamic characteristics of

adsorption of cyclic and cage hydrocarbons in squalane, Russ. Chem. Bull. 59 (2010) 2026–2038.

[123] P. Bielecki, W. Wasiak, Cyclam complexes of Cu(II) and Co(II) as stationary phases for gas chromatography, J. Chromatogr. A 1217 (2010) 4648–4654.

[124] B. Karolat, J. Harynuk, Prediction of gas chromatographic retention time via an additive thermodynamic model, J. Chromatogr. A 1217 (2010) 4862–4867.

[125] L. Liao, D. Qing, J. Li, G. Lei, Structural characterization and Kovats retention index prediction for oxygen-containing organic compounds, J. Mol. Struct. 975 (2010) 389–396.

[126] I. Rykowska, P. Bielecki, W. Wasiak, Retention indices and quantum-chemical descriptors of aromatic compounds on stationary phases with chemically bonded copper complexes, J. Chromatogr. A 1217 (2010) 1971–1976.

[127] V.K. Gupta, H. Khani, B. Ahmadi-Roudi, S. Mirakhorli, E. Fereyduni, S. Agarwal, Prediction of capillary gas chromatographic retention times of fatty acid methyl esters in human blood using MLR, PLS and back-propagation artificial neural networks, Talanta 83 (2011) 1014–1022.

[128] T.M. McGinitie, B.R. Karolat, C. Whale, J.J. Harynuk, Influence of carrier gas on the prediction of gas chromatographic retention times based on thermodynamic parameters, J. Chromatogr. A 1218 (2011) 3241–3246.

[129] Z. Garkani-Nejad, M. Ahmadvand, Comparative QSRR modeling of nitrobenzene derivatives based on original molecular descriptors and multivariate image analysis descriptors, Chromatographia 73 (2011) 733–742.

[130] C. West, Y. Zhang, L. Morin-Allory, Insights into chiral recognition mechanisms in supercritical fluid chromatography. I. Non-enantiospecific interactions contributing to the retention on tris-(3,5-dimethylphenylcarbamate) amylose and cellulose stationary phases, J. Chromatogr. A 1218 (2011) 2019–2032.

[131] I.G. Zenkevich, E.V. Eliseenkov, A.N. Kasatochkin, Z.A. Zhakovskaya, L.O. Khoroshko, Identification of the chlorination products of aliphatic ketones by gas chromatography and gas chromatography/mass spectrometry, J. Anal. Chem. 66 (2011) 396–406. Zh. Anal. Khim. 66 2011; 406–416.

[132] Z. Garkani-Nejad, M. Ahmadvand, Investigation of linear and nonlinear chemometrics methods in modeling of retention time of phenol derivatives based on molecular descriptors, Sep. Sci. Technol. 46 (2011) 1034–1044.

[133] Z. Li, F. Cheng, Z. Xia, Quantitative structure-gas chromatographic retention relationship of polycyclic aromatic sulfur heterocycles using molecular electronegativity-distance vector, Se. Pu. 29 (2011) 63–69 (in Chinese).

[134] H. Noorizadeh, A. Farmany, M. Noorizadeh, Quantitative structure–retention relationship analysis of retention index of essential oils, Quimica Nova 34 (2011) 242–249.

[135] A.A. D'Archivio, A. Incani, F. Ruggieri, Retention modelling of polychlorinated biphenyls in comprehensive two-dimensional gas chromatography, Anal. Bioanal. Chem. 399 (2011) 903–913. DOI: 10.1007/s00216-010-4326-z.

[136] K. Bouharis, M.L. Souici, D. Messadi, Retention indices for programmed-temperature gas chromatography of polycyclic aromatic hydrocarbons: a QSRR study, Asian J. Chem. 23 (2011) 1044–1048.

[137] F.L. Dorman, E.B. Overton, J.J. Whiting, J.W. Cochran, J. Gardea-Torresdey, Gas chromatography, Anal. Chem. 80 (2008) 4487–4497.

CHAPTER

20

Physicochemical Measurements (Inverse Gas Chromatography)

Adam Voelkel

OUTLINE

During the last 40–50 years of evolution, inverse gas chromatography (IGC) has become a widely used, popular, and fruitful technique for physicochemical characterization of various materials, as well as interactions between components in various systems. Several reviews concerning the theoretical background, parameters, interpretation of experimental data, and application have been developed over the last 20 years [1–8]. Therefore, the basic information of IGC "standard" parameters will be reduced here to a necessary minimum. The author intends to focus on the interpretation of the most often used characteristics, analysis of the errors, and comparison of IGC results with those achieved by using other techniques, as well as the most interesting applications.

Newcomers to this field will learn that inverse gas chromatography "was born" in the early stage of the gas chromatography. The first published papers dealt with activity of the catalyst and various adsorbents [9–11]. This kind of scientific activity was soon called IGC and then extended to the examination of polymers, polymer blends, minerals, and other materials.

The characteristic feature of an ICG experiment is the location of the examined material in the column. It performs the role of the stationary phase. Its properties influence the retention of carefully selected test solutes. Retention data (retention time, net retention

Gas Chromatography DOI: 10.1016/B978-0-12-385540-4.00020-1

volume, and specific retention volume) are further converted into the parameters describing the required property of the examined substance.

At the beginning of the second decade of the XXI century, it is worth remembering two statements concerning IGC: i) it is possible to work at finite concentration of test solutes – FC-IGC or in the region of infinitive dilution ID-IGC – and ii) retention of the test solute may be a result of the pure adsorption equilibrium or the mixed retention mechanism, e.g. bulk sorption and adsorption on existing interfaces. The first mechanism is obvious during the examination of solid materials (adsorbents, fibers, and polymers below T_g), while the second one might be important when polymers (above T_g), polymer blends, or oils are characterized.

IGC experiments might also be carried out by using pulse or frontal techniques. In a pulse mode, the given amount of the test solute is injected into the carrier gas (mobile phase) and transported through a column filled with the stationary phase (examined material). In the frontal technique, test solute is continuously added to the mobile phase, which leads to the formation of the breakthrough curve on the chromatogram. The use of the pulse technique is suggested for systems where the equilibrium is quickly established. It happens most often when interactions between the test solute and examined material are relatively weak. The alternative in a system of "slow" equilibrium is the frontal mode [4].

The first important step in the development of IGC investigation is the setting of experimental conditions: temperature, carrier gas flow rate, humidity of the carrier gas, amount of the sample (examined material), and sample preparation. The temperature of the IGC experiment might be imposed either by the interest of the researcher or by the properties of the examined material. During the investigation of polymer systems, it is not unusual for a phase transition to occur within the range of temperature applied in the IGC experiment [12]. In such a case, the $\log V_g$ *vs.* $1/T$ plot will visualize the region of phase transition. The researcher should carefully adopt the appropriate model for transformation of experimental data (retention parameter of test solutes) into the required physicochemical parameter. The application of an inappropriate model will lead, of course, to vague values of the characteristics and inconclusive results. The carrier gas flow rate should be individually established and optimized by using the van Deemter equation [4]. If the flow rate utilized for the experiment is too low, the time required for the experiment will increase, broad peaks will develop, and the retention data will be less precise. Too high a flow rate may prevent the establishment of local equilibrium assumed in the IGC theory; it may also lead to erroneous results. As already mentioned, the method of choice for systems with "slow" equilibrium is frontal analysis [4].

20.1. GAS–SOLID IGC

20.1.1. Nonspecific Interactions

The dispersive component of surface free energy, γ_S^D, is most often used for characterization of the surface layer of examined material by means of IGC. The two most popular procedures used by IGC researchers were proposed by Dorris and Gray [12] and Schultz et al. [13,14].

The dispersive properties of the examined material are calculated from retention data of test solutes determined at infinite dilution in the Henry's law region [2,12]. It is also assumed that interactions between the adsorbed molecules are negligible.

In the infinite dilution regime, the net retention volume, V_N, of adsorbing test solute

$$V_N = j \cdot t'_R \cdot F \tag{20.1}$$

where j is the James—Martin compressibility factor, F is the carrier gas flow rate [cm^3/min] related to the slope of its adsorption isotherm (K):

$$V_N = K \cdot A_{TOT} \tag{20.2}$$

where A_{TOT} is the total area of the examined material in the column. As K is related to the standard free energy of adsorption and

$$-\Delta G_A = RT \ln V_N + C \tag{20.3}$$

where R is the gas constant, T is the — absolute temperature, and C is the constant, leading to

$$RT \ln V_N = 2a_{mol}(\sigma_S^{LW}\sigma_L^{LW})^{1/2} + C \tag{20.4}$$

where σ_S^{LW} denotes the Lifshitz—van der Waals component of the solid (examined material) surface tension, σ_L^{LW} is the LW component of the test solute liquid surface tension, a_{mol} is the the molar area occupied by the adsorbing molecule, and C is a constant that is dependent on the reference state applied.

The σ_S^{LW} value can be determined by plotting $RT \ln V_N$ against $a_{mol} \cdot (\sigma_L^{LW})^{1/2}$. Sun and Berg [15] applied this model for examination of moisture effect on the surface free energy and acid—base properties of mineral oxides. They calculated a_{mol} assuming spherical molecular shape and hexagonal packing for polar test solutes:

$$a_{mol} = 1.33\, N^{1/3} v_{mol}^{2/3} \tag{20.5}$$

with v_{mol} being the liquid molar volume and N is the Avogadro's number.
a_{mol} of nonpolar species were calculated according to the procedure suggested by Dorris and Gray [12].

An equation similar to Eqn (20.4) may be developed by considering the concept of the work of adhesion. For nonpolar probes, only dispersive (Lifshitz—van der Waals) components influence their retention. The free energy of adsorption is related to the work of adhesion by the following equation:

$$-\Delta G_A = aNW_a \tag{20.6}$$

The work of adhesion is described as

$$W_a = W_a^{LW} = 2\left(\gamma_S^{LW}\gamma_L^{LW}\right)^{1/2} \tag{20.7}$$

Eqns (20.3), (20.6) and (20.7) combined lead to

$$RT \ln V_N = 2N\left(\left(\gamma_S^{LW}\right)^{1/2} a \left(\gamma_L^{LW}\right)^{1/2}\right) + C \tag{20.8}$$

In ICG literature, this relationship is commonly presented in the form introduced by Schultz and Lavielle [13,14]:

$$R \cdot T \cdot \ln V_N = 2 \cdot N \cdot a_p \cdot \sqrt{\gamma_S^D \cdot \gamma_L^D} + C \tag{20.9}$$

where symbol γ_S^D is used instead of γ_S^{LW} and denotes the dispersive component of surface free energy of the solid, the symbol γ_L^D used instead of γ_L^{LW} and denotes the dispersive component of surface free energy of the test solute, and the symbol a_p is used instead of a and denotes the area occupied by the adsorbing molecule.

Dorris and Gray calculated γ_S^D according to the following equation:

$$\gamma_S^D = \frac{-R^2 \cdot T^2 \cdot \left[\ln\left[\dfrac{V_N^{(C_{n+1}H_{2n+4})}}{V_N^{(C_nH_{2n+2})}}\right]\right]^2}{4 \cdot N^2 \cdot \left(a_{CH_2}\right)^2 \cdot \gamma_{(CH_2)}}, \tag{20.10}$$

where $a_{(CH_2)}$ is the the surface area of a methylene group, and the value of this parameter is assumed to be equal to 6 Å^2 but sometimes is taken as 5.2 or 5.5 Å^2; any variation of these quantities is even higher, e.g. 3.1 Å^2 was calculated from a spherical model [16] while values up to 7.7 Å^2 have been observed by scanning tunneling microscopy (STM) [17]; N is the Avogadro's number (6.023 10^{23} [1/mol]); $V_N^{(C_{n+1}H_{2n+4})}$ is the net retention volume of alkane

$C_{n+1}H_{2n+4}$; $V_N^{(C_nH_{2n+2})}$ is the net retention volume of alkane C_nH_{2n+2}; and $\gamma_{(CH_2)}$ is the surface energy of the polyethylene-type polymers with a finite molecular weight [mJ/m^2]. The value $\gamma_{(CH_2)}$ is calculated according to the following equation:

$$\gamma_{(CH_2)} = 34.0 - 0.058 \cdot t \tag{20.11}$$

or

$$\gamma_{(CH_2)} = 35.6 + 0.058(293 - T) \tag{20.12}$$

where t is the temperature in °C and T is the temperature in K.

Hamieh [18] summarized the problem of the uncertainty concerning the "a" value. This might be calculated by using various models. The additional limitation is the fact that the "a" value for given test solute varies with the nature of the examined solid, the temperature, and surface coverage. This problem was previously discussed [19]. It is worth noting that the most problematic value is a_p and may be determined with a large error margin, depending on adsorbent properties, temperature, and reference substance. Although the assumption that a_p is constant is most comfortable, under IGC conditions the most realistic approach seems to treat the test solute as ideal gas [20]. Hamieh [18] also pointed out that γ_L^D values are not always available from the literature within the required temperature range.

Schultz et al. [13,14] applied the Fowkes model for the determination. The van Oss et al. [21] approach is most often used in the surface free energy determination by means of contact angle experiments [22] and references cited therein:

$$(1 + \cos\theta)\gamma^{TOT} = 2\left(\sqrt{\gamma_S^{LW} + \gamma_L^{LW}} + \sqrt{\gamma_S^+\gamma_L^-} + \sqrt{\gamma_S^-\gamma_L^+}\right) \tag{20.13}$$

where the surface free energy γ^{TOT} is the sum of nonpolar (LW, Lifshitz–van der Waals; γ_S^{LW} for solid; γ_L^{LW} for liquid) and polar (AB, acid–base; γ_S^+, γ_S^- for solid; γ_L^+, γ_L^- for liquid) components.

Shi et al. [23] compared γ_S^D values obtained by using the Dorris–Gray and Shultz et al. methods. They found that the ratio defined by the following equation,

$$\frac{\gamma^D_{s,\text{Dorris-Gray}}}{\gamma^D_{s,\text{Schultz}}} = \frac{(a_{n+1}\gamma^{0.5}_{L,n+1} - a_n\gamma^{0.5}_{L,n})^2}{\gamma CH_2 \cdot a^2(CH_2)} \tag{20.14}$$

is always higher than 1. It means that the value of γ_s^D will be larger when calculated by the Dorris– Gray procedure. The ratio slightly increases with the temperature of the IGC experiment. A similar conclusion was drawn by Karaoglan and Sakar [25].

A comparison of γ_s^D values found from the contact angle and IGC experiments has shown that the values calculated from the IGC data are higher. Garnier and Glasser [21] suggested that the differences arise from the changes of the "a" value in different states, while others indicated [26,27] that IGC evaluates, primarily, high-energy surface sites. The dispersive component of the surface free energy was often used to characterize various materials [25–36]. However, it is impossible to cite all papers as this review will evolve into the breakdown of bibliographic data.

20.1.2. Influence of the Humidity on IGC Parameters

Thielmann emphasized the importance of carrier gas purity and its dryness [4]. It is clear that any impurity, including moisture, will affect the results of the experiment. However, it is well-known that there is a group of materials containing a "natural" amount of water, which influences their stability, activity, etc. In such a case, during the IGC examination with the use of a standard, i.e., dry, carrier gas will cause a progressive loss of the water and, finally, the "demolition" of the examined substance. This group of materials may consist of cellulose wooden pulp, natural fibers, dental cements, etc. Therefore, the justified use of the wet carrier gas and the influence of the carrier gas humidity on the estimated parameters should be briefly discussed here.

Comte et al. [37] examined the influence of carrier gas relative humidity (RH) on the surface properties of low specific surface area glass beads. They found significant influence on the conditions of the IGC experiment. Specific retention volume and free enthalpy of adsorption of apolar molecules decreased with the increase of the RH of the carrier gas. Progressive coverage of the glass beads by the water molecules hides high-energy sites and the retention times of alkanes decreased. When the whole surface of the glass beads was covered, the alkanes adsorbed on a film of water (Figure 20.1). The specific retention value of the test solute became constant. This effect was achieved at $p/p_0 = 0.15$. The same result, i.e., the value of "critical" RH, was found when changes of the free enthalpy of adsorption with RH were taken into account. Enthalpy and entropy of adsorption also varied with the RH. The authors [24] found that the entropy term went through minima at $p/p_0 = 0.06$ for octane and nonane and $p/p_0 = 0.03$ for decane. Comte et al. explained it by the nonrandom adsorption due to an orientation of the molecules at the surface. With the increase of the RH, the free, nonwetted zones of the bead surface decrease, which imposes orientation of the test solute molecules to reach the nonwetted part of the surface. After formation of a water film, the test solute molecules gain a few degrees of freedom. The decrease of the retention of the test solutes with an increase in RH was also reported by Garcia-Herruzo et al. [38] after an examination of air humidity on the sorption of hydrocarbons on soil.

The adsorption of polar probes show that ΔG_A decrease occurs in two stages [16,37]. The decrease of ΔG_A was attributed to progressive masking by acidic and basic surface groups of water molecules. Stable ΔG_A values indicated the coverage of the surface by water film. The parameter exhibiting surface ability to acidic (K_A) and basic (K_B) interactions decreased when the RH increased.

Sun and Berg [16] indicated that many active surfaces of mineral oxides adsorb water and their surface properties correspond to a clean surface covered by hydroxyl groups and molecular water. Although these are not the characteristics of a "pure" surface, from a practical point of view, "they are a fair representation of the surfaces of everyday use materials."

Boutboul et al. [39] examined the interactions between aroma compounds and native cornstarch under humid and dry conditions. Selective interactions between the system components result from an adsorption phenomenon involving hydrogen-bonding prevailing when dried

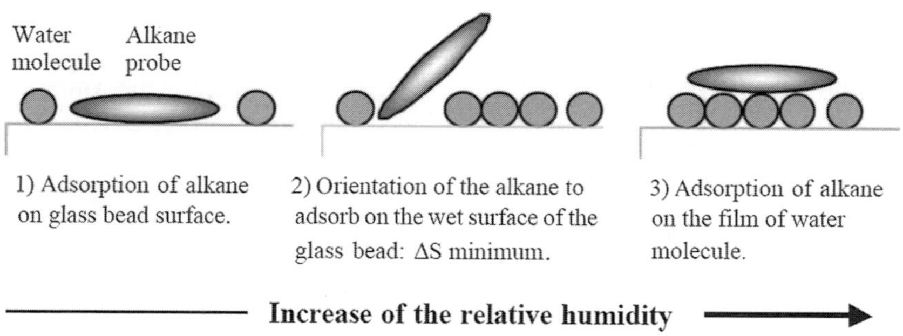

FIGURE 20.1 Representation of the adsorption of alkane probes at different relative humidities *(from Ref. [37] with permission).*

starch (~4.5% water content) was examined. The increase of RH enabled the diffusion of more hydrophilic molecules of test solutes into the native starch matrix. The partial change of retention mechanism is probably an explanation of the increase of retention volume of the polar test solutes used in IGC experiments.

The research group from Valencia (Spain) investigated the effect of moisture on the transport properties of various test solutes across different polymer matrices [40–43]. The sorption of water by these materials, used as stationary phases in PC-IGC, caused the increase of the partition and diffusion coefficients of alcohol with the EVOH (ethylene oxide-vinyl alcohol) copolymer.

Sunkersett et al. [44] studied the influence of the carrier gas humidity upon the dispersive component of the surface free energy (γ_s^D) for two pharmacologically active powders, i.e., carbamazepine and paracetamol. They found that γ_s^D values do not vary significantly with the increase of RH. However, the specific component of the free energy of adsorption ($-\Delta G_A$) remained constant or decreased up to 10%. The authors interpreted these results as an indication of the decrease of surface activity of these materials. This finding corresponds with the findings of Sun and Berg [16] as well as Comte et al. [37]. It is important to note that Sunkersett et al. [31] supported their IGC findings by the GRID analysis of preferential interaction sites of test solutes around paracetamol. They found that the decrease in the adsorption of the test solute might be the result of a competitive sharing of the same chemical site by the probe and the water molecules.

Das et al [45] found that the results of the examination of the influence of humidity on the surface energetic of various powders are unclear. Further reports found that a decrease in the dispersive component of surface free energy with increasing RH was observed [46–48] or it remained constant [44,49]. Das et al. [45] concluded that there is no general relationship between the specific surface energy and the RH of the environment. They found that for salmeterol xinofoate (SX) and α-lactose monohydrate samples, the total surface energies increased after storage at high RH (75%). This increase was attributed to the adsorption of water during storage. The authors pointed out that examination of only the dispersive part of surface energy would result in misleading interpretation of the results. The dispersive component of surface energy of SX decreased after storage, as reported by other research groups. The determination of total surface energy, including the nonspecific and specific contribution, allowed one to properly interpret the changes in the "community" of functional groups onto the powder surface. Das et al. interpreted it as an artifact of shielding by moisture adsorbed at polar sites [37]. However, it is important to remember the differences between the influence of storage at high RH and carrying out the experiments with the use of wet carrier gas.

Kim et al. [50] used IGC for the estimation of thermodynamic characteristics for sorption of water onto three polymer surfaces. Similar values of the molar enthalpy of sorption were found for polystyrene (PS) and poly(methyl methacrylate) (PMMA), while the lower one was found for the poly(2-hydroxyethyl methacrylate) (PHEMA) surface. They found that a larger amount of protein was adsorbed on PS and PMMA in comparison to PHEMA. The authors proved that the strength of water–polymer interactions is an important factor for controlling the amount of nonspecifically adsorbed protein. This finding might be extremely important for evaluation of various polymeric biomaterials where protein deposition may lead to adverse biological consequences such as biofouling and/or activation of immune system.

An interesting paper was published by Slimane et al. [51]. They examined the wetting behavior of materials by polypyrrole. The authors stressed the advantage of IGC over the XPS technique. The analysis depth of XPS is 10 nm (several monolayers), while IGC supplies information on atoms and functional groups at the outermost layers of the materials.

20.1.3. Specific Surface Properties

The ability of the examined surface to create specific interactions by means of IGC might be determined by the injection of the polar probes of known characteristics and the collection of their retention data. The difficulties connected with this approach were presented by Papirer and Balard [52]: (i) a polar test solute will interact with a polar surface through both dispersive and specific interactions; and (ii) the investigator collects only one chromatographic peak and its parameters result from both types of interactions. One has to separate and evaluate both contributions. The term 'specific interactions' denotes all types of interactions except London ones, i.e., bipolar, H-bond type, acid–base, metallic, magnetic, and hydrophobic. Such intermolecular forces are known to dominate over dispersion and dipole–dipole interactions [53]. In the absence of electrostatic, magnetic, or metallic interactions, acid–base interactions prevail over the dispersive interactions. Therefore, special attention was focused on the evaluation of procedures enabling the determination of the acid–base characteristics of the examined surfaces [20].

Acidity and basicity of solid surfaces are deduced from the behavior of polar compounds being injected onto the chromatographic column filled with the studied material [5,54–57]. The ability of the solid surface to interact as a base or acid is expressed in different ways. However, the Gutmann's scale seems to be utilized most often in IGC procedures. It should be noted that *AN** and *DN* numbers of the test solutes express their ability to act as an electron acceptor and electron donor, respectively. Therefore, the final result should be discussed in terms of Lewis acid–base interactions. An interesting discussion of physicochemical aspects of "Gutmann's model" has been presented by Hamieh [58]. Hamieh raised the problem of direct transposition of bulk quantities (*AN*, *AN**, and *DN*) to interfacial interactions. The main contradiction "lies in the discontinuity in the symmetry of the local interactions involving the molecules, when transferring these molecules from the bulk to the rigid surface". Hamieh stated that the dependence of the reference bulk quantities on the nature of interacting species, the local coordination, and physical constraints is properly corrected by the modification of equations used in the specific interaction parameters. This important finding must be carefully verified and discussed, which will be done in one of the following sections.

The ability of the examined surface to cause specific interactions is determined by the procedure consisting of the determination of the specific component of adsorption energy ΔG_A^s, then the specific component of the enthalpy of adsorption ΔH_A^s followed by the solution of the relationship between ΔH_A^s and parameters characterizing the examined material. ΔG_A^s is determined as the difference between the adsorption energy of the polar compound, $\Delta G_{A,\text{polar}}$, hypothetical alkane, $\Delta G_{A,\text{ref}}$ (Eqn (20.15)), having the same selected property as polar test solute (e.g., vapor pressure in the Papirer method [52,58–61] or $a \cdot \sqrt{\gamma_L^D}$ value in the Schultz and Lavielle method):

$$\Delta G_A^s = \Delta G_{A,\text{polar}} - \Delta G_{A,\text{ref}} \tag{20.15}$$

where

$$\Delta G_{A,\text{polar}} = -RT \ln V_{N_{\text{polar}}} + C \quad \text{and}$$
$$\Delta G_{A,\text{ref}} = -RT \ln V_{N_{\text{refr}}} + C$$

Chehimi and Pigois-Landureau [62] compared six methods of evaluating ΔG^S where $RT \ln V_N$ was related to the abscissa coordinates labeled as follows: I, ΔH^d_{vap} (dispersive component of enthalpy of evaporation); II, ΔH_{vap} (enthalpy of evaporation); III, T_b (boiling point); IV, log P^o (saturated vapor pressure); V, $a_F(\gamma_L^D)^{1/2}$ (area occupied by adsorbing molecule and the dispersive component of the liquid solute surface tension, respectively); and VI, $(h\nu)^{1/2}\alpha$ (molecular polarizability). Chehimi and Pigois-Landureau suggested the use of the ΔH^d_{vap} parameter for the description of the reference state for demonstrating the self-association character of polar test solutes earlier postulated by Fowkes [63].

ΔH^s_A is calculated from ΔG^s_A dependence on temperature:

$$\Delta G^s_A = \Delta H^s_A - T\cdot\Delta S^s_A \tag{20.16}$$

where ΔG^s_A is the specific component of the free adsorption energy and ΔS^s_A is the specific component of the free adsorption entropy of the polar compound onto the surface of the investigated solid.

Plotting $\Delta G^s_A/T$ against $1/T$ yields a straight line with the slope of ΔH^s_A. ΔH^s_A should be determined for at least four test compounds, and ΔG^s_A should be determined at at least, three temperatures. However, the calculation of K_A and K_D parameters from ΔH^s_A is time consuming. The estimation of ΔH^s_A from Eqn (20.16) might be the source of the largest error in determination of the specific characteristics of the examined material. Moreover, this procedure might be "dangerous" for examined materials that are labile at the elevated temperature, making the determination of ΔH^s_A impossible.

Specific components of enthalpy of adsorption of the polar compound, ΔH^s_A, are related to acceptor and donor numbers describing the electron acceptor (AN^*) and electron donor (DN) properties of the test compound:

$$\Delta H^s_A = DN\cdot K_A + AN^*\cdot K_D \tag{20.17}$$

where K_A and K_D parameters express the ability of the examined material to act as an electron acceptor and electron donor, respectively.

Plotting $\Delta H^s_A/AN^*$ against DN/AN^*,

$$\frac{\Delta H^s_A}{AN^*} = \frac{DN}{AN^*}\cdot K_A + K_D \tag{20.18}$$

one obtains the straight line with the slope of K_A. As the estimation of K_D from the intercept of Eqn (20.18) may lead to the significant error [64], one should determine this value as the slope of the following relationship:

$$\frac{\Delta H^s_A}{DN} = \frac{AN^*}{DN}\cdot K_D + K_A \tag{20.19}$$

Parameters K_A and K_D were also calculated by using ΔG^s_A [3,58,59,65] instead of ΔH^s_A. This method of determination of K_A and K_D leads to the temperature-dependent values containing entropic factors [5,65].

Hamieh et al. [58,66] proved that the relationship in Eqn (20.18) is, in general, not verified in the case of some metallic oxides. They suggested the use of the corrected version,

$$\begin{aligned}\Delta H^s_A &= K_A\cdot DN + K_D\cdot AN - K(K_A,K_D)AN\cdot DN\\ &= w(K_A\cdot DN + K_D\cdot AN)\end{aligned} \tag{20.20}$$

where w is a parameter expressing a weighing factor for the interactions between adsorbed molecules and solid substrate ($0 \le w \le 1$). $K(K_A, K_D)$ is a constant depending on K_A and K_D and describing the amphoteric nature of the solid substrate.

The very last paper presented by Hamieh [58] is really interesting and is the summary of the earlier papers [66–72]. It combines the series of interesting findings concerning the theory of IGC and the properties of examined materials (poly(α-n-alkyl) methacrylates). However, the proposal expressed by Eqn (20.20) is problematic from a chemometric and mathematical point of view. It is a well-known tendency to construct the complex relationships. It is also known [73,74] that parameters, "constants" used therein, must be independent and any 'intercorrelation' has to be omitted. The approach proposed by Hamieh et al. enhanced, of course, the statistics of ΔH_A^s vs. K_A, K_D relationship, but the meaning of K and/or w parameters is obscure. The authors showed the existing relations between K and K_A, K_D (Eqn (20.18) and Table 11 in [44]). Hopefully, R^2 (regression coefficient) values given therein are relatively low. In my opinion, the deviations in the properties of metallic oxides and other examined materials from those predicted by using Eqn (20.17) should be explained not by overbuilding the appropriate relationships but by looking for another (from Gutman's one) model. Despite any doubts, this model was used many times in the characterization of various species [75–78].

The influence of RH on both dispersive and specific surface properties of various excipients and active materials important for drug delivery was presented [39,47,49,79–86]. The IGC-derived parameters supported by the parameters determined by the use of other techniques (e.g. XPS, DSC, TGA, and DVS) were also used for estimation of the batch-to-batch variations important in the manufacturing of the pharmaceutical species. Various carbohydrates were characterized by using dispersive and specific IGC parameters [87–95]. It is worth noting the examination of chitin and its N-deacetylation derivative [96] by means of IGC. The influence of the nature and treatment of starch on aroma–starch interactions was examined by Boutboul et al. [31,97]. Various minerals, silicas, modified silicas, talc, modified talc, and others were often materials characterized by means of IGC and became the accepted standard [98–110]. Surface energy characteristics of toner articles, ink fillers, printing ink pigments, and components used in coil-coating primers were also achieved by using the methods and procedures of IGC [111–114]. It should be noted that IGC has been successfully applied in the characterization of various zeolites [115–122]. Both surface properties and solubility parameters were determined for two different chewing-gum bases [123]. Chehimi and his co-workers presented a series of interesting papers on the characterization of cement components, cement pastes, and the interactions between structural adhesive components with cement pastes [124–127]. The key role of IGC in the determination of interactions between components of polymeric systems was reviewed by Santos and Guthrie [128,129].

Natural (also vegetable) and synthetic fibers were also characterized by means of IGC [95,130–138], which is an important contribution of IGC research groups to the present-day fiber technology.

Activated carbons were the next group of materials characterized by means of the discussed technique [139–144]. The influence of measurement conditions on the elution peaks of the test solutes was discussed by Wu et al. [140]. The first papers of Kiselev [9–11] were devoted to the examination of catalysts. This is still the current trend in IGC [145].

20.2. BULK PROPERTIES OF POLYMERS AND POLYMER BLENDS

IGC has been used for examination of the bulk properties of polymers and their blends, as well as complex systems consisting of organic and organic components (polymer-filler compositions).

The Flory–Huggins interaction parameter (χ) reflects the interaction between low-molecular-weight solvents and high-molecular-weight polymers, and it has been considered as a Gibbs free energy parameter. According to such an assumption, the interaction parameter χ can be divided into enthalpy χ_H and entropy χ_S components [146]:

$$\chi = \chi_H + \chi_S \tag{20.21}$$

It was found that χ_S is positive and usually should be between 0.2 and 0.6. This term is also sometimes related to the reciprocal of the coordination number in the polymer solution. It has been assumed that in some cases it is between the 0.3 and 0.4 limit [147]. For complete miscibility between polymer and solvent interaction, parameter χ should be less than 0.5. Also, the entropy term is about 0.3; therefore, the enthalpy term χ_H must be very small to meet miscibility's criterion [146].

At infinite dilution of the probe and for high-molecular-weight of the stationary phase, the Flory–Huggins interaction parameter can be determined from [148,149]

$$\chi_{12}^{\infty} = \ln\left(\frac{273.15 \cdot R}{p_1^o \cdot V_g \cdot M_1}\right) - \frac{p_1^o}{R \cdot T} \cdot (B_{11} - V_1^o) + \ln\left(\frac{\rho_1}{\rho_2}\right) - \left(1 - \frac{V_1^o}{V_2^o}\right) \quad (20.22)$$

where V_g is the specific retention volume, *1* denotes the solute and *2* denotes examined material, M_1 is the molecular weight of the solute, p_1^o is the saturated vapor pressure of the solute, B_{11} is the second virial coefficient of the solute, V_i^o is the molar volume, ρ_i is the density, and R is the gas constant.

The simplified form of Eqn (20.36) is also found in the literature [20,150,151]:

$$\chi_{12}^{\infty} = \ln\left(\frac{273.15 \cdot R \cdot v_2}{p_1^o \cdot V_g \cdot V_1^o}\right) - \frac{p_1^o}{R \cdot T} \cdot (B_{11} - V_1^o) - 1 \quad (20.23)$$

where v_2 is the specific volume of the polymer.

Deshpande et al. [152] first suggested the use of IGC for studying polymer blends. Starting from the Flory–Huggins expression for the change of free enthalpy of mixing ΔG_{mix} extended to three-component systems, they proposed a method of elaboration of IGC data collected with the use of polymer blends leading to the polymer–polymer interaction coefficient.

When a mixture of components is used as a stationary phase in a chromatographic column, subscripts 2 and 3 are used to represent the first and second mixtures' components, respectively:

$$\chi_{1m}^{\infty} = \ln\left(\frac{273.15 \cdot R}{p_1^o \cdot V_g \cdot M_1}\right) - \frac{p_1^o}{R^* T} \cdot (B_{11} - V_1^o) + \ln\left(\frac{\rho_1}{\rho m}\right) - \left(1 - \frac{V_1^o}{V_2^o}\right) \cdot \varphi_2 - \left(1 - \frac{V_1^o}{V_3^o}\right) \cdot \varphi_3 \quad (20.24)$$

where φ_2 and φ_3 are the volume fractions of components [146].

The problem of availability and credibility of physical data used in Eqns (20.22)–(20.24) was discussed earlier [20,150]. However, it is worth repeating here that due to uncertainty of the basic physicochemical data, one should take into account the possible error of estimation of the IGC parameter, which may, in several cases, exceed 10%.

Applying the Flory–Huggins equation of polymer solutions to a ternary system with two polymers and one probe, the interaction parameter χ_{1m}^{∞} is related to the probe–polymer interaction parameters and the polymer–polymer interaction parameters by the following equation:

$$\chi_{1m}^{\infty} = \varphi_2 \cdot \chi_{12}^{\infty} + \varphi_3 \cdot \chi_{13}^{\infty} - \varphi_2 \cdot \varphi_3 \cdot \chi_{23}' \quad (20.25)$$

where φ_2 and φ_3 are the volume fractions of the polymers.

The magnitude of interactions between the two components of the polymer blend is expressed in terms of χ_{23}'. Large positive values of χ_{23}' indicates the absence or negligible interactions between components, a low value indicates favorable interactions, while a negative value indicates strong interactions (the polymer pair is miscible).

The interaction parameter χ_{23}' may be calculated from equation [153,154]:

$$\chi_{23}' = \frac{\chi_{23}^{\infty} \cdot V_1^o}{V_2^o} = \frac{1}{\varphi_2 \cdot \varphi_3} \cdot \left(\ln \frac{V_{g,m}}{W_2 \cdot v_2 + W_3 \cdot v_3} - \varphi_2 \cdot \ln \frac{V_{g,2}}{v_2} - \varphi_3 \cdot \ln \frac{V_{g,3}}{v_3}\right) \quad (20.26)$$

Here, the second subscript of V_g identifies the nature of the column:

$$\chi'_{23} = \frac{1}{\varphi_2 \cdot \varphi_3} \cdot (\chi_{12}^{\infty} \cdot \varphi_2 + \chi_{13}^{\infty} \cdot \varphi_3 - \chi_{1m}^{\infty}) \quad (20.27)$$

Values of the Flory–Huggins χ'_{23} parameter depend on the chemical structure of the solute and it is a common phenomenon described by several research groups [155,156]. It has been interpreted as arising for preferential interaction with one of two types of components. Hsu and Prausnitz [157], as well as Su and co-workers [158,159], suggested that the compatibility of polymeric components should reflect not only the interaction between the components themselves, i.e., χ'_{23}, but also the difference in strength of the polymer–probe interactions, i.e., $\Delta\chi = |\chi_{12} - \chi_{13}|$. They called it the $\Delta\chi$ effect, and a large $\Delta\chi$ in addition to a high χ'_{23} value leads to incompatibility. Accordingly, one must select probes that give $\chi_{12}^{\infty} = \chi_{13}^{\infty}$ for studying the blend.

Farooque and Deshpande [156] proposed to rearrange Eqn (20.25) to the following form:

$$\frac{\chi_{1(23)}^{\infty} - \chi_{13}^{\infty}}{V_1} = \varphi_2 \cdot \frac{\chi_{12}^{\infty} - \chi_{13}^{\infty}}{V_1} - \varphi_2 \cdot \varphi_3 \cdot \frac{\chi'_{23}}{V_2} \quad (20.28)$$

By plotting the left-hand side of Eqn (20.28) vs. $\varphi_2 \cdot \frac{\chi_{12}^{\infty} - \chi_{13}^{\infty}}{V_1}$, the interaction parameter can be obtained from the intercept. Although this method gives reliable χ'_{23} values, they are affected by large errors. To reduce the uncertainty in the χ'_{23} values, a possible alternative is an adequate selection of the probes to use. However, the slopes deviated from their theoretical values.

Huang [159–161] proposed to rearrange Eqn (20.28) as follows:

$$\frac{\chi_{1(23)}^{\infty}}{V_1} = \frac{\varphi_2 \cdot \chi_{12}^{\infty} + \varphi_3 \cdot \chi_{13}^{\infty}}{V_1} - \varphi_2 \cdot \varphi_3 \cdot \frac{\chi'_{23}}{V_2} \quad (20.29)$$

A linear plot can be obtained from the left-hand side vs. $(\varphi_2 \cdot \chi_{12}^{\infty} + \varphi_3 \cdot \chi_{13}^{\infty})/V_1$. The polymer– polymer interaction term can be determined from the intercept at $(\varphi_2 \cdot \chi_{12}^{\infty} + \varphi_3 \cdot \chi_{13}^{\infty})/V_1 = 0$. A physical meaning of this procedure is that when $(\varphi_2 \cdot \chi_{12}^{\infty} + \varphi_3 \cdot \chi_{13}^{\infty})/V_1 = 0$, the probe is experiencing a similar environment in the blend as compared to the probe liquid. The disturbance of the liquid polymer structure is expected to be minimal.

Several other papers where IGC was used in examination of the polymer blends should be mentioned [162–164].

Benabdelghani et al. [164] studied the miscibility of poly(styrene-co-methacrylic acide) (PSMA-12) blends containing 12% of methacrylic acid with poly(2,6-dimethyl-1,4-phenylene oxide) (PPO). They rearranged Eqn (20.28) as follows:

$$\frac{1}{V_1} \ln\left(\frac{V_{g,m}}{v_m}\right) = \left[\frac{1}{V_1} \cdot \left(\varphi_2 \ln\left(\frac{V_{g,2}}{v_2}\right) + \varphi_3 \ln\left(\frac{V_{g,3}}{v_3}\right)\right)\right] + \varphi_2 \varphi_3 \frac{\chi'_{23}}{V_2} \quad (20.30)$$

The plot of the left-side term as a function of the expression between brackets of the right-side term of Eqn (20.30) allows us to calculate the true $\frac{\chi'_{23}}{V_2}$ from the intercept. They found that χ'_{23} values obtained for investigated systems are in good agreement with T_g data from the DSC measurements.

Zhao and Choi [165] used nonrandom partitioning solutes in binary polyolefin blends to test whether solutes may cause the solute dependence on the Flory–Huggins parameter χ'_{23} observed in IGC experiments. They concluded that the nonrandom partitioning behavior of solutes is not the real reason to probe the dependence problem. This is mainly attributed to the improper use of the reference volume in calculations of the solute–stationary-phase interaction parameters (χ_{1i}^{∞}). The authors suggested using single common reference volume (V_o) instead of individual molar volumes of the solutes used in IGC experiments for the calculations of the χ_{12}^{∞} parameter. In the reference volume, Zhao and Choi proposed to use the molar volume of the repeated unit of polymer at the experimental temperature [165]:

$$\chi_{1m}^{\infty} = \frac{V_0}{V_1} \cdot \left(\ln \frac{273.15 \cdot R \cdot (W_2 \cdot v_2 + W_3 \cdot v_3)}{V_g \cdot V_1 \cdot p_1^0} - 1 + \frac{V_1}{M_2 \cdot v_2} + \frac{V_1}{M_3 \cdot v_3} - \frac{(B_{11} - V_1)}{R \cdot T} p_1^0 \right) \quad (20.31)$$

Milczewska and Voelkel [166,167] used procedures proposed by Farooque and Deshpande, Zhao and Choi, and Huang to calculate values of χ'_{23} parameters independent from the solute type. The results (values of χ'_{23}) obtained from those procedures minimized the $\Delta\chi$ effect. For the system PU-N2 (a system containing polyetherurethane filled with carbonate– silicate filler modified with octylsilane and stearic acid), they obtained low, negative values, which indicated strong interactions between the polymer and the N2 filler [166,167].

Diffusivity of solvents in polymers, gel permeation parameters, and other transport properties, as well as interaction parameters between system components, were determined for the series of polymeric systems [168–180]. Studies of miscibility of components of polymer blends were presented by several research groups [181–184].

20.2.1. Solubility Parameters

Strong interactions between molecules exist in condensed phases (liquids, solutions, and solid materials), resulting in considerable (negative) potential energy in each molecule. This energy is called the *molar cohesive energy* ($-E$) [185]. Cohesive energy related to a molar volume is called *cohesive energy density c*,

$$c = \frac{-E}{V} \quad (20.32)$$

and the square root of cohesive energy density is called *solubility parameter* δ. This term proposed by Hildebrand for nonpolar systems and used as a measure of intermolecular forces of different solvents is related to the enthalpy of an evaporation ΔH_w:

$$\delta = \sqrt{c} = \sqrt{\frac{E_{\text{coh}}}{V}} = \sqrt{\left(\frac{\Delta H_w - RT}{V}\right)} \quad (20.33)$$

where δ is the solubility parameter, E_{coh} is the cohesive energy, V is the molar volume of a pure liquid, R is the gas constant, and T is the temperature. The solubility parameter expressed by the relation in Eqn (20.32) is called the Hildebrand solubility parameter.

The solubility parameter reflects van der Waals interactions between molecules, forming a liquid. However, such a definition could not be used for other systems where, besides dispersive interactions also the polar ones should be taken into account.

The most widely accepted concept of solubility parameter (total solubility parameter) related to more complex systems has been proposed by Hansen:

$$\delta_T^2 = \delta_d^2 + \delta_p^2 + \delta_h^2 \quad (20.34)$$

where δ_d, δ_p, and δ_h denote dispersive, polar, and hydrogen-bonding contributions, respectively.

The determination of solubility parameters and Hansen's solubility parameters (HSPs) by means of IGC was discussed very recently [20]. The author wishes to focus readers' attention on Hansen's book [186] and the third edition of the current edition of the e-book on practical applications of HSP [187] (also including IGC).

The procedure proposed by Ito and Guillet [188] has been applied by Price [189,190] for the estimation of solubility parameter values for the low-molecular-weight compounds [189] and liquid crystalline systems [190]. Price reported that the experimental relationship between the left-hand side of Eqn (20.35) and solubility parameters of test solutes is different from the linear relationship. The significant curvature of the following relation was observed:

$$\frac{\delta_{1i}^2}{RT} - \frac{\chi_{(12)i}^{\infty}}{V_{1i}} = \frac{2\delta_2}{RT}\delta_{1i} - \left(\frac{\delta_2^2}{RT} + \frac{\chi_s^{\infty}}{V_{1i}}\right) \quad (20.35)$$

The downward curvature for alkanes and upward curvature for other compounds were found. Therefore, the tendency for alkanes leads to underestimation of δ_2, while the tendency for polar compounds may cause the overestimation value of δ_2.

According to the three-component solubility parameter theory of Hansen, Price assumed that different types of intermolecular interactions between examined material and test solute will influence the value of the solubility parameter. Price proposed to express the solubility parameter as the sum of two terms

$$\delta_T^2 = \delta_d^2 + \delta_a^2 \quad (20.36)$$

resulting from dispersive and polar intermolecular interactions and examined material/test solute.

However, in complex systems, additional, hydrogen-bonding interaction should be taken into consideration. The extension of Price's concept, enabling calculation of all components of the total solubility parameter, is the procedure proposed by Voelkel and Janas [191]. They have extended a group of test solutes used in IGC experiments, by the addition of solutes, representing hydrogen-bonding interactions. However, the problems with the correct selection of the test solutes might be the reason for the relatively large error in HSP determination.

Adamska et al. [192] proposed to calculate the HSP data by using the model proposed by Lindvig et al. [193], combining experimental data of the Flory–Huggins interaction parameter χ_{12}^{∞} with components of the solubility parameter for the material:

$$\chi_{12}^{\infty} = \alpha\frac{V_1}{RT}((\delta_{1,d} - \delta_{2,d})^2 + 0,25(\delta_{1,p} - \delta_{2,p})^2 + 0,25(\delta_{1,h} - \delta_{2,h})^2) \quad (20.37)$$

where α, V_1, R, and T are corrective coefficient, molar volume of the test solute, gas constant, and temperature of measurement, respectively. In this relation, literature data for components of different test solutes $(\delta_{1,d}, \delta_{1,p}, \delta_{1,h})$ and the solubility parameters of the material $(\delta_{2,d}, \delta_{2,p}, \delta_{2,h})$ can be used to estimate the Flory–Huggins interaction parameter. For the purposes of IGC, experimentally obtained χ_{12}^{∞} values can be used for the determination of the HSP for the examined material by applying the above relation (Eqn (20.37)) [192].

Choi [194] proposed to calculate the critical value of the interaction parameter by using the following relationship:

$$\chi_{\text{crit}} = \frac{1}{2}\left(1 + \sqrt{\frac{V_1}{V_2}}\right)^2 \quad (20.38)$$

where V_1 is a molar volume of the test solute and V_2 is a molar volume of the polymer. Test solutes for which

$$\chi < \chi_{\text{cri}} \quad (20.39)$$

are further used in the calculation of the solubility parameters. Their $(\delta_{1,d}, \delta_{1,p}, \delta_{1,h})$ values are used to calculate the components of the solubility parameter of the examined material as the average of the respective values for all selected solutes. Therefore, all data for "disqualified" test solutes are omitted and are not taken into consideration. In the author's opinion, one should take into account all interactions between the examined material and test solutes.

The solubility parameter was used in the characterization of various materials [150,195–198].

Regretfully, due to limited space, the author was unable to address the problem of FC-IGC. However, interested readers should refer to the review of Charmas and Leboda [199] and other papers concerning this technique and the DVS technique and published after year 2000 [200–202]. The authors also suggest the careful reading of the reviews and papers on reversed flow-gas chromatography [20,203–213].

In conclusion, it is worth citing the final text from Santos and Guthrie's review [128]: "All these approaches have valuable advantages and relevant drawbacks. Thus, it is recommended the simultaneous use of several alternative approaches in order to corroborate the experimental results, analyses and theoretical predictions."

References

[1] D.R. Lloyd, T.S. Ward, H.P. Schreiber (Eds.), Inverse gas chromatography of polymers and other materials, ACS Symp. Series 391, American Chemical Society, Washington DC, 1989.

[2] M.N. Belgacem, A. Gandini, IGC as a tool to characterize dispersive and acid-base properties of the surface of fibers and powders, in: E. Pefferkorn (Ed.), Interfacial phenomena in chromatography, Marcel Dekker, New York, 1999, p. 41.

[3] A. Voelkel, Crit. Rev. Anal. Chem. 22 (1991) 411.

[4] F. Thielmann, J. Chromatogr. A 1037 (2004) 115.

[5] A. Voelkel, Inverse gas chromatography in examination of acid-base and some other properties of solids, in: A. Dabrowski, V.A. Tertykh (Eds.), Adsorption on new and modified inorganic, Elsevier, Amsterdam, 1996, p. 465.

[6] C. Sun, J.C. Berg, Adv. Coll. Int. Sci. 105 (2003) 151.

[7] A. Voelkel, B. Strzemiecka, K. Adamska, K. Milczewska, K. Batko, Surface and bulk characteristics of polymers by means of inverse gas chromatography, in: A. Nastasovic, S. Jovanovic (Eds.), Polymeric materials, Research Signpost, 2009, pp. 71–102.

[8] R. Wu, D. Que, Z.Y. Al-Saigh, J. Chromatogr. A 1146 (2007) 93.

[9] A.V. Kisielev, V.V. Kulichenko, Dokl. Akad. Nauk. SSSR. 93 (1953) 101.

[10] A.V. Kiselev, B.A. Frolov, Kinetics. Catal. (USSR Eng Transl) 3 (1962) 667.

[11] A.V. Kiselev, B.A. Frolov, Kinetics. Catal. (USSR Eng Transl) 3 (1962) 672.

[12] A.B. Nastasović, A.E. Onjia, J. Chromatogr. A 1195 (2008) 1.

[13] G.M. Dorris, P. Gray, J. Colloid. Interf. Sci. 77 (1980) 353.

[14] J. Schultz, L. Lavielle, C.J. Martin, Adhesion 23 (1987) 45.

[15] J. Schultz, L. Lavielle, Interfacial properties of carbon fibre – epoxy resin matrix composites, in: D.R. Lloyd, T.S. Ward, H.P. Schreiber (Eds.), Inverse gas chromatography of polymers and other materials, ACS Symp. Series 391, American Chemical Society, Washington DS, 1989, pp. 185–202.

[16] C. Sun, J.C. Berg, J. Chromatogr. A 969 (2002) 59.

[17] P.H. Emmett, R.H. Brunauer, J. Am. Chem. Soc. 59 (1937) 155.

[18] G.C. McGonigal, R.H. Bernhardt, Appl. Phys. Lett. 57 (1990) 28.

[19] T. Hamieh, Chromatographia 73 (2011) 705.

[20] A. Voelkel, B. Strzemiecka, K. Adamska, K. Milczewska, J. Chromatogr. A 1216 (2009) 1551–1566.

[21] G. Garnier, W.G. Glasser, J. Adhesion. 46 (1994) 165.

[22] C.J. van Oss, R.J. Good, M.K. Chandhury, Langmuir 4 (1988) 884.

[23] O. Planinšek, A. Trojak, S. Srči&ccaron, Int. J. Pharmaceut. 221 (2001) 211.

[24] B. Shi, Y. Wang, L. Jia, J. Chromatogr. A 1218 (2011) 860.

[25] G.K. Karaoglan, D. Sakar, Chromatographia 73 (2011) 93.

[26] Y. Matsushita, S. Wada, K. Fukushima, S. Yasuda, Ind. Crops. Prod. 23 (2006) 115.

[27] S.K. Papadoulou, G. Dritsas, I. Karapanagiotis, I. Zuburtikudis, Eur. Polym. J. 46 (2010) 202.

[28] A. Onjia, S.K. Milonjić, N.N. Jovanović, S.M. Jovanovi&cacute, React. Funct. Polym. 43 (2000) 269.

[29] A.B. Nastasović, A.E. Onjia, S.K. Miljonić, S.M. Jovanovi&cacute, Eur. Polym. J. 41 (2005) 1234.

[30] F. Dieckmann, C. Klinger, P. Uhlmann, F. Böhme, Polymer 42 (2001) 3463.

[31] A.A. Calhoun, P.D. Nicholson, A.B. Barnes, Polym. Degrad. Stab. 91 (2006) 1964.

[32] N. Rocha, J.A.F. Gamelas, P.M. Gonçalves, M.H. Gill, J.T. Guthrie, Eur. Polym. J. 45 (2009) 3389.

[33] M.M. Chehimi, E. Abdeljalil, Synth. Met. 145 (2004) 15.

[34] Z.Y. Al-Saigh, Polymer 40 (1999) 3479.

[35] M.K. Kozłowska, U. Domańska, M. Lempert, M. Rogalski, J. Chromatogr. A 1068 (2005) 297.

[36] A. Al-Ghamdi, Z.Y. Al-Saigh, J. Chromatogr. A 969 (2002) 229.

[37] S. Comte, R. Calvet, A. Dodds, H. Balard, Powder Technol. 157 (2005) 39.
[38] F. Gracia-Herruzo, J.M. Rodriguez-Maroto, R.A. Garcia-Delgado, C. Gomez-Lahoz, C. Vereda-Alonso, Chemosphere 41 (2000) 1167.
[39] A. Boutboul, P. Giampaoli, A. Feingenbaum, V. Ducruet, Food Chemistry 71 (2000) 387.
[40] D. Cava, J.M. Lagaron, F. Martinez-Gimenez, R. Gavara, J. Chromatogr. A 1175 (2007) 267.
[41] J.M. Lagaron, R. Catala, R. Gavara, Mater. Sci. Technol. Lond. 20 (2004) 1.
[42] G. Lopez-Barballo, D. Cava, J.M. Lagaron, R. Catala, R. Gavara, J. Agric. Food Chem. 53 (2005) 7212.
[43] S. Aucejo, M.J. Pozo, R. Gavara, J. Appl. Polym. Sci. 70 (1998) 711.
[44] M.R. Sunkersett, I.M. Grimsey, S.W. Doughty, J.C. Osborn, P. York, R.C. Rowe, Europ. J. Pharm. Sci. 13 (2001) 219.
[45] S. Das, I. Larson, P. Young, P. Stewart, Eur. J. Pharm. Sci. 38 (2009) 347.
[46] K. Sooben, G. Buckton, J. Newton, Pharm. Sci. 2 (Suppl.) (2000).
[47] H. Newell, G. Buckton, D. Butler, F. Thielmann, D. Williams, Int. J. Pharm. 217 (2001) 45.
[48] H. Balarad, A. Saada, B. Siffert, E. Papirer, Clays Clay. Miner. 45 (1997) 489.
[49] M. Ticehurst, P. York, R. Rowe, S. Dwivedi, Int. J. Pharm. 141 (1996) 93.
[50] J. Kim, W. Qian, Z.Y. Al-Saigh, Surface Sci. 605 (2011) 419.
[51] A.B. Slimane, K. Boukerma, M. Chabut, M.M. Chehimi, Colloids. Surf. A 240 (2004) 45.
[52] E. Papirer, H. Balard, in: E. Pefferkorn (Ed.), Interfacial phenomena in chromatography, Marcel Dekker, New York, 1999, p. 145.
[53] J.M.R.C.A. Santos, K. Fagelman, J.T. Guthrie, J. Chromatogr. A 969 (2002) 119.
[54] J.R. Conder, C.L. Young, Physicochemical measurements by gas chromatography, Wiley, Chichester, 1979.
[55] V. Gutmann, Electrochim Acta. 21 (1976) 661.
[56] U. Mayer, V. Gutmann, W. Gerger, Mh. Chem. 106 (1975) 1235.
[57] F.L. Riddle, F.M. Fowkes, J. Am. Chem. Soc. 112 (1990) 3259.
[58] C.S. Flour, E. Papirer, Ind. Eng. Chem. Prod. Res. Dev. 21 (1982) 666.
[59] C.S. Flour, E. Papirer, J. Colloid. Int. Sci. 91 (1983) 69.
[60] E. Papirer, J.M. Perrin, B. Siffert, G. Philipponneau, J. Colloid. Int. Sci. 144 (1991) 263.
[61] E. Papirer, J.M. Perrin, B. Siffert, G. Philipponneau, J.M. Lamerant, J. Colloid. Int. Sci. 156 (1993) 104.
[62] M.M. Chehimi, E. Pigois-Landureau, J. Mater. Chem. 4 (1994) 741.
[63] F.M. Fowkes, J. Adhesion, Sci. Technol. 4 (1990) 669.
[64] B. Shi, Q. Zhang, J. Lina, L. Yang, L. Bin, J. Chromatogr. A 1149 (2007) 390.
[65] E. Fekete, J. Móczó, B. Pukánszky, J. Colloid. Int. Sci. 269 (2004) 143.
[66] T. Hamieh, M. Nardin, M. Rageul-Lescouët, H. Haïdara, J. Schultz, Colloids. Surf. A 125 (1997) 155.
[67] T. Hamieh, M. Rezzaki, J. Schultz, J. Thermal Anal. 51 (2007) 793.
[68] T. Hamieh, M. Rezzaki, J. Schultz, J. Colloid. Interface Sci. 233 (2001) 339.
[69] T. Hamieh, M. Rezzaki, J. Schultz, J. Colloid Interface Sci. 233 (2001) 343.
[70] T. Hamieh, J. Schultz, J. Chromatogr. A 969 (2002) 17.
[71] T. Hamieh, J. Schultz, J. Chromatogr. A 969 (2002) 27.
[72] T. Hamieh, M.B. Falladah, J. Schultz, J. Chromatogr. A 969 (2002) 37.
[73] R. Kaliszan, Crit. Rev. Anal. Chem. 16 (1986) 323.
[74] R. Kaliszan, HPLC as a source of information about chemical structure of solutes, in: P.R. Brown, A.P. Hartwick (Eds.), High performance liquid chromatography, in: J.D. Winefordner, I.M. Kolthoff, (Ed.), Chemical analysis series of monographs, vol. 98, John Wiley&Sons, New York, 1989.
[75] A. Askin, S.S. Etöz, D.T. Yazici, F. Tümsek, Chromatographia 73 (2011) 109.
[76] J.M.R.C.A. Santos, K. Fagelman, J.T. Guthrie, J. Chromatogr. A 969 (2002) 111.
[77] Q.-C. Zou, S.-L. Zhang, Q.-q. Tang, S.-M. Wang, L.-M. Wu, J. Chromatogr. A 1110 (2006) 140.
[78] Y. Wu, Z. Li, H. Xi, J. Hazard Mater B113 (2004) 131.
[79] G. Buckton, H. Gill, Adv. Drug. Delivery Rev. 59 (2007) 1474.
[80] A.V. Ambarkhane, K. Pincott, G. Buckton, Int. J. Pharm. 294 (2005) 129.
[81] M. Ohta, G. Buckton, Int. J. Pharm. 289 (2005) 31.
[82] A. Columbano, G. Buckton, P. Wikeley, Int. J. Pharm. 253 (2003) 61.
[83] M. Ohta, G. Buckton, Int. J. Pharm. 272 (2004) 121.
[84] M. Ohta, G. Buckton, Int. J. Pharm. 269 (2004) 81.
[85] Y. Yokoi, E. Yonemochi, K. Terada, Int. J. Pharm. 280 (2004) 67.
[86] V. Swaminatham, J. Cobb, I. Saracovan, Int. J. Pharm. 312 (2006) 158.
[87] K.V. Kumar, F. Rocha, J. Chromatogr. A 1216 (2009) 8528.
[88] B.A.P. Ass, M.N. Belgacem, E. Frollini, Carbohydr Polym. 63 (2006) 19.
[89] S. Baumgartner, O. Planinšek, S. Srčič, J. Kristl, Eur. J. Pharm. Sci. 27 (2006) 375.
[90] X. Han, X. Ma, J. Liu, H. Li, Carbohydr Polym. 78 (2009) 533.

[91] Ph. Rousset, P. Sellapan, P. Daoud, J. Chromatogr. A 969 (2002) 97.
[92] G. Buchler-Diller, M.K. Inglesby, Y. Wu, Colloids Surf. A 260 (2005) 63.
[93] G. Czeremuszkin, P. Mukhopadhyay, S. Sapieha, J. Colloid. Int. Sci. 194 (1997) 127.
[94] O. Planinšek, J. Zadnik, Š Rozman, M. Kunaver, R. Dreu, S. Srčič, Int. J. Pharm. 256 (2003) 17.
[95] N. Cordeiro, C. Gouveia, A.G.O. Moraes, S.C. Amico, Carbohydr. Polym. 84 (2011) 110.
[96] B. Shi, S. Zhao, L. Jia, L. Wang, Carbohydr. Polym. 67 (2007) 398.
[97] A. Boutboul, F. Lenfant, P. Giampaoli, A. Feigenbaum, V. Ducruet, J. Chromatogr. A 969 (2002) 9.
[98] W.M. Burry, D.S. Keller, J. Chromatogr. A 972 (2002) 241.
[99] D.S. Keller, P. Luner, Colloids. Surf. A 161 (2000) 401.
[100] M.-P. Comard, R. Calvet, H. Balard, J.A. Dodds, Colloids. Surf. A 238 (2004) 37.
[101] M. Rückriem, A. Inayat, D. Enke, R. Gläser, W.-D. Einicke, R. Roskmann, Colloids. Surf. A 357 (2010) 21.
[102] C. Sun, J.C. Berg, J. Colloid. Interface. Sci. 260 (2003) 443.
[103] Y.-C. Yang, S.-B. Jeong, B.-G. Kim, P.-R. Yoon, Powder Technol. 191 (2009) 117.
[104] S.K. Milonji&ccaron, Colloids. Surf. A 149 (1999) 461.
[105] S. Lazarevi&cacute, Ž Radovanović, Dj . Veljović, A. Onjia, Dj. Janaćković, R. Petrović, Appl. Clay. Sci. 43 (2009) 41.
[106] D.M. Ansari, G.J. Price, Polymer 45 (2004) 1823.
[107] J.-B. Donnet, H. Balard, N. Nedjari, B. Hamdi, H. Barthel, T. Gottschalk-Gaudig, J. Colloid. Int. Sci. 328 (2008) 15.
[108] M.-P. Comard, R. Calvet, H. Balard, J.A. Dodds, Colloids. Surf. A. 232 (2004) 269.
[109] M.P. Comard, R. Calvet, J.A. Dodds, H. Balard, Powder Technol. 128 (2002) 262.
[110] S. Hamdi, B. Hamdi, Z. Kessaissia, H. Barthel, H. Balard, J.B. Donnet, J. Chromatogr. A 969 (2002) 143.
[111] L.G.H.J. Segeren, M.E.L. Wouters, M. Bos, J.W.A. van den Berg, G.J. Vancso, J. Chromatogr. A 969 (2002) 215.
[112] M.N. Belgacem, A. Blayo, A. Gandini, J. Colloid. Interfce. Sci. 182 (1996) 431.
[113] C. Castro, G.M. Dorris, C. Daneault, J. Chromatogr. A 969 (2002) 313.
[114] V. Lavaste, J.F. Watts, M.M. Chehimi, C. Lowe, Int. J. Adhes. Adhes. 20 (2000) 1.
[115] O. Inel, D. Tapaloglu, A. Askin, F. Tümsek, Chem. Eng. J. 88 (2002) 255.
[116] C. Bilgic, A. Askin, J. Chromatogr. A 1006 (2003) 281.
[117] F. Tümsek, O. Inel, Chem. Eng. J. 94 (2003) 57.
[118] A. Askin, C. Bilgiç, Chem. Eng. J. 112 (2005) 159.
[119] E. Diaz, S. Ordonez, A. Vega, J. Coca, J. Chromatogr. A 1049 (2004) 139.
[120] C. Bilgiç, F. Tümsek, J. Chromatogr. A 1162 (2007) 83.
[121] E. Diaz, S. Ordonez, A. Auroux, J. Chromatogr. A 1095 (2005) 131.
[122] E. Diaz, S. Ordonez, A. Vega, J. Coca, J. Chromatogr. A 1049 (2004) 161.
[123] B. Niederer, A. Le, E. Cantergiani, J. Chromatogr. A 996 (2003) 189.
[124] V. Oliva, B. Mrabet, M.I.B. Neves, M.M. Chehimi, K. Benzarti, J. Chromatogr. A 969 (2002) 261.
[125] I.B. Neves, M. Chabut, C. Perruchot, M.M. Chehimi, K. Benzarti, Appl. Surf. Sci. 238 (2004) 523.
[126] C. Perruchot, M.M. Chehimi, M.-J. Vaulay, K. Benzarti, Cem. Concr. Res. 36 (2006) 305.
[127] F. Djouani, C. Connan, M. Delamar, M.M. Chehimi, K. Benzarti, Constr. Bulid. Mater. 25 (2011) 411.
[128] J.M.R.C.A. Santos, J.T. Guthrie, Mater. Sci. Eng. R. 50 (2005) 79.
[129] K.E. Fagelman, J.T. Guthrie, J. Chromatogr. A 1095 (2005) 145.
[130] M. Kazayawoko, J.J. Balatinecz, M. Romansky, J. Colloid. Interface Sci. 190 (1997) 408.
[131] A. van Asten, N. van Veenendaal, S. Koster, J. Chromatogr. A 888 (2000) 175.
[132] E. Cantergiani, D. Benczedi, J. Chromatogr. A 969 (2002) 103.
[133] A. Vega, F.V. Diez, P. Hurtado, J. Coca, J. Chromatogr. A 962 (2002) 153.
[134] P. Jandura, B. Riedl, B.V. Kokta, J. Chromatogr. A 969 (2002) 301.
[135] X. Huang, B. Shi, B. Li, L. Li, X. Zhuang, S. Zhao, Polymer Test 25 (2006) 970.
[136] N. Riba, M. Nardin, J.-Y. Drean, R. Frydrych, J. Colloid. Interface. Sci. 314 (2007) 373.
[137] B. Lindsay, M.-L. Abel, J.F. Watts, Carbon 45 (2007) 2433.
[138] L. Cossaruto, C. Vagner, G. Finqueneisel, J.V. Weber, T. Zimny, Appl. Surf. Sci. 177 (2001) 207.
[139] C. Vagner, G. Finqueneisel, T. Zimny, P. Burg, B. Grzyb, J. Machnikowski, et al., Carbon 41 (2003) 2847.
[140] Y. Wu, Z. Li, H. Xi, Carbon 42 (2004) 3003.
[141] G.S. Singh, D. Lal, V.S. Tripathi, J. Chromatogr. A 1036 (2004) 189.
[142] E. Diaz, S. Ordonez, A. Vega, J. Coca, Microporous Mesoporous Mater. 82 (2005) 173.
[143] M. Perez-Mendoza, M.C. Almazan-Almazan, L. Mendez-Linan, M. Domingo-Garcia, F.J. Lopez-Garzon, J. Chromatogr. A 1214 (2008) 121.

[144] H. Balard, D. Maafa, A. Santini, J.B. Donnet, J. Chromatogr. A 1198-1199 (2008) 173.
[145] J. Xie, Q. Zhang, K.T. Chuang, J. Catal. 191 (2000) 86.
[146] A.F.M. Barton, CRC handbook of solubility parameter and other cohesion parameters, CRC Press, Boca Raton, FL, 2000.
[147] R.F. Blanks, J.M. Prausnitz, Ind. Eng. Chem. Fundam. 3 (1964) 1.
[148] J.M. Barrales-Rienda, J. Vidal Gancedo, Macromolecules 21 (1988) 220.
[149] J. Fall, K. Milczewska, A. Voelkel, J. Mater. Chem. 11 (2001) 1042.
[150] A. Voelkel, J. Fall, J. Chromatogr. A 721 (1995) 139.
[151] D. Topaloğlu Yazici, A. Askin, V. Bütün, J. Chem. Thermodynamics 40 (2008) 353.
[152] D.D. Deshpande, D. Patterson, H.P. Schreiber, C.S. Su, Macromolecules 7 (1974) 530.
[153] M.J. El-Hibri, W. Cheng, P. Hattam, P. Munk. in: D.R. Lloyd, T.C. Ward, H.P. Schreiber, (Eds), Inverse gas chromatography. Characterization of polymers and other materials. ACS symposium series, 391, Washington, 1989, p. 121.
[154] O. Olabisi, Macromolecules 8 (1975) 316.
[155] E. Fernandez-Sanchez, A. Fernandez-Torres, J.A. Garcia-Dominguez, J.M. Santiuste, E. Pertierra-Rimada, J. Chromatogr. 457 (1988) 55.
[156] A.M. Farooque, D.D. Deshpande, Polymer 33 (1992) 5005.
[157] C.C. Hsu, J.M. Prausnitz, Macromolecules 7 (1974) 320.
[158] C.S. Su, D. Patterson, Macromolecules 10 (1977) 708.
[159] J.C. Huang, J. Appl. Polym. Sci. 89 (2003) 1242.
[160] J.C. Huang, J. Appl. Polym. Sci. 90 (2003) 671.
[161] J.C. Huang, Eur. Polym. J. 42 (2006) 1000.
[162] Z. Tan, G.J. Vancso, Macromol Theory Simul. 6 (1997) 467.
[163] F. Feraz, A.S.H. Hamou, S. Djadoun, Eur. Polym. J. 31 (1995) 665.
[164] Z. Benabdelghani, A. Etxeberria, S. Djadoun, J.J. Iruin, C. Uriarte, J. Chromatogr. A 1127 (2006) 237.
[165] L. Zhao, P. Choi, Polimer 43 (2002) 6677.
[166] K. Milczewska, A. Voelkel, J. Appl. Polym. Sci. 107 (2008) 2877.
[167] K. Milczewska, A. Voelkel, J. Polym. Sci. Part B: Polym. Phys. 44 (2006) 1853.
[168] C. Zhao, J. Li, Z. Jiang, C. Chen, Eur. Polym. J. 42 (2006) 615.
[169] S. Zhang, A. Tsuboi, H. Nakata, T. Ishikawa, Fluid. Phase. Equlib. 194—197 (2002) 1179.
[170] J.R. Galdamez, S.J. Kernion, J.L. Duda, R.P. Danner, Polymer 49 (2008) 2873.
[171] G. Overejo, P. Perez, M.D. Romero, I. Diaz, E. Diez, Eur. Polym. J. 45 (2009) 590.
[172] P.M. Budd, N.B. McKeown, B.S. Ghanem, K.J. Msayib, D. Fritsch, L. Starannikova, et al., J. Membr. Sci. 325 (2008) 851.
[173] J.V. Scicolone, P.K. Davis, R.P. Danner, J.L. Duda, Polymer 47 (2006) 5364.
[174] R.Y.M. Huang, P. Shao, G. Nawawi, X. Feng, C.M. Burns, J. Membr. Sci. 188 (2001) 205.
[175] H. Eser, F. Tihminlioglu, Fluid. Phase. Equilib. 237 (2005) 68.
[176] F. Tihminlioglu, R.P. Danner, J. Chromatogr. A 845 (1999) 93.
[177] E. sz. Kovats, G. Foti, A. Dallos, J. Chromatogr. A 1046 (2004) 185.
[178] C. Etxabarren, M. Iriarte, C. Uriarte, A. Etxeberria, J.J. Iruin, J. Chromatogr. A 969 (2002) 245.
[179] D.M. Ansari, G.J. Price, Polymer 45 (2004) 3663.
[180] J.M.R.C.A. Santos, J.T. Guthrie, J. Chromatogr. A 1070 (2005) 147.
[181] S. Ourdani, F. Amrani, J. Chromatogr. A 969 (2002) 287.
[182] I.M. Shillcock, G.J. Price, Polymer 44 (2003) 1027.
[183] E. Fekete, E. Foldes, B. Pukanszky, Eur. Polym. J. 41 (2005) 727.
[184] Q.-C. Zou, S.-L. Zhang, S.-M. Wang, L.-M. Wu, J. Chromatogr. A 1129 (2006) 255.
[185] A.F.M. Barton, Chem. Rev. 75 (1975) 731.
[186] C.M. Hansen, Hansen solubility parameters: a user's handbook, CRC Press, Boca Raton FL, 2007.
[187] S. Abbott, C.M. Hansen, H. Yamamoto, HSPiP: hansen solubility parameters in practice, third ed., (2010), ISBN 9780955122026.
[188] K. Ito, J.E. Guillet, Macromolecules 12 (1979) 1163.
[189] G.J. Price. in: D.R. Lloyd, T.C. Ward, H.P. Schreiber, (Eds). Inverse gas chromatography. Characterization of polymers and other materials, ACS Symposium Series, 391, Washington, 1989, p. 48.
[190] G.J. Price, I.M. Shillcock, J. Chromatogr. A 964 (2002) 199.
[191] A. Voelkel, J. Janas, J. Chromatogr. A 645 (1993) 141.
[192] K. Adamska, R. Bellinghausen, A. Voelkel, J. Chromatogr. A 1195 (2008) 146.
[193] T. Lindvig, M.L. Michelsen, G.M. Kontogeorgis, Fluid Phase Equilib. 203 (2002) 247.
[194] P. Choi, T. Kavassalis, A. Rudin, J. Coll. Int. Sci. 180 (1996) 1.
[195] F. Tümsek, M. Börekçi, D.T. Yazici, A. Askin, Chromatographia 73 (2011) 117.
[196] U. Domańska, Z. Żołek-Tryznowska, J. Chem. Thermodyn. 42 (2010) 363.
[197] N. Scott Bobbitt, J.W. King, J. Chromatogr. A 1217 (2010) 7898.
[198] F. Mutelet, G. Ekulu, M. Rogalski, J. Chromatogr. A 969 (2002) 207.

[199] B. Charmas, R. Leboda, J. Chromatogr. A 886 (2000) 133.

[200] F. Thielmann, D.A. Butler, D.R. Williams, Colloids. Surf. A 187–188 (2001) 267.

[201] N. Marilyn Ahfat, G. Buckton, R. Burrows, M.D. Ticehurst, Eur. J. Pharm. Sci. 9 (2000) 271.

[202] C. Tisserand, R. Calvet, S. Patry, L. Galet, J.A. Dodds, Powder Technol. 190 (2009) 53.

[203] N.A. Katsanos, E. Iliopoulou, V. Plagianakos, H. Mangou, J. Colloid. Interface. Sci. 239 (2001) 10.

[204] A.V. Dremetsika, P.A. Siskos, N.A. Katsanos, J. Hazard, Mat. 149 (2007) 603.

[205] N.A. Katsanos, G. Karaiskakis, Time-resolved inverse gas chromatography and its applications, HNB Publishing, New York, 2004.

[206] D. Gavril, K.A. Rashid, G. Karaiskakis, J. Chromatogr. A 919 (2001) 349.

[207] K.A. Rashid, D. Gavril, N.A. Katsanos, G. Karaiskakis, J. Chromatogr. A 934 (2001) 31.

[208] N.A. Katsanos, J. Chromatogr. A 969 (2002) 3.

[209] N.A. Katsanos, D. Gavril, G. Karaiskakis, J. Chromatogr. A 983 (2003) 177.

[210] N.A. Katsanos, J. Chromatogr. A 1037 (2004) 125.

[211] G. Karaiskakis, D. Gavril, J. Chromatogr. A 1037 (2004) 147.

[212] F. Roubani-Kalantzopoulou, J. Chromatogr. A 1037 (2004) 191.

[213] N.A. Katsanos, J. Kapolos, D. Gavril, N. Bakaoukas, V. Loukopoulos, A. Koliadima, et al., J. Chromatogr. A 1127 (2006) 221.

CHAPTER

21

Separation of Enantiomers

V. Schurig

OUTLINE

21.1. INTRODUCTION

Enantiomers (optical isomers) must be differentiated for the determination of enantiomeric compositions of chiral compounds (*ee, er,* and *ec* [1]). There exist two approaches for enantioseparations by gas chromatography (GC). The *indirect* approach is based on the formation of diastereomeric derivatives by reaction with an enantiomerically pure chiral auxiliary and the subsequent separation of the diastereomers on a conventional achiral stationary phase in the spirit of Pasteur's resolution principles. This method requires the absence of kinetic resolution and racemization of both reaction partners during the derivatization step, the absence of fractionation during sample preparation, as well as an unbiased detection of diastereomers. In the *direct* approach, enantiomers are separated *via* the noncovalent diastereomeric

Gas Chromatography DOI: 10.1016/B978-0-12-385540-4.00021-3

interaction with a nonracemic chiral stationary phase (CSP). The rapid and reversible formation of diastereomeric association complexes of distinct stabilities causes retention differences of the enantiomers resulting in enantioseparation by GC [2]. Here only the *direct* approach will be treated. The enantioseparation *via* diastereomeric derivatives has been reviewed in Chapter 3.5.1 of Ref. [3].

High efficiency, sensitivity, and speed of separation are important advantages of enantioseparations by high-resolution capillary gas chromatography (HRC-GC). Due to the high separation power of HRC-GC, contaminants and impurities are separated from the chiral analytes and the simultaneous analysis of multicomponent mixtures of enantiomers (e.g. derivatized proteinogenic α-amino acids) is straightforward. Ancillary techniques (multidimensional GC and comprehensive GC × GC) and coupling methods (GC-MS) are important tools in chiral analysis. By employing the selected ion-monitoring mode, trace amounts of enantiomers can be detected by GC-MS(SIM). The universal flame ionization detector (FID) is linear over five orders of magnitude, and detection sensitivity can further be increased to the picogram level by electron-capture detection (ECD) and by element-specific detection often aided by special derivatization strategies. In contrast to liquid chromatography or electromigration methods, the delicate choice of solvents (buffers), modifiers, and gradient elution systems is absent in GC. Temperature programming (cf. Chapters 2 and 8) is preferentially employed for complex enantiomeric mixtures. Yet the prerequisites for the use of GC are volatility, thermal stability, and resolvability of the enantiomers, restricting the exclusive use of enantioselective GC. In the first trials on enantioselective GC, deactivated glass capillary columns and packed columns were employed. They have totally been replaced by fused silica capillary columns coated with various CSPs, which are generally commercially available. The state of the art of enantioselective GC has been reviewed in a number of sequential accounts [4–10]. Reviews on enantioselective GC are also included in general treatises on chiral (chromatographic) analysis [3,11–14]. Specific topical reviews are mentioned in the following chapters.

21.2. CHIRAL STATIONARY PHASES BASED ON α-AMINO ACID DERIVATIVES

This topic has been reviewed [2,15–20]. The first direct enantioseparation on a chiral stationary phase (CSP) by GC was discovered in 1966 by Gil-Av, Feibush, and Charles-Sigler [21]. An experimental setup was employed which was not much different from today's use. A homemade 100 m × 0.25 mm i.d. high-resolution glass capillary column was coated with a 20% solution of the CSP *N*-trifluoroacetyl-L-isoleucine lauryl ester in diethyl ether. The 2-propanol, *n*-butanol, and cyclopentanol esters of *N*-trifluoroacetyl-alanine were partially resolved. The design of this enantioselective selector–selectand system was based on the idea of biomimetically imitating the stereoselective peptide enzyme interaction employing simple α-amino acid entities as model substances. In a follow-up paper, eighteen enantiomeric pairs of *N*-trifluoroacetyl-α-amino acid esters were enantioseparated using *N*-trifluoroacetyl-D- or L-isoleucine lauryl ester and *N*-trifluoroacetyl-L-phenylalanine cyclohexyl ester as CSPs [22]. It was found that the derivatives of L-amino acids eluted after the corresponding D-enantiomers on columns coated with a CSP having the L-configuration. Gil-Av et al. noted that the difference in the free energies of solvation Δ(ΔG) of the diastereomeric associates amounted only to 0.006–0.03 kcal/mole at the enantioseparation temperature. The enantioseparation was thought to be due to hydrogen bonding between NH···F and NH···O=C– functions, whereby the latter contribution was considered to be more important. Whereas all previous enantioseparations were performed on an analytical scale, an enantioselective packed column containing the dipeptide *N*-trifluoroacetyl-L-valyl-L-valine cyclohexyl ester as a CSP was used for the semipreparative enantioseparation by GC and a chirooptical detector based on optical rotatory dispersion was employed for the first time to unequivocally identify the enantiomers [23]. Feibush found that only the *N*-terminal amino acid in the *N*-L-Val-*C*-L-Val dipeptide was essential for enantiomer recognition, whereas the *C*-terminal amino acid just provided the second amide bond required for additional hydrogen bonding. The *C*-terminal amino acid was therefore substituted by a *tert*-butyl group to yield the versatile amino acid diamide selector *N*-lauroyl-*L*-valine-*tert*-butylamide [24].

In order to improve the temperature stability of the CSPs, Ôi et al. linked the selectors of Gil-Av et al. to a triazine ring [25]. A breakthrough in enantioselective GC was achieved in 1977 when Frank, Nicholson, and Bayer introduced the L-valine diamide selector of Feibush [24] into to a polysiloxane backbone *via* total synthesis [26]. Thus, dimethylsiloxane and (2-carboxypropyl)methylsiloxane were copolymerized and afterwards reacted with valine *tert*-butylamide. By this pioneering strategy, enantioselectivity was combined with the unique GC properties of fluidic silicones [26]. The *chiral* poly*sil*oxane containing *val*ine diamide was termed *Chirasil-Val* (Figure 21.1 left). It shows a high temperature stability from 70 to 250 °C and a low tendency to column bleeding. The simultaneous enantioseparation of major proteinogenic α-amino acids as *N(O,S)*-trifluoroacetyl-*O*-*n*-propyl esters on Chirasil-L-Val [18] is depicted in Figure 21.2.

The pretreatment of borosilicate glass columns prior to coating with Chirasil-Val as well as immobilization strategies have been described in detail by Nicholson et al. [27,28]. Subsequently, Chirasil-Val-coated borosilicate glass capillaries were replaced by Chirasil-Val-coated fused silica capillaries. The mirror-image CSPs, Chirasil-L-Val and Chirasil-D-Val, were used to verify enantioseparations by switching

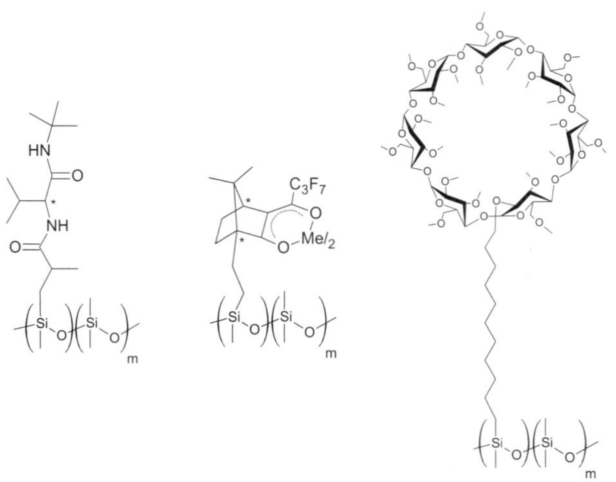

FIGURE 21.1 Structures of the polymeric CSPs Chirasil-Val (left), Chirasil-Metal (middle), and Chirasil-β-Dex (right) used for enantioselective GC.

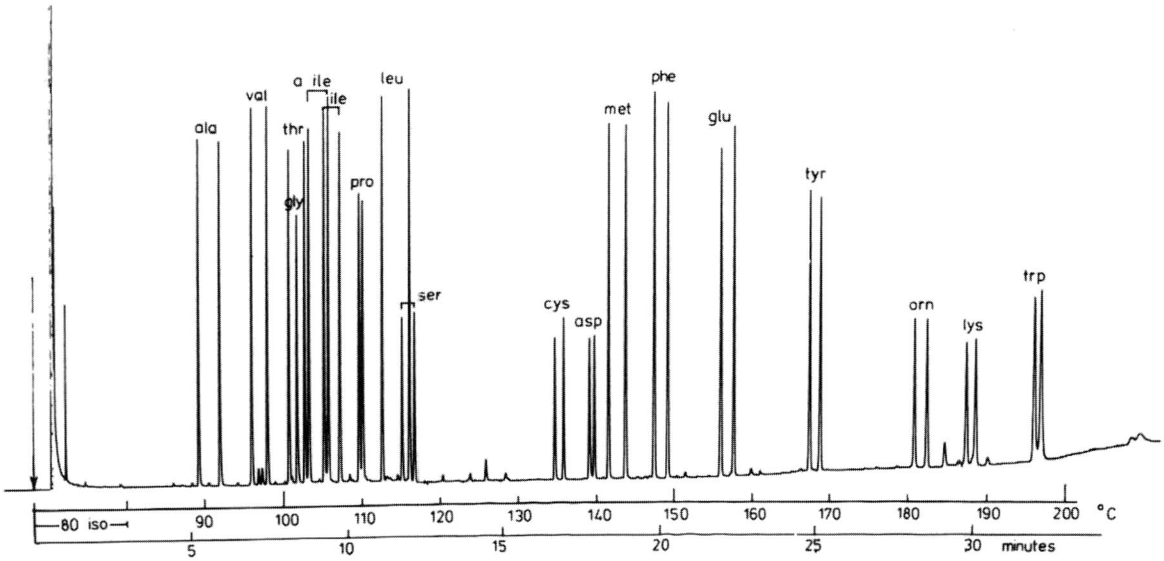

FIGURE 21.2 Simultaneous enantioseparation of proteinogenic *N(O,S)*-TFA α-amino acid *O-n*-propyl esters by GC on a 20 m × 250 μm i.d. glass capillary coated with Chirasil-L-Val. The first eluted peak corresponds to the D-enantiomer. *Source: From Ref. [18] with permission.*

the elution order of enantiomers [29]. A direct straightforward approach to polymeric CSPs is based on the modification of cyanoalkyl-substituted polysiloxanes (XE-60, OV-225) with amino acid selectors [30,31]. In Chirasil-Val-C_{11}, a long undecamethylene spacer separates the valine diamide selector from the polymeric backbone [32]. In Chirasil-Val, the chiral moieties are statistically distributed along the polymer chain. A more ordered Chirasil-type CSP has been obtained by block condensation of 1,5-*bis*-(diethylamino)-hexamethyl-trisiloxane and 2′,2′,2′-trifluoroethyl-(3-dichloromethylsilyl)-2-methylpropionate followed by nucleophilic displacement of the functionalized polysiloxane with chiral amines and amino acids [33]. A highly ordered supramolecular structure has been prepared by linking chiral L-valine-*tert*-butylamide moieties to the eight hydroxyl groups of a resorcine[4]arene basket-type structure obtained from resorcinol and 1-undecanal. The calixarene was subsequently chemically linked *via* four spacer units to a poly(dimethylsiloxane) to give Chirasil-Calix [34]. Detailed mechanistic aspects of GC enantioseparation *via* hydrogen bonding have been addressed in Refs. [15,19].

Enantioseparation of enantiomers by hydrogen bonding CSPs usually requires derivatization of the analytes in order to increase the volatility and thermal stability, to introduce suitable functions for additional hydrogen bonding and for improving detectability of trace amounts of enantiomers (ECD) [15,17]. The derivatization strategy should also assist the simultaneous enantioseparation of numerous α-amino acids without extensive peak overlapping [35]. At the outset of enantioselective GC, Gil-Av et al. used a two-step derivatization strategy for α-amino acids consisting of the formation of *N*-perfluoroacyl-*O*-alkyl esters [21–23] which proceeds without racemization at ambient temperature [36]. For Chirasil-Val, *N*-trifluoroacetyl-*O*-methyl (or *O*-2-propyl and *O*-*n*-pentyl) esters and *N*-pentafluoropropionyl-*O*-2-propyl esters of α-amino acids are routinely employed. The resolution factors R_s of eight representative α-amino acid derivatives as a function of the *N*-perfluoroacyl group (i.e. trifluoroacetyl (TFA), pentafluoropropionyl (PFP), and heptafluorobutyryl (HFB)) and ester alkyl group (*i.e.*, *n*-propyl, 2-propyl, 2-butyl, and isoamyl) on Chirasil-Val have been systematically determined [37]. An automated gas chromatographic chiral analysis system for derivatized α-amino acids has been described [38]. A very fast one-step derivatization procedure of the carboxylic group and all other reactive groups of α-amino acids has been developed by Hušek [39,40]. The use of alkyl chloroformates as derivatizing reagents leads to *N*(*O*)-alkoxycarbonyl alkyl esters of α-amino acids, whereby the intermediate mixed anhydride is decarboxylated to the alkyl ester. The alkyl chloroformate approach bears a number of advantages: (i) the rapid one-step reaction can be carried out in aqueous solution without heating, (ii) the cost of reagent is negligible, (iii) the derivatized amino acids can be easily separated from the mixture using an organic solvent, thus reducing chemical contamination, and (iv) the method can easily be automated. The various derivatization strategies for α-amino acids have been detailed in Ref. [20].

21.3. CHIRAL STATIONARY PHASES BASED ON METAL CHELATES

This topic has been reviewed [41–43]. In the early 1970s, Gil-Av and Schurig conjectured whether abiotic selector–selectand systems displaying metal–organic coordination would display chiral recognition by complexation GC. To this end, attempts were directed toward resolving a chiral olefin on an optically active metal coordination compound by GC. Indeed, dicarbonyl-rhodium(I)-3-(trifluoroacetyl)-(*1R*)-camphorate dissolved in squalane and coated on a long stainless-steel capillary column displayed isotopic selectivity toward deuterated ethenes and enantioselectivity toward the chiral olefin 3-methylcyclopentene [41]. Whereas the scope of enantioseparation of olefins on the rhodium(I)-containing selector was limited, the use of metal(II) *bis* [3-(trifluoroacetyl)-(*1R*)-camphorates] and metal(II) *bis* [3-(heptafluorobutanoyl)-(*1R*)-camphorates] (metal = manganese, cobalt and nickel) as chiral selectors for the enantioseparation of the enantiomers of nitrogen-, oxygen-, and sulfur-containing selectands emerged soon as a routine method for enantioselective GC [44,45]. Most of the racemates were not previously amenable to enantioseparation by the hydrogen bonding CSPs. Significantly, the metal selectors were capable of separating some of the smallest chiral compounds namely alkyl-substituted aliphatic aziridines, oxiranes, and thiiranes [44]. Apart from camphor, eleven different terpene ketones, including 3- and 4-pinanone, methylthujone, carvone, pulegone, menthone, and isomenthone, were utilized as chiral recognition moieties in enantioselective complexation GC [45]. Various coalescence phenomena [46] including the first example of dynamic interconversion of enantiomers (enantiomerization) (Figure 21.3) [44,47] as well as four distinct enantioselective processes [48] were discovered in complexation GC. The enantiomers of chiral insect pheromones were identified by complexation GC [49]. The limited temperature stability of the dissolved metal chelates was improved by anchoring the selectors to poly(dimethylsiloxane) to yield Chirasil-Metal-CSPs [50] (Figure 21.1 middle). With the advent of modified cyclodextrins, the use of enantioselective complexation GC has been discontinued. The method, however, offered important insights into the mechanisms of chirality recognition in the realm of metal organic chemistry [50].

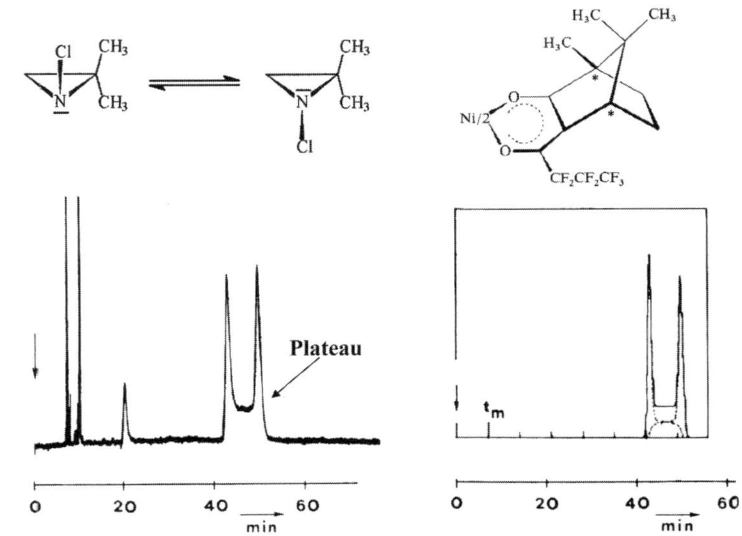

FIGURE 21.3 Plateau formation due to enantiomerization (distorted elution profile caused by molecular inversion at the nitrogen atom) of 1-chloro-2,2-dimethylaziridine upon complexation GC on nickel(II) bis[3-(trifluoroacetyl)-(1R)-camphorate] in squalane at 60 °C. Left: experimental trace, right: simulated trace. *Source: From Ref. [47] with permission.*

21.4. CHIRAL STATIONARY PHASES BASED ON MODIFIED CYCLODEXTRINS (CDS)

This topic has been reviewed [3,51–56]. The first enantioseparation of terpenoic hydrocarbons (α- and β-pinene, *cis*- and *trans*-pinane, and carene) by GC was observed by Kościelski, Sybilska, and Jurczak on native α-cyclodextrin hydrate dissolved in formamide and coated on celite [57]. Unfortunately, peak efficiency of the packed column was low, the column temperature was limited to 70 °C, and the lifetime of the chromatographic system was short due to extensive column bleeding and dehydration of the CD. Yet this work nevertheless started an impressive development of enantioselective GC employing selectively derivatized CDs as CSPs. The high potential of CDs for enantioseparations is due to the different reactivities of the 2-, 3-, and 6-hydroxy groups of the glucose moieties, which can be modified by regioselective alkylation and acylation giving rise to a plethora of possible chemically modified CDs. One disadvantage is the fact that CDs are only available with D-configurated glucose building blocks. The absence of CDs with the unnatural L-configurated glucose prevents the possibility to reverse the sense of enantioselectivity by employing the CSP with opposite configuration, which is essential for validation purposes.

In order to efficiently coat capillary columns (glass, later substituted by fused silica), the CSP should be fluid. Two approaches have been developed independently: Schurig and Nowotny dissolved peralkylated CDs in a moderately polar polysiloxane (e.g. OV 1701), thus combining enantioselectivity with the unique properties of silicones in GC [58,59] (Figure 21.4), whereas König et al. [60] employed low-melting CD derivatives containing *n*-pentyl groups (e.g. per-*O*-pentylated and 3-*O*-acylated-2,6-di-*O*-pentylated α-, β-, and γ-CD [53]) as undiluted liquid stationary phases coated on Pyrex glass capillary

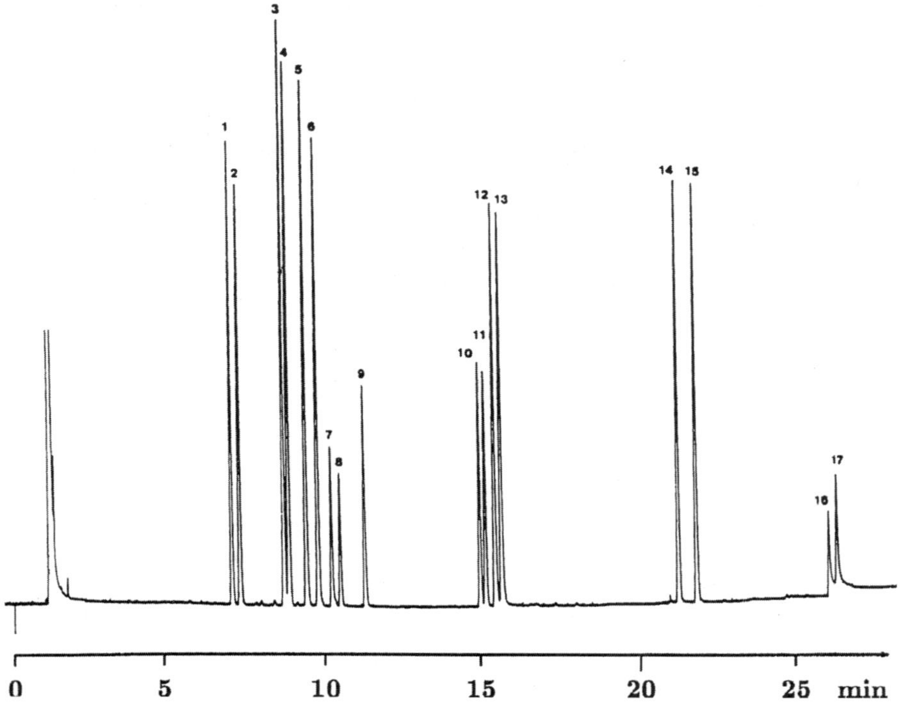

FIGURE 21.4 Simultaneous enantioseparation of various racemic compounds belonging to different classes of chiral compounds ['Schurig test mixture'; α-pinene (1,2), (1R)-(+)-trans-pinane (3), (1S)-(-)-trans-pinane (4), (1S)-(-)-cis-pinane (5), (1R)-(+)-cis-pinane (6), rac-2,3-butanediol (7,8), meso-2,3-butanediol (9), γ-valerolactone (10,11), 1-phenylethylamine (12,13), 1-phenylethanol (14,15), 2-ethylhexanoic acid (16,17)] on permethylated β-cyclodextrin diluted in OV-1701 (70 °C for 5 min followed by 3 °C/min, 0.65 bar H_2, 25 m × 0.25 mm i.d. fused silica capillary column). *(Courtesy, Chrompack International, Middleburg, NL).*

columns. This approach was later extended by Armstrong et al. using permethylated 2-hydroxypropyl and pentylated/acylated CDs coated on fused silica capillary columns [61]. The strategy to dilute modified CDs such as permethylated β-CD and 2,6-di-*O*-methyl-3-*O*-trifluoro-β-CD [58,59] in semipolar polysiloxanes now represents the most frequently used methodology. In addition, the enantioseparations described by König obtained on *undiluted* CDs [53] profited from the dilution approach [5]. The use of permethylated α-, β-, and γ-CD dissolved in polysiloxanes bears a number of merits [52,62]: (i) the excellent coating properties of silicones for high-resolution and high-efficiency capillary columns (HRC-GC) are maintained and combined with the enantioselectivity of CDs; (ii) high melting points or phase transitions of modified cyclodextrins are not detrimental to column performance; and (iii) mixed multicomponent CD-based CSPs can be employed [51].

A comprehensive study involving more than 150 racemates showed the advantages of diluted permethylated β-CD in comparison to the α-CD and γ-CD congeners [52]. A comparison between permethylated α-, β-, and γ-CD showed complementary behavior and in selected cases even a reversal of the elution order of the enantiomers [63]. A collection of enantioseparation factors α of different classes of chiral compounds on (i) octakis (6-*O*-methyl-2,3-di-*O*-pentyl)-γ-cyclodextrin (Lipodex G), (ii) hepakis(2,6-*O*-methyl-3-*O*-pentyl)-β-cyclodextrin, (iii) octakis(2,6-*O*-methyl-3-*O*-pentyl)-γ-cyclodextrin, and (iv) octakis(3-*O*-butanoyl-2, 6-di-*O*-pentyl)-γ-cyclodextrin (Lipodex E) was compiled by König [64]. Second-generation CDs contain the bulky *tert*-butyldimethylsilyl (TBDMS) residue. Per-TBDMS-β-CD diluted in polysiloxane PS-086 and coated on glass capillary columns was used as CSP for GC enantioseparation up to 250 °C [65]. Heptakis(2,3-di-*O*-acetyl-6-*O*-tert-butyldimethylsilyl)-β-cyclodextrin [66] and heptakis(2,3-di-*O*-methyl-6-*O*-tert-butyldimethylsilyl)-β-cyclodextrin [67] represent important CSPs in enantioselective GC. As compared to permethyl-β-CD, the TBDMS-derivatives of β-CD also show improved solubility in polysiloxanes. In diluted systems the enantioseparation factor α is rendered concentration dependent due to the two different retention mechanisms arising (i) from the presence of the achiral solvent and (ii) from the chiral CD selector, both comprising the total stationary phase [68]. A theoretical treatment has shown that α does not linearly increase with the CD concentration but reaches an optimum already at a low concentration [69]. Hence, no gain in enantioselectivity at high concentrations is obtained. Thus, the use of undiluted CDs [53] is being increasingly discontinued.

An extension of the dilution approach represents the chemical linkage of the CD-selector to a poly(dimethylsiloxane) backbone [70–73]. The *chiral* poly*sil*oxane containing cyclodextrins was named Chirasil-Dex (Figure 21.1 right) in analogy to Chirasil-Val. The CSP Chirasil-Dex shows low column bleeding (important for GC-MS) and has a high-temperature range of operation between −25 °C and 250 °C. The CD resides in an apolar environment leading to fast analysis of polar compounds. Chirasil-Dex can be immobilized on the fused silica surface by simple thermal treatment. The same *o*pen-tubular fused silica column (100 cm × 0.05 mm i.d.) coated with Chirasil-Dex has been used for the enantioseparation of, e.g. hexobarbital by all contemporary chromatographic methods: *o*-GC, *o*-SFC, *o*-LC, and *o*-CEC (unified enantioselective approach) [74]. The CSP Chirasil-γ-Dex refers to polysiloxane-linked octakis(3-*O*-butanoyl-2, 6-di-*O*-pentyl)-γ-cyclodextrin (Lipodex E) [75]. Another immobilization strategy to link β-CD to siloxanes has been advanced by Armstrong et al. [76]. The option to mix different cyclodextrin (CD) selectors in one chiral stationary phase (CSP) has been proposed [51] and realized subsequently (see ref. [77] and references cited therein).

The GC enantioseparation of a wide variety of racemic compounds of different classes of compounds on modified cyclodextrins is mostly characterized by very small enantioseparation factors ($1.02 < \alpha < 1.20$) (Figure 21.4) corresponding to a low enantioselectivity ($-\Delta_{D,L}(\Delta G) = RT \ln \alpha$) in the range of 0.014–0.140 kcal/mol at 100 °C. The low enantioselectivities elicited by cyclodextrins GC is sufficient for analytical purposes but is detrimental for reliable mechanistic studies by molecular modeling. The role of inclusion has been challenged by the observation that also modified linear dextrins ('acyclodextrins'), *i.e.*, (heptakis [(2,3-di-*O*,4″-*O*)-acetyl-(1′-*O*,6-*O*)-*tert*-butyldimethylsilyl]-maltoheptaose) are able to enantioseparate various racemic compounds [78]. Even the single derivatized glucose building block, *i.e.*, 2,3,4-tri-*O*-acetyl-1,6-*O*-*tert*-butyldimethylsilyl-glucose, enantioseparates α-amino acids as *N*-TFA methyl esters [78]. One of the main advantages of the use of linear dextrin derivatives resides in the possibility to readily obtain the D- and L-configurated (as well as epimeric) forms of the selectors.

New developments are concerned with ionic cyclodextrin derivatives dissolved in ionic liquids [79] and a new carbohydrate selector based on cyclofructan [80] for enantioselective GC.

21.5. THE TEMPERATURE DEPENDENCE OF ENANTIOSELECTIVITY, ENTHALPY/ENTROPY COMPENSATION, AND THE ISOENANTIOSELECTIVE TEMPERATURE T_{ISO}

Enantioseparation by GC is governed by thermodynamics. The association equilibrium ought to be reversible and kinetics must be fast. The enantioselectivity $-\Delta_{D,L}\Delta G$ is determined by the Gibbs–Helmholtz equation. When an undiluted selector is employed as CSP, enantioselectivity is related to the difference of the retention factors k (the subscripts D and L arbitrarily denote enantiomers, α_{max} is the maximal enantioseparation factor corresponding to the enantiomerically pure selector, R is the gas constant, and T is the absolute temperature):

$$\begin{aligned} -\Delta_{D,L}\Delta G &= -\Delta_{D,L}\Delta H + T\Delta_{D,L}\Delta S \\ &= RT\ln(k_D/k_L) = RT\ln\alpha_{max} \end{aligned}$$

The enantioselectivity $-\Delta_{D,L}\Delta G$ is governed by an enthalpy term $-\Delta_{D,L}\Delta H$ and an entropy term $T\Delta_{D,L}\Delta S$. For a 1:1 association process, both quantities oppose each other in determining $-\Delta_{D,L}\Delta G$. The resulting enthalpy/entropy compensation arises from the fact that the more tightly bonded complex ($\Delta H_D > \Delta H_L$) is more ordered ($\Delta S_D < \Delta S_L$). Since the entropy term increases with the temperature T, an *isoenantioselective temperature* exists [16,48], *i.e.*

$$T_{iso} = \Delta_{D,L}\Delta H/\Delta_{D,L}\Delta S \; (\Delta_{D,L}\Delta G = 0, \alpha = 1)$$

At T_{iso} peak coalescence arises and the coeluted enantiomers cannot be separated due to enthalpy/entropy compensation. Below T_{iso} enantioseparation is governed by the predominant enthalpic contribution to enantiorecognition, whereas above T_{iso} it is governed by the predominant entropic contribution to enantiorecognition. Below and above T_{iso} the enantioselectivity increases either by decreasing or by increasing the temperature, respectively, whereby the association between selectands (enantiomers) and selector (CSP) steadily decreases with increasing temperature T. Very low values for $-\Delta_{D,L}\Delta H$ render the enantioseparation as nearly temperature independent. Enthalpy/entropy compensation must be considered for molecular modeling studies in which the importance of entropy changes should be appreciated. The quantities $-\Delta_{D,L}\Delta H$ and $\Delta_{D,L}\Delta S$ are accessible *via* linear van't Hoff plots when measurements are performed at different temperatures T according to

$$\begin{aligned} R\ln\alpha_{max} &= -\Delta_{D,L}\Delta G/T \\ &= -\Delta_{D,L}\Delta H/T + \Delta_{D,L}\Delta S \end{aligned}$$

Nonlinear plots indicate multimodal mechanisms of enantiorecognition. A characteristic example of peak inversion due to enthalpy/entropy compensation [81] is depicted in Figure 21.5.

T_{iso} is usually high (>100 °C) in enantioselective GC and most GC enantioseparations are consequently governed by the enthalpy term of the Gibbs–Helmholtz equation and the enantioselectivity increases with reducing the temperature. Therefore, the lowest possible enantioseparation temperature should be employed. As involatile racemates usually require a high elution temperature, it is advisable to use short columns (1–5 m × 0.25 mm i.d.) *via* miniaturization [74,82] (Figure 21.6). The loss of efficiency arising from the smaller

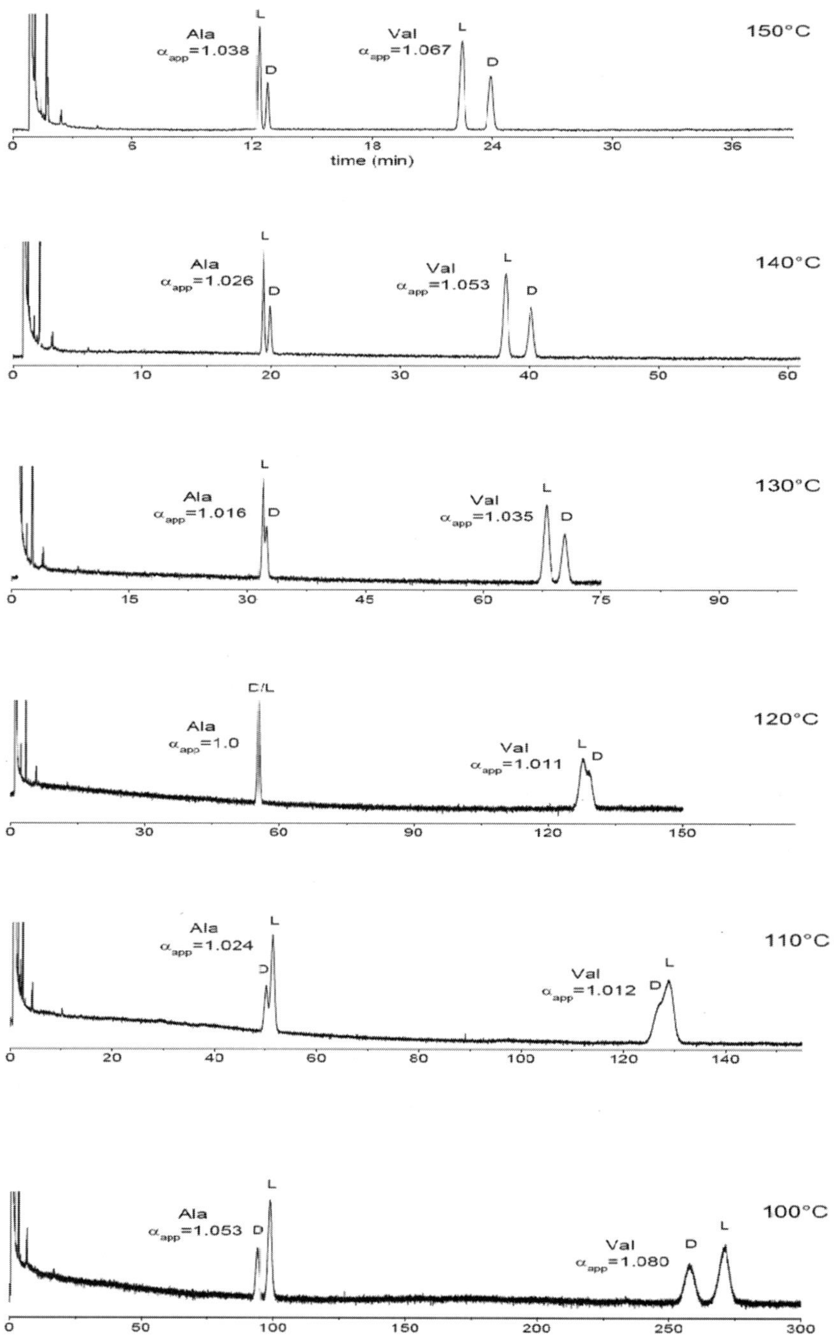

FIGURE 21.5 Temperature-dependent reversal of the elution order of the enantiomer of alanine-ECPA and valine-ECPA (ECPA = N-ethoxycarbonyl *n*-propylamide) on Chirasil-L-Val-C_{11}; the α-amino acids are enriched with the L-enantiomer. Isoenantioselective temperatures, T_{iso}, for alanine-ECPA and valine-ECPA are 120 and 114 °C, respectively. Column: fused silica, 20 m × 250 μm i.d. × 0.25 μm (polymer thickness); carrier gas: H_2; head pressure: 50 kPa (120–170 °C) and 100 kPa (100–110 °C); detector: FID. Time scale in minutes. *Source: From ref. [81] with permission.*

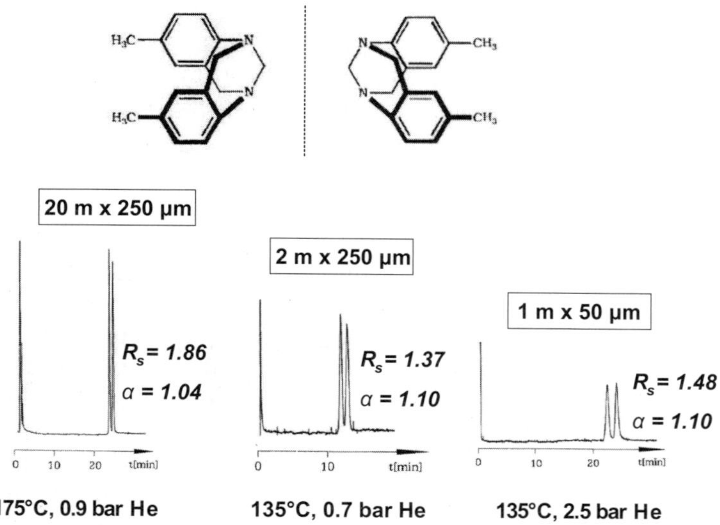

FIGURE 21.6 Column miniaturization for the GC enantioseparation of Tröger's base on Chirasil-β-Dex. *Source: From Ref. [82] with permission.*

theoretical plate number N of a short column is then compensated by the gain of enantioselectivity due to the increased enantioseparation factor α at the low elution temperature.

An alternative treatment of enantioselectivity based on the enthalpy/entropy–compensation–formalism has been advanced [83,84] and a three-phase (pseudo) model for enantioselective GC comprising of permethylated CD/polysiloxane CSP was also developed by Pino et al. [85]. Two separation mechanisms exist in diluted selector systems. Whereas enantiomers are not differentiated by the achiral solvent, they are discriminated by the chiral selector (CSP). Therefore, the two contributions to retention must be separated by the retention-increment (R') formalism in order to get reliable thermodynamic data of enantiorecognition [43,68,83,86].

The retention-increment method has also been applied for mixed CSPs in enantioselective GC [87]. The combination of amino acid diamide selectors and modified cyclodextrins (Chirasil-Val-Dex) [32] in a single column widens the scope of enantioselectivity. Another strategy to combine the enantioselectivities of hydrogen bonding and inclusion type selectors as a single CSP consists of linking a single L-valine diamide moiety in the C6-position of permethylated β-cyclodextrin [88].

21.6. APPLICATIONS

Enantioselective GC is a valuable tool for the determination of enantiomeric compositions (*ee, er,* and *ec* [1]) in various fields of contemporary science (Figure 21.7). In contrast to the indirect method of forming diastereomers, the chiral auxiliary, *i.e.*, the chiral selector, need not be enantiomerically pure. Although an enantiomerically impure selector reduces the maximum enantioseparation factor α_{max}, the enantiomeric ratio of the chiral analyte itself is not affected [48,86]. In order to determine 0.1% of an enantiomeric impurity, the detector and integration facility must be linear by at least four orders of magnitude. For validation purposes, the use of selectors with opposite

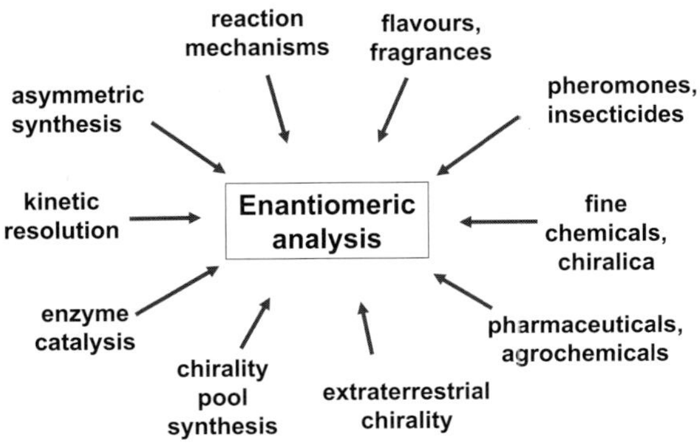

FIGURE 21.7 The importance of enantiomeric analysis in academia and industry.

configuration (when available), e.g. Chirasil-L-Val vs. Chirasil-D-Val, is recommended (Figure 21.8).

The enantiomer of opposite configuration D represents an ideal standard for the quantification of the enantiomer L present in a mixture. This approach, referred to as 'enantiomer labeling' [89,90], exploits the fact that enantiomers possess identical chemical and physical properties in an achiral environment and in nonconcentrated complex mixtures. The enantiomeric composition of sample and standard is not affected by workup,

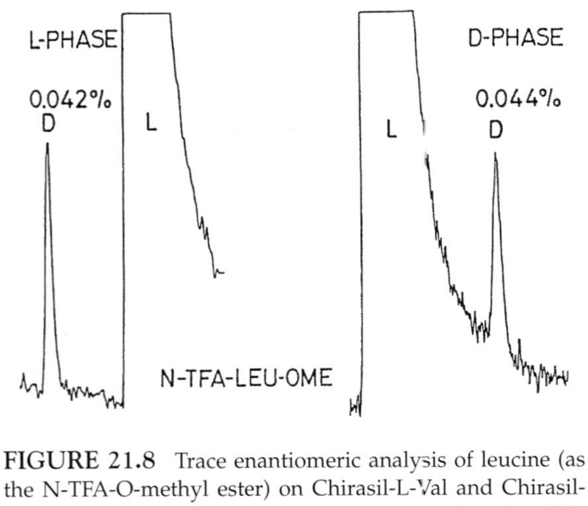

FIGURE 21.8 Trace enantiomeric analysis of leucine (as the N-TFA-O-methyl ester) on Chirasil-L-Val and Chirasil-D-Val (20 m × 250 μm i.d. glass capillary, 95 °C, 0.3 bar H_2). *(Courtesy Prof. B. Koppenhoefer, Habilitation Thesis, 1989, University Tübingen, p. 148).*

sample manipulations, losses, split injection, and detection. Also, the amount of a racemate present in a complex matrix can be quantitatively determined *via* the enantiomer labeling method when a known amount of a single enantiomer (L or D) is added to the mixture and the change of the enantiomeric composition is then determined by enantioselective GC.

HRC-GC columns coated with Chirasil-Val have been used for many applications [15,91] involving amino acid analysis, e.g. the detection of D-amino acids in bacterial cell walls and in peptide antibiotics, the monitoring of amino acid enantiomeric purity in peptide synthesis, the determination of the degree of racemization during peptide hydrolysis (*vide infra*), the dating of paleontological and archeological artifacts by exploiting the time-dependence of amino acid racemization, the search for biogenic amino acids in extraterrestrial material, and the amplification of optical activity under abiotic and prebiotic conditions. Chirasil-Val has also been used for the stereochemical GC analysis of the nerve gas soman [92].

HRC-GC columns coated with modified CDs were applied for enzyme and catalyst screening [93,94], for the enantiomeric analysis of essential oils, flavors, and fragrances [95,96], branched fatty acid esters [97], organochlorine compounds [98], chiral pollutants [99,100], silicon compounds [101], alkyl nitrates as atmospheric constituents [83], inhalational anesthetics [90] (Figure 21.9), clinical compounds [102], and unfunctionalized aliphatic hydrocarbons [103,104] (see also 270 references on pages 222–231 in Ref. [3]). GC enantioseparations of some pharmaceuticals [105] and of (derivatized) stimulant-type drugs of the phenethylamine structure [106] have also been described.

For the determination of extraterrestrial homochirality, three enantioselective GC columns coated with commercially available Chirasil-Val, Chirasil-Dex, and octakis(2,6-di-*O*-pentyl-3-*O*-trifluoroacetyl)-γ-cyclodextrin (G-TA), respectively, were integrated in the COSAC experiment

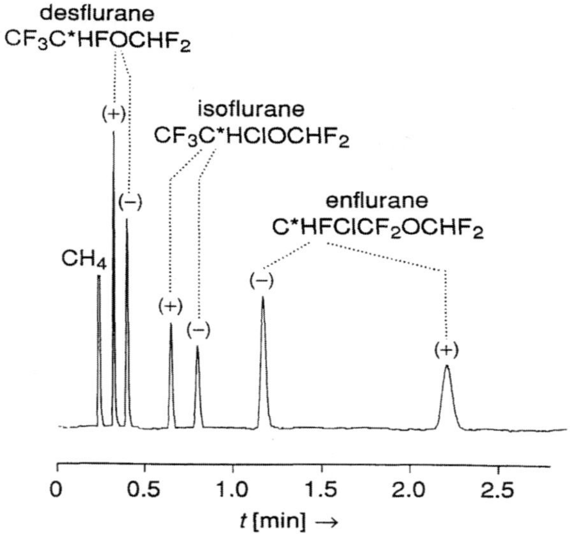

FIGURE 21.9 Gas chromatographic enantioseparation of racemic inhalation anesthetics (28 °C, 10 m × 250 μm i.d. fused silica capillary coated with 0.18 μm (film thickness) of immobilized Chirasil-γ-Dex). *Source: From Ref. [75] with permission.*

as part of the payload of the Rosetta mission of ESA launched in 2004 and scheduled to land at the comet 67P/Churyumov–Gerasimenko in 2014 [107] (Figure 21.10). Enantiomeric bias of extraterrestrial and configurationally stable α-methyl-α-amino acids has been determined in the Murchison meteorite by enantioselective GC on Chirasil-Val [108].

21.7. HYPHENATED APPROACHES IN ENANTIOSELECTIVE GC

When high sensitivity is required, the coupling of enantioselective GC with mass spectrometry (enantio-GC-MS) is the method of choice [8]. GC-MS hyphenation has been considered for enantioseparations of derivatized α-amino acids in extraterrestrial space [109]. Enantio-GC-MS in the selected ion-monitoring mode (enantio-GC-MS(SIM)) represents another refinement. In the SIM mode of the MS system, selected ions are monitored, thus prolonging the detection time of these ions and thereby increasing the signal-to-noise ratio. The four stereoisomers of *E*,*Z*-chalcogran (2-ethyl-1,6-dioxaspiro[4.4]nonane), the aggregation pheromone of the bark beetle *Pityogenes*

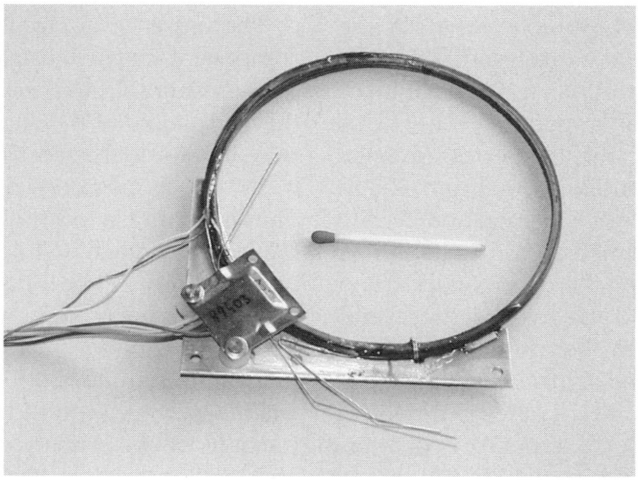

FIGURE 21.10 A prototype of a fused silica capillary column coated with a Chirasil-type CSP connected to micromachined thermal conductivity detector of the COSAC gas chromatographic campaign now present in outer space. *(Courtesy, Prof. U. Meierhenrich).*

chalcographus, were separated at the 20-ppm level in *n*-hexane by complexation GC on nickel(II) *bis* [3-(heptafluorobutanoyl)-(*1R*)-camphorate] in SE-54 (0.2 μm) at the m/z = 29 and m/z = 127.4 molecular ions [42]. Unnatural D-amino acids as *N(O)*-pentafluoropropionyl/2-propyl esters were determined in mammals using Chirasil-Val in the GC-MS(SIM) mode [110]. Glausch et al. detected an enantiomeric bias of the atropisomeric polychlorinated biphenyl PCB 132 in human milk samples using two-dimensional GC in the MS(SIM)-mode [111]. An unexpected small deviation from the administered racemic composition of the inhalational anesthetic isoflurane (2-chloro-2-(difluoromethoxy)-1,1,1-trifluoroethane) during and after surgery in clinical patients was established by headspace GC-MS(SIM) (m/z = 117 and 149) on octakis(3-*O*-butanoyl-2,6-di-*O*-pentyl)-γ-cyclodextrin (Lipodex E) [64] employing a multipurpose headspace sampler combined with a cold injection system for trapping, enrichment, and focusing of the narcotic [112]. The quantitative and stereoisomeric determination of the chiral hydrocarbons 1-methyltetralin, *cis*-1,2- and 1,3-dimethylindane as biomarkers in crude oil and coal samples has been determined by GC-MS(SIM) at the ppt-level [113].

The racemization of α-amino acids occurring during the acid-catalyzed hydrolysis of peptides can falsify the true enantiomeric composition of the building blocks of a peptide [114]. This problem can be overcome by performing the hydrolysis in a fully deuterated medium, e.g. in 6 *N* D_2O/DCl. In this case, racemization during hydrolysis is accompanied by substitution of the hydrogen attached to the stereogenic carbon atom by deuterium and the hydrogenated and deuterated species can be differentiated by mass spectrometry. After hydrolysis, the amino acids are derivatized and enantioseparated on a suitable CSP (Chirasil-Val, Lipodex E, or Chirasil-γ-Dex) by GC and online detected by MS in the SIM mode [115]. For each amino acid, a characteristic ion containing the proton at the stereogenic carbon atom is monitored, whereas the deuterated species formed during the hydrolysis is disregarded. The reliable determination of enantiomeric purities of L-α-amino acids in peptides up to 99.9% is thus possible [115]. This method has also been used to determine the rate constants of configurational inversion at the stereogenic center of α-amino acids under acid hydrolysis conditions (110 °C, 6 *N* DCl) [115]. The time-dependent racemization of α-amino acids has been widely used for dating purposes of archeological artifacts containing α-amino acids [15].

A still rudimentary attempt to hyphenate enantio-GC and ^{1}H-NMR has been undertaken for 2,4-dimethylhexane (C*HMeEt*i*Bu) enantioseparated at 35–50 °C on a 30 m × 250 μm i.d. fused silica capillary coated with 0.5 μm octakis(6-*O*-methyl-2,3-di-*O*-pentyl)-γ-cyclodextrin (Lipodex G) [64] and connected *via* a 2 m transfer capillary (250 μm i.d.) to an NMR spectrometer [116].

21.8. TWO-DIMENSIONAL APPROACHES IN ENANTIOSELECTIVE GC

The presence of chiral compounds in multicomponent matrices causes a doubling of peaks on an enantioselective column, thereby increasing the complexity of the elution pattern. Therefore, two-dimensional approaches are routinely used by introducing a second dimension of separation. The conventional technique involves the heartcut GC–GC approach first demonstrated by Schomburg et al. [117]. Here the first nonenantioselective column coated with an achiral polar stationary phase is used to preseparate components of interest (first dimension), whereas in the second enantioselective column coated with a CSP fractions of chiral analytes are enantioseparated after online transfer through a pneumatic or flow-controlled low-dead-volume heartcut interface (second dimension). There are many applications in the field of terpenes and food chemistry [118,119,120]. From a complex mixture of atropisomeric polychlorinated biphenyls (PCBs) in the formulation Clophen A 60, a single peak was heartcut in the first dimension and resolved into three peaks in the second dimension comprising an enantioselective fused silica capillary column coated with Chirasil-β-Dex. The peaks were assigned to PCB 151 (chiral and containing *two* chlorine substituents in ortho-positions: unresolved) and to PCB 132 (chiral and containing *three* chlorine substituents in ortho-positions: resolved) [121].

Stir-bar sorptive extraction (SBSE) and two-dimensional enantioGC-MS analysis of chiral flavor compounds have been described [122]. The use of miniaturized vs. conventional columns in the two-dimensional mode have been compared [123]. Also, an enantioselective two-dimensional LC-GC approach has been utilized [124].

Comprehensive two-dimensional gas chromatography GC × GC represents an orthogonal separation system of compounds on two columns using cryogenic modulation without the requirement of heartcutting. GC × GC works by transferring a high number of contiguous heartcuts from the first conventional column to the second miniaturized column, thereby generating characteristic high-resolution contour plots. In case an achiral separation system is employed in the first dimension, a very fast separation of enantiomers would be required in the second dimension (GC × enantio-GC). An enantioseparation with $R_S \sim 1.0$ in eight seconds was reported for racemic limonene on a 1 m × 100 μm i.d. fused capillary column coated with a diluted β-cyclodextrin derivative [125]. The column geometry may preferentially be inverted, *i.e.*, the first column consists of an enantioseparation system, whereas the second miniaturized achiral column, e.g. coated with a polar stationary phase, is employed to obtain the two-dimensional contour plot of a multitude of enantioseparations (enantioGC × GC) [126,127].

To differentiate between genuine enantiomers and fortified enantiomers in flavors and fragrances, enantioselective multidimensional gas chromatography has been hyphenated on-line with combustion isotope ratio mass spectrometry (enantio-MDGC-IRMS) [95,128,129]. The enantiomers eluted from the GC column are pyrolyzed in a ceramic oxidation reactor interface and the $^{18}O/^{16}O$ and/or the $^{13}C/^{12}C$ ratio of the carbon dioxide formed is determined by mass spectrometry. Also, biosynthetic or biogenetic pathways may be discerned by MDGC-IRMS. Enantioselective GC-IRMS has been used to determine the carbon isotopic ratio of amino acids present in the Murchison meteorite to prove its extraterrestrial origin [130].

21.9. ENANTIOSELECTIVE STOPPED-FLOW MULTIDIMENSIONAL GAS CHROMATOGRAPHY (SF-MDGC)

By this method, the configurational instability of a chiral compound can be evaluated at elevated temperatures with only a trace of the racemic probe. Three columns are employed in series. The enantiomers are enantioseparated on-line in the first enantioselective column. Either one of the pure enantiomers is then transferred into the second empty reactor column. By stopping the flow for the time t, enantiomerization (inversion of configuration) is allowed to proceed at the elevated temperature T in the gas phase. After the time t the enantiomers are focused by cooling and resuming the flow. The enantiomeric ratio *er* is then determined in the third enantioselective column (or by backflushing into the first column). From T, t, and *er*, the rate constant of enantiomerization can be calculated. This method has been applied for the determination of activation barriers of molecular inversion of atropisomeric polychlorinated biphenyls (chiral PCBs) [131] and of 1-chloro-2,2-dimethylaziridine [132,133]. The determination of enantiomerization kinetics by dynamic GC and stopped-flow GC has been reviewed [134].

21.10. PRACTICAL ASPECTS OF ENANTIOSELECTIVE GC

Chiral test mixtures for the performance of enantioselective GC columns have been devised for hydrogen bonding diamide CSPs [135] and for permethylated β-cyclodextrin in OV-1701 covering a broad spectrum of polarity [62,63] (Figure 21.4). Clearly, the choice of the constituents of those chiral test mixtures is highly arbitrary limiting their general applicability.

Comprehensive information for enantioselective GC is contained in the "Chirbase/GC®" data bank developed by the group of B. Koppenhoefer (University of Tübingen, Germany): http://www.acdlabs.com/products/adh/chrom/chirbase/. It contains method information until the year 2000 for over 24,000 enantioseparations of more than 8,000 chiral molecules. The data bank contains experimental conditions, as well as the structure, the substructure, and structural relations [136]. The "Chirbase/Flavor®" data bank contains 1,000 entries on odor properties of enantiomers. For the development of a specific MS library used for the identification of 134 racemates belonging to flavors and fragrances, 'interactive' linear retention index values I^T were determined on four modified β-cyclodextrin derivatives [137].

Topics such as precision and accuracy of enantioselective gas chromatography, practical hints, and recommendations have been treated in former accounts [17,53,98]. Sources of error have also been discussed in detail [42,98]. In a comprehensive study aimed at detecting small enantiomeric bias (symmetry breaking) in experiments devoted to the amplification of homochirality under prebiotic conditions, it was implied that the limit in determining minute deviations from the true racemic composition of target compounds by enantioselective GC of α-amino acids lies in the range of $50.0 \pm 0.1\%$, corresponding to an apparent $ee = 0.2\%$ [138]. In order to attain such accuracy *and* precision, a number of requirements must be met, including an enantioseparation with a resolution factor $R_S > 1.5$, the exclusion of coeluting impurities, the absence of decomposition of the sample (or its derivative), and a correct peak area integration. Measurement with a precision of $50.0 \pm 0.05\%$ is feasible but the accuracy (trueness) of the value determined must be carefully validated by employing oppositely configurated CSPs leading to the reversal of the elution order of the separated enantiomers [138]. Incidentally, the racemic composition of a chiral compound represents an ideal probe for the accuracy of integration devices because a strict 1:1 ratio should be encountered (statistical fluctuations of enantiomers are well beyond detectability). In the realm of high enantiomeric purities as little as 0.01% of an enantiomeric impurity (after recrystallization) has been inferred for various α-amino acids as *N*-TFA-*O*-methyl esters on Chirasil-Val [139]. Oppositely configurated CSPs were employed to confirm very high enantiomeric ratios as shown for the trace enantiomeric analysis of leucine (as *N*-TFA-*O*-methyl ester) determined on Chirasil-L-Val and Chirasil-D-Val (Figure 21.8). A value of 0.04% was found for a derivatized proline derivative [140]. Clearly, 0.1% of an enantiomeric impurity can routinely be determined in case of baseline enantioseparations by GC. The mode of integration may influence the accuracy of the determination of enantiomeric compositions [138,141].

21.11. (SEMI)PREPARATIVE-SCALE ENANTIOSEPARATIONS BY GC

The state of the art of enantioselective (semi) preparative gas chromatography has been reviewed [142]. This method is restricted to thermally stable and volatile compounds. Contrary to liquid chromatography (LC), the recovery of the isolated enantiomers from the gaseous mobile phase (carrier gas) is straightforward when aerosol and mist formation are prevented by using specially designed collection vessels. The GC approach does not match the overwhelming success of preparative enantioseparation achieved in enantioselective LC. This also holds true for the simulated moving bed (SMB) technology employed in enantioselective GC [143]. It should be mentioned that many CSPs developed for enantioselective GC have also been used as enantioselective sensor devices [144].

References

[1] V. Schurig, Terms for the quantitation of a mixture of stereoisomers, Enantiomer 1 (1996) 139–143.

[2] E. Gil-Av, Present status of enantiomeric analysis by gas chromatography, J. Mol. Evol. 6 (1975) 131–144.

[3] P. Schreier, A. Bernreuther, M. Huffer, Analysis of chiral organic molecules, Walter de Gruyter, Berlin - New York, 1995. [chapter 3.5], pp. 132–233.

[4] C.H. Lochmüller, R.W. Souter, Chromatographic resolution of enantiomers. Selective review, J. Chromatogr. 113 (1975) 283–302.

[5] W.A. König, Enantioselective gas chromatography, Trends Anal. Chem. 12 (1993) 130–137.

[6] V. Schurig, Enantiomer separation by gas chromatography on chiral stationary phases, J. Chromatogr. A 666 (1994) 111–129.

[7] V. Schurig, Separation of enantiomers by gas chromatography, J. Chromatogr. A 906 (2001) 275–299.

[8] V. Schurig, Chiral separations using gas chromatography, Trends Anal. Chem. 21 (2002) 647–661.

[9] L.F. He, T.E. Beesley, Applications of enantiomeric gas chromatography: a review, J. Liq. Chromatogr. Relat. Technol. 28 (2005) 1075–1114.

[10] V. Schurig, Separation of enantiomers by gas chromatography on chiral stationary phases, in: S. Ahuja (Ed.), Chiral separation methods, Wiley, Hoboken, 2011, pp. 251–297. [chapter 9].

[11] D.W. Armstrong, S.M. Han, Enantiomeric separations in chromatography, CRC Crit. Rev. Anal. Chem. 19 (1988) 175–224.

[12] S.G. Allenmark, Chromatographic enantioseparation: methods and applications (Ellis Horwood series in analytical chemistry), second revised ed., Ellis Horwood, New York & London, 1991.

[13] S. Allenmark, V. Schurig, Chromatography on chiral stationary phases, J. Mater. Chem. 7 (1997) 1955–1963.

[14] T.E. Beesley, R.P.W. Scott, Chiral chromatography, John Wiley and Sons, New York, 1998.

[15] V. Schurig, Gas chromatographic separation of enantiomers on optically active metal-complex-free stationary phases, Angew. Chem. Int. Ed. Engl. 23 (1984) 747–765.

[16] B. Koppenhoefer, E. Bayer, Chiral recognition in gas chromatographic analysis of enantiomers on chiral polysiloxanes, J. Chromatogr. Libr. 32 (1985) 1–42.

[17] W.A. König, The practice of enantiomer separation by capillary gas chromatography, Hüthig, Heidelberg, 1987.

[18] H. Frank, Gas chromatography of enantiomers on chiral stationary phases. [chapter 3], in: B. Holmstedt, H. Frank, B. Testa (Eds.), Chirality and biological activity, Alan R. Liss, Inc, New York, 1990, pp. 33–54.

[19] B. Feibush, Chiral separation of enantiomers via selector/selectand hydrogen bondings, Chirality. 10 (1998) 382–395.

[20] V. Schurig, Gas chromatographic enantioseparation of derivatized α-amino acids on chiral stationary phases – past and present, J. Chromatogr. B 897 (2011) 3122–3140.

[21] E. Gil-Av, B. Feibush, R. Charles-Sigler, Separation of enantiomers by gas liquid chromatography with an optically active stationary phase, Tetrahedr. Lett. (1966) 1009–1015.

[22] E. Gil-Av, B. Feibush, R. Charles-Sigler, Separation of enantiomer by gas-liquid chromatography with an optically active stationary phase, in: A.B. Littlewood (Ed.), Gas chromatography 1996, Institute of Petroleum, London, 1967, pp. 227–239, pp. 254–257.

[23] E. Gil-Av, B. Feibush, Resolution of enantiomers by gas liquid chromatography with optically active stationary phases. Separation on packed columns, Tetrahedr. Lett. (1967) 3345–3347.

[24] B. Feibush, Interaction between asymmetric solutes and solvents. N-Lauroyl-L-valyl-t-butylamide as stationary phase in gas liquid partition chromatography, J. Chem. Soc. Chem. Commun. (1971) 544–545.

[25] N. Ôi, M. Horiba, H. Kitahara, H. Shimada, Gas chromatographic separation of enantiomers of some dipeptides on an optically active stationary phase, J. Chromatogr. 202 (1980) 302–304.

[26] H. Frank, G.J. Nicholson, E. Bayer, Rapid gas chromatographic separation of amino acid enantiomers with a novel chiral stationary phase, J. Chromatogr. Sci. 15 (1977) 174–176.

[27] G.J. Nicholson, H. Frank, E. Bayer, Glass capillary gas chromatography of amino acid enantiomers, J. High Resolut. Chromatogr. 2 (1979) 411–415.

[28] G. Lai, G.J. Nicholson, E. Bayer, Immobilization of Chirasil-Val on glass capillaries, Chromatographia 26 (1988) 229–233.

[29] E. Bayer, H. Allmendinger, G. Enderle, B. Koppenhoefer, Anwendung von D-Chirasil-Val bei der gaschromatographischen Analytik von Enantiomeren, Fresenius. Z. Anal. Chem. 321 (1985) 321–324.

[30] T. Saeed, P. Sandra, M. Verzele, Synthesis and properties of a novel chiral stationary phase for the resolution of amino acid enantiomers, J. Chromatogr. 186 (1979) 611–618.

[31] W.A. König, I. Benecke, Gas chromatographic separation of enantiomers of amines and amino alcohols on chiral stationary phases, J. Chromatogr. 209 (1981) 91–95.

[32] P.A. Levkin, A. Levkina, V. Schurig, Combining the enantioselectivities of L-valine diamide and permethylated β-cyclodextrin in one gas chromatographic chiral stationary phase, Anal. Chem. 78 (2006) 5143–5148.

[33] H. Frank, I. Abe, G. Fabian, A versatile approach to the reproducible synthesis of functionalized polysiloxane stationary phases, J. High Resolut. Chromatogr. 15 (1992) 444–448.

[34] A. Ruderisch, J. Pfeiffer, V. Schurig, V, Synthesis of an enantiomerically pure resorcarene with pendant L-valine residues and its attachment to a polysiloxane (Chirasil-Calix), Tetrahedr. Asymm. 12 (2001) 2025–2030.

[35] V. Schurig, M. Juza, M. Preschel, G.J. Nicholson, E. Bayer, Gas-chromatographic enantiomer separation of proteinogenic amino acid derivatives: comparison of Chirasil-Val and Chirasil-γ-Dex used as chiral stationary phases, Enantiomer. 4 (1999) 297–303.

[36] W.A. Bonner, M.A. Van Dort, J.J. Flores, Quantitative gas chromatographic analysis of leucine enantiomers. Comparative study, Anal. Chem. 46 (1974) 2104–2107.

[37] R. Liardon, S. Ledermann, GC behaviour of N(O, S)-perfluoroacyl D, L-amino acid alkyl esters on chirasil-val stationary phase, J. High Resolut. Chromatogr. Chromatogr. Commun. 3 (1980) 475–477.

[38] K. Nokihara, J. Gerhardt, Development of an improved automated gas-chromatographic chiral analysis system: application to non-natural amino acids and natural protein hydrolysates, Chirality. 13 (2001) 431–434.

[39] P. Hušek, Rapid derivatization and gas chromatographic determination of amino acids, J. Chromatogr. 552 (1991) 289–299.

[40] H. Zahradnickova, P. Hušek, P. Simek, GC separation of amino acid enantiomers via derivatization with heptafluorobutyl chloroformate and Chirasil-L-Val column, J. Sep. Sci. 32 (2009) 3919–3924.

[41] V. Schurig, Resolution of enantiomers and isotopic compositions by selective complexation gas chromatography on metal complexes, Chromatographia 13 (1980) 263–270.

[42] V. Schurig, Enantiomer analysis by complexation gas chromatography. Scope, merits and limitations, J. Chromatogr. 441 (1988) 135–153.

[43] V. Schurig, Practice and theory of enantioselective complexation gas chromatography, J. Chromatogr. A 965 (2002) 315–356.

[44] V. Schurig, W. Bürkle, Extending the scope of enantiomer resolution by complexation gas chromatography, J. Amer. Chem. Soc. 104 (1982) 7573–7580.

[45] V. Schurig, W. Bürkle, K. Hintzer, R. Weber, Evaluation of nickel(II) *bis*[α-(heptafluorobutanoyl)-terpeneketonates] as chiral stationary phases for the enantiomer separation of alkyl-substituted cyclic ethers by complexation gas chromatography, J. Chromatogr. 475 (1989) 23–44.

[46] V. Schurig, Peak coalescence phenomena in enantioselective chromatography, Chirality. 10 (1998) 140–146.

[47] W. Bürkle, H. Karfunkel, V. Schurig, Dynamic phenomena during enantiomer resolution by complexation gas chromatography. A kinetic study of enantiomerization, J. Chromatogr. 288 (1984) 1–14.

[48] V. Schurig, Contributions to the theory and practice of the chromatographic separation of enantiomers, Chirality 17 (2005) S205–S226.

[49] R. Weber, V. Schurig, Complexation gas chromatography – A valuable tool for the stereochemical analysis of pheromones, Naturwiss 71 (1984) 408–413.

[50] V. Schurig, K. Betschinger, Metal-mediated enantioselective access to unfunctionalized aliphatic oxiranes: prochiral and chiral recognition, Chem. Rev. 92 (1992) 873–888.

[51] V. Schurig, H.-P. Nowotny, Gas chromatographic separation of enantiomers on cyclodextrin derivatives, Angew. Chem. Int. Ed. Engl. 29 (1990) 939–957.

[52] W. Keim, A. Köhnes, W. Meltzow, H. Römer, Enantiomer separation by gas chromatography on cyclodextrin chiral stationary phases, J. High Resolut. Chromatogr. 14 (1991) 507–529.

[53] W.A. König, Gas chromatographic enantiomer separation with modified cyclodextrins, Hüthig, Heidelberg, 1992.

[54] J. Snopek, E. Smolková-Keulemansová, T. Cserháti, K.H. Gahm, A. Stalcup, Cyclodextrins in analytical separation methods, in: J. Szejtli, T. Osa (Eds.), Comprehensive supramolecular chemistry, vol. 3, Cyclodextrins, Pergamon, 1996, pp. 516–571. [chapter 18].

[55] W. Li, T.M. Rossi, Derivatized cyclodextrins as chiral gas chromatographic stationary phases and their potential applications in the pharmaceutical industry, Chapter 15, in: H.Y. Aboul-Enein, I.W. Wainer (Eds.), The impact of stereochemistry on drug development and use, vol. 142, John Wiley and Sons, New York, 1997, pp. 415–436. in: Chemical Analysis, Winefordner JD (Ed.).

[56] V. Schurig, Use of derivatized cyclodextrins as chiral selectors for the separation of enantiomers by gas chromatography, Ann. Pharm. Fr. 68 (2010) 82–98.

[57] T. Kościelski, D. Sybilska, J. Jurczak, Separation of α- and β-pinene into enantiomers in gas-liquid chromatography systems via α-cyclodextrin inclusion complexes, J. Chromatogr. 280 (1983) 131–134. New chromatographic method for the determination of the enantiomeric purity of terpenoic hydrocarbons, J. Chromatogr. 364(1986) 299–303.

[58] V. Schurig, H.P. Nowotny, Separation of enantiomers on diluted permethylated β-cyclodextrin by high-resolution gas chromatography, in: Proceed. Adv. Chromatogr. 1987, Berlin, 8–10 Sept 1987, A. Zlatkis (Ed.), J. Chromatogr., 441, 1988, pp. 155–163.

[59] H.-P. Nowotny, D. Schmalzing, D. Wistuba, V. Schurig, Extending the scope of enantiomer separation on diluted methylated β-cyclodextrin derivatives by high-resolution gas chromatography, J. High Resolut. Chromatogr. 12 (1989) 383–393.

[60] W.A. König, S. Lutz, P. Mischnick-Lübecke, B. Brassat, G. Wenz, Cyclodextrins as chiral stationary phases in capillary gas chromatography I. Pentylated α-cyclodextrin, J. Chromatogr 447 (447) (1988) 193–197.

[61] D.W. Armstrong, W. Li, C.-D. Chang, Polar-liquid, derivatized cyclodextrin stationary phases for the capillary gas chromatography separation of enantiomers, Anal. Chem. 62 (1990) 914–923.

[62] S. Mayer, D. Schmalzing, M. Jung, M. Schleimer, A chiral test mixture for permethylated β-cyclodextrin-polysiloxane gas–liquid chromatography phases: the Schurig test mixture, LC × GC Intern 5 (4) (April 1992) 58–59.

[63] C. Bicchi, G. Artuffo, A. D'Amato, G.M. Nano, A. Galli, M. Galli, Permethylated cyclodextrins in the GC separation of racemic mixtures of volatiles: part 1, J. High Resolut. Chromatogr. 14 (1991) 301–305.

[64] W.A. König, Forum: collection of enantiomer separation factors obtained by capillary gas chromatography on chiral stationary phases, J. High. Resolut. Chromatogr. 16 (1993) 312–323, 338–352 and 569–586.

[65] W. Blum, P. Aichholz, Gas chromatographic enantiomer separation on *tert*-butylsilylated β-cyclodextrin diluted in PS-086. A simple method to prepare enantioselective glass capillary columns, J. High. Resolut. Chromatogr. 13 (1990) 515–518.

[66] A. Dietrich, B. Maas, V. Karl, P. Kreis, D. Lehmann, B. Weber, et al., Stereoisomeric flavor compounds part. LV: stereodifferentiation of some chiral volatiles on heptakis(2,3-di-O-acetyl-6-O-tert-butyldimethylsilyl)-β-cyclodextrin, J. High. Resolut. Chromatogr. 15 (1992) 176–179.

[67] A. Dietrich, B. Maas, W. Messer, G. Bruche, V. Karl, A. Kaunzinger, et al., Stereoisomeric flavor compounds, part LVIII: The use of heptakis(2,3-di-O-methyl-6-O-tert-butyldimethylsilyl)-β-cyclodextrin as a chiral stationary phase in flavor analysis, J. High Resolut. Chromatogr. 15 (1992) 590–593.

[68] I. Špánic, J. Krupčik, V. Schurig, Comparison of two methods for the gas chromatographic determination of thermodynamic parameters of enantioselectivity, J. Chromatogr. A 843 (1999) 123–128.

[69] M. Jung, D. Schmalzing, V. Schurig, Theoretical approach to the gas chromatographic separation of enantiomers on dissolved cyclodextrin derivatives, J. Chromatogr. 552 (1991) 43–57.

[70] V. Schurig, D. Schmalzing, U. Mühleck, M. Jung, M. Schleimer, P. Mussche, et al., Gas chromatographic enantiomer separation on polysiloxane-anchored permethyl-β-cyclodextrin (Chirasil-Dex), J. High Resolut. Chromatogr. 13 (1990) 713–717.

[71] P. Fischer, R. Aichholz, U. Bölz, M. Juza, S. Krimmer, Permethyl-β-cyclodextrin, chemically bonded to polysiloxane: a chiral stationary phase with wider application range for enantiomer separation by capillary gas chromatography, Angew. Chem. Int. Ed. Engl. 29 (1990) 427–429.

[72] V. Schurig, D. Schmalzing, M. Schleimer, Enantiomer separation on immobilized Chirasil-Metal and Chirasil-Dex by gas chromatography and supercritical fluid chromatography, Angew. Chem. Int. Ed. Engl. 30 (1991) 987–989.

[73] H. Cousin, O. Trapp, V. Peulon-Agasse, X. Pannecoucke, L. Banspach, G. Trapp, et al., Synthesis, NMR spectroscopic characterization and polysiloxane-based immobilization of the three regioisomeric monooctenylpermethyl-ß-cyclodextrins and their application in enantioselective GC, Eur. J. Org. Chem. (2003) 3273–3287.

[74] V. Schurig, M. Jung, S. Mayer, M. Fluck, S. Negura, H. Jakubetz, Unified enantioselective capillary chromatography on a Chirasil-DEX stationary phase. Advantages of column miniaturization, J. Chromatogr. A. 694 (1995) 119–128.

[75] H. Grosenick, V. Schurig, Enantioselective capillary gas chromatography and capillary supercritical fluid chromatography on an immobilized γ-cyclodextrin derivative, J. Chromatogr. A. 761 (1997) 181–193.

[76] D.W. Armstrong, Y. Tang, T. Ward, M. Nichols, Derivatized cyclodextrins immobilized on fused-silica capillaries for enantiomeric separations via capillary electrophoresis, gas chromatography, or supercritical fluid chromatography, Anal. Chem. 65 (1993) 1114–1117.

[77] D. Kreidler, H. Czesla, V. Schurig, A mixed stationary phase containing two versatile cyclodextrin-based selectors for the simultaneous gas chromatographic enantioseparation of racemic alkanes and racemic α-amino-acid derivatives, J. Chromatogr. B. 875 (2008) 208–216.

[78] G. Sicoli, F. Pertici, Z. Jiang, L. Jicsinszky, V. Schurig, Gas-chromatographic approach to probe the absence of molecular inclusion in enantioseparations by carbohydrates. Investigation of linear dextrins ("acyclodextrins") as novel chiral stationary phases, Chirality 19 (2007) 391–400.

[79] K. Huang, X.T. Zhang, D.W. Armstrong, Ionic cyclodextrins in ionic liquid matrices as chiral stationary phases for gas chromatography, J. Chromatogr. A 1217 (2010) 5261–5273.

[80] Y. Zhang, Z.S. Breitbach, C.L. Wang, D.W. Armstrong, The use of cyclofructans as novel chiral selectors for gas chromatography, Analyst. 135 (2010) 1076–1083.

[81] P.A. Levkin, A. Levkina, H. Czesla, V. Schurig, Temperature-induced inversion of the elution order of enantiomers in gas chromatography: N-ethoxycarbonyl propylamides and N-trifluoroacetyl ethyl esters of α-amino acids on Chirasil-Val-C11 and Chirasil-Dex stationary phases, Anal. Chem. 79 (2007) 4401–4409.

[82] V. Schurig, H. Czesla, Miniaturization of enantioselective gas chromatography, Enantiomer. 6 (2001) 107–128.

[83] M. Schneider, K. Ballschmiter, Alkyl nitrates as achiral and chiral solute probes in gas chromatography. Novel properties of a β-cyclodextrin derivative and characterization of its enantioselective forces, J. Chromatogr. A. 852 (1999) 525–534.

[84] A. Berthod, W. Li, D.W. Armstrong, Multiple enantioselective retention mechanisms on derivatized cyclodextrin gas chromatographic chiral stationary phases, Anal. Chem. 64 (1992) 873–879.

[85] V. Pino, A.W. Lantz, J.L. Anderson, A. Berthod, D.W. Armstrong, Theory and use of the pseudophase model in gas-liquid chromatographic enantiomeric separations, Anal. Chem. 78 (2006) 113–119.

[86] V. Schurig, Elaborate treatment of retention in chemoselective chromatography. The retention increment approach and non-linear effects, J. Chromatogr. A. 1216 (2009) 1723–1736.

[87] P.A. Levkin, V. Schurig, Apparent and true enantioselectivity of single- and binary-selector chiral stationary phases in gas chromatography, J. Chromatogr. A 1184 (2008) 309–322.

[88] O. Stephany, F. Dron, S. Tisse, A. Martinez, J.-M. Nuzillard, V. Peulon-Agasse, et al., (L)- or (D)-valine *tert*-butylamide grafted on β-cyclodextrin derivatives as new mixed binary chiral selectors. Versatile tools for capillary gas chromatographic enantioseparation, J. Chromatogr. A 1216 (2009) 4051–4062.

[89] H. Frank, G.J. Nicholson, E. Bayer, Enantiomer labelling, a method for the quantitative analysis of amino acids, J. Chromatogr. 167 (1978) 187–196.

[90] M. Juza, H. Jakubetz, H. Hettesheimer, Quantitative determination of isoflurane enantiomers in blood samples during and after surgery via headspace gas chromatography-mass spectrometry, J. Chromatogr. B 735 (1999) 93–102.

[91] E. Bayer, Chirale Erkennung von Naturstoffen an optisch aktiven Polysiloxanen (Chiral recognition of natural products on optically active polysiloxanes), Z. Naturforsch. 38b (1983) 1281–1291.

[92] H.P. Benschop, C.A.G. Konings, L.P.A. De Jong, Gas chromatographic separation and identification of the four stereoisomers of 1,2,2-trimethylpropyl methylphosphonofluoridate (Soman). Stereospecificity of in vitro detoxification reactions, J. Amer. Chem. Soc. 103 (1981) 4260–4262.

[93] A. Ghanem, V. Schurig, Lipase-catalyzed irreversible transesterification of 1-(2-furyl)-ethanol using isopropenyl acetate, Chirality 13 (2000) 118–123.

[94] M.T. Reetz, K.M. Kuhling, S. Wilensek, H. Husmann, U.W. Hausig, M. Hermes, A GC-based method for high-throughput screening of enantioselective catalysts, Catal. Today. 67 (2001) 389–396.

[95] A. Mosandl, Enantioselective capillary gas chromatography and stable isotope ratio mass spectrometry in the authenticity control of flavours and essential oils, Food Rev. Int. 11 (1995) 597–664.

[96] C. Bicchi, A. D'Amato, P. Rubiolo, Cyclodextrin derivatives as chiral selectors for direct gas chromatographic separation of enantiomers in the essential oil, aroma and flavour field, J. Chromatogr. A 843 (1999) 99–121.

[97] S. Thurnhofer, G. Hottinger, W. Vetter, Enantioselective determination of anteiso fatty acids in food samples, Anal. Chem. 79 (2007) 4696–4701.

[98] W. Vetter, V. Schurig, Enantioselective determination of chiral organochlorine compounds in biota by gas chromatography on modified cyclodextrins, J. Chromatogr. A 774 (1997) 143–175.

[99] A. Imran, H.Y. Aboul-Enein, Chiral pullutants – distribution, toxicity and analysis by chromatography and capillary electrophoresis. Chapter 6: the analysis of chiral pollutants by gas chromatography, Wiley VCH, Weinheim & New York, 2004.

[100] W. Vetter, K. Bester, Gas chromatographic enantioseparation of chiral pollutants – techniques and results. [chapter 6], in: K. Busch, M. Busch (Eds.), Chiral analysis, Elsevier Science, Amsterdam, 2006.

[101] B. Feibush, C.L. Woolley, V. Mani, Separation of chiral silicon compounds using permethylated α-, β-, and γ-cyclodextrin capillary GC columns, Anal. Chem. 65 (1993) 1130–1133.

[102] A. Kaunzinger, A. Rechner, T. Beck, A. Mosandl, A.C. Sewell, H. Böhles, Chiral compounds as indicators of inherited metabolic disease. Simultaneous stereodifferentiation of lactic-, 2-hydroxyglutaric- and glyceric acid by enantioselective cGC, Enantiomer 1 (1996) 177–182.

[103] U.J. Meierhenrich, M.-J. Nguyen, B. Barbier, A. Brack, W.H.-P. Thiemann, Gas chromatographic separation of saturated aliphatic hydrocarbon enantiomers on permethylated β-cyclodextrin, Chirality 15 (2003) S13–S16.

[104] G. Sicoli, D. Kreidler, H. Czesla, H. Hopf, V. Schurig, Gas chromatographic enantioseparation of unfunctionalized chiral alkanes: a challenge in separation science (overview, state of the art, and perspectives), Chirality 21 (2009) 183–198.

[105] Z. Juvancz, K. Grolimund, V. Schurig, Pharmaceutical applications of a bonded cyclodextrin stationary phase, J. Microcol. Sep. 5 (1993) 459–468.

[106] C. Morrison, F.J. Smith, T. Tomaszewski, K. Stawiarska, M. Biziuk, Chiral gas chromatography as a tool for investigations into illicitly manufactured methylamphetamine, Chirality 23 (2011) 519–522.

[107] F. Goesmann, H. Rosenbauer, R. Roll, C. Szopa, F. Raulin, R. Sternberg, et al., COSAC, the cometary sampling and composition experiment on Philae, Space Sci. Rev. 128 (2007) 257–280.

[108] J.R. Cronin, S. Pizzarello, Enantiomeric excesses in meteoritic amino acids, Science 275 (1997) 951–955.

[109] M. Zampolli, G. Basaglia, F. Dondi, R. Sternberg, C. Szopa, M.C. Pietrogrande, Gas chromatography–mass spectrometry analysis of amino acid enantiomers as methyl chloroformate derivatives: Application to space analysis, J. Chromatogr. A 1150 (2007) 162–172.

[110] H. Brückner, A. Schieber, Determination of free D-amino acids in mammalia by chiral gas chromatography-mass spectrometry, J. High. Resolut. Chromatogr. 23 (2000) 576–582.

[111] A. Glausch, J. Hahn, V. Schurig, Enantioselective determination of chiral 2,2′,3,3′,4,6′-hexachlorobiphenyl (PCB 132) in human milk samples by multidimensional gas chromatography/electron capture detection and by mass spectrometry, Chemosphere 30 (1995) 2079–2085.

[112] R. Schmidt, H.G. Wahl, H. Häberle, H.-J. Dieterich, V. Schurig, Headspace gas chromatography-mass spectrometry analysis of isoflurane enantiomers in blood samples after anesthesia with the racemic mixture, Chirality 11 (1999) 206–211.

[113] A. Berthod, X. Wang, K.H. Gahm, D.W. Armstrong, Quantitative and stereoisomeric determination of light biomarkers in crude oil and coal samples, Geochim. et. Cosmochim. Acta. 62 (1998) 1619–1630.

[114] H. Frank, W. Woiwode, G. Nicholson, E. Bayer, Determination of the rate of acidic catalyzed racemization of protein amino acids, Liebigs. Ann. Chem. (1981) 354–365.

[115] R. Liardon, S. Ledermann, U. Ott, Determination of D-amino acids by deuterium labelling and selected ion monitoring, J. Chromatogr. 203 (1981) 385–395.

[116] M. Kühnle, D. Kreidler, K. Holtin, H. Czesla, P. Schuler, V. Schurig, et al., Online coupling of enantioselective capillary gas chromatography with proton nuclear magnetic resonance spectroscopy, Chirality 22 (2010) 808–812.

[117] G. Schomburg, H. Husmann, E. Hübinger, W.A. König, Multidimensional capillary gas chromatography – enantiomeric separations of selected cuts using a chiral second column, J. High Resolut. Chromatogr. 7 (1984) 404–410.

[118] G.P. Blanch, J. Jauch, Enantiomeric composition of filbertone in hazelnuts in relation to extraction conditions. Multidimensional gas chromatography and gas chromatography/mass spectrometry in the single ion monitoring mode of a natural sample, J. Agric. Food Chem. 46 (1998) 4283–4286.

[119] L. Mondello, M. Catalfamo, A.M. Cotroneo, A.,G. Dugo, G. Dugo, H. McNair, Multidimensional capillary GC–GC for the analysis of real complex samples. Part IV. Enantiomeric distribution of monoterpene hydrocarbons and monoterpene alcohols of lemon oils, J. High Resolut. Chromatogr. 22 (1999) 350–356.

[120] P. Rubiolo, B. Sgorbini, E. Liberto, C. Cordero, C. Bicchi, Essential oils and volatiles: sample preparation and analysis. A review, Flavour Fragr. J. 25 (2010) 282–290.

[121] A. Glausch, G.J. Nicholson, M. Fluck, V. Schurig, Separation of the enantiomers of stable atropisomeric polychlorinated biphenyls (PCBs) by multidimensional gas chromatography on Chirasil-Dex, J. High Resolut. Chromatogr. 17 (1994) 347–349.

[122] M. Kreck, A. Scharrer, S. Bilke, A. Mosandl, Enantioselective analysis of monoterpene compounds in essential oils by stir bar sorptive extraction (SBSE)-enantio-MDGC-MS, Flavour. Fragr. J. 17 (2002) 32–40.

[123] C. Bicchi, E. Liberto, C. Cagliero, C. Cordero, B. Sgorbini, P. Rubiolo, Conventional and narrow bore short capillary columns with cyclodextrin derivatives as chiral selectors to speed-up enantioselective gas chromatography and enantioselective gas chromatography-mass spectrometry, J. Chromatogr. A 1212 (2008) 114–123.

[124] G. Dugo, A. Verzera, A. Cotroneo, I.S. d'Alcontres, L. Mondello, K.D. Bartle, Automated HPLC–HRGC: a powerful method for essential oil analysis. Part II. Determination of the enantiomeric distribution of linalol in sweet orange, bitter orange and mandarin essential oils, Flavour Fragr. J. 9 (1994) 99–104.

[125] R. Shellie, P.J. Marriott, Comprehensive two-dimensional gas chromatography with fast enantioseparation, Anal. Chem. 74 (2002) 5426–5430.

[126] P. Shellie, P. Marriott, C. Cornwell, Application of comprehensive two-dimensional gas chromatography (GC × GC) to the enantioselective analysis of essential oils, J. Sep. Sci. 24 (2001) 823–830.

[127] A. Williams, D. Ryan, A.O. Guasca, P. Marriott, Analysis of strawberry volatiles using comprehensive two-dimensional gas chromatography with headspace solid-phase microextraction, J. Chromatogr. B 817 (2005) 97–107.

[128] A. Mosandl, Progress in the authenticity assessment of wines and spirits, Analusis. 25 (1997) M31–M38.

[129] M. Greule, C. Haensel, U. Bauermann, A. Mosandl, Feed additives: authenticity assessment using multicomponent-/multielement-isotope ratio mass spectrometry, Eur. Food. Res. Technol. 227 (2008) 767–776.

[130] S. Pizzarello, Y. Huang, M. Fuller, The carbon isotopic distribution of Murchison amino acids, Geochim. et. Cosmochim. Acta. 68 (2004) 4963–4969.

[131] V. Schurig, S. Reich, Determination of the rotational barriers of atropisomeric polychlorinated biphenyls (PCBs) by a novel stopped-flow multidimensional gas chromatographic technique, Chirality 10 (1998) 316–320.

[132] S. Reich, O. Trapp, V. Schurig, Enantioselective stopped-flow multidimensional gas chromatography – determination of the inversion barrier of 1-chloro-2,2-dimethylaziridine, J. Chromatogr. A. 892 (2000) 487–498.

[133] J. Krupčik, J. Midlová, P. Májek, P. Šimon, D.W. Armstrong, Methods for studying reaction kinetics in gas chromatography, exemplified by using the 1-chloro-2,2-dimethylaziridine interconversion reaction, J. Chromatogr. A. 1186 (2008) 144–160.

[134] O. Trapp, G. Schoetz, V. Schurig, Review: determination of enantiomerization barriers by dynamic and stopped-flow chromatographic methods, Chirality 13 (2001) 403–414.

[135] R. Aichholz, U. Bölz, P. Fischer, A standard test mixture for assessing enantioselectivity of chiral phase capillary columns – CHIRAL-test I for amide phases, J. High. Resolut. Chromatogr. 13 (1990) 234–238.

[136] B. Koppenhoefer, R. Graf, H. Holzschuh, A. Nothdurft, U. Trettin, P. Piras, et al., CHIRBASE, a molecular database for the separation of enantiomers by chromatography, J. Chromatogr. A 666 (1994) 557–563.

[137] E. Liberto, C. Cagliero, B. Sgorbini, C. Bicchi, D. Sciarrone, B. D'Acampora Zellner, et al., Enantiomer identification in the flavour and fragrance fields by "interactive" combination of linear retention indices from enantioselective gas chromatography and mass spectrometry, J. Chromatogr. A 1195 (2008) 117–126.

[138] C. Reiner, G.J. Nicholson, U. Nagel, V. Schurig, Evaluation of enantioselective gas chromatography for the determination of minute deviations from racemic composition of α-amino acids with emphasis on tyrosine: accuracy and precision of the method, Chirality 19 (2007) 401–414.

[139] B. Koppenhoefer, V. Muschalek, M. Hummel, E. Bayer, Determination of the enhancement of the enantiomeric purity during recrystallization of amino acids, J. Chromatogr. 477 (1989) 139–145.

[140] Y. Xiang, G.W. Sluggett, Development and validation of a GC method for quantitative determination of enantiomeric purity of a proline derivative, J. Pharmaceut. Biomed. Anal. 53 (2010) 878–883.

[141] B.J. Asher, L.A. D'Agostino, J.D. Way, C.S. Wong, J.J. Harynuk, Comparison of peak integration methods for the determination of enantiomeric fraction in environmental samples, Chemosphere 75 (2009) 1042–1048.

[142] V. Schurig, Preparative-scale separation of enantiomers on chiral stationary phases by gas chromatography, in: F. Toda (Ed.), Enantiomer separation: fundamentals and practical methods, Kluwer, Dordrecht, 2004, pp. 267–300.

[143] M. Juza, O. Di Giovanni, G. Biressi, V. Schurig, M. Mazzotti, M. Morbidelli, Continuous enantiomer separation of the volatile inhalation anesthetic enflurane with a gas chromatographic simulated moving bed unit, J. Chromatogr. A. 813 (1998) 333–347.

[144] P. Kurzawski, V. Schurig, A. Bogdanski, R. Wimmer, A. Hierlemann, Direct determination of the enantiomeric purity or enantiomeric composition of methylpropionates using a single capacitive microsensor, Anal. Chem. 81 (2009) 1969–1975.

CHAPTER

22

Analysis of Essential Oils and Fragrances by Gas Chromatography

K. Hüsnü Can Başer, Temel Özek

OUTLINE

22.1. DEFINITIONS: WHAT IS ESSENTIAL OIL? WHAT ARE FRAGRANCES?

An essential oil, also known as volatile oil, or ethereal oil, is a liquid containing volatile aroma compounds obtained mainly from plant materials. Essential oils are biosynthesized by living organisms and are generally obtained from their matrix by water, steam and dry distillation, or expression in the case of peel oils from citrus fruits [1–6]. There are some other processes involving extraction with organic solvents, fluidized gasses, or supercritical fluids. Concretes, absolutes, spice oleoresins, etc. obtained by extraction can be classified as aromatic extracts. They are not technically considered as essential oils [1,2,6].

Gas Chromatography DOI: 10.1016/B978-0-12-385540-4.00022-5

Essential oils occur mainly in aromatic plants. However, a few of them are found in animal sources, such as Asian musk deer, North American beaver, civet, and sperm whale, or are produced by microorganisms [1,2,5,7,9]. The Council of Europe describes "*essential oil*" as a product obtained from "*vegetable raw material*." Due to a ban on animal-based essential oils, and flavor and fragrance materials, the essential oil trade is entirely plant based.

In plant materials, essential oils occur in oil cells, glandular hairs, and secretory cells within the epidermis or cavities [2,10]. However, they may also be bound with carbohydrates in glycosidic form. In such cases, hydrolysis should be applied to release the volatile moieties prior to distillation [2].

Essential oils are sources of substances as starting materials for different chemical syntheses. They can be fractionated to isolate aroma chemicals or used as fractions. These fractions and their isolates are widely used in different industrial applications in cosmetics, perfumery, pharmaceuticals, food, flavor, and fragrance sectors. Sometimes, essential oils are also associated with gums and/or resins. They can be liberated from such matrices by distillation [3].

Fragrance compounds are complex combinations of substances with characteristic and usually pleasant odor. They are added to products in order to impart a distinctive smell. Like all other ingredients, they are, therefore, used in a wide variety of products, such as foods, alcoholic and nonalcoholic beverages, cosmetics and toiletries, cleaning products, tobacco products, and a wide variety of pharmaceutical preparations [1–5,7–9]. There are hundreds of fragrances created every year all over the world. Fragrances are thoroughly evaluated for safety prior to marketing [11]. They not only are composed of essential oils but may also contain natural or synthetic aroma chemicals, or are diluted with carrier essential oils.

Essential oils may comprise volatile compounds that can be classified as terpenoids and nonterpenoid Hydrocarbons and their oxygenated derivatives. However, some of them may also contain nitrogen or sulfur derivatives [12–14]. Essential oil compounds may exist in different forms such as alcohols, aldehydes, esters, ketones, acids, etc. The groups of compounds in essential oils are mainly monoterpenes, sesquiterpenes, and even diterpenes including their oxygenated derivatives [1–4,10]. In addition, phenylpropanoids, fatty acids and their esters, or their decomposition products are also encountered as volatiles in essential oils [1,2,15–17]. Essential oils should not be confused with fixed oils or fatty oils, which are composed of a naturally occurring mixture of lipids, which may not necessarily be volatile. Therefore, essential oils differ entirely both in chemical and in physical properties from fatty oils. Essential oil evaporates completely when dropped on filter paper; however, fixed oil leaves a permanent stain which does not evaporate even when heated [2].

22.2. GC PHASES USED IN THE ANALYSIS OF ESSENTIAL OILS AND AROMA CHEMICALS

Gas chromatography (GC) is the most important and common techniques for the separation of essential oils [18]. Separation of components depends on the polarity and volatility of the analytes. Essentially, GC is based on differential partitioning of solutes between the mobile and stationary phase. It is quite simple, fast, reliable, and applicable to the separation of volatile materials which are stable at a temperature up to 350–400 °C. If the sample or the mixture is nonvolatile, then the sample should be derivatized appropriately to make it volatile *e.g.*, silylation, acylation, and alkylation.

GC columns are categorized into two groups as packed and capillary columns. They are further classified according to their construction material, column length and diameter, and type of stationary phase and appropriate film thickness. Usually, preparative columns are made of glass, stainless steel, or fused silica [19]. One of the most important criteria for good GC separations is to select the most suitable stationary phase and column size (Table 22.1).

22.3. SEPARATION CRITERIA AND TECHNIQUES

22.3.1. Chiral Columns

Enantiomers have identical physical properties such as boiling point, melting point, and spectroscopic features. Most of the essential oils contain enantiomeric compounds and these enantiomers may possess different properties of interest in defining the quality of an essential oil. Enantiomers are preferably separated on capillary columns coated with a chiral stationary phases. There is no universal column for the separation of all enantiomeric compounds. Chiral stationary phases can only be applied for the separation of enantiomer pairs according to their structures [20,21]. The three main types of chiral GC stationary phases are chiral amino acid derivatives, chiral metal coordination compounds, and cyclodextrin derivatives (see Chapter 21). The latter have proven to be the most versatile for essential oil analysis.

22.3.2. Multidimensional Techniques

Single-column GC applications are most commonly used for the analysis of essential oils. However, due to the limitations of one-dimensional chromatography for the separation of complex mixtures, interest in multidimensional (MDGC) and comprehensive two-dimensional gas chromatography (2D-GC) methods is increasing (see Chapter 7) [22–24]. In 2D-GC, the separation employs two capillary columns of different polarity in tandem. Each dimension of separation gives specific information about the target compounds. Since an essential oil contains hundreds of compounds, this may be the only approach for complete characterization of complex oils. Comprehensive and multidimensional GC is also used for the efficient separation of enantiomers. In this case, the first column usually contains an achiral stationary phase and the separated components are directed to a chiral column to separate the enantiomers (see Chapter 21).

22.3.3. Headspace Techniques

The gas space above the sample in a closed container is expressed as '*Headspace*' (HS). The headspace technique is used in GC for the analysis of volatiles in solid, liquid, and gas samples (see Chapter 9).

The headspace trapping technique is used to trap the odor of an aromatic material, such as a living flower, fragrance on skin, or any other matrix. It requires a small amount of sample and is solvent-free. Odor can be sampled either directly or trapped on an adsorbent material. The trapped odorous components can be freed by either solvent extraction or thermal desorption prior to analysis by modern instrumental techniques (see Chapter 10). Headspace trapping techniques can be classified as follows [18,25]:

- static headspace sampling,
- vacuum headspace sampling, and
- dynamic headspace sampling

In the **static headspace sampling technique**, The sample is kept in a closed vial and headspace air above the solid or a liquid sample is sampled by a gas syringe or directed on to the GC column. However, in common applications, the components in HS are first concentrated on an adsorbent trap. Heat may be applied to enhance the release of volatiles from the sample. Although a very rapid method, it does not give a comprehensive profile of the volatiles as some important components may not be detected.

TABLE 22.1 GC Column Selection Chart

Column types		• Packed • Capillary
Column materials		• Glass • Stainless steel • Copper • Aluminum • Special alloy • Silica
Column polarity		• Nonpolar • Intermediate polarity • Polar • Highly polar
Stationary phases*	Nonpolar	• 100% dimethyl polysiloxane • 5% phenyl, 95% dimethyl polysiloxane
	Intermediate polarity	• 6% cyanopropyl phenyl, 94% dimethyl polysiloxane • 35% phenyl, 65% dimethyl polysiloxane • 35% phenyl, 65% dimethyl arylenesiloxane • 14% cyanopropyl phenyl, 86% dimethyl polysiloxane
	Polar	• 50% cyanopropyl phenyl, 50% dimethyl polysiloxane • Polyethylene glycol
	High-polar	• Poly(80% biscyanopropyl/20% cyanopropyl phenyl siloxane) • Poly(90% biscyanopropyl/10% cyanopropyl phenyl siloxane) • Poly(biscyanopropyl siloxane)
Column length	Packed	• 1–2 m (max. 10 m for special applications)
	Capillary	• 5–100 m
Column diameter	Packed	• 2–4 mm
	Capillary	• 0.10–0.53 mm
Column film thickness		• 0.1–5 μm

* *Summarized as most common.*

Vacuum headspace sampling technique involves suction of the headspace air *via* a vacuum pump through condensers cooled with liquid nitrogen to condense odorous principles. This technique is also used by some perfumery companies for commercial-scale production of fragrances.

Dynamic headspace sampling involves sweeping the analyte with a stream of air or gas and adsorption of the volatiles from the gas stream on an adsorbent trap. Hydrophobic traps are preferred such as Tenax, Porapak Q, Chromosorb 101-105, or activated charcoal being the most popular. Volatiles from the trap can be removed either by thermal desorption or by solvent extraction. In fragrance analysis, solvent extraction is the preferred method. Dynamic headspace sampling techniques can be applied in one of the following ways:

- ***Closed-loop Stripping Method:*** The analyte is placed in the middle of a closed circuit system in which clean air is continuously pumped through the analyte and odorous components in the headspace air are trapped on an adsorbent material.
- ***Direct Sampling Method:*** The analyte, which may be a living flower, is placed in a glass container and the headspace air is sucked *via* a suction pump through an adsorbent tube in which the odorous components are trapped [25].

22.3.4. Solid-Phase Microextraction (SPME)

Solid-phase microextraction (SPME) is a microsampling technique which has found wide application in flavor and fragrance research. It is a solvent-free method which is used to trap flavors and fragrances either from aqueous samples (immersion SPME) or from the vapor space above a liquid or a solid sample (headspace SPME) [25]. Since it is an equilibrium technique, it does not extract the analytes completely, hence does not disturb the concentration of a liquid. Agitation of the liquid facilitates rapid extraction. For the headspace samples, faster mass transport rates are attained and volatiles are extracted faster than semivolatiles.

Analytes adsorbed on the fiber are thermally desorbed directly in the injection port of a gas chromatograph for separation. The SPME assembly consists of a specially designed injector, also called SPME holder, which enables the coated fiber to move in and out of the needle. During injection into a vial for sampling or injection port for analysis, the fiber is concealed in the needle and exposed during sampling and thermal desorption. Several fiber coatings with different sorption properties are commercially available, such as polydimethylsiloxane (PDMS), polyacrylate (PA), polydimethylsiloxane/divinylbenzene (PDMS/DVB), carbowax (CW), carbowax/divinylbenzene (CW/DVB), and carboxene/polydimethylsiloxane (C/PDMS). SPME has proven itself to be a highly efficient and simple sample preparation technique which may be expected to replace conventional headspace techniques.

22.3.5. Preparative Gas Chromatography

This preparative technique is used when the analytical separation conditions have been established and large amounts of some components in high purity are required for further evaluation (see Chapter 16). Preparative gas chromatography (Prep-GC) is widely used for the isolation of terpenes and other volatiles from essential oils [19].

22.3.6. Gas Chromatography-Mass Spectrometry (GC-MS)

In this commonly used technique for the analysis of essential oils and fragrances, a mass spectrometer coupled to a GC is used as a detector. Mass detection, unlike other common detection techniques, provides structural information for the compound detected including its molecular mass as well as quantitative information. Detection of the compounds by peak matching with compounds in GC-MS Libraries is possible. Full characterization is completed by matching their retention indices with those of authentic compounds.

22.4. RETENTION INDEX

In order to express the retention value of a compound reliably retention indices are used. There are two accepted retention index calculation methods depending on whether isothermal or temperature-programmed conditions are used [26,27]. A homologous series of n-alkanes is generally used as the standards for the calculation of retention index values. These are preferred for low-polarity stationary phases but owing to interfacial adsorption on polar phases ethyl esters are sometimes used as an alternative. The retention index for the standards is expressed as 100 times the number of carbon atoms. Thus, octane has a value of 800 and dodecane 1200 on all stationary phases. However, since the retention index values depend not only on the column temperature but also on the stationary phase used, retention index values change according to the polarity of stationary phase used. If the stationary phases are of the same polarity, the retention indices of the compounds remain the same within an acceptable range of ± 5 units for methyl silicone stationary phases, and ± 10 units for polyethylene glycol phases [27].

22.5. QUALITATIVE AND QUANTITATIVE ASPECTS

The qualitative analysis of essential oils by GC is based on the comparison of the peaks in the chromatogram for the essential oil with those of authentic standards separated with the same chromatographic conditions. GC-MS is used for correct identification of each peak (ideally single compound) separated from the essential oil. Spectroscopic information, in some cases, is an unavoidable alternative together with the use of retention indices and other relative retention parameters. The use of qualitative information alone is not sufficient to correctly characterize an essential oil, and quantitative data are also important. For this reason, selection of the detector is of extreme importance. Apart from the flame ionization detector each detector has associated with it a compound-specific response factor [4] (see Chapter 12). Since the flame ionization detector (FID) has low selectivity an electron-capture detector (ECD) is used to selectively detect compounds with a high electron affinity. Similarly, mass detector and FID do not have the same response, and FID is considered the more reliable and accurate. Therefore, GC-MS is preferred for qualitative identification except for the SIM mode in which only selected ions are quantified. In this mode, GC-MS gives the most accurate quantitative results for individual compounds. In recent years, hyphenated chromatographic and spectroscopic techniques have been used more extensively for the analysis of essential oils. In order to evaluate essential oil composition reliably both qualitative and quantitative data should be obtained from GC and/or GC-MS regarding detector response, column stationary phase (*e.g. some compounds such as thymoquinone decompose on a polar column*), and other retention factors [19].

22.6. GC-MS LIBRARIES

In GC, identification of a compound using retention indices is generally accepted when two successful matches are obtained from known and target compounds on at least two columns of different polarities. On the other hand, identification of the target compounds made on a single column can only be accepted if it is obtained in combination with spectroscopic detection systems. In order to overcome this difficulty, it is necessary to have spectroscopic information of the compound simultaneously. When a suitable reference database (or library) is available, identification of target compounds will be much easier. Mass spectral data from GC-MS are generally considered as the key for component identification.

Three types of libraries are available for essential oil composition identification:

- commercial libraries,
- specific libraries, and
- in-house libraries.

22.6.1. Commercial Libraries

These libraries contain nonspecific collections of spectra mainly taken from the literature. One has to be careful when evaluating data from commercial libraries since they contain mass spectral data taken under different conditions using different instruments. The main drawback of such libraries is the lack of retention data to compare with the retention indices. NIST has recently started providing such a service [28]. Among the commercially available GC-MS libraries for essential oil components and aroma chemicals, Wiley, NIST, NBS, TNO, EPA, NIH, Adams, MassFinder, FFNSC, and Tkachev can be mentioned.

22.6.2. Specific Libraries

Some libraries are dedicated to specific applications or compound groups. Adams, MassFinder, FFNSC, and Joulain and Koenig collections are available for these applications. While the Adams and MassFinder are general essential oil libraries, FFNSC and Joulain and Koenig collections are prepared for flavors and fragrances of natural and synthetic compounds and sesquiterpene hydrocarbons respectively. These libraries contain specific information for the compounds of interest such as retention time, retention indices, and physicochemical information.

22.6.3. In-house Libraries

Researchers dealing mainly with a certain group of products may develop their own libraries in order to facilitate and expedite compound characterization. Such libraries are more reliable since they are created using certified compounds under identical conditions and contain retention data. The BASER library of essential oil constituents is an example of an in-house library. In this library, retention time, retention index, source of mass spectrum, etc. are given for the polar HP-InnoWax column.

22.7. CONCLUSIONS

Gas chromatography has proven its value as a useful, fast, and reliable separation technique for essential oils over the last few decades. Hyphenated techniques have broadened its usefulness. Especially, the gas chromatography-mass spectrometry applications together with GC-MS libraries and retention index databases have revolutionized research into essential oils, flavors, and fragrances. Enantiomeric separations of optically active volatile chemicals through multidimensional techniques are possible and such useful information is highly regarded by the industry. Headspace analysis techniques are well advanced as a nondestructive analytical technique for capturing odor information. Headspace trapping can sample the odor of a living flower and continuous monitoring of diurnal changes of the odor is possible. GC in combination with an olfactometer can simultaneously evaluate the odor of sample components as they leave the column. A perfumer can also sniff the odor of the eluted compounds by attaching a sniffing port to the GC. Preparative GC is useful for isolating new compounds for further spectral analyses and for accumulating enantiomers. GC-MS libraries facilitate and expedite the analyses of essential oils and fragrances. Instruments are only tools in the hands of a well-trained analyst. Therefore, specialization in a certain group of compounds or products is necessary for correct and accurate evaluation of instrumental data.

References

[1] H. Surburg, J. Panten, Common fragrance and flavor materials: preparation, properties and uses, fifth ed., Wiley-VCH Verlag GmbH & Co. KGaA, Weinheim, 2006, p. 318.

[2] K.H.C. Baser, F. Demirci, Chemistry of essential oils, in: R.G. Berger (Ed.), Flavours and fragrances: chemistry, bioprocessing and sustainability, Springer-Verlag, Heidelberg, 2007, pp. 43–86.

[3] G.A. Reineccius, Flavour-isolation techniques, in: R.G. Berger (Ed.), Flavours and fragrances: chemistry, bioprocessing and sustainability, Springer-Verlag, Heidelberg, 2007, pp. 409–426.

[4] K.H.C. Baser, G. Buchbauer (Eds.), Handbook of essential oils, CRC Press Taylor & Francis Group, Boca Raton, FL, 2010, pp. 3–38.

[5] K.H.C. Baser, in: K.T. de-Silva (Ed.), A manual on essential oil industry, UNIDO, Vienna, 1995, p. 155.

[6] K.H.C. Baser, F. Demirci, Essential oils. Kirk-othmer encyclopedia of chemical technology, John Wiley & Sons, Inc, 2011, pp. 1–37.

[7] M. Guentert, The flavour and fragrance industry-past, present, and future, in: R.G. Berger (Ed.), Flavours and fragrances: chemistry, bioprocessing and sustainability, Springer-Verlag, Heidelberg, 2007, pp. 1–14.

[8] D.A. Müller, Flavours: the legal framework, in: R.G. Berger (Ed.), Flavours and fragrances: chemistry, bioprocessing and sustainability, Springer-Verlag, Heidelberg, 2007, pp. 15–24.

[9] Anonymous, Perfumes from animal sources. [updated May 15; cited 2011 May 15]; Available from: Wikimedia Foundation, Inc., 2011 http://en.wikipedia.org/wiki/Perfume#Animal_sources, 2011.

[10] E.J. Bowles, The chemistry of aromatherapeutic oils, Allen & Unwin, Crows Nest, 2003, p. 236.

[11] J.C.R. Demyttenaere, Recent EU legislation on flavors and fragrances and its impact on essential oils, in: K.H.C. Baser, G. Buchbauer (Eds.), Handbook of essential oils, CRC Press Taylor & Francis Group, Boca Raton, FL, 2010, pp. 917–948.

[12] R.A. Clery, C.J. Hammond, A.C. Wright, Nitrogen-containing compounds in black pepper oil (Piper nigrum L.), J. Essential Oil Res. 18 (1) (2006) 1–3.

[13] M. Iranshahi, G. Amin, M.S. Sourmaghi, A. Shafiee, A. Hadjiakhoondi, Sulphur-containing compounds in the essential oil of the root of Ferula persica Willd. var. persica, Flavour Frag. J. 21 (2) (2006) 260–261.

[14] D. Lopes, R.L.O. Godoy, S.L. Goncalves, M. Koketsu, A.M. Oliveira, Sulphur constituents of the essential oil of nira (Allium tuberosum Rottl.) cultivated in Brazil, Flavour Frag. J. 12 (4) (1997) 237–239.

[15] S. Bourgou, I. Bettaieb, M. Saidani, B. Marzouk, Fatty acids, essential oil, and phenolics modifications of black cumin fruit under NaCl stress conditions, J. Agric. Food Chem. 58 (23) (2010) 12399–12406.

[16] C. Messaoud, M. Boussaid, Myrtus communis berry color morphs: a comparative analysis of essential oils, fatty acids, phenolic compounds, and antioxidant activities, Chem. Biodiv. 8 (2) (2011) 300–310.

[17] N. Nasri, N. Tlili, S. Triki, W. Elfalleh, I. Cheraif, A. Khaldi, Volatile constituents of Pinus pinea L. needles, J. Essential Oil Res. 23 (2) (2011) 15–19.

[18] T. Cserhati, Chromatography of aroma compounds and fragrances, Springer-Verlag, Heidelberg, 2010.

[19] T. Özek, F. Demirci, Isolation of natural products by preparative gas chromatography in: Sarker SD, editor. Natural products isolation. 3rd edin-print.

[20] E. Francotte, Chiral stationary phases for preparative enantioselective chromatography, in: G.B. Cox (Ed.), Preparative enantioselective chromatography, Blackwell Publishing Ltd, Oxford, 2005, pp. 48–77.

[21] T.E. Beesley, R.P.W. Scott, in: R.P.W. Scott, C. Simpson, E.D. Katz (Eds.), Chiral chromatography, John Wiley & Sons Ltd, Chichester, 1998, p. 506.

[22] L. Modello, A.C. Lewis, K.D. Bartle (Eds.), Multidimensional chromatography, first ed., John Wiley & Sons, Inc, West Sussex, 2002, pp. 217–250.

[23] P.Q. Tranchida, D. Sciarrone, L. Mondello, Multidimensional gas chromatography, in: E. Grushka, N. Grinberg (Eds.), Advances in chromatography, CRC Press Taylor & Francis Group, Boca Raton, 2010, pp. 289–328.

[24] M.D.R.G.d. Silva, Z. Cardeal, P.J. Marriott, Comprehensive two-dimensional gas chromatography: application to aroma and essential oil analysis, in: H. Tamura, S.E. Ebeler, K. Kubota, G.R. Takeoka (Eds.), Food flavor, American Chemical Society, Washington, DC., 2008, pp. 3–24.

[25] N.C.D. Costa, S. Eri, Identification of aroma chemicals, in: D.J. Rowe (Ed.), Chemistry and technology of flavors and fragrances, Blackwell Publishing Ltd., Boca Raton, FL, 2005, pp. 12–34.

[26] T. Shibamoto, Retention indices in essential oil analysis, in: S. Sandra, C. Bicchi (Eds.), Capillary gas chromatography in essential oil analysis, first ed., Alfred Huethig Verlag, Heidelberg, 1987.

[27] V.I. Babushok, I.G. Zenkevich, Retention indicies for most frequently reported essential oil compounds in gas chromatography, Chromatographia 69 (2009) 257–269.

[28] V.I. Babushok, P.J. Linstrom, J.J. Reed, I.G. Zenkevich, R.L. Brown, W.G. Mallard, et al., Development of a database of gas chromatographic retention properties of organic compounds, J. Chromatogr. A 1157 (2007) 414–421.

CHAPTER

23

Analysis of Lipids by Gas Chromatography

Cristina Cruz-Hernandez, Frédéric Destaillats

OUTLINE

23.1. INTRODUCTION

Gas chromatography (GC) is the standard and certainly the most suitable tool to analyze fatty acid profile of simple and complex samples. The flame ionization detector (FID) is often used as detector since it allows accurate quantification of fatty acid derivatives such as fatty acid methyl esters (FAMEs). The evolution from packed columns to highly polar open-tubular capillary columns facilitated the progressive improvement in the resolution of complex mixtures of positional and geometrical isomers. Nowadays, fatty acid analysis on polar capillary columns is routinely used in the food sector to assess the quality of fats, oils, raw materials, and the fatty acid composition of food products. This type of analysis is widely used to assess the nutritional or health status in human subjects enrolled in nutritional interventions or in prospective epidemiological studies. Current developments in this field are

Gas Chromatography DOI: 10.1016/B978-0-12-385540-4.00023-7

focusing on the reduction of the run time as well as the automation of sample preparation and data management.

GC-FID is used to analyze free fatty acids (FFAs) and partial acylglycerols such as monoacylglycerols (MAGs) or diacylglycerol (DAG), and sterols after derivatization. This type of analysis is very useful to characterize the composition of fats and oils or emulsifiers and can be extended to the profiling of triacylglycerols (TAGs), waxes, or sterol esters using heat-stable stationary phases. Beyond the analysis of fatty acid esters with FID detection, GC is used in tandem with mass spectrometry (MS) for structural analysis of fatty acids. Different techniques have been developed in order to identify and locate double bonds, rings, or other groups. Quantitative or structural analysis of more complex lipids such as glycerophospholipids or TAG is not achievable by GC and requires the use of liquid chromatography-based techniques. These topics have been recently addressed by Christie and Han in their book entitled "Lipid Analysis" and therefore will not be addressed in the present Chapter [1]. Also, applications of emerging multidimensional chromatographic techniques for lipid analysis will not be covered since this topic has been extensively recently reviewed by Tranchida and co-workers [2].

23.2. FATTY ACID ANALYSIS BY GC AS METHYL ESTER DERIVATIVES

23.2.1. Preparation of Fatty Acid Methyl Esters

For many lipids, GC is the primary tool used in the determination of their fatty acid profile as their FAME derivatives. During methylation, *O*-acyl- and *N*-acyl lipids are transesterified in the presence of a catalyst and a short-chain alcohol (i.e. methanol, ethanol, 2-propanol, or butanol) that replaces the glycerol or sphingosine moiety. Derivative formation is extensively used in lipid analysis; many other possibilities, including the preparation of FAME after trans-esterification or trimethylsilyl ethers (TMS) of hydroxyl groups, will be described later in this chapter. FAME can be prepared from isolated lipids, or directly by combining extraction and transesterification in a one-step procedure. The most commonly used catalysts are described below.

Hydrochloric acid (HCl), an acid catalyst, is required when substantial amounts of free fatty acids (FFAs), esters, alk-1-enyl ethers, amides, and glycosides (except ethers) are present in the sample to be analyzed. A solution of HCl/methanol (1−2 M) converts the plasmalogen moiety (alk-1-enyl ethers) in lipids to dimethylacetals (DMAs) and methylations are completed within minutes except for sphingolipids [3]. An alternative method was proposed to be effective with FAME yields > 96% when toluene, methanol, and 8% of HCl solution were added sequentially to the lipid samples [4]. Sulfuric acid (H_2SO_4) has been used as a catalyst for the preparation of isopropyl, butyl, and other esters from dairy fats [5−7]. The longer-chain esters have the advantage of providing FID responses that do not require correction factors for quantification of the FAMEs by GC and less inferences from solvent peaks, although short-chain fatty acids might be lost during aqueous washes and the resolution of closely eluting isomers also appear to be compromised [8−10]. The acidic catalyst boron trifluoride (BF_3) in methanol is suitable for the methylation of all lipid classes, including amides, FFAs, and plasmalogens. However, it is unstable and will produce artifacts if not fresh. In addition, acid-catalyzed methylation is generally not recommended for milk fat analysis since it decreases the content of all the *cis/trans*-conjugated linoleic acid (*c/t*-CLA) isomers, producing *tt*-CLA isomers and methoxy artifacts [11].

For oil samples containing primarily TAG, alkaline transesterification is recommended for direct formation of FAME and is usually performed using hexane and 2 M methanolic potassium or sodium hydroxide [12,13]. For samples containing PL, a method was recommended which allows the direct preparation of FAME and subsequent quantification of long-chain polyunsaturated fatty acids (LC-PUFAs, 4), but does not apply to crude oil with acid values higher than 1.5%. Methylation using sodium methoxide ($NaOCH_3$) is recommended for the analyses of CLA containing lipids. FAMEs are produced with the base (alkoxide), form an anionic intermediate, which is transformed in the presence of large excess of the alcohol (i.e. methanol, ethanol, 2-propanol, and butanol) into a new ester [3,14]. The most common reagent is 1−2 M sodium or potassium hydroxide in anhydrous methanol, which converts all esters to FAME within 10−15 min and is recommended to avoid isomerizations [9]. An aqueous-free system is preferred for the methylation of milk fat lipids [8,13].

Preparation of FAME can also be performed by direct transmethylation of plant materials, biological fluids (including blood and milk), and cultured cells [15,16]. Comparisons between direct methylation and extraction followed by derivatization methods have been done in egg and plasma where fat extraction was preferred over direct derivatization [17,18]. For GC analysis of total blood different methods were compared and the best conditions selected were by using HCl/MeOH 1.2% (w/v) at 45 °C; this was also recommended for other matrices such as blood lipids, fish, and vegetable oils, although a previous solid-phase extraction (SPE) or thin-layer chromatography (TLC) fractionation steps were necessary for lipid classes [4,19]. Additional sample preparation step, based on SPE, was also included in a method for lipids from plant sources to obtain clean samples. This method used 5% HCl/methanol, 2 h at 70 °C, and was preferred over several different methods of derivatization evaluated [20]. Different methods have been reported for fatty acid quantification in microalgae and for lipid-producing bacteria or microheterotrophs [21,22]. An efficient method was reported for oils, waxes, feedstuffs, or fresh, frozen, or lyophilized tissue samples, although no method is mentioned to inhibit the action of degradative enzymes when a mechanical grinder is used at room temperature. Methylation was performed in the presence of up to 33% water and KOH in MeOH and H_2SO_4 as catalyst. Artifact formation or the isomerization of sensitive polyunsaturated fatty acids such as conjugated linoleic acid (CLA) isomers was not mentioned by the authors [23].

23.2.1.1. Analysis on Conventional Columns

Flexible fused silica columns coated with a highly polar cyanoalkyl polysiloxane stationary phase are recommended for the analyses of fats that contain complex mixtures of geometric and positional isomers of monounsaturated fatty acids and CLA isomers as well as a range of FAs from butyric acid (4:0) to LC-PUFAs (e.g. milk fat). These columns are generally available in 100-m length from different suppliers and give improved separations of samples containing fatty acids with different chain lengths [14,24−32]. The 100% cyanopropyl polysiloxane stationary phases are available as CP-Sil 88 (Varian Inc.), SP-2560 (Supelco Inc.), and BPX-70 columns (SGE). The CP-Sil 88 and SP-2560 columns provide similar elution orders of FAME [11,25,27,28,30,31,33−36]. The BPX-70 column [37−40] produces a similar elution pattern for some fatty acids (e.g. CLA isomers) to those obtained using the CP-Sil 88 and the SP-2560 columns [11,14,25,27,29,35,41−43], although for other FAMEs the order is different [1,44]. Some differences have been found also in the order of elution as well as interferences of some fatty acids (21:0 and some 20:2 isomers), when the CP-Sil 88 and SP-2560 columns are used [34,41,44]. Differences have also been found when different temperature program rates are used as well as batch-to-batch variation for a given supplier [3]. Some variability due to differences in the construction of the columns can be found, for example, some suppliers produce continuous 100-m capillary columns (i.e. SP-2560; RTX-2560; HP-88), while others join two 50-m capillary columns (i.e. CP-Sil 88), or two 100-m capillary columns (200 m Select FAME, Agilent). Retention times of certain fatty acids can change depending on the column, the temperature program used, and the age of the column as it is often found between the 21:0 eluted with 9*c*11*c*- [34], with 10*t*12*c*- [44], or between 11*c*13*t*- and 10*t*12*c*-CLA [40]. For routine analysis standards are recommended as well as GC-MS for a definitive identification. The 100-m ionic liquid capillary column SLB-1L111 was recently evaluated and compared with the SP-2560 and CP-Sil 88 columns [45]. SLB-1L111 showed better separations for some *cis* and *trans* CLA isomers when operated at 168 °C. Improvements were also observed for the 14:1, 16:1 18:1, 20:1, and 18:3 isomers, as well as branched chain fatty acids, although the saturated fatty acids eluted between the *cis*- and *trans*-monounsaturated FAMEs. An example using this column is shown in Figure 23.1 [45]. The TC-70-m, 60-m column (70% cyanopropylpolysilphenylene−siloxane liquid phase) has been compared with other columns (e.g. SP-2560, 100 m, 100% cyanopropyl polysiloxane) for the separation of the *trans* isomers. In this example, the *trans* fatty acid content was reported to be quantitatively the same with both columns [46], although other reports suggested that these columns appear to be less efficient [44,47]. Alternatively, although not as reliable, one can compare the results of different GC columns with different polarities. For example, the 60-m Supelcowax 10 capillary column was suggested as a complementary GC column mainly because of its intermediate polarity with a better resolution of the short-chain fatty acids and their monounsaturated fatty acids analogs, as well as a complete separation of the 18:3 n-3 and 20:1 FAME isomers. It has been shown to provide confirmation of the identity of individual FAME by comparison with a 100-m CP-Sil 88 column [29]. Columns with intermediate polarity (200-m Select FAME) have been used lately, although the increased length of this capillary column does not improve separations of the 18:1 isomer region [14,48]. A thorough systematic evaluation of the separation of these columns is needed to establish the elution pattern of most of the common FAMEs including the positional and geometrical 18:1, 18:2, and 18:3 isomers, and the identity of possible interfering FAMEs in the CLA region. Partial separations have been obtained on 120-m BPX-70 columns [44] as well as for the *trans*-18:1 isomers [49,50].

23.2.1.2. High-Resolution Analysis of Positional Isomers

Despite the improvement in the resolution using the 100-m highly polar columns, the separation of some fatty acids such as the geometric 18:1 isomers remains a challenge for the overlapping peaks. Some FAs are not separated or resolved, even when a good resolution of many FAs is accomplished by using different columns and chromatographic conditions; therefore, having an optimum GC program to determine all the *cis* and *trans* isomers in lipid matrices has its limitations [27,44]. For example, the *trans*-18:1 isomers from 6 t- to 11*t*-18:1 are not well resolved by GC using a typical temperature program, and the resolution of the remaining *trans*-18:1 isomers 12*t*- to 16*t*-18:1 may overlap extensively depending on the relative concentration of the major 9*c*-18:1 isomer, sample load applied onto the GC, and temperature program used. Underestimation of these peaks is sometimes reported due to wrong or incomplete analysis [14,44]. Isothermal GC operations at 172, 160, and 150 °C enable the resolution of almost all the *trans*-18:1 isomers by comparing the separations

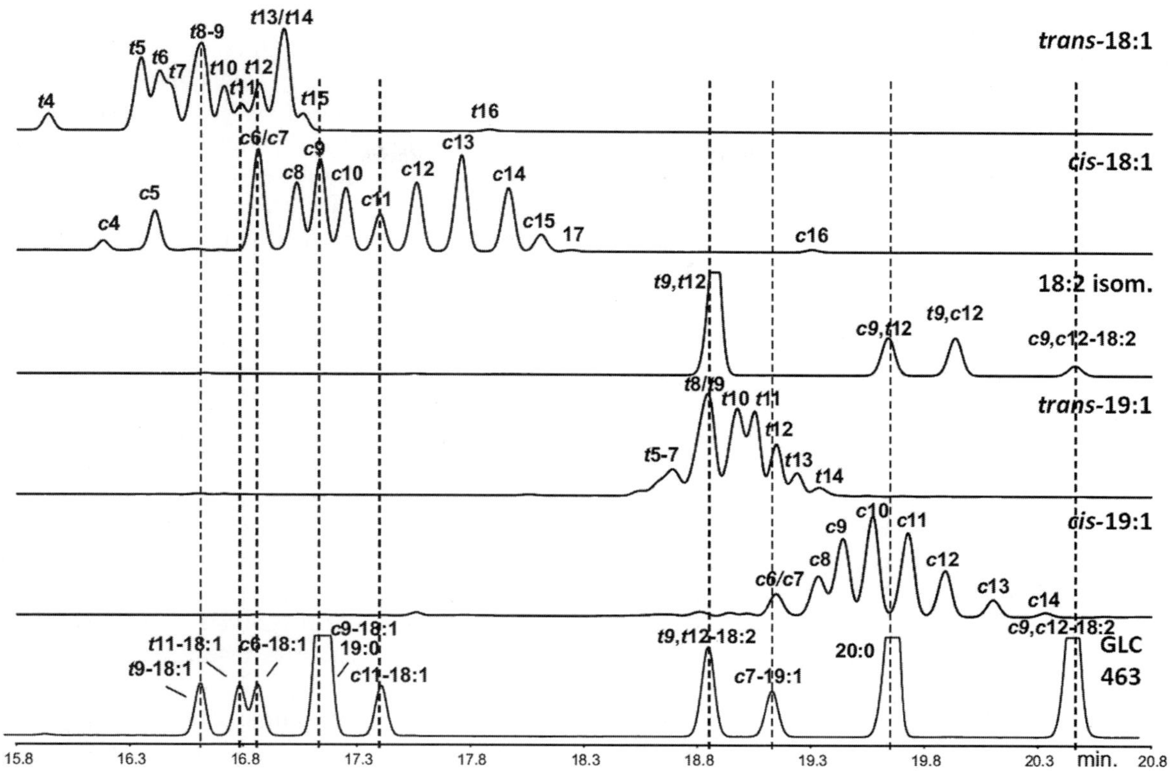

FIGURE 23.1 From P. Delmonte [45]. Partial GC chromatogram of the 18:1, 19:1 and 18:2 region. From top: *trans* and *cis* fractions of 18:1 obtained from a mixture of *cis* 6-, *cis* 9-, and *cis* 13–18:1 FAMEs after two successive brominations and debrominations, linoleic acid FAME isomerized by PTSA, *trans* and *cis* fractions of 19:1 obtained by fractionating *cis* 10-19:1 FAME after six reactions, and reference FAME mixture GLC 463.

of total milk fat FAME without previous separations [3,33]. This requires 3 separate GC analyses but all the *trans*-18:1 isomers could be resolved except for 13*t*/14*t*-18:1 and 6*t*-8*t*-18:1. The identification of individual *trans*-18:1 isomers is possible when the isomers are present at similar concentrations, but this becomes impossible when adjacent isomers are present at greatly different concentrations. The extent of the overlap of the 18:1 geometrical isomers is evident after a prior fractionation using silver-ion chromatography followed by GC separations on long and polar columns [14,51].

The method based on silver-ion thin-layer chromatography (Ag^+-TLC) in tandem with GC is the best way to obtain a complete analysis of the 18:1 isomers. Silver-ion-impregnated TLC plates (Ag^+-TLC) can be easily prepared and have the advantage of fractionation of fatty acid esters according to the number and geometry of the double bonds in FAME moieties [52–55]. Hexane and diethyl ether are used as the mobile phase for the separation of saturated from monounsaturated *trans*- and *cis*-FAMEs and PUFAs. Although less frequently used, separations can also be performed using Ag^+-HPLC columns [56,57]. Eleven *trans*-18:1 isomers can be separated by such techniques and, after the Ag^+-TLC, a longer GC temperature program is recommended to complete the separation and resolution of *trans*-18:1 isomers [27,28,58]. Typical separation of the FAMEs in the 18:0 to 18:2 n-6 region contains overlapping peaks associated with the *cis*-18:1, *trans*-18:1, and *c*/*t*-18:2 isomers as shown in Figure 23.2. Of the *trans*-18:1 isomers, 4*t*-, 5*t*-, 6-8*t*-, 9*t*-, 10*t*-, 11*t*-, and 12*t*-18:1 can generally be resolved, while the 13*t*- and 14*t*-18:1 isomer coelutes with 6*c*-8*c*-18:1 are generally co-eluted with the major 9*c*-18:1 isomer and is generally not resolved. On the other hand, 15*t*-18:1 coelutes with 9c-/10*c*-18:1, and 16*t*-18:1 with 14*c*-18:1 [14]. By comparing the results of Ag^+-TLC-GC analysis, it is possible to analyze the extent of the overlap of the *cis* and *trans*-18:1 isomers in the temperature-programmed analysis of total milk fat FAME [14]. The resolution of some *trans*-18:1 isomers is only partial (6*t*-8*t*- to 11*t*-18:1), and there is a lack of separation of the 6*t*-8*t*-18:1 and the 13*t*-/14*t*-18:1 isomers. Without a prior Ag^+-TLC separation the *trans*-18:1 isomers can be often misidentified. For example, the isomer 6*t*-18:1 was independently reported in a number of publications [59–62] even though its separation from 7*t*- and 8*t*-18:1 by GC is not possible.

After isolation by Ag^+-TLC, all the *trans*-18:1 isomers can be resolved except for the 6*t*-8*t*-18:1. Precht and Molketin [27] suggested that the lack of separation of the 6 t-8*t*-18:1 was due to the small content of the 6*t* and 7*t* isomers. Using the combination of Ag^+-TLC and GC difficult pair of *trans*-18:1 isomers are separated, such as 13*t*-/14*t*-18:1, and 10*t*- and 11*t*-18:1 in samples high in either one of these isomers [14]. The analysis of the *trans*- isomers of *trans*-16:1, *trans*-20:1, and *trans*-22:1 requires the application of higher sample loads onto the GC, as demonstrated for *trans*-16:1 and the *trans*-20:1 isomers as previously described [44]. Using the isothermal low-temperature program starting at 120 °C better resolutions of all the *cis*-18:1 isomers are obtained, but even then 6*c*-8*c*-18:1 remains unresolved and most often 10*c*-18:1 remains as an unresolved peak on the slope of 9*c*-18:1 because of the large difference in their relative amounts. There is evidence that several of the minor *tt*- and *c*/*t*-18:2 isomers in milk fat elute starting with 13*c*-18:1 [11,31].

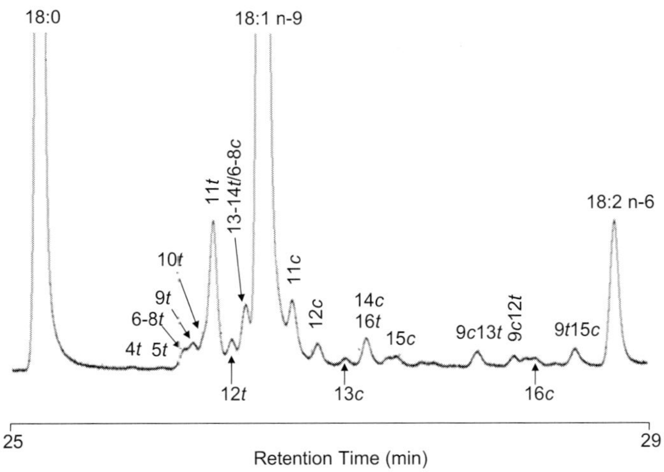

FIGURE 23.2 Gas–liquid chromatograms of butter by GLC on a 100 m capillary column: Enlarged view of the chromatographic region from stearic (18:0) to linoleic (18:2 n-6) acids.

23.2.1.3. *Fast and Ultrafast Analysis*

Faster GC analysis is possible by using the new generation of column and apparatus. This methodology is very efficient for FAME analysis. Reduction of analytical expenses as well as increased laboratory productivity and sample throughput are some of the benefits obtained by GC with the now available instrumentation. In general, fast GC has been defined based on the peak widths obtained [63–65] as well as the analysis of a sample in a shorter time than conventional GC methods [66,67]. Ultrafast-GC belongs to the analytical methods with separation times in the subsecond range and with average peak widths between 5 and 30 ms [63].

Fast GC is run without compromising resolution by using a shorter column, a higher carrier gas velocity, faster program temperatures, flow pressure, or different column film thicknesses. For optimization of the chromatographic system, columns of 100-μm internal diameter are available and there is a wide choice of stationary phases. Wide-bore capillary columns have been of particular interest because it can be used as a direct replacement for a packed column without changing operating parameters or sample preparation. New chromatographs are commercially available and equipped with electronic pressure/flow programming of the carrier gas that can be used to reduce the separation time. For wide-boiling-range samples fast temperature programing is used to minimize the separation time. New instruments employ resistive heating at rates up to 1200 °C/min, and employing cooling from 300 °C to 50 °C in less than 30 s. Different detection systems are compatible with fast GC such as FID, MS, and time-of-flight MS, the latter of which is specifically useful for fast GC-MS. Here we only consider the widely used FID which is characterized by its high sensitivity, fast response, wide dynamic range, and simplicity of construction.

Evaluation of FA composition by fast GC has been applied to different samples such as essential oils [68–77], human plasma, and serum [67,78,79]. For the analysis of plasma lipids, Massod and Salem [80,81] developed an improved method for the preparation and analysis of a large number of samples. The method included the robotic transmethlyation and analysis of the sample, with the main focus on n-3 PUFA, and achieved elution times within 6 minutes. Plasma and plasma phospholipids isolated by SPE from total lipid extracts have been analyzed with short separation times and the identification of 37 fatty acids in 3.2 minutes and 25 FA in 3.8 minutes, respectively [82]. Plasma and red blood cells have been analyzed by fast GC [83] as well as five main polar lipid fractions (phosphatidylethanolamide, phosphatidylcholine, phosphatidylserine, phosphatidylinositol, and sphingomyelin), from brain lipids from rats [84] in less than 5 minutes.

Analysis of menhaden oil, butter, lard, tallow, corn, peanut, olive, colza, sunflower, and soya samples by fast GC have been achieved in about 2.9 min by Mondello et al. [85], significantly faster than previous GC work from the same group having separation times of about 70 and 15 min [76,85]. The use of shorter columns has resulted in shorter times of analysis when cod liver was analyzed with a 0.1 mm i.d. polar column segment of 2 m (1.45 min) or with a 10 m × 0.1 mm i.d. capillary column [73,86]. All methods and examples mentioned in this section can be taken into account to select the specific conditions for fast GC but the final parameters will always depend on the application.

23.3. ANALYSIS OF FREE FATTY ACIDS

Free fatty acids (FFA) occur in different types of samples and various GC-based methodologies have been developed for their analysis. In food products such as cheese, short-chain FFA also named aroma active FFA (4:0 to 10:0) can be analyzed by GC using solid-phase microextraction (SPME) and FID [87] or MS detection [88]. These methods can be used to analyze the evolution of the cheese aroma containing mainly short-chain FFA, but are not suitable for the analysis of the most common medium or long-chain FFA due to the limitation of the SPME technique. For longer FFA such as 12:0 to very-long-chain fatty acids (e.g. 24:0), FFA should be derivatized before analysis. Various types of derivatives such as trimethylsilyl esters TMS, methyl or ethyl esters can be used. Most methods require purification of the FFA from lipid extracts [89]. Methylation of FFA can be achieved using boron trifluoride, methanolic solution of hydrochloric or sulfuric acids, or diazomethane. Methylation using diazomethane produces very clean samples but, due to safety reasons, trimethylsilyldiazomethane is the preferred reagent [90]. Juarez and coworkers developed a method allowing the preparation of FAME without prior purification of the FFA from the lipid extract [91]. This method is based on the preparation of tetramethylammonium soaps (TMA-soaps) which are converted into FAME in the injection port by pyrolysis [91]. Silylation can also be used but TMS esters of FFA do not separate very well compared to FAME and should be analyzed using apolar columns (details regarding analysis of FAME by GC can be found in the previous section of this chapter).

23.4. ANALYSIS OF ACYLGLYCEROLS

Acylglycerols refer to the fatty acid esters of glycerol and comprise TAG, DAG, and MAG. Separation of TAG by GC can be achieved using high-temperature conditions (>300 °C). This technique allows good separations of TAG according to their number of carbons but is limited in terms of resolution due to several reasons such as the temperature required to elute TAG, their thermal stability as well as the thermal stability of the stationary phase under such conditions. Regardless of these limitations, GC analysis of TAG according to their carbon number can be performed to assess, for example, the authenticity of milk fat ([92] – a typical chromatogram of milk fat TAG is shown in Figure 23.3). This type of analysis can be performed using short (<5 m) apolar capillary columns and is very useful for quality control of fats and oils. The use of more polar capillary columns such as RTX-65 (65% diphenyl/35% dimethyl polysiloxane) provides

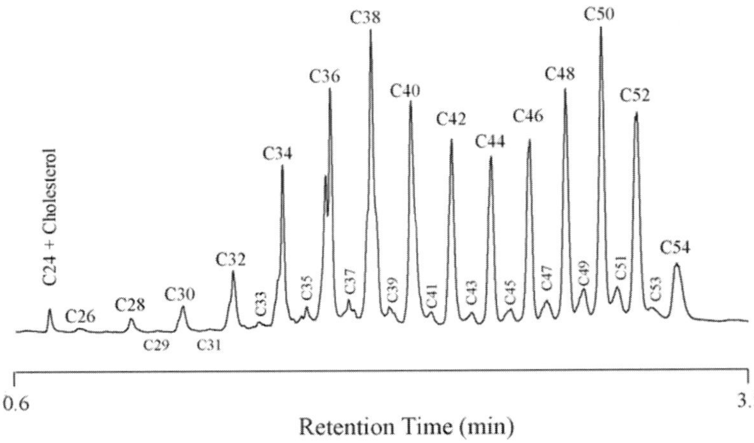

FIGURES 23.3 GC chromatogram of pure milk fat triacylglycerol (TAG) analyzed on a short apolar open-tubular capillary column. *Source: Adapted from Ref. [92].*

improved resolution of TAG in complex samples such as cocoa butter, palm oil, or butter [93]. Separation of TAG regio-isomers such as POP and PPO (P and O standing for palmitic and oleic acid, respectively) cannot be obtained by high-temperature GC. Resolution of TAG regio-isomers containing two butyric acid residues is possible on polar as demonstrated by Angers and Arul in studies of the regio-specific distribution of fatty acids in TAG [94]. Separation of monoacylglycerols (MAG) and diacylglycerols (DAG) in complex samples such as emulsifiers or crude vegetable oils can be achieved by GC after silylation of the samples. FID is often used to detect and quantify TMS esters and ethers, and the use of an appropriate internal standard allows the quantification of the different lipid classes. This method has been standardized by the American Oil Chemist Society [95]. The separation can be achieved using apolar of medium polar columns. Glycerol can be separated simultaneously as its trimethylsilyl ether derivative. The separation of Sn-1,3 and Sn-1,2 DAG isomers with the same fatty acid residues is also possible. This feature is useful in studies of the hydrolysis mechanism of triacylglycerols in plant oisl and milk fat [96,97]. Retention indices of TMS derivatives of MAG and DAG on nonpolar stationary phases (HP-5 and HP-1 types of columns) have been measured by Isodorov and co-workers [98]. Mass spectra of pure MAG and DAG TMS ethers have been reported [99] and it has been demonstrated that GC-MS can be used to identify MAG regio-isomers as their di-TMS derivatives [100].

23.5. ANALYSIS OF STEROLS, STEROL ESTERS, AND STERYL GLYCOSIDES

Cholesterol or plant-derived sterols such as campesterol, stigmasterol, and β-sitosterol are analyzed by GC as their TMS or other derivatives [101–106]. Separation of the unsaponifiable fraction from the lipid extract is required to isolate the sterols; this step also allows the hydrolysis of the steryl esters and the fraction obtained can be analyzed after silylation [103]. Epicoprostanol is often used as an

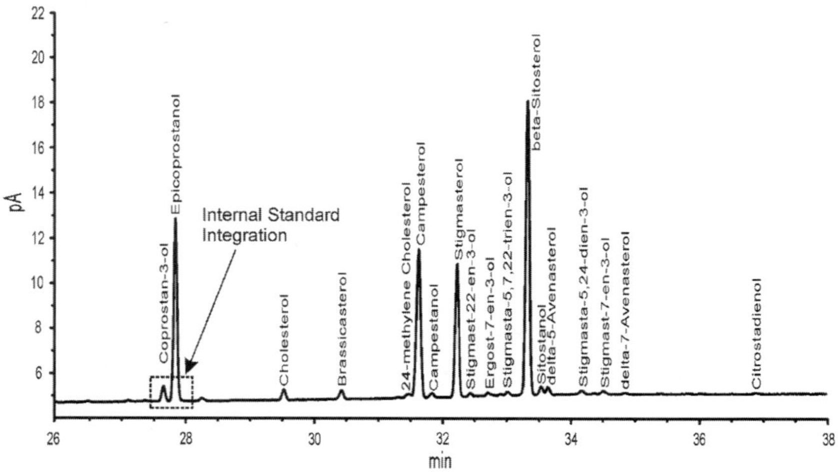

FIGURES 23.4 Gas-chromatogram of TMS derivative of phytosterols obtained from milk formulated with emulsified sterol concentrate. *Source: Adapted from reference [103].*

internal standard and a typical chromatogram of a dietary phytosterol formulation is shown in Figure 23.4. This methodology is used to characterize the sterol profile in different contexts such as characterization of vegetable oils [104], ingredient authenticity [106], quality control of food products fortified with plant sterols [103], clinical nutrition [102], or clinical biology [101,105]. Sterol esters can be analyzed directly without prior hydrolysis of the acyl moiety and TMS derivatization as shown for cholesteryl esters in plasma lipids [107]. GC can also be used to analyze acyl steryl glycosides as proposed by Pieber and co-workers who developed a methodology for their quantitative analysis in biodiesel using MS in the single monitoring ion mode (SIM) as a detector [108]. The method involved a saponification step to release the fatty acid residue but the reaction does not alter the glycosidic bond. The fraction containing steryl glycosides is then silylated using N,O-bis(trimethylsilyl)trifluoroacetamide (BSTFA) with 5% trimethylchlorosilane (TMCS) to ensure complete silylation of the glycoside groups [108].

23.6. ANALYSIS OF WAXES

Waxes are fatty acid and fatty alcohol esters found in almost natural lipid extracts. Analysis of waxes can be done by GC and most recent references report the use of MS as detector [109,110]. The standard procedures for GC analysis of waxes involve as a first step the purification of the waxes from TAG and other lipids by silica gel chromatography, thin-layer chromatography, or solid-phase extraction [1]. This sample preparation step is required to reduce the complexity of the chromatogram and limit coelution of waxes with other analytes. However, Michael-Jubeli et al. [111] reported a method to analyze waxes and other lipids in skin surface lipid samples by GC-MS without prior isolation of the waxes. Alternatively, isolation of the wax fraction from the other lipids can be done on-line using liquid chromatography coupled with GC as reported by Aragon et al. [110].

23.7. ANALYSIS OF LIPID CLASSES

We previously reviewed the methodologies available to analyze FFA, sterols, waxes, MAG, DAG, TAG, or other lipids by either GC-FID or GC-MS after silylation. Several methods have been reported for the simultaneous analysis of these lipid classes in various types of matrices [107,112–114]. A wide range temperature program is required to elute light TMS derivatives such as glycerol tri-TMS and high-molecular weight TAG [95]. The standard method for MAG- and DAG-based emulsifiers [95] can also be used to separate glycerol, FFA, MAG, DAG, and TAG in complex samples such as crude oils. For the quantitative determination of neutral lipid classes after silylation, it is highly recommended to use an internal standard for each lipid class since response factors of the different classes might differ significantly. The chromatograms obtained can be very complex as demonstrated by Torres and co-workers [112] or Verleyen and co-workers who analyzed deodorizer distillates [113]. Therefore, in addition to the most abundant classes such as FFA, MAG, DAG, and TAG, other liposoluble substances such as vitamins E, sterols, and hydrocarbons will also be present in the sample (see, for example, the study by L. Nang Lau et al. [99]). The use of a mixture of pure standards is very useful to identify the different lipid classes and minimize the identification problems [95]. An alternative to the use of standards for identification is to use MS as a detector as described by Lytovchenko and co-workers for plant material [114] or Michael-Jubeli and co-workers for skin surface lipid profiling – Figure 23.5 [111].

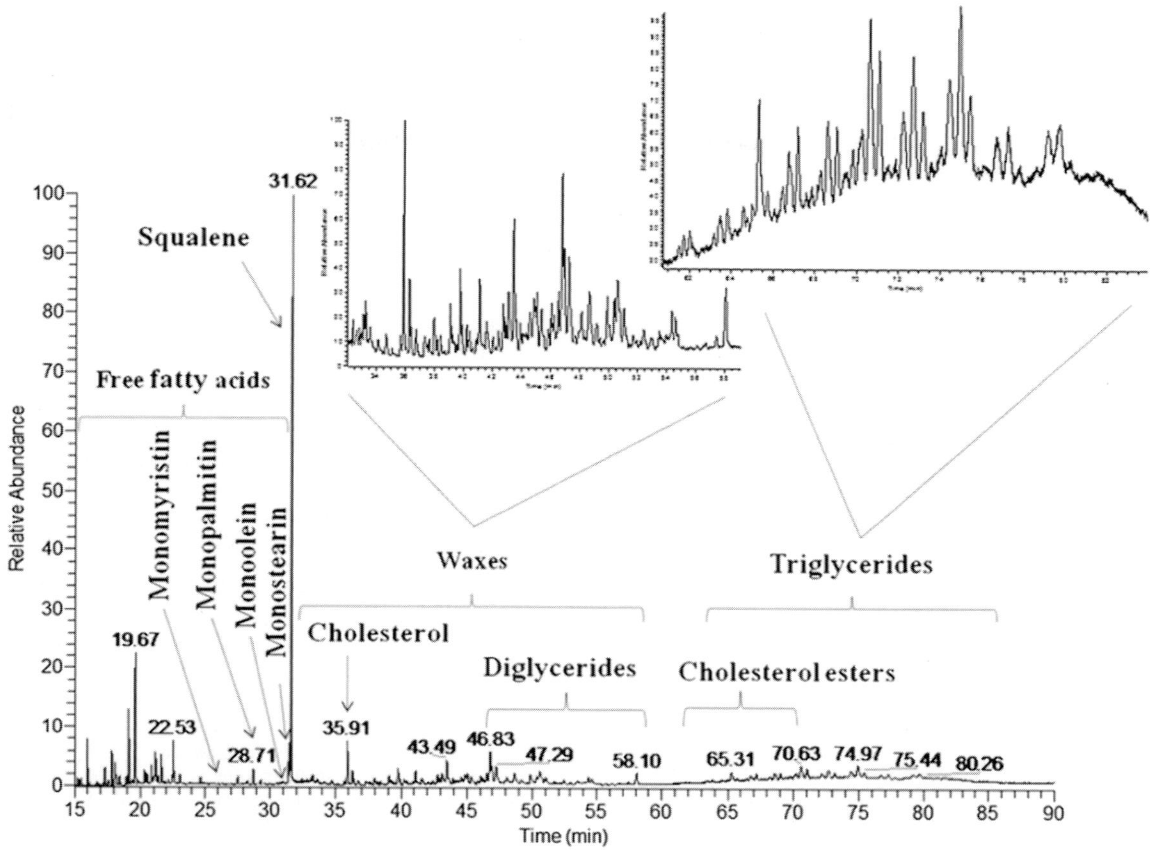

FIGURES 23.5 Gas-chromatogram of TMS derivatives of skin surface lipids obtained by high-temperature gas chromatography-mass spectrometry. *Source: Adapted from Michael-Jubeli et al., [111].*

References

[1] W.W. Christie, Lipid analysis. Isolation, separation, identification and lipidomic analysis, fourth ed., The Oily Press, Bridgwater, UK, 2010.

[2] P.Q. Tranchida, P. Donato, G. Dugo, Comprehensive chromatographic methods for the analysis of lipids, Trends. Anal. Chem. 26 (3) (2007) 191–205.

[3] J.K.G. Kramer, J. Zhou, Conjugated linoleic acids and octadecenoic acids: extraction and isolation of lipids, Eur. J. Lipid. Sci. Technol. 103 (2001) 594–600.

[4] K. Ichihara, Y. Fukubayashi, Preparation of fatty acid methyl esters for gas-liquid chromatography, J. Lipid Res. 51 (2010) 635–640.

[5] R.L. Wolff, C.C. Bayard, Improvement in the resolution of individual *trans*-18:1 isomers by capillary gas-liquid chromatography: use of a 100-m CP-Sil 88 column, J. Am. Oil. Chem. Soc. 72 (1995) 1197–1201.

[6] R.L. Wolff, C.C. Bayard, R.J. Fabien, Evaluation of sequential methods for the determination of butterfat fatty acid composition with emphasis on *trans*-18:1 acids. Application to the study of seasonal variations in French butters, J. Am. Oil Chem. Soc. 72 (1995) 1471–1483.

[7] J.-M. Chardigny, R.L. Wolff, E. Mager, C.C. Bayard, J.-L. Sébédio, L. Martine, et al., Fatty acid composition of French infant formulas with emphasis on the content and detailed profile of *trans* fatty acids, J. Am. Oil Chem. Soc. 73 (1996) 1595–1601.

[8] P.Y. Chouinard, L. Corneau, D.M. Barbano, L.E. Metzger, D.E. Bauman, Conjugated linoleic acids alter milk fatty acid composition and inhibit

milk fat secretion in dairy cows, J. Nutr. 129 (1999) 1579–1584.

[9] W.W. Christie, A simple procedure for rapid transmethylation of glycerolipids and cholesterol esters, J. Lipid Res. 23 (1982) 1072–1075.

[10] F. Ulberth, R.G. Gabernig, F. Schrammel, Flame-ionization detector response to methyl, ethyl, propyl, and butyl esters of fatty acids, J. Am. Oil Chem. Soc. 76 (1999) 263–266.

[11] J.K.G. Kramer, V. Fellner, M.E.R. Dugan, F.D. Sauer, M.M. Mossoba, M.P. Yurawecz, Evaluating acid and base catalysts in the methylation of milk and rumen fatty acids with special emphasis on conjugated dienes and total *trans* fatty acids, Lipids 32 (1997) 1219–1228.

[12] S.W. Christopherson, R.L. Glass, Preparation of milk fat methyl esters by alcoholysis in an essential non-alcoholic solution, J. Dairy Sci. 52 (1969) 1289–1290.

[13] W.W. Christie, Gas chromatography and lipids: a practical guide, The Oily Press, Ayr, 1989, p. 307.

[14] C. Cruz-Hernandez, Z. Deng, J. Zhou, A.R. Hill, M.P. Yurawecz, P. Delmonte, et al., Methods for analysis of conjugated linoleic acids and *trans*-18:1 isomers in dairy fats by using a combination of gas chromatography, silver-ion thin-layer chromatography/gas chromatography, and silver-ion liquid chromatography, J. AOAC Int. 87 (2004) 545–562.

[15] P. Risé, F. Salvetti, C. Galli, Application of a direct transmethylation method to the analysis of fatty acid profile in circulating and cultured cells, Anal. Biochem. 346 (1) (2005) 182–184.

[16] A.I. Carrapiso, C. Garcia, Development in lipid analysis: some new extraction techniques and in situ transesterification, Lipids 35 (11) (2000) 1167–1177.

[17] E. Amusquivar, S. Schiffner, E. Herrera, Evaluation of two methods for plasma fatty acid analysis by GC, Eur. J. Lipid Sci. Technol. 113 (2011).

[18] M.R. Mazalli, N. Bragagnolo, Validation of two methods for fatty acids analysis in eggs, Lipids 42 (5) (2007) 483–490.

[19] K. Ichihara, K. Yoneda, A. Takahashi, N. Hoshino, M. Matsuda, Improved methods for the fatty acid analysis of blood lipid classes, Lipids 46 (2011) 297–306.

[20] S.P. Alves, A.R.J. Cabrita, A.J.M. Fonseca, R.J.B. Bessa, Improved method for fatty acid analysis in herbage based on direct transesterification followed by solid-phase extraction, J. Chromatogr. A 1209 (2008) 212–219.

[21] M.J. Griffiths, R.P. van Hille, S.T. Harrison, Selection of direct transesterification as the preferred method for assay of fatty acid content of microalgae, Lipids 45 (11) (2010) 1053–1060.

[22] T. Lewis, P.D. Nichols, T.A. McMeekin, Evaluation of extraction methods for recovery of fatty acids from lipid-producing microheterotrophs, J. Microbiol. Meth. 43 (2000) 107–116.

[23] J.V. O Fallon, J.R. Busboom, M. L Nelson, T. Gaskins, A direct method for fatty acid methyl ester synthesis: application to wet meat tissues, oils, and feedstuffs, J. Anim. Sci. 85 (2007) 1511–1521.

[24] Stoffel W, Chu F, Ahrens EH. Analysis of long-chain fatty acids by gas-liquid chromatography. Micromethod for preparation of methyl esters. Anal. Chem. 31, 307–308.

[25] K. Eulitz, M.P. Yurawecz, N. Sehat, J. Fritsche, J.A.G. Roach, M.M. Mossoba, et al., Preparation, separation, and confirmation of the eight geometrical cis/trans conjugated linoleic acid isomers 8,10 – through 11,13–18:2, Lipids 34 (1999) 873–877.

[26] R.L. Wolff, N.A. Combe, F. Destaillats, C. Boué, D. Precht, J. Molkentin, et al., Follow-up of the Δ4 to Δ16 Trans-18:1 isomer profile and content in French processed foods containing partially hydrogenated vegetable oils during the period 1995–1999. Analytical and nutritional implications, Lipids 35 (2000) 815–825.

[27] D. Precht, J. Molkentin, C18:1, C18:2 and C18:3 trans and cis fatty acid isomers including conjugated cis Δ9, trans Δ11 linoleic acid (CLA) as well as total fat composition of German human milk lipids, Nahrung 43 (1999) 233–244.

[28] J.K.G. Kramer, C. Cruz-Hernandez, J. Zhou, Conjugated linoleic acids and octadecenoic acids: analysis by GC, Eur. J. Lipid Sci. Technol. 103 (2001) 600–609.

[29] J.K.G. Kramer, C.B. Blackadar, J. Zhou, Evaluation of two GC columns (60-m SUPELCOWAX 10 and 100-m CP Sil 88) for analysis of milkfat with emphasis on CLA, 18:1, 18:2 and 18:3 isomers, and short- and long-chain FA, Lipids 37 (2002) 823–835.

[30] R.L. Wolff, D. Precht, A Critque of 50-m CP Sil 88 capillary columns used alone to assess trans-unsaturated FA in foods: the case of the TRANSFAIR study, Lipids 37 (2002) 627–629.

[31] D. Precht, J. Molkentin, Overestimation of linoleic acid and trans-C18:2 isomers in milk fats with emphasis on Δtrans 9, Δtrans 12-octadecadienoic acid, Milchwissenschaft 58 (2003) 30–34.

[32] R.J.B. Bessa, P.V. Portugal, I.A. Mendes, J. Santos-Silva, Effect of lipid supplementation on growth performance, carcass and meat quality and fatty acid composition of intramuscular lipids of lambs fed dehydrated lucerne or concentrate, Livest Prod. Sci. 96 (2005) 185–194.

[33] R.L. Wolff, Analysis of alpha-linolenic acid geometrical isomers in deodorized oils by capillary gas-liquid chromatography on cyanoalkyl polysiloxane phases: a note of caution, J. Am. Oil Chem. Soc. 71 (1994) 907–909.

[34] J.K.G. Kramer, N. Sehat, M.E.R. Dugan, M.M. Mossoba, M.P. Yurawecz, J.A.G. Roach, et al., Distribution of conjugated linoleic acid (CLA) isomers in tissue lipid classes of pigs fed a commercial CLA mixture determined by gas chromatography and silver ion-high performance liquid chromatography, Lipids 33 (1998) 549–558.

[35] J.K.G. Kramer, N. Sehat, J. Fritsche, M.M. Mossoba, K. Eulitz, M.P. Yurawecz, et al., Separation of conjugated linoleic acid isomers, in: M.P. Yurawecz, M.M. Mossoba, J.K.G. Kramer, M.W. Pariza, G.J. Nelson (Eds.), Advances in conjugated linoleic acid research, vol. 1, AOCS Press, Champaign, IL, USA, 1999, pp. 83–109.

[36] D. Precht, H. Hagemeister, W. Kanitz, J. Voigt, Trans fatty acids and conjugated linoleic acids in milk fat from dairy cows fed a rumen-protected linoleic acid rich diet, Kieler. Milchw. Forsch. 54 (2002) 225–242.

[37] S. Banni, J.–C. Martin, Conjugated linoleic acid and metabolites, in: J.-L. Sébédio, W.W. Christie (Eds.), Trans fatty acids in human nutrition, The Oily Press, Dundee, UK, 1998, pp. 261–302.

[38] F. Lavillonnière, J.C. Martin, P. Bougnoux, J.-L. Sébédio, Analysis of conjugated linoleic acid isomers and content in French cheeses, J. Am. Oil Chem. Soc. 75 (1998) 343–352.

[39] F. Destaillats, P. Angers, Directed sequential synthesis of conjugated linoleic acid isomers from $\Delta^{7,9}$ to $\Delta^{12,14}$, Eur. J. Lipid Sci. Technol. 105 (2003) 3–8.

[40] W.W. Christie, J.L. Sébédio, P. Juanéda, A practical guide to the analysis of conjugated linoleic acid, Inform. 12 (2001) 147–152.

[41] J.A.G. Roach, M.P. Yurawecz, J.K.G. Kramer, M.M. Mossoba, K. Eulitz, Y. Ku, Gas chromatography-high resolution selected-ion mass spectrometric identification of trace 21:0 and 20:2 fatty acids eluting with conjugated linoleic acid isomers, Lipids 35 (2000) 797–802.

[42] AOCS official method Ce 1h-05, Determination of cis-, trans-, saturated, monounsaturated and polyunsaturated fatty acids in non-ruminant animal oils and fats by capillary GLC, in: Sampling and analysis of commercial fats and oils, AOCS, Champaign, IL, USA, 2005.

[43] F. Destaillats, P. Angers, Base-catalyzed derivatization methodology for FA analysis. Application to milk fat and celery seed lipid TAG, Lipids 37 (2002) 527–532.

[44] D. Precht, J. Molkentin, Frequency and distributions of conjugated linoleic acid and trans fatty acid contents in European bovine milk fats, Milchwissenschaft 55 (2000) 687–691.

[45] P. Delmonte, A.R.F. Kia, J.K.G. Kramer, M.M. Mossoba, L. Sidisky, J.I. Rader, Separation characteristics of fatty acid methyl esters using SLB-IL111, a new ionic liquid coated capillary gas chromatographic column, J. Chromatogr. A 1218 (3) (2011) 545–554.

[46] S. Shirasawa, A. Sasaki, Y. Saida, C. Satoh, A rapid method for trans-fatty acid determination using a single capillary GC, J. Oleo. Sci. 56 (2) (2007) 53–58.

[47] A.K. Vickers, M. Hastings, R. Lautamo, R. Davis, S. Watkins, New high polarity bis(cyanopropyl) siloxane stationary phase for GC resolution of positional and geometric isomers of fatty acid methyl esters, Agilent Technologies, November 11, 2004. 5989-181EN.

[48] J. Peene, J. de Zeeuw, F. Biermans, L. Joziasse, CP-Select CB for FAME, A new highly polar bonded stationary phase with a temperature stability up to 290 °C optimized for analyzing cis- and trans FAME isomers with GC, #P-147, Varian Inc, Middelburg, The Netherlands, 2004.

[49] P. Juaneda, S. Brac de la Perriere, J.L. Sebedio, S. Gregoire, Influence of heat and refining on formation of CLA isomers in sunflower oil, J. Am. Oil Chem. Soc. 80 (2003) 937–940.

[50] F. Destaillats, J.P. Trottier, J.M.G. Galvez, P. Angers, Analysis of alpha-linolenic acid biohydrogenation intermediates in milk fat with emphasis on conjugated linolenic acids, J. Dairy Sci. 88 (2005) 3231–3239.

[51] J.,K.G. Kramer M.R. Hernandez, C. Cruz-Hernandez, J. Kraft, M.E.R. Dugan, Combining results of two GC separations partly achieves determination of all cis and trans 16:1, 18:1, 18:2, 18:3 and CLA isomers of milk fat as demonstrated using Ag-ion SPE fractionation, Lipids 43 (2008) 259–273.

[52] G. Dobson, W.W. Christie, B. Nikolova-Damyanov, Silver ion chromatography of lipids and fatty acids, J. Chromatogr. B 671 (1995) 197–222.

[53] B. Nikolova-Damyanova, W.W. Christie, B. Herslöf, Mechanistic aspects of fatty acid retention in silver ion chromatography, J. Chromatogr. A 749 (1996) 47–54.

[54] L.J. Morris, Separations of lipids by silver ion chromatography, J. Lipid Res. 7 (1966) 717–732.

[55] D. Precht, J. Molketin, Rapid analysis of the isomers of trans-octadecenoic acid in milk fat, Int. Dairy J. 6 (1996) 791–809.

[56] B. Nikolova-Damyanova, B.G. Herslof, W.W. Christie, Silver ion high-performance liquid chromatography of derivatives of isomeric fatty acids, J. Chromatogr. A 609 (1992) 133–140.

[57] P. Delmonte, J.K.G. Kramer, S. Banni, M.P. Yurawecz, New developments in silver ion and reverse phase HPLC of CLA, in: M.P. Yurawecz, J.K.G. Kramer, M.M. Mossoba, O. Gudmundsen, M.W. Pariza, S. Banni (Eds.), Advances in conjugated linoleic acid research, vol. 3, AOCS Press, Champaign, IL, USA, 2006, pp. 95–118.

[58] R.L. Wolff, D. Precht, Comments on the resolution of individual trans-18:1 isomers by gas-liquid chromatography, J. Am. Oil Chem. Soc. 75 (1998) 421–422.

[59] A.A. Abu-Ghazaleh, D.J. Schingoethe, A.R. Hippen, Conjugated linoleic acid and other beneficial fatty acids in milk fat from cows fed soybean meal, fish meal, or both, J. Dairy Sci. 84 (2001) 1845–1850.

[60] R.J. Baer, J. Ryali, D.J. Schingoethe, K.M. Kasperson, D.C. Donovan, A.R. Hippen, et al., Composition and properties of milk and butter from cows fed fish oil, J. Dairy Sci. 84 (2001) 345–353.

[61] D.C. Donovan, D.J. Schingoethe, R.J. Baer, J. Ryali, A.R. Hippen, S.T. Franklin, Influence of dietary fish oil on conjugated linoleic acid and other fatty acids in milk fat from lactating dairy cows. J. Dairy Sci. 83 (2000) 2620–2628.

[62] A.A. Abu-Ghazaleh, D.J. Schingoethe, A.R. Hippen, L.A. Whitlock, Feeding fish meal and extruded soybeans enhances the conjugated linoleic acid (CLA) content of milk, J. Dairy Sci. 85 (2002) 624–631.

[63] P. Korytár, H.G. Janssen, E. Matisova, U.A.T. Brinkman, Practical fast gas chromatography: methods, instrumentation and applications, Trends. Anal. Chem. 21 (2002) 558–572.

[64] M. vanDeursen, J. Beens, C.A. Cramers, Possibilities and limitations of fast temperature programming as a route towards fast GC, J. High Resolut. Chromatogr. 22 (1999) 509–513.

[65] L.M. Blumberg, M.S. Klee, Method translation and retention time locking in partition GC, Anal. Chem. 70 (1998) 3828–3839.

[66] J.M. Clemons, T.J. Thomas, in: Fast process gas chromatography, Proceedings of the 45th annual ISA analysis division symposium, West Virginia, 45th Annual ISA analysis division symposium, Charleston, vol. 33, 9–12 April 2000, pp. 115–128.

[67] F. Magni, P.M. Piatti, L.D. Monti, P. Lecchi, A.E. Pontiroli, G. Pozza, et al., Fast gas chromatographic-mass spectrometric method for the evaluation of plasma fatty acid turnover using [1-13C] palmitate, J. Chromatogr. B 657 (1994) 1–7.

[68] P.Q. Tranchida, M.L. Presti, R. Costa, P Dugo, G. Dugo, L. Mondello, High-throughput analysis of bergamot essential oil by fast solid-phase microextraction-capillary gas chromatography-flame ionization detection, J. Chromatogr. A 1103 (2006) 162–165.

[69] P. Rubiolo, F. Belliardo, C. Corsero, E. Liberto, B. Sgorbini, C. Bicchi, Headspace-solid-phase microextraction fast GC in combination with principal component analysis as a tool to classify different chemotypes of chamomile flower-heads (Matricaria recutita l.), Phytochem. Anal. 17 (2006) 217–225.

[70] C. Bicchi, C. Brunelli, C. Cordero, P. Rubiolo, M. Galli, A. Sironi, Direct resistively heated column gas chromatography (ultrafast module-GC) for high-speed analysis of essential oils of differing complexities, J. Chromatogr. A 1024 (2004) 195–207.

[71] C. Bicchi, C. Brunelli, M. Galli, A. Sironi, Conventional inner diameter short capillary columns: an approach to speeding up gas chromatographic analysis of medium complexity samples, J. Chromatogr. A 931 (2001) 129–140.

[72] L. Mondello, A. Casilli, P.Q. Tranchida, R. Costa, P. Dugo, G. Dugo, Fast GC for the analysis of citrus oils, J. Chromatogr. Sci. 42 (2004) 410–416.

[73] L. Mondello, A. Casilli, P.Q. Tranchida, M. Furukawa, K. Komori, K. Miseri, et al., Fast enantiomeric analysis of a complex essential oil with an innovative multidimensional gas chromatographic system, J. Chromatogr. A 1105 (2006) 11–16.

[74] M.J. Bogusz, S. Abu El Hajj, Z. Ehaideb, H. Hassan, M. Al-Tufail, Rapid determination of benzo(a)pyrene in olive oil samples with solid-phase extraction and low-pressure, wide-bore gas chromatography–mass spectrometry and fast liquid chromatography with fluorescence detection, J. Chromatogr. A 1026 (2004) 1–7.

[75] F.L. Godoi, V. Wagner, R.H.M. Godoi, L.V. Vaeck, R.V. Grieken, Application of low-pressure gas chromatography–ion-trap mass spectrometry to the analysis of the essential oil of *Turnera difussa* (Ward.) Urb, J. Chromatogr. A 1027 (2004) 127–130.

[76] L. Mondello, P.Q. Tranchida, R. Costa, A. Casilli, G. Dugo, A. Cotroneo, et al., Fast GC for the analysis of fats and oils, J. Sep. Sci. 26 (2003) 1467–1473.

[77] F. David, D.R. Gere, F. Scanlan, P. Sandra, Instrumentation and applications of fast high-resolution capillary gas chromatography, J. Chromatogr. A 842 (1999) 309–319.

[78] I. Bondia-Pons, C. Moltó-Puigmartí, A.I. Castellote, M.C. López-Sabater, Determination of conjugated linoleic acid in human plasma by fast gas chromatography, J. Chromatogr. A 1157 (2007) 422–429.

[79] Z. Meng, D. Wen, D. Sun, F. Gao, W. Li, Y. Liao, et al., Rapid determination of C12–C26 non-derivatized fatty acids in human serum by fast gas chromatography, J. Sep. Sci. 30 (2007) 1537–1543.

[80] M.A. Masood, K.D. Stark, N. Salem Jr., A simplified and efficient method for the analysis of fatty acid methyl esters suitable for large clinical studies, J. Lipid Res. 46 (2005) 2299–2305.

[81] M.A. Masood, N. Salem Jr., High-throughput analysis of plasma fatty acid methyl esters employing robotic transesterification and fast gas chromatography, Lipids 43 (2008) 171–180.

[82] I. Bondia-Pons, S. Morera-Pons, A.I. Castellote, M.C. López-Sabater, Determination of phospholipid fatty acids in biological samples by solid-phase extraction and fast gas chromatography, J. Chromatogr. A 1116 (2006) 204–208.

[83] F. Destaillats, C. Cruz-Hernandez, Fast analysis by gas-liquid chromatography, J. Chromatogr. A 1169 (1–2) (2007) 175–178.

[84] C. Cruz-Hernandez, F. Destaillats, Recent advances in fast gas-chromatography: application to the separation of fatty acid methyl esters perspective on the resolution of complex fatty acid compositions, J. Liq. Chromatogr. Relat. Technol. 32 (2009) 1672–1688.

[85] L. Mondello, R. Shellie, A. Casilli, P.Q. Tranchida, P. Marriott, G. Dugo, Ultra-fast essential oil characterization by capillary GC on a 50 microm ID column, J. Sep. Sci. 27 (2004) 699–702.

[86] P.Q. Tranchida, M. Mondello, D. Sciarrone, P. Dugo, G. Dugo, L. Mondello, Evaluation of use of a very short polar microbore column segment in high-speed gas chromatography analysis, J. Sep. Sci. 31 (2008) 2634–2639.

[87] N. Noronha, D.A. Cronin, E.D. O'Riordan, Flavouring of imitation cheese with enzyme-modified cheeses (EMCs): sensory impact and measurement of aroma active short chain fatty acids (SCFAs), Food. Chem. 106 (2008) 905–913.

[88] O. Pinho, I. Ferreira, Solid-phase microextraction in combination with GC/MS for quantification of the major volatile free fatty acids in ewe cheese, Anal. Chem. 74 (2002) 5199–5204.

[89] M.C. Perotti, S.M. Bernal, C.A. Meinardi, Free fatty acid profiles of Reggianito Argentino cheese produced with different starters, Int. Dairy J. 15 (2005) 1150–1155.

[90] N. Hashinomoto, T. Aoyama, T. Shioiri, New methods and reagents in organic synthesis. 14. A simple efficient preparation of methy l esters with trimethylsilyldiazomethane (TMSCHN2) and its application to gas chromatographic analysis of fatty acids, Chem. Pharm. Bull. 29 (1981) 1475–1487.

[91] M. Juarez, M.A. De la Fuente, Improved gas chromatographic method for the determination of the individual free fatty acids in cheese using a capillary column and a PTV injector, Chromatographia 33 (7) (1992).

[92] F. Destaillats, M. de Wispelaere, F. Joffre, Authenticity of milk fat by fast analysis of triacylglycerols: application to the detection of partially hydrogenated vegetable oils, J. Chromatogr. A 1131 (2006) 227–234.

[93] R.J. Craven, R.W. Lencki, Rapid analysis of acylglycerols in low molecular weight milk fat fractions, Lipids 42 (2007) 473–482.

[94] P. Angers, J. Arul, A simple method for regiospecific analysis of triacylglycerols by gas chromatography, JAOCS 76 (4) (1999).

[95] AOCS Official Method Cd 11b-91, Determination of mono- and diglycerides by capillary gas chromatography, in: Sampling and analysis of commercial fats and oils, AOCS, Champaign, IL, USA, 2009.

[96] H.L. Nang Lau, C.W. Puah, Y.M. Choo, A.N. Ma, Simultaneous quantification of free fatty acids, free sterols, squalene, and acylglycerol molecular species in palm oil by high-temperature gas chromatography-flame, Lipids 40 (5) (2005) 523–528.

[97] P. Fagan, C. Wijesundera, Determination of mono- and di-acylglycerols in milk lipids, J. Chromatogr. A 1054 (2004).

[98] V.A. Isidorov, Gas chromatographic retention indices of biologically and environmentally important organic compounds on capillary columns with low-polar stationary phases, J. Chromatogr. A 1216 (2009) 8998–9007.

[99] V.A. Isidorov, M. Rusak, L. Szczepaniak, Gas chromatographic retention indices of trimethylsilyl derivatives of mono-and diglycerides on capillary columns with non-polar stationary phases, J. Chromatogr. A 1166 (2007) 207–211.

[100] F. Destaillats, C. Cruz-Hernandez, K. Nagy, Identification of monoacylglycerol regio-isomers by gas chromatography–mass spectrometry, J. Chromatogr. A 1217 (2010).

[101] G. Brufau, R. Codony, M.A. Canela, Rapid and quantitative determination of total sterols of plant and animal origin in liver samples by gas chromatography, Chromatographia 64 (2006) 559–563.

[102] M.L. Forchielli, G. Bersani, S. Tala, G. Grossi, The spectrum of plant and animal sterols in different oil-derived intravenous emulsions, Lipids 45 (2010) 63–71.

[103] L.M. Clement, S.L. Hansen, C.D. Costin, Quantitation of sterols and steryl esters in fortified foods and beverages by GC/FID, JAOCS 87 (2010) 973–980.

[104] J.C. Bada, M. León-Camacho, M. Prieto, Characterization of walnut oils (*Juglans regia* L.) from Asturias, Spain, JAOCS 87 (2010) 1469–1474.

[105] D. Noto, A.B. Cefalù, G. Barraco, E. Martino, Plasma non-cholesterol sterols: a useful diagnostic tool in pediatric hypercholesterolemia, Pediatric. Res. 67 (2) (2010) 200–204.

[106] K.M. Al-Ismail, A.K. Alsaed, R. Ahmad, Detection of olive oil adulteration with some plant oils by GLC analysis of sterols using polar column, Food Chem. 121 (2010) 1255–1259.

[107] A. Kuksis, J.J. Myher, Quantitation of plasma lipids by gas–liquid chromatography on high temperature polarizable capillary columns, J. Lipid Res. 34 (1993) 1029–1038.

[108] B. Pieber, S. Schober, C. Goebl, Novel sensitive determination of steryl glycosides in biodiesel by gas chromatography–mass spectroscopy, J. Chromatogr. A 1217 (2010) 6555–6561.

[109] L. Zhang, Y. Yun, Y. Liang, Discovery of mass spectral characteristics and automatic identification of wax esters from gas chromatography mass spectrometry data, J. Chromatogr. A 1217 (2010) 3695–3701.

[110] A. Aragon, J.M. Cortes, R.M. Toledano, J. Villen, Analysis of wax esters in edible oils by automated on-line coupling liquid chromatography–gas chromatography using the through oven transfer adsorption desorption, J. Chromatogr. A 1218 (2010) 4960–4965.

[111] R. Michael-Jubeli, J. Bleton, High-temperature gas chromatography–mass spectrometry for skin surface lipids profiling, J. Lipid Res. 52 (2011) 143–151.

[112] C.F. Torres, D. Tenllado, F.J. Señorans, A versatile GC method for the analysis of alkylglycerols and other neutral lipid classes, Chromatographia 269 (2009) 729–734.

[113] T. Verleyen, R. Verhé, L. Garcia, K. Dewettinck, Gas chromatographic characterization of vegetable oil deodorization distillate, J. Chromatogr. A 921 (2001) 277–285.

[114] A. Lytovchenko, R. Beleggia, N. Schauer, T. Isaacson, J.E. Leuendoff, H. Hellmann, et al., Application of GC-MS for the detection of lipophilic compounds in diverse plant tissues, Plant Methods 5 (2009) 1–11.

CHAPTER

24

Metabonomics

Eric Chun Yong Chan, Mainak Mal, Kishore Kumar Pasikanti

OUTLINE

24.1. OVERVIEW OF METABONOMICS

Since its inception, the field of metabonomics has grown remarkably in terms of its applications and contributions to system biology research. Metabonomics provides a powerful tool for gaining valuable insight into functional biology, toxicology, pharmacology, diagnosis, and prognosis of diseases. The nonhypothesis-driven global metabolic profiling strategy or metabonomics is defined as the quantitative measurement of the dynamic multiparametric metabolic response of living systems to pathophysiological stimuli or genetic modification [1]. Metabonomics is complementary to genomics and proteomics as it measures the perturbed metabolic endpoints due to environmental, pharmacological, or pathological influences while in genomics and proteomics, more upstream biological events are typically profiled and studied [2]. Metabonomic studies involve the analysis of various biological matrices such as blood, urine, and tissues using suitable analytical platforms. Metabolomics and metabolic fingerprinting are terms used in addition to metabonomics to describe the unbiased and nontargeted analysis of global metabolic profiles in biofluids and tissues [3,4]. In metabonomics, metabolites belonging to diverse metabolic pathways and comprising

Gas Chromatography DOI: 10.1016/B978-0-12-385540-4.00024-9

diverse chemical classes, such as organic acids, amino acids, fatty acids, amines, sugars, sugar alcohols, steroids, and nucleic acid bases, are profiled. Therefore, multiple complementary analytical platforms are often utilized for nontargeted metabonomic studies, in order to cover as large a metabolic space as possible [5]. In this chapter, application of gas chromatography mass spectrometry (GC-MS)-based metabonomics is elaborated for urine and tissue matrices. Urine and tissue matrices were chosen for discussion since the sample preparation procedures for these matrices are distinct and require special attention.

24.2. ANALYTICAL TOOLS IN METABONOMIC RESEARCH

Ideally, an analytical tool for metabonomic research should allow analysis with minimal or no sample preparation, exhibit a high degree of robustness and reproducibility, and be high throughput and highly sensitive. Most importantly, for metabonomics, comprehensive coverage of metabolic space and ease of identification of profiled metabolites are additional desirable attributes of an analytical platform. Analytical platforms that are commonly used in metabonomics include nuclear magnetic resonance (NMR) spectroscopy- and mass spectrometry (MS)-based techniques such as GC-MS, liquid chromatography mass spectrometry (LC-MS), or capillary electrophoresis mass spectrometry (CE-MS) [6]. In addition to these popular techniques, other methods such as Fourier transform infrared (FTIR) spectroscopy [7], LC with ultraviolet [8] or coulometric detection [9], and CE with ultraviolet detection [10] have also been used in metabonomic studies. In addition, hybrid platforms comprising LC, NMR spectroscopy, and MS have been explored in metabonomics. In such systems, the LC eluent is split into two parts and subjected to concomitant analysis by both NMR and MS. The resulting NMR- and MS-based data provide in-depth molecular information and aid in metabolite identification [11,12].

24.3. GC-MS-BASED METABONOMICS

24.3.1. GC-MS Technologies

Among the various analytical methods utilized conventionally for metabonomics, GC-MS has emerged as an essential and complementary analytical technique because of its high sensitivity, reproducibility, and peak resolution. Moreover, the identification of metabolites can be performed straightforwardly using the electron GC-MS impact ionization (EI) spectral library. However, chemical derivatization of the polar functional groups of metabolites is usually required in GC-MS analysis in order to decrease their polarity, and increase their volatility and thermal stability. As a result of the extensive sample preparation process combined with long GC elution times, GC-MS is considered a relatively low-throughput technique when compared to LC-MS or NMR spectroscopy [13]. However, sample throughput is significantly enhanced with the advent of gas chromatography/time-of-flight mass spectrometry (GC-TOFMS). Coupling of GC to the TOF analyzer offers several advantages compared to the coupling to quadrupole MS. Software advances and the fast acquisition rate of TOFMS (up to 500 Hz) facilitate the deconvolution of the mass spectra of closely eluting analytes [14]. In other words, MS spectra of coeluting peaks can be extracted with an incomplete chromatographic resolution of the metabolites. Such an attribute of GC-TOFMS is pertinent in the analysis of the complex biological matrices where coelution of metabolites is prevalent. Importantly, analysis time may be shortened as chromatographic resolution can be compromised moderately due to the peak deconvolution function. In addition, spectral purity and sensitivity are enhanced significantly in GC-TOFMS compared to GC-MS analysis [14]. The advent of two-dimensional gas chromatography time-of-flight mass spectrometry (GC × GC-TOFMS) has comprehensively added on to the metabolic space covered by conventional GC-MS. GC × GC-TOFMS is noted for its ability to analyze complex mixtures and has been successfully applied in metabonomic investigations [15–19]. GC × GC-TOFMS offers several advantages compared to GC-MS in metabonomic analyses. First, peak capacity and peak resolution are significantly enhanced. Approximately 1000 peaks can be detected using GC × GC-TOFMS. Therefore, a larger metabolic space can be covered using GC × GC-TOFMS. Second, artifactual peaks arising from column bleed or derivatizing agent could be chromatographically resolved from the metabolite peaks using GC × GC-TOFMS. Artifactual peaks can be subsequently excluded from the data tables automatically using data preprocessing steps. Third, a five-fold increase in sensitivity was observed using GC × GC-TOFMS compared to GC-MS due to cryogenic focusing. Finally, apart from the increased number of detectable peaks as compared to GC-MS, spectral purity is significantly improved in GC × GC-TOFMS, which in turn aids in mass spectral deconvolution and compound identification [20–23]. In summary, the advantages of GC × GC-TOFMS are (1) increased chromatographic resolution without compromising analytical run time, (2) higher sensitivity through cryogenic focusing, (3) improved spectral match due to increased analytical purity of peaks, and (4) larger coverage of metabolic space through detection of low abundance metabolites. Although two-dimensional GC offered several advantages, it is important to note some of the limitation of GC × GC-TOFMS. Since short and narrow bore columns are commonly used as second-dimension columns in GC × GC analysis, column overloading is commonly observed in metabonomics [24]. Another issue that has not been thoroughly investigated so far is retention time shift in GC × GC-TOFMS analysis. It is also worth noting that although GC × GC offered increased resolution of peaks, coelution of some metabolites remains inevitable due to the complex nature of biological matrices. The principles and instrumentation of comprehensive two-dimensional gas chromatography are reviewed in Chapter 7.

24.3.2. Metabonomic Workflow

The overall workflow adopted in GC-MS-based metabonomics is summarized in Figure 24.1.

24.3.2.1. GC-MS Conditions and Data Acquisition

A sample injection volume of 0.5–2 μL is typically adopted in metabonomic studies, and the injector temperature is maintained at 200–250 °C to facilitate rapid vaporization of injected sample and subsequent mixing with the carrier gas. This is followed by the chromatographic separation of derivatized metabolites on the GC column and subsequent detection by MS [25]. A split injection mode is usually preferred in metabonomic studies where metabolites vary widely in concentrations. As discussed in previous sections, with the advent of TOFMS coupled to deconvolution software, the problem of coeluting analytes has been largely surpassed resulting in considerably reduced GC elution time [13].

Columns with varying polarities (DB-1 to DB-50) [25–27], varying internal diameters (0.25–0.18 mm) [25,28], diverse chemical composition of stationary phases, and different lengths (10–60 m) have been utilized in metabonomic studies [29]. Nevertheless, the use of DB-5MS columns or columns with similar separation properties, are commonly used in metabonomic studies [30,31]. Generally, GC oven temperatures vary from 40 to 325 °C [25,32] and the maximum temperature that can be set depends on the type of GC column

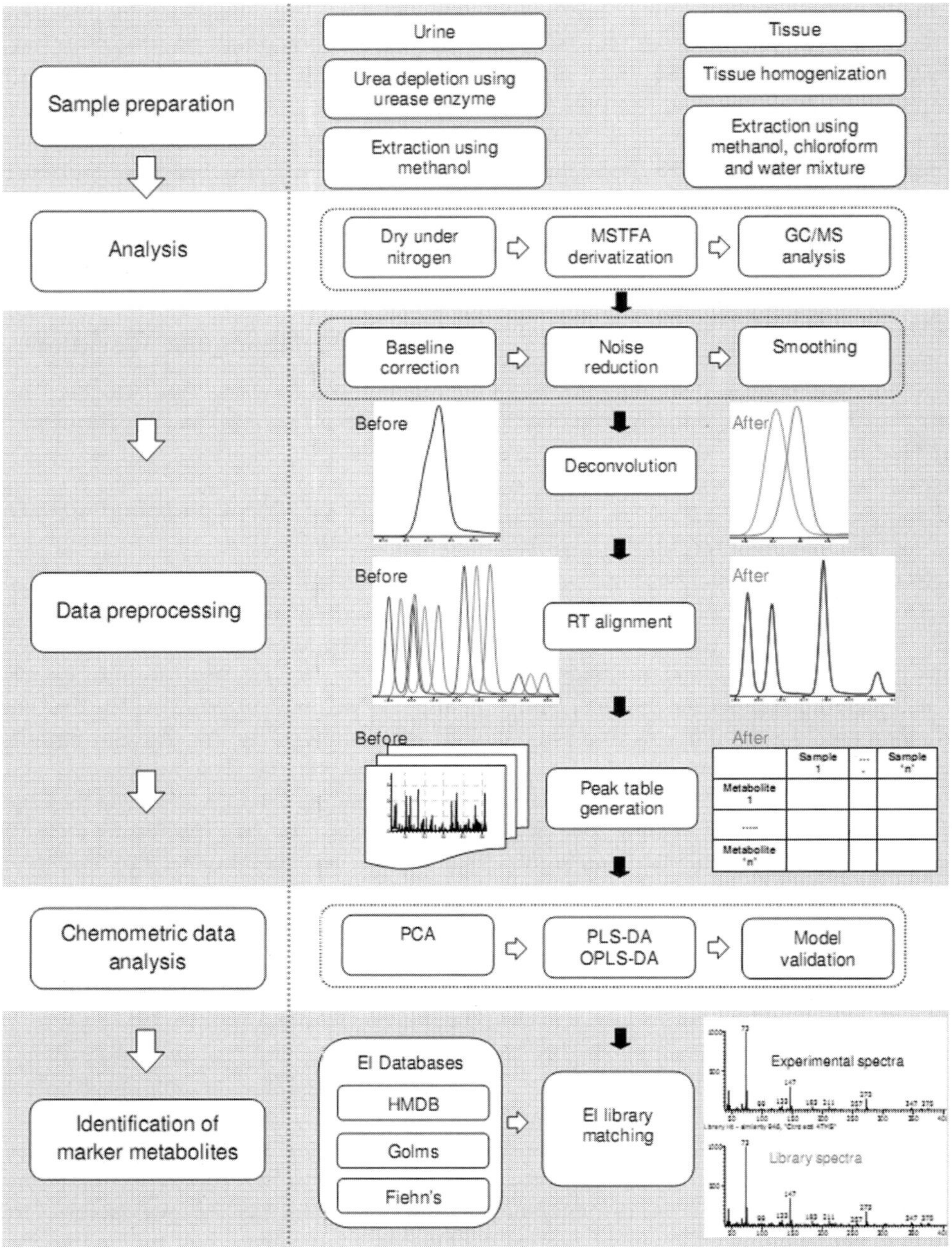

FIGURE 24.1 Summary of the main steps in urine- and tissue-based metabonomics. Subsequent to sample preparation, similar methodologies can be adopted for GC-MS data analysis, data preprocessing, chemometric data analysis, and metabolite identification.

used. Helium is the most commonly used carrier gas and typical column flow rates range from 0.8 to 2 mL/min [26,31].

Although EI is the commonly adopted mode of ionization in metabonomics, chemical ionization (CI) has also been explored particularly for compounds that do not yield molecular ions in EI due to complete fragmentation [33]. In metabonomics, MS is operated typically in full scan mode where a mass range of 50–700 is scanned [26]. As metabolites are commonly subjected to derivatization prior to GC-MS analysis, various by-products are formed concurrently by the derivatizing agent. Such by-products can be present at high concentrations in samples and therefore may cause saturation of the MS detector. Due to their high volatility, these by-products usually elute within the first few minutes of sample injection. Therefore, an acquisition delay is commonly used, during which the MS detector is kept off until these by-products are completely eluted [13].

Data acquisition is carried out by the software provided along with the instrument, and the raw data files are generated in a proprietary data format. However, for subsequent data preprocessing it is often necessary to convert the default data formats to common raw data formats, such as netCDF, ASCII, or mzXML. Some of the software packages provided with the instrument contain features for this purpose [34].

24.3.2.2. Data Preprocessing and Pretreatment

Data preprocessing in metabonomics analysis commonly refers to various steps involved in converting GC-MS raw data into data tables amenable for chemometric analysis [35]. Data preprocessing includes steps such as noise reduction, baseline correction, deconvolution, and retention time alignment (see Chapter 17). Deconvolution is necessary to resolve coeluting peaks [36], and retention time alignment is important to negate variation in retention times as chemometric techniques are inherently sensitive to retention time precision [37]. Commercial software provided by vendors of GC-MS instruments such as ChromaTOF (Leco Corp., USA) [38] as well as freely available software packages such as AMDIS (deconvolution only) [39], MathDAMP [40], MSFACTs [41], MZmine [42], XCMS [43], MetAlign [44], and TagFinder [45] can be used for GC-MS data preprocessing. Although these peak alignment algorithms reduce manual intervention significantly, each software has some limitations such as the mandatory conversion of chromatographic file formats to NetCDF and lack of all data preprocessing steps. For example, MetAlign provides retention time alignment features but does not offer deconvolution of chromatographic peaks [44]. In contrast, AnalyzerPro provides deconvolution but does not offer retention time alignment. In addition, most of the alignment software packages do not provide EI library matching for metabolite identification. Therefore, end-users still have to depend on GC-MS vendor software to perform library matching [46]. Finally, all the software packages require an in-depth understanding of the underlying algorithms, extensive optimization of parameters, and thorough validation before they can be utilized for peak alignment. Recently, Koh et al. evaluated retention time alignment accuracies of various algorithms in the alignment of derivatized peaks of metabolites and standard compounds [38]. Among the peak alignment algorithms evaluated in the study, the calibration feature of the GC-MS software showed superior performance [38]. The use of a calibration feature for peak alignment involves higher manual intervention compared to peak alignment software. This limitation is however acceptable since the focus of metabonomics is not the throughput of analysis but rather the quality of data.

Data pretreatment involves steps such as normalization, centering, scaling, and transformation [35]. Normalization is necessary to counter minor variations in data arising from sample preparation and/or instrument response, while preserving the relevant inherent biological variation [34]. Although different normalization strategies exist, normalization to total area [13] and normalization using single or multiple internal standards [28] are used predominantly in the case of GC-MS-based metabonomics. Centering helps in balancing the difference between high- and low-abundance metabolites and mean centering is typically used for this purpose [47]. Scaling is used to account for the differences in fold changes between the different metabolites by converting the data into differences in concentration relative to a scaling factor. Different scaling methods based on data dispersion such as auto-scaling, unit variance scaling, and pareto scaling are usually used for metabonomic data [48]. Transformation involves nonlinear conversion of the data, for example, log transformation or power transformation to rectify heteroscedastic variation and to improve symmetry of skewed data [48]. Mathematical environments such as MATLAB (Mathworks, Natick, MA, USA) and R (open source GNU project) or commercial chemometric software such as SIMCA-P (Umetrics, Umeå, Sweden) has in-built features to carry out such data pretreatment steps prior to chemometric analysis.

24.3.2.3. *Chemometric Data Analysis*

Metabonomic studies result in the generation of huge and complex data sets containing a large number of observations (samples) and variables (metabolites). Multivariate statistical techniques or chemometric tools are indispensable for the analysis of such data sets. Of the different chemometric methods available such as hierarchical clustering, partitional clustering, artificial neural networks, support vector machine, evolutionary-based algorithms, and regression trees, projection-based methods are extensively used in metabonomics [35]. Projection-based methods include both unsupervised methods such as principal component analysis (PCA) and supervised methods such as partial least squares discriminant analysis (PLS-DA) and orthogonal PLS-DA (OPLS-DA). Projection-based methods are based on the assumption that the system in question is controlled by a few latent variables (LVs) or principal components. As PCA is useful for identifying outliers and inherent clustering trends, it is commonly used as the first step in chemometric analysis to investigate inherent data variability and clustering trends. Outliers can also be detected using the distance to model plot (DModX) based on residual variance of the PCA model [49]. PLS-DA is a supervised method in which class information for observations is included in the analysis to further enhance the investigation of class clustering trends. The OPLS-DA is a modified form of PLS-DA where orthogonal signal correction is used to split the variation in X matrix into two parts, one that is unrelated or orthogonal to Y and the other that is related to Y [29]. This helps in easier model interpretation and identification of important variables contributing to the model as compared to PLS-DA [50]. Thus, PCA (to identify outliers and inherent clustering trends) is conventionally followed by PLS-DA or OPLS-DA to identify potential marker metabolites that are contributing significantly to the model and discriminating one class from another [51]. Chemometric software such as SIMCA-P and Unscrambler (CAMO software, Oslo, Norway) and mathematical environments such as MATLAB and R (open source GNU project) are typically employed for chemometric data analysis in metabonomics.

24.3.2.4. *Model Validation*

A typical chemometric model built in a metabonomic study is characterized by the use of a relatively small number of observations compared to the number of variables [52]. Therefore, in some cases, it is possible that optimistic performance characteristics observed in a PLS-DA or OPLS-DA model could be due to specimen artifacts, overfitting of data, or chance correlation [51,53]. Therefore, validation of each model should be performed before it can be leveraged to predict the unknown observations or identification of marker metabolites. Model validation can be performed in two stages starting with internal validation and followed by external validation. Internal validation of PLS-DA models can be performed using a permutation test and receiver operating characteristic (ROC) analysis [54]. In the permutation test, goodness of fit (R^2 and Q^2) of the original model is compared with the goodness of fit of several PLS-DA models built using the data matrix where the order of the Y observations are randomly permuted, while the X matrix is kept intact [55,56]. Additionally, ROC analysis using the cross-validated predicted Y (predicted class) values can be further performed to verify the robustness of the model. The sensitivity and specificity trade-offs can be summarized for each model using the area under the ROC curve (AUC) and calculated using the trapezoidal rule [54]. ROC analysis can provide a good indication of sensitivity and specificity that can be achieved in predicting unknown samples.

Subsequent to confirmation of validity of each model using internal validation strategies, model validity can be further confirmed using external validation. To perform external validation, a subset of observations is randomly selected for building a training set and classification of the remaining samples is then predicted. The selection of training and tests sets should be defined prior to chemometric analysis to represent actual prediction of unknown samples and to avoid any bias related to data preprocessing and pretreatment. External validation can be performed iteratively by randomly selecting different combinations of training and test sets to estimate the predictive ability of the model. Most software packages (SIMCA-P or MetaboAnalyst) offer in-built features to perform internal and external validation. Once the validity of each model is confirmed using internal and external validation strategies, statistical significance of potential marker metabolites in control and treatment groups should be further verified by univariate statistical tests such as the Welch t-test [13].

24.3.2.5. Biomarker Identification and Pathway Mapping

Subsequent to GC-MS analysis, preliminary identification of metabolites is carried out by library searching against mass spectral library databases provided with the instrument such as the National Institute of Standards and Technology (NIST) database as well as online databases such as the Human Metabolome Database (HMDB) [57], NIST Chemistry Web Book (webbook.nist.gov/chemistry), Madison Metabolomic Consortium Database (mmcd.nmrfam.wisc.edu), SpecInfo (cds.dl.ac.uk/cds/datasets/spec/specinfo), and Spectral Database for Organic Compounds (riodb01.ibase.aist.go.jp/sdbs/cgi-bin/cre_index.cgi). The identities of such significant marker metabolites are further confirmed by using commercially available standard compounds [13]. Subsequent to confirmation, it is pertinent to link the marker metabolites to metabolic pathways. Metabolic pathway mapping is important to identify significantly perturbed metabolic pathways in response to a diseased condition, or genetic modification, or pharmacological intervention. Various databases such as the Kyoto Encyclopedia of Genes and Genomes (KEGG) [58], BioCyc [59], and Reactome [60] and software packages such as KegArray (used with KEGG), Pathway Tools (used with BioCyc), and MetaCore (GeneGo, St Joseph, MI, USA) are available for metabolic pathway analysis.

24.4. GC-MS-BASED TISSUE METABONOMICS

24.4.1. Sample Preparation

The first step in sample preparation for GC-MS-based tissue metabonomics involves effective extraction of metabolites from the tissue matrix. Different extraction strategies such as homogenization alone, or homogenization followed by ultrasonication, or ultrasonication alone (usually applicable for small amount of tissue, ≤20 mg) can be adopted [13,61]. The choice of extraction solvent depends on the nature of the metabonomic study. In the case of global metabonomics, a mixture of different solvents [for instance, chloroform/methanol/water in the ratio of 2:5:2 (v/v/v)] covering a wide polarity range is typically used to cover the extraction of as large a metabolic space as possible [61]. In the case of targeted metabolic profiling, it is relevant to use a solvent or solvent mixture of narrow polarity range to selectively extract metabolites belonging to a specific chemical class. After extraction, the tissue extract is typically dried by using a nitrogen evaporator, followed by addition of a sufficient quantity of anhydrous toluene (about 100 μL) to the extract. The mixture is evaporated to complete dryness using a nitrogen evaporator in order to eliminate any trace of moisture that might interfere with GC-MS analysis. The dried extract thus obtained is subjected to derivatization. The derivatization strategy involves methoximation of metabolites using methoxyamine hydrochloride (MOX) reagent in pyridine (28–37 °C, up to 2 h or 16 h) followed by trimethylsilyl (TMS) derivatization using reagents such as N,O-bis-(trimethylsilyl)trifluoroacetamide (BSTFA) with 1% trimethylchlorosilane (TMCS) or N-methyl-N-trifluoroacetamide (MSTFA) with 1% TMCS at 70 °C for 30 min or 37 °C for 1 h. Methoximation prior to TMS derivatization is considered necessary to prevent cyclization of sugars as well as protection of α-ketoacids against decarboxylation, and fixation of enolizable keto groups. This will help reduce the number of multiple-derivatized metabolites. After derivatization, the samples are injected into a GC-MS system for analysis [13,26,61].

24.4.2. Applications

Tissue-based metabonomics differs from plasma- or urine-based metabonomics as it provides anatomical site-specific information of the endogenous metabolites and is less susceptible to variation as compared to plasma or urine. Therefore, tissues are suitable biomatrices for metabolic profiling where the main aim of the metabonomic study is to reveal the metabolic phenotype and deregulated metabolic pathways associated with a disease or pharmacological intervention rather than to diagnose or prognose a pathology.

Tissue-based metabonomics using GC-MS either alone or in conjunction with other methods such as NMR, CE-MS, or LC-MS has shown promise in identifying metabolite-based biomarkers in different forms of cancer. Denkert et al. found that molecular changes in ovarian tumor tissues can be characterized by quantitative changes in metabolic profiles [62]. Tissue-based metabonomic studies in human colorectal cancer have revealed perturbations in various biochemical processes such as glycolysis, Kreb's cycle, amino acid metabolism, fatty acid biosynthesis, steroid biosynthesis, eicosanoid biosynthesis, bile acid biosynthesis, nucleotide metabolism, and osmoregulation. The majority of these observations were attributed to the high energy demand, tissue hypoxia, and altered synthetic rate of cellular components of rapidly proliferating tumor cells [61,63–65]. Moreover, in the study by Ong et al. it was found that distinct progressive changes in metabolite profiles accompany the transformation of normal colonic mucosa to malignant tumor [65]. GC-MS-based metabonomic studies in human gastric cancer as well as in animal models of gastric cancer have resulted in the identification of several marker metabolites having the potential for diagnosis and staging of gastric cancer as well as for prognosis of metastatic progression [66,67]. Wu et al. utilized GC-MS-based metabolic profiling of biopsied tissue specimens obtained from esophageal cancer patients, to identify several marker metabolites of diagnostic potential [68]. Similar studies have resulted in the identification of metabolite-based biomarkers of brain cancer [69,70].

GC-MS has also been utilized in the metabonomic studies of carbon tetrachloride-induced liver injury in mouse [71] and rat [72], metabolic syndrome using cardiac and adipose tissues of PPAR-alpha null mutant mouse [73], and microbial marker metabolites of endocarditis in human cardiac tissue [74]. The application of GC-MS can also be found for tissue-based targeted metabonomic studies of L-β-methyl-aminoalanine in human brain [75] and prostaglandins in human lung [76] and colorectal cancer [77–79]. GC-MS-based metabolic profiling has also been used to validate the 3-nitropropionic acid-induced early stage Huntington's disease rat model [80].

24.5. GC-MS-BASED URINE METABONOMICS

Although tissue metabonomic studies are important in the mechanistic elucidation of diseases, harvesting tissue is an invasive process, often requiring biopsy. Compared to the analysis of blood and tissue samples, the advantages of urinalysis are numerous: urine sample collection is noninvasive and not limited by volume. In addition, urine can be used for measuring time-resolved, dynamic, or temporal data, an invaluable attribute for the investigation of the pathogenesis, progression, and prognostication of acute and chronic diseases. On the other hand, the use of blood as a matrix has limitations in some cases since the metabolic composition of blood may be affected by multiple organ systems and alterations detected in the serum and plasma may not be specific to a particular disease [81]. The application of GC-MS in urinary metabolic profiling is a long-standing practice [82]. As expected, there were significant improvements in urine sample preparation and data processing over time. Shoemaker et al. contributed to a significant improvement in sample preparation by employing the urease enzyme to deplete urea that caused major chromatographic interference and masked many of the low-intensity metabolite peaks [83]. Remarkable improvement in analytical turnaround time in urinary metabonomics was realized with the coupling of GC to TOFMS, as noted earlier.

The majority of recent research papers have reported the application of GC-MS in metabonomics rather than the advancement of the technique. This indicates a high level of acceptance and maturity of the GC-MS platform. Although the composition of urine may reflect a disease state or a pharmacological effect, it is important to note that the urine metabotype is equally sensitive to other factors such as dietary intake [84] and physiological conditions such as age, gender, and demographic characteristics [85,86]. In some instances, urine may contain xenobiotics and their metabolites, which can introduce additional complexities in downstream data analysis. Furthermore, variation in metabolic composition could be introduced during sample collection and storage. Therefore, proper study design in urinary metabonomics is extremely important to obtain meaningful data. Baseline characteristics of study groups, sample collection, and storage conditions should be carefully controlled and standardized according to the research hypothesis. Variations due to diet and lifestyle factors can be minimized by collection of first-pass urine [84]. However, collection and storage of first-pass urine are challenging due to poor patient compliance. Evaluation of stability of urine samples using GC-MS [85] and other analytical platforms showed that urine samples are stable up to 6 months when stored at −20 or −80 °C. No significant difference in urine metabolite compositions was observed when the urine samples were subjected to repeated freeze−thaw cycles (up to 9 cycles) [85,87]. Lauridsen et al. showed that addition of a preservative is not mandatory provided urine samples are stored at below −20 °C [88]. On the other hand, instability of some of the metabolites was noted when urine was subjected to freeze drying [88].

24.5.1. Sample Preparation

Urine presents a wide dynamic range of metabolite concentrations, the occurrence of urea as a chromatographic interference, and the unpredictable degree of urinary dilution. Therefore, sample preparation in urinary metabolic profiling is distinct from other biomatrices. The urine sample preparation varies according to the derivatizing reagent utilized in a study. Among several derivatizing agents explored in urinary metabolic profiling, TMS derivatizing agents such as BSTFA and MSTFA are used predominantly [89–92]. Recently, derivatization using ethyl chloroformate is gaining popularity particularly in urine metabolic profiling. In contrast to TMS derivatizing agents, which only work in the nonaqueous phase, ethyl chloroformate (ECF) remains reactive in aqueous medium. Briefly, urine preparation for TMS derivatization includes incubation of urine with urease enzyme, extraction using methanol, drying of the metabolic extract, and a two-step derivatization process first using MOX and subsequently by MSTFA, whereas, ECF derivatization includes the addition of anhydrous ethanol and pyridine to urine samples followed by ECF. The mixture is then sonicated, and subsequent extraction is performed using chloroform, with the aqueous layer pH adjusted to 9–10 using NaOH. In some instances, the derivatization procedure is repeated with the addition of ECF to the mixture. After the two successive repetitions of ECF derivatization, the aqueous layer is aspirated off, while the remaining chloroform layer containing the derivatives is isolated, dried with anhydrous sodium sulfate, and analyzed by GC-MS [93].

Due to the wide adoptability of TMS derivatization in metabonomic research, sample preparation employing TMS derivatization is presented in more detail here. Typically, a small volume of urine (200–500 μL) is utilized for studies involving TMS derivatization. Sample preparation is initiated by incubating individual urine samples with urease enzyme to deplete excess urea, which is a major chromatographic interferent. Up to 30–100 U of urease enzyme is used, depending on the volume of urine sample [85,94]. Termination of urease activity and extraction of metabolites are carried out using ethanol [94] or methanol [54]. Similar extraction efficiencies of urinary metabolites by ethanol and methanol were noted [85]. Similar to tissue samples, the extracted metabolites are dried and subjected to two-step derivatization, first with MOX reagent, followed by a TMS derivatizing reagent such as MSTFA. However, it is important to note that silylation can cause conversion reactions for some metabolites, for example, arginine is converted into ornithine by reaction with BSTFA or MSTFA [95].

24.5.2. Applications

Applications of GC-based urinary metabolic profiling are wide ranging, which include studying diseases of the renal system such as bladder, ovarian, and kidney cancers [31,54,96,97] and a host of other nonrenal-related studies [98]. Applications of GC-TOFMS-based metabonomic analyses were reported for biomarker discovery and elucidation of pathogenesis of various diseases such as diabetes [16,99], hepatocellular carcinoma [100], kidney cancer [31], bladder cancer [54], colorectal cancer [101], ovarian cancer [62], osteosarcoma [102], acute coronary syndrome [103], and Crohn's disease [104].

GC-based urinary metabonomics was also successfully applied in toxicological evaluation. One way the biological systems adapt to maintain homeostasis in the face of a toxic challenge is to modulate the composition of biofluids by eliminating substances via the kidney [105]. Even when cellular homeostasis is maintained, subtle responses to toxicity or disease are expressed in altered biofluid composition such as the change in urine metabotypes [105]. Although GC-MS is relatively less utilized in the area of toxicology compared to NMR and LC-MS, recent studies demonstrate the potential of urinary metabolic profiling by GC-MS as a complementary tool in toxicological evaluations, providing a comprehensive understanding of the response of biological system to xenobiotic intervention. Recently, Chen et al. utilized GC-MS-based urinary metabolic profiling to elucidate the toxicity induced by orally administered multiglycosides of *Tripterygium wilfordii* Hook. f. (GTW) in rats [106]. Urine samples at various time points and predose urine were collected from GTW-administered rats, and the samples were analyzed using both GC-MS and LC-MS. The study indicated that GTW caused a time-dependent toxic effect at a high dose as revealed by the perturbed metabolic regulatory network [106]. This multiplatform metabolic profiling approach successfully aided in gaining insights into metabolic alterations associated with the onset and progression of multiorgan toxicity induced by GTW. GC-MS-based urinary metabonomics was applied in nephrotoxicity investigation in rats, where the toxicity was induced by various xenobiotics such as aristolochic acid [30], gentamicin [107,108], cisplatin [108], and tobramycin [108]. Likewise, urinary metabonomics was utilized to understand hepatotoxicity caused by fenofibrate [109], atorvastatin [110], valproic acid [111], and carbon tetrachloride [112]. Other applications of urinary metabonomics comprised nutritional studies [113–115]. Collectively, these studies reinforced the applicability of GC-MS-based urinary metabolic profiling in a large variety of metabonomic investigations.

24.6. FUTURE DIRECTIONS

Of the different sampling and transfer methods available for headspace GC-MS, solid-phase microextraction (SPME) has emerged as a promising technique because of its low sample size requirement, sensitivity, and minimal sample pretreatment as compared to conventional methods. In headspace SPME, an SPME fiber is introduced as adsorbent into the headspace to trap volatile analytes. Various SPME fibers made of materials such as polydimethylsiloxane and polydimethylsiloxane/divinylbenzene are commercially available for this purpose. In order to analyze more polar and nonvolatile analytes by headspace SPME, the direct immersion mode is often necessary in which the SPME device (stir-bars or thin films) is brought into direct contact with the biomatrix of interest [116]. SPME in conjunction with headspace GC-MS has already been successfully utilized in metabonomic studies involving human skin emissions [117–119], breath in human lung cancer [120,121] and lung infections [122], colon cancer cells [123], blood of liver cancer patients [124], and *Helicobacter pylori* infection in human gastric cancer [125]. Recent development of newer SPME devices has further expanded its metabolic space coverage [126]. Therefore, an increasing use of the SPME-based GC-MS method for metabonomic studies is expected in the future.

Gas chromatography tandem mass spectrometry (GC-MS/MS) has emerged as a suitable technique for targeted analysis because of its higher selectivity and unambiguous metabolite identification as compared to conventional GC-MS. Recently, GC-MS/MS has been used for the analysis of marker metabolites such as those related to alcohol abuse in human hair [127–129], 4-aminobiphenyl hemoglobin adducts in smokers and nonsmokers [130], estrone sulfate in human plasma [131], endogenous anabolic steroids in human hair [132], endogenous organic acids in human urine [133], tyrosine metabolites in human urine [134], plasma [135], exhaled breath [136], and pyrimidine metabolites in human urine [137]. Thus, an increasing trend for using GC-MS/MS for targeted profiling and validation of metabolite-based biomarkers is anticipated in the future.

24.7. CONCLUSION

Metabonomics studies utilizing GC-MS alone or as a complementary method to other analytical platforms such as NMR, CE-MS, or LC-MS have a significant impact in the field of biomedical research. The applications of such studies can be found in different spheres of research such as cancer biology, toxicology, nutritional studies, pathogenesis of diseases such as diabetes, osteosarcoma, cardiovascular diseases, and neurodegenerative diseases. Headspace SPME and GC-MS/MS are poised to extend the range of applications of GC-MS-based metabonomics.

References

[1] J.K. Nicholson, J.C. Lindon, E. Holmes, 'Metabonomics': understanding the metabolic responses of living systems to pathophysiological stimuli via multivariate statistical analysis of biological NMR spectroscopic data, Xenobiotica 29 (1999) 1181–1189.

[2] O. Fiehn, Combining genomics, metabolome analysis, and biochemical modelling to understand metabolic networks, Comp. Funct. Genomics 2 (2001) 155–168.

[3] M. Morris, S.M. Watkins, Focused metabolomic profiling in the drug development process: advances from lipid profiling, Curr. Opin. Chem. Biol. 9 (2005) 407–412.

[4] M. Urpi-Sarda, M. Monagas, N. Khan, R. Llorach, R.M. Lamuela-Raventos, O. Jauregui, et al., Targeted metabolic profiling of phenolics in urine and plasma after regular consumption of cocoa by liquid chromatography-tandem mass spectrometry, J. Chromatogr. A 1216 (2009) 7258–7267.

[5] W.B. Dunn, N.J. Bailey, H.E. Johnson, Measuring the metabolome: current analytical technologies, Analyst. 130 (2005) 606–625.

[6] E.M. Lenz, I.D. Wilson, Analytical strategies in metabonomics, J. Proteome. Res. 6 (2007) 443–458.

[7] G.G. Harrigan, R.H. LaPlante, G.N. Cosma, G. Cockerell, R. Goodacre, J.F. Maddox, et al., Application of high-throughput Fourier-transform infrared spectroscopy in toxicology studies: contribution to a study on the development of an animal model for idiosyncratic toxicity, Toxicol. Lett. 146 (2004) 197–205.

[8] H. Pham-Tuan, L. Kaskavelis, C.A. Daykin, H.G. Janssen, Method development in high-performance liquid chromatography for high-throughput profiling and metabonomic studies of biofluid samples, J. Chromatogr. B Analyt. Technol. Biomed. Life. Sci. 789 (2003) 283–301.

[9] K.E. Vigneau-Callahan, A.I. Shestopalov, P.E. Milbury, W.R. Matson, B.S. Kristal, Characterization of diet-dependent metabolic serotypes: analytical and biological variability issues in rats, J. Nutr. 131 (2001) 924S–932S.

[10] S. Zomer, C. Guillo, R.G. Brereton, M. Hanna-Brown, Toxicological classification of urine samples using pattern recognition techniques and capillary electrophoresis, Anal. Bioanal. Chem. 378 (2004) 2008–2020.

[11] J.C. Lindon, E. Holmes, J.K. Nicholson, Metabonomics in pharmaceutical R&D, FEBS J. 274 (2007) 1140–1151.

[12] J.C. Lindon, J.K. Nicholson, I.D. Wilson, Directly coupled HPLC-NMR and HPLC-NMR-MS in pharmaceutical research and development, J. Chromatogr. B Biomed. Sci. Appl. 748 (2000) 233–258.

[13] K.K. Pasikanti, P.C. Ho, E.C. Chan, Gas chromatography/mass spectrometry in metabolic profiling of biological fluids, J. Chromatogr. B Analyt. Technol. Biomed. Life Sci. 871 (2008) 202–211.

[14] J.W. Allwood, A. Erban, S. de Koning, W.B. Dunn, A. Luedemann, A. Lommen, et al., Inter-laboratory reproducibility of fast gas chromatography-electron impact-time of flight mass spectrometry (GC-EI-TOF/MS) based plant metabolomics, Metabolomics 5 (2009) 479–496.

[15] M.F. Almstetter, I.J. Appel, M.A. Gruber, C. Lottaz, B. Timischl, R. Spang, et al., Integrative normalization and comparative analysis for metabolic fingerprinting by comprehensive two-dimensional gas chromatography-time-of-flight mass spectrometry, Anal. Chem. 81 (2009) 5731–5739.

[16] X. Li, Z. Xu, X. Lu, X. Yang, P. Yin, H. Kong, et al., Comprehensive two-dimensional gas chromatography/time-of-flight mass spectrometry for metabonomics: biomarker discovery for diabetes mellitus, Anal. Chim. Acta 633 (2009) 257–262.

[17] K. Ralston-Hooper, A. Hopf, C. Oh, X. Zhang, J. Adamec, M.S. Sepulveda, Development of GCxGC/TOF-MS metabolomics for use in ecotoxicological studies with invertebrates, Aquat. Toxicol. 88 (2008) 48–52.

[18] R.E. Mohler, K.M. Dombek, J.C. Hoggard, K.M. Pierce, E.T. Young, R.E. Synovec, Comprehensive analysis of yeast metabolite GC x GC-TOFMS data: combining discovery-mode and deconvolution chemometric software, Analyst 132 (2007) 756–767.

[19] R.A. Shellie, W. Welthagen, J. Zrostlikova, J. Spranger, M. Ristow, O. Fiehn, et al., Statistical methods for comparing comprehensive two-dimensional gas chromatography-time-of-flight mass spectrometry results: metabolomic analysis of mouse tissue extracts, J. Chromatogr. A 1086 (2005) 83–90.

[20] M. Adahchour, J. Beens, U.A. Brinkman, Recent developments in the application of comprehensive two-dimensional gas chromatography, J. Chromatogr. A 1186 (2008) 67–108.

[21] W. Khummueng, J. Harynuk, P.J. Marriott, Modulation ratio in comprehensive two-dimensional gas chromatography, Anal. Chem. 78 (2006) 4578–4587.

[22] D. Ryan, P. Morrison, P. Marriott, Orthogonality considerations in comprehensive two-dimensional gas chromatography, J. Chromatogr. A 1071 (2005) 47–53.

[23] J.M. Dimandja, G.C. Clouden, I. Colon, J.F. Focant, W.V. Cabey, R.C. Parry, Standardized test mixture for the characterization of comprehensive two-dimensional gas chromatography columns: the Phillips mix, J. Chromatogr. A 1019 (2003) 261–272.

[24] M.M. Koek, B. Muilwijk, L.L. van Stee, T. Hankemeier, Higher mass loadability in comprehensive two-dimensional gas chromatography-mass spectrometry for improved analytical performance in metabolomics analysis, J. Chromatogr. A 1186 (2008) 420–429.

[25] S.A. Fancy, O. Beckonert, G. Darbon, W. Yabsley, R. Walley, D. Baker, et al., Gas chromatography/flame ionisation detection mass spectrometry for the detection of endogenous urine metabolites for metabonomic studies and its use as a complementary tool to nuclear magnetic resonance spectroscopy, Rapid Commun. Mass Spectrom. 20 (2006) 2271–2280.

[26] J. Lisec, N. Schauer, J. Kopka, L. Willmitzer, A.R. Fernie, Gas chromatography mass spectrometry-based metabolite profiling in plants, Nat. Protoc. 1 (2006) 387–396.

[27] Q. Zhang, G. Wang, Y. Du, L. Zhu, A. Jiye, GC-MS analysis of the rat urine for metabonomic research, J. Chromatogr. B Analyt. Technol. Biomed. Life Sci. 854 (2007) 20–25.

[28] S. Bijlsma, I. Bobeldijk, E.R. Verheij, R. Ramaker, S. Kochhar, I.A. Macdonald, et al., Large-scale human metabolomics studies: a strategy for data (pre-) processing and validation, Anal. Chem. 78 (2006) 567–574.

[29] S. Wiklund, E. Johansson, L. Sjostrom, E.J. Mellerowicz, U. Edlund, J.P. Shockcor, et al., Visualization of GC/TOF-MS-based metabolomics data for identification of biochemically interesting compounds using OPLS class models, Anal. Chem. 80 (2008) 115–122.

[30] Y. Ni, M. Su, Y. Qiu, M. Chen, Y. Liu, A. Zhao, et al., Metabolic profiling using combined GC-MS and LC-MS provides a systems understanding of aristolochic acid-induced nephrotoxicity in rat, FEBS Lett. 581 (2007) 707–711.

[31] T. Kind, V. Tolstikov, O. Fiehn, R.H. Weiss, A comprehensive urinary metabolomic approach for identifying kidney cancer, Anal. Biochem. 363 (2007) 185–195.

[32] K. Yuan, H. Kong, Y. Guan, J. Yang, G. Xu, A GC-based metabonomics investigation of type 2 diabetes by organic acids metabolic profile, J. Chromatogr. B Analyt. Technol. Biomed. Life Sci. 850 (2007) 236–240.

[33] H.J. Major, R. Williams, A.J. Wilson, I.D. Wilson, A metabonomic analysis of plasma from Zucker rat strains using gas chromatography/mass spectrometry and pattern recognition, Rapid Commun. Mass Spectrom. 20 (2006) 3295–3302.

[34] M. Katajamaa, M. Oresic, Data processing for mass spectrometry-based metabolomics, J. Chromatogr. A 1158 (2007) 318–328.

[35] R. Goodacre, D. Broadhurst, A.K. Smilde, B.S. Kristal, J.D. Baker, R. Beger, et al., Proposed minimum reporting standards for data analysis in metabolomics, Metabolomics 3 (2007) 231–241.

[36] P. Jonsson, J. Gullberg, A. Nordstrom, M. Kusano, M. Kowalczyk, M. Sjostrom, et al., A strategy for identifying differences in large series of metabolomic samples analyzed by GC-MS, Anal. Chem. 76 (2004) 1738–1745.

[37] K.J. Johnson, B.W. Wright, K.H. Jarman, R.E. Synovec, High-speed peak matching algorithm for retention time alignment of gas chromatographic data for chemometric analysis, J. Chromatogr. A 996 (2003) 141–155.

[38] Y. Koh, K.K. Pasikanti, C.W. Yap, E.C. Chan, Comparative evaluation of software for retention time alignment of gas chromatography/time-of-flight mass spectrometry-based metabonomic data, J. Chromatogr. A 1217 (2010) 8308–8316.

[39] J.M. Halket, A. Przyborowska, S.E. Stein, W.G. Mallard, S. Down, R.A. Chalmers, Deconvolution gas chromatography/mass spectrometry of urinary organic acids—potential for pattern recognition and automated identification of metabolic disorders, Rapid Commun. Mass Spectrom. 13 (1999) 279–284.

[40] R. Baran, H. Kochi, N. Saito, M. Suematsu, T. Soga, T. Nishioka, et al., MathDAMP: a package for differential analysis of metabolite profiles, BMC Bioinformatics 7 (2006) 530.

[41] A.L. Duran, J. Yang, L. Wang, L.W. Sumner, Metabolomics spectral formatting, alignment and conversion tools (MSFACTs), Bioinformatics 19 (2003) 2283–2293.

[42] M. Katajamaa, J. Miettinen, M. Cresic, MZmine: toolbox for processing and visualization of mass spectrometry based molecular profile data, Bioinformatics 22 (2006) 634–636.

[43] C.A. Smith, E.J. Want, G. O'Maille, R. Abagyan, G. Siuzdak, XCMS: processing mass spectrometry data for metabolite profiling using nonlinear peak alignment, matching, and identification, Anal. Chem. 78 (2006) 779–787.

[44] A. Lommen, MetAlign: interface-driven, versatile metabolomics tool for hyphenated full-scan mass spectrometry data preprocessing, Anal. Chem. 81 (2009) 3079–3086.

[45] A. Luedemann, K. Strassburg, A. Erban, J. Kopka, TagFinder for the quantitative analysis of gas chromatography–mass spectrometry (GC-MS)-based metabolite profiling experiments, Bioinformatics 24 (2008) 732–737.

[46] Y. Tikunov, A. Lommen, C.H. de Vos, H.A. Verhoeven, R.J. Bino, R.D. Hall, et al., A novel approach for nontargeted data analysis for metabolomics. Large-scale profiling of tomato fruit volatiles, Plant Physiol. 139 (2005) 1125–1137.

[47] R. Bro, A.K. Smilde, Centering and scaling in component analysis, J. Chemometr. 17 (2003) 16–33.

[48] R.A. van den Berg, H.C. Hoefsloot, J.A. Westerhuis, A.K. Smilde, M.J. van der Werf, Centering, scaling, and transformations: improving the biological information content of metabolomics data, BMC Genomics 7 (2006) 142.

[49] J.C. Lindon, E. Holmes, J.K. Nicholson, Pattern recognition methods and applications in biomedical magnetic resonance, Prog. Nucl. Mag. Res. Sp. 39 (2001) 1–40.

[50] M. Bylesjo, D. Eriksson, A. Sjodin, S. Jansson, T. Moritz, J. Trygg, Orthogonal projections to latent structures as a strategy for microarray data normalization, BMC Bioinformatics 8 (2007) 207.

[51] J. Trygg, E. Holmes, T. Lundstedt, Chemometrics in metabonomics, J. Proteome. Res. 6 (2007) 469–479.

[52] J. Westerhuis, H. Hoefsloot, S. Smit, D. Vis, A. Smilde, E. van Velzen, et al., Assessment of PLSDA cross validation, Metabolomics 4 (2008) 81–89.

[53] S. Smit, M.J. van Breemen, H.C. Hoefsloot, A.K. Smilde, J.M. Aerts, C.G. de Koster, Assessing the statistical validity of proteomics based biomarkers, Anal. Chim. Acta. 592 (2007) 210–217.

[54] K.K. Pasikanti, K. Esuvaranathan, P.C. Ho, R. Mahendran, R. Kamaraj, Q.H. Wu, et al., Noninvasive urinary metabonomic diagnosis of human bladder cancer, J. Proteome. Res. 9 (2010) 2988–2995.

[55] S. Wiklund, D. Nilsson, L. Eriksson, M. Sjostrom, S. Wold, K. Faber, A randomization test for PLS component selection, J. Chemometr. 21 (2007) 427–439.

[56] S. Mahadevan, S.L. Shah, T.J. Marrie, C.M. Slupsky, Analysis of metabolomic data using support vector machines, Anal. Chem. 80 (2008) 7562–7570.

[57] D.S. Wishart, C. Knox, A.C. Guo, R. Eisner, N. Young, B. Gautam, et al., HMDB: a knowledgebase for the human metabolome, Nucleic Acids Res. 37 (2009) D603–610.

[58] M. Kanehisa, S. Goto, KEGG: kyoto encyclopedia of genes and genomes, Nucleic Acids Res. 28 (2000) 27–30.

[59] R. Caspi, H. Foerster, C.A. Fulcher, P. Kaipa, M. Krummenacker, M. Latendresse, et al., The MetaCyc database of metabolic pathways and enzymes and the BioCyc collection of pathway/genome databases, Nucleic Acids Res. 36 (2008) D623–631.

[60] D. Croft, G. O'Kelly, G. Wu, R. Haw, M. Gillespie, L. Matthews, et al., Reactome: a database of reactions, pathways and biological processes, Nucleic Acids Res. 39 (2011) D691–697.

[61] M. Mal, P.K. Koh, P.Y. Cheah, E.C. Chan, Development and validation of a gas chromatography/mass spectrometry method for the metabolic profiling of human colon tissue, Rapid Commun. Mass Spectrom. 23 (2009) 487–494.

[62] C. Denkert, J. Budczies, T. Kind, W. Weichert, P. Tablack, J. Sehouli, et al., Mass spectrometry-based metabolic profiling reveals different metabolite patterns in invasive ovarian carcinomas and ovarian borderline tumors, Cancer Res. 66 (2006) 10795–10804.

[63] C. Denkert, J. Budczies, W. Weichert, G. Wohlgemuth, M. Scholz, T. Kind, et al., Metabolite profiling of human colon carcinoma–deregulation of TCA cycle and amino acid turnover, Mol. Cancer 7 (2008) 72.

[64] E.C. Chan, P.K. Koh, M. Mal, P.Y. Cheah, K.W. Eu, A. Backshall, et al., Metabolic profiling of human colorectal cancer using high-resolution magic angle spinning nuclear magnetic resonance (HR-MAS NMR) spectroscopy and gas chromatography mass spectrometry (GC-MS), J. Proteome. Res. 8 (2009) 352–361.

[65] E.S. Ong, L. Zou, S. Li, P.Y. Cheah, K.W. Eu, C.N. Ong, Metabolic profiling in colorectal cancer reveals signature metabolic shifts during tumorigenesis, Mol. Cell. Proteomics (2010).

[66] H. Wu, R. Xue, Z. Tang, C. Deng, T. Liu, H. Zeng, et al., Metabolomic investigation of gastric cancer tissue using gas chromatography/mass spectrometry, Anal. Bioanal. Chem. 396 (2010) 1385–1395.

[67] J.L. Chen, H.Q. Tang, J.D. Hu, J. Fan, J. Hong, J.Z. Gu, Metabolomics of gastric cancer metastasis detected by gas chromatography and mass spectrometry, World J. Gastroenterol. 16 (2010) 5874–5880.

[68] H. Wu, R. Xue, C. Lu, C. Deng, T. Liu, H. Zeng, et al., Metabolomic study for diagnostic model of oesophageal cancer using gas chromatography/mass spectrometry, J. Chromatogr. B Analyt. Technol. Biomed. Life Sci. 877 (2009) 3111–3117.

[69] S. Wold, E. Johansson, E. Jellum, I. Bjornson, R. Nesbakken, Application of Simca multivariate data-analysis to the classification of gas-chromatographic profiles of human-brain tissues, Anal. Chim. Acta-Comp. 5 (1981) 251–259.

[70] E. Jellum, I. Bjornson, R. Nesbakken, E. Johansson, S. Wold, Classification of human cancer-cells by means of capillary gas-chromatography and pattern-recognition analysis, J. Chromatogr. 217 (1981) 231–237.

[71] G.Y.F.H. Xin, Y. Ke, C. Yi Yu, Gas chromatography-mass spectrometry based on metabonomic study of carbon tetrachloride-induced acute liver injury in mice, Chinese J. Anal. Chem. 27 (2006) 1736–1740.

[72] L. Pan, Y. Qiu, T. Chen, J. Lin, Y. Chi, M. Su, et al., An optimized procedure for metabonomic analysis of rat liver tissue using gas chromatography/time-of-flight mass spectrometry, J. Pharm Biomed. Anal. 52 (2010) 589–596.

[73] H.J. Atherton, N.J. Bailey, W. Zhang, J. Taylor, H. Major, J. Shockcor, et al., A combined H-1-NMR spectroscopy- and mass spectrometry-based metabolomic study of the PPAR-alpha null mutant mouse defines profound systemic changes in metabolism linked to the metabolic syndrome, Physiol. Genomics 27 (2006) 178–186.

[74] O.N. Khabib, N.V. Beloborodova, G.A. Osipov, Detection of bacterial molecular markers in the tissue of cardiac valves in normal and pathological states by gas chromatography and mass spectrometry, Zh Mikrobiol. Epidemiol. Immunobiol. (2004) 62–68.

[75] L.R. Snyder, J.C. Hoggard, T.J. Montine, R.E. Synovec, Development and application of a comprehensive two-dimensional gas chromatography with time-of-flight mass spectrometry method for the analysis of L-beta-methylamino-alanine in human tissue, J. Chromatogr. A 1217 (2010) 4639–4647.

[76] W.C. Hubbard, C.L. Litterst, M.C. Liu, E.R. Bleecker, J.C. Eggleston, T.L. McLemore, et al., Profiling of prostaglandin biosynthesis in biopsy fragments of human lung carcinomas and normal human lung by capillary gas chromatography-negative ion chemical ionization mass spectrometry, Prostaglandins 32 (1986) 889–906.

[77] A. Bennett, A. Civier, C.N. Hensby, P.B. Melhuish, I.F. Stamford, Measurement of arachidonate and its metabolites extracted from human normal and malignant gastrointestinal tissues, Gut. 28 (1987) 315–318.

[78] V.W. Yang, J.M. Shields, S.R. Hamilton, E.W. Spannhake, W.C. Hubbard, L.M. Hylind, et al., Size-dependent increase in prostanoid levels in adenomas of patients with familial adenomatous polyposis, Cancer Res. 58 (1998) 1750–1753.

[79] F.M. Giardiello, E.W. Spannhake, R.N. DuBois, L.M. Hylind, C.R. Robinson, W.C. Hubbard, et al., Prostaglandin levels in human colorectal mucosa: effects of sulindac in patients with familial adenomatous polyposis, Dig. Dis. Sci. 43 (1998) 311–316.

[80] K.L. Chang, L.S. New, M. Mal, C.W. Goh, C.C. Aw, E.R. Browne, et al., Metabolic profiling of 3-nitropropionic acid early-stage Huntington's disease rat model using gas chromatography time-of-flight mass spectrometry, J. Proteome. Res. 10 (2011) 2079–2087.

[81] D. Ng, K. Pasikanti, E. Chan, Trend analysis of metabonomics and systematic review of metabonomics-derived cancer marker metabolites, Metabolomics (2010) 1–24.

[82] E.C. Horning, M.G. Horning, Metabolic profiles: gas-phase methods for analysis of metabolites, Clin. Chem. 17 (1971) 802–809.

[83] J.D. Shoemaker, W.H. Elliott, Automated screening of urine samples for carbohydrates, organic and amino acids after treatment with urease, J. Chromatogr. 562 (1991) 125–138.

[84] M.C. Walsh, L. Brennan, J.P.G. Malthouse, H.M. Roche, M.J. Gibney, Effect of acute dietary standardization on the urinary, plasma, and salivary metabolomic profiles of healthy humans, Am. J. Clin. Nutr. 84 (2006) 531–539.

[85] K.K. Pasikanti, P.C. Ho, E.C. Chan, Development and validation of a gas chromatography/mass spectrometry metabonomic platform for the global profiling of urinary metabolites, Rapid Commun. Mass Spectrom. 22 (2008) 2984–2992.

[86] M. Assfalg, I. Bertini, D. Colangiuli, C. Luchinat, H. Schafer, B. Schutz, et al., Evidence of different metabolic phenotypes in humans, Proc. Natl. Acad. Sci. USA 105 (2008) 1420–1424.

[87] H.G. Gika, G.A. Theodoridis, I.D. Wilson, Liquid chromatography and ultra-performance liquid chromatography-mass spectrometry fingerprinting of human urine: sample stability under different handling and storage conditions for metabonomics studies, J. Chromatogr. A 1189 (2008) 314–322.

[88] M. Lauridsen, S.H. Hansen, J.W. Jaroszewski, C. Cornett, Human urine as test material in 1H NMR-based metabonomics: recommendations for sample preparation and storage, Anal. Chem. 79 (2007) 1181–1186.

[89] K. Dettmer, P.A. Aronov, B.D. Hammock, Mass spectrometry-based metabolomics, Mass Spectrom. Rev. 26 (2007) 51–78.

[90] M. Chen, L. Zhao, W. Jia, Metabonomic study on the biochemical profiles of a hydrocortisone-induced animal model, J. Proteome. Res. 4 (2005) 2391–2396.

[91] S.H. Lee, H.M. Woo, B.H. Jung, J. Lee, O.S. Kwon, H.S. Pyo, et al., Metabolomic apprcach to evaluate the toxicological effects of nonylphenol with rat urine, Anal. Chem. 79 (2007) 6102–6110.

[92] J.L. Little, Artifacts in trimethylsilyl derivatization reactions and ways to avoid them, J. Chromatogr. A 844 (1999) 1–22.

[93] Y. Qiu, M. Su, Y. Liu, M. Chen, J. Gu, J. Zhang, et al., Application of ethyl chloroformate derivatization for gas chromatography-mass spectrometry based metabonomic profiling, Anal. Chim. Acta. 583 (2007) 277–283.

[94] T. Kuhara, Diagnosis and monitoring of inborn errors of metabolism using urease-pretreatment of urine, isotope dilution, and gas chromatography-mass spectrometry, J. Chromatogr. B Analyt. Technol. Biomed. Life. Sci. 781 (2002) 497–517.

[95] K.R. Leimer, R.H. Rice, C.W. Gehrke, Complete mass spectra of N-trifluoroacetyl-n-butyl esters of amino acids, J. Chromatogr. 141 (1977) 121–144.

[96] H.J. Issaq, O. Nativ, T. Waybright, B. Luke, T.D. Veenstra, E.J. Issaq, et al., Detection of bladder cancer in human urine by metabolomic profiling using high performance liquid chromatography/mass spectrometry, J. Urol. 179 (2008) 2422–2426.

[97] K. Kim, P. Aronov, S.O. Zakharkin, D. Anderson, B. Perroud, I.M. Thompson, et al., Urine metabolomics analysis for kidney cancer detection and biomarker discovery, Mol. Cell. Proteomics 8 (2009) 558–570.

[98] J.C. Lindon, J.K. Nicholson, E. Holmes, H. Antti, M.E. Bollard, H. Keun, et al., Contemporary issues in toxicology the role of metabonomics in toxicology and its evaluation by the COMET project, Toxicol. Appl. Pharmacol. 187 (2003) 137–146.

[99] Y. Bao, T. Zhao, X. Wang, Y. Qiu, M. Su, W. Jia, Metabonomic variations in the drug-treated type 2 diabetes mellitus patients and healthy volunteers, J. Proteome. Res. 8 (2009) 1623–1630.

[100] H. Wu, R. Xue, L. Dong, T. Liu, C. Deng, H. Zeng, et al., Metabolomic profiling of human urine in hepatocellular carcinoma patients using gas chromatography/mass spectrometry, Anal. Chim. Acta. 648 (2009) 98–104.

[101] Y. Qiu, G. Cai, M. Su, T. Chen, X. Zheng, Y. Xu, et al., Serum metabolite profiling of human colorectal cancer using GC-TOFMS and UPLC-QTOFMS, J. Proteome. Res. (2009).

[102] Z. Zhang, Y. Qiu, Y. Hua, Y. Wang, T. Chen, A. Zhao, et al., Serum and urinary metabonomic study of human osteosarcoma, J. Proteome. Res. 9 (2010) 4861–4868.

[103] M. Vallejo, A. Garcia, J. Tunon, D. Garcia-Martinez, S. Angulo, J.L. Martin-Ventura, et al., Plasma fingerprinting with GC-MS in acute coronary syndrome, Anal. Bioanal. Chem. 394 (2009) 1517–1524.

[104] H.M. Lin, S.I. Edmunds, N.A. Helsby, L.R. Ferguson, D.D. Rowan, Nontargeted urinary metabolite profiling of a mouse model of Crohn's disease, J. Proteome. Res. 8 (2009) 2045–2057.

[105] J.C. Lindon, E. Holmes, J.K. Nicholson, So what's the deal with metabonomics? Anal. Chem. 75 (2003) 384A–391A.

[106] M. Chen, Y. Ni, H. Duan, Y. Qiu, C. Guo, Y. Jiao, et al., Mass spectrometry-based metabolic profiling of rat urine associated with general toxicity induced by the multiglycoside of Tripterygium wilfordii Hook. f, Chem. Res. Toxicol. 21 (2008) 288–294.

[107] M. Sieber, D. Hoffmann, M. Adler, V.S. Vaidya, M. Clement, J.V. Bonventre, et al., Comparative analysis of novel noninvasive renal biomarkers and metabonomic changes in a rat model of gentamicin nephrotoxicity, Toxicol. Sci. 109 (2009) 336–349.

[108] K.J. Boudonck, M.W. Mitchell, L. Nemet, L. Keresztes, A. Nyska, D. Shinar, et al., Discovery of metabolomics biomarkers for early detection of nephrotoxicity, Toxicol. Pathol. 37 (2009) 280–292.

[109] T. Ohta, N. Masutomi, N. Tsutsui, T. Sakairi, M. Mitchell, M.V. Milburn, et al., Untargeted metabolomic profiling as an evaluative tool of fenofibrate-induced toxicology in Fischer 344 male rats, Toxicol. Pathol. 37 (2009) 521–535.

[110] B.S. Kumar, Y.J. Lee, H.J. Yi, B.C. Chung, B.H. Jung, Discovery of safety biomarkers for atorvastatin in rat urine using mass spectrometry based metabolomics combined with global and targeted approach, Anal. Chim. Acta. 661 (2010) 47–59.

[111] M.S. Lee, B.H. Jung, B.C. Chung, S.H. Cho, K.Y. Kim, O.S. Kwon, et al., Metabolomics study with gas chromatography-mass spectrometry for predicting valproic acid-induced hepatotoxicity and discovery of novel biomarkers in rat urine, Int. J. Toxicol. 28 (2009) 392–404.

[112] X. Huang, L. Shao, Y. Gong, Y. Mao, C. Liu, H. Qu, et al., A metabonomic characterization of CCl4-induced acute liver failure using partial least square regression based on the GC-MS metabolic profiles of plasma in mice, J. Chromatogr. B Analyt. Technol. Biomed. Life. Sci. 870 (2008) 178–185.

[113] W.S. Law, P.Y. Huang, E.S. Ong, C.N. Ong, S.F. Li, K.K. Pasikanti, et al., Metabonomics investigation of human urine after ingestion of green tea with gas chromatography/mass spectrometry, liquid chromatography/mass spectrometry and (1)H NMR spectroscopy, Rapid Commun. Mass Spectrom. 22 (2008) 2436–2446.

[114] Z. Wu, M. Li, C. Zhao, J. Zhou, Y. Chang, X. Li, et al., Urinary metabonomics study in a rat model in response to protein-energy malnutrition by using gas chromatography-mass spectrometry and liquid chromatography-mass spectrometry, Mol. Biosyst. 6 (2010) 2157–2163.

[115] F.A. van Dorsten, C.H. Grun, E.J. van Velzen, D.M. Jacobs, R. Draijer, J.P. van Duynhoven, The metabolic fate of red wine and grape juice polyphenols in humans assessed by metabolomics, Mol. Nutr. Food Res. 54 (2010) 897–908.

[116] D. Vuckovic, X. Zhang, E. Cudjoe, J. Pawliszyn, Solid-phase microextraction in bioanalysis: new devices and directions, J. Chromatogr. A 1217 (2010) 4041–4060.

[117] Y. Xu, S.J. Dixon, R.G. Brereton, H.A. Soini, M.V. Novotny, K. Trebesius, et al., Comparison of human axillary odour profiles obtained by gas chromatography/mass spectrometry and skin microbial profiles obtained by denaturing gradient gel electrophoresis using multivariate pattern recognition, Metabolomics 3 (2007) 427–437.

[118] Z.M. Zhang, J.J. Cai, G.H. Ruan, G.K. Li, The study of fingerprint characteristics of the emanations from human arm skin using the original sampling system by SPME-GC-MS, J. Chromatogr. B Analyt. Technol. Biomed. Life. Sci. 822 (2005) 244–252.

[119] M. Gallagher, C.J. Wysocki, J.J. Leyden, A.I. Spielman, X. Sun, G. Preti, Analyses of volatile organic compounds from human skin, Br. J. Dermatol. 159 (2008) 780–791.

[120] H. Yu, L. Xu, P. Wang, Solid phase microextraction for analysis of alkanes and aromatic hydrocarbons in human breath, J. Chromatogr. B Analyt. Technol. Biomed. Life Sci. 826 (2005) 69–74.

[121] X. Chen, F. Xu, Y. Wang, Y. Pan, D. Lu, P. Wang, et al., A study of the volatile organic compounds exhaled by lung cancer cells in vitro for breath diagnosis, Cancer 110 (2007) 835–844.

[122] M. Syhre, J.M. Scotter, S.T. Chambers, Investigation into the production of 2-Pentylfuran by Aspergillus fumigatus and other respiratory pathogens in vitro and human breath samples, Med. Mycol. 46 (2008) 209–215.

[123] D. Zimmermann, M. Hartmann, M.P. Moyer, J. Nolte, J.I. Baumbach, Determination of volatile products of human colon cell line metabolism by GC-MS analysis, Metabolomics 3 (2007) 13–17.

[124] R. Xue, L. Dong, S. Zhang, C. Deng, T. Liu, J. Wang, et al., Investigation of volatile biomarkers in liver cancer blood using solid-phase microextraction and gas chromatography/mass spectrometry, Rapid Commun. Mass Spectrom. 22 (2008) 1181–1186.

[125] B. Buszewski, A. Ulanowska, T. Ligor, M. Jackowski, E. Klodzinska, J. Szeliga, Identification of volatile organic compounds secreted from cancer tissues and bacterial cultures, J. Chromatogr. B Analyt. Technol. Biomed. Life Sci. 868 (2008) 88–94.

[126] D. Vuckovic, J. Pawliszyn, Systematic evaluation of solid-phase microextraction coatings for untargeted metabolomic profiling of biological fluids by liquid chromatography-mass spectrometry, Anal. Chem. (2011).

[127] C.M. Zimmermann, G.P. Jackson, Gas chromatography tandem mass spectrometry for biomarkers of alcohol abuse in human hair, Ther. Drug. Monit. 32 (2010) 216–223.

[128] Y. Shi, B. Shen, P. Xiang, H. Yan, M. Shen, Determination of ethyl glucuronide in hair samples of Chinese people by protein precipitation (PPT) and large volume injection-gas chromatography-tandem mass spectrometry (LVI-GC-MS/MS), J. Chromatogr. B Analyt. Technol. Biomed. Life. Sci. 878 (2010) 3161–3166.

[129] H. Kharbouche, F. Sporkert, S. Troxler, M. Augsburger, P. Mangin, C. Staub, Development and validation of a gas chromatography-negative chemical ionization tandem mass spectrometry method for the determination of ethyl glucuronide in hair and its application to forensic toxicology, J. Chromatogr. B Analyt. Technol. Biomed. Life Sci. 877 (2009) 2337–2343.

[130] T.H. Seyler, L.R. Reyes, J.T. Bernert, Analysis of 4-aminobiphenyl hemoglobin adducts in smokers and nonsmokers by pseudo capillary on-column gas chromatography- tandem mass spectrometry, J. Anal. Toxicol. 34 (2010) 304–311.

[131] F. Giton, P. Caron, R. Berube, A. Belanger, O. Barbier, J. Fiet, Plasma estrone sulfate assay in men: comparison of radioimmunoassay, mass spectrometry coupled to gas chromatography (GC-MS), and liquid chromatography-tandem mass spectrometry (LC-MS/MS), Clin. Chim. Acta. 411 (2010) 1208–1213.

[132] M. Shen, P. Xiang, B. Shen, M. Wang, Determination of endogenous anabolic steroids in hair using gas chromatography-tandem mass spectrometry, Se Pu 26 (2008) 454–459.

[133] M. Pacenti, S. Dugheri, F. Villanelli, G. Bartolucci, L. Calamai, P. Boccalon, et al., Determination of organic acids in urine by solid-phase microextraction and gas chromatography-ion trap tandem mass spectrometry previous 'in sample' derivatization with trimethyloxonium tetrafluoroborate, Biomed. Chromatogr. 22 (2008) 1155–1163.

[134] D. Tsikas, A. Mitschke, M.T. Suchy, F.M. Gutzki, D.O. Stichtenoth, Determination of 3-nitrotyrosine in human urine at the basal state by gas chromatography-tandem mass spectrometry and evaluation of the excretion after oral intake, J. Chromatogr. B Analyt. Technol. Biomed. Life Sci. 827 (2005) 146–156.

[135] J.P. Gaut, J. Byun, H.D. Tran, J.W. Heinecke, Artifact-free quantification of free 3-chlorotyrosine, 3-bromotyrosine, and 3-nitrotyrosine in human plasma by electron capture-negative chemical ionization gas chromatography mass spectrometry and liquid chromatography-electrospray ionization tandem mass spectrometry, Anal. Biochem. 300 (2002) 252–259.

[136] M. Larstad, A.S. Soderling, K. Caidahl, A.C. Olin, Selective quantification of free 3-nitrotyrosine in exhaled breath condensate in asthma using gas chromatography/tandem mass spectrometry, Nitric Oxide 13 (2005) 134–144.

[137] U. Hofmann, M. Schwab, S. Seefried, C. Marx, U.M. Zanger, M. Eichelbaum, et al., Sensitive method for the quantification of urinary pyrimidine metabolites in healthy adults by gas chromatography-tandem mass spectrometry, J. Chromatogr. B Analyt. Technol. Biomed. Life Sci. 791 (2003) 371–380.

CHAPTER

25

Applications of Gas Chromatography in Forensic Science

Abuzar Kabir, Kenneth G. Furton

OUTLINE

Gas Chromatography DOI: 10.1016/B978-0-12-385540-4.00025-0

25.1. INTRODUCTION AND SCOPE

Forensic science applies scientific principles, tools, and methodologies to resolve legal issues and disputes. Forensic chemists not only analyze a wide variety of forensic samples, but also extract and interpret information from the analytical data that may potentially have to withstand rigorous challenges when presented in civil and/or criminal judicial proceedings. As such, it is imperative that any analytical methodology developed for solving forensic problems should meet, at a bare minimum, the required standard set forth by the court of law.

Among all analytical instruments currently being used in routine forensic analyses as well as in forensic research, gas chromatography (GC) is one of the most widely used analytical tools. High sensitivity, selectivity, resolution, speed, good accuracy and precision, wide dynamic concentration range, simple and robust instrument design, and its ability to be interfaced with many established and emerging detection systems have made GC the instrument of choice in many facets of forensic science. In addition to its numerous advantageous features, the basic principle and the theory of GC has been well studied and understood since its inception more than half a century ago. As such, new GC instruments (hardware) along with their operating systems (software) are so simple and user friendly that even a novice operator can operate it with confidence.

Due to the inherent advantages of GC, applications of this reliable analytical instrument are on a continuous surge. Major application areas in forensic science include bulk seized drug analysis, drug screening from biological specimens, postmortem toxicology, trace evidence analysis, manmade environmental pollution investigation, human odor profiling, explosive analysis, analysis of ignitable liquid residues from fire debris, etc. The application areas of GC in forensic science, commonly used sample preparation techniques that precede GC analysis, are illustrated in Figure 25.1.

Sampling and sample preparation impact the integrity of GC analysis of forensic samples, especially when dealing with trace and ultra-trace levels of the target analyte(s) present in various complex matrices (e.g., biological, environmental, fire debris, and explosive residues). In addition, in the majority of cases, the volume of available samples to the forensic investigators is limited. Therefore, a valid sampling and sample preparation strategy should be adopted prior to beginning the analytical process in order to ensure that the analyzed samples are truly representative of the evidence matrix. Due to the complex nature of the sample matrix where the analyte(s) of interest are present, most often, forensic samples cannot be introduced directly into the GC inlet. This incompatibility stems from two factors. First, the complex sample matrix, if introduced directly into the GC inlet without employing any sample

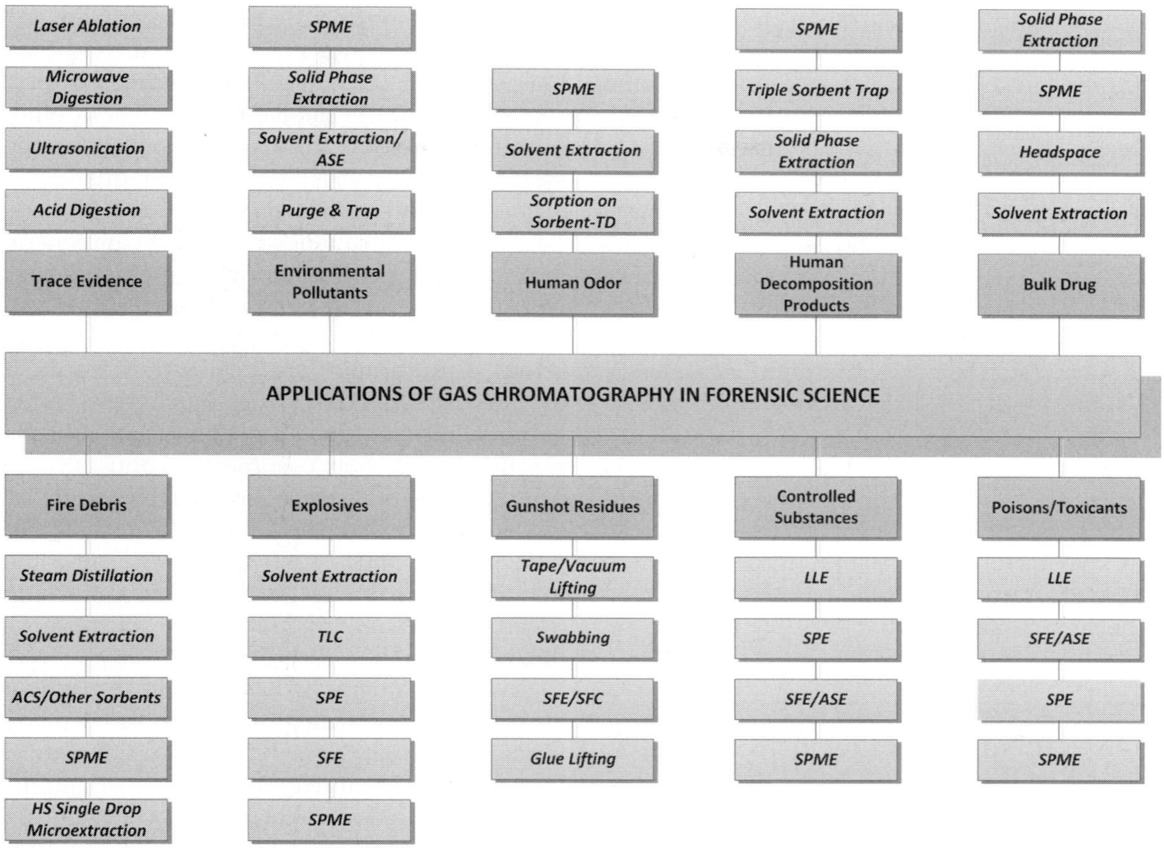

FIGURE 25.1 Major application areas of gas chromatography in forensic science with principal sample preparation techniques.

treatment/cleanup procedure, may exert a detrimental impact on the performance of the GC by contaminating the inlet, as well as by compromising the sensitive stationary phase of the GC column. Second, if the concentration of the target analyte in the sample matrix is very low so that it may fall below the detection limit of the GC, no usable chromatographic data would be garnered. Since every forensic case is unique, standardization of the sampling and sample preparation techniques for the forensic samples is difficult and often dependent upon the knowledge, experience, and judgment of the analyst.

Sample preparation techniques frequently employed in processing forensic samples prior to GC analysis include solvent extraction, solid-phase extraction (SPE), purge and trap, liquid–liquid extraction (LLE), supercritical fluid extraction (SFE), steam distillation, accelerated solvent extraction (ASE), microwave-assisted extraction (MAE), solid-phase microextraction (SPME), liquid-phase microextraction (LPME), stir-bar sorptive extraction (SBSE), solid-phase dynamic extraction (SPDE), etc.

After GC separation, a variety of detectors may be employed for the detection, quantification, and/or identification of the analyte(s) which include flame ionization detector (FID), nitrogen phosphorus detector (NPD), sulfur and nitrogen chemiluminescence detector, flame photometric detector (FPD), atomic emission detector (AED), thermal energy analyzer (TEA), electron capture detector (ECD), ion mobility mass spectrometry (IMMS), time-of-flight mass spectrometry (TOFMS), and isotope ratio mass spectrometry (IRMS). However, the most popular is the mass spectrometer (MS) as it offers both identification and quantification of an unknown substance with high confidence. In some cases, MS/MS is also used as the detector.

Application areas of GC within forensic science are growing rapidly. However, due to the limited scope of this chapter, we will discuss only the major areas within forensic science where gas chromatography has already proved to be an invaluable tool.

25.2. ANALYSIS OF BULK DRUG FOR IDENTIFICATION, IMPURITY PROFILING, AND DRUG INTELLIGENCE PURPOSE

The scope of bulk drug analysis includes seized illicit drugs, as well as counterfeit and traditional medicines in which synthetic medicinal products are occasionally added to enhance their pharmacological efficacy. Being illicit or inadequately regulated, these products often pose an enormous risk to public health. The following section discusses the applications of gas chromatography for analyzing illicit drugs, counterfeit drugs, as well as traditional medicines.

25.2.1. Analysis of Confiscated Illicit Drugs

Law enforcement agencies pose a series of questions to forensic chemists regarding the analysis of confiscated drugs, such as the following: (1) whether the submitted sample contains any controlled substance(s) in it; (2) if yes, then how much of the controlled substance does it contain; and (3) is the sample similar to other seized samples? To address these questions, the seized drug samples require both qualitative (screening) and quantitative analyses (confirmation and quantification) of the illicit component. Federal and state drug trafficking penalties vary with the type and quantity of the drug seized. Both identification and quantification data of seized drug samples are frequently demanded by the prosecution. Drug intelligence often requires impurity profiling, identification of the precursors, reaction by-products, and isotopic profile analysis to establish possible origin of the substance, sources of the raw materials, precursors, and solvents used in the manufacturing process that may lead to tracking down clandestine manufacturing facilities of illicit drugs.

Gas chromatography has been the instrument of choice for confiscated drug analysis for many years. Although GC-FID is still the most frequently used instrument, chromatographic profiling procedures have been continuously shifting toward GC-MS due to its higher analytical performance [1] and gradually decreasing cost. Among others, one major advantage of GC-MS is its ability to use the *target ion* quantification functionality that helps in situations where coelution of structurally related compounds is a problem. Table 25.1 depicts a list and chemical structures of Drug Enforcement Administration (DEA) scheduled drugs [2a].

Cannabis is one of the most widely cultivated and abused illicit drug that produces a wide range of illicit drugs including marijuana, sinsemilla, thai sticks, dichweed, hashish, hash oil, etc. Although Δ^9-tetrahydrocannabinol (Δ^9-THC) is the main psychoactive ingredient in cannabis, other cannabinoids present in cannabis also demonstrate different pharmacological activities. Cannabinoids are extracted from confiscated cannabis using solvent

TABLE 25.1 Classification of DEA Scheduled Drugs and Their Chemical Structures

Drug class	Commercial name	DEA schedule	Chemical structure
Cannabinoids	Marijuana	I	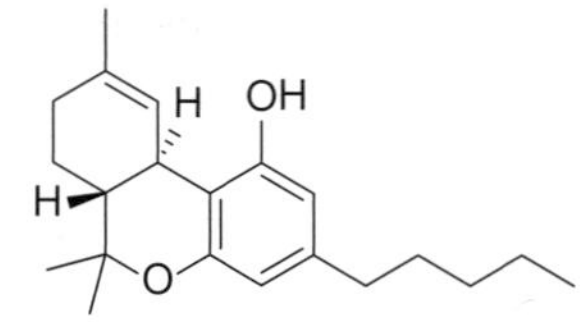
	Hashish	I	
Opioids	Heroin	I	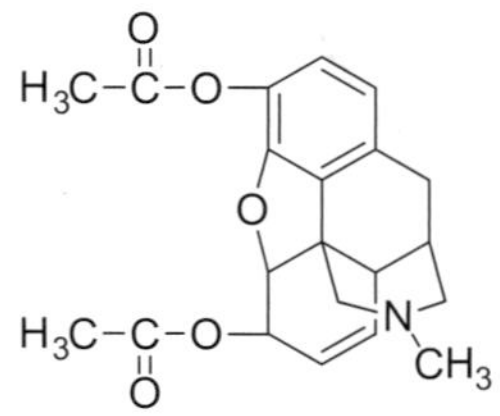
	Opium	II, III, V	Morphine 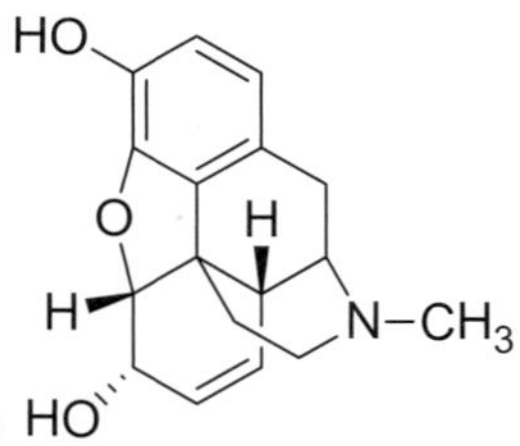 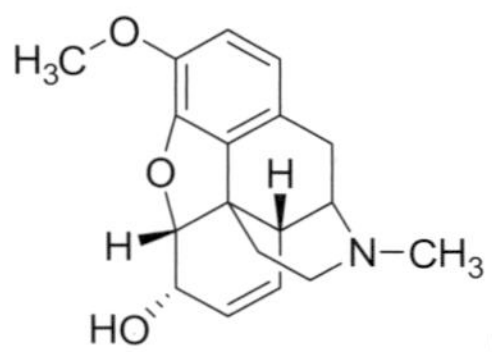 Codeine
Stimulants	Cocaine	II	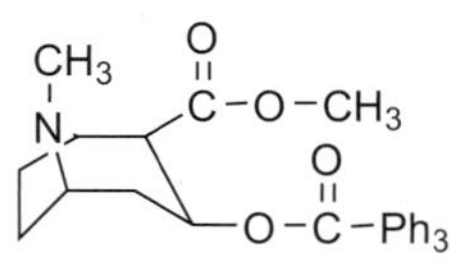
	Amphetamine	II	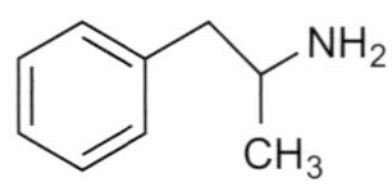
	Methamphetamine	II	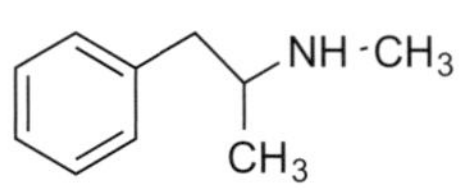
Club drugs	MDMA	I	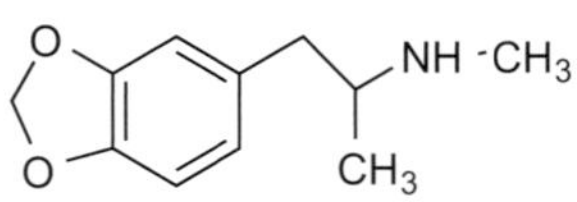
	Flunitrazepam	IV	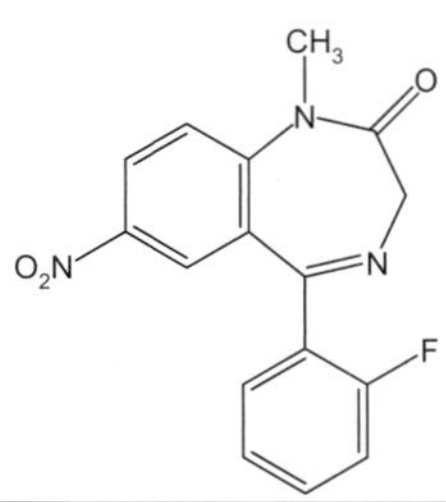

(*Continued*)

TABLE 25.1 Classification of DEA Scheduled Drugs and Their Chemical Structures (*cont'd*)

Drug class	Commercial name	DEA schedule	Chemical structure
	GHB	I	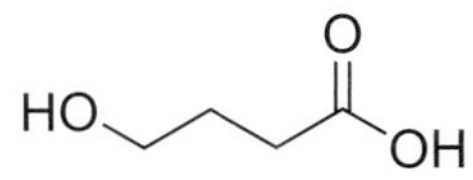
Dissociative drugs	Ketamine	III	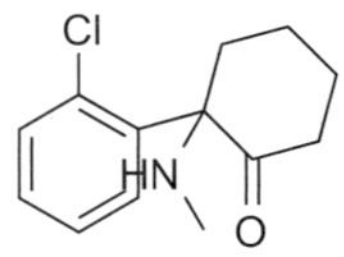
	PCP and Analogs	I, II	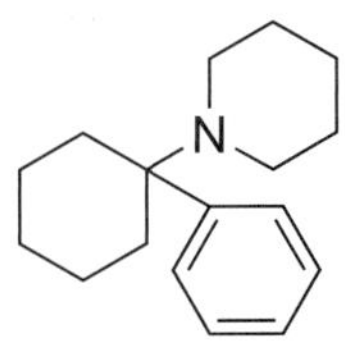
Hallucinogens	LSD	I	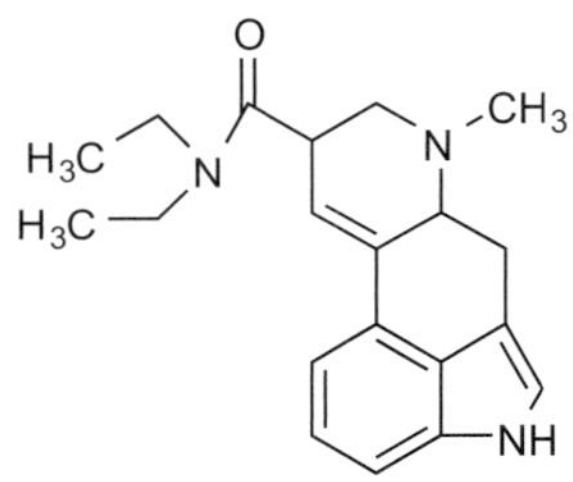
	Mescaline	I	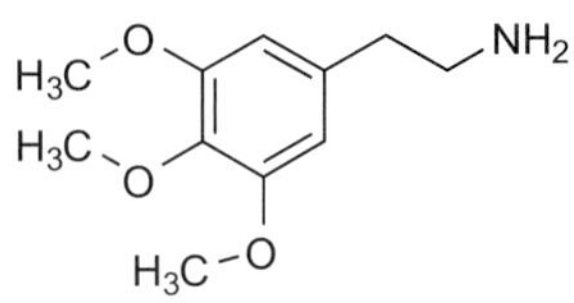
	Psilocybin	I	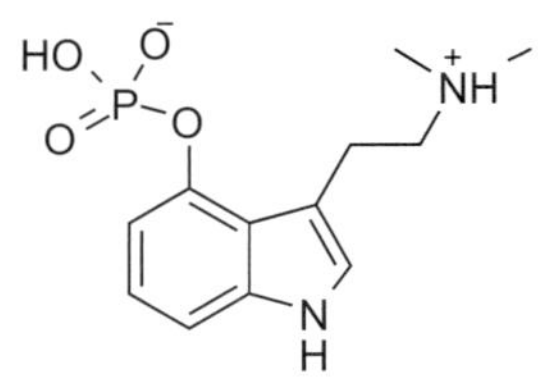

extraction. The extracted samples can be analyzed by GC-FID [2b] or GC-MS [3].

Heroin is a commonly abused drug. It is a semisynthetic product derived from morphine. Due to the differences in agricultural and manufacturing procedures, there is a significant difference in the type and concentration of opium alkaloid content in heroin samples. In addition to that, insertion of diluents and adulterants makes the sample matrix even more complicated. However, profiling of the seized drugs provide information regarding origin, as well as adulterants and diluents used in the manufacturing process. Although FID continues to be a major detector for screening and quantitation of the opium alkaloids and adulterants, due to the higher selectivity and its ability to identify unknown albeit important impurities, MS is used frequently with GC [4,5].

Opium contains more than twenty different kinds of alkaloids that constitute from a small amount to as high as 10% of its weight [6]. Analysis of these alkaloids is carried out using GC-FID or GC-MS. Alkaloid components are extracted from opium samples by solvent extraction. The solution is then treated with 20% trifluoroacetic anhydride (TFAA) in order to transform alkaloids into trifluoroacetylated derivatives before injecting into the GC. A relatively simpler way is to use thermal desorption (TD) in combination with GC to analyze volatile and semivolatile alkaloids, impurities, synthetic by-products, and residual solvents simultaneously. This approach reduces the use of toxic organic solvents and eliminates the tedious sample preparation process.

Bulk cocaine samples, in the form of hydrochloride and free base, are routinely analyzed by GC-MS or GC-FID. For drug intelligence purposes, cocaine samples are profiled to determine major cutting agents and cocaine alkaloids present in the sample that include ecgonine methyl ester, ecgonine, tropacocaine, benzoylecgonine, norcocaine, *cis*- and *trans*-cinnamoylcocaine, and 3,4,5-trimethoxycocaine. Profiling of cocaine requires derivatization and *N*-methyl-*N*-trimethylsilyltrifluoroacetamide (MSTFA), *N,O*-bis (trimethylsilyl) acetamide (BSA), trimethylsilylchloride (TMSCL), *N,O*-bis(trimethylsilyl)trifluoroacetamide (BSTFA), and + 1% trimethylchlorosilane (TMCS) are among the commonly used derivatization reagents used for this purpose. However, if the cocaine sample contains lactose or mannitol as the cutting agent, derivatization with MSTFA is advised [7]. Analysis of solvent residues in illicit cocaine may also provide valuable information regarding the clandestine manufacturing process and can be used for source discrimination purposes. Residual solvent is analyzed by extracting it onto a SPME fiber followed by analysis by GC-MS [8]. If identification, profiling, or residual solvent analysis alone or in combination does not offer conclusive information about the source of the drug, GC-MS-IRMS has been used to obtain $\delta\ ^{13}C$ and/or $\delta\ ^{15}N$ values that can be used, independently or in combination with other relevant information, as an efficient tool for source discrimination [9].

Methamphetamine (MA) can be manufactured using a number of synthesis routes. As a consequence, it exhibits different impurity profiles based on the synthesis route, as well as the reaction conditions employed in the manufacturing process. Impurity profiling of seized MA is carried out by GC-FID or GC-MS. Impurities present in MA are extracted by LLE under alkaline conditions or by SPME. Comparison between LLE and SPME reveals [10] that the relative intensity of impurities in MA appear to be much higher if SPME is employed for sample preparation, thus establishing it as an efficient solvent-less sampling tool for impurity profiling. In addition, larger number of volatiles can be seen in SPME-GC-MS. TD/GC-MS can be another green alternative for impurity profiling of MA [11]. Kuwayama et al. [11] compared the performance of thermal desorption and liquid–liquid extraction in MA impurity profiling and demonstrated that both the chromatographic signal intensity and the number of detected peaks are higher when TD is used.

Amphetamine-type stimulants (ATSs), e.g. ecstasy, have been a global problem due to its ubiquitous availability and abuse around the world. The majority of the methods used for ecstasy profiling are based on GC-FID or GC-MS. Sample preparation for ecstasy seizures includes LLE, SPE, and SPME. Due to its operational simplicity and solvent-free nature, SPME has gained much interest in recent years among forensic chemists.

Designer drugs are synthetic versions of drugs that are designated as controlled substance as per the DEA schedule. They are similar in structure and efficacy to controlled substances. Common designer drugs include fentanyl and meperidine analogs, phenylcyclidine (PCP), amphetamines, and methamphetamine. Designer drugs are screened, identified, and quantified using GC-EI-MS. Prior to chromatographic analysis, the drug sample is often derivatized by trifluoroacetic acid (TFA). Comparison of the chromatographic separation between TFA derivatives of phenethylamine-type drugs and their free base demonstrates that derivatization facilitates the complete separation of some critical components which are not separated in their free-base state [12].

Methaqualone (MTQ) is a sedative–hypnotic drug commonly sold in South Africa as *Mandrax*. GC-MS is routinely used to identify and quantify the concentration of MTQ in seized drugs. Studies show that only chromatographic profiles of multiple batches of seized samples (some of them are similar in appearance and tablet weight), obtained by GC-MS analysis, are not sufficient to associate them with the same manufacturer; hence, identification of precursors or reaction by-products is recommended to make any definitive conclusion [13].

p-Flurofentalyl (pFF), a potent synthetic narcotic analgesic, is often used illicitly and can be found in tablet and capsule form. For gas chromatographic analysis, it is first solubilized in aqueous solution of potassium hydroxide and sodium sulfate followed by extraction into diethyl ether. The organic extract is then evaporated to dryness and reconstituted in methanol. pFF can be analyzed by GC-MS without derivatization [14].

In addition to the use of conventional one-dimensional gas chromatography with suitable detectors for drug screening, comprehensive two-dimensional gas chromatography coupled with time-of-flight mass spectrometry (GC × GC-TOFMS) [15] and gas chromatography-fourier transform infrared spectroscopy (GC-FTIR) [16] have shown promise of faster screening and confirmation of illicit drugs.

25.2.2. Analysis of Counterfeit Drugs and Traditional Medicinal Products

Both counterfeit drugs and inadequately regulated traditional medicines pose significant risk to the human health and public safety. Counterfeit drugs account for at least 10% of the medicines sold worldwide [17]. Forensic analysis of counterfeit medicines play a key role in determining the presence of potentially toxic ingredients, establishing possible association between different counterfeit medicines, and providing valuable clues to the drug regulatory and law enforcement agencies regarding the potential sources. Although HPLC and other analytical instruments are frequently employed for screening of the suspect samples, GC-MS is also used when the analyte of interest is thermally stable [18].

Traditional medicinal products occasionally contain undisclosed synthetic medicinal compounds to increase their efficacy. They are available in various forms including tablet, capsule, powder, and syrup. Since the addition of potent medicinal compounds is not disclosed, consumers are often not aware of the risk associated with the consumption of these products. GC-MS is generally used for the identification of the presence of synthetic medicinal products in traditional medicines [19].

25.3. GAS CHROMATOGRAPHY IN FORENSIC TOXICOLOGY

Toxicology is the *science of poison*. It deals with chemical and physical properties of poisons, their physiological or behavioral effects on living organisms, their qualitative and quantitative methods of analysis, and the development of procedures for the treatment of poisoning [20]. Forensic toxicology, a subunit of the broader field of toxicology, deals with the application of toxicology to cases and issues where the adverse effects of poison have administrative or medico-legal consequences and in addition, there is a potential likelihood that the results will be used in court.

Forensic toxicology has application in four areas: (1) postmortem toxicology/death investigation toxicology, (2) human performance toxicology, (3) doping control, and (4) forensic workplace drug testing.

25.3.1. Postmortem Toxicology/Death Investigation Toxicology

Analysis of postmortem samples to determine the probable cause of death is always challenging due to intrinsic variability of the concentration of drugs/toxicants within the body, different degrees of corpse degradation before it reaches the morgue, and also the inconsistency in the availability of sample specimens submitted for toxicological analysis. Collection of biological samples should strictly follow relevant protocols developed to minimize site-to-site variability. When possible, peripheral blood and at least one other specimen, e.g. urine, vitreous humor, liver, or gastric content, should be collected.

In order to isolate target analytes from the collected biological specimens, different sample preparation techniques are employed. Commonly used sample preparation techniques include SPE, SPME, single-drop microextraction, and solvent extraction. The selection of an appropriate sample preparation technique normally depends on the type of biological specimen.

The samples are generally screened using gas chromatography equipped with NPD or FID. Confirmatory tests are often carried out using MS or MS/MS. As the effectiveness of gas chromatography is largely dependent upon sampling and sample preparation techniques, as well as chromatographic parameters, e.g. column type and quality, column temperature, and analysis time [21], developing a GC method requires careful consideration of all of the listed and important parameters that can greatly influence the analytical results.

The extent and type of postmortem toxicological analysis are dependent upon jurisdiction of death, age of the deceased, availability of medical history, and many other factors. Depending on the availability of information regarding the deceased, a psychoactive drug screening or a comprehensive drug screening may be carried out [22]. Drug screening determines the type and quantity of the drug present in the biological specimen and often includes alcohol, drugs of abuse, and a variety of toxic substances.

When the screening test provides a positive result for a particular drug/toxicant, a confirmatory test is carried out. Often, a different chromatographic technique is recommended for confirmatory test in order to minimize systematic errors.

The following section discusses typical uses of GC in postmortem toxicology.

Nicotine, a water-soluble alkaloid, is the addictive ingredient in cigarettes. A number of poison cases have been reported every year instigated by accidental ingestion of cigarettes, cigarette end, or their liquid extract. In some cases, these ingestions claim human lives. A gas chromatographic method was developed, allowing for the detection of nicotine and its major metabolite cotinine from blood and urine samples with an LOD 2.1 ng/mL for both analytes [23].

Due to its sedative and euphoric effects, gamma-hydroxybutyric acid (GHB) is popular as a recreational drug. GHB is also an endogenous compound, produced as a by-product of gamma-aminobutyric acid (GABA) metabolism. Therefore, it is important to know the endogenous concentration of GHB in humans. GHB can be extracted from plasma and urine samples in chloroform for analysis by GC-FID and in ethyl acetate for analysis by GC-MS. The extracted GHB can be detected by GC-FID without derivatization; however, derivatization using BSTFA containing 1% TMCS improves the method sensitivity to at least one order of magnitude when analysis is carried out by GC-MS. The analysis provides an LOD of 2.5 mg/L by GC-FID and 0.2 mg/L by GC-MS [24].

Frequently abused drugs, e.g. amphetamine, morphine, codeine, opiates, cocaine, and their metabolites, can be extracted from human postmortem tissues including brain [25]. The analytes are extracted from homogenized brain tissues by solid-phase extraction (SPE). Following SPE, extracted drug samples and their metabolites are derivatized in two steps, first using a mixture of MTBSTFA + 1% *tert*-butyldimethylchlorosilane (TBDMCS) and later a mixture of BSTFA + 1% TMCS. The addition of MTBSTFA helps in reducing the loss of amphetamines, due to volatility, during the sample preparation steps. TBDMS derivatizes primary amines of amphetamines while the sterically hindered analytes are derivatized with BSTFA. As such, this dual derivatization process provides a mixture of a thermally stable derivatized system amenable to GC. The analytes are finally quantified by GC-MS operated in positive chemical ionization (PCI) mode.

Basic drugs, e.g. amitriptyline, citalopram, clozapine, diazepam, dihydrocodeine, methadone, and tramadol, from postmortem blood samples are extracted using liquid–liquid extraction under basic condition (pH 10) with diethyl ether as the organic extractant. These analytes can be injected into GC-MS without derivatization [26].

Acidic and neutral drugs including analgesics, anticoagulants, antidiabetics, antiepileptics, barbiturates, diuretics, hypnotics, and muscle relaxants are also screened from blood samples. Drug analytes are extracted from blood samples in ethyl acetate using LLE. Extracted analytes can be analyzed by GC-FID or GC-MS [27].

Organophosphorus insecticides, a substance commonly used in suicidal attempts, are analyzed from postmortem blood samples [28]. Omethoate, dimethoate, dizinon, chlorpyrifos, parathinon-ethyl, and chlorfenvinphos are some of the members of this insecticide family. Sample preparation for extracting target analytes from blood samples include protein precipitation, SPME, LLE, and SPE. Since organophosphorus pesticides are thermally stable, generally no derivatization is required prior to GC-MS analysis.

25.3.2. Human Performance Toxicology

Human performance toxicology primarily deals with driving under the influence of drugs and alcohol. Upon a law enforcement officer's request, a suspect must take chemical DUI (driving under the influence) tests which include breathalyzers, blood tests, and urine tests to establish his/her sobriety.

Ethanol is the most widely abused drug in the world. However, unlike other common drugs of abuse, ethanol poses enormous analytical challenges due to its high volatility, as well as its ability of being produced *in situ* in poorly handled samples or as a natural product of the postmortem process [29]. Due to its high volatility, ethanol cannot be efficiently isolated from complex biological matrices by classical extraction techniques nor can it be introduced directly, along with the matrix, into the GC system. As a result, headspace sampling remains is the only viable alternative sampling and sample preparation technique. To make headspace sampling more reliable and efficient, a HS-GC-FID method was developed utilizing a dual-role robotic autosampler, one for sample preparation and the other for headspace sample introduction into the GC inlet. Carrying out the chromatographic separation on a DB-624 column, this method successfully baseline-separated all the known analytes that normally appear with ethanol, providing LOQ and LOD values of 17 mg/dL and 5 mg/dL, respectively [29].

Kristoffersen et al. studied the effects of blood storage and headspace conditions on the ethanol concentration in plasma, hemolyzed blood, and nonhemolyzed blood using HS-GC-FID [30]. A decrease in ethanol concentration was observed after a few days of storage at room temperature with nearly equivalent increase in aldehyde concentration. The diminishing value of ethanol concentration at prolonged room temperature storage is attributed to the chemical oxidation of ethanol to acetaldehyde. In addition, high headspace equilibration temperature was found to contribute to a higher ethanol value.

Considering the fact that ethanol can be detected in the body after consumption only for a relatively short period of time, ethyl glucuronide (EtG), a direct metabolite of ethanol, is frequently used to investigate alcohol addiction (as opposed to social consumption) that offers an extended window for assessing alcohol consumption and can be detected in body fluids, tissues, and hair samples. Ethyl glucuronoide is nonvolatile and stable upon storage.

Ethyl glucuronide was extracted from urine samples [31] using microwave-assisted extraction (MAE) in chloroform followed by evaporation to dryness under nitrogen stream. Due to its nonvolatile nature, analysis of ethyl glucuronoide by GC requires derivatization. All hydroxyl functional groups present in EtG can be efficiently silated by treating it with BSTFA and pyridine. Microwave-assisted extraction of EtG, followed by analysis by GC-EI-MS ran in selected ion monitoring (SIM) mode provided LOQ and LOD values of 0.1 μg/mL and 5 ng/mL, respectively, with recovery in the range of 80–92%.

A similar microwave-assisted extraction of EtG was also carried out from a hair matrix [32]. A mixture of n-hexane/water (1:1 v/v) was used as the extractant. Due to the advantage of obtaining a greater retrospective window for the detection of chronic use of alcohol, EtG can be monitored using hair samples rather than bodily fluids (urine, blood, and saliva). Analysis of EtG in hair samples has gained attention in recent years and several analytical methods have been developed using GC-NCI-MS [33], GC-NCI-MS/MS [34], and GC-EI-MS/MS [35].

A recent development in EtG analysis from hair samples is the use of a protein precipitation (PPT) technique. EtG was extracted from hair samples in DI water under sonication for 1 h, incubated overnight, centrifuged, and the supernatant was evaporated to dryness. The residue was reconstituted in a mixture of acetonitrile and DI water (7:1 v/v) then passed through a Sirocco™ protein precipitation plate. EtG residue was finally derivatized using a mixture of pyridine and BSTFA before injecting into the chromatographic system. In addition to using a PPT technique for matrix cleaning, a large-volume injection (LVI) using a programmable temperature vaporization (PTV) GC inlet liquid injections of up to 50 μL were allowed. Detection and quantification of EtG was carried out in EI-MS/MS. Due to the high-volume injection, the sensitivity of the method increased significantly with LOQ and LOD values of 10 pg/mg and 5 pg/mg, respectively [36]. Although BSTFA is the most common derivatization reagent for EtG, other derivatization regents, e.g., pentafluropropionic anhydride (PFPA) and heptafluorobutyric anhydride (HFBA), were also tested [37]. PFPA was shown to offer improved sensitivity in GC-MS analysis [34].

Extraction of EtG from hair samples is generally carried out using water as the extractant, followed by matrix cleanup by solid-phase extraction (SPE), derivatization, and finally the extract containing EtG was introduced into the GC as a liquid. As an alternative to low-volume liquid-phase injection (unless LVI is employed), a new method was developed [38] that utilized solid-phase microextraction (SPME) as a sample preconcentration and introduction tool. Following the typical procedure of EtG extraction in water followed by SPE for matrix cleanup and derivatization using HFBA, the derivatized EtG was exposed to HS-SPME. Carboxen-polydimethylsiloxane (CAR-PDMS, 75 μm) SPME fiber demonstrated the best performance in terms of analyte yield with minimal matrix effects. The HS-SPME-GC-MS/MS method provided LLOQ and LLOD values of 2.8 pg/mL and 0.6 pg/mL, respectively, which were significantly lower than the previously reported values obtained by liquid injection [36].

25.3.3. Doping Control

The use of performance-enhancing drugs in sports is commonly known as doping and is considered to be a widespread problem in the athletic community. These drugs are frequently used by athletes to improve their athletic performance. Consumption of these drugs not only is unethical in terms of healthy and fair competition in sports, but also poses serious health risks to the athletes. World Anti-Doping Agency (WADA) strictly monitors the presence of performance-enhancing drugs in urine samples collected from the athletes and publishes a list of prohibited substances each year. The prohibited substances include (1) anabolic agents; (2) peptide hormones, growth factors, and related substances; (3) β-2 agonists; (4) hormone antagonists and modulators; and (5) diuretics and other masking agents. Among all the prohibited substances, anabolic agents, commonly known as anabolic steroids, have been the most frequently detected performance-enhancing drug for many years [39]. Because of the ever-increasing number of performance-enhancing drugs, the structural similarity of exogenous and endogenous steroids, the inherent complexity of the urine matrix, and the trace level of concentration of these drugs in urine samples, analysis of these drugs is always challenging [40].

For doping control purposes, urine samples are collected from the athletes and are divided into two portions, one is sent for analysis and the other is kept under secure custody so that it can be used in case there is a dispute in the analytical results.

Analyses of the urine samples collected from the athletes are carried out in two levels: (1) screening analysis for all samples and (2) confirmatory analysis of suspicious samples that provide positive test results during screening [40]. For both screening and confirmatory analysis, gas chromatography-mass spectrometry (GC-MS) is frequently used. However, due to the low concentration of these drugs in urine samples and their low thermal stability, sample preparation (preconcentration, derivatization, etc.) always precedes the chromatographic analysis. For sample preparation, aliquots of the urine samples are first enzymatically hydrolyzed using β-glucuronidase for deconjugating the steroids. The deconjugated steroids are then extracted from the urine sample matrix by either liquid–liquid extraction or solid-phase extraction. Prior to analysis by GC-MS, the extracted analytes are subjected to derivatization. Most laboratories use trimethylsilyl-based derivatization approach using MSTFA as the predominant derivatization reagent, although other derivatization approaches such as methoxime, trifluoroacetyl, heptyfluorobutyl derivatives, and mixed modifications have been reported [39–46].

A recent development in doping control analysis is the application of two-dimensional gas chromatography coupled with combustion-isotope ratio mass spectrometry (GC × GCC-IRMS) [47]. GC × GC employs two columns in tandem, where cryogenic slices are collected from the first column in few second intervals and are immediately released and separated on a second column. In addition, hyphenation of this system to molecular MS systems provides a third degree of separation. This system offers a viable alternative to already-established GC-MS methods with better separation of the target analytes and reduced urine cleanup procedure.

25.3.4. Forensic Workplace Drug Testing

Due to the continuous increase in the use of illicit drugs among workers, resulting in poor job performance, tardiness, frequent absence from work, and other behavioral issues, the importance of workplace drug testing (pre- and postemployment) is on the rise. Workplace drug testing includes collection and analysis of a biological specimen, e.g. urine, hair, blood, sweat, and oral fluid, to determine the presence or absence of selected drugs or their metabolites. In addition to drug screening in the workplace, biological specimens for drug screening can also be collected from schools, military, and prisons.

Drug screening using urine samples may include a standard five-panel drug test that includes marijuana (THC), cocaine, PCP, opiates, and amphetamine or a ten-panel drug test that includes methamphetamine, barbiturates, benzodiazepines, MDMA, and methadone in addition to the five-panel drugs. Hair samples are frequently tested using either a standard five-panel or seven-panel drug test. A seven-panel drug test includes barbiturates and benzodiazepines, as well as the five-panel drug test. Figure 25.2 represents a typical chromatogram obtained from a clinical sample testing positive for codeine, morphine, and 6-monoacetylmorphine (6-MAM).

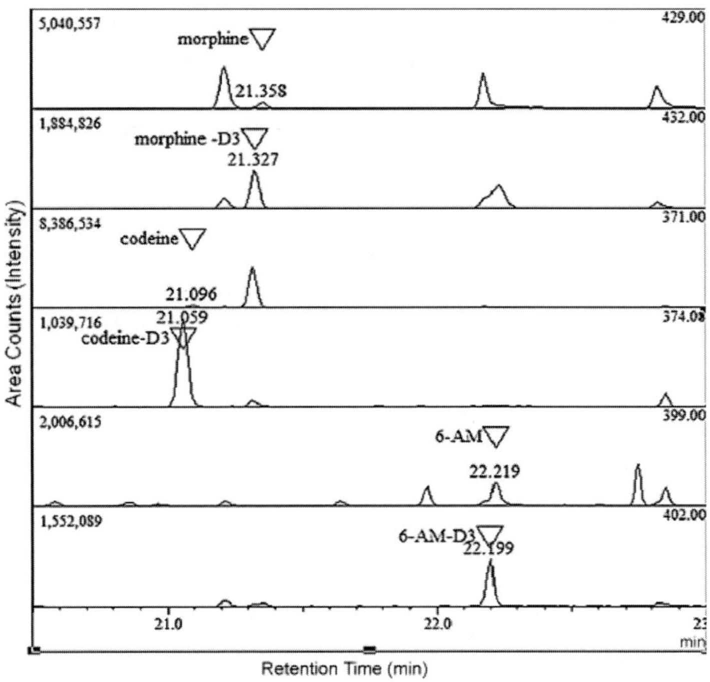

FIGURE 25.2 Chromatogram obtained from a clinical sample testing positive for codeine, morphine, and 6-monoacetylmorphine. *Source: Reproduced with permission from Ref. [48]. Copyright 2010, Elsevier Science.*

25.4. ANALYSIS OF IGNITABLE LIQUID RESIDUES FROM FIRE DEBRIS

Unlike other crime scenes, it is often difficult, if not impossible, to obtain direct physical evidence related to the arsonist (e.g., DNA, fingerprint) from the fire scene as they are most often destroyed by the fire. As such, in a majority of arson cases, the investigators heavily depend on collecting, analyzing, and tracking the potential sources of ignitable liquid residues (ILRs) or accelerants that the arsonist(s) use to aid the fire.

ASTM International has classified ignitable liquids into nine primary classes which include gasoline, petroleum distillates, isoparaffinic products, aromatic products, naphthenic paraffinic products, n-alkanes products, de-aromatized distillates, oxygenated solvents, and others/miscellaneous [49]. Gasoline, kerosene, paint thinners, charcoal lighter fluids, alcohols, mineral spirits, fuel oils, and vegetable oils are the most common accelerants used by arsonists. Detection and identification of ILRs obtained from the fire scene provide the investigators crucial information about the type of accelerants used in the arson case, helping them to track down the suspected arsonist(s). Two recent books *Analysis and Interpretation of Fire Scene Evidence* [50] and *Fire Debris Analysis* [51] extensively covers various important aspects of fire science from a forensic perspective. Some review articles were also published in recent years covering important aspects within this field [52–54].

ASTM International has developed a number of standard practices for screening, isolating and quantifying ILRs, and archiving of extracts recovered from fire debris prior to instrumental analysis.

Due to the complex nature of the sample matrix and high volatility of ignitable liquids, sample collection from the crime scene requires extreme care and attention. On top of that, proper documentation is required in every step to meet any potential future legal challenges. The fire debris samples, as a source of ILRs, are collected from the fire scene in clean, air tight containers and are immediately transported to the laboratory for analysis. Commercial containers, e.g., metal paint cans, glass mason jars, and copolymer bags, are among those most frequently used as fire debris evidence collection and storage containers. Sample preparation techniques prior to injection of ILR samples into GC include dynamic headspace concentration, passive headspace concentration with activated charcoal, solvent extraction, and passive headspace concentration with solid-phase microextraction. Solid-phase microextraction has gained increased attention in recent years as a viable sample preparation technique for fire debris analysis [55–57].

Separation of ILR samples are predominantly carried out by gas chromatography coupled with different detectors, e.g. FID, FTICR-HRMS, PID, ECD, microcells, MS, MS/MS, DMS, and AED, with MS being the most frequently used detection system [58–63]. ASTM has set GC-MS as the preferred analytical tool for fire debris analysis and established a standard test method for investigators (ASTM E1618-11).

Interpretation of the overly complex chromatographic information obtained from ILR analysis is always challenging and most often requires application of chemometric multivariate statistical analysis, such as principal component analysis (PCA), discriminant analysis (DA), and Pearson product moment correlation (PPMC) coefficients [55,64].

In addition to the continuous attempts in developing better gas chromatographic methods, some researchers addressed different aspects of fire debris analysis to gain a better understanding of the fire scenes. J.R. Almirall and K.G. Furton characterized the background and pyrolysis products to assess their potential interferences in fire debris analysis. Many of the compounds identified by combustion/pyrolysis of common substrates, e.g. alkanes and aromatics, are the target compounds commonly found in ILR mixtures [65]. R. Borusiewicz et al. studied the impact of different factors, e.g. type of substrate or burned materials, type of accelerant, length of time between the lighting and extinguishing of fire, and the level of air availability on the possibility of identification of accelerant traces. The study revealed that the type of burned material exerts the greatest influence in identifying the accelerant traces [66]. In an attempt to develop a relatively cheaper and more reliable onsite detection tool for ignitable liquid residues, Y. Lu and P.B. Harrington [67] investigated gas chromatography-differential mobility spectrometry (GC-DMS). Headspace SPME was used as the sampling and sample preconcentration technique. GC-DMS provides unique two-way patterns for each sample. The collected data were classified into one of seven ignitable liquids using a fuzzy rule building expert system (FuRES). The results demonstrated an accuracy of 82.3% for burned samples.

25.5. ANALYSIS OF EXPLOSIVES

Since September 11, 2001 terrorist attacks in the USA, the importance and significance of the identification and quantification of traces of explosives, particularly compounds used in making homemade explosives (HME), have been overwhelmingly recognized among the law enforcement agencies, as well as the forensic scientists working in this field. As such, new instrumental methods, a majority of them based on gas chromatography, have emerged in recent years. Table 25.2 represents a list of common organic explosive compounds with their pertinent properties [68a]. The low vapor pressures of many organic explosive compounds present tremendous challenges in analyzing and detecting them with gas chromatography.

Identification and determination of post-blast organic residues of explosives is an important topic in forensic science as the chemical profile of postblast residues provides valuable information with regard to the explosive type, their tentative origin, as well as an assessment of the severity of environmental pollution.

Organic explosive compounds can be arbitrarily classified as nitro-containing and non-nitro-containing explosives based on the presence or absence of nitro/nitroso functional groups in their chemical structure. Trinitrotoluene (TNT) and nitroglycerine (NG) are examples of nitro-containing explosives, whereas triacetone triperoxide (TATP) and hexamethylene triperoxide diamine (HMTD) are examples of non-nitro-containing explosives.

Over the years, triacetone triperoxide (TATP) has increased in popularity among terrorist groups, as indicated by multiple bombings that have occurred within the United States and throughout the world, where TATP was the explosive compound that was used.

A new method was developed by Sigman et al. [68b] for analyzing TATP in GC-MS (using ammonia positive ion chemical ionization, electron ionization, and methane negative ion chemical ionization). Comparison among different ionization methods revealed that ammonia positive ion chemical ionization yields LLOD in the picogram level compared to 15 ng obtained by GC-FID.

Since homemade explosives are generally made by commercially available ingredients, one possible way to track the origin of the used precursors is to obtain the impurity profile of the precursor and matching the impurity

TABLE 25.2 List of Organic Explosives with Their Pertinent Properties

Explosive class	Name of the explosive	Abbreviation	Molecular weight	Vapor pressure at 25 °C (Torr)
Nitroaliphatic	Nitromethane	NM	61.04	2.8×10^{1}
	2,3-Dimethyl-2,3-dinitrobutane	DMNB	176.17	2.1×10^{-3}
Nitroaromatic	2,4-Dinitrotoluene	2,4-DNT	182.13	2.1×10^{-4}
	2,4,6-Trinitrotoluene	TNT	227.13	5.8×10^{-6}
	2,4,6-Trinitrophenol	Picric Acid	229.10	5.8×10^{-9}
Nitrate ester	Ethylene glycol dinitrate	EGDN	152.06	7.0×10^{-2}
	Nitroglycerine	NG	227.09	3.1×10^{-4}
	Pentaerythritol tetranitrate	PETN	316.14	1.4×10^{-8}
Nitramine	2,4,6,N-Tetranitro-N-methylaniline	Tetryl	287.14	5.7×10^{-9}
	Trinitro triazacyclohexane	RDX	222.12	4.6×10^{-9}
	1,3,5,7-Tetranitro-1,35,7-tetraazacyclooctane	HMX	296.16	1.6×10^{-13}
	Hexatrinitrohexaazaisowurtzitane	CL20	438.18	Not available
Peroxide	Triacetone triperoxide	TATP	222.24	3.7×10^{-1}
	Hexamethylene triperoxide diamine	HMTD	208.17	Not available
By-product	2-Ethyl-1-hexanol		130.23	1.4×10^{-1}
	Hydrogen peroxide		34.03	1.4×10^{0}
	Acetone		58.08	2.3×10^{2}

profile with commercial precursors from different manufacturers. As different manufacturers use different starting materials, solvent, and synthetic routes for synthesizing precursor chemicals, each of the precursors should possess a unique impurity profile that can be easily differentiated from others. Partridge et al. [69] utilized some off-the-shelf precursors to prepare TATP and hexamethylene triperoxide (HMTD). Following detonation, SPME was used to obtain postblast by-products. Analysis of postblast residues by GC-MS demonstrated the persistence of the same impurities present in the precursors, validating the potential application of profiling to trace back the raw materials/precursors used to manufacture the explosive.

A majority of the explosives, except peroxide-based explosives, contain either nitro (NO_2) or nitrate (NO_3) functional groups, and therefore are compatible with a chemiluminescence-based detector known as thermal energy analyzer (TEA). The TEA pyrolyzes the explosive vapors at high temperatures to produce a NO radical, which is later oxidized by ozone to form the excited state of nitrogen dioxide (NO_2*). The excited entity then emits a photon as it moves back to the ground state. The emission of photons can be detected using a photomultiplier tube. Since TEA is only sensitive to nitrogen-containing compounds, the presence of other organic compounds in the sample matrix does not impact the sensitivity of TEA in detecting such explosives in complex environmental matrices [70]. DNT, TNT, NG, and PETN explosives have been analyzed using solvating GC in a packed column with CO_2 as the mobile phase and detected with TEA.

Compared to the detection sensitivity of FID, TEA can detect 10 times more NG, with a linear range from 0.1 to 0.5 ppb [71]. In addition to TEA, other detectors, e.g. flame ionization detector (FID), electron capture detector (ECD), nitrogen-phosphorous detector (NPD), and mass spectrometer (MS), are frequently used for nitrogen-containing explosives, with ECD being the most common. The ECD, commonly used for halogenated compounds, is less selective, but more sensitive for nitroaromatic explosives than TEA or NPD [72].

Routon et al. [73] developed a GC-MS method in order to discriminate Hodgdon Pyrodex® and triple Seven®, two black powder substitutes frequently used as fillers in improvised explosive devices (IEDs), which have claimed thousands of soldiers, as well as civilians, lives. The method utilizes derivatization using BSTFA + 1% TMCS. The derivatization process converts organic fuel compounds (benzoic acid, nitrobenzoic acid, and dicyandiamide) into trimethylsilyl derivatives which can be easily separated and identified by GC-MS.

Emulsion explosives, a new explosive class widely used in China that accounts for 40–50% of Chinese explosive output, consist of an aqueous solution of inorganic oxidizing salt, an organic fuel, and emulsifiers. Postblast residues of these explosives can be analyzed by first reacting the residues with methanolic KOH solution which converts the emulsifiers into methyl esters of fatty acids, followed by derivatization with BSTFA containing 1% TMCS. The GC-MS profile offers sufficient information to recognize the original explosive [74a].

With the rapid increase in the use of organic explosives, it is important to trace the clandestine manufacturing units. As such, Partridge et al. [74b] developed a method to detect impurities in the commercially available precursor chemicals. Detection of those impurities would provide valuable information regarding the potential sources of the precursor chemicals and eventually would be able to retrospectively track down the potential wrongdoer.

Although ion mobility spectrometry (IMS) is considered to be a reliable field detection tool for explosive detection and is routinely used by law enforcement and military personnel, its deficiency to handle chemical mixtures (as opposed to a single pure compound) poses a serious limitation to its effectiveness as a field detection technique. Cook et al. [75] offered a unique approach to boost the capability of IMS by interfacing it with gas chromatography. IMS coupled with GC resolved the problem of separating components from a complex mixture, thus accomplishing enhanced linear dynamic range (LDR). As such, the new system could be able to screen both single- and multicomponent explosive systems with the ability to detect trace to high concentrations of target analytes.

Although commonly used by law enforcement to disperse large groups of protesters, aerosol defense sprays, also known as pepper spray, can be used, with criminal intention as a weapon. Among other irritants, oleoresin capsicum (OC) is one of the major active ingredients used in defense sprays. In forensic investigations, pepper spray residuals have been recovered from the fabrics of the victim/offender by liquid–liquid extraction. However, low and inconsistent recovery in solvent extraction has prompted the development of a new method based on headspace SPME-GC-MS. Both PDMS/DVB and DVB/CAR/PDMS SPME fibers were found to be suitable for extracting capsaicin and its analog dihydrocapsaicin with LODs of 1.08 and 0.73 ng, respectively, and about 70% recovery from spiked samples.

The efficient detection of trace amounts of nitrogen-containing explosives in military training, wartime activities, airports, public meeting places, and other high security risk areas, are of great interest and a high demanding research area. In addition to their security implications, most of the explosives are toxic and pose serious health hazards to people who are exposed to these explosives continuously. As such, it is necessary to develop highly selective and specific analytical methods that can effectively detect and identify explosives from different complex matrices, e.g. air, soil, and water.

Due to the very low concentration of explosives that are present in an overly complex matrix, explosive samples often require sample preparation and/or matrix cleanup prior to analysis in gas chromatography. Water and soil samples are generally prepared by using US Environmental Protection Agency SW-846 Method 8330 [76] which utilizes SPE or salting out. However, solid-phase extraction often involves multiple steps, utilizing toxic organic solvents, and indiscriminately extracting numerous organic interferents that are present in the sample matrix.

Explosive traces can also be recovered from different surfaces in the vicinity of the explosion site using solvent-wetted cotton swabs. Analytes are extracted from the swab by adding additional organic solvent. The volume of the extracting solvent is often reduced by evaporation to increase the analyte concentration.

Being simple, solvent free, and fast, solid-phase microextraction (SPME) has been employed for selective extraction of explosive analytes from different matrices. Both direct immersion SPME [76] and headspace SPME [77] can be used with PDMS/DVB and CW/DVB as the preferred fiber chemistry.

25.6. GAS CHROMATOGRAPHIC ANALYSIS OF ORGANIC GUNSHOT RESIDUES (OGSRs)

In any criminal case involving actual or suspected use of firearms, the detection and identification of residues from the firearm discharge are of prime importance as they provide invaluable information in estimating the firing distance, identifying the bullet holes, and in determining whether or not the suspect was involved in the shooting [78]. Gunshot residue (GSR), also known as cartridge discharge residue (CDR) or firearm discharge residue (FDR), is composed of unburned or partially burned propellant powder, particles from the ammunition primer, grease, lubricants, and tiny metal fragments from the cartridge [79–81]. Gunshot residue comprises both organic and inorganic compounds. However, we would limit our discussion to only organic components of the gunshot residue as GC is only used to detect and identify those compounds that are present in GSR. Table 25.3 provides a list of typical organic compounds commonly encountered in gunshot residue. Some of the compounds in the list are almost obsolete, as they are not being used in current firearm formulations [81]. However, they may still be detected in cases where old ammunitions are used. The major sources of organic compounds in GSR include the propellant powder and the primer mix. Although black powder was the first propellant used in firearms, it has been discontinued with the introduction of a better propellant known as smokeless powder. As the name implies, it produces negligible smoke when fired unlike the black powder. Smokeless powders are classified into (1) single-base powder if only nitrocellulose is used as the explosive; (2) double-base powder if both nitrocellulose and nitroglycerine are used as the explosive components; and (3) triple-base powder when nitroglycerine, nitrocellulose, and nitroguandine are used as the explosive components. In addition to the explosive components, other ingredients that are commonly used in the firearm formulations are additives, stabilizers, plasticizers, flash inhibitors, coolants, surface lubricants, and anti-wear additives, are all contributors to the source of organic compounds in the GSR.

TABLE 25.3 Common Organic Compounds Identified in Gunshot Residues

Compound	Function
2,4-Dinitrophenylamine (2,4-DPA)	Propellant
2,3-Dinitrotoluene (2,3-DNT)	
2,4-Dinitrotoluene (2,4-DNT)	
2,6-Dinitrotoluene (2,6-DNT)	
2-Nitrophenylamine (2-NDPA)	
4-Nitrophenylamine (4-NDPA)	
Carbazole	
Carbanilide	
Camphor	
Butyl centralite	
Butyl phthalate	
Akaridte II (AKII)	
Cresol	
Methyl cellulose	
Methyl centralite	
Methyl phthalate	
Nitroguandine	
Nitrotoluene	
N-Nitrosodiphenylamine (N-NDPA)	
Picric acid	
RDX	
Resorcinol	
Triacetin	
Dextrin	Primer
Diazodinitrophenol	
Diazonitrophenol	
Gum Arabic	
Gum tragacanth	
Karaya gum	
Sodium alginate	
Pentaerythritol tetranitrate (PETN)	Propellant/primer
2,4,6-Trinitrotoluene	
Nitrocellulose (NC)	
Nitroglycerine (NG)	
Tetracene	
Tetryl	

For investigative purposes, gunshot residues are often collected from different areas and surfaces, e.g. skin, hair, body parts, and clothing of the suspect, as well as vehicles, surroundings of the incident, and surfaces of different objects which are in proximity of the incident [82]. Depending upon the objects and the type of surfaces from where GSR is collected, different sampling procedures and strategies are applied, with a common goal of maximizing the amount of GSR collected and minimizing the collection of matrix interferences, as well as to reduce the risk of cross-contamination. The most frequently used sampling procedure includes tape lifts, glue lifts, swabbing, vacuum lifts, combing, nose blowing, etc. [81].

Analysis of the organic compounds present in GSR is generally carried out using gas chromatography coupled with a variety of different detectors, e.g. FID, ECD, MS, and TEA, with TEA being the most frequently used detector. Andrasko et al. [83] investigated various organic compounds and degradation products from smokeless powders in the barrels of firearms after test shooting. Gunshot residues were extracted onto a SPME fiber and introduced into the inlet of GC-TEA for the analysis. The compounds were later identified by GC-MS. As SPME is a nondestructive equilibrium-based sampling and sample preconcentration technique, the same sample can be investigated multiple times in order to validate the analytical results. In another study performed by Zeichner et al. [84], analysis of the organic constituents of GSR was carried out by GC-TEA, GC-MS, and IMS in order to assess their relative performances. GSR was collected on fiberglass and Teflon filters using a portable vacuum sampler. Collected samples were solvent extracted followed by centrifugation or filtration and evaporative concentration. GC-TEA was found to offer a good level of sensitivity for organic compounds in GSR. However, the sensitivity for nitroglycerine

(NG) compared to dinitrotoluene (DNT) was considerably low which could be attributed to the fact that NG undergoes thermal decomposition in the GC column. The researchers found two responses for NG on a longer GC column, proving the thermal instability of NG during GC analysis. The second peak was identified as 1,2-GDN, a decomposition product of NG. The sensitivity obtained from GC-MS was at least one or two orders of magnitude lower than GC-TEA. On the other hand, IMS has shown comparable sensitivity for organic GSR.

In a separate study, Zeichner et al. [85] investigated the potential of analyzing both inorganic and organic components of GSR by first utilizing scanning electron microscopy/energy dispersive x-ray spectroscopy (SEM-EDX) for inorganic GSR and then employing GC-TEA and IMS for organic GSR. The GSR samples were collected from the suspect's body or clothing using double-sided adhesive stubs. The collected GSR on the adhesive stub were extracted in a solution containing 80% v/v aqueous solution of 0.1% w/v sodium azide and 20% v/v ethanol using sonication at 80 °C for 15 min. The residues obtained from the first extraction were then further extracted with methylene chloride followed by evaporative concentration and instrumental analysis. One major drawback of this combined analysis was its low extraction efficiencies (30–90%) for NG and 2,4-dinitrotoluene.

Although GC is compatible with a majority of the compounds listed in Table 25.3, some compounds cannot be analyzed by this system for various reasons. For example, nitrocellulose (NC) is not compatible with GC because it is not sufficiently volatile. Similarly, nitrate esters, a common GSR constituent, are incompatible with GC due to their poor thermal stability.

Another interesting application of GC with forensic implication in GSR analysis is the estimation of time since discharge from the firearm and spent cartridges. This application utilizes SPME to sample for VOCs inside the barrel of the firearm and cartridge case and then analyzing the extract by GC-TEA, GC-FID, or GC-MS. Estimation of time since last discharge is based on the assumption that the dissapation rate of VOCs from the firearm and the cartridge case is a function of time [86–90].

25.7. ANALYSIS OF FORENSIC TRACE EVIDENCE

Forensic trace evidence includes both microscopic and macroscopic physical evidence that is left behind at a crime scene by the perpetrator which can associate him/her linking them to the scene. Hairs, fibers, paint, soil, polymeric materials, glass, and impressions are among commonly encountered trace evidence. Trace evidence is often evaluated by visual and microscopical means to assess their physical characteristics to obtain a potential match with the suspect. When chemical analysis of the trace evidence is inevitable, gas chromatography is among the analytical instruments used for extracting valuable information out of the evidence to aid in establishing a possible associative relationship with the suspect. Although pyrolysis-gas chromatography/mass spectrometry (py-GC-MS) has been considered to be the gold standard for organic trace analysis, GC-MS is also used quite frequently. Synthetic polymeric materials, e.g. polyurethane foam [91], automobile body fillers and paints [92–94], and spray paints on plasters [95], are routinely analyzed by py-GC-MS. Trace evidence, e.g. fragments of photocopied paper [96], amino acids in fingerprint residues [97], and condom lubricants for sexual assault cases [98], are also analyzed by py-GC-MS. Verification of the authenticity and integrity of written documents is carried out by analyzing the ink from the questioned document. This analysis can be performed

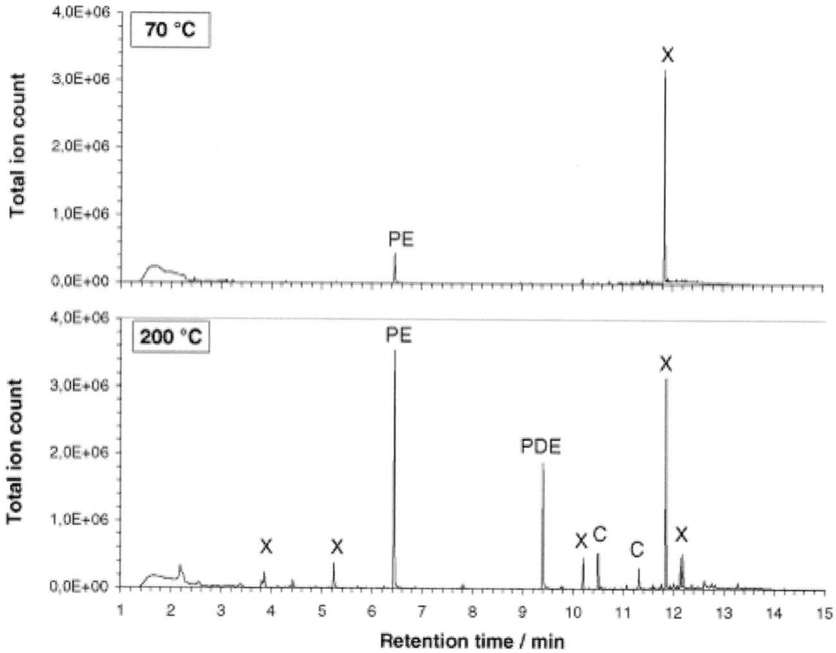

FIGURE 25.3 Chromatogram obtained for age determination of ballpoint pen ink by thermal desorption gas chromatography-mass spectrometry. *Source: Reprinted with permission from Ref. [100]. Copyright 2011, Elsevier Science.*

either by thermal desorption of ink volatiles followed by GC-MS (Figure 25.3) [99,100] or by solid-phase microextraction of extractable volatiles from the ink sample and subsequent analysis by GC-MS [101].

25.8. FORENSIC ENVIRONMENTAL ANALYSIS

The primary motivation behind developing and applying environmental forensic methodologies in the United States originated from the need to determine the environmental liability of a suspected individual/group or business entity in response to a specific law. Most common laws passed in order to safeguard the environment include Comprehensive Environmental Response Compensation and Liability Act (CERLA) and Resource Conservation and Recovery Act (RCRA). Environmental forensics is a very broad field and is growing very rapidly with the advent of numerous sophisticated instrumental and mathematical techniques that help in determining the extent of environmental pollution, as well as to ascertain the source(s) for potential legal action. Gas chromatography is an invaluable tool in many areas of environmental forensics, including analysis of pesticides, polychlorinated biphenyls (PCBs), dioxins and furans, polycyclic aromatic hydrocarbons (PAHs), and automobile gasoline. Fingerprinting of crude oil and refined products is another major area that relies upon gas chromatography.

25.8.1. Analysis of Pesticides

Pesticides are chemical agents used to destroy or control pests. Common pesticides include organochlorines, organophosphates, carbamates, pyrethroids, photosynthesis inhibitors, dithiocarbamates, and benzimidazoles. Pesticides and their degradation products are analyzed from different complex matrices, e.g. soil, water, and biological tissue, to determine their concentration and fate in the environment, as well as to study the exposure and ecological effects [102]. Forensic samples for pesticide analysis often undergo liquid–liquid extraction and liquid–solid extraction. Solid samples are generally treated with sodium sulfate to eliminate moisture followed by extraction with methylene chloride using Soxhlet extraction, ultrasonic extraction, or other suitable extraction techniques. Extracted samples are then subjected to cleanup procedures to reduce matrix interferences, which eventually helps with detecting and quantifying the target analyte(s). Cleanup processes include gel permeation chromatography, adsorption on solid adsorbents, and adsorption on charcoal. Depending upon the target pesticides, different gas chromatographic detectors are employed. For example, organochlorine pesticides and phenoxy acid herbicide are analyzed on GC-ECD, organophosphate pesticides are analyzed on GC-FPD, and organonitrogen pesticides are analyzed using GC-NPD. In addition to that, mass spectrometry is widely used for its capability to detect pesticides and also provides valuable structural information [103].

25.8.2. Analysis of Polychlorinated Biphenyls

Polychlorinated biphenyls (PCBs) belong to the list of stable organic contaminants which, once released into the environment, remain intact for an exceptionally long period of time by resisting photolytic, chemical, and biological degradation. Due to their widespread use in transformers and condensers as dielectrics, as well as in glues, dyes, and construction materials as additives, along with other industrial applications, their presence in the environment is ubiquitous. PCBs are highly toxic. The danger of PCBs is synergized with the presence of polychlorinated dibenzodioxines (PCDDs) and polychlorinated dibenzofurans (PCDFs). About 209 congeners of PCBs have been enormously challenging to efficiently detect and identify. Gas chromatography is routinely used for analyzing PCBs, with ECD and MS being the preferred detectors. Although the use of FID has been reported for characterizing commercial PCB mixtures, its use for trace PCB analysis from environmental samples is not recommended. For identification of individual PCB congeners, MS is used [104]. Due to the high number of PCB congeners, as well as matrix interferents, coelution is always a problem. Tandem mass spectrometry (MS/MS) can also be used for separating coeluting congener pairs which are not separated by GC-MS. Parallel GC can be used to accomplish better separation among closely eluted PCB congeners. Two chromatographic columns with different stationary phases (DB-1 and DB-XLB) are attached in parallel with the GC inlet. The analytes are detected by two ECD detectors connected to the other ends. Such a technique reduces the analysis time significantly with separation of critical congeners [105]. Since the same sample splits up into two plugs and moves along the individual column, separation of the mixture does not exploit both the column chemistries simultaneously and, therefore, much separation cannot be expected. Multidimensional GC is a viable alternative where separation of closely eluted PCB congeners is required. Multidimensional gas chromatography has the capability of separating an order of magnitude more compounds than conventional single-dimensional gas chromatography [106]. Multidimensional GC utilizes two capillary columns connected in series with a modulator between two columns. The modulator periodically injects the analyte mass eluting from the first column (typically, nonpolar) into the second column (generally, polar). The difference in selectivity of the two column chemistries augments the separation efficiency [104].

25.8.3. Analysis of Dioxins and Furans

Dioxins (polychlorinated dibenzo-*p*-dioxines) and furans (polychlorinated dibenzofurans) are among the most toxic group of compounds reported and are in the list of twelve compounds and/or compound classes that were decided to be reduced or eliminated in *Stockholm Conventions on Persistent Organic Pollutants* which came into force in May 2004. Dioxins and furans have structural similarities; both are planar compounds containing chlorine-substituted benzene rings connected by one (furan) or two (dioxins) oxygens [107]. Furan and dioxin may have as many as 75 and 135 congeners, respectively. Due to the complex nature of the environmental samples, different sample preparation techniques are employed based on the availability and the nature of the sample in order to reduce matrix interferences and preconcentrate the analyte. Major sample preparation techniques include Soxhlet extraction, accelerated solvent extraction (ASE), microwave-assisted extraction (MAE), and solid-phase extraction (SPE).

US EPA has developed a number of analytical methods (e.g. US EPA Method 1613 and 23) based on GC. Gas chromatography-high resolution mass spectrometry (GC-HRMS) is considered the gold standard being the only acceptable detector, per several US EPA methods [108]. GC-MS is also used frequently for furan and dioxin analysis. Other gas chromatographic techniques commonly used in dioxin and furan analysis are PTV-LV-GC-MS/MS [109], fast GC-TOFMS [110], GC × GC-TOFMS [111], and GC × GC-ECD [112].

25.8.4. Analysis of Polycyclic Aromatic Hydrocarbons

Polycyclic aromatic hydrocarbons (PAHs) represent a large class of organic compounds containing two or more fused aromatic rings. PAHs are ubiquitous environmental pollutants and have been present in air, soil, sediment, and water samples [113]. PAHs can originate from either natural sources (diagenesis of organic material and biosynthesis by plants and animals) or anthropological sources (combustion of fossil fuels, refuse burning, and automobile exhaustion). Hundreds of PAHs have been identified in nature, many of which are either carcinogens or carcinogen suspects. PAHs can enter into the environment on local, regional, and a global scale. Environmental forensic scientists often deal with the challenging tasks of (1) chemical profiling of PAHs, (2) determining the potential source types, (3) determining the specific source(s) of PAHs from many possible candidates, and (4) allocating fractions of the PAHs to the suspected source(s).

Since PAHs can originate from both natural and anthropological sources, establishing a scientifically sound association between the PAHs extracted and identified from a study area and a suspected source requires a combination of robust analytical techniques as well as advanced statistical tools.

Taking into consideration the enormous challenges involved in isolating PAHs from different overly complicated environmental matrices, the US Environmental Pollution Agency (US EPA) has developed multiple analytical methods for different environmental matrices, e.g. drinking water, municipal and industrial discharges, soils, sludges, solid waste, and ambient air [114]. Each method has been designed to provide particular information regarding the PAHs or volatile/semivolatile organic compounds present in the sample matrix, as well as to address sample preparation issues, as oftentimes sample preparation is needed to preconcentrate the PAHs, eliminate, or reduce the matrix interferences before it can be introduced into the GC inlet. Among all the EPA PAHs methods, EPA Method 610 applies to 16 priority pollutants, which are known to be of high risk for human health [115]. Due to the combined advantages of high selectivity, resolution, and sensitivity for PAH analysis, GC finds itself favorable over LC. As such, GC-FID and GC-MS are considered to be the workhorse in PAHs analysis. Due to the presence of a very high number of PAHs structurally similar to each other in environmental samples, careful selection of a GC column is of high importance if complete separation of each PAH is a goal. Methyl and phenyl-substituted polysiloxane stationary phases are the most common stationary phases used in PAH analysis [116]. However, liquid-crystalline columns seem to have been more effective in separating PAHs which are coeluted in 5% phenylmethylpolysiloxane columns. GC-TOFMS is recommended for complex samples as TOF-MS offers better structural conformations, as well as signal-to-noise ratios [117]. When separation of all PAHs is critical for the investigation, GC × GC with FID, MS, or TOF-MS should be considered [118]. Sources of the origin of environmental PAHs can be established employing gas chromatography-isotope ratio mass spectrometry (GC-IRMS) which enables the calculation of compound-specific carbon isotope ratios ($^{13}C/^{12}C$) of individual PAHs to trace back the origin of that particular PAH [119].

In addition to the techniques mentioned above, large-volume injection using programmed temperature vaporizer (PTV) for increased sensitivity [120], fast GC (approximately 3 min total analysis time) [121], and LC-GC-MS [122] have also been used to analyze PAHs from environmental samples.

25.8.5. Fingerprinting of Crude and Refined Petroleum

Crude and refined petroleum fingerprinting is a globally adopted technique aiming at determining the oil sources by matching the collected samples with the suspected candidate sources and includes differentiating the background hydrocarbons with the alleged spilled oil, monitoring the changes in composition of the spilled oil over time, and assessing the impact on the ecosystem in order to impose monetary liability to the responsible entity. Due to the chemical makeup of the crude and refined petroleum (predominantly hydrocarbons), gas chromatography is considered to be the gold standard in petroleum fingerprinting. GC-FID is frequently used in preliminary screenings of the environmental samples. Figure 25.4 represents a decision tree that is often used in a typical environmental lab to process suspected oil-spill samples.

Identification and/or quantification of an individual compound in the sample matrix are carried out by GC-MS [124]. Oftentimes, GC-IRMS is employed to obtain information regarding geographical origin of the spilled oil by calculating isotope ratios of selected elements. When environmental samples are found overly complex with many coeluted compounds, GC × GC in combination with different detectors, e.g. FID [125,126], TOFMS [106], nitrogen chemiluminescence detector (NCD), sulfur chemiluminescence detector (SCD) [127], atomic emission detector (AED) [128], and mass spectrometry (MS) [129], may be employed.

25.9. ANALYSIS OF HUMAN ODOR PROFILE

In recent years, analysis of human odor profile has garnered a great deal of interest among forensic chemists. Human odor has been reported [130] as being characteristic to an individual and an individualistic parameter that can be potentially used as a biometric identifier similar to that of a fingerprint.

Human odor is believed to be produced from a combination of the body's metabolism, hormones, gland secretions, and bacterial interactions [131]. Human odor is said to consist of

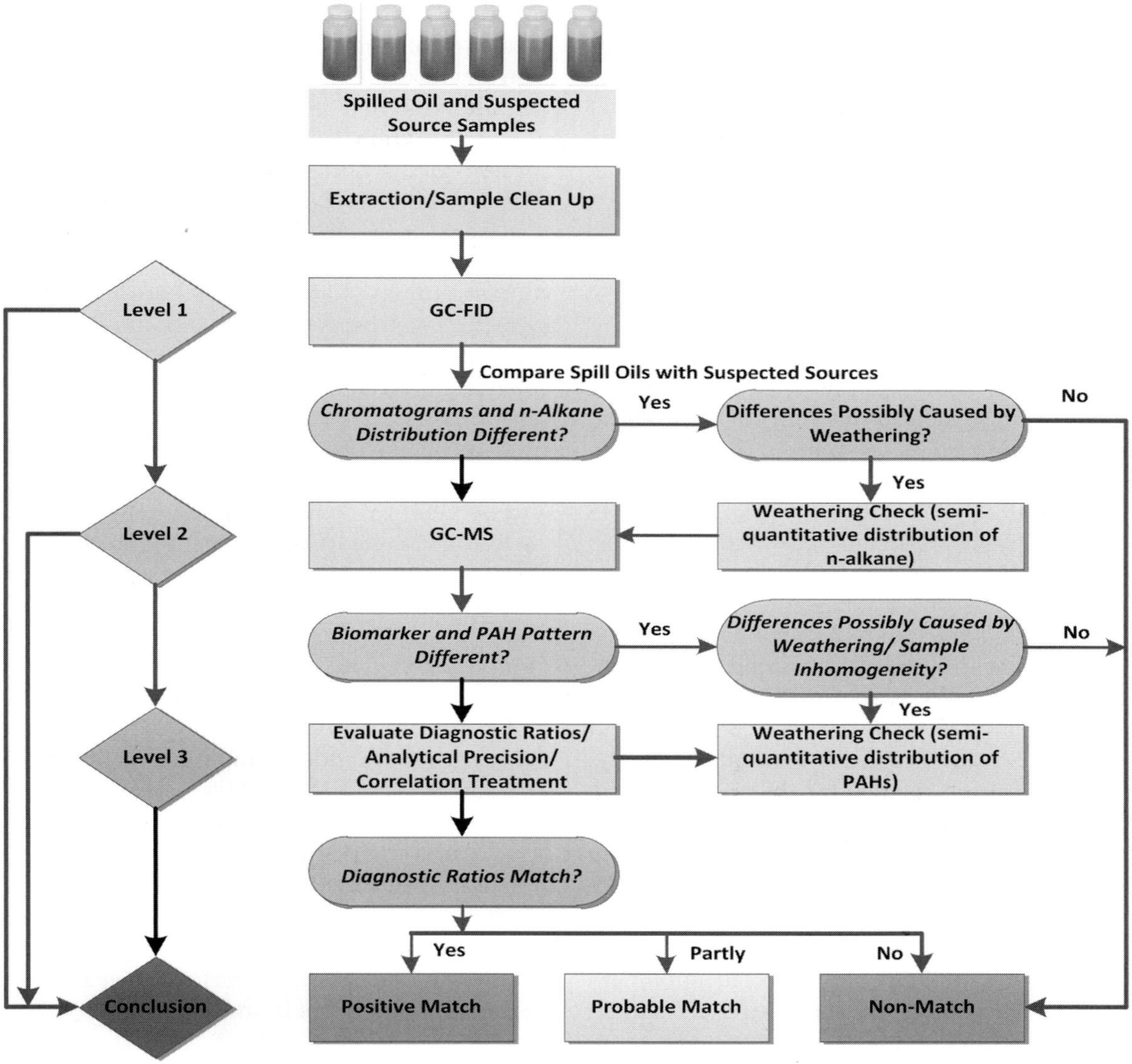

FIGURE 25.4 Decision chart used in oil-spill identification. *Source: Adopted with permission from Ref. [123]. Copyright 2011, Elsevier Science.*

a range of volatile organic compounds (VOCs) and vary in compound classes, e.g. aldehydes, ketones, aliphatic and aromatic hydrocarbons, fatty acids, and carboxylic acid methyl esters.

An individual odor profile is dependent upon several factors, e.g. genetic makeup, environmental and physiological conditions, age, and sex, and can be classified as primary, secondary, and tertiary odor [130]. Primary odor contains VOCs that are characteristic to an individual and is stable over time regardless of diet or environmental factors (surroundings). Secondary odor consists of VOCs that originate from within the body similar, to primary odor, but is functionally related to diet and various environmental factors. Tertiary odor contains VOCs that originate from outside sources (e.g. cosmetics, lotions, perfumes, and smoke).

From a forensic viewpoint, human odor collection is of primary interest to law enforcement officers because it can be collected from evidence and then subsequently provided to scent-discriminating canines where they can compare the collected scent to scent left at a crime scene [132,133].

Gas chromatography is the most predominantly used analytical instrument in human odor research. Due to the very low concentration of compounds present in human odor, a sample preparation technique is required to preconcentrate the analyte(s) prior to GC-MS analysis. Solid-phase microextraction (SPME) [130,134,135], stir-bar sorptive extraction (SBSE) [136], sorption on solid sorbents [137], and solvent extraction [138] have been reported to be used as a sample preparation/preconcentration techniques, with SPME being the most predominant one. Some researchers [133] collect human odor on selected sorbent materials (e.g. cotton, polyester, rayon, and blend fabrics) by placing the collection material in between the palms of the hands allowing the material to remain in contact with the palms for a predefined amount of time. The collection materials are then immediately enclosed in a glass vial in order to equilibrate the collected odor with the headspace inside the enclosed vial. Human odor compounds are then extracted onto a SPME fiber and introduced into the inlet of the GC where the VOCs are thermally desorbed and subsequent chromatographic separation occurs. Some researchers [132,139] have implemented a specially designed sampling tool known as a Scent Transfer Unit 100 (STU-100). STU-100 is a device that produces a dynamic airflow and is comprised of a vacuum pump and a scent collection head to hold the sorbent media (most often, a piece of cotton gauze) on the surface so that odor can be collected without coming into direct contact with the evidence (or palms of the hands). The airflow speed of this instrument can be adjusted which will impact the accumulation of VOCs on the sorbent media. Additionally, time for odor collection, sorbent media, and the distance between the STU-100 and the object of interest can be optimized in order to achieve the desired odor collection efficiency. Upon scent collection the sorbent is then stored in a vial so that SPME extraction can be performed and subsequently analyzed using GC-MS. In addition, instead of collecting odors on the surface of sorbents, solid-phase microextraction of VOCs emanating from the hand can be directly carried out by placing a sealed glass globe on the palm to enclose the extraction surface [138], or by enclosing the hand into a sampling chamber [135] where a continuous and steady nitrogen gas flow carries the human odor to the SPME fiber for preconcentration [135]. Another approach for hand odor collection is to wash the hands with a solvent (water) to transfer the VOCs into the solvent. The solvent containing the washed-hand odor compounds is then subjected to SPME-GC-MS analysis. This approach is recommended for extracting more polar and less volatile compounds (Figure 25.5.) [142].

Human scent evidence collected from objects at crime scenes is often quickly presented to human scent canines in order to identify a criminal from a pool of suspects Hudson et al. [141] studied the stability of human scent collected on a pretreated cotton sorbent. Cotton material retaining human scent volatiles was subjected to moderate to extreme environmental conditions (room temperature, −80 °C, dark, UV light) for a period of time. Solid-phase microextraction using 50/30 μm divinylbenzene/carboxen/ polydimethylsiloxane (DVB/CAR/PDMS), followed by GC-MS analysis was used to monitor the change in retaining human scent. The results revealed that the scent profile change with time, with the maximum changes occurring within the first 3 weeks.

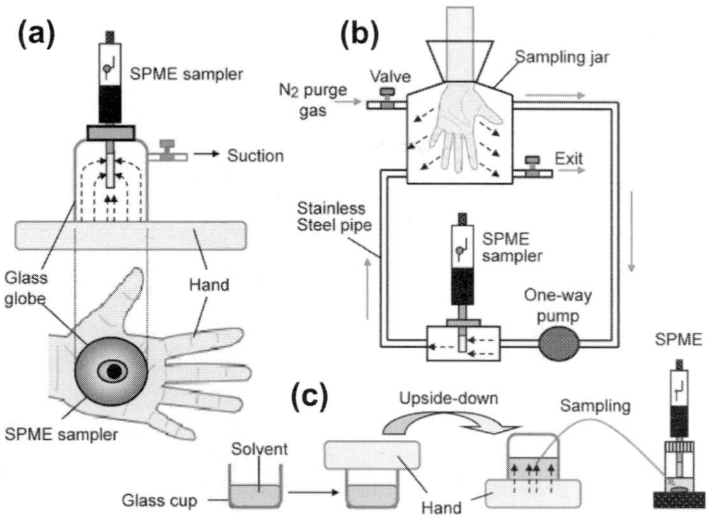

FIGURE 25.5 Illustration of hand odor sampling and preconcentration using solid-phase microextraction (SPME). (a) Direct SPME in sealed glass globe, (b) direct SPME in flow sampling chambers, and (c) liquid sampling in glass cup. *Source: Reproduced with permission from Ref. [142]. Copyright 2011, Elsevier Science.*

Different human scent collection media (cotton, cotton blend, and natural and synthetic composite materials) have also been tested to evaluate their capability and performance in retaining human scent volatile compounds [132,140,141]. Commercial sorbent materials were found to contain volatile compounds that were recognized as being human scent compounds. Therefore, these materials require effective cleaning prior to human scent collection. Supercritical fluid extraction, subcritical water extraction, Soxhlet extractions, and steam sterilization were investigated to identify the best extraction method to analytically clean the sorbent media. Headspace SPME-GC-MS of the sorbent media before and after the cleaning demonstrated that supercritical fluid extraction outperformed all the other extraction methods [140].

25.10. ANALYSIS OF HUMAN DECOMPOSITION PRODUCTS

Within the last several years, there has been a heightened interest in the decomposition process, in particular the evolution of VOCs, in both human and animal. Research has been conducted on the chemical processes that occur during the decomposition process, as well as identifying biomarker(s) that may aid in estimating the postmortem interval with high precision. Some research groups have also engaged in developing advanced methodologies or techniques to uncover buried human remains from clandestine graves.

Human remains and their analogues (often used as an investigational sample matrix,) are extremely complex and when they decompose the proteins, lipids, and carbohydrate macromolecules that are found within the body breakdown producing volatile organic compounds (VOCs) over a range of functional groups, such as aldehydes, ketones, sulfur-containing compounds, nitrogen-containing compounds, hydrocarbons, carboxylic acids, just to name a few.

The human body differs in composition (different protein/fat ratio) in comparison to human cadaver analogues. As a result, obtaining a representative sample from a decaying body is quite challenging. Sampling and sample preparation for human remains are mostly dependent upon broad objectives of study which are, in general, classified into the following categories: (1) estimating the post-mortem interval, (2) understanding the complex decomposition pathways, including the impact of different environmental and geochemical factors on the decomposition process, as well as the decomposition products at various stages, (3) developing human remains detector canine training aids to train them without handling hazardous materials, and (4) assessing the environmental impacts on the decomposition process.

Research on human-remains volatiles is largely reliant on gas chromatography-mass spectrometry, [143].

25.10.1. Estimation of PostMortem Interval (PMI)

There are research groups that are focusing on identifying suitable biomarker(s) that could be used to estimate the postmortem interval of a deceased individual with high precision. Such a biomarker would not only minimize the possibility of human error, but aid in providing a more reliable and precise means for the estimation of the Postmortem Interval.

Among other analytes, amino acids, putrescine, and cadaverine were reported in some human remains studies [143]. Target analytes were extracted from human-remains sample in an EDTA buffer solution. After extracting the analytes from the human remains samples, it was subjected to drying followed by derivatization. Analysis of amino acids was carried out by derivatizing a fraction of dried extract using pyridine and N-methyl-N-[t-butyldimethylsilyl]trifluoroacetamide (MTBSTFA). On the other hand, cadaverine and putrescine were analyzed after derivatization with pyridine and methyl-8® (DMF dimethyl acetal). Both the samples were then injected into the GC-MS.

25.10.2. Characterization of Decomposition Products

The decomposition process begins immediately after death and may continue for years depending upon the environmental conditions and surroundings in which the body is placed [143]. One of the major decomposition products is adipocere, a white soap-like material consisting of saturated fatty acids, unsaturated fatty acids, triglycerides, hydroxyl fatty acids, and other minor ingredients [144]. Formation of adipocere can be affected by environmental conditions, including: temperature, pH, and microbial activity. Characterization of adipocere provides valuable information pertaining to the state of decomposition of the body, as well as to assess the impact of the burial environment on the decomposition process. Characterization of adipocere involves both identification and quantification of its individual constituents, and GC-MS is most often used as the analytical instrument. Oftentimes, adipocere is extracted from soil samples by solvent extraction using chloroform, hexane, or other suitable organic solvents. Due to the high polarity of the majority of the adipocere constituents, derivatization is required prior to injecting into the GC-MS. Hexamethydisilazane (HMDS) [145] and N,O-Bis(trimethylsilyl)trifluoroacetamide BSTFA [144] are commonly used derivatization reagents. Free fatty acids (FFAs) can be isolated from adipocere by solid-phase extraction using disposable cartridges [146]. The extracted FFAs are then derivatized and injected into the GC-MS for analysis. In addition to characterizing decomposition products in GC-MS, GC-IRMS can be used to obtain $\delta^{13}C$ values. The $\delta^{13}C$ values ($\Delta^{13}C_{18.0-16.0}$) are characteristic of an individual and can be used in combination with other information to aid in identifying decomposing remains [147].

Formation of volatile organic compounds (VOCs) is an integral part of the decomposition process. During the course of decomposition,

FIGURE 25.6 A noncontact sampling device used for collecting VOCs from human remains. *Source: Reprinted with permission from Ref. [152]. Copyright 2011, Elsevier Publications.*

different macromolecules (proteins, carbohydrates, and lipids) breakdown to produce a variety of VOCs. For example, carbohydrates produce oxygen-rich compounds, including alcohols, aldehydes, ketones, acids, esters, and ethers; proteins yield nitrogen, sulfur, and phosphorous containing compounds; lipids break down into hydrocarbons, nitrogen, phosphorous, and oxygenated containing compounds. VOCs are often collected from decomposing remains using different solid sorbents (Carbograph, Carbosieve, Carbotrap, Carbosieve S-III, and Carbotrap-C) [148,149]. After collecting the volatiles on the sorbent tubes for a predetermined period of time, the analytes are thermally desorbed and analyzed using GC-MS.

Human-remains detection canines are often trained using tissue, blood, bone, and decomposition fluids which are biohazardous and difficult to obtain. Attempts have been made to identify key odor-producing constituents from decomposing remains to produce nonhazardous canine training aids [132,150]. A noncontact sampling instrument is shown in Figure 25.6. Solid-phase microextraction (SPME) is employed to extract VOCs from cotton gauze pads that were used to collect odor from decomposing remains and then analyzed using GC-MS. Capillary open-tubular columns are predominantly used in human-remains volatile analysis, alumina-coated porous-layer open-tubular (PLOT) columns have also been used to analyze ninhydrin-reactive nitrogen (NRN) collected from the headspace air above gravesoil [151].

25.11. FIELD-PORTABLE GAS CHROMATOGRAPH FOR ONSITE SAMPLE ANALYSIS

Following September 11, 2001, the demand for portable and rapid field-deployable gas chromatographic systems that are capable of detecting lethal, hazardous, and toxic chemicals, such as chemical warfare agents (CWAs) and toxic industrial chemicals (TICs), has increased significantly. Since then, a number of portable and rapid field-deployable gas chromatographic systems have emerged, each of which has its own unique manifestation. The following section discusses portable GCs.

A.C. Lewis et al. [153] developed a microfabricated planar gas chromatograph using two chemically etched 95×95 mm glass substrates sandwiched together resulting in a 7.5 m-long capillary channel. The channel was coated with a nonpolar polydimethylsiloxane (PDMS) stationary phase. A standard flame ionization detector (FID) and a modified lightweight photoionization detector (PID) served as the detector for the miniaturized planar GC. Probe test mixtures containing selected VOCs demonstrated detection sensitivity at the subnanogram level for monoaromatics, indicating its potential as a field-portable GC device.

Man-portable fast gas chromatograph coupled with a time-of-flight mass spectrometer (GC-TOF-MS) was developed. This instrument has low-power consumption, high-speed, a quadrupole ion trap time-of-flight mass spectrometer, and it is lightweight (30 lbs.). This field-portable system is capable of screening a sample rapidly under photoionization MS or full-scale confirmatory analysis using GC-MS in EI mode. Such a system can be conveniently deployed in the field for detecting hazardous chemicals including CWAs [154].

In an attempt to augment the detection capability of ion mobility mass spectrometry (IMMS), a popular explosive detection system already deployed worldwide for explosive detection in key strategic points, has been coupled to a GC. The combined GC-IMS-MS has remarkably decreased the false-positive rates for explosives in comparison to the standalone IMS. In addition, GC-IMS-MS has demonstrated the capability of detecting TNT, RDX, HMTD, and TATP even in the presence of highly concentrated of interferences, allowing it to be a good candidate for field deployment [75].

Modifying commercially available HAPSITE® Smart GC-MS system, J.D. Fair et al. [155] enhanced its capability for rapid sampling and onsite analysis of VOCs at low parts per billion (ppb). The modifications include the addition of a Tenax microtrap concentrator between the GC inlet and the column (1 m capillary GC column). The new system is suitable for air-quality monitoring.

A GC-μFID system has been developed [156] to cater to identification of environmental contaminants from polluted water in conjunction with SPME.

Focusing on the capability of detecting CWAs and TICs, a hand-portable GC coupled with toroidal ion trap mass spectrometer (GC-TMS) that runs on batteries has been developed (Figure 25.7). The system uses SPME for sample collection and introduction into the GC inlet, thereby enabling it to handle target analytes from both air and water matrices [157].

25.12. GAS CHROMATOGRAPHY IN FOOD FORENSICS

Food forensics is an emerging discipline that deals with determining authenticity, adulteration, and safety of foodstuffs to protect consumer's safety and well-being and to enforce food-related laws. Unlike other forensic fields, food forensics has broader implications on the personal level as it is designed to safeguard what we eat or drink.

Although food authenticity is primarily determined by various methods based on DNA analysis, proteomics, and metabolomics, the use of gas chromatography is also a common practice.

In response to the report of melamine-tainted Chinese infant formula, which allegedly inflicted severe adverse health effects in babies, a new confirmatory GC-MS method was developed to confirm the presence of melamine in cow milk (CM) and milk-based powdered infant

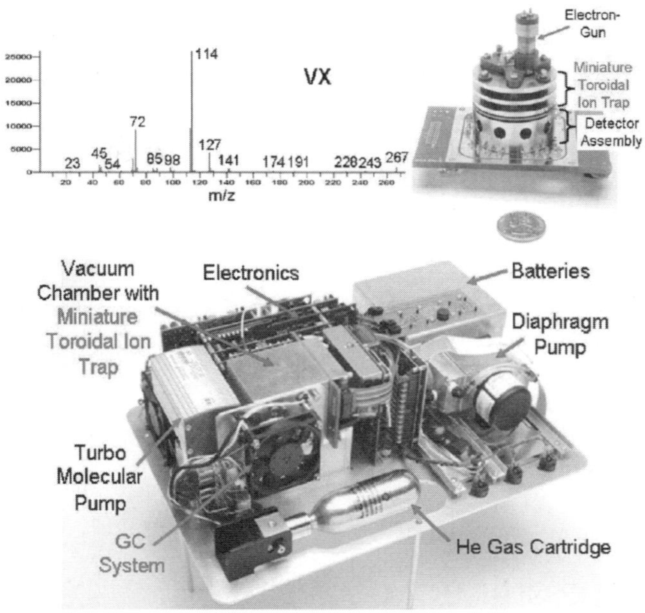

FIGURE 25.7 Internal components of portable Guardion-7 gas chromatograph-toroidal ion trap mass spectrometer (GC-TMS). *Source: Reproduced with permission from Ref. [157]. Copyright 2011, Elsevier Science.*

formula (MBPIF) [158]. Melamine was extracted from MBPIF using solid-phase extraction. The eluent from the SPE cartridge was subjected to derivatization using BSTFA-TMCS (99:1 v/v). The derivatized solution was filtered and introduced into the GC-MS. A low LOD (0.009 mg/kg), high recovery (95–101%), and fair reproducibility of the method was indicative of the method's performance for confirming melamine in the selected matrices.

Olive oils are often subjected to adulteration and if the adulterant is similar in nature (e.g. hazelnut oil in virgin olive oils), then determining authenticity of the claimed substance becomes very challenging. L. Cercaci et al. [159] developed a method by combining solid-phase extraction, thin-layer chromatography, and gas chromatography to confirm the authenticity of virgin olive oil by recognizing the presence of hazelnut in the tested sample. The method was built on the principle that authentic virgin olive oils possess a unique value range of a esterified sterol fraction known as the Mariani ratio (R_{MAR}), which is significantly different in adulterated samples. A SPE silica cartridge was used to separate esterified sterols from the oil sample. The adsorbed analytes were eluted first in n-hexane and then by a mixture of n-hexane and diethyl ether (8:2 v/v). Both the fractions were then combined, dried, and subjected to cold saponification. The unsaponifiable fraction was fractionated on a silica TLC plate, which extracted the sterol band, and derivatized with a mixture of pyridine, hexamethyldisilazane (HMDS), and trimethylchlorosilane (TMCS) prior to GC-MS analysis. The study suggested that calculating the Mariani ratio of any suspected oil by GC-MS may provide reliable information about its authenticity.

U.S. Food and Drug Administration (FDA)'s Forensic Chemistry Center (FCC) received 25 cases during 1992–2004 involving the tampering of beverage products, infant formula, and raw meat, with bleach. The exposure to bleach may cause serious physical damage to the consumer. As such, D.S. Jackson et al. [160] developed a method based on headspace GC-FID capable of detecting bleach in beverages *via* chloroform. Chloroform is produced in aqueous solution of sodium hypochlorite and can be easily detected by GC-FID.

Other interesting applications of GC with forensic significance include recognition of beer brand [161], determination of THE geographic origin of Emmental cheese [162], and investigating wine adulteration by GC-IRMS [163].

25.13. ANALYSIS OF CHEMICAL WARFARE AGENTS (CWAs)

A chemical warfare/weapon agent (CWA) is a substance intended for use in military operations to kill, seriously injure, or incapacitate the enemy because of its physiological effects. Chemical weapon agents are classified into four major classes based on their functional characteristics: (1) nerve agents, (2) blister agents, (3) blood agents, and (4) choking agents. Table 25.4 describes the classification, example, and characteristics of major CWAs.

TABLE 25.4 List of Chemical Warfare Agents (CWAs) and Their Characteristics

List of chemical warfare agents		
Agent classes	**Example**	**Characteristics**
Nerve agents	Tabun (GA) Sarin (GB) Soman (GD) Cyclosarin VX	Attack nervous system. Can enter body through inhalation or skin.
Blister agents	Mustard gas Lewisite	Attack skin. Rapidly absorbed into skin.
Choking Agents	Phosgene Chloropicrin Chlorine	Attack respiratory track.
Blood Agents	Hydrogen cyanide Cyanogen chloride	Attack blood circulatory system.

Nerve agents have a short life span and can readily hydrolyze under environmental conditions to stable degradation products, e.g. alkylphosphonic acids (APAs). A rapid and reliable GC method is necessary to screen and identify the presence of APAs in a short period of time. Although LC/MS can be used for this purpose, it is not compatible with mobile labs for onsite testing. R. Subramaniam et al. [164] addressed this problem by developing a new GC-MS method that requires only 5 min of sample preparation before it can be introduced into the GC-MS NCI SIM for screening. Sample preparation included the derivatization of a sample containing APAs by a novel derivatization reagent, 1-(diazomethyl)-3,5-bis(trifluoromethyl) benzene, under sonication. For identification purposes, the same derivatized solution was quenched with acetic acid for 10 min at 40 °C, evaporated to dryness, and then reconstituted in hexane. The solution was then injected into GC-MS EI and PICI.

Plausible forensic signatures can be obtained by impurity profiling of chemical weapon precursors using GC × GC-TOF-MS which is beneficial in locating the perpetrator(s) [165].

Due to the ongoing risk of chemical attack, a fast and reliable analytical method to identify the presence of chemical warfare agents is of high forensic interest. As such, a number of man-portable and rapid field-deployable GC systems coupled with different detection systems have been introduced recently. Additional information can be found in Section 25.11.

25.14. NEW DEVELOPMENTS IN GAS CHROMATOGRAPHY WITH FORENSIC IMPLICATIONS

In addition to the continuous refinements in the GC hardware, pneumatics, electronics, control software, as well as GC column chemistries, the addition of fast GC, GC × GC, and GC-IRMS has significantly augmented the overall analytical power that GC offers to the forensic community.

Capillary gas chromatography is considered to be the workhorse in forensic laboratories due to it's ability at separating and detecting volatile and semivolatile organic compounds of forensic interests; however, it often suffers from long analysis time which may span from 10–60 min. With the growing number of samples in forensic laboratories, faster analysis has always been a pressing issue. To address such a demand, fast GC has emerged. The primary goal of the fast GC is to maintain resolving power of the column by manipulating a number of chromatographic parameters, e.g. column length, internal diameter of the column, stationary-phase chemistry and thickness, carrier gas and its linear velocity, and GC oven temperature program. Consequently, a 3–10-times faster analysis can be achieved without compromising the quality of separation. Fast GC generally uses 100 μm I.D. column and hydrogen as carrier gas.

Another remarkable development in the field of gas chromatography is the advent of multidimensional GC. Multidimensional GC utilizes two dimensions for separating analytes in a complex mixture by employing two columns of different polarities in a series which resolves a group of peaks with narrow retention time ranges. Sample mixture are first separated on a nonpolar column and then small sections of the primary eluent are introduced into a second short and narrow polar column using a high-precision modulator. Separation achieved in the first column is further augmented in the second column; thus, difficult-to-separate critical pairs of analytes are well resolved in GC × GC. When MS is used as the detector, analytes experience additional dimension in the separation process.

Gas chromatography-isotope ratio mass spectrometry (GC-IRMS), a significant addition to chromatographic separation, made a tremendous impact on forensic science. GC-IRMS provides unique information regarding geographic, chemical, and biological origin of the analyzed sample. The underlying principle of GC-IRMS is that based on the geographic, chemical, or biological origin, samples collected can be characterized by it's C, O, H, N, and S isotope ratio. As a result, even though two samples are apparently identical, they can be discriminated from each other if they are sourced from two different geographic locations.

Recently, GC × GC has been interfaced with IRMS which would allow forensic scientists to extract information about the samples that cannot be obtained by individual systems alone. Figure 25.8 represents the schematic of a GC × GCC-IRMS.

25.15. CONCLUSIONS

Ever since gas chromatography was commercially available, its application has gradually increased by embracing new fields and directions. Forensic science, like other disciplines, is heavily dependent on gas chromatography. A continuous influx of new column chemistries, development of high-precision thermal and pneumatic controlling systems, introduction to programmed temperature vaporizer (PTV), advancement in control electronics, and a large variety of detection systems have positioned gas chromatography as a formidable foe to other competing

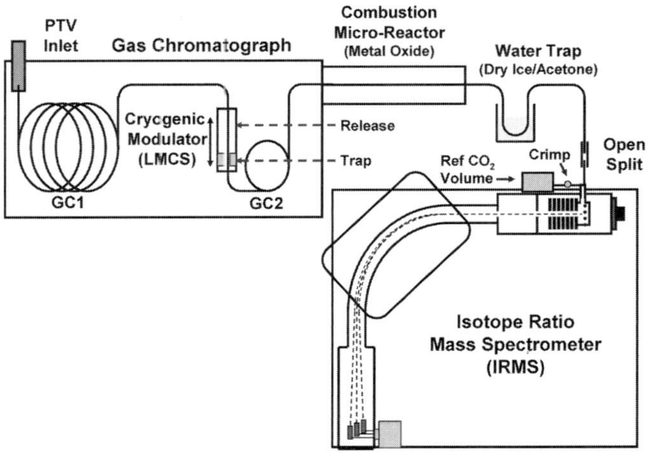

FIGURE 25.8 Schematic representation of a GC × GCC-IRMS system. *Source: Reproduced with permission from Ref. [166]. Copyright 2011, American Chemical Society.*

analytical instruments. Although the applications of gas chromatography are limited to volatile and semivolatile organic compounds, rapid development in derivatization chemistry has allowed for new organic compounds that have a forensic significance amenable to GC extending its horizon.

The introduction of two-dimensional gas chromatography (GC × GC), gas chromatography-isotope ratio mass spectrometry, and fast gas chromatography has contributed significantly to forensic applications by extracting additional information from the collected sample that can aid forensic scientists.

Miniaturization in the gas chromatographic system is another long-sought demand for the forensic field investigators. Such miniaturization would allow forensic investigators to take the analytical instrument to the field to analyze their samples immediately upon collection of evidence and will minimize the possibility of sample contamination, loss, and degradation.

It can be speculated with high confidence that gas chromatography will remain a strong player in the forensic field and will continue to offer new and unique attributes to solve more challenging forensic problems with relative ease.

Abbreviations

AED Atomic emission detection
ASE Accelerated solvent extraction
ASTM American Society for Testing and Materials
BSA N,O-Bis(trimethylsilyl)acetamide
BSTFA N,O-Bis(trimethylsilyl)trifluoroacetamide
BZP N-Benzylpiperazine
CAR-PDMS Carboxen-polydimethylsiloxane
CDR Cartridge discharge residues
CI Chemical ionization
CERLA Comprehensive environmental response compensation, and liability act
CWA Chemical warfare agent
DA Discriminant analysis
DEA Drug Enforcement Administration
DMS Differential mobility spectrometry
DNT Dinitrotoluene
ECD Electron capture detector
EDTA Ethylenediaminetetraacetic acid
EtG Ethyl glucuronide
EPA Environmental Protection Agency
FID Flame ionization detector
FDR Firearm discharge residues
FPD Flame photometric detector
FTICR-HRMS Fourier transform ion cyclotron resonance-high resolution mass spectrometry
GABA Gamma-aminobutyric acid
GC Gas chromatography
GHB Gamma-hydroxybutyrate
GC-DMS Gas chromatography-differential mass spectrometry

GC-EI-MS Gas chromatography-electron ionization-mass spectrometry
GC-FTIR Gas chromatography-fourier transform infrared spectroscopy
GC-NCI-MS Gas chromatography-negative chemical ionization-mass spectrometry
GC-PCI-MS Gas chromatography-positive chemical ionization-mass spectrometry
GC-IRMS Gas chromatography-isotope ratio mass spectrometry
GC-TMS Gas chromatography-toroidal ion trap mass spectrometry
GC-TOFMS Gas chromatography-time-of-flight mass spectrometry
HFBA Heptafluorobutyric anhydride
HMDS Hexamethyldisilazane
HMTD Hexamethylene triperoxide diamine
ILR Ignitable liquid residues
IED Improvised explosive device
IMS Ion mobility spectrometry
LLE Liquid–liquid extraction
LOD Limit of detection
LOQ Limit of quantification
LPME Liquid-phase microextraction
MAE Microwave-assisted extraction
MDMA Methylenedioxymethamphetamine
MSTFA N-methyl-N-trimethylsilyl-trifluoro acetamide
MTBSTFA N-methyl-N-[t-butyldimethylsilyl] trifluoro acetamide
NG Nitroglycerine
NPD Nitrogen phosphorus detector
OC Oleoresin capsicum
OGSR Organic gunshot residues
PAH Polycyclic aromatic hydrocarbon
PCA Principal component analysis
PCB Polychlorinated biphenyls
PCP Phenylcyclidine
PDMS Poly(dimethylsiloxane)
PID Photoionization detector
PPMC Pearson product moment correlation coefficient
PPT Protein precipitation
PTV Program temperature vaporizer
RCRA Resource conservation and recovery act
SBSE Stir-bar sorptive extraction
SEM-EDX Scanning electron microscopy-energy dispersive X-ray spectroscopy
SFE Supercritical fluid extraction
SPDE Solid-phase dynamic extraction
SPE Solid-phase extraction
SPME Solid-phase microextraction
TATP Triacetone triperoxide
TEMPP 1-(3-trifluoromethyl phenyl) piperazine
TEA Thermal energy analyzer
TFAA Trifluroacetic anhydride
TIC Toxic industrial chemical
TMCS Trimethylchlorosilane
TMSCL Trimethylsilylchloride
WADA World anti-doping agency

References

[1] P. Esseiva, L. Gaste, D. Alvarez, F. Anglada, Illicit drug profiling, reflection on statistical comparisons, Forensic Sci. Int. 207 (2011) 27–34.

[2a] List of Scheduling Actions, Controlled Substances, Regulated Chemicals, Drug Enforcement Administration, U.S. Department of Justice, Springfield, VA, 2011.

[2b] Z. Mehmedic, S. Chandra, D. Slade, H. Denham, S. Foster, A.S. Patel, et al., Potency trends of Delta 9-THC and other cannabinoids in confiscated cannabis preparations from 1993 to 2008, J. Forensic Sci. 55 (2008) 1209–1217.

[3] J. Broseus, F. Anglada, P. Esseiva, The differentiation of fibre- and drug type Cannabis seedlings by gas chromatography/mass spectrometry and chemometric tools, Forensic Sci. Int. 200 (2010) 87–92.

[4] S. Klemenc, In common batch searching of illicit heroin samples — evaluation of data by chemometrics methods, Forensic Sci. Int. 115 (2001) 43–52.

[5] R. Dams, T. Benijts, W.E. Lambert, D.L. Massart, A.P. De Leenheer, Heroin impurity profiling: trends throughout a decade of experimenting, Forensic Sci. Int. 123 (2001) 81–88.

[6] M. Hida, T. Mitsui, S. Tsuge, H. Ohtani, Rapid and sensitive determination of morphine in street opium samples by thermal desorption gas chromatography using a microfurnace pyrolyzer, J. Separation Sci. 27 (2004) 1030–1032.

[7] S. Lociciro, P. Hayoz, P. Esseiva, L. Dujourdy, F. Besacier, P. Margot, Cocaine protiling for strategic intelligence purposes, a cross-border project between France and Switzerland - part I. Optimisation and harmonisation of the protiling method, Forensic Sci. Int. 167 (2007) 220–228.

[8] M. Chiarotti, R. Marsili, A. Moreda-Pineiro, Gas chromatographic-mass spectrometric analysis of residual solvent trapped into illicit cocaine exhibits using head-space solid-phase microextraction, J. Chromatogr. B-Anal. Technol. Biomed. Life Sci. 772 (2002) 249–256.

[9] Z. Muccio, G.P. Jackson, Simultaneous identification and delta 13C classification of drugs using GC with concurrent single quadrupole and isotope ratio mass spectrometers, J. Forensic Sci. 56 (2011) S203–S209.

[10] H. Inoue, Y.T. Iwata, K. Kuwayama, Characterization and profiling of methamphetamine seizures, J. Health Sci. 54 (2008) 615–622.

[11] K. Kuwayama, H. Inoue, T. Kanamori, K. Tsujikawa, H. Miyaguchi, Y. Iwata, et al., Contribution of thermal desorption and liquid-liquid extraction for identification and profiling of impurities in methamphetamine by gas chromatography-mass spectrometry, Forensic Sci. Int. 171 (2007) 9–15.

[12] K. Kanai, K. Takekawa, T. Kumamoto, T. Ishikawa, T. Ohmori, Simultaneous analysis of six phenethylamine-type designer drugs by TLC, LC-MS, and GC-MS, Forensic Toxicol. 26 (2008) 6–12.

[13] A. Grove, E. Rohwer, J. Laurens, B. Vorster, The analysis of illicit methaqualone containing preparations by gas chromatography-mass spectrometry for forensic purposes, J. Forensic Sci. 51 (2006) 376–380.

[14] D. de Boer, W. Goemans, V. Ghezavat, R. van Ooijen, R. Maes, Seizure of illicitly produced para-fluorofentanyl: quantitative analysis of the content of capsules and tablets, J. Pharm. Biomed. Anal. 31 (2003) 557–562.

[15] S. Song, P. Marriott, A. Kotsos, O. Drummer, P. Wynne, Comprehensive two-dimensional gas chromatography with time-of-flight mass spectrometry (GC x GC-TOFMS) for drug screening and confirmation RID E-7462-2010, Forensic Sci. Int. 143 (2004) 87–101.

[16] M. Praisler, I. Dirinck, J. Van Bocxlaer, A. De Leenheer, D. Massart, Pattern recognition techniques screening for drugs of abuse with gas chromatography-Fourier transform infrared spectroscopy, Talanta. 53 (2000) 177–193.

[17] R. Martino, M. Malet-Martino, V. Gilard, S. Balayssac, Counterfeit drugs: analytical techniques for their identification, Anal. Bioanal. Chem. 398 (2010) 77–92.

[18] K. Soltaninejad, M. Faryadi, M. Akhgari, L. Bahmanabadi, Chemical profile of counterfeited buprenorphine vials seized in Tehran, Iran. Forensic. Sci. Int. 172 (2007) E4–E5.

[19] S. Liu, S. Woo, H. Koh, HPLC and GC-MS screening of Chinese proprietary medicine for undeclared therapeutic substances, J. Pharm. Biomed. Anal. 24 (2001) 983–992.

[20] L. Langman, B. Kapur, Toxicology then and now, Clin. Biochem. 39 (2006) 498–510.

[21] H. Schütz, F. Erdmann, M.A. Verhoff, G. Weiler, Pitfalls of toxicological analysis, Leg. Med. 5 (Suppl.) (2003) S6–S19.

[22] O.H. Drummer, Requirements for bioanalytical procedures in postmortem toxicology, Anal. Bioanal. Chem. 388 (2007) 1495–1503.

[23] F. Moriya, Y. Hashimoto, Nicotine and cotinine levels in blood and urine from forensic autopsy cases, Leg. Med. 6 (2004) 164–169.

[24] P. Simon, E., Gamma hydroxybutyric acid (GHB) concentrations in humans and factors affecting endogenous production, Forensic Sci. Int. 133 (2003) 9–16.

[25] R. Lowe, A. Barnes, E. Lehrmann, W. Freed, J. Kleinman, T. Hyde, et al., A validated positive chemical ionization GC/MS method for the identification and quantification of amphetamine, opiates, cocaine, and metabolites in human postmortem brain, J. Mass Spectrom 41 (2006) 175–184.

[26] S. Paterson, R. Cordero, S. Burlinson, Screening and semi-quantitative analysis of post mortem blood for basic drugs using gas chromatography/ion trap mass spectrometry, J. Chromatogr. B 813 (2004) 323–330.

[27] D. Lo, T. Chao, S. Ng-Ong, Y. Yao, T. Koh, Acidic and neutral drugs screen in blood with quantitation using microbore high-performance liquid chromatography diode array detection and capillary gas chromatography flame ionization detection, Forensic Sci. Int. 90 (1997) 205–214.

[28] R. Raposo, M. Barroso, S. Fonseca, S. Costa, J.A. Queiroz, E. Gallardo, et al., Determination of eight selected organophosphorus insecticides in postmortem blood samples using solid-phase extraction and gas chromatography/mass spectrometry, Rapid Commun. Mass Spectrom 24 (2010) 3187–3194.

[29] C.L. Morris-Kukoski, E. Jagerdeo, J.E. Schaff, M.A. LeBeau, Ethanol analysis from biological samples by dual rail robotic autosampler, J. Chromatogr. B-Anal Technol. Biomed. Life Sci. 850 (2007) 230–235.

[30] L. Kristoffersen, L. Stormyhr, A. Smith-Kielland, Headspace gas chromatographic determination of ethanol: the use of factorial design to study effects of blood storage and headspace conditions on ethanol stability and acetaldehyde formation in whole blood and plasma, Forensic Sci. Int. 161 (2006) 151–157.

[31] I.A. Freire, A.M. Bermejo Barrera, P. Cid Silva, M.J. Tabernero Duque, P. Fernandez Gomez, P. Lopez Eijo, Microwave assisted extraction for the determination of ethyl glucuronide in urine by gas chromatography-mass spectrometry, J. Appl. Toxicol. 28 (2008) 773–778.

[32] I. Alvarez, A. Maria Bermejo, M. Jesus, P. Fernandez, P. Cabarcos, P. Lopez, Microwave-assisted extraction: a simpler and faster method for the determination of ethyl glucuronide in hair by gas chromatography-mass spectrometry, Anal. Bioanal. Chem. 393 (2009) 1345–1350.

[33] M. Yegles, A. Labarthe, V. Auwärter, S. Hartwig, H. Vater, R. Wennig, et al., Comparison of ethyl glucuronide and fatty acid ethyl ester concentrations in hair of alcoholics, social drinkers and teetotallers, Forensic Sci. Int. 145 (2004) 167–173.

[34] H. Kharbouche, F. Sporkert, S. Troxier, M. Augsburger, P. Mangin, C. Staub, Development and validation of a gas chromatography-negative chemical ionization tandem mass spectrometry method for the determination of ethyl glucuronide in hair and its application to forensic toxicology, J. Chromatogr. B-Anal. Technol. Biomed. Life Sci. 877 (2009) 2337–2343.

[35] R. Paul, R. Kingston, L. Tsanaclis, A. Berry, A. Guwy, Do drug users use less alcohol than non-drug users? A comparison of ethyl glucuronide concentrations in hair between the two groups in medico-legal cases, Forensic Sci. Int. 176 (2008) 82–86.

[36] Y. Shi, B. Shen, P. Xiang, H. Yan, M. Shen, Determination of ethyl glucuronide in hair samples of Chinese people by protein precipitation (PPT) and large volume injection-gas chromatography-tandem mass spectrometry (LVI-GC/MS/MS), J. Chromatogr. B-Anal. Technol. Biomed. Life Sci. 878 (2010) 3161–3166.

[37] C. Jurado, T. Soriano, M.P. Giménez, M. Menéndez, Diagnosis of chronic alcohol consumption: hair analysis for ethyl-glucuronide, Forensic Sci. Int. 145 (2004) 161–166.

[38] R. Agius, T. Nadulski, H. Kahl, J. Schraeder, B. Dufaux, M. Yegles, et al., Validation of a head-space solid-phase microextraction-GC-MS/MS for the determination of ethyl glucuronide in hair according to forensic guidelines, Forensic Sci. Int. 196 (2010) 3–9.

[39] M.K. Parr, W. Schaenzer, Detection of the misuse of steroids in doping control, J. Steroid Biochem. Mol. Biol. 121 (2010) 528–537.

[40] E.M. Brun, R. Puchades, A. Maquieira, Analytical methods for anti-doping control in sport: anabolic steroids with 4,9,11-triene structure in urine, Trac-Trends Anal. Chem. 30 (2011) 771–783.

[41] J. Marcos, J. Pascual, X. de la Torre, J. Segura, Fast screening of anabolic steroids and other banned doping substances in human urine by gas chromatography/tandem mass spectrometry, J. Mass Spectrom 37 (2002) 1059–1073.

[42] H.H. Maurer, Hyphenated mass spectrometric techniques - indispensable tools in clinical and forensic toxicology and in doping control, J. Mass Spectrom 41 (2006) 1399–1413.

[43] H. Maurer, Position of chromatographic techniques in screening for detection of drugs or poisons in clinical and forensic toxicology and/or doping control, Clin. Chem. Lab. Med. 42 (2004) 1310–1324.

[44] H. Maurer, Role of gas chromatography-mass spectrometry with negative ion chemical ionization in clinical and forensic toxicology, doping control, and biomonitoring, Ther. Drug. Monit. 24 (2002) 247–254.

[45] M. Mazzarino, M. Orengia, F. Botre, Application of fast gas chromatography/mass spectrometry for the rapid screening of synthetic anabolic steroids and other drugs in anti-doping analysis, Rapid. Commun. Mass. Spectrom. 21 (2007) 4117–4124.

[46] H. Maurer, Systematic toxicological analysis procedures for acidic drugs and/or metabolites relevant to clinical and forensic toxicology and/or doping control, J. Chromatogr. B 733 (1999) 3–25.

[47] H.J. Tobias, Y. Zhang, R.J. Auchus, J.T. Brenna, Detection of synthetic testosterone use by novel comprehensive two-dimensional gas chromatography combustionâ "isotope ratio mass spectrometry, Anal. Chem. 83 (2011) 7158–7165.

[48] M. Moller, K. Aleksa, P. Walasek, T. Karaskov, G. Koren, Solid-phase microextraction for the detection of codeine, morphine and 6-monoacetylmorphine in human hair by gas chromatography-mass spectrometry, Forensic Sci. Int. 196 (2010) 64–69.

[49] Anonymous, in: ASTM volume 14.02; general test methods; forensic psychophysiology; forensic sciences; terminology; conformity assessment; statistical methods; nanotechnology; forensic engineering, vol. 14.02, Manufacture of Pharmaceutical Products; ASTM International, Pennsylvania, USA, 2011, p. 1496.

[50] J.R. Almirall, K.G. Furton (Eds.), Analysis and interpretation of fire scene evidence, CRC Press LLC, 2004, p. 262.

[51] Eric Stauffer, J. Dolan, R. Newman, in: Fire debris analysis, Academic Press, Boston, Mass, 2008, p. 634.

[52] P.M.L. Sandercock, Fire investigation and ignitable liquid residue analysis - a review: 2001–2007, Forensic Sci. Int. 176 (2008) 93–110.

[53] A.D. Pert, M.G. Baron, J.W. Birkett, Review of analytical techniques for arson residues RID C-6915-2009, J. Forensic Sci. 51 (2006) 1033–1049.

[54] E. Stauffer, A review of the analysis of vegetable oil residues from fire debris samples: analytical scheme, interpretation of the results, and future needs, J. Forensic Sci. 51 (2006) 1016–1032.

[55] M. Monfreda, A. Gregori, Differentiation of unevaporated gasoline samples according to their brands, by SPME-GC-MS and multivariate statistical analysis, J. Forensic Sci. 56 (2011) 372–380.

[56] A. Harris, J. Wheeler, GC-MS of ignitable liquids using solvent-desorbed SPME for automated analysis, J. Forensic Sci. 48 (2003) 41–46.

[57] Z. Deng, G. Tian, Z. Fan, Identification of inflammable liquids (ethanol, ether and acetone) in debris of fire scene by hyphenated GC-MS with SPME, Lihua Jianyan, Huaxue Fence 44 (2008) 624–626.

[58] A. Choodum, N.N. Daeid, Evaluating the performance of three GC columns commonly used for the analysis of ignitable liquid mixtures encountered in fire debris, Anal. Methods 3 (2011) 1525–1534.

[59] Z. Deng, S. Zhao, Z. Lu, Z. Fan, G. Tian, Determination of arson residual of ignitable liquid with solid-phase micro-extraction (SPME)/gas chromatography-mass spectrometry, Xiaofang. Kexue. Yu. Jishu. 28 (2009) 695–697.

[60] R. Qiu, Separation of ignitable liquid residues from fire debris samples and their analysis by gas chromatography-mass spectrometry, Zhongguo Anquan Shengchan Kexue Jishu 3 (2007) 35–39.

[61] D. Sutherland, J. Perr, J.R. Almirall, in: Identification of ignitable liquid residues in fire debris by GC/MS/MS, CRC Press LLC, 2004, pp. 181–230.

[62] B. de Vos, M. Froneman, E. Rohwer, D. Sutherland, Detection of petrol (gasoline) in fire debris by gas chromatography/mass spectrometry/mass spectrometry (GC/MS/MS), J. Forensic Sci. 47 (2002) 736–756.

[63] M. Sittidech, Part I. Analysis of aromatic amines in banned azo dyes by SPME and GC/MS/MS. Part II. Analysis of residual ignitable fluids in fire debris by GC/MS/MS and development of an expert system (2002).

[64] J.M. Baerncopf, V.L. McGuffin, R.W. Smith, Association of ignitable liquid residues to neat ignitable liquids in the presence of matrix interferences using chemometric procedures, J. Forensic Sci. 56 (2011) 70–81.

[65] J. Almirall, K. Furton, Characterization of background and pyrolysis products that may interfere with the forensic analysis of fire debris RID D-1280-2010, J. Anal. Appl. Pyrolysis 71 (2004) 51–67.

[66] R. Borusiewicz, J. Zieba-Palus, G. Zadora, The influence of the type of accelerant, type of burned material, time of burning and availability of air on the possibility of detection of accelerants traces, Forensic Sci. Int. 160 (2006) 115–126.

[67] Y. Lu, P.B. Harrington, Forensic application of gas chromatography - differential mobility spectrometry with two-way classification of ignitable liquids from fire debris, Anal. Chem. 79 (2007) 6752–6759.

[68a] R. J. Harper and K.G. Furton, Biological Detection of Explosives. In: J. Yinon (Ed.), Counterterrorist Detection Techniques of Explosives, 1st Edn, Elsevier, 2007, pp. 395–431.

[68b] M.E. Sigman, C.D. Clark, R. Fidler, C.L. Geiger, C.A. Clausen, Analysis of triacetone triperoxide by gas chromatography/mass spectrometry and gas chromatography/tandem mass spectrometry by electron and chemical ionization, Rapid Commun. Mass Spectrom 20 (2006) 2851–2857.

[69] A. Partridge, S. Walker, D. Armitt, Detection of impurities in organic peroxide explosives from precursor chemicals, Aust. J. Chem. 63 (2010) 30–37.

[70] M.S. Meaney, V.L. McGuffin, Luminescence-based methods for sensing and detection of explosives, Anal. Bioanal. Chem. 391 (2008) 2557–2576.

[71] C.R. Bowerbank, P.A. Smith, D.D. Fetterolf, M.L. Lee, Solvating gas chromatography with chemiluminescence detection of nitroglycerine and other explosives, J. Chromatogr. A 902 (2000) 413–419.

[72] M. Walsh, Determination of nitroaromatic, nitramine, and nitrate ester explosives in soil by gas chromatography and an electron capture detector, Talanta 54 (2001) 427–438.

[73] B.J. Routon, B.B. Kocher, J.V. Goodpaster, Discriminating hodgdon pyrodex (R) and triple seven (R) using gas chromatography-mass spectrometry, J. Forensic Sci. 56 (2011) 194–199.

[74a] F. Tian, J. Yu, J. Hu, Y. Zhang, M. Xie, Y. Liu, et al., Determination of emulsion explosives with Span-80 as emulsifier by gas chromatography–mass spectrometry, J. Chromatogr. A. 1218 (2011) 3521–3528.

[74b] A. Partridge, S. Walker, D. Armitt, Detection of Impurities in Organic Peroxide Explosives from Precursor Chemicals, Aust. J. Chem. 63 (2010) 30–37.

[75] G.W. Cook, P.T. LaPuma, G.L. Hook, B.A. Eckenrode, Using gas chromatography with ion mobility spectrometry to resolve explosive compounds in the presence of interferents, J. Forensic Sci. 55 (2010) 1582–1591.

[76] F. Monteil-Rivera, C. Beaulieu, J. Hawari, Use of solid-phase microextraction/gas chromatography–electron capture detection for the determination of energetic chemicals in marine samples, J. Chromatogr. A 1066 (2005) 177–187.

[77] S. Calderara, D. Gardebas, F. Martinez, Solid phase micro extraction coupled with on-column GC/ECD for the post-blast analysis of organic explosives, Forensic Sci. Int. 137 (2003) 6–12.

[78] R.F. Saverio, P. Margot, Identification of gunshot residue: a critical review, Forensic Sci. Int. 119 (2001) 195–211.

[79] E.B. Morales, A.L.R. Vazquez, Simultaneous determination of inorganic and organic gunshot residues by capillary electrophoresis, J. Chromatogr. A 1061 (2004) 225–233.

[80] F.S. Romolo, P. Margot, Identification of gunshot residue: a critical review, Forensic Sci. Int. 119 (2001) 195–211.

[81] O. Dalby, D. Butler, J.W. Birkett, Analysis of gunshot residue and associated materials-a review, J. Forensic Sci. 55 (2010) 924–943.

[82] A.J. Schwoeble, D.L. Exline. electronic resource, in: Current methods in forensic gunshot residue analysis, CRC Press, Boca Raton, 2000, p. 169.

[83] J. Andrasko, J. Oskarsson, S. Stahling, Ammunition used in the latest shooting, Forensic Sci. Int. 136 (2003). 146–146.

[84] A. Zeichner, B. Eldar, B. Glattstein, A. Koffman, T. Tamiri, D. Muller, Vacuum collection of gunpowder residues from clothing worn by shooting suspects, and their analysis by GC/TEA, IMS, and GC/MS, J. Forensic Sci. 48 (2003) 961–972.

[85] A. Zeichner, B. Eldar, A novel method for extraction and analysis of gunpowder residues on double-side adhesive coated stubs, J. Forensic Sci. 49 (2004) 1194–1206.

[86] J. Andrasko, T. Norberg, S. Stahling, Time since discharge of shotguns, J. Forensic Sci. 43 (1998) 1005–1015.

[87] J. Andrasko, S. Stahling, Time since discharge of spent cartridges, J. Forensic Sci. 44 (1999) 487–495.

[88] J. Andrasko, S. Stahling, Time since discharge of rifles, J. Forensic Sci. 45 (2000) 1250–1255.

[89] J.D. Wilson, J.D. Tebow, K.W. Moline, Time since discharge of shotgun shells, J. Forensic Sci. 48 (2003) 1298–1301.

[90] J. Andrasko, S. Stahling, Time since discharge of pistols and revolvers, J. Forensic Sci. 48 (2003) 307–311.

[91] N.S. Parsons, M.H.W. Lam, S.E. Hamilton, F. Hui, A preliminary investigation into the comparison of dissolution/digestion techniques for the chemical characterization of polyurethane foam, Sci. Justice 50 (2010) 177–181.

[92] S.C. McNorton, G.W. Nutter, J.A. Siegel, The characterization of automobile body fillers, J. Forensic Sci. 53 (2008) 116–124.

[93] J. Zieba-Palus, G. Zadora, J.M. Milczarek, P. Koscieniak, Pyrolysis-gas chromatography/mass spectrometry analysis as a useful tool in forensic examination of automotive paint traces, J. Chromatogr. A 1179 (2008) 41–46.

[94] B. Kochanowski, S. Morgan, Forensic discrimination of automotive paint samples using pyrolysis-gas chromatography-mass spectrometry with multivariate statistics, J. Chromatogr. Sci. 38 (2000) 100–108.

[95] J.M. Milczarek, J. Zieba-Palus, Examination of spray paints on plasters by the use of pyrolysis-gas chromatography/mass spectrometry for forensic purposes, J. Anal. Appl. Pyrolysis 86 (2009) 252–259.

[96] Y. Higashikawa, H. Kanno, T. Kaneko, S. Suzuki, Forensic discrimination of trace fragments of photocopied materials, Bunseki Kagaku 56 (2007) 1147–1152.

[97] A. Richmond-Aylor, S. Bell, P. Callery, K. Morris, Thermal degradation analysis of amino acids in fingerprint residue by pyrolysis GC-MS to develop new latent fingerprint developing reagents, J. Forensic Sci. 52 (2007) 380–382.

[98] G.P. Campbell, A.L. Gordon, Analysis of condom lubricants for forensic casework, J. Forensic Sci. 52 (2007) 630–642.

[99] J. Bugler, H. Buchner, A. Dallmayer, Characterization of ballpoint pen inks by thermal desorption and gas chromatography-mass spectrometry, J. Forensic Sci. 50 (2005) 1209–1214.

[100] J.H. Buegler, H. Buchner, A. Dallmayer, Age determination of ballpoint pen ink by thermal desorption and gas chromatography-mass spectrometry, J. Forensic Sci. 53 (2008) 982–988.

[101] L. Brazeau, M. Gaudreau, Ballpoint pen inks: the quantitative analysis of ink solvents on paper by solid-phase microextraction, J. Forensic Sci. 52 (2007) 209–215.

[102] E. Scribner, E. Thurman, L. Zimmermann, Analysis of selected herbicide metabolites in surface and ground water of the United States RID B-5131-2011, Sci. Total Environ. 248 (2000) 157–167.

[103] S.M. Mudge, I. Ebrary, Methods in environmental forensics [electronic resource], CRC Press, Boca Raton, 2009, p. 386.

[104] O.N. Zabelina, V.I. Saloutin, O.N. Chupakhin, Analysis of polychlorinated biphenyl mixtures by

gas chromatography, J. Anal. Chem. 65 (2010) 1098–1108.

[105] J. Cochran, G. Frame, Recent developments in the high-resolution gas chromatography of polychlorinated biphenyls, J. Chromatogr. A 843 (1999) 323–368.

[106] G.S. Frysinger, R.B. Gaines, C.M. Reddy, GC x GC - A new analytical tool for environmental forensics, Environ. Forens. 3 (2002) 27–34.

[107] E.J. Reiner, The analysis of dioxins and related compounds, Mass Spectrom. Rev. 29 (2010) 526–559.

[108] J. Focant, G. Eppe, M. Scippo, A. Massart, C. Pirard, G. Maghuin-Rogister, et al., Comprehensive two-dimensional gas chromatography with isotope dilution time-of-flight mass spectrometry for the measurement of dioxins and polychlorinated biphenyls in foodstuffs - comparison with other methods, J. Chromatogr. A 1086 (2005) 45–60.

[109] K. Saito, A. Ohmura, M. Takekuma, R. Sasano, Y. Matsuki, H. Nakazawa, Application of a novel large-volume injection method using a stomach-shaped inlet liner in capillary gas chromatographic trace analysis of dioxins in human milk and plasma, Anal. Sci. 23 (2007) 661–666.

[110] E. Matisova, M. Domotorova, Fast gas chromatography and its use in trace analysis, J. Chromatogr. A 1000 (2003) 199–221.

[111] J. Focant, A. Sjodin, D. Patterson, Improved separation of the 209 polychlorinated biphenyl congeners using comprehensive two-dimensional gas chromatography-time-of-flight mass spectrometry RID F-2464-2010, J. Chromatogr. A 1040 (2004) 227–238.

[112] P. Haglund, P. Korytar, C. Danielsson, J. Diaz, K. Wiberg, P. Leonards, et al., GCxGC-ECD: a promising method for the determination of dioxins and dioxin-like PCBs in food and feed, Anal. Bioanal. Chem. 390 (2008) 1815–1827.

[113] S. Dahle, V. Savinov, G. Matishov, A. Evenset, K. Naes, Polycyclic aromatic hydrocarbons (PAHs) in bottom sediments of the Kara Sea shelf, Gulf of Ob and Yenisei Bay, Sci. Total Environ. 306 (2003) 57–71.

[114] D.L. Poster, M.M. Schantz, L.C. Sander, S.A. Wise, Analysis of polycyclic aromatic hydrocarbons (PAHs) in environmental samples: a critical review of gas chromatographic (GC) methods, Anal. Bioanal. Chem. 386 (2006) 859–881.

[115] W.A. Lopes, G.O. da Rocha, P.A.d.P. Pereira, F.S. Oliveira, L.S. Carvalho, N.d.C. Bahia, et al., Multivariate optimization of a GC-MS method for determination of sixteen priority polycyclic aromatic hydrocarbons in environmental samples, J. Separation Sci. 31 (2008) 1787–1796.

[116] S. Wu, S. Tao, F. Xu, R. Dawson, T. Lan, B. Li, et al., Polycyclic aromatic hydrocarbons in dustfall in Tianjin, China RID G-3968-2010, Sci. Total Environ. 345 (2005) 115–126.

[117] D. Zou, K. Liu, W. Pan, J. Riley, Y. Xu, Rapid analysis of PAHs in fly ash using thermal desorption and fast GC-TOF-MS, J. Chromatogr. Sci. 41 (2003) 245–250.

[118] O. Panic, T. Gorecki, Comprehensive two-dimensional gas chromatography (GC × GC) in environmental analysis and monitoring, Anal. Bioanal. Chem. 386 (2006) 1013–1023.

[119] A. Stark, T. Abrajano, J. Hellou, J. Metcalf-Smith, Molecular and isotopic characterization of polycyclic aromatic hydrocarbon distribution and sources at the international segment of the St. Lawrence River, Org. Geochem. 34 (2003) 225–237.

[120] F. Norlock, J. Jang, Q. Zou, T. Schoonover, A. Li, Large-volume injection PTV-GC-MS analysis of polycyclic aromatic hydrocarbons in air and sediment samples RID A-3395-2008, J. Air. Waste. Manage. Assoc. 52 (2002) 19–26.

[121] G. Reed, K. Clark-Baker, H. McNair, Fast gas chromatography of various sample types using fast oven temperature programming, J. Chromatogr. Sci. 37 (1999) 300–305.

[122] A. Christensen, C. Ostman, R. Westerholm, Ultrasound-assisted extraction and on-line LC-GC-MS for determination of polycyclic aromatic hydrocarbons (PAH) in urban dust and diesel particulate matter, Anal. Bioanal. Chem. 381 (2005) 1206–1216.

[123] D. Wang, J. Xing, C. Wu, Application of the novel benzo-15-crown-5 sol-gel coating for solid-phase microextraction in rapid determination of trace PAHs in water at nonequilibrium situations, Fenxi Kexue Xuebao 19 (2003) 109–112.

[124] Z. Wang, M. Fingas, L. Sigouin, Using multiple criteria for fingerprinting unknown oil samples having very similar chemical composition, Environ. Forens 3 (2002) 251–262.

[125] R. Gaines, G. Frysinger, M. Hendrick-Smith, J. Stuart, Oil spill source identification by comprehensive two-dimensional gas chromatography, Environ. Sci. Technol. 33 (1999) 2106–2112.

[126] G.T. Ventura, G.J. Hall, R.K. Nelson, G.S. Frysinger, B. Raghuraman, A.E. Pomerantz, et al., Analysis of petroleum compositional similarity using multiway principal components analysis (MPCA) with comprehensive two-dimensional gas chromatographic data, J. Chromatogr. A 1218 (2011) 2584–2592.

[127] R. Hua, Y. Li, W. Liu, J. Zheng, H. Wei, J. Wang, et al., Determination of sulfur-containing compounds in diesel oils by comprehensive two-dimensional gas

chromatography with a sulfur chemiluminescence detector, J. Chromatogr. A 1019 (2003) 101–109.
[128] L. van Stee, J. Beens, R. Vreuls, U. Brinkman, Comprehensive two-dimensional gas chromatography with atomic emission detection and correlation with mass spectrometric detection: principles and application in petrochemical analysis, J. Chromatogr. A 1019 (2003) 89–99.
[129] G.T. Ventura, B. Raghuraman, R.K. Nelson, O.C. Mullins, C.M. Reddy, Compound class oil fingerprinting techniques using comprehensive two-dimensional gas chromatography (GC × GC), Org. Geochem. 41 (2010) 1026–1035.
[130] A.M. Curran, P.A. Prada, K.G. Furton, The differentiation of the volatile organic signatures of individuals through SPME-GC/MS of characteristic human scent compounds, J. Forensic Sci. 55 (2010) 50–57.
[131] S.K. Pandey, K. Kim, Human body-odor components and their determination, TrAC Trends. Anal. Chem. 30 (2011) 784–796.
[132] L.E. DeGreeff, A.M. Curran, K.G. Furton, Evaluation of selected sorbent materials for the collection of volatile organic compounds related to human scent using non-contact sampling mode, Forensic Sci. Int. 209 (2011) 133–142.
[133] A.M. Curran, C.F. Ramirez, A.A. Schoon, K.G. Furton, The frequency of occurrence and discriminatory power of compounds found in human scent across a population determined by SPME-GEMS, J. Chromatogr. B-Anal. Technol. Biomed. Life Sci. 846 (2007) 86–97.
[134] A. Curran, S. Rabin, P. Prada, K. Furton, Comparison of the volatile organic compounds present in human odor using SPME-GC/MS, J. Chem. Ecol. 31 (2005) 1607–1619.
[135] Z. Zhang, J. Cai, G. Ruan, G. Li, The study of fingerprint characteristics of the emanations from human arm skin using the original sampling system by SPME-GC/MS, J. Chromatogr. B 822 (2005) 244–252.
[136] D.J. Penn, E. Oberzaucher, K. Grammer, G. Fischer, H.A. Soini, D. Wiesler, et al., Individual and gender fingerprints in human body odour, J. R. Soc. Interface. 4 (2007) 331–340.
[137] E. Woolfenden, Sorbent-based sampling methods for volatile and semi-volatile organic compounds in air part 1: sorbent-based air monitoring options, J. Chromatogr. A 1217 (2010) 2674–2684.
[138] M. Gallagher, C.J. Wysocki, J.J. Leyden, A.I. Spielman, X. Sun, G. Preti, Analyses of volatile organic compounds from human skin, Br. J. Dermatol. 159 (2008) 780–791.
[139] B. Eckenrode, S. Ramsey, R. Stockham, G. Van Berkel, K. Asano, D. Wolf, Performance evaluation of the scent transfer unit (TM) (STU-100) for organic compound collection and release, J. Forensic Sci. 51 (2006) 780–789.
[140] P.A. Prada, A.M. Curran, K.G. Furton, Comparison of extraction methods for the removal of volatile organic compounds (VOCs) present in sorbents used for human scent evidence collection, Anal. Methods 2 (2010) 470–478.
[141] D.T. Hudson, A.M. Curran, K.G. Furton, The stability of collected human scent under various environmental conditions, J. Forensic Sci. 54 (2009) 1270–1277.
[142] H. Kataoka, K. Saito, Recent advances in SPME techniques in biomedical analysis, J. Pharm. Biomed. Anal. 54 (2011) 926–950.
[143] A.A. Vass, S. Barshick, G. Sega, J. Caton, J.T. Skeen, J.C. Love, et al., Decomposition chemistry of human remains: a new methodology for determining the postmortem interval, J. Forensic Sci. 47 (2002) 542–553.
[144] S. Forbes, J. Keegan, B. Stuart, B. Dent, A gas chromatography-mass spectrometry method for the detection of adipocere in grave soils, Eur. J. Lipid Sci. Technol. 105 (2003) 761–768.
[145] S. Forbes, B. Stuart, B. Dent, The identification of adipocere in grave soils, Forensic Sci. Int. 127 (2002) 225–230.
[146] S.J. Notter, B.H. Stuart, B.B. Dent, J. Keegan, Solid-phase extraction in combination with GC/MS for the quantification of free fatty acids in adipocere, Eur. J. Lipid Sci. Technol. 110 (2008) 73–80.
[147] I.D. Bull, R. Berstan, A. Vass, R.P. Evershed, Identification of a disinterred grave by molecular and stable isotope analysis, Sci. Justice 49 (2009) 142–149.
[148] M. Statheropoulos, C. Spiliopouiou, A. Agapiou, A study of volatile organic compounds evolved from the decaying human body, Forensic Sci. Int. 153 (2005) 147–155.
[149] A.A. Vass, R.R. Smith, C.V. Thompson, M.N. Burnett, N. Dulgerian, B.A. Eckenrode, Odor analysis of decomposing buried human remains, J. Forensic Sci. 53 (2008) 384–391.
[150] L.E. DeGreeff, K.G. Furton, Collection and identification of human remains volatiles by non-contact, dynamic airflow sampling and SPME-GC/MS using various sorbent materials, Anal. Bioanal. Chem. 401 (2011) 1295–1307.
[151] T.M. Lovestead, T.J. Bruno, Detecting gravesoil with headspace analysis with adsorption on short porous layer open tubular (PLOT) columns, Forensic Sci. Int. 204 (2011) 156–161.
[152] P.A. Prada, A.M. Curran, K.G. Furton, The evaluation of human hand odor volatiles on various

textiles: a comparison between contact and noncontact sampling methods, J. Forensic Sci. 56 (2011) 866–881.

[153] A.C. Lewis, J.F. Hamilton, C.N. Rhodes, J. Halliday, K.D. Bartle, P. Homewood, Sci Justice. Microfabricated planar glass gas chromatography with photoionization detection, J. Chromatogr. A 1217 (2010) 768–774.

[154] J. Syage, B. Nies, M. Evans, K. Hanold, Field-portable, high-speed GC/TOFMS, J. Am. Soc. Mass Spectrom. 12 (2001) 648–655.

[155] J.D. Fair, W.F. Bailey, R.A. Felty, A.E. Gifford, B. Shultes, L.H. Volles, Method for rapid on-site identification of VOCs, J. Environ. Sci. 21 (2009) 1005–1008.

[156] H. Li, C. Deng, X. Zhang, Fast field analysis of short-chain aliphatic amines in water using solid-phase microextraction and a portable gas chrornatograph, J. Separation Sc. 31 (2008) 3225–3230.

[157] J.A. Contreras, J.A. Murray, S.E. Tolley, J.L. Oliphant, H.D. Tolley, S.A. Lammert, Sci Justice. Hand-portable gas chromatograph-toroidal ion trap mass spectrometer (GC-TMS) for detection of hazardous compounds, J. Am. Soc. Mass Spectrom. 19 (2008) 1425–1434.

[158] P. Lutter, M. Savoy-Perroud, E. Campos-Gimenez, L. Meyer, T. Goldmann, M. Bertholet, Sci Justice. Screening and confirmatory methods for the determination of melamine in cow's milk and milk-based powdered infant formula: validation and proficiency-tests of ELISA, HPLC-UV, GC-MS and LC-MS/MS, Food Control 22 (2011) 903–913.

[159] L. Cercaci, M. Rodriguez-Estrada, G. Lercker, Solid-phase extraction-thin-layer chromatography-gas chromatography method for the detection of hazelnut oil in olive oils by determination of esterified sterols, J. Chromatogr. A 985 (2003) 211–220.

[160] D. Jackson, D. Crockett, K. Wolnik, The indirect detection of bleach (Sodium hypochlorite) in beverages as evidence of product tampering, J. Forensic Sci. 51 (2006) 827–831.

[161] T. Cajka, K. Riddellova, M. Tomaniova, J. Hajslova, Recognition of beer brand based on multivariate analysis of volatile fingerprint RID A-8914-2008, J. Chromatogr. A 1217 (2010) 4195–4203.

[162] L. Pillonel, S. Ampuero, R. Tabacchi, J. Bosset, Analytical methods for the determination of the geographic origin of emmental cheese: volatile compounds by GC/MS-FID and electronic nose, EurFood Res. Technol. 216 (2003) 179–183.

[163] A.I. Cabanero, J.L. Recio, M. Ruperez, Isotope ratio mass spectrometry coupled to liquid and gas chromatography for wine ethanol characterization, Rapid Commun. Mass Spectrom. 22 (2008) 3111–3118.

[164] R. Subramaniam, C. Astot, L. Juhlin, C. Nilsson, A. Ostin, Direct derivatization and rapid GC-MS screening of nerve agent markers in aqueous samples, Anal. Chem. 82 (2010) 7452–7459.

[165] J.C. Hoggard, J.H. Wahl, R.E. Synovec, G.M. Mong, C.G. Fraga, Impurity profiling of a chemical weapon precursor for possible forensic signatures by comprehensive two-dimensional gas chromatography/mass spectrometry and chemometrics, Anal. Chem. 82 (2010) 689–698.

[166] H.J. Tobias, G.L. Sacks, Y. Zhang, J.T. Brenna, Comprehensive two-dimensional gas chromatography combustion isotope ratio mass spectrometry RID B-2686-2009 RID G-4123-2010, Anal. Chem. 80 (2008) 8613–8621.

CHAPTER

26

Application of Gas Chromatography to Multiresidue Methods for Pesticides and Related Compounds in Food

Milagros Mezcua, M. Angeles Martinez-Uroz, Amadeo R. Fernandez-Alba

OUTLINE

26.1. INTRODUCTION

Plant production yield is being continually affected by harmful organisms. It is essential to protect plants and plant products against such organisms in order to prevent a reduction in yield or damage to the plants or plant products and to ensure both the quality of the products harvested and high agricultural productivity. The use of active substances in plant protection products is one of the most common methods of protecting plants and plant products from the effects of harmful organisms. In some cases, these products act by confusing

Gas Chromatography DOI: 10.1016/B978-0-12-385540-4.00026-2

insects or making crops less palatable for pests. However, more commonly, the damaging insects, weeds, or fungi are killed by chemicals. Such pesticides have potentially severe undesirable effects if they are not strictly regulated. A possible consequence of their use may be the presence of residues in the treated products, in animals feeding on those products, and in products of animal origin.

Maximum residue levels (MRLs) are the upper legal concentration levels acceptable for pesticide residues in or on food or feed based on Good Agricultural Practices (GAP) and which ensure the lowest consumer exposure possible.

Regulatory and public concern over pesticide residues in food has been increasing due to the potential health hazards. Measuring trace levels of pesticides in the presence of large amounts of sample matrix components that occur naturally is a challenging task. There is growing interest in developing simple, rapid, cost-effective, and reliable analytical methods to ensure that the levels of toxic pesticides incurred in produce are below tolerance levels.

Multiresidue methods (MRMs) are undoubtedly one way of addressing the problem of pesticide determination, and are worth evaluating given the great diversity of these groups of compounds. However, the complex sample matrix may contain abundant quantities of components that can interfere with good sample analysis.

For the development of an MRM, two important aspects must be taken into account. The first issue to consider is the extraction method employed. Next, compounds with different physicochemical properties and different structural characteristics are included in the scope of an MRM, making it necessary to employ extraction methods with a wide versatility.

The first notable MRM was the Mills method developed in the 1960s for the determinations of nonpolar organochlorine pesticides in nonfatty food. The Mills method was based on acetonitrile extraction, the extract was then diluted with water, and the pesticides were partitioned into a nonpolar solvent. The follow-up research was oriented toward extending the analytical polarity range to cover a wider range of polarity of pesticides analyzed in a single procedure. New solvents for initial extraction and addition of sodium chloride for the partitioning step were used, and were tested to reach higher recoveries of the more polar analytes. In the 1980s, environmental and health concerns led to the avoidance of dangerous solvents; later, solid-phase extraction (SPE) was established and to avoid liquid–liquid partitioning and as a cleanup step. Increased urgency to further reduce solvent usage and manual labor led to the introduction of several alternative extraction approaches: matrix solid-phase dispersion (MSPD), supercritical fluid extraction (SFE), and solid-phase microextraction (SPME). In 2003, Anastassiades et al. developed a quick, easy, inexpensive, effective, rugged, and safe method (QuEChERS) based on solid–liquid extraction (SLE) using AcN as the extraction solvent, aimed to overcome critical flaws and practical limitations of existing methods [1].

On the other hand, the instrumental methods need to be able to identify a multiclass group of compounds; in this sense, mass spectrometry detectors offer a greater advantage than traditional detectors. To ensure the desired level of analytical selectivity and sensibility by gaschromatography–mass spectrometry (GC–MS), the regulatory and commercial testing laboratories usually adopt selective techniques like tandem mass spectrometry (MS/MS) or selected ion monitoring (SIM) to get better sensitivity as compared to full scan techniques. Due to wide diversity in the natural and indirect sources of contamination in food, target-oriented residue monitoring by MS/MS or SIM often fails to provide holistic assessment of the contamination status of any food sample.

This overview focuses on a revision of MRMs based on GC–MS published for the determination of pesticides in different groups of food: crops, animal origin food, processed food, and baby food. Within these groups of food, published work from 2005 until recently has been reviewed. Due to the large amount of published MRMs for pesticides in food, especially for crops, not all published work has been referenced in this chapter; a selection has been made detailing the most representative techniques.

26.2. MULTIRESIDUE METHODS FOR PESTICIDES IN CROPS

Plant food can be contaminated by pesticides under a great variety of circumstances and at different times preceding their consumption. Many factors can reduce such contamination, (e.g., rainfall, wind, chemical reactions induced by oxygen, moisture, light, or plant enzymes) The number of publications that describe MRMs for the determination of pesticides in crops is very high; from 2009 to now there are more than 30 publications describing MRMs for determination of pesticides in crops by GC–MS, and most of them are for fruits and vegetables. The past few years have been reviewed in this work. An overview of these methods is shown in Table 26.1.

The scope of multiresidue revised methods varies between 14 and 346 pesticides, the limits of detection published for all authors meet with the actual legislation, and very good recoveries are achieved for all employed methods. In this section, we will describe the more critical aspects that can affect the obtained results after the application of an MRM.

Most of extraction methods employed for determination of pesticides based on solid–liquid extraction; acetonitrile is the solvent of choice for most authors [2–13]. SLE methods have been developed in general miniaturized methods with a small amount of sample and extraction solvent, with the exception of Chen et al. [14], SLE allows recoveries between 70% and 120%, as can be seen in Table 26.1.

Stir bar sorptive extraction is employed for the determination of 20 pesticides in different vegetables [15]; this technique provides very good limits of detection but recoveries are in the range of 10%–110%, and recoveries were even lower than 20% for matrices such as green beans, onions, tomatoes, and green peppers.

Other extraction techniques such as MAE [16], SPME [17], and MSPD [18,19] have been employed for the determination of pesticides in crops. Cleanup steps are employed in some cases; the technique of choice is dispersive SPE using PSA as a dispersive agent.

Regarding sample injection, 1 μL is, in general, injected in a splitless mode, but various authors select a large volume injection mode (20 μL) using a temperature program in the injection ports when programmed temperature vaporising injectors are used; however, no significant differences in limits of detection can be assigned to the different volumes of injection, as can be observed in Table 26.1.

Regarding chromatographic separation, a nonpolar analytical column has been employed in most methods with a stationary phase composition of 5% phenyl/95% dimethylpolysiloxane.

The selection of the instrumental method, together with the selection of the extraction methods, are the most important choice for the correct identification and quantification of target compounds; MRMs for determination of pesticides in crops in past years have been focused on the use of mass spectrometry as a detection system. The ionization mode in mass spectrometry is one of the aspects to consider when a method must be selected. The most current methods for multiresidue analysis in crops are based on electron impact ionization; however, the compounds with highly electronegative elements, such as halogen, oxygen, etc., afforded high sensitivity

TABLE 26.1 Overview of Multiresidue Methods for Determination of Pesticides in Crops

Matrix	Analytical technique	Injection	Column	Cleanup	Extraction method	Scope	LOD/LOQ	Recoveries %	Ref.
Grapes	GC–MS–SIM (EI) quadrupole	1 μL splitless	DB-1701	Dispersive SPE (PSA)	Solid–liquid with acetonitrile	346	1.7–266/Nd μg/kg	60–120	[2]
Green vegetables	GC–MS–SIM (EI) ion trap	1 μL on column injection	ZB-5MS	Dispersive SPE (PSA)	Solid–liquid with acetonitrile	16	0.03–0.1/ 0.1–0.5 μg/kg	85–112	[3]
Lettuce, orange, strawberry, plum	GC–MS–SIM (NCI) quadrupole	2 μL solvent vent	CP-Sil 8 CB	Dispersive SPE (PSA)	Solid–liquid with acetonitrile	26	0.00015–0.619/ 0.0005–2.356 μg/L	70–110	[4]
Orange, grape, pear, and apple	GC–MS–(EI) quadrupole SCAN and SIM simultaneously	1 μL splitless	ZB-5MS		MSPD	31	9–250/N.d μg/kg	62–116	[18]
Corn muffin and cocoa beans	GC–MS–(SIM) (EI) quadrupole	5 μL solvent vent mode	RTX5MS	Disposable pipette extraction (DPX)	Solid–liquid with acetonitrile	27	0.90–12.48/ 2.73–37.8 μg/kg	87–135	[5]
Grape, pomegranate, mango	GC–MS–MRM (EI) ion trap	20 μL PTV LVI	VF 5MS	Dispersive SPE (PSA and GCB)	Solid–liquid with acetonitrile	50	Nd/10–20 μg/kg	70–120	[6]
Lettuce, spinach, green bean, green pepper, tomato, broccoli, potato, carrot, and onion	GC–MS–(SIM) (EI) quadrupole	20 μL solvent vent mode	TRB 5MS		Stir bar sorptive extraction followed by liquid desorption	20	0.01–10 μg/kg	10–110	[15]
Leeks	GC–MS/MS (EI) (triple quadrupole)	1 μL in splitless mode.	TR-5MS	Dispersive SPE (PSA and GCB)	Solid–liquid with acetonitrile (pretreatment with microwave)	20	0.07–1.5/ 0.25–5 μg/kg	81–109	[7]

Tomato, strawberry, potato, orange, and lettuce	LP–GC–MS (EI) (time-of-flight)	10 μL in solvent vent mode	Rti-5MS	Dispersive SPE (PSA) and DPX	Solid–liquid with acetonitrile	150	25 μg/kg	70–120	[56]
Rice, soybean, potato, spinach, cabbage, apple, orange, cacao pumpkin, and green tea			DB-17			10	10 μg/kg	72–120	[8]
Mix of vegetables	GC–MS/MS (EI) (triple quadrupole)	1 μL in splitless mode	TR-5MS		MSPD	15	0.7–32 μg/kg	73–111	[9]
Garlic, onion, spring onion, and chili	GC–MS–(SIM) (NCI) quadrupole	1 μL in pulsed splitless mode	HP-5MS	Dispersive SPE (PSA and GCB) Additionally C18 for chili	Solid–liquid with acetonitrile-hexane	17	0.02–6 μg/kg	54–129.8	[19]
Wheat grains, flour, and bran	GC–MS–(SIM) (NCI) triple quadrupole	2 μL no data mode	VF-5MS	Dispersive SPE (C18)	Solid–liquid with acetonitrile	24	2.5/5–10 μg/kg	70–120	[18]
Fruits and vegetables (one-year survey)	GC–MS/MS (EI) triple quadrupole	6 μL large volume injection technique	VF-5MS	Dispersive SPE (PSA)	Solid–liquid with acetonitrile	121	1–3/10 μg/kg	80–116	[10]
Pachoi, cabbage, legumes, leaf mustard (three-year survey)	GC–MS–(SIM) (EI) quadrupole (only for confirmation)	1 μL no data mode	DB-1701		Solid–liquid with ethyl acetate	22	10 μg/kg	80–120	[11]
Berries (raspberry, strawberry, blueberry, and grape)	GC–MS–(SIM) (EI) quadrupole	1 μL in splitless mode	DB-1701	SPE	Solid–liquid with acetonitrile	88	6–50/20–150 μg/kg	63–137	[12]
Grape	GC–MS–(SIM) (EI) quadrupole	splitless mode (no data about volume)	AT.RPA-1	Extraction and cleanup in one step with SiO_2 fiber	25	0.48–1.4/1.6–28 μg/L	61–108	24	[14]

(*Continued*)

TABLE 26.1 Overview of Multiresidue Methods for Determination of Pesticides in Crops (*cont'd*)

Matrix	Analytical technique	Injection	Column	Cleanup	Extraction method	Scope	LOD/LOQ	Recoveries %	Ref.
Fruits and vegetables	GC–MS–(SIM) for quantitation GC–MS–(full scan) for screening (EI) quadrupole	20 μL in solvent vent mode	HP-5MS	Dispersive SPE	MAE with acetone and acetonitrile	72	2–20/ 25–100 μg/kg	72–114	[57]
Carrot and orange	GC–MS–(SIM) (EI) quadrupole	2 μL in splitless mode	DB-17	Disposable pipette extraction	Solid–liquid with acetonitrile	36	0.2–28.6/ 0.4–96.2 μg/kg	72–116	[58]
Grapes	GC–MS/MS (EI) Ion trap	20 μL PTV-LVI program	TR-5MS	Dispersive SPE (PSA)	Solid–liquid with ethyl acetate	21	0.5–3.1/ 2.5–10 μg/kg	77–115	[59]
Mangoes	GC–MS–(SIM) (EI) quadrupole	splitless mode, injection by desorption of fiber	RTX-1		SPME	14	1.0–3.3/ 3.3–33.3 μg/kg	71.6–117.5	[60]
Cabbage and apples	GC–MS/MS (NCI) Triple quadrupole	10 μL PTV LVI program	RTX-5MS	Dispersive SPE (PSA and C18)	Solid–liquid with acetonitrile	82	0.01–1.82/ 0.03–6.46 μg/kg	58.7–124.4	[17]

in the pesticide analysis with the negative chemical ionization mode. If the chemical ionization mode is used, the chromatograms obtained are cleaner due to the minimization of background interferences from ions derived from the sample matrix than when using electron ionization. NCI is specially recognized for improved selectivity and sensitivity for organochlorine and organophosphorus compounds. The mass spectrometry methods developed in NCI [4,10,11,20] show very low limits of detection and quantification.

Different analyzers have been used, such as MRMs, quadrupole, ion trap, triple quadrupole, and time-of-flight (TOF). When a TOF analyzer was used [9], 150 pesticides where analyzed in less than 10 minutes. The selectivity which provides the measurement of exact mass in full scan mode without the need to create a retention time window makes this analyzer an excellent tool for high scope MRMs. However, the absence of fragments by collision induced in some pesticides make necessary the confirmation of pesticides with another technique; this is a limitation in the use of these detectors in target methods. In recent years, they are used for the development of screening methods [21].

The use of a quadrupole analyzer in the SIM mode is an excellent tool for the target method because of the good selectivity in comparison to full scan. The problem with this configuration is that when the scope of the required method is high; it is more convenient to develop various SIM methods in order to assure good selectivity [2], which involves increased processing time since it involves several injections per sample.

Therefore, the best tool for MRMs is the use of triple quadrupole or ion trap with the capacity to work in MS–MS, which provides good sensitivity and selectivity.

It is well known [22,23] that suppression in the signal in GC is due to a gradual accumulation of nonvolatile matrix components in the GC system, resulting in formation of new active sites and a gradual decrease in analyte responses. The presence of matrix effects and their extent are simultaneously influenced by several factors (matrix concentration, pesticide concentration, matrix type, and analytical range). The matrix sample diversity and the different possibilities of interactions that can occur within the sample/pesticide/chromatographic system make it difficult to establish a trend for the matrix effect of each pesticide in each matrix. Therefore, the quantitation based on the use of analytical standards prepared in a blank matrix extract (matrix-matched calibration) to compensate for the matrix effect and to obtain more accurate results is commonly used in MRMs for pesticides in crops and in the other groups of food commodities considered in this chapter.

The average in the reported limits of quantification is 10 μg/kg; the lowest limits of quantification are achieved when negative chemical ionization is employed.

26.3. MULTIRESIDUE METHODS FOR PESTICIDES IN ANIMAL ORIGIN PRODUCTS

The diversity of matrices in animal origin products make it difficult to establish general trends in MRMs, especially in the extraction stage of the method.

In this section, MRMs reviewed will be commented upon while focusing on the class of pesticides in order to give a homogeneous point of view.

A nonpolar analytical column has been used in all cases with a stationary phase composition of 5% phenyl/95% dimethylpolysiloxane.

Concerning organochlorinated pesticides, they have been analyzed in eggs [24], honey [25], and fish [26]; the extraction methods employed were MSPD, ASE, and SEME respectively; better limits of detection were achieved in honey. In all cases, good recoveries were achieved; ion trap or quadrupole, in electron impact ionization and SIM mode, has been selected for all methods.

Organophosphorus pesticides have been analyzed in cow milk by HS–SPME [27]; limits of detection between 2.2 and 10.9 and limits of quantification between 6.5 and 32.9 μg/kg have been reported. A quadrupole mass spectrometer operated in the electron ionization SIM mode was employed. A group of organophosphate pesticides has been analyzed in honey [28], the extraction method employed is coacervative microextraction ultrasound-assisted back-extraction procedure, with limits of detection between 0.03 and 0.47 μg/kg and recoveries higher than 90%. An ion trap has been used in the electron ionization mode.

Different MRMs have been developed for the determination of phyrethrins and pyrethroids in fish [29], porcine muscle, and pasteurized milk [30]. For fish, SLE with acetonitrile was performed followed by DSPE with PSA and C18 for cleanup, the injection was in splitless mode, and analysis was made by GC-MS in NCI and SIM modes. Recoveries were in the range 70%–115%, limits of detection achieved were between 0.3 and 0.5 μg/kg. For porcine muscle and pasteurized milk, a liquid–liquid extraction (LLE) was employed, the identification was made by GC-MS with electron ionization in the SIM mode, recoveries were in the range of 83%–109%, and very high limits of detection and quantification were reported, between 3 and 9 mg/kg and between 10 and 24.6 mg/kg, respectively.

Multiclass MRMs have also been developed for the determination of pesticides in animal origin products. For determination of pesticides in honey, single drop microextraction, LLE [31], and LLE and low temperature purification [32] were used. In animal fat, SLE with AcN and n-hexane followed by dispersive solid-phase extraction [33] has been developed. In honey, mass spectrometry was used only for confirmation; 1 μL was injected in the splitless mode, and analyses were performed in full scan and electron impact ionization modes. For animal fat, 4 μL was injected in a PTV inlet in a solvent vent mode, a quadrupole was used for identification and quantification in electron impact ionization in the SIM mode.

26.4. MULTIRESIDUE METHODS FOR PESTICIDES IN PROCESSED FOOD

26.4.1. Juice

Fruit juices are low-fat and nutritious beverages, and their consumption can help to fulfill the recommendation to eat more fruits and vegetables. Fruit juice sales in the two major markets, Europe (EU) and United States of America (USA), show a steady increase with sales volumes in both regions at some 11 billion liters per year. Simultaneously, safety concerns have increased in consumers and authorities about these products. Pesticide residues resulting from the treatment of raw fruits or from the water used in the processing are undoubtedly one of the major problems to human health due to their toxicity. Despite the progressive introduction of good practices in production, regulation, and scrutiny in the fruit juice industry to prove that the fruit juices are safe for consumption, the presence of pesticide residues remains a real problem.

In EU and other countries in the world, the MRLs of pesticides are established for raw agricultural commodities and not for processed products. However, pesticide residue levels found in a fruit juice depend on the various factors such as the type of pesticide, process applied, commodity, and degradation process involved. Thus far, EU and USA have prepared recommendations on principles and practices to the establishment of MRLs for fruit juice and other processed food [34].

Different extraction methods have been developed for the determination of pesticides in juices. In Zuin et al. [35], SBSE and MASE are compared for the determination of 18 pesticides in sugar cane juice. SBSE provides better LOD values, precision, and linearity, whereas MASE resulted in better recoveries, being faster, simpler, and fully automated.

Other extraction techniques employed for pesticides in juices are: dispersive LLE for analysis of 25 pesticides in apple [36], in which an efficient and high reproducible extraction was achieved for all pesticides assayed; SPE [37] for determination of 52 pesticides in carrot, peach, grape, orange, pineapple, and apple; and extraction with acetonitrile for the determination of 141 pesticides in apple juice [38].

The scope of the reviewed methods is between 18 and 141 pesticides, electron impact ionization is the technique of choice for the determination of pesticides in juice, and most of authors select the SIM mode for method development in order to achieve a good sensitivity. However, for a high number of pesticides, this mode of work does not allow the selectivity required for MRMs in food commodities [39]; similar LODs are achieved working in full scan and SIM mode, as shown in Figure 26.1 [internal communication]. In a method developed for the determination of 141 pesticides in the SIM mode, the method was divided in two methods, which include 62 and 79 pesticides, respectively, in order to achieve good results [38].

One of the major obstacles in the effective GC–MS analysis is the complexity of food matrixes and the presence of interferences that increase the need for cleanup steps, limit the ruggedness of the instrumental methods, and make low-level pesticide identification and quantification difficult. The use of multidimensional GC (with two different mechanisms of separation) can help solve this problem due to the effective increase in the selectivity and, consequently, in the resolution of the analytes. Cunha et al. [36] have developed and validated a fast analytical MD–GC–MS method for evaluation of multiple pesticide residues in apple juice; the potential of a dual GC column system connected by a Deans switch device was tested in combination with fast GC–MS.

Albero et al. [37] describe a method based on SPE followed by GC–MS (SIM) in which many compounds presented an increase in their chromatographic response, some of them from two- to five-fold, although organochlorine pesticides were the compounds that presented the lowest matrix effect. Sample components may compete for the active sites of the glass liner, decreasing the interaction between the active sites of the glass liner and the analyte, and thus a larger amount of analyte is transferred to the chromatographic column.

In general, limits of detection achieved by authors for determination of pesticides in juices are between 0.06 and 5 μg/L; the method that allows better sensitivity is DLLME followed by MD–GC–MS [36].

26.4.2. Wine

Fungicides, insecticides, and herbicides are commonly used in viticulture. Vineyards are treated with these products in the final stage of vegetation to prevent attack, which may occur shortly before the harvest.

Different MRMs have been developed for the determination of pesticides in wine; most of research describes development of methods for wine, grapes, and must.

The reviewed bibliography involves very different extraction methods: extraction with solvent, SPE, and solvent bar microextraction.

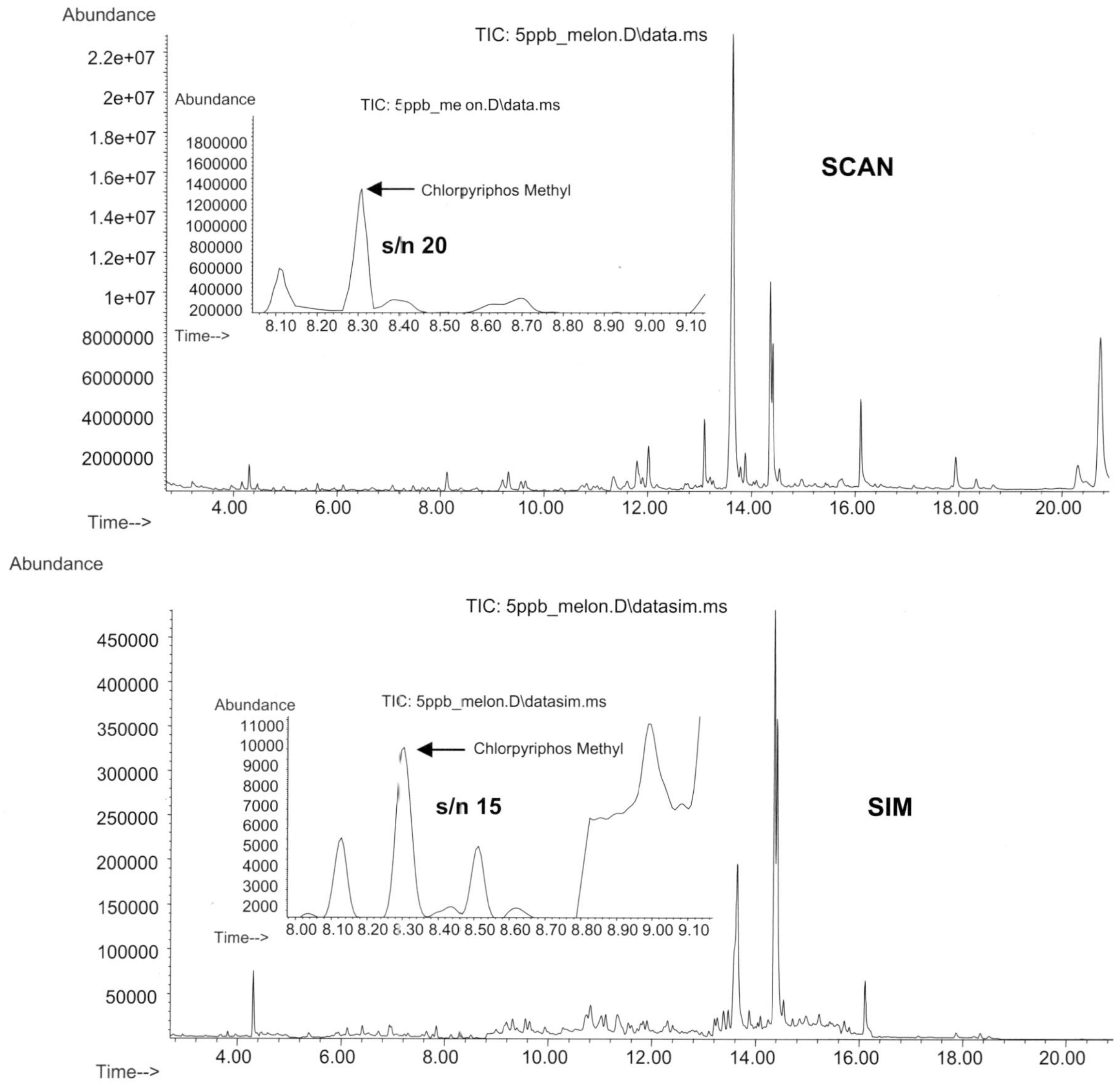

FIGURE 26.1 Total ion chromatograms in full scan and SIM of a melon extract spiked at 5 μg/kg with a mixture of 95 pesticides. The areas of the chromatograms where chlorpyriphos methyl elutes are magnified are shown. Signal to noise ratios of chlorpyriphos methyl are shown in the chromatograms.

Dasgupta et al. developed an extraction with ethyl acetate for the determination of 160 pesticides followed by cleanup by dispersive SPE with PSA [40]; Cunha et al. employed acetonitrile as a solvent for the extraction of 16 pesticides; and the cleanup is performed by dispersive solid extraction as well, but with PSA and C18 [41].

SPE [42] has been optimized for the determination of 16 organochlorinated pesticides in wine; the sorbent and the percentage of ethanol in the sample were optimized for a quantitative extraction of the selected compounds, 30% of ethanol and HySphere Resin SH Cartridges were selected as optima.

Solvent bar microextraction was developed for the determination of only six pesticides in wine [43]; different aspects of the extraction were optimized: hollow fiber, extraction solvents, extraction time, temperature, and finally the effect of ionic strength and ethanol addition. Once optimized, the limits of detection at the fg/ml level were achieved. Obviously, this extraction method allows better sensitivity than others, but the number of compounds analyzed is low.

Residue monitoring by unit-resolution GC−MS can be suitably performed on either a quadrupole or TOF instrument. For a complex mixture of analytes, a time-of-flight mass spectrometer, when coupled with a fast GC system, can perform simultaneous analyses of a large number of compounds within a reasonably short time period with sufficient accuracy, which is otherwise not possible with slower, scanning type mass detector (MS) like quadrupoles or ion traps. In the case of multiresidue screening by GC−MS full scan, the separation of a large component mixture is often a challenging task due to limited peak capacity and the resulting coelution of interfering matrix compounds. Since peak deconvolution algorithms require sufficient sampling to resolve complex peaks, correct identification of residues at trace levels becomes less likely as more compounds coelute at a single location in the chromatogram. Such analytical problems can however be resolved with comprehensive two-dimensional gas chromatography (GC × GC).

Two-dimensional gas chromatography TOF mass spectrometry has been developed for the determination of 160 pesticides in wine [40]; it provides advantages in terms of sensitivity, precision, and accuracy. In addition, the possibility of false negatives is reduced with the lower detection limits offered. Matrix effects can be solved by using GC × GC since baselines of compounds with common ions with matrix components can be separated.

A fast low-pressure gas chromatography-mass spectrometry method [41] was optimized for the determination of 27 pesticides in wine; in this method, using a column of 15-m length and 0.32 mm internal diameter, the carrier gas was set at 2.6 ml/min constant flow; with this condition, a shorter run time can be obtained than with conventional GC. The acquisition was performed in the SIM mode and matrix effects were resolved by the use of matrix match solutions.

Gas chromatography mass spectrometry in tandem is the most commonly used MRM method for the determination of pesticides in wine, such as heptachlor, aldrin, endrin, endosulfan, and dieldrin [42,43].

26.4.3. Oil

The development of sample-treatment procedures for the isolation of pesticides in samples with relatively high fat content (>15%), such as olive samples, by chromatographic techniques, requires the complete removal of the high-molecular mass fat from the sample to maintain the chromatographic system in working order. The main problem associated when working with these kinds of matrices is that dirty extracts with even small amounts of fat may harm the whole system (columns and detectors). For this reason, an additional cleanup step is usually included in MRMs for the determination of pesticides in processed food, such as oil. In addition, the development of MRMs to control a large number of compounds is highly desirable. However, the different natures and physicochemical properties of the classes of compounds to be studied (e.g., organochlorines, organophosphorus compounds, and triazines) makes it even more complex to establish extraction methodologies.

For the analysis of pesticides in the oil samples, several methods were used, such as SPE, gel permeation chromatography (GPC), SPME, MSPD, and SFE.

A total of 95 pesticides including organophosphate, organochlorines, carbamates, triazines, triadiazines, and pyrethroid compounds were extracted from soybean oil by LLE (oil solved in n-hexane was extracted with acetonitrile followed by centrifugation, freezing, and dispersive SPE as a cleanup step, using florisil as a sorbent [44]. Determination of 33 pesticides in peanut oil used low-temperature cleanup at −20 °C for oil precipitation after extraction with ACN and dispersive SPE using PSA and C18 as a sorbent [45].

Eight pesticides in virgin olive oil were determined by carbon nanotube-based SPE [46] without additional cleanup steps.

Extraction with n-hexane and acetonitrile followed by a cleanup step of gel permeation chromatography was applied for the determination of 26 pesticides from olive oil [47,48]. An MSPD method was validated for the determination of 9 pesticides in olive oil; aminopropyl was used to prepare the dispersion with the oil, and florisil was employed for the cleanup in one unique step [49]. A nonpolar analytical column has been used with a stationary phase composition of 5% phenyl/95% dimethylpolysiloxane.

For analytical determination, most authors develop MRMs using quadrupole in electron impact ionization in the SIM mode; in general, 1 μL is injected in the splitless mode [44−46,49]; however, there are some applications in which an ion trap is employed in the determination of 30 pesticides using electron impact ionization in the MS/MS mode [48]. Other applications have been developed using electron impact ionization and chemical ionization simultaneously [47].

26.5. MULTIRESIDUE METHODS FOR PESTICIDES IN BABY FOOD

Baby food is any food that is made specifically for infants, roughly between the ages of 6 months and 2 years. Baby foods combine a wide range of different matrices: nonfatty baby foods based on fruits and vegetables, fatty food based on meat/egg/cheese, and cereal-based foods. Moreover, breast milk and infant formulas are also included. The EU Baby Food Directive 2006/125/EC on processed cereal-based foods and processed foods for infants and young children places emphasis on the control of pesticides or transformation products (including metabolites) of pesticides with a maximum acceptable daily intake of 0.0005 mg/kg body weight. Pesticides are either designated as prohibited, and considered not to have been used if their residue does not exceed 3 μg/kg, or have MRLs set between 4 and 8 μg/kg. Twelve of the pesticides and breakdown products listed in the Directive (cadusafos, ethoprophos, fipronil, fiproni-desulfinyl, heptachlor, trans-heptachlor epoxide, hexachlorobenzene, nitrofen, aldrin, dieldrin, endrin, and dimethoate) are suitable for multiresidue analysis by GC−MS. The remaining compounds specified in the Directive, because of their physicochemical properties, require analysis either by multiresidue liquid chromatography−mass spectrometry (LC−MS) or by specific single residue methods.

MRMs applied to analysis of baby food need to be sensitive enough to meet the current directives mentioned above. In this chapter, MRMs based on GC−MS have been reviewed; the most common matrices analyzed are processed fruits such as purees, milk infant formulas, meat, and vegetables [50−52].

The scope of the MRM for baby food is between twelve and seventeen; nonpolar columns are employed in most of the cases.

Special attention is focused to development of methods for pesticides listed in the Directive 2006/125/EC [53].

Regarding the extraction techniques, SLE with acetonitrile followed by different variations concerning the cleanup step in the extraction is the extraction method of choice for most authors when meat, fruits and vegetables are analyzed.

However, other different extraction methods have been developed for multiresidue determination of pesticides in baby food. DI–SPME–GC–MS has been employed for analysis of carrots, vegetables, chicken with rice, chicken with vegetables, and lamb with vegetables. SPME is an environmentally nonharmful preconcentration system that avoids the use of organic solvents and cleanup steps, thus reducing the sample preparation time and allowing low detection limits [54].

SLE as a modification of the QuEChERS method by using a triple partitioning extraction between water, can, and hexane to considerably reduce lipophilic co-extracts generally present at high levels in baby food is used for the multiresidue analysis of pesticides in meat-based baby food. A wide scope is achieved and 236 pesticides are analyzed [52]; this method allows a preconcentration factor of 10 and very low limits of detection and quantification are achieved, 0.03 and 0.1, respectively.

For apple and apple purée, four sample preparation methods were tested [50]: a modified QuEChERS method utilizing column-based SPE cleanup instead of dispersive SPE provides lower recoveries, but slightly better LOQs. The modified Schenck's method with the best cleaning efficiency provides the lowest LOQs, while MSPD provides the worst cleaning related to the highest LOQs. The QuEChERs method, without any evaporation step, offers extreme improvement in rapidity and simplicity in comparison with the other methods tested. This makes the QuEChERS method the most attractive and useful in the ultratrace analysis of pesticides.

An SLE method (modified QuEChERS citrate) has been evaluated and applied to samples of fruits intended to be baby food [55]; the method has proven to be fast, easy to conduct, and used only small quantities of reagents. The results showed quite good analytical performance in terms of good repeatability, and recovery was achieved in almost all cases for the studied matrices.

When analyzing milk infant formulas, one of the main problems associated with the analysis is the high lipid content, lipids often co-extracted with the analytes of interest. It is well known that, in gas chromatography, large amounts of injected fat may cause problems in the injector and at the top of the column. To avoid some of the main disadvantages of the clean-up steps, such as reducing sample treatment time and minimizing the organic solvents consumed, modern extraction techniques are being applied. Such is the case with the pressurized liquid extraction (PLE) technique, which allows the combination of selective extraction with integrated cleanup strategies, by using fat-retainer compounds placed in the PLE extraction cell [51].

Recoveries achieved in the employed method are generally between 70% and 120%.

The analytical determinations in multiresidue methods for pesticides in baby food have been performed by using different analyzers; Ion Trap (IT) in full scan mode has been employed by Cieslik et al. [55] and Przybyloski et al. [52]. Quadruple in the SIM mode has been used by Viñas et al. [54] and Hercegová et al. [50]. A more selective and sensitive technique has been used by Leandro et al. [53], who employ TSQ in the MS/MS mode and Mezcua et al. [51] who employ IT in the MS/MS mode.

Few pesticides have been found in baby food and concentrations are lower than the established limits; the highest concentration in milk infant formula was 5.03 μg/kg of alpha endosulfan.

26.6. CONCLUSIONS

Attending to the extraction procedure in the selected MRMs for determination of pesticides in food commodities, for crops the extraction technique of choice is solid– liquid extraction with acetonitrile as a solvent, followed in some cases by dispersive solid extraction as cleanup; MRMs in products of animal origin are characterized by the diversity in extraction strategies, depending on the specific matrix.

In oil matrices, the main characteristic in the extraction procedure is the need to remove the high molecular compounds that can interfere in the determination of pesticides; this is mainly performed by GPC in the selected MRMs for determination of pesticides in food commodities.

Low-volume injection (1 μL) in the splitless mode is generally the injection technique of choice for most authors. The chromatographic separation is performed by the use of a nonpolar analytical column with a stationary phase composition of 5% phenyl/95% dimethylpolysiloxane. In all reviewed papers, a higher number of methods describe the analytical determination using quadrupole as an analyzer in the SIM mode, followed by a lower number of publications that employ an ion trap analyzer in the MS/MS mode; electron impact ionization is the ionization mode selected by most authors. In general, there are a scarce number of methods that employ chemical ionization as the ionization mode.

Recoveries achieved are generally in the range 70%–120%. Limits of quantification range between 10 and 25 μg/kg.

Acknowledgments

The authors acknowledge funding support from the Junta de Andalucía-FEDER (Project ref. AGR-4047) and from the EU, DG SANCO Specific Agreement No. 5 to Framework Partnership Agreement No. SANCO/2005/FOOD SAFETY/0025-Pesticides in Fruit and Vegetables.

Milagros Mezcua acknowledges the research contract to Junta de Andalucía-FEDER.

References

[1] A. Hercegová, M. Domotorava, E. Matisová, J. Chromatogr. A 1153 (2007) 54–73.
[2] Y.-J. Lian, G.-F. Pang, H.-R. Shu, C.-L. Fan, Y.-M. Liu, J. Feng, et al., J. Agric. Food Chem. 58 (2010) 9428–9453.
[3] C. Prybyiski, V. Bonnet, Anal. Bioanal. Chem. 394 (2009) 1147–1159.
[4] R. Huskova, E. Matisova, S. Hrouzkava, L. Svorc, J. Chromatogr. A 1216 (2009) 6326–6334.
[5] H. Guan, W.E. Brewer, S.L. Morgan, J. Agric. Chem. 57 (2009) 10531–10538.
[6] R.H. Savant, K. Banerjee, S.C. Utture, S.H. Patil, S. Dasgupta, M.S. Ghaste, et al., J. Agric. Chem. 58 (2010) 1447–1454.
[7] J.W. Wong, K. Zhang, K. Tech, D.G. Hayward, C.M. Malovi, A.J. Krynitsky, et al., J. Agric. Food Chem. 58 (2010) 5868–5883.
[8] L.-J. Qu, H. Zhang, J.-H. Zhu, G.-S. Yang, H.Y. Aboul-Enein, Food Chem. 122 (2010) 327–332.
[9] U. Koesukwiwat, S.J. Lehotay, S. Miao, N. Leepipatpiboon, J. Chromatogr. A 1217 (2010) 6692–6703.
[10] C-yu Shen, Xiao-wen Cao, Wi-jian Shen, Yuang Jiang, Zeng-yun Zhao, Bin Wu, et al., Talanta. 84 (2010) 141–147.
[11] D.I. Kolberg, O.D. Prestes, M.B. Adaime, R. Zanella, Food Chem. 125 (2010) 1436–1442.
[12] F.J. Camino Sanchez, A. Zafra-Gomez, J. Ruiz-Garcia, R. Bermudez-Peinado, O. Ballesteros, A. Navalon, et al., J. Food Compos. Anal. 24 (2010) 427–440.
[13] X. Yang, H. Zhang, Y. Liu, J. Wang, Y.C. Zhang, A.J. Dong, et al., Food Chem. 127 (2011) 855–865.
[14] Chen Chen, Yongzhong Qian, Qiong Chen, Chuanjiang Tao, Chuanyong Li, Yun Li, Food Control. 22 (2011) 114–1120.
[15] M. Barriada-Pereira, P. Serodio, M.J. Gonzalez-Castro, J.M.F. Nogueira, J. Chromatogr. A 1217 (2010) 119–126.
[16] Gouri Satpathy, Yogest Kumar Tyagi, Rajinder Kumar Gupta, Food Chem. 127 (2011) 1300–1308.
[17] Adalberto Menezes Filho, Fabio Neves dos Santos, Pedro Alfonso de Paula Pereira, Talanta. 81 (2010) 346–354.
[18] J.J. Ramos, M.J. Gonzalez, L. Ramos, J. Chromatogr. A 1216 (2009) 7307–7313.

[19] Qin-Bao lin, Hui-Juan Shi, Ping Xue, Chromatographia 72 (2010) 1143–1148.
[20] Jing Dong, Yu Xiang Pan, Jian-Xia Lv, Jun Sun, Xiao-Ming Gong, Kai li, Chromatographia 74 (2011) 109–119.
[21] M. Mezcua, O. Malato, J.F. Garcia-Reyes, A. Molina-Diaz, A.R. Fernandez-Alba, Anal. Chem. 81 (2009) 913–929.
[22] D.R. Erny, A.M. Gillespie, D.M. Gilvydis, J. Chromatogr. 57 (1993) 638.
[23] Jana Hajsloeva, Tomas Cajka. Food Toxicant Analysis-Y, in: Picó (Ed.), Gas chromatography-mass spectrometry (GC-MS), Elsevier, 2007. Chapter 12.
[24] V.I. Valsamaki, V.I. Boti, V.A. Sakkas, Triantafyllos A. Albanis, Anal. Chim. Acta. 573 (2006) 195–201.
[25] Jun Wang, Michael M. Klisks, Soojin Jun, Qing X. Li]
[26] Kamiesh Shrivas, Hui-Fen Wu, J. Sep. Sci. 38 (2008) 380–386.
[27] Frederico de M. Rodrigues, Paulo R.R. Mesquita, Lidia S. De Oliveira, Fabio S. De Oliveira, Adalberto Menezes Filho, Pedro A. De P. Pereira, et al., Microchem. J. 98 (2011) 56–61.
[28] Ariel R. Fontana, Alejandra B. Camargo, Jorgelina C. Altamirano, J. Chromatogr. A. 1217 (2010) 6334–6341.
[29] Dorthea F.K. Rawn, Judy Judge, Veronica Roscoe, Anal. Bional. Chem. 397 (2010) 2525–2531.
[30] Sathya Khay, A.M. AbdEl-Aty, Jeong-Heui Choi, Eun-Ho Shin, Ho-Chul Shin, Jin-Suk Kim, et al., J. Sep. Sci. 32 (2009) 244–251.
[31] Nikolaos G. Tsiropoulos, Elpiniki G. Amvrazi, J. AOAC Int. 94 (2011) 634–644.
[32] Gevany Paulino de Pinho, Antonio Augusto Neves, Maria Eliana Lopes Ribeiro de Queiroz, Flaviano Oliveira Silverio, Food Control 21 (2010) 1307–1311.
[33] Mercedes Castillo, Carmen González, Ana Miralles, Anal. Bioanal. Chem. 400 (2011) 1315–1328.
[34] Fernandez-Alba AR, Beijing, 2009, http://www.crl.pesticides.eu/library/docs/fv/Beijing2009pdf.
[35] Vania Gomes Zuin, Manuela Schellin, Larise Montero, Janete H. Yariwake, Fabio Augusto, Peter Popp, J. Chromatogr. A 1114 (2006) 180–187.
[36] S.C. Cunha, J.O. Fernandez, M.B.P.P. Oliveira, J. Chromatogr. A 1216 (2009) 8835–8844.
[37] Beatriz Albero, Consuelo Sánchez-Brunete, José L. Tadeo, Talanta 66 (2005) 917–924.
[38] Jian-Hua Wang, Yi-Bing Zhang, Xiu-Lin Wang, J. Sep. Sci. 29 (2006) 2330–2337.
[39] Milagros Mezcua, Maria A. Martinez-Uroz, Philip L. Wylie, Amadeo R. Fernande-Alba, J. AOAC Int. 92 (2009) 1790–1860.
[40] Soma Dasgupta, Kaushik Banerjee, Sangram H. Patil, Manoj Ghaste, K.N. Dhumal, Pandurang G. Adsule, J. Chromatogr. A 1217 (2010) 3881–3889.
[41] S.C. Cunha, J.O. Fernandes, A. Alves, M.B.P.P. Oliveira, J. Chromatogr. A 1216 (2009) 119–126.
[42] Jose Antonio Perez-Serradilla, José María Mata-Granados, María Dolores Duque de Castro, Chromatographia 71 (2010) 899–905.
[43] Kan-Jung Chia, Shang-Da Huang, Rapid Comm. Mass Spectrom. 20 (2006) 118–124.
[44] Thang Dong Nguyen, Myoung Hee Lee, Gae Ho Lee, Microchem. J. 95 (2010) 113–119.
[45] Li Li, Hongyan Zhang, Canping Pan, Zhiqiang Zhou, Shuren Jiang, Fengmao Liu, J. Sep. Sci. 30 (2007) 2097–2104.
[46] S. López-Feria, S. Cárdenas, M. Valcarcel, J. Chromatogr. A. 1216 (2009) 7346–7350.
[47] E. Ballesteros, A. Garcia Sánchez, N. Ramos Martos, J. Chromatogr. A 1111 (2006) 89–96.
[48] María Guardi-rubio, María Luisa Fernández-De Córdova, María José Ayora-Cañada, Antonio Ruiz-Medina, J. Chromatogr. A 1108 (2006) 231–239.
[49] Carmen Ferrer, M. Jose Gómez, M. Jose Gómez, Juan F. García-Reyes, Imma Ferrer, E. Michael Thurman, Amadeo R. Fernández-Alba, J. Chromatogr. A 1069 (2005) 183–194.
[50] Andrea Hercegova, Milena Domotorova, Svetlana Hrouzkova, Eva Matisova, Int. J. Anal. Chem. 87 (2007) 957–969.
[51] M. Mezcua, M.R. Repetti, A. Aguera, C. Ferrer, J.F. García-Reyes, A.R. Fernández-Alba, Anal. Bioanal. Chem. 389 (2007) 1833–1840.
[52] Cedric Przbylski, Christophe Segard, J. Sep. Sci. 32 (2009) 1858–1867.
[53] Cristina C. Leandro, Richard J. Fussell, Brenda J. Keely, J. Chromatogr. A 1085 (2005) 207–212.
[54] Pilar Viñas, Natalia Campillo, Nelson Martinez-Castillo, Manuel Hernandez-Cordoba, J. Chromatogr. A 1216 (2009) 140–146.
[55] Ewa Cieslik, Anna Sadowska, Juan Manuel Molina Ruiz, Magdalena Surma-Zamdora, Food Chem. 125 (2011) 773–778.
[56] M.I. Cervera, C. Medina, T. Portolés, E. Pitarch, J. Beltran, E. Serrahima, et al., Anal. Bioanal. Chem. 397 (2010) 2873–2891.
[57] Munetomo Nakamura, Satoko Noda, Masaki Kosugi, Noriko Ishiduka, Kazushi Mizukoshi, Makoto Taniguchi, et al., Food Hyg. Saf. Sci. 51 (2010) 213–219.
[58] Hans Ragnar Norli, Agnethe Christiansen, Borge Holen, J. Chromatogr. A 1217 (2010) 2056–2064.
[59] Hongxia Guan, Willian E. Brewer, Sherry T. Garris, Stephen L. Morgan, J. Chromatogr. A 1217 (2010) 1867–1874.
[60] Kaushik Banerjee, Rhaul H. Savant, Soma Dasgupta, Sangram H. Patil, Dasharath P. Oulkar, Pandurang G. Adsule, J. AOAC Int. 93 (2010) 369–379.

CHAPTER

27

Chemical Warfare Agents

Philip A. Smith

OUTLINE

27.1. INTRODUCTION AND BACKGROUND

27.1.1. The Use of Gas Chromatography for Analysis of CWA Materials

When James and Martin first performed gas–liquid chromatography to separate a series of *n*-alkyl fatty acids, their packed column stationary phase consisted of diatomaceous earth coated with stearic acid dissolved in silicone oil. The column temperature was controlled by passing an isothermally heated liquid through a jacket surrounding the column. For detection, column effluent was passed through a pH indicator solution and base was dispensed from a dropper when the operators noted a color change. Elution time was manually recorded, along with the amount of titration

Gas Chromatography DOI: 10.1016/B978-0-12-385540-4.00027-4

reagent needed to bring the pH indicator solution back to its initial state [1,2].

Important developments in gas chromatography (GC) theory and column design followed in succession, leading by the early 1980s to the current state-of-the-art open-tubular fused silica GC column with, for example, a cross-linked covalently bonded siloxane-based liquid stationary phase. These advances have allowed GC to become a tool routinely used for many applications, both in and out of the laboratory, including for analysis of chemical warfare agent (CWA) compounds. Broadly defined, CWA materials are chemical compounds used historically, or designed and created to kill or injure members of an opposing military force. Several instances have also occurred where CWA materials have been used by governments or terrorists against civilian populations. Specific substances historically used or created for chemical warfare use are the subjects of an international treaty that requires declaration and elimination of existing CWA stockpiles, and prohibits the creation of certain listed chemical compounds [3].

Due to the speed and relatively simple analysis procedures inherent to GC, this is now one of the most used methods for CWA analysis. Even in high-concern situations where detection of any CWA-related analyte is imperative, initial screening by GC "shows you which samples are interesting, and should be further investigated" [4]. The use of selective GC detectors is possible due to the presence of either sulfur or phosphorus in many CWA compounds. Mass spectrometric detection is desirable for GC analysis of CWA materials due to the need for certainty in identification, and mass spectrometric detectors are widely available at reasonable cost to meet this need.

One of the earliest reports of GC analysis for a CWA analyte in the peer-reviewed literature appeared in 1970, as Albro and Fishbein reported both isothermal and temperature program analysis of bis(2-chloroethyl sulfide) (sulfur mustard, or HD) and several related analytes [5]. They used a 0.2-cm I.D., 1.5-m glass column packed with Gas-Chrom Q which had been coated with 3% cyclohexanedimethanol, and a flame ionization detector (FID).

Writing in 1972 of the need for rapid detection of sulfur mustard during permeability testing of chemical protective clothing, Erickson et al. [6] described temperature program analytical performance in gas chromatography that seemed impossible to attain at that time "A total elution time of 2 min was allowed per sample injection... Consequently, it was not possible to use such time-consuming techniques as temperature programming."

A chromatogram produced by these researchers is shown in Figure 27.1, with relatively hot isothermal conditions selected to allow their required sample throughput. Thirty years

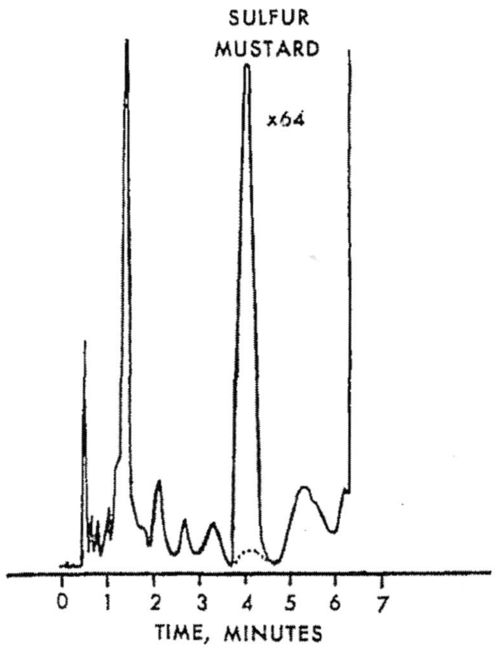

FIGURE 27.1 Gas chromatogram resulting from analysis of sulfur mustard using packed column GC and isothermal column temperature (125 °C) to obtain required speed of analysis. Reprinted from [6]. Copyright (1972) American Chemical Society.

later, advances in GC column design and column heating have produced small, lightweight column modules capable of temperature programming at rates up to several hundred °C per minute with resistive heating of a low thermal mass (LTM) open-tubular fused silica column [7]. In 2003, this column heating approach demonstrated that GC performance unimagined by Erickson et al. in the early 1970s is now possible, both in the laboratory and for use in field analysis. A standard open-tubular fused silica column with bonded liquid stationary phase was used with LTM resistive heating in a small field-portable gas chromatography-mass spectrometry (GC-MS) instrument to separate CWA analytes sampled from water by solid phase microextraction (SPME). These included *O*-isopropylmethylphosphonofluoridate (sarin or GB), *O*-pinacolylmethylphosphonofluoridate (soman or GD), ethyl-*N*,*N*-dimethylphosphoramidocyanidate (tabun or GA), HD, *O*-ethyl *S*-[2-(diisopropylamino)ethyl] methyl phosphonothiolate (VX,) and T-2 mycotoxin (466 u) and the GC-MS analysis was completed in <4 min. The first four of these (including sulfur mustard) were eluted in <1.5 min (Figure 27.2) [8].

This chapter summarizes important developments in GC for analysis of CWA compounds, the GC detectors often used, as well as the development of field-portable GC instrumentation largely driven by a demand to detect CWA analytes in near real time to protect deployed military forces, first responder personnel, and civilian populations.

27.1.2. Chemical Weapons Convention

The convention on the prohibition of the development, production, stockpiling, and use of chemical weapons and of their destruction (chemical weapons convention or CWC) became operative in 1997. The various state parties bound by this multilateral treaty have agreed to declare and destroy CWA materials previously stockpiled, and related production facilities, and to create a means to verify that compounds controlled under the CWC are not used in a prohibited fashion.

To complete the verification tasks defined by the CWC, the Organisation for the Prohibition of Chemical Weapons (OPCW) has been established. Substantial work has been done to define the analytical capabilities required to support

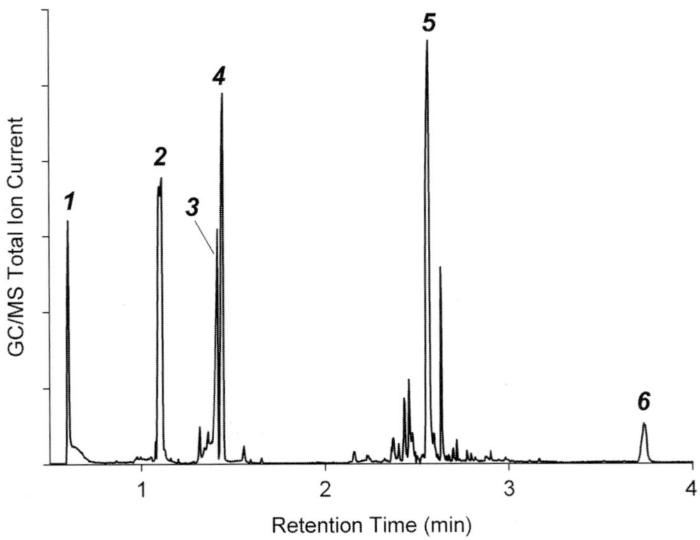

FIGURE 27.2 Direct 5 min SPME sampling of water spiked with (1) sarin, (2) soman, (3) tabun, (4) sulfur mustard, (5) VX, and (6) T-2 toxin. A 100% polydimethylsilixane stationary-phase GC column was used, having a length of 15 m, 0.25 mm I.D., and 25 μm film thickness. Column temperature program: 40 °C for 5 s, 80 °C/min to 100 °C, 20 °C/min to 115 °C, then 200 °C/min to 300 °C which was maintained until the run was completed. Carrier gas was H_2 at constant pressure with initial linear velocity of 100 cm/s. Reprinted from [8], Copyright (2005), with permission of Elsevier.

verification efforts. A major part of these capabilities includes laboratory GC, including GC-MS [4,9], as well as transportable GC-MS [10].

27.1.3. Types of CWA and Related Material

27.1.3.1. Nerve Agents

Nerve agent CWA compounds typically contain organophosphorus functional groups. Nerve agents bind to the mammalian enzyme acetylcholinesterase, deactivating this enzyme. In neurons where acetylcholine (Figure 27.3) is the neurotransmitter, transmission of a nerve signal between two neurons occurs with the release of this compound from the axon of one of the neurons. Diffusion of acetylcholine across a synapse to the dendrite structure of a second neuron may initiate an electrochemical signal that travels down the length of that neuron. In the case of nerve signals initiated to stimulate the activity of muscles (e.g. for breathing) acetylcholine also signals between the final neuron and the muscle tissue. If the acetylcholinesterase enzyme is deactivated, fundamental and necessary activities of the body can be severely impaired as nerve impulses will tend to continue in an uncontrolled fashion at affected synapses.

Figure 27.4 provides the structure for the nerve agent VX, and the similarities to acetylcholine are readily apparent. In normal function, a serine residue in the acetylcholinesterase enzyme forms a transient covalent bond with acetylcholine to cleave the acetyl group from the neurotransmitter molecule. With nerve agent poisoning, a permanent covalent bond

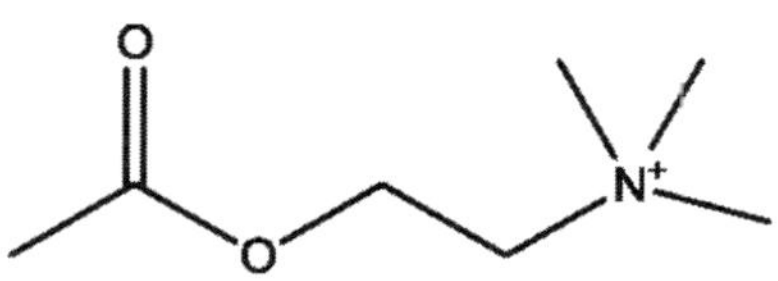

FIGURE 27.3 Acetylcholine.

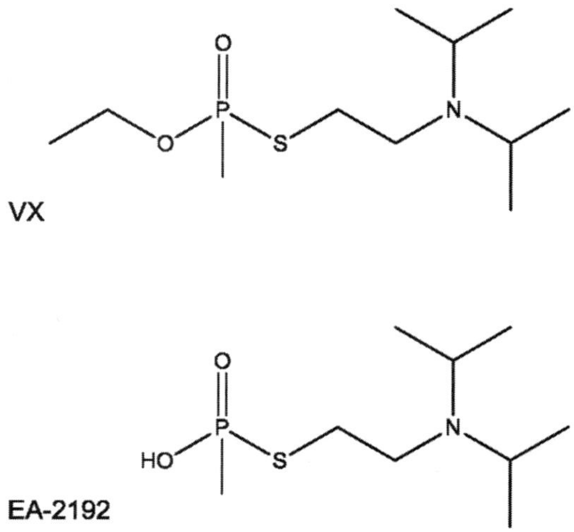

FIGURE 27.4 VX, and the VX degradation product EA-2192.

between the serine residue and the nerve agent causes loss of enzymatic function.

In pursuit of effective insecticides, the German chemist Gerhard Schroeder is reported to have synthesized the first nerve agent tabun (Figure 27.5) in 1936 [11]. This chemical was discovered to have unacceptable mammalian toxicity, and its military potential was recognized. A report was sent to the German Army in 1937, and this resulted in related patents being made secret. German efforts to develop additional nerve agents resulted in the discovery of other compounds with anticholinesterase activity, including sarin and soman (Figure 27.5). During World War II, thousands of tons of tabun and hundreds of tons of sarin were produced by the German military [11], although it is widely held that these stockpiles were not used during the war.

The nerve agents tabun, sarin, cyclohexyl sarin, soman, and VX are all suitable for analysis by GC without the need for derivatization, and GC has been used for analysis of these types of compounds since at least the early 1960s. However, as development of GC occurred during the height of the cold war years the early

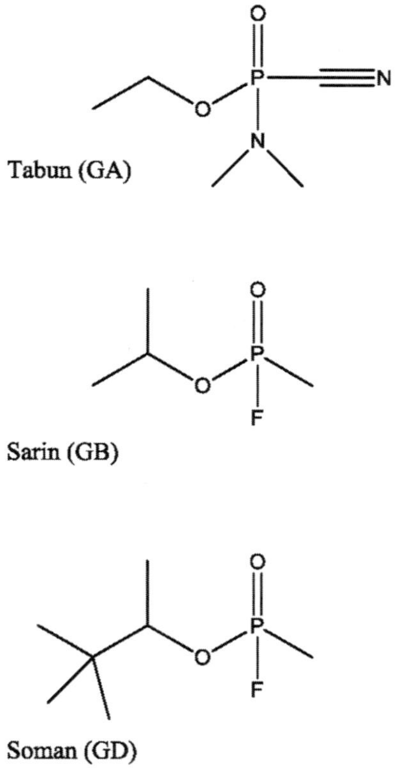

FIGURE 27.5 G Agents.

GC analyses of nerve agents completed at military research facilities were not well-documented in peer-reviewed literature sources. As an example of typical GC instrumentation and methods used in the 1960s for CWA analysis, Baier and Seller describe the use of packed column GC with FID and thermal conductivity detector (TCD) to identify thermal degradation of sarin with and without catalysis [12].

Exemplary of more current approaches, D'Agostino et al. describe the use of capillary column GC with mass spectrometric detection for separation and identification of numerous VX degradation products as well as the parent material [13]. For detection by GC under field conditions, early person-portable GC-MS systems capable of self-contained (i.e. battery powered) operation were able to detect the more volatile G agents directly from air. However, detection of VX required conversion to a more volatile species by reaction of gas-phase VX with AgF. Without the need for a so-called V-to-G conversion step, fast GC separation of degraded VX compounds and the parent material from solid-phase microextraction (SPME) with mass spectrometric detection was recently described using a small person-portable GC-MS instrument [14] (Figure 27.6). The second-generation person-portable GC-MS instrument used is capable of stand-alone operation on battery power for several hours, with GC separation by a well-insulated resistively heated 5-m metal capillary column with liquid film stationary phase (0.10 mm I.D., 1 μm d_f).

27.1.3.2. Vesicants

The prototypical CWA vesicant is sulfur mustard (Figure 27.7). This compound was reportedly first created by Despretz in 1822 by mixing sulfur chloride and ethylene [11]. Despretz did not recognize the toxic properties of the resulting compound, but noted a horse-radish or mustard smell. The synthesis of purified sulfur mustard was reported by Meyer in 1866 [11].

Sulfur mustard was used during armed conflict in 1917, and unlike the permanent gases used as CWA materials in the war prior to this (e.g. Cl_2 deployed from compressed gas cylinders), the effects of sulfur mustard were not limited to pulmonary exposure [11], and thus a chemical protective mask alone no longer offered adequate protection against CWA exposures. Sulfur mustard produced a large number of injuries from pulmonary, dermal, and ocular exposure, and treatment of these casualties required the expenditure of significant logistical and medical efforts. Analysis of sulfur mustard by GC has been routine for some time [5]. Analysis of the primary hydrolysis product of this compound, thiodiglycol, is usually completed by GC analysis following derivatization.

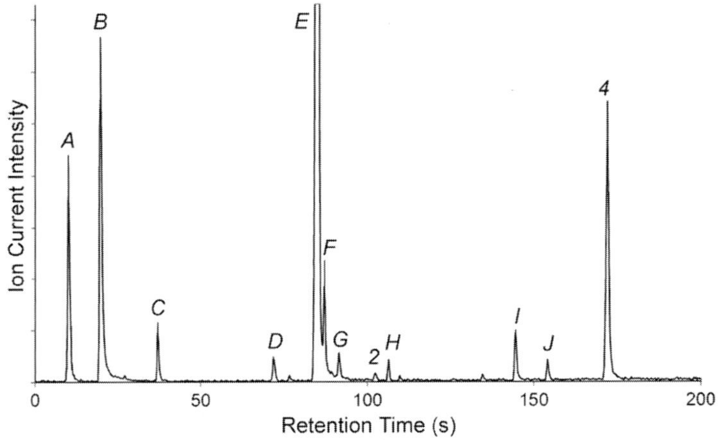

FIGURE 27.6 Chromatogram produced by analysis of sample collected after VX was added to AgF and maintained at 70 °C overnight. The person-portable GC-MS system described in reference [14] was used for analysis of SPME samples collected from the headspace of a vial containing the VX material. Peak identities in order of elution: ***A*,** thiirane; ***B*,** diisopropylamine; ***C*,** *O*-ethyl methylphosphonofluoridate; ***D*,** diethyl methylphosphonate; ***E*,** 2-(diisopropylamino)ethanethiol; ***F*,** unknown analyte, probable $M^{+\bullet}$ 159 *m/z*; ***G*,** *O,S*-diethyl methylphosphonothioate; **2,** 2-(diisopropylaminoethyl)ethyl sulfide; ***H*,** unknown analyte, probable $M^{+\bullet}$ 157 *m/z*; ***I*,** VX; ***J*,** unknown analyte, likely bis(diisopropylaminoethyl)sulfide from presence of 114 *m/z* base peak and elution order; **4,** bis(diisopropylaminoethy)disulfide. Reprinted from [14], Copyright (2011), with permission from Elsevier.

The organoarsenical vesicant 2-chloroethenyldichloroarsine (lewisite 1, Figure 27.7) was produced near the end of the First World War but was not used in that conflict. Lewisite is a fast-acting blister agent, and has been produced for inclusion in a mixture with sulfur mustard to cause more rapid onset of blister formation, and also for use in cold environments where sulfur mustard alone would remain a solid. As produced, lewisite consists of a mixture with three major components, of which lewisite 1 is the dominant species. In contrast to the nerve agents and sulfur mustard, derivatization is required for analysis of lewisite 1 by GC, and Muir et al. described thermal desorption GC-MS analysis for derivatives of these compounds and for underivatized sulfur mustard from sorbent tubes. Derivatization of lewisite 1 and lewisite 2 was completed by reaction with either butanethiol or 3,4-dimercaptotoluene which had been preloaded onto a sorbent tube (Figure 27.8) [15]. When using

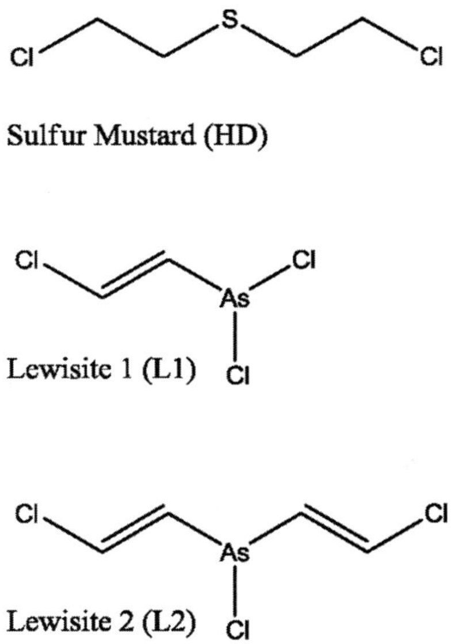

FIGURE 27.7 Vesicants sulfur mustard, lewisite 1, and lewisite 2.

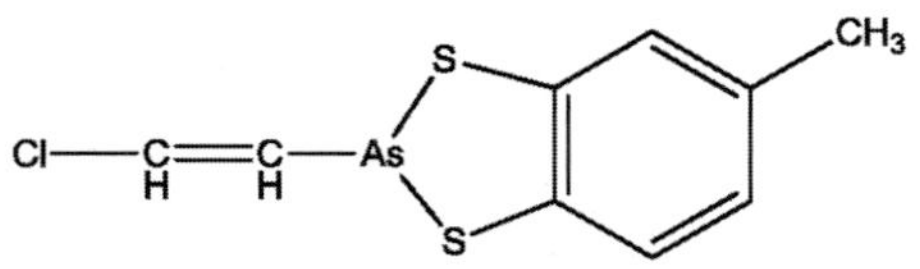

FIGURE 27.8 3,4-Dimercaptotoluene derivative of lewisite 1.

the dimercaptotoluene reagent both lewisite 1 and lewisite 2 yielded the same reaction product.

27.1.3.3. Blood and Pulmonary Agents

Mentioned here for completeness, the blood agents include systemic metabolic poisons, such as hydrogen cyanide (HCN), and the pulmonary (choking) agents include Cl_2 and phosgene, which damage pulmonary tissues. While these agents are highly dangerous in certain circumstances, they are also used extensively for legitimate industrial applications and furthermore are extremely volatile and thus nonpersistent. This makes these types of compounds less useful in armed conflict where opposing forces are found in close proximity to each other, and also tends to limit severe health effects to those who are heavily exposed over a brief period. The low environmental persistence of blood and pulmonary agents also lessens the analytical demands associated with their detection and identification.

27.1.3.4. Toxins

Toxins are harmful compounds produced by biological organisms. In the case of toxins produced by a microorganism such as trichothecene mycotoxins, a state wishing to illicitly use a toxin as a CWA could dismiss allegations of deliberate use due to the possibility that a recovered toxin material could have been produced by natural processes. Illustrative of the political and diplomatic ramifications that may be attendant with analysis of samples related to CWA use or production is the reported use of CWA materials against anticommunist resistance fighters in Southeast Asia in the 1970s. This allegation surfaced and gained credence when it was officially disclosed by the US Secretary of State. At that time the cold war was an important focus for many governments, and the alleged use of trichothecene mycotoxins (eventually termed "yellow rain") against the backdrop of the cold war struggle subjected the allegations to intense scrutiny, and the topic remains controversial.

A sample allegedly collected from an area in Laos where numerous animal deaths had reportedly occurred was analyzed by Rosen and Rosen [16]. These researchers used packed column GC with mass spectrometric detection to identify the presence of the mycotoxins T-2, diacetoxyscirpenol, 4-deoxynivalenol, and zearalenone in the sample as trimethylsilyl esters. A separate sample was analyzed by Mirocha, who also detected the presence of mycotoxins using GC-MS [17].

Referring to mycotoxins as putative agents causing the reported CWA incidents, Watson et al. [18] summarized the following questions and answered them in the affirmative:

1. Were the chemical and physical properties of these compounds suited for their use as warfare agents?
2. Could the toxins be produced in the large quantities that would be needed for such operations?
3. Was there any evidence that these toxins had been the subjects of classified research projects at institutes involved in chemical or biological warfare research?

An alternative explanation for the "yellow rain" material found in Southeast Asia was put forward by Nowicke and Messelson [19], who argued that the material was likely fecal material produced by honeybees.

Gas chromatographic analysis for tricothecene mycotoxins had been reported as early as 1971 [20]. The methods described by Ikediobi et al. involved trimethylsilyl (TMS) derivatization, and both isothermal and temperature program methods using packed pyrex glass columns and flame ionization detection. The favored liquid film for the larger toxins (e.g. T-2 and HT-2) was SE-30, due to a relatively high upper temperature limit [20].

Following the initial yellow rain papers published by Rosen and Rosen [16] and Mirocha et al. [17], considerable interest in GC analysis of trichothecene mycotoxins was raised in the chemical defense community as shown by the work published in 1986. D'Agostino et al. [21] used capillary column GC (DB-1 and DB-5 liquid films) with both FID and MS for analysis of six underivatized mycotoxins. Peak shape was initially poor when the toxins were dissolved in methanol, and the substitution of acetone provided improved chromatographic performance. Electron ionization allowed detection of mycotoxins spiked in human blood at μg/g concentrations, although the mass spectra lacked diagnostic high-mass ions. Using selected ion monitoring, and ammonia chemical ionization, T-2 toxin and diacetoxyscirpenol were detected at levels as low as 2 ng/g. Begley et al. [22] also used capillary column GC (SE-54) with single ion monitoring mass spectrometric detection to detect trichothecenes in the same spiked human blood sample set analyzed by D'Agostino et al., observing similar detection limits. Negative ion chemical ionization was employed, with sensitivity aided by pentafluoropropionyl esterification prior to GC analysis. Development of an SPME method for sampling underivatized T-2 toxin from water for subsequent GC analysis with flame ionization detection was described by Lee et al. [23]. Detection was possible at levels as low as 10 ppb (v/v).

Demonstrating the potential use of SPME and a field-portable GC-MS instrument for rapid sampling and analysis of a range of CWA materials under field conditions, Smith et al. [8] completed SPME sampling for T-2 toxin and several CWA compounds from water, with GC-MS analysis in <4 min using high-velocity H_2 carrier gas and a rapidly heated LTM GC column.

27.1.3.5. Riot Control/Incapacitating Agents

The characteristics of an ideal incapacitating agent include rapid onset of physiological effects that render targeted individuals incapable of performing routine functions, with rapid reversibility when exposure to the agent ceases, and lack of short- and long-term health effects from exposure. Two broad classes of incapacitating agents include those that are routinely used by civil authorities for riot and crowd control, and compounds developed through military research for use on the battlefield. The effects of the latter category (intense nausea or psychological disturbance) are somewhat morally objectionable, and thus this type of incapacitating agent is not used for civilian riot-control situations. Several readily available incapacitating agents that produce intense pain for a brief period are routinely used by law enforcement personnel in many countries, including o-chlorobenzylidenemalononitrile (CS) and phenacyl chloride (CN), shown in Figure 27.9. Beswick [24] describes the use of chemical incapacitating agents in both military conflicts and civil disturbances.

Some controversy exists concerning the categorization of incapacitating agents such as CS and CN among CWA materials. Both CS and CN are considered to be relatively safe, nonlethal agents routinely used in civil disturbances for crowd dispersal. Such use is typically judged to be moral, as alternative means of crowd dispersal would cause a greater risk for harm or loss of life. Wils and Hulst described GC-MS methods for analysis of CS and related compounds [25], and provided relevant mass spectra.

The dispersal of CS and CN is often accomplished by heating. For example, a small thermal CS canister "grenade" contains lactose fuel, permanganate oxidizer, and CS. In a commercially available CS canister, when the mixture was lit by a fuse mechanism,

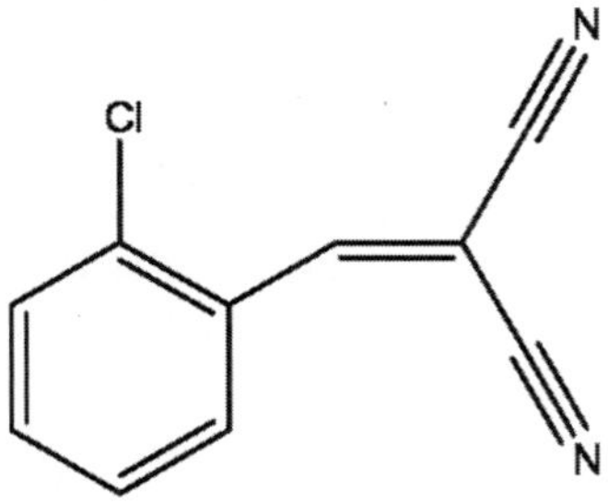

o-Chlorobenzylidenemalononitrile (CS)

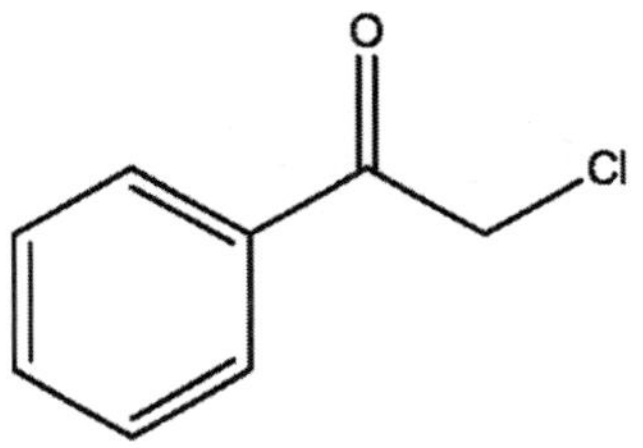

o-Chlorobenzylidenemalononitrile (CN)

FIGURE 27.9 Riot-control agents.

temperatures of about 700 °C were measured inside the canister [26]. Kluchinsky et al. recognized the potential for thermal production of organic degradation products, and characterized a number of these recovered as airborne contaminants produced by incendiary-type CS grenades used in riot control [27]. One of the principal degradation products recovered suggested the loss of hydrogen cyanide (HCN) from the parent CS material, and further work confirmed the presence of airborne HCN produced by high-temperature dispersion of CS [28].

27.1.3.6. Environmental Degradation Products of CWA Compounds

Most analytical methods for detection of CWA materials would deal with either analysis of bulk chemicals (i.e. from a suspected CWA process stream), or with environmental samples collected from air, soil, or water sources. A comprehensive discussion regarding the environmental fate of nerve and blister agents has been provided by Munro et al. [29], who listed the well-known degradation products. The literature cited by these authors refers to numerous papers where GC analysis was used for CWA-related products produced through hydrolysis. Many of these CWA degradation products are not suitable for direct analysis by GC, but, in most cases, this may be accomplished following derivatization.

Many of the nerve agent hydrolysis products contain acidic phosphorus functional groups, while sulfur mustard produces thiodiglycol, and a principal degradation product of lewisite 1 is chlorovinyl arsenous acid. Production of thiodiglycol from sulfur mustard occurs *via* hydrolytic dehydrochlorination, and a number of related sulfoxide and sulfone compounds are also known to result from hydrolysis of the parent material. Militarized vesicants may include both sulfur mustard and longer-chain-length compounds with biological effects similar to sulfur mustard, such as bis(2-chloroethylthio)ethane (sesquimustard) and bis[(2-chloroethylthio)ethyl] ether. D'Agostino and Provost demonstrated the usefulness of GC for these analytes as well by subjecting samples of HQ (a mixture of sulfur mustard and sesquimustard) and HT (a mixture of sulfur mustard and the ether) to hydrolysis. This was followed by GC-MS analyses using both electron ionization (EI) and ammonia chemical ionization (CI) to directly identify a number of the resulting degradation products, and numerous TMS derivatives [30].

Most of the nerve agent hydrolysis products are considerably less dangerous than the intact parent CWA materials. The VX degradation product *S*-2-(N,N-diisopropylaminoethyl) methylphosphonothiolate (EA-2192, Figure 27.4) is a zwitterion, and is not directly extractable as at least one of the two possible EA-2192 ionization sites remains substantially ionized at all pH conditions. The ability to identify this product in waste streams from declared VX destruction processes is important as this compound is a potent cholinesterase inhibitor in its own right [29]. Derivatization approaches for GC analysis of CWA-related compounds are discussed later, under "Analytical Considerations."

27.1.4. CWA Detection Needs as Drivers for Field-Portable GC Instrumentation

An impetus for the development of early field-portable GC systems was the need to analyze CWA materials in near real time. Early instruments (see Chapter 15) used for analyses in the field employed technology relevant to the early years of GC such as packed columns. Due to the power requirements for temperature program operation, the early field-portable GC instruments used low temperatures, and often isothermal column temperature. At the other extreme, the Viking 572 and Bruker EM 640S GC-MS designs developed in the 1990s are exemplary of high-capability field-portable GC instrumentation. Both of these instruments borrowed from typical laboratory-based GC designs, although with miniaturized components when possible. Each design employed a small air bath oven for column heating, for example, and each employed an ion beam quadrupole detector with two-stage vacuum pumping. The Bruker instrument was adopted by the OPCW for official on-site analyses [10]. In addition to the needs for orthogonal analysis driven by forensic [31] and CWC treaty compliance concerns [10], substantial resources have been applied to the development of field-portable GC-based methods with element-selective detectors for analysis of CWA in field settings to protect the health of workers involved in destruction of declared CWA stockpiles, and the public.

Development of a second generation of commercial field-portable GC instruments has primarily been driven by the need to detect and identify CWA materials in the field. Arguably, the most important improvement in this area is the use of LTM column heating. Several low power consumptive approaches to this end have been described in the literature and have engendered commercial ventures [7,32]. Since the events of September 11, 2001, increased interest in field-portable GC-MS for detection of CWA compounds or other dangerous chemicals has led to commercialization of at least four GC-MS instruments that use the LTM heating approach first described by Sloan et al. [7] for control of GC column temperature, and the general approach for this is discussed further below.

27.1.4.1. Minicams

The MINICAMS is a commercially available GC instrument designed specifically for field detection of CWA materials at very low levels during operations to destroy declared CWA stockpiles. This system may automatically pass ambient air through a sorbent trap for subsequent thermal desorption, or through a sample loop if preconcentration is not required. In addition to analysis of airborne nerve or vesicant CWA compounds that contain either sulfur or phosphorus, the instrument may be set up to monitor lewisite using gas-phase dithiol derivatization prior to analysis [33]. The MINICAMS is compact, and completes analyses quickly, but access to stable external power is required. Several detectors are available, but flame photometric or pulsed flame photometric types are the logical choices for detection of CWA analytes.

27.1.4.2. Low Thermal Mass GC Column Heating

The movement away from packed GC columns toward the open-tubular design for use in a laboratory setting was driven by the improved chromatographic performance made possible by the improved design. However, the modern fused-silica, open-tubular GC column that has resulted is also much smaller than the typical packed column. As the open-tubular GC column design (which happens coincidentally to also have a low thermal mass) became accepted and widely used, this also opened up the potential to move away from convection oven heating to quickly change the temperature of a standard open-tubular column using relatively little power.

Several research groups independently demonstrated rapid heating and cooling of a typical fused-silica, open-tubular GC column using very little power [7,32]. With the approach of Sloan et al. [7] thermal control is provided by measuring the temperature-dependent resistance in a thin platinum wire threaded within a small circular column bundle. Column heating is provided by several additional insulated wires intertwined with the coiled GC column (Figure 27.10), which are resistively heated using electrical current under microprocessor feedback control. The commercial availability of high-performance resistively heated LTM GC column modules beginning in the early 2000s has led to adoption of this column heating method in several GC-MS systems designed for both field transportability and person portability. In most of these cases, funding from U.S. military organizations spurred the development of these due to the need for small, fast field-portable instruments to protect the health of deployed forces.

The LTM column heating approach replaced a small convection oven present in earlier versions of the person-portable Hapsite™ GC-MS instrument that was first marketed in the 1990s. Later adoption of this heating approach resulted in lower power consumption. This first-generation person-portable GC-MS instrument uses an ion beam quadrupole detector, with primary mass spectrometer vacuum pumping provided by a nonevaporative getter (NEG) pump and an ion sputter pump to remove residual noble gases. Due to the need

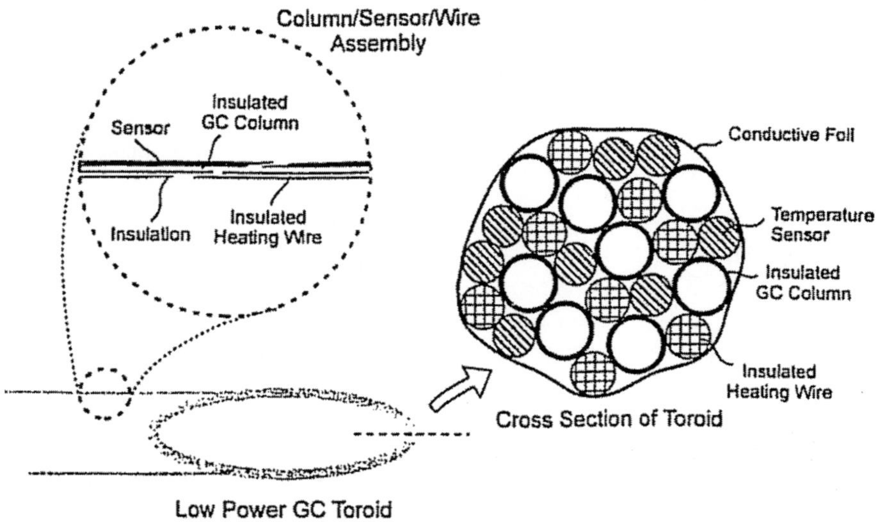

FIGURE 27.10 Diagram illustrating the low thermal mass (LTM) resistive heating design for a standard open-tubular capillary column. Reprinted from [7], © 2002 Wiley Periodicals, Inc., with permission from Wiley Periodicals.

for a reasonable NEG service life, a polymer membrane GC-MS interface is used in this instrument. This challenges the instrument's chromatographic performance, and the use of an approximately 1-m air sample probe heated to only about 40 °C limits this instrument to direct air sampling and analysis of airborne analytes with *n*-alkane linear program temperature retention index values less than about 1300. For air sampling, an onboard sorbent tube may be used to trap analytes when the air sample probe inlet is used. Additional modules to allow desorption of an SPME fiber or an externally collected sorbent tube sample have been added to this instrument's capabilities recently. A chromatogram produced from analysis of volatile CWA analytes by a Hapsite™ instrument is shown in Figure 27.11. The GC-MS membrane interface used in this instrument is the primary cause of the GC peak tailing seen in the chromatogram.

Smith et al. [8] described a transportable GC-MS capable of very fast analysis of CWA compounds (similar to Figure 27.2). Construction of this instrument was funded by the U.S. military with the specific intention that it be built with an LTM column assembly mounted to a standard quadrupole mass filter heavily used for laboratory GC-MS analyses. A refined LTM GC-MS design based on this approach was commercialized in 2010 by Agilent Technologies, using the 5975 mass spectrometric detector, with the resulting instrument designated as the 5975T ("transportable") GC-MS system. In the early 2000s, an additional GC-MS instrument designed primarily for field use incorporated the same basic LTM GC column design as the separation method for a transportable cylindrical ion trap detector GC-MS manufactured by Griffin Analytical (now part of FLIR Systems Inc.).

Beginning in 2008, a GC-MS instrument designed for person portability incorporating LTM GC has been commercially produced by Torion Technologies [35]. The current version of this instrument weighs 14.5 kg, and is small enough to travel onboard commercial aircraft

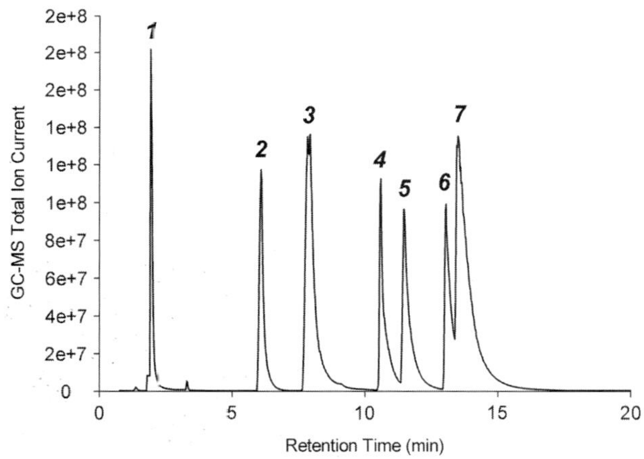

FIGURE 27.11 Hapsite™ sampling/analysis GC−MS chromatogram: 5.0 mg/m^3 air concentration for each of four volatile CWAs. Sample time was 1.0 min, nominal sample rate was 250 ml/min with Tenax concentrator module used. Initial column temperature was 70 °C, ramped to 180 °C at 30 °C/min. 1: Air, methylene chloride; 2: Sarin; 3: N,N-dimethylacetamide (artifact present in clean Tedlar bags); 4: phenol (artifact present in clean Tedlar bags); 5: soman (two diastereomers not resolved here); 6: sulfur mustard; and 7: cyclohexylmethylphosphonofluoridate. Reprinted from [34], Copyright (2004), with permission from Elsevier.

as a carry-on luggage item (after removing the onboard high-pressure He cylinder for transportation safety). Due to the use of a small, well-insulated injector and transfer line components, and a 5-m GC column with 0.10-mm I.D. that is resistively heated as per Sloan et al. [7], the rechargeable battery used in this instrument is adequate to complete about 20 analysis cycles. Vacuum in the mass spectrometer is maintained by a small turbomolecular pump, backed by an onboard membrane roughing pump. As initially designed, sample introduction was limited to desorption from an SPME fiber. The small GC column diameter limits carrier gas flow into the toroidal ion trap detector, allowing direct interface of the GC to the mass spectrometer. A chromatogram produced by this instrument from analysis of a degraded VX sample is shown in Figure 27.6.

27.1.4.3. Volatility Constraints for Field-Portable GC

A significant challenge exists for gas-phase sampling of the nerve agent VX, a compound with limited volatility. For qualitative screening using SPME, Hook et al. showed that with gentle heating of cloth material contaminated with a drop of VX liquid, adequate analyte loading may be rapidly obtained from the headspace of a sealed vial [36]. The use of a sealed vial contributes to increased analyst safety, although it would be wise to use this sampling method with full personal protective measures, and a scrubber-equipped fume hood. For quantitative sampling of airborne VX for GC analysis, a different approach has been used for years where airborne VX is passed through a porous material coated with silver fluoride. The reaction shown in Figure 27.12 occurs readily as demonstrated by Fowler and Smith [37], producing a much more volatile (and still quite dangerous) "G analog" that differs from the G agent sarin only in the presence of an *O*-ethyl group instead of an *O*-isopropyl group. The reaction to produce the G analog is used for field GC analysis methods where the capability to quantify VX vapor concentration is desired. The removal of the diisopropylamine functional group and the sulfur atom from the VX molecule results in not only a more volatile analyte, but also one that is also less susceptible to interactions with active sites [38]. Prior to the work documented by Fowler and Smith, quantitative VX measurements were often completed using liquid impinger sampling and spectrophotometric measurements, or involved wet chemistry titration of cholinesterase activity.

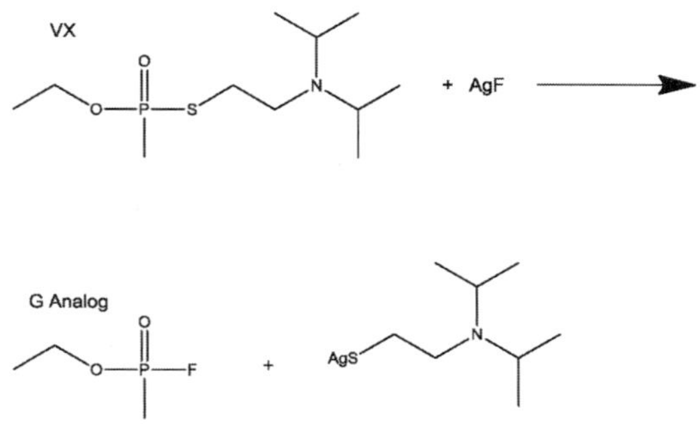

FIGURE 27.12 Conversion of airborne VX to a more volatile and less reactive G analog by reacting VX vapor with AgF, initially described by Fowler and Smith [37].

27.2. ANALYTICAL CONSIDERATIONS FOR SAMPLING AND GAS CHROMATOGRAPHIC ANALYSIS OF CWA-RELATED COMPOUNDS

27.2.1. Derivatization

As discussed above for analysis of CWA degradation products, hydrolysis and metabolism of these generally produce degradation products and metabolites that are quite polar. Many of these compounds may be derivatized for GC analysis following routine procedures to add trimethylsilyl (TMS) or *tert*-butyldimethylsilyl (TBDMS) groups to mask amine or hydroxyl sites. A comprehensive review of derivatization for analysis of CWA-related materials was completed by Black and Muir in 2003, who covered derivatization for GC as well as for liquid chromatography analysis [38].

27.2.1.1. G Agents

The suitability of alkylphosphonic and alkyl methylphosphonic acids for TBDMS derivatization and quantitative GC analysis was investigated by Purdon et al. in 1989 [39]. Trimethylsilylation is also a possibility for analysis of G agent degradation products, and both approaches are discussed by Kuitunen for use in OPCW analytical procedures [40]. Recovery of alkylphosphonic acid compounds is problematic in aqueous samples or soil matrices where inorganic cations are present unless a cation-exchange cleanup is included [38]. Other derivatization approaches (e.g. methylation and pentafluorobenzyl esterification) for G agent degradation products are summarized by Black and Muir [38].

For confirmation of exposure, GC analysis of the sarin metabolite *O*-isopropyl methylphosphonic acid, the *O*-ethyl methylphosphonic acid sarin analog, and methylphosphonic acid was completed for the TMS derivatives by Minami et al. [41]. Urine to be tested was first passed through an ion-exchange column to remove metal ions, followed by drying of the eluate under vacuum. The acid metabolites were derivatized as trimethylsilyl esters for analysis using flame photometric detection. Limits of detection as low as 25 parts per billion for the isopropyl and ethyl phosphonic acid species were reported.

27.2.1.2. VX

Many degradation products of VX do not require derivatization for GC analysis [13,42]. Hydrolytic degradation of VX can produce several acidic phosphorus compounds, including ethyl methylphosphonic acid, ethyl methylthiophosphonic acid, and EA-2192 (Figure 27.4) [43]. Creasy et al. [44] showed that TMS derivatization of alkyl methylphosphonic acids and alkyl methylphosphonothioic acids may be routinely completed for GC analysis. Analysis of EA-2192 as the TMS derivative was problematic, although methylation with trimethylphenylammonium hydroxide (TMPAH) allowed GC analysis of this analyte [45]. Pardasani et al. [43] showed recently that the TMS derivative of EA-2192 may be successfully analyzed by GC if column temperatures that favor gas-phase activity of the derivative are maintained throughout an entire analytical run. These researchers hypothesized that decomposition of the derivative occurred on the column after initial condensation at the relatively cool column temperatures typically used at the beginning of a linear temperature program.

27.2.1.3. Sulfur Mustard

As is the case for all of the nerve and vesicant compounds except for lewisite 1, the parent sulfur mustard compound is well suited for GC analysis. However, virtually all of the mustard degradation products are best analyzed following derivatization. Wils and Hulst [46] reported electron ionization mass spectra for numerous TMS derivatives of analytes related to sulfur mustard, as well for many of the underivatized degradation products.

27.2.1.4. Lewisite

Muir et al. reported that GC analysis of underivatized lewisite 1 and lewisite 2 (both of which contain reactive As−Cl bonds) quickly leads to column degradation [15]. Derivatization of lewisite 1, lewisite 2, and the lewisite hydrolysis products such as chlorovinyl arsenous acid is usually accomplished using a thiol or dithiol reagent. Early work by Fowler et al.

used 1,2,-ethanedithiol to derivatize chlorovinyl arsenous acid in water for GC analysis with flame photometric detection. Excess reagent was precipitated by treating the aqueous sample with $AgNO_3$ prior to solvent extraction [47]. Butanethiol and 3,4-dimercaptotoluene were used by Muir and co-workers who spiked derivatizing reagent into an air stream passing over tenax packed in a thermal desorption tube prior to sampling air that contained lewisite 1 and lewisite 2. While the dithiol reagent provided better detection limits, its use resulted in the production of the same derivative for both lewisite 1 and lewisite 2, restricting the use of this reagent to the simultaneous quantification of the total combined airborne concentration of both analytes [15]. Due to the confusing production of degradation product derivatives with the same identity as those produced from the parent CWA compounds, Hanaoka et al. analyzed samples containing lewisite and sulfur mustard without derivatization using GC with either atomic emission or mass spectrometric detection. This approach is unusual for lewisite, and the authors used a guard column and on-column injection with frequent column solvent washes to allow analysis of the underivatized lewisite compounds [48].

27.2.2. Thermal Desorption

Desorption of air sampling sorbent media is often carried out with liquid solvents [40]. However, when very low airborne concentrations are to be sampled, thermal desorption becomes an attractive alternative as this avoids dilution of the relevant analytes. In 1979, Fowler et al. reported thermal desorption for analysis of CWA-related analytes after sampling sulfur mustard onto a packed tube that was later placed into a heated GC injector, followed by flow of carrier gas through the sorbent to an isothermally heated packed column [49]. Commercially available instrumentation was not an option for thermal desorption of analytes trapped on sorbent tubes when Fowler et al. completed this work.

The apparatus for the thermal desorption of sampling tubes directly into the analytical column of a gas chromatograph can be purchased from commercial sources or fabricated in-house (see Chapter 10). However, commercially available equipment is often prohibitively expensive for those who wish merely to engage in limited experimentation or who expect to use the method only occasionally [49].

Later work by Steinhanses and Schoene demonstrated the usefulness of a commercially available thermal desorption inlet for GC analysis of sulfur mustard and several organophosphorus compounds, including sarin and soman, using flame photometric detection [50]. Black et al. used a commercially available thermal desorption inlet interfaced to laboratory GC-MS to successfully sample sulfur mustard from the headspace above soil collected by an investigative journalist where chemical warfare agents had allegedly been used by the government of Iraq against civilians [51]. Hancock and Peters used a custom-built thermal desorption inlet for GC analysis of compounds "of chemical defence interest," sampled from the gas phase by purging spiked water, and by sampling the headspace above spiked soil. Simultaneous flame ionization and flame photometric detection were used [52].

Several field-portable GC systems available today include an integrated sorbent tube air sampler to preconcentrate airborne analytes for thermal desorption to introduce analytes into GC instrumentation, including the MINICAMS™ fixed-location instrument as well as the Hapsite™ person-portable GC-MS instrument. In both cases, the use of thermal desorption and compact thermal desorption components allows for detection of trace contaminant levels using relatively little power.

27.2.3. SPME Sampling/Sample Introduction for GC Analysis

In 1990, Arthur and Pawliszyn described solid-phase microextraction (SPME) [53]. Analysis of samples collected onto a typical SPME fiber is most often completed by gas chromatography, and thousands of papers have described the use of SPME sampling from a wide range of matrices for GC analysis, including a number with a focus on detection and identification of CWA materials. The use of SPME for sampling and sample introduction relevant to GC analysis of CWA-related compounds may follow two broad approaches: (1) quantitative GC analysis and (2) qualitative screening. Zygmunt et al. [54] reviewed the use of SPME for CWA sampling and GC analysis in 2007, summarizing numerous sample matrices that have been addressed. Updating their list, CWA compounds that have been sampled from soil or sediment or their extracts by SPME for GC analysis include sulfur mustard [55], lewisite degradation products [56,57], and VX degradation products [42]. Those sampled from aqueous systems include sulfur mustard [8,58], nerve agents [8,59,60], T-2 toxin [8,61], degradation products of sulfur mustard and nerve agents [58,62], and lewisite degradation products [57]. Those sampled from air include G nerve agents [34,60,63] and sulfur mustard [34]. Hook et al. also demonstrated the potential to detect VX on contaminated cloth material by short-duration SPME sampling from a closed vial kept at 50 °C, with GC-MS analysis completed in the field [36].

Quantitative GC analysis of SPME samples may employ either passive equilibrium or dynamic air sampling. The former approach is most typically used, although for some analytes attainment of equilibrium between the SPME fiber coating and the matrix sampled can be lengthy. Sampling may stop before equilibrium is attained as long as adequate analyte is available on the fiber and sample duration is consistent from one sampling event to another. It is advisable to avoid termination of sampling in the area of an SPME uptake curve where the curve is steep (sample duration on x-axis, mass loaded to fiber on y-axis), as small errors in sample timing can cause relatively large errors in quantification. The dynamic quantitative air sampling approach uses an adsorptive SPME fiber coating as described by Koziel et al. [64]. An example of equilibrium sampling followed by quantitative analysis of CWA-related materials in water was provided by Lakso and Ng, who used SPME with GC-MS (selected ion monitoring). They obtained detection limits for sarin, soman, and tabun of about 0.05 μg/mL, while the detection limit for VX was reported to be about 0.5 μg/mL [59]. With the exception of the value for VX, these are below or slightly above the respective short-term exposure limits promulgated by the US Army for their presence in water to be consumed by deployed troops. In another example, Kimm et al. used passive headspace SPME sampling to demonstrate that sulfur mustard spiked in soil at several hundred ng/g soil could be detected with GC-MS analysis [55]. In the soil system, equilibrium sampling was approached at room temperature, with a sampling time of 20 min. For dynamic quantitative air sampling using SPME, Hook et al. used a carboxen/polydimethylsiloxane SPME fiber coating and GC-MS analysis to quantify airborne sarin concentrations as low as about 20 ppb (v/v) [63].

The speed and simplicity of SPME, and the ability to desorb GC analytes from an SPME fiber within the heated injector of an unmodified GC system, make SPME useful for rapidly screening large numbers of potentially contaminated items and environmental samples for the presence of relatively concentrated (mg quantities) CWA materials. Additionally, the use of SPME avoids the need for solvent extraction to obtain target analytes from various matrices, and also avoids extensive sample handling. Both of these attributes lessen the likelihood for exposure to technicians using SPME and field-portable GC instrumentation to complete qualitative screening for CWA contaminants. For qualitative CWA screening, GC-MS is usually used to identify analytes sampled quickly and safely using SPME [8,34,42,57]. The U.S. Marine Corps Chemical Biological Incident Response Force (CBIRF), tasked with counterterrorism responsibilities for detection and mitigation of chemical, biological, and radiological attack, uses SPME and GC analysis with fast resistive column heating and mass spectrometric detection to quickly screen samples potentially contaminated with CWA materials. In the field, a limited amount of a liquid sample suspected of being a CWA may be collected by a properly protected individual using a cotton-tipped swab to be sealed inside a vial with a septum top. The exterior of the vial is then decontaminated, and insertion of an SPME fiber through the septum under a portable fume hood allows headspace sampling with low potential for exposure to the GC operator. This approach is attractive for field use in a mobile laboratory or other transportable platform equipped with GC-MS capability, and sample times of 1 min or less are possible if mg quantities of CWA materials such as the G agents, sulfur mustard [58], or even VX [36] are present.

27.2.4. GC Detectors for CWA Analyses

27.2.4.1. Flame Ionization Detector (FID)

Numerous papers describe GC analysis of CWA materials with detection by FID. In situations where samples are to be screened for the possible presence of CWA compounds, the general usefulness of this detector for virtually all hydrocarbon-containing analytes necessitates the consideration of relative retention information such as the linear temperature program retention index (LTPRI) system proposed by Van den Dool and Kratz [65]. D'Agostino and Provost used GC with flame ionization detection to obtain this type of retention index information relative to a homologous series of *n*-alkanes for organophosphorus compounds (including sarin, soman, tabun, and VX), vesicants, irritants, simulants, and precursors [66]. With little information on analyte identity provided by the detector, the flame ionization detector is useful only for screening samples where CWA compounds, precursors, or degradation products are expected, or to specifically rule out their presence. As GC with mass spectrometric detection has become more widely used, this has reduced the need to rely on broadly responding types of detectors for general screening.

Extensive tables of LTPRI data for CWA-related GC analytes have been compiled for use in the work of the OPCW [4]. The usefulness of such information is not limited to GC analysis using a nonorthogonal detector such as the FID. LTPRI information is also useful in those cases where electron ionization mass spectra fail to provide an unambiguous identification [13].

27.2.4.2. Detectors with Selectivity Toward Phosphorus, Sulfur, and Arsenic

The presence of phosphorus and sulfur in nerve agents, sulfur mustard, precursors, and degradation products allows the use of GC detectors with selectivity toward these elements. In addition to the use of a homologous *n*-alkane series, LTPRI information relative to a homologous series of alkyl bis(trifluoromethyl)phosphine sulfides (the M-series) has also been tabulated for many CWA-related GC analytes [4,67]. The use of the M-series for LTPRI measurement of CWA analytes supports the use of detectors with element-specific selectivity.

Lakso and Ng used GC with both mass spectrometric and nitrogen–phosphorus detection to detect nerve agents sampled from water by SPME with detection limits of 0.05 μg/L for the G agents and 0.5 μg/L for VX [59]. Flame photometric detection has been used extensively to detect compounds related to both nerve agents and sulfur mustard. Sass and Parker reported the use of flame photometric detection for GC analyses of a number of nerve agents and other organophosphorus compounds in 1980. In that work, they cited an early defense community technical report on the use of this GC detector as early as 1969 [68], only a few years after it was described by Brody and Chaney [69]. A more recent publication details detection of methyl phosphonic acid metabolites for sarin and the ethyl sarin analog in the urine of patients exposed during the 1995 Tokyo subway chemical terrorism incident [41]. Analysis with flame photometric detection followed TMS derivatization of these analytes.

Derivatization of lewisite compounds with thiol reagents conveniently produces analytes that are suitable for analysis using flame photometric detection [56]. While offering excellent sensitivity for both sulfur and phosphorus compounds (better for phosphorus than for sulfur), it is well-known that the flame photometric detector response is not linear for sulfur [69].

Atar et al. described pulsed flame photometric detection in 1991 [70]. This detector improves on the classical flame photometric detector in several ways, primarily through separation of emission information in time with the use of a pulsed flame, as heteroatoms tend to emit following carbon. In a continuous flame detector, coeluting hydrocarbon compounds can lead to quenching of the desired signal derived from sulfur or phosphorus. In addition, the pulsed flame photometric detector uses less hydrogen than a continuous flame detector, a plus for use in a field-portable detection system [71]. Jing and Amirav [72] discussed the ability of the pulsed flame photometric detector to selectively detect a range of heteroatoms (including arsenic) as well as carbon.

27.2.4.3. Atomic Emission Detection

The atomic emission detector has been used for GC analysis of CWA-related compounds on numerous occasions due to its ability to provide information on the empirical formula of an unknown analyte. Since this detector was described in 1989 [73], it has been used repeatedly to assist in the identification of CWA-related compounds separated by GC. In combination with mass spectrometry, Mazurek et al. used atomic emission detection to identify a number of compounds related to the presence of sulfur mustard in an item caught in the nets of fishermen in the Baltic Sea [74].

27.2.4.4. Mass Spectrometric Detection

As in other fields where correct analyte identification is important, the mass spectrometer is commonly acknowledged to be the most useful detector for GC analysis of CWA-related compounds. As the interface of GC with the ion beam quadrupole mass spectrometer was attaining commercial significance in the 1960s, widespread recognition of the need to control environmental pollution was taking hold in the US and other developed nations. In the US, this led to the creation of the Environmental Protection Agency in 1970, and the quadrupole mass filter rapidly became the most important GC detector for applications where both detection and identification of organic pollutants were required. As recounted by Finnigan, "the combination of GC retention time and MS spectrum gave unambiguous proof of the presence of pollutants. Any technique that left ambiguity in the analytical results was likely to lead to continual controversy and litigation" [75].

Heller et al. described the usefulness of the newly commercialized GC-MS systems available at that time: "The identification of pollutants at the part-per billion level with a high degree of confidence in the result has become nearly routine in several EPA laboratories. What was once an impossible task for a staff of

100 working six months sometimes can be accomplished by a skilled individual in a few hours" [76].

Confirmation of analyte identity where OPCW treaty compliance is in question necessarily follows a conservative approach. A positive identification is confirmed with analysis by two independent methods [9]. Often, this may be obtained with mass spectrometric detection using (sequentially) both electron and chemical ionization (EI and CI), requiring pure analytes as optimally provided by the use of a gas chromatographic inlet.

27.2.4.4.1. EI

The majority of GC-MS analyses for CWA-related materials has used an ion beam quadrupole detector, and 70 eV EI conditions. The quadrupole mass filter and EI produce reasonably standard mass spectra that may be compared to large-mass spectral databases. In some cases, EI data alone are inconclusive, e.g. for identification of VX and degradation products of VX having the diisopropylaminoethyl functional group [13]. The presence of this structure typically imparts a base peak at *m/z* 114, and, in the case of VX and related compounds, signal for $M^{+\cdot}$ and other high-mass ions is either completely absent or very weak, resulting in a number of very similar EI mass spectra that may not be easily differentiated either by automated mass spectral searching or by manual examination.

27.2.4.4.2. CI

The use of CI for GC-MS is important to confirm EI results in forensic and OPCW analyses. Sass and Fisher reported the use of EI, as well as methane, isobutane, and ethylene CI reagents for GC-MS detection of nerve agents in 1979 [77]. When this work was carried out, packed GC columns were still commonly used and a GC-MS interface required the diversion of most of the column flow away from the high vacuum region of a mass spectrometric detector. In the cited work, a membrane interface was used to accomplish this, while later work has been predominantly carried out using capillary columns where a direct interface is possible. D'Agostino et al. demonstrated the use of CI detection for GC-MS using a capillary column and ammonia reagent gas to supplement EI data for the successful identification of VX and a number of its degradation products possessing the diisopropylaminoethyl functional group. The use of ammonia reagent provided soft ionization of the targeted amine compounds, producing mass spectra with abundant $[M+H]^+$ pseudomolecular ions and little fragmentation [13]. The high proton affinity of CI reagent ions produced from ammonia relative to ions produced from other typical CI reagent gases provides some selectivity against the ionization of uninteresting analytes such as hydrocarbon compounds that may also be present in a CWA-related sample. Rohrbaugh used methanol as CI reagent for GC-MS analyses of VX and related degradation products and discussed the relative merits of this liquid reagent for use in a field-portable system to avoid the need to transport compressed gas reagents [78]. Methanol CI generally produced more intense signals for $[M+H]^+$ and less fragmentation compared to spectra obtained using methane or ammonia reagent.

27.2.4.4.3. SELF-CI

The phenomenon of self-CI is commonly seen when an ion trap mass spectrometric detector is used and ionization occurs within the trapping region (internal ionization) [79]. At least two such ion trap GC-MS instruments with internal ionization have been commercialized for use in the field, driven in large part by the need for defensive detection of CWA analytes by military forces [35,80]. While ion beam instruments operated with EI at typically low pressures produce unimolecular decomposition, the simultaneous presence of ions and neutral species within an ion trap using internal ionization may lead to an additional dimension of information related to specific ion/molecule chemistry. When analyzed using an internal ionization ion trap GC-MS detector, numerous CWA analytes produce either protonated pseudomolecular ions or protonated dimer ions [14,81]. When this occurs the resulting mass spectra are not directly comparable to those obtained from large-mass spectral databases mostly produced using ion beam instruments. Nevertheless, the addition of ion/molecule interactions to the EI process may be useful in identifying unknown analytes, as long as the basis for the ion/molecule reactions is understood.

The formation of protonated dimers has been observed when ionization occurs at a phosphoryl or carbonyl oxygen atom [81]. Protonation at this location is thought to occur through self-CI interaction between $M^{+\cdot}$ and neutral molecules. This is then followed by reaction of a resulting electrophilic phosphorus or carbon atom with the nucleophilic neutral species [81]. Even when a phosphoryl oxygen atom is present, if a different site on the molecule is more readily ionized (e.g. the diisopropylamino functional group of VX), the formation of a dimer ion is not observed, presumably as the phosphoryl oxygen remains uncharged and thus unactivated for reaction with the neutral molecule. Self-CI protonation at the amine group, without the formation of a dimer may be observed in this situation, and is also possible for amine compounds that lack a phosphoryl oxygen as well [14]. Further work is needed to verify the reactivity of additional functional groups or elements as well as to incorporate this information into automated algorithms for identification of unknown chemicals using this information combined with existing mass spectral libraries.

27.2.4.4.4. TANDEM MASS SPECTROMETRY

Selective detection with tandem mass spectrometry using either a triple quadrupole or an ion trap mass spectrometer is available to many of the chemical defense community laboratories, and has been used for high-certainty detection of targeted compounds present at low levels in matrices with high concentrations of interferents. D'Agostino et al. described early efforts using GC with a highly specialized triple quadrupole mass spectrometer to selectively detect targeted CWA analytes at pg levels in an extract of charcoal that had been used to sample a diesel exhaust environment [82]. The use of ion trap instrumentation allows for similar MS/MS detection with lower overall instrumentation cost compared to the more specialized triple quadrupole detector. Riches et al. described the use of a benchtop ion trap mass spectrometric GC detector operated in the negative ion chemical ionization (NICI) MS/MS mode to detect pentafluorobenzyl derivatives of nerve agent alkyl alkylphosphonic acid metabolites in urine [83]. The primary negative ion from an alkyl alkylphosphonic acid pentafluorobenzyl derivative results from loss of the pentafluorobenzyl group and thus structural information relevant to the remaining alkyl groups is retained. Full scan and selected ion monitoring NICI data provided detection limits in the low ng/mL range, while the use of selected reaction monitoring MS–MS mode improved the sensitivity of the method by about an additional order of magnitude.

27.3. GC APPLICATIONS FOR BIOMEDICAL CWA ANALYSES

In 1994, Black et al. described the use of GC-MS for "the first documented unequivocal identification of nerve agent residues in environmental samples collected after a chemical attack" [84]. In addition to the need for unequivocal detection of CWA-related compounds in environmental matrices, a similar need exists with regard to biological matrices, for both forensic and clinical purposes. Two instances are amply demonstrated in the literature: detection of sulfur mustard hydrolysis products in those reportedly exposed to this CWA material during the Iran–Iraq conflict of the 1980s and detection of hydrolysis products related to the G agent sarin, found in various tissues of individuals exposed to this compound during the Tokyo subway terrorism incident of 1995, as well as the less-well-known 1994 incident in Matsumoto Japan.

In 1984, Wils et al. found thiodiglycol through GC-MS analyses of urine collected from Iranian soldiers allegedly attacked with the CWA sulfur mustard in 1984 during the Iran–Iraq war [85]. However, "thiodiglycol concentrations from 10 to 100 ng/mL in the urine of both the Iranian patients and the controls precluded an unambiguous verification of the use of mustard gas against the Iranian patients" [86]. In 1985, Vycudilik reported the GC-MS detection of sulfur mustard in the urine of two patients one week after they were reportedly exposed to this CWA material in the Iran–Iraq war [87]. In a subsequent paper, GC with high-resolution mass spectrometry was used to again identify this analyte in the urine of six out of twelve patients reporting exposure [88]. However, Vycudilik noted that the methods used did not specifically differentiate between thiodiglycol and the nonhydrolyzed agent, as "this compound is also synthesized *via* a nucleophilic substitution from thiodiglycol and chloride ions in the course of the extraction procedure" [88]. Hard evidence for the use of CWA materials in this conflict was a goal for numerous chemical defense laboratories, and additional work was performed to examine the usefulness of thiodiglycol as a marker for exposure to sulfur mustard. Black and Read noted that "the detection of free sulfur mustard in the body fluids of hospitalized casualties is unlikely, due to its chemical reactivity and extensive metabolism" [89]. These researchers used pentafluorobenzyl chloride derivatization, followed by NICI GC-MS analysis to detect thiodiglycol in spiked blood and urine samples, allowing detection at levels as low as 1 ng/mL. Thiodiglycol was found at concentrations up to 16 ng/mL and <1 ng/mL in the blood and urine, respectively, of healthy nonexposed control subjects, allowing Black and Read to hypothesize that the reported analytical method could be useful to differentiate exposed and nonexposed individuals with the caveat that additional work was needed to carefully examine the incidence and magnitude of endogenously produced thiodiglycol… "clearly a much larger number of control subjects will need to be analysed for thiodiglycol before any firm conclusions can be drawn about endogenous levels" [89].

Minami et al. extracted alkyl methylphosphonic acid metabolites present in the urine of patients exposed to sarin and related impurities in the Tokyo subway incident of 1995 [41]. Ion exchange cleanup was required, and this was followed by TMS derivatization and analysis by GC with flame photometric detection. The time course for the presence of isopropyl methylphosphonic acid in the urine of two exposed patients was followed, demonstrating relatively high concentrations at 12 h following exposure and a rapid decline thereafter.

Nagao et al. found that for four victims they examined from the Tokyo subway incident "postmortem examinations revealed no macroscopic and microscopic findings specific to sarin poisoning and sarin and its hydrolysis products were almost undetectable in their blood" [90]. To provide information of use to future forensic or clinical work, these researchers described the recovery of isopropyl methylphosphonic acid from sarin-bound acetylcholinesterase enzyme present in peripheral blood of the victims. The sarin-bound enzyme was released by trypsin and alkaline phosphatase digestion, and the free acid was then subjected to TMS derivatization for GC-MS analysis [90].

The less-well-known terrorist release of sarin in the Japanese city of Matsumoto caused seven deaths, compared to the 12 deaths attributed to the Tokyo incident the following year. Nakajiima et al. described GC analysis methods that were similar to those reported by Minami et al. [41] to follow the exponential decay of isopropyl methylphosphonic acid and methylphosphonic acid excreted in the urine of a single case [91]. This person lived in a third floor apartment said to be 50 m away from the sarin release point. He recalled "blurriness of vision immediately after opening the window at ~2300 h on the 27th of June, 1994, and then he went to bed. At 0100 h the next day, he was found unconscious by a rescuer team, and was transferred to a hospital" [91]. The victim's total sarin dose was estimated by extrapolating the decay curves obtained for the urinary metabolites, arriving at a value of ~0.05 mg/kg, slightly above the accepted lethal dose for humans. Noting this information and the clinical findings, Nakajiima et al. stated that this victim "...fortunately had a narrow escape from death."

27.4. CONCLUSION

Gas chromatography rapidly became an important method for detection and identification of chemical compounds related to chemical warfare agents in the decade immediately following the initial experiments completed by James and Martin. With the development of the modern fused-silica, open-tubular GC column and the widespread availability of mass spectrometric detectors, GC-based detection approaches have assumed increased importance to the OPCW treaty compliance laboratories, the chemical defense research community, and to technician-level users of field-portable GC instrumentation. The availability of selective detectors well suited to the CWA-related analytes and the ability to analyze many of the intact CWA compounds by GC without derivatization further add to the usefulness of GC for a variety of applications ranging from OPCW treaty compliance verification of process stream or environmental samples, to clinical efforts focused on the protection of human health. Where derivatization is required for GC analysis, substantial well-documented efforts have resulted in sensitive methods that are suitable for use in many circumstances.

While the ability to complete GC-MS analyses using person-portable instrumentation can extend the capabilities of frontline users who need highly definitive answers in high-stakes situations, the supporting systems, methods, and identification algorithms must be improved to fully realize the potential of the faster, smaller, and more capable instruments that continue to be developed to meet this need.

References

[1] A.T. James, A.J.P. Martin, Gas–liquid partition chromatography: the separation and micro-estimation of volatile fatty acids from formic to dodecanoic acid, Biochem. J. 50 (1951) 679–690.

[2] L.S. Ettre, A. Zlatkis, Archer J.P. Martin (Eds.), 75 years of chromatography, a historical dialogue, Elsevier, Amsterdam, 1979, pp. 285–296.

[3] Convention on the prohibition of the development, production, stockpiling and use of chemical weapons and on their destruction, Technical Secretariat of the Organisation for Prohibition of Chemical Weapons, The Hague, 1997.

[4] O. Kostiainen, Gas chromatography in screening of chemicals related to the chemical weapons convention, in: M. Mesilaakso (Ed.), Chemical weapons convention chemical analysis, sample collection, preparation, and analytical methods, John Wiley and Sons, Chichester, UK, 2005.

[5] P.W. Albro, L. Fishbein, Gas chromatography of sulfur mustard and its analogs, J Chromatogr 46 (1970) 202–203.

[6] R.L. Erickson, R.N. Macnair, R.H. Brown, H.D. Hogan, Determination of bis(2-chloroethyl) sulfide in a Dawson apparatus by gas chromatography, Anal. Chem. 44 (1972) 1040–1041.

[7] K.M. Sloan, R.V. Mustacich, B.A. Eckenrode, Development and evaluation of a low thermal mass gas chromatograph for rapid forensic GC-MS analyses, Field Anal. Chem. Technol. 5 (2001) 288–301.

[8] P.A. Smith, M.T. Sng, B.A. Eckenrode, S.Y. Leow, D. Koch, R.P. Erickson, et al., Towards smaller and faster gas chromatography–mass spectrometry systems for field chemical detection, J. Chromatogr. A 1067 (2005) 285–294.

[9] E.R.J. Wils, Gas chromatography/mass spectrometry in analysis of chemicals related to the chemical weapons convention, in: M. Mesilaakso (Ed.), Chemical weapons convention chemical analysis, sample collection, preparation, and analytical methods, John Wiley and Sons, Chichester, UK, 2005.

[10] M. Sokolowski, The OPCW gas chromatograph/mass spectrometer for on-site analysis. Instrumentation, AMDIS software, and preparations for use, in: M. Mesilaakso (Ed.), Chemical weapons convention chemical analysis, sample collection, preparation, and analytical methods, John Wiley and Sons, Chichester, UK, 2005.

[11] L. Szinicz, History of chemical and biological warfare agents, Toxicology 214 (2005) 167–181.

[12] R.W. Baier, S.W. Weller, Catalytic and thermal decomposition of isopropyl methyl fluorophosphonate, Ind. Eng. Chem. Process Des. Dev. 6 (1967) 380–385.

[13] P.A. D'Agostino, L.R. Provost, J. Visentini, Analysis of *O*-ethyl S-[2-(diisopropylamino)ethyl] methylphosphonothiolate (VX) by capillary column gas chromatography-mass spectrometry, J. Chromatogr. 402 (1987) 221–232.

[14] P.A. Smith, C.R. Jackson Lepage, B. Savage, C.R. Bowerbank, E.D. Lee, M.J. Lukacs, Use of a Hand-Portable Gas Chromatograph-toroidal ion trap mass spectrometer for self-CI identification of degradation products related to *O*-ethyl *S*-2-diisopropylaminoethyl methyl phosphonothiolate (VX), Anal. Chim. Acta 690 (2011) 215–220.

[15] B. Muir, S. Quick, B.J. Slater, D.B. Cooper, M.C. Moran, C.M. Timperly, et al., Analysis of chemical warfare agents II. Use of thiols and statistical experimental design for the trace level determination of vesicant compounds in air samples, J. Chromatogr. A 1068 (2005) 315–326.

[16] R.T. Rosen, J.D. Rosen, Presence of four Fusarium mycotoxins and synthetic material in "yellow rain", Biomed. Mass Spectrom. 9 (1982) 443–450.

[17] C.J. Mirocha, R.A. Pawlosky, K. Chatterjee, S. Watson, W. Hayes, Analysis for *Fusarium* toxins implicated in biological warfare in Southeast Asia, J. Assoc. Off. Anal. Chem. 66 (1983) 1485–1499.

[18] S.A. Watson, C.J. Mirocha, A.W. Hayes, Analysis for trichothecenes in samples from Southeast Asia associated with "yellow rain." Fund. Appl. Toxicol. 4 (1984) 700–717.

[19] J.W. Nowicke, M. Messelson, Yellow rain – a palynological analysis, Nature 309 (1984) 205–206.

[20] C.O. Ikediobi, I.C. Hsu, J.R. Bamburg, F.M. Strong, Gas-liquid chromatography of mycotoxins of the trichothecene group, Anal. Biochem. 43 (1971) 327–340.

[21] P.A. D'Agostino, L.R. Provost, D.R. Drover, Analysis of trichothecene mycotoxins in human blood by capillary column gas chromatography-ammonia chemical ionization mass spectrometry, J. Chromatogr. 367 (1986) 77–86.

[22] P. Begley, B.E. Foulger, P.D. Jeffery, R.M. Black, R.W. Read, Detection of trace levels of trichothecenes in human blood using capillary gas chromatography-electron-capture negative chemical ionization mass spectrometry, J. Chromatogr. 367 (1986) 87–101.

[23] P.K. Lee, S.Y.K. Kee, W. Ng, P. Gopalakrishnakone, Determination of trichothecene toxin (T2 mycotoxin) in aqueous sample with solid phase microextraction technique followed by gas chromatography with flame ionization detection, J. High Resol. Chromatogr. 22 (1999) 424–426.

[24] F.W. Beswick, Chemical agents used in riot control and warfare, Human Toxicol. 2 (1983) 247–256.

[25] E.R.J. Wils, A.G. Hulst, Mass spectra of some derivatives of the irritant o-chlorobenzylidenemalononitrile (CS), Fresenius Z Anal. Chem. 320 (1985) 357–360.

[26] T.A. Kluchinsky Jr., M.V. Sheely, P.B. Savage, P.A. Smith, Formation of 2-chlorobenzylidenemalononitrile (CS riot control agent) thermal degradation products at elevated temperatures, J. Chromatogr. A 952 (2002) 205–213.

[27] T.A. Kluchinsky Jr., P.B. Savage, M.V. Sheely, R.J. Thomas, P.A. Smith, Identification of CS-derived compounds formed during heat-dispersion of CS riot control agent, J. Microcolumn Sep. 13 (2001) 186–190.

[28] P.A. Smith, M.V. Sheely, T.A. Kluchinsky Jr., Solid phase microextraction with analysis by gas chromatography to determine short term hydrogen cyanide concentrations in a field setting, J. Sep. Sci. 25 (2002) 917–921.

[29] N.B. Munro, S.S. Talmage, G.D. Griffin, L.C. Waters, A.P. Watson, J.F. King, et al., The sources, fate, and toxicity of chemical warfare agent degradation products, Env. Health Perspect 107 (1999) 933–974.

[30] P.A. D'Agostino, L.R. Provost, Capillary column electron impact and ammonia chemical ionization gas chromatographic-mass spectrometric and gas chromatographic-tandem mass spectrometric analysis of mustard hydrolysis products, J. Chromatogr. 645 (1993) 283–292.

[31] B.A. Eckenrode, Environmental and forensic applications of field-portable GC-MS: an overview, J. Am. Soc. Mass Spectrom. 12 (2001) 683–693.

[32] E.U. Ehrmann, H.P. Dharmasena, K. Carney, E.B. Overton, Novel column heater for fast capillary gas chromatography, J. Chromatogr. Sci. 34 (1996) 533–539.
[33] J. Padayhag, Vapor validation of monitoring systems for detection of trace levels of chemical warfare agents in air, in: V.M., k Kolodkin (Ed.), NATO security through science series, ecological risks associated with the destruction of chemical weapons, Springer, The Netherlands, 2006.
[34] P. Smith, C. Jackson Lepage, D. Koch, H. Wyatt, B. Eckenrode, G. Hook, et al., Detection of gas phase chemical warfare agents using field-portable gas chromatography-mass spectrometry systems: instrument and sampling strategy considerations, Trends Anal. Chem. 23 (2004) 296–306.
[35] J.A. Contreras, J.A. Murray, S.E. Tolley, J.L. Oliphant, H.D. Tolley, S.A. Lammert, et al., Hand-portable gas chromatograph-toroidal ion trap mass spectrometer (GC-TMS) for detection of hazardous compounds, J. Am. Soc. Mass Spectrom. 19 (2008) 1425–1434.
[36] G.L. Hook, G. Kimm, G. Betsinger, P.B. Savage, A. Swift, T. Logan, et al., Solid phase microextraction sampling and gas chromatography/mass spectrometry for field detection of the chemical warfare agent O-ethyl S-(2-diisopropylaminoethyl) methylphosphonothiolate (VX), J. Sep. Sci. 26 (2003) 1091–1096.
[37] W.K. Fowler, J.E. Smith, Indirect determination of O-ethyl S-(2-diisopropylaminoethyl) methylphosphonothioate in air at low concentrations, J. Chromatogr. 478 (1989) 51–61.
[38] R.M. Black, B. Muir, Derivatisation reactions in the chromatographic analysis of chemical warfare agents and their degradation products, J. Chromatogr. A 1000 (2003) 253–281.
[39] J.G. Purdon, J.G. Pagotto, R.K. Miller, Preparation, stability and quantitative analysis by gas chromatography and gas chromatography -electron impact mass spectrometry of *tert*.-butyldimethylsilyl derivatives of some alkylphosphonic and alkyl methylphoshonic acids, J. Chromatogr. 475 (1989) 261–272.
[40] M.-L. Kuitunen, Sample preparation for analysis of chemicals related to the chemical weapons convention in an off-site laboratory, in: M. Mesilaakso (Ed.), Chemical weapons convention chemical analysis, sample collection, preparation, and analytical methods, John Wiley and Sons, Chichester, UK, 2005.
[41] M. Minami, D.-M. Hui, M. Katsumata, H. Inagaki, C.A. Boulet, Method for the analysis of the methylphosphonic acid metabolites of sarin and its ethanol-substituted analogue in urine as applied to the victims of the Tokyo sarin disaster, J. Chromatogr. B 695 (1997) 237–244.
[42] G.L. Hook, G. Kimm, D. Koch, P.B. Savage, B. Ding, P.A. Smith, Detection of VX in soil through solid-phase microextraction sampling and gas chromatography/mass spectrometry of the VX degradation product bis(diisopropylaminoethyl)disulfide, J. Chromatogr. A 992 (2003) 1–9.
[43] D. Pardasani, A. Purohit, A. Mazumder, D.K. Dubey, Gas chromatography-mass spectrometric analysis of toxic hydrolyzed products of nerve agent VX and its analogues for verification of chemical weapons convention, Anal. Methods 2 (2010) 661–667.
[44] W.R. Creasy, A.A. Rodriguez, J.R. Stuff, R.W. Warren, Atomic emission detection for the quantitation of trimethylsilyl derivatives of chemical-warfare-agent related compounds in environmental samples, J. Chromatogr. A 709 (1995) 333–344.
[45] W.R. Creasy, J.R. Stuff, B. Williams, K. Morrissey, J. Mays, R. Duevel, et al., Identification of chemical-weapons-related compounds in decontamination solutions and other matrices by multiple chromatographic techniques, J. Chromatogr. A 774 (1997) 253–263.
[46] E.R.J. Wils, A.G. Hulst, Mass spectra of some derivatives of 2,2′-dichlorodiethyl sulphide (mustard gas), Fresenius Z Anal. Chem. 321 (1985) 471–474.
[47] W.K. Fowler, D.C. Stewart, D.S. Weinberg, E.W. Sarver, Gas chromatographic determination of the lewisite hydrolysate, 2-chlorovinylarsonous acid, after derivatization with 1,2-ethanedithiol, J. Chromatogr. 558 (1991) 235–246.
[48] S. Hanaoka, K. Nomura, T. Wada, Determination of mustard and lewisite related compounds in abandoned chemical weapons (Yellow shells) from sources in China and Japan, J. Chromatogr. A 1101 (2006) 268–277.
[49] W.K. Fowler, C.H. Duffey, H.C. Miller, Modification of a gas chromatographic inlet for thermal desorption of adsorbent-filled sampling tubes, Anal. Chem. 51 (1979) 2333–2336.
[50] J. Steinhanses, K. Schoene, Thermal desorption-gas chromatography of some organophosphates and S-mustard after trapping on Tenax, J. Chromatogr. 514 (1990) 273–278.
[51] R.M. Black, R. Claarke, D.B. Cooper, R.W. Read, D. Utley, Application of headspace analysis, solvent extraction, thermal desorption and gas chromatography-mass spectrometry to the analysis of chemical warfare samples containing sulphur mustard and related compounds, J. Chromatogr. 673 (1993) 71–80.
[52] J.R. Hancock, G.W. Peters, Retention index monitoring of compounds of chemical defence interest using thermal desorption gas chromatography, J. Chromatogr. 538 (1991) 249–257.

[53] C.L. Arthur, J. Pawliszyn, Solid phase microextraction with thermal desorption using fused silica optical fibers, Anal. Chem. 62 (1990) 2145–2148.

[54] B. Zygmunt, A. Zaborowska, J. Świattlowska, J. Namieśnik, Solid phase microextraction combined with gas chromatography - a powerful tool for the determination of chemical warfare agents and related compounds, Current Org. Chem. 11 (2007) 241–253.

[55] G.L. Kimm, G.L. Hook, P.A. Smith, Application of headspace solid-phase microextraction and gas chromatography-mass spectrometry for detection of the chemical warfare agent bis(2-chloroethyl) sulfide in soil, J. Chromatogr. A 971 (2002) 185–191.

[56] B.A. Tomkins, G.A. Sega, C.-H. Ho, Determination of lewisite oxide in soil using solid phase microextraction followed by gas chromatography with flame photometric or mass spectrometric detection, J. Chromatogr. A 909 (2001) 13–28.

[57] B. Szostek, J.H. Aldstadt, Determination of organoarsenicals in the environment by solid-phase microextraction-gas chromatography-mass spectrometry, J. Chromatogr. A 807 (1998) 253–263.

[58] J.-A.M. Creek, A.M. McAnoy, C.S. Brinkworth, Rapid monitoring of sulfur mustard degradation in solution by headspace solid-phase microextraction and gas chromatography mass spectrometry, Rapid Comm. Mass Spectrom. 24 (2010) 3419–3424.

[59] H.A. Lakso, W.F. Ng, Determination of chemical warfare agents in natural water samples by solid-phase microextraction, Anal. Chem. 69 (1997) 1866–1872.

[60] J.F. Schneider, A.S. Boparai, L.L. Reed, Screening for sarin in air and water by solid-phase microextraction-gas chromatography-mass spectrometry, J. Chromatogr. Sci. 39 (2001) 420–424.

[61] P.K. Lee, S.Y.K. Kee, W. Ng, P. Gopalakrishnakone, Determination of trichothecene toxin (T2 mycotoxin) in aqueous sample with solid phase microextraction technique followed by gas chromatography with flame ionization detection, J. High Res. Chromatogr. 22 (1999) 424–426.

[62] M.T. Sng, W.F. Ng, In-situ derivatisation of degradation products of chemical warfare agents in water by solid-phase microectraction and gas chromatographic-mass spectrometric analysis, J. Chromatogr. A 832 (1999) 173–182.

[63] G.L. Hook, C.J. Lepage, S.I. Miller, P.A. Smith, Dynamic solid phase microextraction for sampling of airborne sarin with gas chromatography-mass spectrometry for rapid field detection and quantification, J. Sep. Sci. 27 (2004) 1017–1022.

[64] J. Koziel, M. Jia, J. Pawliszyn, Air sampling with porous solid-phase microextraction fibers, Anal. Chem. 72 (2000) 5178–5186.

[65] H. van Den Dool, P.D. Kratz, A generalization of the retention index system including linear temperature programmed gas-liquid partition chromatography, J. Chromatogr. 11 (1963) 463–471.

[66] P.A. D'Agostino, L.R. Provost, Gas chromatographic retention indices of chemical warfare agents and simulants, J. Chromatogr. 331 (1985) 47–54.

[67] A. Manninen, M-L. Kuitunen, L. Julin. Gas chromatographic properties of the M-series of universal retention index standards and their application to pesticide analysis. J. Chromatogr. 394 (1987) 465–471.

[68] S. Sass, R.J. Steger, Gas chromatographic differentiation and estimation of some sulfur and nitrogen mustards using a multidetector technique, J. Chromatogr. 238 (1982) 121–132.

[69] S.S. Brody, J.E. Chaney, Flame photometric detector. The application of a specific detector for phosphorous and for sulfur compounds-sensitive to subnanogram quantities, J. Gas Chromatogr. 4 (1966) 42–46.

[70] E. Atar, S. Cheskis, A. Amirav, Pulsed flame - a novel concept for molecular detection, Anal. Chem. 63 (1991) 2064–2068.

[71] G. Frishman, A. Amirav, Fast GC-PFPD system for field analysis of chemical warfare agents, Field Anal. Chem. Technol. 4 (2000) 170–194.

[72] H. Jing, A. Amirav, Pulsed flame photometric detector - a step forward towards universal heteroatom selective detection, J. Chromatogr. A 805 (1998) 177–215.

[73] J.J. Sullivan, B.D. Quimby, Detection of C, H, N, and O in capillary gas chromatography by atomic emission, J. High Res. Chromatogr. 12 (1989) 282–286.

[74] M. Mazurek, Z. Witkiewicz, S. Popiel, M. Śliwakowski, Capillary gas chromatography-atomic emission spectroscopy-mass spectrometry analysis of sulphur mustard and transformation products in a block recovered from the Baltic Sea, J. Chromatogr. A 919 (2001) 133–145.

[75] R.E. Finnigan, Quadrupole mass spectrometers, from development to commercialization, Anal. Chem. 66 (1994) 969A–975A.

[76] S.R. Heller, J.M. McGuire, W.L. W.L. Budde, Trace organics by GC/MS, Env. Sci. Technol. 9 (1975) 210–213.

[77] S. Sass, T.L. Fisher, Chemical ionization and electron impact mass spectrometry of some organophosphate compounds, Org. Mass Spectrom. 14 (1979) 257–264.

[78] D.K. Rohrbaugh, Methanol chemical ionization quadrupole ion trap mass spectrometry of *O*-ethyl *S*-[2-(diisopropylamino)ethyl]

methylphosphonothiolate (VX) and its degradation products, J. Chromatogr. A 893 (2000) 393–400.

[79] S.A. McLuckey, G.L. Glish, K.G. Asano, G.J. Van Berkel, Self chemical ionization in an ion trap mass spectrometer, Anal. Chem. 60 (1988) 2312–2314.

[80] G.E. Patterson, A.J. Guymon, L.S. Riter, M. Everly, J. Griep-Raming, B.C. Laughlin, et al., Miniature cylindrical ion trap mass spectrometer, Anal. Chem. 74 (2002) 6145–6153.

[81] P.A. Smith, C. Jackson Lepage, M. Lukacs, N. Martin, A. Shufutinsky, P.B. Savage, Field-portable gas chromatography with transmission quadrupole and cylindrical ion trap mass spectrometric detection: chromatographic retention index data and ion/molecule interactions for chemical warfare agent identification, Int. J. Mass Spectrom. (2010).

[82] P.A. D'Agostino, L.R. Provost, J.F. Anacleto, P.W. Brooks, Capillary column gas chromatography–tandem mass spectrometry detection of chemical warfare agents in a complex airborne matrix, J. Chromatogr. 504 (1990) 259–268.

[83] J. Riches, I. Morton, R.W. Read, R.M. Black, The trace analysis of alkyl alkylphosphonic acids in urine using gas chromatography-ion trap negative ion tandem mass spectrometry, J. Chromatogr. B 816 (2005) 251–258.

[84] R.M. Black, R.J. Clarke, R.W. Read, M.T.J. Reid, Application of gas chromatography–mass spectrometry and gas chromatography–tandem mass spectrometry to the analysis of chemical warfare samples, found to contain residues of the nerve agent sarin, sulphur mustard and their degradation products, J. Chromatogr. A 662 (1994) 301–321.

[85] E.R.J. Wils, A.G. Hulst, A.L. de Jong, A. Verweij, H.L. Boter, Analysis of thiodiglycol in urine of victims of an alleged attack with mustard gas, J. Anal. Toxicol. 9 (1985) 254–257.

[86] E.R.J. Wils, A.G. Hulst, J. van Laar, Analysis of thiodiglycol in urine of victims of an alleged attack with mustard gas, part II, J. Anal. Toxicol. 12 (1988) 15–19.

[87] W. Vycudilik, Detection of mustard gas bis(2-chloroethyl)-sulfide in urine, Forensic Sci. Int. 28 (1985) 131–136.

[88] W. Vycudilik, Detection of bis(2-chlorethyl)-sulfide (Yperite) in urine by high resolution gas chromatography-mass spectrometry, Forensic Sci. Int. 35 (1987) 67–71.

[89] R.M. Black, R.W. Read, Detection of trace levels of thiodiglycol in blood, plasma and urine using gas chromatography–electron-capture negative-ion chemical ionisation mass spectrometry, J. Chromatogr. 449 (1988) 261–270.

[90] M. Nagao, T. Takatori, Y. Matsuda, M. Nakajima, H. Iwase, K. Iwadate, Definitive evidence for the acute sarin poisoning diagnosis in the Tokyo subway, Toxicol. Appl. Pharmacol. 144 (1997) 198–203.

[91] T. Nakajiima, K. Sasaki, H. Ozawa, Y. Sekijima, H. Morita, Y. Fukushima, et al., Urinary metabolites of sarin in a patient of the Matsumoto sarin incident, Arc. Toxicol. 72 (1998) 601–603.

CHAPTER

28

Emerging and Persistent Environmental Compound Analysis

Frank L. Dorman, Eric J. Reiner

OUTLINE

28.1. INTRODUCTION

The discovery of gas chromatography in the early 1950s [6] was a very important development as it enabled the separation and detection of the many components present in environmental samples. Early detectors such as the thermal conductivity detector were not very sensitive or selective. The electron-capture detector (ECD), very sensitive and selective to organohalogen compounds, and the mass spectrometer, a universal detector, were both first used in the late 1950s. The ECD [7] is still used in many applications, but the development of cheaper, more rugged, and user-friendly mass spectrometers has made them the GC detector of choice for most applications [8]. Packed GC columns were used from the

Gas Chromatography DOI: 10.1016/B978-0-12-385540-4.00028-6

1950s to the 1970s. For these types of columns, the stationary phase was coated onto a support material that was packed in a column (see Chapter 4). The separating power was quite poor as compared to today's standards with peak capacities of less than 10. The wall-coated open-tubular column (WCOT) or capillary column was developed in the early 1960s, but not commonly used until the late 1970 or early 1980s following the development of fused silica which enabled long columns (30 m and longer) to be wound on a support cage that could be placed in an oven [9,10]. The separating power of capillary columns is significantly greater than packed columns with peak capacities ranging from 50 to 100 (see chapter 3). Standard configurations are 0.25 mm id, 0.25 um d_{f}., and are used for most applications. By decreasing the inner diameter and reducing the film thickness while keeping the phase ratio (inner diameter to film thickness) constant, the relative retention times remain the same. Narrower bore columns have a greater number of theoretical plates per meter enabling shorter columns to be used resulting in shorter analysis times. The GC peaks become taller and narrower which increases sample detectability. This technique is termed "fast GC" and chromatographic run times can be reduced by up to 80%. The main challenge with the narrow peaks produced using fast GC is to obtain the necessary 7–10 sampling points across a GC peak to ensure accurate determination of peak area. Comprehensive two-dimensional gas chromatography (GC × GC) [11,12] is a relatively new technique where two GC columns of different phases are linked using a device called a modulator (see chapter 7). The modulator traps the eluent from the first (dimension) column and reinjects it into the second (dimension) column. The peaks eluting from the second column are very narrow (a few hundred milliseconds wide) resulting in significant improvement in separations. A schematic of a GC × GC system is shown in Figure 28.1. Peak capacities of GC × GC analyses can be over 1000.

Although many thousands of chemicals are used, only a small number are regulated on an international scale. The Stockholm Agreement [13], first ratified in 2001, included only 12 halogenated compounds or compounds groups (see Table 28.1). An additional nine were added in 2009 and three more are currently under review. All of the original and recently added compounds are persistent, bioaccumulative, and toxic organic compounds. To analyze these compounds, samples must be quantitatively extracted and the extract must be simplified to remove matrix coextractables so that a portion of the cleaned extract can be injected into the analytical instrument without affecting or biasing the results or damaging the instrument. Compounds such as dioxin, polychlorinated biphenyls (PCBs), PCNs, polychlorinated diphenylethers (PCDEs), or polybrominated diphenylethers (PBDEs) are multicomponent mixtures comprised of congeners (a series of related compounds where hydrogen atoms are serially replaced by chlorine or bromine atoms – see Figure 28.2). The toxicity of the specific congeners can vary significantly [14]. For example, the toxicity of the two major dioxins present in Agent Orange, 1,3,6,8-tetrachlorodibenzo-p-dioxin (TCDD) and 2,3,7,8-TCDD, can vary by 6 orders of magnitude, with the 2,3,7,8 congener being the most toxic of all dioxin congeners. Mass spectrometry is considered the most selective of all detectors. A major drawback in the analysis of isomers (and congeners if they fragment to common ions) is that they cannot be distinguished if their mass spectra are identical or very similar. This is frequently the case with the persistent environmental pollutants described in this chapter and therefore requires a separation technique such as gas chromatography to be used in conjunction with MS detection. In general, the combination of gas chromatography and mass spectrometry is considered the most sensitive

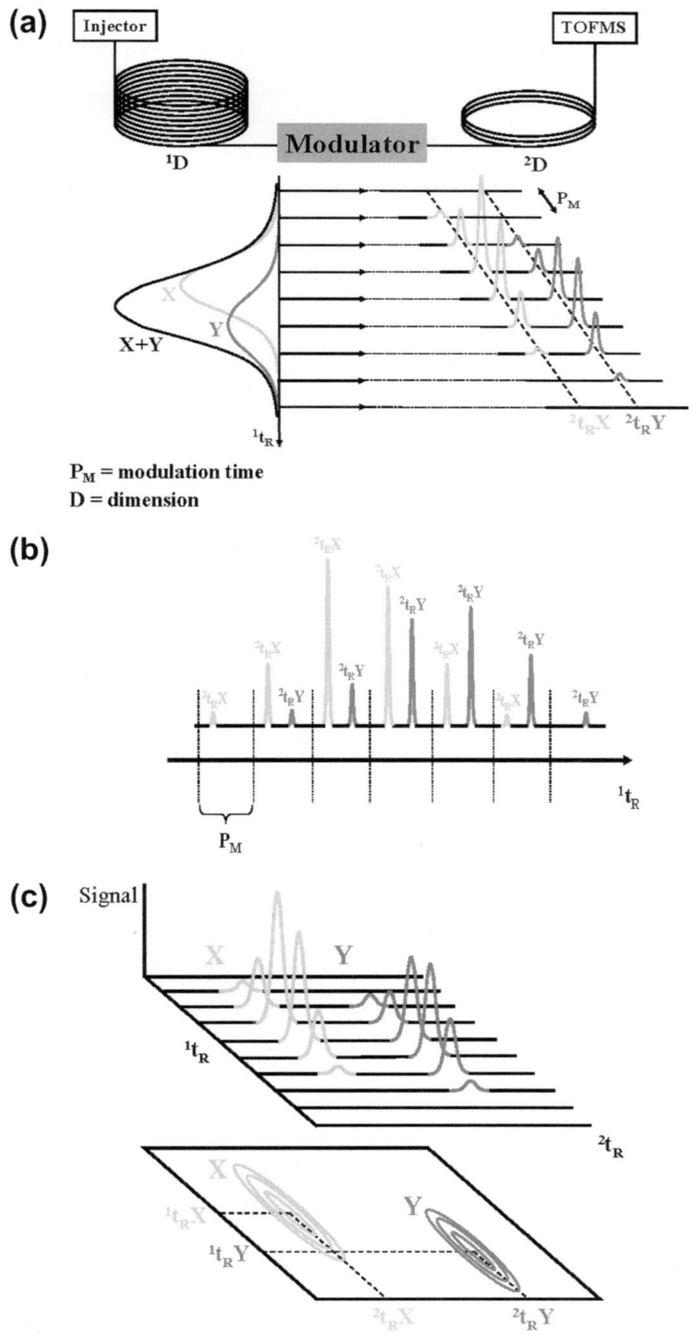

FIGURE 28.1 Schematic of the GC × GC-TOFMS apparatus and how chromatograms are produced. (a) The modulator allows sampling of the analytes eluting from the first-dimension GC (1D) and reinjects them into the second-dimension column (2D). The modulation process is illustrated for two coeluting compounds in 1D (X and Y) retention time ^1t_R in the first dimension. As the modulation process occurs during the modulation period P_m, narrow bands of analytes enter 2D and elute with different second-dimension retention times $^2t_R(X)$ and $^2t_R(Y)$. (b) Raw data signals are recorded by the TOFMS throughout the entire separation process. (c) Construction of a two-dimensional contour plot from the high-speed secondary chromatograms obtained in (b), in which similar signal intensities are connected by the contour lines. *Source: Modified from Focant et al.,* Talanta *63, 1231 (2004).*

TABLE 28.1 Stockholm Convention POPs

Category	Compounds
Pesticides	• Aldrin • Chlordane • Dieldrin • DDT • Endrin • Heptachlor • Hexachlorobenzene (HCB) • Mirex • Toxaphene
Industrial chemicals	• Polychlorinated Biphenyls (PCBs)
Unintentional production	• Polychlorinated Dibenzodioxins (PCDDs) and Dibenzofurans (PCDF) • PCBs • HCB
Added, May 2009	• Chlordecone • α-hexachlorocyclohexane • β-hexachlorocyclohexane • Hexabromobiphenyl • Hexabromodiphenyl ether and heptabromodiphenyl ether • Lindane (gamma-hexachlorocyclohexane) • Pentachlorobenzene • Perfluorooctanesulfonic acid (PFOS), its salts and perfluorooctanesulfonyl fluoride • Tetrabromodiphenyl ether and pentabromodiphenyl ether
Compounds under review nominated for addition, Oct 2009	• Short-chain chlorinated paraffin (SCCPs) • Endosulfan • Hexabromocyclododecane (HBCD)

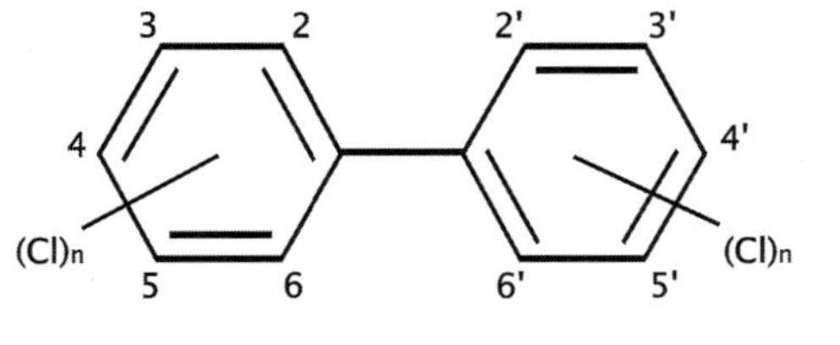

FIGURE 28.2 Structure and naming for PCBs.

and selective analytical technique for these types of compounds.

28.2. POLYCHLORINATED BIPHENYLS

Polychlorinated biphenyls (PCBs) were first synthesized in 1881. Commercial production began in 1929, by the Anniston Ordnance Company (Anniston, Alabama) whose name was later changed to the Theodore Swann Company after the founder Theodore Swann. In 1935, this facility was purchased by Monsanto who became one of the largest manufacturers of PCBs with production levels peaking in 1970. Manufacturing continued until 1977 when production was halted. During this time, estimates are that over 1 million tons of PCBs were produced, a considerable amount of which unfortunately has entered the environment through a variety of pathways [15]. PCBs had a wide variety of potential uses, including extenders in insecticide production, paint ingredients, insulators for electrical devices such as transformers and capacitors, and heat exchange fluids [16].

PCBs are a family of chlorinated organic compounds which consist of two benzene rings linked by a carbon–carbon bond. Chlorine is substituted on the two rings from 1 to 10 available positions (Figure 28.2), which accounts for 209 possible congeners in the family of PCBs. A PCB is then commonly referred to either by its specific congener structure (e.g. 2,2′,4,4′ tetrachlorobiphenyl) as in the case of a specific compound or by the commercial product itself.

Since individual congeners were not manufactured and isolated in the commercial process, many PCB analytical methods are based around the identification and quantification of the commercial mixture. Commercial mixtures were sold under a variety of trade names, as listed in Table 28.2. Among these commercial names, "Aroclor" is one of the most prevalent, and, within each family of commercial mixtures, there are typically several different products [17]. These products were developed for a variety of uses and range both in overall distribution of congeners and in the degree of chlorination. For the Aroclor family, each mixture is given a numerical identifier that denotes its chlorination range. Specifically the last two digits of the Aroclor number denote the percent chlorine substitution by weight (except for Aroclor 1016). Table 28.3 lists the common Aroclor mixtures.

28.2.1. Analytical Considerations

When developing an analytical method for PCBs it is important to first determine the reason for performing the work: identification and quantification of the specific congeners, or determination of the commercial mixture. Since PCBs were used as commercial mixtures, not as specific congeners, many methods have been focused on the identification of the commercial mixture or mixtures present in a sample, followed by their quantification [18]. Methods employing this approach use commercial mixtures as calibration standards and typically identify several of the highest responding components which are characteristic for each of the mixtures. Several difficulties arise from these so-called "Aroclor" methods. Most notably, weathering of the PCB mixture since its introduction into the environment can cause significant distortion of the pattern of the individual congeners

TABLE 28.2 PCB Commercial Mixtures and Primary Country of Use

APIROLIO	(Italy)
AROCLOR	(U.K., U.S.A.)
ASBESTOL	(U.S.A)
ASKAREL	(U.K., U.S.A.)
BAKOLA 131	(U.S.A.)
CHLOREXTOL	(U.S.A.)
CLOPHEN	(Germany)
DELOR	(Czechoslovakia)
DK	(Italy)
DIACLOR	(U.S.A.)
DYKANOL	(U.S.A.)
ELEMEX	(U.S.A.)
FENCLOR	(Italy)
HYDOL	(U.S.A.)
INTERTEEN	(U.S.A.)
KANECLOR	(Japan)
NOFLAMOL	(U.S.A.)
PHENOCLOR	(France)
PYRALENE	(France)
PYRANOL	(U.S.A.)
PYROCLOR	(U.K.)
SAFT-KUHL	(U.S.A.)
SOVOL	(U.S.S.R.)
SOVTOL	(U.S.S.R.)

Adapted from Ref. [15].

TABLE 28.3 Common Aroclor Formulations

CAS #	Formulation	Approximate weight % Cl
12674-11-2	Aroclor 1016	40
11104-28-2	Aroclor 1221	21
11141-16-5	Aroclor 1232	32
53469-21-9	Aroclor 1242	42
12672-29-6	Aroclor 1248	48
11097-69-1	Aroclor 1254	54
11096-82-5	Aroclor 1260	60
37324-23-5	Aroclor 1262	62
11100-14-4	Aroclor 1268	68

Table adapted from Ref. [17].

relative to the original commercial mixture. It is easily possible to determine the most significant congeners in the original Aroclor from the calibration standards, and a calibration curve can be developed for each of these. If the pattern or distribution of these congeners changes as a result of weathering of the samples, however, the Aroclor may be misidentified or incorrectly quantified.

The individual identification and quantification of the PCB congeners is a possible solution to issues of weathering; however, this approach is arguably more difficult. Since there are 209 possible congeners, all of which are chemically similar, the complete separation of these has been a goal of the analytical community for quite some time. Even though there are some very sophisticated methods for analysis, to date, nobody has separated all 209 PCB congeners in a single separation, even when employing dual-column separations and/or mass spectrometric detection. One significant advantage of the congener-based analysis is that the individual PCBs range considerably in toxicological effect. Many of the PCBs are considered relatively nontoxic, but some are of considerable concern. PCB congeners without chlorine substitution at the ortho-positions on the two rings are able, via rotation, to align in a planar geometry. This causes these specific congeners to adopt a structure more similar to the toxic dioxin and furan congeners, discussed in another section. These congeners may be

TABLE 28.4 PCB Congeners with TEF Values as Assigned by the World Health Organization (WHO) [19]

Non-ortho-PCB congener	IUPAC name	WHO-TEF
PCB77	3,4,4′,5-TeCB	0.0001
PCB 81	3,3′,4,4′-TeCB	0.0003
PCB 126	3,3′,4,4′,5-PeCB	0.1
PCB 169	3,3′,4,4′,5,5′-HxCB	0.03
MONO-ORTHO PCB CONGENER		
PCB 105	2′,3,4,4′,5-PeCB	0.00003
PCB 114	2,3′,4,4′,5-PeCB	0.00003
PCB 118	2,3,4,4′,5-PeCB	0.00003
PCB 123	2,3,3′,4,4′-PeCB	0.00003
PCB 156	2,3,3′,4,4′,5′-HxCB	0.00003
PCB 157	2,3′,4,4′,5,5′-HxCB	0.00003
PCB 167	2,3,3′,4,4′,5-HxCB	0.00003
PCB 189	2,3,3′,4,4′,5,5′-HpCB	0.00003

termed "dioxin-like" or "coplanar" congeners. A detailed congener analysis is, therefore, the method of choice if the goal of the analysis is to measure PCBs based on their toxic qualities to arrive at a toxic-equivalent (TEQ) value. There are generally 12 PCB congeners that are considered to have these properties as listed in Table 28.4.

28.2.2. Analysis of PCBs Based on Commercial Mixtures

Each of the commercial mixtures (Aroclors, Kaneclors, etc.) contains a distribution of individual congeners. Several studies have been conducted to determine the individual congener composition of these mixtures using either GC-ECD or GC-MS. Current regulatory methods simplify the analytical methodology either by specifying quantification as a combination of one or more commercial-product congener distributions or by specifying short lists of target congeners for individual quantification. There have been studies to detail the congener composition of the various commercial mixtures so that, in theory, identification of the actual commercial mixture would not be necessary [20]. Even so, "Aroclor" methods still abound. These methods rely on the analysis of chemical reference materials which are the commercial mixtures. These standards are analyzed over a range of concentrations with enough chromatographic resolution to allow for the identification of the most significant congeners in the distribution.

For calibration, standards of the possible commercial mixtures are analyzed to allow for pattern identification as shown in Figure 28.3. Aroclors identified in the samples are then quantified against a multipoint calibration curve for each Aroclor based on the 3–5 of the most significant congeners for each. Each congener chosen must be characteristic for each Aroclor; however, in general, the later-eluting congeners are subject to less weathering or breakdown.

Unknowns must first then be qualitatively identified as to which commercial mixture or mixtures represent the PCB distribution in the sample. Provided this is successful, the same significant congeners are used to provide 3–5 separate estimates of concentration through the use of the previously mentioned calibrations. The lowest value (assumed to have the least bias from possible interference) is then used for final quantification of the sample. Many methods employ some sort of QA/QC criterion in terms of the variance of the quantified values. Quantification is considered to be more valid if the 3–5 chosen congeners all yield similar results for a sample.

Methods employing ECD detection are generally subject to a second-column conformational analysis. This may be done at the time of initial sample analysis through the use of a dual-column

Aroclor PCB's
Rtx-CLPesticides

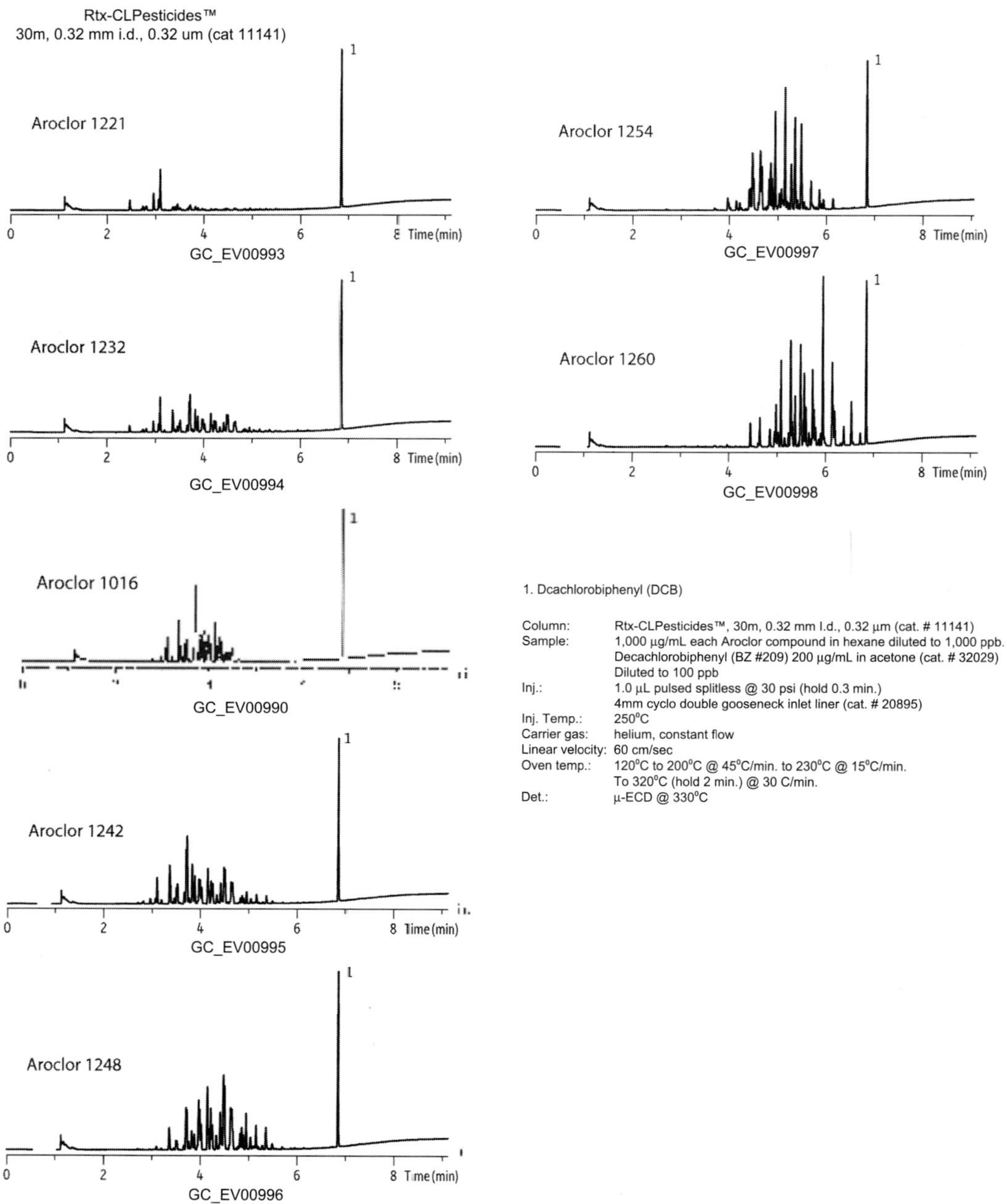

Restek Corporation 110 Benner Circle Bellefonte, PA 16823
814-353-1300 800-356-1688 Fax 814-353-1309 www.restek.com

FIGURE 28.3 Common Aroclor standard chromatograms showing the distribution of PCB congeners.

instrumental setup, or sample with positive analyses for PCBs may be reanalyzed using a secondary, or confirmation, column. The confirmatory column must have a dissimilar elution order to minimize the possibility of interference from pesticides, and other compounds commonly found in some PCB-containing samples.

The quantitation of PCBs as Aroclors is appropriate for many regulatory compliance determinations, but is particularly difficult when the Aroclors have been weathered by long exposure in the environment. Since PCBs were eliminated from production many years ago, most samples have undergone some amount of weathering. In addition, when mixtures of PCBs are in a single sample, identification of the Aroclors may be difficult and, if identification is incorrect, then the quantification will be inaccurate as the sample may be compared to the wrong standard. These reasons, in addition to the common need to determine toxicological-based data, have led to the ever-increasing adoption of congener-based analytical methods.

28.2.3. Analysis of PCBs Based on Individual Congeners

The individual PCB congeners are named according to the convention of their single-ring chlorine substitution. This yields their IUPAC name. Due to the complexity of these names, there have been PCB numbering systems based on the positional substitution of chlorine on the two rings of the molecule [21,22]. It should be noted that the two listed naming systems differ for congeners 107, 108, 109, 199, 100, and 201.

What must first be determined in the development of a congener-specific analysis is the intended target compound list. For all the Aroclors, approximately 150 congeners may be present [20]. Numerous GC column combinations were evaluated in order to determine the chromatographic conditions that resulted in the best overall separation, but in general "XLB-type" phases are considered the best as far as single-column separations using GC-MS. For two-column chromatographic separations using electron-capture detection, numerous column combinations have been evaluated, but in general 5% diphenyl polydimethylsiloxane columns, or XLB columns as the primary column and either carborane phases (e.g. HT-8 or SGE) [23], or other functionalized polysiloxane phases as confirmatory columns are preferred. In the case of congener-specific analysis, each compound is treated as an individual. If a total PCB quantity is required, the congeners are merely summed following individual quantification. If the identification of specific Aroclor is desired from a congener-specific analysis, the congener distribution as a function of Aroclor must be considered (Figure 28.4) [20]. While this reconstitution is possible, it can be difficult especially in complex samples and in cases of weathering.

28.2.4. Recent Improvements to Chromatographic Separation of PCBs

As previously mentioned, comprehensive gas chromatography (GC $\times$ GC) is a technique that has been used to increase the peak capacity of a GC separation. This technique has allowed for a further increase in the total number of PCB congeners resolved in a single analysis [24]. Using two GC columns with considerably different selectivity (ca. HT-8 and BPX-50) it is possible to separate 188 of the 209 PCB congeners. Further, by using time-of-flight mass spectrometry (TOFMS) as a detector, it is currently possible to separate 192 of the 209 PCB congeners. This required an analysis time of 146 minutes, but represents the current state of the art at the time of writing for PCB congener analysis. It may be only a matter of time before the complete separation of all 209 congeners becomes possible – a feat that has been attempted by analytical chemists for quite a long time.

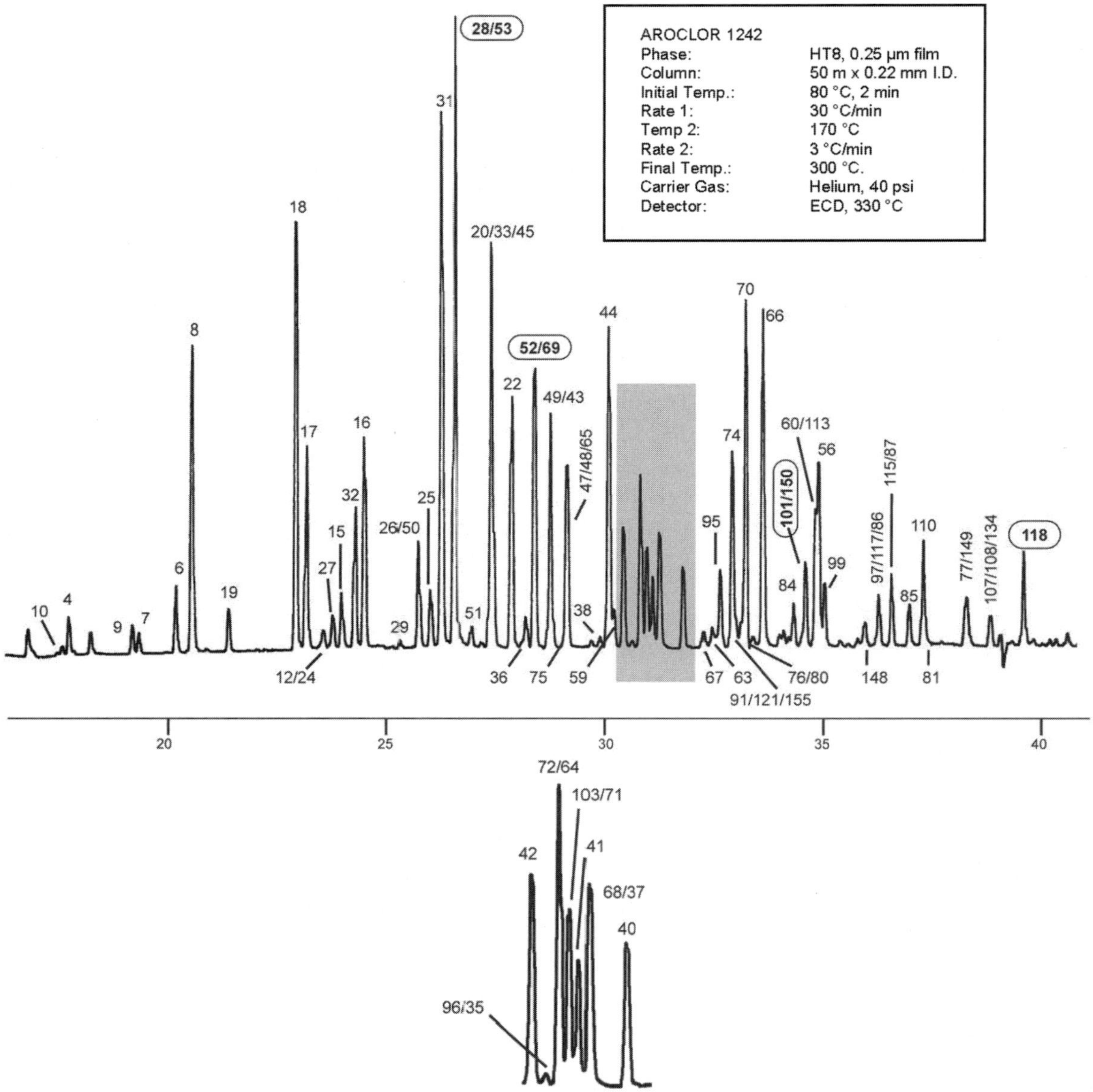

FIGURE 28.4 Aroclor 1242 separation. *Source: Adapted from Ref. [23].* The bottom chromatogram is expansion of shaded area in the chromatogram above.

28.2.5. PCB Summary

PCB analysis is still one of the most common environmental analyses performed in US commercial laboratories, despite the fact that PCB manufacturing voluntarily ceased in the US in the 1970s. Due to the fact that there is no "consensus method" for the analysis of PCBs, they can be especially demanding. Numerous congener and Aroclor methods have been reported, but analytically the congener methods

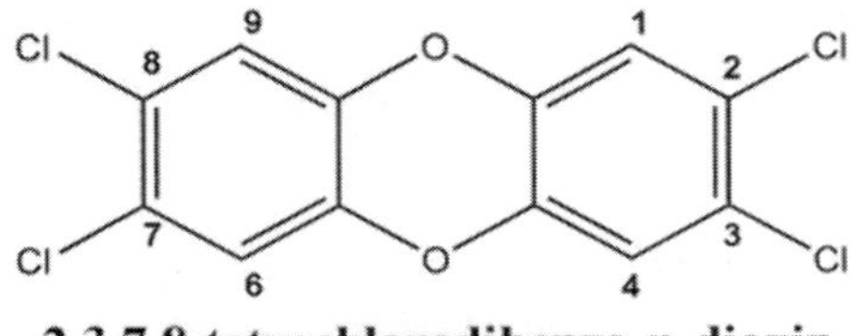

2,3,7,8-tetrachlorodibenzo-p-dioxin

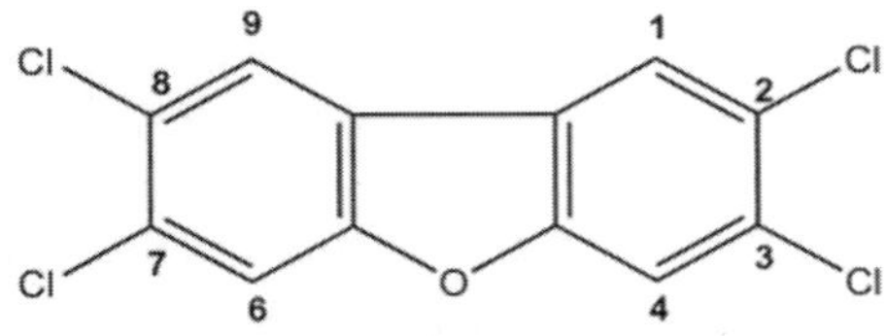

2,3,7,8-tetrachlorodibenzofuran

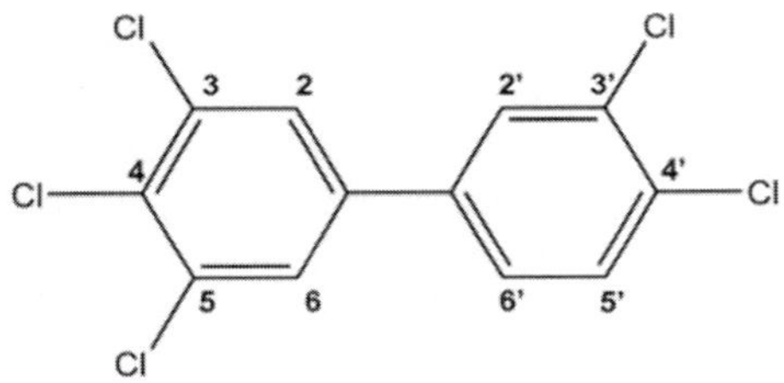

3,3',4,4',5-pentachlorobiphenyl (PCB-126)

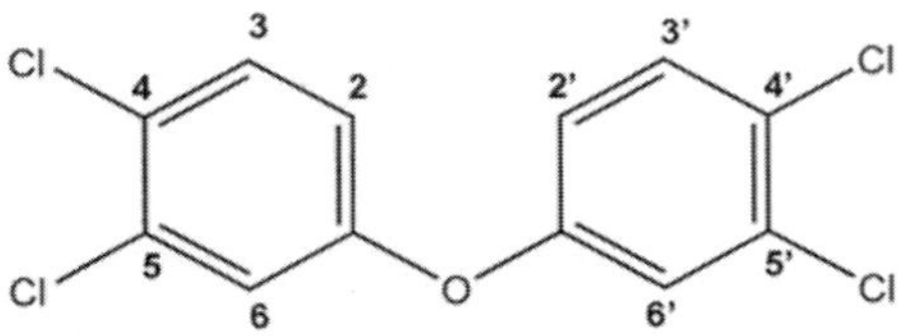

3,3',4,4'-tetrachlorodiphenlyether

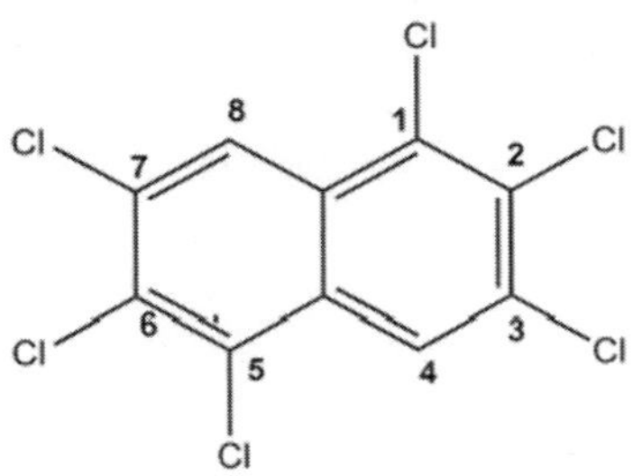

1,2,3,5,6,7-tetrachloronaphthalene (PCN-67)

FIGURE 28.5 Structures of a number of dioxin-like compounds.

are preferred. Considering the total number of possible congeners this analysis is complex, but unless sample fractionation is being performed (to isolate the coplanar PCBs, for example) it is best that the entire list of environmentally found PCBs be the minimum target compound list, and possibly the entire list of 209 congeners.

28.3. DIOXINS

Polychlorinated dibenzo-p-dioxins (PCDDs) and polychlorinated dibenzofurans (PCDFs) are structurally related planar compounds with two chlorine-substituted benzene rings connected by one (furan) or two oxygen atoms (dioxin). There are 75 possible dioxin congeners and 135 possible furan congeners where up to 8 chlorines can replace hydrogen atoms in the three-ring structure. Congeners with chlorines in the 2,3,7,8 positions can bind with the aryl hydrocarbon receptor (AhR) which can promote a number of toxic effects including weight loss, immune impairment, reproductive disorders, development toxicity, and cancer [25]. Dioxin-like compounds (DLCs) all act through a common mechanism (examples – Figure 28.5); therefore, the degree of their toxicological potency can be determined and compared using a relative toxicity scheme normalized to 2,3,7,8-TCDD, the most toxic congener. The toxic equivalent factor (TEF) is a value assigned to each 2,3,7,8- substituted congener representative of its relative potency compared to 2,3,7,8-tetrachlorodibenzo-p-dioxin. The sum of the concentration of each individual dioxin-like compound multiplied by its TEF can be expressed as the toxic equivalent quantity (TEQ) of 2,3,7,8-TCDD in the sample:

$$
\begin{aligned}
\mathrm{TEQ} = {} & \sum[\mathrm{PCDD}_i \times \mathrm{TEF}_i] \\
& + \sum[\mathrm{PCDF}_j \times \mathrm{TEF}_j] \\
& + \sum[\mathrm{PCB}_k \times \mathrm{TEF}_k] + \sum \mathrm{DLC}_l \times \mathrm{TEF}_l
\end{aligned}
$$

The analysis of dioxins is one of the most challenging in analytical chemistry [14,26]. Due to very high toxicity of these compounds, complex sample preparation procedures and very sensitive and selective instrumentation are required. Isotope dilution with $^{13}C_{12}$-labeled 2,3,7,8-congeners is used to account for losses due to the extensive sample cleanup, act as chromatographic time markers for the toxic congeners, increase linearity, and help increase sensitivity for low-level samples. The analytical steps for PCDD/F are similar to those for analytical methods of most compounds. These steps include the following:

- Subsampling representative aliquot for analysis.
- Addition of isotopically labeled PCDD/F internal standards to the sample aliquot.
- Quantitative extraction of PCDD/Fs and other matrix coextractives from sample by one or more extraction methods (see below).
- Removal of interfering coextractives from the sample extracts using a series of sample cleanup procedures (see below).
- Concentration of cleaned extract (in suitable solvent) to amount needed to meet required detection limits for analytical determinations. Final extract volumes typically range from 10 μL to 50 μL.
- Injection of 0.5–5 μL (or more if a large-volume injector is used) into gas chromatography-mass spectrometry detection system.
- Review of data and determination of areas of analytical peaks that meet positive identification criteria in chromatograms (see below).
- Correction of raw data based on recovery of isotopically labeled PCDD/F internal standards added to samples before extraction.
- Use of mass spectrometry calibration curves to determine concentrations of all seventeen 2,3,7,8-substituted PCDD/Fs.
- Review of all available quality control data to ensure that analytical run passes required quality assurance criteria.

Samples can be extracted using standard procedures. Solid samples including soils and sediments can be extracted using Soxhlet extraction, sonication, microwave extraction, or pressurized liquid extraction (PLE). Biological samples (tissue and vegetation) can be extracted using the same techniques after drying with sodium sulfate or diatomaceous earth or acid digested followed by liquid-liquid extraction. Liquid samples can be extracted using solid-phase extraction, liquid-liquid extraction, or using passive samplers. Air samples are typically extracted using high-volume samplers with a glass fiber filter to capture the particulates and a polyurethane foam plug (PUFF) to capture the vapor phase.

Due to the high sample concentration factors (10^6 or more) required to meet the very low detection limits for PCDD/Fs, extensive sample cleanup is required to remove matrix coextractable and interfering compounds. The classical cleanup procedure is a three-stage cleanup based on the Smith–Stallings procedure which involves an acid- and base-impregnated layered silica column to remove polar compounds, and alumina column to remove the remaining compounds of lower polarity and begin to separate compounds such as ortho-substituted PCBs from the other dioxin-like compounds (DLCs). The third stage is an activated carbon column that is used to separate the planar compounds (all DLCs are planar) from the nonplanar ones. Nonplanar compounds (PBDEs, ortho-substituted PCBS) are eluted through the carbon column in the forward direction and the planar dioxin-like compounds (PCDD/Fs, non-ortho-PCBs, and PCNs) are strongly retained and must be removed with a strong elution solvent such as toluene, typically in the reverse direction. Biological samples and other samples with significant amounts of coextractable matrix compounds are often treated by acid and/or base wash or gel permeation chromatography (GPC) prior to the three-stage cleanup.

Chromatographic separation of isomers is a very important step in dioxin analysis due to the very large difference in toxicity [27,28]. There is currently no single GC column that can uniquely resolve all of the 17 toxic congeners from all others. The classical method involves initial analysis on a 60 m × 0.25 mm id 0.25 d_f 5% diphenyl polydimethylsiloxane column with confirmation using a polar phase (e.g., DB-225). In most cases, phases such as arylene-based 5% diphenyl polydimethylsiloxane, DB-Dioxin, or Rtx-Dioxin-2 (Chapter 3) can provide results that that are not significantly different when confirmation on a secondary column is done and many laboratories no longer routinely perform confirmation analysis, especially on biological and human samples where only 2,3,7,8-isomers bioaccumulate. Table 28.5 shows various combinations of chromatographic columns that have been used to analyze PCDD/Fs and their ability to resolve the specific 2,3,7,8-congeners. Shorter columns, e.g. 40 m, 0.18 mm id, and 0.15 um film thickness, have been used to reduce run times by 30–40%. Shorter columns tend to compromise GC peak capacity because often the dioxin-like PCBs are analyzed with the dioxins, and up to 20 ions are monitored in the tetra-dioxin/furan window precluding the ability to obtain a minimum of 7–10 data points across the narrower GC peaks obtained with microbore columns. Comprehensive two-dimensional gas chromatography (GC × GC) [29,30] has been used to enhance chromatographic resolution. Very fast detectors such as TOFMS are required for GC × GC due to the narrow GC peaks (<1 s wide) that are produced. Unfortunately, current TOF mass spectrometers are less selective (nominal mass resolution) and at least an order of magnitude less sensitive than high-resolution mass spectrometers (HRMS).

Magnetic sector instruments running at 10,000 resolution are still the gold standard for PCDD/F analysis. Modern sector instruments are able to detect PCDD/Fs at subfemtogram levels. The carrier effect of $^{13}C_{12}$-labeled surrogates help to minimize adsorption and carry the 2,3,7,8-native congeners through the gas chromatograph. A resolving power of 10,000 is used because this is the best compromise between sensitivity and selectivity. At resolutions greater than 10,000, sensitivity drops off exponentially. Electron ionization (EI) using a reduced electron energy of about 35 eV is typically used. Dioxins are very stable compounds showing little fragmentation and therefore the molecular ion is monitored. Negative chemical ionization (NCI) is not used for PCDD/F analysis. The vast majority of dioxins are more sensitive in NCI than EI, except for 2,3,7,8-TCDD which is significantly less sensitive because the molecular ion readily fragments with the charge located on the stable chloride anion. Single ion monitoring (SIM) is typically used to enhance sensitivity, because the detector is set only to scan for masses (mass to charge ratios of the dioxin and furan ions of interest, see Table 28.5) where PCDD/Fs can be detected. Magnetic sector instruments can scan by either varying the magnetic field or accelerating voltage to switch between ions of interest. In order to scan as fast as possible, accelerating voltage is scanned. In order to obtain the greatest sensitivity, the analytical run is broken into groups or windows consisting of congeners of the same degree of chlorination. If too wide a mass range is scanned, sensitivity will be lost with reduced accelerating voltage. Table 28.6 lists the specific ions of interest for PCDD/Fs and dl-PCBs. Typically, only the tetra- to octa-PCDD/Fs and 12 WHO PCBs are analyzed because they are the only compounds with TEFs.

Tandem quadruple mass spectrometry [31] or ion trap mass spectrometry can be used with single reaction monitoring (MS/MS) for PCDD/F analysis and has been shown to be very selective. The dioxin molecular ion uniquely fragments by the loss of COCl (63 Daltons) resulting in a very clean mass chromatogram with very few interferences. dl-PCBs can also be analyzed using SRM; however, the loss of Cl_2 (70 Da) is

TABLE 28.5 Isomeric Specific Separation of 2,3,7,8-Substituted Dioxins and Furans Using Various Gas Chromatographic Phases

	DB-5, Rtx-5MS HP5-MS, Equity-5	CP-Sil 8 CB/MS	DB-5MS ZB-5MS	Rtx-Dioxin2	ZB5UMS	DB-XLB	DB-225	SP-2331
PCDDs								
2,3,7,8-TCDD	+ +	+ −	+ +	+ +	+ +	+ −	+ −	+ −
1,2,3,7,8-PeCDD	+ +	+ −	+ −	− −	− −	− −	− −	− −
1,2,3,4,7,8-HxCDD	+ +	+ −	+ +	+ +	+ +	+ +	+ +	+ +
1,2,3,6,7,8-HxCDD	+ +	+ +	+ +	+ +	+ +	+ +	+ +	+ +
1,2,3,7,8,9-HxCDD	− −	+ −	+ −	+ +	+ +	+ +	+ +	+ +
1,2,3,4,6,7,8-HpCDD	+ +	+ +	+ +	+ +	+ +	+ +	+ +	+ +
1,2,3,4,6,7,8,9-OCDD	+ +	+ +	+ +	+ +	+ +	+ +	+ +	+ +
PCDFs								
2,3,7,8-TCDF	− −	+ −	+ −	++	++	+ +	+ +	+ −
1,2,3,7,8-PeCDF	+ +	+ +	+ +	++	++	+ +	− −	− −
2,3,4,7,8-PeCDF	− −	− −	− −	− −	− −	− −	+ +	+ +
1,2,3,4,7,8-HxCDF	− −	+ +	+ +	+ +	+ +	+ +	+ +	− −
1,2,3,6,7,8-HxCDF	+ +	+ +	+ +	+ +	+ +	+ +	− −	+ +
1,2,3,7,8,9-HxCDF	+ +	− −	− −	− −	− −	+ +	++	+ +
2,3,4,6,7,8-HxCDF	+ −	− −	− −	− −	− −	+ −	− −	+ +
1,2,3,4,6,7,8-HpCDF	+ +	+ +	+ +	+ +	+ +	+ +	++	+ +
1,2,3,4,7,8,9-HpCDF	+ +	+ +	+ +	+ +	+ +	+ +	++	+ +
1,2,3,4,6,7,8,9-OCDF	+ +	+ +	+ +	+ +	+ +	+ +	++	+ +

+ +: Baseline separation or at least 10% valley. Peak resolution: R > 1
+ −: Quantifiable result (separation that allows peak resolution of R ~ 0.8)
− −: Coelution or interference present. Maximum possible concentration.
Reproduced with permission from Fishman, J. Chromatogr. A, 1139, 285 (2007).

TABLE 28.6 HRMS Multiple Ion Monitoring Program for CDDs/CDFs (DPE = Chlorinated Diphenyl Ethers)

Ion group	m/z (quantification ions)	Compound	Dwell (ms)	Delay (ms)	Theoretical isotope ratio	Acceptable range
1	303.9016, 305.8987	TCDF	50	10	0.77	0.65–0.89
	315.9419, 317.9389	$^{13}C_{12}$-TCDF	25	10	0.77	0.65–0.89
	316.9824	PFK LockMass	30	10		
	319.8965, 321.8936	TCDD	50	10	0.77	0.65–0.89
	327.8847	$^{37}C_{12}$-TCDD	25	10		
	331.9368, 3339339	$^{13}C_{12}$-TCDD	25	10	0.78	0.65–0.89
	375.8364	Hexa-DPE	25	10		
2	339.8597, 341.8567	Penta-CDF	50	10	1.61	1.32–1.78
	351.9000, 353.8970	$^{13}C_{12}$-Penta-CDF	25	10	1.56	1.33–1.79
	354.9792	PFK LockMass	30	10		
	353.8576, 355.8546, 357.8517	Penta-CDD	50	10	1.55	1.31–1.78
	367.8949, 369.8919	$^{13}C_{12}$-Penta-CDD	25	10	1.56	1.32–1.79
	409.7974	Hexa-DPE	25	10		
3	373.8208, 375.8178	Hexa-CDF	50	10	1.24	1.06–1.43
4	383.8369, 385.8610	$^{13}C_{12}$ - Hexa-CDF	25	10	0.52	0.44–0.60
	389.8157, 391.8127	Hexa-CDD	50	10	1.24	1.05–1.43
	392.9760	PFK LockMass	30	10		
	401.8559, 403.8529	$^{13}C_{12}$ - Hexa-CDD	25	10	1.25	1.06–1.43
	445.7555	Octa-DPE	25	10		
	407.7818, 409.7789	Hepta-CDF	50	10	1.04	0.88–1.19
5	417.8250, 419.8220	$^{13}C_{12}$ - Hepta-CDF	25	10	0.45	0.38–0.51
	423.7766, 425.7737	Hepta-CDD	50	10	1.03	0.88–1.19
	435.8169, 437.8140	$^{13}C_{12}$ - Hepta-CDD	25	10	1.04	0.88–1.20
	442.9729	PFK LockMass	30	10		
	479.7165	Nona-DPE	25	10		
	441.7428, 443.7400	Octa-CDF	50	10	0.89	0.76–1.02
6	457.7377, 459.7348	Octa-CDD	50	10	0.89	0.75–1.02
	469.7779, 471.7750	$^{13}C_{12}$ - Octa-CDD	25	10	0.89	0.76–1.03
	480.9697	PFK LockMass	30	10		

not unique for PCBs, therefore interferences can be observed in the SRM traces for PCB analysis.

Dioxin analysis involves a significant amount of quality control. Instruments are tuned to 10,000 resolution (10% valley at $M/\Delta M$) and group windows are determined using a window setting standard containing the first- and last-eluting PCDD/F for each congener group. In addition, a column performance mixture containing the closest eluting congeners for 2,3,7,8-TCDD and 1,2,3,7/1,2,3,8-TCDD are analyzed to ensure that 2,3,7,8-TCDD can be separated from 1,2,3,7/1,2,3,8-TCDD with at least a 25% valley [26]. Following a multipoint calibration typically containing 5 points or more, sample analysis can be carried out. Normally isotopic peaks of the molecular ion cluster are monitored, such that the most abundant ion becomes the target mass for quantification, and the next most abundant ion becomes a qualifier ion. A positive identification must include the elution of the two isotopic peaks within $\pm$ 2 sec of each other as well as the corresponding $^{13}C_{12}$-labeled surrogate standard within the correct time window. The peak shapes must be Gaussian and have a signal-to-noise ratio of at least 3 to 1. Details are listed in EPA method 1613, MOE method 3418, or ISO 18703/ISO 17858 (see Table 28.7)

28.4. ORGANOCHLORINE PESTICIDES

Also listed as compounds in the Stockholm Convention, many of the organochlorine pesticides (OCs) are persistent and bioaccumulative. Compounds such as DDT, endrin, dieldrin, and others are also mentioned by name in Silent Spring. Most of these types of compounds are now banned from use, but due to their nature are frequently found in many solid matrices from plant and animal tissue through the range of soils commonly analyzed. There are still some locations around the globe that use some of these compounds (South Africa for malaria control, for example); so discussion of their analysis is still relevant, unfortunately.

Similar sample preparation strategies as mentioned above are also amenable to OCs. As hydrophobic compounds they are more commonly found in solid matrices, or organic matrices, as opposed to aqueous ones. Following extraction, there are many extract cleanup techniques, details of which can be found in the cited methodology in Table 28.7. Preparatory gel permeation chromatography (prep-GPC) can be extremely helpful in cleanup of these extracts prior to analysis. Prep-GPC is a relatively simple technique that easily removes both sulfur and larger hydrocarbons typically found in biota and samples taken from locations that are high in concentration of various tar-like materials.

Analysis using either MS or ECD is common, but, if using ECD, then a confirmatory analysis must be performed on a GC column exhibiting different selectivity so as to change the elution profile of the compounds; this helps to minimize quantification bias as well as misidentification. Due to the relatively recent increase in MS sensitivity and ruggedness, MS-based methods seem to be increasing in popularity. Even with this increase in sensitivity, it can still be challenging to reach desired detection limits without additional sample extract concentration (as in done with dioxin samples) or without using SIM, or mass spectrometers that are designed to be especially sensitive (TOFMS or high-resolution sector instruments).

For GC columns, many manufacturers now offer primary and confirmation columns designed for this separation that employ proprietary stationary-phase chemistries. The Rtx-CLPesticides and Rtx-CLPesticides2 columns (Restek Corporation) were developed in 1997 and are used frequently for this analysis. Since this time, there have been a number of other options including the MR-1 and MR-2 (Phenomenex). These columns all have relatively short

TABLE 28.7 Selected Methods for Analysis of POPs

Method	Analytes/comments	Reference
USEPA1613	Seventeen 2,3,7,8-substituted dioxins and furans with congener group totals in water and wastewater. Uses isotope dilution – GC-HRMS	USEPA 1994b
USEPA1614	Brominated diphenyl ethers in water soil, Sediment and tissue by HRGC/HRMS	USEPA 2007
USEPA1668a	209 PCB congeners. 12 WHO dioxin-like PCBs by GC-HRMS, the remaining 197 by GC–MS	USEPA 1999
USEPA23	Seventeen 2,3,7,8-substituted dioxins and furans with congener group totals in incinerator stack gasses. Uses isotope dilution – GC-HRMS	USEPA 1995
USEPA 8290 (SW-846)	Seventeen 2,3,7,8-substituted dioxins and furans with congener group totals in materials and waste. Uses isotope dilution – GC-HRMS	USEPA 1994a
ISO 18073	Seventeen 2,3,7,8-substituted dioxins and furans with congener group totals in water and wastewater. Uses isotope dilution – GC-HRMS. Allows GC–MS as an alternate detection method	ISO 2004
ISO 17858	12 WHO dioxin-like PCBs in environmental matrices by GC-HRMS. Allows GC–MS as an alternate detection method	ISO 2006
ISO 22032	Water quality – determination of selected polybrominated diphenyl ethers in sediment and sewage sludge – Method using extraction and gas chromatography/mass spectrometry	ISO 2006
EN 1948	Seventeen 2,3,7,8-substituted dioxins and furans and congener group totals in Stationary Sources by isotope dilution – GC-HRMS	European Standard 1997
MOE3418	Seventeen 2,3,7,8-substituted dioxins and furans including congener group totals and 12 WHO dioxin-like PCBs by GC-HRMS. Uses isotope dilution – GC-HRMS	Ontario Ministry of the Environment, 2010
MOE 3430	Polybrominated diphenyl ethers in environmental matrices using GC-HRMS	Ontario Ministry of the Environment, 2010
MOE3136	The determination of polychlorinated biphenyls (PCBs), organochlorines (OCs) and chlorobenzenes (CBs) in fish , clams and mussels by Fast GC-ECD	Ontario Ministry of the Environment, 2009
MOE3487	The determination of polychlorinated biphenyls (PCBs), organochlorines (OCs) and chlorobenzenes (CBs) in solids by GC × GC- μECD.	Ontario Ministry of the Environment, 2010
ENVCAN 1/RM/19	Seventeen 2,3,7,8-substituted dioxins and furans and congener group totals in pulp and paper effluents by isotope dilution – GC-HRMS	Environment Canada 1992

(Continued)

TABLE 28.7 Selected Methods for Analysis of POPs (*cont'd*)

Method	Analytes/comments	Reference
JIS K0311	Seventeen 2,3,7,8-substituted dioxins and furans including congener group totals in incinerator stack gasses by isotope dilution – GC-HRMS	JIS 1999a
JIS K0312	Seventeen 2,3,7,8-substituted dioxins and furans including congener group totals in wastewater by isotope dilution – GC-HRMS	JIS 1999b

References
Environment Canada, (1992) Reference Method for the Determination of PCDDs and PCDFs in Pulp and Paper Mill Effluents, Report EPS 1/RM/19.
European Standard EN 1948 (1997) Stationary Source Emissions, Determination of the Mass Concentration of PCDDs/PCDFs, CEN, Brussels, Belgium.
ISO 18073 International Organization for Standardization (ISO) (2004) Water quality – Determination of tetra- to octa-chlorinated dioxins and furans – Method using isotope dilution HRGC/HRMS, Geneva, Switzerland.
ISO 17585 International Organization for Standardization (ISO) (2006) Water quality – Determination of Dioxin-like polychlorinated biphenyls – Method using gas chromatography and mass spectrometry, Geneva, Switzerland.
ISO 22032. International Organization for Standardization (ISO) (2006) Water Quality – Determination of selected polybrominated diphenyl ethers in sediment and sewage sludge – Method using extraction and gas chromatography/mass spectrometry.
JIS K0311 Japanese Industrial Standard (1999) Method for determination of tetra- through octa-chlorodibenzo-p-doxins, tetra- through octa-chlorinatedfurans and coplanar polychlorinated biphenyls in stationary source emissions.
JIS K0312 Japanese Industrial Standard (1999) Method for determination of tetra- through octa-chlorodibenzo-p-doxins, tetra- through octa-chlorinated furans and coplanar polychlorinated biphenyls in industrial water and wastewater.
Ontario Ministry of the Environment, (2009), The determination of polychlorinated dibenzo-p-dioxins, polychlorinated furans and dioxin-like PCBs in environmental matrices by GC-HRMS. Environment Ontario Laboratory Services Branch Method DFPCB-E3418. Toronto, ON, Canada.
Ontario Ministry of the Environment, (2009), The determination Polybrominated Diphenyl Ethers in Environmental Matrices using GC-HRMS. Environment Ontario Laboratory Services Branch Method BDE-E3418. Toronto, ON, Canada.
Ontario Ministry of the Environment, (2009), The determination of polychlorinated biphenyls (PCBs), organochlorines (OCs) and chlorobenzenes (CBs) in fish , clams and mussels by gas liquid chromatography electron capture detection (GLC-ECD). Environment Ontario Laboratory Services Branch Method PFAOC-3136. Toronto, ON, Canada.
Ontario Ministry of the Environment, (2010), The determination of polychlorinated biphenyls (PCBs), organochlorines (OCs) and chlorobenzenes (CBs) in solids by two dimensional gas liquid chromatography - micro electron capture detection (GC × GC- μECD). Environment Ontario Laboratory Services Branch Method PSAHOC-3487. Toronto, ON, Canada.
U.S. EPA SW-846 Method 8290 1994a Polychlorinated Dibenzodioxins and Polychlorinated Dibenzofurans by High-Resolution Gas Chromatography / High-Resolution Mass Spectrometry, Revision 0.
U.S. EPA Method 1613, 1994b, Revision B: Tetra- Through Octa-Chlorinated Dioxins and Furans by Isotope Dilution HRGC/HRMS, EPA 821-B94-0059 Office of Water.
U.S. EPA Method 23, 1995, Determination of Polychlorinated Dibenzo-p-dioxins and Polychlorinated Dibenzofurans from Municipal Waste Combustors.
U.S. EPA Method 1668, Revision A: 1999, Chlorinated Biphenyl Congeners in Water, Soil, Sediment, and Tissue by HRGC/HRMS, EPA-821-R-00-002 Office of Water.

analysis times, with complete separation of the common OCs. They may also be used for GC-MS analysis of the OCs, but, in addition, many users also use more common 5% diphenyl polydimethylsiloxane, or XLB-type stationary phases. An example of this separation can be found in Figure 28.6 coupled with a fast GC option using resistive heating. As observed from the figure, this analysis can be conducted in a short time (less than 5 minutes) provided

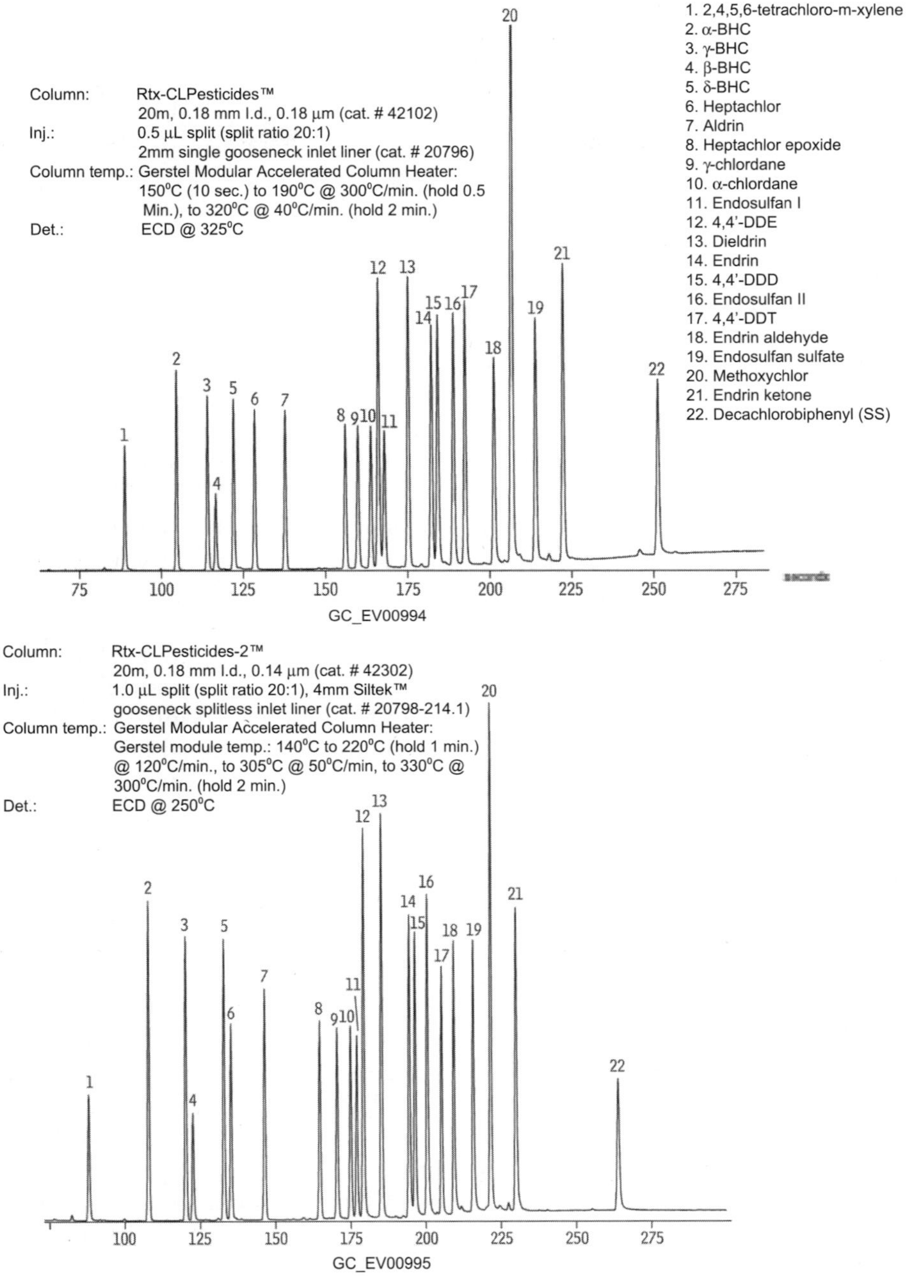

Sample: Organochlorine Pesticide Mix AB #2 (8-80 μg/mL each component In hexane/toluene 1:1, cat.# 32292), Pesticide Surrogate Mix (200 μg/mL each component in acetone, cat.# 32000)
Inj.: 0.5μL, split (split ratio 20:1), 2mm single gooseneck inlet liner (cat. # 20796)
Inj. Temp.: 250°C
Carrier gas: helium, constant flow
Flow rate: 1.5 mL/min.
Oven temp.: 275°C

FIGURE 28.6 OC's separated on Rtx-CL Pesticides and Rtx-Cl Pesticides2 GC columns using Gerstel Mach Fast GC system.

the samples have been prepared in a manner that removes many of the common interferents.

28.5. HALOGENATED FLAME RETARDANTS

Flame retardants have been used for thousands of years [32]. Halogenated flame retardants (HFRs) have been used since the 1960s mainly because they are significantly more compatible with the many polymeric materials we use today and are much better at causing charring and reducing smoke which allows more time for escape [33,34]. Hundreds of different brominated and chlorinated flame retardants have been developed and are being detected in the environment. Polybrominated diphenyl ethers (PBDEs) are the most commonly known and widely used HFRs and along with the hexabromobiphenyls are the only HFRs currently on the Stockholm Convention list. Hexabromocyclododecane (HBCD) and tetrabromobisphenyl-A (TBBPA) are also high-production-volume brominated flame retardants (BFRs) used in numerous applications. There are also a number of chlorinated flame retardants including dechlorane plus (DP) and other dechloranes developed for specific polymer uses such as electrical cabling. A number of the more common halogenated flame retardants and their applications are shown in Table 28.8.

The properties that make halogenated flame retardants very good at retarding flames also make them a real challenge to analyze. Most HFRs readily decompose during the analytical process as they have been designed to easily release a radical halogen that combines with the radicals formed in the pre-ignition stage of combustion. Analytical methods must therefore be designed to minimize the exposure to elevated temperatures, which can be a real problem for GC analysis. Surprisingly enough, most HFRs have a significant vapor pressure and can be analyzed using lower temperatures (<300 °C) and shorter GC columns. Most HFRs are also light sensitive and care must be taken to reduce exposure to light during the analytical process to minimize dehalogenation. In addition, many of the HFRs are present in building materials, furniture, and electronic equipment present in the laboratory. Great care must be taken to minimize exposure to dust, as levels of HFRs in dust can be high and a significant source of contamination.

28.6. POLYBROMINATED DIPHENYL ETHERS

Sample extraction methods are similar to dioxins, PCBs, and other halogenated organics and can include PLE, Soxhlet, microwave, and sonification for solid samples, acid digestion, or extraction by PLE/Soxhlet for biological samples and SPE or liquid-liquid extraction for liquid samples. Sample preparation also involves conditions similar to PCDD/F analysis. In fact, sample preparation for PBDEs, dl-PCBs, PCDD/Fs, and other organic compounds can be combined. Silica is typically used as a sample preparation material with alumina or Florisil® as an optional cleanup step depending on the amount of matrix coextractable materials in the sample. In a combined analyte method (PBDEs, dl-PCBs, PCDD/Fs, and other HFRs), the silica eluent can be loaded onto activated carbon. The PBDEs, ortho-PCBs, and other nonplanar compounds are eluted in the forward direction in one fraction and the PCDD/Fs, non-ortho-PCBs, and other planar compounds such as PCNs can be eluted in the reverse direction. Most methods analyze at least 7–10 congeners, including 17, 28, **47,** 49, **99, 100,** 119, **153, 154, 189,** and **209 (key congeners in bold),** and some methods analyze for more than 30 congeners. The majority of the concentration comes from the ones listed above. Earlier methods did not analyze for BDE209 because analyses

TABLE 28.8 Various Halogenated Flame Retardants

Compound	Chemical formula	Uses [3,4,27]	Analytical method
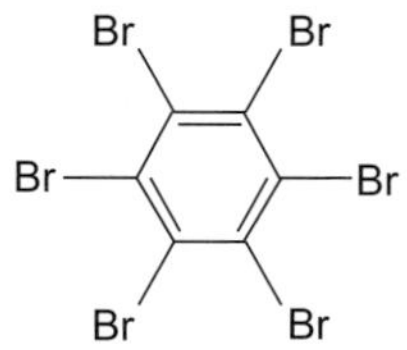 hexabromobenzene (HBB)	C_6Br_6	Paper; electrical goods; polyamides; polypropylene	GC-MS
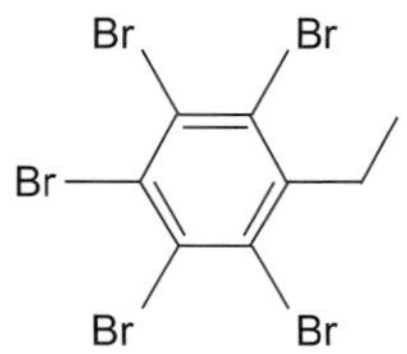 pentabromoethylbenzene (PBEB)	$C_8H_5Br_5$	Unsaturated polyesters; polyethylene; polypropylenes; polystyrene; SBR-latex; textiles, rubbers, ABS	GC-MS
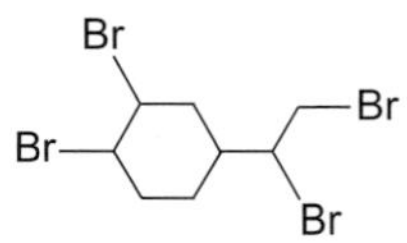 tetrabromoethylcyclohexane (TBECH)	$C_8H_{12}Br_4$	Expandable polystyrene beads	GC-MS LC-MSMS
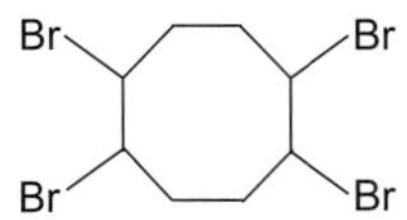 1,2,5,6-tetrabromocyclooctane (TBCO)	$C_8H_{12}Br_4$	Polystyrene	GC-MS LC-MSMS
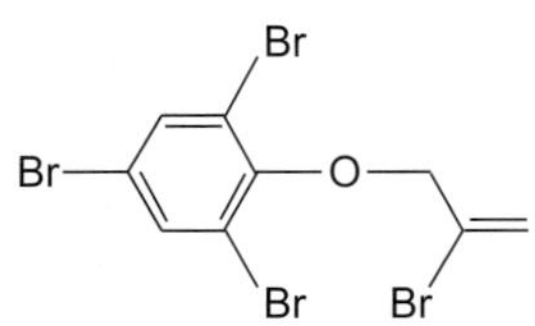 2-bromoallyl 2,4,6-tribromophenyl ether (BATE)	$C_9H_6Br_4O$	High impact plastic	GC-MS LC-MSMS

(Continued)

TABLE 28.8 Various Halogenated Flame Retardants (*cont'd*)

Compound	Chemical formula	Uses [3,4,27]	Analytical method
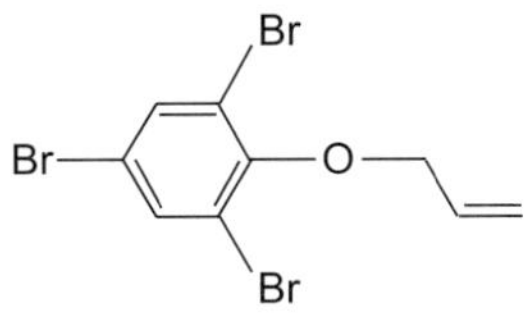 allyl 2,4,6-tribromophenyl ether (ATE)	$C_9H_7Br_3O$	Polyamide; polyester; polyethylene; polypropylene; polystyrene; polycarbonates	GC-MS LC-MSMS
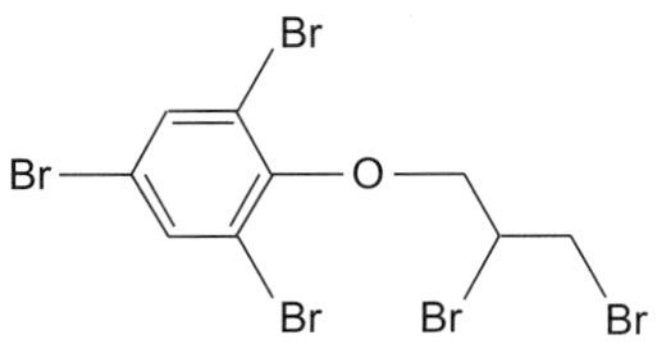 2,3-dibromopropyl 2,4,6-tribromophenyl ether (DPTE)	$C_9H_7Br_5O$	Polypropylene	GC-MS LC-MSMS
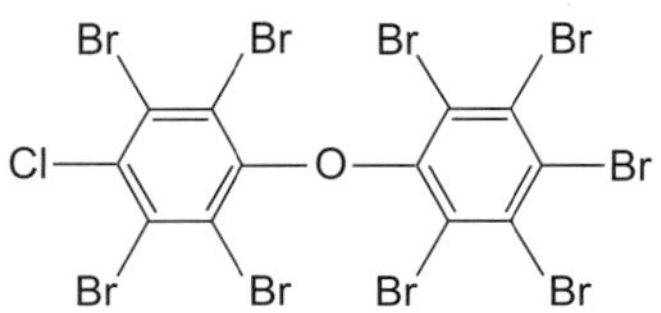 2,2′,3,3′,4,5,5′,6,6′-nonabromo-4′-chlorodiphenyl ether (4PC-BDE208)	$C_{12}Br_9ClO$	Potential instrument injection Standard for BDE209, DBDPE	GC-MS LC-MSMS
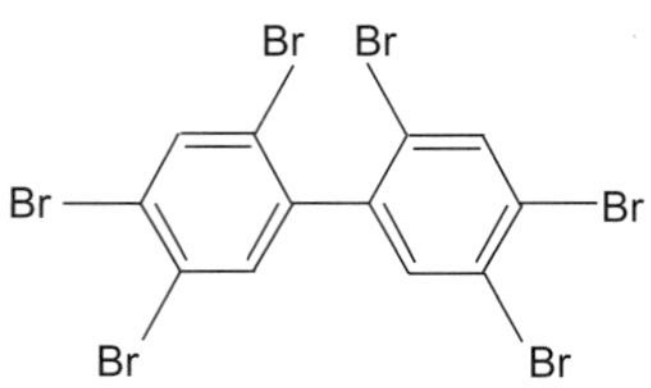 2,2′,4,4′,5,5′-hexabromobiphenyl (BB-153)	$C_{12}H_4Br_6$	Molded plastics and synthetic fibers	GC-MS

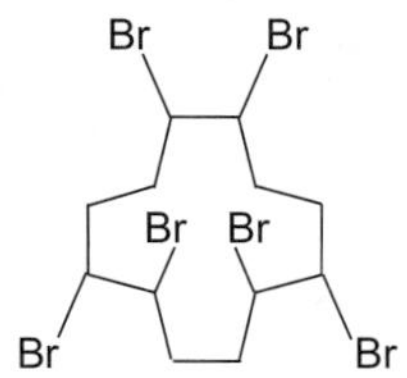 hexabromocyclododecane (HBCD)	$C_{12}H_{18}Br_6$	Polystyrene; latex; textiles; adhesives; coatings; polyesters	LC-MSMS
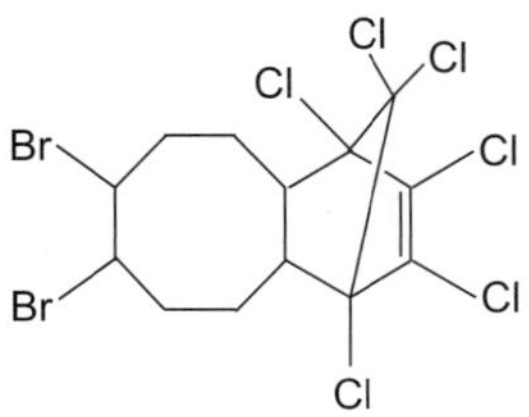 hexachlorocyclopentadienyl-dibromocyclooctane (HCDBCO)	$C_{13}H_{12}Br_2Cl_6$	Styrenic polymer	GC-MS
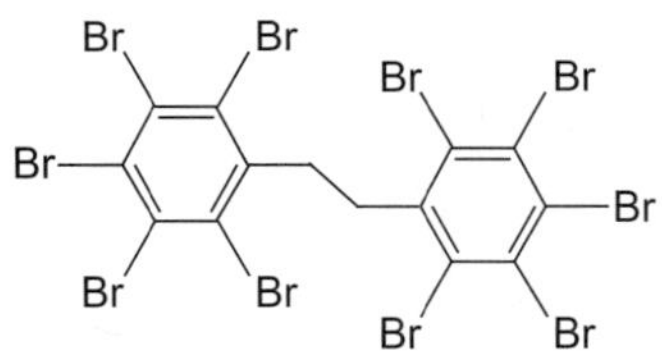 decabromodiphenylethane (DBDPE)	$C_{14}H_4Br_{10}$	High impact plastic; polyamide; polypropylenes; polystyrene; polyester/cotton	GC-MS LC-MSMS
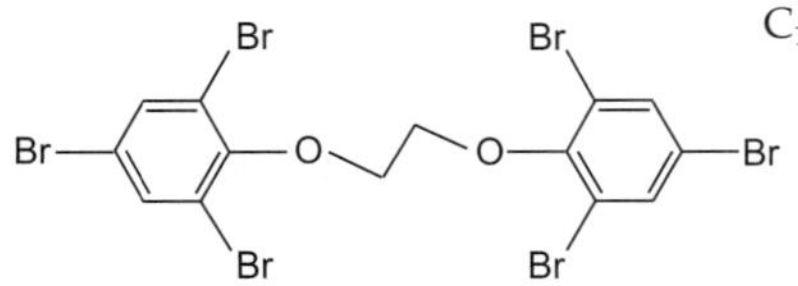 1,2-bis(2,4,6-tribromophenoxy) ethane (BTBPE)	$C_{14}H_8Br_6O_2$	Thermoplastics; ABS polymer systems high impact polystyrene	GC-MS LC-MSMS

(*Continued*)

TABLE 28.8 Various Halogenated Flame Retardants (*cont'd*)

Compound	Chemical formula	Uses [3,4,27]	Analytical method
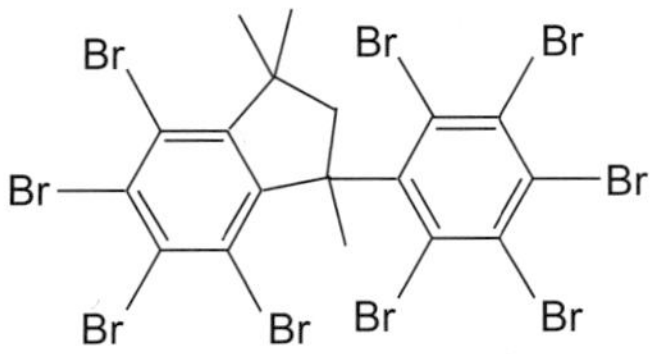 octabromotrimethylphenylindane (OBIND)	$C_{18}H_{12}Br_8$	Hi impact polystyrene (HIPS); Acrylonitrile butadiene styrene (ABS); Polyethylene; polyamides	GC-MS
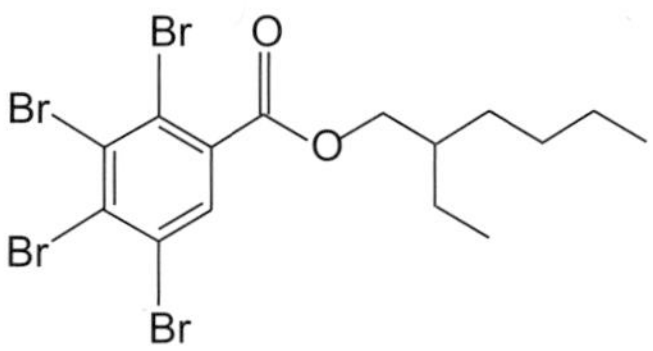 2-ethylhexyl-2,3,4,5-tetrabromobenzoate (EHTeBB)	$C_{15}H_{18}Br_4O_2$	Thermoplastics; PVC; rubber	GC-MS LC-MSMS
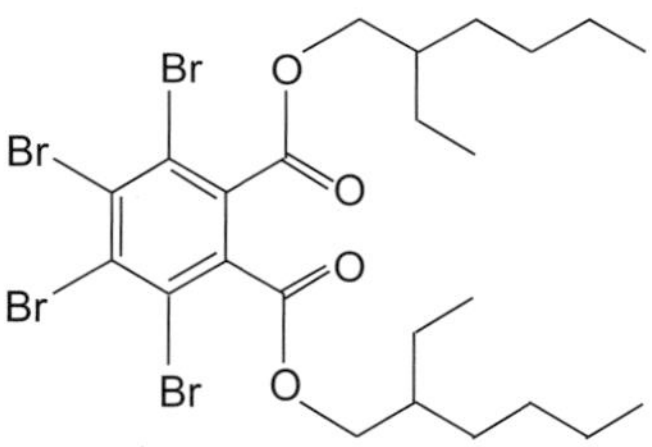 bis(2-ethyl-1-hexyl) tetrabromophthalate (BEHTBP)	$C_{24}H_{34}Br_4O_4$	Thermoplastics; PVC; rubber	GC-MS LC-MSMS
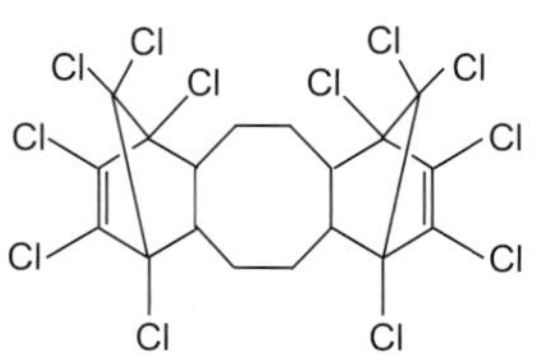 dechlorane plus (DP)	$C_{18}H_{12}Cl_{12}$	Polyamides; polystyrene	GC-MS

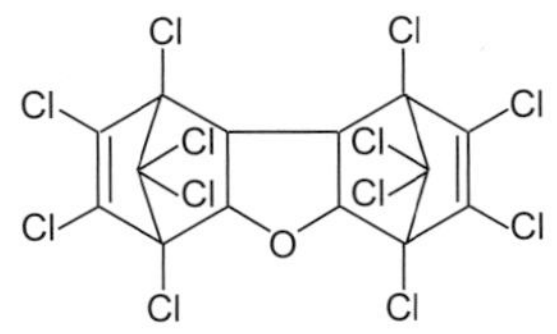 Dechlorane 602	$C_{13}H_2Cl_{12}O$	Thermoplastics; PVC, high-voltage electrical cabling	GC-MS
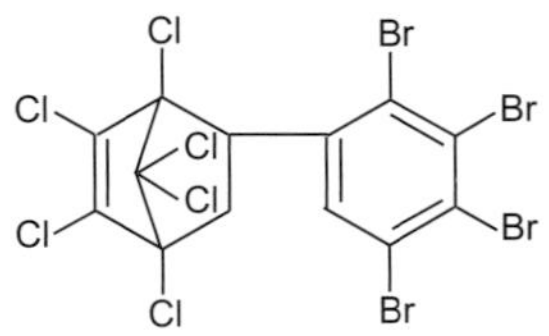 Dechlorane 604	$C_{12}H_4Br_4Cl_6$	Thermoplastics; PVC; high-voltage electrical cabling, silicon grease	GC-MS

were carried out on 30 m × 0.25 mm id film thickness 0.25 μm columns and BDE209 degrades on these columns due to the prolonged residence at elevated temperatures. In addition, unless extreme care is taken, significant degradation could occur resulting in large biases because labeled standards were not available. With the development of labeled internal BDE standards, the use of shorter, thinner film columns, and a better understanding of the conditions required, results for BDE209 are more accurate. The major problem is still extensive contamination which can be random and often no reason can be found. Levels of 10 ng or more in a blank can be randomly detected after obtaining multiple blanks at levels below 100 pg.

The standard column for PBDE analysis is a 5% diphenyl polydimethylsiloxane, 15 m × 0.25 mm id and 0.10 μm film thickness. Short thin-film columns are used to enable high-molecular-weight PBDEs such as decaBDE (BDE209) to pass through the injector to detector. Most laboratories now only use the 15 m column for analysis. It is able to resolve most major interferences and runs can be completed in less than 20 min. A variety of different methods including HRMS, NCI, and MS/MS are used, with HRMS and NCI being the most common. In HRMS methods, isotope dilution using the $^{13}C_{12}$-labeled surrogates for the main congeners is the method of choice where the $M^{+}-Br_2$ ion is monitored for hexa and higher substituted congeners and M^{+} for mono- to penta-congeners. In the NCI method, the bromine anion is monitored. This method is much less selective and possibly less accurate depending on the amount of matrix interferences that are present because isotope dilution cannot be used. It is however less expensive as a much simpler low-resolution mass spectrometer can be used. More details are in EPA 1614, MOE 3430, or ISO 22032 (Table 28.7).

28.7. OTHER HALOGENATED FLAME RETARDANTS

There are many other halogenated flame retardants with many different physical and chemical properties [35]. A number of the more common ones are shown in Table 28.8. These compounds range from polar to nonpolar and have a significant range in molecular weights ranging from just over 100 Da to about 1000. Many of them can be analyzed by GC-MS, but due to their polar nature, e.g. TBBPA or tendency for isomer interconversion at GC temperatures (e.g. HBCD), a number of them must be analyzed using liquid chromatography-tandem mass spectrometry (LC-MS/MS).

Extraction methods are similar to the other halogenated compounds discussed above. Sample preparation can be slightly different for each compound, or compounds being analyzed. Some compounds, bis(2-ethyl-1-hexyl)tetrabromophthalate (BEHTBP) and 2-ethylhexyl-2,3,4,5-tetrabromobenzoate (EHTeBB), for example, are acid sensitive and HBCD isomers are strongly retained on alumina and therefore neutral silica and/or Florisil® are used for sample cleanup. It is important to perform recovery studies to ensure that compounds of interest are recovered quantitatively for the method and procedures being used.

In most cases, the same columns as for PBDEs are used for the brominated and chlorinated flame retardants — 5% diphenyl polydimethylsiloxane, 15 m × 0.25 mm id, film thickness 0.10 (μ)m. Longer columns are also used for the more volatile compounds, e.g. HBB, PBEB, BATE, and ATE, as they are not retained as strongly as some of the higher-molecular-weight compounds DPBPE and BTBPE. Most methods use GC-ECNI-MS for analysis for the non-BDE brominated flame retardants as many of them fragment extensively under GC-EI-MS conditions. The chlorinated dechlorane-type flame retardants exhibit strong retro-Diels—Alder fragments at m/z = 272 and m/z = 237 characteristic of this group of compounds. Many of the organochlorine (OC) pesticides also exhibit these peaks. Most OC pesticides were produced by forming an adduct of hexaclorocyclopentadiene (HCCPD) with another nonhalogenated cyclodiene. Adding a second HCCPD to the diene through a Diels—Alder addition produced a much less toxic compound with excellent flame-retarding properties. Both GC-EI-HRMS and GC-NCI-MS are used to analyze the dechlorane compounds.

28.8. PERFLUORINATED COMPOUNDS

Perfluorinated compounds (PFCs) are a broad range of compounds used in numerous applications including stain repellents for textiles, additive to paper products, and in aqueous film forming foams used to fight electrical fires [36—39]. They have been added to the Stockholm list because they are persistent, bioaccumulative, and toxic, and have been detected globally. The most common are the perfluorinated carboxylic acids (PFCAs) and perfluorocarbonsulfonic acids (PFSAs) of which perflurooctonoic acid (PFOA) and perfluorooctane sulfonate (PFOS) are the most well known. Other PFC compounds include fluorotelomer alcohols (FTOHs), fluorotelomer methacrylates (FTMACs), fluorotelomer acrylates (FTACs), perfluorooctane sulfonamides (FOSAs), perfluorooctane sulfonamidoethanols (FOSEs), polyfluoroalkyl phosphoric acid diesters (diPAPs), and perfluorinated phosphonic acids (PFPAs).

The method of choice for PFCs is LC-MS/MS as the majority of these compounds are polar and require derivatization to be analyzed by GC. Except for the PFCAs, obtaining a stable derivative is a significant challenge. Methyl or 2,4 difluoroanalide derivatives may be used to analyze PFCA on a 30 m × 0.25 mm id, film thickness 0.25 (μ)m 5% diphenyl polydimethylsiloxane column. The less polar PFCs, FTMAC, FTACs, FTOHs, FOSAs, and FOSEs, can be analyzed underivatized on a poly(ethylene glycol) phase — a 30 m × 0.25 mm id, film thickness 0.25 μm. GC-MS is typically used only if specific isomer information is required or compounds such as PFOA where background contamination from

Teflon® lines in liquid chromatographs can cause issues with accurate quantification.

28.9. POLYCYCLIC AROMATIC HYDROCARBONS

Polycyclic aromatic hydrocarbons (PAHs) are widespread environmental contaminants present in fossil fuels and formed through combustion of biomass [40]. They are made of multifused benzene rings up to 1000 in molecular weight or more. PAHs with less than 10 benzene rings are known carcinogens and are toxic to humans and wildlife. Early methods used high-performance liquid chromatography with fluorescence detection because of the strong fluorescent properties of these compounds. With the development of capillary columns, GC-MS is often the method of choice. Most methods include 16–20 compounds containing at least the 16 priority PAHs (see Table 28.9). There are many different alkyl-substituted PAHs and hetero-PAHs (PAH with S, N, or O). Hetero-PAHs are typically not regulated except if they are halo substituted e.g. PCDD/F or PCNs because there are too many compounds and a lack of analytical standards.

Extraction methods are similar to those used for PCDD/Fs, PBDEs, and other organic compounds. Because they are typically present in the environment at levels higher than those discussed above, a simple cleanup, typically a silica SPE cartridge cleanup, is used. PAHs contain no halogens and therefore are mass sufficient (accurate mass is greater than nominal mass). There are fewer interfering compounds at these higher masses of these highly unsaturated hydrocarbons. As a result, low-resolution mass spectrometry (single quadrupole or ion trap) mass spectrometry is used. In order to increase sensitivity, single ion monitoring (SIM) is used. Deuterium-labeled internal standards and in some cases a full isotope dilution method is used. ^{13}C-labeled internal standards are rarely used because of cost. The most common column is a 5% diphenyl polydimethylsiloxane phase with a 30 m x 0.25 mm id, film thickness 0.25 μm column. Fast GC with 20 and 10 m columns has been used to reduce run times [41]. Other stationary phases, most commonly 50% diphenyl polydimethylsiloxane and their arylene equivalents (Chapter 3), are also used. Electron ionization is used. PAHs are very stable and for most compounds the molecular ion is the base peak and also the quantifier ion. The major problem is selection of a second ion that can be used for a qualifier ion. It is typically the $M - H^+$ or $M + 1^+$ (^{13}C isotope) ion. In some cases, the doubly charged molecular ion at half of the M^+ is also used. Challenging separations include the anthracene/phenanthrene pair and the benzo[b]fluoranthene/benzo[k]fluoranthene pair. Benzo[j]fluoranthene can elute between the benzo[b]fluoranthene/benzo[k]fluoranthene pair and often bias the results of these two PAHs. An alternate phase like DB-17 can separate the benzofluoranthenes.

28.10. OTHER COMPOUNDS NOT SPECIFICALLY DISCUSSED

Obviously, there are many other compounds of environmental importance. The ones discussed are those identified by the Stockholm Convention, and therefore constitute those with the highest importance. In addition to these, however, there are many other important compound classes such as organophosphate pesticides, hydrocarbons, and herbicides that are of equal interest to commercial laboratories; however, because these compounds are not considered bioaccumulative, they were deliberately not discussed here. Finally, many analytical methods are based not on compound class (PCBs, dioxins, etc.) but on general chemical properties. Volatile and semivolatile compound methods are also quite common, and typically performed with

TABLE 28.9 Selected PAHs Reported for Environmental and Health-Effects Studies

WHO EHC[a]	GENO[a,b]	CARC[a,b]	US EPA[c]	U.S. ATSDR[d]	MOE 3418	ISO 17993[e]
Acenaphthene	(?)	(?)	X	X	X	X
Acenaphthylene	(?)	No data	X	X	X	X
Anthracene	–	–	X	X	X	X
Benz[a]anthracene	+	+	X	X	X	X
Benzo[b]fluoranthene	+	+	X	X	X	X
Benzo[j]fluoranthene	+	+		X		
Benzo[ghi]fluoranthene	(+)	(–)			X	X
Benzo[k]fluoranthene	+	+	X	X	X	X
Benzo[ghi]perylene	+	–	X	X	X	X
Benzo[a]pyrene	+	+	X	X	X	X
Benzo[e]pyrene	+	?	X*	X	X	X
Chrysene	+	+	X	X	X	X
Coronene	(+)	(?)	X*			
Dibenz[a,h]anthracene	+	+	X	X	X	X
Fluoranthene	+	(+)	X	X	X	X
Fluorene	–	–	X	X	X	X
Indeno[1,2,3-cd]pyrene	+	+	X	X	X	X
Naphthalene	–	(?)	X		X	X
Perylene	+	(–)				
Phenanthrene	(?)	(?)	X	X	X	X
Pyrene	(?)	(?)	X	X	X	X
Triphenylene	+	(–)				

[a] *Reviewed in World Health Organization (WHO) Environmental Health Criteria Monograph on PAHs.*
[b] *GENO = genotoxicity; CARC = carcinogenicity; +, positive; –, negative; ?, questionable; parentheses, result derived from small database.*
[c] *US Environmental Protection Agency (EPA) Method 610 PAHs; PAHs noted with asterisk (*) included in Method TO-13A for PAHs in air.*
[d] *US Agency for Toxic Substances and Disease Registry (ATSDR) Toxicological Profile for PAHs.*
[e] *International Organization for Standardization (http://www.iso.org) Method 17993:2002 Water quality – determination of 15 PAH in water by HPLC–FL after liquid–liquid extraction.*
Modified from D.L. Poster, M.M. Schantz, L.C. Sander, S.A. Wise, Analysis of polycyclic aromatic hydrocarbons (PAHs) in environmental samples: a critical review of gas chromatographic (GC) methods, Analytical and Bioanalytical Chemistry, 386 (2006) 859–881.

general-purpose GC columns using MS detection to allow for the identification of target compounds without the need for multiple confirmatory columns.

28.11. SUMMARY

This chapter discusses several different classes of environmentally significant chemicals

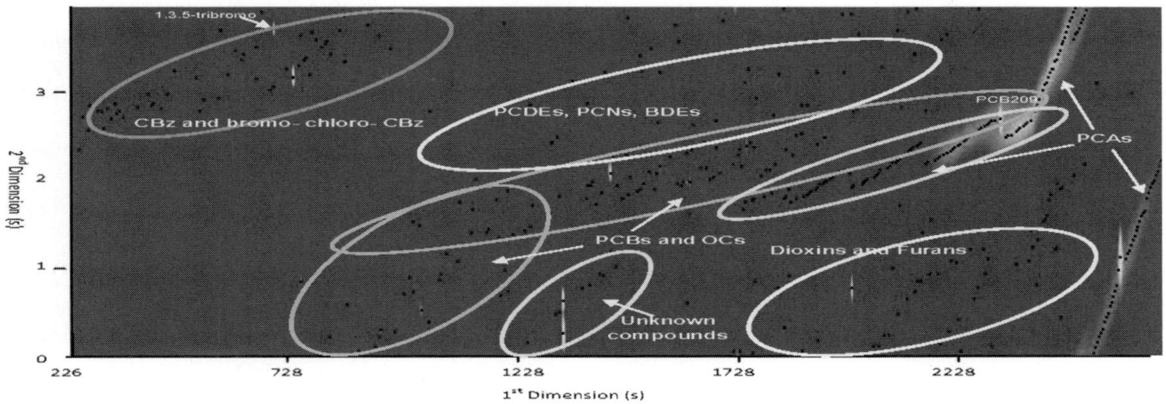

FIGURE 28.7 Two-dimensional chromatogram of a sediment sample showing the presence of a variety of halogenated organic compounds. *Source: From A.M. Musalu, E.J. Reiner, S.N. Liss, T. Chen, G. Ladwig, D. Morse, A Routine Method for the Analysis of Polychlorinated Biphenyls, Organochlorine Pesticides, Chlorobenzenes and screening of other halogenated organics in Soil, sediment and Sludge by GC × GC-μECD, Analytical and Bioanalytical Chemistry, in press (2011) DOI: 10:1007/s00216-011—5114-0.*

and their analysis. It must be noted that there are entire texts devoted to each of these compound classes, and this information is intended to be a starting point for the interested reader. Even with GC and GC-MS being considered mature, the rate of advancement of analytical methods is still fairly fast. Finally, GC × GC shows considerable promise for the possible combination of these individual methods into one or two analyses. Figure 28.7 is the GC × GC-ECD chromatogram of most of the compound classes mentioned in this chapter, resolved in a single injection. This will likely continue to be explored as a way of increased resolution and sensitivity for many analyses. There has been considerable refinement in the methods discussed here relative to where a lot of this first began in the 1970s. It is certain that this will continue to evolve.

References

[1] D.C.G. Muir, P.H. Howard, Are there other persistent organic pollutants? A challenge for environmental chemists, Environ. Sci. Technol. 40 (2006) 7157–7166.

[2] D. Hayward, Identification of bioaccumulating polychlorinated naphthalenes and their toxicological significance, Environ. Res. 76 (1998) 1–18.

[3] E.J. Reiner, A.R. Boden, T. Chen, K.A. MacPherson, A.M. Muscalu, Advances in the analysis of persistent halogenated organic compounds. LC GC Europe, February, 23 (2) (2010) 60–70.

[4] N. Yamashita, S. Taniyasu, N. Hanari, Y. Horii, J. Falandysz, Polychlorinated naphthalene contamination of some recently manufactured industrial products and commercial goods in Japan, J. Environ. Sci. Health - Part A Toxic/Hazardous Substances Environ. Eng. 38 (2003) 1745–1759.

[5] R. Carson, Silent spring, Houghton Miffin, Boston USA, 1962.

[6] A.T. James, A.J.P. Martin, Gas-liquid partition chromatography; the separation and micro-estimation of volatile fatty acids from formic acid to dodecanoic acid, Biochem. J. 35 (1952) 679–690.

[7] J.E. Lovelock, Ionization methods for the analysis of gases and vapors, Anal. Chem. 33 (1961) 162–178.

[8] K.D. Bartle, P. Myers, History of gas chromatography, Trends. Anal. Chem. 21 (2002) 547–557.

[9] P. Korytar, H.-G. Janssen, E. Matisova, U.A.T. Brinkman, Practical fast gas chromatography: methods, instrumentation and applications, Trends. Anal. Chem. 21 (2002) 558–572.

[10] K. Mastovska, S.J. Lehotay, A practical approaches to fast gas chromatography-mass spectrometry, J. Chormatogr. 1000 (2003) 153–180.

[11] M. Adahchour, J. Beens, J.J. Vreuls, U.A.T. Brinkman, Recent developments in the application of comprehensive two-dimensional gas chromatography, J. Chromatogr. A 1186 (2008) 67–108.

[12] O. Panic, T. Gorecki Comprehensive, two-dimensional gas chromatography (GC × GC) in environmental analysis and monitoring, Anal. Bioannal. Chem. 386 (2006) 1013–1023.

[13] Stockholm Convention Secretariat, 2001, UNEP, http://chm.pops.int/

[14] E.J. Reiner, R.E. Clement, A.B. Okey, C.H. Marvin, Advances in analytical techniques for polychlorinated dibenzo-p-dioxins, polychlorinated dibenzofurans and dioxin-like PCBs Anal, Bioanal. Chem. 386 (2006) 791–806.

[15] "PCB transformers and capacitors: from management to reclassification and disposal", UNEP Chemicals, Geneva, Switzerland, www.chem.unep.ch/pops/pdf/PCBtrnscap.pdf

[16] M.D. Erickson, Analytical chemistry of PCB's, Butterworth Pub., Stoneham, MA, 1986.

[17] Aroclor and other PCB mixtures, USEPA web document, http://www.epa.gov/osw/hazard/tsd/pcbs/pubs/aroclortable.pdf

[18] USEPA SW-846, Method 8082A, Revision 1, Feb, 2007.

[19] World Health Organization, Re-evaluation of human and mammalian toxic equivalency factors for dioxins and dioxin-like compounds (2005). http://www.who.int/ipcs/assessment/tef_update/en/

[20] G.M. Frame, J.W. Cochran, et al., Complete PCB congener distributions for 17 aroclor mixtures determined by 3 HRGC systems optimized for comprehensive, quantitative, congener-specific analysis, Hrc-J. High Resol. Chromatogr. 19 (12) (1996) 657–668.

[21] R. Guitart, P. Puig, et al., Requirement for a standardized nomenclature criterium for PCB's – computer assisted assignment of correct congener denomination and numberings, Chemosphere 27 (8) (1993) 1451–1459.

[22] K. Ballschmiter, M. Zell, Analysis of polychlorinated-biphenyls (PCB) by glass-capillary gas-chromatography - composition of technical Aroclor-PCB and Clophen-PCB mixtures, Fresenius Zeitschrift Fur Analytische. Chemie. 302 (1) (1980) 20–31.

[23] SGE Application note, http://www.sge.com/uploads/eb/ce/ebcee7536aa32ddab2c44fcbaac67f33/AP-0040-C_RevB.pdf

[24] J.F. Focant, A. Sjodin, D.G. Patterson Jr., Improved separation of the 209 polychlorinated biphenyl congeners using comprehensive two-dimensional gas chromatography-time-of-flight mass spectrometry, J. Chromatogr. A 1040 (2004) 227–238.

[25] A.B. Okey, An aryl hydrocarbon receptor odyssey to the shores of toxicology: the Deichmann lecture, International Congress of Toxicology-XI, Toxicol. Sci. 98 (2007) 5–38.

[26] E.J. Reiner, The analysis of dioxins and related compounds, Mass Spectrom Rev 29 (2010) 526–559.

[27] J. Boer, Capillary gas chromatography for the determination of halogenated micro-contaminants, J. Chromatogr. A 843 (1999) 179–198.

[28] S.P.J. Van Leeuwen, J. de Boer, Advances in the gas chromatographic determination of persistent organic pollutants in the aquatic environment, J. Chrom. A 1186 (2008) 116–182.

[29] J. DeVos, P. Gorst-Allman, E. Rowher, Establishing an alternative method for the quantitative analysis of polychlorinated dibenzo-p-dioxins and polychlorinated dibenzofurans by comprehensive two dimensional gas chromatography-time-of-flight mass spectrometry for developing countries, J. Chrom. A 1218 (2011) 3282–3290.

[30] J.F. Focant, E.J. Reiner, K. MacPherson, T. Kolic, A. Sjodin, D.G. Patterson Jr., et al., Measurement of PCDD, PCDFs, and non-ortho-PCBs by comprehensive multidimensional gas chromatography-isotope dilution time-of-flight mass spectrometry (GcxGC-IDTOFMS)", Talanta 63 (2004) 1231–1240.

[31] E.J. Reiner, D.H. Schellenberg, V.Y. Taguchi, Environmental applications for the analysis of chlorinated dibenzo-p-dioxins and dibenzofurans using mass spectrometry/mass spectrometry, Environ. Sci. Technol. 25 (1991) 110–117.

[32] M. Alaee, P. Arias, A. Sjodin, A. Bergman, An overview of commercially used flame retardants, their applications, their use patterns in different countries/regions and possible modes of release, Environ. Int. 29 (2003) 683–689.

[33] T.M. Kolic, L. Shen, K. MacPherson, L. Fayez, T. Gobran, P. Helm, et al., The analysis of halogenated flame retardants by GC-HRMS in environmental samples, J. Chromatogr. Sci. 47 (2009) 83–91.

[34] A. Covaci, S. Harrad, M.A.-E. Abdallah, N. Ali, R.J. Law, D. Herzke, et al., Novel brominated flame retardants: a review of their analysis, environmental fate and behaviour, Environ. Int. 37 (2011) 532–556.

[35] A. Covaci, S. Voorspoels, J. de Boer, Determination of brominated flame retardants, with emphasis on polybrominated diphenyl ethers (PBDEs) in environmental and human samples – a review, Environ. Int. 29 (2003) 735–756.

[36] S.P.J. van Leeuwen, J. de Boer, Extraction and clean-up strategies for the analysis of poly- and perfluoroalkyl substances in environmental and human matrices, J. Chromatogr. A 1153 (2007) 172–185.

[37] M. Villagrasa, M. López De Alda, D. Barceló, Environmental analysis of fluorinated alkyl substances by liquid chromatography-(tandem) mass spectrometry: a review, Anal. Bioanal. Chem. 386 (2006) 953–972.

[38] J.C. D'Eon, P.W. Crozier, V.I. Furdui, V.I.E.J. Reiner, E.L. Libelo, S.A. Mabury, Observation of a commercial fluorinated material, the polyfluoroalkyl phosphoric acid diesters, in human sera, wastewater treatment plant sludge, and paper fibers, Environ. Sci. Technol. 43 (2009) 4589–4594.

[39] J.P. Benskin, A.O. De Silva, J.W. Martin, Isomer profiling of perfluorinated substances as a tool for source tracking: a review of early findings and future applications, Rev. Environ. Contam. Toxicol. 208 (2010) 111–160.

[40] D.L. Poster, M.M. Schantz, L.C. Sander, S.A. Wise, Analysis of polycyclic aromatic hydrocarbons (PAHs) in environmental samples: a critical review of gas chromatographic (GC) methods, Anal. Bioanal. Chem. 386 (2006) 859–881.

[41] A.R. Boden, E.J. Reiner, Development of an isotope-dilution gas chromatographic-mass spectrometric method for the analysis of polycyclic aromatic compounds in environmental matrices, Polycyclic Aromatic Comp. 24 (2004) 309–323.

CHAPTER

29

Role of Gas Chromatography in the Identification of Pheromones and Related Semiochemicals

Jocelyn G. Millar

OUTLINE

29.1. INTRODUCTION

Scientists have known that insects use volatile chemical signals (pheromones) for communication between individuals since the early 1900s, but it was not until 1959 that the first sex attractant pheromone, bombykol (10*E*,12*Z*-hexadecadien-1-ol), was identified from the silk moth, *Bombyx mori* [1]. The identification was done before the advent of gas chromatography as a mainstream method, and took several decades of work and 500,000 female moths to obtain the 6 mg of a crystalline derivative of bombykol from which the structure was determined. To place this in perspective in terms of how far we have come, with modern gas chromatography-mass spectrometry (GC-MS), aided by coupled GC-electroantennogram detection (GC-EAD) and other techniques, the same identification now could

 DOI: 10.1016/B978-0-12-385540-4.00029-8

be carried out in a few days with extracts from a few moths or even a single female moth. It is no exaggeration to say that gas chromatography revolutionized and invigorated the discipline of chemical ecology, as evidenced by the exponential growth in the identifications of semiochemicals and the resulting publications through the 1960s and 70s. Today, due in large part to gas chromatography, several thousand insect pheromones have been identified [2], along with numerous other semiochemicals that mediate insect behaviors or insect physiology, such as the volatile cues produced by plant and animal hosts upon which insects feed and lay eggs. Even the relatively nonvolatile contact pheromones that are constituents of the hydrocarbon layer of insect cuticles, and that insects use for short-range recognition of sex and species, have proven amenable to gas chromatographic analysis [3]. A number of microscale analytical techniques have been specifically developed or adapted for use with semiochemical research. These include aeration apparatus for collection of nanogram amounts of bioactive compounds from live organisms of all types, development of GC-EAD as a tool for rapidly locating likely semiochemical components in crude extracts, and development of a variety of microchemical methods to aid in structure determination [4,5].

Gas chromatography is an essential tool for semiochemical research for several reasons. First and most obvious, pheromones and other signaling molecules are usually only produced in minute amounts (approximately nanograms to femtograms per individual), and are often only trace components of extracts, depending to some extent on how the extract was made. Thus, a key characteristic that differentiates many semiochemical identification projects from other natural product identifications is the fact that researchers are inherently limited in the amount of material that can be obtained, either because samples are difficult to obtain or because the compounds are produced in minute amounts, or a combination of the two. Thus, both the sensitivity and the excellent separation power of gas chromatography are major advantages in comparison to other separation methods. The latter characteristic is particularly important for identification of pheromones of lepidopteran insects, which frequently consist of precise blends of positional or geometric isomers that would be difficult to separate and quantify by other methods [2]. GC separation power is further increased by the variety of stationary-phase chemistries that are available. These can be exploited to produce quite different separations and retention orders of the compounds in an extract, based on the combination of analyte vapor pressure and polarity. This not only allows the separation of compounds that may be overlapped on one column, but also provides cross-referenced retention time data indicative of specific functional groups, methyl branches, or other structural features (*vide infra*).

The value of the excellent separating power of GC is further enhanced by the coupling of GC to electroantennography (GC-EAD) and to mass spectrometry. Analysis of an aliquot of a crude semiochemical extract with GC-EAD locates the likely bioactive components, and subsequent analysis by GC-MS then provides mass spectra of each of these components. Thus, two straightforward analyses of a single extract can provide both the number of likely bioactive components in the extract and spectral and retention time data to assist in their identifications.

Another and possibly less obvious advantage of GC and GC-MS over other separation and analysis methods for chemical ecology research is that the flame ionization detector and MS analysis with 70-eV electron impact or chemical ionization are essentially universal and highly sensitive (picogram) detection methods that are relatively insensitive to molecular structure. In contrast, detectors used with HPLC, such as UV-visible diode array detectors, are highly dependent on molecular structure, with compounds lacking a chromophore being largely invisible to the detector. The detection problem carries through to LC-MS to some extent, because almost all ionization methods used with LC-MS require that a functional group be present in an analyte in order for it to be ionized and thus become visible to the mass spectrometer. Thus, LC-MS may fail to detect alkanes altogether, and may detect alkenes only poorly.

The GC columns and methodology used in chemical ecology research from the 1960s to the 1990s have been summarized [6], and the interested reader is referred to these references. The pioneers of chemical ecology were quick to realize the enormous potential of using gas chromatography for semiochemical identification and analysis, initially with packed columns, and then with capillary columns as they became available. Even with packed columns, nanogram amounts could be detected, representing an increase in sensitivity of orders of magnitude over previously available LC methods. The introduction of capillary GC further decreased detection limits, to approximately the picogram range. However, even this level of sensitivity is still inadequate for insects that produce subpicogram amounts of pheromones, below the detection limits of GC or MS detectors. In such cases, it has still been possible to fully identify new structures by using an insect's antenna as a living detector with GC-EAD analyses. Because insect antennae can often detect pheromones into the femtogram range, GC-EAD can be several orders of magnitude more sensitive (and more selective) than an FID detector. Then, careful use of comparative retention index data on several different columns can provide sufficient information about the structure, such as chain length and functional groups, that a tentative identification can be made. The details of this process of identification of vanishingly small quantities of pheromones are discussed below.

29.2. COUPLED GAS CHROMATOGRAPHY-ELECTROANTENNOGRAM DETECTION (GC-EAD)

The electroantennogram assay, in which an insect antenna is used as a living detector, was developed by Dietrich Schneider in the 1950s during the identification of bombykol described above [7], and it rapidly became one of the most useful tools for insect chemical ecology. The method is essentially unique to insects and related arthropods because unlike most other animals, insects' sensory organs, such as the antennae and mouthparts, are located on the body surface and so are readily accessible. In addition, because insects' respiration occurs by passive diffusion of oxygen into their tissues through spiracles, a detached insect antenna can remain alive and functional for many hours. Furthermore, insects are particularly suitable subjects for chemical ecology research because so much of their biology and ecology is mediated by chemical signals rather than other forms of communication.

Moorhouse and co-workers were the first to couple GC with EAD [8]. Because they used packed columns, with peak widths of a minute or more, the continuous flow of the GC effluent over the antennal preparation was found to be unsatisfactory. Thus, they developed a system for collecting the GC effluent in 15-second aliquots, and then pulsed each aliquot over the antenna so that the antenna was stimulated with a concentrated burst of pheromone rather than a continuous flow. With the advent of capillary GC, with much narrower peak widths of only a few seconds, continuous flow of the column effluent over the antenna was found to work well [9]. The impact of this technique in advancing chemical ecology research with arthropods cannot be overstated, because it freed researchers from the drudgery of using bioassay-driven fractionation to locate and isolate bioactive compounds in extracts. Instead, one simple GC-EAD analysis of a crude extract is often sufficient to pinpoint the active compounds in the extract, so that further efforts can be immediately focused on isolating and identifying those compounds. Another major advantage from the streamlining effect of GC-EAD is that far less starting material is required. That is, the multiple sequential fractionation steps usually needed in traditional bioassay-driven fractionation, multiplied by the replication needed for reliable bioassay results at each step, required that hundreds or thousands of insects be extracted to obtain sufficient material to carry a project through to completion. This problem is compounded if the pheromone consists of several components, none of which is active alone, so that subsets of fractions have to be recombined in bioassays in order to locate those fractions that contain active components. All of this is rendered unnecessary with GC-EAD to guide one directly to the bioactive compounds in an extract.

An example is shown in Figure 29.1, in which GC-EAD was used to clearly pinpoint the pheromone component in a crude extract from female grape mealybugs [10]. In many cases, isolation of individual compounds for identification may not be necessary; it is often possible to identify new insect semiochemicals by a combination of mass spectral interpretation, information from comparative retention times on different GC columns, and microchemical tests for functional groups [11]. This is particularly true when working with insects such as moths, where hundreds of pheromones from a wide range of lepidopteran families are known. Because related species usually use similar, if not the same, pheromone components (usually in different ratios) [2], likely structures are often predictable, and can be verified by comparisons of the mass spectrum and retention times (on several different GC columns) of the unknown with those of standards. The construction of GC-EAD amplifiers and apparatus has been reviewed [12,13], and commercial EAD amplifiers and associated hardware and software are available from Syntech (Kirchzarten, Germany) and Grass Technologies (West Warwick RI, USA).

GC-EAD works best for pheromones because a large fraction of an insect's antennal receptors is specifically tuned to its pheromones. However, the technique is by no means restricted

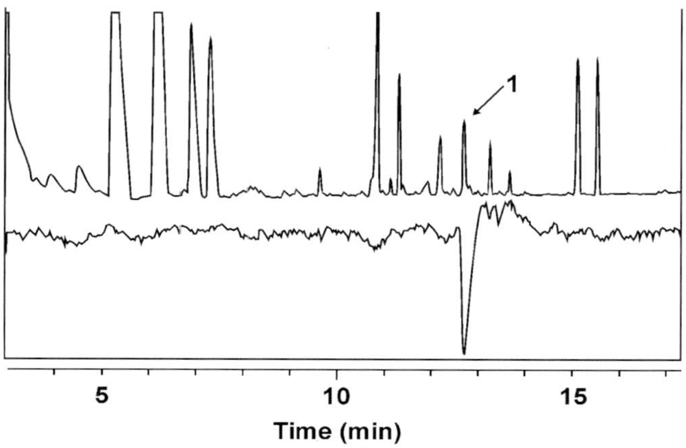

FIGURE 29.1 Coupled gas chromatography-electroantennogram analysis of an extract of headspace volatiles from virgin female grape mealybugs. Top trace shows the chromatogram; bottom, inverted trace shows the male's antennal response. *Source: Reprinted with Permission from [10].*

to pheromones. It also has been widely used to study topics such as host plant selection by herbivorous insects, host location by blood-feeding insects such as mosquitoes or biting flies, or oviposition site selection. However, it must be emphasized that a strong EAD signal only indicates that the insect has receptors for and can detect a stimulus; it provides no information as to the biological function of the stimulus, which could be an attractant, a repellent, a behavioral inhibitor, a defensive compound, or some other behavior-modifying chemical. In addition, molecules that are very similar in structure to the natural ligands of antennal receptors can elicit quite strong responses.

The GC-EAD technique has been further refined by using microelectrodes to record from only a single receptor cell instead of from a whole antennal preparation [14,15]. Because the individual cells are usually quite specific for one or a few related compounds, this enables a specific cell or cell type to be directly correlated with a particular stimulus.

29.3. USE OF COMPARATIVE GC RETENTION INDICES IN STRUCTURE IDENTIFICATION

One of the most useful characteristics of gas chromatography for identification of trace amounts of analytes is that the retention times of compounds on a particular stationary phase *relative to a defined set of standards* (usually straight-chain alkanes, but methyl esters of alkanoic acids have also been used) are highly reproducible. This allows the calculation of retention indices by a simple algebraic formula for programmed temperature runs [16,17]. These retention indices are in essence analogous to other physical constants such as boiling point that are part of the chemical and physical properties of a particular compound. The retention indices of a particular compound are different on different stationary phases because of the interactions of the analytes with the chemistries of the stationary phases. Thus, exact retention time (=retention index) matches between an unknown and a standard of known structure on two or more different columns provide compelling evidence for an identification.

In addition, a substantial amount of information about the structure of an unknown can be gleaned from retention indices because of the diagnostic shifts in retention indices that result from the presence of specific functional groups in an analyte. It may even be possible to determine the position and geometry of double bonds. For example, hexadecan-1-ol has a retention index (RI) of 1882 on a nonpolar DB-5 column [18], an increment of 282 RI units vs. hexadecane (RI = 1600 by definition). In contrast, on a polar DB-WAX column, the RI of hexadecan-1-ol is 2385, reflecting the strong interactions between the polar alcohol group and the polar stationary phase. Furthermore, the *differential between the retention indices* on any two phases (here, 2385−1882 = 503 RI units) is also robust. That is, in its simplest form, a retention index differential of about 500 units for an unknown that is analyzed on DB-5 and DB-WAX suggests that the unknown may have a primary alcohol group. Adding even more power to the system, the increments caused by different functional groups or other structural features are approximately additive. Thus, analysis of an unknown on two or more phases of different polarities followed by careful use of the retention indices and retention index differentials between columns allows both an estimate of the chain length or size of the molecule and an estimation of the number and type of functional groups that might be present. For example, retention indices and differentials provided crucial data in the identification of a dimethyl branched carboxylic acid pheromone for a cerambycid beetle [19]. Because this pheromone was the first ever identified from an insect in this subfamily, and the first female-produced pheromone from within the whole family of more than 35,000 identified species, there was no *a priori* information about the structure. Furthermore, it was available in only nanogram quantities, so that the retention times and mass spectrum constituted all the data available for its identification. In a slightly different context, retention indices allow estimation of the number and placement of methyl groups in the branched chain alkanes that are ubiquitous in insect cuticular lipids, components of which are often contact pheromones [20,21].

Whereas retention indices can provide valuable data for the identification of any unknown, the method is most powerful in cases where the unknown pheromone is present in such low amounts as to be undetectable by GC or GC-MS. In such cases, the analyst can take advantage of the femtogram-level sensitivity of insect antennae for pheromone components, and use GC-EAD to obtain retention indices and differentials of the unknown on several columns. With only these data to work with, it has been possible to identify otherwise undetectable amounts of a number of novel pheromones (e.g. midges [22–24] and Lepidoptera [25,26]).

29.4. DETERMINATION OF ENANTIOMERIC PURITY AND ABSOLUTE CONFIGURATION

For insects such as scarab beetles [27] or some Lepidoptera [28] that produce chiral pheromones, enantiomeric purity can be critically important to biological activity, with strong inhibition of behaviors by even trace amounts of the unnatural enantiomer. The use of one enantiomer in combination with strong inhibition by the other enantiomer reinforces the uniqueness of the pheromone signal and prevents energetically wasteful cross-attraction to females (or males, as the case may be) of the wrong species. Still other insects have gone a step further by using scalemic mixtures in their pheromone blends [28,29,30], analogous to other insects using specific mixtures of positional or geometric isomers to form unique pheromone signals. Before the introduction of chiral GC stationary phases, determination of the enantiomeric purity or ratio of enantiomers in a pheromone required relatively large amounts of material (tens of nanograms or more) and careful derivatization with a chiral derivatizing agent to form diastereomeric derivatives that were hopefully separable on a normal GC column or by HPLC [31]. In the worst cases, for example, with chiral alkanes that have no functional groups that can be derivatized, microscale resolution and determination of enantiomeric purity by any means were not possible. Similarly, determination of the enantiomeric purity of typical components of host plant extracts such as mono- and sesquiterpenes was laborious and tedious at best.

With the advent of chiral GC stationary phases, and particularly the cyclodextrin-based phases that do not require analytes to have polar functional groups in order to achieve resolution, what was once a thorny problem has been reduced in most cases to a straightforward GC analysis that can be done with subnanogram amounts. As with the enhancement of separating power achieved by the variety of normal GC stationary-phase chemistries, the breadth of different chiral phase chemistries and different resolving mechanisms provides additional options for enantiomeric analysis. With these powerful tools now available, pheromone identifications now routinely include determination of the absolute configuration for those compounds that are chiral. Chiral stationary-phase columns have also been employed in GC-EAD analyses of crude pheromone extracts, using the insect antenna as a detector to show unequivocally which enantiomer the insects produce [22]. GC-based and other methods for determination of the absolute configuration and enantiomeric purities of pheromones [31] and the full range of biological activities of enantiomers of pheromones from synergists to powerful inhibitors of behavior [28] have been reviewed.

The determination of the enantiomeric composition of bioactive compounds in the context of chemical ecology has also carried over into the analysis of plant volatiles, with many studies now reporting both the identifications and the enantiomeric compositions of the individual bioactive compounds. In the course of this work, and the analysis of essential oils in general, one of the surprising things that has emerged is the plasticity in the production of enantiomers by plants. It is not unusual for plants to produce scalemic mixtures, and changes in the enantiomeric composition of plant volatiles can occur in response to herbivory or other stresses [32–34].

One lingering problem with completely novel compounds is that although it is possible to resolve enantiomers of most volatile semiochemicals on a chiral GC stationary phase, there is no *a priori* method of knowing which enantiomer is which. Thus, in most cases, it is still necessary to stereoselectively synthesize at least one enantiomer as a standard. To date, the only other solution to this problem is to compare the theoretically calculated vibrational circular dichroism (VCD) spectrum of each enantiomer with the experimentally measured spectrum of the isolated pheromone, or of one of the enantiomers resolved from the synthetic racemate by HPLC on a chiral stationary phase or an enzyme-based kinetic resolution. However, milligram quantities may be required to obtain a VCD spectrum of quality, sufficient for comparison with the theoretically calculated spectra [35,36].

29.5. MICROSCALE PREPARATIVE GAS CHROMATOGRAPHY

Preparative gas chromatography has been and remains enormously useful in the isolation and identification of volatile semiochemicals for several reasons. First, even packed column GC has considerable separation power, with a typical packed column having several thousand theoretical plates, while also being capable of separating nanogram to milligram quantities of material. Separation power is enhanced by an order of magnitude by using megabore columns for separation of a few micrograms or less, or even capillary columns for separation of submicrogram quantities [37]. With modern microprobe NMR instruments, one microgram is enough to obtain basic proton and COSY spectra sufficient to identify most compounds [38,39]. Overall, preparative GC has been widely used in chemical ecology, from preliminary fractionation of crude extracts through to isolation of pure products.

The second major advantage of preparative gas chromatography is that it provides pure samples of compounds that are free of solvent and other volatile impurities, ready for NMR analysis. The importance of this should not be underestimated because it is virtually impossible to obtain microgram quantities of pure samples of volatile compounds, free of traces of solvent that complicate NMR analysis, by any other method. From bitter personal experience, attempts to remove traces of solvent from extracts or liquid chromatography fractions containing microgram amounts of a volatile semiochemical, for example, by concentrating under a gentle stream of nitrogen, often result in loss of most or all of the sample.

Preparative GC fractions are typically collected in dry-ice-cooled capillary tubes [6] or, more recently, in sections of uncoated megabore GC column [40]. Apparatus for collection of microscale preparative GC fractions has been described by several authors [6,38–40]. Whereas commercial preparative GCs are available, it is straightforward to adapt an analytical GC for preparative work, for example, by using a GC-MS interface as a heated outlet port [6].

Alternatively, an unused injector or detector block can be readily modified to form a heated outlet port [6]. Because general-purpose GCs usually are equipped with flame ionization detectors (FIDs) that destroy the sample, for preparative work the column effluent is split with a Y connector, with a small percentage of the flow going to the FID detector, and the bulk being directed to the fraction collector.

29.6. SUMMARY

Gas chromatography is an indispensible tool for chemical ecology research, having been used in the identifications of nearly all volatile semiochemicals found to date. As a result, our understanding of intra- and interspecific communication within and between organisms varying from microbes to elephants has been profoundly increased. This in turn has allowed us to use chemical ecology in the manipulation and control of both beneficial and deleterious organisms, and particularly, in the detection, monitoring, and control of pest insects.

References

[1] E. Hecker, A. Butenandt, Bombykol revisited – reflections on a pioneering period and on some of its consequences, in: H.E. Hummel, T.A. Miller (Eds.), Techniques in pheromone research, Springer-Verlag, New York, NY, 1984, pp. 1–44.

[2] A.M. El-Sayed, The pherobase: database of insect pheromones and semiochemicals, (2011), http://www.pherobase.com.

[3] G.J. Blomquist, A.-G. Bagneres, Insect hydrocarbons. Biology, biochemistry, and chemical ecology, Cambridge University Press, Cambridge, UK, 2010.

[4] H.E. Hummel, T.A. Miller (Eds.), Techniques in pheromone research, Springer-Verlag, New York, NY, 1984.

[5] J.G. Millar, K.F. Haynes (Eds.), Methods in chemical ecology, Chemical methods, vol. 1, Chapman & Hall, Norwell MA, 1998.

[6] R.R. Heath, B.D. Deuben, Analytical and preparative gas chromatography, in: J.G. Millar, K.F. Haynes (Eds.), Methods in chemical ecology, Chemical methods, vol. 1, Chapman & Hall, Norwell MA, 1998, pp. 85–125.

[7] D. Schneider, Elektrophysiologische Untersuchungen von Chemo- und Mechanorezeptoren der antenne des Seidenspinners *Bombyx mori* L, Z. Vergl. Physiol. 40 (1957) 8–41.

[8] J.E. Moorhouse, R. Yeadon, P.S. Beevor, B.F. Nesbitt, Method for use in studies of insect communication, Nature 223 (1969) 1174–1175.

[9] H. Arn, E. Städler, S. Rauscher, The electroantennographic detector – a selective and sensitive tool in the gas chromatographic analysis of insect pheromones, Z. Naturforsch. 30c (1975) 722–725.

[10] B.A. Figadère, J.S. McElfresh, D. Borchardt, K.M. Daane, W. Bentley, J.G. Millar, Trans-α-necrodyl isobutyrate, the sex pheromone of the grape mealybug, *Pseudococcus maritimus*, Tetrahedron Lett. 48 (2007) 8434–8437.

[11] A.B. Attygalle, Microchemical techniques, in: J.G. Millar, K.F. Haynes (Eds.), Methods in chemical ecology, Chemical methods, vol. 1, Chapman & Hall, Norwell MA, 1998, pp. 207–294.

[12] D.L. Struble, H. Arn, Combined gas chromatography and electroantennogram recording of insect olfactory responses, in: H.E. Hummel, T.A. Miller (Eds.), Techniques in pheromone research, 1984, pp. 161–178.

[13] L.B. Bjostad, Electrophysiological methods, in: J.G. Millar, K.F. Haynes (Eds.), Methods in chemical ecology, Chemical methods, vol. 1, Chapman & Hall, Norwell MA, 1998, pp. 339–375.

[14] L.J. Wadhams, The coupled gas chromatography-single cell recording technique, in: H.E. Hummel, T.A. Miller (Eds.), Techniques in pheromone research, 1984, pp. 179–190.

[15] M.C. Stensmyr, M.C. Larsson, S. Bice, B.S. Hansson, Detection of fruit- and flower-emitted volatiles by olfactory receptor neurons in the polyphagous fruit chafer *Pachnoda marginata* (Coleoptera: Cetoniinae), J. Comp. Physiol. A Sens. Neural. Behav. Physiol. 187 (2001) 509–519.

[16] H. Van den Dool, P.D. Kratz, A generalization of the retention index system including temperature programmed gas-liquid partition chromatography, J. Chromatogr A. 11 (1963) 463–471.

[17] E. Kováts, Gas chromatographic characterization of organic substances in the retention index system, Adv. Chromatogr. 1 (1965) 229–247.

[18] F.A. Marques, J.S. McElfresh, J.G. Millar, Kovats retention indexes of monounsaturated C12, C14, and C16 alcohols, acetates, and aldehydes commonly found in lepidopteran pheromone blends, J. Braz. Chem. Soc. 11 (2000) 592–599.

[19] J. Rodstein, J.S. McElfresh, J.D. Barbour, A.M. Ray, L.M. Hanks, J.G. Millar, Identification and synthesis of a female-produced sex pheromone for the cerambycid beetle *Prionus californicus*, J. Chem. Ecol. 35 (2009) 590–600.

[20] D.A. Carlson, U.R. Bernier, B.D. Sutton, Elution patterns from capillary GC for methyl-branched alkanes, J. Chem. Ecol. 24 (1998) 1845–1865.

[21] S. Schulz, Composition of the silk lipids of the spider *Nephila clavipes*, Lipids 36 (2001) 637–647.

[22] R. Gries, G. Khaskin, G. Gries, R.G. Bennett, G.G.S. King, P. Morewood, et al., (Z,Z)-4,7-tridecadien-(*S*)-2-yl acetate: sex pheromone of Douglas-fir cone gall midge, *Contarinia oregonensis*, J. Chem. Ecol. 28 (2002) 2283–2297.

[23] R. Gries, G. Khaskin, R.G. Bennett, A. Miroshnychenko, K. Burden, G. Gries, (*S*,*S*)-2,12-, (*S*,*S*)-2,13-, and (*S*,*S*)-2,14-diacetoxyheptadecanes: Sex pheromone components of red cedar cone midge, *Mayetiola thujae*, J. Chem. Ecol. 31 (2005) 2933–2946.

[24] M.-Y. Choi, G. Khaskin, R. Gries, G. Gries, B.D. Roitberg, D.A. Raworth, et al., (2*R*,7*S*)-diacetoxytridecane: Sex pheromone of the aphidophagous gall midge, *Aphidoletes aphidimyza*, J. Chem. Ecol. 30 (2004) 659–670.

[25] J.G. Millar, A.E. Knudson, J.S. McElfresh, R. Gries, G. Gries, J.H. Davis, Sex attractant pheromone of the pecan nut casebearer, Bioorg. Med. Chem. 4 (1996) 331–339.

[26] M.L. Evenden, B.A. Mori, R. Gries, J. Otani, Sex pheromone of the red clover casebearer moth, *Coleophora deauratella*, an invasive pest of clover in Canada, Entomol. Exp. Appl. 137 (2010) 255–261.

[27] W.S. Leal, Scarab beetles, in: J. Hardie, A.K. Minks (Eds.), Pheromones of non-lepidopteran insects associated with agricultural plants, CABI Publishing, Wallingford, UK, 1999, pp. 51–68.

[28] K. Mori, Significance of chirality in pheromone science, Bioorg. Med. Chem. 15 (2007) 7505–7523.

[29] A.C. Oehlschlager, G.G.S. King, H.D. Pierce Jr., A.M. Pierce, K.N. Slessor, J.G. Millar, et al., Chirality of macrolide pheromones of grain beetles in the genera *Oryzaephilus* and *Cryptolestes* and its implications for species specificity, J. Chem. Ecol. 13 (1987) 1543–1554.

[30] D.R. Miller, J.H. Borden, K.N. Slessor, Inter- and intrapopulation variation of the pheromone, ipsdienol produced by male pine engravers, *Ips pini* (Say) (Coleoptera: Scolytidae), J. Chem. Ecol. 15 (1989) 233–247.

[31] K. Mori, Separation of enantiomers and determination of absolute configuration, in: J.G. Millar, K.F. Haynes (Eds.), Methods in chemical ecology, Chemical methods, vol. 1, Chapman & Hall, Norwell MA, 1998, pp. 295–338.

[32] J.F. Tooker, W.A. Koenig, L.M. Hanks, Altered host plant volatiles are proxies for sex pheromones in the gall wasp *Antistrophus rufus*, Proc. Natl. Acad. Sci. USA 99 (2002) 15486–15491.

[33] C.E. Reisenman, J.A. Riffel, E.A. Bernays, J.G. Hildebrand, Antagonistic effects of floral scent in an insect-plant interaction, Proc. Royal. Soc. B Biol. Sci. 277 (2010) 2371–2379.

[34] H.P. Bais, T.S. Walker, F.R. Stermitz, R.A. Hufbauer, J.M. Vivanco, Enantiomeric-dependent phytotoxic and antimicrobial activity of (±)-catechin. A rhizosecreted racemic mixture from spotted knapweed, Plant Physiol. 128 (2002) 1173–1179.

[35] B. Figadère, F.J. Devlin, J.G. Millar, P.J. Stephens, Determination of the absolute configuration of the sex pheromone of the obscure mealybug by vibrational circular dichroism analysis, Chem. Comm. (2008) 1106–1108.

[36] P.S. Stephens, F.J. Devlin, J.-J. Pan, The determination of the absolute configurations of chiral molecules using vibrational circular dichroism (VCD) spectroscopy, Chirality 20 (2008) 643–663.

[37] P.J. McCall, R.R. Heath, B.D. Dueben, M.D. Wilson, Oviposition pheromone in the *Simulium* damnosum complex: biological activity of chemical fractions from gravid ovaries, Physiol. Entomol. 22 (1997) 224–230.

[38] S. Nojima, D.J. Kiemle, F.X. Webster, W.L. Roelofs, Submicro scale NMR sample preparation for volatile chemicals, J. Chem. Ecol. 30 (2004) 2153–2161.

[39] S. Nojima, D.J. Kiemle, F.X. Webster, C.S. Apperson, C. Schal, Nanogram-scale preparation and NMR analysis for mass-limited small volatile compounds, PLoS One 6 (2011) e18178.

[40] S. Nojima, C.S. Apperson, C. Schal, A simple, convenient, and efficient preparative GC system that uses a short megabore capillary column as a trap, J. Chem. Ecol. 34 (2008) 418–428.

CHAPTER

30

Gas Chromatographic Analysis of Wines: Current Applications and Future Trends

Susan E. Ebeler

OUTLINE

Dedicated to Dr. Walt Jennings – a wonderful mentor and friend.

30.1. INTRODUCTION

Following the development of gas chromatography (GC) described in the seminal works of James and Martin in the early 1950s [1–7], among the early applications of GC were studies on the volatile composition of wines [8–12]. GC has continued to be important for understanding wine chemistry, and currently, the applications of gas chromatography for analysis of wine are vast. As with other applications, GC analysis of wines has benefited from advances in column technology, improved stationary phases, and new detection methods. Early GC analyses with packed columns and flame ionization detectors provided insights into the alcohol, ester, and aldehyde components that were present in relatively high concentrations and that influenced wine aroma [8–12]. In the mid- to late-1980s, the introduction of capillary columns with high resolving power and efficiencies resulted in a flurry of research activity and identification of hundreds

Gas Chromatography DOI: 10.1016/B978-0-12-385540-4.00030-4

of new volatile compounds in grapes and wines, including many that are present at trace levels [13,14]. At the current time, there is a resurgence of activity as multidimensional separations and new hyphenated detection techniques are allowing even greater resolution and sensitivity with more than 200 compounds being detected in a single run [15–17]. Several articles reviewing the gas chromatographic analysis of wine have been published e.g. [14,18–20]. This chapter will focus on recent applications of GC analysis of wines and alcoholic beverages with a focus on new stationary phases, detection methods, multidimensional separations, and sample preparation methodologies. The goal of this review is to highlight some selected and emerging trends and to provide insight into some of the future opportunities and challenges associated with gas chromatographic analysis of wines.

30.2. COLUMNS

The majority of GC separations for wine analysis continue to use polydimethylsiloxane-based and polyethylene glycol-based stationary phases, as reviewed in other chapters of this volume. However, two areas are currently receiving much attention for their potential as GC stationary phases: carbon-based nanotubes/nanofibers and ionic liquids.

Carbon-based sorbents have been used for isolation and separation of volatiles for many years [21]. Recently, the development of ***nanostructured carbon-based sorbents*** with fullerene-like structures has received attention as stationary phases due to their excellent adsorption and mass transfer properties for a wide range of compounds [22,23]. The nanotubes typically have diameters of <1 to 50 nm and their structures are generally single-walled or multiwalled with concentric layers (Figure 30.1) [24,25]. As discussed by Kwon and Park [26], the curved internal surface of nanotubes provides for stronger binding affinities for molecules, as compared to planar carbon surfaces.

In a recent application, Merli et al. [23] used functionalized multiwalled carbon nanotubes to separate alcohols and esters in distilled spirits. The columns gave good separation and quantification of ethyl acetate, methanol, ethanol, 1-propanol, 2-butanol, 2-methyl-1-propanol, and 3-methyl-1-butanol in these beverages. Analysis of methanol and ethanol can be challenging due to their poor retention on many GC stationary phases; however, with the functionalized nanotube phases, excellent retention and separation of these compounds were obtained. The columns are easy to prepare, reproducible, and cheaper than traditional stationary phases. However, the functionalized materials have a limited operational temperature range (<200 °C) and are currently limited to analysis of low-boiling analytes. In a novel application, Reid et al. [22] designed a microfabricated chip containing a carbon nanotube stationary phase for high-speed micro-GC analyses. Using the microchip, a series of C_6–C_{11} hydrocarbons was separated in less than 2.5 seconds. Although much more work is required before these columns can be used for more complex separations, such as are required for wine analysis, the nanotube stationary phases show much promise for rapid chromatographic analysis of volatile compounds.

Applications of ***ionic liquids*** as GC stationary phases are also of increasing interest. As recently reviewed [27–29], ionic liquids are salts with melting points below 100 °C; for GC applications, salts with melting points between −40 °C and 50 °C provide the best characteristics for GC separations. Cross-linking of the ionic liquids provides stability for both low- and high-temperature separations (up to 350 °C). A distinct advantage of many ionic liquid stationary phases is their ability to separate both polar and nonpolar analytes and they have also been applied to chiral separations.

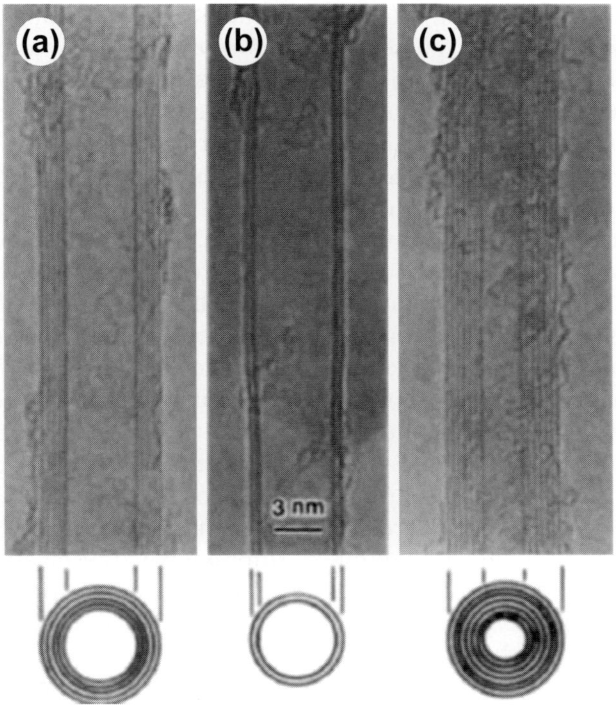

FIGURE 30.1 Electron micrographs of multiwall coaxial nanotubes. Parallel dark lines correspond to the lattice images of graphite. A cross-section of each tubule is illustrated. (a) Tube consisting of five graphitic sheets, diameter 6.7 nm; (b) two-sheet tube, diameter 5.5 nm; and (c) seven-sheet tube, diameter 6.5 nm, which has the smallest hollow diameter (2.2 nm). *Source: With Permission from [24].*

With simple modifications to the salt (i.e. by changing the cation, anion, or substituents), a high degree of analyte selectivity can be obtained; in some cases, the ionic liquid may provide novel selectivities not possible with current GC stationary phases, making these ionic liquid phases particularly promising [29].

In a recent application, cationic imidazolium-based ionic liquids were able to separate a complex mixture of hydrocarbons, alcohols, aldehydes, esters, terpenes, and ketones, typical of those found in wines [30]. Analysis of fatty acid esters (including *cis*- and *trans*-fatty acid isomers) and pesticides in a variety of food and beverage matrices has also been demonstrated [29–31]. Several ionic liquid GC stationary phases are commercially available providing a range of selectivities; however, further studies are needed to fully characterize the ionic liquid stationary phases and to fully evaluate their applications for complex matrices such as wines.

30.3. MULTIDIMENSIONAL SEPARATIONS

As recently reviewed by Ebeler and Thorngate [20], two- or multidimensional separations and the development of comprehensive GC × GC analyses have provided the opportunity for improved resolution of the hundreds of volatile compounds that are found in grapes and wines. In the most common application of

multidimensional GC, a chromatographic peak (or region) from one column is passed to a second column with a different stationary phase. Components that are not separated on the first column are separated on the second column by choosing a stationary phase for the second column that is different from and complementary to that of the first column. By using a chiral stationary phase in the second column, stereoisomers of selected compounds can be effectively separated in complex mixtures without prior isolation. Analysis of chiral compounds with aroma-active properties has been an important application of multidimensional GC for wine analysis [32–34]. In a recent application, Barba et al. [35] used multidimensional GC-MS to separate the stereoisomers of 2,3-butanediol and linalool in white wines. Since the different isomers of linalool have different sensory properties, this separation may provide an improved chemical characterization of wine aroma. Bertsch [36] and Mondello et al. [37] have provided excellent overviews of the theory and applications of multidimensional GC. Other chapters of this volume also provide more detailed discussions of multidimensional GC.

Comprehensive GC (or GC × GC) is a variation of multidimensional GC where the entire effluent from the first analytical column is placed on the second column and separated in the second dimension. Pulses of effluent from the first column are trapped over a defined time and then injected onto the second column. Modulation of the pulses requires careful optimization of instrumental parameters and, as with conventional two-dimensional GC, the two separation columns are chosen to be complementary to each other. The second column for comprehensive GC must also provide a rapid separation so that pulses of effluent remain focused and separated from each other; therefore, the second column is often shorter and has a narrower diameter and thinner film thickness than the first column. Ionic liquid stationary phases are of particular interest for multidimensional separations because the ionic liquids can be designed to provide a wide range of polarities for optimal separation of specific analyte functionalities [29]. Excellent reviews of comprehensive GC are provided by Bertsch [38] and Mondello et al. [37]. The advantages and disadvantages of both conventional and comprehensive GC have also been reviewed [36–39].

Comprehensive GC is a relatively new technique and so applications to wine analysis are somewhat limited. In general, the applications can be classified as targeted analyses focusing on quantification of specific known compounds and nontargeted analyses aimed at profiling large numbers of peaks in a sample without prior knowledge of analyte identity. An early application of comprehensive GC was the targeted analysis of 3-isobutyl-2-methoxypyrazine (IBMP) in wines [40–43]. IBMP is usually the most predominant methoxypyrazine found in grapes and wine, although other methoxypyrazines (e.g. 3-isopropyl-2-methoxypyrazine and 3-*sec*-butyl-2-methoxypyrazine) are also often present (reviewed in [44]). IBMP is found only in selected grape varieties (e.g. Cabernet Sauvignon, Sauvignon blanc, and Merlot) [45] and contributes a bell-pepper aroma when present at concentrations above the sensory threshold (~2–10 ng L^{-1} in water; [46]). The low sensory threshold requires very sensitive analyses which have been achieved by extensive solvent extraction combined with chemical ionization mass spectrometry [47] or, more recently, by solid-phase microextraction (SPME) of the headspace followed by GC-electron impact (EI)-MS in the selected ion mode (SIM) [48]. The one-dimensional GC-(EI)MS-SIM analysis often suffers from interferences with the methoxypyrazine peaks, however, making quantification difficult [40,43]. Comprehensive GC can improve the separation of IBMP from matrix interferences and using this technique, highly sensitive methods for analysis of IBMP have been reported in wines and grape berries, with limits of detection of 0.5 ng L^{-1} and 0.6–1.8 pg g^{-1}, respectively [40,42].

Schmarr et al. [43] have observed, however, that matrix interferences can also occur with the comprehensive GC analysis of IBMP in grapes and wines. These authors proposed that additional sample cleanup, combined with on-line high-performance liquid chromatography (HPLC) followed by multidimensional GC-MS, provided the best approach for removing matrix interferences in the analysis of IBMP. As the authors note, this hyphenated LC-GC approach is expensive and complex and may not be feasible for many applications. However, these results point to the need to carefully evaluate peak purity in such analyses, particularly for trace analytes such as IBMP. The development of standard protocols for monitoring analyte ion ratios and internal standard responses is necessary for ensuring that valid and accurate data are being obtained.

Recently, comprehensive GC × GC, combined with SPME, was used for analysis of ethyl carbamate (urethane) in wines [49]. Ethyl carbamate is a potential carcinogen formed in wines from urea and ethanol [50]. Voluntary limits for ethyl carbamate in wines have been established in the US (weighted average <15 μg/L; [51]), while in Canada, maximum levels (30 μg/L) have been set (limits are higher for fortified wines, distilled spirits, fruit brandies, and liquors [52]). The high resolving power of the GC × GC separation allowed for direct analysis of ethyl carbamate in Spanish Madeira wines without solvent extraction or derivatization as has been used in most previous methods. Limits of detection were well below the maximum allowed levels (LOD = 2.75–4.31 μg/L for dry and sweet model wine solutions, respectively).

Dasgupta et al. [53] separated an impressive 185 pesticides and organic pollutants in grape and wine extracts in less than 38 min using GC × GC-TOFMS. The authors report that the method provided improved sensitivity for a wider range of targeted analytes compared to previous reported procedures. While the high resolving power of GC × GC minimized many matrix interferences observed with other methods, some matrix enhancement or suppression was still observed for some analytes, requiring the use of matrix-matched standards for the most accurate quantification. In addition, deconvolution algorithms could not separate two pairs of coeluting analytes (diazinon and fluchloralin; pyremethanil and etrimphos) when these compounds were present at low concentrations (<25 μg L^{-1}).

Using GC × GC with either time-of-flight (TOF) or quadrupole MS detection, the nontargeted, simultaneous analysis of over 200 volatile compounds in wines has very recently been reported [17,54,55]. Approximately 2–4 times more compounds were detected in these studies compared to typical one-dimensional analyses [17,56]. The limitation of these analyses is the extensive time required for peak identification and data analysis and interpretation [54,55]. In an interesting application, Rocha et al. [56] profiled only the monoterpene compounds in grapes using GC × GC-TOFMS analysis and extracting ions that were characteristic of the terpenes (*m/z* 93, 121, and 136). When retention times of the extracted ion peaks from both separation columns were plotted against each other, the terpenes with similar structures clustered together; this clustering aided in compound identification. Schmarr et al. [54] used an isotopically labeled internal standard to provide quantitative information on volatiles measured in their study. Although internal standards were used in the studies of Robinson et al. [17] and Rocha et al. [56], no quantitative data were reported, albeit relative quantification could be possible.

The extensive characterization of volatile composition that is possible with comprehensive GC analysis now provides the opportunity for a much more detailed understanding of wine components that may contribute to sensory properties, as well as the ability to better discriminate among wines based on variety, growing region, vintage, or processing variables [17,54,55]. However, critical challenges remain in the availability of software and chemometric tools for analysis of the complex GC × GC data that are obtained. Research into the development of new data analysis and statistical tools is critical in order to efficiently process these large data sets, to allow for automated and accurate peak identifications, and to provide for sophisticated pattern recognition for sample classification.

30.4. DETECTORS AND HYPHENATED TECHNIQUES

Among the most common GC detectors for wine analysis are the flame ionization (FID) and mass spectrometric (MS) detectors. Other detectors can provide improved sensitivity and/or selectivity for some analytes relative to these "universal" detectors. For example, flame photometric (FPD) and chemiluminescence detectors have been widely used for analysis of sulfur-containing volatiles that may contribute to off-aromas of wines [57–59]. Electron-capture detectors (ECDs) have been used for analysis of trace levels of haloanisoles (e.g. 2,4,6-trichloroanisole responsible for cork-taint aromas; [60]), pesticide residues [61–65], and volatile aldehydes (as their pentafluorobenzyl hydroxylamine oximes) [66]. Nitrogen–phosphorus detectors (NPDs) have also seen applications for analysis of pesticide residues [65,67] as well as for analysis of methoxypyrazines that contribute a bell-pepper aroma to some wine varieties such as Cabernet Sauvignon [40,68]. The NPD detector has also been used for analysis of thiazolidine derivatives of aldehydes, including acetaldehyde, which contribute oxidative aromas to wines [69–71]. In an interesting application, an atomic emission detector (AED) was used to identify and quantify eight organic sulfur and two organic selenium species (dimethyl selenide and dimethyl diselenide) in wines [72]. Finally, coupling the GC effluent to an olfactory port and using the human nose as a detector has also been widely used to identify important odor-active compounds in wines (reviewed in [19,20]). Detailed discussions of GC detectors are also provided in other chapters of this volume.

Development and application of newer detector technologies that provide the potential for faster analyses with improved selectivity and sensitivity compared to traditional GC detectors are currently an active area of research. These new detector technologies include time-of-flight (TOF) mass spectrometry and hyphenated techniques, such as tandem mass spectrometry (MS/MS), for the analysis of a wide range of wine components.

Time-of-flight (TOF) MS detectors offer very fast acquisition rates yielding larger numbers of spectra across chromatographic peaks compared to quadrupole instruments. This aids in analyte resolution, quantification, and identification. Many commercial GCs with TOF detectors provide nominal mass accuracy (unit mass resolution), although new high-resolution instruments with accurate mass capabilities are now becoming more widely available. Recent studies by Setkova et al. [15,16] highlight the potential for GC-TOFMS analysis to provide rapid analysis of large numbers of analytes. In these studies ~201 volatile components in ice wine samples were identified and quantified in <5 min; the volatile profile was used to classify a range of ice wines according to their origin, grape variety, and processing conditions.

GC-TOFMS analysis is also widely used for metabolomics applications, often employing a two-step methoximation–silylation protocol to derivatize metabolites including amino acids, organic acids, sugars, sugar alcohols, sugar acids, and fatty acids contained in biological samples [73–76]. Skogerson et al. [77] recently identified 108 metabolites in several white wines

(Chardonnay, Pinot gris, Riesling, Sauvignon blanc, and Viognier) using GC-TOFMS. Differences in metabolite profiles were apparent among the different wines and could be used to predict differences in the sensory perception of wine "body" or viscous mouthfeel. Ding et al. [78] used GC-TOFMS profiling to characterize metabolic changes occurring during *Saccharomyces cerevisiae* fermentations. Atanassov et al. [79] have recently described the potential for metabolomics research to improve winemaking practices and grapevine breeding programs for producing high-quality wines.

The fast acquisition rates associated with the TOF detector are ideally suited for analysis of the narrow peaks associated with comprehensive GC analysis. Applications of GC × GC using TOF detectors for analysis of wines were reviewed above.

The hyphenated technique, ***tandem GC-MS (GC-MS/MS)***, is gaining increasing acceptance for targeted quantification of trace analytes in wines. In tandem MS, effluent from the GC column is ionized and enters the first MS where precursor ions are selected and transferred to a collision or reaction cell. In the collision cell, the precursor ions are bombarded with energy causing them to further fragment into product ions which are then detected. The GC-MS/MS experiment can be performed in several ways: (a) all precursor ion masses are scanned in MS1 and, following fragmentation in the collision cell, only selected product ions are monitored in MS2; (b) only a selected precursor ion from MS1 is allowed to pass into the collision cell and all product ions from the reaction are monitored; (c) similar to (a) above, all precursor masses are scanned in MS2 and fragmented in the collision cell – however, MS2 scans for masses associated with a constant neutral loss from the precursor ions; and (d) both MS1 and MS2 are set to select only for specific preidentified masses. Two types of GC-MS/MS instruments are available: quadrupole instruments, where the ions travel through the mass separators and collision cell sequentially in space, and ion-trap instruments, which contain an ion trap that manipulates the precursor ions in time to sequentially perform the functions of MS1, the collision cell, and MS2. With GC-MS/MS analyses, the high degree of selectivity for both precursor and product ions typically provides very sensitive analysis of trace compounds. In addition, analyte structural information can be obtained by monitoring the product ions produced from specific precursor ions. Theory and applications of tandem GC-MS have been reviewed [80–83].

One of the most widely used applications of GC-MS/MS for wine analysis is for multiresidue pesticide screening [84–88]. These methods provide rapid and simultaneous screening for a large number of pesticides, and, depending on the sample preparation procedures, limits of detection in the low pg mL^{-1} range have been reported [85].

GC-MS/MS has also been used for quantification of trace levels of aroma and off-aroma compounds. Recent applications include analysis of (a) compounds responsible for cork taint (2,4,6-trichloranisole, 2,3,4,6-tetrachloroanisole, 2,4,6-tribromoanisole, and pentachloroanisole) [89–95]; (b) *Brettanomyces*-related off-aromas (4-ethylguaiacol, 4-ethylphenol, 4-vinylguaiacol, and 4-vinylphenol) [94,95]; (c) methoxypyrazines associated with bell-pepper varietal aromas (2-methoxy-3-isobutylpyrazine, 2-methoxy-3-isopropylpyrazine, 2-methoxy-3-*sec*-butylpyrazine, and 2-methoxy-3-ethylpyrazine) [96]; (d) 1-octen-3-one which has a mushroom-like aroma [97]; and (e) rotundone, an oxygenated sesquiterpene which is associated with a peppery aroma in some wine varieties [98]. With the exception of the *Brettanomyces*-related aroma compounds, all of these compounds have aroma thresholds in the ng L^{-1} range and therefore require highly sensitive and selective methods for their analysis.

In an interesting application, the diastereoisomers of *S*-3-(hexan-1-ol)cysteine in grape juice were analyzed by ion-trap GC-MS/MS following isolation by affinity chromatography and derivatization with heptafluorobutyric anhydride and heptaflourobutanol [99]. Using this method, the authors observed that the diastereomeric ratio of this cysteinylated compound changed as a result of infection of the juice with the noble rot fungus, *Botrytis cinerea*. During fermentation, the cysteinylated precursor is enzymatically cleaved to form the important aroma compound 3-sulfanylhexan-1-ol. Thibon et al. [99] observed that the diastereomeric ratio of the precursor did not change during fermentation and the aglycone that was released also had a nearly identical isomeric ratio to that of the precursor. The diastereomers of the aglycon 3-sulfanylhexan-1-ol have similar and very low aroma thresholds of 50–60 ng L^{-1}; however, the aroma quality is dependent on the enantiomeric form. The R-form has a grapefruit-like character, while the aroma of the S-form is described as passion fruit [100]. The results of this study demonstrate the application of GC-MS/MS as an important tool for accurately measuring low levels of diastereomeric precursors that can impact wine sensory properties.

30.5. SAMPLE PREPARATION

No discussion of gas chromatographic analysis of wines would be complete without the mention of sample preparation. Jennings and Filsoof [101] used a model solution containing compounds with a range of functional groups and boiling points to elegantly demonstrate the dramatic qualitative and quantitative effects of different sample preparation techniques on GC responses. The importance of sample preparation was further emphasized by Flath [102]:

> "Numerous approaches have been developed for the concentration and isolation of volatile materials, but all have shortcomings which limit their universal applicability. It is necessary to be aware of these limitations and reduce their impact if possible by modifying and improving the most appropriate method, or by applying complementary techniques to compensate for shortcomings of any one method."

Sample preparation techniques for wine analysis were briefly reviewed by Polaskova et al. [19]. In many cases, sample preparations that include derivatization reactions can be used to enhance the volatility and/or stability of analytes that are not normally amenable to GC (due to their strong polarity, high boiling point, and/or thermal instability). Recent applications of derivatization reactions for analysis of wine components include analysis of amines [103,104], organic and inorganic metals such as arsenic [105], phenols and polyphenols, including resveratrol [106–108], amino acids [109], and sugars [110]. Derivatization reactions can also be used to add functional groups to analytes to enhance selectivity (and sensitivity) with detectors such as the NPD and ECD as discussed previously. While not a comprehensive listing, these applications and others discussed throughout this chapter show the wide range of analytes in wines that can be analyzed by GC.

Currently, most sample preparation/extraction approaches for wine analysis are aimed at providing rapid analysis times with minimal or no solvent usage. Two areas in particular have received much attention over the past several years: development of sorptive extraction methods (e.g. solid-phase microextraction or SPME, solid-phase dynamic extraction or SPDE, stir-bar sorptive extraction or SBSE, and solid-phase extraction or SPE) and liquid microextraction methods (e.g. single-drop microextraction, membrane extraction, and dispersive liquid–liquid microextraction). Kloskowski et al. [111] have provided an excellent general overview of sample preparation techniques. Sorptive sample preparation techniques are the focus of reviews by Baltussen et al. and Nongonierma et al. [112,113], and single-drop liquid microextractions have been reviewed by Xu et al. [114].

30.5.1. Sorptive Extraction Methods

Since its introduction in the 1990s, ***solid-phase microextraction (SPME)*** has become one of the most widely used sample preparation techniques for GC, and SPME extraction from the headspace (HS) is commonly used for analysis of wine volatiles. HS-SPME extractions are easy to perform, sensitive, and rapid (most extractions are completed in less than one hour). Several fiber coatings are available offering some selectivity in analyte extraction. However, with coating materials containing carboxen, where competition for sorption sites can occur, matrix effects can impact extraction efficiencies; this matrix effect has particularly been observed for analysis of volatile organic sulfur compounds [57,59,115] and for direct analysis of aldehydes in wines [116].

In recent SPME developments, sample extraction, concentration, and derivatization are accomplished simultaneously with on-fiber derivatization. Here, a derivatizing agent is deposited on the SPME fiber, the fiber is exposed to the sample headspace, and analytes sorb to the fiber and react directly with the derivatizing reagent, or, after sorption the fiber is inserted into the hot GC inlet where the derivatization reaction occurs; the derivatized analyte is then thermally desorbed from the fiber, enters the analytical column, and is analyzed in the normal manner. Headspace-SPME analysis combined with on-fiber derivatization (e.g. *O*-(2,3,4,5,6-pentafluorobenzyl)hydroxylamine (PFBHA) hydrochloride) has been utilized for sensitive and selective analysis of aldehydes (e.g. alkanals, including acetaldehyde, (*E*)-2-alkenals, (*E,E*)-2,4-alkadienals, and furfural) and ketones (e.g. 2,3-butanedione and 3-hydroxy-2-butanone), in wines and distilled beverages [117–120]. As noted previously, these aldehydes can impact wine flavor and, in some cases, their levels can be associated with oxidative conditions during processing or storage. On-fiber derivatization of polyphenols including resveratrol has been accomplished using silylating derivatizing reagents (e.g. bis(trimethylsilyl)trifluoroacetamide, BSTFA) [107,108]. Polyphenols impact wine bitterness and astringency and have antioxidant properties that affect wine stability and the health effects of wine.

Zapata et al. [121] compared on-fiber derivatization of wine carbonyls with a method where the carbonyls were first sorbed to a solid-phase extraction (SPE) cartridge (LiChrolut-EN), derivatized with PFBHA, and eluted with solvent for GC analysis. While the SPME method was overall relatively fast and easy to perform, the SPE isolation/derivatization procedure was overall more reproducible (repeatability relative standard deviations of $<10\%$ for SPE vs. $<20\%$ for SPME). Sensitivity (limits of detection) for the two methods depended on the type of detector used and the analyte studied.

Risticevic et al. [122,123] have provided excellent reviews of HS-SPME measurements of volatile and semivolatile constituents in wine samples. As they note, there is increasing interest in faster, high-throughput sample preparation methods, and they describe an HS-SPME protocol using preequilibrium, short extraction times (2–5 min extraction time with agitation at 500 rpm), combined with preloading of a known amount of internal standard onto the fiber, for accurate and reproducible analysis of wine volatiles. Using this procedure, combined with GC-TOFMS analysis and detection, the authors state they are able to analyze 200–500 analytes in wines in a total analysis time of 10–15 min per sample.

Solid-phase dynamic extraction (SPDE) is a variation of SPME, where the polymeric extracting material is coated on the inside of a needle, the sample (liquid or gas phase) is repeatedly drawn into the needle core for analyte sorption to occur, and then the needle is placed into the GC inlet for desorption and GC separation and detection. SPDE offers increased sorbent capacity compared to SPME, thereby providing better extraction efficiencies and improved sensitivity (reviewed in [124]). SPDE needles are often made of stainless steel, rather than fused silica, hence are less fragile than SPME fibers. SPDE has been used for analysis of short-chain, volatile alcohols (fusel oils) and esters in wines and other alcoholic beverages, to profile volatiles in fermenting musts in order to compare formation of volatiles in normal and problem fermentations, and to classify white wines [124,125].

Stir-bar sorptive extraction or SBSE (also known as Twister®), like SPME, uses a polymeric material for analyte sorption (currently only polydimethylsiloxane and ethylene glycol-modified silicone phases are commercially available); however, in the case of SBSE, the polymeric material is coated on the outside of a magnetic stir bar. The amount of SBSE sorbent is significantly greater, compared to SPME; hence extraction capacities are greater [111]. Either the liquid or headspace/vapor phase of the sample can be extracted. Following extraction, the stir bar is removed from the sample (rinsed with water if extraction was from the liquid phase), placed into a special thermal desorption GC inlet for analyte desorption, and the analytes are cryofocused at the front of the column for GC analysis. While the desorption and cryofocusing steps are fully automated, insertion and removal of the stir bar into the sample vials for sampling must still be done manually. SBSE has been widely adopted for analysis of grape and wine volatiles associated with wine aroma [126–134].

Rather than thermally desorbing and cryofocusing sorbed analytes from the stir bar, Coelho et al. [135] have used liquid/solvent desorption of the stir bar, combined with large-volume injection, to profile and quantify 71 volatiles in sparkling wines. Vestner et al. [136] combined microwave-assisted extraction of cork stoppers with SBSE to quantify 2,4,6-trichloroanisole from the cork. Solvent extraction from the cork matrix is typically slow (24 h); however, Vestner et al. reported total extraction times of ~4½ hours (3½ h for microwave-assisted extraction, 1 h for SBSE) and up to 10 samples could be extracted simultaneously.

Solid-phase extraction (SPE) has been widely used for isolation of volatiles and nonvolatiles in wines. A variety of SPE stationary phases is available providing a range of analyte selectivities and, typically, only small amounts of solvent are needed for extractions. Reversed-phase (C_{18}) SPE extraction cartridges are widely used to extract glycosidically bound aroma precursors from grapes and wine. Following isolation of the glycoside fraction, the volatile aglycone aroma compounds are released by acid or enzyme hydrolysis and the free aglycones analyzed by GC [137–139]. Other recent applications include isolation of terpenes in wines followed by GC analysis of the extract [140], isolation of methoxypyrazines [141], complexing trace levels (ng L^{-1}) of polyfunctional thiol aroma compounds on cartridges containing *p*-hydroxymercurybenzoate [142], and isolation followed by direct GC analysis of 10 different organic acids in red and white wines (the method detects up to 29 acids in foods; [143]). Although extensively used for many years, new SPE stationary phases and applications continue to be developed [144].

30.5.2. Liquid Microextractions

Liquid microextraction procedures that use very small amounts of solvents that are amenable to injection into the GC are receiving growing interest as replacements for traditional liquid-liquid extractions. As reviewed by Kloskowski [111] and Pena-Pereira [145], these methods generally fall into one of the following three categories: (1) ***single-drop microextraction*** where a drop of extracting solvent is suspended into the sample (or the headspace above the sample) from the tip of a rod (often made of polytetrafluoroethylene, PTFE) or a syringe needle and extraction from the sample occurs under equilibrium or nonequilibrium conditions (Figure 30.2), i.e. sample flows past the solvent drop or the solvent drop and sample are repeatedly withdrawn into the needle where the solvent forms a thin layer with increased extraction surface area along the sides of the needle (Figure 30.2b); (2) ***membrane extraction*** where the sample and extracting solvent are separated by a porous membrane – the membranes are typically hydrophobic polymeric materials and/or are filled with extracting solvent and are typically assembled using a hollow fiber, and these extractions can be either static or dynamic; and (3) ***dispersive liquid–liquid microextraction*** where a high-density organic extracting solvent and a dispersing solvent that is miscible in both the organic solvent and the aqueous sample phase are rapidly added to the sample forming small dispersed, emulsified droplets of the organic solvent within the sample. The dispersed droplets have a high surface area and rapidly extract the analytes from the aqueous phase; following centrifugation, the heavier organic layer, containing the extracted analytes, sinks and is isolated for GC analysis. Dispersive liquid–liquid extractions were introduced by Rezaee et al. in 2006 for analysis of hydrophobic pollutants and pesticides in water [146]. Further modifications of the dispersive liquid–liquid extraction method by Regueiro et al. [147] eliminated the emulsifier phase and used ultrasound to assist in droplet formation. General advantages and disadvantages of these microextraction methods are reviewed by Kloskowski [111] and Pena-Pereira [145]. Using these methods, high analyte enrichment factors can typically be obtained, particularly with the dispersive liquid extractions and membrane extractions; however, some sample is typically lost in the membrane when performing membrane extractions and extraction kinetics for the single droplet and membrane extractions can be slow depending on the analytes, the solvents, and the configurations used.

All of these new liquid microextraction techniques have gained rapid acceptance and have

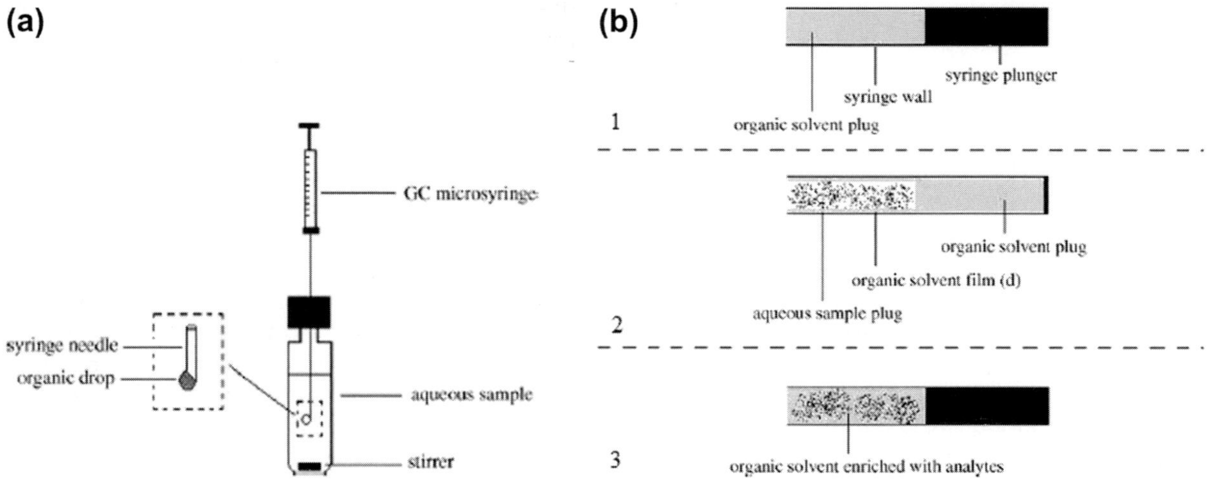

FIGURE 30.2 Schematic for single-drop liquid microextractions. (a) Equilibrium single-drop extraction in liquid phase. (b) Dynamic, nonequilibrium single-drop extraction performed in syringe with repeated retraction of plunger. **Step 1**: Syringe is filled with organic extracting solvent. **Step 2**: Plunger is retracted to withdraw aqueous sample into syringe; extracting solvent forms thin film on inner surface of the syringe. **Step 3**: Expulsion of sample by pushing plunger back into syringe; solvent remains in syringe. Steps 1–3 repeated several times to extract analytes from sample. *Source: With Permission from [114].*

been applied to wine analysis. Droplet microextraction has been used for analysis of pesticides in wines [148] and polyphenols in grapes [149] and membrane extraction has been used to test for leaching of 2-ethylhexyl 4-(dimethylamino) benzoate from sterile tetra pak packaging materials into wines [150]. Dispersive liquid–liquid extraction appears to be among the most widely tested of these new methods with applications for analysis of fungicides in wines [151]; phenols, halophenols, and haloanisoles in wines and corks [93,94,152,153]; varietal-sulfur aroma compounds in wines (methylmercaptoacetate, methyl(methylthio)acetate, 2-methylthioethanol, 3-methylthiopropanol, 3-methylthiohexanol, 4-methylthio-4-methyl-2-pentanone, and hexanethiol) [154]; and the aroma compounds geosmin and methylisoborneol in wines [155]. All of the liquid microextraction methods appear promising as rapid and simple sample preparation procedures; however, in some cases the complex wine matrices can result in significant interferences such as precipitation of matrix components in the dispersive liquid–liquid extractions of fungicides from red wines as observed by Montes et al. [151].

30.5.3. Other Sample Preparation Methods and Comparisons of Methods

A relatively new sample technique that combines both liquid–liquid extraction and solid-phase extraction is the so-called QuEChERS (quick, easy, cheap, effective, rugged, and safe) procedure [156]. As recently reviewed [157–159], the sample is first extracted with a solvent (e.g. acetonitrile), salt ($MgSO_4$ or NaCl) is added to separate the phases, and a sorbent (e.g. primary secondary amine (PSA), C_{18}) is then added to remove matrix interferences and the supernatant is analyzed by GC-MS (or LC-MS). QuEChERS has been most widely used for analysis of pesticide residues in foods and has been applied to multiresidue pesticide analysis of grapes and wines [160–162]. The method is rapid, simple to perform, requires minimal solvent, is easily transferable among different laboratories, and yields high analyte recoveries; however, acetonitrile solvent is not ideal for GC and a programmable temperature vaporization-large-volume injection (PTV-LVI) inlet is recommended for improved sensitivity [158]. Application of QuEChERS for mulitresidue pesticide analysis of foods has currently been recognized as an official method by the American Association of Official Analytical Chemists (AOAC 2007.01; [163,164]) and by the European Committee for Standardization (CEN Standard Method EN 15662; [165]).

All of these emerging sample preparation methods are highly versatile, with applications to many types of analytes and matrices. However, these different sample preparation methods have been directly compared in relatively few cases. For analysis of halophenols and haloanisols in wines, Maggi et al. [92] observed that SBSE (extracting from the liquid) gave lower limits of detection for most of the tested compounds compared to both HS-SPME and direct immersion SPME. Perestrelo et al. [166] also observed that SBSE (sampling from the liquid) had lower limits of detection and quantitation for a range of target analytes compared to SPME (sampling from the headspace); analysis of volatiles in red and white wines showed that more volatile esters were identified and quantified using the SBSE method compared to HS-SPME (25 vs. 16). Andujar-Ortiz et al. [167] recently compared SPME and SPE with traditional liquid–liquid extraction for the analysis of wine volatiles. Based on results from 30 volatile compounds, SPME showed the poorest recovery among the three methods for polar volatiles such as 1-hexanol, furfural, and geraniol. However, the SPME method used significantly less sample (8 mL vs. 50 mL), was faster (extraction times of 20 min), and required no organic solvents. As new sample preparation methods continue to evolve, comparisons of these types will be critical so that appropriate sample preparation methods can be chosen based on both the experimental objectives and an understanding of how the analytes of interest respond to various sample preparation procedures.

For all of these methods, accurate quantitative analysis requires careful consideration of analyte recoveries and matrix effects (e.g. matrix components interfering with the chromatographic separations or causing ion suppression or enhancement in the MS). As reviewed by Polaskova et al. [19], stable isotope dilution analysis (SIDA) offers the most effective way to minimize matrix effects on analyte recovery and reproducibility during analysis. An internal standard (IS) that structurally matches the analyte of interest but contains one or more stable isotope atoms (typically ^{2}H or ^{13}C) is added to the sample at the beginning of the analysis. The IS and analyte theoretically respond similarly to any matrix interaction and so the response will be comparable for both the analyte and the IS. The analyte and IS can be measured by MS and the response ratios used to calculate concentrations. One of the most comprehensive applications of SIDA was reported by Siebert et al. [168] where 31 compounds in wine (including fatty acids, alcohols, acetate esters, and ethyl esters) were quantified by SPME-GC-MS using 29 different deuterated compounds as internal standards. However, recent reports by Koch et al. [45] have indicated that even with an isotopically matched IS changes in the grape matrix during ripening influenced recovery and reproducibility of the analysis of IBMP using HS-SPME-GC-MS. In these cases, standard addition calibration methods may provide the most accurate method of quantification, even though large numbers of analyses are required [45]. Lavignini et al. [169] have recently published an interesting comparison of analysis of sulfur compounds in wines using both isotopically labeled and nonisotopically labeled internal standards. They report application of a statistical variance component model to effectively account for matrix effects when using nonisotopically labeled internal standards that are typically cheaper and more readily available than isotopically labeled compounds. Although the use of one or sometimes two internal standards is common for most analyses, Rebiere et al. [170] also report that reproducibility of SPME methods can be enhanced by application of four internal standards (nonisotopically labeled) to monitor fiber performance as well analyte extraction and quantitation.

As noted initially, no sample preparation technique is without bias, sample preparation and quantitation methods should be chosen to meet experimental objectives, and, in some cases, more than one sample preparation method may be needed to yield the most complete information. Although there have been many recent advances, additional research on sample preparation techniques and statistical approaches that will enhance data analysis and interpretation are needed.

30.6. SUMMARY

Since its development in the 1950s, GC has proven to be a critical tool for the identification and quantification of many wine analytes. This review has largely focused on research and applications that have occurred in the past ~10 years and, from this overview, several trends become apparent: (1) There is increasing focus on development of sustainable and environmentally friendly stationary-phase materials, such as ionic liquids and carbon nanotubes, that give the same performance as today's siloxane- and polyethylene glycol-based phases. (2) Development of multidimensional separations and sensitive new detectors and hyphenated techniques allows for the analysis of more compounds in a given chromatographic run than was possible even 10 years ago. (3) The new separation modes and detectors also allow for faster analysis times; along with research focused on faster, high-throughput sample preparation methods, total analysis times of only a few seconds or minutes are now often possible. (4) There is increasing interest in miniaturization of columns and sample preparation techniques; this miniaturization also contributes to faster analysis times in some cases, and also reduces solvent and chemical use and costs (there is also an increasing focus on miniaturization of GC instruments, but this was not a focus of the current review).

While these active areas of research are providing an improved understanding of wine composition and chemistry, many challenges remain: (1) While the new stationary phases being developed may be cheaper and offer unique selectivities compared to existing stationary phases, many of the new column materials do not have the thermal stability and chromatographic efficiencies of traditional columns; therefore, much research is still needed before these stationary phases will be widely used for analysis of complex matrices such as wines. (2) The development of software tools and approaches for identifying, aligning, deconvoluting, and integrating the hundreds of peaks in multidimensional GC analyses remains challenging; new multivariate statistical approaches are needed to analyze the large amounts of data that are obtained from the profiling and metabolomic analyses. (3) Comparisons and optimization of sample preparation procedures for different analytes are still needed, and for the most accurate quantification, the ability to understand and account for variable matrix effects remains difficult for some grape and wine matrices.

As analytical chemists strive to meet these challenges, new advances in the GC analysis of wines will most surely occur. Over the past 60 years, GC analysis has provided critical information about grape and wine composition and the changes that occur as grapes are transformed into wine. However, much remains unknown and future improvements in GC separations, detection, sample preparation, and data handling processes will be needed as scientists continue to push the boundaries of our knowledge of grape and wine chemistry.

Acknowledgments

A special thank you to John Thorngate for his review of this manuscript.

References

[1] A.T. James, Gas−liquid partition chromatography: the separation of volatile aliphatic amines and of the homologues of pyridine, Biochem. J. 52 (1952) 242−247.

[2] A.T. James, A.J.P. Martin, Gas−liquid partition chromatography: the separation and micro-estimation of volatile fatty acids from formic acid to dodecanoic acid, Biochem. J. 50 (1952) 679−690.

[3] A.T. James, A.J.P. Martin, G.H. Smith, Gas−liquid partition chromatography: the separation and micro-estimation of ammonia and the methylamines, Biochem. J. 52 (1952) 238−242.

[4] A.T. James, A.J.P. Martin, Gas−liquid chromatography: a technique for the analysis and identification of volatile materials, Br. Med. Bull. 10 (1954) 170−176.

[5] A.J.P. Martin, A.T. James, Gas−liquid chromatography: the gas-density meter, a new apparatus for the detection of vapours in flowing gas streams, Biochem. J. 63 (1956) 138−143.

[6] A.T. James, A.J.P. Martin, Gas−liquid chromatography: the separation and identification of the methyl esters of saturated and unsaturated acids from formic acid to *n*-octadecanoic acid, Biochem. J. 63 (1956) 144−152.

[7] A.T. James, A.J.P. Martin, The separation and identification of some volatile paraffinic, naphthenic, olefinic, and aromatic hydrocarbons, J. App. Chem. 6 (1956) 105−115.

[8] E. Bayer, Anwendung Chromatographischer Methoden zur Qaulitätsbeurteilung von Weien and Mosten, Vitis. 1 (1958) 298−312.

[9] E. Bayer, L. Bässler, Systematische Identifizierung von Estern im Weinaroma. II. Mitteilung zur systematischen Identifizierung verdampfbarer organischer Substanzen, Z. Anal. Chem. 181 (1961) 418−424.

[10] A. Castille, Methanol in spirits, An. Bromatologia. 12 (1960) 335−340.

[11] A.D. Webb, R.E. Kepner, Fusel oil analysis by means of gas−liquid partition chromatography, Am. J. Enol. Vitic. 12 (1961) 51−59.

[12] A.D. Webb, R.E. Kepner, The aroma of flor sherry, Am. J. Enol. Vitic. 13 (1962) 1−14.

[13] P. Schreier, Flavor compositions of wines: a review, CRC Crit. Rev. Food Sci. Nutr. 12 (1979) 59–111.

[14] A. Rapp, Wine aroma substances from gas chromatographic analysis, in: H.F. Linskens, J.F. Jackson (Eds.), Modern methods of plant analysis, vol. 6, Springer-Verlag, Berlin, Germany, 1988, pp. 29–66. Wine Analysis.

[15] L. Setkova, S. Risticevic, J. Pawliszyn, Rapid headspace solid-phase microextraction-gas chromatographic-time-of flight mass spectrometric method for qualitative profiling of ice wine volatile fraction I. Method development and optimization, J. Chromatogr. A 1147 (2007) 213–223.

[16] L. Setkova, S. Risticevic, J. Pawliszyn, Rapid headspace solid-phase microextraction-gas chromatographic-time-of-flight mass spectrometric method for qualitative profiling of ice wine volatile fraction II: classification of Canadian and Czech ice wines using statistical evaluation of the data, J. Chromatogr. A 1147 (2007) 224–240.

[17] A.L. Robinson, P.K. Boss, H. Heymann, P.S. Solomon, R.D. Trengove, Development of a sensitive non-targeted method for characterizing the wine volatile profile using headspace solid-phase microextraction comprehensive two-dimensional gas chromatography time-of-flight mass spectrometry, J. Chromatogr. A 1218 (2011) 505–517.

[18] I.L. Francis, J.L. Newton, Determining wine aroma from compositional data, Austr. J. Grape Wine Res. 11 (2005) 114–126.

[19] P. Polaskova, J. Herszage, S.E. Ebeler, Wine flavor: chemistry in a glass, Chem. Soc. Rev. 37 (2008) 2478–2489.

[20] S.E. Ebeler, J.H. Thorngate, Wine chemistry and flavor: looking into the crystal glass, J. Agric. Food Chem. 57 (2009) 8098–8108.

[21] T. Cserháti, Carbon-based sorbents in chromatography. New achievements, Biomed. Chromatogr. 23 (2009) 111–118.

[22] V.R. Reid, M. Stadermann, O. Bakajin, R.E. Synovec, High-speed, temperature programmable gas chromatography utilizing a microfabricated chip with an improved carbon nanotube stationary phase, Talanta 77 (2009) 1420–1425.

[23] D. Merli, A. Speltini, D. Ravelli, E. Quartarone, L. Costa, A. Profumo, Multi-walled carbon nanotubes as the gas chromatographic stationary phase: role of their functionalization in the analysis of aliphatic alcohols and esters, J. Chromatogr. A 1217 (2010) 7275–7281.

[24] S. Iijima, Helical microtubules of graphitic carbon, Nature 354 (1991) 56–58.

[25] M.S. Dresselhaus, G. Dresselhaus, Ph. Avouris, Carbon nanotubes. synthesis, structure, properties, and application, Springer, New York, 2001.

[26] S.H. Kwon, J.H. Park, Intermolecular interactions on multiwalled carbon nanotubes in reversed-phase liquid chromatography, J. Sep. Sci. 29 (2006) 945–952.

[27] S.A. Shamsi, N.D. Danielson, Utility of ionic liquids in analytical separations, J. Sep. Sci. 30 (2007) 1729–1750.

[28] B. Buszewski, S. Studzinska, A review of ionic liquids in chromatographic and electromigration techniques, Chromatographia 68 (2008) 1–10.

[29] D.W. Armstrong, T. Payagala, L.M. Sidisky, The advent and potential impact of ionic liquid stationary phases in GC and GC × GC, LCGC North America 27 (2009) 596–605.

[30] J. González Álvarez, D. Blanco Gomis, P. Aria Abrodo, D. Díaz Llorente, E. Busto, N. Rios Lombardía, et al., Evaluation of new ionic liquids as high stability selective stationary phases in gas chromatography, Anal. Bioanal. Chem. 400 (2011) 1209–1216.

[31] Y. Meng, V. Pino, J.L. Anderson, Exploiting the versatility of ionic liquids in separation science: determination of low-volatility aliphatic hydrocarbons and fatty acid methyl esters using headspace solid-phase microextraction coupled to gas chromatography, Anal. Chem. 81 (2009) 7107–7112.

[32] P. Bouchilloux, P. Darriet, D. Dubourdieu, R. Henry, S. Reichert, A. Mosandl, Stereodifferentiation of 3-mercapto-2-methylpropanol in wine, Eur. Food Res. Technol. 210 (2000) 349–352.

[33] P. Darriet, S. Lamy, S. La Guerche, M. Pons, D. Dubourdieu, D. Blancard, et al., Stereodifferentiation of geosmin in wine, Eur. Food Res. Technol. 213 (2001) 122–125.

[34] L. Fernandes, A.M. Relva, M.D.R. Gomes da Silva, Costa Freitas, different multidimensional chromatographic approaches applied to the study of wine malolactic fermentation, J. Chromatogr. A 995 (2003) 161–169.

[35] C. Barba, G. Flores, M. Herraiz, Stereodifferentiation of some chiral aroma compounds in wine using solid phase microextractions and multidimensional gas chromatography, Food Chem. 123 (2010) 846–861.

[36] W. Bertsch, Two-dimensional gas chromatography. Concepts, instrumentation, and applications—Part 1: fundamentals, conventional two-dimensional gas chromatography, selected applications, J. High Resolut. Chromatogr. 22 (1999) 647–665.

[37] L. Mondello, P.Q. Tranchida, P. Dugo, G. Dugo, Comprehensive two-dimensional gas chromatography-mass spectrometry: a review, Mass Spectrom. Rev. 27 (2008) 101–124.

[38] W. Bertsch, Two-dimensional gas chromatography. Concepts, instrumentation, and applications—Part 2: comprehensive two-dimensional gas chromatography, J. High Resolut. Chromatogr. 23 (2000) 167–181.

[39] P.J. Marriott, P.D. Morrison, R.A. Shellie, M.S. Dunn, E. Sari, D. Ryan, Multidimensional and comprehensive two-dimensional gas chromatography. LCGC Europe, Dec. (2003). 2–10.

[40] D. Ryan, P. Watkins, J. Smith, M. Allen, P. Marriott, Analysis of methoxypyrazines in wine using headspace solid phase microextraction with isotope dilution and comprehensive two-dimensional gas chromatography, J. Sep. Sci. 28 (2005) 1075–1082.

[41] I. Ryona, B.S. Pan, D.S. Intrigliolo, A.N. Lakso, Effects of cluster light exposure on 3-isobutyl-2-methoxypyrazine accumulation and degradation patterns in red wine grapes (*Vitis vinifera* L. cv. Cabernet Franc), J. Agric. Food Chem. 56 (2008) 10838–10846.

[42] I. Ryona, B.S. Pan, G.L. Sacks, Rapid measurement of 3-alkyl-2-methoxypyrazine content of winegrapes to predict levels in resultant wines, J. Agric. Food Chem. 57 (2009) 8250–8257.

[43] H.-G. Schmarr, S. Ganβ, S. Koschinski, U. Fischer, C. Riehle, J. Kinnart, et al., Pitfalls encountered during quantitative determination of 3-alkyl-2-methoxypyrazines in grape must and wine using gas chromatography-mass spectrometry with stable isotope dilution analysis comprehensive two-dimensional gas chromatography-mass spectrometry and on-line liquid chromatography-multidimensional gas chromatography-mass spectrometry as potential loopholes, J. Chromatogr. A 1217 (2010) 6769–6777.

[44] M.S. Allen, M.J. Lacy, S. Boyd, Methoxypyrazines of grapes and wines, in: A.L. Waterhouse, S.E. Ebeler (Eds.), Chemistry of Wine Flavor, ACS Symposium Series No. 714, American Chemical Society, Washington, DC, 1999, pp. 31–38.

[45] A. Koch, C.L. Doyle, M.A. Matthews, L.E. Williams, S.E. Ebeler, 2-Methoxy-3-isobutylpyrazine in grape berries and its dependence on genotype, Phytochemistry 71 (2010) 2190–2198.

[46] R.G. Buttery, R.M. Seifert, D.G. Guadagni, L.C. Ling, Characterization of some volatile constituents of bell peppers, J. Agric. Food Chem. 17 (1969) 1322–1327.

[47] R.L.N. Harris, M.J. Lacey, W.V. Brown, M.S. Allen, Determination of 2-methoxy-3-alkylpyrazines in wine by gas chromatography/mass spectrometry, Vitis 26 (1987) 201–207.

[48] D.M. Chapman, J.H. Thorngate, M.A. Matthews, J.-X. Guinard, S.E. Ebeler, Yield effects on 2-methoxy-3-isobutylpyrazine concentration in Cabernet Sauvignon using a solid phase microextraction gas chromatography/mass spectrometry method, J. Agric. Chem. 52 (2004) 5431–5435.

[49] R. Perestrelo, S. Petronilho, J.S. Camara, S.M. Rocha, Comprehensive two-dimensional gas chromatography with time-of-flight mass spectrometry combined with solid phase microextraction as a powerful tool for quantification of ethyl carbamate in fortified wines. The case study of Madeira wine, J. Chromatogr. A 1217 (2010) 3441–3445.

[50] C.E. Butzke, L.F. Bisson, Ethyl carbamate preventative action manual (1997) available on-line.http://www.fda.gov/Food/FoodSafety/FoodContaminants Adulteration/ChemicalContaminants/EthylCarbamate Urethane/ucm078546.htm (accessed on-line 15 July, 2011).

[51] Wine Institute, Ethyl carbamate in wine, technical committee summary background document (2008). http://www.calwinexport.com/files/Ethyl%20Carbamate%20in%20Wine%202008_02_12%20Tech Comm%20FINAL.PDF (accessed on-line 15 July, 2011).

[52] Health Canada, Canadian standards ("Maximum Levels") for various chemical contaminants in foods, http://www.hc-sc.gc.ca/fn-an/securit/chem-chim/contaminants-guidelines-directives-eng.php (accessed on-line 15 July, 2011).

[53] S. Dasgupta, K. Banerjee, S.H. Patil, M. Ghaste, K.N. Dhumal, P.G. Adsule, Optimization of two-dimensional gas chromatography time-of-flight mass spectrometry for separation and estimation of the residues of 160 pesticides and 25 persistent organic pollutants in grape and wine, J. Chromatogr. A 1217 (2010) 3881–3889.

[54] H.-G. Schmarr, J. Bernhardt, U. Fischer, A. Stephan, P. Müller, D. Durner, Two-dimensional gas chromatographic profiling as a tool for a rapid screening of the changes in volatile composition occurring due to microoxygenation of red wines, Anal. Chim. Acta. 672 (2010) 114–123.

[55] B.T. Weldegergis, A. de Villiers, C. McNeish, S. Seethapathy, A. Mostafa, T. Górecki, et al., Characterisation of volatile components of Pinotage wines using comprehensive two-dimensional gas chromatography coupled to time-of-flight mass spectrometry (GC × GC-TOFMS), Food Chem. 129 (2011) 188–199.

[56] S.M. Rocha, E. Coelho, J. Zrostlíková, I. Delgadillo, M.A. Coimbra, Comprehensive two-dimensional gas chromatography with time-of-flight mass spectrometry of monoterpenoids as a powerful tool for grape origin traceability, J. Chromatogr. A 1161 (2007) 292–299.

[57] R. López, A.C. Lapeña, J. Cacho, V. Ferreira, Quantitative determination of wine highly volatile sulfur compounds by using automated headspace solid-phase microextraction and gas chromatography-pulsed flame photometric detection. Critical study and optimization of a new procedure, J. Chromatogr. A 1143 (2007) 8–15.

[58] Y. Fang, M.C. Qian, Sensitive quantification of sulfur compounds in wine by headspace solid-phase microextraction technique, J. Chromatogr. A 1080 (2005) 177–185.

[59] J. Herszage, S.E. Ebeler, Analysis of volatile organic sulfur compounds in wine using headspace solid-phase microextraction gas chromatography with sulfur chemiluminescence detection, Am. J. Enol. Vitic. 62 (2011) 1–8.

[60] D. Özhan, R.E. Anli, N. Vural, M. Bayram, Determination of chloroanisoles and chlorophenols in cork and wine by using HS-SPME and GC-ECD detection, J. Inst. Brew. 115 (2009) 71–77.

[61] S. Navarro, A. Barba, G. Navarro, N. Vela, J. Oliva, Multiresidue method for the rapid determination—in grape, must and wine—of fungicides frequently used on vineyards, J. Chromatogr. A 882 (2000) 221–229.

[62] J. Oliva, A. Barba, N. Vela, F. Melendreras, S. Navarro, Multiresidue method for the rapid determination of organophosphorus insecticides in grapes, must and wine, J. Chromatogr. A 882 (2000) 213–220.

[63] M. Correia, C. Delerue-Matos, A. Alves, Multiresidue methodology for pesticide screening in wines, J. Chromatogr. A 889 (2000) 59–67.

[64] M. Correia, C. Delerue-Matos, A. Alves, Development of a SPME-GC-ECD methodology for selected pesticides in must and wine samples, Fresinius' J. Anal. Chem. 369 (2001) 647–651.

[65] J.J. Jiménez, J.L. Bernal, M.J. del Nozal, L. Toribio, E. Arias, Analysis of pesticide residues in wine by solid-phase extraction and gas chromatography with electron capture and nitrogen-phosphorus detection, J. Chromatogr. A 919 (2001) 147–156.

[66] V. Ferreira, L. Culleré, N. Loscos, J. Cacho, Critical aspects of the determination of pentafluorobenzyl derivatives of aldehydes by gas chromatography with electron-capture or mass spectrometric detection: validation of an optimized strategy for the determination of oxygen-related odor-active aldehydes in wine, J. Chromatogr. A 1122 (2006) 255–265.

[67] L. Vaquero-Fernández, J. Sanz-Asensio, P. Fernández-Zurbano, M. Sainz-Ramírez, M. López-Alonso, S.-I. Epifanio-Fernández, et al., Development of a liquid–liquid extraction method for the determination of pyrimethanil, metalaxyl, dichlofluanid, and penconazol during the fermentative process of must by GC-NPD, J. Sci. Food Agric. 89 (2009) 750–757.

[68] P.J. Hartmann, H.M. McNair, B.W. Zoecklein, Measurement of 3-alkyl-2-methoxypyrazine by headspace solid-phase microextraction in spiked model wines, Am. J. Enol. Vitic. 53 (2002) 285–288.

[69] S.E. Ebeler, R.S. Spaulding, Characterization and measurement of aldehydes in wine, in: A.L. Waterhouse, S.E. Ebeler (Eds.), Chemistry of wine flavor, ACS symposium series no. 714, American Chemical Society, Washington, DC, 1999, pp. 166–179.

[70] M. Lau, J.D. Ebeler, S.E. Ebeler, Gas chromatographic analysis of aldehydes in alcoholic beverages using a cysteamine derivatization procedure, Am. J. Enol. Vitic. 50 (1999) 324–333.

[71] S.K. Frivik, S.E. Ebeler, Influence of sulfur dioxide on the formation of aldehydes in white wine, Am. J. Enol. Vitic. 54 (2003) 31–38.

[72] N. Campillo, R. Peñalver, I. López-García, M. Hernández-Córdoba, Headspace solid-phase microextraction for the determination of volatile organic sulphur and selenium compounds in beers, wines and spirits using gas chromatography and atomic emission detection, J. Chromatogr. A 1216 (2009) 6735–6740.

[73] J. Lisec, N. Schauer, J. Kopka, L. Willmitzer, A.R. Fernie, Gas chromatography mass spectrometry-based metabolite profiling in plants, Nature Protocols 1 (2006) 387–396.

[74] D. Steinhauser, J. Kopka, Methods, applications and concepts of metabolite profiling: primary metabolism, vol. 97, EXS, 2007. 171–194.

[75] O. Fiehn, G. Wohlgemuth, M. Scholz, T. Kind, D.Y. Lee, Y. Lu, et al., Quality control for plant metabolomics: reporting MSI-compliant studies, Plant J. 53 (2008) 691–704.

[76] J.W. Allwood, A. Erban, S. de Konig, W.B. Dunn, A. Luedemann, A. Lommen, et al., Inter-laboratory reproducibility of fast gas chromatography-electron impact-time of flight mass spectrometry (GC-EI-TOF/MS) based plant metabolomics, Metabolomics 5 (2009) 479–496.

[77] K. Skogerson, R. Runnebaum, G. Wohlgemuth, J. de Ropp, H. Heymann, O. Fiehn, Comparison of gas chromatography-coupled time-of-flight mass spectrometry and ^{1}H nuclear magnetic resonance spectroscopy metabolite identification in white wines from a sensory study investigating wine body, J. Agric. Food Chem. 57 (2009) 6899–6907.

[78] M.-Z. Ding, J.S. Cheng, W.-H. Xiao, B. Qiao, Y.-J. Yuan, Comparative metabolomic analysis on industrial continuous and batch ethanol fermentation processes by GC-TOF-MS, Metabolomics 5 (2009) 229–238.

[79] I. Atanassov, Tz. Hvarleva, K. Rusanov, I. Tsvetkov, A. Atanassov, Wine metabolite profiling: possible application in winemaking and grapevine breeding in Bulgaria, Biotechnol. Biotechnol. Eq. 23 (2009) 1449–1452.

[80] J.H. Gross, Mass spectrometry. A textbook, Springer-Verlag, Berlin, 2004.

[81] E. de Hoffmann, V. Stroobant, Mass spectrometry: principles and applications, third ed., Wiley and Sons, West Sussex England, 2007.

[82] M.C. McMaster, GC/MS. A practical user's guide, second ed., Wiley and Sons, NJ, 2008.

[83] H.-J. Hübschmann, Handbook of GC/MS. Fundamentals and applications, second ed., Wiley-VCH Verlag GmbH & Co, Weinheim, 2009.

[84] M. Nantangelo, S. Tavazzi, E. Benfenati, Evaluation of solid phase microextraction-gas chromatography in the analysis of some pesticides with different mass spectrometric techniques: application to environmental waters and food samples, Anal. Lett. 35 (2002) 327–338.

[85] K.J. Chia, S.D. Huang, Analysis of organochlorine pesticides in wine by solvent bar microextraction coupled with gas chromatography with tandem mass spectrometric detection, Rapid Commun. Mass Spectrom. 20 (2006) 118–124.

[86] P. Paya, M. Anastassiades, D. Mack, I. Sigalova, B. Tasdelen, J. Oliva, et al., Analysis of pesticide residues using the Quick Easy Cheap Effective Rugged and Safe (QuEChERS) pesticide multiresidue method in combination with gas and liquid chromatography and tandem mass spectrometric detection, Anal. Bioanal. Chem. 389 (2007) 1697–1714.

[87] J.A. Perez-Serradilla, J.M. Mata-Granados, M.D.L. de Castro, Low-level determination of organochlorine pesticides in wines by automatic preconcentration and GC-MS-MS detection, Chromatographia 71 (2010) 899–905.

[88] J. Martins, C. Esteves, A. Limpo-Faria, P. Barros, N. Ribeiro, T. Simoes, et al., Multiresidue method for the determination of organophosphorus pesticides in still wine and fortified wine using solid-phase microextraction and gas chromatography-tandem mass spectrometry, Anal. Lett. 44 (2011) 1021–1035.

[89] J.L. Gomez-Ariza, T. Garcia-Barrera, F. Lorenzo, Analysis of anisoles in wines using pervaporation coupled to gas chromatography-mass spectrometry, J. Chromatogr. A 1049 (2004) 147–153.

[90] J.L. Gomez-Ariza, T. Garcia-Barrera, F. Lorenzo, Simultaneous separation, clean-up and analysis of musty odorous compounds in wines by on-line coupling of a pervaporation unit to gas chromatography-tandem mass spectrometry, Anal. Chim. Acta. 516 (2004) 165–170.

[91] J.L. Gomez-Ariza, T. Garcia-Barrera, F. Lorenzo, Dynamic headspace coupled to perevaporation for the analysis of anisoles in wine by gas chromatography-ion-trap tandem mass spectrometry, J. Chromatogr. A 1056 (2004) 243–247.

[92] L. Maggi, A. Zalacain, V. Mazzoleni, G.L. Alonso, M.R. Salinas, Comparison of stir bar sorptive extraction and solid-phase microextraction to determine halophenols and haloanisoles by gas chromatography-ion trap tandem mass spectrometry, Talanta 75 (2008) 753–759.

[93] A.R. Fontana, S.H. Patil, K. Banerjee, J.C. Altamirano, Ultrasound-assisted emulsification microextraction for determination of 2,4,6-trichloroanisole in wine samples by gas chromatography tandem mass spectrometry, J. Agric. Food Chem. 58 (2010) 4576–4581.

[94] C. Pizarro, C. Saenz-Gonzalez, N. Perez-del-Notario, J.M. Gonzalez-Saiz, Development of a dispersive liquid–liquid microextraction method for the simultaneous determination of the main compounds causing cork taint and Brett character in wines using gas chromatography-tandem mass spectrometry, J. Chromatogr. A 1218 (2011) 1576–1584.

[95] C. Pizarro, N. Perez-del-Notario, J.M. Gonzalez-Saiz, Multiple headspace solid-phase microextraction for eliminating matrix effect in the simultaneous determination of haloanisoles and volatile phenols in wines, J. Chromatogr. A 1166 (2007) 1–8.

[96] R. Godelman, S. Limmert, T. Kuballa, Implementation of headspace solid-phase-microextraction-GC-MS/MS methodology for determination of 3-alkyl-2-methoxypyrazines in wine, Eur. Food Res. Technol. 227 (2008) 449–461.

[97] L. Cullere, J. Cacho, V. Ferreira, Validation of an analytical method for the solid phase extraction, in cartridge derivatization and subsequent gas chromatographic-ion trap tandem mass spectrometric determination of 1-ocen-3-one in wines at ng L^{-1} level, Anal. Chim. Acta. 563 (2006) 51–57.

[98] F. Mattivi, L. Caputi, S. Carlin, T. Lanza, M. Minozzi, D. Nanni, et al., Effective analysis of rotundone at below-threshold levels in red and white wines using solid-phase microextraction gas chromatography/tandem mass spectrometry, Rapid Comm. Mass Spectrom. 25 (2011) 483–488.

[99] C. Thibon, S. Shinkaruk, T. Tominaga, B.E. Bennetau, D. Dubourdieu, Analysis of the diastereoisomers of the cysteinylated aroma precursor of 3-sulfanylhexanol in *Vitis vinifera* grape must by gas chromatography coupled with ion trap tandem mass spectrometry, J. Chromatogr. A 1183 (2008) 150–157.

[100] D. Dubourdieu, T. Tominaga, Polyfunctional thiol compounds, in: M.V. Moreno-Arribas, M.C. Polo (Eds.), Wine chemistry and biochemistry, Springer, New York, 2009, pp. 275–293. Chapter 8B.

[101] W.G. Jennings, M. Filsoof, Comparison of sample preparation techniques for gas chromatographic analysis, J. Agric. Food Chem. 25 (1977) 440–445.

[102] R. Flath, Symposium on methods for the isolation of trace volatile constituents, J. Agric. Food Chem. 25 (1977) 439.

[103] K.K. Ngim, S.E. Ebeler, M.E. Lew, D.G. Crosby, J.W. Wong, Optimized procedures for analyzing primary alkylamines in wines by pentafluorobenzaldehyde derivatization and GC-MS, J. Agric. Food Chem. 48 (2000) 3311–3316.

[104] A. Pendem, V.M. Pawar, S. Jayaraman, Development of a gas chromatography method for the estimation of alkylamines in foods, J. Agric. Food Chem. 58 (2010) 8904–8910.

[105] N. Campillo, R. Peñalver, P. Viñas, I. López-García, M. Hernández-Cordoba, Speciation of arsenic using capillary gas chromatography with atomic emission detection, Talanta 77 (2008) 793–799.

[106] R. Montes, M. García-López, I. Rodríguez, R. Cela, Mixed-mode solid-phase extraction followed by acetylation and gas chromatography mass spectrometry for the reliable determination of *trans*-resveratrol in wine samples, Anal. Chim. Acta. 673 (2010) 47–53.

[107] L. Cai, J.A. Koziel, M. Dharmadhikari, J.H. van Leeuwen, Rapid determination of trans-resveratrol in red wine by solid-phase microextraction with on-fiber derivatization and multidimensional gas chromatography-mass spectrometry, J. Chromatogr. A 1216 (2009) 281–287.

[108] P. Vinas, N. Campillo, N. Martinez-Castillo, M. Hernandez-Cordoba, Solid-phase microextraction on-fiber derivatization for the analysis of some polyphenols in wine and grapes using gas chromatography-mass spectrometry, J. Chromatogr. A 1216 (2009) 1279–1284.

[109] H.S.M. Ali, R. Patzold, H. Bruckner, Gas chromatographic determination of amino acid enantiomers in bottled and aged wines, Amino Acids 38 (2010) 951–968.

[110] A.I. Ruiz-Matute, M.L. Sanz, M.V. Moreno-Arribas, I. Martínez-Castro, Identification of free disaccharides and other glycosides in wine, J. Chromatogr. A 1216 (2009) 7296–7300.

[111] A. Kloskowski, W. Chrzanowski, M. Pilarczyk, J. Namiesnik, Modern techniques of sample preparation for determination of organic analytes by gas chromatography, Crit. Rev. Anal. Chem. 37 (2007) 15–38.

[112] E. Baltussen, C.A. Cramers, P.J.F. Sandra, Sorptive sample preparation-a review, Anal. Bioanal. Chem. 373 (2002) 3–22.

[113] A. Nongonierma, A. Voilley, P. Cayot, J.-L. Le Quéré, M. Springett, Mechanisms of extraction of aroma compounds from foods, using adsorbents. Effect of various parameters, Food Rev. Internat. 22 (2006) 51–94.

[114] L. Xu, C. Basheer, H.K. Lee, Developments in single-drop microextraction, J. Chromatogr. A 1152 (2007) 184–192.

[115] R.A. Murray, Limitations to the use of solid-phase microextraction for quantitation of mixtures of volatile organic sulfur compounds, Anal. Chem. 73 (2001) 1646–1649.

[116] S.J. Pérez Olivero, J.P. Pérez Trujillo, A new method for the determination of carbonyl compounds in wines by headspace solid-phase microextraction coupled to gas chromatography-ion trap mass spectrometry, J. Agric. Food Chem. 58 (2010) 12976–12985.

[117] R. Flamini, A. Dalla Vedova, A. Panighel, N. Perchiazzi, S. Ongarato, Monitoring of the principal carbonyl compounds involved in malolactic fermentation of wine by solid-phase microextraction and positive ion chemical ionization GC/MS analysis, J. Mass Spectrom. 40 (2005) 1558–1564.

[118] W.K. Carlton, B. Gump, K. Fugelsang, A.S. Hasson, Monitoring acetaldehyde concentrations during micro-oxygenation of red wine by headspace solid-phase microextraction with on-fiber derivatization, J. Agric. Food Chem. 55 (2007) 5620–5625.

[119] H.-G. Schmarr, W. Sang, S. Ganß, U. Fischer, B. Köpp, C. Schulz, et al., Analysis of aldehydes *via* headspace SPME with on-fiber derivatization to their *O*-(2,3,4,5,6-pentafluorobenzyl)oxime derivatives and comprehensive 2D-GC-MS, J. Sep. Sci. 31 (2008) 3458–3465.

[120] G.M. Valtierra, R.J. Ciprés, A.P. Álvarez, Identification and quantification of aldehydes in mezcal by solid phase microextractions with on-fiber derivatization-gas chromatography, J. Mex. Chem. Soc. 55 (2011) 84–88.
[121] J. Zapata, L. Mateo-Vivaraco, J. Cacho, V. Ferreira, Comparison of extraction techniques and mass spectrometric ionization modes in the analysis of wine volatile carbonyls, Anal. Chim. Acta. 660 (2010) 197–205.
[122] S. Risticevic, H. Lord, T. Górecki, C.L. Arthur, J. Pawliszyn, Protocol for solid-phase microextraction method development, Nature Protoc. 5 (2010) 122–139.
[123] S. Risticevic, Y. Chen, L. Kudlejova, R. Vatinno, B. Baltensperger, J.R. Stuff, et al., Protocol for the development of automated high-throughput SPME-GC methods for the analysis of volatile and semi-volatile constituents in wine samples, Nature Protoc. 5 (2010) 162–176.
[124] M.A. Jochman, M.P. Kmiecik, T.C. Schmidt, Solid-phase dynamic extraction for the enrichment of polar volatile organic compounds from water, J. Chromatogr. A 1115 (2006) 208–216.
[125] S. Malherbe, V. Watts, H.H. Nieuwoudt, F.F. Bauer, M. du Toit, Analysis of volatile profiles of fermenting grape must by headspace solid-phase dynamic extraction coupled with gas chromatography-mass spectrometry (HS-SPDE GC-MS): Novel application to investigate problem fermentations, J. Agric. Food Chem. 57 (2009) 5161–5166.
[126] Y. Hayasaka, K. MacNamara, G.A. Baldock, R.L. Taylor, A.P. Pollnitz, Application of stir bar sorptive extraction for wine analysis, Anal. Bioanal. Chem. 375 (2003) 948–955.
[127] J. Diez, C. Dominguez, D.A. Guillen, R. Veas, C.G. Barroso, Optimisation of stir bar sorptive extraction for the analysis of volatile phenols in wines, J. Chromatogr. A 1025 (2004) 263–267.
[128] R.F. Alves, A.M.D. Nascimento, J.M.F. Nogueira, Characterization of the aroma profile of Madeira wine by sorptive extraction techniques, Anal. Chim. Acta. 546 (2005) 11–21.
[129] A. Zalacain, J. Marin, G.L. Alonso, M.R. Salinas, Analysis of wine primary aroma compounds by stir bar sorptive extraction, Talanta 71 (2007) 1610–1615.
[130] Y. Fang, M.C. Qian, Quantification of selected aroma-active compounds in Pinot noir wines from different grape maturities, J. Agric. Food Chem. 54 (2006) 8567–8573.
[131] T. Kosmerl, E. Zlatic, Determination of 2-aminoacetophenone in wines using the stir bar sorptive extraction method coupled with GC-MS and GC-NPD, Mitteilungen Klosterneuburg 59 (2009) 121–126.
[132] R. Delgado, E. Duran, R. Castro, R. Natera, C.G. Barroso, Development of a stir bar sorptive extraction method coupled to gas chromatography-mass spectrometry for the analysis of volatile compounds in Sherry brandy, Anal. Chim. Acta. 672 (2010) 130–136.
[133] C. Franc, F. David, G. de Revel, Multi-residue off-flavour profiling in wine using stir bar sorptive extraction-thermal desorption-gas chromatography-mass spectrometry, J. Chromatogr. A 1216 (2009) 3318–3327.
[134] D.J. Caven-Quantrill, A.J. Buglass, Comparison of volatile constituents extracted from model grape juice and model wine by stir bar sorptive extraction-gas chromatography-mass spectrometry, J. Chromatogr. A 1218 (2011) 875–881.
[135] E. Coelho, M.A. Coimbra, J.M.F. Nogueira, S.M. Rocha, Quantification approach for assessment of sparkling wine volatiles from different soils, ripening stages, and varieties by stir bar sorptive extraction with liquid desorption, Anal. Chim. Acta. 635 (2009) 214–221.
[136] J. Vestner, S. Fritsch, D. Rauhut, Development of a microwave assisted extraction method for the analysis of 2,4,6-trichloroanisole in cork stoppers by SIDA-SBSE-GC-MS, Anal. Chim. Acta. 660 (2010) 76–80.
[137] M.J. Ibarz, V. Ferreira, P. Hernández-Orte, N. Loscos, J. Cacho, Optimization and evaluation of a procedure for the gas chromatographic-mass spectrometric analysis of the aromas generated by fast acid hydrolysis of flavor precursors extracted from grapes, J. Chromatogr. A 1116 (2006) 217–229.
[138] N. Loscos, P. Hernández-Orte, J. Cacho, V. Ferreira, Release and formation of varietal aroma compounds during alcoholic fermentation from nonfloral grape odorless flavor precursors fractions, J. Agric. Food Chem. 55 (2007) 6674–6684.
[139] N. Loscos, P. Hernandez-Orte, J. Cacho, V. Ferreira, Comparison of the suitability of different hydrolytic strategies to predict aroma potential of different grape varieties, J. Agric. Food Chem. 57 (2009) 2468–2480.
[140] Z. Piñiero, M. Palma, C.G. Barroso, Determination of terpenoids in wines by solid phase extraction and gas chromatography, Anal. Chim. Acta. 513 (2004) 209–214.
[141] R. López, E. Gracia-Moreno, J. Cacho, V. Ferreira, Development of a mixed-mode solid phase extraction method and further gas chromatography mass spectrometry for the analysis of 3-alkyl-2-methoxypyrazines in wine, J. Chromatogr. A. 1218 (2011) 842–848.

[142] L. Mateo-Vivaracho, J. Cacho, V. Ferreira, Selective preconcentration of volatile mercaptans in small SPE cartridges: quantitative determination of trace odor-active polyfunctional mercaptans in wine, J. Sep. Sci. 32 (2009) 3845–3853.

[143] B. Jurado-Sánchez, E. Ballesteros, M. Gallego, Gas chromatographic determination of 29 organic acids in foodstuffs after continuous solid-phase extractions, Talanta 84 (2011) 924–930.

[144] R.E. Majors, New chromatography columns and accessories at Pittcon 2011: part II, LC-GC North America 29 (2011) 301–316.

[145] F. Pena-Pereira, I. Lavilla, C. Bendicho, Miniaturized preconcentration methods based on liquid-liquid extraction and their application in inorganic ultra-trace analysis and speciation: a review, Spectrochim. Acta B 54 (2009) 1–15.

[146] M. Rezaee, Y. Assadi, M.-R.M. Hosseini, E. Aghaee, F. Ahmadi, S. Berijani, Determination of organic compounds in water using dispersive liquid-liquid microextraction, J. Chromatogr. A 1116 (2006) 1–9.

[147] J. Regueiro, M. Llompart, C. Garcia-Jares, J.C. Garcia-Monteagudo, R. Cela, Ultrasound-assisted emulsification-microextraction of emergent contaminants and pesticides in environmental waters, J. Chromatogr. A 1190 (2008) 27–38.

[148] A. Garbi, V. Sakkas, Y.C. Fiamegos, C.D. Stalikas, T. Albanis, Sensitive determination of pesticides residues in wine samples with the aid of single-drop microextraction and response surface methodology, Talanta 82 (2010) 1286–1291.

[149] P. Viñas, N. Martínez-Castillo, N. Campillo, M. Hernández-Córdoba, Directly suspended droplet microextraction with in injection-port derivatization coupled to gas chromatography-mass spectrometry for the analysis of polyphenols in herbal infusions, fruits and functional foods, J. Chromatogr. A 1218 (2011) 639–646.

[150] J.G. March, C. Genestar, B.M. Simonet, Determination of 2-ethylhexyl 4-(dimethylamino) benzoate using membrane-assisted liquid-liquid extraction and gas chromatography-mass spectrometric detection, Anal. Bioanal. Chem. 394 (2009) 883–891.

[151] R. Montes, I. Rodríguez, M. Ramil, E. Rubi, R. Cela, Solid-phase extraction followed by dispersive liquid-liquid microextraction for the sensitive determination of selected fungicides in wine, J. Chromatogr. A 1216 (2009) 5459–5466.

[152] N. Campillo, P. Viñas, J.I. Cacho, R. Peñalver, M. Hernández-Córdoba, Evaluation of dispersive liquid-liquid microextraction for the simultaneous determination of chlorophenols and haloanisoles in wines and cork stoppers using gas chromatography-mass spectrometry, J. Chromatogr. A 1217 (2010) 7323–7330.

[153] C. Pizarro, C. Sáenz-González, N. Perez-del-Notario, J.M. González-Sálz, Optimisation of a dispersive liquid-liquid microextraction method for the simultaneous determination of halophenols and haloanisoles in wines, J. Chromatogr. A 1217 (2010) 7630–7637.

[154] V.P. Jofré, M.V. Assof, M.L. Fanzone, H.C. Goicoechea, L.D. Martínez, M.F. Silva, Optimization of ultrasound assisted-emulsification-dispersive liquid-liquid microextraction by experimental design methodologies for the determination of sulfur compounds in wines by gas chromatography-mass spectrometry, Anal. Chim. Acta. 683 (2010) 126–135.

[155] C. Cortada, L. Vidal, A. Canals, Determination of geosmin and 2-methylisoborneol in water and wine samples by ultrasound-assisted dispersive liquid-liquid microextraction coupled to gas chromatography-mass spectrometry, J. Chromatogr. A 1218 (2011) 17–22.

[156] M. Anastassiades, S.J. Lehotay, D. Stajnbaher, F.J. Schenck, Fast and easy multiresidue method employing acetonitrile extraction/partitioning and "dispersive solid-phase extraction" for the determination of pesticide residues in produce, J. AOAC Int. 86 (2003) 412–431.

[157] R.E. Majors, QuEChERS—A new sample preparation technique for multiresidue analysis of pesticides in foods and agricultural samples, LCGC 25 (2007) 436–446.

[158] S.J. Lehotay, M. Anastassiades, R.E. Majors, QuEChERS, a sample preparation technique that is "catching on": an up-to-date interview with the inventors, LCGC 28 (2010) 504–516.

[159] J. Fenik, M. Tankiewicz, M. Biziuk, Properties and determination of pesticides in fruits and vegetables, Tr. Anal. Chem. 30 (2011) 814–826.

[160] S.C. Cunha, J.O. Fernandes, A. Alves, M.B.P.P. Oliveira, Fast low-pressure gas chromatography-mass spectrometry method for the determination of multiple pesticides in grapes, musts, and wines, J. Chromatogr. A 1216 (2009) 110–126.

[161] Y. Jiang, X. Li, J. Xu, C. Pan, J. Zhang, W. Niu, Multiresidue method for the determination of 77 pesticides in wine using QuEChERS sample preparation and gas chromatography with mass spectrometry, Food Addit. Contam. A 26 (2009) 859–866.

[162] Y.-J. Lian, G.-F. Pang, H.-R. Shu, C.-L. Fan, Y.-M. Liu, J. Feng, et al., Simultaneous determination of 346 multiresidue pesticides in grapes by PSA-MSPD and GC-MS-SIM, J. Agric. Food Chem. 58 (2010) 9428–9453.

[163] S.J. Lehotay, M. O'Neil, J. Tully, A.V. García, M. Contreras, H. Moi, et al., Determination of pesticide residues in foods by acetonitrile extraction and partitioning with magnesium sulfate: collaborative study, J. AOAC Int. 90 (2007) 485–520.

[164] Association of Official Analytical Chemists, AOAC Official Method 2007.01. Pesticide Residues in Foods by Acetonitrile Extraction and Partitioning with Magnesium Sulfate. AOAC. Gaithersburg, MD. http://www.weber.hu/PDFs/QuEChERS/AOAC_2007_01.pdf (accessed on-line 17 April, 2012). (accessed on-line 15 July, 2011).

[165] EN 15662, Foods of plant origin-determination of pesticide residues using GC-MS and/or LC-MS/MS following acetonitrile extraction/partitioning and clean-up by dispersive SPE-QuEChERS-method, European Committee for Standardization, 2008. CEN/TC 276-Food Analysis-Horizontal Methods, ISC 67.050-General Methods of Tests and Analysis for Food Products (accessed via www.cen.eu, July 24, 2011).

[166] R. Perestrelo, J.M.F. Nogueira, J.S. Camara, Potentialities of two solventless extraction approaches-stir bar sorptive extraction and headspace solid-phase microextraction for determination of higher alcohol acetates, isoamyl esters and ethyl esters in wines, Talanta 80 (2009) 622–630.

[167] I. Andujar-Ortiz, M.V. Moreno-Arribas, P.J. Martin-Alvarez, M.A. Pozo-Bayón, Analytical performance of three commonly used extraction methods for the gas chromatography-mass spectrometry analysis of wine volatile compounds, J. Chromatogr. A 1216 (2009) 7351–7357.

[168] T.E. Siebert, H.E. Smyth, D.L. Capone, C. Neuwöhner, K.H. Pardon, G.K. Skouroumounis, et al., Stable isotope dilution analysis of wine fermentation products by HS-SPME-GC-MS, Anal. Bioanal. Chem. 382 (2005) 937–947.

[169] I. Lavignini, B. Fedrizzi, G. Versini, F. Magno, Effectiveness of isotopically labeled and non-isotopically labeled internal standards in the gas chromatography/mass spectrometry analysis of sulfur compounds in wines: use of a statistically based matrix comprehensive approach, Rapid Comm. Mass Spec. 23 (2009) 1167–1172.

[170] L. Rebiere, A.C. Clark, L.M. Schmidtke, P.D. Prenzler, G.R. Scollary, A robust method for quantification of volatile compounds within and between vintages using headspace-solid-phase micro-extraction coupled with GC-MS-application on Semillon wines, Anal. Chim. Acta. 660 (2010) 149–157.

CHAPTER

31

Gas Chromatography in Space Exploration

Maria Chiara Pietrogrande

Department of Chemistry, University of Ferrara, Ferrara, Italy

OUTLINE

31.1. INTRODUCTION

Space missions are designed to study the physics and chemistry of extraterrestrial environments in the hopes of shedding light on the origins, evolution, distribution, and future of life in the Universe; more specifically, such missions seek to understand the origins of life and to test the hypothesis that life does indeed exist elsewhere and not just on the Earth (this is the realm of astrobiology) [1–3]. With this aim, planetary atmospheres and surfaces are directly investigated using remote-sensing techniques (usually spectroscopy) from orbiting observatories and *in situ* chemical analysis using atmospheric or surface-landing probes [4,5].

Gas chromatography (GC) has been, and remains, one of the most frequently used techniques for such *in situ* chemical analyses. This is because it offers great efficiency at high speed, it takes advantage of the enormous separation power provided by capillary GC columns, its formats are sensitive and straightforward, and it can be hyphenated with spectrometric and spectroscopic devices as well as with multicolumn operations. In combination with laboratory simulation, experiments, and computation of theoretical models, *in situ* GC analyses

Gas Chromatography DOI: 10.1016/B978-0-12-385540-4.00031-6

provide accurate information on the elemental-isotopic and chemical composition of the atmospheres and soils of various bodies within the solar system, useful in identifying the chemical signature of extinct or extant life (or proto-life) processes [6–14].

This chapter reviews the relevant applications of onboard GC instrumentation used in space missions to Titan, with its complex atmosphere resembling that of the primitive Earth [10,13,14], to comets that retain traces of the Earth's early evolution [7,8,9], and to Mars as the planet which most closely resembles the Earth [11,12].

31.2. TECHNOLOGICAL AND OPERATING CONSTRAINTS IN SPACE GC

Flight conditions that restrict mass, size, and energy consumption do indeed require specially designed gas chromatographic instruments and specific operative conditions if they are to meet space-mission-imposed constraints [6–10]. An overview of the GC instruments installed on probes of *in situ* space missions is reported in Table 31.1.

It must be underlined that the comprehensive definition of instrument design – i.e. the selection of chromatographic columns and detectors, as well as the definition of the sampling system – is vitally important in space analysis, as hardware configuration cannot be changed after the probe is launched and no significant modification of its working conditions can be implemented.

To fulfill the important criterion of the robustness – resistance to vibrations and temperature variations – metallic capillary columns with cross-linked and bonded phases are the selection of choice because they are highly efficient and require short analysis times

TABLE 31.1 Overview of the GC Instruments Installed on Probes of *in situ* Space Missions

Mission, launch, arrival year	Experiment, sample type	GC columns	Detectors
NASA Viking to Mars, 1975, 1976	GEX, gas	Pairs of packed Porapack Q	Thermistore TCD
	GC-MS, soil	Tenax coated with polymetaphenoxylene	MS
NASA and ESA Cassini–Huygens to Titan 1997, 2004	GC-MS; gas or aerosol	Six columns in parallel: packed columns (carbon molecular sieve, glassy carbon) and three capillary WCOTs	Five MS sources in parallel
ESA Rosetta mission to Comet 2003, 2011	GC-MS; comet nucleus	Six WCOTs and two PLOTs in parallel, three chiral columns	One time-of-flight MS and eight nanoTCDs in parallel
Phoenix mission, 2007, 2008	TEGA; Mars soil	six GC capillary columns	
NASA Mars Science Laboratory mission, 2011, 2012	Sample Analysis at Mars (SAM); Mars soil	six GC capillary columns for C1-C15 hydrocarbons, including a Chirasil-Dex for separation of chiral compounds	TCD, QMS
ESA ExoMars Mission to Mars biological environment of the Martian surface, 2013	Sample preparation and distribution system (SPDS); Mars soil	Mars Organic Analyzer (MOA), a portable microfabricated capillary electrophoresis (CE) instrument	

(Table 31.1, 3rd column). Shock and vibration resistance is one of the main criteria for the selection of a detector: thermal conductivity (TCD) and electron-capture (ECD) detectors are the detectors of choice for space GC [6–10]. In addition, over the last 10 years, progress in mass spectrometer (MS) design has resulted in miniaturized, onboard MS detectors that provide additional information relevant to peak identification (Table 31.1, 4th column).

Another limitation encountered when operating in space is the restricted amounts of power and carrier gas available for the instrument. Consequently, short analysis runs – i.e. less than 30 min – and low operating temperatures – historically in the 20–70 °C range – are used in isothermal conditions, since temperature-programming operations increase power consumption [7,9,10]. These operating conditions strongly limit the selection of a column's dimensions and stationary phases – i.e. column length and polymer film thickness. To overcome the power supply limitation, space probes are fitted with solar cells which provide greater amounts of energy than the electric batteries used in the past.

In addition, data storage and transmission in space is another problem for space GC analyses since onboard electronic and data handling systems suffer from analysis time (generally limited to 10–20 min) and data acquisition rate restrictions, factors of primary importance if one has to define and interpret a chromatogram [8,13].

All these constraints present challenges for the development of new GC equipment for future explorations.

31.3. PREBIOTIC CHEMISTRY IN TITAN'S ATMOSPHERE: THE CASSINI–HUYGENS MISSION

It is assumed that the atmosphere on Titan, the largest moon of the planet Saturn, strongly resembles that of the primitive Earth, before the appearance of life and this makes it one of the most interesting places to search for extinct, or extant life in extraterrestrial environments. The joint NASA-ESA Cassini mission was launched on October 15, 1997 to explore Saturn and Titan: after the Huygens probe landed on Titan on January 14, 2005 (two distinct atmospheric samples were collected at separate altitudes as the probe descended), the Cassini spacecraft completed its initial four-year mission to explore the Saturn System in June 2008 and a Cassini Equinox Mission was extended through 2010 [10,13–19]. A GC-MS instrument was installed on the Cassini probe: three capillary columns operated in parallel to detect and identify organic compounds by connecting each column to an independent MS ion source (Table 31.1, 3rd row). It was coupled to an aerosol and collector pyrolyzer (ACP) to investigate the composition (including isotope ratio) of refractory materials such as aerosols and soil [6,10].

The GC-MS analyses on the Huygens probe found evidence of a moist surface, with ethane (among other volatiles) evaporating from the surface heated by the warm bottom of the probe itself. Taken together with remote sensing from a visual and infrared mapping spectrometer and from radar instruments on the Cassini orbiter, these observations of surface features suggest that a "methalogical" cycle exists on Titan, with methane playing a role similar to that of water in the hydrological cycle seen on the Earth. The ACP result is the first evidence of the presence of complex macromolecular organic matter constituting the solid organic refractory core of the aerosol particles in Titan's atmosphere (tholins) [18,19]. These results were predicted by theoretical models and supported by simulation studies on laboratory analogs representative of the hydrocarbon and nitrile gases in Titan's aerosols, i.e. tholins generated either by photochemistry or by cold plasma discharges [14–17].

31.4. PREBIOTIC CHEMISTRY IN COMET ENVIRONMENTS: ROSETTA MISSION

Being among the most primitive objects in the solar system, comets are assumed to conserve the solar system's average composition and to contain a great abundance of extraterrestrial organic material, the prebiotic building blocks for the emergence of life on the Earth. The European Space Agency (ESA) Rosetta space mission was launched on March 2, 2004 and its main objective was a rendezvous with Comet 67P/Churyumov–Gerasimenko. Over a period of nearly two years, it studied the comet's nucleus and environment in great detail and sent a probe to the comet's surface to gain vital insight into our origins, just as the Rosetta stone enabled us to decipher hieroglyphics [7–9,20].

Among the eight instruments aboard the Rosetta lander, the cometary sampling and composition (COSAC) experiment was dedicated to the *in situ* analysis of compounds obtained by thermal volatilization of the material in the comet's nucleus by a pyrolyzer (maximum heating temperature: 800 °C). GC detection was achieved by a miniaturized TCD connected to each column and a miniaturized (<1 kg), high-resolution TOF mass spectrometer with a mass range of 12–1500 amu [7,9]. The GC system contained eight capillary columns for chemical characterization, mounted in parallel: five capillary columns with different polarities were selected to unambiguously identify the broad range of compounds expected in comets. In addition, three chiral columns were installed for enantiomeric separation of chiral aliphatic hydrocarbons and cometary amino acids (Table 31.1, 4th row).

Chromatograms recovered from space missions are interpreted by comparing them with laboratory calibration measurements under operating conditions simulating those present in the probe mission, i.e. isothermal condition or a slow temperature-increasing program [8,9,10]. As an example, Figure 31.1 shows a GC-MS chromatogram of a standard mixture containing 36 organic compounds representative of extraterrestrial atmospheres, i.e. 36 hydrocarbons and oxygenated compounds with carbon numbers ranging from 2 to 8. The column used was the MXT-1 silanized stainless-steel column selected for space missions (Huygens probe, COSAC experiment) and the operating temperature was the same isothermal condition at 30 °C planned for use in the COSAC experiments [8].

31.5. SEARCH FOR KEY CHEMICAL BIOMARKERS: MARS EXPLORATION

Mars is of great interest to astrobiologists because it is the planet that most closely resembles the Earth and may still retain the original prebiotic organic compounds that led to life. Mars exploration started in 1976 with the first NASA Viking mission when, for the first time ever, a pyrolysis-based GC-MS instrument was installed on the Viking space probe (Table 31.1, 1st, 2nd rows). However, at that time, GC-MS instrumentation lacked the detection sensitivity needed to catch low levels of organics on Mars and only detected CO_2 and water [21,22]. Later, Mars exploration continued with the detailed data from rovers such as Opportunity and Spirit [12,21].

The upcoming Mars Science Laboratory (MSL) mission will be the most comprehensive search thus far for organic molecules in Martian rocks and soil. The MSL spacecraft was launched on November, 26 2011 and the rover Curiosity is expected to land on Mars in August 2012 and operate for two years [11,12,23–28]. Curiosity's scientific payload includes the Sample Analysis at Mars (SAM, scheme reported in Figure 31.2) instrument package composed of three instruments: a GC, an MS, and a tunable laser

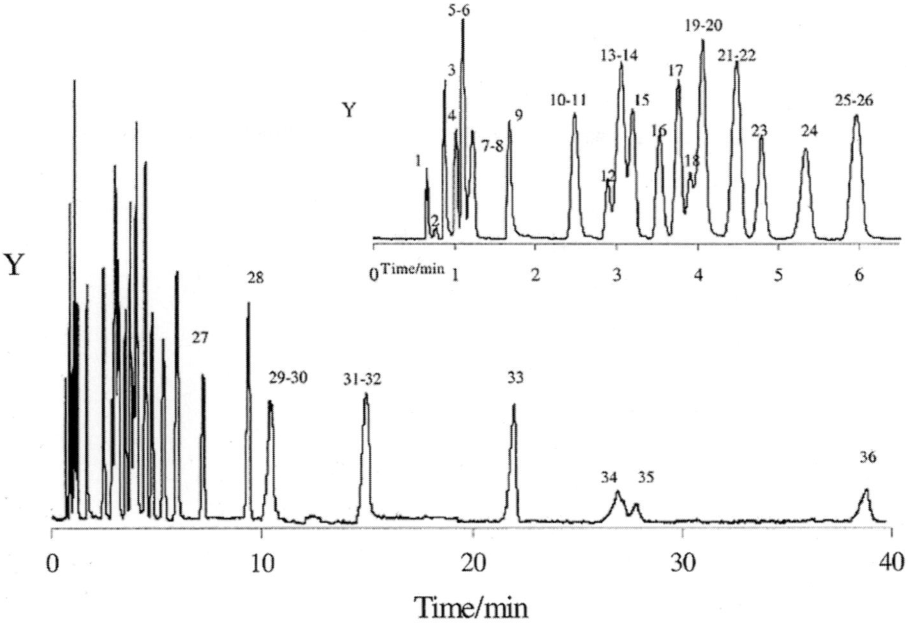

FIGURE 31.1 Chromatogram of a complex mixture of VOCs representative of the targeted chemical species in space *in situ* analysis. Column: MXT-1 column. Operating conditions: isothermal at 30 °C and flow rate 25.6 cm s^{-1}. FID detector. 1: methanol, 2: methyl formate, 3: ethanol, 4: acetone, 5: 1-pentene, 6: isopropanol, 7: n-pentane, 8: ethyl formate, 9: 1-propanol, 10: n-hexane, 11: ethyl acetate, 12: methyl propionate, 13: 2,2 dimethylpentane, 14: methylcyclopentane, 15: 2,4-dimethylpentane, 16: isopropyl methyl ketone, 17: benzene, 18: isopropyl acetate, 19: cyclohexane, 20: 1-butanol, 2: 2,3-dimethylpentane, 22: 2-methyl hexane, 23: 3-methyl hexane, 24: isooctane, 25: n-heptane, 26: propyl acetate, 27: methyl butyl ketone, 28: toluene, 29: isobutyl acetate, 30: 1-pentanol, 31: n-octane, 32: butyl acetate, 33: ethylbenzene, 34: 1-hexanol, 35: isoamyl acetate, 36: amyl acetate. *Source: Reprinted with Permission from Ref. [8]. Copyright (2003) Wiley-VCH Verlag GmbH & Co. KGaA.*

spectrometer (TLS), i.e. to measure carbon isotope ratios ($^{13}C/^{12}C$) and to identify biotic or abiotic organic compounds. The GC and MS systems allow high mass and volume availability using six GC columns to identify a broader range of organic species, taking advantage of the significant advances in throughput and sensitivity over what the instruments on the Viking were able to provide (Table 31.1, 6th row). Moreover, to gain better access to samples, the SAM will be housed on a rover rather than inside a lander as the Viking GC-MS was.

In addition, the SAM instruments are supported by a sample manipulation system (SMS) and a Chemical Separation and Processing Laboratory (CSPL) that include different sample handling and pretreatment devices, i.e. high conductance and microvalves, gas manifolds with heaters and temperature monitors, chemical and mechanical pumps, pyrolysis ovens, and chemical scrubbers and getters [11,12,24–26]. CSPL subsystems will also be implemented to analyze nonvolatile compounds in the Martian soil, i.e. carboxylic acids and amino acids, as such compounds play an important role in terrestrial biochemistry. For GC *in-situ* analysis of such molecules, a chemical reactor has been developed based on one-pot, one-step extraction (either by solvent extraction with isopropanol or by pyrolysis) and derivatization reaction using the

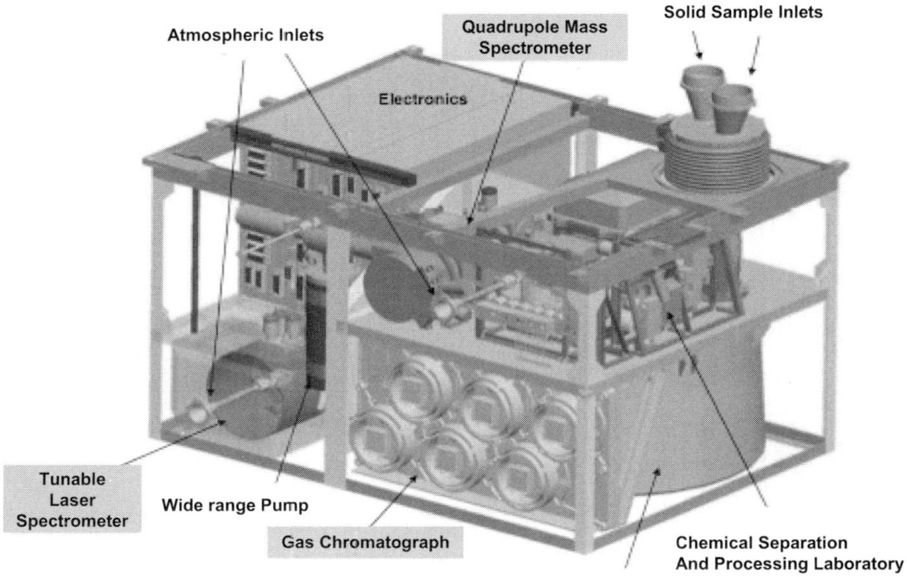

FIGURE 31.2 Outline of the sample analysis at mars (SAM) included in curiosity's scientific payload [23].

chemical derivatizing reagent already developed in the Rosetta mission, i.e. MTBSTFA (N-methyl-N-(tert-thylsilyl)trifluoroacetamide) [24–26]. As an example, Figure 31.3 shows the GC-MS signal obtained by analyzing an analog material of the Mars surface – Atacama Desert soil, from the aridest part of the desert located in Chile – spiked with a standard mixture of carboxylic and amino acids (10^{-3} M each): 15 target compounds can be separated under representative space operating conditions (miniaturization, automation, and low energy consumption) [26]. Further developments are currently under investigation to design a GC-MS instrument for future Mars missions [27,28].

31.6. SEARCH FOR CHIRALITY IN SPACE

The next step in the search for the signature of the prebiotic/biotic materials and the occurrence of life will be the detection of enantiomeric excess of amino acids or sugars in extraterrestrial environments. In fact, these molecules are known to be present in one enantiomer in living macromolecules (L for amino acids and D for sugars), whereas they are found as racemic mixtures (equal parts of L and D) in abiotic systems [29,30]. Chirality discrimination requires amino acid derivatization using a one-step derivatization reaction able to preserve the enantiomeric configuration and prevent racemization [31–33].

A single-step derivatization strategy using a DMF-DMA (N,N-dimethylformamide dimethylacetal) reagent has proved useful in *in situ* space analysis to search for homochirality in extraterrestrial space: it is a rapid, one-step reaction (without any cofactors) that can occur at relatively low temperatures (140 °C); the reaction can easily be automated and it yields low-molecular-weight products compatible with space MS performance (limited mass range for detection).

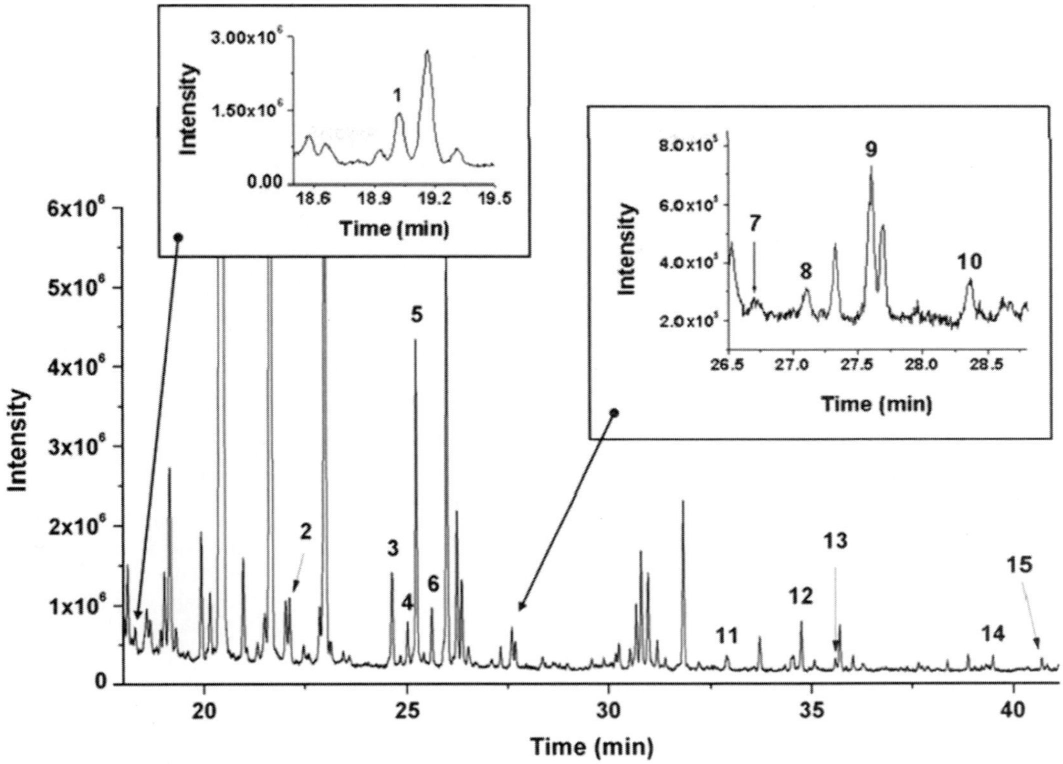

FIGURE 31.3 GC-MS analysis of Atacama soil sample spiked with a mixture of standard acids (10^{-3} M each), under representative space operating conditions. The sample was heated at 500 °C for 5 min. and derivatized by MTBSTFA reagent. GC column: RTX 5MS capillary column; programmed ramp temperature from 100 °C to 270 °C at 4 °C min^{-1}; MS detection in EI mode. Standard compounds: 1: 4-methyl pentanoic acid, 2: heptanoic acid, 3: benzoic acid, 4: octanoic acid, 5: hydroxyethanoic acid, 6: 2-hydroxypropanoic acid, 7: alanine, 8: glycine, 9: nonanoic acid, 10: valine, 11: dodecanoic acid, 12: phosphoric acid, 13: 1,2 benzendicarboxylic acid, 14: pentandecanoic acid, 15: 1,2 benzendicarboxylic acid bis(2-methylpropyl)ester. *Source: Reprinted from Ref. [26]. Copyright (2009) with Permission from Elsevier.*

Enantioseparation and characterization of DMF-DMA derivatives can be achieved using the Chirasil-Dex chiral capillary column, the most effective chiral capillary column used in the SAM experiments. As an example, Figure 31.4 reports the enantiomeric separation of 20 proteinic amino acids with good enantiomer resolution (R_S values $\geq$0.8) and high sensitivity (LOD values $\approx$1 pmol) [32]. Coupled with miniaturized GC-MS (SIM) formats, the DMF-DMA procedure has been incorporated in the Rosetta mission's COSAC experiment and the SAM Mars experiments [6,7,24,25].

31.7. CONCLUSIONS AND PERSPECTIVES

The experience of GC instruments used in space missions has proven the feasibility of the GC technique for the *in situ* analysis of extraterrestrial environments, providing good chemical characterization of extraterrestrial environments and thus making a fundamental contribution to our understanding of the Earth and planetary systems.

Space research has shown positive synergism with technological developments in GC

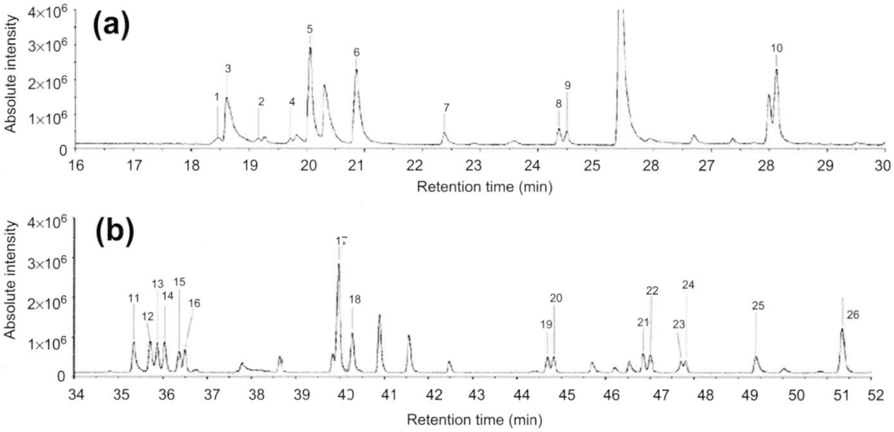

FIGURE 31.4 Chromatogram of 20 separated amino acids (1—10 10^{-7} mol of each enantiomer). Derivatization with 20 μL DMF-DMA, 3 min at 140 °C. GC column: CP-Chirasil-DEX CB capillary column; temperature program: 70 °C for 5 min, increasing temperature rate 3 °C min^{-1} up to 190 °C, isotherm for 10 min at 190 °C. MS detection in EI mode (m/z range from 40 to 350 uma). 1: l-Thr, 2: d-Thr, 3: l-Ser, 4: d-Ser, 5: d-Ala, 6: l-Ala, 7: Gly, 8: d-Val, 9: l-Val, 10: d + l Ile, 11: l-Pro, 12: d-Pro, 13: d-Asp, 14: l-Asp, 15: d-Cys, 16: l-Cys, 17: d + l Met, 18: d + l Glu, 19: d-Phe, 20: l-Phe, 21: d + l-Lys, 22: d + l Lys, 23: d or l Tyr, 24: l or d Tyr, 25: d + l His, 26: d + l Arg. *Source: Reprinted from Ref. [32]. Copyright (2010) with Permission from Elsevier.*

instrumentation. Dramatic improvements in sophisticated space instruments have taken advantage of all the technological advances made in conventional GC used in terrestrial laboratories. Likewise, these improvements – mainly in terms of miniaturization, automation, and energy saving – have, in turn, led to developments in benchtop scale GC equipment. In the future, the experiments performed on onboard space landers should become increasingly similar to what analytical chemists generally perform in terrestrial laboratories.

It is, however, true that new breakthroughs are required to further enhance space GC techniques. The most pressing demands are to: i) extend the techniques to higher temperatures and temperature programming and ii) shorten elution times for most compounds analyzed. Moreover, new stationary phases must be tailored to broaden the range of compounds targeted in such missions. Last, but not least, detector sensitivity must be further enhanced: improving the dynamic MS analysis range, miniaturization, and making detection limits as low as possible are the main goals for mass spectrometers i.e., currently, the detectors of choice because they can provide more structural information than traditional detectors. In addition, new sample-enrichment procedures should be explored to develop reliable, fully automated sample pretreatment methods that meet space requirements, i.e. trapping and thermal desorption, as well as liquid extraction techniques.

Other new breakthroughs in GC space application concern the development of miniaturized equipment and methods: lab-on-a-chip systems are actively being studied because they may be able to provide high-performance chemical analysis in a miniaturized format. As an alternative, complementary techniques such as high-performance liquid chromatography, capillary electrophoresis, and supercritical fluid chromatography should also be adapted to space applications.

For these reasons, in the next space exploratory missions, sophisticated GC instruments will be specially designed for integration into space instrument subsystems so they can gather a wide range of chemical signatures of present or extinct life in extraterrestrial environments.

References

[1] S.A. Benner, Defining life, Astrobiol. 10 (2010) 1021–1030.

[2] B.S. Blumberg, Astrobiology, space and the future age of discovery, Philos. Transact. A Math. Phys. Eng. Sci. 1936 (2011) 508–515.

[3] C.H. Gibson, R.E. Schild, N.C. Wickramasinghe, The origin of life from primordial planets, Int. J. Astrobiol. 10 (2011) 83–98.

[4] M. Fridlund, Extra-terrestrial life in the European space agency's cosmic vision plan and beyond, Philos. Transact. A Math. Phys. Eng. Sci. 1936 (2011) 582–593.

[5] M. Carr, Astronomy: Martian illusions, Nature 470 (2011) 172–173.

[6] C. Szopa, R. Sternberg, C. Rodier, D. Coscia, F. Raulin, Development and analytical aspects of gas chromatography for space exploration, LC•GC Europe (February 2001) 1–6.

[7] R. Sternberg, C. Szopa, C. Rodier, Analyzing a comet nucleus by capillary GC, Anal. Chem. 74 (17) (2002) 481A–487A.

[8] M.C. Pietrogrande, I. Tellini, A. Felinger, F. Dondi, C. Szopa, R. Sternberg, et al., Decoding of complex isothermal chromatograms: application to chromatograms recovered from space missions, J. Sep. Sci. 26 (2003) 569–577.

[9] C. Szopa, M. De Pra, I. Tellini, R. Sternberg, M.C. Pietrogrande, C. Vidal-Madjar, et al., Dual column capillary gas chromatographic system for the in situ analysis of volatile organic compounds on a cometary nucleus, J. Sep. Sci. 27 (2004) 495–503.

[10] C. Szopa, G. Freguglia, R. Sternberg, M.J. Nguyen, P. Coll, F. Raulin, et al., Performances under representative pressure and temperature conditions of the gas chromatography–mass spectrometry space experiment to investigate Titan's atmospheric composition, J. Chromatogr. A 1131 (2006) 215–226.

[11] M.N. Heinrich, B.N. Khare, C.P. McKay, Prebiotic organic synthesis in early Earth and Mars atmospheres: laboratory experiments with quantitative determination of products formed in a cold plasma flow reactor, Icarus 191 (2007) 765–778.

[12] R. Navarro-Gonzalez, E. Iniguez, J. de la Rosa, C.P. McKay, Characterization of organics, microorganisms, desert soils, and Mars-like soils by thermal volatilization coupled to mass spectrometry and their implications for the search for organics on Mars by Phoenix and future space missions, Astrobiol. 8 (2009) 703–715.

[13] M.C. Pietrogrande, P. Coll, R. Sternberg, C. Szopa, R. Navarro-Gonzalez, C. Vidal-Madjar, et al., Analysis of complex mixtures recovered from space missions: statistical approach to the study of Titan atmosphere analogues (tholins), J. Chromatogr. A 939 (2001) 69–77.

[14] N.T. Buu, C.J. Jeffrey, M. Force, R.G. Briggs, V. Vuitton, J.P. Ferris, Photochemical processes on Titan: irradiation of mixtures of gases that simulate Titan's atmosphere, Icarus 177 (2005) 106–115.

[15] M. McGuigan, J.H. Waite, H. Imanaka, R.D. Sacks, Analysis of Titan tholin pyrolysis products by comprehensive two-dimensional gas chromatography–time-of-flight mass spectrometry, J. Chromatogr. A. 1132 (2006) 280–288.

[16] M. Ruiz-Bermejo, C. Menor-Salván, E. Mateo-Martí, S. Osuna-Esteban, J.Á Martín-Gago, S. Veintemillas-Verdaguer, $CH_4/N_2/H_2$-spark hydrophobic tholins: a systematic approach to the characterisation of tholins, Icarus 198 (2008) 232–241.

[17] M. Ruiz-Bermejo, C. Menor-Salván, J.L. de la Fuente, E. Mateo-Martí, S. Osuna-Esteban, J.Á Martín-Gago, et al., $CH_4/N_2/H_2$-spark hydrophobic tholins: a systematic approach to the characterisation of tholins, Part II, Icarus 204 (2009) 672–680.

[18] E.H. Wilson, S.K. Atreya, Titan's carbon budget and the case of the missing ethane, J. Phys. Chem. A 113 (2009) 11221–11226.

[19] E. Sciamma-O'Brien, N. Carrasco, C. Szopa, A. Buch, G. Cernogora, Titan's atmosphere: an optimal gas mixture for aerosol production? Icarus 209 (2010) 704–714.

[20] J.F.J. Todd, S.J. Barber, I.P. Wright, G.H. Morgan, A.D. Morse, S. Sheridan, et al., Ion trap mass spectrometry on a comet nucleus: the Ptolemy instrument and the Rosetta space mission, J. Mass. Spectrom. 42 (2007) 1–10.

[21] B.E. DiGregorio, The search for organic molecules on Mars, Anal. Chem. 77 (2005) 348A–353A.

[22] R. Mukhopadhyay, The Viking GC-MS and the search for organics on Mars, Anal. Chem. 79 (2007) 7249–7256.

[23] http://msl-scicorner.jpl.nasa.gov/Instruments/SAM/.

[24] M.C. Pietrogrande, M.G. Zampolli, F. Dondi, C. Szopa, R. Sternberg, A. Buch, et al., *In situ* analysis of the Martian soil by gas chromatography: decoding of complex chromatograms of organic molecules of exobiological interest, J. Chromatogr. A 1071 (2005) 255–261.

[25] D. Meunier, R. Sternberg, F. Mettetal, A. Buch, D. Coscia, C. Szopa, et al., A laboratory pilot for in situ analysis of refractory organic matter in Martian soil by gas chromatography–mass spectrometry, Adv. Space Res. 39 (2007) 337–344.

[26] A. Buch, R. Sternberg, C. Szopa, C. Freissinet, C. Garnier, El J. Bekri, et al., Development of a gas chromatography compatible sample processing system (SPS) for the in-situ analysis of refractory organic matter in Martian soil: preliminary results, Adv. Space Res. 43 (2009) 143–151.
[27] L.M. Pratt, C. Allen, A. Allwood, A. Anbar, S. Atreya, M. Carr, et al., The Mars Astrobiology Explorer-Cacher (MAX-C): a potential rover mission for 2018. Final report of the Mars Mid-Range Rover Science Analysis Group (MRR-SAG) October 14, 2009, Astrobiol. 10 (2010) 127–163.
[28] S.A. Getty, L.F. Inge, L. Feng, W.B. Brinckerhoff, E.H. Cardiff, V.E. Holmes, et al., Development of an evolved gas-time-of-flight mass spectrometer for the volatile analysis by Pyrolysis of Regolith (VAPoR) instrument, Int. J. Mass. Spectr. 295 (2010) 124–132.
[29] T.J. Ward, Chiral separations, Anal. Chem. 74 (2002) 2863–2872.
[30] L Caglioti, O. Holczknecht, N. Fujii, C. Zucchi, G. Palyi, Astrobiology and biological chirality, Orig Life Evol. Biosph. 36 (2006) 459–466.
[31] M.G. Zampolli, G. Basaglia, F. Dondi, R. Sternberg, C. Szopa, M.C. Pietrogrande, Gas chromatography–mass spectrometry analysis of amino acid enantiomers as methyl chloroformate derivatives: application to space analysis, J. Chromatogr. A 1150 (2007) 162–172.
[32] C. Freissinet, A. Buch, R. Sternberg, C. Szopa, C. Geffroy-Rodier, C. Jelinek, et al., Search for evidence of life in space: analysis of enantiomeric organic molecules by N, N dimethylformamide dimethylacetal derivative dependant gas chromatography–mass spectrometry, J. Chromatogr. A 1217 (2010) 731–740.
[33] M.C. Pietrogrande, G. Basaglia, Enantiomeric resolution of biomarkers in space analysis: chemical derivatization and signal processing for gas chromatography–mass spectrometry analysis of chiral amino acids, J. Chromatogr. A 1217 (2010) 1126–1133.

Index

Note: Page numbers followed by "f" indicate figures and "t" indicate tables.

D

E

F

G

H

I

J

K

L

M